Biological Consequences of Plate Tectonics

Vertebrate Paleobiology and Paleoanthropology Series

Edited by

Eric Delson
Vertebrate Paleontology, American Museum of Natural History
New York, NY, USA

Eric J. Sargis
Yale University
Department of Anthropology, New Haven, CT, USA

Focal topics for volumes in the series will include systematic paleontology of all vertebrates (from agnathans to humans), phylogeny reconstruction, functional morphology, Paleolithic archaeology, taphonomy, geochronology, historical biogeography, and biostratigraphy. Other fields (e.g., paleoclimatology, paleoecology, ancient DNA, total organismal community structure) may be considered if the volume theme emphasizes paleobiology (or archaeology). Fields such as modeling of physical processes, genetic methodology, nonvertebrates or neontology are out of our scope.

Volumes in the series may either be monographic treatments (including unpublished but fully revised dissertations) or edited collections, especially those focusing on problem-oriented issues, with multidisciplinary coverage where possible.

More information about this series at http://www.springer.com/series/6978

Biological Consequences of Plate Tectonics

New Perspectives on Post-Gondwana Break-up—A Tribute to Ashok Sahni

Edited by

Guntupalli V. R. Prasad

Department of Geology, Center for Advanced Studies, University of Delhi, Delhi, India

and

Rajeev Patnaik

Center of Advanced Study in Geology, Panjab University, Chandigarh, India

Professor Ashok Sahni

Springer

Editors
Guntupalli V. R. Prasad
Department of Geology
Center for Advanced Studies
University of Delhi
Delhi, India

Rajeev Patnaik
Center of Advanced Study in Geology
Panjab University
Chandigarh, India

ISSN 1877-9077 ISSN 1877-9085 (electronic)
Vertebrate Paleobiology and Paleoanthropology Series
ISBN 978-3-030-49752-1 ISBN 978-3-030-49753-8 (eBook)
https://doi.org/10.1007/978-3-030-49753-8

Cover image: Cretaceous-Paleogene boundary paleogeographic map in the background (from The PALEOMAP PaleoAtlas [Scotese 2016], modified from Figure 10.1 of Rage and Gheerbrant [this volume]); Reconstruction of the stem proboscidean *Phosphatherium* from the earliest Eocene (55 Ma) of Morocco (top centre) (Fig. 10.6. of Rage and Gheerbrant [this volume]); Ornithischian dinosaur tooth (top right) (Fig. 1.3B of Prasad and Parmar [this volume]) from the Middle Jurassic Kota Formation, India; Left mandible (holotype) of *Bishops whitmorei* (in labial view), a tribosphenic mammal from the Early Cretaceous of Australia (bottom center) (Fig. 3.4B of Rich et al. [this volume]); Skull of *Indohyus* (left center) (Fig. 14.1C of Thewissen et al. [this volume]) from the Middle Eocene Subathu Formation, India.

This Springer imprint is published by the registered company Springer Nature Switzerland AG
The registered company address is: Gewerbestrasse 11, 6330 Cham, Switzerland

Dedication to Jean-Claude Rage (1943–2018)

It is with great emotion and respect that we dedicate this volume to the memory of our French colleague Jean-Claude Rage, a researcher at the CNRS (MNHN-CR2P) paleontological team, Paris, who passed away recently. Jean-Claude contributed to this volume as a leading author of two chapters on two of his most important research specializations, paleoherpetology and continental paleobiogeography, for which his work is well-known globally. He worked for more than 50 years in collaboration with many colleagues from around the world, identifying remains of squamates and amphibians from various Mesozoic and Cenozoic microvertebrate localities of Europe, Africa, India, South and North Americas. In particular, Jean-Claude had a long association with Ashok Sahni and worked on the Late Cretaceous herpetofaunas of India since 1980 in collaboration with the Indian colleagues.

Jean-Claude graduated in 1967 from the University of Lyon, France and received his Ph.D. under the mentorship of Prof. Robert Hoffstetter in 1976. He started his CNRS career at University Pierre et Marie Curie (Paris) in 1975, where he stayed until 1997, and then moved to the MNHN Paleontology Lab (current CR2P team). Jean-Claude retired from MNHN in 2008 but continued as Emeritus Research Director until his demise on March 30, 2018. Jean-Claude published 226 research papers, 3 books, and 29 chapters in books.

Jean-Claude was a very simple, endearing, and warm personality, and was a source of inspiration to the work and research of many students and colleagues in France and abroad. We will miss our great friend, colleague, and mentor, Jean-Claude.

Foreword

This volume has two themes. It presents the results of new research on the evolution of terrestrial biotas during and after the fragmentation of Gondwana, which included parts of what are now India, South America, Antarctica, Africa, Madagascar, and Australia. Also, it is a well-deserved recognition of the contributions of Prof. Ashok Sahni to our understanding of the complex evolutionary history of the biota of the Gondwana landmass. He, his former students, and other colleagues have and continue to make significant contributions to this area of study. Many are presented in this volume.

This collection of papers is a worthy successor to the issue of the *Journal of Biological Sciences* (vol. 34, no. 5, 2009) that was edited by Profs. Bajpai and Sahni. Through a series of studies of the fossil record, that issue of the journal recognized the bicentennial of the birth of Charles Darwin. In the intervening decade, many significant advances have been made. These reflect the increasing interest of Indian and other paleontologists and geologists in advancing field and laboratory studies of the fossil records of India and other parts of Gondwana. Here the focus is on India with articles contributing to our knowledge of its history from the early Mesozoic, when it was part of the greater Gondwanan continent, through the later Mesozoic and Cenozoic during its northward movement. In addition, the net is cast more widely with analyses of the evolution of African terrestrial faunas, evaluation of changing faunal similarities among the Gondwanan continents, and descriptions of new Australian Early Cretaceous mammaliaforms.

Two palynological studies give new insights on floral evolution. Mishra and Prasad (2020) trace the evolution of the Jurassic to Early Cretaceous Gondwanan floras of India and their relationships to contemporaneous floras of other areas. On a different scale, Prasad et al. (2020) relate the evolution of angiosperms to patterns of plate tectonic movement and climate change. Using biomarkers (plant terpenoids found in coal-bearing sediments) Paul and Dutta (2020) trace the effects on floral evolution of the northward movement of India during the Cretaceous and Early Paleogene.

Knowledge of the taxonomic diversity of the Early Jurassic fauna of the Kota Formation is enhanced by Prasad and Parmar (2020). They describe teeth of theropods and ornithischians, the first records of these dinosaurs from this formation. In a temporally wider ranging study, Chatterjee (2020) summarizes and evaluates the Indian fossil record of dinosaurs from the Late Triassic to the end of the Cretaceous. Occurrences of other reptiles as well as amphibians from the Late Cretaceous intertrappean beds are analyzed in a study of lissamphibian and squamate diversity by Rage et al. (2020). In a review of the Cretaceous turtle faunas of India, Lapparent de Broin and Prasad (2020) describe new Late Cretaceous Pelomedusoides turtles. They also review the paleobiogeography of Indian turtles in the context of what is known of the turtles of Africa and Madagascar.

Solé et al. (2020) describe new material of the enigmatic mammal *Frugivastodon* from the Eocene Cambay Shale. Their analysis confirms its membership in the Apatemyidae and points to a phylogenetic relationship to the North American Uintan *Aethomylos*. The Cambay Shale has yielded records of a variety of non-volant mammals. Kapur (2020) analyzes the variation in size of these Early Eocene mammals and points to the possibility of dispersal of Laurasian mammals into India in the Paleocene. Considering Eocene records from both India and Pakistan, Thewissen et al. (2020) highlight an endemic radiation of raoellid artiodactyls. Network analyses, which provide a very informative basis for determining faunal similarity, are applied by Halliday et al. (2020) in a paleogeographic study of the evolution of the Indian fauna. They compare the Late Cretaceous fauna of India to those of Madagascar and South America and then trace the evolution of Laurasian affinities by the Eocene.

Two papers report the results of research on the evolution of rodents on the Indian subcontinent from the earliest Miocene onward. Patnaik (2020) provides new information on murines, rhizomyines, and ctenodactylines. Flynn et al. (2020), noting the influence of topographic and oceanic barriers, trace the origins of extant mice and rats.

Rage and Gheerbrant (2020) note that from the mid-Cretaceous to the Early Miocene Africa was geographically isolated as an island continent with only intermittent connections to Eurasia. A few groups of Eurasian vertebrates entered Africa where they radiated. Later hominoid apes, which originated in Africa, dispersed into Eurasia. Gilbert et al. (2020) suggest that there may have been several dispersal events and discuss the role of climate change influencing their evolution.

In two chapters, Rich et al. (2020a and b) add to our knowledge of the Mesozoic fauna of another Gondwanan continent, Australia. One addition is a remarkably small form with a tribosphenic dentition, which might be referable to the Ausktribosphenidae. The other is based on a very large upper premolar interpreted as belonging to an early monotreme.

The diversity of topics covered in the articles in this volume reflects the wide range of Prof. Ashok Sahni's research interests. He obtained his Ph.D. at the University of Minnesota in 1968. His dissertation was based on a study of the Campanian vertebrate fauna of the Judith River Formation, Montana. His fieldwork produced a large sample of the then poorly known mammalian fauna. On his return to India, he led in the development of vertebrate paleontological research, particularly expanding field and laboratory research on its fossil record. As the articles in this volume testify, he promoted diversification of this research to include studies of Indian paleobotany, formation of the Deccan Traps and its faunas, as well as Early Eocene vertebrates, particularly those found at the Vastan Mine in Gujarat. Also, Prof. Sahni has fostered interdisciplinary research in the area of biomineralization. His and his colleagues' research ranged widely from studies of the structure of molluscan shells, enamel of mammalian teeth, and calcified shells of amniote eggs. Now, Dr. Sahni is a Professor Emeritus at the Panjab University, Chandigarh.

In sum, the articles in this volume testify to the increasing international interest in the evolutionary history of the Gondwanan biota. Also, they pay tribute to Prof. Sahni's many contributions to increasing the tempo and intellectual scope of research based on the growing samples of the fossil records of India and other areas of Gondwana.

William A. Clemens
Curator, University of California Museum of Paleontology
Berkeley, USA

Professor Emeritus, Department of Integrative Biology
University of California Berkeley
Berkeley, USA

References

Chatterjee, S. (2020). The age of dinosaurs in the land of Gonds. In G. V. R. Prasad & R. Patnaik (Eds.), *Biological consequences of plate tectonics: New perspectives on post-Gondwana break-up* (pp. 181–226). Cham: Springer.

Flynn, L. J., Kimura, Y., & Jacobs, L. L. (2020). The murine cradle. In G. V. R. Prasad & R. Patnaik (Eds.), *Biological consequences of plate tectonics: New perspectives on post-Gondwana break-up* (pp. 347–362). Cham: Springer.

Gilbert, C. C., Pugh, K. D., & Fleagle, J. G. (2020). Dispersal of Miocene hominoids (and pliopithecoids) from Africa to Eurasia in light of changing tectonics and climate. In G. V. R. Prasad & R. Patnaik (Eds.), *Biological consequences of plate tectonics: New perspectives on post-Gondwana break-up* (pp. 393–412). Cham: Springer.

Halliday, T. J. D., Holroyd, P. A., Gheerbrant, E., Prasad, G. V. R., Scanferla, A., Beck, R.M.D., et al. (2020). Leaving Gondwana: The changing position of the Indian subcontinent in the global faunal network. In G. V. R. Prasad & R. Patnaik (Eds.), *Biological consequences of plate tectonics: New perspectives on post-Gondwana break-up* (pp. 227–249). Cham: Springer.

Kapur, V. (2020). Size variation amongst the non-volant mammals from the Early Eocene Cambay Shale deposits of western India: Paleobiogeographic implications. In G. V. R. Prasad & R. Patnaik (Eds.), *Biological consequences of plate tectonics: New perspectives on post-Gondwana break-up* (pp. 305–335). Cham: Springer.

Lapparent de Broin, F. de, & Prasad, G. V. R. (2020). Chelonian Pelomedusoides remains from the Late Cretaceous of Upparhatti (Southwestern India): Systematics and paleobiogeographical implications. In G. V. R. Prasad & R. Patnaik (Eds.), *Biological consequences of plate tectonics: New perspectives on post-Gondwana break-up* (pp. 123–180). Cham: Springer.

Mishra, S., & Prasad, V. (2020). Upper Gondwana (Jurassic to Early Cretaceous) palynoflora of India: Its correlation with other Gondwana continents and phytogeographical implications. In G. V. R. Prasad & R. Patnaik (Eds.), *Biological consequences of plate tectonics: New perspectives on post-Gondwana break-up* (pp. 31–66). Cham: Springer.

Patnaik, R. (2020). New data on the Siwalik murines, rhizomyines and ctenodactylines (Rodentia) from the Indian subcontinent. In G. V. R. Prasad & R. Patnaik (Eds.), *Biological consequences of plate tectonics: New perspectives on post-Gondwana break-up* (pp. 363–391). Cham: Springer.

Paul, S., & Dutta, S. (2020). Contrasting biomarker signatures of Cretaceous and Early Paleogene coal-bearing sediments during the northward flight of India. In G. V. R. Prasad & R. Patnaik (Eds.), *Biological consequences of plate tectonics: New perspectives on post-Gondwana break-up* (pp. 265–278). Cham: Springer.

Prasad, G. V. R., & Parmar, V. (2020). First ornithischian and theropod dinosaur teeth from the Middle Jurassic Kota Formation of India: Paleobiogeographic relationships. In G. V. R. Prasad & R. Patnaik (Eds.), *Biological consequences of plate tectonics: New perspectives on post-Gondwana break-up* (pp. 1–30). Cham: Springer.

Prasad, V., Mishra, S., Mishra, A. K., & Bansal, M. (2020). Role of plate tectonics and global climate change in the evolution of angiosperms. In G. V. R. Prasad & R. Patnaik (Eds.), *Biological consequences of plate tectonics: New perspectives on post-Gondwana break-up* (pp. 83-98). Cham: Springer.

Rage, J.-C., & Gheerbrant, E. (2020). Island Africa and vertebrate evolution: A review of data and working hypotheses. In G. V. R. Prasad & R. Patnaik (Eds.), *Biological consequences of plate tectonics: New perspectives on post-Gondwana break-up* (pp. 251–264). Cham: Springer.

Rage, J.-C., Prasad, G. V. R., Verma, O., Khosla, A., & Parmar, V. (2020). Anuran lissamphibian and squamate reptiles from the Upper Cretaceous (Maastrichtian) Deccan intertrappean sites in Central India, with a review of lissamphibian and squamate diversity in the northward drifting Indian plate. In G. V. R. Prasad & R. Patnaik (Eds.), *Biological consequences of plate tectonics: New perspectives on post-Gondwana break-up* (pp. 99–121). Cham: Springer.

Rich, T. H., Flannery, T. F., & Vickers-Rich, P. (2020a). Evidence for a remarkably large toothed-monotreme from the Early Cretaceous of Lightning Ridge, NSW, Australia. In G. V. R. Prasad & R. Patnaik (Eds.), *Biological consequences of plate tectonics: New perspectives on post-Gondwana break-up* (pp. 77–81). Cham: Springer.

Rich, T. H., Trusler, P., Kool, L., Pickering, D., Evans, A., Siu, K., et al. (2020b). A third, remarkably small, tribosphenic mammal from the Mesozoic of Australia. In G. V. R. Prasad & R. Patnaik (Eds.), *Biological consequences of plate tectonics: New perspectives on post-Gondwana break-up* (pp. 67–75). Cham: Springer.

Solé, F., De Bast, E., Legendre, H., Rana, R. S., Kumar, K., Rose, K.D., et al. (2020). New specimens of *Frugivastodon* (Mammalia: Apatotheria) from the Early Eocene of India confirm its apatemyid status and elucidate dispersal of Apatemyidae. In G. V. R. Prasad & R. Patnaik (Eds.), *Biological consequences of plate tectonics: New perspectives on post-Gondwana break-up* (pp. 279–304). Cham: Springer.

Thewissen, J. G. M., Nanda, A. C., & Bajpai, S. (2020). *Indohyus*, endemic radiation of raoellid artiodactyls in the Eocene of India and Pakistan. In G. V. R. Prasad & R. Patnaik (Eds.), *Biological consequences of plate tectonics: New perspectives on post-Gondwana break-up* (pp. 337–346). Cham: Springer.

Preface

This volume is a tribute to the professional life of Prof. Ashok Sahni on his 80th birthday. Prof. Sahni has had a remarkably full and diverse career spanning 50 years. Throughout these 50 years, a period of exciting advancement and achievements in Indian vertebrate paleontology and geology, Prof. Sahni has contributed substantially as both researcher and teacher.

Professor Sahni, a celebrated vertebrate paleontologist of India, has inspired several generations of students, young paleontologists as well as colleagues with his great breadth of knowledge and down to earth demeanor. He leaves a lasting impression irrespective of how briefly one interacts with him. His contributions to Indian paleontology, whether it is through his scientific findings or popularizing the subject, are immense.

Professor Ashok Sahni was born on August 19, 1941, into an illustrious family of academicians that include his grandfather Ruchi Ram Sahni, a Professor of Chemistry at Panjab University, Lahore in undivided India, currently in Pakistan; uncle Birbal Sahni, trained at Cambridge University in Paleobotany under the tutelage of A. C. Seward and a known authority on Gondwana flora, and founder of the unique Institute of Palaeobotany (rechristened as Birbal Sahni Institute of Palaeosciences) at Lucknow; and Prof. Ashok Sahni's father Mulk Raj Sahni, a geologist by training and a specialist in invertebrate paleontology, especially brachiopods, who was instrumental in establishing the Department of Geology at Panjab University, Chandigarh. The Sahnis were the inhabitants of Bhera town located on the eastern bank of Jhelum River prior to the partition of India. After partition, the family has moved over to India. The Sahni family's history of exploration to remote parts of the Himalaya for adventure and scientific enquiry greatly influenced the young Ashok Sahni in choosing natural science for his graduation. Ashok Sahni completed his B.Sc. (Hons.) from Lucknow University in 1961 with Zoology, Botany, and Geology as science subjects. After completing M.Sc. (Hons) in Geology from Lucknow University in 1963, Ashok Sahni went to the University of Minnesota, in Minneapolis, the USA to pursue his Ph.D. He served as a Research Assistant to Prof. Preston Cloud at the Department of Geology and Geophysics and worked on paleoatmospheres and community structure of the Precambrian biota. Prof. Sahni completed his Ph.D. on the Late Cretaceous (Campanian) vertebrates of the Judith River Formation in 1968 under the supervision of Robert E. Sloan. His pioneering work on Campanian mammals of the Judith River Formation was published in the Bulletin of the American Museum of Natural History in 1972. During his tenure at the University of Minnesota, he had the privilege of working with leading mammalian paleontologists Malcolm McKenna, William Clemens, and Jason Lillegraven.

After returning from Minnesota, Ashok Sahni joined the Faculty of the Department of Geology, Lucknow University as a Lecturer in 1968 and was promoted to Reader in 1976. Later, he joined as Professor of Paleobiology in the Centre of Advanced Study in Geology, Panjab University, Chandigarh in 1979 where he continues as an Emeritus Professor and Oil and Natural Gas Corporation (ONGC) Chair Professor after retirement in 2003.

With a small bunch of determined research students, Prof. Sahni, on his return from the USA in 1968, began his journey as a vertebrate paleontologist in India and tried to address some of the longstanding research problems in Indian paleontology employing vertebrate fossils and evidence from allied fields. He was instrumental in the establishment of a school of vertebrate paleontology in the country and as many as 25 students were awarded their Ph.D. under his guidance. Prof. Sahni has more than 250 research articles to his credit, 10 of which were published in prestigious journals like *Nature, Nature Communications*, and *Science*. Some of the important contributions Prof. Sahni and his students made to Indian paleontology include

- Cetacean fossils of the Paleogene strata of Kachchh, western India and Jammu and Kashmir in the Lesser Himalaya shed light on the oldest record of fossil whales and their transition from a terrestrial ancestor to obligate marine swimmer (Sahni and Mishra, 1972; Kumar and Sahni 1986; Bajpai et al. 1996; Thewissen et al. 1996; Roe et al. 1998);
- Diversity and evolution of post-India/Asia collision vertebrate faunas from the Middle Eocene Subathu Formation of northwestern Himalaya, their intercontinental dispersals, and the relevance of these faunas for dating the India/Asia collision event (Sahni and Khare 1972, 1975; Sahni and Kumar 1980; Sahni et al. 1981a, b; Kumar and Sahni 1985, 2000; Sahni and Jolly 1993; Kumar et al. 1997);
- Arvicolid rodents of the Karewa Group, which recorded paleoclimatic fluctuations during the Neogene-Quaternary transition in Kashmir valley (Sahni and Kotlia 1983, 1993; Kusumgar et al. 1986; Agarwal et al. 1989);
- Report of the first adapid fossils from the continental Siwalik deposits of the Sub-Himalayan hills in Himachal Pradesh and Uttarakhand (Gingerich and Sahni 1979, 1984); in recognition of his contribution as the first to identify *Sivaladapis* and *Indraloris* as adapids (Gingerich and Sahni 1979), a new adapid species from the Lower Siwalik deposits of Jammu and Kashmir, India was named in Prof. Sahni's honor (*Ramadapis sahnii* Gilbert et al., 2017);
- Pioneered the work on enamel ultrastructure of Paleogene and Neogene mammalian teeth in India for better understanding evolutionary relationships, biomechanical properties, and biomineralization processes (Sahni 1979, 1980, 1981, 1986, 1987; Sahni and Lester 1988; Kotlia and Sahni 1993; Kobayashi et al. 1993);
- Discovery of a rich mammalian fauna with more than 10 orders from the poorly understood Early Eocene time interval of the Cambay Shale of the Cambay Basin (Vastan Lignite Mine), western India in collaboration with R. S. Rana, Kishor Kumar (India), Thierry Smith (Belgium), and Kenneth D. Rose (USA). The global significance of this work includes some of the oldest occurrences of perissodactyls, anthropoids, bats, birds, and a variety of insectivorous mammals, as well as amber inclusions of insects (wasps, bees, ants, and termites) and plants (Sahni et al 2006; Rose et al. 2006, 2007, 2008, 2009a, 2012, 2014; Smith et al. 2006, 2007, 2016; Rust et al. 2010; Kumar et al. 2010; Folie et al. 2012; Dun et al. 2016);
- Collaboration in the early 1980s with French colleagues Jean-Jacques Jaeger, Jean-Louis Hartenberger, Monique Vianey-Liaud, Eric Buffetaut, Jean-Claude Rage, Mireille Gayet, Henri Cappetta, Vincent Courtillot, and Didier Vandamme on the Deccan volcanic province, and integration of vertebrate paleontology, geochronology, and paleomagnetism to understand the biodiversity and faunal changes during the Deccan volcanism (Sahni et al. 1982; Besse et al. 1986; Vianey-Liaud et al. 1987, 1997). The research has dramatically changed our perception on the age of initiation and duration of Deccan volcanism and its role in Cretaceous-Paleogene boundary mass extinction (Keller et al. 2009). This work has ultimately led to our current understanding that the Deccan Traps of peninsular India had erupted in the Late Cretaceous (Maastrichtian) and lasted over a narrow time span of 5 million years;

- Demonstration of lack of endemism during the northward drift of Indian plate, based on the fossil evidence from the sedimentary beds associated with the Deccan lava flows (Sahni 1984). Subsequent works have confirmed this view of lack of pronounced endemism in Late Cretaceous biotas of India during its northward journey;
- Discovery of the primitive eutherian mammals from the Cretaceous of India with his students and colleagues has thrown light on the evolution of basal eutherian mammals in the Gondwana (Prasad and Sahni 1988; Prasad et al. 1994, 2007a, b, 2010);
- Pioneered the research on dinosaur eggs and eggshells in India; the research was later carried on by his research students and geologists from the Geological Survey of India. Prof. Sahni and his team were able to record extensive dinosaur nesting sites spreading over 800 km in an east-west direction in the Narmada valley. He has integrated eggshell histology, sedimentology, isotope geochemistry, and taphonomy to interpret the nesting behavior, reproductive strategies, and pathological conditions in titanosaurid sauropod dinosaurs of India. The parataxonomic classification of Late Cretaceous dinosaur eggshells of India revealed the presence of at least nine oospecies representing three oogenera (Sahni et al. 1984; Srivastava et al. 1985; Vianey-Liaud et al. 1987, 1997; Sahni et al. 1994; Khosla and Sahni 1995);
- Investigations of the dietary habits of titanosaurid sauropod dinosaurs in collaboration with paleobotanists of the Birbal Sahni Institute of Palaeosciences from dinosaur coprolites using palynological proxies resulted in the discovery of one of the oldest records of grass phytoliths and shed light on the origin and evolution of the family Poaceae (Prasad et al 2005, 2011).

Besides his research contributions, Prof. Sahni is also keenly interested in the popularization of science, especially paleontology, among the common public. He participated in several interactive outreach programs for school and college students under the auspices of the Department of Science and Technology's (DST) program Innovation in Science Pursuit for Inspired Research (INSPIRE) or TV shows and interviews. At the request of the then Lieutenant Governor of Chandigarh administration, he helped the Government Museum in Sector 10, Chandigarh in establishing a Gallery on Dinosaurs of India in 2005 which has become a popular destination for children as well as adults interested in natural history and evolution. He is also a consultant to the Punjab Government's Science City at Kapurthala for the Evolution Park and Gallery. His book on the Dinosaurs of India (National Book Trust of India) is written in a lucid style and is the only popular book on dinosaurs of India.

In recognition of his contributions to Earth Science teaching and research in India, Prof. Sahni was conferred with many national and international honors and awards including the prestigious National Geoscience Life Time Achievement Award (Ministry of Mines, Govt. of India), Life Time Excellence Award (Ministry of Earth Sciences, Govt. of India), D. N. Wadia Medal (Indian National Science Academy), and Life Time Achievement Award (Electron Microscope Society of India). The three Indian Academies of Science and the Third World Academy of Sciences, Trieste, Italy elected him as Fellow of their respective academies in recognition of his outstanding research contributions to Indian paleontology and capacity building in the country through training of young researchers. He was also elected as Foreign Associate (Associé Étranger) of the Geological Society of France and Honorary Member of the Society of Vertebrate Paleontology.

We are greatly honored to have known him, worked under his tutelage and learned the fundamentals of vertebrate paleontological methods and concepts from him. We are equally privileged to be able to contribute to this volume that honors his scholarship. Though a subtle taskmaster, he is known for sharing his knowledge and affable guidance, and his generous ways of helping young students, colleagues, as well as amateurs.

Contents of the Volume

The 17 chapters comprising this volume in honor of Prof. Ashok Sahni were contributed by 50 leading paleontologists and geologists. Sixteen contributors are from India, three from Belgium, four each from France and the UK, ten each from the USA and Australia, and one each from Japan, Slovak Republic, and Argentina. The chapters were presented by his friends, colleagues, collaborators, and students. The title of the *book Biological Consequences of Plate Tectonics: New Perspectives on Post-Gondwana Break-up* aptly highlights Prof. Ashok Sahni's passion for research on the evolution of life and biogeographic links during the northward drift phase of Indian plate from Gondwana to its collision with Asia. Prof. Sahni was the first to demonstrate that the Late Cretaceous continental biota of India representing India's rapid drift phase is not distinctly endemic, but rather included both Gondwanan and Laurasian elements (Sahni 1984). This stirred up an intensive debate on the historical biogeography of India in later decades and a plethora of paleobiogeographic models have been proposed to explain the co-occurrence of both Gondwanan and Laurasian taxa (Jaeger et al. 1989; Briggs 1989, 2003; Prasad and Rage 1991, 1995; Thewissen and McKenna 1991; Krause et al. 1997; Prasad and Sahni 1999, 2009; Rage 2003; Chatterjee et al. 2013). Several chapters of the current volume track how plate tectonic events shaped the evolution of life with special reference to the break-up of Gondwana and cover the fossil record spanning pre-break-up, syn-drift, and post-collision phases. A brief summary of each chapter is presented below.

The first chapter by Prasad and Parmar (2020) describes new theropod and ornithischian dinosaur remains from the Middle Jurassic Kota Formation of India representing Indian plate's pre-drift phase. This work not only improves the diversity of continental Jurassic vertebrate fauna of India but also expands the poorly known dinosaur fossil record of Gondwana. Their analysis of biogeographic affinities of the Kota vertebrate fauna points to limited biogeographic links with other Jurassic faunas of Gondwana. The predominant occurrence of Laurasian taxa as compared to Gondwanan forms with close similarities to the Middle Jurassic vertebrate fauna of England underscores a close biogeographic link with Laurasia and restricted contacts with other Gondwanan continents.

In the following chapter, Mishra and Prasad (2020) deal with floral responses to changing plate configurations and eustatic sea level changes. According to their narrative, the Early Jurassic flora of India was cosmopolitan in nature, whereas post-Middle Jurassic flora was closely related to that of Australia and Antarctica than to that of Africa and South America and there was a progressive strengthening of provincialism concomitant with the disjunction of continental masses. It is further emphasized that by the Late Cretaceous with drifting apart of continents, the flora of India became more similar to that of Africa than to that of Australia and Antarctica.

In the recent past, several new discoveries of mammals with tribosphenic dentition have been made from the Jurassic and Cretaceous deposits of Gondwana (Rich et al. 1997, 2001; Flynn et al. 1999; Rauhut et al. 2002; Rougier et al. 2007). Because of their great age, these discoveries challenged the traditional view that mammals with tribosphenic molars originated in Laurasian continents in the Early Cretaceous (Lillegraven 1974). However, Luo et al. (2001) proposed a dual origin for tribosphenic mammals with forms endemic to Gondwana representing the infraclass Australosphenida which is survived by extant monotremes and Laurasian clade of placentals, marsupials, and their relatives included in the infraclass Boreosphenida. Rich et al. (1998, 2002) disputed the dual origin hypothesis and on the contrary suggested that the Early Cretaceous *Ausktribosphenos* and *Bishops* from Australia are unquestioned placentals and are not related to monotremes. In the third chapter, Rich and his co-authors (2020a) report a new genus and species of tribosphenic mammal, *Kryoparvus gerriti,* from the Lower Cretaceous Flat Rock locality in Victoria, Australia. *Kryoparvus gerriti* is the third tribosphenic mammal, added to already known *Ausktribosphenos nyktos* and

Bishops whitmorei, that represent Ausktribosphenidae, a Gondwanan clade of tribosphenic mammals. With an estimated body mass of 2 g, the Early Cretaceous *Kryoparvus gerriti* is one of the most diminutive mammals known so far and inhabited the frigid polar latitudes of southeastern Australia. In the following chapter, Rich and colleagues (2020b) describe an isolated upper premolar of a new genus and species of monotreme, *Stirtodon elizabethae,* from an Early Cretaceous locality at Lightning Ridge, New South Wales, Australia. *Stirtodon* further enhances the diversity of Mesozoic monotremes from Australia and supports the idea that Australia was the center of monotreme radiation. These new mammalian finds come from a time interval when Australia along with Antarctica and the Indo-Madagascar block was breaking away from the western Gondwana.

In Chapter 5, Prasad and her co-authors (2020) highlight the role of plate tectonics and climate in the origin and rapid radiation of angiosperms. They also emphasize that angiosperm radiation was accelerated as most arid regions had transformed into wetter climatic belts from the Barremian to the Cenomanian and during the Late Cretaceous distinct floral provinces emerged with the fragmentation of continents. India with its shifting latitudinal position from high to low latitudes hosted the constituents of all miofloral provinces and formed an important linkage between different floral provinces.

As early as in 1982, Prof. Sahni and colleagues described fossil remains of Pelobatidae, an amphibian family of Laurasian origin, from the Upper Cretaceous Deccan intertrappean beds of Central India and suggested that faunal dispersals were possible between India and Laurasia during the former's northward drift as a Noah's Ark (Sahni et al. 1982). This notion was further supported by the later discoveries of gobiatid frogs and anguimorph lizards from the Late Cretaceous of India (Prasad and Rage, 1991, 1995). In Chapter 6, Rage and his colleagues (2020) describe new fossil material of amphibians and squamates from new Deccan intertrappean beds of Central India and also review the fossil record of amphibians and squamates from the Late Cretaceous (Maastrichtian) of India. This study indicates that the Late Cretaceous amphibian diversity of India is at par with that of North America and Europe. On the other hand, the lizard taxa of Laurasia are more numerous than snake taxa in contrast to the snake taxa being more numerous than the lizard taxa in India and other Gondwanan landmasses. Further, it also demonstrated that the Late Cretaceous anuran and squamate fauna of India has mixed Laurasian and Gondwanan affinities pointing to some possible biogeographic links between India and other landmasses during the former's extended physical isolation.

Based on a close phylogenetic relationship between *Kinkonychelys rogersi* (Gaffney et al., 2009), a bothremydid turtle from the Late Cretaceous of Madagascar and Indian bothremydid turtles *Kurmademys* and *Sankuchemys* (tribe Kurmademydini), Gaffney et al. (2009) hypothesized that a biogeographic connection existed between India and Madagascar in the latest Cretaceous. In Chapter 7, Lapparent de Broin and Prasad (2020) describe new turtle remains from the Upper Cretaceous intertrappean beds of Upparhatti, southwestern part of the Deccan volcanic province and also review the taxonomic status of known Cretaceous turtle fossils from India. Until now, all the turtle records from the Cretaceous of India are known either by carapace or skull remains. The authors report for the first time postcranial bones in addition to skull and shell elements representing at least three new taxa (one long-necked and one or two short-necked), belonging to the family Bothremydidae. This study suggests that the Cretaceous turtle fauna from India is endemic in nature and does not support any connection between India and Madagascar in the Late Cretaceous, contrary to the view held by Gaffney et al. (2009).

A comprehensive review of the Mesozoic dinosaur fossil record of India is presented in Chapter 8 by Chatterjee (2020). He offers a historical perspective of dinosaur research in India with some interesting anecdotal accounts of field prospecting in peninsular India. He discussed the distribution of dinosaur fauna of India in space and time, recent discoveries from the Triassic and Jurassic formations, taxonomic revisions of the Cretaceous dinosaurs and how their intercontinental affinities changed with the break-up of Gondwana. A discussion on sauropod dinosaur nesting sites and dietary preferences of sauropod dinosaurs of India highlights the key role played by Prof. Ashok Sahni and his students in identifying the extensive hatcheries in Central India and in providing parataxonomic classification of Indian sauropod dinosaur eggshells. Chatterjee strongly argues that the Late Cretaceous Oman-Kohistan-Dras Island Arc biotic link facilitated the dispersal of dinosaurs between India and Africa-Europe at a time when India was physically isolated from other landmasses. He further advocates that the combined effects of twin asteroid impacts (Shiva impact off the coast of Bombay and Chicxulub crater of Mexico) and eruption of Deccan volcanism over an extended period were responsible for the extinction of dinosaurs and other organisms at the Cretaceous-Paleogene boundary.

Indian subcontinent has a unique geological history of being part of Gondwana for a major part of the Phanerozoic, passing through a period of isolation from the Cretaceous to Early Eocene, and finally colliding with Asia in the Middle Eocene. During its northward journey, the subcontinent remained isolated from other landmasses. As a consequence, it carried many relictual Gondwanan taxa, supported some endemic radiations, and received some immigrants following the collision with Asia. In his classic paper on the non-endemic nature of the biota, Sahni (1984) had demonstrated that Late Cretaceous biota of India exhibit mixed Gondwanan and Laurasian affinities. In Chapter 9, Halliday and others (2020) show how vertebrate faunas of Indian subcontinent changed from having close faunal relationships with Madagascar and South America in the Cretaceous to sharing more similarities with Eurasian fauna in the Eocene. This study reveals that mixed biogeographic affinities of Late Cretaceous faunas of India become lessened in the Eocene with the mixing of faunas from other continents and strengthening the signal for Laurasia-dominated faunas. It also highlights the importance of intercontinental migrations in the development of unique Late Cretaceous-Eocene faunas of the Indian subcontinent.

Africa remained geographically isolated for a considerable time interval ($\sim$90 million years) from the middle of the Cretaceous to the Early Miocene, leading to endemic evolution of terrestrial vertebrates. In Chapter 10, Rage and Gheerbrant (2020) discuss African isolation and its consequences for the evolution of low-diversity native faunas. They argue that intermittent terrestrial connections between Africa and Laurasia across Tethys allowed the immigration of taxa of Laurasian origin into Africa which subsequently evolved into distinct African clades. These immigrants gave rise to endemic adaptive radiations that repopulated Africa along with the native fauna and represent important components of the extant African fauna. According to the authors, island Africa served as a cradle of evolution for Eurasian stem groups and when Africa came in contact with Europe in the Early Miocene, the endemic African fauna did not suffer major extinctions, rather some of them dispersed out of Africa and became successful in their newly adapted landmasses.

In Chapter 11, Paul and Dutta (2020) used biomarkers to document the evolution of plants during the northward journey of Indian plate from the Early Cretaceous to Early Eocene. Their study revealed that the Lower Cretaceous lignite samples yielded sesquiterpenoids and diterpenoids of high abundance and lack angiosperm-dominated triterpenoids implying that in the Early Cretaceous conifer (Araucariaceae and Podocarpaceae) dominated gymnosperm forests were in existence in western India. In contrast, the Lower Eocene lignites of this region were dominated by triterpenoids of angiosperm derivation and completely lack diterpenoids. This is in conformity with earlier documentation of biomarkers diagnostic of tropical angiosperm family Dipterocarpaceae in the amber sampled from the Lower Eocene lignite deposits of Kachchh, western India.

We recall that when we were working in Prof. Ashok Sahni's lab for our Ph.D.s, he used to reiterate why Late Cretaceous and Paleocene–Early Eocene mammals of India are crucial for understanding the geographic origins of the Middle Eocene mammals from the Subathu Formation of the Lesser Himalaya. He was convinced that the Deccan intertrappean beds and the Eocene lignite mines of western India hold great promise for Cretaceous and Early Eocene mammals, respectively. His words proved to be prophetic and in 1988 as a result of focused exploration in the Deccan Volcanic Province, the first Cretaceous mammals of India were discovered in the intertrappean beds of Naskal, near Hyderabad (Prasad and Sahni 1988). Later, prospecting in the lignite mines of western India led to the identification of vertebrate yielding horizons intercalated with the lignite seams within the Cambay Shale, Vastan Lignite Mine in Surat District, Gujarat. In the past decade, the Lower Eocene Cambay Shale has remained a source of highly diversified vertebrate fauna comprising of fish, ranid and bombinatorid frogs, agamid lizards, terrestrial and aquatic snakes, birds, proteutherians, adapisoriculids, didelphomorphs, lagomorphs, rodents, tillodonts, artiodactyls, perissodactyls, bats, hyaenodontids, and primates (see references in Kapur 2020; Solé et al. 2020). This fauna is highly significant from the point of mammalian evolution and intercontinental dispersals at a time when India was still isolated but fast approaching the Asian mainland. Prof. Sahni has been closely associated with the study of the Vastan vertebrate fauna until recently. In Chapters 12 and 13, some recent results of the studies on the vertebrate fauna of Vastan Lignite Mine are presented. The phylogenetic position of Apatemyidae, an insectivorous group from North America and Europe, is not clear but recent works preferred the placement of these mammals in their own order. This mammalian group was considered to have originated in North America with wide geographic distribution in North America and Europe. Outside North America and Europe, apatemyid mammals are known only from the Lower Eocene Cambay Shale of Vastan Lignite Mine as *Frugivastodon cristatus* Bajpai et al., 2005. In Chapter 12, Solé and his co-workers (2020) present its phylogenetic analysis based on additional dental material from the Cambay Shale. According to the authors, similarities between *Frugivastodon* and *Aethomylos* of North America suggest that these two taxa represent a separate apatemyid clade which possibly originated in the late Paleocene or at about the Paleocene-Eocene boundary, though the place of origin remains problematic. In Chapter 13, Kapur (2020) discusses possible modes of dispersal for mammals of Vastan Lignite Mine. Based on a statistical analysis of size variation in the lower dentition of non-volant mammals, he has categorized them into small, medium, and large size groups. According to him, the small and medium-sized mammals may have dispersed across the shallow Tethys through chance dispersal or island hopping. On the other hand, the large-sized mammals such as perissodactyls with close evolutionary relationships with faunas from Eurasia may have used short-lived dispersal corridors between India and Eurasia prior to 54.5 Ma.

In the early Middle Eocene, as the Tethys Sea was shallowing and Indian subcontinent was on the verge of collision with Asian mainland, an important macroevolutionary event took place. This was the origination of whales from a terrestrial raoellid artiodactyl ancestor. The oldest fossil records of whales come from Pakistan and India, and *Indohyus*, considered as the sister taxon for cetaceans, is known from the lower Middle Eocene Subathu Formation of Kalakot area in Jammu and Kashmir. In fact, Prof. Sahni began his vertebrate paleontological research in India after returning from the USA with projects on the Eocene whales of Kachchh (Sahni and Mishra 1972, 1975) and the Middle Eocene mammals of the Subathu Formation (Sahni and Khare 1972, 1973) and continued to work on these vertebrates until recently. In Chapter 14, Thewissen, Nanda, and Bajpai (2020), describe new dental material of *Indohyus indirae* from the early Middle Eocene Sindkhatudi locality, Kalakot district, Jammu and Kashmir. The dental morphology of *Indohyus* points to the fact that it used its teeth for processing food differently from that of generalized artiodactyls, and in some ways more similar to that of Eocene cetaceans.

The Siwalik Group of the Indian subcontinent preserves a rich continental vertebrate fauna of post-India/Asia collision time interval representing the past 20 million years. Prof. Sahni also made significant contributions to our understanding of the evolution of the Siwalik vertebrate fauna, especially the rodent fauna and adapid primates, and functional adaptations in mammalian teeth in response to changing ecosystems (Gingerich and Sahni 1979, 1984; Sahni 1989; Patnaik and Sahni 1996, 2000; Cameron et al. 2000). Chapters 15 and 16 deal with the evolution and range expansion of the Siwalik rodents and how tectonics and climate affected these events. In Chapter 15, Flynn, Kimura, and Jacobs (2020) discuss the evolution of murine rodents in the Siwalik ecosystem following the docking of India with Asia. These authors stress that endemic evolution of continental vertebrates that prevailed during the isolation phase of the Indian plate during the Late Cretaceous and Early Paleogene is also reflected in Siwalik mammals in the Miocene, as mountain ranges had evolved to the north, large rivers and subtropical habitat to the east, and desert conditions to the west. The fossil record of murid rodents from the Siwalik Hills of Potwar Plateau captures the splitting event of extant tribes Murini and Arvicanthini, a speciation event that took place before 10.5 Ma and was probably underway prior to 11.2 Ma. In related work, Patnaik (2020, Chapter 16) revises the taxonomy of murid, rhizomyid, and ctenodactylid rodents ranging in age from Late Miocene to Early Pleistocene from the Siwalik deposits of India and also describes some new specimens from these deposits. He discusses how plate tectonics played a major role in the range expansion and contraction of early murid, rhizomyid, and ctenodactylid rodents. The range expansion of these rodent groups took place in the subtropical wet forests during the Early Miocene collision of Afro-Arabian and Eurasian plates. On the other hand, as arid climate and grasslands expanded in the Late Miocene, these rodent groups adapted to more open conditions and faunal dispersals could take place in both directions. In the final chapter, Gilbert, Pugh, and Fleagle (2020) discuss the dispersal of Miocene hominoids from Africa to Eurasia and the role of tectonics and climate in these dispersals. According to them, the Early Miocene ($\sim$22–19 Ma) land bridge between Europe and Africa via the Arabian Peninsula enabled dispersal of pliopithecoids and later hominoids such as *Griphopithecus* and *Kenyapithecus* into Europe and Anatolia. In Asia, *Sivapithecus*, *Khoratpithecus*, and *Ankarapithecus* appeared in the Middle Miocene, whereas other hypothesized pongines emerged in the Late Miocene through Pleistocene. According to these authors, the pongine clade did not evolve in Asia, rather it was a Middle Miocene immigrant. They attribute the Late Miocene ($\sim$9-8 Ma) extinction of hominoids in several parts of Asia to the climatic changes associated with the uplift of Himalaya and Tibet and suggest a similar extinction of apes in Europe as a result of climatic changes linked to changing oceanic currents.

Delhi, India — Guntupalli V. R. Prasad
Chandigarh, India — Rajeev Patnaik

References (those listed in Prof. Ashok Sahni's bibliography below are not included here)

Bajpai, S., Kapur, V. V., Das, D. P., Tiwari, B. N., Saravanan, N., & Sharma, R. (2005). Early Eocene land mammals from Vastan Lignite Mine, district Surat (Gujarat), western India. *Journal of the Palaeontological Society of India*, *50*(1), 101–113.

Briggs, J. C. (1989). The historic biogeography of India: Isolation or contact? *Systematic Zoology*, *38*, 322–332.

Briggs, J. C. (2003). The biogeographic and tectonic history of India. *Journal of Biogeography*, *30*(3), 381–388.

Chatterjee, S. (2020). The age of dinosaurs in the land of Gonds. In G. V. R. Prasad & R. Patnaik (Eds.), *Biological consequences of plate tectonics: New perspectives on post-Gondwana break-up* (pp. 181–226). Cham: Springer.

Chatterjee, S., Goswami, A., & Scotese, C. R. (2013). The longest voyage: Tectonic, magmatic, and paleoclimatic evolution of the Indian plate during its northward flight from Gondwana to Asia. *Gondwana Research*, *23*(1), 238–267.

Flynn, L. J., Kimura, Y., & Jacobs, L. L. (2020). The murine cradle. In G. V. R. Prasad & R. Patnaik (Eds.), *Biological consequences of plate tectonics: New perspectives on post-Gondwana break-up* (pp. 347–362). Cham: Springer.

Flynn, J. J., Parrish, J. M., Rakotosamimanana, B., Simpson, W. F., & Wyss, A. R. (1999). A Middle Jurassic mammal from Madagascar. *Nature*, *401*, 57–60.

Gaffney, E. S., Krause, D. W., & Zalmout, I. S. (2009). *Kinkonychelys*, a New Side-Necked Turtle (Pelomedusoides: Bothremydidae) from the Late Cretaceous of Madagascar. *American Museum Novitates, 3662*, 1–25.

Gilbert, C. C., Patel, B. A., Singh, N. P., Campisano, C. J., Fleagle, J. G., et al. (2017). New sivaladapid primate from Lower Siwalik deposits surrounding Ramnagar (Jammu and Kashmir State), India. *Journal of Human Evolution, 102*, 21–41.

Gilbert, C. C., Pugh, K. D., & Fleagle, J. G. (2020). Dispersal of Miocene hominoids (and pliopithecoids) from Africa to Eurasia in light of changing tectonics and climate. In G. V. R. Prasad & R. Patnaik (Eds.), *Biological consequences of plate tectonics: New perspectives on post-Gondwana break-up* (pp. 393-412). Cham: Springer.

Halliday, T. J. D., Holroyd, P. A., Gheerbrant, E., Prasad, G. V. R., Scanferla, A., Beck, R. M. D., et al. (2020). Leaving Gondwana: The changing position of the Indian subcontinent in the global faunal network. In G. V. R. Prasad & R. Patnaik (Eds.), *Biological consequences of plate tectonics: New perspectives on post-Gondwana break-up* (pp. 227–249). Cham: Springer.

Jaeger, J.-J., Courtillot, V., & Tapponnier, P. (1989). Paleontological view of the ages of the Deccan Traps, the Cretaceous/Tertiary boundary, and the India-Asia collision. *Geology, 17*, 316–319.

Kapur, V. (2020). Size variation amongst the non-volant mammals from the Early Eocene Cambay Shale deposits of western India: Paleobiogeographic implications. In G. V. R. Prasad & R. Patnaik (Eds.), *Biological consequences of plate tectonics: New perspectives on post-Gondwana break-up* (pp. 305–335). Cham: Springer.

Lapparent de Broin, F. de, & Prasad, G. V. R. (2020). Chelonian Pelomedusoides remains from the Late Cretaceous of Upparhatti (Southwestern India): Systematics and paleobiogeographical implications. In G. V. R. Prasad & R. Patnaik (Eds.), *Biological consequences of plate tectonics: New perspectives on post-Gondwana break-up* (pp. 123–180). Cham: Springer.

Lillegraven, J. A. (1974). Biogeographical considerations of the marsupial-placental dichotomy. *Annual Review of Ecology and Systematics, 5*, 263–283.

Luo, Z.-X., Cifelli, R. L., & Kielan-Jaworowska, Z. (2001). Dual origin of tribosphenic mammals. *Nature, 409*, 53–57.

Mishra, S., & Prasad, V. (2020). Upper Gondwana (Jurassic to Early Cretaceous) palynoflora of India: Its correlation with other Gondwana continents and phytogeographical implications. In G. V. R. Prasad & R. Patnaik (Eds.), *Biological consequences of plate tectonics: New perspectives on post-Gondwana break-up* (pp. 31–66). Cham: Springer.

Patnaik, R. (2020). New data on the Siwalik murines, rhizomyines and ctenodactylines (Rodentia) from the Indian subcontinent. In G. V. R. Prasad & R. Patnaik (Eds.), *Biological consequences of plate tectonics: New perspectives on post-Gondwana break-up* (pp. 363–391). Cham: Springer.

Paul, S., & Dutta, S. (2020). Contrasting biomarker signatures of Cretaceous and Early Paleogene coal-bearing sediments during the northward flight of India. In G. V. R. Prasad & R. Patnaik (Eds.), *Biological consequences of plate tectonics: New perspectives on post-Gondwana break-up* (pp. 265–278). Cham: Springer.

Prasad, G. V. R., & Parmar, V. (2020). First ornithischian and theropod dinosaur teeth from the Middle Jurassic Kota Formation of India: Paleobiogeographic relationships. In G. V. R. Prasad & R. Patnaik (Eds.), *Biological consequences of plate tectonics: New perspectives on post-Gondwana break-up* (pp. 1–30). Cham: Springer.

Prasad, G. V. R., & Rage, J.-C. (1991). A discoglossid frog in the latest Cretaceous (Maastrichtian) of India. Further evidence for a terrestrial route between India and Laurasia in the latest Cretaceous. *Comptes Rendus de l'Académie des Sciences (série II), 313*, 273–278.

Prasad, G. V. R., & Rage, J. C. (1995). Amphibians and squamates from the Maastrichtian of Naskal, India. *Cretaceous Research, 16*, 95–107.

Prasad, V., Mishra, S., Mishra, A. K., & Bansal, M. (2020). Role of plate tectonics and global climate change in the evolution of angiosperms. In G. V. R. Prasad & R. Patnaik (Eds.), *Biological consequences of plate tectonics: New perspectives on post- Gondwana break-up* (pp. 83–98). Cham: Springer.

Rage, J. C. (2003). Relationships of the Malagasy fauna during the Late Cretaceous: Northern or Southern routes? *Acta Palaeontologica Polonica, 48*, 661–662.

Rage, J.-C., & Gheerbrant, E. (2020). Island Africa and vertebrate evolution: A review of data and working hypotheses. In G. V. R. Prasad & R. Patnaik (Eds.), *Biological consequences of plate tectonics: New perspectives on post-Gondwana break-up* (pp. 251–264). Cham: Springer.

Rage, J.-C., Prasad, G. V. R., Verma, O., Khosla, A., & Parmar, V. (2020). Anuran lissamphibian and squamate reptiles from the Upper Cretaceous (Maastrichtian) Deccan intertrappean sites in Central India, with a review of a lissamphibian and squamate diversity in the northward drifting Indian plate. In G. V. R. Prasad & R. Patnaik (Eds.), *Biological consequences of plate tectonics: New perspectives on post-Gondwana break-up* (pp. 99–121). Cham: Springer.

Rauhut, O. W. M., Martin, T., Ortiz-Jaureguizar, E., & Puerta, P. (2002). A Jurassic mammal from South America. *Nature, 416*, 165–168.

Rich, T. H., Flannery, T. F., Trusler, P., Kool, L., van Klaveren, N. A., & Vickers-Rich, P. (2001). A second tribosphenic mammal from the Mesozoic of Australia. *Records of the Queen Victoria Museum, 110*, 1–9.

Rich, T. H., Flannery, T. F., Trusler, P., Kool, L., van Klaveren, N. A., & Vickers-Rich, P. (2002). Evidence that monotremes and ausktrobosphenids are not sister groups. *Journal of Vertebrate Paleontology, 22*(2), 466–469.

Rich, T. H., Flannery, N. A., & Vickers-Rich, P. (1998). Alleged Cretaceous placental from down under: Reply. *Lethaia, 31*, 346–348.

Rich, T. H., Flannery, T. F., & Vickers-Rich, P. (2020b). Evidence for a remarkably large toothed-monotreme from the Early Cretaceous of Lightning Ridge, NSW, Australia. In G. V. R. Prasad & R. Patnaik (Eds.), *Biological consequences of plate tectonics: New perspectives on post-Gondwana break-up* (pp. 77–81). Cham: Springer.

Rich, T. H., Trusler, P., Kool, L., Pickering, D., Evans, A., Siu, K., et al. (2020a). A third, remarkably small, tribosphenic mammal from the Mesozoic of Australia. In G. V. R. Prasad & R. Patnaik (Eds.), *Biological consequences of plate tectonics: New perspectives on post-Gondwana break-up* (pp. 67–75). Cham: Springer.

Rich, T. H., Vickers-Rich, P., Constantine, A., Flannery, T. F., Kool, L., & van Klaveren, N. (1997). A tribosphenic mammal from the Mesozoic of Australia. *Science, 278*, 1438–1442.

Rougier, G., Martinelli, A. G., Foraisepi, A. M., & Novacek, M. J. (2007). New Jurassic mammals from Patagonia, Argentina: A reappraisal of australosphenidan morphology and interrelationships. *American Museum Novitates, 3566*, 1–54.

Solé, F., De Bast, E., Legendre, H., Rana, R. S., Kumar, K., Rose, K. D., et al. (2020). New specimens of *Frugivastodon* (Mammalia: Apatotheria) from the Early Eocene of India confirm its apatemyid status and elucidate dispersal of Apatemyidae. In G. V. R. Prasad & R. Patnaik (Eds.), *Biological consequences of plate tectonics: New perspectives on post-Gondwana break-up* (pp. 279–304). Cham: Springer.

Thewissen, J. G. M., & McKenna, M. C. (1992). Paleobiogeography of Indo-Pakistan: A response to Briggs, Patterson, and Owen. *Systematic Biology, 41*, 248–251.

Thewissen, J. G. M., Nanda, A. C., & Bajpai, S. (2020). *Indohyus*, endemic radiation of raoellid artiodactyls in the Eocene of India and Pakistan. In G. V. R. Prasad & R. Patnaik (Eds.), *Biological consequences of plate tectonics: New perspectives on post- Gondwana break-up* (pp. 337–346). Cham: Springer.

Acknowledgments

We express our sincere thanks to Federico Agnolin, Racquel Lopez Antoñanzas, Alexander O. Averianov, Sunil Bajpai, Paul M. Barrett, Kenneth Christopher Beard, Achim Bechtel, John Briggs, David Button, Isaac Casanova, William A. Clemens, Sitindra Dirhangi, Susan E. Evans, Lawrence J. Flynn, Peter M. Galton, Adán Pérez Garcia, James D. Gardner, Philip D. Gingerich, Marc Godinot, Annette E Götz, Terry Harrison, Simon Hills, Ren Hirayama, Jeremy Hooker, David W. Krause, Nao Kusuhashi, Thomas Martin, Ralph E. Molnar, Gustavo R. Oliveira, Maeva Orliac, Khum Paudyal, Mercedes Beatriz Pramparo, Li Quang, Todd C. Rae, Thomas H. Rich, Zbyněk Roček, Samir Sarkar, Vijay Sathe, Anju Saxena, Sevket Sen, Dhurjati Sengupta, Mary T. Silcox, Salvador Moyà-Solà, Ksenjia Stojanovic, Jessica M. Theodor, J. G. M. Thewissen, Takehisa Tsubamoto, Paul Upchurch, Michael O. Woodburne, and Anna Maria Zavattieri for reviewing the chapters of this book. Special thanks are due to William Clemens, Kenneth Christopher Beard, and Paul Upchurch for reviewing multiple chapters. We express our deep sense of gratitude to Prof. William Clemens for writing the foreword to this book. We are highly thankful to the Series Editors, Eric Delson and Eric Sargis, for responding promptly and patiently to our innumerable queries and guiding us from time to time in completing all editorial tasks effectively. Thanks are due to Emmanuel Gheerbrant, Thomas H. Rich, and J. G. M. Thewissen for allowing us to use some of their figures for the cover page. We acknowledge the help from Lourembam Ranjit Singh and Harsha Dhiman in formatting the chapters and figures.

Bibliography of Ashok Sahni

1968

1. Sahni, A. A review of Cretaceous mammals. *Journal of the Palaeontological Society of India, 12,* 21–24.

2. Sahni, A. Community structure of Campanian (Upper Cretaceous) vertebrates from the Judith River Formation in north-central Montana. *Proceedings of International Geological Congress, 23rd session, Prague, Czechoslovakia,* 329–334.

1969

3. Sahni, A. Techniques in prospecting for terrestrial microvertebrates. *Journal of the Palaeontological Society of India, 13,* 38–43.

1970

4. Sahni, A. *A laboratory manual of practical geology.* Luckow: Pustak Kendra Publication.

1971

5. Kapoor, H. M., & Sahni, A. *Ctenacanthus ishii* sp. nov., a shark from the Zewan Series of Guryul Ravine, Kashmir. *Memoirs of the Faculty of Science, 38*(1), 163–166.

1972

6. Sahni, A., & Mishra, V. P. A new species of *Protocetus* (Cetacea) from the Middle Eocene of Kutch, western India. *Palaeontology, 15*(3), 490–495.

7. Sahni, A., & Choudhary, N. K. Lower Eocene fishes from Barmer District, Rajasthan. *Proceeding of the Indian National Science Academy, 38*(A), 97–102.

8. Sahni, A., & Khare, S. K. Three new mammals from the Eocene of Kalakot, District Rajauri, Jammu and Kashmir. *Journal of the Palaeontological Society of India, 16,* 41–53.

9. Sahni, A. Palaeoecology of Lameta Formation at Jabalpur (M.P.). *Current Science, 47*(17), 652.

10. Sahni, A. The vertebrate fauna of the Judith River Formation. *Bulletin of the American Museum of Natural History, 147*(6), 323–412.

1973

11. Misra, R. C., Sahni, A., & Chhabra, N. L. Triassic conodonts and associated fish remains from the Niti Pass region, Kumaun, Himalaya. *Himalayan Geology, 3*, 48–161.

12. Sahni, A., & Khare, S. K. Additional mammals from the Eocene Subathu Formation, Jammu and Kashmir. *Journal of the Palaeontological Society of India, 17,* 31–49.

13. Sahni, A., Chhabra, N. L., & Kumar, S. Middle Triassic conodonts from Malla Johar, Kumaun Himalaya. *Current Science, 42*(22), 806–807.

14. Sahni, A., & Prakash, I. Rhaetic conodonts and associated fish remains from the Niti Pass region, Kumaun Himalayas. *Current Science, 42*(6), 218.

1974

15. Sahni, A., & Chhabra, N. L. Middle Triassic conodonts from Kalapani Limestone, Malla Johar, Kumaun Himalaya. *Himalayan Geology, 4,* 259–295.

16. Sahni, A., & Chhabra, N. L. A record of Triassic ostracodes from Kashmir Himalaya. *Current Science, 43*(10), 325.

17. Sahni, A., & Kumar, V. Palaeogene palaeobiogeography of the Indian subcontinent. *Palaeogeography, Palaeoclimatology, Palaeoecology, 15*(3), 209–226.

18. Sahni, A., Kumar, V., & Srivastava, V. C. *Dryopithecus* (Subgenus *Sivapithecus*) and associated vertebrates from the Lower Siwaliks of Uttar Pradesh. *Bulletin of the Indian. Geologists Association, 7,* 54.

19. Sahni, A., & Mehrotra, D. K. Turonian terrestrial communities of India. *Geophytology, 4* (l), 102–105.

1975

20. Sahni, A., & Mishra, V. P. Lower Tertiary vertebrates from western India, *Monograph of the Palaeontological Society of India, 3,* 1–48.

21. Sahni, A., & Khare, S. K. Siwalik insectivora. *Journal of the Geological Society of India, 17,* 114–116.

22. Nautiyal, A. C., & Sahni, A. Lower Triassic argillaceous sequence of Pahalgam (Kashmir Himalaya) and paleoecological observations. *Proceeding of the Indian National Science Academy, 42*(4), 364–371.

1976

23. Sahni, A., & Chhabra, N. L. Microfish remains from certain Triassic sections in the Kashmir and Kumaun Himalayas. Proceedings of VIth Indian Colloquim on Micropalaeontology and Stratigraphy, Varanasi, 218–224.

24. Sahni, A., & Srivastava, V. C. Eocene rodents and associated reptiles from the Subathu Formation, Jammu and Kashmir. *Journal of Paleontology, 50*(5), 922–928.

1977

25. Sahni, A., & Srivastava, H. C. Eocene rodents of India: Their paleogeographic significance. *Géobios Special Memoir, 1,* 87–95.

26. Sahni, A., & Khare, S. K. A Middle Siwalik fish fauna from Ladhyani (Haritalyangar), Himachal Pradesh. *Biological Memoirs, Vertebrate Paleontology Series, 2*(1&2), 187–221.

27. Sahni, A., & Choudhary, N. K. A new Eocene louvar from Barmer, south-western Rajasthan, India. *Journal of the Palaeontological Society of India, 20,* 391–395.

1978

28. Sahni, A., Srivastava, M. C., & Souza, R. D. Eocene ziphodont Crocodilia from northwestern India. *Géobios, 11*(5), 779–785.

1979

29. Sahni, A. Enamel ultrastructure of certain North American Cretaceous mammals. *Palaeontographica, 166,* 37–49.

30. Sahni, A. An Eocene mammal from the Subathu/Dagshai transition zone, Dharampur, Simla Himalaya. *Bulletin of the Indian Geologists Association, 12*(2), 259–262.

31. Sahni, A., & Tewari, B. N. Lower Siwalik vertebrates from Kalagarh, District Pauri-Garhwal, U.P. *Himalayan Geology, 9,* 339–347.

32. Chhabra, N. L., Kumar, S., & Sahni, A. A record of holothurian sclerites from the Triassic of Malla Johar, Kumaun Himalaya. *Current Science, 48,* 1038–1039.

33. Gingerich, P., & Sahni, A. *Indraloris* and *Sivaladapis*, Miocene adapid primates from the Siwaliks of India and Pakistan. *Nature, 279*(5712), 415–416.

34. Sahni, A. Miocene vertebrates from the coastal Tertiaries of peninsular India. *Geological Survey of India Miscellaneous Publication, 45,* 197–205.

1980

35. Sahni, A., & Kumar, K. Lower Eocene mammal, *Ishatherium subathuensis* gen. et. sp.nov. from the type area, Subathu Formation, Subathu, Simla Himalaya, H.P. *Journal of the Palaeontological Society of India, 23 & 24, 132*–135.

36. Sahni, A., & Mitra, H. C. Neogene paleobiogeography of the Indian subcontinent with special reference to vertebrates. *Palaeogeography, Palaeoclimatology, Palaeoecology, 31,* 39–62.

37. Sahni, A. SEM studies of Indian Eocene and Siwalik rodent enamels. *Geosciences Journal, 1*(2), 21–30.

38. Agarwal, P. N., Singh, S. N., & Sahni, A. Sycthian ostracodes from Kashmir Himalayas. *Journal of the Palaeontological Society India, 23 & 24,* 110–114.

39. Sahni, A., Tewari, B. N., & Kumar, K. A report on the occurrence of *Ramapithecus punjabicus* (Hominoidea) from the Uttar Pradesh Siwaliks. *Himalayan Geology, 11,* 193–197.

40. Sahni, A., & Mitra, H. C. Lower Miocene (Aquitanian-Burdigalian) paleobiogeography of the Indian subcontinent. *Geologische Rundschau, 69*(3), 824–848.

41. Sahni, A., Tewari, B. N., & Kumar. K. Lower Eocene marine mammal (Sirenia) from Dharampur, Simla Himalayas, H.P. *Bulletin of the Indian Geologists Association,* 7, 271–272.

42. Budurov, K. G., Sahni, A., Gupta, V. J., Kachroo, R. K., Chhabra, N. L., Ahluwalia, A., & Azmi, R. J. On the validity of the name *Neospathodus gondolelloides* (Bender). *Bulletin of the Indian Geologists Association, 13*(2), 145–146.

43. Sahni, A., Tewari, B. N., & Kumar, K. Additional Lower Siwalik vertebrates from Kalagarh District, Pauri-Garhwal, U.P. Proceeding of the Indian Geological Congress, Pune, 81–90.

1981

44. Sahni, A. Enamel ultrastructure of fossil Mammalia: Eocene archaeocete from Kutch. *Journal of the Palaeontological Society of India, 25,* 33–37.

45. Chhabra, N. L., & Sahni. A. Late Lower Triassic and early Middle Triassic conodont faunas from Kashmir and Kumaun Himalayas. *Journal of the Palaeontological Society India, 25,* 135–147.

46. Uppal, S., Sahni, A., & Gupta, V. J. New Permian fish locality from the northern Pir Panjal flank, Kashmir. *Bulletin of the Indian Geologists Association, 14*(2), 155–158.

47. Sahni, A., Bhatia, S. B., Jaeger, J.-J., Hartenberger, J.-L., Kumar, K., Sudre, J., et al. Vertebrates from the type section of the Subathu Formation and comments on the paleobiogeography of the Indian subcontinent during the Early Palaeogene. *Bulletin of the Indian Geologists Association, 14*(2), 89–100.

48. Gupta, V. J., & Sahni, A. *Theropithecus delsoni*, a new cercopithecine species from the upper Siwaliks of India. *Bulletin of the Indian Geologists Association, 14*(l), 69–71.

49. Sahni, A., Bhatia, S. B., Hartenberger, J.-L., Jaeger, J.-J., Kumar, K., Sudre, J., et al. Vertebrates from the Subathu Formation and comments on the biogeography of the Indian subcontinent during the Early Paleogene. *Bulletin de la Société géologique de France, 23*(6), 689–695.

50. Sahni, A., & Mehrotra, D. K. The elasmobranch fauna of coastal Miocene sediments of peninsular India. *Biological Memoirs, 5*(2), 83–121.

51. Sahni, A. Timing of fragmentation and collision of the Indian plate. Proceeding of the Himalayan Geology Seminar, Ministry of Steel and Mines, New Delhi, Sect. 2 (Tectonics), pp. 235–242.

52. Sahni, A. Dental and skeletal microstructure and ultrastructure of Indian Tertiary vertebrates. Proceeding of IX Indian Colloquium on Micropalaeontology & Stratigraphy, Udaipur, pp. 109–122.

53. Sahni, A. A report on a small collection of Karewa vertebrates investigated under the Kashmir paleoclimate project. *Man and Environment, 5,* 87.

1982

54. Sahni, A. Karewa Vertebrates; Biostratigraphy, palaeohistology and palaeoecology. *Man and Environment, 6,* 16–20.

55. Sahni, A. The structure, sedimentation and evolution of Indian continental margins. Indian Ocean. In A. E. M. Nairn, & F. G. Stehli (Eds), *The ocean basins and margins: The Indian Ocean* (pp 353–398). New York: Springer.

56. Kotlia, B. S., Sahni, A., Aggarwal, D. P., & Pant, R. K. New vertebrate evidence for the age of the Karewa sediments, Kashmir. *Man and Environment, 6,* 13–15.

57. Uppal, S., Saigal, S. K., Sahni, A., & Gupta, V. J. Lower Carboniferous plant fossils from near Banihal, District Doda, Jammu and Kashmir, *Contributions to Himalayan Geology, 2,* 83–86.

58. Sahni, A., Kumar, K., & Rana, R. S. On the occurrence of a Pleistocene ossiferous gravel at Nagpur, Central India. *Current Science, 51*(12), 610.

59. Sahni, A. Upper Cretaceous paleogeography of peninsular India and the Cretaceous-Paleocene transition - the vertebrate evidence. Proceeding of the Indian Association Palynostratigraphers, 128–140.

60. Sahni, A., & Kotlia, B. S. Karewa microvertebrates: Biostratigraphical and paleoecological implications. Proceeding International Workshop on Late Cenozoic Palaeoclimates, PRL, Ahmedabad, Current Trends in Geology 6 (29–43). Delhi: Today and Tomorrow.

61. Sahni, A., Batra, R. S., & Bhatia, S. B. Vertebrate assemblage from the Upper Subathu (Middle Eocene) of the Bilaspur area, Himachal Pradesh, India. Proceeding of Xth Indian Colloquim on Micropalaeontology & Stratigraphy, pp. 357–368.

62. Sahni, A., & Gupta, V. J. *Deinotherium indicum* from the Middle Siwaliks of the Kangra Foothills, Himachal Pradesh, India. *Recent Researches in Geology, 8,* 351–355.

63. Sahni, A., & Saxena, R. K. Middle Eocene otoliths from Jhadwa, northwestern Kutch. *Journal of the Palaeontological Society of India, 27,* 64–67.

64. Sahni, A., Kumar, K., Hartenberger, J.-L., Jaeger, J.-J., Rage, J.-C., Sudre, J., et al. Microvertébrés nouveaux des Trapps du Deccan (Inde): Mise en évidence d'une voie de communication terrestre probable entre la Laurasie et l'Inde à la limite Crétacé-Tertiaire. *Bulletin de la Société géologique de France*, *24*(5–6), 1093–1099.

65. Sahni, A. The present status of Hora's Satpura Hypothesis. In K. S. Valdiya, V. K. Gaur &. S. B. Bhatia (Eds.). *Geology of Vindhyachal* (pp. 143–147). Delhi: Hindustan Publcation Corporation.

66. Sahni, A., & Gupta, V. J. Cretaceous eggshell fragments from the Lameta Formation of India. *Bulletin of the Indian Geologists Association, 15*(l), 85–86.

1983

67. Sahni, A., and Kotlia, B. S. Micromammals from the Karewa Group of Kashmir. *Man and Environment, 7,* 157–158.

68. Sahni, A., Bhatia, S. B., & Kumar, K. Faunal evidence for the withdrawal of the Tethys in the Lesser Himalaya, northwestern India. *Bollettino Della Societa Paleontologica Italiana, 22* (1&2), 77–86.

69. Uppal, S., Sahni, A., & Gupta, V. J. First record of fossil ophiuroids from the Upper Triassic rocks of Kashmir Himalaya. *Contributions to Himalayan Geology, 2,* 299–302.

70. Sahni, A., Nagpaul, K. K., Prasad, A. K., Suri, P. S., Sharma, P. K., & Bal, K. D. Behaviour of Uranyl-ion during the fossilization process of vertebrate in exogenic cycle. Proceeding of IIIrd National Seminar on Solid State Nuclear Track Detector, Physics Dept. GND University, Amritsar, pp. 89–94.

1984

71. Gingerich, P. D., & Sahni, A. Dentition of *Sivaladapis nagri* (Adapidae) from the Late Miocene of India. *International Journal of Primatology, 5*(1), 63–79.

72. Sahni, A., Srikantia, S. V., Ganesan, T. H., & Wangdus, C. Early Tertiary fishes and molluscs from the Kuksho Formation of the Indus Group near Nyoma, Ladakh. *Journal of the Geological Society of India, 25*(11), 742–747.

73. Sahni, A., Rana, R. S., Kumar K., & Loyal, R. S. New stratigraphic nomenclature for the intertrappan beds of Nagpur. *Geosciences Journal, 5*(l), 55–58.

74. Sahni, A., Rana, R. S., & Prasad, G. V. R. SEM studies of thin egg shell fragments, from the intertrappeans (Cretaceous-Tertiary transition) at Nagpur and Asifabad, Peninsular India. *Journal of the Palaeontological Society of India, 29,* 26–33.

75. Sahni, A. Upper Cretaceous-Early Paleogene paleobiogeography of India based on terrestrial vertebrate faunas. *Mémoires de la Société géologique de France*, 147, 125–138.

76. Sahni, A. Cretaceous-Paleocene terrestrial faunas of India: Lack of faunal evidence during drifting. *Science, 226,* 441–443.

1985

77. Kumar, K., & Sahni, A. Eocene mammals from the Upper Subathu Group, Kashmir Himalaya, India. *Journal of Vertebrate Paleontology, 5*(2), 153–168.

78. Jain, S. L., & Sahni, A. Dinosaurian eggshell fragments from the Lameta Formation of Pisdura, Chandrapur District, Maharashtra. *Geosciences Journal, 6*(2), 211–220.

79. Srivastava, S., Mohabey, D. M., Sahni, A., & Pant, S. C. Upper Cretaceous dinosaur egg clutches from Kheda District, Gujarat, India: Their distribution, shell ultrastructure and paleoecology. *Palaeontographica A, 193,* 219–233.

80. Sahni, A., Rana R. S., & Prasad, G. V. R. Late Cretaceous - Lowermost Paleocene coastal faunas from peninsular India. *Proceedings of the Sixth International Gondwana Symposium, American Geophysical Union, Geophysical Monograph, 41,* 207–214.

81. Sahni, A. Cretaceous-Palaeocene microfossil assemblages from infra and intertrappeans of peninsular India. Proceeding of XIth Indian Colloquium on Micropalaeontology and Stratigraphy, Calcutta, 91–100.

82. Sahni, A., & Kakar, R. K. The Krol-Tal Field Meeting. Editors Note. *Bulletin of the Indian Geologists Association, 18*(l), FM I–Fm 5.

83. Sahni, A. *Mammalian Paleofaunas of the World.* Book Review. *Bulletin of the Indian Geologists Association, 18*(2), 145.

1986

84. Kumar, K., & Sahni, A. *Remingtonocetus harudiensis*, new combination, a Middle Eocene archaeocete (Mammalia, Cetacea) from western Kutch, India. *Journal of Vertebrate Paleontology, 6*(4), 326–349.

85. Sahni, A. Application of electron microscopy to problems of palaeontology and biomineralization. Presidential Address of XIIth Indian Colloquium on Micropalaeontology and Stratigraphy, University of Delhi, pp. 1–28.

86. Sahni, A. Enamel structure of early mammals and its role in evolutionary relationships among rodents. In W. P. Luckett & J.-L. Hartenberger. (Eds), *Evolutionary Relationships among Rodents.* NATO Advanced Science Institutes (ASI) Series (Series A: Life Sciences), 92 (pp.133–150). Boston: Springer.

87. Sahni, A. Antecedants of early man in northwestern India: Paleontological and paleoecological evidences. Book Review *Journal of the Geological Society of India, 28*(l), 63.

88. Sahni, A., Prasad, G. V. R., & Rana, R. S. New paleontological evidence for the age and initiation of the Deccan volcanics, central peninsular India. *Gondwana Geological Magazine*, l, 13–24.

89. Sahni, A., Gupta, V. J., Prakash, B., & Kotlia. B. S. Fossil Bird egg shell fragments from the Karewas of Kashmir valley (J & K) India: A scanning electron microscope study. *Journal of the Bombay Natural History Society, 83*(3), 623–630.

90. Prasad, G. V. R., Sahni A., & Gupta, V. J. Fossil assemblage from the infra and intertrappean beds of Asifabad, Andhra Pradesh and their geological implications. *Geosciences Journal, 7*(2), 623–630.

91. Kusumgar, S., Kotlia, B. S., Agarwal, D. P., & Sahni, A. Biochronologie de fossiles de vertébrés de formations de Karwas du Cachemire, Inde. *L'Anthropologie, 90*(2), 151–164.

92. Besse, J., Buffetaut, E., Capetta, H., Courtillot, V., Jaeger, J.-J., Montigny, R., et al. The Deccan Traps (India) and the Cretaceous-Tertiary boundary events. In O. H. Walliser, (Ed.), *Global bioevents: A critical approach. Lecture Notes in Earth Sciences* (pp 365–370). Berlin: Springer.

1987

93. Sahni, A. 1987. Evolutionary Aspects of reptilian and mammalian enamel structure. *Scanning Microscopy International, 1*(14), 1903–1912.

94. Vianey-Liaud, M., Jain, S. L., & Sahni. A. Dinosaur assemblage (Saurischia) from the Late Cretaceous Intertrappean and Lameta Formations (Deccan, India). *Journal. of Vertebrate Paleontology*, *7*(4), 408–424.

95. Brookfield, M., & Sahni, A. Paleoenvironments of the Lameta beds (Late Cretaceous) at Jabalpur, Madhya Pradesh, soils and biotas of a semi-arid alluvial plain. *Cretaceous Research, 8*(1), 1–14.

1988

96. Sahni, A., Kaul, S., Singh, M. P., & Sharma, S. K. Further evidence of hominoid remains from the Pinjor Formation. *Proceeding of the Indian National Science Academy*, *54A*(3), 553–562.

97. Sahni, A., & Bajpai, S. Cretaceous - Tertiary boundary events: The fossil vertebrate, paleomagnetic and radiometric evidence from peninsular India. *Journal of the Geological Society of India, 32*(5), 382–396.

98. Sahni, A., & Lester, K. S. The nature and significance of enamel tubules in multituberculates and marsupials. In J.-P. Santoro, D. Sigogneau-Russell & D. E. Russell (Eds.), *Teeth Revisited.* Proceeding of the VIIth Symposium on Dental Morphology. *Mémoires du Muséum d'Histoire naturelle (Ser.C), 53,* 85–99.

99. Prasad, G. V. R., & Sahni. A. First Cretaceous mammal from India. *Nature, 332*(6164), 638–640.

100. Sahni, A. Cretaceous-Tertiary boundary events: Mass extinctions, iridium enrichment and Deccan volcanism. *Current Science, 57*(10), 513–519.

101. Sahni, A., Jaeger, J.-J., Rana, R. S., & Prasad, G. V. R. Takli Formation and coeval Deccan intertrappeans. H. K. Maheshwari (Ed.), Proceedings of the Symposium on Paleocene of India: Limits and Subdivisions (pp. 28–33). Lucknow: BSIP.

102. Sahni, A. Application of electron microscopy to problems in paleontology and biomineralisation. In P. Kalia (Ed.), *Micropaleontology of shelf sequences of India* (pp. 1–11). Delhi: Papyrus Publication House.

103. Sahni, A. Cretaceous - Tertiary boundary events: The fossil vertebrate paleomagnetic and radiometric evidence from Peninsular India. In National Seminar on Stratigraphic Boundary Problems in India (Key Paper preprint, University of Jammu), 36–57.

1989

104. Sahni, A. Eocene rodents of South Asia: Origins, dispersals, and paleoecology. In C. C. Black and M. Dawson (Eds.), *Papers on fossil rodents in honor of Albert Elmer Wood* (pp. 147–156). Los Angeles: Los Angeles County Museum.

105. Sahni, A. Sivalik adapid primates: Taxonomy, distribution and enamel ultrastructure. In A. Sahni & R. Gaur, R.(Eds.), *Perspectives in human evolution* (pp. 173–193). New Dehi: Renaissance Publishing House.

106. Aggarwal, D. P., Dodia, R., Kotlia, B. S., Razdan, H., & Sahni, A. The Plio-Pleistocene geologic and climatic record of the Kashmir Valley, India: A review and new data. *Palaeogeography, Palaeoclimatology Palaeoecology, 73,* 267–286.

107. Sahni, A. Paleoecology and paleoenvironments of Late Cretaceous dinosaur eggshell sites from peninsular India. In D. D. Gillette & M. G. Lockley (Eds.), *Dinosaur tracks and traces* (pp. 179–185). New York: Cambridge University Press.

108. Sahni, A., & Gaur, R. *Perspectives in Human Evolution*. Delhi: Renaissance Publishing House.

109. Sahni. A., Kumar, G., Bajpai, S., & Srinivasan, S. Ultrastructure and taxonomy of ostrich eggshells from Upper Paleolithic sites of India. *Journal of the Palaeontological Society of India, 34,* 91–98.

110. Rana, R. S., & Sahni, A. Fish otoliths from the Takli Formation (intertrappean beds) of Nagpur, India. *Geosciences Journal, 10*(1–2), 145–164.

1990

111. Prakash, T., Singh, R. Y., & Sahni, A. Palynofloral assemblage from Padwar Deccan intertrappean (Jabalpur) M.P. In A. Sahni, & A. Jolly (Eds.), *Cretaceous event stratigraphy and correlation of the Indian non-marine strata*. Contribution to Seminar cum Workshop IGCP 216 and 245 (pp. 68–69). Chandigarh.

112. Sahni, A., & Jolly, A. Cretaceous event stratigraphy and the correlation of the Indian nonmarine strata. Contribution to Seminar cum Workshop IGCP 216 & 245 (pp. 1–125), Chandigarh.

113. Sahni, A., & Tripathi, A. Age implication of the Jabalpur Lameta Formation and intertrappean biotas. In A. Sahni & A. Jolly (Eds.), *Cretaceous event stratigraphy and correlation of the Indian non-marine strata*. Contribution to Seminar cum Workshop IGCP 216 and 245 (pp. 35–37). Chandigarh.

114. Sahni, A. Recent advances and shortcomings in understanding non-marine Cretaceous basins of India (Conveners Report). In A. Sahni & A. Jolly (Eds.), *Cretaceous event stratigraphy and correlation of the Indian non-marine strata*. Contribution to Seminar cum Workshop IGCP 216 and 245 (pp. 18–21). Chandigarh.

115. Bajpai, S., Sahni A., Jolly, A., & Srinivasan, S. Kachchh intertrappean biotas: Affinities and correlation. In A. Sahni & A. Jolly (Eds.), *Cretaceous event stratigraphy and correlation of the Indian non-marine strata*. Contribution to Seminar cum Workshop IGCP 216 and 245 (pp. 101–105). Chandigarh.

116. Sahni, A. Deccan volcanism as the terminal Cretaceous catastrophic event: An analysis. In A. Sahni & A. Jolly (Eds.), *Cretaceous event stratigraphy and correlation of the Indian non-marine strata*. Contribution to Seminar cum Workshop IGCP 216 and 245 (pp. 110–111). Chandigarh.

117. Nanda, A. C., & Sahni, A. Oligocene vertebrates from the Ladakh Molasse Group, Ladakh Himalayas. *Himalayan Geology, 21,* 1–10.

118. Flynn, L. J., Sahni, A., Jaeger, J.-J., Singh, B., & Bhatia, S. B. Additional fossil rodents from the Siwalik beds of India. *Proceedings of the Koninklijke Nedelandse Akademie van Wetenschappen, 93*(l), 7–20.

119. Kumar, G., Sahni, A., Pancholi, R. K., & Narvare, G. Archaeological discoveries and a study of Late Pleistocene ostrich eggshells and egg shell objects in India. *Man and Environment, 16*(1), 29–40.

120. Sahni, A., Kumar, G., Bajpai, S., & Srinivasan, S. A review of Late Pleistocene ostriches (*Struthio* sp.) in India. *Man and Environment, 15*(1), 41–47.

1991

121. Sahni, A., & Bajpai, S. Eurasiatic elements in the Upper Cretaceous nonmarine biotas of Peninsular India. *Cretaceous Research, 12,* 177–183.

122. Sahni, A. Uncle Birbal – Reminiscences of a nephew. *Current Science, 61*(9/10), 570–571.

123. Sahni, A. The Deccan Traps. In extinct plants, evolution and Earth's history. *Current Science, 61*(9/10), 654–656.

124. Sahni, A. Faunal assemblage, correlation and paleontology of Late Cretaceous nonmarine sequences of peninsular India. In N. A. Mateer & C. Peiji (Eds.), Proceeding of Ist International Symposium IGCP 245 (pp.148–163). Beijing: Ocean Press.

125. Sahni, A. Marginal and deep-sea sedimentation of the India subcontinent: Basinal evolution in space and time. In S. K. Tandon, C. Pant & S. M. Casshyap (Eds.), *Sedimentary Basins in India: Tectonic context* (pp. 131–146). Nainital: Gyanodaya Prakashan.

1992

126. Bajpai, S., Sahni, A., & Srinivasan, S. Ornithoid eggshells from Deccan intertrappean beds near Anjar (Kachch), Western India. *Current Science, 64*(1), 42–45.

127. Srinivasan, S., Sahni, A., & Bajpai, S. Fossil charophyta from the Deccan intertrappean beds of Gurmatkal, Gulbarga district, Karnataka. *Current Science, 63*(7), 396–398.

128. Prithiraj, C., Mishra, P. K., & Sahni, A. Late and terminal Cretaceous foraminiferal assemblages from Ukhrul Melange Zone, Manipur. *Current Science, 62*(6), 478–481.

1993

129. Sahni, A., & Jolly, A. Eocene mammals from Kalakot, Kashmir Himalaya: Community structure, taphonomy and paleobiogeographical Implications, *Darmstadt Beitrage zur Naturgechichte, 3,* 200–222.

130. Patnaik, R., Bahadur, M., Sharma, T., & Sahni, A. A comparative analysis of the molars of *Mus booduga, Mus dunni* and fossil *Mus* of the Indian Subcontinent: Phylogenetic and paleobiogeographic implications. *Current Science, 65*(10), 782–786.

131. Kobayashi, I., Mutvei, H., & Sahni, A. Fossil hard tissues: Structure., formation and evolution. 29th IGC symposium volume, Kyoto. Tokyo: Tokai University Press.

132. Kotlia, B. S., & Sahni, A. Enamel ultrastructure of fossil and extant arvicolids. In I. Kobayashi, H. Mutvei & A. Sahni (Eds.), *Fossil hard tissues: Structure, formation and evolution* (pp.159–172). Tokyo: Tokai University Press

133. Sahni, A., & Kotlia, B.S. Upper Pliocene-Quaternary vertebrate communities of the Karewas and Siwaliks. *Current Science, 64*(11/12), 893–898.

1994

134. Prasad, G. V. R., Jaeger, J.-J., Sahni, A., Gheerbrant, E., & Khajuria. C. K. Eutherian mammals from the Upper Cretaceous (Maastrichtian) intertrappean beds of Naskal. Andhra Pradesh, India. *Journal of Vertebrate Paleontology, 14*(2), 260–277.

135. Sahni, A., & Bardhan, S. Palaeontology. In A. Mookherjee (Ed.), *A profile of geoscience research in India* (pp. 32–37). New Delhi: Diamond Jubilee Publication of Indian National Science Academy.

136. Sahni, A., Tandon, S. K., Jolly, A., Bajpai, S., Sood, A., & Srinivasan, S. Upper Cretaceous dinosaur eggs and nesting sites from the Deccan volcano-sedimentary province of Peninsular India. In K. Carpenter, K. F. Hirsch, & J. R. Horner (Eds.), *Dinosaur eggs and Babies* (pp. 204–228). Cambridge: Cambridge University Press.

137. Patnaik, R., & Sahni, A. Record of a bird humerus from Upper Pleistocene Narmada Valley sediments. *Journal of the Palaeontological Society of India, 39,* 77–79.

138. Sahni, A. Cretaceous Basins of India. *Chiyoda Japan, 16*(5), 290–294.

139. Sahni, A., & Khosla, A. A Maastrichtian ostracod assemblage (Lameta Formation) from Jabalpur Cantonment, Madhya Pradesh, India. *Current Science, 67*(6), 456–460.

140. Sahni, A., & Khosla, A. Paleobiological, taphonomical and paleoenvironmental aspects of Indian Cretaceous sauropod nesting sites. *GAIA, 8,* 1–8.

1995

141. Sakae, T., Mishima, H., Suzuki, K., Kozawa, Y., & Sahni, A. Crystallographic and chemical analyses of eggshell of dinosaur (sauropod, titanosaurids). *Journal of Fossil Research, 27,* 50–54.

142. Khosla, A., & Sahni, A. Parataxonomic classification of Late Cretaceous dinosaur eggshells from India. *Journal of the Palaeontological Society of India, 40,* 87–102.

143. Sahni, A., & Jolly, A. Status of taphonomical studies in India: A review. *The Palaeobotanist, 44,* 29–37.

144. Sahni, A., & Kakar, R. K. Paleoenvironment during the Indian Cretaceous: Geodynamic evolution of contrasting basinal tectonic settings. *Proceedings of the 15th International Symposium Kyungpook National University*, Daegu, South Korea, pp.131–147.

1996

145. Sahni, A., Venkatachala, B. S., Kar. R. K., Rajanikanth, A., Prakash, T., Prasad, G. V. R., et al. New palynological data from the intertrappean beds: Implications for the latest record of dinosaurs and synchronous initiation of volcanic activity in India. *Memoir of the Geological Society of India, 67,* 267–203.

146. Patnaik, R., Auffray J.-C., Jaegar, J.-J., & Sahni, A. House mouse ancestor from Late Pliocene Siwalik sediments of India. *Comptes Rendus de l Académie des Sciences Paris, Science de la vie/Life Science, 319,* 431–434.

147. Patnaik, R., & Sahni, A. Siwalik rodent biostratigraphy: implications for intra-subcontinental correlation. In J. Pandey, R. J. Azmi, A. Bhandari & A. Dave (Eds.), Contribution to XV Indian Colloquium on Micropalaeontology and Stratigraphy (pp. 509–512). Dehradun

148. Sahni, A., & Sahni. N. Onset of biomineralization: New insights using scanning electron microscopy and other modern techniques. *Indian Minerals, 50*(1&2), 9–18.

149. Bajpai, S., Thewissen, J. G. M., & Sahni, A. *Indocetus* (Cetacea, Mammalia) endocasts from Kachchh (India). *Journal of Vertebrate Paleontology, 16*(3), 582–584.

150. Loyal, R. S., Khosla, A., & Sahni, A. Gondwana dinosaurs of India: Affinities and paleobiogeography. *Memoirs of the Queensland Museum, 39*(3), 627–638.

151. Thewissen, J. G. M., Roe, L. J., O'Neil, J. R., Hussain, S. T., Sahni, A., & Bajpai. S. Evolution of cetacean osmoregulation. *Nature, 301,* 379–380.

152. Sahni, A. (Ed.) *Cretaceous stratigraphy and paleoenvironments. Geological Society of India, Memoir, 37.*

1997

153. Sahni, A., & von Koenigswald, W. The enamel structure of some fossil and recent whales from the Indian subcontinent. In W. von Koenigswald & P. M. Sander (Eds.), *Tooth enamel microstructure* (pp. 177–191). Rotterdam: Balkema.

154. Sahni, A. Paleoenviromnents of Late Cretaceous dinosaur nesting sites of India. In D. L. Wolberg, E. Stump & G. D. Rosenberg (Eds.), *Proceeding of Symposium Dinofest International*, Tucson. Special Publication, Academy Natural Sciences, Philadelphia, pp. 291–300.

155. Sahni, A. Indian dinosaurs. In P. J. Currie & K. Padian (Eds.), *Encyclopedia of dinosaurs* (pp.361–368). Cambridge: Academic Press

156. Bajpai, S., Srinivasan, S., & Sahni, A. Fossil turtle eggshells from intertrappean beds of Duddukuru, Andhra Pradesh. *Journal of the Geological Society of India, 49,* 209–213.

157. Krause, D. W., Prasad, G. V. R., von Koenigswald, W., Sahni, A., & Grine, F. G. Cosmopolitanism amongst Gondwana Late Cretaceous mammals. *Nature, 390,* 504–507.

158. Kumar, K., Srivastava, R., & Sahni, A. Middle Eocene Rodents from the Subathu Group, Northwest Himalaya. *Palaeovertebrata, 26*(1–4), 83–128.

159. Vianey-Liaud, M., Hirsch, K. F., Sahni, A., & Sigé, B. Late Cretaceous Peruvian eggshells and their relationships with Laurasian and Eastern Gondwanian material. *Géobios, 30*(1), 75–90.

1998

160. Bajpai, S., Sahni, A., & Schleich, H. H. Late Cretaceous gekkonid eggshells from the Deccan intertrappeans of Kutch (India). *Veroffentlichungen aus dem Fuhlrott-Museum, 4,* 301–306.

161. Singh, S. D., Sahni, A., Gaffney, A., & Schleich, H. H. Reptilia from the intertrappean beds of Bombay (India). *Veroffentlichungen aus dem Fuhlrott-Museum, 4,* 307–320.

162. Roe, L. J., Thewissen, J. G. M., Quade, J., O'Neil, J. R., Bajpai, S., Sahni, A., et al. Isotopic approaches to understanding the terrestrial-to-marine transition of the earliest cetaceans. In J. G. M Thewissen (Ed.), *The emergence of whales* (pp. 399–422). New York: Plenum Press.

163. Sahni, A. Nursing scientific talent in India: Changing perspectives. In brain drain, *Indian National Science Academy,* pp. 23–27.

164. Kar, R. K., Sahni, A., Ambwani, K., & Singh, R. S. Palynology of Indian onshore-offshore Maastrichtian sequences in India: Implication for correlation and paleobiogeography. *Indian Journal of Petroleum Geology, 7*(2), 39–49.

165. Nanda, A. C., & Sahni, A. Ctenodactyloid rodent assemblage from Kargil Formation, Ladakh Molasse Group: Age and paleogeographic implications for the Indian subcontinent in the Oligo-Miocene. *Géobios, 31*(4), 533–544.

166. Sahni, A., & Kad, S. Evidence of the tectonic events in the foreland basin: phase relationships with Himalayan and Tibetan uplifts. *Himalayan Geology, 19*(2), 49–53.

167. Kar, R. K., Sahni, A., Ambwani, K., & Sharma, P. A new fossiliferous intertrappean locality at Amajiri, Chhindwara District, Madhya Pradesh. *Geophytology, 27*(1&2), 115–117.

168. Loyal, R. S., Mohabey, D. M., Khosla, A., & Sahni, A. Status and paleobiology of the Late Cretaceous Indian theropods with description of a new theropod eggshell oospecies, *Ellipsoolithus khedaensis*, from the Lameta Formation (District Kheda, Gujarat), Western India. *GAIA, 15,* 1–9.

1999

169. Sahni, A., Khosla, A., & Sahni, N. Fossil seeds from the Lameta Formation (Late Cretaceous), Jabalpur, India. *Journal of the Palaeontological Society of India, 44,* 15–23.

170. Sahni, A. India-Asia collision: Ecosystem changes. J. B. Auden Lecture. Wadia Institute of Himalayan Geology, Dehradun, pp. 1–27.

171. Prasad, G. V. R., & Sahni, A. Were there size constraints on biotic exchanges during the northward drift of the Indian plate?. In A. Sahni & R. S. Loyal (Eds.), *Special Issue on "Gondwana Assembly: New Issues and Perspectives"* (pp. 169–188). New Delhi: Proceedings of the Indian National Science Academy.

172. Sahni, A. & R. S. Loyal (Eds.). Special issue on Gondwana assembly: New issues and perspectives. *Proceedings of the Indian National Science Academy, 65A*(3), 1–229.

2000

173. Khosla, A., & Sahni A. Late Cretaceous (Maastrichtian) ostracodes from the Lameta Formation, Jabalpur cantonment area, Madhya Pradesh, India. *Journal of the Palaeontological Society of India, 45,* 57–78

174. Patnaik, R., & Sahni, A. Paleobiogeography of Pliocene Siwalik murid and gerbillid rodents. *Geological Survey of India Miscellaneous Publication, 64,* 43–47.

175. Sahni, A. Late Cretaceous dinosaur nesting sites of India: Stratigraphical, paleoenvironmental and paleoecological controls. A. M. Bravo & T. Reyes (Eds.), First International Symposium of Dinosaur Eggs and Embryos (pp. 161–165). Imprenta Provincial: Lleida, Isona, Spain

176. Cameron, D., Patnaik, R., & Sahni, A. News and views *Sivapithecus* dental specimens from Dhara locality, Kalagarh Distict, Uttar Pradesh, Siwaliks, Northern India. *Journal of Human Evolution, 37*(6), 861–868.

177. Kumar, K., & Sahni, A. Himalayan Foreland Basin with special reference to pre-Siwalik Tertiaries. *Special Volume Himalayan Geology, 21*(1&2), 1–214.

2001

178. Ambwani, K., Kar, R. K., & Sahni, A. Re-investigation on *Sahnipushpam shukla* from the Deccan intertrappean sediments of Madhya Pradesh. *Ameghiniana, 38*(4),393–398.

179. Patnaik, R., Sahni, A., & Prasad, G. V. R. Tooth Enamel microstructure of a Late Cretaceous Gondwanatherian mammal from India. *Journal of the Palaeontological Society of India, 46,* 15–23.

2002

180. Sahni, A. Deccan volcano-sedimentary sequences: Their status in hydrocarbon exploration, in Keynote Addresses, Proceedings of Conference on Strategic Challenges and Paradigm Shift in Hydrocarbon Exploration with Special Reference to Frontier Basins, 1, 153.

2003

181. Khosla. A., & Sahni, A. Biodiversity during the Deccan volcanic eruptive episode. *Journal of Asian Earth Sciences, 21, 895*–908.

182. Sahni ,A., Patnaik, R., & Alimohamadian, H. Tephra stratigraphy in India with reference to regional Quaternary volcanic activity. Proceeding of the IV South Asia Geological Congress (GEOSAS IV), Geological Survey of India, pp. 1–5.

183. Ashraf, A. R., & Sahni, A. Cretaceous environments of Afghanistan: A synthesis based on selected sections. *Journal of Geoscientific Research in Northeast Asia, 6*(1), 1–13.

184. Gaffney, E. S., Sahni, A., Schleich, H., Singh, S. D., & Srivastava, R. *Sankuchemys*, a new side-necked turtle (Pelomedusoides: Bothremydidae) from the Late Cretaceous of India. *American Museum Novitates, 3405,* 1–10.

185. Sahni, A. Indian dinosaurs revisited. *Current Science, 85*(7), 101–107.

186. Khosla, A., Kapur, V. V., Sereno, P. C., Wilson, J. A., Dutheil, D., Sahni, A., et al. First dinosaur remains from the Cenomanian-Turonian of the Nimar Sandstone (Bagh Beds), District Dhar, Madhya Pradesh, India. *Journal of the Palaeontological Society of India, 48,* 115–127.

187. Wilson, J. A., Sereno, P. C., Srivastava, S., Bhatt, D. K., Khosla, A., & Sahni, A. A new abelisaurid (Dinosauria, Theropoda) from the Lameta Formation, (Cretaceous, Maastrichtian) of India. *Contribution from the Museum of Paleontoloy, University of Michigan, 31*(1), 1–42.

188. Ambwani, K., Sahni, A., Kar, R. K., & Dutta, D. Oldest known non-marine diatoms (*Aulacoseira*) from the Deccan intertrappean beds and Lameta Formation (Upper Cretaceous of India). *Revue de Micropaléontologie, 46*, 67–71.

189. Ghosh, P., Bhattacharya, S. K., Sahni, A., Kar, R. K., Mohabey, D. M., & Ambwani, K. Dinosaur coprolites from the Late Cretaceous (Maastrichtian) Lameta Formation of India: Isotopic and other markers suggesting a C3 plant diet. *Cretaceous Research, 24,* 743–750.

190. Kar, R. K., Ambwani, K., Sahni, A., & Sharma, P. Unisexual flowers from the Deccan intertrappean bed of Madhya Pradesh, India. *The Palaeobotanist, 52,* 73–79.

2004

191. Sahni, A, Rana, R. S., Loyal, R. S, Saraswati, P. K., Mathur, S. K., Rose, K. D., et al. Western margin Paleocene-Lower Eocene lignites: Biostratigraphic and paleoecological constraints. Proceeding II Association Petroleum Geology Conference, Khajuraho, 1–18

192. Weishampel, D. B., Barret, P. M., Coria, R. A., Loeuff, J. L., Xing, Z., Sahni, A., et al. Dinosaur distribution and biology. In D. B. Weishampel, P. Dodson & H. Osmolska (Eds.), *The Dinosauria* (pp. 517–607). California: University of California Press.

193. Patnaik, R., Cameron, D., & Sahni, A. The phylogenetic significance of the Middle Pleistocene Narmada hominin cranium from central India. *Osteoarchaeology, 14,* 419–447.

194. Arya, R., Ambwani, K., Sahni, N., & Sahni, A. First mammal and additional fossil flowers from the Kasauli Formation, Kasauli, Himachal Pradesh. *Journal of the Geological Society of India, 64,* 317–324.

195. Khosla, A., Prasad, G. V. R., Verma, O., Jain, A. K., & Sahni, A. Discovery of a micromammal-yielding Deccan intertrappean site near Kisalpuri, Dindori District, Madhya Pradesh. *Current Science, 87*(3), 380–383

196. Sahni, A. The boyhood of Birbal Sahni - taking to science in 19th Century Panjab. *Resonance, 9*(4), 42–49.

197. Cameron, D., Patnaik, R., & Sahni, A. The systematic status of the Middle Pleistocene Narmada hominin from central India with comments regarding the paleoecology of the locality. *International Journal of Osteoarchaeology, 14*(6), 419–447.

2005

198. Prasad, V., Strömberg, C. A. E., Alimohammadian, H., & Sahni, A. Dinosaur coprolites shed light on the early evolution of grasses and grazers. *Science, 310,* 1177–1180.

199. Rana, R. S., Singh, H., Sahni, A., Rose, K. D., & Saraswati, P. K. Early Eocene chiropterans from the Vastan lignites, Gujarat, western peninsular margin: Oldest record of bats from Asia. *Journal of the Palaeontological Society of India, 50*(1), 93–100

200. Srivastava, R., Sahni, A., Jafar, S. A., & Mishra, S. 2005. Microstructure-dictated resistance properties of some Indian dinosaur eggshells: Finite element modeling. *Paleobiology, 31*(2), 317–325.

201. Pillans, B., Williams, M., Cameron, D., Patnaik, R., Hogarth, J., & Sahni, A. Revised correlation of the Haritalyangar magnetostratigraphy, Indian Siwaliks: Implications for the age of the Miocene hominids *Indopithecus* and *Sivapithecus*, with a note on a new hominid tooth. *Journal of Human Evolution, 48,* 507–515.

202. Kar, R. K., Sahni, A., Ambwani, K., & Dutta, D. Fossil flora (*Costatheca* and *Spermatites*) from the Upper Maastrichtian Deccan intertrappean beds of India. *Geologica Carpathica, 56*(2), 149–154.

203. Alimohammadian, H., Sahni, A., Patnaik, R., Rana, R. S., & Singh, H. First Record of an exceptionally diverse and well preserved amber-embedded biota from the Lower Eocene (~52Ma) lignites, Vastan, Gujarat. *Current Science, 89*(8), 1328–1330.

2006

204. Sahni, A., Saraswati, P. K., Rana, R. S., Kumar, K., Singh, H., Alimohammadian, H., et al. Temporal constraints and depositional paleoenvironments of the Vastan Lignite sequence, Gujarat: Analogy for the Cambay Shale hydrocarbon source rock. *Indian Journal of the Petroleum Geology, 15*(1), 1–20.

205. Bajpai, S., Thewissen, J. G. M., Kapur, V. V., Tiwari, B. N., & Sahni, A. Eocene and Oligocene sirenians (Mammalia) from Kachchh, India. *Journal of Vertebrate Paleontology, 26* (2), 400–410.

206. Rose, K. D., Smith, T., Rana, R. S., Sahni, A., & Singh, H. First Early Eocene (Ypresian) continental mammal assemblage form India, with description of a new anthracobunid (Mammalia, Tethytheria). *Journal of Vertebrate Paleontology, 26*(1), 1–7.

207. Sahni, A. Biotic response to the India-Asia collision: Changing paleoenvironments and vertebrate faunal relationships. *Palaeontographica, abt. A. 278*(1), 15–26.

208. Rana, R. S., Kumar, K., Loyal, R. S., Sahni, A., Rose, K. D., Mussell, J., et al. from the Early Eocene Kapurdi Formation (Fullers Earth), Barmer District, Rajasthan, India. *Journal of the Geological Society of India, 67,* 509–522.

209. Smith, T., Rana, R. S., Rose, K. D., & Sahni, A. Earliest bats from India. *Society of Vertebrate Paleontology, Abstract. 127A.*

210. Sahni, A. The future of paleontology and basic science research: An armchair view. Reminisces, Lucknow: BSIP, Diamond Jubilee Volume.

2007

211. Prasad, G. V. R., Verma, O., Sahni, A., Parmar, V., & Khosla, A. A Cretaceous hoofed mammal from India. *Science, 318,* 937.

212. Mayr, G., Rana, R. S., Sahni, A., & Smith, T. Oldest fossil avian remains from the Indian plate. *Current Science, 92*(9), 1266–1269.

213. Sahni, A. The drift of India and its collision with Asia: Biotic response and evolutionary implications. *Bulletin Indian Geological Congress, 1*(2), 123–130

214. Smith, T., Rana, R. S., Missiaen, P., Rose, K. D., Sahni, A., Singh, H., & Singh, L. High bat diversity in the Early Eocene of India. *Naturwissenschaften, 94,* 1003–1009.

215. Prasad, G. V. R., Verma, O., Sahni, A., Krause, D. W., Khosla, A., & Parmar, V. A new Late Cretaceous Gondwanatherian Mammal from Central India. *Proceedings of the Indian National Science Academy, 73*(1), 17–24.

216. Rose, K. D., Rana, R. S., Sahni, A., & Smith, T. A new adapoid primate from the Early Eocene of India. *Contribution from the Museum Paleontology, University of Michigan, 31* (14), 379–385.

2008

217. Rose, K. D., DeLeon, V. B., Missiaen, P., Rana, R. S., Sahni, A., Singh, L., et al., Early Eocene lagomorph (mammalian) from western India and the early diversification of Lagomorpha. *Proceedings of the Royal Society London B, 275,* 1203–1208.

218. Sahni, A. Paleontology: In Search of an identity in the 21st Century. *The Palaeobotanist, 57,* 1–5.

219. Sahni, A., & Prasad, G. V. R. Geodynamic evolution of the Indian plate: Consequences for the dispersal and distribution of biota. *Golden Jubilee Memoir of the Geological Society of India, 66,* 203–225.

220. Rana, R. S., Kumar, K., Escarguel, G., Sahni, A., Rose, K. D., Smith, T., et al., Ailuravine rodents (Mammalia) from the Lower Eocene lignites of India: Paleobiogeographic implications. *Acta Palaeontologica Polonica, 53*(1), 1–14.

221. Garcia, G., Khosla, A., Jafar, S. A., Sahni, A., & Vianey-Liaud, M. Eggshell Microstructure and porosity of the Nicobar scrubfowl (*Megapodius nicobariensis,* Great Nicobar Island, India). *Palaeovertebrata, 36*(1–4), 75–88.

2009

222. Patnaik, R., Chauhan, P. R., Rao, M. R., Blackwell, B. A. B., Skinner, A. R., Sahni, A., et al. New geochronological, paleoclimatological, and archaeological data from the Narmada Valley hominin locality, central India. *Journal of Human Evolution, 56,* 114–133.

223. Rose, K. D., Rana, R. S., Sahni, A., Kumar, K., Missiaen P., Singh, L., et al. Early Eocene primates from Gujarat, India. *Journal of Human Evolution, 56,* 366–404.

224. Keller, G., Adatte, T., Bajpai, S., Mohabey, D. M., Widdowson, M., Khosla, A., et al. K-T transition in Deccan Traps of central India marks major marine seaway across India. *Earth and Planetary Science Letters, 282,* 10–23.

225. Keller, G., Sahni, A., & Bajpai, S. Deccan Volcanism, the K-T mass extinction and dinosaurs, *Journal of Biosciences, 34*(5), 709–728.

226. Bajpai, S., Thewissen, J. G. M., & Sahni, A. The origin and early evolution of whales: Macroevolution documented on the Indian subcontinent. *Journal of Biosciences, 34*(5), 673–686.

227. Rose, K. D., Rana, R. S., Sahni, A., Kumar, K., Singh, L., & Smith, T. First tillodont from India: Additional evidence for an early Eocene faunal connection between Europe and India?. *Acta Palaeontologica Polonica, 54*(2), 351–355.

228. Patnaik, R., Sahni, A., Cameron , D., Pillans, B., Chatrath, P., Simons, E., et al. Ostrich-like eggshells from a 10.1 million year old Miocene ape locality, Haritalyangar (Himachal Pradesh) India. *Current Science, 96*(11), 1485–1495.

2010

229. Sahni, A. Dinosaur nesting sites of India: A review. *Science and Culture, 81*(5–6), 123–127.

230. Bera, M. K., Bhattacharya, K., Sarkar, A., Samanta, A., Kumar, K., & Sahni, A. Oxygen isotope analysis of bone and tooth enamel phosphate from Palaeogene sediments: Experimental techniques and initial results. *Journal of the Geological Society of India, 76,* 275–282.

231. Sahni, A. Planetary scale events and evolution: The Indian raft. *Journal of Biosciences, 35*(4), 1–8

232. Chungkham, P., & Sahni, A. Biostratigraphy of the exotics, Ukhrul melange zone (Manipur): Unravelling the closure of the eastern Tethys. *Memoirs of the Geological Society of India, 75,* 115–122.

233. Rust, J., Singh, H., Rana, R. S., McCann, T., Singh, L., Anderson, K., et al. Biogeographic and evolutionary implications of a diverse paleobiota in amber from the early Eocene of India. *Proceeding of the National Academy of Sciences, 107*(43), 1–6.

234. Prasad, G. V. R., Verma, O., Gheerbrant, E., Goswami, A., Khosla, A., Parmar, V., et al. First mammal evidence from the Late Cretaceous of India for biotic dispersal between India and Africa at the K-T transition. *Comptes Rendus Palevol, 9,* 63–71.

235. Archibald, J. D., Clemens, W. A., Padian, K., Rowe, T., MacLeod, N., Barrett, P. M. A., et al. Multiple causes, not a single cause for terminal Cretaceous extinctions. *Science, 328,* 973.

236. Mayr, G., Rana, R. S., Rose, K. D., Sahni, A., Kumar, K., Singh, L., et al. *Quercypsitta*-like birds from the Early Eocene of India (Aves, ?Psittaciformes); *Journal of Vertebrate Paleontology, 30*(2), 467–478.

237. Kumar, K., Rose, K. D., Rana, R. S., Singh, L. Smith, T., and Sahni, A. Early Eocene artiodactyls (Mammalia) from India. *Journal of Vertebrate Paleontology, 30*(4), 1245–1274.

238. Sahni A. Indian Cretaceous terrestrial vertebrates: Cosmopolitanism and endemism in a geodynamic plate tectonic framework. In S. Bandopadhyay (Ed.), *New aspects of Mesozoic biodiversity. Lecture Notes in Earth Sciences* 132 (pp. 91–104). Heidelberg: Springer Verlag.

2011

239. Prasad, V., Strömberg, C. A. E., Leaché, A. D., Samant, B., Patnaik, R., Tang, L., et al. Late Cretaceous origin of the rice tribe provides evidence for early diversification in Poaceae. *Nature Communications, 2*(480), 1–8.

240. Sahni, A. Krishna Rajaram Surange 1920-2010 (biography). *Current Science, 100,* 1089.

241. Sahni, A. Digging for Dino. *BBC Knowledge, 1*(4), 65–66.

2012

242. Folie, A., Rana, R. S., Rose, K. D., Sahni, A., Kumar, K., & Singh, L. Early Eocene frogs from Vastan Lignite Mine, Gujarat, India. *Acta Palaeontologica Polonica, 58*(3), 511–524.

243. Sahni, A. Biomineralization: Some complex crystallite-oriented skeletal structures. *Journal of Biosciences, 38,* 1–11.

244. Rose, K. D., Kumar, K., Rana, R. S., Sahni, A., & Smith, T. New hypsodont tillodont (Mammalia, Tillodontia) from the Early Eocene of India. *Journal of Paleontology, 87*(5), 842–853.

245. Mayr, G., Rana, R. S., Rose, K. D., Sahni, A., Kumar, K., & Smith, T. New specimens of the Early Eocene bird *Vastanavis* and the interrelationships of stem group Psittaciformes. *Paleontological Journal, 47*(11), 1308–1314.

2014

246. Prasad, G. V. R., & Sahni, A. Vertebrate fauna from the Deccan volcanic province: Response to volcanic activity. In G. Keller and A. C. Kerr (Eds.), *Volcanism, Impacts, and Mass Extinctions: Causes and Effects, Geological Society of America Special Paper, 505,* 193–211.

247. Rose, K. D., Holbrook, L. T., Rana, R. S., Kumar, K., Jones, K. E., Ahrens, H. E., et al., Early Eocene fossils suggest that the mammalian Order Perissodactyla originated in India. *Nature Communications, 5*(5570), 1–9.

2015

248. Singh, I. B., Sahni, A., Jain, A. K., Upadhyay, R., Parcha, S. K., Parmar, V. Post-Collision sedimentation in the Indus basin (Ladakh, India): Implications for the evolution of the northern margin of the Indian plate. *Journal of the Palaeontological Society of India, 60* (2), 97–146

249. Srivastava, R., Patnaik, R., Shukla, U. K., & Sahni, A. Crocodilian nest in a Late Cretaceous sauropod hatchery from the Type Lameta Ghat Locality, Jabalpur, India. *PLoS One* Doi:10.1371/journal.pone.0144369.

250. Smith, T., Solé. F., Missiaen, P., Rana, R. S., Kumar, K., Sahni, A., et al. First Early Eocene tapiroid from India and its implication for the paleobiogeographic origin of perissodactyls. *Palaeovertebrata, 39*(2), 1–9.

2016

251. Dunn, R. H., Rose, K. D., Rana, R. S., Kumar, K., Sahni, A., & Smith, T. New euprimate postcrania from the early Eocene of Gujarat, India, and the strepsirrhine-haplorhine divergence. *Journal of Human Evolution, 99, 25*–51.

252. Prasad, G. V. R., Verma, V., Grover, P., Priyadarshini, R., Sahni, A., & Lourembam, R. S. Isolated archosaur teeth from the green sandstone capping the coralline limestone (Bagh Group) of the Narmada valley: Evidence for the presence of pre-Late to Late Maastrichtian abelisaurids in India. *Island Arc, 25,* 410–420.

253. Sahni, A. *Dinosaurs of India (2nd Edition).* New Delhi: National Book Trust of India.

254. Smith, T., Kumar, K., Rana, R. S., Folie, A., Solé, F., Noiret, C., et al. New Early Eocene vertebrate assemblage from western India reveals a mixed fauna of European and Gondwana affinities. *Geoscience Frontiers, 7,* 969–1001.

2017

255. Prasad G. V. R., Verma, V., Sahni, A., Lourembam R. S., & Rajkumari, P. Elasmobranch fauna from the upper most part of the Cretaceous Bagh Group, Narmada valley, India. *Island Arc.* e12200. https://doi.org/10.1111/iar.12200.

256. Sahni, A. How climate changing activities may have positively impacted the human race. The Indian Economist (Science and Technology) online May 22nd 2017.

257. Sensarma, S., Singh, H., Rana, R. S., Paul, D. & Sahni, A. Nature and composition of interbedded marine basaltic pumice in the ~52–50 Ma Vastan lignite sequence, western India: Implication for Early Eocene MORB volcanism offshore Arabian Sea. *Journal of Earth System Science, 126*(17), 1–19.

2018

258. Sahni, A. Birbal Sahni and his father Ruchi Ram: Science emerging from the Shadows of the Raj. In A. R. Chaudhary & D. Kumar (Eds.), *Emergence of Modern Science in Colonial India. Indian Journal of History of Science, 53*(4), 160–166. DOI: 10.16943/ijhs/2018/v53i4/49539

2020

259. Dutta, S., Kumar, S., Singh, H., Khan, M. A., Barai, A., Tewari, A., et al. Chemical evidence of preserved collagen in 54 million year old fish vertebrae. *Palaeontology, 63*(2), 195–202.

Contents

Contributors

Sunil Bajpai Department of Earth Sciences, Indian Institute of Technology Roorkee, Roorkee, India

Mahi Bansal Birbal Sahni Institute of Palaeosciences, Lucknow, India

Robin M. D. Beck School of Environment and Life Sciences, University of Salford, The Crescent, Salford, UK

Sankar Chatterjee Department of Geosciences, Museum of Texas Tech University, Lubbock, Texas, USA

Eric De Bast Directorate Earth and History of Life, Paleobiosphere Evolution, Royal Belgian Institute of Natural Sciences, Brussels, Belgium

France de Lapparent de Broin Centre de Recherches sur la Paléobiodiversité et les Paléoenvironnements, CR2P, MNHN, CNRS, UPMC, Sorbonne Universités, Muséum national d'Histoire naturelle, Paris, France

Suryendu Dutta Department of Earth Sciences, Indian Institute of Technology Bombay, Mumbai, India

Alistair Evans Museums Victoria, Melbourne, VIC, Australia
School of Biological Sciences, Monash University, Melbourne, VIC, Australia

Timothy F. Flannery Melbourne Sustainable Society Institute, University of Melbourne, Parkville, VIC, Australia

John G. Fleagle Department of Anatomical Sciences, Stony Brook University, Stony Brook, NY, USA

Lawrence J. Flynn Department of Human Evolutionary Biology, Harvard University, Cambridge, MA, USA

Emmanuel Gheerbrant Centre de Recherches CR2P UMR CNRS 7207, Sorbonne Université, MNHN/CNRS/UPMC, Paris, France

Christopher C. Gilbert Department of Anthropology, Hunter College of the City University of New York, New York, NY, USA
Graduate Center of the City University of New York, New York, NY, USA
New York Consortium in Evolutionary Primatology, New York, NY, USA
Division of Paleontology, American Museum of Natural History, New York, NY, USA

Neil J. Gostling Biological Sciences, The University of Southampton, Southampton, UK

Anjali Goswami Department of Earth Sciences, UCL, London, UK
Department of Life Sciences, Natural History Museum, London, UK

Thomas J. D. Halliday School of Geography, Earth and Environmental Sciences, University of Birmingham, Edgbaston, UK
Department of Genetics, Evolution, and Environment, UCL, London, UK

Patricia A. Holroyd Museum of Paleontology, University of California, Berkeley, CA, USA

Louis L. Jacobs Roy M. Huffington Department of Earth Sciences, Southern Methodist University, Dallas, TX, USA

Vivesh V. Kapur Birbal Sahni Institute of Palaeosciences, Lucknow, Uttar Pradesh, India

Ashu Khosla Centre of Advanced Study in Geology, Panjab University, Chandigarh, India

Yuri Kimura Department of Geology and Paleontology, National Museum of Nature and Science, Tsukuba, Ibaraki, Japan

Lesley Kool School of Earth, Atmosphere and Environment, Monash University, Melbourne, VIC, Australia

David W. Krause Department of Earth Sciences, Denver Museum of Nature and Science, Denver, CO, USA

Kishor Kumar Wadia Institute of Himalayan Geology, Dehradun, Uttarakhand, India

Martin Kundrat Center for Interdisciplinary Biosciences, Technological and Innovative Park, Pavol Jozef Safarik University, Kosice, Slovak Republic

Hélène Legendre Institut Polytechnique LaSalle-Beauvais, Beauvais, France

Anton Maksimenko Australian Synchrotron, Clayton, VIC, Australia

Ashish K. Mishra Birbal Sahni Institute of Palaeosciences, Lucknow, India

Shreya Mishra Department of Geology, Centre for Advanced Studies, University of Delhi, Delhi, India

Steven Morton School of Physics and Astronomy, Monash University, Melbourne, VIC, Australia

A. C. Nanda Wadia Institute of Himalayan Geology, Dehradun, Uttarakhand, India

Varun Parmar Department of Geology, University of Jammu, Jammu, India

Rajeev Patnaik Centre of Advanced Study in Geology, Panjab University, Chandigarh, India

Swagata Paul Department of Earth Sciences, Indian Institute of Technology Bombay, Mumbai, India

David Pickering Deceased

Guntupalli V. R. Prasad Department of Geology, Centre for Advanced Studies, University of Delhi, Delhi, India

Vandana Prasad Birbal Sahni Institute of Palaeosciences, Lucknow, India

Kelsey D. Pugh New York Consortium in Evolutionary Primatology, New York, NY, USA; Division of Anthropology, American Museum of Natural History, New York, NY, USA

Jean-Claude Rage Centre de Recherches sur la Paléobiodiversité et les Paléoenvironnements CR2P, UMR CNRS 7207, Sorbonne Universités, MNHN/CNRS/UPMC, Muséum national d'Histoire naturelle, Paris, France

Rajendra S. Rana Department of Geology, H.N.B. Garhwal University, Srinagar, Uttarakhand, India

Thomas H. Rich Department of Chemistry and Biotechnology, Faculty of Science, Engineering & Technology, Swinburne University of Technology, Hawthorn, VIC, Australia
Museums Victoria, Melbourne, VIC, Australia
School of Physics and Astronomy, Monash University, Melbourne, VIC, Australia
School of Earth, Atmosphere and Environment, Monash University, Clayton, VIC, Australia

Kenneth D. Rose Center for Functional Anatomy and Evolution, Johns Hopkins University School of Medicine, Baltimore, MD, USA

Agustín Scanferla CONICET-Instituto de Bio y Geociencias, del NOA (IBIGEO), Rosario de Lerma, Salta, Argentina

Karen Siu School of Physics and Astronomy, Monash University, Melbourne, VIC, Australia

Thierry Smith Directorate Earth and History of Life, Paleobiosphere Evolution, Royal Belgian Institute of Natural Sciences, Brussels, Belgium

Floréal Solé Directorate Earth and History of Life, Paleobiosphere Evolution, Royal Belgian Institute of Natural Sciences, Brussels, Belgium

J. G. M. Thewissen Department of Anatomy and Neurobiology, Northeast Ohio Medical University, Rootstown, Ohio, USA

Peter Trusler School of Earth, Atmosphere and Environment, Monash University, Melbourne, VIC, Australia

Omkar Verma Geology Discipline Group, School of Sciences, Indira Gandhi National Open University, New Delhi, India

Patricia Vickers-Rich Department of Chemistry and Biotechnology, Faculty of Science, Engineering & Technology, Swinburne University of Technology, Hawthorn, VIC, Australia
School of Earth, Atmosphere and Environment, Monash University, Melbourne, VIC, Australia
Museums Victoria, Melbourne, VIC, Australia;
Faculty of Science, Deakin University, Burwood and Warun Ponds, VIC, Australia

Chapter 1
First Ornithischian and Theropod Dinosaur Teeth from the Middle Jurassic Kota Formation of India: Paleobiogeographic Relationships

Guntupalli V. R. Prasad and Varun Parmar

Abstract The Middle Jurassic Kota Formation of the Pranhita-Godavari Valley in peninsular India is well known for its vertebrate fauna comprising fishes, sphenodontians, iguanian lizards, cryptodire turtle, crocodilians, pterosaurs, sauropod dinosaurs and early mammals. However, no theropod and undoubted ornithischian dinosaur remains have been reported from the Jurassic of India until now. Here we describe the first theropod dinosaur teeth representing five morphotypes of Dromaeosauridae, one *Richardoestesia*-like form, and one Theropoda indet. The ornithischian dinosaur teeth are described under five morphotypes of Ornithischia indet. The new dinosaur fauna improves the diversity of the Jurassic vertebrate fauna of India significantly. It also improves the impoverished Jurassic record of dromaeosaurid and primitive ornithischian dinosaurs of the Gondwana. At higher taxonomic levels, the Kota fauna demonstrates close compositional similarities with Laurasian Jurassic faunas, such as the Middle Jurassic fauna of England, and limited Gondwanan affinities, which may suggest closer connection with the Laurasian continents and existence of some biogeographic partitioning within the Gondwana in the Jurassic. However, this interpretation is subject to further verification in the least explored areas of Gondwana.

Keywords Ornithischia • Dromaeosauridae • Vertebrate • Jurassic • Pranhita-Godavari • Gondwana • Laurasia

Introduction

In India, continental Jurassic sequences are represented by the Lower Jurassic Lathi Formation of western India and the Middle Jurassic Kota Formation of the Pranhita-Godavari Valley, southcentral India. Plant, invertebrate, vertebrate, and trace fossils including theropod footprints have been reported from the Lathi Formation (Pieńkowski et al. 2015). Conversely, the Kota Formation, has been an important source of vertebrate fossils over the past three decades and since the first reports of fish fossils by Egerton (1851, 1878), many groups of vertebrates have been documented. These include: fish, such as *Lissodus indicus* (Yadagiri 1986; Prasad et al. 2004), ?*Polyacrodus* sp. (Prasad et al. 2004), *Tetragonolepis oldhami* (Egerton 1878; Jain 1973), *Paradapedium egertoni* (Jain 1973), *Indocoelacanthus robustus* (Jain 1974a), *Lepidotes deccanensis* (Sykes 1851; Jain 1983), *Pholidophorus kingi*, *P. indicus* (Yadagiri and Prasad 1977), reptiles, including the sphenodontians *Rebbanasaurus jaini* and *Godavrisaurus lateefi* (Evans et al. 2001), the iguanian lizard *Bharatagama rebbanensis* (Evans et al. 2002), teleosaurid crocodiles (Owen 1852; Nath et al. 2002), the cryptodire turtle *Indochelys spatulata* (Datta et al. 2000), atoposaurid crocodiles, the pterosaurs *Campylognathoides indicus* (Jain 1974b) and *Rhamphorhynchus* sp. (Rao and Shah 1963), and the sauropod dinosaurs *Barapasaurus tagorei* (Jain et al. 1975; Bandyopadhyay et al. 2010) and *Kotasaurus yamanapalliensis* (Yadagiri et al. 1979; Yadagiri 2001), and mammals *Kotatherium haldeni* (Datta 1981), *Trishulotherium kotaensis*, *Indotherium pranhitai* (Yadagiri 1984; Prasad and Manhas 2002 [= *Indozostrodon simpsoni* Datta and Das 2001]), *Nakunodon paikasiensis* (Yadagiri 1985), *Paikasigudodon yadagirii* (Prasad and Manhas 1997, 2002), *Dyskritodon indicus* (Prasad and Manhas 2002), *Gondtherium dattai* (Prasad and Manhas 2001, 2007), *Indobaatar zofiae* (Parmar et al. 2013) and an australosphenidan (Parmar et al. 2015; Prasad et al.

G. V. R. Prasad (✉)
Department of Geology, Centre for Advanced Studies, University of Delhi, Delhi, 110007, India
e-mail: guntupalli.vrprasad@gmail.com

V. Parmar
Department of Geology, University of Jammu, Jammu, 180006, India

G. V. R. Prasad and R. Patnaik (eds.), *Biological Consequences of Plate Tectonics: New Perspectives on Post-Gondwana Break-up—A Tribute to Ashok Sahni*, Vertebrate Paleobiology and Paleoanthropology Series, https://doi.org/10.1007/978-3-030-49753-8_1

2015). It is apparent from this faunal list that the Kota Formation dinosaurs have been represented only by basal sauropods thus far. Theropods and ornithischians have been conspicuously absent from the Kota vertebrate fauna. In recent years, the Kota Formation has been explored extensively for Jurassic mammals (Prasad and Manhas 1997, 2002, 2007; Datta and Das 2001; Parmar et al. 2013). Bulk samples (7000 kg) from mammal yielding horizons were screen-washed in search of micromammals, which led to the recovery of many isolated teeth of crocodiles, pterosaurs, ornithischians and theropods. In view of the rarity of archosaur fossils from the Indian Jurassic, any new find is of great significance in terms of faunal diversity. In this paper, the ornithischian and theropod teeth are described and their paleobiogeographic significance is discussed.

Geological Setting

The Pranhita-Godavari (P-G) Valley in Telangana state, India preserves an almost uninterrupted sequence of continental Gondwana Supergroup of rocks ranging in age from Early Permian to Early Cretaceous. The Upper Gondwana Group exposed in this valley are of special significance as they provide an excellent opportunity to document the rare Mesozoic terrestrial ecosystems from South Asia. These rocks are exposed in a NW-SE trending fault bounded linear belts. In the beginning, vertebrate yielding Upper Triassic – Lower Jurassic strata of the P-G Valley were referred to as the Kota-Maleri beds (Hughes 1876). Later, King (1881) separated the Kota-Maleri beds into two distinct stratigraphic units, the lower Maleri beds containing a Late Triassic fauna and the overlying Kota beds yielding an Early Jurassic fauna. Following extensive field mapping, Kutty (1969) revised the Upper Gondwana stratigraphy of the P-G Valley and subdivided the Upper Gondwana Group into the lower Middle Triassic Yerrapalli Formation, upper Middle Triassic Bhimaram Sandstone, lower Upper Triassic Maleri Formation, Upper Triassic – Lower Jurassic Dharmaram Formation, Lower Jurassic Kota Formation, and Lower Cretaceous Gangapur Formation (Kutty et al. 1987). The Triassic Yerrapalli, Maleri and Dharmaram formations are red clay dominated sequences with subordinate sandstones and lime-pellet rocks, and the Bhimaram Sandstone Formation is predominantly represented by coarse sandstones. The Gangapur Formation, which overlies the Kota Formation with an angular unconformity, is a sequence of coarse sandstones and alternating sandstones, mudstones and silty mudstones (Kutty et al. 1987).

The Kota Formation succeeds the topmost red clay band of the Dharmaram Formation with a coarse pebbly sandstone that becomes thick and conglomeratic towards the eastern part of the basin. The conglomerate bed, together with an abrupt change in the fauna from that of the underlying Dharmaram Formation, was taken as evidence for a minor break in sedimentation (Kutty et al. 1987). The Kota Formation, with its type section exposed near the village Kota in Sironcha district, Maharashtra on the left bank of the Pranhita river, approximately 8 km north of its confluence with the Godavari river, is lithologically divided into lower and upper units (Rudra 1982). The lower unit comprises coarse, hard, compact and pebbly sandstone that grades upwards into fine-grained white sandstone, red and green clays, mudstones and siltstones. The dinosaur and mammal-yielding mudstones and siltstones of the lower unit are succeeded by the calcareous shale and limestones of the upper unit. The limestone is overlain by mudstones and ferruginous shale interbedded with sandstone. The clay - mudstone - siltstone sequence of the lower unit produced sauropod dinosaurs, mammals, ostracods and charophytes, whereas the limestone-bearing upper unit yielded articulated fish skeletons, a teleosaurid crocodylomorph, sphenodontians, iguanian lizards, a cryptodiran turtle, mammals and ostracods (Bandyopadhyay and Roychowdhury 1996; Parmar et al. 2013).

Rudra (1982) interpreted the lower unit of the Kota Formation as a fining upward sequence deposited by a laterally shifting meandering river system, while the upper unit is a coarsening upward sequence deposited in a braided river system. The limestone facies of the upper part was considered to have been deposited in a distal playa-type system of interconnected lakes (Rudra 1982; Rudra and Maulik 1987). The faunal and floral elements of the Kota Formation also indicate freshwater lacustrine depositional environment (King 1881; Robinson 1970; Tasch et al. 1973; Govindan 1975; Jain 1980, 1983; Jain and Roy Chowdhury 1987; Rudra and Maulik 1987; Feist et al.1991; Bhattacharya et al. 1994).

Age

Based on the fishes *Paradapedium egertoni*, which was considered similar to Early Jurassic *Dapedium* of Europe, *Tetragonolepis oldhami* (considered morphologically close to *Tetragonolepis* of Upper Liassic of Europe [Jain 1973]), and *Lepidotes deccanensis* (that shares affinities with Upper Liassic [Toarcian] *Lepidotes elevensis* from Europe), the Kota Formation has been assigned an Early Jurassic (Liassic) age (Jain 1980, 1983). A similar age was suggested by Yadagiri and Prasad (1977) on the basis of the pholidophorid fishes, *Pholidophorus kingi* and *P. indicus*. Palynofossils (Prabhakar 1989) and charophytes (Bhattacharya et al. 1994) also indicate an Early Jurassic age. However, freshwater ostracods, represented by *Darwinula* cf. *D. sarytirmensis* and *Timiriasevia*, have been used to infer a Middle Jurassic age for the Kota Formation (Govindan 1975). Vijaya and Prasad (2001) assigned a younger Middle – Upper Jurassic age based on a palynological study. Contrary to these later, younger, age assignments, Evans et al. (2001) regarded the Early Jurassic age as more likely because the dental morphology of the two Kota sphenodontian taxa, *Rebbanasaurus* and *Godavarisaurus*, is more derived than those of the basal rhynchocephalians *Diphydontosaurus* and *Planocephalosaurus* of Late Triassic age, and *Gephyrosaurus* of Early Jurassic age. The current study reveals the presence of ornithischian and theropod teeth in association with a vertebrate microfossil assemblage compositionally similar to that of Middle Jurassic of UK pointing to a potentially younger Middle Jurassic age for the Kota Formation.

Methods and Materials

Locality Information

The Kota Formation outcrop yielding the dinosaur teeth described herein is a localized section with a short lateral extent as is the case with most known sections of this formation. The outcrop is a small cliff exposed along a stream channel 300 m southwest of Paikasigudem village, Rebbana Mandalam, Adilabad district, Telangana, India (Fig. 1.1). The measured lithological section consists of 20 m of sandstones, clays, mudstones and siltstones with two distinctive limestone and two impure thin limestone bands (Fig. 1.1, 1.2). The dinosaur teeth are derived from a 27 cm thick greenish-grey nodular siltstone situated about 9.5 m above the base of the section representing Rudra's (1982) upper unit of the Kota Formation. Between 1997 and 2013 about 7000 kg of this siltstone were subjected to screen-washing in search of early mammals. As the siltstones are relatively soft, the combined use of kerosene and water was employed to disaggregate the rock matrix.

Fossils Collected

The residue left after screen-washing was sorted under the microscope for vertebrate microfossils which led to the recovery of more than 5000 specimens of vertebrate fossils predominantly represented by fish teeth and scales in addition to jaw fragments of sphenodontians and iguanian lizards, crocodilian, few pterosaur and dinosaur teeth, and 50 mammalian teeth. The fossil-bearing horizon is also highly enriched in ostracod carapaces. The vertebrate microfossils are very well-preserved except for damage related to mechanical breakage which could be pre-or post-burial or due to breakage during screen-washing. No substantial rounding of bones and teeth is observed which implies that the specimens have not undergone extensive transportation. The dinosaur teeth are represented mainly by crowns possibly due to resorption of the root during tooth replacement as is evident from the presence of resorption pits at the crown base. The presence of wear facets on tooth apices and loss of denticles also attest to tooth use. The dinosaur fauna described here is represented by 71 theropod and 35 ornithischian teeth. Photographs of the described teeth were taken with Leica S8APO Stereoscopic Binocular Microscope attached with Leica MC 120 HD Digital Camera and Zeiss Scanning Electron Microscope and Zeiss EVOMA 10 Scanning Electron Microscope.

The utility of isolated dinosaur teeth in identification, at least at family level and above, has been demonstrated by Currie et al. (1990), Farlow et al. (1991), Basso (1997), Smith et al. (2005), Larson and Currie (2013), Hendrickx and Mateus (2014) and Gerke and Wings (2016) for theropod and Coombs (1990), Parkar et al. (2005), Butler et al. (2006), Irmis et al. (2007) for ornithischian dinosaurs. Tooth measurements were made using the methods of Currie et al. (1990), Farlow et al. (1991) and Smith et al. (2005) for theropod teeth. In addition to fore-apt basal length (FABL), basal labiolingual width (BW), total apico-basal height of the crown (TCH) perpendicular to FABL proposed by Currie et al. (1990), the denticle size difference index (DSDI) of Rauhut and Werner (1995), involving the measurement of the ratio of the number of denticles per given length unit of the mesial and distal carinae was also measured. The nomenclature used for the description of teeth is mesial (towards the snout), distal (away from this joint), lingual (towards the tongue), labial (towards the lip), apical (towards the tip of the tooth), and basal (towards the base of the tooth). For ornithischian teeth, maximum height is measured from the base of the crown to the tip of the apical cusp and the maximum mesiodistal length is measured between the flanks of mesial and distal most denticles. Isolated ornithischian teeth are difficult to assign to their original position in the jaw. Barrett (2001) has observed that tooth wear is predominantly present on the lingual cingula of maxillary teeth and labial cingula of dentary teeth. In the Kota teeth, no wear facets are observed with the exception of two specimens. However, distally offset apex is used as one criterion for orienting the specimens of both theropod and ornithischian teeth. Secondly, the convex face of the crown is interpreted as the labial side and the concave or flat face as the lingual side in both groups of dinosaurs. No attempt is made to assign the teeth to the maxillary or dentary bones as isolated teeth without wear facets are difficult to identify with these bones. In view of disarticulated nature of fossil material described, open nomenclature is followed for their systematic identification and description.

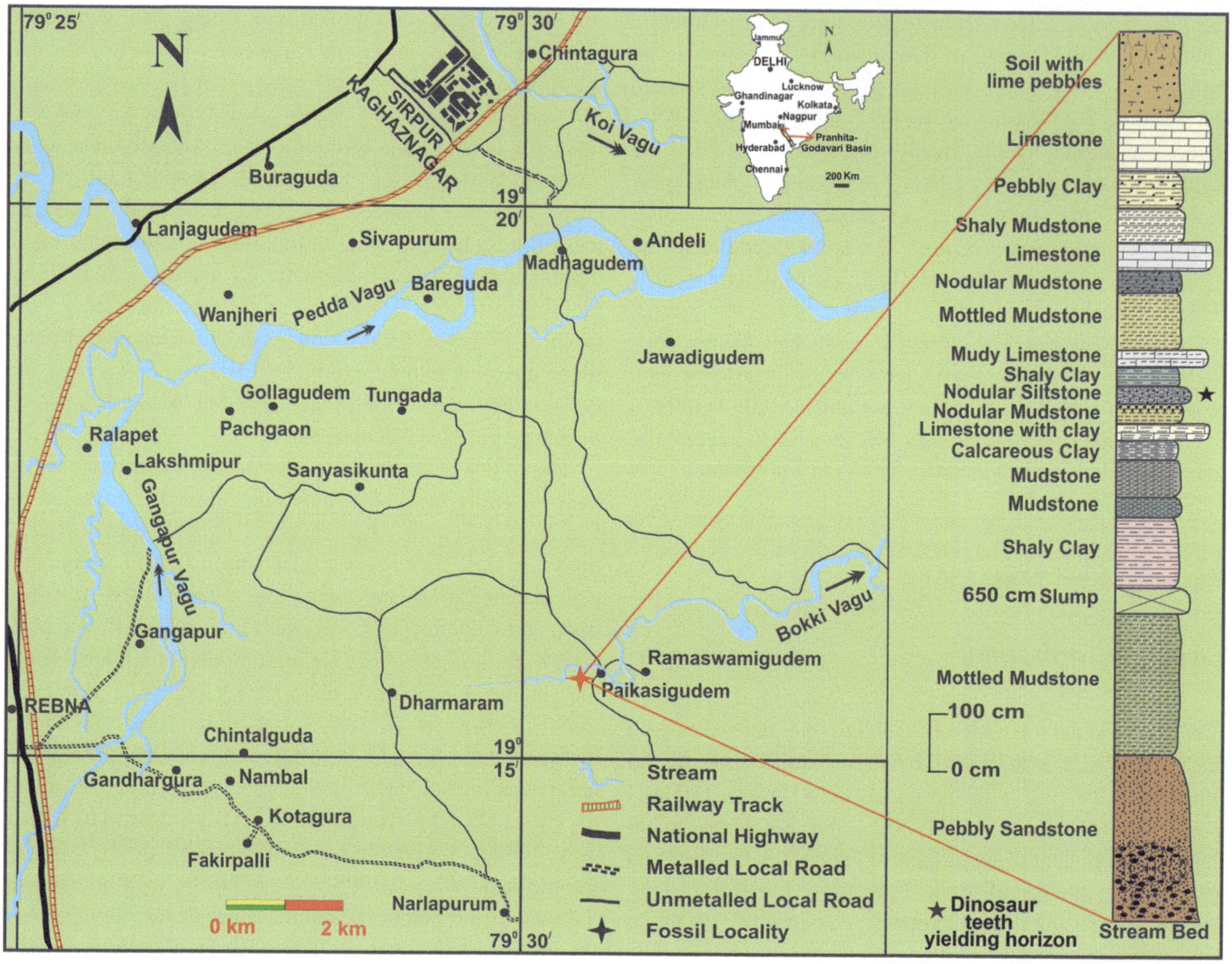

Fig. 1.1 Location map and lithocolumn of the dinosaur teeth yielding section of the Kota Formation exposed at Paikasigudem village, Rebbana Mandalam, Adilabad district, Telangana, India

Institutional Abbreviations

DUGF/J – Delhi University, Geology Department Fossil Collection/Jurassic), Delhi, India.

Systematic Paleontology

Class Dinosauria Owen, 1842
Order Ornithischia Seeley, 1887
Ornithischia indet.

Morphotype 1

(Figs. 1.3A–H, 1.4A–S, 1.5A–H)

Referred Material: Five fairly well preserved cheek tooth (DUGF/J1-4, 12).

Description: DUGF/J1 is a well-preserved tooth lacking the root. It is as long mesiodistally as apicobasally high. The tooth is slightly asymmetrical in form, labiolingually compressed and appears broadly rounded in both labial and lingual views. The crown bears well-developed denticulated carinae mesially and distally. The occlusal tip of the crown has a relatively large and blunt cusp that is split into a smaller less individualized denticle on its mesial margin. The apex of the central cusp is slightly tilted distally. There are four major marginal denticles mesial to the apical cusp and three individualized denticles on its distal flank (Fig. 1.3A–B, E–F). Except for the mesial-most one, the denticles on the mesial carina are better developed than those situated distally. The denticles are bluntly rounded and conical or leaf-shaped in form. The median part of each denticle is strongly convex which makes its cross-section triangular in shape. The second mesial denticle from the apex is next to the apical cusp in size. In addition to the three regular denticles, a minute denticle is present at the base of the third

Fig. 1.2 Field photograph of the fossiliferous section at Paikasigudem village, Rebbana Mandalam, Adilabad district, Telangana, India

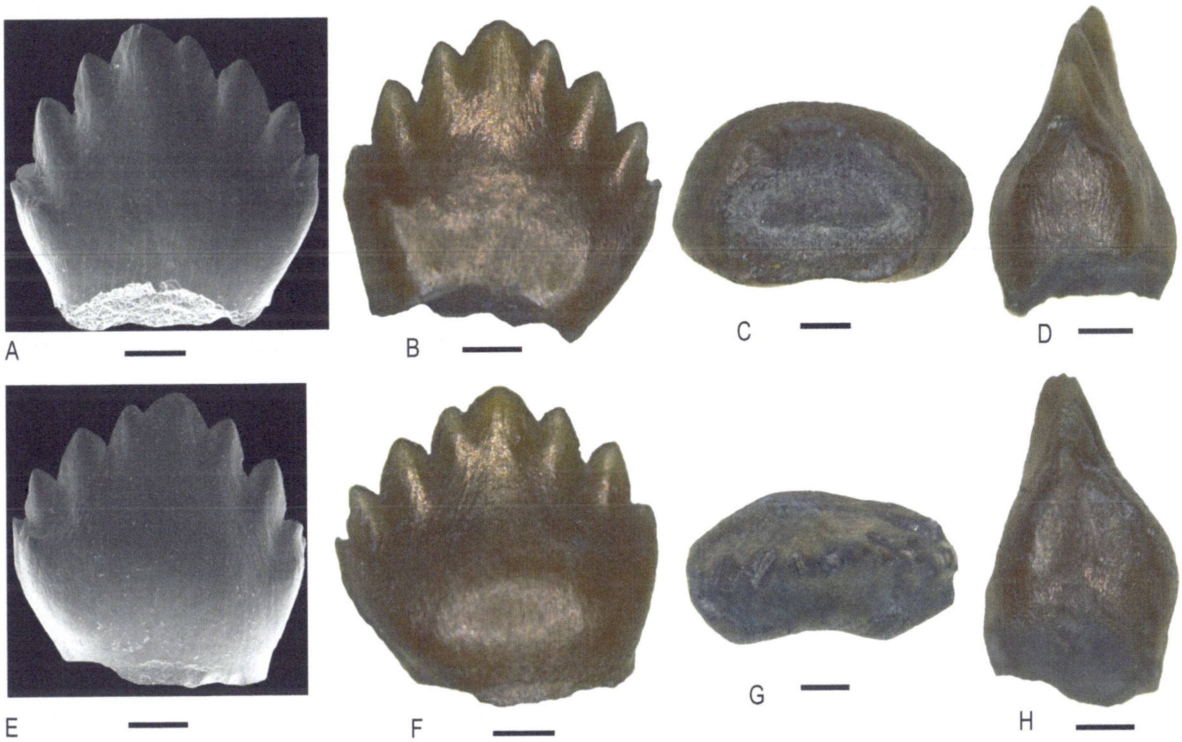

Fig. 1.3 Ornithischia indet. **A–H**. Morphotype 1, cheek tooth (DUGF/J1), **A**. lingual view (SEM photomicrograph), **B**. lingual view, **C**. basal view, **D**. view of mesial flank, **E**. labial view (SEM photomicrograph), **F**. labial view, **G**. occlusal view, **H**. view of distal flank. Scale bar equals 500 μm

distal denticle, which is slightly lingually placed from the distal margin and terminates in a short rudimentary lingual 'cingulum'. The marginal denticles though variable show a general tendency towards decrease in size towards the base. The surface of the crown is covered by very faint striations that are more prominent labially. The labial crown surface is

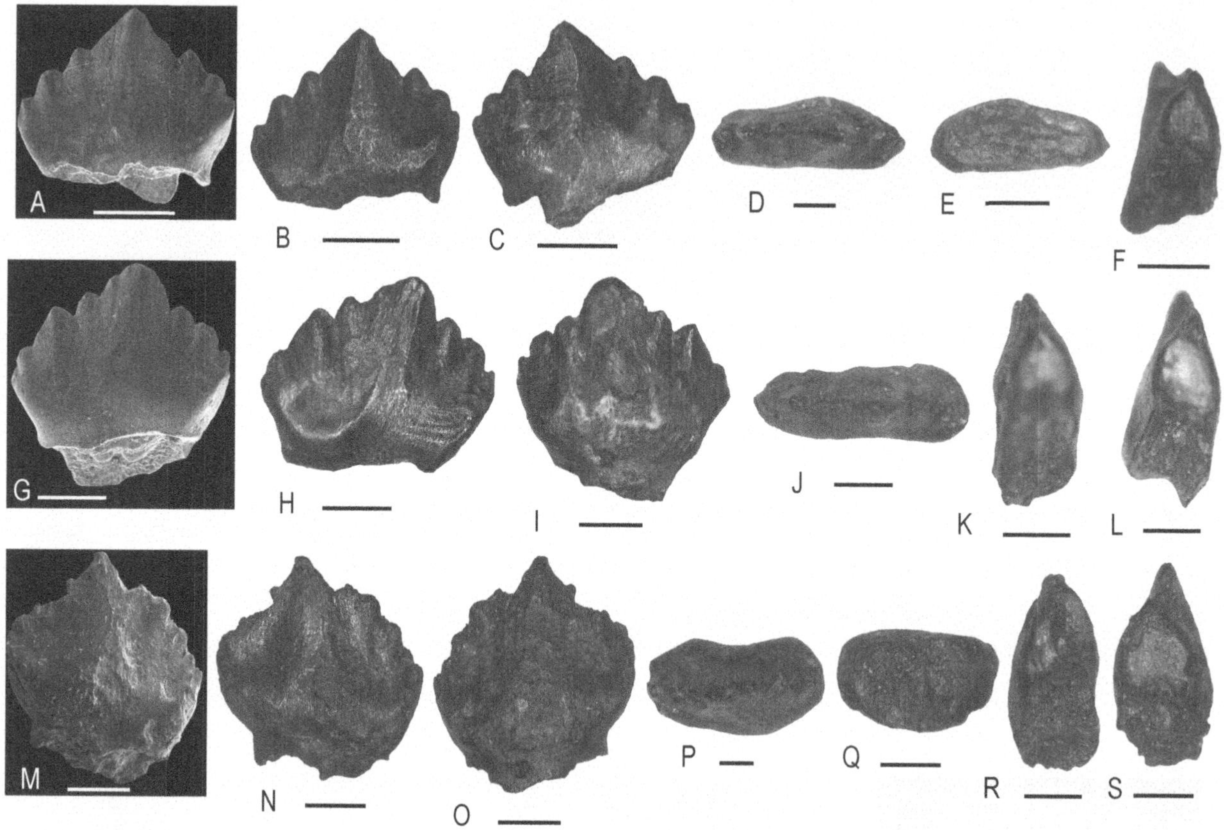

Fig. 1.4 Ornithischia indet. **A–H**. Morphotype 1, cheek tooth (DUGF/J4), **A**. lingual view (SEM photomicrograph), **B**. lingual view, **C**. labial view, **D**. occlusal view, **E**. basal view, **F**. view of distal flank; **G–L**. Morphotype 1, cheek tooth (DUGF/J2), **G**. lingual view (SEM photomicrograph), **H**. lingual view, **I**. labial view, **J**. occlusal view, **K**. view of distal flank, **L**. view of mesial flank; **M–S**. Morphotype 1, cheek tooth (DUGF/J3), **M**. lingual view (SEM photomicrograph), **N**. lingual view, **O**. labial view, **P**. occlusal view, **Q**. basal view, **R**. view of distal flank, **S**. view of mesial flank. Scale bar equals 500 μm

strongly inflated at its base. As on the distolingual margin, a short rudimentary mesiolabial 'cingulum' is also present on the labial crown surface. The basal cross-section is bean-shaped with a concave lingual and convex labial margins (Fig. 1.3C). Flat distal and slightly concave mesial 'interdental pressure facets' are present. As there are resorption pits in the basal view of the crown, DUGF/J1 is considered as a shed tooth.

DUGF/J2-4 are relatively small in size (almost 50% smaller than DUGF/J1, Table 1.1) and have more asymmetrical crown with the apical cusp offset more distally than in DUGF/J1 (Fig. 1.4A–S). They are more labiolingually compressed, asymmetrically triangular in outline, as high as long and have an elliptical or lenticular basal cross-section. The apical cusp is broadly triangular in shape and relatively larger than the marginal denticles unlike in DUGF/J1 and has a convex lingual surface with a broad central ridge extending from the apex to the base of the crown. The crown is slightly concave at the lingual base of this ridge as in DUGF/J1. In DUGF/J2 (Fig. 1.4G–I), there are three denticles on the distal carina, of which the distal-most one is the smallest whereas the remaining two are nearly of the same size. On the mesial carina, the first one from the apical cusp is very small and twinned with the latter, the second mesial denticle is slightly larger in size, the third one being the largest of all the mesial denticles, the fourth one is much smaller in size and is flanked by a miniscule denticle at the mesial extremity. The 'interdental pressure facet' is flat distally and concave mesially (Fig. 1.4K–L). The crown is convex labially and slightly concave lingually. The base of the crown, although swollen, is not as inflated as in DUGF/J1. The crown is laterally compressed rendering a lenticular outline to basal cross-section. The enamel surface displays very faint striations. In DUGF/J 3 (Fig. 1.4M–S), the apex of the tooth, particularly the carinae are broken but in general, the morphology is similar to that of DUGF/J2. DUGF/J4 (Fig. 1.4A–F), as compared to DUGF/J3, is mesiodistally longer than high and labiolingually compressed. The apical cusp is triangular in shape and relatively large and mesiodistally broader than in DUGF/J2 and 3. Both carinae bear three denticles each. The base of the crown in DUGF/J 4 is less swollen than in DUGF/J2 and 3.

Fig. 1.5 Ornithischia indet. **A–E**. Morphotype 1, cheek tooth (DUGF/J12), **A**. labial view, **B**. lingual view, **C**. basal view, **D**. view of mesial flank, **E**. view of distal flank; **F–H**. Morphotype 2, cheek tooth (DUGF/J17), **F**.?labial view, **G**. basal view, **H**. occlusal view. Scale bar equals 500 μm

The crown of DUGF/J12 is subtriangular to subtrapezoidal in shape with a large apical cusp and worn but relatively coarse marginal denticles (Fig. 1.5A–E). The marginal denticles extend some distance as cylindrical ridges separated by long grooves onto the body of the crown, particularly on the lingual crown face. There are four marginal denticles on either side of the apical cusp (Fig. 1.5A–B). The first distal marginal denticle is smaller with respect to the apical cusp and is more or less at the same level as the latter and twins with it. Distal to this, three subequal denticles are present. At its distal extremity the distal carina curves lingually and extends to a short distance along the crown base before terminating in a minute denticle. The mesial carina bears three subequal denticles followed by a minute denticle

Table 1.1 Measurements of ornithischian teeth

Specimen No.	Apicobasal height of the crown in mm	Mesiodistal length in mm
DUGF/J1	3.0	3.0
DUGF/J2	1.0	2.0
DUGF/J3	2.0	2.0
DUGF/J4	1.0	1.0
DUGF/J5	2.0	2.0
DUGF/J6	2.0	3.0
DUGF/J7	2.0	3.0
DUGF/J8	1.0	2.0
DUGF/J9	1.0	2.0
DUGF/J10	1.0	2.0
DUGF/J11	2.0	3.0
DUGF/J12	2.0	3.0
DUGF/J16	3.0 (in lingual view) 2.0 (in labial view)	2.0 (in lingual view) 2.0 (in labial view)
DUGF/J17	3.0	3.0

at its mesial extremity. The lingual face of the crown is slightly concave (Fig. 1.5B), whereas the labial face is slightly convex (Fig. 1.5A), particularly in the middle part. The crown is swollen at its base to form a 'cingulum' which is more prominent on the lingual side. The lingual face of the crown is less high than the labial face and as a consequence the lingual swelling of the crown at its base is at a higher level than that of the labial face. On the lingual face of the crown, the swollen base is medially arched and connects to the apical cusp by a broad central eminence. Faint vertical grooves and ridges are present on this central eminence and on the basal swelling of the crown on this face (Fig. 1.5B). On the labial face, the crown is flanked by two narrow and obliquely oriented, elongated ridges that connect to the mesial and distal most denticles (Fig. 1.5A). The base of the crown is swollen both labially and lingually forming a distinct 'cingulum'. The 'interdental pressure facet' is broad and flat distally, while it is broader than its distal counterpart and slightly concave mesially. At the labial base of the crown, a small part of the root is preserved which is narrower than the crown and is constricted from it. The basal cross-section is bean-shaped with a slightly indented lingual margin and a convex labial margin (Fig. 1.5C). All of the denticles exhibit wear facets at their apices. The wear facet on the mesial denticle next to the apical cusp extends down to its labial base. Another very broad and comma-shaped wear facet with its narrow and long axis paralleling the basal 'cingulum' occurs on the mesial half of the labial crown surface. Fine striae oriented parallel to the long axis of the crown are present on this wear facet.

Remarks Several characters of the maxillary/dentary teeth, such as low, sub-triangular shape in labial view, crowns mesiodistally expanded above the root such that the maximum mesiodistal length is greater than the maximum mesiodistal length of the root, labiolingual expansion of the crowns above the root forming a "cingulum", asymmetrical crowns in mesial and distal views, and crowns with enlarged denticles on the mesial and distal margins often associated with 'interdental pressure facet' were considered as synapomorphies of Ornithischia (Sereno 1986, 1991; Hunt and Lucas 1994; Heckert 2002, 2004; Norman et al. 2004). However, recent reviews on Late Triassic dinosaurs have shown that some of these features are also present in non-dinosaurian Triassic archosaurs (Dzik 2003; Parkar et al. 2005; Butler et al. 2006; Irmis et al. 2007). Barrett et al (2008) were of the opinion that the majority of these features are not known in non-ornithischian Jurassic archosaurs. However, the labiolingual expansion of the crown at its base or presence of a basal "cingulum" connected to the mesial and distal most denticles by ridges has been considered as an unambiguous synapomorphy for Ornthischia (Parkar et al. 2005; Butler et al. 2006; Irmis et al. 2007). Because of the presence of asymmetric basal swelling of the crown in all the teeth described here, they are identified here as belonging to ornithischian dinosaurs.

In some symplesiomorphic characters such as labiolingually compressed, sub-triangular to triangular crowns with many denticles on mesial and distal carinae, smooth crown surfaces, these teeth compare well with those of basal ornithischian *Lesothosaurus diagnosticus* Galton (1978), basal thyreophorans *Scelidosaurus harrisonii* Owen (1861), *Scutellosaurus lawleri* Colbert (1981), *Emausaurus ernsti* Haubold (1990), *Laquintasaura venezuelae* Barrett et al. (2014) and the basal neornithischians *Agilisaurus louderbacki* Peng (1990) and *Hexinlusaurus multidens* (He and Cai 1984). However some of these taxa (*Scelidosaurus*, *Agilisaurus*) have tall, triangular to rhombic crowns which are apicobasally high in comparison to mesiodistally long and sub-triangular crowns of Kota teeth. In *Scelidosaurus*, the number of denticles on each margin vary from six to nine in comparison to three to four in the Kota teeth. *Emausaurus* differs from the Kota teeth in having slightly spatulate crowns. The maxillary and dentary teeth of *Scutellosaurus* and *Lesothosaurus* are identical to each other in sub-equal dimensions of mesiodistal length and apicobasal height, sub-triangular crown outlines, and bulbous crown base. But they differ from each other in the number of denticles on mesial and distal carinae. The Kota teeth although subtriangular in outline, their mesiodistal length is equal or greater than apicobasal height, the crown base is not as bulbous as in *Lesothosaurus* and *Scutellosaurus*, and have three to four denticles on each crown margin as compared six to nine denticles in *Lesothosaurus*. In contrast to subequal apicobasal height and mesiodistal length and smooth crowns of the Kota teeth, *Laquintasaura* and *Hexinlusaurus* have crowns that are apicobasally higher than mesiodistally long and apicobasal ridges on the crown surface. The maxillary/dentary teeth of heterodontosaurids also differ from the Kota teeth in having chisel-shaped crowns with denticles restricted to the apical most third of the crown (Butler et al. 2006). Although maxillary/dentary teeth of stegosaurs have sub-triangular crowns as in Kota teeth and some basal ornithischians, the "cingulum" is strongly developed and in the form of a horizontal ring around the crown base, the denticles are usually rounded at their tips as compared to sharply pointed denticles in other ornithischians, and vertical grooves extend from the apex of the denticles down to the 'cingulum' at the base (Blows and Honeysett 2014). The Kota teeth are also distinct from the teeth of iguanodontian and ceratopsid ornithischians in having sub-triangular crowns in labial/lingual view, coarse denticles on mesial and distal margins and asymmetry of the crowns in mesial and distal views (Sereno 1999; Butler et al. 2008).

In characters that are common to many ornithischian taxa such as leaf-like crown, labiolingual compression, twinned apical cusp with adjacent mesial denticle, and asymmetrical basal swelling of the crown, the Kota teeth are broadly similar to those of an indeterminate ornithischian from the Middle Jurassic (Bathonian) of France (Kriwet et al. 1997), neornithischian *Nanosaurus agilis* Marsh (1877) known from the Upper Jurassic Morrison Formation (Carpenter and Galton 2018) and *Alocodon kuehnei* and *Phyllodon henkeli* from the Upper Jurassic Guimarota lignite deposits of Portugal (Thulborn 1973; Rauhut 2001), and the upper Bathonian Forest Marble Formation of southern England (Evans and Milner 1994). The Middle Jurassic teeth of France differ from those of Kota in the presence of well-developed secondary ridges connecting the denticles to the swollen base. In the cheek teeth of *Nanosaurus*, each denticle of posterior cheek teeth is divided into a central cusplet and adjacent blades reinforced with ridges that render a pointed triangular shape to its apex (Carpenter and Galton 2018, fig. 10H), whereas mid-dentary teeth have no such multi-cusped denticles (Carpenter and Galton 2018: fig. 10P). Moreover, in the Kota teeth, the ridges supporting the marginal denticles and associated grooves do not extend farther on to the crown as in *Nanosaurus*. The teeth of *Phyllodon* differ from DUGF/J1-4 and 12 in their lozenge shape, having apicobasal height greater than the mesiodistal length, and the labial side of the crown significantly higher than the lingual side. Similarly, *A. kuehnei*, referred variously to Fabrosauridae, Hypsilophodontidae, Ornithischia *incertae sedis* or Thyreophora (Thulborn 1973; Sereno 1991; Ruiz-Omeñaca 2001; Knoll 2002) or *nomen dubium* (Knoll 2002; Weishampel et al. 2004; Godefroit and Knoll 2003), differs from the Kota teeth in the presence of ridges on the lingual surface of the cheek teeth.

DUGF/J2-4 differ from DUGF/J1 in their smaller size, less swollen crown base, greater labiolingual compression, lenticular basal cross-section and a relatively broad and triangular apical cusp. Besides the size disparity, the morphological differences are sufficient enough to consider DUGF/J2-4 as representing different morphotype from that of DUGF/J1. These differences in morphology may be attributed either to their position on a different bone (dentary vs maxillary) or a different part of the tooth row (posterior vs anterior). DUGF/J4, previously described from the Kota Formation exposed at Gorlapalli village in Adilabad district, Telangana and situated about 40 km southeast of Paikasigudem site, was erroneously referred to Hypsilophodontidae (Prasad 1986).

DUGF/J12 is morphologically distinct from DUGF/J1-4 in having deep grooves separating cylindrical ridges of the denticles that extend some distance onto the crown and better expressed asymmetrical basal swelling of the crown or 'cingulum' which is arched below the apical cusp lingually. In size also it is intermediate to DUGF/J1 and DUGF/J2-4. The presence of a well-developed wear facet on the labial 'cingulum' indicates that DUGF/J 12 is a possible dentary tooth. Although its phylliform, labiolingually compressed and mesiodistally expanded crown with bulbous base, and well-developed constriction between the crown and root recall the crown morphology of ankylosaurian ornithischians, because of the poor preservation of DUGF/J12 and lack of synapomorphies of ankyosaurians, it is assigned to Ornithischia indet. Although there are significant morphological differences between these three sets of teeth, due to lack of knowledge on variation in dental morphology related to heterodonty, ontogeny, and intraspecific variation (Coombs 1990), they are lumped together in Ornithischia indet. Mophotype 1.

Morphotype 2

(Fig. 1.5F–H)

Referred Material: One partially preserved tooth (DUGF/J17).

Description: DUGF/J17 is a partially complete crown. It is lozenge or sub-rhombic in shape with a high and large apical cusp flanked by two marginal denticles at the mesial and distal ends of the crown. The apex of the prominent apical cusp is slightly titled distally. On one side of the crown, the apical cusp is connected to the base of the crown by a broad central eminence that extends from the apex to the base. The mesial and distal denticles are relatively small with respect to the apical cusp but salient and are connected to the crown base by narrow and low longitudinal ridges. The base of the crown is partly broken but from the preserved mesial part it appears to have a slightly swollen basal rim rather than a distinct 'cingulum'. The mesial carina is convexly curved while the distal one is straight or less curved. There are no denticles on the mesial and distal carinae. The enamel of the lateral ridge connecting the distal denticle with the crown base is chipped off. Between these lateral ridges and the central eminence, the crown is longitudinally depressed. Five faint ridges are present distal to the central eminence (Fig. 1.5F). A smooth and flat vertical surface is present on the other side of the crown. This surface displays some fine striations oriented obliquely to the long axis of the crown. The mesial face of the tooth bears an acutely-triangled, concave 'interdental pressure facet'. This facet is flat and relatively narrow on the distal flank. The basal cross-section is sub-trapezoidal in outline. In occlusal view, the crown presents a sub-rectangular outline with a broader mesial end than the distal one.

Remarks DUGF/J17 is a poorly preserved tooth but displays a steeply inclined wear facet with fine striae on one side of the crown and one prominent subcentral vertical ridge and two marginal crests that descend basally to merge with the basal crown rim on the opposite side. Teeth with a prominent central ridge and two lateral ridges merging with basal crown rim are known to occur in heterodontosaurids such as *Echinodon*, basal thyreophorans and neornithischian *Agilisaurus*, which have triangular to rhomboidal crowns with denticulated margins (Galton 1986). DUGF/J17 differs from the basal ornithischian *Scelidosaurus* in lacking several small marginal denticles on the mesial and distal crown margins and in the presence of a prominent central ridge and fine longitudinal flutings on the crown surface. Contrary to the presence of a prominent central ridge in DUGF/J17, in *Echinodon*, there are no ridges on lingual or labial crown surfaces except for a rounded central eminence and the mesial and distal margins bear denticles. *Agilisaurus* also bears several marginal denticles both mesially and distally contrary to the condition of one denticle each at the mesial and distal flanks of the crown in DUGF/J17. The crown morphology of DUGF/J17 is somewhat similar to that of *Heterodontosaurus* in which the crown bears curved ridges running down the mesial and distal margins and meeting at the base forming a rim and enclosing a shield area. A central ridge as in DUGF/J17 divides the shield area of *Heterodontosaurus* teeth into two longitudinal depressions. However, in *Heterodontosaurus*, the occlusal surface displays characteristic 'w' shape wear facet and the crown merges into the root without forming a 'cingulum' (Norman et al. 2011). The root is not preserved and crown is heavily worn in DUGF/J17 making its identification difficult. In the absence of diagnostic crown features allying it with any known basal ornithischians, the present tooth is assigned to Ornithischia indet.

Morphotype 3

(Fig. 1.6A–Q)

Referred Material: Seven fairly well-preserved teeth (DUGF/J5-11) and three fragmentary teeth.

Description: DUGF/J5-11 appear to belong to the cheek region. Only the crown is preserved in all the recovered teeth. DUGF/J5 (Fig. 1.6B–E) is higher than long and asymmetrical triangle in outline with a distally offset apex. The distal carina of the tooth is nearly straight while its mesial carina is slightly arched. The apical cusp is relatively broad in comparison to the marginal denticles and in the lingual view it extends basally in a broad and convex central eminence. On either side of the central eminence, the crown is flat on the lingual face. The mesial and distal carinae are finely denticulated with the denticles being smaller on the mesial carina than on the distal one. There are 7–8 denticles on each carina. The mesial-most and distal-most denticles are salient and diverging away from the crown and form a short and faint 'cingulum' at the mesial and distal extremities. Except for these two denticles, the remaining denticles are minute and less conspicuous. The labial face of the crown is strongly convex. The surface of the crown is largely smooth. The crown is swollen at its base near its junction with the root. The basal cross-section of the crown presents an oval outline with a convex labial margin and nearly straight lingual border.

DUGF/J6-10 (Fig. 1.6A, F–Q) are more or less similar to DUGF/J5 in general crown morphology except for the fact that their mesiodistal length is nearly equal to their apicobasal height or slightly longer than high and they are not as asymmetrical as DUGF/J5. These teeth possibly represent middle cheek region. However, it is difficult to differentiate the maxillary and dentary teeth. In some of the teeth (DUGF/J5, 6 and 7), the crown shows faint vertical striations. The marginal denticles are not well preserved in all teeth, but in one tooth (DUGF/J7), there are at least seven denticles (Fig. 1.6A). In this tooth, the mesial carina is slightly longer than the distal one and is straight whereas the distal one is slightly concave. In all other morphological features, these teeth are similar to DUGF/J5. In all these teeth, the mesial and distal carinae intersect at right angles and the denticles decrease in size apically.

Remarks Although DUGF/J5 and DUGF/J6-10 differ in overall shape of the crown, salient and strongly diverging mesial most and distal most denticles and marginal denticles decreasing in size towards the apex are the characters that unite them in a single taxon. This variation in crown morphology may be due to variation in their position on the jaw. Thulborn (1973) described triangular, labiolingually compressed teeth that are as long as high and with apex canted distally, smooth enamel surfaces, straight lateral edges intersecting at right angles, marginal denticles decreasing in size towards the apex, and salient distal most and mesial most denticles diverging away from the crown from the Upper Jurassic lignite deposits of Guimarota at Porto Pinheiro, near Lourinhã, Provincia do Estremadura, Portugal. These teeth were assigned to a new ornithischian taxon *Trimucrodon cuneatus* within the Hypsilophodontidae. Thulborn (1973) suggested that *Alocodon kuehnei* and *T. cuneatus* were closely related to *Fabrosaurus australis* and might be referred to the Fabrosauridae. Later, Galton (1994) described additional teeth of *Trimucrodon* from the Upper Jurassic Guimarota lignite deposits. Morphologically, the Kota teeth are indistinguishable from those of Guimarota. Thulborn (1973) opined that *Trimucrodon cuneatus* is closely related to *Echinodon becklesii* Owen (1861), a heterodontosaurid (Norman and Barrett 2002) known from the Middle Purbeck Beds, Durleston Bay, U.K. in many of

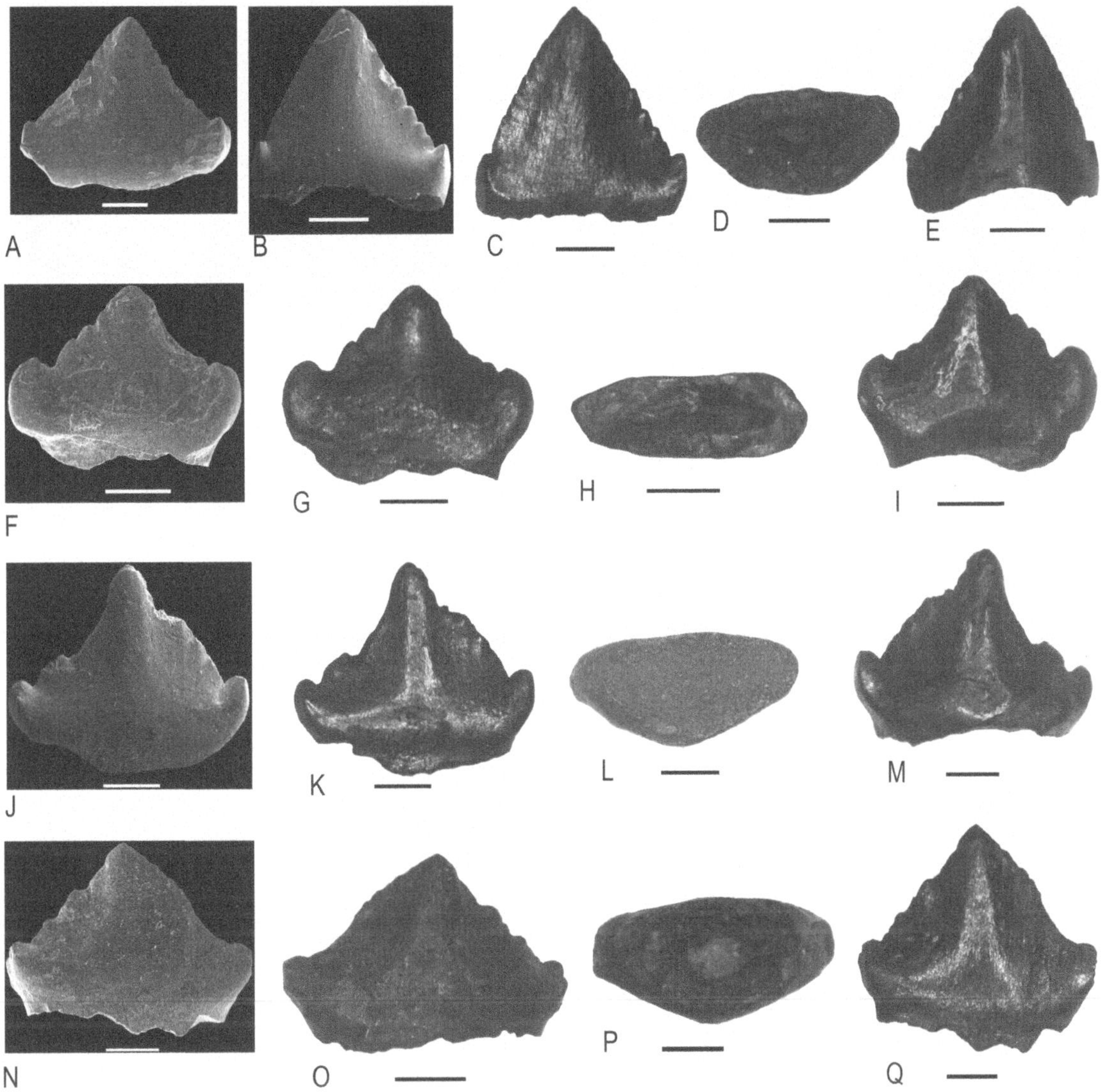

Fig. 1.6 Ornithischia indet. **A–S**. Morphotype 3. **A**. cheek tooth (DUGF/J7), labial view (SEM photomicrograph); **B–E**. ?mesial cheek tooth (DUGF/J5), **B**. lingual view (SEM photomicrograph), **C**. lingual view, **D**. basal view, **E**. labial view; **F–I**. cheek tooth (DUGF/J9), **F**. lingual view (SEM photomicrograph), **G**. lingual view, **H**. basal view, **I**. labial view; **J–M**. cheek tooth (DUGF/J8), **J**. lingual view (SEM photomicrograph), **K**. lingual view, **L**. basal view, **M**. labial view; **N–Q**. cheek tooth (DUGF/J6), **N**. lingual view (SEM photomicrograph), **O**. labial view, **P**. basal view, **Q**. lingual view. Scale bar equals 500 μm

the above cited morphological features. However, *Trimucrodon* teeth differ from those of *Echinodon* in having fewer marginal denticles that disappear well below the apex of the crown and more strongly divergent mesial most and distal most denticles. Galton (1978) also felt that there are more differences than similarities between *Echinodon* and *Trimucrodon*. Though Weishampel and Witmer (1990) considered both *T. cuneatus* and *A. kuehnei* as *nomina dubia*, Sereno (1991) placed them in Ornithischia *incertae sedis*. In light of these taxonomic uncertainties, the teeth from the Kota Formation are referred here to Ornithischia indet.

Morphotype 4

(Fig. 1.7A–D)

Referred Material: One fairly well-preserved tooth (DUGF/J16).

Description: In DUGF/J16, the crown is asymmetrical, laterally compressed, and lozenge-shaped with a high central cusp with its tip slightly tilted distally. In the lingual view, the crown is slightly concave or flat and slightly higher than mesiodistally long. On both labial and lingual surfaces, the middle of the crown is enlarged into a broad central

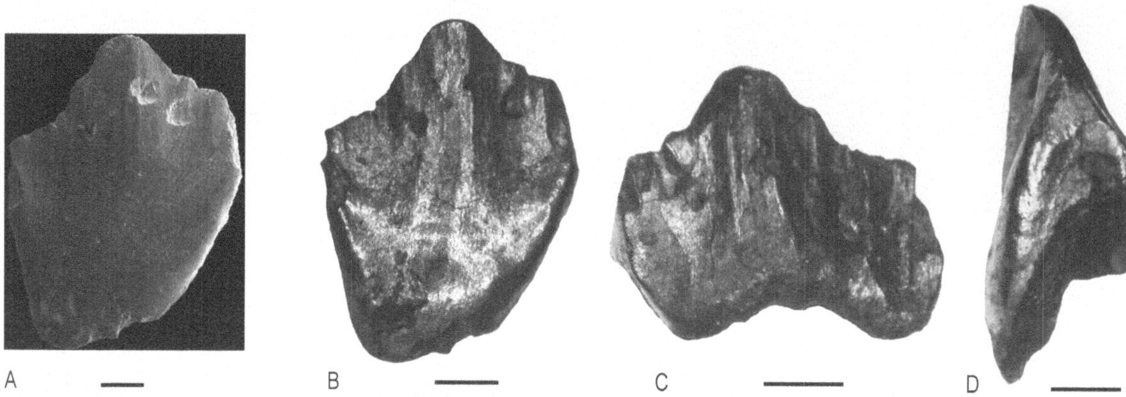

Fig. 1.7 Ornithischia indet. **A–D**. Morphotype 4, cheek tooth (DUGF/J16), **A**. lingual view (SEM photomicrograph), **B**. lingual view, **C**. labial; view, **D**. view of distal flank. Scale bar equals 500 μm

eminence. There are two narrow lateral ridges that flank the crown mesially and distally. Broad longitudinal depressions separate these ridges and the central eminence. The mesial denticulated margin is relatively longer than the distal one and both these carinae are slightly concave. Two small denticles separated by narrow longitudinal grooves are present on the mesial carina, whereas there are three such minute denticles on the distal carina. The base of the crown is produced into a slightly swollen rim. The crown is broader near the apical region than at its base which is tapering down gradually. The crown surface shows faint striations on the lingual crown surface (Fig. 1.7A–B) but these are more salient on the labial face (Fig. 1.7C). The labial crown surface is slightly convex, less high than the lingual one and the basal enamel rim is located at the level of mid lingual crown height. This could be an artifact of breakage. It has a slightly arched, swollen base. The crown is laterally compressed and has a basal sub-elliptical to sub-quadrangular cross-section. The tips of the marginal cusps are slightly worn labially. Wear facets are also observed on the mesial and distal flanks of the crown labially. The mesial and distal flanks of the crown base are slightly concave and flat, respectively.

Remarks The basal thyreophoran *Laquintasaura* and basal neornithischian *Huxinlusaurus* have apicobasally high crowns as in DUGF/J16, the crown is triangular in shape in *Laquintasaura* and the crown surface is ornamented with well-developed apicobasal ridges in both *Laquintasaura* and *Hexinlusaurus*. The crown morphology of DUGF/J16 is more close to that of heterodontosaurids. Heterodontosaurids are primarily known by *Lycorhinus angustidens* Haughton (1924), *Abrictosaurus consors* (Thulborn 1974), *Heterodontosaurus tucki* Crompton and Charig (1962), *Pegomastax africanus* Sereno (2012) from the Lower Jurassic (Hettangian-?Sinemurian) Upper Elliot Formation, South Africa, *Manidens condorensis* a small-bodied heterodontosaurid from the Middle Jurassic Cañadón Asfalto Formation in central Patagonia, Argentina (Pol et al. 2011), *Tianyulong confuciusi* from the latest Middle Jurassic (Callovian) upper horizon of the Lanqi (= Tiaojishan) Formation in Liaoning Province, northern China (Zheng et al. 2009) and *Fruitadens haagarorum* from the Upper Jurassic (Tithonian) Morrison Formation (Butler et al. 2010, 2012). DUGF/J16 compares well with the cheek teeth of heterodontosaurids, particularly with *L. angustidens* and *A. consors* in being slightly higher than long, spatulate in form, apex slightly deflected distally, crown expanding in mesiodistal length towards the apex, two to three denticles occurring towards the apex of the crown, the middle part of the crown both labially and lingually forming a broad central eminence, and less robust mesial and distal ridges converging at the base forming a swollen rim or weak 'cingulum'. In *Heterodontosaurus*, there is no basal rim, rather the crown merges with the root. Further, in *Heterodontosaurus* and possibly in *Lycorhinu*s as well, the broad central eminence is in the form of a strong ridge in contrast to its weak development in DUGF/J16 and *Abrictosaurus*. DUGF/J16 displays longitudinal enamel ridges on the labial crown surface, but the labial crown face is less high than its lingual counterpart possibly due to breakage in this region. DUGF/J16 is also somewhat similar to the teeth described as Iguanodontia gen. et sp. indet. from the Upper Jurassic strata of Guimarota lignite mine, Portugal (Rauhut 2001: fig. 5A–B) in the crown shape, development of central eminence and lateral ridges and absence of prominent secondary ridges. Because of absence of secondary ridges, the specimens from Portugal were not assigned definitively to *Iguanodon*. DUGF/J16 is also comparable to dentary/maxillary teeth described under Ceratopsia indet. from the Lower Cretaceous Shahai and Fuxin formations, northeastern China (Amiot et al. 2010: fig. 7, V16748.10) on account of laterally compressed, lozenze-shaped crown with a central primary ridge and the mesial and distal secondary ridges forming a rim at the base and the lingual crown higher than its labial counterpart. However, in the Chinese specimens, the labial crown face seems to be smooth as compared to ridged labial crown face with an arched swollen base in DUGF/J16. All known heterodontosaurid taxa share one synapomorphy i.e., dentary and maxillary cheek teeth are chisel-shaped with denticles restricted to the apical most third of the crown (Weishampel and Witmer 1990; Norman and Barrett 2002; Galton 2002; Norman et al. 2004). The presence of 'cingulum' or basal swelling of the crown was considered as a valid synapomorphy used for the identification of ornithischians (Parkar et al. 2005; Butler et al. 2006). However, 'cingulum' is absent or weakly developed in several ornithischian taxa, such as *Heterodontosaurus tucki* and *Abrictosaurus consors* (Heterodontosauridae), and an unusual heterodontosaurid from the Late Triassic of Argentina (cf. *Heterodontosaurus* sp., Baez and Marsicano 2001) and derived members of the clade Ornithopoda (Butler et al. 2006). In most other heterodontosaurids, the 'cingulum' is weakly developed (Hopson 1980; Norman and Barrett 2002). Though the Kota specimen appears to be similar to maxillary/dentary teeth of heterodontosaurids in general shape (chisel-shape) and weak development of 'cingulum', because of poor preservation it is regarded as an indeterminate ornithischian.

Morphotype 5

(Fig. 1.8A–D)

Referred Material: Three fragmentary teeth (DUGF/J13-15).

Description: There are three fragmentary teeth (DUGF/J13-15, Fig. 1.8A–D) with only a small longitudinal part preserved. A couple of marginal denticles are preserved in DUGF/J13-14 (Fig. 1.8C, D). The vertical grooves between adjacent marginal denticles do not extend far onto the crown flank. A distinct cingular ridge extends from the mesial or distal most denticle towards the middle of the crown on the inferred lingual face. In one specimen (DUGF/J 14), the 'cingulum' is finely crenulated (Fig. 1.8 D). In DUGF/15 (Fig. 1.8A, B), the crown is more elongated apicobasally than in other specimens (possibly leaf-like), and the 'cingulum' is thick extending onto the middle part and the crown surface is longitudinally grooved. However, the marginal denticles are small and not separated by deep grooves as in DUGF/J13-14.

Remarks The two fragmentary teeth DUGF/J13-14 have similar development of 'cingulum', coarse marginal denticles and the grooves between them. The fragmentary nature of these teeth do not permit any taxonomic assignment beyond Ornithischia indet. DUGF/J15 is closer to ankylosaurid teeth on account of elongated crown (leaf-like) and the widely spaced marginal denticles with shallow grooves between them. DUGF/J15 with a more elongated (higher than long) crown and marginal denticles not separated by deep grooves and a 'cingulum' extending on to the middle part resembles the teeth described under *Gargoyleosaurus* and *Ankylosaurus* by Blows and Honeysett (2014: figs. 4D, 5B) from the Early Cretaceous of England. In possessing apicobasally extending ridges on its enamel, DUGF/J15 equally compares with the crown morphology of maxillary or dentary teeth described as *Laquintasaura* from the Lower or Middle Jurassic La Quinta Formation of western Venezuela (Barrett et al. 2008: fig. 4) and *Hexinlusaurus* from the Middle Jurassic of China (He and Cai 1984; Barrett et al. 2005). However, in *Hexinlusaurus*, the crown has a rhombic outline compared to the triangular outline of *Laquintasaura*

Fig. 1.8 Ornithischia indet. Morphotype 5, **A–B**. cheek tooth fragment (DUGF/J15), **A**. lingual view; **B**. labial view; **C**. cheek tooth fragment (DUGF/J13), ?lingual view; **D**. cheek tooth fragment (DUGF/J14), ?lingual view; Scale bar equals 500 μm

and the inferred triangular outline of DUGF/J15. Though DUGF/J15 exhibits dental characters similar to those of the basal ornithischian *Laquintasaura*, in view of limited fossil material at hand, the fragmentary tooth from the Kota Formation is conservatively referred to an indeterminate ornithischian.

Order Saurischia Seeley, 1887
Suborder Theropoda Marsh, 1881
Family Dromaeosauridae Matthew and Brown, 1922
Dromaeosauridae indet.

(Fig. 1.9A–AE)

Referred Material: Twenty six fairly well preserved (DUGF/J19-38, 43-48) and six fragmentary teeth.

Description: Based on morphological differences, the teeth are described below under five different morphotypes.

Morphotype 1

DUGF/J19-25 (Fig. 1.9A–R) have relatively broad-based crowns, slightly recurved with a nearly straight distal margin and convex mesial margin. In all these teeth, the root is not preserved. The labial face of the crown is strongly convex, while the lingual face is flat to slightly convex in the middle. A convex and broad central eminence extends from the base towards the apex where it becomes narrow (DUGF/J19). In some of these teeth (DUGF/J19, 21, 22-24), the middle part of the convex lingual face is flat (DUGF/J23-25). The tip of these teeth is slightly lingually canted. The mesial and distal carinae bear denticles, but the distal denticles are comparatively larger than the mesial ones. On both carinae, the denticles are larger in the middle and decrease in size apically and basally. The denticles on the distal carina are oriented perpendicular to the longitudinal axis at the base of the crown but towards the apex their tips get slightly apically curved. Denticles are almost as high as they are long. In DUGF/J22, the mesial carina does not bear any denticles. This face shows a tear-drop shaped wear tapering down the mesial carina from the apex. A small wear facet also occurs just below the apex on the distal carina. The surface is smooth but appears to have very fine longitudinal striations. DUGF/J25 has a moderately deep attritional furrow that extends from the apex to one-third height of the crown on its lingual face and an oval basal cross-section. This tooth is less compressed labiolingually than the other teeth. In DUGF/J19-21, the mesial carina terminates one-third height above the crown base and is slightly lingually twisted. In DUGF/J22-23 and 25, though it terminates before reaching the base, the mesial carina occupies midline of the mesial margin. The basal cross-section is tear-drop shaped in DUGF/J19, 21, 22, 24, 25 and elliptical to oval in DUGF/J20 and 23. A longitudinal groove which tapers towards the apex is present on the central eminence of the lingual face of most of these teeth.

Morphotype 2

DUGF/J/27-30, 32-34 (Fig. 1.9V-AB) are more strongly distally recurved than those of Morphotypes 1 and 3 with a strongly convex mesial carina and concave or straight distal carina. The tip of the crown extends slightly beyond the base distally in DUGF/J28, whereas it does not extend beyond distal base in rest of the teeth referred to this morphotype. In one of the better preserved larger specimens (DUGF/J29), the mesial carina twists lingually about one-fourth the distance from the apex and terminates before reaching the base. The mesial carina is not denticulated in these teeth. The distal margin is concave and the carina bears relatively larger denticles in the middle and the denticles decrease in size apically and basally. The distal carina, occupying a middle position along the distal margin, terminates at the base of the tooth and its denticles are perpendicular to the long axis at the base and have slightly apically tilted tips towards the apex. The mesial margin of the crown is rounded as compared to labiolingually compressed distal margin. On the lingual face, a broad convex eminence extends from the apex to the base of the crown medially, but at the base it becomes flattened with a central shallow groove. The labial surface is convex in the middle and mesially but a shallow longitudinal depression exists between the central broad ridge and the distal margin. The crown base is elliptical in outline in DUGF/J29 whereas DUGF/J27, 28, 30 and 33 have oval cross-sections.

Morphotype 3

DUGF/J26 (Fig. 1.9S–U) is comparatively larger in size than the teeth referred to Morphotype 1 (Table 1.2). In this tooth, the lingual face of the tooth is less flattened than in Morphotype 1, rather the median part is convex and in the form of a broad eminence with laterally depressed areas near the carinae. The mesial carina is convex and the distal one is rectilinear as in Morphotype 1. The apex is slightly lingually canted and blunt. It is also tilted distally. The distal carina, which is broken at its base, bears small, chisel-shaped denticles which are perpendicular to the long axis of the crown. On the mesial margin of the crown, the enamel is heavily etched and probably had minute denticles. The mesial carina is better preserved in the apical half of the crown and appears to have been slightly twisted lingually. The labial face of the crown is uniformly convex with a slight longitudinal depression between the distal carina and the broad median

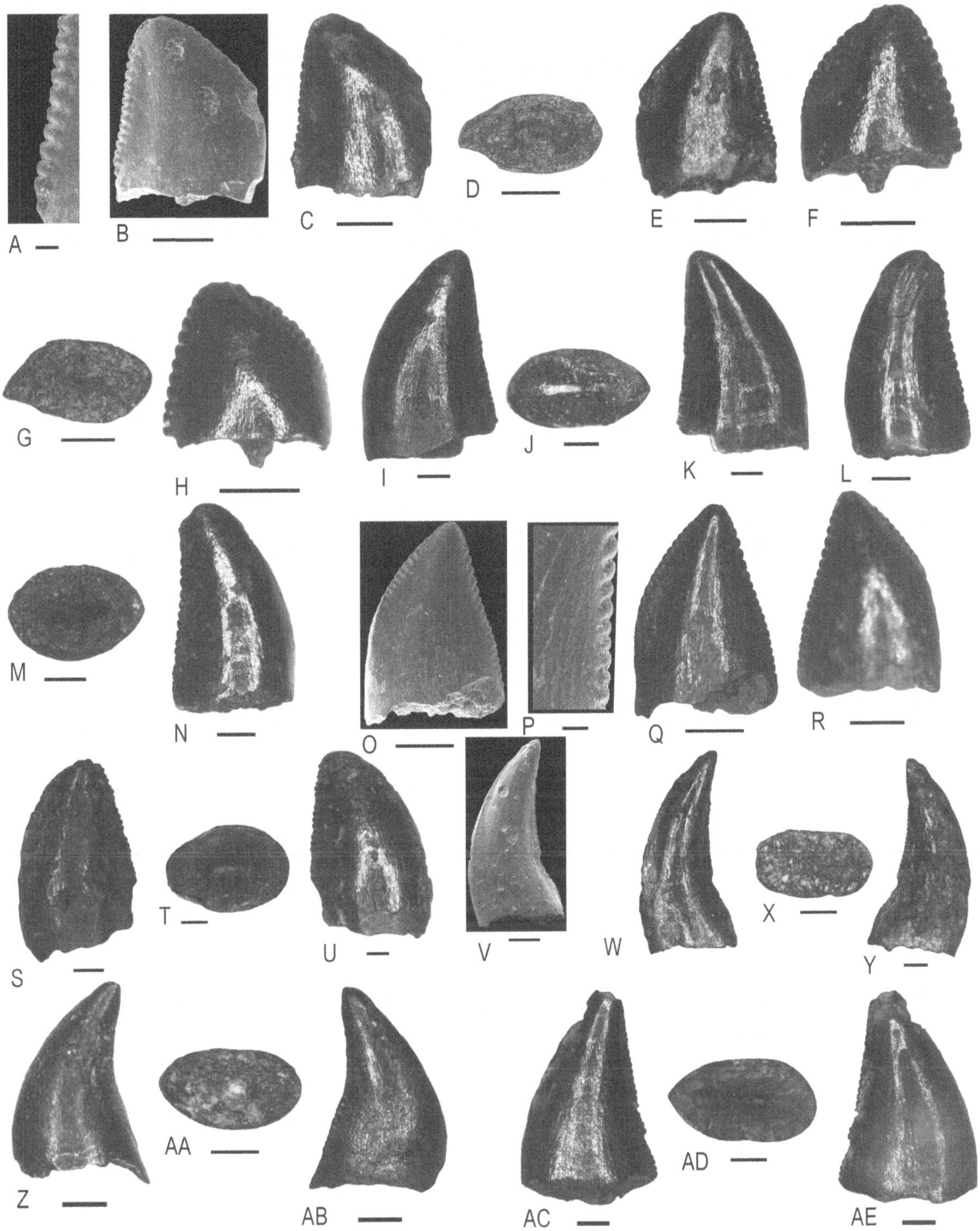

Fig. 1.9 A–AE. Isolated teeth of Doromaeosauridae indet. **A–R**. Morphotype 1, **A–E** (DUGF/J21), **A**. enlarged view of distal carina (SEM photomicrograph), **B**. labial view (SEM photomicrograph), **C**. labial view, **D**. basal view, **E**. lingual view; **F–H** (DUGF/J20), **F**. lingual view, **G**. basal view, **H**. labial view; **I–K** (DUGF/J22), **I**. labial view, **J**. basal view, **K**. lingual view; **L–N** (DUGF/J25), **L**. lingual view, **M**. basal view, **N**. labial view; **O–R**. (DUGF/J19), **O**. lingual view (SEM photomicrograph), **P**. enlarged view of distal carina (SEM photomicrograph), **Q**. lingual view, **R**. labial view; **S-U**. Morphotype 3 (DUGF/J26), **S**. lingual view, **T**. basal view, **U**. labial view; **V–AB**. Morphotype 2, **V–Y** (DUGF/J29), **V**. lingual view (SEM photomicrograph), **W**. lingual view, **X**. basal view, **Y**. labial view; **Z–AB** (DUGF/J32), **Z**. lingual view, **AA**. Basal view, **AB**. labial view; **AC–AE**. Morphotype 4 (DUGF/J36), **AC**. lingual view, **AD**. basal view, **AE**. labial view. Scale bar equals 500 μm for **B-O**, **Q-AE** and 100 μm for **A** and **P**

Table 1.2 Measurements of theropod teeth

Specimen no.	TCH	FABL	BW	CBR (TCH/BW)	MDC	DDC	DSDI
DUGF/J19	2.00	1.00	0.70	1.42	16/1mm	13/1.0 mm	1.23
DUGF/J20	1.00	1.00	0.60	1.66	15/1.0 mm	13/1.0 mm	1.15
DUGF/J21	2.00	1.00	0.70	1.42	Poorly preserved	13/1.0 mm	
DUGF/J22	3.00	2.00	1.00	2.00	No denticles	11/1.0 mm	
DUGF/J23	2.00	1.00	0.90	1.11	16/1.0 mm	14/1.0 mm	1.14
DUGF/J25	3.00	2.00	1.00	2.00	Worn	7.5/1.0 mm	
DUGF/J26	4.00	2.00	2.00	1.00	Heavily worn	11/1.0 mm	
DUGF/J27	3.00	1.00	0.70	1.42	No denticles	11/1.0 mm	
DUGF/J28	2.00	1.00	0.60	1.66	No denticles	7.5/1.0 mm	
DUGF/J29	3.00	2.00	0.90	2.22		13.5/1.0 mm	
DUGF/J30	2.00	1.00	0.80	1.25	No denticles	No denticles	
DUGF/J31	2.00	1.00	0.80	1.25	No denticles	Weakly developed	
DUGF/J32	2.00	2.00	0.80	2.50	No denticles	11/1.0 mm	
DUGF/J33	3.00	1.00	0.80	1.25	No denticles	Not preserved	
DUGF/J34	2.00 tip broken	1.00	0.70	1.42	No denticles	14/1.0 mm	
DUGF/J35	1.00	0.70	0.50	1.40		20/1.0 mm	
DUGF/J36	3.00	2.00	1.00	2.00	No denticles	10.5/1 mm	
DUGF/J37	2.00	1.00	0.80	1.25	9.0/1.0 mm	8.0/1.0 mm	1.25
DUGF/J39	4.00	2.00	1.00	2.00		14/1.0 mm	
DUGF/J45	2.00	2.00	1.00	2.00	14/1.0 mm	11/1.0 mm	1.27
DUGF/J46	3.00	3.00	1.00	3.00	No denticles	10/1.0 mm	
DUGF/J52	2.00	1.00	0.80	1.25	No denticles	12/1.0 mm	
DUGF/J53	3.00	1.00	0.90	1.11	No denticles	16/1.0 mm	
DUGF/J54	2.00	1.00	0.70	1.42		16/1.0 mm	
DUGF/J55	3.00	0.80	0.70	1.14		15/1 mm	
DUGF/J56	3.00	1.00	1.00	1.00	No denticles, only flange	12.5/1 mm	
DUGF/J60	2.00	1.00	1.00	1.00	No denticles	13/1mm	
DUGF/J61	4.00	1.00	0.90	1.11	18/1.0 mm	12/1.0 mm	1.50
DUGF/J62	3.00	1.00	1.00	1.00	No denticles	13.5/1.0 mm	
DUGF/J63					No denticles	18/1.0 mm	
DUGF/J64	3.00	0.90	0.70	1.28	22/1.0 mm	18/1.0 mm	1.22
DUGF/J65	3.00	0.90	0.70	1.28	19/1.0 mm	14/1.0 mm	1.35
DUGF/J68	Apex broken	1.0	1.0	1.0	16/1.0 mm	12/1.0 mm	1.33

BW = basal width, CBR = crown base ratio, DSDI = denticle size difference index, FABL = fore-apt basal length, MDC = mesial carina denticle count per mm, DDC = distal carina denticle count per mm, TCH = tooth crown height

longitudinal eminence. The basal cross-section is oval in form with a slight indentation at the distal margin of the lingual face. DUGF/J26 is less labiolingually compressed than the teeth of Morphotypes 1 and 2.

Morphotype 4

DUGF/J36 (Fig. 1.9AC–AE) has a broad base, triangular outline, denticulated distal and non-denticulated mesial margin, flat labial and lingual surfaces in the middle of the crown, chisel-like denticles perpendicular to the long axis of the crown at the base and inclined towards the apex in the apical region, and lenticular cross-section. The longitudinal groove present on the central eminence of the lingual face of teeth belonging to morphotype 1 and 2 is not distinct on this tooth.

Morphotype 5

DUGF/J43-48 (Fig. 1.10A–Q) are incompletely preserved teeth characterized by triangular crowns and strong labiolingual compression. DUGF/J43, broken at its base, is a better preserved specimen of the six referred teeth (Fig. 1.10A–E). This tooth is recurved distally with a convex mesial margin and slightly concave distal margin and its apex is slightly canted lingually. The mesial and distal carinae are denticulated and the distal denticles are larger than the mesial ones. On both carinae, the denticle tips are pointing towards the apex. In DUGF/J43 and 44, however, the denticles of the distal carina are perpendicular to the long axis at the base. The lingual face of the crown in DUGF/J43 is in the form of a convex broad eminence in the middle which slopes laterally towards the mesial and distal carinae

(Fig. 1.10C). Whereas in DUGF/J 44, the lingual face of the crown is covered by a broad, flat to slightly convex longitudinal eminence that extends from the apex to the base (Fig. 1.10I). This median eminence is separated from the mesial and distal carinae by shallow and narrow longitudinal depressions. On the labial surface of all teeth, a medial convex eminence extends from the apex to the base which gently slopes to the mesial margin. This central eminence is separated from the distal carina by a shallow longitudinal groove and the distal margin is more expanded laterally in the basal half forming a labiolingually compressed flange. Along this flange the denticles become comparatively larger than in the apical half. Because of the labiolingual compression, the teeth have a lenticular cross-section. DUGF/J37-38 also display a crown morphology very similar to that of DUGF/J43-48, but DUGF/J37 has a tear drop-shaped basal cross-section (Fig. 1.10G) in comparison to oval or lenticular cross-section of DUGF/J43-48 (Fig. 1.10D, J, M, P).

Remarks Isolated theropod dinosaur teeth are difficult to distinguish at genus and species level as many of their morphological features, such as crown basal length and width, crown height, presence or absence of denticles on mesial and distal carinae, denticle shape, density and orientation on the carina, are similar in many theropod dinosaur groups (Hendrickx and Mateus 2014; Gerke and Wings 2016). Besides this, differences in dental morphology due to position in the jaw, intraspecific and ontogentic variations are poorly understood at present. However, many recent works have adapted morphometric methods to assign isolated teeth at least to higher taxonomic levels (Smith et al. 2005; Fanti and Therrien 2007; Larson 2008; Larson and Currie 2013; Hendrickx and Mateus 2014; Gerke and Wings 2016). Because of limited number of theropod teeth recovered from the Kota Formation, only qualitative morphological features are used in the present study for identifying them at higher taxonomic levels.

The labiolingual compression and recurvature of the Kota teeth with mesial and distal serrations are plesiomorphic characters observed in theropod dinosaurs. In view of their diminutive size, the taxonomic affinity of the Kota teeth with Tyrannosauroidea, Allosauroidea, Carcharodontosauridae, Ceratosauridae and Megalosauridae is ruled out. The small-sized Kota teeth may have come from juvenile individuals or small-sized theropod taxa. Han et al. (2011) while describing similar-sized theropod teeth from the Middle-Upper Jurassic Shishugou Formation, China discussed their possible derivation from juveniles or adults of dromaeosaurids (Osborn 1924; Xu et al. 2000; Norell et al. 2006), compsognathids (Currie and Chen 2001), troodontids (Sankey et al. 2002), basal ornithomimosaurs (Perez-Moreno et al. 1994), and primitive alvarezsauroids (Choiniere et al. 2010). Lateral teeth of Compsognathidae differ from those of the Kota Formation in lacking mesial denticles, distal denticles disappearing well beneath the apex of the crown and in the presence of a constriction between the crown and root (Currie and Chen 2001; Peyer 2006; Dal Sasso and Maganuco 2011; Hendrickx and Mateus 2014). Among troodontids, *Troodon* differs from all the Kota morphotypes in its less recurved teeth with large, apically inclined distal denticles and constriction between the crown and the root (Currie et al. 1990), whereas *Byronosaurus* (Makovicky et al. 2003), *Mei* (Xu and Norell 2004), *Urbacodon* (Averianov and Sues 2007) and *Anchiornis* (Hu et al. 2009) are distinct from the Kota teeth in having non-serrated magins and a constriction between crown and root. In the presence of unserrated and basally constricted teeth, basal ornithomimosaurs such as *Pelicanimimus* (Perez-Moreno et al. 1994) and *Shenzhousaurus* (Ji et al. 2003) are distinctly different from the Kota teeth. Basal alvarezsauroids are also distinguished from the Kota teeth in having several vertical ridges on the labial surface of the crown (Han et al. 2011).

Dromaeosaurid teeth are relatively small in size, moderately labiolingually compressed and strongly curved distally, and bear mesial and distal carinae with prominent denticles at least on the distal carina and have a sigmoidal carina in mesial view with some lingual twisting (Currie et al. 1990; Sankey et al. 2002). Lingual spiraling or twisting of mesial carina is also known in basal tyrannosauroid *Guanlong* (Han et al. 2011), some tyrannosaurids (Sankey et al. 2002) and allosauroids (Currie and Carpenter 2000). In *Dromaeosaurus* and basal tyrannosauroid *Guanlong*, the lingual twisting of the mesial carina begins almost at the apex (Han et al. 2011), whereas in Kota Morphotypes 1–3, the mesial carina descends a short distance from the apex before twisting lingually as in isolated dromaeosaurid teeth from Barremian of Uña, Spain (Rauhut 2002). The relatively small and moderately laterally compressed teeth described here under Morphotypes 1–3 exhibit distally curved crowns with a twisted mesial carina, oval cross-section (with a rounded mesial margin and a more acute distal margin), fore-aft basal length twice basal width, lack of constriction between the crown and root, denticles either on both carinae or restricted to the distal one, when both carinae are denticulated, distal ones slightly larger than the mesial ones, and chisel-shaped distal denticles placed perpendicular to the long axis of the carina at the base and slightly tilted apically near the tip that conform to the dental morphology described for Dromaeosauridae (Currie et al. 1990; Basso 1997; Han et al. 2011; Sankey et al. 2002; Hendrickx and Mateus 2014; Gerke and Wings 2016). The tooth described under Morphotype 4 differs from those of Morphotypes 1–3 in lacking the lingual twist of mesial carina and in having less labiolingually compressed crowns. Rather they have straight mesial carina. Teeth similar to those of *Dromaeosaurus* with

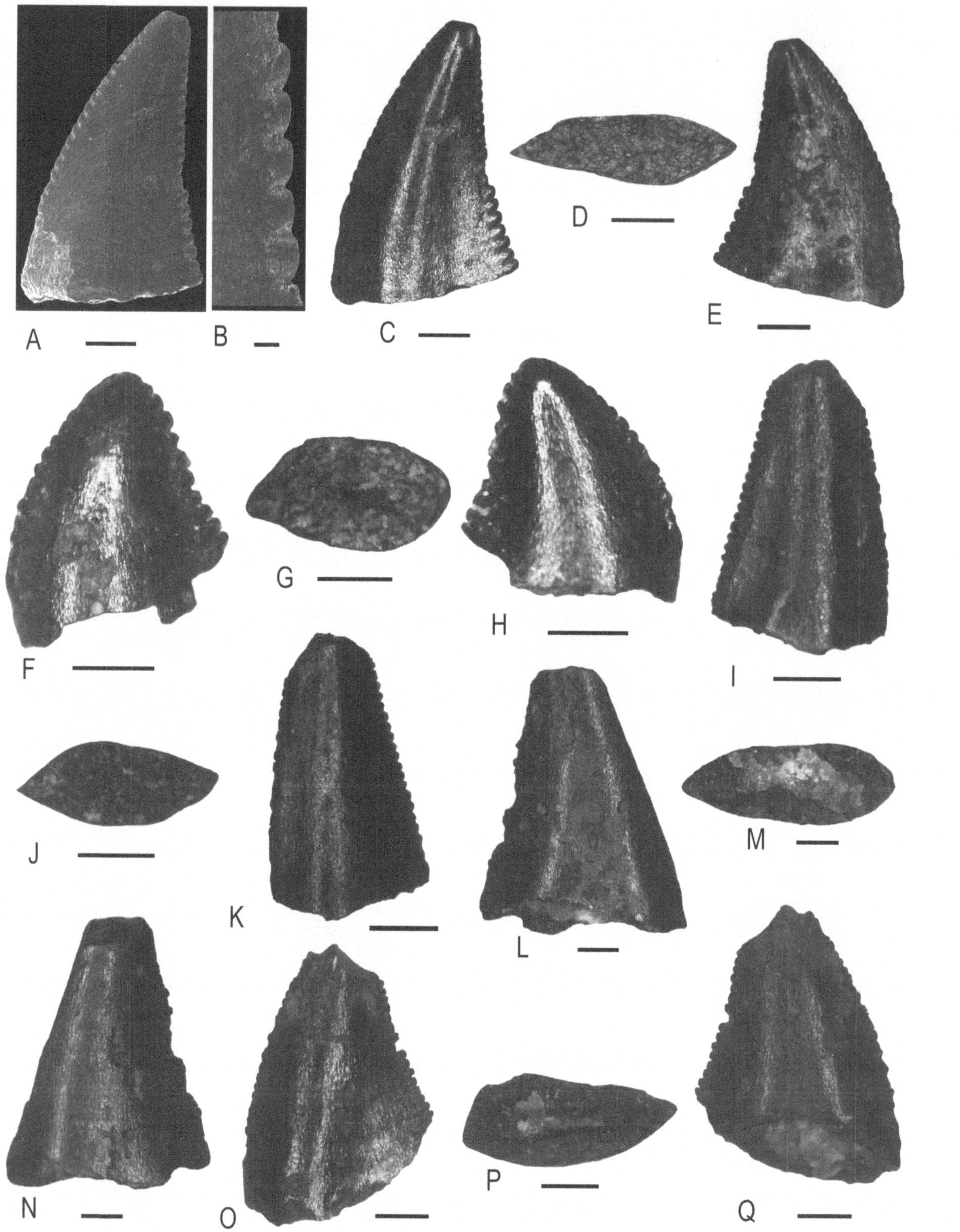

Fig. 1.10 A-Q. Isolated teeth of Dromaeosauridae indet., Morphotype 5. **A–E** (DUJF/J43), **A**. lingual view (SEM photomicrograph), **B**. enlarged view of distal carina at its base (SEM photomicrograph), **C**. lingual view, **D**. basal view, **E**. labial view; **F–H** (DUGF/J37), **F**. labial, **G**. basal, **H**. lingual; **I–K** (DUGF/J44), **I**. lingual, **J**. basal, **K**. labial; **L–N** (DUGF/J46), **L**. lingual, **M**. basal, **N**. labial; **O–Q** (DUGF/J45), **O**. labial, **P**. basal, **Q**. lingual. Scale bar equals 500 μm for **A**, **C–Q** and 100 μm for **B**

a straight mesial carina were reported earlier also (Currie et al. 1990). Keeping this in view, Morphotype 4 is tentatively referred to Dromaeosauridae indet. until more taxonomically diagnostic specimens are found. Dromaeosaurids teeth are known to exhibit considerable variation in shape, some being long and fang-like or resembling curved daggers, some being short and strongly hooked (Longrich 2008), while others have a figure of eight in cross-section caused by concavities on labial and lingual surfaces of the crown (e.g. *Buitreraptor* [Makovicky et al. 2005; Gianechini et al. 2011], *Austroraptor* [Novas et al. 2009]). The denticles are also variably present or absent in premaxillary, maxillary and dentary teeth and on mesial and distal carinae in *Microraptor*, *Tsaagan*, *Bambiraptor*, *Shanag* (Xu et al. 2000; Hwang et al. 2002; Burnham 2004; Hwang 2005; Norell et al. 2006) and *Sinornithosaurus* (Xu and Wu 2001). Lateral teeth of dromaeosaurids *Buitreraptor*, *Dromaeosaurus* and *Saurornitholestes* are characterized by a crown morphology distinct from other dromaeosaurids. *Buitreraptor* lacks denticles on both mesial and distal carinae (Gianechini et al. 2011). In *Dromaeosaurus*, the carinae bear sub-quadrangular denticles with convex margins (Currie et al. 1990; Currie 1995), whereas in *Saurornitholestes*, the teeth bear large and apically hooked denticles. In light of these observed variations in dental morphology of Dromaeosauridae, though the teeth referred to Morphotype 5 approach the crown morphology of velociraptorine dromaeosaurids (Currie et al. 1990) in their strong lateral compression, distal curvature of crown apex, location of mesial carina on the midline of the crown, shape of the denticles and the size difference observed between the denticles of mesial and distal carinae (DSDI 1.25–1.27), it is conservatively included in Dromaeosauridae.

Prior to this report, Jurassic dromaeosaurid teeth have been documented from the Middle Jurassic (Bathonian) Isalo Group (Isalo IIIb) of the Mahajanga Basin, northern Madagascar (Maganuco et al. 2005), the Middle Jurassic (Bathonian) of the UK (Evans and Milner 1994), the Middle–Upper Jurassic Shishugou Formation of northwest Xinjiang, China (Han et al. 2011), the Upper Jurassic Guimarota lignite deposits of Portugal (Zinke 1998), the Upper Jurassic of Germany (van der Lubbe et al. 2009; Gerke and Wings 2016), and the Upper Jurassic – Lower Cretaceous succession of Galve area, NE Spain (Sanchez-Hernandez et al. 2007). The teeth belonging to Morphotypes 1–5 of the Kota Formation fall within the size range of those from UK, Guimarota (Portugal), Shishugou (China) and Galve (Spain). The DSDI measured for a few of the Kota teeth with denticles on both carinae falls within the range of 1.15–1.23. This value is slightly higher than 0.86–1.17 and 1.0 measured for dromaeosaurid teeth from the Lower Cretaceous of Uña, Spain (Rauhut 2002) and the Upper Jurassic of Portugal (Zinke 1998), respectively, but is within the range of 1.2 measured for dromaeosaurid teeth from the Late Jurassic – Early Cretaceous of Galve, Spain (Sanchez-Hernandez et al. 2007).

Richardoestesia-like teeth

(Fig. 1.11A–AJ)

Referred Material: 22 fairly well-preserved (DUGF/J39, 49–69) and four fragmentary teeth (DUGF/J70-73).

Description: The teeth are moderately labiolingually compressed, lanceolate or triangular with apicobasally elongated crowns and have suboval basal cross-section. The tip of the crown is slightly distally recurved and lingually canted. The mesial margin of the tooth is convexly rounded, whereas the distal margin is nearly straight. The lingual face of the teeth is moderately convex in the middle and in the form of a broad eminence extending from the apex to the base. The labial face of the crown is more convex than the lingual face. A shallow, medial longitudinal groove is present on both labial and lingual faces of the crown which gradually disappears towards the apex. Some of these teeth are slightly constricted at the crown-root junction due to rounding of the tooth at the base of mesial and distal carinae (Fig. 1.11A, K–M, X–Z). A few teeth (DUGF/J56, 60, 61-63, and 71) have slightly broader bases (Fig. 1.11 T–W, AA, AC–AE) as compared to narrow bases of majority of the teeth. In one set of teeth, only the distal carina is denticulated (DUGF/J49-61) (Fig. 1.11C, E, G, Q, S, X, Z, AA–AC, AE), while the second set of teeth have mesial denticles as well which are relatively small in size as compared to the distal ones (DUGF/J62, 64-65,68, 72) (Fig. 1.11H–J, K, M, AF–AI, AJ). These teeth have broad, mesiodistally short minute denticles with tips that are square-shaped or rounded and oriented perpendicular to the long axis. The denticles decrease in size towards the apex and also towards the base. Interdenticle spaces are minute and less discernible. In one specimen (DUGF/J54; Fig. 1.11Q–S), the lingual face at one-third height of the base is constricted with two shallow and elongated depressions on either side of the median ridge. The relevance of this constriction is not known. Faint longitudinal (apicobasal) striations are present on both labial and lingual faces of the crown in some specimens (e.g., DUGF/J61). The basal cross-section of these teeth is oval or rounded oval.

DUGF/J39 (Fig. 1.11N–P) is an elongated tooth with a broken apex. This tooth is morphologically similar to those described above except for the presence of longitudinal flutings on its labial and lingual faces. The distal carina bears minute denticles oriented perpendicular to the long axis of the crown at its basal one-fourth, whereas rest of this carina is worn flat. The mesial carina is also worn flat, but shows traces of worn denticles and terminates before reaching the base at least

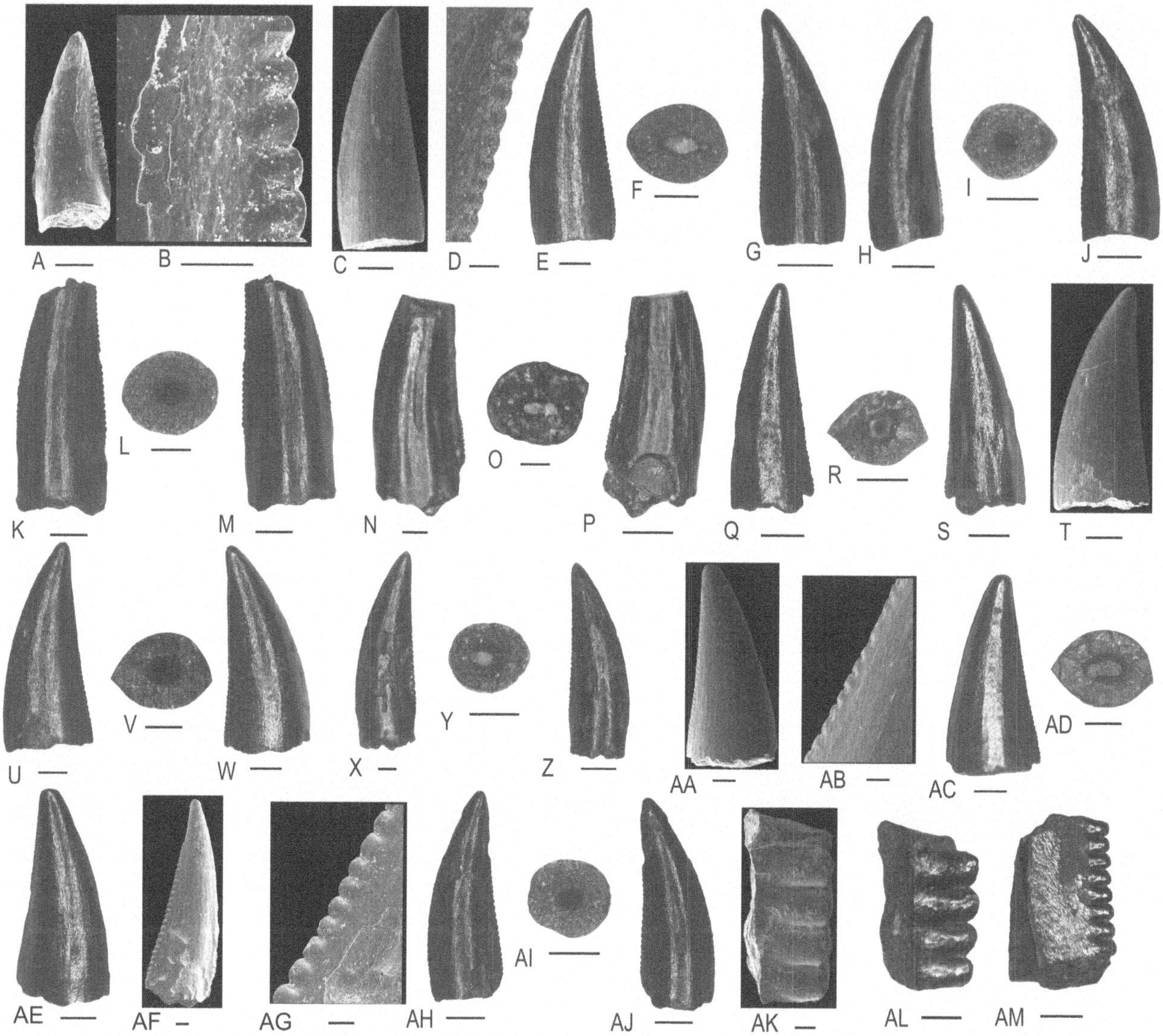

Fig. 1.11 **A–AJ**. Isolated teeth of *Richardoestesia*-like teeth. **A–B** (DUGF/J50), **A**. lingual view (SEM photomicrograph), **B**. enlarged view of distal carina (SEM photomicrograph); **C–G** (DUGF/J61), **C**. lingual view (SEM photomicrograph), **D**. enlarged view of distal carina (SEM photomicrograph), **E**. lingual view, **F**. basal view, **G**. labial view; **H–J** (DUGF/J64), **H**. labial view, **I**. basal view, **J**. lingual view; **K–M** (DUGF/J68), **K**. lingual view, **L**. basal view, **M**. labial view; **N–P** (DUGF/J39), **N**. labial view, **O**. basal view, **P**. lingual view; **Q–S** (DUGF/J54), **Q**. lingual view, **R**. basal view, **S**. labial view; **T–W**. (DUGF/J56), **T**. lingual view (SEM photomicrograph), **U**. lingual view, **V**. basal view, **W**. labial view; **X–Z** (DUGF/J55), **X**. lingual view, **Y**. basal view, **Z**. labial view; **AA–AE** (DUGF/J62), **AA**. labial view (SEM photomicrograph), **AB**. enlarged view of distal carina (SEM photomicrograph), **AC**. lingual view, **AD**. basal view, **AE**. labial view; **AF–AJ** (DIUGF/J65), **AF**. labial view (SEM photomicrograph), **AG**. enlarged view of distal carina (SEM photomicrograph), **AH**. lingual view, **AI**. basal view, **AJ**. labial view; **AK–AM**. Theropoda indet. **AK–AL** (DUGF/J41), **AK**. in labial or lingual view (SEM photomicrograph), **AL**. in labial or lingual view; **AM** (DUGF/J40) in labial or lingual view. Scale bar equals 100 μm for **B**, **D**, **AB**, **AG**, and **AK**, 300 μm for **A**, and 500 μm for **E–AF**, **AH–AJ**, **AL**, and **AM**

one-fourth the height of the crown from the base. The tip is broken perpendicular to the long axis and a wear facet extends slightly labially at its middle region. The basal cross-section is oval to elliptical in outline with a broadly rounded mesial margin and laterally compressed distal margin.

Remarks In having lanceolate shape and minute denticles as compared to the tooth size, these teeth are comparable to those of *Richardoestesia* described from the Upper Cretaceous strata of North America (Currie et al. 1990; Basso 1997; Longrich 2008; Sankey 2008; Larson 2008; Larson et al. 2010; Larson and Currie 2013). *Richardoestesia* has a long stratigraphic range from Santonian to Maastrichtian and a wide geographic distribution in North America, extending from West Texas to southern Alberta (Larson and Currie 2013; Sankey 2001; Sankey et al. 2002, 2005; Larson 2008). At present, *Richardoestesia* is known by three named species, *R. gilmorei* from the Upper

Cretaceous (Campanian) Dinosaur Park Formation of southern Alberta (Currie et al. 1990), *R. isosceles* from the Upper Campanian Aguja Formation, Texas (Sankey 2001), and *R. asiatica* from the Upper Cretaceous (Turonian) Bissekty Formation, central Kyzylkum Desert, Uzbekistan (Sues and Averianov 2013). According to Sankey et al. (2002), the teeth of *R. gilmorei* are strongly recurved resembling those of *Saurornitholestes*, whereas the teeth of *R. isosceles* have a shape of isosceles triangle. Moreover, the denticles of *R. gilmorei* are pointed as compared to their squarish appearance in *R. isosceles*. The teeth of *R. asiatica* are characterized by straight mesial and distal margins, labiolingually flattened crown and the presence of triangular central depression at the base, both labially and lingually. Besides the occurrences from Canada, USA and Uzbekistan, *Richardoestesia*-like teeth have also been documented from the Upper Jurassic Guimarota coal mine, Portugal (Zinke 1998), the Lower Cretaceous localities at Uña (Rauhut and Zinke 1995) and Galve, Spain (Rauhut 2002), the Upper Cretaceous deposits of southcentral Pyrenees (Prieto-Márquez et al. 2000) and the Upper Cretaceous (Maastrichtian) Ocozocoautla Formation of Chiaps, southern Mexico (Carbot-Chanona and Rivera-Sylva 2011).

The teeth of *R. gilmorei* differ from the Kota specimens in the curvature of the proximal part and mesiodistally broad crown bases, pointed denticles and flattened oval to oval cross-section. In *R. isosceles*, the teeth are not constricted at the crown base and have a trapezoidal cross-section at the base (Sankey et al. 2002). Furthermore, the teeth of *R. gilmorei* and *R. isosceles* are characterised by subequal mesial and distal denticles as in the teeth from the Upper Jurassic Guimarota lignite mine (Zinke 1998) and the Lower Cretaceous of Uña (Rauhut and Zinke 1995). This is contrary to the smaller mesial denticles of the present teeth and the Upper Cretaceous teeth from Uzbekistan (Sues and Averianov 2013), Mexico (Carbot-Chanona and Rivera-Sylva 2011) and the San Juan Basin, northwestern New Mexico (Williamson and Brusatte 2014). The longitudinal groove, generally present on the labial and lingual faces of *R. gilmorei*, *R. isosceles* (Currie et al. 1990; Sankey et al. 2005), *R. asiatica* (Sues and Averianov 2013) and cf. *Richardoestesia* sp. from the Upper Jurassic of Guimarota (Zinke 1998) and the Lower Cretaceous of Galve (Rauhut 2002), are also present on many of the Kota teeth. *Richardoestesia*-like teeth from the Late Cretaceous of the south-central Pyrenees (Prieto-Márquez et al. 2000), Chiaps, southern Mexico (Carbot-Chanona and Rivera-Sylva 2011), and New Mexico (Williamson and Brusatte 2014) have denticles on the distal carina only as in some teeth from the Kota Formation. The Mexican teeth, however, differ from the Kota specimens in having mesiodistally expanded crowns with a less flattened oval cross-section. DUGF/J 39 with longitudinal flutings and minute denticles on the distal carina with respect to the size of the tooth is comparable to *Richardoestesia* sp. described from the Upper Cretaceous Lance Formation, USA (Longrich 2008: fig. 9.8F). The basal constriction observed in the teeth of *R. gilmorei* and in the teeth from San Juan Basin, New Mexico (Williamson and Brusatte 2014), is also present in some of the Kota teeth (DUGF/J 55, 68). Though the denticles are chisel-shaped in the Guimarota, Uña and Kota teeth, the mesial denticles in the Kota teeth are smaller than the distal ones. The DSDI of the Kota teeth with denticles on both carinae is 1.22 for DUGF/J 64, 1.33 for DUGF/J 68, 1. 35 for DUGF/J 65, and 1.50 for DUGF/J 61. The DSDI ranges between 1.08 and 1.5 for cf. *Richardoestesia* sp. from the Lower Cretaceous of Uña, Spain (Rauhut 2002) and from 0.80 to 1.33 for cf. *Richardoestesia* sp. from the Upper Jurassic of Portugal (Zinke 1998). Hence the DSDI of the Kota conforms to the values documented for *Richardoestesia* from Jurassic and Cretaceous sites.

The teeth from the Kota Formation of India, the Upper Jurassic of Portugal and Lower Cretaceous of Spain closely resemble the North American species *R. isosceles* in having elongated crowns with an isosceles triangular shape that might indicate a long temporal range for this clade. It may also be the result of convergent evolution as suggested by Zinke (1998). In the past, *Richardoestesia* has not been identified with any known families below the clade Maniraptora within the order Theropoda as its phylogenetic relationships were not clear (Currie et al. 1990). Some considered *Richardoestesia* as a possible spinosaurid or as a product of convergent evolution with spinosaurids (Sankey et al. 2002), whereas others regarded them as troodontids (Hwang 2005). On re-examining *R. gilmorei*, Longrich (2008) suggested that it may be referred to Paraves. A piscivorous diet was suggested for *Richardoestesia* because of the lanceolate shape of the teeth and high abundance in near coastal deposits such as Hell Creek and Lance formations (Basso 1997; Sankey 2001; Longrich 2008). However, based on cladistic and quantitative analyses of isolated theropod teeth from the Upper Jurassic of Portugal Hendrickx and Mateus (2014) identified a tooth (ML 939) as *Richardoestesia* aff. to *R. gilmorei* and placed *Richardoestesia* in the Dromaeosauridae. Following Hendrickx and Mateus (2014), *Richardoestesia*-like teeth from the Kota Formation are assigned to the Dromaeosauridae.

Theropoda indet.

(Fig. 1.11AK-AM)

Referred Material: Two tooth fragments (DUGF/J/40-41).

Description: There are two fragments (DUGF/J 40-41) representing one of the dental carinae in the Kota microvertebrate fossils. Of these, the denticles of DUGF/J40

(Fig. 1.11AM) are smaller than those of DUGF/J41 (Fig. 1.11AK–AL). The denticles are cylindrical, chisel-like in form with their mesiodistal length greater than their apicobasal height. The denticles are oriented perpendicular to the long axis of the carinae. The blood grooves are prominent and deeply developed extending up to the base of the denticles. However, they do not bend basally at their contact with the crown surface as in allosaurids or tyrannosaurids.

Remarks These specimens are too fragmentary to comment on their taxonomic affinity and are thus left in open nomenclature.

Discussion

Prior to the present finds from the Kota Formation, a partial skull and postcranial skeleton, attributed to a stegosaur *Dravidosaurus blanfordi*, was described from the Coniacian Trichinopoly Group within the *Kosmaticeras theobaldianum* Zone exposed near Siranattam village, Cauvery Basin, South India (Yadagiri and Ayyasami 1979). However, its identification as a stegosaur was questioned by some (Chatterjee and Rudra 1996; Wilson et al. 2011), while others considered it as *nomen dubium* (Maidment et al. 2008) or ?Stegosauria indet (Pereda Suberbiola et al. 2015). It is the first record of an ornithischian dinosaur from India. Presence of an ornithischian tooth in the Kota Formation was noted by Prasad (1986) though it was wrongly identified as a hypsilophodontid. Following a detailed study of supposed ornithischian dinosaur bone identified as a stegosaur dermal plate from the Upper Cretaceous Kallamedu Formation (Anonymous 1978), Galton and Ayyasami (2017) concluded that this bone probably belongs to a sauropod. They, however, agreed with the referal by Mateus et al. (2011) of a right pes impression from the Upper Cretaceous (Maastrichtian) Lameta Formation of the Jetholi area, Balasinor, Kheda district, Gujarat, western India (Mohabey 1986) to stegosaurian ichnotaxon *Deltapodus* sp. Ankylosaur bones comprising spines, limbs and girdle elements are also known to occur in the Lameta Formation at Rahioli in Gujarat, western India (Chatterjee 2020). The presence of ornithischian remains in the Upper Triassic Upper Dharmaram Formation of the Pranhita-Godavari Valley was also mentioned by Kutty et al. (1987), but these fossils have not yet been described or illustrated.

In comparison to well documented early sauropod skeletons from the Kota Formation (*Barapasaurus tagorei* Jain et al. 1975; *Kotasaurus yamanapalliensis* Yadagiri et al. 1979), no theropod and ornithischian dinosaur remains have been reported so far except for one report of a supposed hypsilophodontid tooth (Prasad 1986) and alleged carnosaur (Yadagiri 1982) and ankylosaur (Nath et al. 2002). Yadagiri (1982) reported the occurrence of fragmentary ischium, dorsal and caudal vertebrae and isolated tooth representing a supposed carnosaur in the Kota Formation under a new taxon *Dandakosaurus indicus* (Theropoda *incertae sedis*) in a progress report submitted to the Geological Survey of India. This remained inaccessible in the classified unpublished field reports of the Geological Survey of India. Neither a full description nor illustrations of this find have been published so far and thus regarded as *nomen nudum*. Purported ankylosaur bones consisting of parts of the skull, maxillae and dentaries, vertebrae, 30 specimens of body armor, fragments of ilium and scapula and teeth were reported from the red clays of the Kota Formation occurring below the limestone (Nath et al. 2002). However, the illustrated tooth (Nath et al. 2002: plate 1, fig. e) appears little like an ankylosaurid tooth, rather it compares well with a crocodilian tooth. The bones reported by Nath et al. (2002) were regarded as those of a crocodylomorph (Wilson and Mohabey 2006: 471). However, Galton and Carpenter (2016: 203) identified the dermal armor of Nath et al. (2002) as Thyreophora indet. The present study provides evidence for greater diversity in the dinosaur fauna of the Kota Formation, particularly for the small dinosaur fauna. The occurrence of a scute and fragmentary limb bone of possible scelidosaurid affinity in the Kota Formation has also been noted (Chatterjee 2020).

The Paikasigudem section of the Kota Formation has not yielded any large vertebrates so far and the vertebrate microfossils recovered range in size from 1 to 8 mm. Although the teeth described are limited in number and have relatively few diagnostic features, we can make reasonably acceptable taxonomic inferences on the presence of several ornithischian and theropod morphotypes. The dinosaur fauna described here is clearly represented by small individuals <10 kg body weight as is the case with the dinosaur fauna from the Upper Jurassic Guimarota Coal Mine (Rauhut 2001) and many other such microfaunas. The most commonly present teeth in the Kota assemblage are those of *Richardoestesia*. The predominance of theropod teeth in the assemblage is similar to the Guimarota fauna. Isolated teeth with resorption pits indicate that they are shed teeth. The microvertebrate assemblage from the Paikasigudem section is primarily represented by small-sized fauna as in the cases of Guimarota and Uña sites (Zinke 1998; Rauhut 2001, 2002), and most of the Middle Jurassic vertebrate microsites of UK (Evans and Milner 1994; Sweetman 2004). Apparently it could be an artifact of size sorting of the fauna before its burial. But there are no visible evidences for long distance transport in the vertebrate microfossils recovered from the studied section. Alternatively, the environmental niche may have favored small bodied taxa.

Although the fossil material described is fragmentary in nature, it improves our current understanding of vertebrate diversity of the Kota Formation. It also adds to our knowledge on limited fossil data available for ornithischian and theropod dinosaurs of the Gondwana. The large number of morphotypes present in the ornithischian and theropod dinosaur teeth assemblage of the Kota Formation indicates that the taxonomic diversity is relatively high though identification at lower taxonomic levels is not possible because of the absence of taxon specific synapomorphies and poor preservatioin of some of the teeth. On a conservative note, we can conclude that there are at least five distinct taxonomic units of early ornithischians within the seven morphological variants of indeterminate ornithischians described. Similarly, there are at least a minimum of three distinct taxonomic units (one dromaeosaurid, one *Richardoestesia*-like, one indeterminate) among the seven theropod morphotypes. However, more well-preserved and unambiguously associated fossil materiali is needed to know the real diversity of Kota dinosaur fauna.

Some of the earliest records of dromaeosaurid dinosaurs came from the Middle Jurassic Isalo Group (Isalo IIIb) of Mahajanga Basin, Madagascar (Maganuco et al. 2005; Flynn et al. 2006), the Middle Jurassic (Bathonian) Chipping Norton Limestone Formation at Hornsleasow Quarry, Gloucestershire, England (Metcalf et al. 1992), the Middle Jurassic (Upper Bathonian) Forest Marble, Kirtlington Quarry, UK (Evans and Milner 1994), the Middle-Upper Jurassic Shishugou Formation, northwest Xinjiang, China (Han et al. 2011), the Middle Jurassic Balabansai Svita in the northern Fergana depression, Kyrgyzstan (Averianov et al. 2005), the Upper Jurassic (Tithonian) Mugher Mudstone Formation of Ethiopia (Goodwin et al. 1999; Hall and Goodwin 2011), the Upper Jurassic of Boulonnais, northern France (Buffetaut and Martin 1993), and the Upper Jurassic (Kimmeridgian) of Portugal (Zinke 1998; Hendrickx and Mateus 2014), and the Upper Jurassic of Germany (van der Lubbe et al. 2009; Gerke and Wings 2016). All these reports were based on tooth morphotypes. It is evident from this list that the geographic distribution of dromaeosaurid dinosaur is skewed towards the Laurasian continents and the apparent scarcity of coelurosaurian clades in the former Gondwana is an artifact of poor sampling in these localities.

The age of the Kota Formation has remained problematic as conflicting ages were suggested on the basis of different fossil groups. In general, the Early Jurassic age was favored based on weak biostratigraphic correlations of different vertebrate groups (Jain 1973, 1980, 1983; Yadagiri and Prasad 1977; Evans et al. 2001), whereas the Middle Jurassic age was suggested using ostracod and palynological data (Govindan 1975; Vijaya and Prasad 2001). As there are no marine intercalations within the Kota Formation, no biostratigraphic constraints are available from marine microfossils. Likewise, no radiometric dates are available for the Kota Formation as it contains no volcanic ash beds or associated radiometrically datable horizons. The new fossils described here provide some constraints on the age of this formation. Until now, some of the oldest records of *Phyllodon* and *Alocodon*-like (Ornithischia incertae sedis) teeth came from the Lower-Middle Bathonian Chipping Norton Limestone Formation (Metcalf et al. 1992; Metcalf and Walker 1994), Upper Bathonian Stonesfield Slate (Galton 1980) and Kirtlington (Evans and Milner 1994). The oldest dromaeosaurids are known from the Middle Jurassic (Bathonian) of UK (Evans and Milner 1994). The Kota sauropod dinosaurs *Barapasaurus* and *Kotasaurus* are of similar evolutionary grade as *Cetiosaurus* of UK. As the ornithischian and dromaeosaurid taxa as well as other vertebrate groups from the Middle Jurassic of UK are compositionally similar to those of Kota Formation, we conclude that Middle Jurassic age is a more reasonable estimate for the Kota Formation.

Biogeographic Links of the Kota Vertebrate Fauna

Some of the well-known Jurassic continental vertebrate localities in the world are the Lower Jurassic Kayenta Formation, USA, the Middle Jurassic Kota Formation of India, the Middle Jurassic sites of U.K., the Middle Jurassic Cañadón Asfsalto Formation of Argentina, the Middle Jurassic Isalo Group III of Madagascar, the Middle Jurassic Guelb el Ahmar site of Morocco (Haddoumi et al. 2016), the Upper Jurassic Morrison Formation of USA, the Upper Jurassic Tendaguru Beds of Tanzania, the Upper Jurassic Guimarota Lignite deposits of Portugal, and the Upper Jurassic Mugher Mudstone of Ethiopia (Goodwin et al. 1999). The Kota Formation has produced one of the richest continental Jurassic vertebrate faunas in Gondwana and with >42 taxa it represents a taxonomically diverse assemblage. All major groups of vertebrates are represented. Ostracods occur in great abundance and charophytes are also present though in limited numbers. The fauna is characterized by predominantly Laurasian taxa along with a few Gondwanan taxa. The Kota vertebrate fauna includes three taxa, viz., *Barapasaurus*, *Dyskritodon* and an australosphenidan which are restricted to the Gondwana in their distribution. *Barapasaurus tagorei*, a sauropod dinosaur from the Kota Formation is considered to be phylogenetically related to *Vulcanodon* of Zimbabwe. *Dyskritodon amazighi*, a eutriconodontid, was described from the Early Cretaceous Anoual Ksar Metlili site of Morocco, NW Africa (Sigogneau-Russell 1995). Later, Prasad and Manhas (2002) described a new species of this genus, *Dyskritodon indicus*, from the Kota Formation of India. Australosphenidan mammals are also restricted to the Middle Jurassic of Madagascar (Flynn et al.

1999) and South America (Rauhut et al. 2002; Martin and Rauhut 2005; Rougier et al. 2007), and Early Cretaceous of Australia (Rich et al. 1997, 2001). More recently, a definite lower molar and a possible upper molar of an australosphenidan mammal was recorded from the Kota Formation (Parmar et al. 2015). On the other hand, semionotid fishes, primitive ornithischian and dromaeosaurid dinosaurs, kuehneotheriid and amphidontid symmetrodonts, morganucodontids, docodonts, possible dryolestids, multituberculates and yet to be described eleutherodontid haramiyidans are typical Laurasian taxa. None of these taxa are known from the Middle Jurassic Cañadón Asfalto Formation of Argentina and the Isalo Group III of Madagascar, and the Upper Jurassic Tendaguru Beds.

In the presence of primitive ornithischians (*Alocodon*-like, *Trimucrodon*-like), Dromaeosauridae indet., *Richardoestesia*-like, and multituberculate mammals, the fauna from the Upper Jurassic Guimarota Coal Mine is closely comparable to that of the Kota Formation. However, large theropod dinosaurs such as allosaurids and tyrannosaurids of the Guimarota fauna are missing in the Kota fauna. In fact at higher taxonomic levels, the vertebrate faunal composition of the Kota Formation comprising of freshwater hybodont sharks, semionotid fishes, amphibians, testudines, sphenodontids, crocodylians, pterosaurs, sauropod, primitive ornithischian and dromaeosaurid dinosaurs, symmetrodont, morganucodontid, eutriconodont, docodont, multituberculate mammals and eleutherodontid haramiyidans is more similar to that of English Bathonian localities such as the Upper Bathonian Forest Marble, Kirtlington Quarry (Evans and Milner 1994), the Middle Jurassic of Skye (Scotland) (Evans and Waldman 1996; Evans et al. 2006) and the Chipping Norton Limestone Formation at Hornsleasow Quarry, Gloucestershire (Metcalf et al. 1992; Metcalf and Walker 1994; Evans and Milner 1994) than to any other localities. Hence the Kota fauna broadly displays greater taxonomic affinities to the Laurasian British faunas rather than to Gondwanan faunas.

Among the Gondwanan faunas, the Moroccan Guelb el Ahmar site shares *Lepidotes*, Testudinata, rhynchocephalians, Atoposauridae crocodiles, pterosaur, theropod and ornithischian dinosaur remains with the Kota Formation (Haddoumi et al. 2016). Slightly younger ?Berriasian Ksar Metlili site also shares eutriconodont (*Dyskritodon*), multituberculates, and symmetrodonts, in addition to dromaeosaurids, ornithischians, pterosaurs, cryptodiran turtle, freshwater hybodont sharks, *Lepidotes* with the Kota Fauna. Even in the case of the Moroccan sites, the vertebrate fauna was considered to have close affinities with Laurasian British Bathonian faunas.

The predominantly Laurasian taxa in the Kota fauna implies that either these are the relict taxa with formerly Pangaean distribution or dispersed along a trans-Tethyan route across the Mediterranean Tethyan Sill as inferred for the Laurasian elements of Guelb el Ahmar fauna (Haddoumi et al. 2016). Haddoumi et al. (2016), while discussing the biogeographic relationships of supposedly Laurasian affinities of the Middle Jurassic Guelb el Ahmar biota, pondered over two possible biogeographic scenarios: 1) the taxa had Pangaean distribution in the geological past and there are significant gaps in the fossil record of Gondwana; 2) the absence of Laurasian taxa reported from Guelb Al Ahmar site in other Gondwanan Jurassic sites may represent real absences. The relatively restricted occurrence of the Kota fauna in other sites in Gondwana may also suggest provincialism or biogeographic partitioning within Gondwana. However, this lack of common fauna in the Gondwanan continents may also be the result of poor sampling in these areas as compared to Laurasia. Therefore, a clear picture would emerge as and when all the Jurassic sites of the southern hemisphere are extensively sampled for microvertebrate faunas.

Conclusions

The Indian subcontinent has a unique history of being part of Gondwana that broke-up in the Late Jurassic and having subsequently drifted rapidly towards the north to collide with Asia in the Middle Eocene. The fauna from the Kota Formation, therefore, represents the fauna of pre-drift India and offers some insights into the biodiversity prior to the drift phase. Based on qualitative morphological observations of isolated teeth, five morphotypes of ornithischians, five morphotypes of Dromaeosauridae indet., one *Richardoestesia*-like form, and one morphotype of Theropoda indet. are identified in the vertebrate fauna of the Kota Formation. It adds at least five possible ornithischian and three theropod taxonomic groups to the existing record of two sauropod dinosaur taxa. Prior to this study no theropod dinosaurs have been reported from this formation and only the occurrence of ankylosaurids among the ornithischians have been documented so far. Although the morphotypes are identified at higher taxonomic levels, they point to greater diversity in the Jurassic dinosaur fauna of India. The close similarities between the vertebrate fauna of the Kota Formation and the Middle Jurassic deposits of UK favours a Middle Jurassic age for the Kota Formation. The predominant occurrence of Laurasian taxa over Gondwanan taxa underscores close biogeographic links between India and Laurasia with a restricted faunal interchanges between India and other Gondwanan landmasses in the Middle Jurassic. However, it may also be argued that many of these vertebrate groups had a Pangaean distribution and limited sampling in the Gondwanan continents obscures true intra-Gondwanan biogeographic relationships. A clear picture will emerge only when the Gondwanan Jurassic sites are extensively and intensively sampled.

Acknowledgements This contribution is dedicated to Prof. Ashok Sahni who has been a source of inspiration for a generation of Indian vertebrate paleontologists. The authors are thankful to Peter Galton, Paul Barrett, and Alexander O. Averianov for their insightful comments and suggestions on the original manuscript. Lourembam Ranjit Singh is thanked for his help in making the figures. GVRP acknowledges grants for this work from J. C. Bose National Fellowship of Science and Engineering Research Board, New Delhi. The research was also supported by funds from the Department of Science and Technology (DST), New Delhi (SR/FTP/ES-46/2009) to VP in the form of Fast Track Research Project.

References

Amiot, R., Kusuhashi, N., Xu, X., & Wang, Y. (2010). Isolated dinosaur teeth from the Lower Cretaceous Shahai and Fuxin formations of northeastern China. *Journal of Asian Earth Sciences, 39,* 347–358.

Anonymous, (1978). New dinosaurian remains. *Geological Survey of India, News, 9*(5), 4.

Averianov, A. O., Martin, T., & Bakirov, A. A. (2005). Pterosaur and dinosaur remains from the Middle Jurassic Balabansai Svita in the northern Fergana Depression, Kyrgyzstan (Central Asia). *Palaeontology, 48,* 135–155.

Averianov, A. O., & Sues, H. D. (2007). A new troodontid (Dinosauria: Theropoda) from the Cenomanian of Uzbekistan, with a review of troodontid records from the territories of the former Soviet Union. *Journal of Vertebrate Paleontology, 27,* 87–98.

Baez, A. M., & Marsicano, C. A. (2001). A heterodontosaurid ornithischian dinosaur from the Upper Triassic of Patagonia. *Ameghiniana, 38,* 271–279.

Bandyopadhyay, S., Gillette, D. D., Ray, S., & Sengupta, D. P. (2010). Osteology of *Barapasaurus tagorei* (Dinosauria: Sauropoda) from the Early Jurassic of India. *Palaeontology, 53*(3), 533–569.

Bandyopadhyay, S., & Roychowdhury, T. K. (1996). Beginning of the continental Jurassic in India: A paleontological approach. In M. Morales (Ed.), *The continental Jurassic, Museum of Northern Arizona Bulletin, 60* (pp. 371–378). Arizona: Museum of Northern Arizona Press.

Barrett, P. M. (2001). Tooth wear and possible jaw action of *Scelidosaurus harrisonii* Owen and a review of feeding mechanisms in other thyreophoran dinosaurs. In K. Carpenter (Ed.), *The armored dinosaurs* (pp. 25–52). Bloomington: Indiana University Press.

Barrett, P. M., Butler, R. J., & Knoll, F. (2005). Small-bodied ornithischian dinosaur from the Middle Jurassic of Sichuan, China. *Journal of the Vertebrate Paleontology, 25*(4), 823–834.

Barrett, P. M., Butler, R. J., Mundil, R., Scheyer, T. M., Irmis, R. B., & Sánchez-Villagra, M. R. (2014). A paleoequatorial ornithischian and new constraints on early dinosaur diversification. *Proceedings of the Royal Society B, 281*(1791), 20141147. https://dx.doi.org/10.1098/rspb.2014.1147.

Barrett, P. M., Butler, R. J., Novas, F. E., Moore-Fay, S. C., Moody, J. M., Clark, J. M., et al. (2008). Dinosaur remains from the La Quinta Formation (Lower or Middle Jurassic) of the Venezuelan Andes. *Paläontologische Zeitschrift, 82*(2), 163–177.

Basso, S. (1997). Investigations on Canadian dinosaurs: Systematic paleontology of isolated dinosaur teeth from the latest Cretaceous of south Alberta, Canada. *Courier Forschunginstitut Senckenberg, 196,* 33–37.

Bhattacharya, A., Ray, S., Datta, P.M., & Maulick, P. (1994). Fossil Charophyta from the Kota Formation of the Pranhita-Godavari valley, Andhra Pradesh, India. *Proceedings of Ninth International Gondwana Symposium 1* (pp. 471–475). New Delhi: Oxford and IBH Publishing Co. Pvt. Ltd.

Blows, W. T., & Honeysett, K. (2014). New teeth of nodosaurid ankylosaurs from the lower Cretaceous of southern England. *Acta Paleontologica Polonica, 59*(4), 835–841.

Buffetaut, E., & Martin, M. (1993). Late Jurassic dinosaurs from the Boulonnais (Northern France): A review. *Revue de Paléobiologie, 7,* 17–28.

Burnham, D.A. (2004). New information on *Bambiraptor feinbergi* (Theropoda: Dromaeosauridae) from the Late Cretaceous of Montana. In P. J. Currie, E. B. Koppelhus, M. A. Shugar & J. L. Wright (Eds.), *Feathered Dragons: Studies on the transition from dinosaurs to birds* (pp. 67–111), Bloomington: Indiana University Press.

Butler, R. J., Galton, P. M., Porro, L. B., Chiappe, L. M., Henderson, D. M., & Erickson, G. M. (2010). Lower limits of ornithischian dinosaur body size inferred from a new Upper Jurassic heterodontosaurid from North America. *Proceedings of Royal Society B, 277,* 375–381.

Butler, R. J., Porro, L. B., Galton, P. M., & Chiappe, L. M. (2012). Anatomy and cranial functional morphology of the small-bodied dinosaur *Fruitadens haagarorum* from the Upper Jurassic of the USA. *PLoS One, 7,* e31556. https://doi.org/10.1098/rspb.2009.1494
.

Butler, R. J., Porro, L. B., & Heckert, A. B. (2006). A supposed heterodontosaurid tooth from the Rhaetian of Switzerland and a reassessment of the European Late Triassic record of Ornithischia (Dinosauria). *Neues Jahrbuch für Geologie und Paläontologie, Abhandlungen, 2006*(10), 613–633.

Butler, R. J., Upchurch, P., & Norman, D. B. (2008). The phylogeny of the ornithischian dinosaurs. *Journal of Systematic Palaeontology, 6,* 1–40.

Carbot-Chanona, G., & Rivera-Sylva, H. E. (2011). Presence of a maniraptoriform dinosaur in the Late Cretaceous (Maastrichtian) of Chiapas, southern Mexico. *Boletín de la Sociedad Geológica Mexicana, 63*(3), 393–398.

Carpenter, K., & Galton, P. M. (2018). A photo documentation of bipedal ornithischian dinosaurs from the Upper Jurassic Morrison Formation, USA. *Geology of the Intermontane West, 5,* 167–207.

Chatterjee, S. (2020). The age of dinosaurs in the land of Gonds. In G. V. R. Prasad & R. Patnaik (Eds.), *Biological consequences of plate tectonics: New perspectives on post-Gondwana break-up* (pp. 181–226). Cham: Springer.

Chatterjee, S., & Rudra, D. K. (1996). KT events in India: Impact, rifting, volcanism and dinosaur extinction. *Memoirs of Queensland Museum, 39,* 489–532.

Choiniere, J. N., Xu, X., Clark, J. M., Forster, C. A., Guo, Y., & Fan, H. (2010). A basal alvarezsauroid theropod from the early Late Jurassic of Xinjiang, China. *Science, 327,* 571–574.

Colbert, E. H. (1981). A primitive ornithischian dinosaur from the Kayenta Formation of Arizona. *Bulletin Museum of Northern Arizona, 53,* 1–61.

Coombs, W. P. (1990). Teeth and taxonomy in ankylosaurs. In K. Karpenter & P. J. Currie (Eds.), *Dinosaur systematics, approaches and perspectives* (pp. 269–279). Cambridge: Cambridge University Press.

Crompton, A. W., & Charig, A. J. (1962). A new ornithischian from the Upper Triassic of South Africa. *Nature, 196,* 1074–1077.

Currie, P. J. (1995). New information on the anatomy and relationships of *Dromaeosaurus albertensis* (Dinosauria: Theropoda). *Journal of Vertebrate Paleontology, 15,* 576–591.

Currie, P. J., & Carpenter, K. (2000). A new specimen of *Acrocanthosaurus atokensis* (Theropoda, Dinosauria) from the Lower Cretaceous Antlers Formation (Lower Cretaceous, Aptian) of Oklahoma, USA. *Geodiversitas, 22,* 207–246.

Currie, P. J., & Chen, P.-J. (2001). Anatomy of *Sinosauropteryx prima* from Lioaning, northeastern China. *Canadian Journal of Earth Sciences, 38,* 1705–1727.

Currie, P. J., Rigby, J. K., & Sloan, R. E. (1990). Theropod teeth from the Judith River Formation of southern Alberta, Canada. In K. Carpenter & P. J. Currie (Eds.), *Dinosaur systematics. approaches and perspectives* (pp. 107–125). Cambridge: Cambridge University Press.

Dal Sasso, C., & Maganuco, S. (2011). *Scipionyx samniticus* (Theropoda: Compsognathidae) from the Lower Cretaceous of Italy: osteology, ontogenetic assessment, phylogeny, soft tissue anatomy, taphonomy and paleobiology. *Memoire della Società Italiana di Scienze Naturali e del Museo Civico di Storia Naturale di Milano, 37,* 1–281.

Datta, P. M. (1981). The first Jurassic mammal from India. *Zoological Journal of the Linnean Society of London, 73,* 307–312.

Datta, P. M., & Das, D. P. (2001). *Indozostrodon simpsoni*, gen. et sp. nov., an Early Jurassic megazostrodontid mammal from India. *Journal of Vertebrate Paleon*tology, *21(3),* 528–534.

Datta, P. M., Manna, P., Ghosh, S. C., & Das, D. P. (2000). The first Jurassic turtle from India. *Palaeontology, 43,* 99–109.

Dzik, J. (2003). A beaked herbivorous archosaur with dinosaur affinities from the early Late Triassic of Poland. *Journal of Vertebrate Paleontology, 23,* 556–574.

Egerton, P. M. G. (1851). Description of fossil fishes from the Deccan. *Quarterly Journal of the Geological Society of London, 7,* 273.

Egerton, P. M. G. (1878). On some remains of ganoid fishes from the Deccan. *Palaeontologia Indica, 4*(1–2), 1–8.

Evans, S. E., Barrett, P. M., Hilton, J., Butler, R. J., Jones, M. E. H., Liang, M. M., et al. (2006). The Middle Jurassic vertebrate assemblage of Skye, Scotland. In P. M. Barrett & S. E. Evans (Eds.), *Ninth international symposium on Mesozoic terrestrial ecosystems and biota* (pp. 36–39). London: Natural History Museum Publications.

Evans, S. E., & Milner, A. R. (1994). Middle Jurassic microvertebrate assemblages from the British Isles. In N. C. Fraser & H.-D. Sues (Eds.), *In the shadow of dinosaurs, Early Mesozoic tetrapods* (pp. 302–321). Cambridge: Cambridge University Press.

Evans, S. E., Prasad, G. V. R., & Manhas, B. K. (2001). Rhynchocephalians (Diapsida: Lepidosauria) from the Jurassic Kota Formation of India. *Zoological Journal of the Linnean Society, 133,* 309–334.

Evans, S. E., Prasad, G. V. R., & Manhas, B. K. (2002). Fossil lizards from the Jurassic Kota Formation of India. *Journal of Vertebrate Paleontology, 22,* 299–312.

Evans, S. E., & Waldman, M. (1996). Small reptiles and amphibians from the Middle Jurassic of Skye, Scotland. In M. Morales (Ed.), *The continental Jurassic, Museum of Northern Arizona, Bulletin, 60* (pp. 219–226). Arizona: Museum of Northern Arizona Press.

Fanti, F., & Therrien, F. (2007). Theropod tooth assemblages from the late Cretaceous Maevarano Formation and the possible presence of dromaeosaurids in Madagascar. *Acta Palaeontologica Polonica, 52,* 155–166.

Farlow, J. O., Brinkman, D. L., Abler, W. L., & Currie, P. J. (1991). Size, shape and serration density of theropod dinosaur lateral teeth. *Modern Geology, 16,* 161–198.

Feist, M., Bhatia, S. B., & Yadagiri, P. (1991). On the oldest representative of the family Characeae and its relationships with the Porocharaceae. *Bulletin de la Société Botanique de France, Actualités Botaniques, 138*(1), 25–32.

Flynn, J. J., Fox, S. R., Parrish, J. M., Ranivoharimanana, L., & Wyss, A. R. (2006). Assessing diversity and paleoecology of a Middle Jurassic microvertebrate assemblage from Madagascar. In J. D. Harris, S. G. Lucas, J. A. Spielmann, M. G. Lockley, A. R. C. Milner, & J. I. Kirkland (Eds.), *The Triassic-Jurassic terrestrial transition, New Mexico Museum of Natural History and Science Bulletin, 37* (pp. 476–489). Albuquerque: New Mexico Museum of Natural History.

Flynn, J. J., Parrish, J. M., Rakotosamimanana, B., Simpson, W. F., & Wyss, A. R. (1999). A Middle Jurassic mammal from Madagascar. *Nature, 401,* 57–60.

Galton, P. M. (1978). Fabrosauridae, the basal family of ornithischian dinosaurs (Reptilia: Ornithopoda). *Paläontologische Zeitschrift, 52,* 138–159.

Galton, P. M. (1980). Armored dinosaurs (Ornithischia: Ankylosauria) from the Middle and Upper Jurassic of England. *Géobios, 13,* 825–837.

Galton, P. M. (1986). Herbivorous adaptations of Late Triassic and Early Jurassic dinosaurs. In K. Padian (Ed.), *The beginning of the age of dinosaurs, faunal change across the Triassic-Jurassic boundary* (pp. 203–221). Cambridge: Cambridge University Press.

Galton, P. M. (1994). Notes on dinosauria and pterodactylid from the Cretaceous of Portugal. *Neues Jahrbuch für Geologie und Paläontologie, Abhandlungen, 194,* 253–267.

Galton, P. M. (2002). New material of ornithischian (?heterodontosaurid) dinosaur *Echinodon* (Early Cretaceous, southern England) from the Late Jurassic of Fruita near Grand Junction, Colorado, USA. *Journal of Vertebrate Paleontology, 22* (suppl.), 55A–56A.

Galton, P. M., & Ayyasami, K. (2017). Purported latest bone of a plated dinosaur (Ornithischia: Stegosauria), a "dermal plate" from the Maastrichtian (Upper Cretaceous) of southern India. *Neues Jahrbuch für Geologie und Paläontologie, Abhandlungen, 285*(1), 241–246.

Galton, P. M., & Carpenter, K. (2016). The plated dinosaur *Stegosaurus longispinus* Gilmore, 1914 (Dinosauria: Ornithischia; Upper Jurassic, western USA), type species of *Alcovasaurus* n. gen. *Neues Jahrbuch für Geologie und Paläontologie, Abhandlungen, 279*(2), 185–208.

Gerke, O., & Wings, O. (2016). Multivariate and cladistic analyses of isolated teeth reveal sympatry of theropod dinosaurs in the Late Jurassic of Northern Germany. *PLoS ONE, 11*(7), e0158334. https://doi.org/10.1371/journal.pone.0158334.

Gianechini, F. A., Makovicky, P. J., & Apesteguía, S. (2011). The teeth of the unenlagiine theropod *Buitreraptor* from the Cretaceous of Patagonia, Argentina, and the unusual dentition of the Gondwanan dromaeosaurids. *Acta Palaeontologica Polonica, 56,* 279–290.

Godefroit, P., & Knoll, F. (2003). Late Triassic dinosaur teeth from southern Belgium. *Comptes Rendus Palevol, 2,* 3–11.

Goodwin, M. B., Clemens, W. A., Hutchison, J. H., Wood, C. B., Zavada, M. S., Kemp, A., et al. (1999). Mesozoic continental vertebrates with associated palynostratigraphic dates from the northwestern Ethiopian plateau. *Journal of Vertebrate Paleontology, 19,* 728–741.

Govindan, A. (1975). Jurassic freshwater ostracods from the Kota limestone of India. *Palaeontology, 18*(1), 207–216.

Haddoumi, H., Allain, R., Meslouh, S., Metais, G., Monbaron, M., Pons, D., et al. (2016). Guelb el Ahmar (Bathonian, Anoual Syncline, eastern Morocco): first continental flora and fauna including mammals from the Middle Jurassic of Africa. *Gondwana Research, 29*(1), 290–319.

Hall, L. E., & Goodwin, M. B. (2011). A diverse dinosaur tooth assemblage from the Upper Jurassic of Ethiopia: Implications for Gondwana dinosaur biogeography. *Society of Vertebrate Paleontology Poster*.

Han, F., Clark, J. M., Xu, X., Sullivan, C., Choiniere, J., & Hone, D. W. E. (2011). Theropod teeth from the Middle-Upper Jurassic Shishugou Formation of northwest Xinjiang, China. *Journal of Vertebrate Paleontology, 31*(1), 111–126.

Haubold, H. (1990). Ein neuer dinosaurier (Ornithischia, Thyreophora) aus dem Unteren Jura des nördlichen Mitteleuropa. *Revue de Paléobiologie, 9*, 149–177.

Haughton, S. H. (1924). The fauna and stratigraphy of the Stormberg Series. *Annals of the South African Museum, 12*, 323–497.

He, X.-L., & Cai, K.-J. (1984). The Middle Jurassic dinosaurian fauna from Dashanpu, Zigong, Sichuan. Vol. 1. *The Ornithopod Dinosaurs*. (Chengdu: Sichuan Scientific and Technical Publishing House [Chinese with English Summary].

Heckert, A. B. (2002). A revision of the Upper Triassic ornithischian dinosaur *Revueltosaurus* with a description of a new species. *New Mexico Museum of Natural History Science Bulletin, 21*, 253–268.

Heckert, A. B. (2004). Late Triassic microvertebrates from the lower Chinle Group (Otischalkian-Adamanian: Carnian), southwestern USA. *New Mexico Museum of Natural History Science Bulletin, 27*, 1–170.

Hendrickx, C., & Mateus, O. (2014). Abelisauridae (Dinosauria: Theropoda) from the Late Jurassic of Portugal and dentition-based phylogeny as a contribution for the identification of isolated theropod teeth. *Zootaxa, 3759*(1), 1–74.

Hopson, J. A. (1980). Tooth function and replacement in early Mesozoic ornithischian dinosaurs: Implications for aestivation. *Lethaia, 13*, 93–105.

Hu, D., Hou, L., Zhang, L., & Xu, X. (2009). A pre-*Archaeopteryx* troodontid theropod from China with long feathers on the metatarsus. *Nature, 461*, 640–643.

Hughes, T. W. H. (1876). On the relation of the fossiliferous strata Maleri and Kota near Sironcha. *Central Palaeontological Records of Geological Survey of India, 9*, 63–114.

Hunt, A. P., & Lucas, S. G. (1994). Ornithischian dinosaurs from the Upper Triassic of the United States. In N. C. Fraser & H.-D. Sues (Eds.), *In the shadow of dinosaurs, early Mesozoic tetrapods* (pp. 227–241). Cambridge: Cambridge University Press.

Hwang, S. H. (2005). Phylogenetic patterns of enamel microstructure in dinosaur teeth. *Journal of Morphology, 266*, 208–240.

Hwang, S. H., Norell, M. A., Qiang, J., & Keqin, G. (2002). New specimens of *Microraptor zhaoianus* (Theropoda: Dromaeosauridae) from North-eastern China. *American Museum Novitates, 3381*, 1–44.

Irmis, R. B., Parker, W. G., Nesbitt, S. J., & Liu, J. (2007). Early ornithischian dinosaurs: the Triassic record. *Historical Biology, 19* (1), 3–22.

Jain, S. L. (1973). New specimens of Lower Jurassic holostean fishes from India. *Palaeontology, 16*, 149–177.

Jain, S. L. (1974a). *Indocoelacanthus robustus* n. gen. n. sp. (Coelacanthidae, Lower Jurassic), the first fossil coelacanth from India. *Journal of Paleontology, 48*, 49–62.

Jain, S. L. (1974b). Jurassic pterosaur from India. *Journal of the Geological Society of India, 15*, 334–335.

Jain, S. L. (1980). The continental Lower Jurassic fauna from Kota Formation, India. In L. L. Jacobs (Ed.), *Aspects of vertebrate history* (pp. 99–123). Flagstaff: Museum of Northern Arizona Press.

Jain, S. L. (1983). A review of the genus *Lepidotes* (Actinopterygii: Semionotiformes) with special reference to the species from the Kota Formation (Lower Jurassic), India. *Journal of the Palaeontological Society of India, 28*, 7–42.

Jain, S. L., Kutty, T. S., Roychowdhury, T., & Chatterjee, S. (1975). The sauropod dinosaur from the Lower Jurassic Kota Formation of India. *Proceedings of the Royal Society, London, 188*, 221–228.

Jain, S. L., & Roy Chowdhury, T. (1987). Fossil vertebrates from the Pranhita-Godavari valley (India) and their stratigraphic correlation. In G. D. McKenzie (Ed.), *Gondwana six: Stratigraphy, sedimentology and paleontology*, Geophysical Monograph 41, (pp. 219–228). Washington, D.C: American Geophysical Union.

Ji, Q., Norrell, M., Makovicky, P. J., Gao, K., Ji, S., & Yuan, C. (2003). An early ostrich dinosaur and implications for ornithomimosaur phylogeny. *American Museum Novitates, 3420*, 1–19.

King, W. (1881). The geology of the Pranhita-Godavari valley. *Memoirs of the Geological Survey of India, 18*, 151–311.

Knoll, F. (2002). Les Fabrosauridae Galton, 1972 (Dinosauria: Ornithischia): Répartition géographique et stratigraphique; systématique et phylogénie. Thèse Dr., Muséum national d'Histoire Naturelle.

Kriwet, J., Rauhut, O. W. M., & Gloy, U. (1997). Microvertebrate remains (Pisces, Archosauria) from the Middle Jurassic (Bathonian) of southern France. *Neues Jahrbuch für Geologie und Paläontologie, Abhandlungen, 206*, 1–28.

Kutty, T. S. (1969). Some contributions to the stratigraphy of the Upper Gondwana formations of the Pranhita-Godavari valley, Central India. *Journal of the Geological Society of India, 10*(1), 33–48.

Kutty, T. S., Jain, S. L., & Roy Chowdhury, T. (1987). Gondwana sequence of the northern Pranhita-Godavari valley: Its stratigraphy and vertebrate faunas. *The Palaeobotanist, 36*, 263–282.

Larson, D. W. (2008). Diversity and variation of theropod dinosaur teeth from the uppermost Santonian Milk River Formation (Upper Cretaceous), Alberta: A quantitative method supporting identification of the oldest dinosaur tooth assemblage in Canada. *Canadian Journal of Earth Sciences, 45*, 1455–1468.

Larson, D. W., Brinkman, D. B., & Bell, R. (2010). Faunal assemblages from the upper Horseshoe Canyon Formation, an early Maastrichtian cool-climate assemblage from Alberta, with special reference to the *Albertosaurus sarcophagus* bone bed. *Canadian Journal of Earth Sciences, 49*, 1159–1181.

Larson, D. W., & Currie, P. J. (2013). Multivariate analyses of small theropod dinosaur teeth and implications for paleoecological turnover through time. *PLoS ONE, 8*(1), e54329. https://doi.org/10.1371/journal.pone.0054329.

Longrich, N. (2008). Small theropod teeth from the Lance Formation of Wyoming, USA. In J. T. Sankey & S. Baszio (Eds.), *Vertebrate microfossil assemblages: Their role in paleoecology and paleobiogeography* (pp. 135–158). Bloomington: Indiana University Press.

Lukose, N. G. (1971). *Palynological evidences on the age of the Lathi Formation, western Rajasthan* (p. 20). Calcutta: Seminar on Paleopalynology and Indian Stratigraphy.

Maganuco, S., Cau, A., & Pasini, G. (2005). First description of theropod remains from the Middle Jurassic (Bathonian) of Madagascar. *Atti della Società Italiana di Scienze Naturali e del Museo civico di Storia Naturale de Milano, 146*, 165–202.

Maidment, S. C. R., Norman, D. B., Barrett, P. M., & Upchurch, P. (2008). Systematics and phylogeny of Stegosauria (Dinosauria: Ornithischia). *Journal of Systematic Palaeontology, 6*, 367–407.

Makovicky, P. J., Apesteguia, S., & Agnolin, F. L. (2005). The earliest dromaeosaurid theropod from South America. *Nature, 437*, 1007–1011.

Makovicky, P. J., Norell, M. A., Clark, J. M., & Rowe, T. (2003). Osteology and relationships of *Byronosaurus jaffei* (Theropoda: Troodontidae). *American Museum Novitates, 3402*, 1–32.

Marsh, O. C. (1877). Notice of some new vertebrate fossils. *American Journal of Science, Series, 3*(14), 253–255.

Marsh, O. C. (1881). Principal characters of American Jurassic dinosaurs, Part V. *American Journal of Science, 21*, 417–423.

Martin, T., & Rauhut, O. W. M. (2005). Mandible and dentition of *Asfaltomylos patagonicus* (Australosphenida, Mammalia) and the evolution of tribosphenic teeth. *Journal of Vertebrate Paleontology, 25,* 414–423.

Mateus, O., Milàn, J., Romano, M., & Whyte, M. A. (2011). New finds of stegosaur tracks from the Upper Jurassic Lourinhã Formation, Portugal. *Acta Palaeontologica Polonica, 56*(3), 651–658.

Matthew, W. D., & Brown, B. (1922). The family Deinodontidae, with notice of a new genus from the Cretaceous of Alberta. *Bulletin of the American Museum of Natural History, 46,* 367–385.

Metcalf, S. J., Vaughan, R. F., Benton, M. J., Cole, J., Simms, M. J., & Dartnall, D. L. (1992). A new Bathonian (Middle Jurassic) microvertebrate site, within the Chipping Norton Limestone Formation at Hornsleasow Quarry, Gloucestershire. *Proceedings of the Geologist's Association, 103,* 321–342.

Metcalf, S. J., & Walker, R. J. (1994). A new Bathonian microvertebrate locality in the English Midlands. In N. C. Fraser & H.-D. Sues (Eds.), *In the shadow of the dinosaurs-Mesozoic small tetrapods* (pp. 322–331). Cambridge: Cambridge University Press.

Mohabey, D. M. (1986). Note on dinosaur foorprint from Kheda District, Gujarat. *Journal of the Geological Society of India, 27,* 456–459.

Nath, T. T., Yadagiri, P., & Moitra, A. K. (2002). First record of armored dinosaur from the Lower Jurassic Kota Formation, Pranhita-Godavari valley, Andhra Pradesh. *Journal of the Geological Society of India, 59,* 575–577.

Norell, M. A., Clark, J. M., Turner, A. H., Makovicky, P. J., Barsbold, R., & Rowe, T. (2006). A new dromaeosaurid theropod from Ukhaa Tolgod (Ömnögov Mongolia). *American Museum Novitates, 3545,* 1–51.

Norman, D. B., & Barrett, P. M. (2002). Ornithischian dinosaurs from the Lower Cretaceous (Berriasian) of England. *Special Papers in Paleontology, 68,* 161–189.

Norman, D. B., Crompton, A. W., Butler, R. J., Porro, L. B., & Charig, A. J. (2011). The Lower Jurassic ornithischian dinosaur *Heterodontosaurus tucki* Crompton & Charig, 1962: Cranial anatomy, functional morphology, taxonomy and relationships. *Zoological Journal of the Linnean Society, 163,* 182–276.

Norman, D. B., Witmer, L. M., & Weishampel, D. B. (2004). Basal Thyreophora. In D. B. Weishampel, P. Dodson & H. Osmólska (Eds.), *The Dinosauria* (pp. 335–342). Berkeley: University of California Press.

Novas, F. E., Pol, D., Canale, J. I., Porfiri, J. D., & Calvo, J. O. (2009). A bizarre Cretaceous theropod dinosaur from Patagonia and the evolution of Gondwanan dromaeosaurids. *Proceedings of Royal Society of London B, 276,* 1101–1107.

Osborn, H. F. (1924). Three new Theropoda, *Protoceratops* Zone, central Mongolia. *American Museum Novitates, 144,* 1–12.

Owen, R. (1842). Report on British fossil reptiles. Part II. *Report of the British Association of the Advancement of Science, 11,* 60–204.

Owen, R. (1852). Note on the crocodilian remains accompanying Dr. T. L. Bell's paper on Kotah. *Proceedings of the Geological Society, London, 7,* 233.

Owen, R. (1861). A monograph of the fossil Reptilia of the Lias formations. I. *Scelidosaurus harrisonii. Palaeontographical Society Monographs, 13,* 1–14.

Parkar, W. G., Irmis, R. B., Nesbitt, S. J., Martz, J. W., & Browne, L. S. (2005). The Late Triassic pseudosuchian *Revueltosaurus callenderi* and its implications for the diversity of early ornithischian dinosaurs. *Proceedings of the Royal Society of London B, 272,* 963–969.

Parmar, V., Prasad, G. V. R., & Kumar, D. (2013). The first multituberculate mammal from India. *Naturwissenschaften, 100,* 515–523.

Parmar, V., Prasad, G. V. R., & Kumar, D. (2015). An overview of Jurassic mammalian fauna of India. In Y. Zhang, S. Z. Wu, & G. Sun (Eds.), *The 12th symposium of Mesozoic terrestrial ecosystems, Extended Abstract* (pp. 18–20). China: Shenyang.

Peng, G.-Z. (1990). A new species of small ornithopod from Zigong, Sichuan. *Journal of the Zigong Dinosaur Museum, 2,* 19–27. (Chinese).

Pereda Suberbiola, X., Díaz-Martínez, I., Salgado, L., & de Valais, S. (2015). Síntesis del registro fósil de dinosaurios tireóforos en Gondwana. *Argentinian Paleontological Association Publication Electronica, 15*(1), 90–107.

Perez-Moreno, B. P., Sanz, J. L., Buscalioni, A. D., Moratalla, J. J., Ortega, F., & Rasskin-Gutman, D. (1994). A unique multi-toothed ornithomimosaur dinosaur from the Lower Cretaceous of Spain. *Nature, 370,* 363–367.

Peyer, K. (2006). A reconsideration of *Compsognathus* from the Upper Tithonian of Canjuers, southeastern France. *Journal of Vertebrate Paleontology, 26,* 879–896.

Pieńkowski, G., Brański, P., Pandey, D. K., Schlögl, J., Alberti, M., & Fürsich, F. T. (2015). Dinosaur footprints from the Thaiat ridge and their paleoenvironmental background, Jaisalmer Basin, Rajasthan, India. *Volumina Jurassica, 13*(1), 17–26.

Pol, D., Rauhut, O. W. M., & Becerra, M. (2011). A Middle Jurassic heterodontosaurid dinosaur from Patagonia and the evolution of heterodontosaurids. *Naturwissenschaften, 98,* 369–379.

Prabhakar, M. (1989). Palynological evidence and its significance for the Kota Formation in the Pranhita-Godavari basin. *Proceedings of 12th Indian Colloquium on Micropalaeontology and Stratigraphy,* pp 59–65.

Prasad, G. V. R. (1986). Microvertebrate assemblage from the Kota Formation (Early Jurassic) of Gorlapalli, Adilabad District, Andhra Pradesh. *Indian Society of Geoscientists Bulletin, 2,* 3–13.

Prasad, G. V. R., & Manhas, B. K. (1997). A new symmetrodont mammal from the Lower Jurassic Kota Formation, Pranhita-Godavari valley, India. *Geobios, 30,* 563–572.

Prasad, G. V. R., & Manhas, B. K. (2001). First docodont mammals of Laurasian affinity from India. *Current Science, 81,* 1235–1238.

Prasad, G. V. R., & Manhas, B. K. (2002). Triconodont mammals from the Jurassic Kota Formation of India. *Geodiversitas, 24,* 445–464.

Prasad, G. V. R., & Manhas, B. K. (2007). A new docodont mammal from the Jurassic Kota Formation of India. *Palaeontologia Electronica, 10* (2), 11.

Prasad, G. V. R., Manhas, B. K., & Arratia, G. (2004). Elasmobranch and actinopterygian remains from the Jurassic and Cretaceous of India. In G. Arratia & A. Tintori (Eds.), *Mesozoic Fishes 3: Systematics, paleoenvironments and biodiversity* (pp. 625–638). München: Verlag Dr. Friedrich Pfeil.

Prasad, G. V. R., Parmar, V., & Kumar, D. (2015). A Jurassic australosphenidan mammal from India: Implications for mammalian evolution and distribution in former Gondwanaland. In Abstract volume (pp. 462–463), *XII International Symposium on Antarctic Earth Science* (ISAES 2015) Goa, India, July 13–17, 2015.

Prieto-Márquez, A., Gaete, R., Galobart, A., & Ardèvol, L. (2000). A *Richardoestesia*-like theropod tooth from the Late Cretaceous foredeep, southcentral Pyrenees, Spain. *Eclogae Geologicae Helvetiae, 93,* 497–501.

Rao, C. N., & Shah, S. C. (1963). On the occurrence of pterosaur from the Kota-Maleri beds of Chanda district, Maharashtra. *Records of the Geological Survey of India, 92,* 315–318.

Rauhut, O. W. M. (2001). Herbivorous dinosaurs from the Late Jurassic (Kimmeridgian) of Guimarota, Portugal. *Proceedings of the Geologist's Association, 112,* 275–283.

Rauhut, O. W. M. (2002). Dinosaur teeth from the Barremian of Uña, Province of Cuenca, Spain. *Cretaceous Research, 23,* 255–263.

Rauhut, O. W. M., Martin, T., Ortiz-Jaureguizar, E., & Puerta, P. (2002). A Jurassic mammal from South America. *Nature, 416,* 165–168.

Rauhut, O. W. M., & Werner, C. (1995). First record of the family Dromaeosauridae (Dinosauria: Theropoda) in the Cretaceous of Gondwana (Wadi Milk Formation, northern Sudan. *Paläontologische Zeitschrift, 69,* 475–489.

Rauhut, O. W. M., & Zinke, J. (1995). A description of the Barremian dinosaur fauna from Uña with a comparison to that of Las Hoyas. *II International Symposium on Lithographic Limestones, Lleida-Cuenca (Spain)* (pp. 123–126). Extended Abstracts: Universidad Autónoma de Madrid, Madrid.

Rich, T. H., Flannery, T. F., Trusler, P., Kool, L., van Klaveren, N., & Vickers-Rich, P. (2001). A second tribosphenic mammal from the Mesozoic of Australia. *Records of the Queen Victoria Museum Launceston, 110,* 1–8.

Rich, T. H., Vickers-Rich, P., Constantine, A., Flannery, T. F., Kool, L., & Klaveren, N. V. (1997). A tribosphenic mammal from the Mesozoic of Australia. *Science, 278,* 1438–1442.

Robinson, P. L. (1970). *The Indian Gondwana Formations: A review* (pp. 201–268). South America: First International Symposium on Gondwana Stratigraphy, IUGS.

Rougier, G. W., Martinelli, A. G., Forasiepi, A. M., & Novacek, M. J. (2007). New Jurassic mammals from Patagonia, Argentina: a reappraisal of australosphenidan morphology and interrelationships. *American Museum Novitates, 3566,* 1–54.

Rudra, D. K. (1982). Upper Gondwana stratigraphy and sedimentation in the Pranhita-Godavari valley, India. *Quarterly Journal of the Geological Mining and Metallurgical Society of India, 54,* 56–79.

Rudra, D. K., & Maulik, P. K. (1987). Stromatolites from Jurassic freshwater limestone, India. *Mesozoic Research, 1,* 135–146.

Ruiz-Omeñaca, J. I. (2001). Dinosaurios hipsilofodóntidos (Ornithischia: Ornithopoda) en la Península Ibérica. In Colectivo Arqueológico-Paleontológico de Salas (Ed.), *Actas de las I Jornadas internacionales sobre Paleontologia de Dinosaurios y su entorno, Salas de los Infantes* (pp. 175–266).

Sanchez-Hernandez, B., Benton, M., & Naish, D. (2007). Dinosaurs and other fossil vertebrates from the Late Jurassic and Early Cretaceous of the Galve area, NE Spain. *Palaeogeography, Palaeoclimatology, Palaeoecology 249,* 180–215.

Sankey, J. T. (2001). Late Campanian southern dinosaurs, Aguja Formation, Big Bend, Texas. *Journal of Paleontology, 75,* 208–215.

Sankey, J. T. (2008). Diversity of latest Cretaceous (Late Maastrichtian) small theropods and birds: Teeth from the Lance and Hell Creek formations, USA. In J. T. Sankey & S. Baszio (Eds.), *Vertebrate microfossils: Their role in paleoecology and paleobiogeography* (pp. 117–134). Bloomington: Indiana University Press.

Sankey, J. T., Brinkman, D. B., Guenther, M., & Currie, P. J. (2002). Small theropod and bird teeth from the Late Cretaceous (Late Campanian) Judith River Group, Alberta. *Journal of Paleontology, 76,* 751–763.

Sankey, J. T., Standhardt, B. R., & Schiebout, J. A. (2005). Theropod teeth from the Upper Cretaceous (Campanian-Maastrichtian), Big Bend National Park, Texas. In K. Carpenter (Ed.), *The carnivorous dinosaurs* (pp. 127–152). Bloomington: Indiana University Press.

Seeley, H. G. (1887). On the classification of the fossil animals commonly named dinosaurs. *Proceedings of the Royal Society of London, 43,* 165–171.

Sereno, P. C. (1986). Phylogeny of the bird-hipped dinosaurs (Order Ornithischia). *National Geographic Research, 2,* 234–256.

Sereno, P. C. (1991). *Lesothosaurus*, "fabrosaurids", and the early evolution of Ornithischia. *Journal of Vertebrate Paleontology, 11* (2), 168–197.

Sereno, P. C. (1999). The evolution of dinosaurs. *Science, 284,* 2137–2147.

Sereno, P. C. (2012). Taxonomy, morphology, masticatory function and phylogeny of heterodontosaurid dinosaurs. *Zookeys, 226,* 1–225.

Sigogneau-Russell, D. (1995). Two possibly aquatic triconodont mammals from the Early Cretaceous of Morocco. *Acta Palaeontologica Polonica, 40,* 149–162.

Smith, J. B., Vann, D. R., & Dodson, P. (2005). Dental morphology and variation in theropod dinosaurs: Implications for the taxonomic identification of isolated teeth. *Anatomical Record Part A, 285,* 699–736.

Sues, H.-D., & Averianov, A. (2013). Enigmatic teeth of small theropod dinosaurs from the Upper Cretaceous (Cenomanian-Turonian) of Uzbekistan. *Canadian Journal of Earth Sciences, 50,* 306–314.

Sweetman, S. C. (2004). The first record of velociraptorine dinosaurs (Saurischia, Theropoda) from the Wealden (Early Cretaceous, Barremian) of southern England. *Cretaceous Research, 25,* 353–364.

Sykes, C. (1851). On a fossil fish from the table land of the Deccan, in the peninsular India, with a description of the specimens by P. M. G. Egerton. *Quarterly Journal of the Geological Society London, 7,* 272–273.

Tasch, P., Sastry, M. V. A., Shah, S. C., Rao, B. R. J., Rao, C. N., & Ghosh, S. C. (1973). Estherids of the Indian Gondwana: Significance for continental drift. In *3rd Symposium on Gondwana Stratigraphy, Australia*, pp. 443–452.

Thulborn, R. A. (1973). Teeth of ornithischian dinosaurs from the Upper Jurassic of Portugal. *Memoire Services Géologiques du Portugal, (Nueva Series), 22,* 89–134.

Thulborn, R. A. (1974). A new heterodontosaurid dionosaur from the Upper Triassic Red Beds of Lesotho. *Zoological Journal of the Linnean Society, 55,* 151–175.

van der Lubbe, T., Richter, U., & Knötschke, N. (2009). Velociraptorine dromaeosaurid teeth from the Kimmeridgian (Late Jurassic) of Germany. *Acta Palaeontologica Polonica, 54,* 401–408.

Vijaya, & Prasad, G. V. R. (2001). Age of the Kota Formation, Pranhita-Godavari valley, India: A palynological approach. *Journal of the Palaeontological Society of India, 46,* 77–93.

Weishampel, D. B., Dodson, P., & Osmólska, H. (2004). *The Dinosauria, First Edition (733 pp).* Berkeley: University of California Press.

Weishampel, D. B., & Witmer, L. M. (1990). Heterodontosauridae. In D. B. Weishampel, P. Dodson, & H. Osmólska (Eds.), *The Dinosauria* (pp. 486–497). Berkeley: University of California Press.

Williamson, T. E., & Brusatte, S. L. (2014). Small theropod teeth from the Late Cretaceous of the San Juan Basin, northwestern New Mexico and their implications for understanding latest Cretaceous dinosaur evolution. *PLoS ONE, 9*(4), e93190. https://doi.org/10.1371/journal.pone.0093190.

Wilson, J. A., Barrett, P. M., & Carrano, M. T. (2011). An associated partial skeleton of *Jainosaurus* cf. *septentrionalis* (Dinosauria: Sauropoda) from the Late Cretaceous of Chhota Simla, central India. *Palaeontology, 54,* 981–998.

Wilson, J. A., & Mohabey, D. M. (2006). A titanosauriform (Dinosauria: Sauropoda) axis from the Lameta Formation (Upper Cretaceous: Maastrichtian) of Nand, central India. *Journal of Vertebrate Paleontology, 26*(2), 471–478.

Xu, X., & Norell, M. A. (2004). A new troodontid dinosaur from China with avian-like sleeping posture. *Nature, 431,* 838–841.

Xu, X., & Wu, X.-C. (2001). Cranial morphology of *Sinornithosaurus millenii* Xu et al., 1999 (Dinosauria: Theropoda: Dromaeosauridae) from the Yixian Formation of Liaoning, China. *Canadian Journal of Earth Sciences, 38,* 1739–1752.

Xu, X., Zhou, Z., & Wang, X. (2000). The smallest known non-avian theropod dinosaur. *Nature, 408,* 706–708.

Yadagiri, P. (1982). Osteological studies of a carnosaurian dinosaur from the Lower Jurassic Kota Formation: Andhra Pradesh. Geological Survey of India Progress Report for Field Season Programme 1981–1982, Regional Paleontological Laboratories, Southern Region.

Yadagiri, P. (1984). New symmetrodonts from Kota Formation (Early Jurassic), India. *Journal of the Geological Society of India, 25,* 514–621.

Yadagiri, P. (1985). An amphidontid symmetrodont from the Early Jurassic Kota Formation, India. *Zoological Journal of the Linnean Society of London, 85,* 411–417.

Yadagiri, P. (1986). Lower Jurassic lower vertebrates from Kota Formation, Pranhita-Godavari valley, India. *Journal of the Palaeontological Society of India, 31,* 89–96.

Yadagiri, P. (2001). The osteology of *Kotasaurus yamanapallensis*, a sauropod dinosaur from the Early Jurassic Kota Formation of India. *Journal of Vertebrate Paleontology, 21*(2), 242–252.

Yadagiri, P., & Ayyasami, K. (1979). A new stegosaurian dinosaur from the Upper Cretaceous sediments of South India. *Journal of the Geological Society of India, 20,* 521–530.

Yadagiri, P., & Prasad, K. N. (1977). On the discovery of *Pholidophorus* fishes from the Kota Formation, Adilabad district, Andhra Pradesh. *Journal of the Geological Society of India, 18*(8), 436–444.

Yadagiri, P., Prasad, K.N., & Satsangi, P. P. (1979). The sauropod dinosaur from Kota Formation of Pranhita-Godavari valley, India. In B. Laskar & C. S. Raja Rao (Eds.), *Proceedings of the Fourth International Gondwana Symposium* (pp. 199–203). Delhi: Hindustan Publishing Corporation.

Zheng, X. T., You, H. L., Xu, X., & Dong, Z. M. (2009). Early Cretaceous heterodontosaurid dinosaur with integumentary structures. *Nature, 458,* 333–336.

Zinke, J. (1998). Small theropod teeth from the Upper Jurassic coal mine of Guimarota (Portugal). *Paläontologische Zeitschrift, 72,* 179–189.

Chapter 2
Upper Gondwana (Jurassic to Early Cretaceous) Palynoflora of India: Its Correlation with Other Gondwana Continents and Phytogeographical Implications

Shreya Mishra and Vandana Prasad

Abstract Upper Gondwana deposits occur extensively across peninsular and extra-peninsular parts of India. The present study provides a critical palynological review of the Indian Upper Gondwana sequence, mainly focusing on the paleobotanical distribution, diversity, and transitions. The current problems and stratigraphic lacuna in dating and correlating the Upper Gondwana sequence of India are also discussed. The palynological data pertaining to the Jurassic and Cretaceous periods of Gondwana reveals the presence of four phases of floral transition, i.e., (I) the Triassic-Jurassic boundary flora (replacement of corystosperms by cheirolepidacean conifers); (II) the post-Toarcian flora (replacement of cheirolepidacean conifers by araucarian conifers); (III) the Tithonian-Valanginian flora (rise of araucarians and podocarps; (IV) the post-Hauterivian flora (appearance of angiosperms). These floral transitions were greatly affected by tectonic movements and eustatic changes. The palynostratigraphic correlation of Indian flora with other contemporaneous strata across the Gondwana continents shows its close similarity with eastern Antarctica and Australia constituting a single phytogeographic province.

Keywords Spores and Pollen • Palynostratigraphy • Gondwana • Continental drift • Paleovegetation • Paleoclimate • Phytogeography

S. Mishra (✉)
Department of Geology, Centre for Advanced Studies, University of Delhi, Delhi, 110007, India
e-mail: shreyamishra1005@gmail.com

V. Prasad
Birbal Sahni Institute of Paleosciences, 53-University Road, Lucknow, 226007, India

Introduction

The Supercontinent Pangaea (Golonka 2004), started to fragment into two landmasses during mid-Triassic, Laurasia (northern) and Gondwana (southern). Present-day India, along with Antarctica, Australia, South America, Africa and Madagascar were the part of Gondwana and remained as a unified landmass till Early Cretaceous. However, slight drifting of Gondwana has already started by the end of the Triassic, but significant tectonic events occurred between 180 and 35 Ma (Early Jurassic to Eocene; McLoughlin 2001; Geiger et al. 2004; Golonka 2004). These paleogeographic changes in Gondwana took place during Middle Jurassic times resulting in the breakup of East Gondwana from West Gondwana followed by the separation of India from Antarctica and Australia during the Early Cretaceous (McLoughlin 2001) (Fig. 2.1A–C). The subsequent drifting of these continents resulted in the breakup of one single unified landmass, and hence marking the closure of Gondwana period by the end of Early Cretaceous.

The Supercontinent Pangaea provided land-based pathways for biotic exchange although the high-latitudinal Gondwanan flora that had maintained a distinctly different composition from that of the paleoequatorial and boreal palynoflora of Laurasia throughout the Paleozoic and Mesozoic (McLoughlin 2001). The Gondwanan flora had experienced radical turn-overs during end Permian and end Triassic which can be linked to the mass extinction events but appears to be non-correlatable with the tectonic movements (McLoughlin 2001). The mass extinction events transformed the Permian *Glossopteris* dominated flora to the Triassic *Dicroidium* flora which was again replaced by the *Ptilophyllum* flora by the Jurassic and Cretaceous (Lele 1964; McLoughlin 2001). In spite of active latitudinal shifting during the Jurassic and Cretaceous, the flora was almost similar and represented by cheirolepidacean, araucarian and podocarp conifers along with a wide variety of

G. V. R. Prasad and R. Patnaik (eds.), *Biological Consequences of Plate Tectonics: New Perspectives on Post-Gondwana Break-up—A Tribute to Ashok Sahni*,
Vertebrate Paleobiology and Paleoanthropology Series, https://doi.org/10.1007/978-3-030-49753-8_2

A

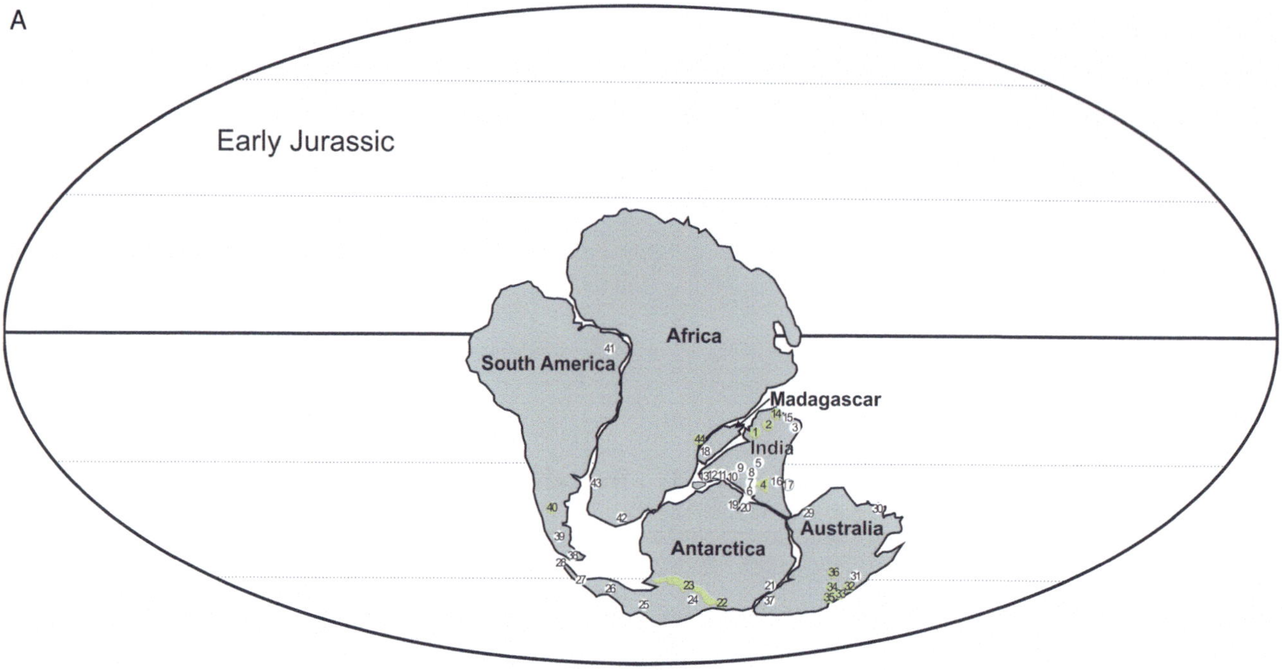

Fig. 2.1 Paleogeographic positions of major Upper Gondwana basins of Gondwana (modified after Scotese 1997; green circles indicate presence of palynoflora of respective age). **India**: 1-Kachchh, 2-Rajasthan, 3-Niti, Spiti Valley and Lamayuru Complex, Ladakh, 4-Rajmahal Basin, Domra sub-basin, Panagarh area and Birbhum Coalfield, 5-Jabalpur area, 6-Athgarh Basin, 7-Mahanadi Basin, 8-South Rewa Basin, 9-Satpura Basin, 10-Pranhita Godavari Basin, 11-Krishna-Godavari Basin, 12-Palar Basin, 13-Cauvery Basin, **Pakistan**: 14-Salt Range, 15-Kala Chitta Range, **Tethyan Margin**: 16-Gamba, 17-Xixang, Southern China, **Malagasy**: 18-Morondavan Basin, **Antarctica**: 19-McRobertson shelf, 20-Prydz Bay, 21-George V Coast, 22-South Victoria Land, 23-Transantarctic Mountain, 24-Nielson Plateau, 25-Ross Province (Marie Byrd Land), 26-Falkland Plateau, 27-Palmer Basin, 28-Byers Peninsula, **Australia**: 29-Perth Basin, 30-Bonaparte Basin, 31-Mulgildie Basin, 32-Bowen Basin, 33-Nambour Basin, 34-Surat Basin, 35-Clearence Moreton Basin, 36-Euromanga Basin, 37-Otway Basin, **South America**: 38-Austral Basin, 39-Cañadón Asfalto Basin, 40-Neuquén Basin, 41-Araripe Basin, **Africa**: 42-Algoa Basin, 43-Orange Basin, 44-Tanzania. **A** Paleogeographic position of Gondwana continents and floral distribution during Early Jurassic (green circles display presence of Early Jurassic palynological records). **B** Paleogeographic position of Gondwana continents and floral distribution during Late Jurassic (green circles display presence of Middle to Late Jurassic palynological records). **C** Paleogeographic position of Gondwana continents and floral distribution during Early Cretaceous (green circles display presence of Early Cretaceous palynological records)

seed ferns (McLoughlin 2001) (Fig. 2.1A–C). The emergence of the modern angiosperm dominated flora occurred in the Early Cretaceous. The angiosperm flora was established during the Aptian, replacing gymnosperms followed by their abundance and diversification by the end-Cretaceous, after the Gondwanan breakup (McLoughlin 2001). These transitions were almost coeval across the entire Gondwana along with few latitudinal differences. The Gondwana floral transition across the extinction events is widely associated and correlated whereas its relation with tectonic movements is still intriguing and needs further research.

India occupied more or less a central position within Gondwana and experienced a complex tectonic history with the longest voyage among all the continents which was initiated by its separation from Gondwana to its final collision with Eurasian plate during the Eocene (Chatterjee et al. 2013). The tectonic history of India as part of Upper Gondwana spans from the Jurassic to the Early Cretaceous. The Jurassic and Early Cretaceous deposits in India are referred as Upper Gondwana based on the three fold classification (Feistmantel 1882; Wadia 1957; Lele 1964). The Upper Gondwana deposits occur as a thick sedimentary succession in several basins along the central, northern, eastern and southern part of the country (Fig. 2.2).

The Indian Upper Gondwana deposits have different stratigraphic names in different basins which create an ambiguity in identifying their stratigraphic status and age (Sastry et al. 1977) (Fig. 2.3). Moreover, the available

B

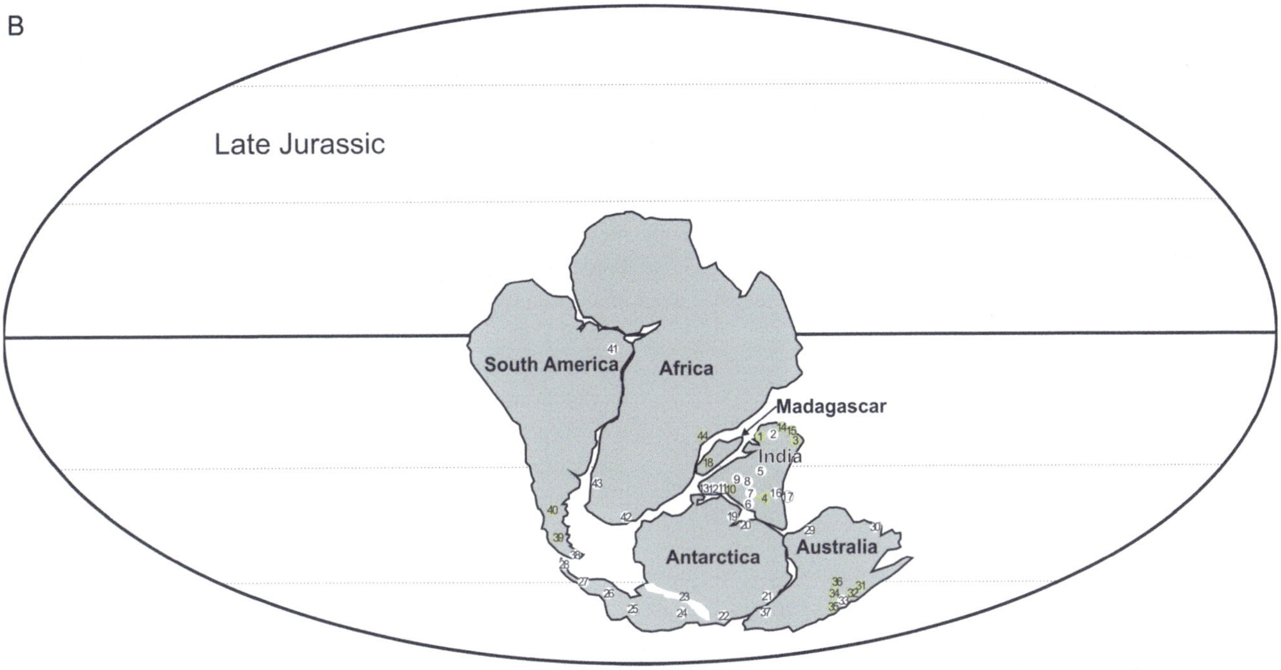

C

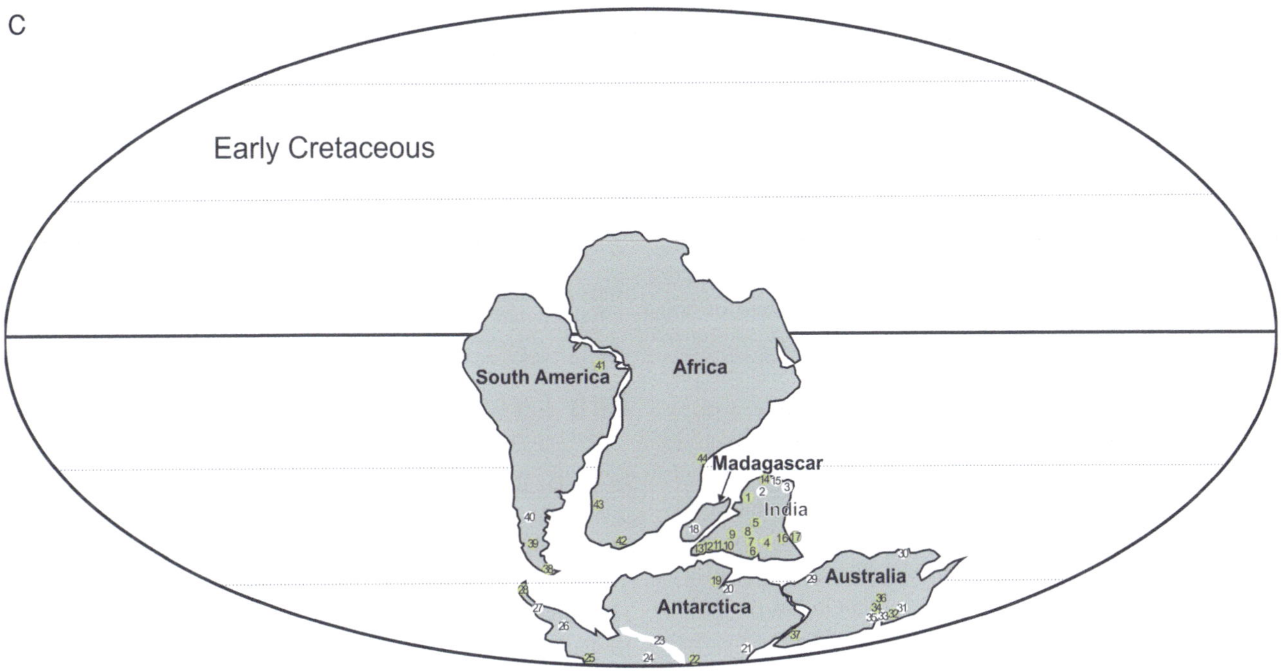

Fig. 2.1 (continued)

palynological data have not been precisely correlated with the international time-scale thus adding on to problems in recognising the ages. Therefore, the present chapter aims to analyse and correlate Indian and Gondwanan counterparts based on available palynological data of the Indian Upper Gondwana sequence for age assignment and paleoenvironment interpretations. The study also addresses the gaps in palynological data in dating and correlating the Indian Upper

Gondwana sequence and future prospects. The emphasis has also been given to analyse the impact of the tectonic evolution of the Indian subcontinent on the paleofloral distribution. The intra- and inter-basinal correlation amongst the Gondwanan continents helps to understand the paleovegetational and paleoclimatic relation as well as phytogeographic provincialism.

Geological Setting and Lithostratigraphy

Gondwana basins occur within the suture zones of Precambrian cratonic blocks of Peninsular India. These basins are bordered by boundary faults having graben or half-graben geometry. The early Permian sediments were deposited in initial sag basins, the so-called "Master Basin" which later on evolved as rift basins due to pre-break-up distension of Gondwana and sedimentation took place as syn-rift fills by the Late Permian-Early Triassic (Veevers and Tewari 1995; Biswas 1999). By the end of the Triassic, the rift cycle was completed and followed by the Jurassic to Cretaceous Upper Gondwana rift cycle (Casshyap and Khan 2000; Dutta 2002; Bastia 2006). In India, the Gondwana deposits occur in peninsular as well as extra-peninsular regions covering an area of 50,000 km^2, and are estimated to have an average thickness of 600–700 m (Sastry et al. 1979).

The Upper Gondwana rocks are deposited in varied depositional environments ranging from marine, estuarine to continental while few occur as intertrap deposits (i.e., Rajmahal Formation). They are mostly deposited under continental facies in the central and eastern part of the country (Satpura and South Rewa basins; Sastry et al. 1979) (Fig. 2.2). The Upper Gondwana deposits of the western part of the country are deposited under continental as well as estuarine facies (Kachchh, Kathiawar also known as Saurashtra and Jaisalmer basins; Singh and Venkatachala 1987) (Fig. 2.2). The well-developed Jurassic and Early Cretaceous marine deposits occur in the form of isolated patches along the east coast of the Mahanadi, Cauvery, Krishna-Godavari, Palar and Pranhita-Godavari basins (Singh and Venkatachala 1987) (Fig. 2.2). The northeastern Rajmahal Basin preserves Early Cretaceous strata as intertrap deposits within fifteen basalt flows (Tripathi 2008). The Upper Gondwana deposits have also been reported from the extra-peninsular parts, i.e., Niti and Spiti valley, Lamayuru Complex of Ladakh area (Fig. 2.2).

Jurassic

The major basins which preserve the Jurassic aged deposits include the Pranhita-Godavari, Krishna-Godavari, Satpura, South Rewa, Cauvery, Domra and Rajmahal basins (Figs. 2.2 and 2.3, Table 2.1). Rocks of this age also occur in the central and western part of the country (Madhya Pradesh, Kachchh, Gujarat, and Rajasthan) and the regions of east-coast basins. Most of the Jurassic deposits are terrestrial (Sastry et al. 1979). The Lathi Formation represents the Jurassic deposits in Rajasthan state, the Kota Formation in the Pranhita-Godavari Basin and the Jhumara, Jhuran, Katrol formations and Kuar Bet Beds in Kachchh area (Jana and Hilton 2007) (Figs. 2.2 and 2.3). While there are some diachronous units such as the Dubrajpur Formation of Rajmahal Basin and Kota Formation of Pranhita-Godavari Basin (Tables 2.1 and 2.2). The Dubrajpur Formation includes strata from the Early Triassic to the Early Cretaceous, indicating a striking diachroneity (Tables 2.1 and 2.2). It has been correlated with the Panchet (Early Triassic), Supra-Panchet (Late Triassic) and Kamthi (Early Triassic to Late Triassic) formations of the Pranhita-Godavari Basin based on lithological grounds whereas, sedimentation of the Kota Formation of the Pranhita-Godavari Basin extends from late Middle Jurassic to the Early Cretaceous (Jha and Srivastava 1996; Vijaya and Prasad 2001) (Tables 2.1 and 2.2). In the extra-Peninsular parts, the Jurassic sediments occur at Spiti (as Spiti Shale B Formation) and Lamayuru Complex in Ladakh area (Figs. 2.2 and 2.3). The presence of Jurassic age rocks has been witnessed in the Panagarh area of West Bengal between the Panchet and Rajmahal formations through palynological studies (Vijaya and Bhattacharji 2003) (Table 2.1).

Early Cretaceous

In India, the major formations which are identified as the lower Cretaceous deposits are more widely distributed in comparison to the Jurassic deposits (Figs. 2.2 and 2.3, Table 2.2). The thick deposits of this age occur as Rajmahal Formation in the Rajmahal Basin, as Bansa Formation in Rewa Basin, as Jabalpur Formation in Satpura Basin, as Chikiala/Gangapur Formation in the Pranhita-Godavari Basin, as Gollapalli Sandstone and Raghavapuram Shale in the Krishna-Godavari Basin, as Sriperumbudur Formation in the Palar Basin, as Karaikal sediments in the Cauvery Basin,

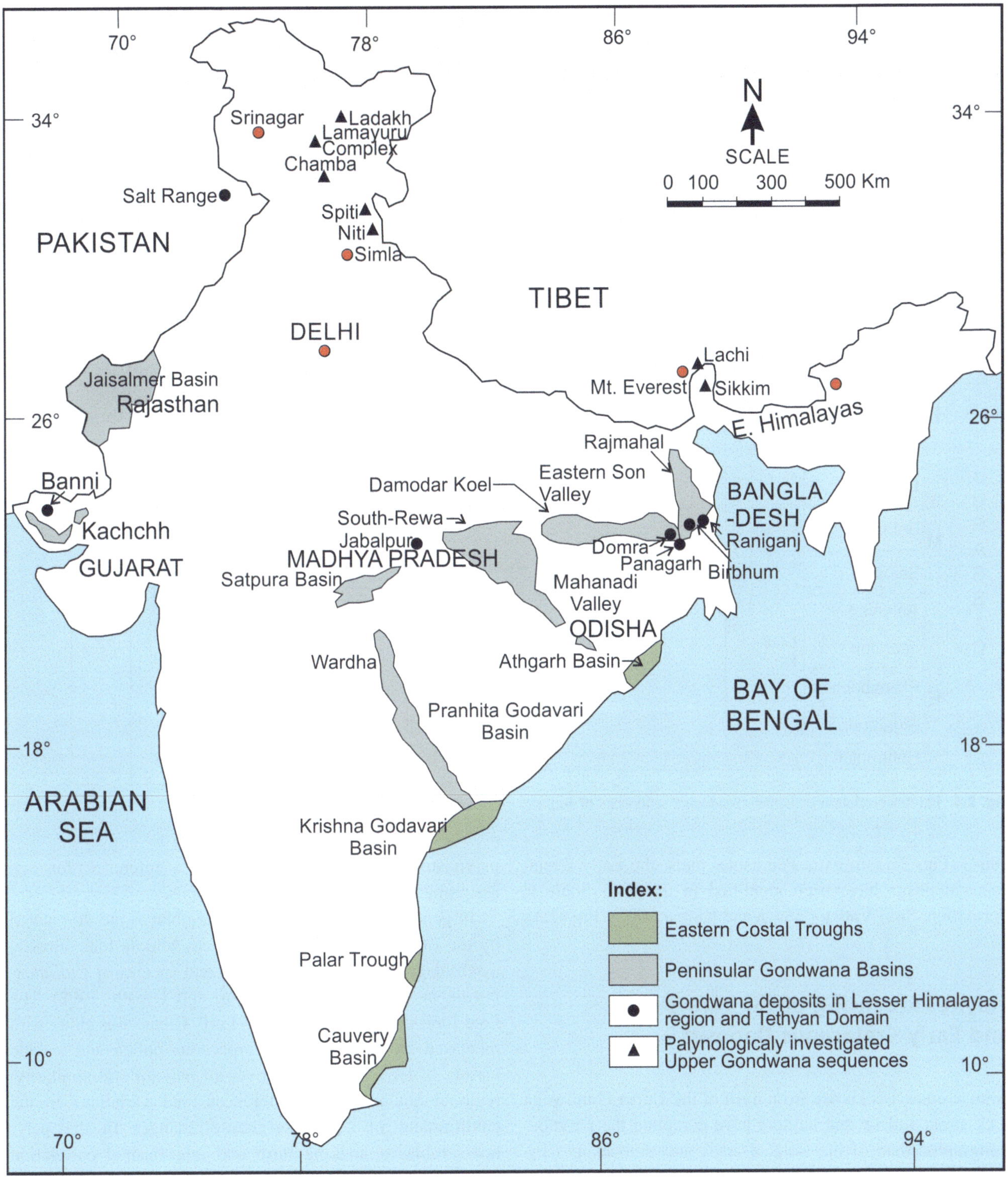

Fig. 2.2 Present day location of Upper Gondwana basins and areas in India (after Geological Survey of India, 1996)

as Athgarh Sandstone in the Athgarh Basin and as Bhuj Formation in Kachchh Basin (Table 2.2). The Bhuj Formation represents the topmost beds of the Gondwana succession in India and is known as Umia plant-bearing beds (Venkatachala and Kar 1969). The upper part of the Dubrajpur Formation is also regarded as Early Cretaceous in age (Tripathi

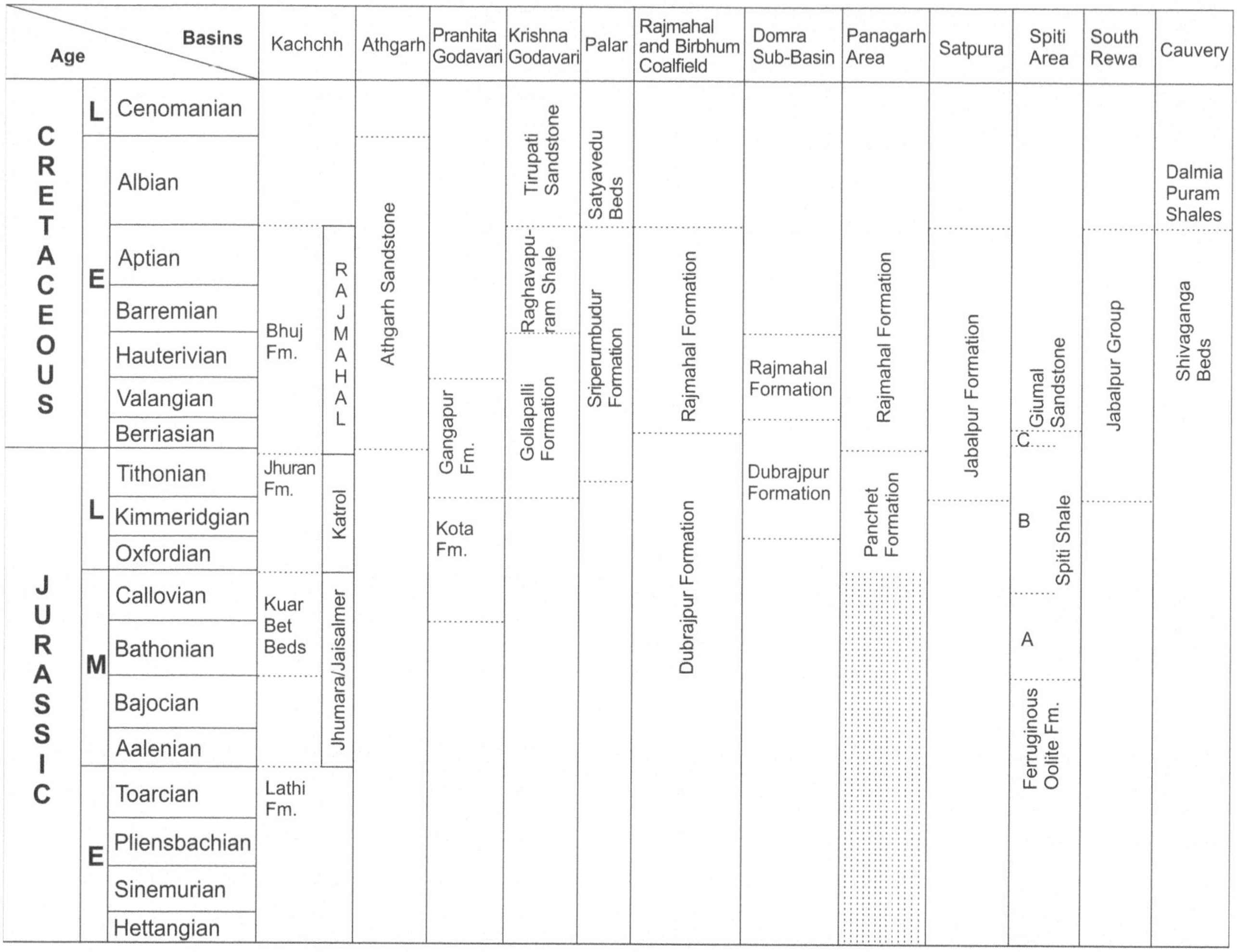

Fig. 2.3 Palynocorrelation of Upper Gondwana sequences of India

2004) (Fig. 2.3). In extra-Peninsular parts, the Early Cretaceous deposits have been identified in the Spiti Shale C Formation, Spiti Valley (Vijaya and Kumar 2002) (Fig. 2.3).

Palynosequencing of the Jurassic and Early Cretaceous Deposits

A huge dataset is known from most of the Upper Gondwana rock strata and several authors have compiled the palynostratigraphic data of the isolated areas and time-spans (Srivastava 1966; Venkatachala 1974, 1977; Venkatachala and Kar 1969; Koshal 1975; Tripathi and Tiwari 1991; Ramanujam 1993; Prasad et al. 1995; Prasad 1997; Vijaya 1997a, b, 1999; Goswami et al. 2008; Dogra et al. 2010). However, there are very few palynostratigraphic schemes which cover the entire Gondwana sequence (Bharadwaj et al. 1970; Tiwari and Tripathi 1992; Prasad et al. 1995) (Fig. 2.4). These complete palynostratigraphic schemes act as a reference for age assignment and inter-basinal correlations.

In general, the Early Jurassic assemblages are dominated by the *Classopollis* complex, whereas Middle-Late Jurassic assemblages are dominated by several species of *Callialasporites* and *Araucariacites* (Tiwari and Tripathi 1992). The Late Jurassic flora extends up to Early Cretaceous along with increased incidence of podocarpaceous pollen and a wide variety of ferns. In spite of over-all palynofloral similarity, regional environmental differences, and variations in the environment of deposition cause changes in the palynoassemblages, making intra and inter-basinal correlation difficult. The present chapter describes palynodata from the Jurassic and Early Cretaceous sequences to synthesize inter-basinal stratigraphic correlation (Tables 2.1 and 2.2). The Indian palynozones have been correlated with the well-dated palynostratigraphic scheme of Australia for age assignment (Burger 1990a, b, 1995a, b) (Fig. 2.4). The stratigraphic range of marker palynotaxa and list of various

Table 2.1 Jurassic palynoassemblage records from different parts of India

Formation and area		References	Abundant and stratigraphically important taxa	Abundant palynotaxa	Age
India					
Lathi Formation, Jaisalmer		Lukose (1972)	*Callialasporites, Podocarpidites*	*Classopollis*	Early Jurassic
Kachchh area	Banni area	Koshal (1975)	*Callialasporites, Podocarpidites*	*Classopollis*	Early Jurassic
	Kuar Bet Beds	Jana and Hilton (2007)	*Concavissimisporites* sp. cf. *Concavissimisporites variverrucatus, Klukisporites, Callialasporites dampieri*	*Araucariacites*	Palynoflora: Bathonian/Callovian
	Katrol Formation	Venkatachala and Kar (1969)	*Concavissimisporites, Boseisporites,* Various species of *Callialasporites*	*Callialasporites, Podocarpidites, Contignisporites*	Late Jurassic
Birbhum Coalfield	Dubrajpur Formation	Vijaya and Roy (2000)	*Araucariacies australis, Duplexisporites probiemensis,*	*Callialasporites*	Early Jurassic
		Vijaya and Sen (2005)	*Cicatricosisporites australiensis, Aequitriaradites, Ruffordiaspora, Crybelosporites, Microcachryidites antarcticus, Foraminisporis, dailyi, Neoraistrickia*	*Callialasporites*	Middle-Late Jurassic
		Vijaya et al. (2010)	*Podocarpidites, Ceratosporites, Concavissimisporites, Klukisporites, Lycopodiacidites, Foramninisporis, Microcachryidites antarcticus*	*Callialasporites, Araucariacites*	Middle Jurassic
Rajmahal Basin	Dubrajpur Formation	Tripathi (2000)	*Foveosporites, Nevesisporites. Enzonalasporites, Classopollis, Playfordiaspora, Arcutipollenites, Infernopollenites, Podocarpidites*	*Araucariacites, Callialasporites*	Early Jurassic
		Tripathi (2001)	*Callialasporites trilobatus, C. turbatus, Araucariacites australis, Boseisporites indicus, Biformaesporites baculosus*	*Podocarpidites, Araucariacites*	Early-early Middle Jurassic (Toarcian-Bajocian)
		Tripathi (2004)	Various species of *Callialasporites,* and presence of *Biformaesporites, Boseisporites*	*Callialasporites, Podocarpidites*	Early-early Middle Jurassic
			Contignisporites, Klukisporites, Microreticulatisporites	*Callialasporites, Concavissimisporites*	Middle Jurassic
			Ruffordiaspora, Concavissimisporites, Klukisporites, Microreticulatisporites	*Callialasporites*	Late Jurassic
		Tripathi and Ray (2006)	*Arcuatipollenites, Infernopollenites, Nevesisporites, Boseisporites, Phallocysta, Informaesporites baculosus, Contignisporites*	*Callialasporites, Classopollis*	Early Jurassic-early Middle Jurassic
Domra Sub-basin	Dubrajpur Formation	Vijaya (1997a)	*Baculatisporites, Biformaesporites, Concavissimisporites, Foraminisporis, Foveotriletes, Osmundacidites, Murospora forida, Contignisporites cooksoniae*	*Callialasporites, Araucariacites*	Late Jurassic (Kimmeridgian-Tithonian)
Panagarh area	Panchet Formation, West Bengal	Vijaya and Bhattacharjii (2003)	*Contignisporites cooksoniae, Murospora florida, Cicatricosisporites australiensis. Lycopodiacidites asperatus, Impardecispora indica, Concavissimisporites kutchensis*	*Callialasporites*	Late Jurassic
P-G Basin	Kota Formation	Vijaya and Prasad (2001)	*Murospora florida, Contignisporites cooksoniae, Crybelosporites stylosus, Aequitriaradites*	*Callialasporites*	Middle Jurassic-Late Jurassic (Callovian-Tithonian)
Ladakh Area	Spiti Shale B, Spiti	Vijaya and Kumar (2002)	*Concavissimisporites kutchensis, Contignisporites cooksoniae, C. multimuratus, Klukisporites scabratus, Podocarpidites ellipticus, Alisporites.*	*Callialasporites*	Middle Jurassic-Early Tithonian
	Lamayuru Complex	Upadhyay et al. (2005)	*Aequitriradites, Araucariacites australis, Alisporites grandis, Coptospora, Impardecispora apiverrucata, Murospora florida, Microcachryidites antarcticus, Perinopollenites elatoides, Podocarpidites grandis, Podosporites tripakshi, Podosporites*	*Callialasporites*	Middle-Late Jurassic

Table 2.2 Early Cretaceous palynoassemblage records from different parts of India

Formation and Area		References	Stratigraphically Important taxa	Abundant palynotaxa	Age
India					
Katrol Formation, Kachchh area		Venkatachala and Kar (1969)	*Cicatricosisporites, Polycingulatisporites, Ceratosporites, Staplinisporites, Trilobosporites, Caoronatispora, Foraminisporis, Microcachryidites*	*Callialasporites, Podosporites, Classopollis*	Late Jurassic
Jabalpur Formation, Madhya Pradesh		Dogra et al. (1994, 2010)	*Aequitriradites spinosus, Neoraistrickia pallida, Contignisporites dettmanii, Coptospora mesozoica, Crybelosporites, Cicatricosisporites australis, Schizosporis minutus, Tricolpites psilatus*	*Narbadapollis, Araucariacites, Podocarpidites*	Early Cretaceous (Tithonian-Neocomian)
Rajmahal Basin	Dubrajpur Formation	Tripathi (2001)	*Klukisporites, Retitriletes, Ruffordiaspora, Santhalisporites, Cingutriletes, Nevesisporites, Coptospora, Araucariacites, Classopollis, Aequitriradites, Ruffordiaspora australiensis, Klukisporites venkatachalae, Trilites tuberculiformis, Coptospora verrucosa*	*Callialasporites, Podocarpidites, Araucariacites*	Berriasian
		Tripathi (2004)	*Aequitriradites,* Species diversity in costate and hilate spores, presence of *Foraminisporis, Januasporites, Triporoletes*	*Araucariacites, Podocarpidites, Callialasporites*	Early Cretaceous
		Tripathi and Ray (2006)	*Contignisporites, Aequitriradites verrucosus, Coptospora, Klukisporites, Ruffordiaspora australiensis*	*Callialasporites*	Late Jurassic-Early Cretaceous
	Rajmahal Formation	Tripathi et al. (1990)	*Murospora, Cicatricosisporites ausraliensis, Contignisporites stylosus, Aequitriradites spinulosus*	*Callialasporites, Podocarpidites*	Late Jurassic-Early Cretaceous
		Tiwari and Tripathi (1995)	*Coptospora, Aequitriradites verrucosus, Contignisporites, Cooksonites, Cicatricosisporites australiensis, Microchachryidites antarcticus*	*Araucariacites, Callialasporites Podocarpidites*	Early Cretaceous (Aptian)
		Tripathi et al. (2013)	*Araucariacites australis, Podocarpidites ellipticus, P. grandis, P. multesimus, Callialasporites trilobatus, Coptospora kutchensis, Balmeisporites*	Poor assemblage	Early Cretaceous
Domra Sub-basin	Dubrajpur Formation	Vijaya (1997a)	*Alsophilidites, Cicatricosisporites, Concavissimisporites, Contignisporites, Crybelosporites, Duplexisporites, Foveotriletes, Klukisporites, Leptolepidites, Lycopodiacidites, Murospora florida, Undulatisporites, Crybelosporites stylosus, Aequitriradites spinulosus, Cooksonites*	*Callialasporites, Araucariacites*	Tithonian-Berriasian
	Rajmahal Formation	Vijaya (1997b)	*Microcachryidites antarcticus, Crybelosporites stylosus, Cicatricosisporites hughesii, C. ludbrookiae, Foraminisporis, Retitrilets, Contignisporites multimuratus*	*Callialasporites, Araucariacites, Podocarpidites*	Early Cretaceous Valangian-Hauterivian

(continued)

Table 2.2 (continued)

Formation and Area		References	Stratigraphically Important taxa	Abundant palynotaxa	Age
Birbhum Coalfield	Rajmahal Formation	Vijaya et al. (2010)	*Ruffordiaspora ludbrookiae, Leptolepidites, Aequitriradites verrucosus, A. spinolosus, A. hispidus, Boseisporites praeclarus, Murospora florida, Impardecispora apiverrucata, Concavissimisporites, Crybelosporites stylosus*	*Callialasporites, Araucariacites Podocarpidites*	Early Cretaceous (Berriasian)
Raniganj Coalfield		Vijaya (2011)	*Callialasporites, Araucariacites*, and *Podocarpidites, Foraminisporis wonthaggiensis, Cicatricosisporites, Contignisporites fornicatus, C. multimuratus*	Poor assemblage	Valangian-Barremian
Panagarh area		Vijaya and Bhattacharji (2002)	*Cicatricosisporites australiensis, Aquitriradites spinulosus, Foraminisporis wonthaggiensis. C. australiensis, Aequitriaradites spinulosus*	*Callialasporites, Araucariacites, Podocarpidites*	Early Cretaceous (Berriasian-Valangian)
Orissa area	Talchir Formation,	Tripathi (2009)	*Contignisporites multimuratus, Klukisporites lacunus, Leptolepidites rimatus, Murospora florida, Ruffordiaspora australiensis, Santhalisporites blosus, Todisporites minor*	*Podocarpidites, Araucariacites*	Late Jurassic-Early Cretaceous
	Athgarh Sandstone, Athgarh Basin,	Goswami et al. (2008)	*Coptospora cauveriana, C. kutchensis, C. microgranulosa, C. verrucosa, Podosporites tripakshii*, No angiosperm pollen reported	*Araucariacites*	Early Cretaceous (Neocomian-Albian)
Krishna Godavari Basin	Gollapalli Formation,	Prasad and Pundeer (1999)	*Contignisporites cooksoniae, Foraminisporis, Contignidporites, Concavissimisporites, Crybelosporites, Cooksonites, Podosporites, Ischyosporites, Ceratosporites*	*Microcachryidites antartcticus, Cicatricosisporites, Callialasporites*	Early Cretaceous (Neocomian)
	Raghavapuram Shale		*Gleichnidites, Asteropollis, Foraminisporis, Callialasporites, Liliacidites, Execispollenites*	*Callialasporites, Microcachryidites, Classopollis*	Barremian-Aptian to Early Albian
Sriperumbudur Formation, Tamil Nadu		Tripathi and Vijaya (1997)	*Microcachryidites antarcticus, Cicatricosisporites ausraliensis, Crybelosporites stylosus, Aequitriradites spinulosus*	*Callialasporites, Araucariacites*	Tithonian-Valangian
Karaikal sediments, Cauvery Basin		Venkatachala and Jain (1970)	*Aequitriradites, Cooksonites, Coronatispora, Sestrosporites, Microcachyidites, Classopollis*	*Callialasporites*	Neocomian-Aptian
Pranhita-Godavari Basin (P-G)	Kota Formation	Vijaya and Prasad (2001)	*Araucariacites, Klukisporites, Crybelosporites, Foraminisporis, Appendicisporites, Cooksonites, Cicatricosisporites*	*Callialasporites*	Late Hauterivian-Early Barremian
	Gangapur Formation	Jha et al. (2017)	*Coptospora, Ceratosporites equalis, Appendicisporites erdtmanii, Biretisporitesspectabilis, Cicatricosisporites hughesii, Concavissimisporites, Contignisporites, Converrucosisporites, Klukisporites scabratus, Plicifera delicata, Balmeiopsis limbatus, Araucariacites australis, A. cooksonii, Callialasporites, Dacrycarpites australiensis, Podocarpidites*	*Araucariacites, Callialasporites*	Late Jurassic – Early Cretaceous (Tithonian-Valangian)

(continued)

Table 2.2 (continued)

Formation and Area	References	Stratigraphically Important taxa	Abundant palynotaxa	Age
		ellipticus, *Podosporites variabilis*, *Crybelosporites stylosus*, *Foraminisporis wonthaggiensis*		
Kota/Gangapur Formation, Asifabad area	Jha et al. (2017)	*Alsophyllidites kerguelensis*, *Biretisporites eneabbaensis*, *Cicatricosisporites ludbrooki*, *Classopollis classoides*, *Foveosporites* cf. *canalis*, *Latosporites*, *Podocarpidites ellipticus*, *P. rarus*, *P. typicus*	*Araucariacites*, *Callialasporites*	Late Jurassic-Early Cretaceous
Spiti Shale C, Spiti area, Ladakh area	Vijaya and Kumar (2002)	*Contignisporites cooksoniae*, *Ruffordiaspora australiensis*, *Ruffordiaspora ludbrookiae*, *Cicatricosisporites hughesii*, *C. annulatus*, *Aequitriradites spinulosus*, *Podosporites tripakshii*, *Cooksonites variabilis*	*Callialasporites*	Late Tithonian-Early Cretaceous

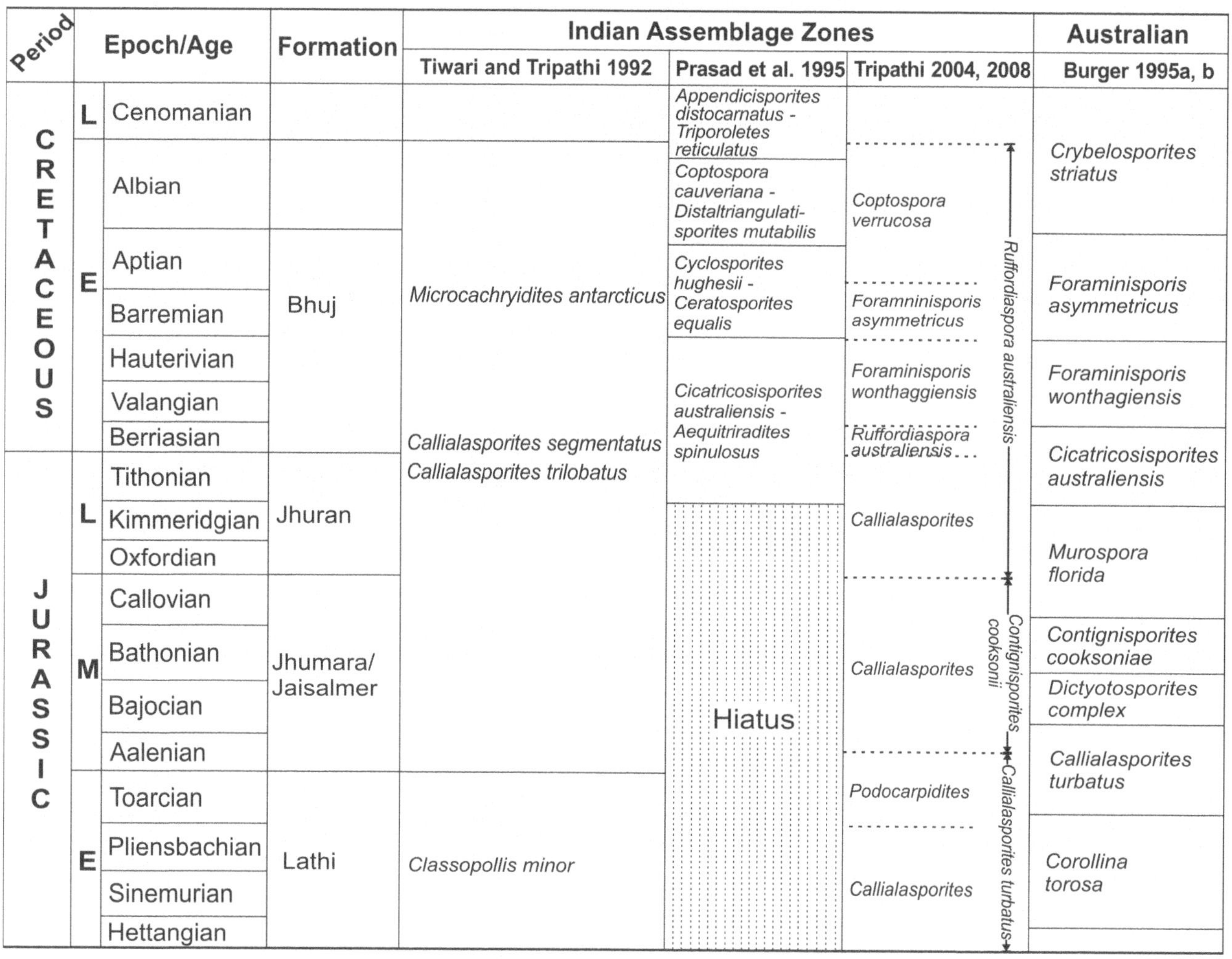

Fig. 2.4 Correlation of different Upper Gondwana palynostratigraphic schemes of India and their comparison with Australian palynozones

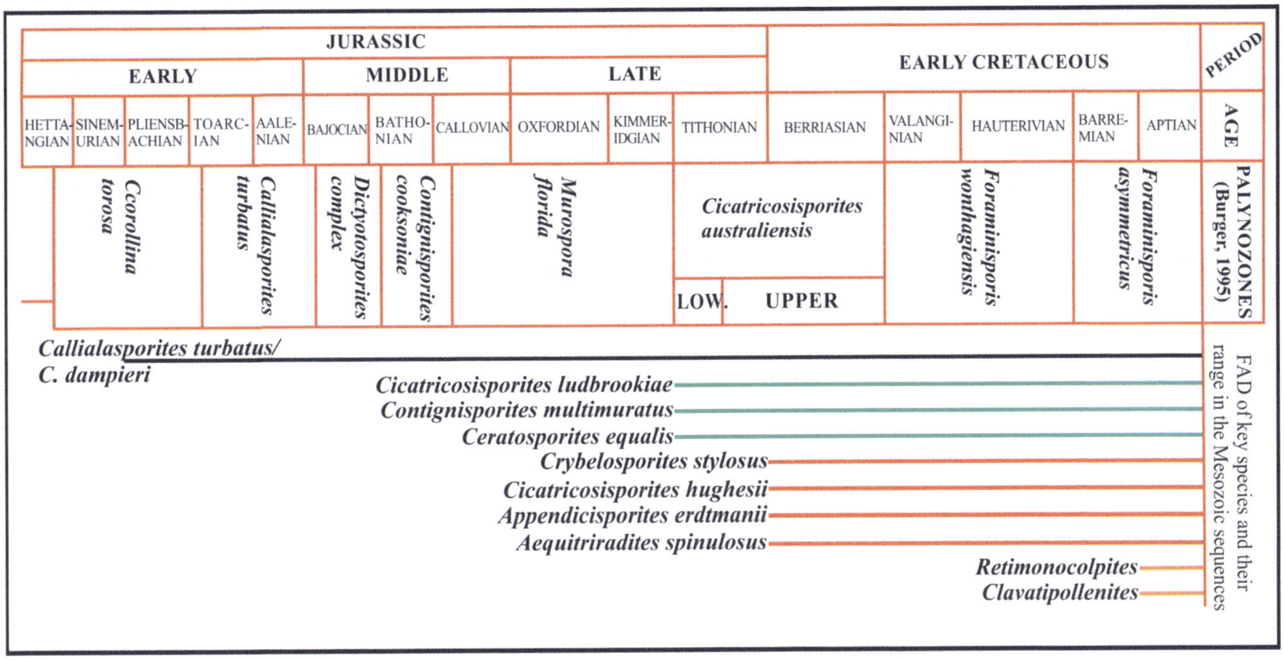

Fig. 2.5 Range chart of stratigraphically significant palynotaxa (after Singh, 1964, 1971; Venkatachala 1974; Helby et al. 1987; Mohr 1990; Dettmann et al. 1992; Mohr and Gee 1992; Burger 1995a, b; Tiwari and Tripathi 1995; Burger and Shafik 1996a, b; Vijaya 1999; Tripathi 2001, 2002; Partridge 2006; Volkheimer et al. 2009)

palynotaxa from the Indian Upper Gondwana sequence along with their paleobotanical affinities are given in Fig. 2.5 and Table 2.3, respectively.

Jurassic

The Jurassic palynoflora is characterized by the abundance and diversification of *Classopollis, Callialasporites,* and *Araucariacites* along with few long-ranging Triassic taxa such as *Densoisporites, Cyathidites, Osmundacidites, Verrucosisporites* and *Podocarpidites* among many others (Table 2.3). There are few other palynomorphs that appear in the Jurassic including *Ceratosporites, Klukisporites, Ischyosporites, Foveotriletes, Leptolepidites, Murospora, Contignisporites* which are long-ranging and extend up to Early Cretaceous. Tiwari and Tripathi (1992) (Fig. 2.3) had identified three palynozones in the Jurassic which are described below:

i. *Classopollis minor* Assemblage Zone (Early Jurassic): This palynoassemblage is characterized by the abundance of *Classopollis* and *Spheripollenites* species. The key elements include spores such as *Cyathidites, Staplinisporites, Gleichenidites, Matonisporites cooksonii, Verrucosisporites* and pollen of *Classopollis minor, C. classoides, Callialasporites trilobatus. C. triletus, C. segmentatus,* and *Podocarpidites* spp. This assemblage has been dated as Early Jurassic in age. The *Callialasporites turbatus/dampieri* pollen is an age-diagnostic species of the Early Jurassic, and its FAD marks the Pliensbachian (Vijaya 2000a).
ii. *Callialasporites trilobatus* Assemblage Zone (Middle to Late Jurassic): This assemblage is dated as Middle to Late Jurassic and recovered from Banni Well, belonging to the Jhumara and Jhuran formations (Koshal 1975). The assemblage is characterized by the presence of several species of *Callialasporites* along with *Densoisporites, Klukisporites, Gleicheniidites, Araucariacites, Coptospora, Classopollis minor, C. classoides, C. itunensis* and *Cicatricosisporites.* The FAD of *Murospora florida* marks the Callovian age (Vijaya 2000a) (Fig. 2.4).
iii. *Callialasporites segmentatus* Assemblage Zone (Late Jurassic to Berriasian): This palynozone is characterized by the presence of palynomorphs such as *Araucariacites australis, Stereisporites antiquasporites, Classopollis classoides, Podocarpidites ellipticus, P. multesimus, Ceratosporites equalis, C. acutus, Cicatricosisporites ludbrooki, C. australiensis, Contignisporites glebulentus, C. multimuratus, Baculatisporites comaumensis, Retitriletes austroclavatidites, R. reticulumsporites* and several species of *Callialasporites* such as *C. dampieri, C. monoalasporus, C. segmentatus, C. trilobatus. C triletus,* etc. This zone was described by Venkatachala et al. (1972) from the Cauvery Basin. *Cicatricosisporites australiensis* is a characteristic pollen of Late Jurassic

Table 2.3 List of stratigraphically significant palynomorphs of Jurassic (J) and Early Cretaceous (C) recovered from different parts of India and their affinities based on the compiled data of Balme (1995) and Brugman et al. (1994)

Thallophyta	**J**	**C**
Schizosporis rugulatus Cookson and Dettmann, 1959	*	*
Schizosporis minutus		*
Schizosporis sp.	*	*
Bryophyta		
Aequitriradites spinulosus (Cookson and Dettmann) Cookson and Dettmann, 1966		*
Aequitriradites verrucosus (Cookson and Dettmann) Cookson and Dettmann, 1966		*
Aequitriradites sp.		*
Cooksonites variabilis Pocock, 1962		*
Coptospora cauveriana Venkatachala et al., 1912	*	*
Coptospora kutchensis Venkatachala, 1969	*	*
Coptospora mesozoica Couper, 1958	*	*
Coptospora verrucosa Tripathi et al., 1990		*
Cyclosporites hughesii (Cookson and Dettmann) Cookson and Dettmann, 1959		*
Foraminisporis asymmetricus (Cookson and Dettmann) Dettmann, 1963		*
Foraminisporis dailyii (Cookson and Dettmann) Dettmann, 1963		*
Foraminisporis tribulosus Playford and Dettmann, 1965	*	*
Foraminisporis wonthaggiensis Dettmann, 1963		*
Foraminisporis sp.	*	*
Stereisporites antiquasporites (Wilson and Webster) Dettmann, 1963	*	*
Stereisporites sp.	*	*
Polycingulatisporites crenulatus Playford and Dettmann, 1965	*	*
Polycingulatisporites reduncus (Bolkhovitina) Playford and Dettmann, 1965		*
Taurocusporites minor Singh, 1964		*
Taurocusporites segmentatus Stover, 1962		*
Triporoletes levigatus (Pocock) Playford, 1971		*
Triporoletes radiatus (Dettmann) Playford, 1971		*
Triporoletes reticulatus (Pocock) Playford, 1971	*	*
Triporoletes simplex (Cookson and Dettmann 1958) Playford, 1971		*
?Ceratosporites equalis Cookson and Dettmann, 1958		*
?Ceratosporites helidonensis de Jersey, 1971		*
Neoraistrickia neozealandica (Couper) Bharadwaj and Kumar, 1984	*	*
Neoraistrickia pallida Kumar, 1973	*	*
Neoraistrickia truncate Kumar, 1973	*	*
Tracheophyta		
Order – Lycopodiales		
Densoisporites sp.	*	*
Foveosporites canalis Balme, 1957		*
Foveosporites triassicus Kumaran and Maheshwari, 1980	*	
Foveosporites sp.	*	
Foveotriletes parviretus (Balme) Dettmann, 1963	*	*
Foveotriletes foveolatus Venkatchala et al., 1969		*
Foveotriletes sp.	*	*
Leptolepidites major Couper, 1958	*	*
Leptolepidites rimatus Tripathi et al., 1990	*	*
Leptolepidites verrucatus Couper, 1953	*	*
Lycopodiacidites dettmannae Burger, 1980	*	*
Lycopodiacidites subtriangulus Venkatachala, Kar and Raza, 1969	*	*
Lycopodiacidites sp.	*	*
Lycopodiumsporites agathoecus (Potonié) Thiergart, 1937		*
Lycopodiumsporites austroclavatidites (Cookson) Potonie, 1956	*	*
Lycopodiumsporites circolumenus Cookson and Dettmann, 1958	*	*
Lycopodiumsporites pallidus Kumar, 1973	*	*
Lycopodiumsporites sinuosus Kumar, 1973	*	*

(continued)

Table 2.3 (continued)

Lycopodiumsporites sp.	*	*
Retitriletes austroclavatidites (Cookson) Doring, Krutzsch, Mai and Schulz, 1963	*	*
Retitriletes circolumenus (Cookson and Dettmann) Backhouse, 1978		*
Retitriletes reticulumsporites (Rouse) Doring, Krutzsch, Mai and Schulz., 1963	*	*
Sestrosporites alveolatus (Couper) Dettmann, 1985	*	*
Sestrosporites irregulatus (Couper) Dettmann, 1985	*	*
Sestrosporites pseudoalveolatus Dettmann, 1963	*	*
?Staplinisporites caminus (Balme) Pocock, 1962		*
Class – Filicopsida		
Order – Filicales		
Family – Dicksoniaceae/Cyathaceae		
Alsophilidites sp.	*	*
Concavissimiporites crassatus (Delcourt and Sprumont) Delcourt et al., 1963		*
Concavissimisporites penolaensis Dettmann, 1963		*
Concavissimisporites punctatus (Delcourt and Sprumont) Brenner, 1964	*	*
Concavissimisporites subverrucosus Venkatachala, 1969		*
Concavissimisporites verrucosus (Delcourt and Sprumont) Delcourt et al., 1963	*	*
Concavissimisporites variverrucatus Brenner, 1963	*	*
Concavissimisporites sp.	*	
Cyathidites australis Couper, 1953		*
Cyathidites concavus (Bolkhovitina) Dettmann, 1963		*
Cyathidites cutchensis Singh, Srivastava and Roy, 1964		*
Cyathidites densus Kumar, 1973		*
Cyathidites ghuneriensis Singh, Srivastava and Roy, 1964		*
Cyathidites grandis Singh et al., 1964		*
Cyathidites sp. cf. *C. asper* Couper, 1953		*
Cyathidites minor Couper, 1953		*
Cyathidites pseudopunctatus Singh et al. 1964		*
Cyathidites sp.		*
Deltoidospora hallii Miner, 1935		*
Deltoidospora sp.		*
Family – Dipteridaceae/Matoniaceae/Cheiropleuriaceae		
Boseisporites indicus Kumar, 1973	*	*
Boseisporites insignitus Venkatachala, 1969	*	*
Boseisporites jabalpurensis Kumar, 1973	*	*
Boseisporites minutus Venkatachala, Kar and Raza, 1969	*	*
Boseisporites praeclarus (Dev) Maheshwari, 1974	*	*
Boseisporites sehoraensis Kumar, 1973	*	*
Converrucosisporites jenensis Reinhardt, 1984	*	*
Converrucosisporites sp.	*	*
Dictyophyllidites haradaensis Kumar, 1973	*	*
Dictyophyllidites harrisii Couper, 1958	*	*
Dictyophyllidites sp.	*	*
Matonisporites cooksonii Dettmann, 1963	*	*
Matonisporites crassiangulatus (Balme) Dettmann, 1963	*	*
Matonisporites discoidalis Kumar, 1973	*	*
Matonisporites dubius Kumar, 1973	*	*
Matonisporites kutchensis Venkatachala, 1969	*	*
Family – Gleicheniaceae		
Clevifera triplex (Bokhovitina) Bokhovitina, 1966		*
?Concavisporites crassus Venkatachala, Kar and Raza, 1969	*	*
?Concavisporites indicus Venkatachala, 1969	*	*
?Concavisporites novicus Kumar, 1973	*	*
?Concavisporites sp.	*	*
Gleichenidites mundus Sah and Jain, 1965	*	*

(continued)

Table 2.3 (continued)

Gleicheniidites apicus Kumar, 1973	*	*
Gleicheniidites cercinidites (Cookson) Dettmann, 1963	*	*
Gleicheniidites glaucus Kumar, 1973	*	*
Gleicheniidites senonicus Ross, 1949	*	*
Gleicheniidites sp.	*	*
Family – Marsiliaceae		
Crybelosporites punctatus Dettmann, 1963		*
Crybelosporites stylosus Dettmann, 1963	*	*
Crybelosporites striatus (Cookson and Dettmann) Dettmann, 1963		*
Family – Osmundaceae		
Baculatisporites comaumensis (Cookson) Potonié, 1956		*
Baculatisporites rotundus Kumar, 1973	*	*
Baculatisporites sp.	*	*
Osmundacidites wellmanii Couper, 1953	*	*
Todispoites major Couper, 1958	*	
Todisporites minor Couper, 1953	*	*
Todisporites sp.	*	
Verrucosisporites ambiplicatus Kar, 1968	*	*
Verrucosisporites bosei Maheshwari and Banerji, 1975	*	*
Verrucosisporites kazigaonensis Tripathi, Tiwari and Kumar, 1990	*	*
Verrucosisporites narmianus Balme, 1970	*	*
Verrucosisporites racemus (Peppers) Smith, 1971	*	*
Family – Schizeaceae		
Cicatricosisporites angustus Singh, 1970		*
Cicatricosisporites annulatus Archangelsky and Gamerro, 1966		*
Cicatricosisporites australiensis (Cookson) Potonié, 1956		*
Cicatricosisporites hughesii Dettmann, 1963		*
Cicatricosisporites ludbrooki Dettmann, 1963		*
Cicatricosisporites minor (Bolkhovitina) Pocock, 1962		*
Cicatricosisporites purbeckensis Norris, 1969.		*
Cicatricosisporites sp.		*
Ruffordiaspora australiensis (Cookson) Dettmann and Clifford, 1992		*
Ruffordiaspora ludbrookiae (Dettmann) Dettmann and Clifford, 1992		*
Undifferentiated Filicopsids		
Appendicisporites distocarinatus Dettmann and Playford, 1968		*
Appendicisporites erdtmanii Pocock, 1964		*
Aquitriradites spinulosus (Cookson and Dettmann) Cookson and Dettmann, 1961		*
Dictyotriletes pseudoreticulatus (Couper) Pocock, 1962	*	*
Biformaesporites baculosus Singh and Kumar, 1971	*	*
Biformaesporites sp.	*	*
Biretisporites eneabbaensis Backhouse, 1978		*
Biretisporites spectabilis Dettmann, 1963	*	*
Biretisporites sp.	*	*
Contignispoites cooksoniae (Balme) Dettmann, 1963	*	*
Contignispoites dettmanii Singh and Kumar, 1966		*
Contignispoites fornicatus Dettmann, 1963		*
Contignispoites glebulentus Dettmann, 1963		*
Contignisporites fornicatus Dettmann, 1963		*
Contignisporites glebulentus Dettmann, 1963		*
Contignisporites multimuratus Dettmann, 1963		*
Contignisporites psilatus Singh and Kumar, 1966	*	*
Contignisporites sp.	*	*
Crassimonoletes surangei Singh, Srivastava and Roy, 1964		*
Crassimonoletes sp.		*
Didecitriletes ericianus (Balme and Hennelly) Venkatchala and Kar, 1985	*	*

(continued)

Table 2.3 (continued)

Distaltriangulatisporites mutabilis Singh, 1971		*
Duplexisporites problematicus (Couper) Playford and Dettmann, 1965	*	*
Duplexisporites sp.	*	*
Impardecispora apiverrucata (Couper) Venkatachala, Kar and Raza, 1969	*	*
Impardecispora indica Venkatachala, Kar and Raza, 1969	*	*
Impardecispora nebulosus Balme, 1970	*	*
Impardecispora uralensis (Bolkhovitina) Venkatachala, Kar and Raza, 1969	*	*
Ischyosporites crateris Balme, 1957	*	*
Ischyosporites haradensis Kumar, 1973	*	*
Ischyosporites punctatus Cookson and Dettmann, 1958	*	*
Ischyosporites sp.	*	*
Januasporites sp.	*	*
Klukisporites apunctus Venkatachala, Kar and Raza, 1969	*	*
Klukisporites areolatus Singh, 1970	*	*
Klukisporites haradensis Kumar, 1973	*	*
Klukisporites pseudoreticulatus Couper, 1958	*	*
Klukisporites scaberis (Cookson and Dettmann) Dettmann, 1963	*	*
Klukisporites variegatus Couper, 1958	*	*
Klukisporites venkatachalae Tripathi, Tiwari and Kumar, 1990	*	*
Klukisporites sp.	*	*
Laevigatosporites sp.	*	*
Metamonoletes haradensis Singh and Kumar, 1972	*	*
Microreticulatisporites fusus (Nilson) Morbey, 1975	*	*
Murospora florida (Balme) Pocock, 1961	*	*
Murospora truncata Singh, 1971	*	*
Pilosisporites notensis Cookson and Dettmann, 1958	*	*
Pilosisporites trichopapillosus Cookson and Dettmann, 1958		*
Plicifera delicata Bolkhovitina, 1973		*
Rajmahalispora rugulata Tiwari, Tripathi and Kumar, 1984	*	*
Rajmahalispora triassica Tiwari, Tripathi and Kumar, 1984	*	*
Triletes fusus Kumar, 1973	*	*
Trilites tuberculiformis Cookson, 1947	*	*
Triplexisporites playfordii (de Jersey and Hamilton) Foster, 1979		*
Undulatisporites fossulatus Singh, 1971		*
Undulatisporites microcutis Thomson and Pflug, 1953		*
Undulatisporites undulapolus Brenner, 1963		*
Verrucatosporites sp.		*
Unknown Affinity		
Callispora potoniei Dev em. Bharadwaj and Kumar, 1972	*	*
Coronatispora perforate (Dettmann) Evans, 1966	*	*
Coronatispora telata (Balme) Dettmann, 1963	*	*
Dubrajisporites bulbosus Tiwari and Tripathi, 1987	*	
Dubrajisporites isolatus Tiwari and Tripathi, 1987		*
Dubrajisporites triassicus Tiwari and Tripathi, 1987	*	
Dubrajisporites unicus Tiwari and Tripathi, 1987	*	*
Lakhnavitriletes bansaensis Maheshwari, 1974		*
Lametatriletes indicus Singh and Kumar, 1972	*	*
Lametatriletes tenuis Kumar, 1973	*	*
Leschikisporis rudis Kar and Sah, 1970	*	*
Leschikisporis sp.	*	*
Trilobosporites hannonicus (Delcourt and Sprumont) Potonié, 1956		*
Trilobosporites trioreticulatus Cookson and Dettmann, 1958		*
Class – Gymnospermopsida		
Order – Caytoniales (Pteridosperms)		
Family – Corystospermaceae		

(continued)

Table 2.3 (continued)

Alisporites asansoliensis Maheshwari and Banerji, 1975	*	*
Alisporites grandis (Cookson) Dettmann, 1963	*	*
Alisporites haradensis Kumar, 1973	*	*
Alisporites landianus Balme, 1970	*	*
Alisporites ovalis Kumar, 1973	*	*
Alisporites sehoraensis Kumar, 1973	*	*
Alisporites sp.	*	*
Order – Ephedrales		
Family – Ephedraceae		
Ephedripites sp.	*	*
Ginkgoretectina sp.	*	*
Family – Caytoniaceae		
Vitreisporites pallidus (Reissinger) Nilsson, 1958	*	*
Vitreisporites signatus Leschik emend. Jansonius, 1962	*	*
Vitreisporites sp.	*	*
Family – Peltaspermaceae		
Cycadopites couperi Kumar, 1973	*	*
Cycadopites follicularis Wilson and Webster, 1946		*
Cycadopites grandis de Jersey and Hamilton, 1967	*	
Cycadopites sp.	*	*
Ginkgocycadophytus sp.	*	*
Monosulcites ellipticus Kumar, 1973	*	*
Monosulcites sp.	*	*
Platysaccus indicus Sah and Jain, 1965	*	
Order – Pteridosperms *(incertae sedis)*		
Order – Coniferales		
Family – Araucariaceae		
Araucariacites australis Cookson, 1947	*	*
Araucariacites cooksoni Singh, Srivastava and Roy, 1964	*	*
Araucariacites ghuneriensis Singh, Srivastava and Roy, 1964	*	*
Araucariacites limbatus (Balme) Habib, 1969	*	*
Balmeiopsis limbatus (Balme) Archangelsky, 1977		*
Callialasporites baculosus (Dev) Maheshwari, 1974	*	*
Callialasporites barragaoensis Srivastava, 1969	*	
Callialasporites circumplectus Kumar, 1973	*	*
Callialasporites crenulatus Pocock, 1970	*	
Callialasporites dampieri (Balme) Dev, 1961	*	*
Callialasporites discoidalis (Doring) Bharadwaj and Kumar, 1972	*	*
Callialasporites doering Kumar, 1973	*	*
Callialasporites enigmatus (Singh and Kumar) Kumar, 1973	*	*
Callialasporites grandis Sah and Jain, 1965	*	*
Callialasporites granulatus Venkatachala and Kar, 1972	*	*
Callialasporites indicus (Singh and Kumar) Kumar, 1973	*	*
Callialasporites lametaensis Kumar, 1973	*	*
Callialasporites lenticularis (Doering) Venkatachala and Kar, 1969	*	*
Callialasporites limbatus (Singh and Kumar) Kumar, 1973	*	*
Callialasporites lucidus (Pocock) Maheshwari, 1974	*	*
Callialasporites microvelatus Schulz, 1967		*
Callialasporites monoalasporus Dev, 1961	*	*
Callialasporites plicatus (Singh and Kumar) Kumar, 1973	*	*
Callialasporites primus (Singh and Kumar) Kumar, 1973	*	*
Callialasporites rudisacus Maheshwari, 1974	*	*
Callialasporites segmentatus (Balme) Dev, 1961		*
Callialasporites triletus Singh, Srivastava and Roy, 1964		*
Callialasporites trilobatus (Balme) Dev, 1961	*	*

(continued)

Table 2.3 (continued)

Callialasporites turbatus (Balme) Schulz, 1967	*	*
Callialasporites velatus Singh and Kumar, 1972		*
Callialasporites sp.	*	*
Family – Cheirolepidiaceae		
Classopollis chateaunovi Reyre, 1970	*	*
Classopollis classoides (Pflug) Pocock and Jansonius, 1961	*	*
Classopollis indicus Maheshwari, 1974	*	*
Classopollis itunensis Pocock, 1962	*	*
Classopollis meyeriana (Klaus) de Jersey, 1973	*	*
Classopollis minor Pocock and Jansonius, 1961	*	*
Classopollis torosus (Reiss) Couper, 1958	*	*
Classopollis sp.	*	*
Family – Podocarpaceae		
Dacrycarpites australiensis Cookson and Pike, 1953		*
Microcachryidites antarcticus Cookson, 1947		*
Podocarpidites cristiexinus Sah and Jain, 1965	*	*
Podocarpidites ellipticus Cookson, 1947	*	*
Podocarpidites grandis Sah and Jain, 1965	*	*
Podocarpidites magnus Maheshwari, 1974	*	*
Podocarpidites multesimus (Bolkhovitina) Pocock, 1962	*	*
Podocarpidites novus Sah and Jain, 1965	*	*
Podocarpidites vermiculatus Kumar, 1973	*	*
Podocarpidites sp.	*	*
Podosporites microsaccatus (Couper) Dettmann, 1963	*	*
Podosporites raoi Singh, Srivastava and Roy, 1964	*	*
Podosporites tripakshii Rao, 1943	*	*
Podosporites variabilis Dev, 1961	*	*
Family – Taxodiaceae		
?*Inaperturopollenites* sp.	*	*
Family – Pinaceae		
Abiespollenites sp.		*
Unknown Gymnosperm pollen		
Cedripites nudis Kar and Sah, 1970	*	*
Enzonalasporites densus Dolby and Balme, 1976	*	*
Exesipollenites tunuis Balme, 1957	*	*
Exesipollenites sp.	*	*
Lygestipollenites florinii		*
Narbadapollis indicus		*
Nidipollenites sp.	*	*
Satsangisaccites sp.	*	*
Spheripollenites scabratus Couper, 1958	*	*
Class – Angiospermopsida		
Asteropollis asteroids Hedlund and Norris, 1968		*
Clavatipollenites hughesii Couper, 1958		*
Liliacidites reticulatus (Brenner) Singh, 1971		*
Retimonocolpites peroreticulatus (Brenner) Doyle, 1975		*
Retimonocolpites sp.		*
Santhalisporites baskoensis Sah and Jain, 2004		*
Santhalisporites bulbosus Tripathi, Tiwari and Kumar, 1990		*
Santhalisporites sp.		*
Turonipollis helmighii Van Amerom, 1965		*
Tricolpites micromurus (Groot and Penny) Singh, 1971		*
Tricolpites psilatus Singh, 1971		*

and marks the Kimmeridgian-Tithonian boundary (Vijaya 2000a). The *Cicatricosisporites australiensis* Zone was earlier dated as Tithonian, but the presence of this zone has been interpreted as an indication of an Early Cretaceous age (Singh and Venkatachala 1987) (Fig. 2.3).

Inter-basinal Correlation

In India, deposits of the Jurassic age are not widely distributed as the Early Cretaceous deposits (Tables 2.1 and 2.2). The Jurassic sedimentation in most of the basins is interrupted by several hiatuses, or Jurassic deposits are absent as in the Damodar Basin (Fig. 2.3). The isolated distribution of Jurassic deposits and lack of palynological data hampers proper correlation of the different Jurassic rock sequences. The Triassic-Jurassic boundary succession in India is represented by the Dubrajpur Formation in different areas such as the Rajmahal Basin, Domra sub-basin, Birbhum Coalfield and Panagarh area (Table 2.1). The *Arcuatipollenites tethyensis* Assemblage Zone, ranging from the Late Triassic to earliest Jurassic records a gradual transition of palynoflora across the Triassic/Jurassic boundary (Tripathi 2004). This palynoassemblage is characterized by dominance of *Arcuatipollenites, Striatopodocarpites* and *Satsangiasaccites* along with the first appearances of *Callialasporites* and *Classopollis meyerina.* The Dubrajpur Formation shows a hiatus during the Early Jurassic and thus the dominant phase of cheirolepidacean pollen is not reported. The flora with a high percentage of *Classopollis minor* is recovered from the Banni area, Kachchh (Koshal 1975) and the Lathi Formation from the south-east of Jaisalmer (Lukose 1972) (Table 2.1). Both of these assemblages show a high proportion of cheirolepidacean pollen, *Classopollis* spp. and extend up to late Early Jurassic (i.e., Toarcian).

The Dubrajpur Formation records a phase of decline in the abundance of *Classopollis* spp. from the Toarcian onwards and a coeval rise in the abundance and diversifications of *Callialasporites* and *Podocarpidites* in early Middle Jurassic within the *Callialasporites turbatus* Assemblage Zone (Tripathi 2000, 2001, 2003, 2004, 2008; Tripathi and Ray 2006) (Table 2.1). The other forms present in this zone include *Araucariacites, Biformaesporites, Nidipollenites, Boseisporites, Lundbladispora, Goubinispora indica, Foraminisporis tribulosus, Aracariacites australis, Klukisporites* and *Satsangisaccites* along with *Phallocysta* dinoflagellate (Tripathi 2001) (Table 2.1). These palynotaxa and similar palynofloral transitions are also reported from the Middle to Late Jurassic of Australia (Filatoff 1975; Helby et al. 1987; Burger 1995a, b; Grant-Mackie et al. 2000). A similar palynoassemblage has been recovered from the Dubrajpur Formation of the Birbhum Coalfield with a high proportion of diversified *Callialasporites* and *Araucariacites* (Vijaya and Roy 2000) (Table 2.1).

The Indian paleoflora changed during the Middle Jurassic as the dominance of cheirolepidacean pollen is replaced by the dominance of conifer pollen of the Araucariaceae and Podocarpaceae families (Table 2.1). Their dominance continues till the Early Cretaceous and becomes stratigraphically insignificant (Tables 2.1 and 2.2). Thus, age of the palynoassemblages have been marked on the basis of FADs of stratigraphically significant palynotaxa. The palynoassemblages recovered from the Kuar Bet Beds of Kachchh show dominance of conifer pollen, *Araucariacites* along with the presence of stratigraphically significant palynotaxa such as *Cyathidites australis, C. concavus, Dictyophyllidites harrisii, Todisporites major, Concavissimisporites variverrucatus* (Jana and Hilton 2007) (Table 2.1).

Apart from Kachchh, the Middle Jurassic palynoassemblage is also reported from the Dubrajpur Formation of the Rajmahal Basin as the *Contignisporites cooksoniae* Assemblage Zone (Tripathi 2004). This zone is dominated by *Callialasporites* and marks FAD of *Contignisporites cooksoniae* and diversification of *Concavissimisporites* and *Klukisporites* (Table 2.1). A similar palynoassemblage is recovered from the Dubrajpur Formation of the Pachami area of West Bengal (Vijaya 2004, 2006; Vijaya et al. 2010) (Table 2.1). A detailed study of the Dubrajpur Formation area provided a late Middle Jurassic to Early Cretaceous age for this formation (Callovian to Berriasian; Vijaya 2004, 2006, 2009).

The Late Jurassic palynoflora of India is represented by a uniform dominance of *Callialasporites* along with *Araucariacites* and *Podocarpidites* sub-dominance. Equivalent deposits have also been recovered from the Dubrajpur Formation of the Domra sub-basin having dominance of *Callialasporites* and *Araucariacites* along with significant presence of *Crybelosporites, Klukisporites, Murospora florida* (Vijaya 1997a) (Table 2.1).

A Late Jurassic palynoassemblage has been recovered from the Barakar Formation of the Birbhum Coalfield (Vijaya 2000b). Similarly, the Late Jurassic palynoflora has been also identified within the Panchet Formation of the Panagarh area. Here, the Panchet Formation extends from the Early Triassic to Late Jurassic with a large sedimentation gap spanning the Late Triassic to Middle Jurassic. The Early Triassic palynoflora is directly followed by the Late Jurassic palynoflora comprising several species of *Callialasporites* spp. (*C. turbatus, C. dampieri*), *Contignisporites cooksoniae, Cicatricosisporites australensis* and *Murospora florida* indicating an Oxfordian to Tithonian age (Vijaya 2000b; Vijaya and Bhattacharji 2003).

The Late Jurassic deposits are also reported from the western (Kachchh) and central part (Madhya Pradesh) of the country. In Kachchh, the Katrol and Jhuran formations represent a Late Jurassic sequence (Venkatachala and Kar 1969, 1970). The low diversity Palynozone-1 from the Jhuran Formation is characterised by the restricted occurrence of *Katrolites* sp., the dominance of conifer pollen (*Callialasporites, Podocarpidites, Microcachryidites*) and low proportion of cheirolepidacean pollen (*Classopollis*). The palynoassemblages from the Chaugaon beds (Kumar 1993) and Nimar Formation of the Narbada Valley (Kumar 1994a) are less diverse and show a high percentage of *Callialasporites* followed by *Araucariacites* and *Podocarpidites*. However, dominance of these pollen also occurs in the Early Cretaceous. Thus, the age assignment is still controversial and needs further investigation.

There are several Indian palynoassemblages which lack a precise age assignment (mainly due to taphonomic factors or lack of stratigraphically significant palynotaxa and marine markers) and extend from the Middle to Late Jurassic and up to the Early Cretaceous. A detailed study is needed to assign precise ages to such formations. The Dubrajpur Formation of the Birbhum Coalfield is dominated by *Callialasporites* and extends from the Middle to Late Jurassic (Vijaya and Roy 2000; Vijaya and Sen 2005; Vijaya et al. 2010) (Table 2.1). Similarly, palynoassemblages from the extra-peninsular parts, i.e., Lamayuru Complex of Ladakh and from Spiti area of Ladakh also span from the Middle Jurassic to Late Jurassic (Vijaya and Kumar 2002; Upadhyay et al. 2005). These assemblages are dominated by *Callialasporites* and *Araucariacites*. The age-diagnostic taxa of both the palynoassemblages are listed in Table 2.1.

The age of the Kota Formation, Pranhita-Godavari Basin is still controversial and palynological investigations indicate that it is a diachronous lithological unit deposited within a time period spanning the Middle Jurassic to Early Cretaceous age (Vijaya and Prasad 2001). The main palynotaxa recovered from the formation in four different localities include various species of *Callialasporites* spp., *Murospora florida, Contignisporites cooksoniae, Aequitriradites* sp., *Crybelosporites stylosus, C. punctatus* and *Triporoletes simplex* (Table 2.1).

Jurassic-Cretaceous Boundary

The transition from the Jurassic to the Cretaceous periods was not abrupt and similar palynomorphs extend from the latest Jurassic to the Early Cretaceous (Tithonian-Berriasian). The Late Jurassic palynomorphs such as *Araucariacites, Callialasporites, Podocarpidites, Classopollis, Cycadopites, Microcachryidites* and several others range into the Early Cretaceous (Singh and Venkatachala 1987; Tripathi 2004, 2008) (Table 2.3). Moreover, the lack of a visible lithological change adds on to the problems in placing the Jurassic-Cretaceous boundary, as in other parts of the world (Vijaya 1999). Tiwari and Tripathi (1992) defined the Jurassic/Cretaceous boundary based on the dominance of *Callialasporites* in the Jurassic and *Callialasporites/Podocarpidites/Araucariacites* pollen in the Early Cretaceous. However, stratigraphically it seems insignificant as dominance of these pollen grains also extends to the Late Jurassic (Table 2.1). Thus, Early Cretaceous palynostratigraphy is based on the FADs of stratigraphically significant taxa whereas most of them range from the Tithonian-Early Cretaceous without any visible difference across the boundary (Figs. 2.4 and 2.5, Table 2.2). Palynotaxa such as *Cicatricosisporites australiensis, C. hughesii, Appendicisporites erdtmanii, Crybelosporites stylosus* among others are the characteristic Berriasian marker taxa (Tripathi 2004). There are very few palynoassemblages which are found to document this gradual transitional phase such as Palynozone-2 from Kachchh which is considered to include the Jurassic/Cretaceous boundary based on the first appearance of *Cicatricosisporites australiensis* (Venkatachala and Kar 1970). This is an acme zone of *Impardecispora* and *Araucariacites*. The spore abundance and diversity of *Cyathidites* spp. (*C. cutchensis, C. pseudopunctatus, C. grandis*), *Concavisporites indicus, Murospora punctate, Aequitriradites verrucosus, Coptospora, Trilobosporites, Schizosporis* along with pollen of *Platysaccus indicus* play a key role in defining the Jurassic/Cretaceous boundary (Tiwari and Tripathi 1992). The Jurassic/Cretaceous boundary has been also identified in the transition of the Spiti Shale B to the Spiti Shale C based on the FADs of *Aequitriradites spinulosus, Ruffordiaspora ludbrookiae* and *Cicatricosisporites hughesii* (Vijaya and Kumar 2002) (Fig. 2.3). The Jurassic-Cretaceous transition has been also identified in the Rajmahal Formation of the Domra sub-basin based on the FAD of *Cicatricosisporites australiensis* and *C. angustus* along with the rare occurrence of *Aequitriradites spinulosus* and *Crybelosporites stylosus* (Vijaya 1997a). Similarly, intertrappean bed-II of the Rajmahal Basin marks the first appearances of *Cicatricosisporites australiensis, Aequitriradites spinulosus, Lycopodiacidites dettmannae* and *Crybelosporites stylosus* and is believed to preserve the Jurassic/Cretaceous boundary (Vijaya 2000a). Bera and Banerjee (1997) have identified the four palynostratigraphic palynozones in the western margin of the Bengal Basin. They placed the Jurassic/Cretaceous boundary within the *Podocarpidites ellipticus* sub-zone (PESZ) based on the first appearance of the marker taxa *Cicatricosisporites australiensis* in the Bengal Basin.

Early Cretaceous

Lower Cretaceous rocks are more widely distributed and well-represented by palynofloral assemblages (Fig. 2.1C,

Table 2.2). The palynoassemblage identified in the Lower Cretaceous strata of India is represented by *Microcachryidites antarcticus* Assemblage Zone (Tiwari and Tripathi 1992). This zone has been identified in the Cauvery Basin (Venkatachala et al. 1972). The assemblage zone follows of the *Callialasporites segmentatus* Assemblage Zone and shows a dominance of *Callialasporites, Podocarpidites* and *Araucariacites*. The significant palynomorphs of this palynozone are *Microcachryidites antarcticus, Spheripollenites scabris, Podosporites tripakshi, Cooksonites variabilis, Klukisporites scabratus, Aequitriradites verrucosus, Murospora florida, Staplinisporites caminus, Impardecispora apiverrucata, Appendicisporites distocornitatus, Leptolepidites verrucatus, L. rimatus*, and *Cicatricosisporites ludbrooki*. The palynozone marks the FADs of *Aequitriradites spinulosus, Coptospora kutchensis, Leptolepidites major, Contignisporites dettmanni, Cicatricosisporites australiensis, Podosporites tripakshi, P. microsaccatus, Foraminisporis* sp. and *Baculatisporites comaumensis* (Tiwari and Tripathi 1992) (Fig. 2.4). The Early Cretaceous palynoassemblages can be differentiated from the Late Jurassic assemblages by showing a higher abundance of podocarpaceous pollen, bryophytic and pteridophytic spores and from the Middle to Late Cretaceous by the absence of tricolpate and the diversification of hilate spores and angiosperms (Tiwari and Tripathi 1992; Dogra et al. 1994, 2010).

Prasad et al. (1995) introduced four palynozones spanning the Tithonian to Cenomanian in the Krishna-Godavari Basin (Fig. 2.4). These palynozones also correspond to the marine dinoflagellate zones enabling palynostratigraphic dating on higher resolution:

a. *Cicatricosisporites australiensis-Aequitriradites spinulosus* (?Tithonian-Hauterivian). This zone is characterized by the dominance of *Classopollis, Callialasporites, Microcachryidites, Aequitriaradites* and *Cicatricosisporites*.
b. *Cyclosporites hughesii-Ceratosporites equalis* (Barremian to Aptian). *Classopollis, Callialasporites* and *Microcachryidites* continue to dominate along with increased occurrence of some new forms such as *Crybelosporites, Ceratosporites, Contignisporites* and *Coptospora cauveriana*.
c. *Coptospora cauveriana-Distaltriangulatisporites mutabilis* (Late Aptian to Early Albian). The zone is characterized by the FADs of angiosperm pollen such as *Clavatipollenites hughessi, Lilliacidites reticulatus, Tricolpites micromurus* and *Asteropollis asteroids*.
d. *Appendicisporites distocarnatus-Triporoletes reticulatus* (late Albian to Cenomanian). A marked palynological shift occurs with the decline and termination of previously dominated palynomorph species and appearance of *Lygestipollenites florinii, Turonipollis helmigii, Murospora truncata* and *Clavifera triplex*.

Inter-basinal Correlation

The Late Jurassic and Early Cretaceous epoch was the heyday of pteridophytes and conifers with regard to diversity and abundance, respectively. Similar flora has been reported from most of the Indian peninsular parts (Fig. 2.1C). The complete and well-dated Early Cretaceous sequence (dating based on plant megafossils, pollen and spores, dinoflagellates and isotopic data) is reported from the Dubrajpur Formation and overlying inter-trappean beds of the Rajmahal Formation in eastern India (Table 2.2). The palynomorph assemblage includes pollen of *Araucariacites, Podocarpidites, Callialasporites* and spores such as *Coptospora, Cyathidites* and *Balmeisporites*. The common dinoflagellates observed are *Dissiliodinium* sp. and *Batiacasphaera* sp. The plant megafossils from these beds represent a typical Early Cretaceous *Ptilophyllum* flora containing *Pterophyllum, Ginkgoites, Pachypteris, Araucaria, Elatocladus* and *Sphenopteris* etc. The palynoassemblage is equivalent to the upper part of the *Ruffordiaspora australiensis* Assemblage Zone and is also characterised by the diversification of costate and hilate spores and presence of *Foraminisporis, Januasporites* and *Triporoletes* (Tiwari and Tripathi 1995; Tripathi 2004). The Rajmahal Formation of the Rajmahal Basin preserves a rich Early Cretaceous palynoassemblage within the *Ruffordiaspora australiensis* Assemblage Zone and is characterized by the first occurrence of *Ruffordiaspora* and presence of *Concavissimisporites, Klukisporites, Microreticulatisporites* (Tripathi 2008) (Fig. 2.3). The detailed study of the Rajmahal Formation intertraps enabled to distinguish four palynological associations based on the composition and first appearance datum of age-diagnostic taxa (Tripathi 2008) (Fig. 2.3). Detailed palynological study of these intertraps indicated that the oldest basalt flows have erupted in the Berriasian continuing up to the Aptian (Tiwari and Tripathi 1995).

The Lower Cretaceous deposits are also reported from other eastern India regions. The Rajmahal Formation of the Panagarh area, West Bengal has a diverse spore assemblage of *Cicatricosisporites* (=*Ruffordiaspora*) *australiensis, Aquitriradites spinulosus*, and *Foraminisporis wonthaggiensis* indicating a Berriasian to Valangian age (Vijaya and Bhattacharji 2002). Here, the basalt flow was of different age from that of the Rajmahal Basin as it started during the Berriasian and lasted only up to the Hauterivian (earlier than Rajmahal Basin). A younger palynoflora (Valangian to Hauterivian) equivalent to the *Foraminisporis wonthaggiensis* Assemblage Zone has been also identified from the

Domra sub-basin (Vijaya 1997a). The condensed intertrappean sequence of the Early Cretaceous aged Rajmahal Formation has been also found in the Birbhum Coalfield (Vijaya 2000b). The palynoassemblages referable to intertrappeans of the Rajmahal Formation have also been recovered from the Pachami areas of West Bengal (Vijaya 2004, 2009; Vijaya et al. 2010).

The Early Cretaceous-aged palynomorphs have also been recorded from the Talchir Formation of the Brahmani River Section of the Talcher Coalfield (Tripathi 2009). The palynoassemblage shows a dominance (about 95%) of *Podocarpidites* and *Araucariacites* along with the presence of palynomorphs such as *Aequitriradites, Coptospora, Callialasporites, Contignisporites, Klukisporites, Murospora, Ruffordiaspora (=Cicatricosisporites), Santhalisporites* and *Lametatriletes*. These findings clearly supports the significance of palynological studies for stratigraphic purposes as lithologically the Brahmani river section has a strong similarity with the Talchir sediments, but from palynological studies it can be inferred to be Late Jurassic to Early Cretaceous age. Similarly, a palynological study conducted in the Panchet Formation of the Raniganj Coalfield has revealed the presence of infra-trappean sediments (Vijaya 2011). These intertraps of the Rajmahal Formation are intercalated between the Panchet and Tertiary deposits, which resulted in the mixing of Late Permian, Early Triassic and Early Cretaceous palynoflora due to ongoing volcanic eruptions at that time (Vijaya 2011).

In central India, an Early Cretaceous (Tithonian-Neocomian age) palynoflora is reported from the Jabalpur Formation, Madhya Pradesh (Table 2.2). The recovered palynoassemblage has been named as *Narbadapollis indicus* after its respective dominance (Dogra 1986, 1994; Dogra et al. 1988, 1994, 2003, 2010; Rana et al. 2000) (Table 2.2). The marker palynotaxa of the assemblage are *Neoraistrickia pallida, Tricolpites psilatus, Schizosporis minutus* and *Coptospora mesozoica* (Dogra 1986, 1994; Dogra et al. 1988, 1994, 2003, 2010; Rana et al. 2000).

In southern India, Early Cretaceous deposits are named differently in different basins (Fig. 2.3). The palynological identification of the Athgarh Sandstone records a dominance of araucariacean pollen, basically by *Araucariacites* and *Callialasporites* indicating an Early Cretaceous age but not extending beyond the Albian due to the absence of angiospermic elements (Goswami et al. 2008) (Table 2.2). The Sriperumbudur Formation, Palar Basin is correlatable to the *Microcachrydites antarcticus* Zone (Tiwari and Tripathi 1992) and is Tithonian to Valangian in age (Tripathi and Vijaya 1997; Rajnikanth et al. 2010). The assemblage marks the FAD of *Cicatricosisporites australiensis* and *C. ludbrooki*. The fossil record from the two localities in the Pranhita-Godavari Valley are mega floristically rich in *Ptilophyllum* and *Pterophyllum* whereas conifer pollen dominate the microflora (Jha et al. 2016, 2017). The recovered fossil flora indicates a Tithonian to Neocomian age (Jha et al. 2016, 2017).

The southern Indian Krishna-Godavari Basin also preserves a rich Early Cretaceous palynoassemblage of terrestrial and marine palynomorphs. The palynological analysis of the Gollapalli Formation and Raghavapuram Shale of the Krishna-Godavari Basin yielded a Necomanian-Early Cretaceous and Barremian-Aptian age, respectively (Patra and Sahoo 1995; Prasad 1998; Prasad and Pundeer 1999). The palynomorphs of the Nellore Claystone, Bapatla Sandstone and Pennar Shale of Pennar Basin are almost similar and assignable to a latest Jurassic-Early Cretaceous age (Tithonian to Valanginian age; Prasad 1999). The Gollapalli Formation is coeval to these formations. The succeeding Krishna Formation yielded an assemblage of Valanginian-Hauterivian to Barremian age. The Gollapadu Shale is comparable to the Raghavapuram Shale based on the rich assemblages of dinoflagellates, spores and pollen indicating a Barremian-Aptian to Early Albian and Aptian-Albian to Cenomanian age of the respective formations (Prasad 1999). A younger Aptian-aged pteridophyte rich palynoassemblage assignable to the *Coptospora cauveriana* Zone has been also documented from the coastal Tamil Nadu and Pondicherry areas (Tiwari et al. 1996a, b).

The younger Aptian-Albian-aged palynoassemblage characterized by a dominance of *Araucariacites australis, Callialasporites dampieri* and *Cycadopites grandis* has been also identified from the Jabalpur Formation of the Satpura Basin (Kumar 2011). The Bhuj Formation (also known as Umia Formation) from the Kachchh represents the youngest Upper Gondwana sequence, i.e., post-Aptian age based on the presence of several key palynotaxa such as *Appendicisporites, Baculatisporites comaumensis, Trilobosporites, Coronatispora perforate, C. telata, Neoraistrickia verrucata, Foveosporites canalis* (Venkatachala and Kar 1969; Venkatachala and Kar 1972; Rana et al. 2000). The Dharangadhra Formation of western India, Gujarat has been dated as Early Cretaceous (Neocomian-Aptian) based on mega and micro plant fossils (Kumar 1994b; Jana 2003).

Tectonics Versus Eustasy

The breakup of the supercontinent Gondwana had a great impact on the life, causing extinctions and transitions of existing life forms. The tectonic events occurred in several intervals and the movement of continents into new latitudinal position resulted in climatic shifts followed by extinction and emergence of various life forms in response to adaptation under new environments and climates. The imprints of tectonic movements can be observed in the changing paleoflora which responded to the changing climatic conditions. The major change was the replacement of gymnosperm flora that dominated in the early Mesozoic (Jurassic) by incoming angiosperm flora in the Early Cretaceous of Gondwana and parallel divergence of pteridophytes. This transition was not abrupt and occurred in several intervals in response to changing latitudinal position of the continents. The following section analyses and correlates the Jurassic and Early Cretaceous flora to trace the changes in the vegetation within India and other Gondwana continents.

Intra-Gondwanan Correlation

Palynological data are available in great numbers from the Indian Upper Gondwana. However, most of the data are not calibrated to the international chronostratigraphic time scale making its Intra-Gondwana correlation difficult. Other than India, palynological well-dated Gondwanan records are from Australia and Argentina. The western Australian sequence is biostratigraphically well calibrated to the marine record while eastern Australia lacks such data as they are mainly deposited in terrestrial basins (Turner et al. 2009). Well-dated palynostratigraphic schemes for the Jurassic and Early Cretaceous deposits of Australia are proposed by Evans (1964), Burger (1968), de Jersey (1976), Helby et al. (1987), Price (1997), Sajjadi and Playford (2002a, b) etc. These palynostratigraphic schemes were further modified by several authors incorporating new palynological data (Burger 1995a, b; Price 1997; Grant-Mackie et al. 2000 etc.). Palynostratigraphic scheme by Helby et al. (1987) is widely used for dating and correlating the Upper Gondwana sequence and is divided into two super zones: the lower *Callialasporites dampieri* Superzone (Hettangian to Kimmeridgian) and the upper *Microcachryidites* Superzone ranging from the Tithonian to the Albian. The *Callialasporites dampieri* Superzone is further subdivided in ascending order into the *Corollina torosa* Zone, *Callialasporites turbatus* Zone, *Dictyosporites* complex Zone, *Contignisporites cooksoniae* Zone and *Murospora florida* Zone. The same scheme was further modified by Burger (1995a, b) (Fig. 2.3). The *Callialasporites dampieri* Superzone of Australia (Helby et al. 1987) compares well with the Indian Jurassic sequence from the Jabalpur Formation (Prakash 2008; Kumar and Ram-Awatar 2010).

The Jurassic deposits in most of the Gondwana continents are not well-studied in comparison to the Permian, Triassic and Cretaceous deposits (Fig. 2.1A–B). The Jurassic sedimentation in Australia was not continuous and records several hiatuses during the Triassic-Jurassic boundary interval (during the Hettangian) and the latest Middle-early Late Jurassic and there are different unconformity time periods in different basins (Turner et al. 2009). The *Toripustulatisporites hokonuiensis* Assemblage Zone of Australia and New Zealand (de Jersey and McKeller 2013; Bomfleur et al. 2014) is equivalent to APT5U informal Unit of the Surat and Bowen basins (Price 1997) and represents the youngest Jurassic assemblage (Hettangian-early Sinemurian). The *Corollina torosa* subzone of *Callialasporites dampieri* Superzone is marked by the replacement of corystosperm-dominated Late Triassic flora by cheirolepidiacean flora in the Early Jurassic (McKellar 1996; Grant-Mackie et al. 2000). The other Early Jurassic-aged (up to Toarcian) palynozones include the *Corollina torosa* Abundance Zone from the Surat Basin (Late Hettangian to Early Toarcian; Reiser and Williams 1969; Helby et al. 1987), the *Polycingulatisporites crenulatus* Assemblage Zone of eastern Australia (de Jersey 1975), the Marburg Subgroup from the Clarence-Moreton Basin (Jansson et al. 2008), the Plover Formation of the Bonaparte Basin (Mantle 2009); the *Classopollis* Abundance Zone of Australia (de Jersey and McKeller 2013), the *Retitriletes austroclavatidites* Zone of New Zealand (de Jersey and Raine 1990, 2002) and the APJ unit (Price 1997). Based on the megafossil records from the Clarence-Morenton and Nambour basins of Australia the Pliensbachian-Toarcian assemblages contain a wide range of pteridophytes such as equisetaleans, lycophytes, dipteridaceous, osmundaceous, dicksoniaceous and matoniaceous besides equisetaleans, lycophytes, the pteridosperms *Komlopteris* and *Rintoulia,* caytonialeans and bennettitales (Gould 1968, 1971; Rozefelds and Sobbe 1983; Grant-Mackie et al. 2000; Pattemore and Rigby 2005; Jansson et al. 2008). The similar dominance of cheirolepidacean pollen, i.e., *Classopollis* spp. is also marked in the Indian Early Jurassic palynozone *Classopollis minor* along with a similar presence of palynotaxa such as *Araucariacites, Dictyophyllidites, Callialasporites, Minutosaccus, Nevesisporites, Infernopollenites* and *Staurosaccites* (Tripathi and Ray 2006) (Table 2.1). However, the assemblage differs in showing a low diverse pteridophytic assemblage (Lukose 1972; Koshal 1975).

A comparable palynoassemblage is also recorded from Antarctica, Africa, South America and Pakistan. In Antarctica, the Carapace Sandstone and Kirkpatrick Basalt of the Ferrar Group, South Victoria Land (Ribecai 2007), the ?*Classopollis* Abundance Zone from the Transantarctic (Bomfleur et al. 2014), the Ferrar Large Igneous Province and parts of East Antarctica (Bomfleur et al. 2014) have been palynologically dated as Early Jurassic and show similar dominance of *Classopollis* pollen. The African Pindiro Group from southern coastal Tanzania is coeval to the Upper Triassic and Lower Jurassic deposits based on the occurrences of stratigraphically significant palynotaxa such as *Playfordiaspora* spp., *Classopollis, Ashmoripollis, Ceratosporites* spp. (Hudson and Nicholas 2014). The South American *Classopollis classoides* Assemblage Zone (Late

Sinemurian-Early Toarcian; Quattrocchio et al. 1996) and the palynoassemblage from Cañadón Asfalto Formation, Cañadón Asfalto Basin, Argentina (Early Jurassic to mid Middle Jurassic; Olivera et al. 2015) is also similar to these assemblages in dominance of *Classopollis, Araucariacites* and *Callialasporites*. The palynoassemblages of high *Classopollis* abundance are also reported from the Neuquen Basin, Patagonia (Volkheimer et al. 1978; Zavattieri et al. 2008) and Nestares Formations, Argentina (Zavattieri and Volkheimer 2003).

The Salt Range of Pakistan preserves a continuous sedimentation along the Triassic/Jurassic boundary between the Kingraili and Datta formations. Here, the Triassic/Jurassic boundary is characterized by the presence of marker dinoflagellates like *Rhaetogonyaulax*, *Suessia* and *Baumontella* which indicate Rhaetian age while Early Jurassic is marked by FAD of *Cerebropollenites thiergatii,* marking the base of the Hettangian. This observation is also supported by the evidence of carbon isotope analysis. The Early Jurassic-aged Datta Formation is dominated by *Clossopollis turosus* pollen (Kuerschner et al. 2017).

The Middle and Late Jurassic palynological record is more abundant in comparison to the Early Jurassic, except in Antarctica (Fig. 2.1B). In Australia, the Middle and Late Jurassic flora is represented by the *Callialasporites dampieri* Superzone. An increase of araucarian conifers characterises the Middle Jurassic paleoflora, while the high proportion of podocarp conifer pollen suggest the Late Jurassic (Grant-Mackie et al. 2000). The Australian Middle to Late Jurassic palynozones *Contignisporites cooksoniae* and *Murospora florida* (Burger 1995a, b) are also similar to the Indian palynoassemblages characterised by *Callialasporites trilobatus* and *Callialasporites segmentatus*, respectively (Burger 1990a, b; Tiwari and Tripathi 1992; Burger 1995a, b; Vijaya and Sen 2005) (Fig. 2.3). These zones show a similar palynofloral pattern of araucariaceae and podocarpacean pollen dominance during the Middle and Late Jurassic, respectively.

The transition of Early to Middle Jurassic (Late Toarcian and Early Bajocian) is documented by the Upper Evergreen Formation and Injune Creek Group of the north-eastern Surat Basin, respectively in Australia (McKellar 1974, 1998), spanning the *Araucariacites fissus* (middle-late Toarcian) and *Camarozonosporites ramosus* (late Toarcian to middle-late Bajocian) assemblage zones (McKellar 1998). Other known Middle Jurassic assemblages from Australia occur in the Walloon Coal Measures, the Clarence-Moreton, Surat and Elang formations of the Bonaparte and Mulgildie basins dated as Bathonian to Callovian (Mantle 2009; Turner et al. 2009) (Fig. 2.1B). The Bajocian to late Middle Callovian palynoassemblages have been also recorded from the north-eastern Surat Basin (McKellar 1998) as *Retitriletes circolumenus* (Middle-Late Bajocian to Early Bathonian), *Aequitriradites norrisii* (Bathonian) and *Contignisporites glebulentus* assemblage zones (late Bathonian to late middle Callovian). The Middle Jurassic palynoassemblage from New Zealand show similar dominance of araucarian and podocarp pollens (Thorn 2001). The Kuar Bet Bed assemblage and Middle Jurassic assemblages from the Dubrajpur Formation of India are comparable to these assemblages (Tripathi 2008; Jana and Hilton 2007).

A similar Middle Jurassic palynoassemblage referable to the *Callialasporites dampieri* Superzone has been identified from Africa based on biomarker studies (Killops et al. 2003). The similar *Callialasporites dampieri* zone (late Toarcian-Oxfordian) described from the Neuquèn Basin, Argentina, South America (Quattrocchio et al. 1996, 2001), the Las Chacritas Member of the Cañadón Asfalto Basin and the Lajas Formation of the Neuquén Basin is also Middle Jurassic in age (García et al. 1994; Volkheimer et al. 2001, 2015). However, they show a high proportion of *Classopollis* spp. along with pollen of Araucariaceae (*Callialasporites* and *Araucariacites*) and Podocarpaceae (*Podocarpidites* and *Phrixipollenites*).

Several Middle to Late Jurassic palynoassemblages have been identified in Australia and correspond to assemblages in India. The Yarragadee Formation of the Perth Basin was dated Middle Jurassic (Bathonian) to Late Jurassic (Tithonian) based on palynofloral similarities with *Murospora florida* Palynozone (Backhouse 1984; Burger 1990a, b). A similar palynological association occurs in the *Murospora florida* Oppel zone (middle Callovian to late Oxfordian) and *Retitriletes watherooensis* Oppel zone (late Oxfordian to middle Tithonian) of the Euromanga Basin of Queensland (Sajjadi and Playford 2002a, b). The Indian palynoassemblage from the Lamayuru Complex, Ladakh is dominated by araucaria pollen and dated as late Middle Jurassic to early Late Jurassic (Callovian-Kimmerdigian; Upadhyay et al. 2005). Palynological studies from the two basins of Patagonia, Argentina i.e., the Cañadón Asfalto and Neuquén basins revealed the presence of Middle and Upper Jurassic paleoflora (Volkheimer et al. 2001, 2015). Similarly, a poor palynoflora from the Samana Suk Limestone Formation of the Kala Chitta Range of Pakistan is assigned to the Middle Jurassic, while the Chichali and Lumshiwal formations yielded a Late Jurassic to Lower Cretaceous palynoflora (Massood et al. 2008).

Late Jurassic palynoassemblages characterize the Frigate Formation of the Bonaparte Basin (Mantle 2009), where the *Murospora florida* Assemblage Zone of the Surat and Bowen basins (late Middle Callovian to Early-Middle Oxfordian; Filatoff 1975; Helby et al. 1987) and the *Retitriletes watherooensis* Assemblage Zone of the Surat Basin are distinguished (middle Oxfordian, Backhouse 1988; Helby et al. 1987). The Panchet Formation of the Domra sub-basin and the Dubrajpur Formation of India have been dated as Late Jurassic (Kimmeridgian to Tithonian) due to similarities with the *Murospora florida* palynoassemblage zone (Vijaya 1997b; Tripathi 2004, 2008).

The *Classopollis-Araucariacites-Shanbeipollenites* Assemblage Zone from the Tendaguru Beds of southeast Tanzania, Africa are also dated as mid-Oxfordian to Tithonian (Schrank 2010). The African palynoflora shows great affinity with the South American flora showing a dominance of *Classopollis* pollen. A similar flora in South America is documented by the *Microcachryidites antarcticus* Zone (Tithonian) of the Neuquén Basin, Argentina (Quattrocchio et al. 1996), the Puesto Almada Member of the Cañadón Asfalto Basin and the Auquilco Formation of the Neuquén Basin (Volkheimer et al. 2015). The Morondava Basin of southwest Madagascar also yielded a Late Jurassic palynoflora (Jehowsky and Goubin 1964). The main elements of the assemblage are conifer pollen (*Araucariacites, Podocarpidites*, and *Callialasporites*), cheirolepidacean pollen (*Classopollis*) and many pteridophytic spores.

Early Cretaceous deposits are more widely distributed in all Gondwana continents. In comparison to the Jurassic (Fig. 2.1), a single palynological zone of the Australian Tithonian-Valangian strata is proposed by Helby et al. (1987; *Microcachryidites* Superzone) and Sajjadi and Playford (2002a, b; *Ruffordiaspora* (=*Cicatricosisporites*) *australiensis* Interval Zone). However, the palynostratigraphic scheme by Burger (1995a, b) is more descriptive and categorises the entire Early Cretaceous palynoflora into four zones (Fig. 2.4). A youngest exposed Albian strata in the Otway Basin has been investigated by Korasidis et al. (2016), revealing a diverse palynoassemblage of bryophytic and pteridophytic spores, and angiosperm pollen. Palynozones corresponding to Burger's palynostratigraphic scheme has been well correlated with the Indian Rajmahal Formation (Tripathi 2004, 2008) (Table 2.2). The Early Cretaceous Australian flora resembles Indian flora in showing similar dominance of araucarian pollen (*Podocarpidites and* Araucariacites) along with bryophytic and pteridophytic spores like *Cyathidites, Microcachryidites, Lycopodiumsporites, Gleichniidites, Stereisporites* (Morgan 1980; Helby et al. 1987; Wagstaff et al. 2012). The flora of the Lumshiwal Nala section from the Salt Range shows close resemblance with the Indian palynoflora showing a dominance of *Callialasporites* and *Araucariacites.* It is assigned to the Late Jurassic and Early Cretaceous (Neocomian; Fredrikson et al. 2007).

Apart from the western extremities of the Indian subcontinent, the data analysed from the Tethys Himalayan margin (Gamba, Xixang, Southern China) also shows a close similarity with the flora from Indian Peninsula (Li et al. 2013, 2017). The Gamba flora, from southern China shows a pronounced dominance of conifer pollens (*Podocarpidites, Araucariacites, Callialasporites, Spheripollenites*) and a variety of cryptogramic spores (*Cooksonites, Coptospora, Gleicheniidites, Crybelosporites*). The Gamba flora is also more closely related with the Indian and Australian flora.

Antarctica has a well dated Early Cretaceous palynological record from Alexander, Livingston Island, James Ross Island and Marie Byrd Land. The Byers Peninsula is also an important Late Jurassic-Early Cretaceous sedimentary basin of peninsular Antarctica. Palynological analysis of the President Beaches Formation and Chester Cone Formation of the Byers Group revealed a Tithonian to Valangian age based on its correlation with the *Biretisporites eneabbaensis* Oppel Zone and marine dinoflagellate assemblages (Helby et al. 1987; Duane 1994, 1996). The palynoassemblage of President Beaches Formation has many taxa in common with Early Cretaceous palynoflora of India such as *Cyathidites minor, C. australis, Stereisporites antiquasporites, Osmundacidites wellmannii, Gleicheniidites circinidites, Foramnisporis wonthagginesis, Classopollis* spp., *Cicatricosisporites australiensis, Biretisporites eneabbaensis* etc. The Robertson shelf has also yielded a Jurassic to Early Cretaceous palynoflora chiefly dominated by *Araucariacites* (Truswell et al. 1999) (Fig. 2.1C). Palynological investigations in Marie Byrd Land by Domack et al. (1980) revealed a late Early Cretaceous age based on the correlation with Australian and New Zealand palynozones. The Albian flora of Alexander Island is paleogeographically important as it possess a mixed flora of abundant elements of eastern Gondwana in association with rare elements from the western Gondwana (Cantrill and Nagalingum 2005). The palynological studies in James Ross Island also revealed an Albian-Cenomanian aged palynoflora (Dettmann and Thomson 1987). The Early Albian flora is similar to the Indian, Australian and New Zealand palynoflora showing a high proportion of podocarpaceous pollen and *Microcachryidites* along with araucarean pollen and pteriodspermous spores.

The Early and Late Cretaceous-aged palynoflora is also wide spread in Africa. The Early Cretaceous flora has been investigated from the Orange Basin, South Africa which is dominated by *Classopollis* along with Podocarpaceae, *Cyathidites,* and *Gleicheniidites* as dominant spore taxa (Sandersen et al. 2011). The Tendaguru Beds of southeast Tanzania also contain a similar palynoflora *Classopollis-Cicatricosisporites-Ruffordiaspora* Assemblage Zone, ranging from the late Berriasian to the Hauterivian age (Schrank 2010). Ferns, cycadales, bennettitales and conifers dominated megaflora has been recovered from the Kirkwood Formation of the Algoa Basin, South Africa. The megafloral assemblage is indicative of an Early Cretaceous age (Muir et al. 2015).

The Early Cretaceous sequence has been also investigated from the Cañadón Calcareo locality, Patagonia. The dominant genera of the Early Cretaceous palynoflora include *Classopollis* and *Callialasporites,* with *Cyclosphaera* being present indicating a Valanginian age (Riccardi et al. 1990). The similar Early Cretaceous flora has been recovered from the Rio Mayer and Kachaike formations from the Austral Basin, southwestern Argentina ranging from the late Hauterivian to the earliest Albian, Cariri and Araripe Group from Araripe Basin (Pothe de Baldis and Ramos 1983; Coimbra et al. 2002, Medina et al. 2008; Pérez Loinaze et al. 2012). The Early Cretaceous palynoassemblages are rich in pteridophytic spores (Archangelsky and Archangelsky 2002, 2004, 2005, 2006a, b, 2010; Archangelsky and Llorens 2005, 2009). The assemblage is mainly rich in spores from Anemiaceae, Pteridaceae family and pollens from Cheirolepidiaceae, Podocarpaceae, Araucariaceae and Ephedraceae. The flora is assigned to the *Antulsporites-Clavatipollenites* Zone (Pérez Loinaze et al. 2012). The Valanginian to the Aptian palynoflora of Argentina are dominated by conifer gymnosperm pollens, *Cyclusphaera-Balmeiopsis-Classopollis* and in the Aptian and post-Aptian deposits, angiosperms become important stratigraphic markers (Prámparo 2012).

The oldest angiosperm (Early Aptian) records from the southern South America and in SW Gondwana is reported from Ticó flora of Patagonia having *Clavatipollenites, Retimonocolpites* and *Anacostia* type (Archangelsky and Archangelsky 2013). Few other Pre-Aptian angiosperms have also been reported from the Springhill Formation having *Clavatipollenites* dominance (Archangelsky and Archangelsky 2004; Quattrocchio et al. 2006) and La Cantera Formation (Aptian-lower Albian), San Luis Basin, Argentina (Prámparo 1999). *Clavatipollenites* grains are reported from various Lower Cretaceous (Early Aptian) deposits of Patagonia (Archangelsky and Taylor 1993). The Aptian flora of the Baqueró Group shows a dominance of *Clavatipollenites* and *Retimonocolpites* similar to Argentina, Australia (Loinaze and Llorens 2018). The pollen have affinity with chloranthaceae and their presence indicates that the family coevolved in comparison with Northern Hemisphere. A rich palynoassemblage of Aptian-Albian boundary has been identified in the Río Mayer Formation (latest Aptian) and Piedra Clavada Formation (earliest Albian; Archangelsky et al. 2008; Medina et al. 2008). The Río Mayer Formation is rich in pollens from conifer family mainly Cheirolepidaceae with few pteridophytic spores and saccate pollens of podocarpaceae family. The percentage of pteridophytic spores and podocarpaceae pollens increases considerably in the Piedra Clavada Formation. Angiosperm pollen is consistently present throughout the section, but they become more significant in the Upper Piedra Clavada Formation and lot of new tricolpate, tricolporate forms appear. Archangelsky et al. (2009) did a very exhaustive study of the Lower Cretaceous deposits from the southern South America and divided the evolution of angiosperms into three stages ranging from the late Barremian (128.3 Ma) to the end of the Coniacian (85.8 Ma). Stage-I (late Barremian-Aptian) witnesses appearance of first angiosperms; Stage-II (latest Aptian-earliest Albian) characterized by the appearance of the eudicots; Stage-III (middle Albian-Coniacian) preserves angiosperm diversification and abundance.

Paleovegetation and Paleoenvironment

In general, the Gondwanan flora of India and other continents shows a general pattern of paleoflora from the Permian to Early Cretaceous. However, the initiation of drifting of continents during the Late Jurassic has greatly affected the vegetation due to climatic shifts. The *Glossopteris* flora of the Permian was replaced by *Dicroidium* flora in Triassic due to end-Permian mass extinction. The *Dicroidium* paleoflora also faced a turn over by the end-Triassic, forced by another mass extinction event and resulted in the conifer and bennettitalean dominated floral communities of the Jurassic (McLoughlin 2001). Climate shifts are also reported in the Jurassic to Early Cretaceous floral record, but here the causal agent was the tectonic movement of the continents as no major mass extinction event has been reported during this phase.

The Early Jurassic earth was represented by two huge landmasses, i.e., Laurasia and Gondwana. The Jurassic flora is found to be more cosmopolitan than the Permian, Triassic or Cretaceous (Maeyen 1987). This cosmopolitan flora was a consequence of huge continental assembly into Pangaea resulting due to land-based pathways for biotic exchange and almost uniform global climates. However, the latitudinal differences played an important role in developing a characteristic flora across a landmass as huge as Gondwana. The tectonic movement of eastern Gondwana into lower paleolatitudes in association with falling sea-level during the latest Triassic and Early Jurassic was the main triggering factor for the extinction of the *Dicroidium* flora adapted to warm and humid climate and the subsequent rise of the *Ptilophyllum* flora of the Jurassic and Early Cretaceous. Palynologically, the corystosperm-dominated palynoflora including *Falcisporites* and *Alisporites* was replaced by cheirolepidacean conifer pollen, *Classopollis*. The Triassic-Jurassic boundary is not well exposed in most parts of Gondwana. However there are few records of replacement of the corystosperm-dominated Late Triassic flora by cheirolepidacean (dominated by *Classopollis*) and first appearance of araucarian flora. The radiation of gymnosperms is not comparable to the massive radiation observed after the Permian-Triassic extinction event. However, rise and diversification of cheirolepidacean conifers, caytoniales and seed-ferns is observed (Anderson et al. 1999).

The Jurassic was unambiguously the period of conifers, ranging from low diversity in the Early Jurassic to a highly diversified group in the Late Jurassic (McLoughlin 2001). Other gymnosperm-dominated plant groups were represented by Bennettitales and Pentoxylales (endemic to Gondwana), which appeared in the Triassic and continued through the Jurassic (Herbst and Troncoso 1996). Based on the well-dated megafossil records from the Clarence-Moreton and Nambour basins of Australia the Pliensbachian–Toarcian assemblages contain wide range of pteridophytes like equisetaleans, lycophytes, dipteridaceous, osmundaceous, dicksoniaceous and matoniaceous, the pteridosperms (*Komlopteris* and *Rintoulia*), caytonialeans and bennettitales (Gould 1968, 1971; Rozefelds and Sobbe 1983; Grant-Mackie et al. 2000; Pattemore and Rigby 2005; Jansson et al. 2008). There palynoflora representatives have been reported from various parts of Gondwana including, Kachchh of India (Lukose 1972; Koshal 1975), Australia (Mckellar 1996; Grant-Mackie et al. 2000; Turner et al. 2009; de Jersey and McKeller 2013), Antarctica (Bomfleur et al. 2014), South America (Volkheimer et al. 1978; Quattrocchio et al. 1996; Zavattieri et al. 2008) and Africa (Hudson and Nicholas 2014). The *Classopollis* pollen is produced by *Cheirolepis*, *Pagiophyllum* or *Brachyphyllum* plants which prefer to grow in warmer and dry climates (McKellar 1996, 2002). The phase of *Classopollis* dominance is reported up to the Toarcian and is referred to as Cheirolepidacean Phase by Grant-Mackie et al. (2000).

The Indian and Gondwanan floras transformed over a short transitional phase through the late Early Jurassic (i.e., Toarcian) due to the rise in global sea level, an anoxic event in the Tethys Ocean or combinations of both (McKellar 1998). This second phase of palynofloral transition includes the replacement of the cheirolepidacean flora by the araucarean flora with acme of *Callialasporites* and *Araucariacites* pollen (Grant-Mackie et al. 2000) (Table 2.3). This Early Jurassic flora represents base of the *Callialasporites dampieri* Zone (Helby et al. 1987; Burger 1995a, b). *Callialasporites* and *Araucariacites* are present in the Early Jurassic as subsidiary element and gradually gain dominance in the Middle Jurassic along the Pliensbachian-Toarcian boundary. These araucarian pollen prefer to grow under hydrophilic conditions (McKellar 1996, 2002). A similar late Early to early Middle Jurassic flora along with several pteridophytes has been reported in various parts of Gondwana including India (Vijaya and Roy 2000; Tripathi 2000, 2001, 2004; Tripathi and Ray 2006) (Table 2.1). The Australian megaflora of the Pleinsbachian-Toarcian flora is dominated by a wide range of mega and microfossils of pteridophytes, pteridosperms, caytoniales and bennettitales (Gould 1968, 1971; Rozefelds and Sobbe 1983; Grant-Mackie et al. 2000; Pattemore and Rigby 2005; Jansson et al. 2008). The Antarctica flora from South Victoria Land also shows a dominance of *Callialasporites* (Ribecai 2007). The advent of the *Callialasporites dampieri* flora on the eastern Gondwana indicates the warmest climatic phase during the Paleozoic-Mesozoic interval (McKellar 1998).

The third and final phase of palynofloral transition occurred within the Late Jurassic and marked by the increased frequency of bi- and trisaccate pollen of podocarpaceous conifers (Grant-Mackie et al. 2000). This transition has been assigned to different ages by different authors due to depositional hiatuses in the late Oxfordian by McKellar (1996), in the mid-Oxfordian by Grant-Mackie et al. (2000), and in the middle Kimmeridgian by McKellar (1998). A similar flora ranging from the Middle to Late Jurassic with high proportions of conifer pollen of araucarians and podocarp (*Callialasporites, Araucariacites, Microcachryidites* and *Podocarpidites*; Table 2.3) occurs in various parts of India (Fig. 2.1, Table 2.1), Australia (Grant-Mackie et al. 2000; Mantle 2009), Antarctica (Truswell et al. 1999) and Kala Chitta Range, Pakistan (Masood et al. 2008). The flora of Africa (Killops et al. 2003; Schrank 2010), South America (Quattrocchio et al. 1996; Volkheimer et al. 2001, 2015), Madagasacar (Jehowsky and Goubin 1964; Goubin 1965) shows a different pattern as *Classopollis* pollen continue to dominate along with increasing incidence of various araucarian pollen, witnessing warm and humid environments.

The east Antarctica flora represents a mixed flora of eastern and western Gondwana having dominance of *Classopollis* but differs in lacking *Araucariacites* and having rare occurrence to dominance of *Exesipollenites* pollen in comparison to the later (Truswell et al. 1999; Norris 1965; Tasch and Lammons 1978). The palynoassemblage of Antarctica is also characterised by low percentages of pteridophytic spores in comparison to India, barring the presence of few bryophytic and lycophyte representatives. The pre-Albian assemblage of the Prydz Bay contains palynotaxa which are absent in the high latitudes of the southern region (Domack et al. 1980; Truswell 1991a, b).

The Late Jurassic to Early Cretaceous was a period of enormous floral diversity and gradual vegetational turnovers. The foundation of the present day flora has been laid down in the Early Cretaceous as most of the modern vascular families had their origins within this time and mostly confined to Gondwana regions. Similarly, the bryophytes and pteridophytes also underwent diversification as most of the bryophytic spores became important stratigraphic markers in various parts of Gondwana (Morgan et al. 1995; Archangelsky and Villar de Seoane 1996; Vijaya 1999; Nagalingum et al. 2002; Tripathi 2002) and most of the modern fern families have their origin in the Early Cretaceous

(Dettmann 1994). This floral change was mainly induced by the continental drifting and resultant environmental changes.

These changes shaped the Indian flora and displays almost similar floral transitional pattern with other Gondwana continents. However, Indian flora differs in having more evolved and diverse bennettitales than elsewhere (Prakash 2008). Early Cretaceous palynofloral records are widely distributed in India and show a dominance of araucarian and podocarp pollen along with a wide variety of angiosperms, pteridophytes and bryophytes (Tables 2.2 and 2.3). The advent of this floral composition is related to the reference of cooling climatic conditions (McKellar 1998). The Tithonian-Neocomian transition in India is marked by the FAD of characteristic taxa such as *Cicatricosisporites australiensis, Aequitriradites spinosus, Crybelosporites stylosus, Contignisporites dettmanii* and *Cooksonites reticulatus* (Rana et al. 2000). While, the Hauterivian is marked by the appearance of angiospermous pollen with primitive morphology and *Podosporites microsaccatus* along with diversification of hilates, apiculates, murornates (Tiwari 1996). A similar Early Cretaceous palynoflora were reported from various parts of Gondwana like Australia (McKellar 1974; Helby et al. 1987; Sajjadi and Playford 2002a, b; Wagstaff et al. 2012), Antarctica (Domack et al. 1980; Dettmann and Thomson 1987; Duane 1994, 1996; Truswell 1991a, b; Truswell et al. 1999; Cantrill and Nagalingum 2005), Pakistan (Massod et al. 2008) and Tethys Himalayas (Li et al. 2013, 2017). The flora of western Gondwana is similar in dominance of podocarp pollens but differs in high percentages of *Classopollis* pollen and consistent occurrence of endemic *Cyclusphaera psilata* reported from Africa (Schrank 2010; Sandersen et al. 2011; Muir et al. 2015) and South America (Pothe de Baldis and Ramos 1980, 1983; Coimbra et al. 2002; Medina et al. 2008; Parez Loinaze et al. 2012). A contemporaneous increase in diversity of free sporing plants is documented in Australian, Indian and South American palynofloras. However, there are differences in assemblages at generic and family levels.

Phytogeography

The Late Triassic assemblages show a general Gondwanan uniformity of non-striate bisaccate pollen dominance, mainly *Alisporites australis* and *Falcisporites nuthallensis*. However, the restricted presence of characteristic palynomorphs in the provinces and their mixing in certain regions shows latitudinal dependence of species distribution from north to south (Césari and Colombi 2013). This latitudinal difference is represented by two phytogeographic areas during Late Triassic: the Ipswitch and the Onslow province (Artabe et al. 2003; Spalletti et al. 2003; Dolby and Balme 1976). The Onslow province shows an influence of European flora and is recognized in northwestern Australia, northwest Madagascar, East Africa, in the Indian Godavari, South Rewa and Mahanadi basins and in the East Antarctic Lambert Graben. The Onslow flora reflects a warm temperate climate (Césari and Colombi 2013). While the Ipswitch province surrounds the southern pole and represents cool temperate plant communities encompassing southern and eastern Australia, the Transantarctic Mountains, South Africa and Argentina characterised by Gondwana species.

The floral transition from the Late Triassic to the Early Jurassic was gradual (Cirilli 2010) and is documented by the *Ashmoripollis reducta* to *Classopollis torosus* Zone which marks the Triassic-Jurassic boundary in Western Australia. This zone witnesses the decline in bisaccate pollen such as *Falcisporites*, the dominance of *Classopollis torosus* and the last appearance of Triassic spores such as *Playfordiaspora velata* (Backhouse and Balme 2002; Grice et al. 2005). The dominant forms of the Triassic such as Peltasperms and Corystospermales were replaced by low diverse conifers (Araucariaceae, Cupressaceae and Podocarpaceae) and bennettitales (Llanos and Prez 2013). A similar palynofloral transition is observed within the *Craterisporites rotundus* Oppel Zone of eastern Australia (de Jersey 1975) as well as the *Alisporites* mioflora (de Jersey and McKeller 2013), the *Arcuatipollenites tethyensis* Zone of the Indian Peninsula (Tripathi 2000) and the Late Triassic-Lower Jurassic assemblage of the Dubrajpur Formation (Tripathi 2004). The flora of New Zealand also shows a similar transition and dominance of *Classopollis* from the Rhaetian *Foveosporites moretonensis* Zone to the Hettangian *Retitriletes austroclavatidites* Zone (de Jersey and Raine 1990). While the Argentinian and African microfloras have influence of the Ipswitch flora, containing palynofloras from the northern and southern components. Thus, a uniform palynofloral pattern can be observed on the entire Gondwana with few latitudinal differences (Quattrocchio et al. 1996; Tiwari 1999). The series of events like the end-Triassic mass extinction, a lowstand global sea level, and the movement of eastern Gondwana during the latest Triassic-earliest Jurassic into lower paleolatitudes led to global warming and resulted in wide dispersal of the cheirolepidacean flora (McKellar 2002). The global dominance of cheirolepidacean pollen, (*Classopollis*) indicates a relatively warm and seasonally dry period (McKellar 1996, 2002; Abbink et al. 2001).

The Early Jurassic experienced hothouse conditions with an absence of polar ice caps (Scotese et al. 1999). Then Australia occupied a 60–70° S latitudinal position and experienced mesothermal conditions inferred from wide spread coal deposits and absence of evaporites (Rees et al. 2000; Greb et al. 2006). India was situated in the southern Subtropical Arid Belt (30° S) during the Early Jurassic ($\sim$167 Ma) supporting very warm and dry climatic conditions across the country with mean annual precipitation of less than 2 cm/month (Chatterjee et al. 2013). The onset of warmer conditions favoured the development of the *Classopollis* pollen-dominated flora and lack of coal deposits (Lukose 1972; Koshal 1975). The Early Jurassic cheirolepidacean flora was more cosmopolitan than the Middle and Late Jurassic ones, except by the fact that the flora of middle latitudes (Africa and South America) were more diverse (Rees et al. 2000) (Fig. 2.1A). The percentages of *Classopollis* and *Exesipollenites* are high in the Early Jurassic assemblage from Carapace Nunatak locality of the Transantarctic Mountains of East Antarctica (Truswell et al. 1999). Percentages of these palynotaxa are also high in low diverse assemblages from Western Australia, western India and Madagascar. While in the parts of southern South America, eastern Australia, New Caledonia and New Zealand the frequency of *Classopollis* is higher, and *Exesipollenites* is a rare to absent constituent of the assemblage. These parts contain more diverse cryptogram associations than those of central areas of southern Gondwana.

The tectonic movements took place from the Middle to Late Jurassic and Gondwana occupied a middle to high (30°–80°) latitudinal position in the Jurassic. Although, northern Africa and northern South America were more extended towards the equator (Smith et al. 1981). This latitudinal difference is clearly visible in the palynofloral transition, i.e., araucarian conifers (*Callialasporites* and *Araucariacites*) and a pteridosperm-dominated flora by the end of the Toarcian continued up to the mid-Oxfordian (Balme 2000; Grant-Mackie et al. 2000). The coeval rise in the sea level outcompeted the cheirolepidacean flora by a hydrophilic araucarian flora inferring reoccurrence of consistent humid conditions (McKellar 1996, 2002).

The flora of India, Australia, New Zealand and East Antarctica shows a clear dominance of Araucarian conifer pollen of *Callialasporites* and *Araucariacites*. The Late Jurassic period also supported the growth and diversification of bryophytes and pteridophytes which extended up to the Early Cretaceous in response to the short Ice House condition across the boundary (McLoughlin 2001). The movement of Australia to higher and cooler latitudes is visible in the floral transition of the Araucariaceae phase to the Podocarpaceae phase (Grant-Mackie et al. 2000) supporting the interpretation of cool climate based on the paleoclimatic model proposed by Frakes et al. (1992). The cooler and more balanced global temperatures were also favoured by the development of the proto-Indian Ocean during the Middle Jurassic to Early Cretaceous in the form of the Argo Abyssal Plain which acted as the heat sink (McLoughlin 2001). However, the increased atmospheric CO_2 levels during the Middle to Late Jurassic ameliorated this cooling climatic effect. Floral regionalism became more pronounced during the Late Jurassic and shows a steep environmental gradient across the latitudes. The floras of eastern Australia-western Antarctica were mainly dominated by araucarian and only a few forms of cheirolepidacean and podocarps and a diverse cryptogram association of fern and lycopod spores. India occupied lower latitudes in comparison to Australia and Antarctica (Fig. 2.1B) supporting warmer conditions and a dominance of araucarian along with low percentage of *Classopollis* and podocarpaceous pollen and diverse bryophytes and pteridophyte spores (Tiwari and Tripathi 1992). Southern South American, African and the Falkland Plateau floras were very different and dominated by *Classopollis* with lesser proportions of araucarian, podocarps and pteridosperms together with a dominant bryophytic component (Riccardi et al. 1990). The increased diversity coincided with the opening of the proto South Atlantic Ocean (Dettmann 1986). Thus, during the Late Jurassic the India showed a closer similarity with the Australian and Antarctica ones than with those of Africa and South America (Truswell et al. 1999).

Previously, the Early Cretaceous Gondwana flora was suggested to be relatively uniform forming only one phytogeographic province (Herngreen and Chlonova 1981; Maeyen 1987; Herngreen et al. 1996). However, new data by Vakhrameev (1991) proposed two phytogeographical provinces corresponding to low-middle and high latitudes. Herngreen et al. (1996) distinguished the one large phytogeographic province into two subprovinces: the *Cyclusphaera-Classopollis* subprovince located in western Gondwana (southern South America, Africa and Madagascar) and the *Murospora florida* subprovince of eastern Gondwana (Australia, New Zealand and India). The lack of palynological data from Antarctica in the 1990s made it difficult to incorporate this landmass into the recognised floristic provinces. Antarctica occupied the central position in the Gondwana and played a significant role in establishing the history of paleobiota (McLoughlin 2001). The compilation of recent palynological data from various parts of Antarctica resulted in the recognition of four phytogeographic provinces in Gondwana: the Patagonia-Palmer (southern South America-Antarctic Peninsula), the South African, the Indo-Eromangan (India-northern Australia) and the Tasman (south eastern Australia and New Zealand) provinces based on the climatic differences, geographic boundaries originated due to continental breakup. The tectonic evolution resulted in more pronounced regionalism. The flora of the southern South America-Falkland Plateau assemblage was found to be more associated with the cheirolepidacean and araucariacean pollen together with diverse ferns and hepatics (Dettmann 1986). While the peninsular part of Antarctica was occupied by the forest of podocarp/araucaricean conifers with understorey of ferns. The flora of Peninsular Antarctica is comparable with the flora of New Zealand and eastern Australia except with the fact that lycopods were the dominant flora of the understorey

(Dettmann 1994; Cantrill and Nagalingum 2005). The Indian and eastern Antarctic Peninsula (especially the Nielsen Basin) show a close relation and are dominated by araucarian pollen of *Araucariacites* and *Callialasporites* with few forms of bi- and trisaccate pollen and representing tropical to sub-tropical climate with high rainfall (Goswami et al. 2010). But they differ as the Indian flora contains a more diverse flora of ferns. The Gamba flora from the Tethys Himalayan region shows an important signature of increased provincialism during the Early Cretaceous (Li et al. 2013, 2017). The Gamba palynoflora represents a similar floral evolution with Australia prior to Albian as both the plates were juxtaposed. The Australian flora had to face many changes due to its eastward drifting while the Indian plate remained relatively static and show no remarkable change in the flora. The Indian Gamba flora witnesses the inception of numerous genera and species of African origin led by the northward journey of the subcontinent during Late Cretaceous. The disjunction of Indian plate with Australia during Late Cretaceous is clearly displayed in the flora via inception of African species and increased disparity from the Australian flora. This indicates that tectonic movements had greatly controlled the evolution of the flora and Africa and India were more closely placed during Late Cretaceous in comparison to Early Cretaceous (Li et al. 2013, 2017). The regionalism with in Eastern Australia also became more pronounced and the flora reported from the George V-Addie coast had close floral similarity with the vegetation of southeastern Australia (Domack et al. 1980).

However, within these provinces a latitudinal similarity in the floral composition can be observed. The flora of James Ross Island, Antarctica suggests a strong link with the *Murospora florida* flora of Australia and New Zealand in comparison to South America (Dettmann and Thomson 1987). The Antarctica flora is somewhat mixed due to its paleolatitudinal position and shows strong connection with the eastern and west Gondwana continents. However, relatively it shows more similarity with the eastern Gondwana province than with the western (Raine 1984; Dettmann and Thomson 1987; De Jersey and Grant-Mackie 1989; Duane 1996; Hathway et al. 1999).

The accelerated drifting of the Gondwana continents during the late Early Cretaceous favoured regional climatic gradients. These changes can be observed in the palynofloral differences between western and eastern Antarctica (Dettmann and Thomson 1987). The flora of eastern Antarctica has a strong connection with India and eastern? Australia, while to a lesser extent with those of west Antarctica, southern South America, Madagascar and New Zealand (Raine 1984; Dettmann and Thomson 1987; de Jersey and Grant-Mackie 1989; Duane 1996; Hathway et al. 1999; Truswell et al. 1999; Grant-Mackie et al. 2000; Cantrill and Nagalingum 2005). The Early Albian palynoflora of Antarctica can be distinguished from the mid Albian to Cenomanian flora by more diverse angiosperm pollen (Dettmann and Thomson 1987). The existence of angiosperm pollen *Clavatipollenites, Asteropoliis, Liliacidites, Retimonocolpites,* etc. (Table 2.3) during the Albian suggests synchronous wide distribution of angiosperms in Antarctica and Australia (Dettmann and Thomson 1987) along with India (Tiwari 1996).

Conclusions

The Upper Gondwana palynofloral changes are more consistent with the tectonic events and record four phases of floral transitions related to mass extinction and continental breakup: (I) the Triassic-Jurassic boundary interval: recording the replacement of *Falcisporites* and *Alisporites* (Corystosperms) by clear dominance of *Classopollis* (cheirolepidacean conifers); (II) the post-Toarcian flora documenting the reassembly of the Conifer dominated flora by the replacement of cheirolepidacean conifers by araucarian conifers; (III) Tithonian-Valangian flora showing the coeval rise of araucarian and podocarp canopy flora with various bryophytes and pteridophytes occupying the understorey; (IV) the post-Hauterivian flora with the first appearance of angiosperms along with their parallel diversification of bryophytes (including liverworts or hepatophytes), Bryophytina (Mosses), and diverse families of pteridophytes.

The Early Jurassic flora was more cosmopolitan and widely distributed due to close continental assembly and uniform climatic conditions.

The tectonic drifting was more active during post-Middle Jurassic times and resulted in more provincialized flora. The decline in *Classopollis* pollen during the Middle to Late Jurassic in India, Australia, New Zealand, Antarctica and Salt range was synchronous. In contrast, African and South American floras were still dominated by this pollen up to the Early Cretaceous. Thus, the Indian flora is more closely related to the flora of east Antarctica and Western Australia than those of Africa and South America.

The continental assembly of the Indian Satpura, South-Rewa and East Coastal Gondwana basins with East Antarctica and West-Australia was the cause of development of similar paleoflora.

The Early Cretaceous flora shows a similar dominance of podocarp pollen grains and diversified pteridophyte spores. However, provincialism kept on increasing through Cretaceous as the continents kept on drifting in different directions. The *Murospora florida* flora of Eastern Gondwana is dominated by araucarian pollen and *Cooksonites, Coptospora, Gleichenidites, Cyathidites* and *Microcachryidites* spores while the *Cyclusphaera psilata* flora has high percentages of *Classopollis, Araucariacites, Exesipollenites* pollen and many spores.

The continued disjunction of the continents in the Late Cretaceous and India's northward journey resulted in the increased similarity with African flora in comparison to Australia and Antarctica.

Acknowledgements This research is supported by the National Post-Doctoral Fellowship (PDF/2017/002297) grant from the Science and Engineering Research Board of India (SERB). Prof. Annette Goetz, Prof. Ana Maria Zavattieri and Dr. Anju Saxena are gratefully acknowledged for their valuable suggestions that significantly improved the article.

References

Abbink, O. A., Targarona, J., Brinkhuis, H., & Visscher, H. (2001). Late Jurassic to Early Cretaceous paleoclimate evolution of the Southern North Sea. *Global and Planetary Change, 30,* 231–256.

Anderson, J. M., Anderson, H. M., Archangelsky, S., Bamford, M., Chandra, S., Dettmann, M., et al. (1999). Patterns of Gondwana plant colonisation and diversification. *Journal of African Earth Sciences, 28,* 145–167.

Archangelsky, A., Archangelsky, S., Poire, D. G., & Canessa, N. D. (2008). Registros palinológicos en la Fm. Piedra Clavada (Albiano) en suárea tipo, provincia de Santa Cruz, Argentina. *Revista del Museo Argentino de Ciencias Naturales, Nueva Serie, 10,* 105–111.

Archangelsky, A., & Llorens, M. (2005). Palinología de la Formación Kachaike, Cretácico Inferior de la Cuenca Austral, provincia de Santa Cruz. II. Esporas. *Ameghiniana, 42,* 311–328.

Archangelsky, A., & Llorens, M. (2009). Palinología de la Formación Kachaike, Cretácico Inferior de la Cuenca Austral, provincia de Santa Cruz. Granos de polen de Gimnospermas. *Ameghiniana, 46,* 225–234.

Archangelsky, S., & Archangelsky, A. (2002). Palinología estadística en el Cretácico de la Cuenca Austral, Plataforma Continental Argentina. I. Seis perforaciones delárea de Magallanes. *Revista del Museo Argentino de Ciencias Naturales, Nueva Serie, 4,* 25–34.

Archangelsky, S., & Archangelsky, A. (2004). Palinología estadística en el Cretácico de la Cuenca Austral, Plataforma Continental Argentina. I. Seis perforaciones delárea de Magallanes. III. Discusiones y conclusiones. *Revista del Museo Argentino de Ciencias Naturales, Nueva Serie, 6,* 245–255.

Archangelsky, S., & Archangelsky, A. (2005). *Aequitriradites* Delcourt & Sprumont and *Couperisporites* Pocock liverwort spores from the Early Cretaceous of Patagonia, Argentina. *Revista del Museo Argentino de Ciencias Naturales, Nueva Serie, 7,* 119–138.

Archangelsky, S., & Archangelsky, A. (2006a). Putative Early Cretaceous pteridaceous spores from the offshore Austral Basin in Patagonia, Argentina. *Cretaceous Research, 27,* 473–486.

Archangelsky, S., & Archangelsky, A. (2006b). *Pilunsporites*, a new genus of a Cretaceous spore from Patagonia. *Revista del Museo Argentino de Ciencias Naturales, Nueva Serie, 8,* 41–51.

Archangelsky, S., & Archangelsky, A. (2010). Revisión taxonómica y estratigráfica de esporas cicatricosus del Cretácico Inferior de Patagonia: 2. Gèneros *Cicatricosisporites* Potonié & Gelletich y *Ruffordiaspora* Dettmann & Clifford. *Revista del Museo Argentino de Ciencias Naturales, 12,* 179–201.

Archangelsky, S., & Archangelsky, A. (2013). Aptian angiosperm pollen from the Tico flora, Patagonia Argentina. *International Journal of Plant Sciences, 174,* 559–579.

Archangelsky, S., Barreda, V., Passalia, M. G., Gandolfo, M., Prámparo, M., Romero, E., et al. (2009). Early angiosperm diversification: Evidence from southern South America. *Cretaceous Research, 30,* 1073–1082.

Archangelsky, S., & Taylor, T. N. (1993). The ultrastructure of in situ *Clavatipollenites* pollen from the Early Cretaceous of Patagonia. *American Journal of Botany, 80,* 879–885.

Archangelsky, S., & Villar de Seoane, L. (1996). Estudios palinológicos de la Formación Baqueró (Cretácico), provincia de Santa Cruz, Argentina. VII. *Ameghiniana, 33,* 307–313.

Artabe, A. E., Morel, E. M., & Spalletti, L. A. (2003). Caracterizacióo′n de las provincias fitogeográa′ficas Triáa′sicas del Gondwana extratropical. *Ameghiniana, 40,* 387–405.

Backhouse, J. (1984). Revised Late Jurassic and Early Cretaceous stratigraphy in Perth Basin. *Geological Survey of Western Australia. Report, 12,* 1–6.

Backhouse, J. (1988). Late Jurassic and Early Cretaceous palynology of the Perth Basin, Western Australia. *Geological Survey of Western Australia Bulletin, 135,* 1–233.

Backhouse, J., & Balme, B. (2002). *Late Triassic palynology of the Northern Carnarvon Basin.* Perth, Australia: Minerals and Energy Research Institute of Western Australia.

Balme, B. E. (1995). Fossil in situ spores and pollen grains: An annotated catalogue. *Review of Paleobotany and Palynology, 87,* 81–323.

Balme, B. E. (2000). Flora. In J. A. Grant-Mackie, Y. Aita, B. E. Balme, H. J. Campbell, A. B. Challinor, D. A. B. MacFarlan, et al. (Eds.), *Jurassic paleobiogeography of Australasia.* Association of Australian Palaeontological Memoir, *23,* 311–353.

Bastia, R. (2006). An overview of Indian sedimentary basins with special focus on emerging east coast deepwater frontiers. *The Leading Edge, 27,* 839–845.

Bera, S., & Banerjee, M. (1997). Palynostratigraphy of Mesozoic sediments from western Bengal Basin, India. *Journal of Palynology, 33,* 39–82.

Bharadwaj, D. C., Kumar, A., & Singh, H. P. (1970). Palynostratigraphy of coal deposits in Jabalpur Stage, Upper Gondwana, India. *The Palaeobotanist, 19,* 225–247.

Biswas, S. K. (1999). A review of the evolution of rift basins of India during Gondwana with special reference to Western Indian basins and their hydrocarbons prospects. *Proceedings of the Indian National Science Academy, 65*(3), 261–283.

Bomfleur, B., Schöner, R., Schneider, J. W., Viereck, L., Kerp, H., & McKellar, J. L. (2014). From the Transantarctic Basin to the Ferrar Large Igneous Province-New palynostratigraphic age constraints for Triassic-Jurassic sedimentation and magmatism in East Antarctica. *Review of Palaeobotany and Palynology, 207,* 18–37.

Brugman, W. A., Van Bergen, P. F., & Kerp, J. H. F. (1994). A quantitative approach to Triassic palynology, the Lettenkeuper of the Germanic Basin as an example. In A. Traverse (Ed.), *Sedimentation of organic particles* (pp. 409–429). Cambridge: Cambridge University Press.

Burger, D. (1968). *Relationship of palynology to stratigraphy in the Lower Cretaceous of the Surat Basin in Queensland.* Commonwealth of Australia, Department of National Development, Bureau of Mineral Resources, Geology and Geophysics, Australia, Record 1968/125, 1–26.

Burger, D. (1990a). *Australian Phanerozoic timescales: 8. Jurassic biostratigraphic charts and explanatory notes.* Bureau of Mineral Resources, Australia, Record 1989/38, 38 pp.

Burger, D. (1990b). *Australian Phanerozoic timescales: 9. Cretaceous biostratigraphic charts and explanatory notes.* Bureau of Mineral Resources, Australia, Record 1989/39, 36 pp.

Burger, D. (1995a). *Timescales: 8. Jurassic Australian Phanerozoic timescales, biostratigraphic charts and explanatory notes. Second Series.* Bureau of Mineral Resources, Australia, Record 1995/37, 30 pp.

Burger, D. (1995b). *Timescales: 9. Cretaceous Australian Phanerozoic timescales, biostratigraphic charts and explanatory notes. Second Series.* Bureau of Mineral Resources, Australia, Record 1995/38, 35 pp.

Burger, D., & Shafik, S. (1996a). Jurassic (Charts 8 and 9). In G. C. Young & J. R. Laurie (Eds.), *An Australian Phanerozoic timescale* (pp. 148–159). Melbourne: Oxford University Press.

Burger, D., & Shafik, S. (1996b). Cretaceous (Charts 10 and 11). In G. C. Young & J. R. Laurie (Eds.), *An Australian Phanerozoic timescale* (pp. 160–174). Melbourne: Oxford University Press.

Cantrill, D. J., & Nagalingum, N. S. (2005). Ferns from the Cretaceous of Alexander Island, Antarctica: Implications for Cretaceous phytogeography of the Southern Hemisphere. *Review of Palaeobotany and Palynology, 137,* 83–103.

Casshyap, S. M., & Khan, A. (2000). Tectono-sedimentary evolution of the Gondwana Satpura basin of Central India: Evidence of pre-trap doming, rifting and paleoslope reversal. *Journal of African Earth Sciences, 31,* 65–76.

Césari, S. N., & Colombi, C. E. (2013). A new Late Triassic phytogeographical scenario in westernmost Gondwana. *Nature Communication.* https://doi.org/10.1038/ncomms2917.

Chatterjee, S., Goswami, A., & Scotese, C. R. (2013). The longest voyage: Tectonic, magmatic, and paleoclimatic evolution of the Indian plate during its northward flight from Gondwana to Asia. *Gondwana Research, 23*(1), 238–267.

Cirilli, S. (2010). Upper Triassic lowermost Jurassic palynology and palynostratigraphy: A review. *Geological Society, London, Special Publication, 334,* 285–314.

Coimbra, J. C., Ara, M., & Carreño, A. L. (2002). Biostratigraphy of Lower Cretaceous microfossils from the Araripe basin, northeastern Brazil. *Geobios, 35,* 687–698.

de Jekhowsky, B., & Goubin, N. (1964). Subsurface palynology in Madagascar: A stratigraphic sketch of the Permian, Triassic and Jurassic of the Morondava Basin. *Society of Economic Paleontologists and Mineralogists, Special Publication, 11,* 116–130.

de Jersey, N. J. (1975). Miospore zones in the Lower Mesozoic of southeastern Queensland. In K. S. W Campbell (Ed.), *Gondwana geology.* Papers Presented at the Third Gondwana Symposium, Canberra, Australia, 1973 (pp. 159–172). Canberra: Australian National University Press,

de Jersey, N. J. (1976). *Palynology* and time relationships in the lower Bundamba Group (Moreton Basin). *Queensland Govt. Mineralogy Journal, 77,* 460–465.

de Jersey, N. J., & Grant-Mackie, J. A. (1989). Palynoflora from the Permian, Triassic and Jurassic of New Caledonia. *New Zealand Journal of Geology and Geophysics, 32,* 463–476.

de Jersey, N. J., & McKellar, J. L. (2013). The palynology of the Triassic-Jurassic transition in southeastern Queensland, Australia, and correlation with New Zealand. *Palynology, 37,* 77–114.

de Jersey, N. J., & Raine, J. I. (1990). Triassic and earliest Jurassic miospores from the Murihiku Supergroup, New Zealand. *New Zealand Geological Survey Paleontological Bulletin, 62,* 164.

de Jersey, N. J., & Raine, J. I. (2002). Early to Middle Jurassic miospore zonation, New Zealand. IPC 2002, *Geological Society of Australia, 68,* 41.

Dettmann, M. E. (1963). Upper Mesozoic microfloras from southeastern Australia. *Proceedings of Royal Society of Victoria, 77,* 1–148.

Dettmann, M. E. (1986). Significance of the spore genus *Cyatheacidites* in tracing the origin and migration of *Lophosoria* (Filicopsida). *Special Papers in Palaeontology, 35,* 63–94.

Dettmann, M. E. (1994). Cretaceous vegetation: The microfossil record. In R. S. Hill (Ed.), *History of the Australian vegetation: Cretaceous to recent* (pp. 143–170). Cambridge: Cambridge University Press.

Dettmann, M. E., Molnar, R. E., Douglas, J. G., Burger, D., Fielding, C., Clifford, H. T., et al. (1992). Australian Cretaceous terrestrial faunas and floras: Biostratigraphic and biogeographic implications. *Cretaceous Research, 13,* 207–262.

Dettmann, M. E., & Thomson, M. R. A. (1987). Cretaceous palynomorphs from the James Ross Island area, Antarctica-A pilot study. *Bulletin of the British Antarctic Survey, 77,* 13–59.

Dogra, N. N. (1986). Palynostratigraphy of Jabalpur and Lameta groups of sediments in the type area. Ph.D. Dissertation, Panjab University.

Dogra, N. N. (1994). Palynological evidence in support of marine environment for the deposition of Lameta Formation (Maastrichtian) in the type area. *Quarterly Journal of Geological Association and Research Centre, 2(2),* 1–12.

Dogra, N. N., Singh, R. Y., & Kulshreshtha, S. K. (1994). Palynostratigraphy of infratrappean Jabalpur and Lameta formations (Lower and Upper Cretaceous) in Madhya Pradesh, India. *Cretaceous Research, 15*(2), 205–215.

Dogra, N. N., Singh, R. Y., & Kulshreshtha, S. K. (1988). Palynological evidences on the age of Jabalpur and Lameta formations in the type area. *Current Science, 57,* 954–956.

Dogra, N. N., Singh, R. Y., & Kumar, S. (2010). The lithostratigaphic status of Jabalpur and Lameta formations and their palynological constraints on the age and environment. *Gondwana Geological Magazine, 25*(1), 185–194.

Dogra, N. N., Singh, R. Y., & Singh, Y. R. (2003). Palynology of Jabalpur Formation in type area: Age and Environment. *Gondwana Geological Magazine, 18,* 24–42.

Dolby, J. H., & Balme, B. E. (1976). Triassic palynology of the Carnarvon Basin, Western Australia. *Review of Palaeobotany and Palynology, 22,* 105–168.

Domack, E. W., Fairchild, W. W., & Anderson, J. B. (1980). Lower Cretaceous sediment from the East Antarctic continental shelf. *Nature, 287,* 625–626.

Duane, A. M. (1994). Preliminary palynological investigation of the Byers Group (Late Jurassic-Early Cretaceous), Livingston Island, Antarctic Peninsula. *Review of Palaeobotany and Palynology, 84,* 113–120.

Duane, A. M. (1996). Palynology of the Byers Group (Late Jurassic Early Cretaceous) of Livingstone and Snow Islands, Antarctic peninsula: Its biostratigraphical and paleoenvironmental significance. *Review of Palaeobotany and Palynology, 91,* 241–281.

Dutta, P. K. (2002). Gondwana lithostratigraphy of the peninsular India. *Gondwana Research, 5,* 540–553.

Evans, P. (1964). The tectonic framework of Assam. *Journal of the Geological Society of India, 5,* 80–96.

Feistmantel, O. (1882). The fossil flora of the South Rewa Gondwana Basin. Fossil Flora of the Gondwana System. *Memoirs of the Geological Survey of India, Palaeontologia Indica, Series 12, 4(1),* 1–52.

Filatoff, J. (1975). Jurassic palynology of the Perth Basin. *Palaeontographica Abt. B, 154,* 1–113.

Frakes, L. A., Francis, J. E., & Syktus, J. I. (1992). *Climate modes of the Phanerozoic.* Cambridge: Cambridge University Press.

Fredrikson, N. O., Sheehan, T. P., & Andrle, V. A. S. (2007). Spore-pollen biostratigraphy and paleoecology of Mesozoic and Lower Tertiary samples from the Surghar and Salt Ranges, Northern Pakistan. *U.S. Geological Survey Bulletin, 2078-D,* 1–15.

García, V. M., Zavala, C. A., & Quattrocchio, M. E. (1994). Relación entre análisis palinológico y análisis de facies. Aplicación al Grupo Cuyo (Jurásico Medio) en la Cuenca Neuquina. *Revista de la Asociación Geológica Argentina, 49,* 184–195.

Geiger, M., Clark, D. N., & Mette, W. (2004). Reappraisal of the timing of the breakup of Gondwana based on sedimentological and seismic evidence from the Morondava basin, Madagascar. *Journal of African Earth Sciences, 38,* 363–381.

Golonka, J. (2004). Plate tectonic evolution of the southern margin of Eurasia in the Mesozoic and Cenozoic. *Tectonophysics, 381,* 235–273.

Goswami, S., Das, M., & Guru, B. C. (2010). Paleoenvironment in the Mahanadi Basin: Inferences from Mesozoic plant and ichno fossils diversity. *An International Quarterly Journal of Environmental Sciences, 4*(1), 7–14.

Goswami, S., Meena, K. L., Das, M., & Guru, B. C. (2008). Upper Gondwana palynoflora of Mahanadi Master Basin, Orissa, India. *Acta Palaeobotanica, 48*(2), 171–181.

Goubin, N. (1965). Description et répartition des principaux pollenites permiens, Triasiques, et Jurassiques des sondages du Basin de Morondava (Madagasacar). *Revue Institute France de Petrole, 20,* 1415–1461.

Gould, R. E. (1968). Morphology of *Equisetum laterale* Phillips, 1829, and *E. bryanii* sp. nov. from the Mesozoic of south-eastern Queensland. *Australian Journal of Botany, 16,* 153–176.

Gould, R. E. (1971). Report on plant fossils from Durikai, southeastern Queensland, Appendix 3. In N. F. Exon, R. F. Reiser, D. J. Casey, & R. L. Brunker (Eds.), *The post-Paleozoic rocks of the Warwick 1:250 000 Sheet area, Queensland and New South Wales* (pp. 63–64). Bureau of Mineral Resources, Geology and Geophysics, Australia, Report, 140.

Grant-Mackie, J. A., Aita, Y., Balme, B. E., Campbell, H. J., Crallinor, A. B., MacFarlane, D. A. B., et al. (2000). Jurassic paleobiogeography of Australasia. *Association of Australasian Palaeontologists Memoir, 23,* 311–353.

Greb, S. F., DiMichele, W. A., & Gastaldo, R. A. (2006). Evolution and importance of wetlands in earth history. *Geological Society of America Special Paper, 399,* 1–40.

Grice, K., Cao, C., Love, G. D., Böttcher, M. E., Twitchett, R. J., Grosjean, E., et al. (2005). Photic Zone Euxinia during the Permian-Triassic Superanoxic Event. *Science, 307,* 706–709.

Hathway, B., Duane, A. M., Cantrill, D. J., & Kelley, S. P. (1999). $^{40}Ar/^{39}Ar$ geochronology of the Cerro Negro Formation, South Shetland Island, Antarctica: A new radiometric tie for Cretaceous terrestrial biostratigraphy in the Southern Hemisphere. *Australian Journal of Earth Sciences, 46,* 593–606.

Helby, R., Morgan, R., & Panridge, A. D. (1987). A palynological zonation of the Australian Mesozoic palynology. *Association of Australian Palaeontologists, Memoir, 4,* 1–94.

Herbst, R., & Troncoso, A. (1996). La tafoflora de Juan de Morales del Jurásico Medio (Formación Chacarilla), Región de Tarapacá, Chile. *Andean Geology, 23,* 3–15.

Herngreen, G. F. W., & Chlonova, A. F. (1981). Cretaceous microfloral provinces. *Pollen et Spores, 23,* 441–555.

Herngreen, G. F. W., Kedves, M., Rovina, L. V., & Smirnova, S. B. (1996). Cretaceous palynofloral provinces: A review. In J. Jasonius & D. C. McGregor (Eds.), *Palynology: Principles and Applications* (pp. 1157–1188). Houston 3: American Association of Stratigraphic Palynologists Foundation.

Hudson, W. E., & Nicholas, C. J. (2014). The Pindiro Group (Triassic to Early Jurassic Mandawa Basin, southern coastal Tanzania): Definition, paleoenvironment, and stratigraphy. *Journal of African Earth Sciences, 92,* 55–67.

Jana, B. N. (2003). Diversity in the Lower Cretaceous flora of Dharangadhra Formation, Gujarat. *Geophytology, 33*(1–2), 81–85.

Jana, B. N., & Hilton, J. (2007). Resolving the age of the Mesozoic Kuar Bet Beds (Kachchh, Gujarat, India): A reinvestigation of paleobotanical and palynological assemblages. *Journal of Asian Earth Sciences, 30*(3–4), 457–463.

Jansson, I.-M., McLoughlin, S., Vajda, V., & Pole, M. (2008). An Early Jurassic flora from the Clarence-Moreton basin, Australia. *Review of Palaeobotany and Palynology, 150,* 5–21.

Jha, N., Joshi, H., & Mishra, S. (2016). Record of Gondwana plant mega- and microfossils in Nimugudem area, Telangana, India: Palynodating and paleoenvironmental interpretation. *Current Science, 111,* 416–424.

Jha, N., Prakash, N., & Joshi, H. (2017). Integrated paleobotanical and palynological analysis of subsurface Gondwana sedimentary succession (Jurassic–Cretaceous) in Jangareddygudem area, Chintalapudi Sub-basin, South India: Stratigraphical and phytogeographical implications. *Palaeoworld, 26,* 173–193.

Jha, N., & Srivastava, S. C. (1996). Kamthi Formation palynofloral diversity. In P. K. S. Guha, S. Sengupta, K. Ayyasami, & R. N. Ghosh (Eds.), *Gondwana Nine* (pp. 355–368). Calcutta: Oxford and IBH Publishing Company.

Killops, S., Cook, R., Raine, I., Weston, R., & Woolhouse, T. (2003). A tentative New Zealand chemostratigraphy for the Jurassic-Cretaceous based on terrestrial plant biomarkers. *New Zealand Journal of Geology and Geophysics, 46,* 63–77.

Korasidis, V. A., Wagstaff, B. E., Gallagher, S. J., Duddy, I. R., Tosolini, A.-M. P., Cantrill, D. J., et al. (2016). Early angiosperm diversification in the Albian of southeast Australia: Implications for flowering plant radiation across eastern Gondwana. *Review of Palaeobotany and Palynology, 232,* 61–80.

Koshal, V. N. (1975). Palynozonation of Mesozoic subsurface sediments of Banni, Kutch, Gujarat. *Quarterly Journal of Geology, Mining, Metallurgical Society of India, 4*(2), 79–82.

Kuerschner, W. M., Iqbal, S., Wagreich, M., Jan, I., & Gier, S. (2017). *Palynology and Carbon isotope stratigraphy of the Triassic – Jurassic transition in the Salt Range (Pakistan).* Geophysical Research Abstracts, 19, EGU2017-15291-1, EGU General Assembly 2017.

Kumar, M. (2011). Palynofloral and sedimentary organic matter analysis of Early Cretaceous sediments exposed along Sher River in Satpura Basin, Madhya Pradesh, India. *Geophytology, 41,* 9–23.

Kumar, M., & Ram-Awatar (2010). Early to Late Jurassic palynofossils from South Rewa Gondwana Basin, Madhya Pradesh, India. *Earth Science Frontier, 17,* 215–216.

Kumar, P. (1993). Palynodating of Chaugaon beds exposed near Khatama caves, Hoshangabad District, Madhya Pradesh, India. *Biological Memoirs, 19*(1), 1–16.

Kumar, P. (1994a). Palynology of carbonaceous clays of Nimar Formation, Jhabua District, Madhya Pradesh, India. *Journal of the Geological Society of India, 44,* 671–674.

Kumar, P. (1994b). A note on Pipli palynoassemblage from Dharngadhra Formation, Surendra Nagar district, Gujarat, India. *Biological Memoirs, 20*(2), 109–111.

Lele, K. M. (1964). Studies in the Talchir flora of India-2. Resolution of the spore genus *Nuskoisporites* Pot. & KI. *The Palaeobotanist, 12,* 147–163.

Li, J., Peng, J., & Batten, D. J. (2013). Palynostratigraphy of a Jurassice Cretaceous transitional succession in the Himalayan Tethys, southern Xizang (Tibet), China. *Cretaceous Research, 46,* 123–135.

Li, J., Wub, Y., Peng, J., & Batten, D. J. (2017). Palynofloral evolution on the northern margin of the Indian Plate, southern Xizang, China during the Cretaceous Period and its phytogeographic significance. *Palaeogeography, Palaeoclimatology, Palaeoecology, 515,* 107–122.

Llanos, M. P. I., & Prezz, C. B. (2013). The role of true polar wander on the Jurassic paleoclimate. *International Journal of Earth Sciences, 102,* 745–759.

Loinaze, V. S. P., & Llorens, M. (2018). Palynology of Baqueró Group (upper Aptian), Patagonia Argentina: An integrated study. *Cretaceous Research, 86,* 219–237.

Lukose, N. G. (1972). Palynological evidence on the age of Lathi Formation, western Rajasthan, India. In A. K. Ghosh (Ed.),

Proceedings of the Seminar on Palaeopalynology of Indian Stratigraphy, Calcutta (pp. 155–166).
Maeyen, S. V. (1987). *Fundamentals of paleobotany.* London: Chapman and Hall.
Mantle, D. J. (2009). Palynology, sequence stratigraphy, and paleoenvironments of Middle to Upper Jurassic strata, Bayu-Undan Field, Timor Sea region-Part One. *Palaeontographica Abt. B, 280,* 1–86.
Masood, R. K., Qureshi, K. A., Butt, A. A., & Ghazi, S. (2008). Palynology of the Mesozoic succession of the Kala Chitta Range Pakistan. *Geological Bulletin of Punjab University, 43,* 83–99.
McKellar, J. L. (1974). Jurassic miospores from the upper Evergreen Formation, Hutton Sandstone, and basal Injune Creek Group, north-eastern Surat Basin. *Geological Survey of Queensland, Publication, 361, Palaeontological Paper, 35,* 1–89.
McKellar, J. L. (1996). Palynofloral and megafloral indications of paleoclimate in the Late Triassic, Jurassic, and Early Cretaceous of southeastern Queensland. In *Mesozoic 96. Mesozoic Geology of the Eastern Australia Plate Conference,* Brisbane, Geological Society of Australia, Extended Abstracts, 43, 366–373.
McKellar, J. L. (1998). Late Early to Late Jurassic palynology, biostratigraphy and paleogeography of the Roma Shelf area, northwestern Surat Basin, Queensland, Australia: Including phytogeographic-paleoclimatic implications of the Callialasporites dampieri and Microcachryidites Microfloras in the Jurassic – Early Cretaceous of Australia, based on an overview assessed against a background of floral change and apparent true polar wander in the preceding late Paleozoic – early Mesozoic. Ph.D. Dissertation, The University of Queensland.
McKellar, J. L. (2002). Palynological biostratigraphy and age of the Cooper and Eromanga Basins. In J. J. Draper (Ed.), *Geology of the Cooper and Eromanga Basins, Queensland.* Brisbane: Queensland Minerals and Energy Review Series, Queensland Department of Natural Resources and Mines.
McLoughlin, S. (2001). The breakup history of Gondwana and its impact on pre-Cenozoic floristic provincialism. *Journal of Geological Society of India, 49,* 271–300.
Medina, F. A., Archangelsky, S., Guler, V., Archangelsky, A., & Cardenas, O. (2008). Estudio bioestratigráfico integrado del perfil La Horqueta (límite Aptiano-Albiano), Lago Cardiel, Patagonia, Argentina. *Revista del Museo Argentino de Ciencias Naturales, Nueva Serie, 10,* 141–155.
Mohr, B. A. R. (1990). Early Cretaceous palynomorphs from ODP Sites 692 and 693, the Weddell Sea, Antarctica. In P. F. Barker, J. P. Kennett, S. Ó. Connell, S. Berkowit, W. R. Bryant, L. H. Burckle, et al. (Eds.), *Proceedings of the Ocean Drilling Program,* Scientific Results, 113 (pp. 449–464).
Mohr, B. A. R., & Gee, C. T. (1992). 17. An Early Albian palynoflora from the Kerguelen Plateau, southern Indian Ocean (Leg 120). In S. W. Wise Jr., A. A. P. Julson, R. Schlich, & E. Thomson (Eds.), *Proceedings of the Ocean Drilling Program,* Scientific Results 120 (pp. 255–271).
Morgan, R. (1980). Palynostratigraphy of the Australian Early and middle Cretaceous. *Memoirs of the Geological Survey of New South Wales, Paleontology, 18,* 1–153.
Morgan, R., Alley, N. F., Rowett, A. I., & White, M. R. (1995). Biostratigraphy. In J. G. G. Morton & J. F. Drexel (Eds.), *The petroleum geology of South Australia. Vol. 1: Otway Basin* (pp. 95–101). Mines and Energy South Australia, Report Book 95/12.
Muir, R. A., Bordy, E. M., & Prevec, R. (2015). Lower Cretaceous deposit reveals first evidence of a post-wildfire debris flow in the Kirkwood Formation, Algoa Basin, Eastern Cape, South Africa. *Cretaceous Research, 56,* 161–179.
Nagalingum, N. S., Drinnan, A. N., Lupia, R., & McLoughlin, S. (2002). Fern spore diversity and abundance in Australia during the Cretaceous. *Review of Palaeobotany and Palynology, 119,* 69–92.
Norris, G. (1965). Triassic and Jurassic miospores and acritarchs from the Beacon and Ferrar groups, Victoria Land, Antarctica. *New Zealand Journal of Geology and Geophysics, 8,* 236–277.
Olivera, D. E., Zavattieri, A., & Quattrocchio, M. E. (2015). The Palynology of the Cañadón Asfalto Formation (Jurassic), Cerro Cóndor Depocentre, Cañadón Asfalto Basin, Patagonia, Argentina: Paleoecology and paleoclimate based on ecogroup analysis. *Palynology, 39,* 362–386.
Partridge, A. D. (2006). Jurassic-Early Cretaceous spore-pollen and dinocyst zonations for Australia. In E. Monteil (Ed.), *Australian Mesozoic and Cenozoic Palynology Zonations-updated to the 2004 Geologic Time Scale.* Geoscience Australia Record 2006/23 (on CD-ROM).
Patra, B. P., & Sahoo, N. K. (1995). A reappraisal of geology and paleobotany of the Athgarh Sandstone, Orissa, India. *Geophytology, 25,* 17–26.
Pattemore, G. A., & Rigby, J. F. (2005). Fructifications and foliage from the Mesozoic of southeast Queensland. *Memoirs of the Queensland Museum, 50,* 329–345.
Perez Loinaze, V. S., Archangelsky, S., & Cladera, G. (2012). Palynostratigraphic study of the Early Cretaceous Río Mayer and Kachaike formations at the Quebrada El Moro Section, Austral Basin, southwestern Argentina. *Cretaceous Research, 34,* 161–171.
Pothe de Baldis, E. D. & Ramos, V., (1980). Las microfloras de la Formacion Rio Mayer y su significado estratigratico, Provincia de Santa Cruz, Reptiblica Argentina. Actas II Congreso Argentina de Paleontologic v Bioestratigrafia v I Congreso Latinoamericano dc Paleontologia, (Iluenos Aires) 5, 201–213.
Pothe de Baldis, E. D., & Ramos, Y. (1983). Dinoflagelados del Aptiano inferior de Rio Fossiles, Lago san Martin, Provincia de Santa Cruch, Argentina. *Revista Espanola de Micropaleontologia, 15,* 427–446.
Prakash, N. (2008). Biodiversity and paleoclimatic interpretation of Early Cretaceous flora of Jabalpur Formation, Satpura Basin, India. *Palaeoworld, 17,* 253–263.
Prámparo, M. B. (1999). Granos de Polen de primitivas angiospermas en el Cretácico inferior de la Cuenca de San Luis y su distribuciónen otras cuencas cretácicas de Argentina. Boletim do 5 Simposio sobre o Cretáceo do Brasil, UNESP, 539–543.
Prámparo, M. B. (2012). Non-marine Cretaceous palynomorph biostratigraphy of Argentina, a brief summary. *Journal of Stratigraphy, 36,* 215–228.
Prasad, B. (1997). Palynology of the subsurface Triassic sediments of Krishna-Godavari Basin India. *Palaentographica B, 242,* 91–125.
Prasad, B. (1998). Status and age of red beds in Krishna-Godavari Basin (India) and their relationship with Hydrocarbon occurrences. *Journal of the Geological Society of India, 52,* 579–589.
Prasad, B. (1999). Palynological characterisation, biostratigraphy and depositional environment of pre-Tertiary lithounits of Krishna-Godavari Basin, India. *Geoscience Journal, 20*(2), 143–164.
Prasad, B., Jain, A. K., & Mathur, Y. K. (1995). A standard palynozonation scheme for the Cretaceous and pre-Cretaceous subsurface sediments of Krishna-Godavari Basin, India. *Geoscience, 16*(2), 155–233.
Prasad, B., & Pundeer, B. S. (1999). Biostratigraphy of the exposed Gondwana and Cretaceous rocks of Krishna-Godavari Basin, India. *Journal of the Palaeontological Society of India, 44,* 91–117.
Price, P. L. (1997). Permian to Jurassic palynostratigraphic nomenclature of the Bowen and Surat Basins. In P. M. Green (Ed.), *The Surat and Bowen Basins, South-east Queensland. Queensland Minerals and Energy Review Series* (pp. 137–178). Brisbane: Queensland Department of Mines and Energy.
Quattrocchio, M. E., García, V., Martínez, M., & Zavala, C. (2001). A hypothetic scenario for the Middle Jurassic in the southern part of

the Neuquén Basin, Argentina. 7th International Symposium on Mesozoic Terrestrial Ecosystems. *Asociación Paleontológica Argentina. Publicación Especial, 7,* 163–166.

Quattrocchio, M. E., Martinez, M. A., Carpinelli Pavisich, A., & Volkheimer, W. (2006). Early Cretaceous palynostratigraphy, palynofacies and paleoenvironments of well sections in northeastern Tierra del Fuego, Argentina. *Cretaceous Research, 27,* 584–602.

Quattirocchio, M. E., Sarjeant, W. A. S., & Volkheimer, W. (1996). Marine and terrestrial Jurassic microfloras of the Neuquen Basin (Argentina): Palynological zonation. *Geophysical Research Forum, 1,* 167–178.

Raine, J. I. (1984). Outline of a palynological zonation of Cretaceous to Paleogene terrestrial sediments in the West Coast region, South Island, New Zealand. *New Zealand Geological Survey report, 109,* 82.

Rajnikanth, A., Aggarwal, A., & Stephen, A. (2010). An integrated Inquiry of Early Cretaceous Flora, Palar Basin, India. *Phytomorphology, 60,* 21–28.

Ramanujam, C. G. K. (1993). *Spore-pollen suites of some Upper Gondwana deposits of South India – a critical appraisal* (pp. 462–478). Lucknow, Special Publication: Birbal Sahni Institute of Palaeobotany.

Rana, S., Singh, R. Y., & Dogra, N. N. (2000). Geological history of Early Cretaceous sediments in Kutch and Jabalpur basins, a palynological approach. *Bulletin Indian Geological Association, 33*(1), 47–58.

Rees, P. M., Ziegler, A. M., & Valdes, P. J. (2000). Jurassic phytogeography and climates: New data and model comparisons. In B. T. Huber, K. G. Macleod, & S. L. Wing (Eds.), *Warm climates in earth history* (pp.297–318). Cambridge: Cambridge University Press.

Reiser, R. F., & Williams, A. J. (1969). Palynology of the Lower Jurassic sediments of the northern Surat Basin, Queensland. *Geological Survey of Queensland Publication, 339, Palaeontological Paper, 15,* 1–24.

Ribecai, C. (2007). Early Jurassic miospores from Ferrar Group of Carapace Nunatak, South Victoria Land, Antarctica. *Review of Paleobotany and Palynology, 144,* 3–12.

Riccardi, A. C., Damborenea, S. E., & Manceñido, O. (1990). 3. South America and Antarctica Peninsula. 3.1 Lower Jurassic of South America and Antarctic Peninsula. *Newsletters in Stratigraphy, 21,* 75–103.

Rozefelds, A. C., & Sobbe, I. (1983). Fossil plants of the Durikai plant beds near Warwick, southeast Queensland. *The Fossil Collector Bulletin, 10,* 15–20.

Sajjadi, F., & Playford, G. (2002a). Systematic and stratigraphic palynology of Late Jurassic-earliest Cretaceous strata of the Euromanga basin, Queensland, Australia: Part One. *Palaeontographica, 261B,* 1–97.

Sajjadi, F., & Playford, G. (2002b). Systematic and stratigraphic palynology of Late Jurassic-earliest Cretaceous strata of the Euromanga basin, Queensland, Australia: Part Two. *Palaeontographica, 261B,* 99–165.

Sandersen, A., Scott, L., McLachlan, I. R., & Hancox, P. J. (2011). Cretaceous bio-zonation based on terrestrial palynomorphs from two wells in the offshore Basin of South Africa. *Palaeontologia Africana, 46,* 21–41.

Sastry, M. V. A., Acharyya, S. K., Shah, S. C., Satsangi, P. P., Ghosh, S. C., Raha, P. K., et al. (1977). Stratigraphic Lexicon of Gondwana Formation of India. *Geological Survey of India Miscellaneous Publication No. 36,* 1–170.

Sastry, V. V., Raju, A. T. R., Venkatchala, B. S., & Acharya, S. K. (1979). Sediementary Basins map of India. Part of the Map. In: Stratigraphic Correlation between sedimentary basins of the ESCAP region. *Mineral Resources Development Series*, 45. United Nations, New York.

Schrank, E. (2010). Pollen and spores from the Tendaguru Beds, Upper Jurassic and Lower Cretaceous of southeast Tanzania: Palynostratigraphical and paleoecological implications. *Palynology, 34,* 3–42.

Scotese, C. R. (1997). 'Paleogeographic Atlas'. (Department of Geology, University of Texas at Arlington: Arlington, Texas.) *PALEOMAP Progress Report,* 90- 0497, 1–45.

Scotese, C. R., Boucot, A. J., & McKerrow, W. S. (1999). Gondwana paleogeography and paleoclimatology. *Journal of African Earth Sciences, 28,* 99–144.

Singh, C. (1964). Microflora of the Lower Cretaceous Mannville Group, east-central Alberta. *Alberta Research Council, Bulletin, 15,* 1–239.

Singh, C. (1971). Lower Cretaceous microfloras of the Peace River area, North-western Alberta. *Alberta Research Council, Bulletin, 28,* 1–542.

Singh, H. P., & Venkatachala, B. S. (1987). Upper Jurassic Lower Cretaceous spore-pollen assemblages in the peninsular India. *The Palaeobotanist, 36,* 168–176.

Smith, A. G., Hurley, A. M., & Briden, J. C. (1981). *Phanerozoic paleocontinental world maps*. Cambridge: Cambridge University.

Spalletti, L. A., Artabe, A. E., & Morel, E. M. (2003). Geological factors and evolution of southwestern Gondwana Triassic plants. *Gondwana Research, 6,* 119–134.

Srivastava, S. K. (1966). Jurassic microflora from Rajasthan, India. *Micropaleontology, 12,* 87–103.

Tasch, P., & Lammons, J. M. (1978). Palynology of some lacustrine interbeds of Antarctic Jurassic, Palinologia, numero extraordinary 1. In F. Cramer (Ed.), *Proceedings of the Coloquio International de Palinologia,* Leon (pp.455–460).

Thorn, V. (2001). Vegetation communities of a high paleolatitude Middle Jurassic forest in New Zealand. *Palaeogeography, Palaeoclimatology, Palaeoecology, 168,* 273–289.

Tiwari, R.S. (1996). Palynoevent stratigraphy in Gindwana sequence of India. *IXth International Gondwana Symposium*, Hyderabad, India (pp. 3–19).

Tiwari, R. S. (1999). The palynological succession and spatial relationship of the Indian Gondwana sequence. *Proceedings of Indian National Science Academy, 65*(3), 329–375.

Tiwari, R. S., & Tripathi, A. (1992). Marker assemblage-zones of spores and pollen species through Gondwana Paleozoic and Mesozoic sequence in India. *The Palaeobotanist, 40,* 194–236.

Tiwari, R. S., & Tripathi, A. (1995). Palynological assemblages and absolute age relationship of intertrappean beds in the Rajmahal Basin, India. *Cretaceous Research, 16,* 53–72.

Tiwari, R. S., Vijaya, Mamgain, V. D., & Misra, R. S. (1996a). Palynological studies on a Late Paleozoic–Mesozoic Tethyan sequence in the Niti area of the Central Himalaya, Uttar Pradesh, India. *Review of Palaeobotany and Palynology, 94,* 169–196.

Tiwari, R. S., Vijaya, & Tripathi, A. (1996b). Gondwana palynostratigraphy from drill-core studies in coastal Tamil Nadu and Pondicherry. *Journal of the Geological Society of India, 48(1)*, 65–74.

Tripathi, A. (2000). Palynological events during Late Triassic-Early Jurassic time in India. *The Palaeobotanist, 49*(3), 399–408.

Tripathi, A. (2001). Permian, Jurassic and Early Cretaceous palynofloral assemblages from subsurface sedimentary rocks in Chuperbhita Coalfield, Rajmahal Basin, India. *Review of Palaeobotany and Palynology, 113*(4), 237–259.

Tripathi, A. (2002). Role of pteridophytic spores in Early Cretaceous stratigraphy and demarcating the Jurassic-Cretaceous boundary in India. In P. N. Trivedi (Ed.), *Advances in Pteridology* (pp. 268–279). Lucknow: Pointer Publisher.

Tripathi, A. (2003). Palynology evidences of hitherto unrecognised Jurassic sedimentation in Rajmahal Basin, India. *Rivista Italiana di Paleontologia Stratigrafia, 110*(1), 35–42.

Tripathi, A. (2004). Unique organic remains from Upper Permian coal bearing horizon in Talcher Coalfield, Orissa, India. *Palynology, 28,* 121–128.

Tripathi, A. (2008). Palynochronology of Lower Cretaceous volcano-sedimentary succession of the Rajmahal Formation in the Rajmahal Basin, India. *Cretaceous Research, 29*(5–6), 913–924.

Tripathi, A. (2009). Palynology of the Brahmani River Section, Talcher Coalfield, Orissa, India. *Journal of the Palaeontological Society of India, 54*(2), 179–187.

Tripathi, A., Jana, B. N., Verma, O., Singh, R., & Singh, A. K. (2013). Early Cretaceous palynomorphs, dinoflagellates and plant megafossils from the Rajmahal Basin, Jharkhand, India. *Journal of the Palaeontological Society of India, 58,* 125–134.

Tripathi, A., & Ray, A. (2006). Palynostratigraphy of the Dubrajpur Formation (Early Triassic to Early Cretaceous) of the Rajmahal Basin, India. *Palynology, 30,* 133–149.

Tripathi, A., & Tiwari, R. S. (1991). Early Cretaceous angiospermous pollen from the intenrappean beds of Rajmahal Basin. *The Palaeobotanist, 39,* 50–56.

Tripathi, A., Tiwari, R.S., & Kumar, P. (1990). Sporae dispersae and their distributional pattern in subsurface Mesozoic sediments of Rajmahal Basin, Bihar, India. *Palaeobotanist, 37,* 367–388.

Tripathi, A., & Vijaya. (1997). Biostratigraphic age-determintion of Sriperumbudur Formation in bore hole PBSD-1, Palar Basin, Tamil Nadu: A reassessment. *Journal of the Palaeontological Society of India, 42,* 101–116.

Truswell, E. M. (1991a). Data report: Palynology of sediments from Leg 119 drillsites in Prydz Bay, East Antarctica. In J. Barron, B. Larsen, J. G. Baldalf, & J. Anderson (Eds.), *Proceedings of the Ocean Drilling Program,* Scientific Results 119 (pp. 941–945).

Truswell, E. M. (1991b). Antarctica: A history of terrestrial vegetation. In R. J. Tingey (Ed.), *The geology of Antarctica* (pp. 499–537). Oxford: Clarendon.

Truswell, E. M., Dettmann, M. E., & O'Brien, P. E. (1999). Mesozoic palynofloras from the Mac. Robertson shelf, East Antarctica: Geological and phytogeographic implications. *Antarctic Science, 11,* 239–255.

Turner, S., Bean, L. B., Dettmann, M., McKellar, J. L., McLoughlin, S., & Thulborn, T. (2009). Australian Jurassic sedimentary and fossil successions: Current work and future prospects for marine and non-marine correlation. *GFF, 131*(1–2), 49–70.

Upadhyay, R., Ram-Awatar, Kar, R. K., & Sinha, A. K. (2005). First record of Middle-Late Jurassic palynomorphs from the Lamayuru Complex, Indus Suture Zone, Ladakh, India. *Current Science, 88* (6), 980–986.

Vakhrameev, V. A. (1991). *Jurassic and Cretaceous floras and climates of the Earth.* Cambridge: Cambridge University Press.

Veevers, J. J., & Tewari, R. C. (1995). Gondwana Master Basin of Peninsular India between Tethys and the interior of the Gondwanaland Province of Pangea. *Memoir Geological Society of America, 187,* 1–73.

Venkatachala, B. S. (1974). Palynological zonation of the Mesozoic and Teniary subsurface sediments in the Cauvery Basin. In K. R. Surange, R. N. Lakhanpal & D. C. Bharadwaj (Eds.), *Aspects & appraisal of Indian palaeobotany* (pp. 476-496). Lucknow: Birbal Sahni Institute of Palaeobotany.

Venkatachala, B. S. (1977). Fossil floral assemblages in the east coast Gondwana – a critical review. *Journal of the Geological Society of India, 18,* 378–397.

Venkatchala, B. S., & Jain, A. K. (1970). Fossil spores and pollen from the Lower Cretaceous sub surface sediments near Karaikal, Cauvery Basin. *The Palaeobotanist, 18,* 63–66.

Venkatachala, B. S., & Kar, R. K. (1969). Palynology of the Mesozoic sediments of Kutch, W. India-6. *Three new species of Applanopsis with remarks on the morphology of the genus* (pp. 33–43). Calcutta: Memorial Volume, Botanical Society Bengal.

Venkatachala, B. S., & Kar, R. K. (1970). Palynology of the Mesozoic sediments of Kutch, W. India. Palynological zonation of Katrol (Upper Jurassic) and Bhuj (Lower Cretaceous) sediments in Kutch, Gujarat. *The Palaeobotanist, 18,* 75–86.

Venkatachala, B. S., & Kar, R. K. (1972). Palynology of the Mesozoic sediments of Kutch, W. India-9. Palynological fossils from the Bhuj exposures near Dayapar, Kutch District, Gujarat State. In A. K. Ghosh (Ed.), *Proceedings of the Seminar on Palaeopalynology and Indian Stratigraphy* (pp. 166–171). Calcutta.

Venkatachala, B. S., Sharma, K. D., & Jain, A. K. (1972). Palynological zonation of Jurassic-Lower Cretaceous sediments in the subsurface of Cauvery Basin. In A. K. Ghosh (Ed.), *Proceedings of the Seminar on Palaeopalynology and Indian Stratigraphy* (pp. 172–187) Calcutta.

Vijaya. (1997a). Palynoflora from subsurface Lower Cretaceous intertrappean beds in Domra sub-basin of the Raniganj Gondwana Basin, West Bengal, India. *Cretaceous Research, 18(1),* 37–50.

Vijaya. (1997b). Palynological dating of infra- and intertrappean sediments in the Mesozoic succession of bore hole PGD-6, Domra sub-basin, West Bengal, India. *Cretaceous Research, 18,* 833–847.

Vijaya. (1999). Palynological dating of the Neocomian-Aptian succession in the Indian Peninsula. *Cretaceous Research, 20,* 597–608.

Vijaya (2000a). Palynology of the Jurassic-Cretaceous transition in the Rajmahal Formation, West Bengal, India. *Alcheringa,* 24, 125–133.

Vijaya. (2000b). Search for Jurassic in subsurface Mesozoic sediments Birbhum District, West Bengal. *Proceedings XVI Indian Colloquium on Micropalaentology and Stratigraphy; Bulletin, Oil & Natural Gas Commission, 37(2),* 127–135.

Vijaya. (2004). Palynostratigraphy of the Permian Coal Measures and the Mesozoic succession in bore hole DPD-6, Pachami area, West Bengal, India. *Palaeontographica Abl. B, 267,* 33–55.

Vijaya. (2006). Permian to Jurassic palynoflora in subsurface strata, Pachami area, West Bengal, India. *Gondwana Geological Magazine, 21(1),* 13–26.

Vijaya. (2009). Palynofloral changes in the Upper Paleozoic and Mesozoic of the Deocha-Pachami area, Birbhum Coalfield, West Bengal, India. *Science in China Series D: Earth Sciences* D, *52,* 1932–1952.

Vijaya. (2011). Palynostratigraphy of subsurface Upper Permian and Mesozoic succession, Rakshitpur area, Raniganj Coalfield, West Bengal. *Palaeoworld, 20,* 61–74.

Vijaya, & Bhattacharji, T. K. (2002). An Early Cretaceous age for the Rajmahal traps, Panagarh area, West Bengal: Palynological evidence. *Cretaceous Research, 23,* 789–805.

Vijaya, & Bhattacharji, T. K. (2003). Palynological evidence for Jurassic in the Panagarh area, West Bengal, India. *Journal of Asian Earth Sciences, 21,* 1171–1190.

Vijaya, & Kumar, S. (2002). Palynostratigraphy of the Spiti Shale (Oxfordian-Berriasian) of Kumaon Tethys Himalaya, Malla Johar area, India. *Review of Palaeobotany and Palynology, 122*(3–4), 143–153.

Vijaya, Murthy, S., & Vethanayagam, S. M. (2010). Palynology and age of the subsurface strata in Birbhum Coalfield, West Bengal, India. *Gondwana Geological Magazine, 25*(2), 291–303.

Vijaya, & Prasad, G. V. R. (2001). Age of Kota Formation, Pranhita Godavari Valley, India: A palynological approach. *Journal of the Palaeontological Society of India, 46,* 77–93.

Vijaya, & Roy, A. (2000). Late Lower Jurassic palynomorphs assemblage in the Dubrajpur sediments, Birbhum district, West Bengal, India. *The Palaeobotanist, 49,* 43–49.

Vijaya, & Sen, K. K. (2005). Palynological study of the Dubrajpur Formation in the Mesozoic succession, Pachami area, Birbhum Coalfield, West Bengal. *Journal of the Palaeontological Society of India, 50(1)*, 121–133.

Volkheimer, W., Gallego, W. O., Cabaleri, N. G., Armella, C., Narvez, P. L., Nieto, D. G. S., et al. (2009). Stratigraphy, palynology, and conchostracans of a Lower Cretaceous sequence at the Canadon Calcareo locality, Extra-Andeancentral Patagonia: Age and paleoenvironmental significance. *Cretaceous Research, 30,* 270–282.

Volkheimer, W., Manceñido, M., & Damborenea, S. (1978). Zur Biostratigraphie des Lias in der Hochkordillere von San Juan, Argentinien. *Münstersche Forschungen für Geologie und Paläontologie, 44*(45), 205–235.

Volkheimer, W., Quattrocchio, M. E., & Cabaleri, N. G. (2001). Palynology and paleoenvironment of the lacustrine Cañadón Asfalto Formation, Jurassic of Central Patagonia, Argentina. In *34th Annual Meeting of the American Association of Stratigraphic Palynologists,* Abstract Volume: 43, San Antonio.

Volkheimer, W., Quattrocchio, M. E., Cabaleri, N. G., Narvaez, P. L., Rosenfeld, U., Scafati, L., et al. (2015). Environmental and climatic proxies for the Cañadón Asfalto and Neuquén basins (Patagonia, Argentina): Review of Middle to Upper Jurassic continental and near coastal sequences. *Revue Brassileria Paleontologia, 18,* 71–82.

Wadia, D. N. (1957). *Geology of India* (3rd ed.). Revised, London: MacMillan.

Wagstaff, B. E., Gallagher, S. J., & Trainor, J. K. (2012). A new subdivision of the Albian sporepollen zonation of Australia. *Review of Palaeobotany and Palynology, 171,* 57–72.

Zavattieri, A. M., Rosenfeld, U., & Volkheimer, W. (2008). Palynofacies analysis and sedimentary environments of Early Jurassic coastal sediments at the southern border of the Neuquén Basin, Argentina. *Journal of South American Earth Sciences, 25,* 227–245.

Zavattieri, A. M., & Volkheimer, W. (2003). Palynostratigraphy and paleoenvironments of Early Jurassic strata (Nestares Formation) in northern Patagonia, Argentina. Part 1. Terrestial species. *Ameghiana, 40,* 545–558.

Chapter 3
A Third, Remarkably Small, Tribosphenic Mammal from the Mesozoic of Australia

Thomas H. Rich, Peter Trusler, Lesley Kool, David Pickering*, Alistair Evans, Karen Siu, Anton Maksimenko, Martin Kundrat, Neil J. Gostling, Steven Morton, and Patricia Vickers-Rich

Abstract A third species of tribosphenic mammal is now known from Australia. Owing to the fragmentary nature of the two specimens on which this new, diminutive species is based, its affinities within the tribosphenic Mammalia are uncertain. When better known, they could well prove to belong to the Ausktribosphenidae. In estimated body mass, 2 g, the new species is amongst the smallest mammals known.

*Deceased

T. H. Rich · P. Vickers-Rich (✉)
Faculty of Science Engineering & Technology, Swinburne University of Technology, PO BOX 2183122 Hawthorn, VIC, Australia

T. H. Rich
Museums Victoria, PO Box 666, Melbourne, VIC 3001, Australia
e-mail: trich@museum.vic.gov.au

T. H. Rich · K. Siu · S. Morton
School of Physics and Astronomy, Monash University, 3800 Melbourne, VIC, Australia

P. Trusler · L. Kool · P. Vickers-Rich
School of Earth, Atmosphere and Environment, Monash University, 3800 Melbourne, VIC, Australia

A. Evans · P. Vickers-Rich
Museums Victoria, PO Box 6663001 Melbourne, VIC, Australia

A. Evans
School of Biological Sciences, Monash University, 3800 Melbourne, VIC, Australia

A. Maksimenko
Australian Synchrotron, 800 Blackburn Road, 3168 Clayton, VIC, Australia

M. Kundrat
Center for Interdisciplinary Biosciences, Technological and Innovative Park, Pavol Jozef Safarik University, Jesenna 5, SK-04154 Kosice, Slovak Republic

N. J. Gostling
Biological Sciences, The University of Southampton, Life Sciences Building 85, University Road, Southampton, SO17 1BJ, UK

Keywords Australia • Cretaceous • Ausktribosphenidae • Batodonoides • Samson effect

Introduction

Two tribosphenic mammals have previously been described from the Early Cretaceous Flat Rocks locality, Victoria, Australia: *Ausktribosphenos nyktos* Rich et al. (1997) and *Bishops whitmorei* Rich et al. (2001). Both of these species have been assigned to the Australian endemic family Ausktribosphenidae. The placement of the ausktribosphenids within the Mammalia is contentious. Rich et al. (1997) tentatively assigned them to the Placentalia while Luo, et al. (2001) erected the Infraclass Australosphenida into which they placed the ausktribosphenids and excluded the placentals, which were assigned to their second new Infraclass Boreosphenida that was simultaneously proposed.

A third new species of tribosphenic mammals is recognized here, also from the Flat Rocks locality. Based on two fragmentary jaws, as far as can be determined they could also be ausktribosphenids. Like the undoubted ausktribosphenids, they have three molars, and there is no inflected angle. Those two features shared with ausktribosphenids were two of the three primary ones that initially led Rich et al. (1997) to suggest that *Ausktribosphenos nyktos* might be a placental.

Methods

In order to measure the tiny teeth central to this report, the method used was to work with the length and width of their M_1s. (Images were made at the SPring-8 Synchrotron, Harima Science Park City, Hyogo Prefecture, Japan). On the images, scales were present with a resolution of 0.01 mm.

Bloch et al. (1998) present in their Appendix III an equation relating body masses of insectivorous mammals 1984) of 40 different species of Soricidae, Erinaceidae,

G. V. R. Prasad and R. Patnaik (eds.), *Biological Consequences of Plate Tectonics: New Perspectives on Post-Gondwana Break-up—A Tribute to Ashok Sahni*, Vertebrate Paleobiology and Paleoanthropology Series, https://doi.org/10.1007/978-3-030-49753-8_3

Talpidae, Tupaidae and Macroscelecidae. Substituting the appropriate values for their sample in their two equations, they may be written as follows.

Body mass estimate = $e^{[1.628\ (\ln(\text{m1 tooth length} * \text{m1 tooth width})) + 1.726]}$

95% confidence interval = Body Mass Estimate

$\pm$ 2.024 $\sqrt{}$[0.1795 * (1/40 + ((ln(m1 tooth length * m1 tooth width) – 1.35)^2)/ 39.28890]

The estimates of body mass in the two specimens of *Kryoparvus gerriti* were based on the dimensions of the m2, there being no m1 preserved in either of these fossils.

Type Locality

Flat Rocks fossil site on the marine shore platform in the Bunurong Marine Park near Inverloch, Victoria, Australia (38° 39′ 40±02″ S, 145° 40′ 52±03″ E, World Geodetic System 1984). Four other mammalian taxa are known from this locality: two possible eutherians/australosphenidians, *Ausktribosphenos nyktos* Rich et al. (1997), *Bishops whitmorei* Rich et al. (2001), the monotreme *Teinolophos trusleri* Rich et al. (1999), and the multituberculate *Corriebaatar marywaltersae* Rich et al. (2009a).

Type Horizon, *Rock Unit and Age*

Previously, strata at the Flat Rocks locality have been assigned to the 'Wonthaggi Formation' in the 'Strzelecki Group'. However, VandenBerg et al. (2006) suggested extending the application of the name, Eumeralla Formation (Otway Group) from the Otway Basin to encompass equivalent deposits dominated by feldspathic sandstones in the Gippsland Basin, including those of the Flat Rocks area. Based on palynological evidence, the age of the fossil site is considered to be late Barremian to early Aptian (Wagstaff et al. 2020). A discussion of the taphonomy of the Flat Rocks site can be found in Rich et al. (2016).

Institutional Abbreviation

NMV P, Paleontology collection, Museums Victoria, Melbourne, Australia.

Systematic Paleontology

Mammalia Linnaeus (1758).

Theria Parker and Haswell (1897).

Placentalia? Owen (1837).

Ausktribosphenidae? Rich et al. (1997).

Genus *Kryoparvus* gen. nov.

Type species: *Kryoparvus gerriti.*

Etymology: The *small, cold animal* in allusion to the small size of this mammal that lived in a polar environment.

Diagnosis: Same as for the only known species.

Kryoparvus gerriti sp. nov.

(Figs. 3.1 and 3.2).

Etymology: In honor of Gerrit Kool, who found the holotype in 2001.

Holotype: NMV P210087, partial right mandible with m2-3, talonid region of m1, three premolar alveoli, ascending ramus complete in anterior region, ventral margin of angular process complete to its distal extremity.

Referred Specimen: NMV P212940. Left ramus with m2-3, roots of m1, ramus extends anteriorly far enough to contain alveoli for as many as three single-rooted premolars. Found by Wendy White in 2003.

Diagnosis: Distinguished from the ausktribosphenids *Ausktribosphenos nyktos* and *Bishops whitmorei* by being two-thirds or less their sizes in linear dimensions of the m2-3 (Table 3.1) and having only one double-rooted lower premolar rather than three as in *A. nyktos* and six as in *B. whitmorei.*

Description: As far as *Kryoparvus gerriti* can be compared, except for being significantly smaller in size than the ausktribosphenids *Ausktribosphenos nyktos* and *Bishops whitmorei* from the same fossil locality, it is largely similar to them. The two well preserved lower molars in both specimens, m2-3, are double-rooted tribosphenic molars with the protoconid significantly taller than the paraconid and metaconid and the trigonid somewhat more elongated than the talonid. Neither of the elongated, shallow rami are well enough preserved to determine whether mental foramina were present on the buccal surface. The holotype has a well-preserved Meckelian groove, and the scan of the holotype clearly indicates the position of the mandibular foramen together with the extent of the mandibular canal as far anteriorly as the ramus is preserved (Fig. 3.3). The prominence of these features is more like that of *Ausktribosphenos nyktos* than the more derived condition seen in *Bishops whitmorei* where these features are more subdued if present at all.

On the holotype, the ventral margin of the angular process is preserved to its posterior end. On the perimeter of the internal side can be seen a distinct ridge surrounding the area where a slip of the pterygoideus muscle would have inserted. In referred specimens of both *A. nyktos* and *B. whitmorei*, but not the holotypes, there is a well-developed angular process preserved, but in neither case is it preserved as far posteriorly as on the holotype of *Kryoparvus gerriti.*

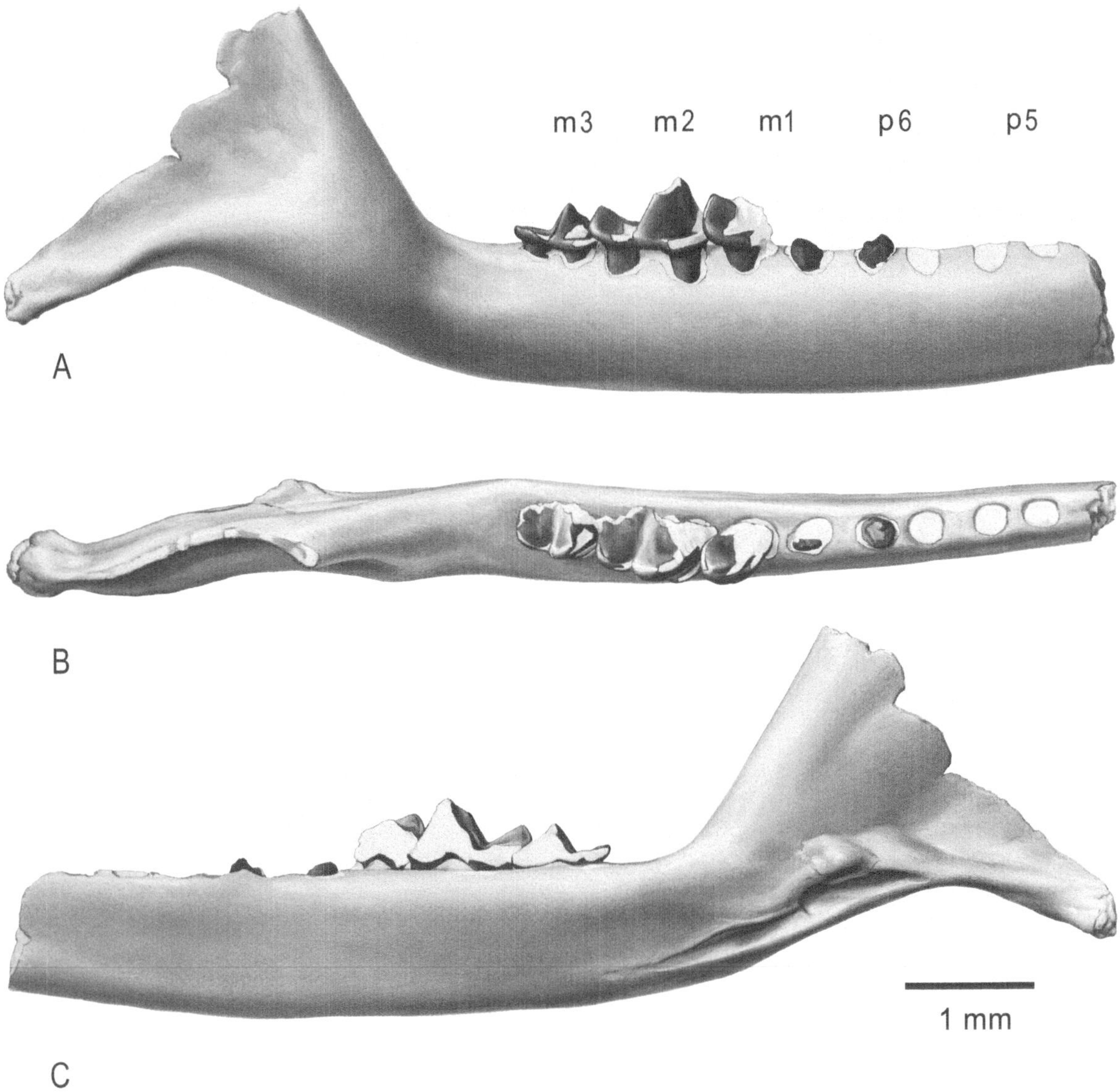

Fig. 3.1 **A**. Buccal view, **B**. Occlusal view, and **C**. Lingual view of right mandible, NMV P210087, holotype of *Kryoparvus gerriti*. Scale bar represents 1.0 mm. (Art by Peter Trusler)

Discussion and Conclusions

While the presence of three lower molars, rather than four, and the lack of an inflected angle on *Kryoparvus gerriti* are features found on ausktribospenids, they also occur in unquestioned eutherians. It is only the fact that as yet, no unquestioned eutherians are known to occur in the Mesozoic of Australia and ausktribosphenids are found at the same locality as the two fragmentary specimens of *K. gerriti* that this new species is tentatively placed there.

Bloch et al. (1998) described and named *Batodonoides vanhouteni* from the Early Eocene Willwood Formation of Wyoming, USA. They characterized this species as possibly the smallest known mammal with an estimated body mass of "*…between 0.93 and 1.82 g, with a best estimate of about 1.30 g*". Their assessment was based on measurements of the length and width of m1 that were correlated with the body masses of small, extant insectivorous mammals.

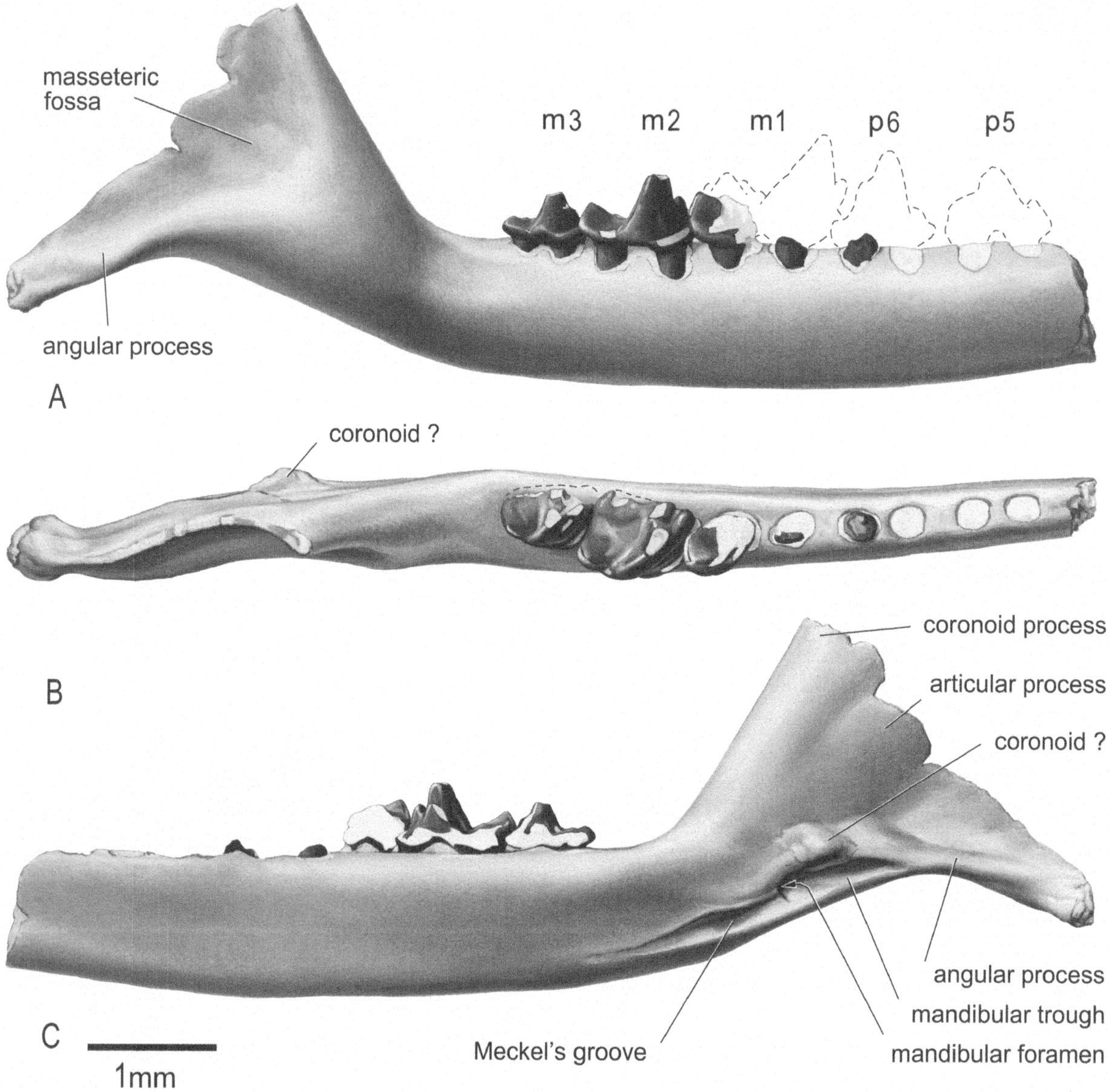

Fig. 3.2 A. Buccal view, B. Occlusal view, and C. Lingual view of a composite reconstruction of the holotype of *Kryoparvus gerriti*, NMV P210087 and referred specimen NMV P212940. Scale bar is based on measurements of the right m2-3 of NMV P210087 and represents 1 mm. (Art by Peter Trusler)

Unfortunately, the m1 of *Kryoparvus gerriti* is unknown. The m2 is present in the two known specimens of *K. gerriti*. Substituting the dimensions of those two teeth in the equations of Bloch et al. (1998), gives the following body masses estimates and ranges of the 95% confidence intervals: NMV P210087, 1.51 g, 1.19 g–1.84 g; NMV P212940, 1.07 g, 0.72 g–1.42 g (see Table 3.2). Although these results suggest that *K. gerriti* was in the general size range of *Batodonoides vanhouteni*, the fact that of necessity the calculations were based on measurements of the m2 rather than m1 as Bloch et al. (1998) utilized, comparisons of the body mass estimates must be made with this in mind. Where *K. gerriti* and *B. vanhouteni* can be directly compared, the dimensions of m2, *K. gerriti* is somewhat larger. The m2 lengths, 0.93 and 0.86

Table 3.1 Measurements of lower dentition of Ausktribosphenidae

	NMV P208090 *Ausktribosphenos nyktos*	NMV P208228 *Ausktribosphenos nyktos*	NMV P210075 *Bishops whitmorei* holotype	*Bishops whitmorei* NMV P208484	*Bishops whitmorei* NMV P209975	*Bishops whitmorei* NMV P210070	NMV P210087 *Kryoparvus gerriti* holotype	*Kryoparvus gerriti* NMV P212940
p1 length			0.9			1.0		
p1 width			0.4					
p2 length			0.8			0.9		
p2 width			0.4					
p3 length			0.8					
p3 width			0.4					
p4 length			0.9			0.9		
p4 width			0.6					
p5 length			1.1			1.0		
p5 width			0.7					
p6 length	1.4	1.2	1.2					
p6 width	1.2	1.1	1.1					
m1 length	1.4	2.3	1.4		1.1 ± 0.1	1.5		
m1 trigonid length	1.6		0.8			0.9		
m1 trigonid width	1.8	1.4	1.2			1.2		
m1 talonid width	1.7	1.4	1.1			1.0		
m2 length	1.8	2.3*	1.7 ± 0.1	1.6	1.8		0.93	0.86
m2 trigonid length			0.9 ± 0.1	1.2	1.2		0.48	0.42
m2 trigonid width		1.7		1.0			0.48	0.79
m2 talonid width		1.3*		1.0			0.52	0.71
m3 length	1.2		1.1 ± 0.1		1.3		0.70	
m3 trigonid length			0.5 ± 0.1			0.5	0.30	
m3 trigonid width	0.90		0.9 ± 0.1			0.9	0.39	0.45
m3 talonid width			0.8			0.9	0.37	

*Measurement of cast

mm, and widths, 0.48 and 0.42 mm of *K. gerriti* are somewhat greater than the length and width of the m2 of *B. vanhouteni*, 0.70 and 0.56 mm.

Visual comparison of the holotypes of *K. gerriti* and *Bishops whitmorei*, the next largest ausktribosphenid known (Table 3.2) certainly leaves no doubt that *K. gerriti* was a remarkably small mammal (see Fig. 3.4).

Because the sizes and proportions of the molars in the three tribosphenic mammals from the Flat Rocks site (Table 3.1) are sufficiently different from one another that, in conjunction with the differences in the estimates of their respective body masses (Table 3.2), it is likely that each occupied a somewhat different ecological niche than the other two, although presumably all were insectivorous in the broadest sense of the term.

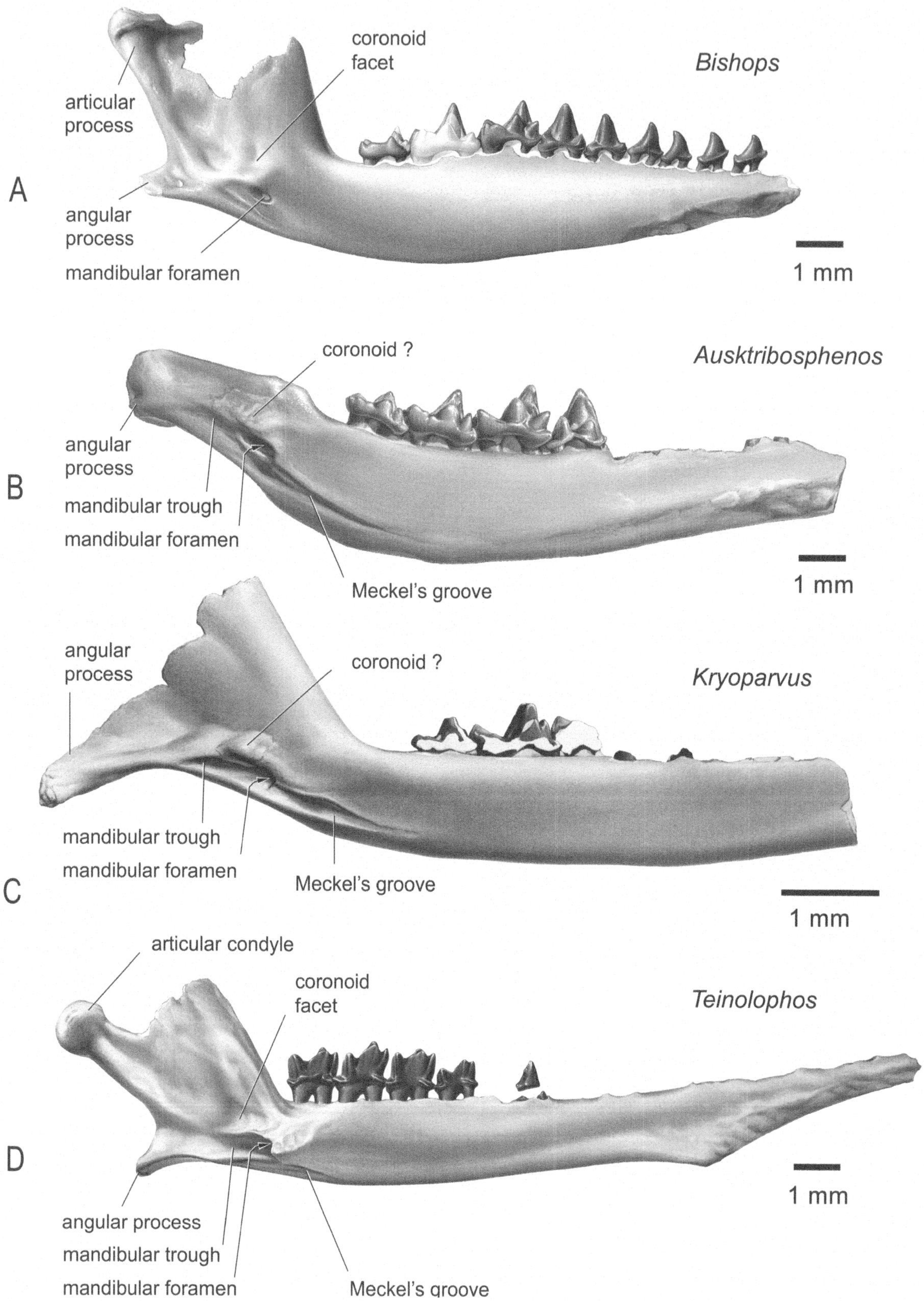

Fig. 3.3 Lingual view of the mammalian mandibles from the Early Cretaceous, Flat Rocks locality, Victoria, Australia, from which this element is known. All specimens depicted as if they are left mandibles. Note the degree of difference in the expression of the angular process, coronoid facet, mandibular groove, mandibular trough, and Meckel's groove. At least one of the primitive mandibular conditions present in some of these mammals also occurs in the unquestioned eutherian *Prokennalestes minor*. Kielan-Jaworowska and Dashzeveg (1989) illustrate a remnant of the Meckelian groove in their Fig. 23. Scale bars represent 1 mm. (Art by Peter Trusler)

Table 3.2 Estimate of body weights of ausktribosphenids

	Ausktribosphenos nyktos Nmvp 208090	*Ausktribosphenos nyktos* NMV P208228	*Bishops whitmorei* holotype NMV P210075	*Bishops whitmorei* NMV P210070	*Kryoparvus gerriti* holotype NMV P210087	*Kryoparvus gerriti* NMV P212940
Estimated body mass	20.88 g	37.70 g	6.76 g	9.26 g	1.51 g	1.07 g
95% confidence interval	20.73–21.04 g	37.57–37.84 g	6.57–6.97 g	8.96–9.36 g	1.19–1.84 g	0.72–1.42 g

Fig. 3.4 A. Holotype of *Kryoparvus gerriti*. **B**. Holotype of *Bishops whitmorei*. Both specimens in buccal view. (Photography by Steve Morton)

That a mammal as small as *Kryoparvus gerriti* lived at polar latitudes at a time when southeastern Australia was rather frigid (Constantine et al. 1998) is not unprecedented. Weighing less than 6 g during the winter, the common shrew *Sorex araneus* is active throughout the year in northern Europe close to the Arctic Circle (Hyvärinen and Heikura 1971).

Two other mammalian groups in addition to aukstribosphenids occur at the Flat Rock locality: multituberculates and monotremes. The multituberculate *Corriebaatar marywaltersae* is represented by a well preserved p4 (Rich et al. 2009). Clearly, *Kryoparvus gerriti* lacks any indication of a plagiaulacoid premolar and thus is clearly not a multituberculate. The monotreme *Teinolophos trusleri* has the typical dilambdodont condition on the lower molars characteristic of all known toothed monotremes. Rather than a basined talonid with wear facets within it on the rear half of the lower molar typical of tribosphenic mammals including *K. gerriti* (Fig. 3.5), the posterior lambda on the rear half of a toothed monotreme lower molar is anteroposteriorly compressed, and wear facets are developed on the external surfaces rather than within the posterior lambda.

The term, "Samson Effect" was coined to describe the pronounced bias in the sample of fossil mammals collected from the Flat Rocks locality where all but one ausktribosphenid specimen has been found (Rich et al. 2009b). The bias in the sample that is with the exception of a single isolated upper premolar of the monotreme *Teinolophos trusleri*, all the fifty-five other mammalian fossils found at that locality have been lower jaws. Evidently, the mode of accumulation of these fossils involved the transport of them for a significant distance when many of the fossils that have been recovered were significantly abraded. The abrasion was so intense that not one other element of the mammalian skeletons in addition to the lower jaws and single isolated upper premolar has been preserved. In the case of the holotype of *Ausktribosphenos nyktos*, the remnant of the angular process was so obscure owing to abrasion that it was not noticed in the original description (Rich et al. 1997: Fig 3.2). A referred specimen had an obvious remnant of the angular process (Rich et al. 1999: Fig. 3.2). Given the severe abrasion that prevailed in the depositional environment in which the mammal fossils accumulated at the Flat Rocks locality, the fact that the holotype of *Kryoparvus gerriti* has the remnants of what was clearly a prominent angular process suggests that ausktribosphenids generally could have them, but they have been even more highly abraded in all other known specimens.

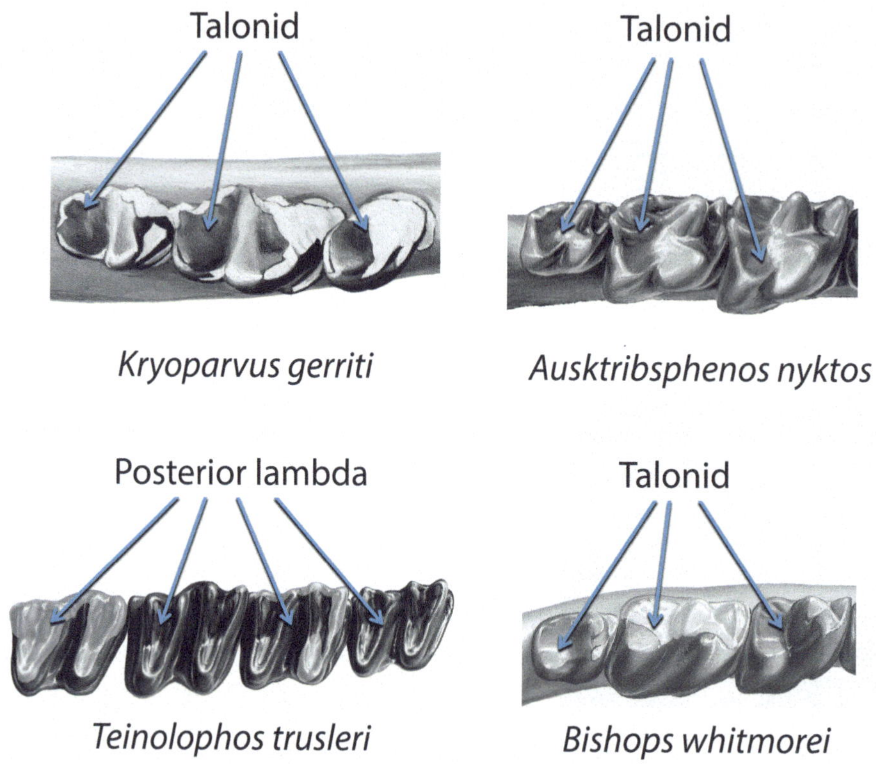

Fig. 3.5 Occlusal views of the lower molars of the various mammals from the Early Cretaceous, Flat Rocks locality, Victoria, Australia from which those teeth are known. Note the development of basins on the talonid into which the protocone of the upper molars occluded in the tribosphenic mammals *Ausktribosphenos nyktos*, *Bishops whitmorei* and *Kryoparvus gerriti*. Compare this with the compressed nature of the posterior lambda of the monotreme *Teinolophos trusleri* which did not have a cusp on the upper molars occluding within the posterior lambdas. Rather, in *T. trusleri* the lambdas of the upper molars occluded with the external surfaces of the posterior lambdas of the lower molars. This fundamental difference in the occlusal pattern between the lower molars of a toothed monotreme and the Tribosphenida was first pointed out by Kielan-Jaworowska et al. (1989). (Art by Peter Trusler)

Acknowledgements First and foremost, the volunteers who participated in the field work at the Flat Rocks site are thanked for their hard work necessary to make the discoveries. Valuable comments were made on the manuscript by William A. Clemens, Thomas Martin, Wendy White, Mary Walters, and an anonymous reviewer. The Committee for Research and Exploration of the National Geographic Society and the Australian Research Council provided much of the funding needed for the fieldwork carried out at the Flat Rocks locality over two decades. Permits to carry out the excavations in the Bunurong Marine Park were issued by Parks Victoria. Travel to Japan was funded by the International Synchrotron Access Programme of the Australian Government, managed by the Australian Synchrotron. Synchrotron Proposal no. 2011A1122 was performed with the approval of the Japan Synchrotron Radiation Research Institute. Assistance with the preparation of Tables 3.1 and 3.2 was provided by Sally Rogers-Davidson.

References

Bloch, J. I., Rose, K. D., & Gingerich, P. D. (1998). New species of *Batodonoides* (Lipotyphla, Geolabididae) from the early Eocene of Wyoming: Smallest known mammal? *Journal of Mammalogy, 79,* 804–827.

Constantine, A., Chinsamy, A., Vickers-Rich, P., & Rich, T. H. (1998). Periglacial environments and polar dinosaurs. *South African Journal of Science, 94,* 137–141.

Hyvärinen, H., & Heikura, K. (1971). Effect of age and seasonal rhythm on the growth patterns of some small mammals in Finland and in Kirkenes, Norway. *Journal of Zoology, 165,* 545–556.

Kielan-Jaworowska, Z., Crompton, A. W., & Jenkins, F. A., Jr. (1989). The origin of egg-laying mammals. *Nature, 326,* 871–873.

Kielan-Jaworowska, Z., & Dashzeveg, D. (1989). Eutherian mammals from the early Cretaceous of Mongolia. *Zoologica Scripta, 18,* 347–355.

Linnaeus, C. (1758). Systema Naturae per regna tria naturae secundum classes, ordines, genera, species cum characteribus, differentiis, synonymis locis. Editio decima, reformata. Stockholm, Laurentii Salvii, vol. 1 [ii] + 824 p.

Luo, Z.-X., Cifelli, R. L., & Kielan-Jaworowska, Z. (2001). Dual origin of tribosphenic mammals. *Nature, 409,* 53–57.

Owen, R. (1837). Teeth. In R. B.Todd (Ed.), *The cyclopedia of anatomy and physiology* (pp. 864–935). London: Sherwood, Gilbert and Piper.

Parker, T. J., & Haswell, W. A. (1897). *A text-book of Zoology*. London: McMillian & Co.

Rich, T. H., Flannery, T. F., Trusler, P., Kool, L., van Klaveren, N. A., & Vickers-Rich, P. (2001). A second tribosphenic mammal from the Mesozoic of Australia. *Records of the Queen Victoria Museum, 110,* 1–9.

Rich, T. H., Hopson, J. A., Gill, P. G., Trusler, P., Rogers-Davidson, S., Morton, S., et al. (2016). The mandible and dentition of the early Cretaceous monotreme *Teinolophos trusleri*. *Alcheringa, 40,* 475–501.

Rich, T. H., Vickers-Rich, P., Constantine, A., Flannery, T. F., Kool, L., & van Klaveren, N. (1997). A tribosphenic mammal from the Mesozoic of Australia. *Science, 278,* 1438–1442.

Rich, T. H., Vickers-Rich, P., Constantine, A., Flannery, T. F., Kool, L., & van Klaveren, N. (1999). Early Cretaceous mammals from Flat Rocks, Victoria, Australia. *Records of the Queen Victoria Museum, 106,* 1–34.

Rich, T. H., Vickers-Rich, P., Flannery, T. F., Kear, B. P., Cantrill, D., Komarower, P., et al. (2009a). An Australian multituberculate and its paleobiogeographic implications. *Acta Palaeontologica Polonica, 54,* 1–5.

Rich, T. H., Vickers-Rich, P., Flannery, T. F., Pickering, D., Kool, L., Tait, A. M., et al. (2009b). A fourth Australian Mesozoic mammal locality. *Bulletin of the Museum of Northern Arizona, 65,* 677–681.

VandenBerg, A. H. M., Cayley, R. A., Willman, C. E., Morand, V. J., Seymon, A. R., Ooborne, C. R., et al. (2006). Walhalla-Woods Point–Tallangallook special map area. *Geological Survey of Victoria Report, 127,* 1–448.

Wagstaff, B. E., Gallagher, S. J., Hall, W. M., Korasidis, V. A., Rich, T. H., Seegets-Villiers, D. E., Vickers-Rich, P. (2020). Palynological-age determination of Early Cretaceous vertebrate-bearing beds along the south Victorian coast of Australia and implications for the spore-pollen biostratigraphy. *Alcheringa.* doi:10.1080/03115518.2020.1754464.

Chapter 4
Evidence for a Remarkably Large Toothed-Monotreme from the Early Cretaceous of Lightning Ridge, NSW, Australia

Thomas H. Rich, Timothy F. Flannery, and Patricia Vickers-Rich

Abstract An isolated upper premolar from the Early Cretaceous (late Albian) Griman Creek Formation of Lightning Ridge, NSW, Australia, represents a new genus and species of a toothed monotreme that is much larger than any known from there or elsewhere.

Keywords Monotreme • Lightning Ridge • New South Wales • Australia • Cretaceous • Opal • *Obdurodon*

Introduction

Clemens et al. (2003) described and illustrated an isolated tooth characterized by them as an "enigmatic (Synapsid?)". Based on the morphological similarity of this fossil to the upper premolar of *Teinolophos trusleri* Rich, Vickers-Rich, Constantine, Flannery, Kool and van Klaveren, 1999 and the P2 of *Obdurodon dicksoni* Archer, Jenkins, Hand, Murray and Godthelp, 1992 (Rich et al. 2016: Fig. 7), it is regarded here as an upper premolar of a monotreme significantly larger than any toothed ones previously discovered.

T. H. Rich (✉)
Museums Victoria, PO Box 666, Melbourne, VIC 3001, Australia
e-mail: trich@museum.vic.gov.au

T. H. Rich · P. Vickers-Rich
Department of Chemistry and Biotechnology, Faculty of Science, Swinburne University of Technology, Hawthorn, VIC 3122, Australia

T. H. Rich · P. Vickers-Rich
School of Earth, Atmosphere and Environment, Monash University, Clayton, VIC 3800, Australia

T. F. Flannery
Melbourne Sustainable Society Institute, University of Melbourne, Parkville, VIC 2052, Australia

P. Vickers-Rich
Faculty of Science, Deakin University, Burwood and Warun Ponds, VIC, Australia

Living monotremes are widely considered to be edentulous. This is true in the case of adults. However, in 1888, Poulton announced the existence of non-functional but true teeth in the young of the platypuses. Nearly a century would pass after that before the discovery in the Oligocene of South Australia on 27 June 1971 by one of us (PVR) of what, in life, had been a functional upper molar of a fossil monotreme, *Obdurodon insignis* Tedford and Woodburne, 1975 (Poulton 1888; Rich et al. 1989, p. 651). Subsequently, functional dentitions of monotremes have been found in the Oligocene of Queensland, *Obdurodon dicksoni* Archer, Jenkins, Hand, Murray, and Godthelp, 1992 and *Obdurodon tharalkooschild* Pian, Archer and Hand, 2013, the Cretaceous of New South Wales, *Steropodon galmani* Archer, Flannery, Ritchie and Molnar, 1985, and quite likely *Kollikodon ritchei* Flannery, Archer, Rich and Jones, 1995, the Cretaceous of Victoria, *Teinolophos trusleri* Rich, Vickers-Rich, Constantine, Flannery, Kool, and van Klaveren, 1999, and the Paleocene of Argentina, *Monotrematum sudamericanum* Pascual, Archer, Ortiz Jaureguizar, Prado, Godthelp, and Hand, 1992.

To this slowly growing list of the past diversity of toothed-monotremes is here added one more genus and species from New South Wales.

The orientation of AMF 118621 assumed in Fig. 1 of Clemens et al. (2003) is as if the tooth is a left upper molar. As interpreted here, the tooth is instead a right upper premolar. This alternative orientation of AMF 118621 is based on comparison with both the right P2 of *Obdurodon dicksoni,* which occurs *in situ* in the skull of the holotype specimen of that species (Vickers-Rich and Rich 1993: Fig. 189) and the isolated right upper premolar of *Teinolophos trusleri* (Fig. 4.1B-B′). Therefore, the buccal and anterior margins as depicted by Clemens et al. (2003) are considered the reverse here. Allowing for this difference, the description of AMF118621 in Clemens et al. (2003) is quite adequate and, therefore, need not be repeated here. The descriptive terminology employed here follows that in Rich et al. (2016: Fig. 7) except to replace "labial" with "buccal".

G. V. R. Prasad and R. Patnaik (eds.), *Biological Consequences of Plate Tectonics: New Perspectives on Post-Gondwana Break-up—A Tribute to Ashok Sahni*,
Vertebrate Paleobiology and Paleoanthropology Series, https://doi.org/10.1007/978-3-030-49753-8_4

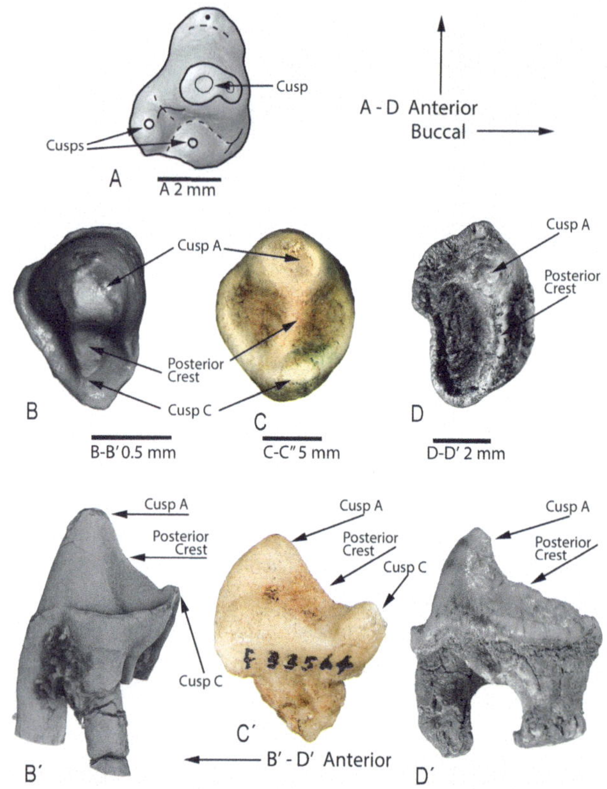

Fig. 4.1 Right upper premolars. **A** *Kollikodon ritchei*, occlusal view, ultimate premolar, based on sketch of a maxilla of that species held in the National Opal Collection, Melbourne and Sydney, Australia, a cast of which is deposited in the Australian Museum, Sydney, AMF140201. **B** Occlusal, **B′** labial view, *Teinolophos trusleri*, NMV P229194. **C** Occlusal, **C′** labial view, *Stirtodon elizabethae*, AMF118621. **D** Occlusal, **D′** labial view *Obdurodon dicksoni*, P2, QM F57736

Institutional Abbreviations

AMF. Fossil Collection of the Australian Museum, Sydney, Australia.

NMV P. Palaeontology Collection of Museums Victoria, Melbourne, Australia.

QM F. Fossil Collection, Queensland Museum, Brisbane, Australia.

Systematic Paleontology

Class Mammalia Linnaeus, 1758
Subclass Monotremata Bonaparte, 1837
Genus Stirtodon gen. nov.

Etymology: In honor of the late Prof. Ruben Arthur Stirton or "Stirt" as he preferred to be known, noted pioneer of Australian paleomammalogy, who wrote in his diary on his first day in Australia on 14 February 1953, "At last I had gotten to Australia in my quest for the ancestry of the monotremes and the marsupials." He did much to advance knowledge in that area in the decade that followed (Rich et al. 2019).

Genus and Species: *Stirtodon elizabethae* gen. and sp. nov.

Etymology: In recognition of the tireless effort by Dr. Elizabeth T. Smith of Lightning Ridge to bring to light the fossil history of that opal field.

Holotype: AMF 118621, natural cast of a right upper monotreme premolar. This specimen was not available for study. Therefore a cast of it, QM F33564, mentioned in Clemens et al. (2003) and provided by Kristen Spring of the Queensland Museum, was examined instead.

Diagnosis: Distinguished from the known upper premolars of toothed monotremes by its significantly greater size: more than eleven times as long anteroposteriorly as the upper premolar of *Teinolophos trusleri,* three times as long as the ultimate upper premolar *Kollikodon ritchei*, and almost twice as long as the P2 of *Obdurodon dicksoni*. Distinguished from *K. ritchei* by the presence of two major cusps rather than three and a quite different outline in occlusal view. Distinguished from *O. dicksoni* by the lack of a well-developed cingulum on the anterior margin of the crown and the presence of a low cusp on the posterior margin of the tooth, Cusp C.

Type Locality: "...exposed in a mine at 'Vertical Bill's'," near, "…the Three-Mile Field, Lightning Ridge, New South Wales…," (Clemens et al. 2003). "The specimen was recovered during mining by opal miners Ed Long and Hennie Kunzelman at 29° 27′ 19″ S:147° 58′ 02″ E. from a depth of approximately 14–15 m. It was found associated with fossils of plants, molluscs, fish, pliosaurs and theropod dinosaurs; this assemblage of specimens is held at the Australian Museum, Sydney," (E. T. Smith, *pers. comm*).

Type Horizon: Finch Clay facies (E. T. Smith, *pers. comm.*), Griman Creek Formation, Albian (Raza et al. 2009).

Discussion: In attempting to allocate AMF 118621, Clemens et al. (2003) considered at length several different tetrapods as possibilities: notosuchian crocodilians; trirachodont, diademodont, and traversodont gomphodonts; and the Australian Cretaceous mammals then known. Having considered those possibilities, Clemens et al. (2003) concluded that the most similar fossils were upper molars of some dryolestids. In particular, they noted the resemblance of the upper molars of *Laolestes* and *Groebertherium.* Examination of Fig. 4.2 illustrates the plausibility of this hypothesis. Where they do differ most noticeably, however, is that the dryolestids have stylar cusps unlike AMF 118621. The border of AMF 118621 that Clemens et al. (2003) considered to be the buccal margin, regarded as the anterior margin here, is rounded in occlusal view in contrast to the shape of the buccal margin of the dryolestid molars.

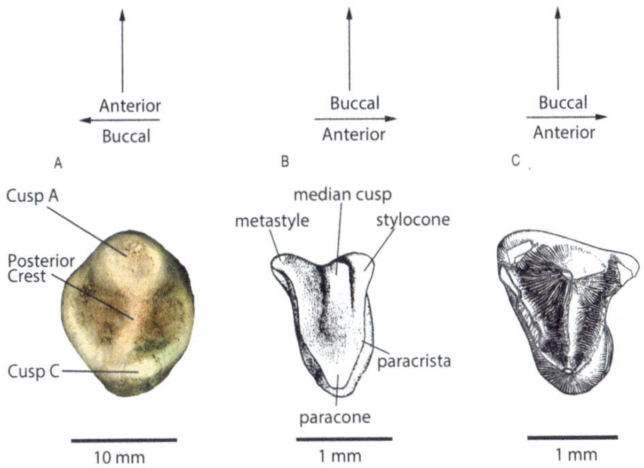

Fig. 4.2 Comparison in occlusal view of the holotype of *Stirtodon elizabethae* with two dryolestids. **A** *S. elizabethae*. **B** *Groebertherium stipanicici* (Kielan-Jaworowska et al. 2004: Fig. 10.6 A_1). **C** *Laolestes andresi* (Martin 1999: mirror image of Fig. 28A)

In 2007, an upper premolar of the monotreme *Teinolophos trusleri* NMV P229194 was discovered at the Early Cretaceous Flat Rocks site in Victoria, Australia (Rich et al. 2016). The tooth assigned to *Stirtodon elizabethae* is similar to the upper premolar of *T. trusleri* in that it has a prominent cusp at its anterior end (Cusp A), and a weaker cusp (Cusp C) at its posterior end, the two being linked by a posterior crest that in occlusal view is just slightly lingually convex. In outline, the teeth are similar in having a broadly rounded anterior end and a somewhat more angular posterior end. Posterior to Cusp A, there are buccal and lingual cingula on both, the buccal cingula being just slightly wider than the lingual ones. Up to this point, these comparisons apply equally to the P2 of *Obdurodon dicksoni*. However, that species differs in having a well-developed anterior cingulum in front of Cusp A, a feature which, if developed at all on *S. elizabethae* and *T. trusleri,* is not prominent, and in lacking Cusp C at the posterior end of the tooth (Fig. 4.1).

The holotype specimen of *Stirtodon elizabethae* differs markedly from the upper premolar of *Teinolophos trusleri* and the right P2 of *Obdurodon dicksoni* in that its features are quite subdued. In contrast to the upper premolar of *T. trusleri,* which AMF 118621 most closely resembles, *Stirtodon elizabethae* has low and rounded A and C cusps, rather than having tall and sharp ones. Likewise the posterior crest is low and broad on *S. elizabethae* rather than tall, narrow and sharp. These differences are almost certainly in large part owing to the way in which the two specimens were collected.

AMF 118621 was processed in a "puddler", a wire mesh basket with rotating rubber paddles, notorious for causing heavy damage and abrasion of opal items (E. T. Smith *pers. comm.*). If a fossil happened to be preserved embedded in a hard rock that does not break down in this process, it may survive in a pristine condition, as did the holotype of *Steropodon galmani* Archer, Flannery, Ritchie and Molnar, 1985. If not enclosed in a resistant rock, a fossil subjected to this same procedure can be abraded. This appears to have been the case with AMF 118621, for as noted in Clemens et al. (2003), "…the fossil might have suffered slight abrasion during collection for it was recovered after being separated from the clay matrix in a tumbling tank."

In contrast, the upper premolar of *Teinolophos trusleri,* NMV P229194, was found on 8 March 2007 by Nicola Sanderson *née* Barton when searching for fossils by deliberately breaking indurated sandstone from the Early Cretaceous Flat Rocks locality in Victoria. Having broken open a rock with a blow from a small hammer, she carefully examined the newly exposed surface and spotted the tiny black fossil still embedded in the rock. How she did that is quite incredible, because the tooth was not much larger than the surrounding sand grains. The fossil was subsequently carefully removed from the rock by David Pickering using a pin vice with a needle inserted in it to dislodge individual sand grains one at a time from around the tiny fossil. That careful approach did not abrade the tiny premolar at all.

Thus, the different history of collection and preparation is probably the primary reason for the differences seen in the morphology of the two fossils.

Once the likely abraded nature of the *Stirtodon elizabethae* tooth is taken into account, the close morphological similarity of it to the isolated upper premolar of *Teinolophos trusleri* is truly remarkable, particularly in light of the fact that it is about fourteen times as large in its linear dimensions (Fig. 4.3).

In attempting to assess the relative dimensions of the animal that bore AMF 118621, a key piece of evidence is the known dimensions of the m1 in three species of *Obdurodon* (Table 4.1). The m1 of *O. tharalkooschild* is one-third larger in both length and width than the two known m1·s of *O. dicksoni.* Likewise, it is three-fifths larger in these two dimensions than the m1 of *Obdurodon insignis*. The constancy of these ratios between the three species of *Obdurodon* suggests that within the genus there is a linear relationship between the dimensions of the m1. From this it may reasonably be expected that within the genus, there is a linear relationship between the various dimensions of the dentition as a whole and thus the upper premolars as well. On that basis, it is plausible that the length and width of the as yet unknown P2 of *O. tharalkooschild,* the largest species of the genus, were 7.5 mm and 4.2 mm respectively. These estimates imply that AMF118621 would be expected to be one and four fifths as long and slightly over twice as wide as the as yet unknown P2 of *Obdurodon tharalkooschild* Pian, Archer and Hand, 2013.

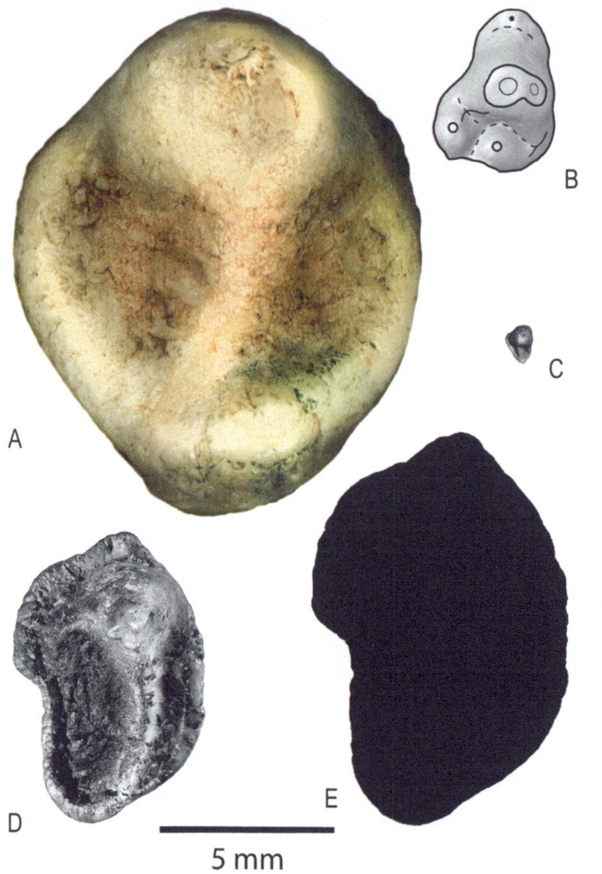

Fig. 4.3 Right upper premolars at the identical scale. **A** *Stirtodon elizabethae*, AMF118621. **B** *Kollikodon ritchei*. **C** *Teinolophos trusleri*. **D** *Obdurodon dicksoni*. **E** Hypothetical premolar of *Obdurodon tharalkooschild*

Table 4.1 Upper premolar dimensions of toothed monotremes

	Length (mm)	Width (mm)	Length/Width
Teinolophos trusleri	1.2	0.9	1.33
Kollikodon ritchei {ultimate}	4.49	3.58	1.25
Obdurodon dicksoni {P2}	7.5	4.2	1.79
Obdurodon tharalkooschild[a]	10.2	5.7	1.79
Stirtodon elizabethae	14.0	12.2	1.15

[a]Estimate based on the length and width of the P2 relative to those dimensions of the m1 of *Obdurodon dicksoni*

Although this observation suggests that *S. elizabethae* was a larger living animal than *O. tharalkooschild,* caution is warranted for such a hypothesis. This is because AMF118621, as previously pointed out, is more similar to the upper premolar of *Teinolophos trusleri* than to *Obdurodon dicksoni*. Those two species, as far as is known, are quite unlike one another in the conformation of their lower jaws (Archer et al. 1993; Rich et al. 2016). Thus, it is not warranted to estimate the size of a living *S. elizabethae* based on a single isolated tooth, large though it may be when compared to all other known upper premolars of monotremes.

The m1 of the holotype of *Steropodon galmani* Archer et al., 1985 is only about three-quarters the length and width of the same tooth in *Obdurodon dicksoni* (Pian et al. 2013) (Table 1). It is, therefore, quite likely that the unknown upper premolar or premolars of *S. galmani* are even smaller than the P2 of *O. dicksoni* relative to the upper premolar of AMF 118621.

Not only is the ultimate premolar of *Kollikodon ritchei* Flannery, Archer, Rich and Jones, 1995 far smaller than AMF118621, but it is so different in morphology that no meaningful estimate can be made of the relative sizes of that species and *Stirtodon elizabethae* on the basis of the upper premolar.

The position of AMF118621 in the tooth row is not certain. Compared with the holotype of *Obdurodon dicksoni*, the specimen clearly resembles the upper ultimate premolar, P2. However, that AMF118621 has a closer resemblance to the upper premolar of *Teinolophos trusleri* renders the assumption that it is an ultimate premolar uncertain. While *O. dicksoni* is known to have only two upper premolars, the P1 being a simple, mediolaterally compressed tooth unlike the more complex P2 and the equally complex AMF 18621. The number and morphology of the various upper premolars of *T. trusleri* is not certain. In the lower various jaw of *T. trusleri* there are alveoli for four widely spaced premolars. Therefore, it is not certain if the known upper premolar of that species is the ultimate premolar or one further forward.

That AMF 118621 is a monotreme at all is based on a logical chain of observations ultimately linking it with the P2 of the unquestioned ornithorhynchid, *Obdurodon dicksoni*. First, *Teinolophos trusleri* was recognized to be a monotreme based on the similarity of its then single known lower molar to lower molars of *O. dicksoni* (Rich et al. 2001). Subsequently, NMV P229194 was identified as an upper premolar of *T. trusleri* on the basis of evidence regarding its morphological similarity to the P2 of *O. dicksoni,* its provenance at the same locality where all known *T. trusleri* material has been found and finally, the ratio of its size relative to the lower molars of *T. trusleri* being similar to the ratio between the P2 and lower molars of *O. dicksoni* (Rich et al. 2016). This logical chain did not exist when Clemens et al. (2003) was written, because the single upper premolar of *T. trusleri* was only discovered four years after that paper was published.

Conclusions

Stirtodon elizabethae, based solely on an isolated tooth AMF118621 from the Early Cretaceous Griman Creek Formation at Lightning Ridge, NSW, Australia, is the largest toothed monotreme known. Although fourteen times larger in linear dimensions than the upper premolar of *Teinolophos trusleri,* the close morphological resemblance of AMF118621 to that tooth, rather than the P2 of the undoubted ornithorhynchid *Obdurodon dicksoni*, suggests that whatever *S. elizabethae* was in life, it was an animal quite unlike a platypus.

Stirtodon elizabethae is either the second or third monotreme known from Lightning Ridge. While the monotreme nature of *Steropodon galmani* has never been questioned, the same cannot be said for *Kollikodon ritchei*. In a recent publication focused on a maxilla referred to *Kollikodon ritchei*, the authors concluded that *K. ritchei* was a monotreme (Pian et al. 2016). However, in a review of monotremes generally, Musser (2013) expressed doubts as to that placement.

No matter whether there are two or three monotremes now known from Lightning Ridge, the addition of *Stirtodon elizabethae* to the list of Mesozoic monotremes adds further support for the idea posed by Darlington more than half a century ago that there had been into a wide variety of monotremes in Australia that was then completely unknown (Darlington 1957).

Acknowledgements First and foremost, TR would like to thank Guillermo Rougier and Alexander Averianov for completely independent of one another suggesting that NMV P P229194 should be compared with the upper premolars of *Obdurodon dicksoni*. This after TR had spent a number of years futilely attempting to identify this tooth with no success. Determining that this tooth is an upper premolar of *Teinolophos trusleri* was an essential step in identifying the holotype of *Stirtodon elizabethae* as a monotreme upper premolar. Sally Rogers-Davidson assisted with the layout of the three figures. Together, James Rule, Suzanna Hand, Steve Poropat and Peter Trusler produced the illustration of the ultimate upper premolar of *Kollikodon ritchei*. William Clemens, Steve Poropat, Ralph Molnar, Mary Walters, Wendy White, and Elizabeth T. Smith together with two anonymous reviewers read and commented upon the manuscript.

References

Archer, M., Flannery, T. F., Ritchie, A., & Molnar, R. E. (1985). First Mesozoic mammal from Australia-an early Cretaceous monotreme. *Nature, 318*, 363–366.

Archer, M., Jenkins, F. A., Hand, S. J., Murray, P., & Godthelp, H. (1992). Description of the skull and non-vestigial dentition of a Miocene platypus (*Obdurodon dicksoni* n. sp.) from Riversleigh, Australia and the problem of monotreme origins. In M. L. Augee (Ed.), *Platypus and echidnas* (pp. 15–27). Sydney: The Royal Zoological Society of New South Wales.

Archer, M., Murray, P., Hand, S., & Godthelp, H. (1993). Reconsideration of monotreme relationships based on the skull and dentition of the Miocene *Obdurodon dicksoni*. In F. S. Szalay, M. J. Novacek, & M. C. McKenna (Eds.), *Mammal phylogeny* (pp. 75–94). New York: Springer.

Clemens, W. A., Wilson, G. P., & Molnar, R. E. (2003). An enigmatic (synapsid?) tooth from the Early Cretaceous of New South Wales, Australia. *Journal of Vertebrate Paleontology, 23*, 232–237.

Darlington, P. J. (1957). *Zoogeography: The geographical distribution of animals*. New York: Wiley.

Flannery, T. F., Archer, M., Rich, T. H., & Jones, R. (1995). A new family of monotremes from the Cretaceous of Australia. *Nature, 377*, 418–420.

Kielan-Jaworowska, Z., Cifelli, R. L., & Luo, Z. X. (2004). *Mammals in the age of dinosaurs*. New York: Columbia University Press.

Martin, T. (1999). Dryolestidae (Dryolestoidea, Mammalia) aus dem Oberen Jura von Portugal. *Abhandlungen der Senckenbergischen Naturforschenden Gesellschaft, 550*, 1–119.

Musser, A. M. (2013). Classification and evolution of the monotremes. In K. Ashwell (Ed.), *Neurobiology of monotremes: Brain evolution in our distant mammalian cousins* (pp. 1–16). Collingwood: CSIRO Publishing.

Pian, R., Archer, M., & Hand, S. J. (2013). A new, giant platypus, *Obdurodon tharalkooschild*, sp. nov. (Monotremata, Ornithorhynchidae), from the Riversleigh World Heritage Area, Australia. *Journal of Vertebrate Paleontology, 33*, 1255–1259.

Pian, R., Archer, M., Hand, S. J., Beck, R. M. D., & Cody, A. (2016). The upper dentition and relationships of the enigmatic Australian Cretaceous mammal *Kollikodon ritchiei*. *Memoirs of Museum Victoria, 74*, 97–105.

Poulton, E. B. (1888). True teeth in the young *Ornithorhynchus paradoxes*. *Proceedings of the Royal Society, London, 43*, 353–356.

Raza, A., Hill, K. C., & Korsch, R. J. (2009). Mid-Cretaceous uplift and denudation of the Bowen and Surat Basins eastern Australia: Relationship to Tasman Sea rifting from apatite fission-track and vitrinite-reflectance data. *Australian Journal of Earth Sciences, 56*, 501–531.

Rich, P. V., Rich, T. H., & Fenton, M. A. (1989). *The fossil book*. New York: Doubleday.

Rich, T. H., Hopson, J. A., Gill, P. G., Trusler, P., Rogers-Davidson, S., Morton, S., et al. (2016). The mandible and dentition of the Early Cretaceous monotreme *Teinolophos trusleri*. *Alcheringa, 40*, 475–501.

Rich, T. H., Lawson, P. F., Vickers-Rich, P., & Tedford, R. H. (2019). RA Stirton: pioneer of Australian mammalian palaeontology. *Transactions of the Royal Society of South Australia, 143*, 244–282.

Rich, T. H., Vickers-Rich, P., Constantine, A., Flannery, T. F., Kool, L., & van Klaveren, N. (1999). Early Cretaceous mammals from Flat Rocks, Victoria, Australia. *Records of the Queen Victoria Museum and Art Gallery, 106*, 1–34.

Rich, T. H., Vickers-Rich, P., Trusler, P., Flannery, T. F., Cifelli, R., Constantine, A., et al. (2001). Monotreme nature of the Australian Early Cretaceous mammal *Teinolophos*. *Acta Paleontologica Polonica, 46*, 113–118.

Vickers-Rich, P., & Rich, T. H. (1993). *Wildlife of Gondwana: Dinosaurs and other vertebrates from the ancient supercontinent*. New South Wales: Reed, Chatswood.

Chapter 5
Role of Plate Tectonics and Global Climate Change in the Evolution of Angiosperms

Vandana Prasad, Shreya Mishra, Ashish K. Mishra, and Mahi Bansal

Abstract Angiosperms being the most dominant plant community in the terrestrial landscape, naturally invite a look in to their evolution. Angiosperms originated during the Middle Triassic in fern and gymnosperm dominated ecosystems, diversified and completely covered the terrestrial landscape by the Late Cretaceous. In this paper, consistent fossil angiosperm pollen grain evidences from across the globe are being assessed to unravel the evolutionary history of angiosperms. The bulk of fossil evidences suggest Early Cretaceous origin of angiosperms while some of the fossil evidences do point to the presence of angiosperms in the Triassic. Diversification of angiosperms in the low latitudinal regions during mid Cretaceous period provides evidence that the warm and humid climate during Cretaceous must have played an important role in the proliferation and diversification of tropical angiosperms. Biogeography of most of the tropical angiosperm families can be linked with the continental plate tectonic history. This review is an attempt to bring together the palynological evidences that place angiosperm evolution in perspective.

Keywords Early angiosperms • Pollen • Palynology • Cretaceous • Warm climate • Plate tectonics

V. Prasad (✉) · A. K. Mishra · M. Bansal
Birbal Sahni Institute of Palaeosciences, 53-University Road, Lucknow, 226 007, India
e-mail: prasad.van@gmail.com

S. Mishra
Department of Geology, Centre for Advanced Studies, University of Delhi, Delhi, 110 007, India

Introduction

Angiosperms are the most dominant plant community and are represented in the majority of the ecosystems of the world. They evolved gradually during the Early Cretaceous, rapidly diversified and dominated the Earth's terrestrial ecosystem by the Late Cretaceous (Vakhrameev 1991; Crane 1987; Crane and Lidgard 1989; Lidgard and Crane 1990; Crane et al. 1995; McLoughlin 2001; Crepet et al. 2004; Heimhofer et al. 2005, 2007 and references therein). Tracing the evolution of angiosperms is one of the most challenging and relevant issue in plant evolution. Several hypotheses have been propounded to explain the evolution and diversification of angiosperms such as radiation of pollinating insects, eco-physiological and macroevolutionary innovations in plants (Chaboureau et al. 2014). One of the theories proposes that angiosperms originated in the Permian or Triassic tropical uplands and invaded the lowland Cretaceous habitats during equable climates (Axelrod 1970). Based on the high abundance of basal angiosperm families in Burma, Thailand, Indo-China, and Malaysia, south-central Asia was considered the 'center of origin' for the evolution of angiosperms (Takhtajan 1967). However, a basic flaw in this hypothesis was the lack of early fossil records from these regions and it was considered that the present-day regions of basal angiosperms may represent 'center of survival' rather than 'center of origin' (Takhtajan 1969). The disparity in the location of early angiosperm fossils and extant regions of basal angiosperms suggests that the descendants of earliest angiosperms must have dispersed far from their point of origin (Takhtajan 1969). It also suggests that the changes in the continental configurations due to more active plate tectonic movements played their role in the evolution and diversification of angiosperms. Global climate change, particularly the humid climate state during the Cretaceous, has been considered as the driver for the evolution and dispersal of early angiosperms. Different criteria

G. V. R. Prasad and R. Patnaik (eds.), *Biological Consequences of Plate Tectonics: New Perspectives on Post-Gondwana Break-up—A Tribute to Ashok Sahni*, Vertebrate Paleobiology and Paleoanthropology Series, https://doi.org/10.1007/978-3-030-49753-8_5

were taken into account in determining the earliest angiosperm fossil records, as some workers only considered the reproductive structures (flowers or parts of flower such as carpels and stamens with pollen sacs) as undoubted angiosperm fossils (Friis et al. 2000, 2006, 2010a, b, 2011) while others relied on the continuous pollen grain records in the palynoassemblages in the lower Cretaceous successions (Cornet 1989; Doyle 1973, 1992; Friis et al. 2000). Due to their wide geographic and stratigraphic coverage, fossil pollen grain can provide more or less continuous records of early angiosperms. Hence in the present review global angiosperm fossil pollen grain records have been analyzed and the role of plate tectonics and climate has been assessed. Although the consistent fossil records of early angiosperms are from Early Cretaceous onwards, there are few reports of angiosperm like fossil pollen grain from pre-Cretaceous sedimentary successions.

Palynological Records

The rise and diversification of angiosperm has been regarded as the "Darwin's abominable mystery" for being the most important transformation of the vegetation and several events of co-evolution in pollinators and dispersal agents that occurred in synchronization (Darwin and Seward 1903; Labandeira et al. 1994; Hu et al. 2008). The oldest undoubted records of angiosperms are from Hauterivian and Barremian (Penny 1986, 1991; Hughes 1994; Brenner 1996). However, still the complete picture of this important event is elusive (Archangelsky et al. 2009). Here we document and discuss all the major palynological findings related with angiosperm origin, diversity and distribution.

Pre-cretaceous Records

The oldest record of angiosperm pollen taxon, *Daminites* relatable to *Cornetipollis* Pocock and Vasanthy (1988) and *Monocrinopollis* Cornet (1989) with infrareticulate-infrabaculate exine and two to three sulci were reported from late Permian sediments of Talcher Coalfield, Orissa, India (Tripathi 1998). However, this area needs to be reinvestigated in detail for its validation. Palynomorphs with angiosperm-like features were also recorded from Triassic sediments from the Norwegian Arctic, Barents Sea region (Hochuli and Feist-Burkhardt 2004, 2013) and the Late Triassic Chinle Formation of Arizona, Utah and New Mexico (Pocock and Vasanthy 1988; Zaavada 1990) and the Richmond Basin, Virginia, USA (Cornet 1989; Vasanthy et al. 2004; Fig. 5.1). The paleo-latitudinal positioning of both Barents Sea region and Arizona and the Richmond Basin, USA were low latitudinal during the Triassic period and suggested Laurasian origin of angiosperms (Fig. 5.1). The Middle Triassic fossil palynomorphs consisted of reticulate and gemmate exine pattern in their pollen grains wall but lacked definitive angiospermous endexine

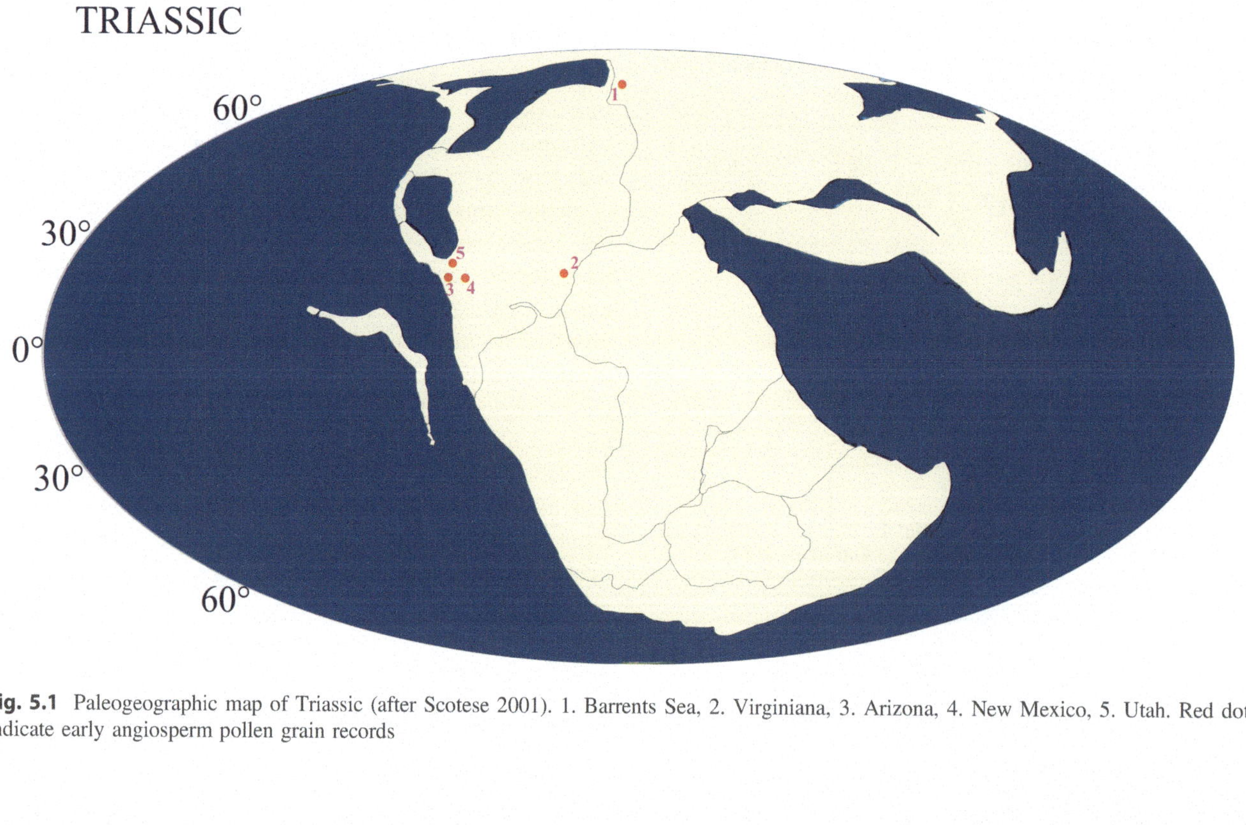

Fig. 5.1 Paleogeographic map of Triassic (after Scotese 2001). 1. Barrents Sea, 2. Virginiana, 3. Arizona, 4. New Mexico, 5. Utah. Red dots indicate early angiosperm pollen grain records

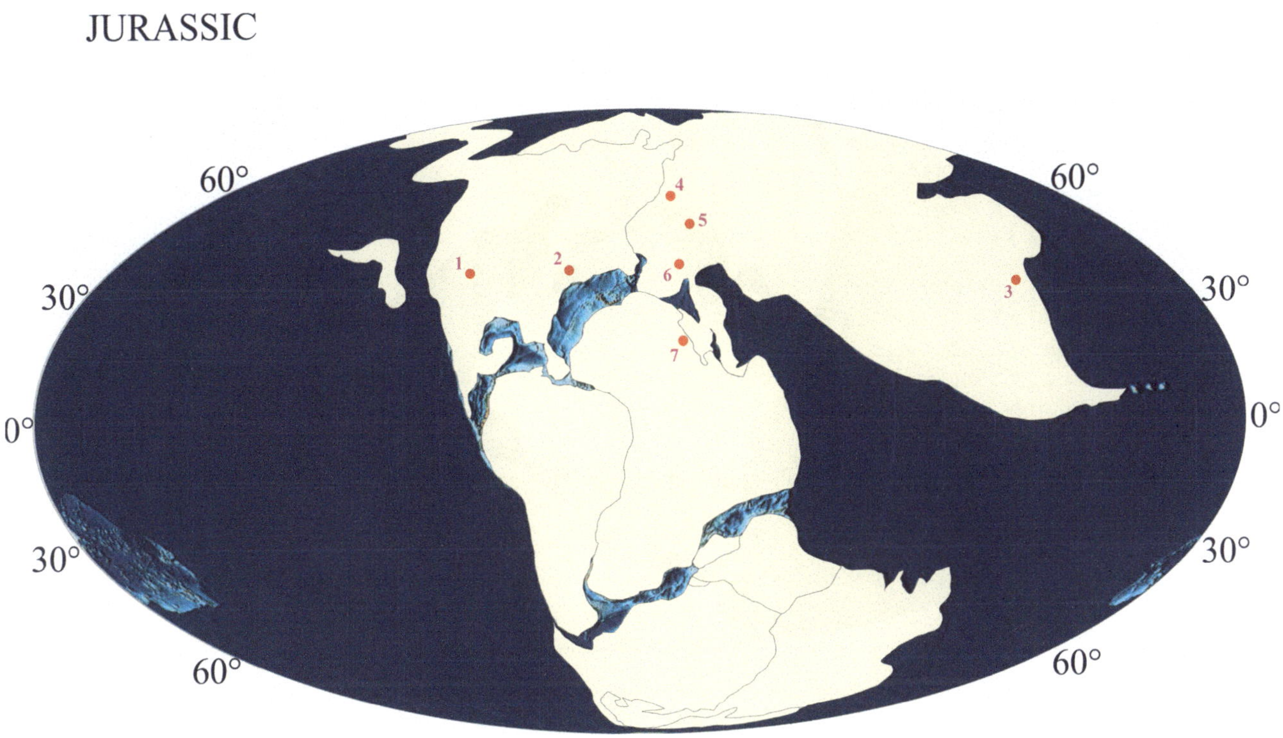

Fig. 5.2 Paleogeographic map of Jurassic (after Scotese 2001): 1. Utah, 2. Virginia, 3. Japan, 4. Nova Scotia, 5. Europe, 6. England, 7. Israel. Red dots indicate early angiosperm pollen grain records

and aperture. These pollen grains with primitive angiosperm-like morphological features were considered proto-angiosperms known as the *Crinopolles* group. The affinity for pollen grains of *Crinopolles* group cannot be linked to any of the extant angiosperm families. Hickey and Doyle (1977) mentioned that these taxon which were recognized as angiosperms were not truly angiosperms and could be ancestor of some other groups.

Subsequently a new angiosperm taxon, *Stellatapollis pocokii*, was reported from the upper Jurassic outcrop of Oxford Clay from Normandy, France (Cornet and Habib 1992). Some monocotyledons from the Jurassic of Portugal were also discovered (Doyle 1973) (Fig. 5.2). The Jurassic data provided evidence that the early angiosperms were restricted to understory habitats of wet forests and did not radiate until the Cretaceous (McLoughlin 2001).

Early Cretaceous Records

Most convincing angiosperm fossil pollen grains records were from the Early Cretaceous and were essentially obtained from the Barremian sediments from southern Laurasia and northern Gondwana (Buerki et al. 2014) (Fig. 5.3). Sporadic occurrences of the angiosperm pollen grains have been reported from Barremian of Israel (Brenner 1996), England (Hughes 1994), Morocco (Gübeli et al. 1984) and Egypt (Schrank and Mahmoud 2002). However, monosulcate angiosperm fossil palynomorphs with columellate exine and radial rods were also reported from late Valanginian to late Hauterivian of China, Israel, Italy and England (Muller 1984; Crane and Lidgard 1989; Friis et al. 2011).

The earliest recognizable angiosperm pollen grain, *Clavatipollenites* (monosulcate, tectate–columellate form) was established by Couper (1958) and has been primarily reported from the Lower Cretaceous sediments of Patagonia by Archangelsky and Gamerro (1966). The Springhill Formation of the Austral Basin yielded the earliest record of this pollen grain from the Barremian aged sediments (Archangelsky and Archangelsky 2002, 2004; Quattrocchio et al. 2006). *Clavatipollenites* type pollen grains emerged as the common element of Early Cretaceous palynoassemblages all along the globe (Archangelsky and Gamerro 1966; Playford 1971; Volkheimer and Salas 1975a, b; Walker and Walker 1984; Burger 1990, 1993; Archangelsky and Taylor 1993; Dettmann 1994; Hughes 1994; Brenner 1996; Eklund et al. 2004) (Fig. 5.3).

Another pollen grain, *Asteropollis* was the next most common element of angiosperm assemblage during the Barremian (Couper 1958; Walker and Walker 1984; Nichols et al. 2006; Friis et al. 2011; Martínez et al. 2013). In addition, *Afropollis* fossil pollen grains were recorded from the Barremian sediments of Nova Scotia, England, Morocco,

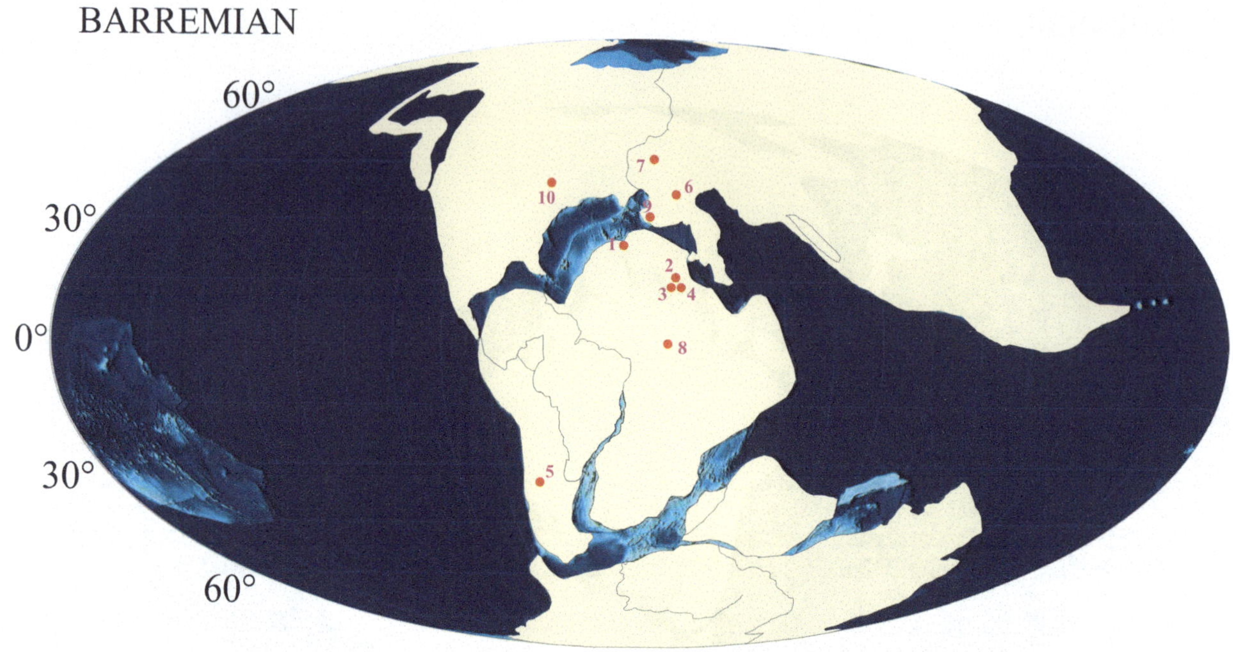

Fig. 5.3 Paleogeographic map of Barremian (after Scotese 2001): 1. Morocco, 2. Israel, 3. Libya, 4. Egypt, 5. Argentina, 6. England, 7. Nova Scotia, 8. Equatorial Africa, 9. Portugal, 10. Eastern U.S.A. Red dots indicate early angiosperm pollen grain records

Libya, Israel, Argentina and South America (Prámparo 1990, 1994, 1999; Doyle 1992; Vallati 1995, 2006; Medina et al. 2008) (Fig. 5.3). Later, many studies were carried out to show several monosulcate fossils with angiosperm affinity described as: *Brenneripollis*, *Stellatopollis*, *Similopollis*, *Retimonocolpites* and *Liliacidites* from the Upper Barremian-Lower Aptian of England Egypt, Patagonia (Hickey and Doyle 1977; Romero and Archangelsky 1986). These monosulcate and trichotomosulcate pollen taxa such as *Clavatipolleites*, *Pennipollis* and *Asteropollis* were found to be well associated with the Chloranthaceae, which was considered a basal angiosperm due to the presence of primitive floral characters (Walker and Walker 1984).

Some monocotyledons from the Lower Cretaceous of Portugal were also discovered (Doyle 1973) (Fig. 5.3). The late Barremian monosulcate angiosperm pollen grains were considered a mixture of ANITA lines, Chloranthaceae, and Magnoliids which were more common and diverse in the northern Gondwana than in Laurasia (Doyle et al. 1977, 1982; Schrank 1983). The three main groups of angiosperms, magnoliids, monocots and eudicots (e.g., Ranunculales) were globally common during the Hauterivian-Aptian interval, providing evidence for angiosperm diversification during the Early Cretaceous (Friis et al. 2011; Buerki et al. 2014). The stratigraphic succession of these fossils is well-represented in eastern North America (Potomac sequence; Doyle 1992), Portugal (Lusitanian Basin; Heimhofer et al. 2007; Friis et al. 2010a, b) and from Yixian Formation of China (Friis et al. 2011).

The appearance of eudicots with tricolpate morphological character of pollen grains started appearing in the fossil assemblages of latest Barremian deposits of England while they appeared a bit late, i.e., during Aptian in Africa and South America (Doyle et al. 1977; Regali and Viana 1989; Hughes and McDougall 1990; Brenner and Bickoff 1992; Doyle 1992; Friis et al. 2000; Archangelsky et al. 2009; Heimhofer and Hochuli 2010). These primary eudicot pollen grains were tricolpate and microreticulate, e.g., *Tricolpites* and *Retitricolpoites* (Volkheimer and Salas 1975a, b; Medina et al. 2008). The tricolpate pollen grains were more in numbers in the palynoassemblage of the Albian of southern Laurasia and Gondwana (Friis et al. 2011) (Fig. 5.4). In equatorial regions, the scenario was different, and eudicots appeared earlier by the early Aptian in Israel (Brenner and Bickoff 1992) and Brazil (Regali and Viana 1989).

Monosulcate pollen grains (pollen with one single slit like aperture) described as *Pennipollis*, *Peromonolites*, *Liliacidites*, *Retimonocolpites* and *Brenneripollis* were the most common angiosperm pollen taxa of Aptian-Albian sediments of Portugal (Penny 1988) (Fig. 5.4) and South America (Brazil) (Muller 1981; Prámparo 1990, 1999; Friis et al. 2000, 2004) (Fig. 5.4), while *Tucanopollis*, having circular aperture with reticulate exine was the most common palynotaxon of the Barremian and Aptian of Brazil and Africa (Regali et al. 1974a, b; Doyle et al. 1977). Similar palynomorph named as *Transitoripollis* was reported from Albian (Spain and England) and Cenomanian of Europe (Muller 1981; Góczán and Júhász 1984). It was considered

Fig. 5.4 Paleogeographic map of Albian (after Scotese 2001): 1. Morocco, 2. Israel, 3. Libya, 4. Egypt, 5. Spain, 6. England, 7. Nova Scotia, 8. Equatorial Africa, 9. Portugal, 10. Eastern U.S.A, 11. Brazil, 12. Australia, 13. New Zealand, 14. Gabon, 15. West Canada, 16. India, 17. Alexander Island of Antarctica. Red dots indicate early angiosperm pollen grain records

that these taxa were supposed to be derived from the eudicot clade which appeared in the late Barremian and Aptian of northern Gondwana (South America, Africa) and the Middle East (Brenner 1976; Doyle et al. 1977; Doyle 1992). Another important morphological feature in the pollen grains evolution was appearance of tricolporate pollen taxa (pollen grains with three slit like aperture along with a pore within each slit) that appeared and diversified in late Albian (Furness et al. 2007). In southern Gondwana, early angiosperm pollen taxa records were few and confined to Albian deposits of Australia, Antarctica and South America (Dettmann 1973; Volkheimer and Salas 1975a, b; Dettmann and Thomson 1987; Burger 1990; Aguirre-Urreta 2002; Guler and Archangelsky 2006; Archangelsky et al. 2008; Medina et al. 2008) (Fig. 5.4). Albian to Cenomanian times were very important for angiosperm diversification and dominance which resulted in the development of pollen grains with varied and complex apertures and sculptures (Barreda and Archangelsky 2006). Similar pattern has been observed all over the world (Drinnan and Chambers 1986; Barreda and Archangelsky 2006; Hochuli et al. 2006; Narváez et al. 2014). Early angiosperm pollen grain records from Antarctica belong to Albian sediments from the Alexander Island (Cantrill and Nichols 1996) (Fig. 5.4).

In India, Hauterivian has been regarded as the time period of appearance of angiosperm pollen grains with primitive morphologies (Tiwari 1996). There are several palynoassemblages of Hauterivian to Early Cretaceous in age, but they are devoid of angiosperm pollen grains (Vijaya 1997; Vijaya and Prasad 2001; Vijaya and Kumar 2002; Goswami et al. 2008; Tripathi 2009; Vijaya 2011). There are very few detailed palynological studies on the Early Cretaceous sediments of India (Tiwari and Tripathi 1995; Prasad and Pundeer 1999; Jha et al. 2016; Mishra and Prasad 2020). Jha et al. (2016) did a detailed study of the Gangapur Formation of Godavari Graben where they have identified few monocolpate, tricolpate and tetracolpate pollen grains of unknown affinity in the Tithonian-Valangian aged deposits. This could be the earliest record of angiosperm pollen grains from the Gondwana continents. In another study from southern India, Prasad and Pundeer (1999) have identified pollen taxa of *Clavatipollenites hughessii, Tricolpites variabilis*, *Stephanocolpites* sp., *Stellatopollis* sp. and *Asteropollis asteroids* from the Raghavapuram Shale (Barremian to Early Albian) of Krishna-Godavari Basin. The well-dated angiosperm pollen taxa records with monosulcate angiospermous morphology has been reported from the intertrappean sediments of Albian age in the Rajmahal Basin

(Tripathi 1993). Based on the 40 Ar/39 Ar analysis the sediments have been dated to span from 118 Ma to 116 Ma (Tiwari and Tripathi 1995).

Late Cretaceous Records

Cenomanian witnessed the diversification of eudicots having numerous tricolpates, tricolporates and periporate forms (Crane and Lidgard 1989; Friis et al. 2006; Prámparo et al. 2007). The diversification of angiosperm led to the domination of flowering plants in varied habitats supplanting Mesozoic flora of conifers, Bennettitales and seed ferns (Friis et al. 2011). During Turonian, Santonian and Campanian, angiosperm diversity and abundance increased in both low and high latitudes of Gondwanan (mainly Morocco, Libya, Egypt, Equatorial Africa, Brazil, Australia, New Zealand, Gabon, France, Peru and India) and Laurasian continents (mainly Israel, Spain, England, Nova Scotia, Portugal, Eastern U.S.A., West Canada, Northern Alaska, Siberia, Japan and Borneo) (Muller 1981, 1984; Askin 1994; Hofmann and Zetter 2010) (Figs. 5.5 and 5.6). In India pre-Maastrichtian palynological records are very less, therefore it's very tough to assess anything regarding the gradual transition and evolutionary development in the angiosperm flora on the Indian subcontinent. The only example of such palynological study is by Prasad and Pundeer (1999) on the Tirupati Sandstone Formation of Krishna Godavari Basin. This Campanian to Maastrichtian aged flora includes pollen grains with varied apertures including monosulcate (*Clavatipollenites* sp., *Liliacidites* sp., *Spinizonocolpites echinatus*, *S. bulosus*), trichotomosulcates (*Asteropollis asteroids*), colpates (*Racemonocolpites ramosus*, *R. fissilis*), tricolpates (*Rousea georgensis*), polycolpate (*Retistephanocolpites* sp., *Polycolpites* sp.), tricolporates (*Tricolporites* sp., *Tubuliflouridites* sp.), tetracolporates (*Tetracol porites* sp.), and polyporates (*Erdtmanipollis* sp.).

The angiosperm diversity reached its acme during the Maastrichtian and the four major palynofloristic provinces can be observed (Zaklinskaya 1981; Herngreen 1996; Vajda and Bercovici 2014 and references therein) (Fig. 5.7). The four main provinces named after the abundant pollen grains/group present in various geographical regions are *Aquilapollenites* Province, *Normapolles* Province, Palmae Province, *Proteacidites/Nothofagidites* Province (Fig. 5.7). The *Aquilapollenites* group pollen grains were commonly distributed in the northern high-latitudinal areas during the Late Cretaceous, including Greenland, northern and eastern Asia and North America (Herngreen 1996). The group reached its maximum expansion during the Maastrichtian and relatively high abundances were reported from the Palmae Province of Bolivia (Pérez Leytón 1987; Vajda-Santivanez 1999) and Brazil (Regali et al. 1974a, b; Prámparo 1990, 1994). The *Normapolles* Province was

Fig. 5.5 Paleogeographic map of Turonian (after Scotese 2001): 1. Morocco, 2. Israel, 3. Libya, 4. Egypt, 5. Spain, 6. England, 7. Nova Scotia, 8. Equatorial Africa, 9. Portugal, 10. Eastern U.S.A, 11. Brazil, 12. Australia, 13. New Zealand, 14. Gabon, 15. West Canada, 16. France, 17. Northern Alaska, 18. Peru, 19. Siberia, 20. India. Red dots indicate early angiosperm pollen grain records

SATONIAN-CAMPANIAN

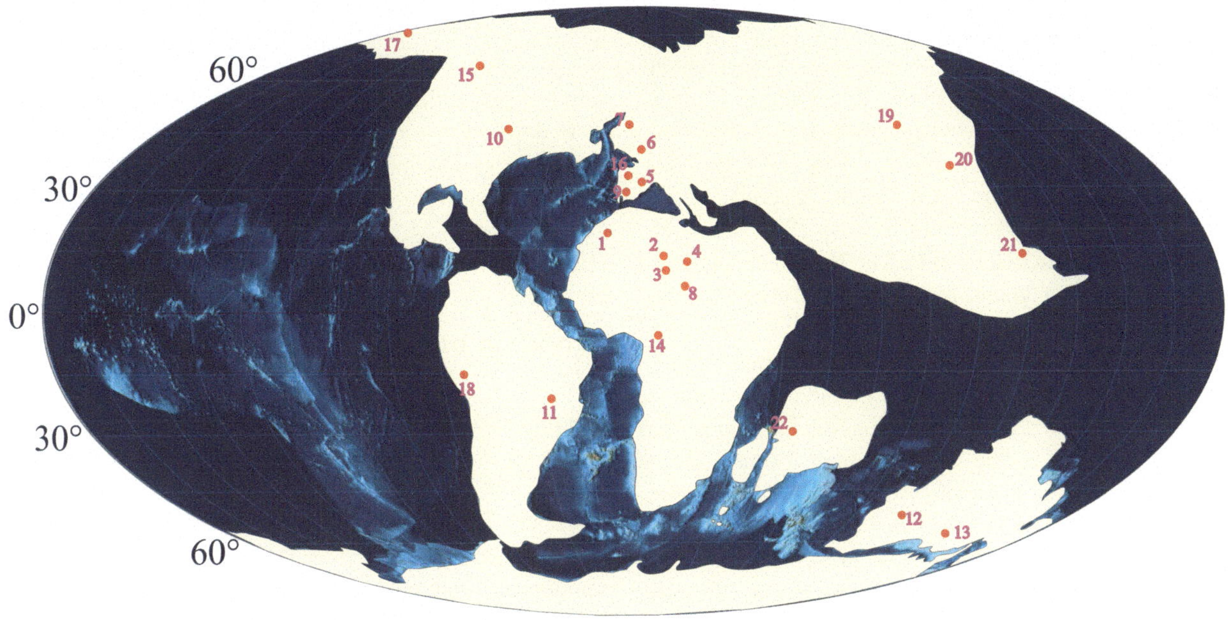

Fig. 5.6 Paleogeographic map of Santonian-Campanian (Scotese 2001): 1. Morocco, 2. Israel, 3. Libya, 4. Egypt, 5. Spain, 6. England, 7. Nova Scotia, 8. Equatorial Africa, 9. Portugal, 10. Eastern U.S.A, 11. Brazil, 12. Australia, 13. New Zealand, 14. Gabon, 15. West Canada, 16. France, 17. Northern Alaska, 18. Peru, 19. Siberia, 20. Japan, 21. Northwest Borneo, 22. India. Red dots indicate early angiosperm pollen grain records

MAASTRICHTIAN

Fig. 5.7 Paleogeographic map of Maastrichtian showing various floristic provinces modified after Herngreen (1996) and Vajda and Bercovici 2014

characterized by triaperturate and oblate pollen grains having an affinity with Fagales (Friis et al. 2006). The province extended from eastern North America, Europe and to western Siberia, also including the Middle East (Herngreen and Chlonova 1981; Herngreen 1996). The Palmae Province consists of dominant palm pollen grains (up to 50%) and represents the areas of low-latitudes during the Late Cretaceous extending from northern South America, central Africa and India (Herngreen and Chlonova 1981; Herngreen 1996; Vajda and Bercovici 2014). The high abundance and diverse palm pollen grains represent relatively warm and ever-wet conditions. The megafossils of palm-like fossil flowers and leaves are very common in the intertrappean beds of India (Friis et al. 2011). Contrary to the Palmae Province, the *Proteacidites/Nothofagidites* Province was mainly represented in the areas of high latitude in the Southern Hemisphere including Australia, Antarctica, New Zealand and southern South America (Herngreen 1996).

In India, Maastrichtian deposits occur as infratrappean or intertrappean deposits of Deccan Volcanic region of central India (Samant and Mohabey 2014). Based on the provinces described by Vajda and Bercovici (2014), India bears three palynological provinces, i.e., *Normapolles* Province (in the lower Northern region); Palmae Province (widely spread in the central to western part of India); mixed flora of Palmae and *Proteacidites/Nothofagidites* Province (Southern region). There are several records of Maastrichtian palynofloral assemblages and most common pollen grains of the assemblage are *Gabonisporites vigouoxii*, *Azolla cretacea* and *Aquilapollenites bengalensis* (Samant and Mohabey 2014; Prasad et al. 2018) The Indian species of *Aquilapollenites* is fairly represented in the late Maastrichtian records, extending from various parts of central India to the south in Godavari and Cauvery basins (Baksi and Deb 1981; Venkatchala and Sharma 1974a, b, 1984; Sahni et al. 1996; Samant and Mohabey 2014; Prasad et al. 2018). Hence, the proposed provincialism seems vague for Indian palynoflora (Fig. 5.7). The flora of the world changed considerably after the Maastrichtian and lot of modern families differentiated and old ancient groups died out (*Normapolles*, *Aquilapollenites*, pollen taxa with elateroids; Romero 1993). They acquired the level of prominence compared to current vegetation (Friis et al. 2006).

Discussion

There are many theories on the evolution of angiosperms. Arguably, angiosperm evolution is driven by a multitude of characters in plants including both morphological and physiological as well as the influence of environment and climate. There are two points of views regarding the origin of angiosperms: 1. paleoequatorial origin due to high diversity at lower latitudes during the early radiation of angiosperms (Raven and Axelrod 1972; Muller 1981; Crane and Lidgard 1989; Brenner 1996); 2. Higher latitudinal origin suggests that the structural characters of angiosperms were developed in response to the strong selective pressure and harsh environmental conditions at higher latitudes and later migrated to low latitudinal regions (Hickey and Doyle 1977; Crane and Lidgard 1989; Drinnan and Crane 1989; Hill and Scriven 1995).

Evolution of Angiosperm

The pro-angiosperm like fossils were recorded from Triassic sediments from various regions (Pocock and Vasanthy 1988; Cornet 1989; Hochuli and Feist-Burkhardt 2004, 2013) (Fig. 5.1). The angiosperm-like pollen grains recovered from Triassic deposits have monosulcate aperture along with semitectate, collumellate and reticulate ornamentation on their thin nexine. However, the uniformly thick alveolar endexine associated with the Triassic pollen grains were more gymnosperm like. The Triassic forms were described as *Crinopolles* group, *Cornetipollis* and *Retisulcites*. The affinity of the Triassic fossil forms with the angiosperms continues to be a subject of controversy as views were given for and against them. It was suggested that the angiosperm like fossils from Triassic probably represented plants of angiosperm stem lineage (Crane 1985; Doyle and Donoghue 1986). The paleolatitudinal position of pro-angiosperm Triassic fossil localities suggested that the evolution of pro-angiosperms like vegetation took place in the low latitude of Laurasia post Pangea break up (McLoughlin 2001) (Fig. 5.1). Most likely the prevalence of humid climatic conditions of low latitudinal regions would have prompted the evolution of angiosperm like vegetation in the Laurasia. However, at the Germanic Basin and northern Switzerland the pro-angiosperm fossils were associated with the evaporites as well as xerophytic spore forms and hence suggested arid environmental conditions (Hochuli and Feist-Burkhardt 2013).

Jurassic records of pro-angiosperm fossils are few and controversial with respect to stratigraphy and age and may indicate contamination. However, few records of monosulcate pollen grains from the Jurassic of western Canada, North Sea indicated presence of pro-angiosperms in Laurasia (Abbink 1998). It was also speculated that probably the prevalence of dry climate during Jurassic restricted the evolution and diversification of angiosperms. During the Late Jurassic and Early Cretaceous, global climatic conditions started becoming wetter. The period coincides with the rifting of Laurasian and Gondwanan supercontinental plates (McLoughlin 2001).

The Early Cretaceous records were the most accepted angiosperm fossil records and were spread all over the low latitudinal belts of northern Gondwana and southern Laurasia. This suggests that probably the wetter climatic conditions of low latitudes were ideal for the proliferation of early angiosperms in Laurasian and Gondwanan regions. Based on fossil records angiosperm invasion of Gondwana took place in the Early Cretaceous (Prámparo et al. 2007). Based on the studies done by Archangelsky et al. (2004) angiosperm evolution in South America took under three major stages. Stage-I includes time between late Barremian to Aptian age, a time when first angiosperm pollen grains appeared within the gymnosperm-fern dominated flora (Archangelsky 2001, 2003; Quattrocchio et al. 2006). *Clavatipollenites* (a monosulcate, tectate- columellate form of genus) was the first recognizable angiosperm pollen grains type found in South American Barremian strata. Other than this it has also been reported from the Barremian deposits of England, equatorial Africa, U.S.A, western North Atlantic, North American Potomac Group and Portugal (Brenner 1963; Doyle and Robbins 1977; Doyle et al. 1977; Hughes et al. 1979; Hochuli and Kelts 1980; Schrank and Mahmoud 2002) (Fig. 5.3). Thus, this pollen taxon has a worldwide synchronous emergence (Kemp 1968; Heimhofer et al. 2005). There are few records of this pollen grain as old as Late Hauterivian sediments from Israel (Brenner and Bickoff 1992; Brenner 1996). However, this needs to be validated. The diversity of *Clavatipollenites* was much higher in the Early Cretaceous sediments of Portugal (Pais and Reyre 1981; Trincão 1990; Mendes et al. 2014) (Fig. 5.4). By the end of this phase several other pollen grains like, *Retimonocolpites* sp. A, *Retimonocolpites.* sp. B, *Reticolpites* sp., *Asteropollis asteroids* and *Monocolpopollenites* sp. also emerged, which provide evidence of widespread appearance of Angiosperm by the end of Aptian (Archangelsky et al. 2004). The Laurasian angiosperm fossil records as observed in England and Egypt bear more diversified late Barremian-early Aptian flora showing monosulcate pollen grains of *Brenneripollis*, *Stellatopollis*, *Similopollis*, *Retimonocolpites* and *Liliacidites* (Hickey and Doyle 1977). *Afropollis* sp. (monosulcate finely or coarsely reticulate) was also recorded from Barremian deposits of Nova, Scotia, England, Morocco, Libya and Israel (Doyle 1992; Archangelsky et al. 2009). Stage-II includes latest Aptian-earliest Albian flora, a phase of eudicots emergence (Archangelsky et al. 2004). The oldest angiosperms with clear morphological features comparable to extant angiosperms were found in the late Aptian and Albian fossil records. By this time the angiosperms were significantly diversified in South America and many new forms of basal angiosperms such as *Liliacidites* sp., *Pennipollis proreticulatus*, *Stellapollis* and various species of *Afropollis* appeared. The tricolpate pollen grains appeared during earliest Albian of South America (Volkheimer and Salas 1975a, b; Medina et al. 2008) and in mid-high latitude regions of North America (Nova Scotia, West Canada; Doyle and Hickey 1976; Muller 1981), Europe (Heimhofer et al. 2005, 2007), Middle East (Israel; Muller 1984) as well as in Australia (Dettmann 1973; Burger 1990). Tricolpate pollen grains appeared during early Aptian in equatorial regions (Regali and Viana 1989; Brenner and Bickoff 1992). However, despite considerable divergence in the angiosperm group, vegetation was still gymnosperm-fern dominated during Early and mid-Cretaceous (Archangelsky et al. 2009). Stage-III was the most important phase of angiosperm divergence and dispersal including sediments of middle Albian-Coniacian age. The tricolpate forms of early Albian evolved during this phase and new reticulate tricolpate and psilate pollen taxa morphological features appeared along with few tricolporoid ones like, *Rousea* sp., *Verrutricolpites* sp., *Dryadopollis* sp. and *Nyssapollenites* sp. (Archangelsky et al. 2009). The development of endoapertures may had a harmomegathic functions and provided a selective advantage to eudicots to colonize drier niches (Furness et al. 2007). Beside these, pollen grains of *Afropollis?/Schrankipollis* and *Walkeripollis* sp. have also been reported from the middle-late Albian of Patagonia (Barreda and Archangelsky 2006; Archangelsky et al. 2008). By the end of Albian, the sculptural modification in the external wall can be seen in the pollen grains with gemmate or verrucate ornamentations. Cenomanian pollen grains show reduction in the polar axis and shortening of the colpi, which may have resulted into the development of triporate forms.

The stages of evolution and dispersal of angiosperms in South America were more or less synchronous in other regions such as Australia and in the Potomac Group, Western Portugese, Algarve Basin, Northeast China and Siberian basins of Northern Hemispheres (Archangelsky et al. 2009). However, due to the rarity of palynological records from India and Antarctica, evolutionary history of angiosperm from Barremian till Early Maastrichtian from these continents is still vague. Angiosperms have considerably diversified and reached relative abundance as today by the end of Cretaceous (Vajda and Bercovici 2014). The development of new bioclimatic continents due to tectonic activities and geological isolation of many archipelagos due to sea-level rise favored angiosperm diversification and ecological radiation (Chaboureau et al. 2014). The diversification and evolutionary success of the angiosperm was greatly regulated by the allopatric speciation due to climate based morphological modification for successful reproduction and ecological adaptation (Farrel 1998; Boyce et al. 2009; Busch and Zachgo 2009; de Boer et al. 2012). This resulted in the enhanced provincialism during Maastrichtian (Vajda and Bercovici 2014).

Angiosperm Evolution versus Tectonics

Angiosperm evolution was greatly induced by the tectonic history of the Pangea (Chaboureau et al. 2014). Angiosperm diversified and evolved in a sequential manner from Barremian to Cenomanian as depicted by pollen taxa and leaf records from South America (Archangelsky et al. 2004), Australia (Dettmann 1973; Pole 2000), North America (Doyle and Hickey 1976), Europe (Coiffard et al. 2012), North Eastern China (Sun and Dilcher 2002; Sun et al. 2000, 2008) and Siberian region (Krassilov 1997).

The emergence and spread of early angiosperms can be linked with the break-up of Gondwana and major sea-level fluctuations during Early and mid-Cretaceous. Coastal areas were flooded by shallow seas as the sea level was 100–200 m higher than today (Haq et al. 1987; Hallam 1992). The disturbed environmental conditions (high and low relief, variable soil conditions and rapidly changing topography) in the newly developed epicontinental seaways must have provided a habitable land for the dispersal of angiosperms towards higher latitudes (Dettmann 1989). It was considered that the early herbaceous angiosperms were the opportunistic species and exploited the newly formed rift corridors and seaways (Retallack and Dilcher 1981). This assumption is supported by the presence of characteristic Northern Gonadwanan pollen grains (*Schrankipollis, Brennipollis, Pennipollis*) at the Austral tip of South America (Barreda and Archangelsky 2006). *Nothofagus*, having an early Gondwana origin and its distribution was considered regulated by the disturbances in the area (Veblen et al. 1980).

During the Early Cretaceous, monosulcate pollen grains were more abundant and diversely dispersed all over the Northern Gondwana (Doyle et al. 1977), while tricolpates pollen grains had a poleward spread in the arid environments in the Northern Gondwana (Brenner 1976). The increased number of apertures (three) would have provided the benefit of extra germination sites to get adapted in a drier condition (Furness and Rudall 2004). This suggested that tricolpate pollen grains were adapted easily to arid climatic conditions. Tricolpate pollen grains were very rare in the latest Barremian in Southern Laurasia as well as Northern Gondwana (Doyle 1992). They expanded considerably throughout the Aptian and Albian in southern part of Gondwana (Heimhofer and Hochuli 2010). The rarity of tricolpates during Barremian (Fig. 5.3) and their wide dispersal during Albian was due to equable climatic conditions during the latter period (Burger 1980; Friis et al. 2006; Smith et al. 2010) (Fig. 5.4). Most of the fossil angiosperm sites during Aptian were situated close to low-mid latitudinal regions (i.e., Gabon, Congo, Tunisia, Egypt and Brazil; Friis et al. 2011). The angiosperm radiation accelerated as most of the arid regions transformed into wetter temperate biomes from Barremian to Cenomanian globally (Chaboureau et al. 2014). Thus emerging temperate humid conditions due to global paleogeogeographic changes set the stage for angiosperm expansion. The start of breakup of large latitudinal dessert belts from Early Cretaceous to their final discontinuation during the Cenomanian also played a vital role in the angiosperm evolution (Chaboureau et al. 2014).

The Kerguelen Plateau may have a minor role in the dispersal of early angiosperms to southern high latitudes by segregating the oceanic barriers between India and Australia-Antarctica due to its large size and its paleoposition off north-eastern India (Crane and Lidgard 1989). As a consequence of major continental fragmentation due to plate tectonic movement, floristic provincialism was seen in the Late Cretaceous successions of different continents (Vajda and Bercovici 2014) (Fig. 5.7). While the Upper Cretaceous successions of high and low latitude continents were characterized by specific types of microfloral provinces, the Upper Cretaceous successions of the Indian subcontinent show the constituents of all the microfloral provinces (Prasad et al 2018). Hence the Indian subcontinent provided a link between the different palynofloral provinces that can be explained by its geodynamic history. India's connections with other Gondwana continents during the Cretaceous and its latitudinal shifting from mid-high to low latitudes could possibly have provided a major platform for the evolution of tropical biota on this landmass (Prasad et al. 2018).

One of the relevant physiological aspects in the angiosperms that also evolved is their high capacity of transpiration rate, due to higher leaf vein and stomatal densities (Boyce et al. 2009, 2010; Boyce and Lee 2017). Both these characters enabled angiosperms to suitably adapt to wetter climatic conditions of tropical region (Boyce et al. 2010). The angiosperms dominance in the rainforest of tropical region is well known. One of the major facts is that most of the early angiosperm fossil records were obtained from low latitudinal zones. This is suggestive of a role of humid climate of tropical region in the evolution and diversification of angiosperms. Early Cretaceous palynological records suggested that early angiosperms first become floristically prominent in low paleolatitude areas ($\sim 20°$ N to 20° S) (Crane and Lidgard 1989). Hence latitudinal climate gradient must have played an important role in the evolution and dispersal of angiosperms. Despite being separated with each other, different continents share common tropical plant families/genera/species. Fossil data revealed that the floral dispersal took place during periods of equable climate. It was suggested that warm climate of the Cretaceous and early Paleogene facilitated dispersal of tropical angiosperms in far and wide regions via northern high latitudinal land bridges (Morley 2000). Many representatives of tropical plant family appeared in mid and high latitudinal regions of Europe, North America and Asia during warm and humid climate of early Paleogene (Morley 2000). The tectonic movements of continental plates in deep geological times resulted in the restriction of dispersal by creating barriers or formation of effective land bridges that might have resulted in the dispersal and diversification of angiosperms. This way the globally warm climate and tectonic evolution of the continents during early Paleogene played an important role in the dispersal of tropical angiosperms in the widely separated regions.

Paleobotanical Records of Angiosperm Evolution

Despite several paleobotanical and phylogenetic attempts, angiosperm evolution is still a "abominable mystery" as stated by Darwin (Darwin and Seward 1903). The paleobotanical history of angiosperms dates back to Hauterivian, which records the first valid records of angiosperm from China, Israel and England (Friis et al. 2011). However, molecular analysis indicate that angiosperm crown group extend back to Jurassic (Sanderson and Doyle 2001; Bell et al. 2010) or even up to Triassic and Permian (Smith et al. 2010; Clarke et al. 2011). The evidence that support this view is the restricted presence of current species with lineages from ANA grade (Basal angiosperms like, Amborellales, Nymphaeales and Austrobaileyales) in the Southern Hemisphere (e.g., New Caledonia, Australia, South-East Asia). This view contradicts the paleobotanical interpretations as no identifiable angiosperm pollen taxa or megafossil records has been identified in the Triassic, Jurassic or even Early Cretaceous deposits in spite of worldwide studies (Friis et al. 2011). However, this paradox can be explained by the complex paleogeographic history of New Caledonia, Australia, South-East Asia and recent volcanic activities due to which fossil records are either under water or destroyed due to tectonic activities (Hall 2009). Another possible explanation could be the "xerophobic" nature of the terrestrial ANITA lines. The early angiosperms were restricted to wet and shaded forest understorey habitats, which were rare during Triassic and Jurassic and escaped detection due to low diversity and abundance (Ziegler et al. 2003; Doyle and Endress 2014).

Clavatipollenites, a monocolpate pollen taxon with reticulate exine is the first recognized angiosperm pollen grain having its affinity to plant family Magnoliaceae and to the monocots. In one of the studies, the affinity of *Clavatipollenites* was pointed out with the *Ascarina* of family Chloranthaceae and it was believed that *Clavatipollenites-Ascarina* lineage formed one of the ancestral angiosperms (Archangelsy and Archangelsky 2013; Doyle and Endress 2014). Thus, the early angiosperm fossil records from Barremian-Aptian indicated affinity with the extant basal angiosperm families. This suggested gradual evolution and diversification of angiosperms from the basal clades. The tricolpate reticulate pollen taxon type from Aptian which is smaller than monosulcate *Clavatipollenites* pollen taxon type and showed its affinity to non-magnoliid dicotyledons (Ranunculidae, Hamamelideae and Dilleniidae).

Tucanopollis, a common pollen taxon of Barremian and Aptian deposit of Northern Gondwana considered a relative of *Amborella* of western Gondwana (Brazil and Africa; Regali et al. 1974a, b; Doyle et al. 1977). *Amborella* and other woody groups were considered restricted to forest's wet understorey habitat (Feild et al. 2004, 2009), contradictory to the view that during the Early Cretaceous most of the Northern Gondwana was semiarid or arid (Brenner 1976; Ziegler et al. 2003). This suggested that during Aptian-Albian time period climate oscillated between wet and dry conditions in Gondwana region. The most common families of Chloranthaceae produced monosulcate pollen taxa that showed predominance during late Aptian and Albian (Feild et al. 2004, 2009). Buxales marked by striate tricolpate pollen taxa appeared during Aptian-Albian in northern Gondwana and during Cenomanian in Laurasian flora. Angiosperm families Lauraceae, Araceae and Nymphaeaceae were abundant during the Late Albian. Palm pollen grains have also been associated with Albian and Cenomanian of Africa (Schrank and Awad 1990) while there are earlier records back in the Barremian/Albian from South America (Martínez et al. 2016). However, most of the modern plant families radiated during the Late Cretaceous.

Future Prospects

In order to define the evolutionary history of angiosperms, fossil sampling must be carried out at a global scale since there is a dearth of information from many geographical areas (e.g., India and Antarctica) regarding their place of origin and diversification history. A comprehensive fossil data set, together with plate tectonic and paleoclimatic histories are necessary to determine the geographic origin, timing of radiation, and biogeography of angiosperms. Lack of consistent fossil records in one particular geographical region and vegetation turnover in deep geological times are linked with global climate change and thus indicate less adaptive abilities of in-situ vegetation. The knowledge about the origin, radiation and dispersal of angiosperms on Indian continent is minimal and more focused studies are needed in this aspect as India occupies a central position in the Gondwana and could have played an important role in the evolution of dispersal of flora. The present knowledge about the evolution of angiosperms is limited and hopefully in the coming years more advanced and focused studies could solve this puzzle into a clear picture which Darwin defined as the "Abominable Mystery".

Acknowledgements Authors are grateful to Director BSIP for encouragement during this study. Mahi Bansal and Ashish K. Misra are thankful to the MoES, New Delhi (Grant No MoES/P.O.(GeoSci.)/36/2014) for financial assistance. Shreya Mishra thanks Science and Engineering Research Board, New Delhi (PDF/2017/002297) for financial support. We are also thankful to the two anonymous reviewers and Dr. Mercedes B. Prámparo for critically reading the manuscript and suggesting substantial improvements.

References

Abbink, O. A. (1998). *Palynological investigations in the Jurassic of the North Sea Region*. Ph.D. thesis. Utrecht, Universiteit Utrecht, 192.

Aguirre-Urreta, M. B. (2002). Invertebrados del Cretácico inferior. In M. J. Haller (Ed.), *Geología y Recursos Naturales de Santa Cruz* (pp. 439–459). Relatorio del 15_Congreso Geológico Argentino, Calafate.

Archangelsky, A., Archangelsky, S., Poire, D. G., & Canessa, N. D. (2008). Registros palinológicos en la Fm. Piedra Clavada (Albiano) en su área tipo, provincia de Santa Cruz, Argentina. *Revista del Museo Argentino de Ciencias Naturales, Nueva Serie, 10,* 105–111.

Archangelsky, S. (2001). Evidences of an Early Cretaceous floristic change in Patagonia, Argentina. 7th international symposium on Mesozoic Terrestrial Ecosystem, Buenos Aires, 1999. *Asociación Paleontológica Argentina, Publicación Especial, 7,* 15–19.

Archangelsky, S. (2003). La flora cretácica del Grupo Baqueró, Santa Cruz, Argentina. *Monografías del Museo Argentino de Ciencias Naturales, 4,* 1–14.

Archangelsky, S., & Archangelsky, A. (2002). Palinología estadística en el Cretácico de la Cuenca Austral, Plataforma Continental Argentina. I. Seis perforaciones del área de Magallanes. *Revista del Museo Argentino de Ciencias Naturales, Nueva Serie, 4,* 25–34.

Archangelsky, S., & Archangelsky, A. (2004). Palinología estadística en el Cretácico de laCuenca Austral, Plataforma Continental Argentina. II. Seis perforaciones del área Gallegos. III. Discusión y conclusiones. *Rev Mus Argent Cienc Nat, NS, 6,* 245–255.

Archangelsy, S., & Archangelsky, A. (2013). Aptian Angiosperm pollen from the Tico Flora Patagonia, Argentina. *International Journal of Plant Science, 174,* 559–571.

Archangelsky, S., Barreda, V., Passalía, M. G., Gandolfo, M., Prámparo, M., Romero, E., et al. (2009). Early angiosperm diversification: Evidence from southern South America. *Cretaceous Research, 30,* 1073–1082.

Archangelsky, S., Cesari, S. N., & Cunéo, R. (2004). On some similar Patagonian and Indian Permian ferns. In P. C. Srivastava (Ed.), *Vistas in palaeobotany and plant morphology: Evolutionary and environmental perspective* (pp. 71–81). Lucknow.

Archangelsky, S., & Gamerro, J. C. (1966). Spore and pollen types of the lower Cretaceous in Patagonia (Argentina). *Review of Palaeobotany and Palynology, 1,* 211–217.

Archangelsky, S., & Taylor, T. N. (1993). The ultrastructure of in situ Clavatipollenites pollen from the Early Cretaceous of Patagonia. *American Journal of Botany, 80,* 879–885.

Askin, R. A. (1994). Monosulcate angiosperm pollen from the López de Bertodano formation (upper Campanian–Maastrichtian–Danian) of Seymour Island, Antarctica. *Review of Palaeobotany and Palynology, 81,* 151–164.

Axelrod, D. I. (1970). Mesozoic paleogeography and early angiosperm history. *Botanical Review, 36,* 277–319.

Baksi, S. K., & Deb, U. (1981). Palynology of the upper Cretaceous of the Bengal Basin, India. *Review of Palaeobotany and Palynology, 31,* 335–365.

Barreda, V., & Archangelsky, S. (2006). The southernmost record of tropical pollen grains in the mid-Cretaceous of Patagonia, Argentina. *Cretaceous Research, 27,* 778–787.

Bell, C.D., Soltis, D.E., & Soltis, P.S. (2010). The age and diversification of the angiosperms re-revisited. *American Journal of Botany, 97,* 1296–1303.

Boyce, C. K., Brodribb, T., Field, T. S., & Zwieniecki, M. A. (2009). Angiosper leaf vein evolution was physiologically and environmentally transformative. *Proceedings of Royal Society of London B, 276,* 1771–1776.

Boyce, C. K., & Lee, J. E. (2017). Plant evolution and climate over geological timescales. *Annual Review of Earth and Planetary Sciences, 45,* 61–87.

Boyce, C. K., Lee, J. E., Feild, T. S., Brodribb, T. J., & Zwieniecki, M. A. (2010). Angiosperms helped put the rain in the rainforests: the impact of plant physiological evolution on tropical biodiversity. *Annals of the Missouri Botanical Garden, 97*(4), 527–540.

Brenner, G. J. (1963). The spores and pollen of the Potomac group of Maryland. *Bulletin Department of Geology, Mines and Water Resources, 27,* 1–215.

Brenner, G. J. (1976). Midlle Cretaceous floral provinces and early migration of angiosperms. In C. B. Beck (Ed.), *Origin and early evolution of angiosperms* (pp. 23–47). New York, Columbia University Press,

Brenner, G. J. (1996). Evidence for the earliest stage of angiosperm pollen evolution: a paleoequatorial section from Israel. In D. W. Taylor & L. J. Hickey (Eds.), *Flowering plant origin, evolution and phylogeny* (pp. 91–115). New York: Chapman & Hall, New York.

Brenner, G. J., & Bickoff, I. S. (1992). Palynology and age of the lower Cretaceous basal Kurnub Group from the coastal plain to the northern Negev of Israel. *Palynology, 16,* 137–185.

Buerki, S., Forest, F., & Alvarez, N. (2014). Proto-South-East Asia as a trigger of early angiosperm diversification. *Botanical Journal of the Linnean Society, 174,* 326–333.

Burger, D. (1980). *Palynological studies in the lower Cretaceous of the Surat Basin, Australia*. Bureau of Mineral Resources, Geology and Geophysics, Australian Government Publishing Service Canberra.

Burger, D. (1990). Early Cretaceous angiosperms from Queensland, Australia. *Review of Palaeobotany and Palynology, 65,* 153–163.

Burger, D. (1993). Early and middle Cretaceous angiosperm pollen grains from Australia. *Review of Palaeobotany and Palynology, 78,* 183–234.

Busch, A., & Zachgo, S. (2009). Flower symmetry evolution: Towards understanding the abominable mystery of angiosperm radiation. *BioEssays, 31,* 1181–1190.

Cantrill, D. J., & Nichols, G. J. (1996). Taxonomy and paleoecology of Early Cretaceous (late Albian) angiosperm leaves from Alexander Island, Antarctica. *Review of Palaeobotany and Palynology, 92,* 1–28.

Chaboureau, A. C., Sepulchre, P., Donnadieu, Y., & Franc, A. (2014). Tectonic-driven climate change and the diversification of angiosperms. *Proceedings of the National Academy of Sciences of the USA, 111,* 14066–14070.

Clarke, J., Warnock, R., & Donghue, P. (2011). Establishing a time-scale for plant evolution. *New Phytologists, 192,* 266–301.

Coiffard, C., Gomez, B., Daviero-Gomez, V., & Dilcher, D. L. (2012). Rise to dominance of angiosperm pioneers in European Cretaceous environments. *Proceedings of National Academy of Sciences of the USA, 109,* 20955–20959.

Cornet, B. (1989). Late Triassic angiosperm-like pollen from the Richmond rift basin of Virginia, USA. *Palaeontographica, Abteilung B, 213,* 37–87.

Cornet, B., & Habib, D. (1992). Angiosperm-like pollen from the ammonite-dated Oxfordian (Upper Jurassic) of France. *Review of Palaeobotany and Palynology, 71,* 269–294.

Couper, R. A. (1958). British Mesozoic microspores and pollen grains, a systematic and stratigraphie study. *Palaeontographica, Abteilung B, 103,* 75–179.

Crane, P. R. (1985). Phylogenetic relationship in seed plants. *Cladistics, 1,* 329–348.

Crane, P. R. (1987). Paleobotanical evidence on the early radiation of non-magnoliid dicotyledons. *Plant Systematics and Evolution, 162,* 165–191.

Crane, P. R., Friis, E. M., & Pedersen, K. R. (1995). The origin and early diversification of angiosperms. *Nature, 374,* 27–33.

Crane, P. R., & Lidgard, S. (1989). Angiosperm diversification and paleolatitudinal gradients in Cretaceous floristic diversity. *Science, 246,* 675–678.

Crepet, W. L., Nixon, K. C., & Gandolfo, M. A. (2004). Fossil evidence and phylogeny: The age of major angiosperm clades based on mesofossil and macrofossil evidence from Cretaceous deposits. *American Journal of Botany, 91,* 1666–1682.

Darwin, F., & Seward, A. C. (1903). *More letters of Charles Darwin. A record of his work in a series of hitherto unpublished letters,* vol. 1. London: John Murray.

de Boer, H. J., Eppinga, M. B., Wassen, M. J., & Dekker, S. C. (2012). A critical transition in leaf evolution facilitated the Cretaceous angiosperm revolution. *Nature Communications*. https://doi.org/10.1038/ncomms2217.

Dettmann, M. E. (1973). Angiosperm pollen from Albian to Turonian sediments of Eastern Australia. *Geological Society of Australia. Special Publication, 4,* 3–34.

Dettmann, M. E. (1989). Antarctica: Cretaceous cradle of austral temperate rainforests? *Geological Society, London, Special Publications, 47,* 89–105.

Dettmann, M. E. (1994). Cretaceous vegetation: the microfossil record. In R. S. Hill (Ed.), *History of the Australian vegetation: Cretaceous to recent* (pp. 143–170). Cambridge: Cambridge University Press.

Dettmann, M. E., & Thomson, M. R. A. (1987). Cretaceous palynomorphs from the James Ross Island area, Antarctica: a pilot study. *British Antarctica Survey Bulletin, 77,* 13–59.

Doyle, J. A. (1973). Fossil evidence on early evolution of the monocotyledons. *The Quarterly Review of Biology, 48,* 399–413.

Doyle, J. A. (1992). Revised palynological correlations of the lower Potomac Group (USA) and the cocobeach sequence of Gabon (Barremian-Aptian). *Cretaceous Research, 13,* 337–349.

Doyle, J. A., Biens, P., Doerenkamp, A., & Jardine, S. (1977). Angiosperm pollen from the pre-Albian lower Cretaceous of equatorial Africa. *Bulletin of Central Research and Exploration-Prod Elf-Aquitaine, 1,* 451–473.

Doyle, J. A., & Donoghue, M. J. (1986). Seed plant phylogeny and the origin of angiosperms: An experimental cladistics approach. *Botanical Review, 52,* 321–431.

Doyle, J. A., & Endress, P. K. (2014). Integrating Early Cretaceous fossils into the phylogeny of living angiosperms: ANITA lines and relatives of chloranthaceae. *International Journal of Plant Sciences, 175,* 555–600.

Doyle, J. A., & Hickey, L. J. (1976). Pollen and leaves from the mid-Cretaceous potomac group and their bearing on the early angiosperm evolution. In C. B. Beck (Ed.), *Origin and early evolution of angiosperms* (pp. 139–206). New York: Columbia University Press.

Doyle, J. A., Jardine, S., & Doerenkamp, A. (1982). Afropollis, a new genus of early angiosperm pollen, with notes on the Cretaceous palynostratigraphy and paleoenvironments of Northern Gondwana. *Bulletin des Centres de Recherches Exploration-Production Elf-Aquitaine, 6,* 39–117.

Doyle, J. A., & Robbins, E. I. (1977). Angiosperm pollen zonation of the continental Cretaceous of the Atlantic Coastal plain and its application to deep wells in the salisbury embayment. *Palynology, 1,* 43–78.

Drinnan, A. N., & Chambers, T. C. (1986). Flora of the lower Cretaceous Koonwarra fossil bed (Korumburra Group) South Gippsland, Victoria. *Association of Australasian Palaeontologists Memoir, 3,* 24–37.

Drinnan, A. N., & Crane, P. R. (1989). Cretaceous paleoboatny and its bearing on the biostratigraphy of Austral angiosperms. In T. N. Tayloy & E. L. Taylor (Eds.), *Antarctic paleobiology* (pp. 192–219). New York: Springer.

Eklund, H., Doyle, J. A., & Herendeen, P. S. (2004). Morphological phylogenetic analysis of living and fossil Cloranthaceae. *International Journal of Plant Science, 165,* 107–151.

Farrel, B. D. (1998). "Inordinate Fondness" explained: Why are there so many beetles? *Science, 281,* 555–559.

Feild, T. S., Arens, N. C., Doyle, J. A., Dawson, T. E., & Donoghue, M. J. (2004). Dark and disturbed: A new image of early angiosperm ecology. *Paleobiology, 30,* 82–107.

Field, R., Hawkins, B. A., Cornell, H. V., Currie, D. J., DinizFilho, J. A. F., Guégan, J.-F., et al. (2009). Spatial species-richness gradients across scales: A meta-analysis. *Journal of Biogeography, 36,* 132–147.

Friis, E. M., Crane, P. R., & Pedersen, K. R. (2011). *Early flowers and angiosperm evolution*. Cambridge: Cambridge University Press.

Friis, E. M., Pedersen, K. R., & Crane, P. R. (2000). Fossil floral structures of a basal angiosperm with monocolpate, reticulate-acolumellate pollen from the Early Cretaceous of Portugal. *Grana, 39,* 226–239.

Friis, E. M., Pedersen, K. R., & Crane, P. R. (2004). Araceae from the Early Cretaceous of Portugal: Evidence on the emergence of mono-cotyledons. *Proceedings of the National Academy of Sciences of the USA, 101,* 16565–16570.

Friis, E. M., Pedersen, K. R., & Crane, P. R. (2010a). Cretaceous diversification of angiosperms in the western part of the Iberian Peninsula. *Review of Palaeobotany and Palynology, 162,* 341–361.

Friis, E. M., Pedersen, K. R., & Crane, P. R. (2010b). Diversity in obscurity: Fossil flowers and the early history of angiosperms. *Philosophical Transactions of Royal Society B, 365,* 369–382.

Friis, E. M., Pedersen, K. R., & Schönenberger, J. (2006). *Normapolles* plants: A complex of extinct fagalean lineages. *Plant Systematics and Evolution, 260,* 107–140.

Furness, C. A., Magallon, S., & Rudall, P. J. (2007). Evolution of endoapertures in early-divergent eudicots, with particular reference to pollen morphology in Sabiaceae. *Plant Systematics and Evolution, 263,* 77–92.

Furness, C. A., & Rudall, P. J. (2004). Pollen aperture evolution: A crucial factor for eudicot success? *Trends in Plant Science, 9,* 154–158.

Góczán, F. M., & Júhász, F. (1984). Monosulcate pollen grains of angiosperms from Hungarian Albian sediments I. *Acta Botanica Hungaria, 30,* 289–319.

Goswami, S., Meena, K. L., Das, M., & Guru, B. C. (2008). Upper Gondwana palynoflora of Mahanadi Master Basin, Orissa, India. *Acta Palaeobotanica, 48,* 171–181.

Gübeli, A. A., Hochuli, P. A., & Wildi, W. (1984). Lower Cretaceous turbiditic sediments from the Rift chain (northern Morocco): palynology, stratigraphy and paleogeographic setting. *Geologische Rundschau, 73,* 1081–1114.

Guler, M. V., & Archangelsky, S. (2006). Albian dinoflagellate cysts from the Kachaike Formation, Austral Basin, southwest Argentina. *Revista del Museo Argentino de Ciencias Naturales, Nueva Serie, 8,* 179–184.

Hall, R. (2009). Hydrocarbon basins in SE Asia: Understanding why they are there. *Petroleum Geoscience, 15,* 131–146.

Hallam, A. (1992). *Phanerozoic sea-level changes.* New York: Columbia University Press.

Haq, B. U., Hardenbol, J., & Vail, P. R. (1987). Chronology of fluctuating sea levels since the Triassic (250 million years ago to present). *Science, 235,* 1156–1167.

Heimhofer, U., & Hochuli, P. A. (2010). Early Cretaceous angiosperm pollen from a low-latitude succession (Araripe Basin, NE Brazil). *Review of Palaeobotany and Palynology, 161,* 105–126.

Heimhofer, U., Hochuli, P. A., Burla, S., Dinis, J. M. L., & Weissert, H. (2005). Timing of Early Cretaceous angiosperm diversification and possible links to major paleoenvironmental change. *Geology, 33,* 141–144.

Heimhofer, U., Hochuli, P. A., Burla, S., & Weissert, H. (2007). New records of Early Cretaceous angiosperm pollen from Portuguese coastal deposits: Implications for the timing of the Early angiosperm radiation. *Review of Palaeobotany and Palynology, 144,* 39–76.

Herngreen, G. F. W. (1996). Cretaceous palynofloral provinces: a review. *Palynology: principles and applications. American Association of Stratigraphic Palynologists Foundation, 3,* 1157–1188.

Herngreen, G. F. W., & Chlonova, A. F. (1981). Cretaceous microfloral provinces. *Pollen et Spores, 23,* 441–557.

Hickey, L. J., & Doyle, J. A. (1977). Early Cretaceous fossil evidence for angiosperm evolution. *The Botanical Review, 43,* 3–104.

Hill, R. S., & Scriven, L. J. (1995). The angiosperm-dominated woody vegetation of Antarctica: A review. *Review of Palaeobotany and Palynology, 86,* 175–198.

Hochuli, P. A., & Feist-Burkhardt, S. (2004). A boreal Early cradle of Angiosperms? Angiosperm-like pollen from the Middle Triassic of the Barents Sea (Norway). *Journal of Micropalaeontology, 23,* 97–104.

Hochuli, P. A., & Feist-Burkhardt, S. (2013). Angiosperm-like pollen and Afropollis from the Middle Triassic (Anisian) of the Germanic Basin (northern Switzerland). *Frontiers in Plant Science, 4,* 344.

Hochuli, P. A., Heimhofer, U., & Weissert, H. (2006). Timing of Early angiosperm radiation: Recalibrating the classical succession. *Journal of the Geological Society, 163,* 587–594.

Hochuli, P. A., & Kelts, K. (1980). Palynology of middle Cretaceous black clay facies from DSDP sites 417 and 418 of the Western North Atlantic. *Initial Reports of the Deep Sea Drilling Project, 51–53,* 897–955.

Hofmann, C. C., & Zetter, R. (2010). Upper Cretaceous sulcate pollen from the Timerdyakh formation, Vilui Basin (Siberia). *Grana, 49,* 170–193.

Hu, X. M., Jansa, L., & Wang, C. S. (2008). Upper Jurassic—lower Cretaceous stratigraphy in south-eastern Tibet: A comparison with the western Himalayas. *Cretaceous Research, 29,* 301–315.

Hughes, N. F. (1994). *The enigma of angiosperm origins.* Cambridge: Cambridge University Press.

Hughes, N. F., Drewry, G. E., & Laing, J. F. (1979). Barremian earliest angiosperm pollen. *Palaeontology, 22,* 513–553.

Hughes, N. F., & McDougall, A. B. (1990). Barremian-Aptian angiospermid pollen records from southern England. *Review of Palaeobotany and Palynology, 65,* 145–151.

Jha, N., Prakash, N., & Joshi, H. (2016). Integrated paleobotanical and palynological analysis of subsurface Gondwana sedimentary succession (Jurassic–Cretaceous) in Jangareddygudem area, Chintalapudi Sub-basin, South India: Stratigraphical and phytogeographical implications. *Palaeoworld, 26,* 173–193.

Kemp, E. M. (1968). Probable angiosperm pollen from British Barremian to Albian strata. *Palaeontology, 11,* 421–434.

Krassilov, V. A. (1997). *Angiosperm origins: Morphological and ecological aspects.* Sophia: Pensoft.

Labandeira, C. C., Dilcher, D. L., Davis, D. R., & Wagner, D. L. (1994). Evolution Ninety-seven million years of angiosperm-insect association: Paleobiological insights into the meaning of coevolution. *Proceedings of National Academy of Sciences of the USA, 91,* 12278–12282.

Lidgard, S., & Crane, P. R. (1990). Angiosperm diversification and Cretaceous floritic trends: a comparison of palynofloras and leaf macrofloras. *Paleobiology, 16,* 77–93.

Martínez, C., Madriñán, S., Zavada, M., & Jaramillo, A. C. (2013). Tracing the fossil pollen record of Hedyosmum (Chloranthaceae), an old lineage with recent Neotropical diversification. *Grana, 52,* 161–180.

Martínez, G., Panda, K., Köhler, C., & Slotkin, R. K. (2016). Silencing in sperm cells is directed by RNA movement from the surrounding nurse cell. *Nature Plants, 2,* 16030.

McLoughlin, S. (2001). The breakup history of Gondwana and its impact on pre-Cenozoic floristic provincialism. *Australian Journal of Botany, 49,* 271–300.

Medina, F., Archangelsky, S., Guler, V., Archangelsky, A., & Cárdenas, O. (2008). Estudio bioestratigráfico integrado del perfil La Horqueta (límite Aptiano-Albiano), lago Cardiel, Patagonia, Argentina. *Revista del Museo Argentino de Ciencias Naturales, NS, 10,* 273–289.

Mendes, M. M., Grimm, G. W., Pais, J., & Friis, E. M. (2014). Fossil *Kajanthuslusitanicus* gen. et sp. nov. from Portugal: Floral evidence for Early Cretaceous Lardizabalaceae (Ranunculales, basal eudicot). *Grana, 53,* 283–301.

Mishra, S., & Prasad, V. (2020). Upper Gondwana (Jurassic to Early Cretaceous) palynoflora of India: Its correlation with other Gondwana continents andphytogeographical implications. In G. V. R. Prasad & R. Patnaik (Eds.), *Biological consequences of plate tectonics: New perspectives on post-Gondwanaland break-up* (pp. 31–66). Cham: Springer.

Morley, R. J. (2000). *Origin and evolution of tropical rain forests.* Wiley.

Muller, J. (1981). Fossil pollen records of extant angiosperms. *Botanical Review, 47,* 1–142.

Muller, J. (1984). Significance of fossil pollen for angiosperm history. *Annals of the Missouri Botanical Garden, 71,* 419–443.

Narváez, P. L., Prámparo, M. B., & Sabino, I. F. (2014). First palynologic record of the Cretaceous La Yesera Formation (Salta Group), Northwestern Argentina. *Revista Brasileria de Paleontologia, 17,* 141–156.

Nichols, D. J., Matsukawa, M., & Ito, M. (2006). Palynology and age of some Cretaceous nonmarine deposits in Mongolia and China. *Cretaceous Research, 27,* 241–251.

Pais, J., & Reyre, Y. (1981). Problèmes poses par la population sporopollinique d'un niveauiplantes de laserie de Buarcos (Portugal). *Sociedade Geologica de Portugal, Boletim, 22,* 35–40.

Penny, J. H. J. (1986). An Early Cretaceous angiosperm pollen assemblage from Egypt. *Special Papers in Palaeontology, 35,* 121–134.

Penny, J. H. J. (1988). Early Cretaceous striate tricolpate pollen from the borehole Mersa Matruh 1, north West Desert, Egypt. *Journal of Micropalaeontology, 7,* 201–215.

Penny, J. H. J. (1991). Early Cretaceous angiosperm pollen from the borehole Mersa Matruh 1, North West Desert, Egypt. *Palaeontographica, Abteilung B, 222,* 31–88.

Pérez Leytón, M. (1987). Datospalinológicosdel Cretácico Superior de la secciónestratigráfica de Carata. In *Potosi, Bolivia. IV Congreso Latinoamericano de Paleontologia* (pp. 739–756).

Playford, G. (1971). Palynology of lower Cretaceous (Swan River) strata of Saskatchewan and Manitoba. *Palaeontology, 14,* 533–565.

Pocock, S. A., & Vasanthy, G. (1988). *Cornetipollisreticulata*, a new pollen with angiospermid features from Upper Triassic (Carnian) sediments of Arizona (USA), with notes on Equisetosporites. *Review of Palaeobotany and Palynology, 55,* 337–356.

Pole, M. (2000). Dicotyledonous leaf macrofossils from the latest Albian-earliest Cenomanian of the Euromanga Basin, Queensland, Australia. *Paleontological Research, 4,* 39–52.

Prámparo, M. B. (1990). Palynostratigraphy of the Lower Cretaceous of the San Luis Basin, Argentina. Its place in the Lower Cretaceous flora provinces pattern. *Neues Jahrbuch für Geologie und Paläontologie, Abhandlungen, 181,* 255–266.

Prámparo, M. B. (1994). Lower Cretaceous palynoflora of the La Cantera Formation, San Luis Basin: Correlation with other Cretaceous palynofloras of Argentina. *Cretaceous Research, 15,* 193–203.

Prámparo, M. B. (1999). *Granos de polen de primitivas angiospermas en el Cretácico Inferior de la Cuenca de San Luis y su distribución en otras cuencas cretácicas de Argentina.* 5th Symposium on the Cretaceous of Brazil, Sao Paulo, (pp. 539–543).

Prámparo, M. B., Quattrocchio, M., Gandolfo, M. A., Zamaloa, M. del C., & Romero, E. (2007). Historia evolutiva de las angiospermas (Cretácico-Paleógeno) en Argentina a través de los registros paleoflorísticos. *Publicación Especial 1, Ameghiniana 50° Aniversario,* 157–172.

Prasad, B., & Pundeer, B. S. (1999). Biostratigraphy of the exposed Gondwana and Cretaceous rocks of Krishna-Godavari Basin, India. *Journal of the Palaeontological Society of India, 44,* 91–117.

Prasad, V., Farooqui, A., Murthy, S., Sarate, O. S., & Bajpai, S. (2018). Palynological assemblage from the Deccan Volcanic Province, central India provides insights into Early history of angiosperms and the terminal Cretaceous paleogeography of peninsular India. *Cretaceous Research, 86,* 186–198.

Quattrocchio, M. E., Martínez, M. A., Carpinelli Pavisich, A., & Volkheimer, W. (2006). Early Cretaceous palynostratigraphy, palynofacies and paleoenvironments of well sections in northeastern Tierra del Fuego, Argentina. *Cretaceous Research, 27,* 584–602.

Raven, P. H., & Axelrod, D. I. (1972). Plate tectonics and Australasian paleobiogeography. *Science, 186,* 1379–1386.

Regali, M. S. P., Uesugui, N., & Santos, A. A. (1974a). Palinologia dos sedimentos mesQ-cenozóicos do Brasil. I. *Boletim Técnico da Petrobras, 17,* 177–190.

Regali, M. S. P., Uesugui, N., & Santos, A. S. (1974b). Palinologia dos sedimentos mesQ-cenozóicos do Brasil. II.*Boletim Técnico da Petrobras, 17,* 263–301.

Regali, M. S. P., & Viana, C. F. (1989). Sedimentos do Neojurassico-Eocretaceo do Brasil: Idade e Correlacao com a Escala Internacional. *Petrobrás,* 95.

Retallack, G., & Dilcher, D. L. (1981). A coastal hypothesis for the dispersal and rise to dominance of flowering plants. In K. J. Niklas (Ed.), *Paleobotany, paleoecology, and evolution, 2* (pp. 27–77). New York: Praeger.

Romero, E. J. (1993). South American paleofloras. In P. Goldblatt (Ed.), *Relationships between Africa and South America* (pp. 62–85). New Haven and London: Yale University Press.

Romero, E. J., & Archangelsky, S. (1986). Early Cretaceous angiosperm leaves from southern South America. *Science, 234* (4783), 1580–1582.

Sahni, A., Venkatchala, B. S., Kar, R. K., Ranikanth, A., Prakash, T., Prasad, G. V. R., et al. (1996). New palynological data from the Deccan Intertrappean Beds: Implications for the latest records of Dinosaurs and Synchronous initiation of volcanic activity in India. *Memoir Geological Society of India, 37,* 267–283.

Samant, B., & Mohabey, D. M. (2014). Deccan volcanic eruptions and their impact on flora: Palynological evidence. *Geological Society of America, Special Paper, 505,* 171–191.

Sanderson, M. J., & Doyle, J. A. (2001). Sources of error and confidence intervals in estimating the age of angiosperms from rbcL and 18s rDNA data. *American Journal of Botany, 88,* 1499–1516.

Schrank, E. (1983). Scanning electron and light microscopic investigations of angiosperm pollen from the Lower Cretaceous of Egypt. *Pollen et Spores, 25,* 213–242.

Schrank, E., & Awad, M. Z. (1990). Palynological evidence for the age and depositional environment of the Cretaceous Omdurman Formation in the Khartoum area, Sudan. *Berliner Geowiss. Abh., A, 120,* 169–182.

Schrank, E., & Mahmoud, M. S. (2002). Barremian angiosperm pollen and associated palynomorphs from the Dakhla Oasis area, Egypt. *Palaeontology, 45,* 33–56.

Scotese, C. R. (2001). *Atlas of Earth History* (vol. 1, pp. 1–52). Paleogeography. Arlington, Texas: PALEOMAP Project.

Smith, S. A., Beaulieu, J. M., & Donoghue, M. J. (2010). An uncorrelated relaxed-clock analysis suggests an earlier origin for flowering plants. *Proceedings of the National Academy of Sciences of the USA, 107,* 5897–5902.

Sun, G., & Dilcher, D. L. (2002). Early angiosperms from the Lower Cretaceous of Jixi, eastern Heilongjiang, China. *Review of Palaeobotany and Palynology, 121,* 91–112.

Sun, G., Dilcher, D. L., & Zheng, S. (2008). A review of recent advances in the study of early angiosperms from northeastern China. *Palaeoworld, 17,* 166–171.

Sun, G., Zheng, S., Wang, X., Mei, S., & Liu, Y. (2000). Subdivision of developmental stages of early angiosperms from NE China. *Acta Palaeontologica Sinica, 39,* 186–199.

Takhtajan, A. (1969). Flowering plants: origin and dispersal. 310pp. (transl. from Russian by C. Edinburgh.: Jeffrey) Oliver & Boyd. *Gen_monocots, Gen_dicots, Evolution, Phylogeny (PMBD, 185405557).*

Takhtajan, A. L. (1966/1967). *System and phylogeny of flowering plants.* Nauka, Moskva-Leningrad (in Russian).

Tiwari, R. S. (1996). Palynoevent stratigraphy in Gondwana sequence of India. In *IXth Int. Gond. Symp., Hyderabad, India* (pp. 3–19).

Tiwari, R. S., & Tripathi, A. (1995). Palynological assemblages and absolute age relationship of intertrappean beds in the Rajmahal Basin, India. *Cretaceous Research, 16,* 53–72.

Trincão, P. R. P. (1990). *Esporos e polenes do Cretácico Inferior (Berriasiano-Aptiano) de Portugal: Paleontologia e biostratigrafia.* Ph.D. Thesis, Universidale Nova de Lisboa, Lisboa, Potugal.

Tripathi, A. (1993). Advent of angiospermous pollen in India and its spatial relationship in Gondwanaland. *Proceedings of 8th International Gondwana Symposium, Tasmania (Australia)*, pp. 277–289.

Tripathi, A. (1998). Pollen showing angiospermoid character from Late Permian of Peninsular India. *Geophytology, 27,* 7–15.

Tripathi, A. (2009). Palynology of the Brahmani River Section, Talcher Coalfield, Orissa, India. *Journal of the Palaeontological Society of India, 54,* 179–187.

Vajda, V., & Bercovici, A. (2014). The global vegetation pattern across the Cretaceous-Paleogene mass extinction interval: A template for other extinction events. *Global and Planetary Change, 122,* 29–49.

Vajda-Santivanez, V. (1999). Miospores from upper Cretaceous-Paleocene strata in northwestern Bolivia. *Palynology, 23,* 181–196.

Vakhrameev, V. A. (1991). *Jurassic and Cretaceous floras and climates of the earth.* Cambridge: Cambridge University Press.

Vallati, P. (1995). Una microflora con Afropollis (pollen de angiosperma) en el Cretácico Inferior de la Cuenca Neuquina. VI Congreso Argentino de Paleontología y Bioestratigrafía. *Trelew, 1994,* 277–290.

Vallati, P. (2006). Las primeras angiospermas en el Cretácico de la Cuenca Neuquina (centrooeste de Argentina): Aspectos geológicos relacionados. *Revista Brasileira de Paleontologia, 9,* 83–92.

Vasanthy, G., Cornet, B., & Pocock, S. A. J. (2004). Evolution of proangiosperms during Late Triassic: Pre-Cretaceous pollen trends towards mono-and dicotyledonous taxa diversification. *Geophytology, 33*(1–2), 99–113.

Veblen, T. T., Schlegel, F. M., & Escobar, B. R. (1980). Structure and dynamics of old-growth Nothofagus forests in the Valdivian Andes, Chile. *The Journal of Ecology, 68,* 1–31.

Venkatchala, B. S., & Sharma, K. D. (1974a). Palynology of the Cretaceous sediments from the subsurface of Pondicherry area (Cauvery Basin). *New Botanist, 1,* 170–200.

Venkatchala, B. S., & Sharma, K. D. (1974b). Palynology of the Cretaceous sediments from the subsurface of Vridhachalam area (Cauvery Basin). *Geophytology, 4,* 153–183.

Venkatchala, B. S., & Sharma, K. D. (1984). Palynological zonation in subsurface sediments in Narsapur Well No. 1, Godavari-Krishna Basin, India. In *Proceedings of the Xth Indian Colloquium on Micropalaeontology and Stratigraphy* (pp. 445–466).

Vijaya, (1997). Palynoflora from subsurface lower Cretaceous intertrappean beds in Domra Sub-basin of the Raniganj Gondwana Basin, West Bengal, India. *Cretaceous Research, 18*(1), 37–50.

Vijaya, (2011). Palynostratigraphy of subsurface Upper Permian and Mesozoic Succession, Rakshitpur area, Raniganj Coalfield, West Bengal. *Palaeoworld, 20,* 61–74.

Vijaya, & Kumar, S. (2002). Palynostratigraphy of the Spiti Shale (Oxfordian-Berriasian) of Kumaon Tethys Himalaya, Malla Johar area, India. *Review of Palaeobotany and Palynology, 122*(3–4), 143–153.

Vijaya & Prasad, G. V. R. (2001). Age of Kota Formation, Pranhita Godavari Valley, India: a palynological approach. *Journal of the Palaeontological Society of India, 46,* 77–93.

Volkheimer, W., & Salas, A. (1975a). Die älteste Angiospermen-Palynoflora Argentiniens von der Typuslokalität der unterkretazischen Huitrín-Folge des Neuquén Beckens. Mikrofloristische Assoziation und biostratigraphische Bedeutung. *Neues Jahrbuch Geologie Paläontologie, Monatshefte, 7,* 424–436.

Volkheimer, W., & Salas, A. (1975b). Estudio palinológico de la Formación Huitrín, Cretácico de la Cuenca Neuquina, en su localidad tipo. *Congreso Geológico Argentino, Bahía Blanca, 6,* 433–456.

Walker, J. W., & Walker, A. G. (1984). Ultrastructure of Lower Cretaceous angiosperm pollen and the origin and early evolution of flowering plants. *Annals of the Missouri Botanical Garden, 71,* 464–521.

Zaklinskaya, E. D. (1981). Phylogeny and classification of the *Normapolles. Review of Palaeobotany and Palynology, 35,* 139–147.

Zaavada, M. S. (1990). The ultrastructure of three monosulcate pollen grains from the Triassic Chinle Formation, western United States. *Palynology, 14,* 41–45.

Ziegler, A. M., Eshel, G., Rees, P. M., Rothfus, T. A., Rowley, D. B., & Sunderlin, D. (2003). Tracing the tropics across land and sea: Permian to present. *Lethaia, 36,* 227–254.

Chapter 6
Anuran Lissamphibian and Squamate Reptiles from the Upper Cretaceous (Maastrichtian) Deccan Intertrappean Sites in Central India, with a Review of Lissamphibian and Squamate Diversity in the Northward Drifting Indian Plate

Jean-Claude Rage, Guntupalli V. R. Prasad, Omkar Verma, Ashu Khosla, and Varun Parmar

Abstract The Maastrichtian intertrappean beds of Kisalpuri and Kelapur, India, have yielded new amphibian and squamate reptile fossils that increase our knowledge of these groups. In addition to focusing on these two localities, we review the faunas of amphibians and squamates from other Maastrichtian intertrappean beds. In the Maastrichtian sites of India, aside from a possible salamander, amphibians are all anurans. Costata (formerly Discoglossidae) and Gobiatidae may be present and Neobatrachia are represented by four or five morphotypes. Importantly, one of the anurans (?Gobiatidae) appears identical to a form from the Late Cretaceous of Central Asia. Squamates are represented by lizards and snakes. Lizards comprise an Anguidae and five or six indeterminate taxa, including possible scincomorphs. Snakes include a primitive (possible stem) form, three Madtsoiidae, one Nigerophiidae, perhaps a booid, and a peculiar, unnamed taxon. These assemblages from Maastrichtian sites in India include taxa of both Laurasian and Gondwanan origins, but most taxa are still unidentifiable at the familial level and below.

Keywords Frogs • Lizards • Snakes • Kelapur • Kisalpuri • Paleobiogeography • Gondwana • Laurasia

Introduction

Late Cretaceous continental vertebrate faunas of India assume great significance because they provide critical information on the biodiversity and evolution of vertebrates and the paleobiogeography of the Indian plate during its northward journey. In India, continental Upper Cretaceous deposits are known from two geographically separate regions: the Cauvery basin in southern India and the Deccan Volcanic Province that occupies about two-thirds of peninsular India. The Upper Cretaceous (Maastrichtian) Kallamedu Formation is the only continental sequence within the predominantly shallow marine depositional setting of the Cauvery basin and, until now, it has yielded few vertebrate fossils (Yadagiri and Ayyasami 1989; Gaffney et al. 2001; Goswami et al. 2013; Prasad et al. 2013; Halliday et al. 2016). By comparison, the Deccan volcano-sedimentary sequences (Lameta Formation [=infratrappean] and intertrappean beds) have been extensively studied in recent years and a good diversity of fish, amphibians, turtles, crocodiles, snakes, lizards, dinosaurs, and mammals is now known (e.g., Khosla and Sahni 2003; Prasad 2012; Prasad and Sahni 2014).

As far as amphibians are concerned, Owen (1847) was the first to report complete anuran skeletons described as *Rana pusilla* from the intertrappean beds of Bombay. Subsequently, Noble (1930) re-assigned the species to his new

Jean-Claude Rage—*Deceased after submission on March 30, 2018*

J.-C. Rage
Centre de Recherches sur la Paléobiodiversité et les Paléoenvironnements CR2P, UMR CNRS 7207, Sorbonne Universités, MNHN/CNRS/UPMC, Muséum national d'Histoire naturelle, 57 rue Cuvier, CP 38, 75005 Paris, France

G. V. R. Prasad (✉)
Department of Geology, Centre for Advanced Studies, University of Delhi, Delhi, 110007, India
e-mail: guntupalli.vrprasad@gmail.com

O. Verma
Geology Discipline Group, School of Sciences, Indira Gandhi National Open University, New Delhi, 110068, India

A. Khosla
Centre of Advanced Study in Geology, Panjab University, Chandigarh, 160014, India

V. Parmar
Department of Geology, University of Jammu, Jammu, 180006, India

G. V. R. Prasad and R. Patnaik (eds.), *Biological Consequences of Plate Tectonics: New Perspectives on Post-Gondwana Break-up—A Tribute to Ashok Sahni*,
Vertebrate Paleobiology and Paleoanthropology Series, https://doi.org/10.1007/978-3-030-49753-8_6

genus *Indobatrachus* as *I. pusillus*. Chiplonkar (1940) and Verma (1965) described additional fossils and each created a new species of *Indobatrachus*. Špinar and Hodrová (1985), however, lumped all the species in *Indobatrachus pusillus* and referred the genus to Myobatrachidae (as Myobatrachinae, subfamily of Leptodactyidae); Roček and Rage (2000) regarded this family assignment as not settled definitely, but they did not reject it. Sahni et al. (1982) reported the presence of a pelobatid frog from the intertrappean beds of Nagpur. Prasad and Rage (1991, 1995, 2004) described cranial and postcranial bones belonging to Gobiatinae (included in Discoglossidae at that time), ?Hylidae, Ranoidea, and ?Leptodactylidae or Hemisotidae from the Upper Cretaceous (Maastrichtian) intertrappean beds of Naskal, a richly fossiliferous site situated on the southern fringe of the Deccan Traps.

Squamate reptiles are less frequently known from the Late Cretaceous of India, and are mainly comprised of snakes (Rage and Prasad 1992; Prasad and Rage 1995; Rage et al. 2004; Wilson et al. 2010; Mohabey et al. 2011). Lizards have been mostly reported as indeterminate taxa, except for Anguidae and an indeterminate scincomorph (Gayet et al. 1984; Prasad and Khajuria 1990; Prasad and Rage 1995).

Here we report isolated bones of anurans and lizards from microvertebrate-yielding intertrappean localities exposed at Kisalpuri and Kelapur villages in central India. The richer locality, Kisalpuri, is within a section exposed 1.5 km southwest of Kisalpuri (Fig. 6.1) in Dindori District, Madhya Pradesh (northeastern Deccan Volcanic Province). The Kelapur site, is located 3 km south of Kelapur village in Yeotmal District, Maharashtra (central Deccan Volcanic Province, Fig. 6.1). The Kisalpuri intertrappean section is 4.5 m thick and consists of clays, mudstones, and siltstones. The amphibian and squamate fossils described here come from a 70 cm thick siltstone bed. The Kisalpuri intertrappean site occurs about 700 km north of the Naskal intertrappean beds, which have produced a diverse assemblage of terrestrial and lacustrine vertebrates. The Kisalpuri intertrappean beds also yielded a diversified continental to brackish water microvertebrate assemblage consisting of fish: *Igdabatis indicus* Prasad and Cappetta, 1993, *Rhombodus* sp., *Lepisosteus* cf. *L. indicus* Woodward, 1908, Pycnodontidae gen. et sp. indet., Osteoglossidae gen. et sp. indet., Siluriformes indet.; anurans; lizards; turtles: Bothremydidae indet.; Crocodiles: Crocodilia indet., Dyrosauridae gen. et sp. indet.; turtle and crocodilian eggshells (Khosla et al. 2004; 2009; Prasad et al. 2015; Verma et al. 2017); dinosaur oospecies: *Megaloolithus baghensis* Khosla and Sahni, 1995 and *Subtiliolithus kachchhensis* Khosla and Sahni, 1995; and mammals: *Deccanolestes hislopi* Prasad and Sahni, 1988, *D. robustus* Prasad et al., 1994, *D. narmadensis* Prasad et al., 2010, *Kharmerungulatum vanvaleni* Prasad et al., 2007a, and *Bharattherium bonapartei* Prasad et al., 2007b. The anuran and lizard remains reported here were recovered from the intertrappean beds of Kisalpuri by screen-washing 3000 kg of sediments. The intertrappean beds of Kisalpuri have been assigned a Maastrichtian age based on the occurrence of Maastrichtian fish *Igdabatis indicus* Prasad and Cappetta, 1993, and ostracods *Paracypretta bhatiai* Khosla and Sahni, 2000 and *Paracandona jabalpurensis* Sahni and Khosla, 1994 (Khosla et al. 2004).

The intertrappean beds of Kelapur are located about 260 km north and 440 km south of Naskal and Kisalpuri intertrappean beds, respectively (Fig. 6.1). The intertrappean section of Kelapur is about 2 m in thickness and comprises clays, hard marl, and limestone. Screen-washing of 1000 kg of the gastropod shell-rich, basal clay led to the recovery of a moderately rich microvertebrate assemblage including fish (*Igdabatis*, Lepisosteidae gen. et sp. indet., Osteoglossidae gen. et sp. indet., Siluriformes indet., and Pycnodontiformes indet.), snakes (Madtsoiidae gen. et sp. indet. and the nigerophiid *Indophis sahnii*, Rage et al. 2004), and dinosaur egg shells, along with invertebrates (ostracods and gastropods). Additionally, anurans and lizards also have been recovered from the Kelapur intertrappean section. These intertrappean beds have been dated as Maastrichtian based on ostracods and the close similarity of the microvertebrate fauna to those of Naskal (Prasad and Rage 2004) and Kisalpuri (Khosla et al. 2004) sites.

Repository

The fossil specimens described here are deposited in the Vertebrate Paleontology Laboratory of the Department of Geology, University of Delhi, Delhi, India and bear catalogue numbers with the acronym DUGF. The acronym VPL/JU refers to the vertebrate paleontological collections of Jammu University, India.

Systematic Paleontology

Ambhibia Linnaeus, 1758
Lissamphibia Haeckel, 1866
Anura Fischer von Waldheim, 1813

The only amphibians found at Kisalpuri and Kelapur are anurans. All taxa are represented by disarticulated bones that cannot be reliably associated. Ilia are the most commonly

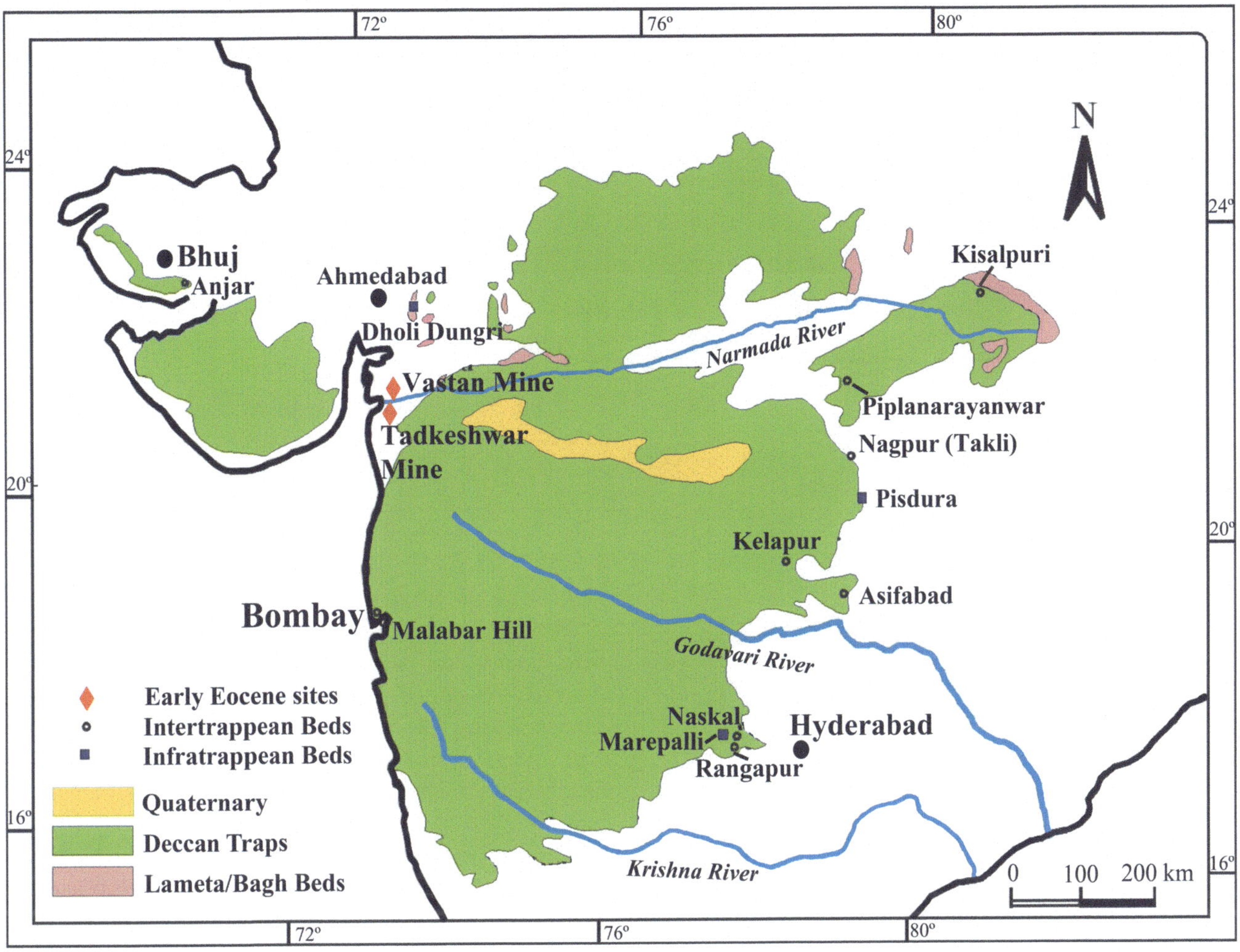

Fig. 6.1 Location map of amphibian and squamate fossil yielding Upper Cretaceous (Maastrichtian) and Lower Eocene localities in Central and Western India

recovered bones and provide the main source of taxonomic information (Roček et al. 2013). We also used ilia to estimate minimum taxonomic diversity. However, other bones should not be neglected. Some show peculiar morphologies and those may be of interest for future studies.

It is worth noting that in a previous study, dealing with the anurans of the Maastrichtian of Naskal, we used ilia to make tentative identifications at the family level (e.g., "?Leptodactylidae or Hemisotidae"; Prasad and Rage 2004). However, in view of our poor knowledge of Mesozoic frogs and of the osteology of taxa from the southern continents, here we refrain from assigning any fossils to families and instead we distinguish them as "morphotypes", as did Roček et al. (2010, 2013), and, where necessary, "forms" as subdivisions of morphotypes, it being understood that each morph and/or form presumably represents one that remains to be identified precisely.

?Costata Lataste, 1879
(Fig. 6.2)

Frost et al. (2006) divided Discoglossidae Günther, 1858 into two families, Alytidae Fitzinger, 1843 and Bombinatoridae Gray, 1825 and they replaced the name Discoglossidae with Costata Lataste, 1879. Costata were formerly reported from the Cretaceous of India as Discoglossidae (Prasad and Rage 1991, 1995).

The morphology of the ilia of Costata is not homogenous. A dorsal crest is either present or absent. Where the crest is present, its posterior border is thickened and forms the *tuber superius* as is the case in all anurans reported from Kisalpuri and Kelapur. However, the ilia of Costata are characterized by the elongation of their ischiadic process.

Horizon and Locality: The Upper Cretaceous Deccan intertrappean beds of Kisalpuri, Dindori District (Madhya Pradesh) and Kelapur, Yeotmal District (Maharashtra).

Referred Material: Three isolated ilia (DUGF/484 from Kisalpuri, DUGF/526 and 527 from Kelapur), and likely two poorly preserved ilia (DUGF/528 and 529 from Kelapur).

Description: None of these specimens is complete, but they collectively document the structure of the ilium. A low,

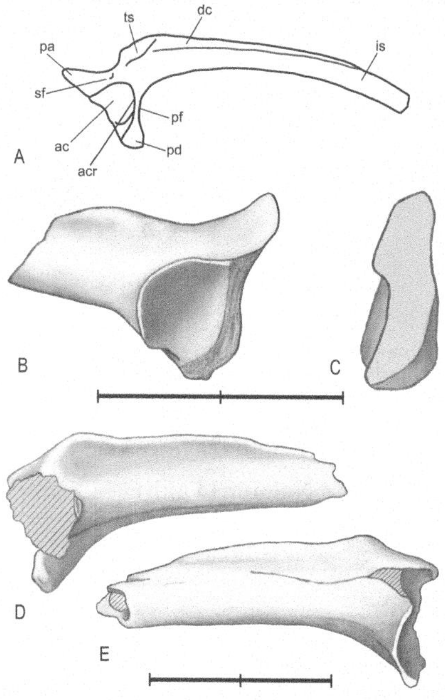

Fig. 6.2 Descriptive terminology of generalized anuran right ilium (not to scale) in lateral **A** view (Adapted from fig. 1A of Folie et al. 2013). **B–E.** Ilia of ?Costata from the Maastrichtian of India left ilium (DUGF/527) from Kelapur in lateral (**B**) and posterior (**C**) views; **D–E.** right ilium (DUGF/484) from Kisalpuri in medial (**D**) and lateral (**E**) views. Abbreviations: ac, acetabulum; acr, acetabular rim; dc, dorsal crest; is, iliac shaft; pa, *pars ascendens*; pd, *pars descendens*; pf, preacetabular fossa; sf, supraacetabular fossa; ts, *tuber superius*. Scale bars represent 2 mm for **B–E**

but distinct dorsal crest is present on the iliac shaft; it is recurved medially. The border of the posterior most part of the dorsal crest is thickened, forming the *tuber superius* (i.e., dorsal prominence + dorsal protuberance, sensu Gómez and Turazzini 2016). The *tuber* either slightly protrudes dorsally (e.g., DUGF/484) or is level with the dorsal border of the crest (e.g., DUGF/527). The iliac shaft has an ovoid cross-section. The *pars ascendens* (supraacetabular expansion) is well developed. It ends posteriorly as an elongate ischiadic process that is relatively well preserved only in DUGF/527 (Fig. 6.2 B); its extremity is broken and its full posterior length remains unknown, but it clearly was relatively long and directed posterodorsally. The dorsal border of the *pars ascendens* (including the ischiadic process) is clearly inclined posterodorsally. A shallow but relatively broad supraacetabular fossa is present on the *pars ascendens*, just posterior to the *tuber superius*. The *pars descendens* (subacetabular expansion) is reduced to such an extent that, in lateral view, it is concealed by the overhanging anteroventral part of the acetabular rim. The medial face of the acetabular area is largely concealed by matrix in DUGF/484 (Fig. 6.2D), but it may be stated with confidence that an interiliac tubercle (interiliac synchondrosis) is lacking; however, the posterior sutural surface for the ischium is relatively thick. The ilia are not strongly waisted between the shaft and acetabular part.

Discussion: A dorsal crest is present on the ilia of all pipids and ranoids, as well as in several members of the paraphyletic "leptodactylids" (including hemiphractines that were formerly assigned to Hylidae), some alytids, and myobatrachoids. In pipids, the *tuber superius* is a distinct process posterior to and separate from the dorsal crest; this morphology is quite different from that of the fossils under consideration. On the other hand, as in the fossils from Kisalpuri and Kelapur, in ranoids, "leptodactylids", and alytids that have a dorsal crest, the *tuber superius* is part of the crest, i.e., it is formed by the thickening of the posterior termination of the crest (Prasad and Rage 2004; Gómez and Turazzini 2016). Within these listed extant groups, there is no clear distinction between taxonomically significant morphologies of the ilia. However, the fossils have a relatively elongated ischiadic process pointing posterodorsally; that is a derived feature characteristic of Costata, more precisely of Alytidae. Today, among costatans, only *Discoglossus* (Alytidae) is provided with a dorsal crest, but several extinct costatans also display this feature. Sanchiz (1998) regarded these extinct taxa as all closely related to *Discoglossus*. However, Wang et al. (2000) considered one of them (*Eodiscoglossus*) as the sister group to living Costata, the more inclusive group that includes *Discoglossus* but also taxa that lack a dorsal crest. If Wang et al. (2000) are right, then the presence of the crest is perhaps not a reliable indication of relationships to *Discoglossus* and Alytidae. By contrast, the strong reduction of the *pars descendens* and the presence of a shallow supraacetabular fossa point to the costatan family Bombinatoridae, not to Alytidae. Another important feature is the absence of an interiliac tubercle. Among confirmed Costata, only the Jurassic-Cretaceous *Eodiscoglossus* (Evans et al. 1990; Venczel and Csiki 2003) appears to lack an interiliac tubercle. The significance of the presence of an interiliac tubercle is unknown: it may be either a plesiomorphic character or an adaptation to aquatic life (Vergnaud-Grazzini and Hoffstetter 1972). Therefore, it is of no help for the evaluation of relationships. A final problem is that the posterodorsal elongation of the ischiadic process is perhaps not as marked as in other alytids, which may cast further doubts on assignment to Alytidae.

On the basis of the evidence at hand, referral to Costata appears possible, but the combination of characters shown by these ilia renders assignment within Costata uncertain. The fossils are regarded as possible Costata *incertae sedis*.

Neobatrachia Reig, 1958
Leptodactylid-Hemisotid morphotype
(Fig. 6.3)

Ilia of these frogs bear a dorsal crest; the thickened posterior border of the crest forms the *tuber superius*. The latter is higher than the dorsal crest. A *tuber superius* projecting

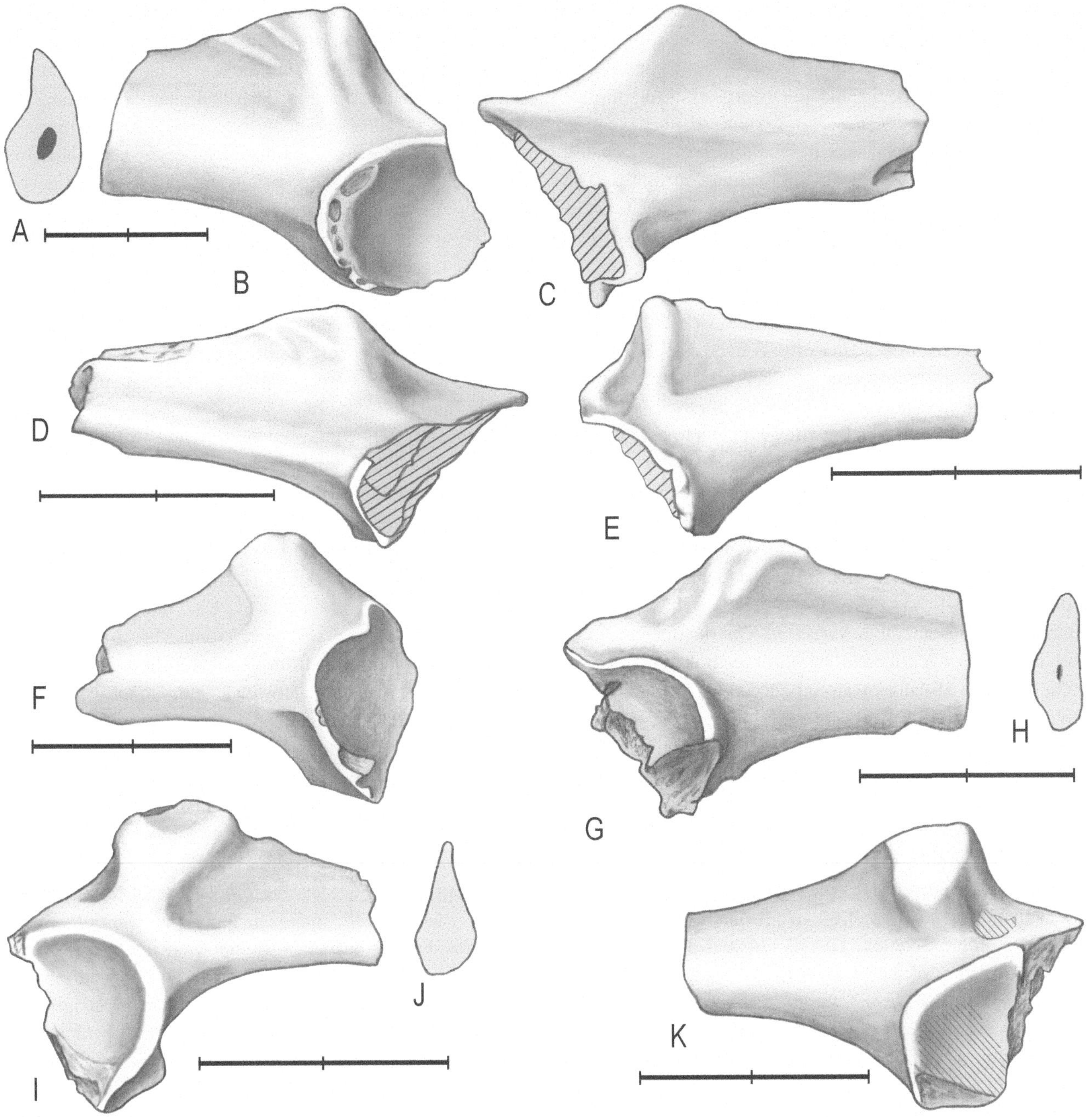

Fig. 6.3 Ilia of leptodactylid-hemisotid morphotype from the Maastrichtian of India; all specimens from Kisalpuri. Form 1: **A–B**. left ilium (DUGF/453), cross-section of the shaft, medial side to the left (**A**) and lateral view (**B**); **C–D**. left ilium (DUGF/454) in medial (**C**) and lateral (**D**) views; **E**. right ilium (DUGF/458) in lateral view. Intermediate morphology: **F**. left ilium (DUGF/455) in lateral view. Form 2: **G–H**. right ilium (DUGF/456) in lateral view (**G**) and cross-section of the shaft, lateral side to the left (**H**); **I–J**. right ilium (DUGF/462) in lateral view (**I**) and cross-section of the shaft, medial side to the left (**J**); **K**. left ilium (DUGF/469) in lateral view. Scale bars represent 2 mm

higher than the dorsal crest is found in leptodactylids and hemisotids.

pars, ?Leptodactylidae or Hemisotidae, indeterminate genus 1: Prasad and Rage 2004: 101–103, Fig. 1E–G.

Horizon and Locality: The Upper Cretaceous Deccan intertrappean beds of Kisalpuri, Dindori District (Madhya Pradesh).

Referred Material: Isolated ilia DUGF/453-459, 461, 462, 469, 470, 478, and perhaps DUGF/463-467 and DUGF/473-477; all from Kisalpuri.

Description: These ilia are all characterized by the presence of a dorsal crest. As in the above described possible Costata, the posterior border of the crest is thickened and forms the *tuber superius*. However, these ilia differ from those referred to Costata in having the *tuber superius* markedly higher than the dorsal crest. From the high *tuber*, the dorsal crest descends at a shallow angle anteriorly. Unfortunately, none of the specimens preserves a complete shaft; consequently, the anterior extent of the crest is unknown. The cross-section of the shaft itself is more or less oval and the crest is vertical, not inclined medially. The acetabular area is damaged in all specimens, but DUGF/453 suggests that the acetabulum was ovoid, more or less extended anteroposteriorly. The ventral portion of the acetabular rim is markedly raised in comparison to the dorsal part. The *pars ascendens* is small and the ischiadic process does not project strongly posteriorly. A depression is present dorsal to the acetabulum, but contrary to typical supraacetabular fossae it is not restricted to the *pars ascendens*; it looks more like a short, shallow furrow than a true fossa, and from the anterior part of the *pars ascendens* it extends prominently onto the posterior face of the *tuber superius*; it may be a peculiar supraacetabular fossa. The *pars descendens* is never completely preserved, but in DUGF/461 a remaining part suggests that it was not clearly reduced. A preacetabular fossa occurs against the acetabular rim; its depth shows considerable variation. On the medial face, there is no interiliac tubercle, but the acetabular surface and the medial face of the area strongly diverge ventrally, which renders the ventral part very thick.

Two forms may be distinguished on the basis of the *tuber superius*; they are here referred to as forms 1 and 2. Form 1 includes DUGF/453, 454, 458, 459 and perhaps DUGF/463-466, 470, 473-475, and 478. Form 2 comprises DUGF/456, 457, 461, 469, and perhaps DUGF/462 and 467. It should be noted that DUGF/455 displays a morphology intermediate between that of the two forms. Because of their poor state of preservation, DUGF/476 and 477 cannot be allocated to either of the two forms.

In form 1 (Fig. 6.3A–E), the dorsal margin of the *tuber superius* is more or less pointed to rounded in lateral view, whereas in form 2 it appears as a more or less straight and horizontal margin. In addition, in form 1 the apex of the *tuber* is generally located slightly anterior to the anterior limit of the acetabular rim. The posterolateral part of the *tuber* appears as a triangular surface that grades into the *pars ascendens* posteroventrally. The triangular surface is limited laterally by a low, indistinct ridge. More anteriorly, the area transitional between the *tuber* and dorsal crest, laterally bears two short, slightly anteriorly inclined ridges in the larger ilia (DUGF/453, 454); smaller specimens (DUGF/458, 459) lack these ridges.

In form 2 (Fig. 6.3G–K), the posterior part of the elongated upper margin of the *tuber superius* is located above the anterior part of the acetabular rim. There is no well demarcated triangular surface on the posterior face of the *tuber* and the anteriorly inclined ridges are lacking. The *tuber* appears to be more massive than in form 1 and it shows variation. It is thick in DUGF/456, 461, and 469, but remains comparatively thin in DUGF/457; moreover, in the last specimen, the supraacetabular fossa is deep and sharply limited. The *tuber* appears rather globulous in DUGF/461, whereas a nearly flat area facing laterodorsally occurs in DUGF/469 and apparently DUGF/456, but in the last specimen a vertical groove extends along the flat area.

In the sole example of the intermediate form, DUGF/455 (Fig. 6.3F), the upper margin of the *tuber superius* is more obtusely rounded than in form 1; as this margin is indented by a tiny depression it does not form a straight border. It lacks the short anteriorly inclined ridges and the posterior triangular surface is poorly limited. Its preacetabular fossa is strongly marked.

Discussion: The morphology of the *tuber superius* forming a dorsal thickening of the posterior end of the dorsal crest and projecting clearly above it is observed in Leptodactylidae and Hemisotidae. This similarity between Leptodactylidae and Hemisotidae results from convergence. Leptodactylids and hemisotids are members of two distinct assemblages, the Hyloides and Ranoides, respectively. These two assemblages are sister groups (Frost et al. 2006) and the large majority of taxa included in both assemblages lack this peculiar feature. However, although lacking phylogenetic significance, this character may be useful for taxonomic purposes. As the *pars descendens*, which is reduced in Hemisotidae, is poorly or not preserved in the Kisalpuri ilia, it is difficult to assign them to either of the two groups on this basis and, instead, only the morphology of the *tuber superius* may be used.

Anteroposteriorly elongated *tubers* are known in leptodactylids, whereas the *tuber* is more pointed dorsally in hemisotids. On the basis of this character, ilia of form 1 are reminiscent of Hemisotidae, whereas those assigned to form 2 resemble Leptodactylidae. However, there is an intermediate morphology in one specimen (DUGF/455), which casts doubts on the presence of two distinct taxa within this assemblage. Nevertheless, in view of the distinct morphological differences between forms 1 and 2, the presence of two distinct taxa appears to be more probable than the presence of a single, highly variable taxon. The intermediate ilium may simply be a variant of form 1. Despite resemblances to leptodactylids and hemisotids, and taking into account the fact that the osteology of Gondwanan frogs remains poorly known, we prefer not to assign these ilia to any family.

Form 1 corresponds to the taxon previously labelled "Indeterminate genus 1" in the Maastrichtian fauna of Naskal, India (Prasad and Rage 2004). The morphology of the *tuber superius*, the peculiar shape and position of the supraacetabular fossa, as well as the ventral divergence between the acetabular and medial surfaces clearly confirm that these ilia from Kisalpuri and Naskal belong to the same, single taxon (likely the same species).

At Naskal, there are no ilia corresponding to form 2 from Kisalpuri. This might be taken as indirect evidence that form 1 and 2 from Kisalpuri do not belong to a single taxon.

Ranoid-Myobatrachoid Morphotype
(Fig. 6.4D)

Ilia assigned to this assemblage have a dorsal crest, a *tuber superius* formed by a thickening of the posterior border of the crest, and the *tuber* is not higher than the crest. Such a morphology occurs in ranoids (i.e., Ranoides sensu Frost et al. 2006) and in some Myobatrachidae and Limnodynastidae, i.e., myobatrachoids (e.g., Tyler 1976), and Hemiphractidae.

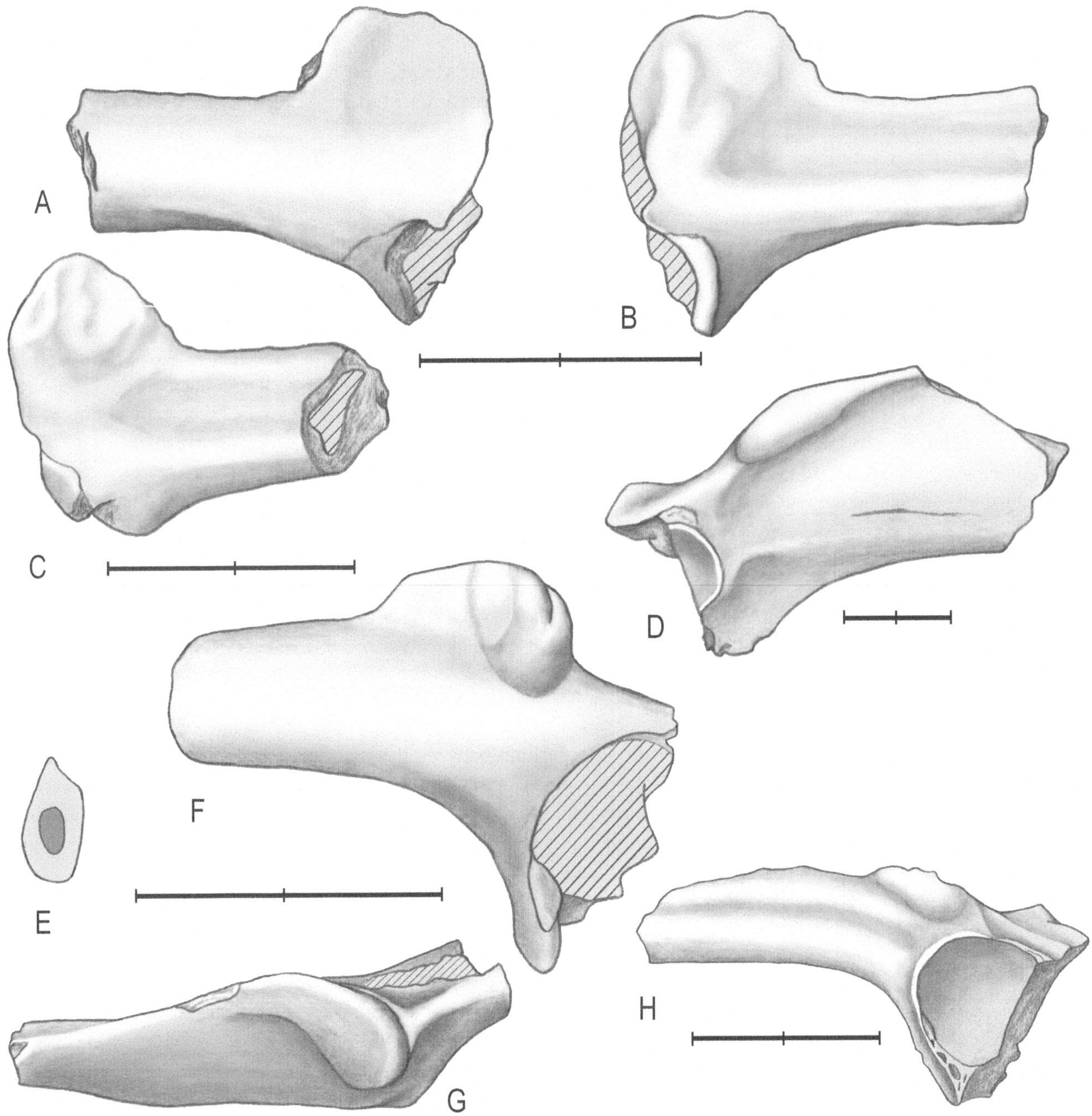

Fig. 6.4 Ilia of ranoid-myobatrachoid, ?ranoid, and unknown morphotypes from the Maastrichtian of India; all specimens from Kisalpuri. **A–B**. ?ranoid morphotype, right ilium (DUGF/479) in medial (**A**) and lateral (**B**) views; **C**. ?ranoid morphotype, right ilium (DUGF/480) in lateral view; **D**. ranoid-myobatrachoid morphotype, right ilium (DUGF/450) in lateral view; **E–G**. unknown morphotype, Form A left ilium (DUGF/482) in cross-section of the shaft anterior to the dorsal crest, medial side to the left (**E**); lateral (**F**) and medial (**G**) views; **H**. unknown morphotype, Form B left ilium (DUGF/483) in lateral view. Scale bars represent 2 mm

Horizon and Locality: The Upper Cretaceous Deccan intertrappean beds of Kisalpuri, Dindori District (Madhya Pradesh).

Referred Material: Isolated ilia (DUGF/450, 452, 460, and perhaps DUGF/451, 471, 472); all from Kisalpuri.

Description: Only DUGF/450 is sufficiently well-preserved to be described (Fig. 6.4D). This is a right ilium that lacks the posterior part of the acetabular area and at least the anterior half of the shaft. It is the largest anuran ilium recovered from Kisalpuri. A comparatively high dorsal crest is present, but it is largely broken away. As in the presumed Costata and in the leptodactylid-hemisotid morphotype from the same locality, the *tuber superius* is a thickening of the posterior border of the dorsal crest. The *tuber superius* is situated well anterior to the anterior part of the acetabular rim, it is not higher than the dorsal crest and perhaps may even be slightly lower than the top of the crest. The *tuber* is in the form of an elongate, ovoid bulge that tapers anteriorly and grades in the dorsal edge of the crest. In lateral view, the *tuber* rises gently and it forms an obtuse angle with the *pars ascendens*. On its lateral face, the *tuber superius* appears clearly demarcated from the crest; consequently, the posterior part of the lateral face of the crest appears as a broad depression between the *tuber superius* and the shaft.

In cross-section the shaft is ovoid. The dorsal crest is vertical along the posterior part of the shaft, but is likely bent slightly medially along the anterior part. The bone lacks a supraacetabular fossa. As shown by its preserved portion, the *pars descendens* was large. A preacetabular fossa is present; it occupies a relatively dorsal position. The rim of the acetabulum does not appear to be strongly projecting. Despite the incompleteness of the acetabular part, it may be inferred that there was no interiliac tubercle on the medial face.

DUGF/452 and 460 do not afford additional information. DUGF/471 and 472 are small; their *tuber superius* is scarcely salient. These two specimens likely represent juveniles; they are tentatively referred to the same taxon.

Discussion: The morphology of the dorsal crest and *tuber superius* of these ilia is similar to that of ilia of Limnodynastidae, Myobatrachidae, Hemiphractidae, ranoids, and even alytid costatans. A *tuber superius* that rises anteriorly rather gradually, thus forming a gentle curve in lateral view, is also a characteristic that occurs in these extant taxa. However, the large size of the *pars descendens* is not consistent with costatans. The observable features do not permit us to discriminate between Limnodynastidae, Myobatrachidae, Hemiphractidae, and ranoids.

Ilia similar to those from Kisalpuri were reported from the Maastrichtian of Naskal (India); they were referred to an indeterminate anuran (Prasad and Rage 1995: Fig. 7) and subsequently to an indeterminate family of Ranoidea (Prasad and Rage 2004: Fig. 2), it being understood that, at that time, Ranoidea represented a group larger than Ranoidea sensu Frost et al. (2006) and approximately correspond with Ranoides of Frost et al. (2006). Although similar, specimens from Kisalpuri and Naskal are not identical. More specifically, the *pars descendens* is larger in the Kisalpuri ilia. However, the specimens from Naskal likely belong to a taxon close to that of Kisalpuri. The similarity between the specimens from Kisalpuri and more complete ilia from Naskal confirms that the ilia described here cannot be referred to alytids. However, it is not possible to refer the fossils from Kisalpuri (and those from Naskal) to any anuran family.

?Ranoid Morphotype
(Fig. 6.4A–C)

Horizon and Locality: The Upper Cretaceous Deccan intertrappean beds of Kisalpuri, Dindori District (Madhya Pradesh).

Referred Material: Two incomplete ilia (DUGF/479 and 480 from Kisalpuri).

Description: These two ilia are very incomplete, but they clearly represent a distinct taxon. They have a dorsal crest that is unfortunately largely broken off; however, its posterior most part remains on both specimens. The dorsal crest was high and vertical, at least in its posterior portion. The posterior border of the crest is not clearly thickened to form a typical *tuber superius*; however, the posterior most part of the crest, including the posterior border, is slightly thickened and its lateral face bears irregular, broad, and low protuberances that were likely for muscle insertions (Fig. 6.4B, C). Therefore, this area is likely an anteroposteriorly broad, poorly defined *tuber superius*. In both specimens, the posterodorsal border of this area is preserved. It was very steeply inclined, contrary to the same border in the ilia above referred to the ranoid-myobatrachoid morph. On the lateral face of the two ilia a marked elongate depression follows the limit between the shaft and the dorsal crest (Fig. 6.4B, C). Posteriorly, it ends against the ventral part of the broad *tuber superius*. Only the anterior most part of the acetabulum area is preserved in both ilia. DUGF/479 shows that the anterior part of the acetabular rim is clearly projecting and close to the anterior border of the *pars descendens*. The lateroventral border of the shaft in DUGF/479 is not rounded as is usual, but instead forms a marked ridge. The ridge grows progressively blunter posteriorly and it vanishes where it approaches the acetabular rim. Apparently, this ridge was less marked in DUGF/480.

Discussion: Within anurans whose *tuber superius* is formed by a thickening of the posterior termination of the dorsal crest, the posterior border of the *tuber* rises more or less gently or steeply. A *tuber superius* whose posterior border rises steeply might be indicative of ranoids because such a morphology appears to be unknown in non-ranoids.

Therefore, on this weak basis, these two ilia may be referred to ranoids. However, the unusual morphology of the *tuber superius* renders this assignment quite tentative. It should be noted that the ridge-like lateroventral border of the shaft is unusual in anuran ilia. The significance of this character is unknown, but it is of interest for further comparisons. These two ilia are regarded as likely, but not certainly, representing a ranoid morphology that is unknown elsewhere.

Unknown Morphotypes
FORM A
(Fig. 6.4E–G)

Horizon and Locality: The Upper Cretaceous Deccan intertrappean beds of Kisalpuri, Dindori District (Madhya Pradesh).

Referred Material: One incomplete left ilium (DUGF/482 from Kisalpuri).

Description: DUGF/482 is an incompletely preserved left ilium whose *tuber superius* and likely dorsal crest display a strange and quite unusual morphology. Except for only a short, superficial trace of break on the dorsal border (Fig. 6.4G), which does not change the shape of the crest, the dorsal crest is intact. The crest is comparatively high, but very short anteroposteriorly. It prolongs the *tuber superius* anteriorly and ends abruptly as a steep anterior border. More anteriorly, a low keel on the shaft may be regarded as the strongly reduced anterior continuation of the dorsal crest (Fig. 6.4E). In lateral view, an oblique, curved line approximately marks the limit between the crest and the *tuber*. The *tuber superius* appears as a globulous thickening of the posterior part of the dorsal crest and it is not higher than the latter (Fig. 6.4F). The *tuber superius* is situated just anterior to the anterior acetabular rim. There is no clear boundary between the *tuber* and the crest; the *tuber* becomes thinner anteriorly, grading into the thin, vertical dorsal crest. Whereas the dorsal crest is in line with the long axis of the shaft, the transition between the *tuber* and crest, and the *tuber* itself are markedly oblique in dorsal aspect (Fig. 6.4 G), which is unusual. Posterodorsally, the *tuber superius* bears a marked groove. The shaft is somewhat compressed laterally. The *pars ascendens* and *pars descendens* are incomplete, but it may be stated that supra- and preacetabular fossae are absent. The acetabulum was likely broad. Apparently, there was no interiliac tubercle.

Discussion: This specimen is very peculiar and, to our knowledge, its morphology does not match any known anuran ilium. Therefore, it cannot be referred to a known taxon. On the other hand, this single and incomplete specimen cannot be the basis for naming a new taxon.

Form B
(Fig. 6.4H)

Horizon and Locality: The Upper Cretaceous Deccan intertrappean beds of Kisalpuri, Dindori District (Madhya Pradesh).

Referred Material: One incomplete left ilium (DUGF/483 from Kisalpuri).

Description: This ilium (Fig. 6.4H) has a dorsal crest and its *tuber superius* is a thickening of the posterior border of the dorsal crest. Unfortunately, the dorsal part of the dorsal crest is broken off; it is not possible to estimate the height or shape of the crest. Similarly, the dorsal margin of the *tuber superius* is damaged. However, from the preserved part, it clearly appears that the *tuber* is inclined gently, not steeply, and that it is well demarcated from the dorsal crest in lateral aspect. The posterior limit of the *tuber* is located posterior to the anterior acetabular rim. A marked supraacetabular fossa is present, but there is no preacetabular one. The acetabulum is large. The *pars ascendens*, which is largely broken away, likely extended markedly posterodorsally. The *pars descendens* is apparently reduced. A marked groove runs on the lateral face of the shaft and, posteriorly, the groove contacts the acetabular rim. The medial face of the acetabular area lacks an interiliac tubercle.

Discussion: The presence of a dorsal crest, as well as the nature (thickening of the posterior border of the dorsal crest) and gentle inclination of the *tuber superius* invites comparisons with and assignment to the above described ranoid-myobatrachoid morphotype. However, DUGF/483 is clearly different in bearing a marked longitudinal groove on the lateral face of the shaft. A similar groove was reported by Tyler (1976, 1986) in species of the extant *Litoria* from Australasia and in the extinct *Australobatrachus ilius* from the Miocene of Australia. Both genera were assigned to Hylidae by Tyler (1976), but Sanchiz (1998) referred *Australobatrachus* to the Microhylidae. The ilia of those southern hemisphere genera lack dorsal crests. To our knowledge, the specimen from Kisalpuri is the only anuran whose ilium has both a dorsal crest and a lateral groove on the shaft.

Indeterminate Anurans
(Fig. 6.5)

The above identifications are based on ilia. Other skeletal elements are available but are less informative. In addition, we cannot confidently associate these isolated bones with the above described ilia. Two categories of bones may be distinguished here. Some provide taxonomic information more limited than the above described ilia, either because they display relatively limited morphological variation or they are too incomplete (maxillae, presacral and sacral vertebrae, urostyles, coracoids, scapulae, humeri); they are briefly described and discussed below. Several other bones provide no information; they cannot be identified within anurans either because they are taxonomically uninformative bones

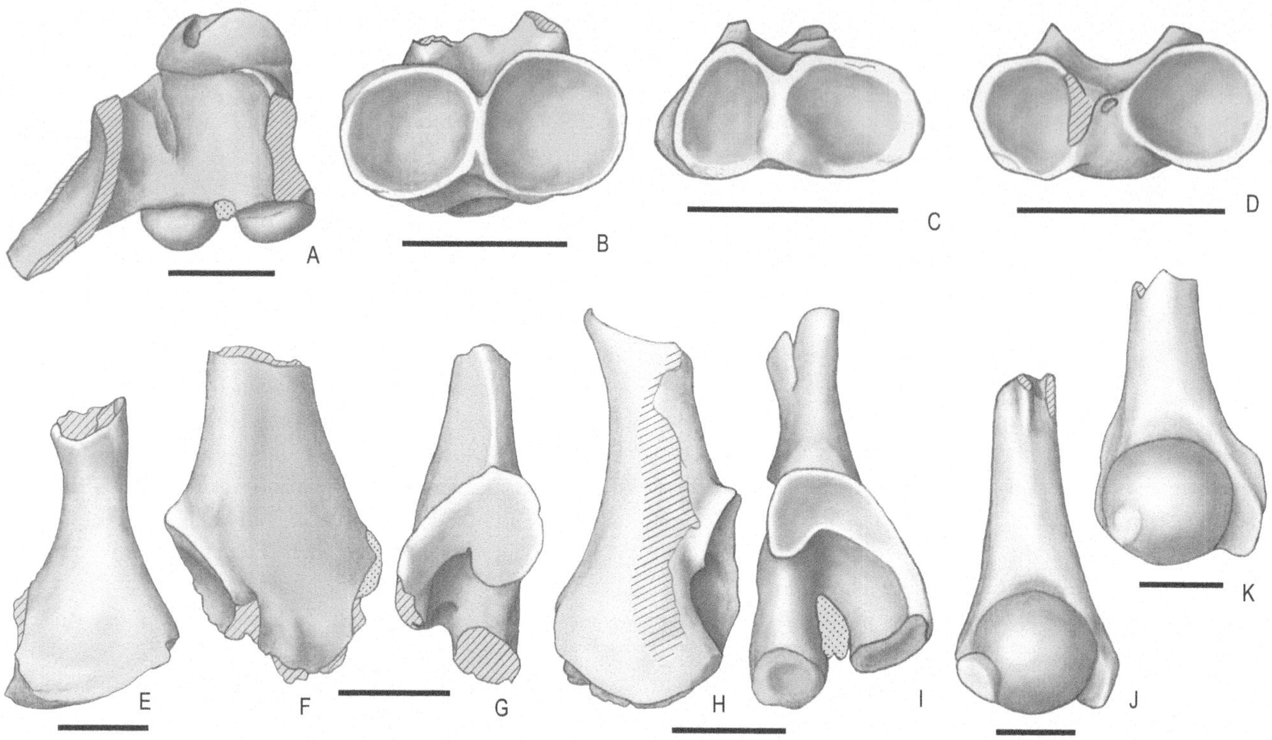

Fig. 6.5 Non-ilia bones of indeterminate anurans from the Maastrichtian of India; all specimens from Kisalpuri. **A**. Sacral vertebra (DUGF/513) in dorsal view; **B**. urostyle (DUGF/516) in anterior view; **C**. urostyle (DUGF/517) in anterior view; **D**. urostyle (DUGF/518) in anterior view; **E**. right coracoid (DUGF/519) in medial view; **F–G**. right scapula (DUGF/524) in lateral (**F**) and posterior (**G**) views; **H–I**. left scapula (DUGF/525) in lateral (**H**) and posterior (**I**) views; **J**. right humerus (DUGF/485) in ventral view; **K**. right humerus (DUGF/486) in ventral view. Scale bars represent 1 mm

(radioulnae and tibiofibulae) or because they are too poorly preserved (several ilia and humeri).

Maxillae (not figured)

Four small, poorly preserved fragments of maxillae are available: DUGF/522 and 523 from Kisalpuri and DUGF/535 and 536 from Kelapur. They show two distinct morphotypes. However, all had teeth (not preserved). DUGF/522, 523, and 535 represent one morphotype. Their labial surface displays low and rather poorly defined wrinkles. The largest fragment (DUGF/523) shows that these wrinkles are elongate and longitudinal. On the lingual face, the *lamina horizontalis* is strong, deep and its medial margin is rounded. This morphotype likely corresponds to the "maxillae type 1" from the Maastrichtian of Naskal (Prasad and Rage 2004: Fig. 5A, B). DUGF/536 represents a different morphotype in having a very shallow, thin *lamina horizontalis*. Its labial face is damaged, but apparently it was smooth. It is not possible to state to which taxa these maxillae belong.

Presacral vertebra (not figured)

DUGF/515 is a fragmentary presacral vertebra from Kisalpuri. It is procoelous with only the ventral part preserved. The centrum is depressed, not cylindrical; as a result, the cotyle and condyle are sub-ovoid. The pedicels supporting the neural arch are broken through at their mid-height and are closer to the anterior cotyle. Among taxa identified in the Cretaceous of India, only Costata are not consistent with this vertebra because members of that clade have opisthocoelous vertebrae. The Kisalpuri vertebra may correspond to vertebrae "type 2" reported from the Maastrichtian of Naskal (Prasad and Rage 2004).

Sacral vertebrae (Fig. 6.5A)

DUGF/513 (Fig. 6.5A) from Kisalpuri is a very incomplete sacral vertebra, only the centrum and posterior border of the left sacral apophysis remain. The centrum is opisthocoelous and there are two posterior condyles for articulation with the urostyle. The centrum and condyles are compressed dorsoventrally. The posterior border of the sacral apophysis is directed markedly posterolaterally. From what remains of this apophysis, it is clear that it was at least slightly expanded (i.e., "dilated") and likely flat, but not cylindrical.

Opisthocoelous sacral vertebrae occur in both opisthocoelous and diplasiocoelous vertebral columns. Opisthocoelous columns are known in Costata, Pipidae, and Rhinophrynidae, whereas diplasiocoelous ones (i.e., presacral vertebrae procoelous, except for the last one which is amphicoelous, and sacral vertebra opisthocoelous) occur in ranoids. Pipidae may be discarded because their sacral vertebra is fused to the urostyle. The presence of two posterior condyles excludes referral to bombinatorid Costata, because articulation with the urostyle is monocondylar (but the condyle can be slightly subdivided by a weak vertical groove) in bombinatorids. The expanded or dilated shape of the sacral apophyses is consistent with Rhinophrynidae and also with various costatans and ranoids. Although none of the observable features permits us to discard Rhinophrynidae, the presence of a rhinophrynid in the Cretaceous of India would be surprising because the family is today restricted to Central and North America (Henrici 1991). However, although the latter argument is a geographic one, i.e., a weak argument, this sacral vertebra seems better attributed to either alytid costatans or ranoids.

The centrum of another sacral vertebra (DUGF/514) from Kisalpuri belongs to a distinct taxon. It is not possible to state whether it was pro- or opisthocoelous, but the two posterior condyles are clearly more widely spaced than in DUGF/513. However, DUGF/514 offers practically no taxonomic information.

Previously, only one sacral vertebra was reported from the Cretaceous of India; it comes from the Maastrichian of Naskal (Prasad and Rage 1995). This sacral vertebra differs from those of Kisalpuri in being procoelous and in having closely spaced posterior condyles; it was regarded as a possible Hylidae, a family that also was tentatively identified at Naskal on the basis of an ilium (Prasad and Rage 1995).

Urostyles (Fig. 6.5B–D)

Three anterior parts of urostyles from Kisalpuri are available. In all three specimens the neural arch is broken off and they are all bicotylar. The anterior cotyles display three distinct morphologies.

In DUGF/516 (Fig. 6.5B) the cotyles are clearly circular in outline and are closely spaced, being separated by a thin median crest. DUGF/517 and 518 (Fig 6.5C, D, respectively) are smaller than DUGF/516. DUGF/517 has the two cotyles widely spaced, separated by a rather broad, poorly defined and ventrally swollen crest. The entirely preserved cotyle has a thick rim, is slightly ovoid in outline, and slightly mediolaterally elongated. In DUGF/518, the well preserved cotyle is rather ovoid with some mediolateral elongation. The sagittal area is damaged, but the two cotyles were likely rather closely spaced. The morphologies illustrated by the urostyles from Kisalpuri were previously not reported from the Cretaceous of India.

Coracoid (Fig. 6.5E)

An anuran coracoid (DUGF/519) is reported for the first time from the Cretaceous of India. It comes from Kisalpuri and, although incomplete, it affords some information. Its glenoid extremity is damaged, but when intact would have been approximately circular. The neck of the bone is narrow, short, straight, and approximately cylindrical. The epicoracoidal part is broad and mediolaterally flattened; its anterior projecting part is broken away, but it may be inferred that the epicoracoidal part was markedly asymmetrical. Such an expanded and asymmetrical epicoracoidal part clearly points to ranoids.

Scapulae (Fig. 6.5F–I)

There are two scapulae in the present collection from Kisalpuri. DUGF/524 (Fig. 6.5F, G) is a right scapular fragment. Most of the dorsal (*pars suprascapularis*) and ventral (*pars acromialis* and glenoid process) parts are broken away. Between them, the bone is clearly narrow anteroposteriorly, but thick. The glenoid articular surface is broad and prolonged medially on the remaining part of the glenoid process. The latter process is broken away, but it is clear that a notch (*sinus interglenoidalis*) separated it from the *pars acromialis*. There is no anterior lamina. The narrowness of the bone, dorsal to the glenoid surface, demonstrates that this scapula does not belong to the dorsoventrally short type. Instead, the scapula was perhaps elongated dorsoventrally.

DUGF/525 (Fig. 6.5H, I) is a comparatively better preserved left scapula. Its *pars suprascapularis* is broken off, but its glenoid facet, glenoid process, and *pars acromialis* are well preserved. The narrow part of the bone, just dorsal to the glenoid surface, appears to be slightly narrower anteroposteriorly and is clearly less thick than in DUGF/524. The glenoid facet is less broad than in DUGF/524 and it extends across the entire posterior margin of the glenoid process. The latter process is not directed posteriorly, i.e., it does not diverge from the *pars acromialis*. Consequently, the *sinus interglenoidalis* is narrow. The scapula also lacks an anterior lamina. As with DUGF/524, it was certainly elongated dorsoventrally and perhaps relatively gracile in build.

Although morphologically similar to each other, the different thicknesses of the middle part and differences in size of the glenoid facets show that these two scapulae represent two distinct taxa. They do not belong to the stout, dorsoventrally short type of scapula known in several non-neobatrachian anurans, including the living *Ascaphus* (Leiopelmatidae), most Costata, Pipidae, and the extinct *Neusibatrachus* (Early Cretaceous; Báez and Sanchiz 2007) and Paleobatrachidae (Late Cretaceous-Pleistocene; Špinar 1972). The dorsoventrally and clearly elongate shape of both Kisalpuri scapulae excludes them from all other "archaeobatrachian" anurans, that is to say, considering taxa identified in the Cretaceous of India, these scapulae cannot be allocated to Costata.

Humeri (Fig. 6.5J–K)

There are seven humeral bones in the anuran collection of Kisalpuri, of which two (DUGF/485 and 486; Fig. 6.5J, K, respectively) are better preserved. These humeri preserve the distal part of the bone. In both humeri, the ball or humeral condyle is spherical, large, and very prominent; it occupies about 75% of the humeral width distally. It is in line with the shaft axis.

In DUGF/485, the medial epicondyle is well developed and its distal tip extends to the level of the distal surface of the humeral ball. There is no medial crest. The lateral epicondyle is very distinct; it appears as an ovoid tubercle applied against the ball. A very narrow lateral crest is present. The *fossa cubitalis* is deep, narrow, and crescentic in shape, not limited laterally, and is without a proximal extension. The olecranon scar is poorly marked, large, and with its proximal tip offset laterally. A small distal part of the *crista ventralis* is preserved. In addition, medial to the *crista ventralis*, a small ridge likely corresponds to the distal tip of a *crista paraventralis*. The ovoid shape of the lateral epicondyle is reminiscent of the humerus "type 2" from Naskal (VPL/JU/1028) that was referred to as indeterminate anuran (Prasad and Rage 2004: Fig. 3C, D). However, the latter morphological type differs from the humerus from Kisalpuri in having a relatively smaller humeral ball and a relatively larger medial epicondyle.

DUGF/486 shows a similar overall morphology. However, it is a shorter fragment of a humerus and it is not possible to state whether a *crista ventralis* and *crista paraventralis* were present. It differs from DUGF/485 in that its medial epicondyle is larger and sharp-cornered proximodorsally and that its lateral crest is stronger. The lateral epicondyle is less prominent than in DUGF/485. The *fossa cubitalis* is poorly limited and it opens laterally. There is no delimited olecranon scar. DUGF/486 matches humeri "type 1" previously reported from the Maastrichtian of Naskal (Prasad and Rage 2004: Fig. 3A, B). As stated previously (Prasad and Rage 2004:104–105), it is not possible to confidently assign such humeri to any anuran family.

These two humeri likely belong to distinct taxa. This is mainly demonstrated by the larger size of the medial epicondyle in DUGF/486; this feature likely is not a sexual difference because both humeri appear to belong to females (they each lack a medial crest). If a *crista paraventralis* was really present in DUGF/485, this may be a weak evidence for referral of this specimen to Costata.

Other skeletal elements (not figured)

One incomplete vertebra (DUGF/521 from Kisalpuri); one incomplete scapula (DUGF/481 from Kisalpuri); one incomplete coracoid bone (DUGF/520 from Kisalpuri); two ilia (DUGF/468 from Kisalpuri, DUGF/531 from Kelapur); 29 radioulnae (DUGF 487-512 from Kisalpuri, DUGF/532-534 from Kelapur); five tibiofibulae (DUGF/556-560 from Kisalpuri).

These bones do not provide any taxonomic information, either because they are very incomplete (vertebra, scapula, coracoid, ilia) or they do not bear significant features (radioulnae, tibiofibulae).

Remarks on anurans from Kisalpuri and Kelapur: In the new amphibian material from the Maastrichtian of Kisalpuri and Kelapur, India, only anurans are represented. First, we must clarify that two anuran taxa formerly identified from the Indian Maastrichtian were not found at Kisalpuri and Kelapur. One of these taxa is an indeterminate Pelobatidae reported from Takli (=Gitti Khadan) by Sahni et al. (1982). The second taxon, from Naskal, was first reported as Discoglossidae by Prasad and Rage (1991, 1995) and subsequently as Gobiatinae (as a subfamily of Discoglossidae) by Prasad and Rage (2004).

Based on ilia, five or six taxa are distinguished within the material from Kisalpuri and Kelapur. One taxon may belong to Costata (formerly Discoglossidae), however, this remains to be confirmed. This taxon is present at both Kisalpuri and Kelapur.

The other taxa are likely neobatrachians but, within this large taxonomic group, they are only distinguished as "morphotypes" and "forms". They all come from Kisalpuri. One of them, referred to as "leptodactylid-hemisotid morphotype", shows variability and two forms (form 1 and form 2) may be distinguished within this morphotype. One ilium is intermediate between the two forms and it is therefore not possible to state whether the "leptodactylid-hemisotid morphotype" corresponds to one or two taxa. Form 1 of this morphotype was previously reported from Naskal; it was assigned to either the Leptodactylidae or the Hemisotidae as "indeterminate genus 1" (Prasad and Rage 2004). Assignment to one of these families is possible, but we cannot exclude referral to another, unknown family.

Another taxon is represented by the "ranoid-myobatrachoid morphotype". It was formerly reported from Naskal as indeterminate anuran (Prasad and Rage 1995) and as an indeterminate family of ranoid (Prasad and Rage 2004).

The taxon reported as "?Ranoid morphotype" perhaps belongs to ranoids (see above the meaning of "ranoids"), but the morphology of the referred ilia appears to be unusual by ranoid standards.

It is worth noting that the presence of ranoids is confirmed by a typical coracoid. Unfortunately, it is not possible to determine to which morphotype of ilium ("ranoid-myobatrachoid" or "?ranoid" morphotype?) it should be associated.

Finally, two quite distinct taxa at Kisalpuri are referred to as Form A and Form B. They are each represented by a single, peculiar ilium. In no known frog does the morphology of ilia approach those of the specimens assigned to Forms A and B.

Diapsida Osborn, 1903
Squamata Oppel, 1811

At Kisalpuri and Kelapur, squamates include mostly lizards. Snakes are represented only by three vertebrae from Kelapur, which is somewhat surprising because snakes are relatively numerous and diverse in the Maastrichtian of Naskal (Rage and Prasad 1992; Prasad and Rage 1995; Rage et al. 2004).

Indeterminate Lizards
(Fig. 6.6)

Horizon and Locality: The Upper Cretaceous Deccan intertrappean beds of Kisalpuri, Dindori District (Madhya Pradesh) and Kelapur, Yeotmal District (Maharashtra).

Referred Material: Isolated premaxillae (DUGF/537 from Kelapur, DUGF/540, from Kisalpuri), maxillae (DUGF/538-539 from Kelapur, DUGF/544-555 from Kisalpuri), dentaries (DUGF/541-543 from Kisalpuri).

Premaxillae (Fig. 6.6A–D)

In anterior view, DUGF/537 (Fig. 6.6A, B) from Kelapur, has an approximately triangular shape (Fig. 6.6A). The labial face of the bone is smooth. The nasal process is broad at the base, but becomes parallel-sided dorsally; a strong sagittal keel strengthens the process lingually (Fig. 6.6B). On either side of the base of the nasal process, a foramen perforates the bone. On the lingual side, the supradental shelf is thin. Its lingual edge is broken, therefore the vomerine processes are not preserved. There was likely no incisive process. There are three preserved pleurodont teeth and three tooth slots. The teeth are slender, slightly compressed laterally, and slightly inwardly curved. The parapet is deep and

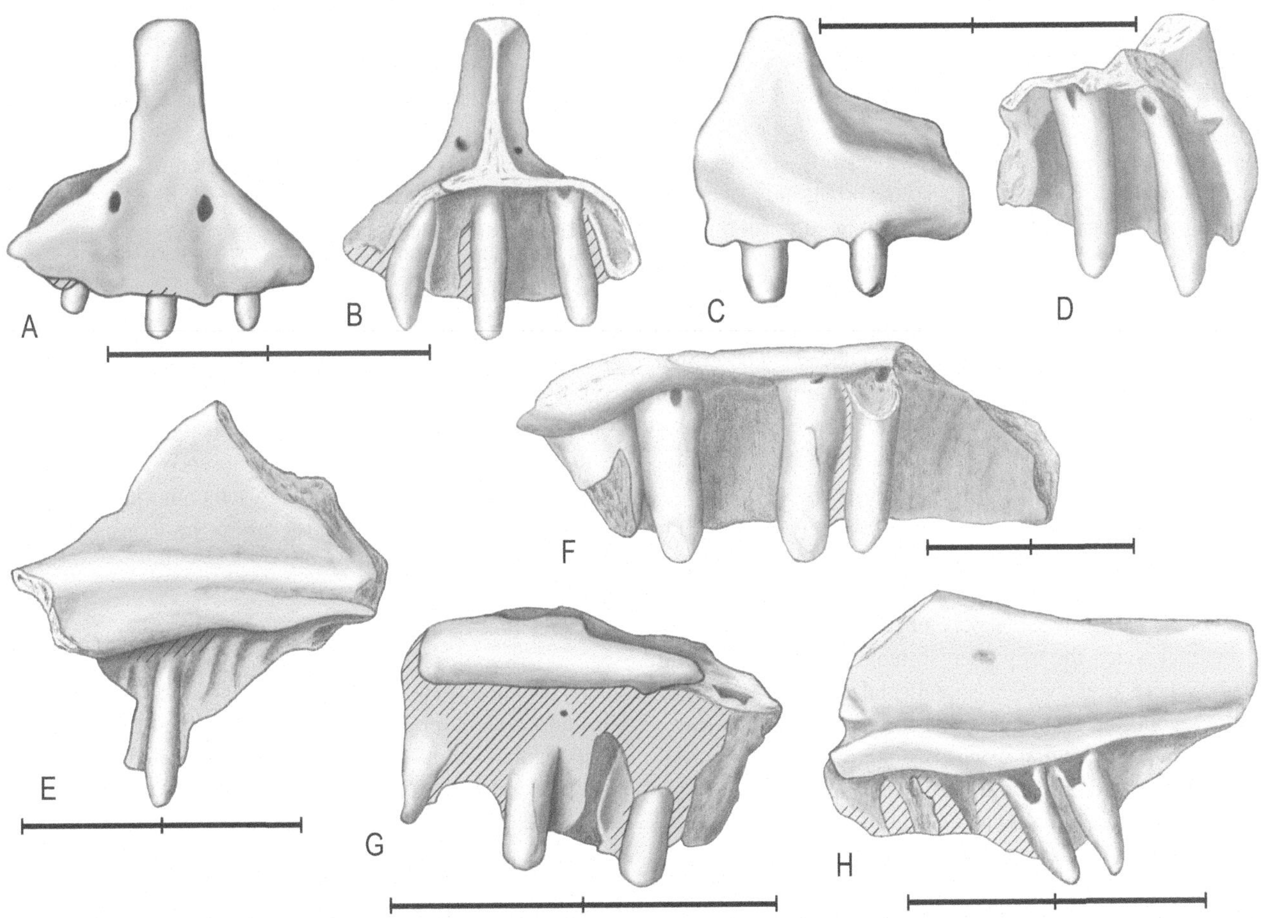

Fig. 6.6 Jaws of indeterminate lizards from the Maastrichtian of India. **A–B**. premaxilla from Kelapur (DUGF/537) in labial (**A**) and lingual (**B**) views; **C–D**. premaxilla from Kisalpuri (DUGF/540) in labial (**C**) and lingual (**D**) views; **E**. fragment of maxilla from Kisalpuri (DUGF/544) bearing a long, slender tooth, in lingual view; **F**. fragmentary maxilla from Kelapur (DUGF/538) bearing blunt teeth, in lingual view; **G**. fragment of maxilla from Kelapur (DUGF/539) in lingual view; **H**. posterior portion of maxilla from Kisalpuri (DUGF/549) in lingual view. Scale bars represent 2 mm

only a quarter of the tooth length projects beyond it. The teeth are unicuspid and no striae are discernible on the crowns. Small resorption pits open in the tooth bases.

DUGF/540 (Fig. 6.6C, D), from Kisalpuri, is a fragment of a premaxilla larger than DUGF/537. It provides little information; however, it clearly lacks the foramina that pierce the base of the nasal process and its teeth are more compressed. Therefore, DUGF/540 and 537 likely belong to distinct taxa.

Maxillae (Fig. 6.6E–H)

Several very fragmentary maxillae show that at least three morphotypes are present at Kisalpuri and Kelapur. These morphotypes likely represent distinct taxa. They are distinguished by details of the teeth and of the supradental table and shelf (sensu Rage and Augé 2010). In all specimens, teeth are pleurodont. Comparisons with premaxillae on the basis of teeth alone may not be reliable, because premaxillary and maxillary teeth may be somewhat different within individuals.

DUGF/544 (Fig. 6.6E) from Kisalpuri is a small fragment of maxilla that includes the anteroventral area of the nasal process (=dorsal process). The labial face is pierced by several foramina and, in the dorsal most area of the fragment, faint traces of osteoderms occur. The fragment bears a single tooth that is slightly inclined anteriorly. The tooth is long and slender. The parapet is so deep that only one-fifth of the tooth length projects beyond it. The mid-portion of the tooth is swollen lingually; as a result, this portion of the tooth shaft appears to be compressed anteroposteriorly. By contrast, the apex is slightly compressed labiolingually; consequently, its labial face is slightly concave. The crown lacks striae. Apparently, there is no resorption pit. The supradental shelf is relatively broad transversely, but there is no *sulcus dentalis*. At the level of the tooth, the shelf is thin and deflected ventrolingually.

Several specimens (DUGF/538 from Kelapur, DUGF/545-548 from Kisalpuri) belong to a morphotype characterized by blunt teeth and broad supradental table and shelf (Fig. 6.6F). The labial face of the bone is smooth. On the lingual side, the supradental table is horizontal and broad. Where preserved, a supradental shelf projects lingually beyond the tooth bases, but it is largely broken away and, thus, its full lingual extent is unknown. A *sulcus dentalis* is present. The teeth are relatively thick, anteroposteriorly compressed, and slightly inwardly curved. Depending on their positions along the tooth row, the tooth bases can extend more or less lingually on the supradental table. Small resorption pits are present lingually. The teeth are moderately elongate and only one-fifth to one-quarter of their length exceeds the parapet. Tooth crowns are blunt, rounded, and no striae are visible. Despite the absence of marked striae, the lingual side of the crown forms a rectangular to ovoid facet, which does not appear to be a wear facet. DUGF/548 (not figured) has only a short erupting tooth; the morphology of the tooth is similar to the larger teeth in other specimens of this morphotype. DUGF/538 is relatively large as compared to all other known maxillae from the inter-trappean beds of Kisalpuri and Kelapur.

DUGF/539 (Fig. 6.6G) is a fragment of maxilla, from Kelapur, with only two teeth. The labial face of the bone is smooth. It has a comparatively thick supradental shelf. The teeth are not well preserved. They resemble those of DUGF/538 and DUGF/545-548, but they are apparently more slender and slightly more elongate. Moreover, they lack the small facet that occurs on the lingual side of the crown in the above-described specimens. DUGF/539 may belong to the same taxon as DUGF/538 and 545-548 or it may represent a distinct species.

DUGF/549 from Kisalpuri is a small posterior portion of maxilla bearing two preserved teeth (Fig. 6.6H). The labial face is smooth. On the lingual face, the supradental table is transversely narrow. A moderately wide and relatively thick supradental shelf is present. There is no *sulcus dentalis*. The teeth are inclined posteriorly, likely reflecting their posterior position on the maxilla. The teeth are not long and about one-quarter of their length projects beyond the parapet. The shaft is slightly bulbous and it narrows toward the apex, which gives a conical shape to the tooth. The crown is unicuspid and pointed, but not acute. Striae are lacking. Resorption pits open lingually.

The other fragments of maxillae (DUGF/550-555) are too incomplete to be assigned to any of the above-mentioned morphotypes.

None of the maxillae offers significant characters. The overall morphology of the teeth and their attachment to the bone is reminiscent of scincomorphs, but such an assignment cannot be clearly demonstrated. The absence of striae on the crown, which is a widespread scincomorphan feature (Richter 1994; Kosma 2004), is perhaps not consistent with referral to this assemblage.

Dentaries (not figured)

There are three dentary fragments (DUGF/541-543) in the Kisalpuri collection. The morphology of their teeth permits comparisons and, potentially, association with the above-described maxillae. In all three specimens, the subdental shelf is comparatively thick and projects lingually for a moderate distance.

DUGF/541 is a fragment that retains three teeth. Tooth bases and the dorsal area of the subdental shelf are concealed by matrix. Teeth are long and slender; the morphology of their exposed portions is similar to that of the sole tooth in maxilla DUGF/544 from the same locality. These two specimens likely belong to the same taxon.

DUGF/542 bears three teeth whose ventral parts are largely concealed by matrix. Their distal portions are clearly conical, similar to those in maxilla DUGF/549 from Kisalpuri. The two specimens are likely from the same taxon.

DUGF/543 preserves two teeth. The teeth are long and slender, but less elongate than in DUGF/541; moreover, no part of the shaft is swollen lingually. The crown is blunt and unicuspid. In addition, a marked *sulcus dentalis* is present, whereas there is no *sulcus* in DUGF/544, the maxilla that is associated to the dentary DUGF/541. DUGF/543 likely belongs to a taxon distinct from that of DUGF/541 and 544 and cannot be associated with any of the above-described maxillae.

Ophidia Brongniart, 1800 sensu Caldwell and Lee, 1997

Ophidia were not recovered from Kisalpuri and they are rare at Kelapur. The latter specimens were previously described by Rage et al. (2004). Snakes from Kelapur belong to Madtsoiidae and Nigerophiidae.

Madtsoiidae Hoffstetter, 1961
Genus and species indeterminate
Madtsoiidae, Indeterminate genus and species: Rage et al. 2004: 427, Fig. 2.

Horizon and Locality: The Upper Cretaceous Deccan intertrappean beds of Kelapur, Yeotmal District (Maharashtra).

Referred Material: One posterior trunk vertebra (DUGF/561 from Kelapur).

Discussion: The single vertebra recovered from Kelapur (Rage et al. 2004: Fig. 2) shows a combination of characters that points to madtsoiids: no trace of prezygapophyseal processes; paradiapohyses projecting laterally beyond the prezygapophyseal facets; and a large parazygantral foramen in a fossa on either side of the zygantrum. The vertebra is small (overall length of neural arch slightly above 2 mm), but it does not display juvenile characters. This vertebra cannot be referred to the large-sized madtsoiid assemblage, which perhaps represents a clade (Mohabey et al. 2011; Rage et al. 2014).

Madtsoiidae are extinct snakes primarily known from Gondwana (Rage 1998; LaDuke et al. 2010). They are relatively rare in India. Aside from Kelapur, confirmed madtsoiids were recovered from the Maastrichtian of Dholi Dungri (Wilson et al. 2010) and Pisdura (Mohabey et al. 2011) and from the early Eocene of Tadkeshwar Lignite Mine (Smith et al. 2016). In addition, it should be noted that the overall vertebral morphology of madtsoiids may be similar to that of booids; if specimens are not well preserved, differentiating between these two taxa may be impossible. Such doubtful assignments (? madtsoiids or booids) apply to fossils from the Maastrichtian of Takli (=Gitti Khadan; Gayet et al. 1984), Pisdura (Jain and Sahni 1983), and Asifabad (Prasad and Sahni 1987) and from the early Eocene of Kutch (Rage et al. 2003) and Vastan Lignite Mine, Surat (Rage et al. 2008).

Serpentes Linnaeus, 1758
Nigerophiidae Rage, 1975
cf. *Indophis sahnii* Rage and Prasad, 1992

Indophis sahnii Rage and Prasad, 1992: Rage et al. 2004: 430, Fig. 3H–F.

Horizon and Locality: The Upper Cretaceous Deccan intertrappean beds of Kelapur, Yeotmal District (Maharashtra).

Referred Material: Two poorly preserved vertebrae (DUGF/562 and 563) from Kelapur.

Discussion: Rage et al. (2004) reported the possible presence of *Indophis sahnii* at Kelapur on the basis of two vertebrae (Rage et al. 2004: Fig. 3F–H). Because of the poor preservation of the specimens, the referral was uncertain. At that time, only one species was assigned to *Indophis, I. sahnii*, which was based on material from the Maastrichtian of Naskal (Rage and Prasad 1992). Subsequently, Pritchard et al. (2014) described a second species, *I. fanambinana* from the Maastrichtian of Madagascar. The vertebrae from Kelapur are distorted, nevertheless the shape of their zygosphene and the relative size of their cotyle show that they cannot belong to the Malagasy species. On the other hand, in view of their state of preservation, the Kelapur vertebrae can only tentatively be referred to *Indophis sahnii*.

As for all nigerophiids, *Indophis* was an aquatic snake. Confirmed *I. sahnii* were recovered from the Maastrichtian of Naskal and Anjar, India (Rage et al. 2004). It is worth noting that assignment of *Indophis* to Nigerophiidae was first regarded as doubtful (Rage and Prasad 1992). However, Pritchard et al. (2014) dismissed those concerns and assigned *Indophis* to Nigerophiidae, which is likely accurate. Nigerophiidae are known from the Late Cretaceous of Africa, Madagascar, and India and from the Paleocene of Africa (Rage 1975; LaDuke et al. 2010; Pritchard et al. 2014). Questionable nigerophiids have been reported from more northern territories: "*Nessovophis*" *zhylga* from the early Eocene of Central Asia (Averianov 1997; Rage et al. 2003) and *Woutersophis novus* from the middle Eocene of Europe (Rage 1980).

Remarks on squamates from Kisalpuri and Kelapur: The squamate assemblage from Kisalpuri and Kelapur includes some lizards and snakes. On the basis of maxillae, three or four taxa are distinguished among lizards. Unfortunately, they cannot be identified confidently, even at higher taxonomic level, although the presence of scincomorph(s) is suspected. In addition, diagnostic vertebrae of two snakes, a madtsoiid and a nigerophiid, were recovered from Kelapur. Kisalpuri did not produce any snake fossils.

Taxonomic Status of Amphibians and Squamates from the Maastrichtian of India: An Overview

Amphibians

A fragment of bone bearing empty tooth slots from the Maastrichtian of Takli was assigned to urodeles, i.e., Caudata, by Sahni et al. (1982). However, no caudatan was recovered from the rich faunas discovered subsequently. Therefore, the presence of this group in the Indian Maastrichtian remains to be confirmed on the basis of more diagnostically reliable remains.

Anuran amphibians make up an important part of Late Cretaceous microvertebrate assemblages. One notable characteristic of the anuran fauna from India is the absence of Pipidae or Pipimorpha. Pipidae are aquatic (freshwater) frogs that occur, as living and extinct forms, only in Africa and South America. In Africa, they are the dominant taxon among extinct frogs and occur in most post-Cenomanian localities (Gardner and Rage 2016). Pipids and pipimorphs may be absent in India and Madagascar, because they appeared in Africa only after its separation from the India-Madagascar block. The earliest pipid from Africa was recovered from the Cenomanian and even the earliest representative of the more inclusive pipimorphs may have occurred after the Africa's isolation (Gardner and Rage 2016).

Another problem is the purported occurrence of Gobiatidae in the Maastrichtian of India. Gobiatinae, as a subfamily of the Costata (i.e., Discoglossidae s. l.), was reported from the Maastrichtian of India (Sanchiz 1998; Prasad and Rage 2004). This taxon is now regarded as a family distinct from Costata (Roček and Nessov 1993; Gubin 1999; Roček 2008). In addition, Gubin (1999) and Wang et al. (2000) recovered gobiatines as a pelobatoid group in their phylogenetic analyses, not closely related to Costata. With regard to the Indian fossils, two cases should be distinguished: a possible gobiatid from Naskal (Prasad and Rage 2004) and those from Asifabad, Pisdura, and Takli (Sanchiz 1998). At Naskal, the frog in question is represented by some ilia that were first assigned to Discoglossidae, i.e., to Costata without more precision (Prasad and Rage 1991). Subsequently, Prasad and Rage (1995) suggested referral to Discoglossidae s. s., i.e., to Alytidae sensu Frost et al. (2006), and later to Gobiatinae (Prasad and Rage 2004). The assignment of the Naskal ilia to gobiatines was based on their similarity to ilia from the Cretaceous of Central Asia that were first referred, erroneously, to Pelobatidae (Nessov 1981, 1988), and subsequently assigned to gobiatids by Roček and Nessov (1993). Sanchiz (1998) included gobiatids in Costata, as Gobiatinae, leading to Prasad and Rage (2004) assigning the Naskal ilia to that subfamily. It is worth noting that the family Gobiatidae is mainly based on articulated specimens (Roček 2013). However, Roček and Nessov (1993) assigned isolated ilia from the Cretaceous of Central Asia to gobiatids, as Gobiatidae indet., but they did not substantiate this referral. The only confirmed gobiatid ilia are those in two articulated specimens from the Late Cretaceous of Mongolia: the holotype of *Cretasalia* (Gubin 1999) from the Maastrichtian and an articulated specimen from the Campanian referred to as *Gobiates* sp. by Roček (2008). In the *Cretasalia* specimen, only the shafts of the ilia are observable; a dorsal crest is absent (Roček 2013). The specimen of *Gobiates* sp. lacks its skull, but the vertebrae and pectoral girdle secure assignment at genus level. The ilia are partly embedded in the matrix but most traits may be observed. They lack a dorsal crest and their *tuber superius* is similar to that of the Naskal specimens. A spiral groove is present just anterior to the *tuber superius* as in various ilia from Naskal (Prasad and Rage 1991: Fig. 2A), but not in all (Prasad and Rage 2004). This character might be an argument for assigning the Naskal specimens to gobiatids but, on the other hand, these specimens show differences from confirmed gobiatid ilia. In *Gobiates* sp., an interiliac tubercle is present (Roček 2008: Fig. 2E1), whereas ilia from Naskal lack it. In addition, the ilium of *Gobiates* sp. appears to lack an elongate ischiadic process, whereas this process is pointed posteriorly in ilia from Naskal. Unfortunately, this character cannot be regarded as ascertained in *Gobiates* sp., because the posterior parts of ilia are partly damaged in the few unequivocally known ilia for that genus. Finally, the distinction between Alytidae and Gobiatidae is not clear if based only on what is presently known for their ilia. Because of their elongated ischiadic process, the ilia from Naskal likely belong to Costata as suggested formerly (as Discoglossidae s. s.) by Prasad and Rage (1995) and by Sanchiz (1998), but assignment to Gobiatidae cannot be definitely discarded. The problem of identification of these ilia (costatans or gobiatids?) is perfectly illustrated by anuran material recently described from the Bostobe Formation (Santonian-Campanian of Kazakhstan) by Skutchas and Kolchanov (2016). They identified both Costata (as Discoglossidae) and Gobiatidae in the locality, but they identified the only available ilium simply to Anura indet.

According to Sanchiz (1998: 147, 155, 168), gobiatids might also be present at Asifabad, Pisdura, and Takli (Gitti Khadan), three Maastrichtian sites in India. However, the ilium from Takli is typical of pelobatids, more specifically it lacks a *tuber superius* but has a spiral groove (Sahni et al.

Table 6.1 Distribution of fossil amphibians and squamates in the Upper Cretaceous Lameta Formation and Deccan intertrappean beds of India

Stratigraphic position and locality	Age	Amphibians	Squamates
Intertrappean Beds of Bombay, Maharashtra	Late Cretaceous – Early Paleocene	Leptodactylidae (Myobatrachinae): *Indobatrachus pusillus* (Noble, 1930) (=*Rana pusilla* Owen, 1947; *Oxyglossus pusillus* Stoliczka, 1869; *Indobatrachus trivialis* Chiplonkar, 1940; *Indobatrachus malabaricus* Verma, 1965)	
Intertrappean Beds of Naskal, Telangana	Late Cretaceous (Maastrichtian)	Discoglossidae gen. et sp. indet. (Prasad and Rage, 1991) Discoglossidae, ?Hylidae, Indeterminate Anura (Prasad and Rage, 1995) Discoglossidae: Gobiatinae indeterminate genus, ?Gobiatinae, ?Leptodactylidae or Hemisotidae: indeterminate genus 1 and ?indeterminate genus 2 Ranoidea family indeterminate, Indeterminate anurans (Prasad and Rage, 2004)	?Nigerophiidae: *Indophis sahnii* Rage and Prasad, 1992 Serpentes incertae sedis (Rage and Prasad, 1992) Anguidae, Indeterminate Lacertilia, Serpentes: ?Cholophidia ?Nigerophiidae: *Indophis sahnii* Rage and Prasad, 1992 (Prasad and Rage, 1995) Aniliidae s. l.: *Coniophis* sp. Indeterminate family, genus and species A and B (Rage et al., 2004)
Intertrappean Beds of Nagpur (=Takli), Maharashtra	Late Cretaceous (Maastrichtian	Pelobatidae indet., Caudata (Sahni et al., 1982)	Indeterminate scincomorph (Gayet et al., 1984) Eolacertilia incertae sedis: *Litakis* sp., Iguanidae: *Pristiguana* sp., Iguanidae gen. et sp. indet., Scincidae: ? *Contogenys* sp., Xenosauridae: *Exostinus estesai* Rana, 2005 (Rana, 2005) Indeterminate boid snake (Sahni et al., 1982)
Intertrappean Beds of Rangapur, Telangana	Late Cretaceous (Maastrichtian)		Agamidae: *Agama* sp., Scincidae: *Eumeces* sp. (Rana, 2005)
Intertrappean Beds of Anjar, Gujarat	Late Cretaceous (Maastrichtian)		?Nigerophiidae: *Indophis sahnii* Rage and Prasad, 1992 (Rage et al., 2004)
Intertrappean Beds of Asifabad, Telangana	Late Cretaceous (Maastrichtian)	Anura incertae sedis (Prasad and Sahni, 1987) or Gobiatidae (Sanchiz, 1998)	Lacertilia indet., Boidae gen. et sp. indet. (Prasad and Sahni, 1987)
Intertrappean Beds of Kisalpuri, Madhya Pradesh	Late Cretaceous (Maastrichtian)	?Costata, Leptodactylid–Hemisotid morphotype, Ranoid–Myobatrachoid morphotype, ?Ranoid morphotype, unknown morphotypes Form A and B, indeterminate anurans (present work)	Indeterminate lizards (present work)
Intertrappean Beds of Kelapur, Maharashtra	Late Cretaceous (Maastrichtian)	?Costata, Indeterminate anurans (present work)	Indeterminate lizards (present work) Madtsoiidae gen. et sp. indet., ?Nigerophiidae: *Indophis sahnii* Rage and Prasad, 1992 (Rage et al., 2004)
Lameta Formation (=Infratrappean Beds) of Pisdura, Maharashtra	Late Cretaceous (Maastrichtian)	Pelobatidae gen. et sp. indet. (Jain and Sahni, 1983) or Gobiatidae (Sanchiz, 1998)	Boidae gen. et sp. indet. (Jain and Sahni, 1983) Madtsoiidae: *Madtsoia pisdurensis* (Mohabey et al., 2011)
Lameta Formation (=Infratrappean Beds) of Dholi Dungri, Gujarat	Late Cretaceous (Maastrichtian)		Madtsoiidae: *Sanajeh indicus* (Wilson et al., 2010)

1982: Fig. 2A). In addition, the associated humerus has a well-developed lateral epicondyle, whereas gobiatids appear to lack it (Roček 2008). Consequently, the frog from Takli cannot be referred to the Gobiatidae. Moreover, the specimens from Pisdura and Asifabad do not appear to be identifiable at family level (Jain and Sahni 1983; Prasad and Sahni 1987).

In India, only the ilia from Naskal may belong to the Gobiatidae, but this is not confirmed, because the specimens lack an interiliac tubercle. Whatever the taxonomic affinities of the ilia in question, the ilia from Naskal are similar to those from the Coniacian-Campanian of Central Asia. More specifically, they appear to be identical to the ilium from the Santonian-Campanian of Kazakhstan in which the acetabulum is significantly large (Skutchas and Kolchanov 2016).

Finally, taking into account all fossiliferous localities, the anuran fauna from the Maastrichtian of India includes the above described six or seven taxa from Kisalpuri and Kelapur, to which should be added *Indobatrachus* from Bombay, the pelobatid from Takli, and two frogs from Naskal (the one labelled "Gobiatinae" by Prasad and Rage (2004) and a possible Hylidae [Prasad and Rage 1995]) (Table 6.1).

In relation to the morphology of ilia, a remark by Roček et al. (2013) deserves attention. They noted that the ilia of anurans from the Mesozoic typically lack dorsal crests and that ilia with dorsal crests are extremely rare. However, on the contrary, the ilia of more than half the anurans from the Mesozoic of India bear such a crest. In India, a dorsal crest occurs on the ilia of the six or seven taxa from Kisalpuri and Kelapur, whereas it is absent on the ilia of the pelobatid from Takli, of *Indobatrachus* from the intertrappean beds of Bombay, and of the ?gobiatid and ?hylid from Naskal. In fact, Roček et al.'s (2013) remark pertains essentially to North American taxa. It is true that, in the Mesozoic of North America, almost all reported anurans have ilia without dorsal crests. For example, among the 65 morphotypes recognized by Roček et al. (2013) in the Late Cretaceous of Utah, only one has a dorsal crest. The dominance of taxa whose ilia lack a dorsal crest also holds true for Laurasian Asia, where only two of the 13 named Mesozoic taxa have a dorsal crest (Dong et al. 2013; Ikeda et al. 2016). In addition, Roček and Nessov (1993) described numerous indeterminate anurans from the Late Cretaceous of Central Asia whose ilia are devoid of dorsal crests. In Europe, as in India, the ratio is inverted. Eight named European taxa have a crest, whereas only two lack it (Szentesi and Venczel 2012; Roček 2013). In Gondwana, out of a dozen or so genera described from the Mesozoic of South America, only three have a crest (Báez et al. 2009; Agnolin 2012). The two genera known from Madagascar both have ilia without crests (Rage and Roček 1989; Evans et al. 2008, 2014). In Africa, where few Mesozoic anuran taxa have been named, ilia with a dorsal crest are more frequent than ilia without crests (JCR, pers. obs.). From this overview, it may be stated that if forms without a dorsal crest on the ilium are overwhelmingly dominant in the Mesozoic of North America, the situation is more variable on other continents. However, it may be noted also, that ilia with dorsal crests are dominant in continents from which the number of known Mesozoic anuran remains is low (i.e., India, Europe, and Africa).

Squamates

Lizards from the Maastrichtian of India are not numerous. Rana (2005) reported a lizard fauna consisting of *Litakis* sp. (as Eolacertilia incertae sedis), *Pristiguana* sp. (Iguanidae), Iguanidae indet. *Agama* sp. (Agamidae), ?*Contogenys* sp., *Eumeces* sp. (Scincidae), and *Exostinus estesai* Rana, 2005 (Xenosauridae) from the intertrappean beds of Nagpur and Rangapur. However, precise identifications based on these specimens need to be confirmed by additional well-preserved specimens (Prasad 2012). Aside from the taxa reported by Rana (2005), an Anguidae from Naskal (Prasad and Rage 1995) can be added to the three or four taxa from Kisalpuri and Kelapur (Table 6.1). Two possible scincomorphs were reported from Takli by Gayet et al. (1984). One of them is represented by a maxilla, but comparison with specimens from other localities is difficult because it is embedded in matrix; it may belong to the same taxon as one of the lizards with blunt teeth from Kisalpuri and Kelapur.

Snakes are more frequent than lizards in the Maastrichtian of India. They include a primitive, perhaps stem snake, three madtsoiids, one nigerophiid, at least one indeterminate and somewhat peculiar taxon, and perhaps booids (if the last are not madtsoiids). The primitive snake was recovered only from Naskal and was assigned to *Coniophis* s. l. (Rage et al. 2004). *Coniophis* is likely an artificial assemblage that may comprise stem snakes and basal alethinophidian snakes (Rage and Augé 2010; Longrich et al. 2012a). Madtsoiidae of the Indian Maastrichtian are *Sanajeh indicus* from Dholi Dungri (Wilson et al. 2010), *Madtsoia pisdurensis* from Pisdura (Mohabey et al. 2011) and the above reported indeterminate taxon from Kelapur (Rage et al. 2004). The last is a small madtsoiid that cannot be a juvenile of either of the two other large-sized taxa. The nigerophiid is *Indophis sahnii*, an aquatic snake that appears to be the most frequent squamate in the Maastrichtian of India. Finally, an indeterminate snake ("indeterminate genus and species A"; Rage et al. 2004) represents a distinct taxon; it is known by a single, incomplete vertebra and cannot be identified more precisely. Other specimens from the Maastrichtian were reported as indeterminate snakes; potentially, some of those may represent additional, distinct taxa. In addition, as stated above, the presence of booids, which may be confused with madtsoiids, cannot be dismissed. It should be noted that Boidae initially were reported from the Indian Maastrichtian by Sahni et al. (1982) and Gayet et al. (1984). However, at that time madtsoiids were included in Boidae as a subfamily and these reports refer to Madtsoiidae, not to Boidae, as they are accepted today.

Comparison with the Anuran and Squamate Diversities in Other Continents

In geographically close Gondwanan landmasses of Africa and Madagascar, the diversity of amphibians is poor in the Maastrichtian. In Africa, only one locality of confirmed Maastrichtian age yielded one frog (a Pipidae; Gardner and Rage 2016). This paucity might be a bias, but it affects all terrestrial groups in the Late Cretaceous of Africa and it appears to be a characteristic of that continent (Gheerbrant and Rage 2006). In the Maastrichtian of Madagascar, two taxa (or more ?) of anurans have been reported (Evans et al. 2014; Gardner and Rage 2016). South America deserves special attention because biogeographical connections during the Mesozoic have been proposed between that continent and the Madagascar-India block (Krause et al. 1997; Rage 2003; Gheerbrant and Rage 2006). Comparison of Cretaceous anuran faunas from India and South America is difficult, because in India frogs are represented by numerous isolated bones, whereas in South America the record consists mostly of a few articulated specimens. It may be stated that neobatrachians are present in both South America (Báez et al. 2012) and India (this study), but whereas South America lacks non-neobatrachians such as costatans and pelobatids, India instead lacks pipimorphs.

In Laurasian continents, the diversity of anurans is generally higher, approximately similar to that of the Indian Maastrichtian. In the Maastrichtian of North America (Gardner and DeMar 2013) and Europe (Roček 2013), the number of taxa does not appear to be significantly different from that in the Maastrichtian of India; however, taxa are likely different at higher taxonomic level, except probably Costata. In Laurasian Asia, the diversity of Maastrichtian frogs is lower than in India (Roček 2013), which is likely a bias because this territory is still poorly explored.

Squamates from the Maastrichtian of India cannot be compared to coeval faunas of Africa; in the latter continent, only highly aquatic, marine squamates were recovered from the Maastricthian: mosasauroids, varanoids, and a paleophiid snake (if the geological age of the last group is confirmed) (Bardet et al. 2010). The Maastrichtian of Madagascar produced a single lizard, a cordyliform (Krause et al. 2003), whereas its snake fauna is diverse (six taxa; LaDuke et al. 2010; Pritchard et al. 2014). The snake assemblage of the Malagasy Maastrichtian is reminiscent of that from India, being composed of madtsoiids and nigerophiids (LaDuke et al. 2010; Pritchard et al. 2014). In addition, the snake reported as "Serpentes incertae sedis" from the Malagasy Maastrichtian (Pritchard et al. 2014) may belong to the same taxon as the "indeterminate genus and species A" from Naskal (Rage et al. 2004). They both show a peculiar character, the relatively anterior position of the ventral most projection of the hemal keel (a hypapophysis?). Precise comparisons of Indian squamates with those from South America are difficult because, in the latter continent, most squamates were recovered from localities whose age is not precisely determined, they are either Campanian or Maastrichtian. However, it may be stated that in the latest Cretaceous of South America, as in other Gondwanan continents, lizards are less numerous than snakes. Lizards are represented by an iguanid, a taxon unknown from India, whereas South American snakes comprise a confirmed anilioid (Gómez et al. 2008), several madtsoiids, and perhaps a booid (Albino 2007).

The overall composition of the squamate fauna from India is different from that of the Maastrichtian of the Laurasian continents in that, contrary to India, lizards are more numerous and more diverse than snakes in Laurasia. In the Maastrichtian of North America, the number and diversity of lizards are high, whereas only three snake taxa have been reported (Longrich et al. 2012b; Nydam 2013). In Europe, lizards are less numerous than in North America, but they still clearly outnumber snakes (Blain et al. 2010; Csiki-Sava et al. 2015; Pereda-Suberbiola et al. 2015). In Laurasian Asia, the latest Cretaceous produced several taxa of lizards (Alifanov 2000; Gao and Norell 2000), but no snakes. In addition, iguanian lizards that have been reported from the Late Cretaceous of all Laurasian continents (Blain et al. 2010; Alifanov 2013; Nydam 2013), are unknown from India. On the other hand, anguids that are present in the Maastrichtian of India are frequent components of latest Cretaceous squamate assemblages on Laurasian continents (Blain et al. 2010; Nydam 2013).

On the whole, the Maastrichtian anuran and squamate fauna of India includes taxa with Laurasian and Gondwanan affinities, but the biogeographic affinities of several taxa from India remain unknown. Taxa indicative of Laurasian relationships are pelobatid frogs, anguid lizards and, if really present, costatan frogs. In addition, the anuran taxon reported from Naskal as a gobiatine indicates Laurasian affinities; whatever its actual affinities, this taxon is present in both the Maastrichtian of India and in the Coniacian-Campanian of Central Asia, i.e., Laurasian Asia (Prasad and Rage 2004; Skutschas and Kolchanov 2016). The problem of the presence of Laurasian forms in the Maastrichtian of India was discussed by, e.g., Prasad et al. (1995, 2010), Prasad and Sahni (1999), Chatterjee and Scotese (2010) and recently by Vérard et al. (2017). Gondwanan taxa are Madtsoiidae and Nigerophiidae. In addition, the presence of the taxa previously used to identify anuran morphotypes (leptodactylids, hemisotids, ranoids, and myobatrachoids) cannot be rejected. These taxa all appear to be of Gondwanan origin. In other words, the amphibian and squamate assemblages confirm the biogeographically mixed nature of faunas from the Maastrichtian of India.

Conclusions

Prior to the present work, Maastrichtian amphibian and squamate remains from India were described mainly from the western (Bombay) and southern (Naskal) parts of the Deccan Volcanic Province, and isolated records were reported from the northwestern and eastern areas. The new fossil remains from the rich intertrappean beds of Kisalpuri and Kelapur significantly extend the geographic distribution of anurans and squamates to widely separated areas in central and northeastern parts of the Deccan Volcanic Province. The composition of the new faunas from Kisalpuri and Kelapur is consistent with a mixed origin (Gondwanan and Laurasian) for the Indian Maastrichtian assemblages. However, in the material from Kisalpuri and Kelapur, taxa of Gondwanan or unknown geographic affinities are largely dominant.

Interestingly, the anuran, lizard, and snake remains occur in relatively large numbers in mammal-bearing Deccan intertrappean horizons. The intertrappean beds of Naskal and Kisalpuri yielded both eutherian and gondwanatherian mammals (Prasad and Sahni 1988; Prasad et al. 1994, 2007a, b, 2010; Verma et al. 2012) and are also a source of a large number of anuran and squamate remains. Though, the intertrappean beds of Kelapur have yielded many anuran, lizard, and snake remains, so far only one mammalian humerus was recovered from this intertrappean section. However, it is a matter of scale. For fossiliferous levels that were prospected for mammals, the Naskal and Kisalpuri sites have been more extensively screen-washed and sorted. Hence the co-occurrence of anuran, lizard, snake, and mammal remains in these intertrappean horizons is possibly an artifact of depositional environments (flood-plain lacustrine environments). In view of the presence of a freshwater lacustrine and terrestrial biota, the Naskal intertrappean beds were considered to have been deposited in a freshwater floodplain lake distal in position to the sea coast (Prasad and Khajuria 1996). The Kelapur intertrappean beds with a similar faunal and floral composition are inferred to represent analogous depositional environment. The Kisalpuri intertrappean beds yielding marine/brackish water elements like myliobatid (*Igdabatis indicus*) and pycnodontid fishes were deposited in freshwater lacustrine basin proximal to the sea coast with intermittent connections with the sea. On the other hand, the well-studied infratrappean beds of Marepalli (Prasad and Cappetta 1993) and Pisdura (Jain and Sahni 1983), and the intertrappean beds of Asifabad (Prasad and Sahni 1987; Prasad and Cappetta 1993), Nagpur (Gayet et al. 1984), and Piplanarayanwar (Lourembam et al. 2017) yielded predominantly marine/brackish water fauna and limited anuran and squamate remains. Therefore, future prospecting for Cretaceous mammals in Deccan volcano-sedimentary sequences in freshwater lacustrine horizons rich in anuran and squamate remains and lacking or low in marine/brackish water elements may result in the discovery of mammals.

Acknowledgements This paper is dedicated to Prof. Ashok Sahni, who was instrumental in training a new generation of vertebrate paleontologists in India. We are indebted to Susan E. Evans, James D. Gardner and Zbyněk Roček for their reviews and constructive remarks. This research was supported by J. C. Bose National Fellowship, by the Science & Engineering Research Board (SERB), New Delhi to G. V. R. Prasad. The comparative study of recent and fossil anuran material was undertaken by G. V. R. Prasad at the Muséum national d'Histoire naturelle (MNHN), Paris as Maître de Conferences of MNHN.

References

Agnolin, F. (2012). A new Calyptocephalellidae (Anura, Neobatrachia) from the Upper Cretaceous of Patagonia, Argentina, with comments on its systematic position. *Studia Geologica Salmanticensia, 48,* 129–178.

Albino, A. M. (2007). Lepidosauromorpha. In Z. Gasparini, L. Salgado, & R. A. Coria (Eds.), *Patagonian Mesozoic Reptiles* (pp. 87–115). Bloomington and Indianapolis: Indiana University Press.

Alifanov, V. R. (2000). The fossil record of Cretaceous lizards from Mongolia. In M. J. Benton, M. A. Shishkin, D. M. Unwin, & E. N. Kurochkin (Eds.), *The age of dinosaurs in Russia and Mongolia* (pp. 368–389). Cambridge: Cambridge University Press.

Alifanov, V. R. (2013). *Desertiguana gobiensis* gen. et sp. nov., a new lizard (Phrynosomatidae, Iguanomorpha) from the Upper Cretaceous of Mongolia. *Paleontological Journal, 47,* 417–424.

Averianov, A. O. (1997). Paleogene sea snakes from the eastern part of the Tethys. *Russian Journal of Herpetology, 4,* 128–142.

Báez, A. M., Gómez, R. O., Ribeiro, L. C. B., Martinelli, A. G., Teixeira, V. P. A., & Ferraz, M. L. F. (2012). The diverse Cretaceous neobatrachian fauna of South America: *Uberabatrachus carvalhoi*, a new frog from the Maastrichtian Marília Formation, Minas Gerais, Brazil. *Gondwana Research, 22,* 1141–1150.

Báez, A. M., Moura, G. J. B., & Gómez, R. O. (2009). Anurans from the Lower Cretaceous Crato Formation of northeastern Brazil: Implications for the early divergence of neobatrachians. *Cretaceous Research, 30,* 829–846.

Báez, A. M., & Sanchiz, B. (2007). A review of *Neusibatrachus wilferti*, an early Cretaceous frog from the Montsec Range, northeastern Spain. *Acta Palaeontologica Polonica, 52,* 477–487.

Bardet, N., Pereda Suberbiola, X., Jouve, S., Bourdon, E., Vincent, P., Houssaye, A., et al. (2010). Reptilian assemblages from the latest Cretaceous—Paleogene phosphates of Morocco: From Arambourg to present time. *Historical Biology, 22,* 186–199.

Blain, H. A., Canudo, J. I., Cuenca-Bescós, G., & López-Martínez, N. (2010). Amphibians and squamate reptiles from the latest Maastrichtian (Upper Cretaceous) of Blasi 2 (Huesca, Spain). *Cretaceous Research, 31,* 433–446.

Brongniart, A. (1800). Essai d'une classification naturelle des reptiles. 1e partie. Etablissement des ordres. *Bulletin des Sciences, Société Philomatique de Paris, 2,* 81–82.

Caldwell, M. W., & Lee, M. S. Y. (1997). A snake with legs from the marine Cretaceous of the Middle East. *Nature, 386,* 705–709.

Chatterjee, S., & Scotese, C. (2010). The wandering Indian Plate and its changing biogeography during the Late Cretaceous-Early Tertiary Period. In S. Bandyopadhyay (Ed.), *New aspects of Mesozoic biodiversity* (pp. 105–126). Lecture Notes in Earth Sciences Dordrecht: Springer.

Chiplonker, G. W. (1940). A new species of frog from the intertrappean beds of Worli Hill, Bombay. *Journal of Bombay Natural History Society, 51,* 799–804.

Csiki-Sava, Z., Buffetaut, E., Ősi, A., Pereda-Suberbiola, X., & Brusatte, S. L. (2015). Island life in the Cretaceous—Faunal composition, biogeography, evolution, and extinction of land-living vertebrates on the Late Cretaceous European archipelago. *ZooKeys, 469,* 1–161.

Dong, L., Roček, Z., Wang, Y., & Jones, M. E. H. (2013). Anurans from the Lower Cretaceous Jehol Group of Western Liaoning. *China. PLoS ONE, 8,* e69723.

Evans, S. E., Groenke, J. R., Jones, M. E. H., Turner, A. H., & Krause, D. W. (2014). New material of *Beelzebufo*, a hyperossified frog (Amphibia: Anura) from the Late Cretaceous of Madagascar. *PLoS ONE, 9,* e87236.

Evans, S. E., Jones, M. E. H., & Krause, D. W. (2008). A giant frog with South American affinities from the Late Cretaceous of Madagascar. *Proceedings of the National Academy of Sciences, 105,* 2951–2956.

Evans, S. E., Milner, A. R., & Mussett, F. (1990). A discoglossid frog from the Middle Jurassic of England. *Palaeontology, 33,* 299–311.

Fischer von Waldheim, G. (1813). *Zoognosia tabulis synopticis illustrata, in usum prælectionum Academiæ Imperialis Medico-Chirurgicæ Mosquensis edita.* (3rd ed., Vol. 1). Moscow: N.S. Vsevolozsky.

Fitzinger, L. J. F. J. (1843). *Systema reptilium. Fasciculus primus.* Wien: Braumüller und Seidel.

Folie, A., Rana, R. S., Rose, K. D., Sahni, A., Kumar, K., Singh, L., et al. (2013). Early Eocene frogs from Vastan Lignite Mine, Gujarat, India. *Acta Palaeontologica Polonica, 58*(3), 511–524.

Frost, D., Grant, T., Faivovich, J., Bain, R. H., Haas, A., Haddad, C. F. B., et al. (2006). The amphibian tree of life. *Bulletin of the American Museum of Natural History, 297,* 1–370.

Gaffney, E. S., Chatterjee, S., & Rudra, D. K. (2001). *Kurmademys*, a new side-necked turtle (Pelomedusoides: Bothremydidae) from the Late Cretaceous of India. *American Museum Novitates, 3321,* 1–16.

Gao, K., & Norell, M. A. (2000). Taxonomic composition and systematics of Late Cretaceous lizard assemblages from Ukhaa Tolgod and adjacent localities, Mongolian Gobi Desert. *Bulletin of the American Museum of Natural History, 249,* 1–118.

Gardner, J. D., & DeMar, D. G., Jr (2013). Mesozoic and Paleocene lissamphibian assemblages of North America: A comprehensive review. In J. D. Gardner & R. L. Nydam (Eds.), *Mesozoic and Cenozoic lissamphibian and squamate assemblages of Laurasia. Palaeobiodiversity and Palaeoenvironments, 93,* 459–515.

Gardner, J. D., & Rage, J. C. (2016). The fossil record of lissamphibians from Africa, Madagascar, and the Arabian Plate. In J. D. Gardner & T. Přikryl (Eds.), *Contributioons in honour of Zbyněk Roček. Palaeobiodiversity and Palaeoenvironments, 96,* 169–220.

Gayet, M., Rage, J. C., & Rana, R. S. (1984). Nouvelles ichthyofaune et herpétofaune de Gitti Khadan, le plus ancien gisement connu du Deccan (Crétacé/Paléocène) à microvertébrés. Implications paléogéographiques. In E. Buffetaut, J. J. Jaeger & J. C. Rage (Eds.), *Paléogéographie de l'Inde, du Tibet et du Sud-Est asiatique. Mémoire de la Société géologique de France (1985), 147,* 55–65.

Gheerbrant, E., & Rage, J. C. (2006). Palaeobiogeography of Africa: How distinct from Gondwana and Laurasia? *Palaeogeography, Palaeoclimatology, Palaeoecology, 241,* 224–246.

Gómez, R. O., Báez, A. M., & Rougier, G. W. (2008). An anilioid snake from the Upper Cretaceous of northern Patagonia. *Cretaceous Research, 29,* 481–488.

Gómez, R. O., & Turazzini, G. F. (2016). An overview of the ilium of anurans (Lissamphibia, Salientia), with a critical appraisal of the terminology and primary homology of main ilial features. *Journal of Vertebrate Paleontology, 36,* e1030023.

Goswami, A., Prasad, G. V. R., Verma, O., Flynn, J. J., & Benson, R. B. J. (2013). A troodontid dinosaur from the latest Cretaceous of India. *Nature Communications, 4,* 1703.

Gray, J. E. (1825). A synopsis of the genera of Reptiles and Amphibia, with a description of some new species. *Annals of Phylosophy, 10,* 193–217.

Gubin, Y. M. (1999). Gobiatids (Anura) from the Upper Cretaceous locality Khermeen-Tsav (Gobi Desert, Mongolia). *Paleontological Journal, 33,* 77–87.

Günther, A. (1858). On the systematic arrangement of the tailless batrachians and the structure of *Rhinophrynus dorsalis*. *Proceedings of the Zoological Society of London, 1858,* 339–352.

Haeckel, E. (1866). *Generelle Morphologie der Organismen.* Berlin: Reimer.

Halliday, T. J. D., Cuff, A. R., Prasad, G. V. R., Thanglemmoi, M. S., & Goswami, A. (2016). New record of *Egertonia* (Elopiformes, Phyllodontidae) from the Late Cretaceous of South India. *Papers in Palaeontology, 2,* 1–8.

Henrici, A. C. (1991). *Chelomophrynus bayi* (Amphibia, Anura, Rhinophrynidae), a new genus and species from the middle Eocene of Wyoming: Ontogeny and relationships. *Annals of Carnegie Museum, 60,* 97–144.

Hoffstetter, R. (1961). Nouveaux restes d'un serpent boïdé (*Madtsoia madagascariensis* nov. sp.) dans le Crétacé supérieur de Madagascar. *Bulletin du Muséum national d'Histoire naturelle, 33,* 152–160.

Ikeda, T., Ota, H., & Matsui, M. (2016). New fossil anurans from the Lower Cretaceous Sasayama Group of Hyogo Prefecture, Western Honshu, Japan. *Cretaceous Research, 61,* 108–123.

Jain, S. L., & Sahni, A. (1983). Some Upper Cretaceous vertebrates from Central India and their paleogeographic implications. In H. K. Maheshwari (Ed.), *Cretaceous of India* (pp. 66–83). Lucknow: Indian Association of Palynostratigrapher's Symposium.

Khosla, A., Prasad, G. V. R., Verma, O., Jain, A. K., & Sahni, A. (2004). Discovery of a micromammal-yielding Deccan intertrappean site near Kisalpuri, Dindori District, Madhya Pradesh. *Current Science, 87,* 380–383.

Khosla, A., & Sahni, A. (1995). Parataxonomic classification of the Late Cretaceous dinosaur eggshells from India. *Journal of the Palaeontological Society of India, 40,* 87–102.

Khosla, A., & Sahni, A. (2000). Late Cretaceous (Maastrichtian) ostracodes from the Lameta Formation, Jabalpur Cantonment area, Madhya Pradesh, India. *Journal of the Palaeontological Society of India, 45,* 57–78.

Khosla, A., & Sahni, A. (2003). Biodiversity during the Deccan volcanic eruptive episode. *Journal of Asian Earth Sciences, 21,* 895–908.

Khosla, A., Sertich, J. J. W., Prasad, G.V.R., & Verma, O. (2009). Dyrosaurid remains from the intertrappean beds of India and the Late Cretaceous distribution of the Dyrosauridae. *Journal of Vertebrate Paleontology, 29*(4), 1321–1326.

Kosma, R. (2004). *The dentitions of recent and fossil scincomorphan lizards (Lacertilia, Squamata). Systematics, functional morphology, paleoecology.* Ph.D. dissertation, University of Hannover.

Krause, D. W., Evans, S. E., & Gao, K. Q. (2003). First definitive record of Mesozoic lizards from Madagascar. *Journal of Vertebrate Paleontology, 23,* 842–856.

Krause, D. W., Prasad, G. V. R., von Koenigswald, W., Sahni, A., & Grine, F. E. (1997). Cosmopolitanism among Gondwanan Late Cretaceous mammals. *Nature, 390,* 504–507.

Laduke, T. C., Krause, D. W., Scanlon, J. D., & Kley, N. J. (2010). Late Cretaceous (Maastrichtian) snake assemblage from the Maevarano Formation, Mahajanga Basin, Madagascar. *Journal of Vertebrate Paleontology, 30,* 109–138.

Lataste, F. (1879). Etude sur le *Discoglossus pictus* Otth. *Actes de la Société Linnéenne de Bordeaux, Série, 4,* 275–342.

Linnaeus, C. (1758). *Systema naturae* (10th ed.). Stockholm: Laurentius Salvius.

Longrich, N. R., Bhullar, B.-A. S., & Gauthier, J. A. (2012a). A transitional snake from the Late Cretaceous period of North America. *Nature, 488,* 205–208.

Longrich, N. R., Bhullar, B.-A. S., & Gauthier, J. A. (2012b). Mass extinction of lizards and snakes at the Cretaceous-Paleogene boundary. *Proceedings of the National Academy of Sciences, 109,* 21396–21401.

Lourembam, R. S., Prasad, G. V. R., & Grover, P. (2017). Ichthyofauna (Chondrichthyes, Osteichthyes) from the Upper Cretaceous intertrappean beds of Piplanarayanwar, Chhindwara District, Madhya Pradesh, India. *Island Arc, 26,* e12180.

Mohabey, D. M., Head, J. J., & Wilson, J. A. (2011). A new species of the snake *Madtsoia* from the Upper Cretaceous of India and its paleobiogeographic implications. *Journal of Vertebrate Paleontology, 31,* 588–595.

Nessov, L. A. (1981). Cretaceous salamanders and frogs of Kyzylkum Desert. In N. B. Ananjeva & L. J. Borkin (Eds.), *The fauna and ecology of amphibians and reptiles of the Palaearctic Asia. Proceedings of the Zoological Institute of the USSR Academy of Science, 101,* 57–88 (in Russian).

Nessov, L. A. (1988). Late Mesozoic amphibians and lizards of Soviet Middle Asia. In Z. Szyndlar (Ed.), Festschrift for M. Młynarski. *Acta Zoologica Cracoviensia, 31,* 475–486.

Noble, G. K. (1930). The fossil frogs of the intertrappean beds of Bombay, India. *American Museum Novitates, 401,* 1–13.

Nydam, R. L. (2013). Squamates from the Jurassic and Cretaceous of North America. In J. D. Gardner & R. L. Nydam (Eds.), *Mesozoic and Cenozoic lissamphibian and squamate assemblages of Laurasia. Palaeobiodiversity and Palaeoenvironments, 93,* 535–565.

Oppel, M. (1811). *Die Ordnungen, Familien und Gattungen der Reptilien als prodrom einer Naturgeschichte derselben.* Munich: J. Lindauer.

Osborn, H. F. (1903). On the primary division of the Reptilia into two sub-classes, Synapsida and Diapsida. *Science, 17,* 275–276.

Owen, R. (1847). On the batracholites, indicative of a small species of frogs (*Rana pusilla,* Ow.). *Quarterly Journal of the Geological Society of London, 3,* 224–225.

Pereda-Suberbiola, X., Corral, J. C., Astibia, H., Badiola, A., Bardet, N., Berreteaga, A., et al. (2015). Late Cretaceous continental and marine vertebrate assemblages of the Laño Quarry (Basque-Cantabrian Region, Iberian Peninsula): an update. *Journal of Iberian Geology, 41,* 101–124.

Prasad, G. V. R. (2012). Vertebrate biodiversity of the Deccan volcanic province of India: A comprehensive review. *Bulletin de la Société géologique de France, 183,* 597–610.

Prasad, G. V. R., & Cappetta, H. (1993). Late Cretaceous selachians from India and the age of the Deccan Traps. *Palaeontology, 36,* 231–248.

Prasad, G. V. R., Jaeger, J.-J., Sahni, A., Gheerbrant, E., & Khajuria, C. K. (1994). Eutherian mammals from the Upper Cretaceous (Maastrichtian) intertrappean beds of Naskal, Andhra Pradesh, India. *Journal of Vertebrate Palaeontoogy, 14,* 260–277.

Prasad, G. V. R., & Khajuria, C. K. (1990). A record of microvertebrate fauna from the intertrappean beds of Naskal, Andhra Pradesh. *Journal of the Palaeontological Society of India, 35,* 151–161.

Prasad, G. V. R., & Khajuria, C. K. (1996). Paleoenvironment of the Late Cretaceous mammal-bearing intertrappean beds of Naskal, Andhra Pradesh, India. In A. Sahni (Ed.), *Cretaceous Stratigraphy and Palaeoenvironments. Geological Society of India Memoir, 37,* 337–362.

Prasad, G. V. R., Khajuria, C. K., & Manhas, B. K. (1995). Paleobiogeographic significance of the Deccan infra- and inter-trappean biota from peninsular India. *Historical Biology, 9,* 319–334.

Prasad, G. V. R., & Rage, J.-C. (1991). A discoglossid frog in the latest Cretaceous (Maastrichtian) of India. Further evidence for a terrestrial route between India and Laurasia in the latest Cretaceous. *Comptes Rendus de l'Académie des Sciences (série II), 313,* 273–278.

Prasad, G. V. R., & Rage, J.-C. (1995). Amphibians and squamates from the Maastrichtian of Naskal, India. *Cretaceous Research, 16,* 95–107.

Prasad, G. V. R., & Rage, J.-C. (2004). Fossil frogs (Amphibia: Anura) from the Upper Cretaceous intertrappean beds of Naskal, Andhra Pradesh. *Revue de Paléobiologie, 23,* 99–116.

Prasad, G. V. R., & Sahni, A. (1987). Costal-plain microvertebrate assemblage from the terminal Cretaceous of Asifabad, Andhra Pradesh. *Journal of the Palaeontological Society of India, 32,* 5–19.

Prasad, G. V. R., & Sahni, A. (1988). First Cretaceous mammal from India. *Nature, 332,* 638–640.

Prasad, G. V. R., & Sahni, A. (1999). Were there size constraints on biotic exchanges during the northward drift of the Indian plate? *Proceedings of the Indian National Science Academy, 65 A,* 377–396.

Prasad, G. V. R., & Sahni, A. (2014). Vertebrate fauna from the Deccan volcanic province: Response to volcanic activity. In G. Keller & A. C. Kerr (Eds.), *Volcanism, impacts, and mass extinctions: Causes and effects. Geological Society of America Special Paper, 505,* 1–18.

Prasad, G. V. R., Sharma, A., Verma, O., Khosla, A., Singh, L. R., & Priyadarshini, R. K. (2015). Testudoid and crocodiloid eggshells from the Upper Cretaceous Deccan Intertrappean Beds of Central India. *Comptes Rendus Palevol, 14,* 513–526.

Prasad, G. V. R., Verma, O., Flynn, J. J., & Goswami, A. (2013). A new Late Cretaceous vertebrate fauna from the Cauvery basin, South India: Implications for Gondwanan paleobiogeography. *Journal of Vertebrate Paleontology, 33,* 1260–1268.

Prasad, G. V. R., Verma, O., Gheerbrant, E., Goswami, A., Khosla, A., Parmar, V., et al. (2010). First mammal evidence from the Late Cretaceous of India for biotic dispersal between India and Africa at the KT transition. *Comptes Rendus Palevol, 9,* 63–71.

Prasad, G. V. R., Verma, O., Sahni, A., Krause, D. W., Khosla, A., & Parmar, V. (2007b). A new Late Cretaceous gondwanatherian mammal from central India. *Proceedings of the Indian National Science Academy, 73,* 17–24.

Prasad, G. V. R., Verma, O., Sahni, A., Parmar, V., & Khosla, A. (2007a). A Cretaceous hoofed mammal from India. *Science, 318,* 937.

Pritchard, A. C., McCartney, J. A., Krause, D. W., & Kley, N. J. (2014). New snakes from the Upper Cretaceous (Maastrichtian) Maevarano Formation, Mahajanga Basin, Madagascar. *Journal of Vertebrate Paleontology, 34,* 1080–1093.

Rage, J.-C. (1975). Un serpent du Paléocène du Niger. Etude préliminaire sur l'origine des Caenophidiens (Reptilia, Serpentes). *Comptes Rendus de l'Académie des Sciences de Paris D, 281,* 515–518.

Rage, J.-C. (1980). Un serpent marin nouveau de l'Eocène de Belgique. Le problème des serpents marins du Paléogène. *Comptes Rendus de l'Académie des Sciences de Paris D, 291,* 469–471.

Rage, J.-C. (1998). Fossil snakes from the Paleocene of São José de Itaboraí, Brazil. Part I. Madtsoiidae. *Aniliidae. Palaeovertebrata, 27,* 109–144.

Rage, J.-C., & Augé, M. (2010). Squamate reptiles from the middle Eocene of Lissieu (France). A landmark in the middle Eocene of Europe. *Geobios, 43,* 243–268.

Rage, J.-C., Bajpai, S., Thewissen, J. G. M., & Tiwari, B. N. (2003). Early Eocene snakes from Kutch, Western India, with a review of the Paleophiidae. *Geodiversitas, 25,* 695–716.

Rage, J.-C., Folie, A., Rana, R. S., Singh, H., Rose, K. D., & Smith, T. (2008). A diverse snake fauna from the early Eocene of Vastan

Lignite Mine, Gujarat, India. *Acta Palaeontologica Polonica, 53,* 391–403.

Rage, J.-C., Métais, G., Bartolini, A., Brohi, I. A., Lashari, R. A., Marivaux, L., et al. (2014). First report of the giant snake *Gigantophis* (Madtsoiidae) from the Paleocene of Pakistan: Paleobiogeographic implications. *Geobios, 47,* 147–153.

Rage, J.-C., & Prasad, G. V. R. (1992). New snakes from the late Cretaceous (Maastrichtian) of Naskal, India. *Neues Jahrbuch für Geologie und Paläontologie Abhandlungen, 187,* 83–97.

Rage, J.-C., Prasad, G. V. R., & Bajpai, S. (2004). Additional snakes from the Uppermost Cretaceous (Maastrichtian) of India. *Cretaceous Research, 25,* 425–434.

Rage, J.-C., & Roček, Z. (1989). Redescription of *Triadobatrachus massinoti* (Piveteau, 1936) an Anuran Amphibian from the early Triassic. *Palaeontographica A, 206,* 1–16.

Rana, R. S. (2005). Lizard fauna from intertrappean (Late Cretaceous-Early Paleocene) beds of peninsular India. In D. M. Mohabey (Ed.), *Indian non-marine Late Cretaceous: Advances and challenges. Gondwana Geological Magazine, 8 Supplement,* 123–132.

Reig, O. A. (1958). Proposiciones para una neuva macrosistemática de los anuros (nota preliminar). *Physis, 21,* 109–118.

Richter, A. (1994). Lacertilia aus der Unteren Kreide von Uña und Galve (Spanien) und Anoual (Marokko). *Berliner geowissenschaftliche Abhandlungen E, 14,* 1–147.

Roček, Z. (2008). The late Cretaceous frog *Gobiates* from Central Asia: its evolutionary status and possible phylogenetic relationships. *Cretaceous Research, 29,* 577–591.

Roček, Z. (2013). Mesozoic and Tertiary Anura of Laurasia. In J. D. Gardner & R. L. Nydam (Eds.), *Mesozoic and Cenozoic lissamphibian and squamate assemblages of Laurasia. Palaeobiodiversity and Palaeoenvironments, 93,* 397–439.

Roček, Z., Eaton, J. G., Gardner, J. D., & Přikryl, T. (2010). Evolution of anuran assemblages in the Late Cretaceous of Utah, USA. *Palaeobiodiversity and Palaeoenvironments, 90,* 341–393.

Roček, Z., Gardner, J. D., Eaton, J. G., & Přikryl, T. (2013). Anuran ilia from the Upper Cretaceous of Utah—Diversity and stratigraphic patterns. In A. L. Titus & M. A. Loewen (Eds.), *At the top of the grand staircase: The Late Cretaceous of southern Utah* (pp. 273–294). Bloomington and Indianapolis: Indiana University Press.

Roček, Z., & Nessov, L. A. (1993). Cretaceous anurans from Central Asia. *Palaeontographica A, 226,* 1–54.

Roček, Z., & Rage, J.-C. (2000). Tertiary Anura of Europe, Africa, Asia, and North America. In H. Heatwole & R. L. Carroll (Eds.), *Amphibian Biology* (Vol. 4, pp. 1332–1387). Paleontology. Chipping Norton: Surrey Beatty and Sons.

Sahni, A., & Khosla, A. (1994). A Maastrichtian ostracode assemblage (Lameta Formation) from Jabalpur Cantonment, Madhya Pradesh, India. *Current Science, 67,* 456–460.

Sahni, A., Kumar, K., Hartenberger, J.-L., Jaeger, J.-J., Rage, J.-C., Sudre, J., et al. (1982). Microvertébrés nouveaux des Trapps du Deccan (Inde): Mise en évidence d'une voie de communication terrestre probable entre la Laurasie et l'Inde à la limite Crétacé-Tertiaire. *Bulletin de la Société Géologique de France, 24,* 1093–1099.

Sanchíz, B. (1998). Salientia. In P. Wellnhofer (Ed.), *Encyclopedia of Paleoherpetology, Part 4* (pp. 1–276). Munich: Friedrich Pfeil.

Skutschas, P. P., & Kolchanov, V. V. (2016). Anurans (Amphibia: Anura) from the Upper Cretaceous (Santonian–lower Campanian) Bostobe Formation of Northeastern Aral Sea Region, Kazakhstan. *Palaeoworld, 26,* 202–208.

Smith, T., Kumar, K., Rana, R. S., Folie, A., Solé, F., Noiret, C., et al. (2016). New early Eocene vertebrate assemblage from western India reveals a mixed fauna of European and Gondwana affinities. *Geoscience Frontiers, 7,* 969–1001.

Špinar, Z. V. (1972). *Tertiary frogs from Central Europe*. The Hague: W. Junk.

Špinar, Z. V., & Hodrová, M. (1985). New knowledge of the genus *Indobatrachus* (Anura) from the Lower Eocene of India. *Amphibia-Reptilia, 6,* 363–376.

Szentesi, Z., & Venczel, M. (2012). A new discoglossid frog from the Upper Cretaceous (Santonian) of Hungary. *Cretaceous Research, 34,* 327–333.

Tyler, M. J. (1976). Comparative osteology of the pelvic girdle of Australian frogs and description of a new fossil genus. *Transactions of the Royal Society of South Australia, 100,* 3–14.

Tyler, M. J. (1986). Additional records of *Australobatrachus ilius* (Anura: Hylidae) from the Tertiary of South Australia. *Alcheringa, 10,* 401–402.

Venczel, M., & Csiki, Z. (2003). New frogs from the latest Cretaceous of Hateg Basin, Romania. *Acta Palaeontologica Polonica, 48,* 609–616.

Vérard, C., Stampfli, G., Borel, G., & Hochard, C. (2017). The Indian Promontory: A bridge between plate tectonics and life evolution models. *Universal Journal of Geoscience, 5,* 25–32.

Vergnaud-Grazzini, C., & Hoffstetter, R. (1972). Présence de Palaeobatrachidae (Anura) dans des gisements tertiaires français. Caractérisation, distribution et affinités de la famille. *Palaeovertebrata, 5,* 157–177.

Verma, K. K. (1965). Note on a new species of fossil frog from the intertrappean beds of Malabar Hill, Bombay. *Current Science, 34,* 182–184.

Verma, O., Khosla, A., Kaur, J., & Prashanth, M. (2017). Myliobatid and pycnodontid fish from the Late Cretaceous of Central India and their palaeobiogeographic implications. *Historical Biology, 29,* 253–265.

Verma, O., Prasad, G. V. R., Khosla, A., & Parmar, V. (2012). Late Cretaceous gondwanatherian mammals of India: Distribution, inter-relationships and biogeographic implications. *Journal of the Palaeontological Society of India, 57*(2), 1–10.

Wang, Y., Gao, K., & Xu, X. (2000). Early evolution of discoglossid frogs: New evidence from the Mesozoic of China. *Naturwissenschaften, 87,* 417–420.

Wilson, J. A., Mohabey, D. M., Peters, S. E., & Head, J. J. (2010). Predation upon hatchling dinosaurs by a new snake from the late Cretaceous of India. *PLoS Biology, 8,* 1–10.

Woodward, A. S. (1908). On some fish remains from the Lameta Beds at Dongargaon, Central Province. *Memoirs of Geological Survey of India, Palaeontologia Indica, New Series, 3,* 1–6.

Yadagiri, P., & Ayyasami, K. (1989). A carnosaurian dinosaur from the Kallamedu Formation (Maastrichtian horizon), Tamil Nadu. In M. V. A. Sastry, V. V. Sastry, C. G. K. Ramanujam, H. M. Kapoor, B. R. J. Rao, P. P. Satsangi et al. (Eds.), *Symposium on three decades of development in palaeontology and stratigraphy in India. Volume 1. Precambrian to Mesozoic. Geological Society of India Special Publications, 11,* 523–528.

Chapter 7
Chelonian Pelomedusoides Remains from the Late Cretaceous of Upparhatti (Southwestern India): Systematics and Paleobiogeographical Implications

France de Lapparent de Broin and Guntupalli V. R. Prasad

Abstract We report here Pelomedusoides turtle remains found in the Late Cretaceous deposits of peninsular India. Previously reported from the Upper Cretaceous (Maastrichtian) strata such as the intertrappean beds of Bombay and Kachchh (western India), Raibasa and Kisalpuri (central India), infratrappean Lameta Formation (Pisdura) (central India) and the Kallamedu Formation (southeastern India), and also Early Albian – Middle Turonian beds of the Karai Formation (southeastern India), the new material from the intertrappean beds of Upparhatti (Karnataka state), represents the first record of turtles from southwestern India. It is compared with the previous Cretaceous discoveries from India, which are also reviewed here. The Maastrichtian turtle fauna of India is rich with about nine taxa of Bothremydidae but lacks members of Podocnemididae. The turtle fauna of Upparhatti, consisting of shell, limb, few skull and neck elements, represents about three indeterminate taxa at genus and species level. An analysis of all the Cretaceous turtle taxa of India provided us with an opportunity to review the turtle paleobiogeography in the context of Africa – Madagascar – India land connections.

Keywords Chelonians • India • Cretaceous • Revisions • New data • Bothremydidae • Paleobiogeography

F. de Lapparent de Broin (✉)
Centre de Recherches sur la Paléobiodiversité et les Paléoenvironnements, (CR2P: MNHN, CNRS, UPMC-Paris, Sorbonne Universités, 6, Muséum national d'Histoire naturelle), 57 rue Cuvier, CP 38, 75231 Paris Cedex 05, France
e-mail: france.delapparent123@orange.fr

G. V. R. Prasad
Department of Geology, Centre for Advanced Studies, University of Delhi, Delhi, 110007, India

Introduction

Carapace material of pleurodiran Pelomedusoides turtles were first recognized in India by Carter (1852) from the intertrappean beds of Bombay as *Testudo leithii*, which is a member of Cryptodira but then considered as a Pleurodiran turtle (as a *Hydraspis* member [chelid] in Gray 1871) (Lydekker 1890). Later Williams (1953) erected *Carteremys* for this taxon and thus became *Carteremys leithii* (Carter 1852), within the family "Pelomedusidae" at a time individualized Pelomedusidae, Podocnemididae and Bothremydidae were not much accepted in the zoological nomenclature from a neontological perspective. But Williams (1953) while revising *Carteremys* compared *C. leithii* with genera such as *Elochelys perfecta* Nopcsa, 1931a, *Palaeaspis* Gray, 1870, and *Rosasia* Carrington da Costa, 1940 which are now placed in Bothremydidae (Broin 1977, 1988; Lapparent de Broin et al. 2018; Pérez-García 2018b). Subsequent discoveries have demonstrated that Pelomedusoides (including Pelomedusidae and Podocnemidoidea (family Bothremydidae + epifamily Podocnemidoidae, in Broin [1988]) first appear in India in the Early Albian to Middle Turonian (Ayyasami and Das 1990). Both families were considered as present in the Upper Cretaceous deposits of India, when Jain (1986) attributed "*Carteremys*" *pisdurensis* Jain, 1977 of the Lameta Formation to the podocnemidid genus *Shweboemys* Swinton, 1939. Here we refute the presence of Podocnemididae in the Cretaceous of India and recognize the presence of Bothremydidae (Podocnemidoidea). Podocnemidoidea are represented in the Upper Cretaceous (Maastrichtian) Lameta Formation (=infratrappean beds) of Pisdura (Jain 1977, 1986) and the intertrappean beds. Relatively complete specimens of chelonians (including skull or carapace or both) have been first reported from the intertrappean beds of "Island of Bombay" and Lakshmipur (Kachchh) (Carter 1852; Lydekker 1890; Bajpai et al. 1990; Singh et al. 1998; Gaffney et al. 2003, 2006) and

G. V. R. Prasad and R. Patnaik (eds.), *Biological Consequences of Plate Tectonics: New Perspectives on Post-Gondwana Break-up—A Tribute to Ashok Sahni*,
Vertebrate Paleobiology and Paleoanthropology Series, https://doi.org/10.1007/978-3-030-49753-8_7

from the Upper Cretaceous (Late Mastrichtian) Kallamedu Formation of southeastern India (Gaffney et al. 2001a, 2006). Other infra- and inter-trappean beds have yielded poorly preserved and isolated remains (mainly carapace fragments) (Kelkar and Gupte 1943; Williams 1953; Prasad and Sahni 1987; Prasad and Khajuria 1990; Prasad and Singh 1991).

The present chapter gives the first detailed account of chelonian remains from the Upper Cretaceous intertrappean beds of Upparhatti, Belgaum District, Karnataka State, southwestern India (Fig. 7.1G). Although these Pelomedusoides remains are indeterminate at the genus and species levels, the material offers an opportunity to more fully document the diversity and distribution of the Pelomedusoides Podocnemidoidea from the Cretaceous of India and their intercontinental affinities. Bothremydidae is the major and possibly the only component of the Cretaceous chelonian fauna of India. "*Carteremys*" *pisdurensis*, previously attributed to Podocnemididae (Jain 1986) and *Carteremys leithii* (Carter 1852) are confirmed as Bothremydidae and the latter is affirmed as Maastrichtian instead of Paleocene or Eocene in age. This fauna from the Cretaceous of India indicated an origin in the Afro-South American northern part of the Gondwana during the earliest Cretaceous, and close biogeographic relations with Africa and Madagascar. But by the Late Cretaceous time, the Bothremydidae dispersed littorally to all continents, i.e. Americas, Europe and Middle East (but not Asia), facilitated by the opening of the western Tethys part of the young Atlantic. However, in the present state of knowledge, the continental Late Cretaceous bothremydid fauna of India is interpreted as endemic at generic level pointing to a certain physical isolation of the Indian subcontinent.

This chapter is a partial and revised version of the one presented in the "*Gaffney Turtle Symposium*" held at the Royal Tyrrell Museum (Drumheller, Alberta, 17–18 October 2009) (Lapparent de Broin et al. 2009).

Geology

The Cretaceous fossil record of India has been extensively documented from the Deccan volcanic province of peninsular India. The sediments associated with the Deccan lava flows were informally designated as the infratrappean beds (occurring below the oldest volcanic flow) and the intertrappean beds interbedded with the volcanic flows. The infratrappean beds are primarily non-marine sediments and were formally designated as the Lameta Formation. The vertebrate fauna from the infratrappean beds is represented by fish (Woodward 1908; Jain and Sahni 1983; Sahni and Tripathi 1990; Prasad and Singh 1991; Prasad and Cappetta 1993; Mohabey and Udhoji 1996), turtles (Jain 1977, 1986), snakes (Wilson et al. 2010; Mohabey et al. 2011), crocodiles (Rana and Sati 2000; Prasad and Singh 1991), and dinosaurs (Khosla and Sahni 2003 and references therein, Wilson et al. 2003). Paleontological data based on fish fauna (Jain and Sahni 1983; Sahni and Tripathi 1990; Prasad and Cappetta 1993; Mohabey 1996), dinosaurs (Vianey-Liaud et al. 1987; Khosla and Sahni 1995; Loyal et al. 1998; Sahni and Khosla 1994; Sahni et al. 1994; Khosla 2001), and ostracods (Khosla and Sahni 2000) favored a Maastrichtian age for the infratrappean beds.

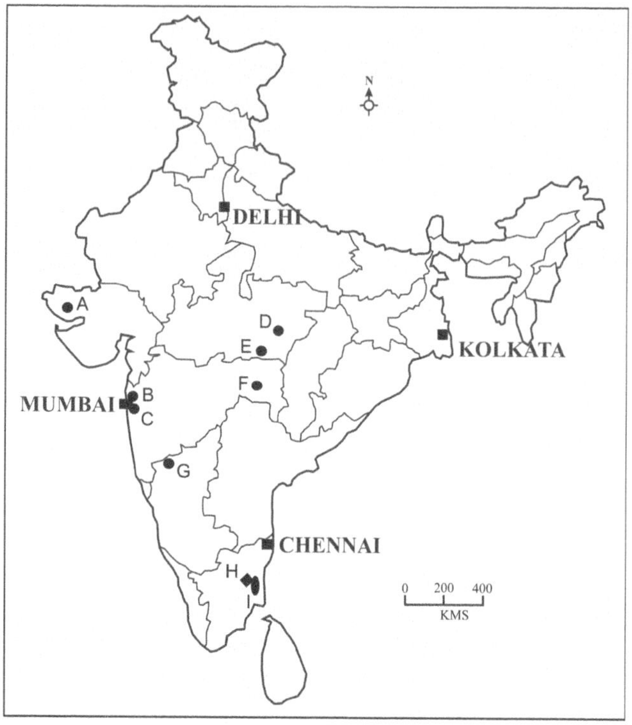

Fig. 7.1 Map of India showing Late Cretaceous localities with turtles. Deccan intertrappean beds: A to E and G. Maastrichtian. Circles: Western India: A, Lakshmipur (Kachchh); B, Amboli Quarry, Bombay island (Mumbai); C, Worli Hill, Bombay island (Mumbai). Central India: D, Kisalpuri; E, Raibasa; F, Lameta Formation (infratrappean beds): Pisdura area, Pisdura and Dongargaon, Maastrichtian. Southwestern India: G, Upparhatti. Southeastern India: H, rhomb, Kunnam, Karai Formation, Early Albian – Middle Turonian; I, oval, Kallamedu Formation, Maastrichtian

The intertrappean beds are fluvio-lacustrine sediments preserved between the Deccan basaltic flows. The intertrappean sediments, which are present at different stratigraphic levels within the Deccan volcanic sequence, were deposited during periods of quiescence in the volcanic activity. The vertebrate groups recorded from the intertrappean beds include fishes (Hora 1938; Gayet et al. 1984; Prasad and Cappetta 1993; Arratia et al. 2004; Verma et al. 2016; Lourembam et al. 2017); amphibians (Noble 1930;

Chiplonkar 1940; Verma 1965; Sahni et al. 1982; Špinar and Hodrová 1985; Prasad and Rage 1991, 1995, 2004; Rage et al. 2020); chelonians (Carter 1852; Williams 1953; Bajpai et al. 1990; Singh et al. 1998; Gaffney et al. 2001a, 2003; Khosla et al. 2004; Gaffney et al. 2006); snakes (Rage and Prasad 1992; Rage et al. 2004; Rage et al. 2020); lizards (Prasad and Rage 1995; Rana 2005; Rage et al. 2020); crocodiles (Rana 1989; Prasad and Lapparent de Broin 2002; Khosla et al. 2009); dinosaurs (Sahni et al. 1984; Vianey-Liaud et al. 1987; Ghevariya 1988; Khosla and Sahni 2003 for supplementary references); and mammals (Prasad and Sahni 1988; Das Sarma et al. 1995; Godinot and Prasad 1994; Prasad and Godinot 1994; Prasad et al. 1994; Krause et al. 1997; Rana and Wilson 2003; Khosla et al. 2004; Anantharaman et al. 2006; Prasad et al. 2007a, b; Wilson et al. 2007; Boyer et al. 2010; Prasad et al. 2010; Goswami et al. 2011, 2012). The Deccan intertrappean outcrops, with the exception of those of Rajahmundry, Jhilmili and Papro, have been considered as Late Cretaceous (Maastrichtian) in age based on pollen and spores and other paleontological studies (see Khajuria et al. 1994; Khosla and Sahni 2003; Prasad and Sahni 2014). In addition to faunal and floral evidences, paleomagnetic and radiometric data of the volcanic flows provide further constraints on the age of infra- and inter-trappean beds (Courtillot et al. 1986, 1988; Duncan and Pyle 1988). The $^{40}Ar/^{39}Ar$ age data for the basaltic dykes exposed along the western region reveal Paleocene (Danian) ages (i.e. 62.8 ± 0.2 Ma) (Widdowson et al. 2000) but no well-defined Cretaceous/Paleogene boundary section has yet been delineated in any of the intertrappean outcrops of peninsular India. The intertrappean beds of Rajahmundry on the southeastern coast, Jhilmili in central India and Papro in the northern Deccan volcanic province are considered as Early Paleocene (Danian) in age based on paleontological (Raju et al. 1991; Singh and Kar 2002; Keller et al. 2009) and geochronological criteria (Knight et al. 2003).

In addition to the intratrappean beds, remains of a Bothremydidae were recovered from the Early Albian to Middle Turonian Karai Formation of Uttatur Group (Ayyasami and Das 1990) and the Upper Cretaceous (Late Maastrichitan) Kallamedu Formation of the Ariyalur Group, Cauvery Basin in southeastern India (Gaffney et al. 2001a, 2006). The Karai Formation was deposited in a shallow marine environment. It has been first dated as early Turonian on the basis of ammonites (Sundaram et al. 2001) but it has been recently revised as Early Albian to Middle Turonian (Nagendra et al. 2013; Bragina and Bragin 2013; Gautam et al. 2015). The formation preserves a shallowing upwards sequence.

Based on sedimentology and invertebrate macrofossils, it has been suggested that the lower part of the Kallamedu Formation was deposited in shallow marine conditions, whereas the deposition of vertebrate yielding upper part took place in a continental fluvio-lacustrine environment (Sundaram et al. 2001). A suite of terrestrial vertebrates has been recovered from the Kallamedu Formation, including anuran and fish remains, turtle skulls and postcranial bones of sauropod dinosaurs, teeth of abelisaurid and troodontid dinosaurs, and a gondwanthere mammal tooth (Blanford 1862; Matley 1929; Gaffney et al. 2001a, 2006; Goswami et al. 2012; Goswami et al. 2013; Prasad et al. 2013). A Late Maastrichtian age was assigned to the lower marine horizons of this formation based on nannofossils (Rai et al. 2013).

Turtle Yielding Cretaceous Localities of India

The distribution of the Cretaceous Pelomedusoides of India is given in Fig. 7.1 and Table 7.1A–B by their geographic position (westcentral, western, southwestern and southeastern India). Most of these sites belong to the Maastricthian age and only one is attributed to the Early Albian to Middle Turonian.

Southwestern India

Upparhatti, Belgaum District, Karnataka State (Fig. 7.1G)

It is the only turtle-bearing locality that is known from southwestern India. The fossil-bearing Upper Cretaceous intertrappean beds of Upparhatti are exposed in a stream section north of Upparhatti village in Belgaum district, Karnataka state. The intertrappean section of Upparhatti consists of a Deccan Trap flow at the base followed upwards by red to dark brown sandy marl with three to four sandstone partings. The red marl is overlain by a weathered basaltic flow. Foote (1876) was the first to identify this site as an intertrappean lacustrine deposit yielding *Unio* shells. Prashad (1918) identified the unionid shells as representing a species of *Indonaia*. Later Kelkar and Gupte (1943) described shells of *Unio*, *Physa*, *Lymnaea*, and *Paludina* from this site. These authors have also noted the presence of some chelonian bones in their collection, but no description was given. During the field season of 1990, one basisphenoid, jaw elements, cervical vertebrae, several first dorsal vertebrae, isolated carapace and plastral fragments and girdles and limb bones were recovered from the red sandy marl of the Upparhatti intertrappean section by one of us (GVRP). This material is described below. The fragmentary material is not sufficient for naming at genus and species levels.

Table 7.1 A, B. distribution of Cretaceous Pelomedusoides bothremydid taxa of India by regions. Localities shown in Fig. 7.1 by the number of each locality. Part A, western India. Part B, central, southwestern and southeastern India

Part A. Western India

Map Fig. 7.1	Taxon	Systematics	Western India, Worli Hill Locality	Western India, Kachchh Locality
			Late Cretaceous, Maastrichtian	
Fig. 7.1 C	*Carteremys leithii* (Carter, 1852) type species *Testudo leithii* Carter, 1852, Plates X–XI, Williams, 1953, figs. 1 and 2	Bothremydidae, *Carteremys* group	Worli Hill, "Island of Bombay", Mumbai district, Maharashtra state Intertrappean	
Fig. 7.1 C	*Carteremys leithii* (Carter, 1852). Kolkata, GSI N° 20337: proposed neotype, Williams 1953, pl. 3	Bothremydidae, *Carteremys* group	Worli Hill, "Island of Bombay", Mumbai district, Maharashtra state Intertrappean	
Fig. 7.1 C	?cf. *Carteremys leithii* (Carter, 1852), (fig. 1: 3) Kolkata GSI collections. Williams, 1953: 2 specimens, not seen	Bothremydidae, *Carteremys* group	Worli Hill, "Island of Bombay", Mumbai district, Maharashtra state Intertrappean	
Fig. 7.1 B	*Sankuchemys sethnai*, Gaffney et al., 2003	Bothremydidae, Kurmademydini in Gaffney et al. 2006.	Amboli Quarry, Jogeshwari, "Island of Bombay", Mumbai district, Maharashtra state Intertrappean	
Fig. 7.1 A	*Carteremys* aff. *leithii*, Fig. 7.2.A VPL/IB/3866	Bothremydidae, *Carteremys* group		Lakshmipur, Kachchh, Gujarat state Intertrappean

Part B. Central, southwestern, souteastern India

Map Fig. 7.1	Systematics		Central India Locality	Southwestern India Locality	Southeastern India Locality
			Late Cretaceous, Maastrichtian		
Fig. 7.1D	cf. *Carteremys* sp., Fig. 7.2. B	Bothremydidae *Carteremys* group	Kisalpuri, Dindori district, Madhya Pradesh state Intertrappean		
Fig. 7.1E	Indeterminate gen. sp., cf. *C. leithii*	?Bothremydidae doubtfully *Carteremys* group	Raibasa, Chhindwara district, Madhya Pradesh state Intertrappean		
	Indeterminate gen. sp., Williams (1953) ?*C. leithii*	Pelomedusoides indet.? cf. Bothremydidae	"Central India" "intertrappean carbonaceous shales" Intertrappean		
Fig. 7.1F	(1) « *Carteremys* » *pisdurensis* (Jain, 1977) skull figs, + (2) "*Shweboemys*" *pisdurensis* Jain, 1986, figs of skull, postcranial.	Bothremydidae: « *Carteremys* » *pisdurensis* incert. gen. (nov.) incert. group	Pisdura (1) and Dongargaon (2), about 16 km southeast of Pisdura, Chandrapur district, Maharashtra state, Lameta Formation Infratrappean		
Fig. 7.1G	Indeterminate gen. Fig. 7.2.C-I, Figs. 7.3–7.12.	Bothremydidae, two to three taxa: indeterminate short-necked Bothremydidae + 1 to 2 indeterminate long-necked Bothremydidae		Upparhatti, Belgaum district, Karnataka State Intertrappean	
Fig. 7.1I	*Kurmademys kallamedensis* Gaffney et al. 2001	Bothremydidae Kurmademydini in Gaffney et al. 2006.			Kallamedu Ariyalur district, Tamil Nadu state, Kallamedu Formation
			Late Cretaceous, Aptian-Albian (formerly early Turonian)		
Fig. 7.1H	Indeterminate gen. sp. Ayyasami and Das 1990	cf. Bothremydidae			Kunnam, Karai Formation, Uttatur Group, Perambalur District, Tamil nadu state

Comparisons with members of the Pelomedusoides families Bothremydidae and Podocnemididae allow a general understanding of the diversity and relationships of the taxa present. We recognize a minimum of three taxa in the Upparhatti assemblage of Bothremydidae indet: (1) One or two long necked forms; (2) Two other forms with at least one short-necked one. The *Carteremys leithii* group of Central-Western India (Mumbai, Kachchh) (Fig. 7.1A–D and E?) is not recognized in the postcranial elements, as they lack the characteristic shell marked ornamentation (strong decoration) of the group. Some of the carapace elements may belong to *Kurmademys* (Fig. 7.1I), known from the Upper Cretaceous Kallamedu Formation (Gaffney et al. 2006), or to another form either close to it or to "*Carteremys*" *pisdurensis* from the Upper Cretaceous Lameta Formation (Fig. 7.1F). This is recognized as a bothremydid instead of a podocnemidid, but in the absence of strong ornamentation it does not belong to the *Carteremys* group. The presence of Podocnemididae is unlikely because of the lack of elements showing definitive derived podocnemid features. This turtle fauna from Upparhatti is described below and comparisons are made with the turtles of the other regions.

Western India

Fossil turtle remains have been found from three localities in western India (Fig. 7.1A–C). Their taxonomic status is reconsidered below. Two of these localities, Worli Hill and Amboli Quarry, are in the city of Mumbai, and the third one, Lakshmipur is in Gujarat State. All of these are exposures of the Upper Cretaceous intertrappean beds.

Worli Hill (Fig. 7.1C)

Located in southern Mumbai, it is one of the original seven islands that formed Mumbai (formerly Bombay). Prior to the expansion of the city, exposures of the intertrappean beds were present along the flanks of the Worli Hill. In addition to the turtle material, these exposures have yielded many frog remains (Chiplonkar 1940). Until a few decades ago, the intertrappean beds of Bombay were regarded as early Eocene in age (Carter 1852; Chiplonkar 1940; Verma 1965). Recent paleontological investigations have indicated a Late Cretaceous (Maastrichtian) age for these intertrappean beds (Singh and Sahni 1996; Singh et al. 1998). In Mumbai, the intertrappean beds were deposited in a sheltered, peripheral marine setting in an increasingly humid climate (Cripps et al. 2005). Cretaceous turtles from India were first found in the Worli Hill locality, and they were described by Carter (1852) as *Testudo leithii*. The syntypes of *Testudo leithii* (nine specimens according to Carter [1852]), which included a figured carapace, plastron, skull and partial mandible, were possibly deposited in the Bombay branch of Asiatic Society (Jain 1977), but are now considered as lost. The only other specimen from the Worli Hill locality that was described and figured is a carapace (GSI (Kolkata) No. 20337) (Williams 1953, pl. 3). This specimen is here considered as the neotype of *C. leithii*. It was described by Singh et al. (1998) as Pelomedusidae indet. Singh et al. (1998) reported that the specimen is housed in the "Sarketi" (sic, for Saketi) Fossil Park near Chandigarh (North India) and a cast was deposited at the Institute of Paleontology, University of Munich. The Saketi Fossil Park was established for the Siwalik fossils, whereas the specimen of *C. lethii* is from the intertrappean beds, and it is the specimen that Williams (1953) illustrated. Two more specimens collected later by Sukheswala (1947) (reported in West 1950, see Williams [1953]) from the intertrappean beds of Worli Hill are in GSI, Kolkata collections. These specimens were neither described nor illustrated. Contrary to Gaffney et al. (2006) *Carteremys leithii* is not considered here as *nomen dubium* but as a definable bothremydid taxon of a *Carteremys* group. *Carteremys leithii* is related to the material from Lakshmipur (see below).

Amboli (Fig. 7.1B)

Located in Mumbai, it contains an exposure of approximately 10 m intertrappean beds overlain by basalts. As with the Worli Hill, this site is currently largely obscured by housing construction. The only vertebrate fossil that has been recovered from this locality is a dorsoventrally flattened/crushed skull of the turtle (SDS/VPL 1125) *Sankuchemys sethnai* Gaffney et al., 2003. Its possible relationship to *Carteremys* (whose skull is barely known) is difficult to establish. Its kurmademydine affinity (Gaffney et al. 2009) is discussed here and its possible presence in Upparhatti is examined.

Lakshmipur, Kachchh, Gujarat State (Fig. 7.1A)

Previously, palynomorphs and ostracods have been described from this intertrappean section (Whatley and Bajpai 2000; Samant and Mohabey 2009). A mid-anterior plastron (VPL/IB/3866), was collected by S. Bajpai from the Upper Cretaceous (Maastrichtian) intertrappean beds of this locality (Bajpai 1990), although it was incorrectly described by Singh et al. (1998) as coming from the intertrappean beds of Bombay. This incorrect locality information was repeated in Gaffney et al. (2006), who attributed the specimen to "*Carteremys leithii* (nomen dubium)". We confirm its attribution to *Carteremys* and provisionally assign it to *C.* aff. *leithii.* Presented in the frame work of the comparative ornamentation of the shells (Fig. 7.2A) for its impact on the nomenclature, the specimen will be dealt in detail later with the study of related Kisalpuri material (decoration, Fig. 7.2B). It is not related to the material from Upparhatti in shell ornamentation (Fig. 7.2C–I).

Central India

On the eastern edge of the Deccan volcanic province, fossil turtle remains have been found at three localities, Kisalpuri in the state of Madhya Pradesh, Pisdura and Dongargaon in the state of Maharashtra. The Kisalpuri locality represents an intertrappean section and the Pisdura and Dongargaon localities contain exposures of infratrappean beds. All these localities are of Late Maastrichtian age.

Fig. 7.2 *Carteremys* aff. *leithii*, Bothremydidae, Upper Cretaceous (Maastrichtian) intertrappean beds of Lakshmipur, Kachchh (Kutch), western India: **A1–A3**. VPL/IB/3866, half anterior plastron, ventral view (**A1**), magnified left ventral part showing the typical strong decoration (**A2**), and dorsal view (**A3**). Decoration of external face of shell plates of Bothremydidae in Upper Cretaceous (Maastrichtian) intertrappean beds. DUGF/T coll. (Work catalog U numbers on the specimens): **B.** Kisalpuri, Central India, DUGF/T271, *Carteremys* sp., fragmentatry entoplastron. C–I. Upparhatti, southwestern India, Bothremydidae indet., **C.** DUGF/T110, neural 2; **D.** DUGF/T57, peripheral 1; **E.** DUGF/T66 peripheral 9 or 10; **F.** DUGF/T125, neural 3 or 5; **G.** DUGF/T111, neural 6; **H.** DUGF/T87, costal 7; **I.** DUGF/T154, fragmentary xiphiplastron. Scale bars equals 2 cm for **A1**, **A3**, **B–I**, and 3 cm for **A2**. Fig. 7.2**A–C**. photographs (MNHNF) by courtesy of Sunil Bajpai

Kisalpuri, *Dindori District, Madhya Pradesh State (Fig. 7.1D)*

Exposures of the intertrappean beds near Kisalpuri village in Madhya Pradesh have yielded a rich microvertebrate assemblage consisting of fish, amphibians, lizards, snakes, turtles, crocodiles, dinosaurs, and mammals, although only the gondwanatherian and eutherian mammals, dyrosaurid crocodiles, squamates, amphibians, and fishes have been formally described (Khosla et al. 2004, 2009; Prasad et al. 2007a, b, 2010; Verma et al. 2012, 2016; Rage et al. 2020). All the fossil material from the site was recovered from a 70 cm thick siltstone horizon. The Kisalpuri site, though predominantly represented by terrestrial and freshwater faunas, also contained some brackish water and marine forms like the myliobatid *Igadabatis indicus*. This indicates that the Kisalpuri site might have been subjected to some marine incursion or it was in close proximity of the sea. The turtle fauna of this site is distinct from that of Upparhatti in having strong decoration of the *Carteremys* group (Fig. 7.2B) of Bothremydidae. The Kisalpuri turtle specimens will be described separately.

Pisdura Area, Chandrapur District, Maharashtra State (Fig. 7.1F).

At Pisdura, the Upper Cretaceous (Maastrichtian) Lameta Formation has yeilded a diverse assemblage of fossil vertebrates including dinosaur bones and eggshells. Turtles are represented by a nearly complete skull (ISI R200), which is the holotype of *Carteremys pisdurensis* Jain, 1977. It was later attributed to the podocnemidid "*Shweboemys*" (Jain 1986). However, the latter assignment is refuted on comparison with the skull of *Shweboemys* s.l. (*Shweboemys* subgroup sensu Zouhri et al. 2017). From the Lameta red clays at Pisdura, presence of partial carapaces is mentioned, and a dorsal vertebra articulated with the shell (R. 194) was figured in Jain (1986, fig. 1) but was not described.

Dongargaon, Chandrapur District, Maharashtra State (Fig. 7.1F)

At Dongargaon, exposures of the infratrappean beds, located about 16 km southeast of Pisdura, have also yielded turtle remains (Jain 1986). Material is represented by numerous elements of carapace and axial skeleton. Elements that Jain (1986) figured include (ISI coll. Calcutta [now Kolkata], India): ISI R. 185 to 194: carapaces with plastron (R. 185 and R. 186); plastron (R. 187); various isolated elements of anterior girdle with humerus and ulna (R. 192), scapulae (R. 190, R. 191), coracoid (R. 189), femur (R. 188), and ilium (R. 193). All this carapacial material is referred, as the skull from Pisdura, to the podocnemidid "*Swheboemys*" *pisdurensis* in Jain (1986), an attribution which is not retained here. The whole material of Jain (1977) and Jain (1986) represent a bothremydid as that of *Carteremys* although it does not belong to the latter genus. It is provisionally attributed to "*Carteremys*" *pisdurensis* as originally suggested by Jain (1977), pending a revision of the genus. Except for an unclear nuchal part, all the figures (principally schematic and pictures in a low quality) clearly favour the attribution of the skull from Pisdura and postcranials from Dongargaon to Bothremydidae on comparison with other Cretaceous bothremydids (Table 7.5).

Southeastern India

The Early Albian – Middle Turonian Karai Formation (Fig. 7.1H), and the Upper Cretaceous (Late Maastrichtian) Kallamedu Formation of Ariyalur District, Tamil Nadu state on the southeastern coast (Fig. 7.1I) have yielded some turtle remains. In addition, an early report of "parts of what appeared to be of two of fossil tortoises" was made by Muzzy (1856) (in Ayyasami and Das 1990) from the Cretaceous rocks of South India but there are few references to these specimens in later works and the location of the specimens is unknown at present.

Kunnam Locality, Perambalur District., Tamil Nadu State (Fig. 7.1H)

Exposures of the Early Albian – Middle Turonian Karai Formation ("Lower Turonian Kunnam Formation" in Ayyasami and Das [1990]), 1 km south of Kunnam village in the Cauvery basin, have yielded the inner mold of an indeterminate carapace and plastron (Ayyasami and Das 1990, figs. 1–3). The specimen is a bothremydid that is possibly as well in the line of the Cretaceous bothremydini members of North America (Zangerl 1948; Gaffney and Zangerl 1968; Gaffney et al. 2006) and Palestine (Haas 1978a, b), as the indeterminate form from the Cenomanian of Madagascar (Lapparent de Broin and Werner 1998; Gaffney et al. 2006). But, being an inner mold, it is not comparable to the Upparhatti disarticulated bones.

Kallamedu Locality

A vertebrate assemblage comprising of fish, amphibian, crocodilian, turtle, abelisaurid and titanosaurid dinosaur and gondwanathere mammal remains has been documented from the Upper Cretaceous Kallamedu Formation of the Ariyalur Group, Ariyalur district, Tamil Nadu state (Fig. 7.1I) (Blanford 1862; Matley 1929; Yadagiri and Ayyasami 1989;

Gaffney et al. 2001a, 2006; Prasad et al. 2013; Goswami et al. 2013; Halliday et al. 2016). The bothremydid *Kurmademys* described from this locality includes shell, skull and lower jaw elements. The cranial elements of Kallamedu are not similar to the Upparhatti material such as the basisphenoid and jaw. But some Upparhatti shell parts (in particular of the bothremydid form b) present features similar to some shell elements of *Kurmademys*.

Material and Methods

The bone plates overlapping the thoracic ribs are named "costals" and the horny scutes covering the costals are named "pleurals" according to Hay (1908) and Gaffney's works (for example) and others, as opposed to the terminology used in zoology for living species, Młynarski (1976 in part), and most of previous works of Lapparent de Broin and others.

Abbreviations

BMNH, British Museum, Natural History (old catalog), NHMUK, Natural History Museum, London; DAMR, Department of Antiquities and Museums, Riyadh; CMN, Canadian Museum of Nature, Ottawa; DUGF/T, Delhi University Geology Department Fossil Turtles; GSI, Geological Survey of India, Kolkata; ICZN, International Code of Zoological Nomenclature; ISI, Indian Satistical Institute, Kolkata; MHNM, Muséum d'Histoire Naturelle de Marseille; MNHN (ZA AC, F), Muséum national d'Histoire naturelle, Paris (Zoologie des reptiles et amphibiens et Anatomie Comparée, Paléontologie); MEAS, Department of Marine Earth and Atmospheric Sciences of NCSU, National State of North Carolina; MZUSP, Museu de Zoologia da Universidade de São Paulo; OCP, Office Chérifien des Phosphates, Khouribga, Morocco; SDS/VPL, Panjab University, Chandigarh, India, Vertebrate Paleontology Laboratory; SMNS, Staatliches Museum für Naturkunde-Stuttgart; ULR, UNLaR, Universidad Nacional de La Rioja, Argentina; UM II, Universtité Montpellier II; USP, Universidade de São Paulo, Brazil; VPL/IB, Vertebrate Paleontology Laboratory, Department of Earth Sciences, Indian Institute of Technology, Roorkee.

Systematic Paleontology

Order Testudines Linnaeus, 1758
Infraorder Pleurodira Cope, 1864
Hyperfamily Pelomedusoides Cope, 1868
Superfamily Podocnemidoidea Cope, 1868
Family Bothremydidae Baur, 1891
Bothemydidae indet.

Stratigraphic Horizon and Locality: The red sandy clay horizon of the Upper Cretaceous (Maastrichtian) intertrappean beds exposed north of Upparhatti village, close to Gokak town, Belgaum district, Karnataka State (South India).

Referred material: Around 206 specimens (complete to partial). DUGF/T47 to DUGF/T255 (work catalog UI to U209, numbers indicated on figured specimens). They consist of: skull: one basiphenoid; lower jaw: 3 dentaries; postcranial bones: 24 cervical and dorsal vertebrae; shell: 91 dorsal elements, 14 plastral fragments, 12 buttresses; girdle: 12 scapulae, 17 pelvic fragments; limb bones: 7 humeri and 18 femora, various metapodials and phalanges, and fragments of bones with indeterminate position on the body.

Description: Turtle remains are abundant in the Upparhatti locality but no articulated specimens are present. Isolated elements pose a serious challenge for taxonomic studies because of the problem of associating different elements. Despite this, isolated elements can provide useful information about the minimum level of diversity and relationships of taxa present. Thus, an element-by-element approach is taken in the description of the turtle material from Upparhatti, with each kind of element being described and compared with the corresponding element in other Pelomedusoides. This is followed by a discussion of the minimum level of diversity of the assemblage, the kinds of Pelomedusoides present, and the possible associations of elements.

Skull: Basisphenoid (Fig. 7.3). The only skull element recovered is a basisphenoid DUGF/T167. This element is complete except for a very short anterior tip of the flat rostrum basiphenoidale. The main body of basiphenoid is wide and short both dorsally and ventrally, measuring 1.9 cm × 1.7 cm (Fig. 7.3) (Table 7.2). In ventral view (Fig. 7.3A1), it is trapezoidal and distinctly pointed anteriorly. Ventrally, the anterior pointed part is partly obliquely covered by the pterygoids, which were united and sutured with the anterolateral borders of the basisphenoid. This part of the basisphenoid is too short to have borne pterygoid wings anteriorly covering an enlarged podocnemidid carotid canal ("pterygoid channel" and "enlarged carotid canal" of Gaffney [1979], cavum pterygoidei of Gaffney et al. 2006) (Broin 1988; Broin 1991; Lapparent de Broin 2000b; Lapparent de Broin et al. 2007; Lapparent de Broin and Werner 1998). Ventrally and at mid-length of the lateral border, there is an arc of circle forming an acute border with the flat ventral surface, which overhangs a short vertical wall and limits the fossa pterygoidea (pf). At the lateral corner, there is no indication of a suture for a ventral participation of the prootic. The two short, longitudinally straight, posterolateral basisphenoid borders are sutured with the quadrate which dorsally covered the basisphenoid obliquely along its length. The wide posterior border was completely sutured with the basioccipital resulting in a closed fenestra postotica medially, which is a podocnemidoid character. The suture with the basioccipital is nearly transverse and little convex ventrally. The vertical short wall borders a small depression (the fossa pterygoidea), and the basisphenoid was just sutured to the pterygoid below the acute border without horizontal or oblique lateral expansion. On the ventral surface of the bone, this acute border is framed by two scars of muscle insertion (ms, Fig. 7.3A1). In DUGF/T167, the foramen posterior canalis caroticus internus (car, Fig. 7.3A1, A2) is situated in the basisphenoid, in a thin canal along the smooth free border of the bone, above the ventral surface. This is located posterior to the suture of the pterygoid with the basisphenoid, close to the junction of the suture for the quadrate and pterygoid. The carotid lies in a thin canal which ends in the ventral surface of the pterygoid just posterior to the free basiphenoid border, approximately at the angle with the quadrate. Dorsally (Fig. 7.3A2), the ossified lateral trabeculae of the rostrum basiphenoidale of DUGF/T167 are well separated posteriorly, and united in a relatively long and flat rostrum anteriorly (ros, Fig. 7.3A2), which is not rounded at its extremity. The anterior outline of the DUGF/T167 dorsum sellae is wide and concave, rounded, slightly surrounding the sella turcica where the well separated anterior foramina of the carotid canal enter on each side. The foramina of the abducens nerve (VI) are located behind the processi clinoidei (Fig. 7.3A2).

Discussion of Basisphenoid: On the whole, the short Upparhatti basiphenoid does not correspond to any pre-podocnemidid or podocnemidid. It corresponds to a bothremydid with a relatively short basicranium and a fossa pterygoidea that is probably weak and, anyway, limited by a very short and vertical border at the suture between the pterygoid and basiphenoid, and a narrow carotid canal and foramen ending at the corner of the basisphenoid when suturing with the pterygoid and the quadrate. In its outline, proportions and the sharp border of the fossa pterygoidea, and the position of the carotid foramen in a relatively posterior and medial position at the boundary with the pterygoid, it does not correspond to any described bothremydid basisphenoid knowing that this bone is not always clearly described and figured or not well preserved (Broin 1988; Lapparent de Broin and Werner 1998; Gaffney et al. 2001a, b, 2003, 2006, 2009; Gaffney and Krause 2011).

The depression seen above is the "podocnemidoid" fossa (pf, Fig. 7.3) of Lapparent de Broin (2000b), identified as the fossa pterygoidea of Gaffney et al. (2006). It is present in most Podocnemidoidea but very variably excavated in the pterygoid ventral face and it includes the lateral border of the basisphenoid on a variable width up to the pterygoid suture laterally when present. It is present in pre-podocnemidids such as Lower Cretaceous *Brasilemys* Lapparent de Broin, 2000b from Araripe (Brazil), and the mid-Cretaceous (prior to the Cenomanian transgression) of Kem Kem beds *Hamadachelys* Tong and Buffetaut, 1996 from Morocco, and in Podocnemididae and in most Bothremydidae (Table 7.2). In the Upparhatti basiphenoid, the fossa pterygoidea is known by its acute border limiting a vertical and short wall overhanging the fossa. An acute fossa border exists in the Cretaceous bothremydid *Cearachelys* from Araripe, *Galianemys whitei* Gaffney et al., 2006, fig. 91b) and *Galianemys* cf. *whitei* (Lapparent de Broin and Werner 1998: fig. 4a) both from Kem Kem beds. Whereas the fossa border is generally rounded in other bothremydids. However, DUGF/T167 has no oblique lateral basisphenoid expansion in the fossa, contrary to the condition in other bothremydids, particularly *Kurmademys kallamedensis* Gaffney et al., 2001a from the Upper Cretaceous Kallamedu Formation (Fig. 7.1I), *Kinkonychelys* Gaffney et al., 2009 from the Mastrichtian of Madagascar and others (Figs. in Lapparent de Broin and Werner 1998; Gaffney et al. 2006, 2009, 2011). Thus, the holotype of *Kurmademys* (in Gaffney et al. 2006, fig. 56) shares with DUGF/T167 the acute medial border of the fossa that is rounded in other genera, but its basisphenoid is expanded laterally up to the suture with the pterygoid in contrast to DUGF/T167. The fossa pterygoidea is wide and deep in *Kurmademys* and narrow and much deeper in *Kinkonychelys*, including at mid-width of the basisphenoid lateral suture with the pterygoid and the quadrate, and with an anterior position of the carotid foramen. In the crushed skull of *Sankuchemys sethnai* Gaffney et al., 2003 from the intertrappean beds of Amboli Quarry (Fig. 7.1B), the palate is flat, seeming to have (on the photographs) primitively either a slight pterygoid fossa or without one, as in *Sokatra antitra* Gaffney and Krause, 2011 from Maastrichtian Berivotra site of Madagascar. But there appears to be an apparent prootic exposition in this place (a primitive feature not being present in the three others, but also known in *Galianemys* [see Gaffney et al. 2006: fig. 81]). A flat area without acute or rounded crest exists in *Zolhafah bella* Lapparent de Broin and Werner, 1998 from the Maastrichtian of Ammonite Hill (Egypt) and other bothremydines, and as well as taphrosphyines (Gaffney et al. 2006). Therefore, the acute border in Upparhatti taxon differentiates it from all these taxa. The muscle scars of DUGF/T167 (Table 7.2) are limited to the vertical border as in the pre-podocnemidid *Hamadachelys*, but in the latter they are wider and closer medially. The position of these scars is rarely represented on published bothremydid schematic figures. They may be present in a narrow depression, more medially situated in *Kinkonychelys* pictures, for example, and in primitive pelomedusoides forms such as *Euraxemys essweini* Gaffney et al., 2006 and in Podocnemididae.

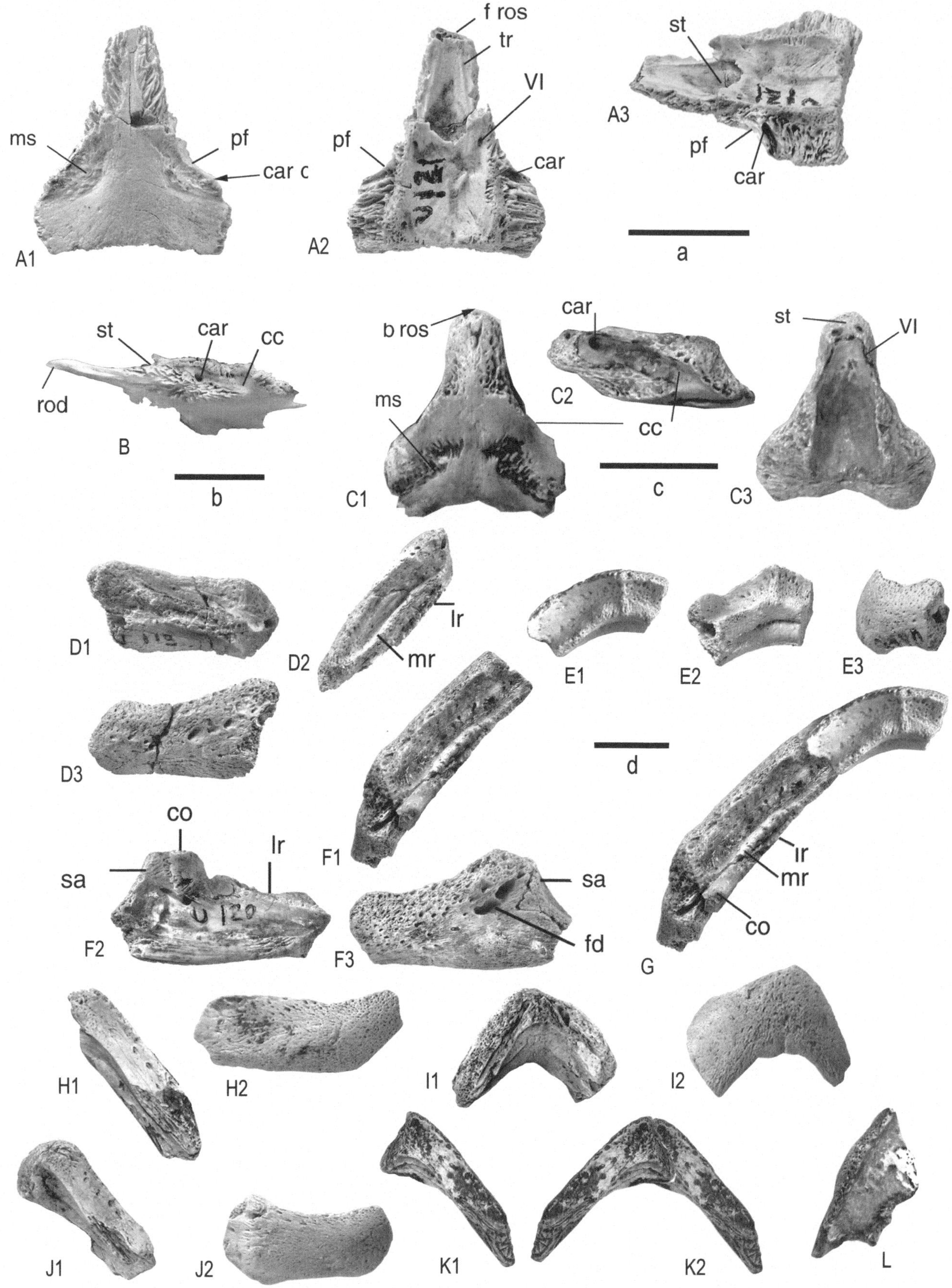
f ros
tr
VI
ms
pf
car c
pf
car
st
A3
pf
car
A1
A2
a
st
car
cc
rod
B
b
b ros
ms
C1
car
C2
cc
c
st
VI
C3
D1
D2
lr
mr
E1
E2
E3
D3
d
F1
co
sa
lr
F2
sa
fd
F3
lr
mr
co
G
H1
H2
I1
I2
J1
J2
K1
K2
L

◀ **Fig. 7.3** Comparison of basisphenoids: Bothremydidae gen. et sp. indet., Upparhatti, southwestern India, intertrappean beds, Maastrichtian, DUGF/T coll. (Work catalog U numbers on the specimens): DUGF/T167, **A1–A3**. ventral (**A1**), dorsal (**A2**), and left latero-dorsal (**A3**) views; *Podocnemis cayennensis*, extant, Guyanne, MNHN (ZA AC), B. left latero-ventral view. *Erymnochelys* group gen. et sp. indet., continental Senonian, Ibeceten, Niger, MNHNF IBC 1861; **C1–C3**. ventral (**C1**), left lateral (**C2**) and dorsal (**C3**) views; Jaws: Bothremydidae indet. gen. sp., Upparhatti, Southwestern India, intertrappean beds, Maastrichtian, DUGF/T coll. Fragmentary dentaries: **D1–D3**. DUGF/T164, left dentary part, lingual (**D1**), dorsal (**D2**), and left lateral (**D3**) views; **E1–E3**. DUGF/T165, symphysial dentary part, dorsal (**E1**), posterodorsal (**E2**), and left lateral (labial) (**E3**) views; **F1–F3**. DUGF/T166, left lower jaw part, dorsal (**F1**), lingual (**F2**) and left lateral (**F3**) views; **G**. associated dentaries DUGF/T165 and DUGF/T166, dorsal view; *Erymnochelys* group indet. gen. sp., continental Senonian, Ibeceten, Niger, MNHNF coll.: **H1–H2**. IBC 1869, right dentary, dorsal (**H1**) and right lateral (**H2**) views; **I1–I2**. IBC 1875, symphysial dentary part, posterodorsal (**I1**) and ventral (**I2**) views; **J1–J2**. IBC 1870, right dentary part, dorsal (**J1**) and right lateral (**J2**) views; **K1–K2**. IBC 594, fragmentary dentary, right part with symphysis (**K1**) and reconstructed by symmetry (**K2**); **L**. IBC 600, right maxilla, posterior fragment, ventral view. Abbreviations: b ros., broken rostrum sphenoidale at the sella turcica; car, foramen canalis carotici interni for the entrance within the basiphenoid toward the sella turcica; car c (arrow), carotid canal ventral extremity; cc, podocnemidid enlarged carotid canal; co, processus coronoideus; fd, foramen dentofaciale majus; f ros, flat rostrum basiphenoidale; lr, lingual ridge; mr, median ridge; ms, muscle insertion; pf, podocnemidoid fossa (fossa prerygoidea) border; rod, rounded rod-like rostrum basisphenoidale; sa, surangular; st, sella turcica; tr, lateral ossified part of the trabeculae; VI, foramen nervi abducentis. Scale bar equals 1 cm for **A1–A3**, **C1–L**, and 2 cm for **B**, a for **A1–A3**, b for **B**, c for **C1–C3**, d for **D1–L**

The position of the carotid foramen (Table 7.2) is much variable in bothremydid species (Lapparent de Broin and Werner 1998; Lapparent de Broin et al. 2007). In most named Bothremydidae, the carotid foramen is ventrally well visible because the fossa border is not acute at a right angle in relation to the basisphenoid surface but inclined at a wider angle and because the ventral basisphenoid participation in the fossa is extended laterally. Among Indian bothremydids, in *Kurmademys*, the ventro-lateral surface of the basisphenoid, which participates in the wide fossa, is wider and the carotid foramen is more anteriorly positioned, although well medial to the basisphenoid-pterygoid suture. In *Sankuchemys*, the carotid foramen is most anteriorly positioned, on the basisphenoid-pterygoid boundary (as figured in Gaffney et al. 2006: fig. 65), while it is posteriorly positioned in *Zolhafah* at the junction of the basiphenoid with the quadrate (Lapparent de Broin and Werner 1998: fig. 3a, pl. 1: 3) and also posteriorly, at the edge of the deep fossa podocnemidoidea, medial to the junction in *Kurmademys*. In the Upparhatti specimen, the foramen is more posterior and hidden at the basisphenoid-pterygoid boundary. In other Podocnemidoidea, the posterior position of the carotid as seen in DUGF/T167 is similar to that in *Hamadachelys* (podocnemidid by the confluent incisura columellae auris and Eustachian tube but pre-podocnemidid by the absence of an enlarged carotid canal covered by pterygoid wings) and *Brasilemys* Lapparent de Broin, 2000b. But in *Brasilemys*, where the ventral fossa border is acute, the carotid foramen is hidden by a very thin plate of basisphenoid bone bordering a small "enlarged podocnemidid canal". In *Hamadachelys*, the fossa is weak ventrally, with a border inclined towards the carotid foramen, as in bothremydids, well visible laterally and inside the basisphenoid in contrast to Upparhatti basiphenoid. In Podocnemididae such as extant *Podocnemis* (Fig. 7.3B) (Williams 1954a) and *Erymnochelys madagascariensis* (Grandidier, 1867 [see Broin 1977: pl. 3, 1], Gaffney 1979), the basisphenoid is longer in relation to its width, the fossa, plunged much deeper in the

Table 7.2 Comparison of bothremydid and podocnemidid basisphenoids. Extant podocnemidids *Erymnochelys* and *Podocnemis* and Senonian Ibeceten (IBC) elements; two known bothremydid skulls of India, *Kurmademys* and *Sankuchemys*, and *Foxemys* from the late Cretaceous of France. Abbreviations: ant, anterior; bsph, basisphenoid; car. canal, enlarged carotid canal; lat, lateral, laterally; Mod, moderately; pf, podocnemidoid fossa; pter, pterygoid; pter-bsph suture, pterygoid-basisphenoid anterior suture; post, posterior. Brackets: (inferred)

Bsph characters	Upparhatti DUGF/T167	*Kurmademys* Gaffney et al. (2006)	*Sankuchemys* Gaffney et al. (2006)	*Foxemys* Gaffney et al. (2006)	*Erymnochelys madagascariensis* MNHN (ZA AC) coll.	Aff. *Erymnochelys* sp. MNHNF IBC 1861	*Podocnemis cayennensis* MNHN (ZA ACH) coll.
Muscle scar	Antero-lat	Medio-lat?	?	?	Medio-post inner	Medio-post inner	Medio-post inner
Basicranium dorsal	Short wide	Short wide	(Short)	?	Long	Long narrow	Long
Pter wings	(No place)	No	No	No	Yes	(Yes)	Yes
Foramen VI	Mod ant	?	?	?	Antero-lat border	Antero-lat border	Postero-medial
Carotid foramen position	Post to pter–bsph suture in pf. fossa	Post to pter–bsph suture in pf. fossa	Ventral midway on flat pter–bsph suture	Post to pter-bsph suture in pf. fossa	Ant to pter–bsph suture deep in car. -canal	Ant to pter–bsph suture deep in car. canal	Ant to pter–bsph suture deep in car. canal

skull, becomes the enlarged podocnemidid canal or cavum, half covered by the posteriorly prolonged pterygoid wings. The posterior foramen of the carotid canal is not located ventrally in the lateral border of the basisphenoid. Rather it is located more anteriorly in the cavum, hidden closer to the rostrum basiphenoidale than in the Upparhatti basiphenoid, and overall ventrally hidden by the pterygoid wing, whose insertion is just in front of the free visible lateral border of the cavum (Lapparent de Broin and Werner 1998: Fig. 5).

The shape of the rostrum basiphenoidale is variably developed in Pelomedusoides (Gaffney et al. 2006; Lapparent de Broin and Werner 1998), particularly in Podocnemididae (Broin 1991). In these, the dorsal surface of the basisphenoid is more elongated, and the rostrum ends anteriorly either in a rounded rod or in a flat double tip. The rostrum is a rounded rod in *Erymnochelys*, *Peltocephalus* and some Podocnemidine species (extant *Podocnemis expansa, P. cayennensis* [*unifilis* of authors], and *Bauruemys elegans* [Suárez 1969] from the Late Cretaceous of Brazil). But it is flatter in other Podocnemidines such as the extant small species *P. lewyana*, and *Lapparentemys vilavilensis* (Broin 1971), a much larger form than the former, from the Paleocene of Bolivia. It is also flat in the Erymnochelyine (*Shweboemys* subgroup member) *Cordichelys antiqua* (Andrews 1903) (and see Andrews 1906) from the Priabonian of the Fayum (Egypt) (NHMUK [BMNH] R11998) and in some extant species of *Pelusios* (Pelomedusidae). In Bothremydidae, the dorsal surfaces of the ossified trabeculae may be well separated as in the Upparhatti specimen, and the central body of the basisphenoid is dorsally short. The rostrum may be short in some taxa, such as *Kurmademys* (Gaffney et al. 2006: fig. 61), *Zolhafah* (Lapparent de Broin and Werner 1998: pl. 2) and *Rosasia soutoi* Carrington da Costa, 1940 from the Late Cretaceous of Portugal (Broin in Antunes and Broin 1988). The basisphenoid central body and the rostrum are longer and form also a rounded rod in North American Late Cretaceous bothremydid *Bothremys cooki* Leidy, 1865 (in Gaffney et al. 2006: fig. 134), and *Chedighaii* sp. (in Gaffney et al. 2006: fig. 167). The Upparhatti specimen falls between the two extremes. The rostrum is rather long, but the end is flattened, not rounded and rod-like and the trabeculae are well separated.

The dorsum sellae of DUGF/T167 conforms to the condition in Pleurodira, as present in Bothremydidae (Gaffney et al. 2006) and Podocnemididae. However, the paired foramina of the abducens nerve (VI) (Fig. 7.3A2, Table 7.2) are in a possible intermediate position, not passing into a slight long notch at the lateral border of the processus clinoideus as in *Erymnochelys* and an indeterminate Ibeceten specimen (Senonian of Niger) (MNHNF IBC 1861) of the *Erymnochelys* group (Fig. 7.3C3), and not crossing the bone as far posteriorly as in bothremydid *Rosasia*, podocnemidids *Peltocephalus* and *Podocnemis* and in living Pelomedusidae. In *Bothremys cooki* and *Chedighaii* sp., the foramen nervi abducentis (paired) is absent, the nerve must have passed fully laterally and hence in a position that is different from that of the Upparhatti specimen.

One undescribed basisphenoid (MNHNF, IBC 1861) is recovered from the continental Senonian of Ibeceten (Niger) a locality which also yielded *Nigeremys* from the overlying Maastrichtian marine deposit (Broin et al. 1974; Broin 1988; Lapparent de Broin 2000a, b). As the relationship of India with Africa in the Late Cretaceous is important from a paleobiogeographic perspective, this basisphenoid has been illustrated here (Fig. 7.3C). It shows diagnostic features of the basiphenoid in pleurodires. The free border, delimiting laterally a fossa, has an acute border, like in Upparhatti specimen, but there is an enlarged carotid canal (cc, Fig. 7.3C). The carotid foramen is anterior in the fossa and the basisphenoid has a longer bone body (Fig. 7.3C), sufficiently longer anteriorly to accommodate pterygoid wings as in Podocnemididae. In addition, the position of the muscle scars is more posterior than in the Upparhatti specimen (Fig. 7.3A1, C1) and more medial not covering the acute border of the fossa as in *Hamadachelys* and extant Podocnemididae including *Erymnochelys*. Although the position of this scar is variable in Podocnemidoidea, the completely lateral position of the scars in the Upparhatti basisphenoid does not fit with the Ibeceten Senonian basiphenoid or with a podocnemidid skull presenting pterygoid wings which might hide a deep fossa above an enlarged canal. The Ibeceten fauna contains numerous other elements of several Pelomedusoides most of which belong to a form representing the *Erymnochelys* group in the typical pattern of short and narrow intergular posteriorly enclosed by the joined gulars (Fig. 7.11A1, A2; Lapparent de Broin 2008b: Fig. 9; Pérez-García et al. 2017b). As it presents some differences with the type species *Erymnochelys madagascariensis* (see also below in the dentary and vertebrae sections), the Ibeceten basisphenoid is attributed to an indeterminate gen. and sp. in the *Erymnochelys* group of Podocnemididae (Lapparent de Broin 2000b; Pérez-García and Lapparent de Broin 2013, 2015; Pérez-García et al. 2017b; Zouhri et al. 2017). It does not correspond to the bothremydid basisphenoid from Upparhatti.

"*Carteremys*" *pisdurensis* Jain, 1977 (Fig. 7.1F), a bothremydid from the Upper Cretaceous Lameta Formation of Pisdura, is based on a skull considered as related to *Carteremys leithii* (Carter 1852) of the intertrappean beds of Worli Hill, Mumbai (Fig. 7.1C). It was latter named as "*Shweboemys*" *pisdurensis* by Jain (1986) while adding postcranial material from the Lameta Formation of Dongargaon, a nearby infratrappean locality, as Jain (1986) felt that it was distinct from that of *C. leithii* (see figs in Williams 1953) and that it shared skull features with those of the species of *Shweboemys* (s.l.). "*Carteremys*" *pisdurensis* is a

Podocnemidoid in the contact of basisphenoid-quadrate hiding the fenestra postotica. It has rather an elongated palate, a feature considered as a characteristic of the genus *Shweboemys* Swinton, 1939. Divided into several new genera by Gaffney et al. (2011) (see below and comparative geographic and stratigraphic distribution in Pérez-García et al. 2017b), the middle Eocene-Pliocene *Shweboemys* (s.l.) represents part of the *Shweboemys* group of Lapparent de Broin (1988), i.e., of the subtribe Stereogenyina Gaffney et al., 2011. It includes two subgroups: (1) the littoral forms: *Shweboemys* subgroup of Zouhri et al. (2017) that includes all forms with a "secondary palate" morphology as in the holotype of *Shweboemys*, and (2) *Stereogenys* subgroup which presents its own distinguishing features of "secondary palate". However, both are podocnemidids by having the fossa pterygoidea with an enlarged carotid canal, and both have much widened and elongated maxillae and palatines in "a kind" of secondary palate. The latter reference is made as it remains longitudinally narrowly open and not medially closed as in the "true" secondary palate of chelonioids. In "*S" pisdurensis* ("*C*" *pisdurensis* as reviewed above) unlike in *Shweboemys* s.l. (Zouhri et al. 2017), there is no long and medially open "secondary palate", but just wide triturating surfaces of the maxillae, as in Bothremydini and *Kurmademys* in particular, and pterygoids anteriorly prolonged into a point nearly up to the vomer; the basisphenoid area is not precise enough but is long, narrow and anteriorly pointed. On each side and well separated from the basisphenoid, there is a schematized oval shape (without any relief) which has been interpreted by Jain (1986) as the "podocnemidid enlarged carotid canal area" (as in *Shweboemys* s.l.). But we do not agree with this interpretation, like Gaffney et al. (2011) because of the following characters: the foramen area, interpreted as carotid foramen, is figured posteriorly and laterally at the posterior end of this oval-shaped area, at the pterygoid-quadrate suture not being hidden in the depth of the skull anteromedially and not half-covered by pterygoid wings as in podocnemidids. The oval-shaped area is only a podocnemidoid fossa pterygoidea, as present in bothremydids. This fossa is weak as in older bothremydid forms *Cearachelys* and *Galianemys*, while it is wide in *Kurmademys* (with the preserved primitive prootic presence as in *Galianemys*) and particularly narrow and deeper in *Kinkonychelys* (without prootic presence) (Gaffney et al. 2006), with a congruent carotid foramen.

In the type species of the genus *Shweboemys*, *S. pilgrimi* Swinton, 1939 from the Pliocene of Burma, a long and anteriorly pointed basisphenoid, as in "*S*". *pisdurensis* (in Jain 1986), contacts the palatines ventrally, preventing the pterygoids from contacting each other at the midline (Wood 1970). However, among *Shweboemys* s.l., in *Cordichelys antiqua* (BMNH R11998) from the Fayum (Egypt), and other Mio-Pliocene Egyptian species *Lemurchelys diasphax* and *Latentemys plowdeni* Gaffney et al., 2011 and the early Miocene *Brontochelys gaffneyi* (Wood 1970) from Bugti Hills (Pakistan), the pterygoids contact each other medially and the basiphenoid is triangular with a wide base, shorter, and even wider than the Upparhatti basisphenoid. In *C. antiqua* and *B. gaffneyi,* the muscle scars are posterior and more medial than in Upparhatti bone, as in other podocnemidids including Senonian Ibeceten specimen although the area is neither preserved nor figured in other "*Shweboemys*" s.l. species. In *Stereogenys*, the secondary palate of which is much longer than in previous genera, the lateral borders of the palatines in the temporal inferior fossa are parallel for most of their length, the wide basisphennoid is trapezoid in outline, widely separating the pterygoids all along the length of the latter, with parallel lateral borders and pointed anterior ones, filling the gap beween the prolonged palatine secondary palate borders. But, if "*C.*" *pisdurensis* shares with "*Shweboemys*" the rather long palate and if it shares the long pointed basisphenoid of *S. pilgrimi*, it cannot be a podocnemidid: the absence of a true enlarged carotid canal (as figured in the diagram and not appearing in the picture), absence of posteriorly prolonged pterygoid wings, and of a *Shweboemys* secondary palate morphotype, justifies its attribution to new genus other than "*Shweboemys*" within Bothremydidae (see **section on** shell). Many Euro-Afro-American Bothremydini have posteriorly widened maxillary triturating surfaces as in "*Carteremys*" *pisdurensis* (Jain 1977: pl. 1b), *Kurmademys* and *Carteremys leithii*. Within them, there are particularly the Late Cretaceous *Polysternon provinciale* (Matheron 1869) and *Foxemys mechinorum* Tong et al., 1998, from the Maastrichtian of France, and *Chedighaii* from USA (Gaffney et al. 2006: Figs. 103, 110, 157). All of them are devoid of the particular deep rounded fossa that is posterior in the maxillary surface (maxillary pit in Gaffney et al. 2006) and characterizes the bothremydine group of *Bothremys*, *Rosasia*, *Araiochelys* and *Zolhafah* (Broin 1977; Broin in Antunes and Broin 1988; Gaffney et al. 2006; Lapparent de Broin and Werner 1998). On the other hand, in all these genera, the maxillae are posteriorly widened without intermediate maxillary crest in contrast to *Sankuchemys* Gaffney et al., 2003 (Fig. 2).

As far as *Carteremys leithii* (Carter 1852) is concerned, the basiphenoid is unknown. Its skull is known only by a dorsal diagram of an anteriorly wide incomplete snout and showing wide posterior temporal fossae, as in "*Carteremys*" *pisdurensis*. Its ventral side shows the corresponding incomplete part of the lower jaw with a preserved dentary posterior part sufficiently wide for receiving important triturating surface. This morphology fits with a bothremydid with wide crushing maxillae and mandible as in Bothremydini. Furthermore, the shell of *Carteremys leithii* is typically bothremydid, as the postcranial remains of

"*Carteremys*" *pisdurensis* of Jain (1986). However, the latter differs particularly in the absence of visible ornamentation, contrasting with the strong decoration of *Carteremys*. *Sokatra*, another Podocnemidoid, in its skull outline and maxilla might correspond to the dentaries of Upparhatti (see below). As mentioned earlier, the figure of "*Carteremys*" *pisdurensis* represents a ventral basicranium partly crushed and broken in the seemingly flat area, but a possible fossa pterygoidea may be present as indicated by the presence of a depression in the preserved part beside the quadrate; a sutured contact of the primitively exposed prootic with the anterolateral border of a triangular basisphenoid is drawn but not confirmed by the photograph. Anyway, it is a configuration that is different from that of the Upparhatti basisphenoid by the absence of a vertical basisphenoid border on the fossa.

To sum up, taking into account all the variabilities between species for the basisphenoid elements, and all the congruencies, the Upparhatti basisphenoid differs from *Shweboemys* s.l. in the absence of an evidently enlarged podocnemidid carotid canal. In all its features, it appears that it is not a podocnemidid. Although appearing as a bothremydid, DUGF/T167 differs from "*C.*" *pisdurensis* by its morphology and proportions, including the position of the podocnemidoid fossa and of the carotid foramen. DUGF/T167 differs from all known bothremydids of the Cretaceous of India, although perhaps not from *C. leithii*, the basisphenoid of which is unknown. It also differs from the two forms of Madagascan bothremydid.

Lower Jaw: *Dentaries* (Fig. 7.3D–G): Three jaw elements from Upparhatti conform in size and appearance with the basisphenoid. DUGF/T164 a left dentary without the anterior extremity, 2.7 cm long × 1.3 cm high (posteriorly); DUGF/T165, symphysial part of sutured left and right dentaries, 1.1 cm long at symphysis; DUGF/T166, left dentary (without symphysial part), coronoid and anterior surangular, 3.1 cm long × 1.55 cm high (at coronoid process). (Fig. 7.3D–G). These jaw bones are similar in size and morphology and allow the reconstruction of the dentary (Fig. 7.3G). From comparison with the podocnemidid *Erymnochelys*, DUGF/T166 corresponds to a lower jaw about 4.2 cm long and a skull about 5.5 to 6 cm long from the anterior tip to the occipital condyle. The anterior outline is rounded in dorsal view, and the dentary rami were widely divergent, the angle between the two dentary rami being about 110–120°. The symphysis is very short both dorsally at the triturating surface and ventrally below this surface (Fig. 7.3E1 in dorsal and Fig. 7.3E2 in posterodorsal views). A symphyseal ridge is present separating the two triturating surfaces. The symphysis is slightly pointed anteriorly forming a very short vertical hook, visible laterally (Fig. 7.3E2, E3). The dentary is relatively deep (Fig. 7.3D3, E3, F2, F3). The triturating surface is narrow, showing almost no widening posteriorly (Fig. 7.3D2, F1, G). Each triturating surface forms a longitudinal groove between the labial and lingual borders and that lingual border extends from a point slightly lateral to the symphyseal midline crest up to the coronoid process. At about mid-length of the dentary, the lingual border is very narrowly divided into a double lingual crest, the intermedial part on the surface (mr) being rounded, stronger and higher, but the lingual (lr) part on the surface being lower and thinner (Fig. 7.3D2, E1, F1, G). The two lingual crests become reunited anterior to the coronoid process. The lingual and labial crests isolate a short and narrow lenticular sulcus, barely like a groove. The labial border is sharp in posterior and lateral views (Fig. 7.3E2, E3) and its dorsal external surface is covered by nutritive foramina for the ramphoteca. Externally, its outline appears to be concave between the short and pointed hook and the moderately elevated and rounded coronoid process (Fig. 7.3D3, E3, F3). At the latero-posterior limit of the ramphoteca, the foramen dentofaciale majus is well developed and doubled (Fig. 7.3F3). Internally, the sulcus meckelii runs longitudinally (Fig. 7.3F2).

Based on the Upparhatti lower jaw elements, the shape of the corresponding skull must have been short and wide with a rounded outline anteriorly and a weak premaxillary fossa for the short dentary hook. The premaxillo-maxillary masticatory surface was short and narrow, not or very slightly widened posteriorly, and possibly would have had an intermediate posterior, short and narrow medial maxillary crest.

Discussion on Dentaries: The dentaries of Upparhatti present a similarity with Meiolaniidae but also with *Proganochelys* among others. The similar features are: - a not triangularly pointed short symphysis, - dentaries with a similar triturating surface consisting of a concave surface all along, relatively narrow and not posteriorly widened, between two crests (labial and lingual) in Meiolaniidae and *Proganochelys*, - and division of the lingual crest with its lingual border less elevated than its median (intermediate in the surface) upper border, - and both framing a groove (or a simple sulcus in Upparhatti dentaries) more or less long (short in Upparhatti dentaries). Such features are present in many other turtles (among them testudinids, which are terrestrial as *Proganochelys* and Meiolaniidae), with similar narrow concave and not posteriorly widened surface). The Upparhatti dentaries have to be examined in conjunction with all the Upparhatti turtle body elements, not independently: the associated basisphenoid, neck vertebrae, limb bones, shell elements are all pleurodiran Pelomedusoides. After comparison, these bones appear to represent a bothremydid rather than a podocnemidid and certainly not chelid or meiolaniid which are from the southern Gondwana. As far as Upparhatti dentaries are concerned, they differ from *Proganochelys* and Meiolaniidae (and other taxa): by a derived shorter lower symphysis part and less long upper symphysis part with a clearly derived elevated symphyseal longitudinal crest; by the way the lingual crest (which has neither a sharp nor sharp but rounded border) begins after a smooth border point on each side of the symphyseal crest; by the way this lingual crest is divided, from the mid-dentary length, into two parts (a rounded one [mr] and a thin lingual one [lr] framing an intermedial short and narrow simple sulcus); and by the way the division ends before the coronoid process (co), beside a long coronoid part which is integrated next to the dentary. This coronoid process is much higher than in *Proganochelys* and *Meiolania* (i.e., it is more derived as in various Pelomedusoides). Morphology of lower jaw is much variable among Pelomedusoides, Bothremydidae as well as Podocnemididae. Few lower jaws are known for Bothremydidae (Table 7.3) among which, those from the Paleogene Posphates of Morocco represented by numerous species (Gaffney et al. 2006, stratigraphic distribution summarized in Bardet et al. 2017) provided the bulk of studied specimens. They differ according to the group to which they belong and to their dietary preferences, as in Cheloniidae (Parham and Pyenson 2010; Bardet et al. 2013; Lapparent de Broin et al. 2014, 2018). The primitive Pelomedusoides features of the dentaries are, notably: no supernumerary intermediary ridge, no longitudinal crest on the symphysis which is primitively a little elongated and smooth, no posteriorly or all along widened triturating surface, no double lingual crest, no medially inclined triturating surface. Each bothremydid form presents one or several derived states of these features, according to the phylogeny and to the feeding pattern. All of them have a symphysis more elongated than in the Upparhatti specimens (according to Gaffney et al. 2006 and MNHNF coll.), and lack derived sagittal sympyseal crest. As seen above, some podocnemidoids are adapted to a crushing diet, with triturating surfaces posteriorly widened such as cited for above bothremydids and podocnemidid Stereogenyini and like in many Cheloniidae. Others have a rather narrow triturating surface but inclined medially such as the taphrosphyine *Rhothonemys brinkmani* Gaffney et al., 2006, which has a cutter-snatching or shearing beak as in some Cheloniidae (Lapparent de Broin et al. 2014, 2018); it is one of the few species without posteriorly widened triturating surface as in Upparhatti specimens, but the surface is medially inclined and without lingual crest and there is no symphyseal crest. *Foxemys* is the only bothremydid known by lower jaws with a slight double lingual crest as in Upparhatti specimen (Fig. 7.3G). But its triturating surface is posteriorly clearly

Table 7.3 Comparison of dentaries. Bothremydidae indet. From Upparhatti compared with two known bothremydid jaws of India, *Kurmademys* and *Sankuchemys*, with *Foxemys*, the other known bothremydid with dentary ridge, and, in podocnemidids, two extant (*Erymnochelys* and *Podocnemis*) and aff. *Erymnochelys* sp. from Ibeceten Senonian (IBC). Abbreviations: ant, anterior; antly, anteriorly; corr., corresponding; dent. dentary; elong-mod, moderately elongated; H, depth; L, length; lat, lateral, laterally; mx, maxilla; Mod, moderately; trit surf, triturating surface; post, posterior, posteriorly; W, width. Brackets: inferred

Dentary characters	Upparhatti DGUF164, 165, 166	*Kurmademys* Gaffney et al. (2006)	*Sankuchemys* Gaffney et al. (2006)	*Foxemys* Gaffney et al. (2006)	*Erymnochelys madagascariensis* MNHN ZA AC coll.	Aff. *Erymnochelys* sp. MNHNF IBC 17 dent. + 4 mx.
Dentary symphysis below trit surface	Very short	Long	?	Long	Long	Long
Symphysis full L	Short	Long	?	Long	Long	Long, mod
Dentary hook	A small upward point	No	(No)	No	Yes	Yes, small
Symphysis medial crest	Yes	No	?	No	No	No
Dentary trit surf posteriorly	Narrow. A groove	Post widened. Mod weak	?	Post widened. Weak	Mod post widened. Weak	Narrow. Weak
Dentary ridge	Yes post, short	No	?	Yes long	No	No
Dentary lat H	Deep	? (ant short)	Mod deep	Low	Deep	Deep
Corr. Maxillary ridge	(No)	No	Yes, long	Yes, ant	No	No
Corr. snout ant border	(Rounded slightly pointed)	Trapezoid slightly pointed	Rounded	Rounded antly pointed	Triangular pointed	(Pointed)
Skull L and W	Short, wide	Short, mod wide	Short, mod wide	Short, wide	Slightly elong, mod wide	? (Short)

widened (corresponding to widened maxillae, as in *Polysternon*). Its surface is well individualized into two triturating longitudinal grooved surfaces by the intermediary crest, which extends up to the anterior part of the symphysis close to the sagittal part, that is free of sagittal crest (Gaffney et al. 2006: fig. 235). Furthermore, in this taxon, the inferior part of the symphysis is longer and its anterior angle is narrower than in Upparhatti dentaries. An intermediary ridge, similar to that of *Foxemys*, unnamed but figured, has been reported for the lower jaw of a bothremydid from the Late Cretaceous of Hungary (Rabi and Botfalvai 2006). However, this ridge is possibly not exactly homologous to that of *Foxemys*. It appears that this ridge constitutes the labial border and the structure identified as the labial border might be a depressed expansion of the external dentary surface, lateral to the border of triturating surface. In any case, this structure differs from the Upparhatti postero-medial ridge, which does not isolate a supplementary lingual groove and is positioned on a narrow surface.

Among Bothremydidae from India, *Kurmademys* differs from the Upparhatti dentary in that its lower jaw has a posteriorly widened, concave and smooth triturating surface and no symphyseal ridge as in *Cearachelys* and bothremydine spp. (Gaffney et al. 2006). Also, the angle of the symphysis is not as wide as that of the Upparhatti species (corresponding to a less rounded snout) and the lower symphysis part is longer. The upper frontal triturating part of the symphysis is smooth and very short, and the lower part with the meckelian canal extremity is much longer. The Upparhatti form has relatively few derived features among those examined above. But contrary to *Kurmademys* (among others) it has the symphyseal crest, the small hook and the deep narrow groove while, no widened triturating surface and no long lower symphysis part. The lower jaw of *Sankuchemys* is unknown, but the shape of the triturating surface can be inferred from the maxillae. As in the Upparhatti specimens (Fig. 7.3D2, F1), the posterior end of the triturating surface on the maxilla is not expanded relative to its anterior end, but in *Sankuchemys* the surface is too much wide in relation to its entire length to correspond to the Upparhatti dentaries. Also, the maxilla has a paralingual ridge that extends along the entire surface. This does not correspond to the short posterior ridge of the Upparhatti dentaries, rather corresponds to a long intermediate dentary ridge.

As seen above, *Carteremys leithii* had widened maxillae and widened lower jaws (posteriorly known only). Its lower jaw is represented only by a part lacking the tip. It has been figured by Carter (1852) (diagrams, Williams 1953). The posterior part of the symphysis is represented ventrally, and this is much longer than in Upparhatti specimens, and the preserved parts of triturating surfaces are posteriorly widened. The partial remains of *Carteremys* do not compare with the Upparhatti elements.

In "*Carteremys*" *pisdurensis*, the lower jaw is not preserved but the skull (Jain 1977) allows its differentiation from the podocnemidids and, in particular, from "*Schweboemys*" (s.l.) to which Jain (1986) referred this species. In this group, the lower jaw has, in agreement with the secondary palate of the skull, a triangular and long symphysis covering the lower part of the symphysis, i.e. a flat and posteriorly wide triturating surface and a triangular hook that is little elevated, as shown in *Bairdemys* Gaffney and Wood, 2002, the South American branch of this littoral subgroup (Sánchez-Villagra and Winkler 2006; Weems and Knight 2013). This is confirmed in a littoral, indeterminate species of "*Shweboemys*" (s.l.) with a lower jaw similar to that of *Bairdemys* from the As Sarrar (late early Miocene) locality, Saudi Arabia (Thomas et al. 1982). The As Sarrar lower jaw (as that of *Bairdemys*) has a long symphysis corresponding to the anteriorly triangular skull with a long symphysis in "*Shweboemys*" (length 75% of the width at coronoid processes and 50 to 65% in *Bairdemys*) (in Weems and Knight 2013) with a weak anterior hook that would have corresponded to the weak premaxillary concavity. The As Sarrar lower jaw surface is weakly sagitally domed, rather than being flat or crested, with two lateral depressions. It has a rounded and protruding coronoid process corresponding in shape to the anterior part of the fossa temporalis inferior. This jaw cannot fit with that of "*C.*" *pisdurensis* and it does not fit either with the Upparhatti dentaries, where the symphysis is not enlarged to correspond to a secondary palate.

In Podocnemididae (Stereogenyini being excluded), there is variability according to the species. The dorsal part of the symphysis is short and its lower part is longer, being a double of the upper part or a little more, but the symphysis full length is no more than a third of the jaw width at coronoid processes (against 13% in DUGF/T165-166) and the angle between the jaw branches is acute in relation to Upparhatti jaws. In Podocnemidinae, a doubled lingual crest is present (Table 7.3) in the lower jaw branches of extant *Podocnemis cayennensis*, *P. lewyana* and *P. vogli*, anteromedially extending up to the sagittal symphyseal ridge which separates the two (left and right) triturating surfaces. There are variations, for example, in the intermediary crest of the double lingual crest which is more or less visible. *P. expansa* has a clear doubled lingual crest but no symphyseal median ridge. *P. sextuberculata* has barely perceptible, very low intermediary lingual crest and a clear symphyseal sagittal crest. Among *Podocnemis* fossil relatives, *Cerrejonemys* Cadena et al., 2010 from the Paleocene of Colombia and *Lapparentemys vilavilensis* (Broin 1971, 1991) from the early Paleocene of Vila and Tiupampa (Bolivia) have neither lower jaw lingual ridge division nor symphyseal ridge. In Erymnochelyinae, a long intermediate crest is present in the lower jaw of the African *Turkanemys pattersoni* Wood, 2003 from the late Miocene Mio-Pliocene of Lothagam, Kenya. It is a close relative of *E. madagascariensis*, a member of *Erymnochelys* group by its long gulars / short intergular pattern present. The podocnemidid *Turkanemys* lower jaw is rather similar to that of the bothremydid *Foxemys* but it is dorsally wider with wide triturating grooves. It is anterioly more rounded dorsally but with a less long inferior part of the symphysis and is elevated in a high hook. The symphyseal crest figured by Wood (2003) does not appear on the photograph of *Turkanemys* (Wood 2003: fig. 4.6) as in the other forms of *Erymnochelys* group. A lower jaw of "cf. *Erymnochelys* sp." is known from the Upper Cretaceous of Madagascar (Gaffney and Forster 2003). From the continental Senonian Ibeceten level (Niger) (Broin et al. 1974), there are 17 fragmentary dentaries and four corresponding maxillary parts, known representing an indeterminate gen. and sp. of the *Erymnochelys* group (MNHNF IBC coll.) (Fig. 7.3H–L) (Table 7.3). The Ibeceten specimens conform to the *Erymnochelys* jaw morphotype (Broin 1977: pl. 3; Gaffney and Forster 2003) in having a high dentary symphysis without a median crest, a moderately pointed hook, a not posteriorly widened triturating surface, lacking a double lingual crest, and the lower part of the symphysis barely longer than its upper surface (Fig. 7.3H1, K, L1 in dorsal view); less long than in aff. *Erymnochelys* from Madagascar and still less than in living *E. madagascariensis*. The angle at the symphysis of the Ibeceten specimens is approximately 90°, slightly more than in *E. madagascariensis* (75°), indicating a more acute triangular snout in the latter. These angles differ from that of the Upparhatti species which is much wider (110–120°), indicating that this had a shorter snout, and less pointed hook than in *Erymnochelys* and its relatives. The Ibeceten species, *E. madagascariensis* and the Cretaceous Madagascan jaw share the absence of a double lingual crest. A difference between the jaw of the Ibeceten species on the one hand and that of *E. madagascariensis* and the Cretaceous Madagascan jaw on the oher hand is that the triturating surfaces of Ibeceten jaws remain primitively narrower up to the coronoid process, as in the Upparhatti species. Further, the symphysis of Ibeceten specimens has its lower part relatively shorter, nearly as in the Upparhatti specimens. Whereas it is much longer than the upper part in the Cretaceous Madagascan jaw and *E. madagascariensis* and as in above observed bothremydids and podocnemidines. Triturating surfaces are slightly posteriorly widened in the two taxa from Madagascar (although not as much as in the crushing jaws cited for bothremydini and of *Carteremys*) and the whole surface is widened (moderately) in *Turkanemys* due to its double crest. With a more acute anterior angle, the symphyses of all Ibeceten jaws are clearly longer than in the Upparhatti specimens and have a longer hook. However, it is short in Ibeceten specimens in relation to stratigraphically younger erymnochelyines. On the whole, the anterior skull was more triangular and shorter in the Ibeceten species than in *E. madagascariensis*, *Turkanemys* and (but less) the Cretaceous Madagascan taxon. The skull of Upparhatti species was also short, but it was more rounded and wider with a skull outline as in bothremydids *Zolhafah* and *Sokatra.* By contrast, *Zolhafah* had a crushing diet pattern of the *Bothremys* morphotype with wide posterior maxillae bearing the typical posterior rounded pit of the group. The lower jaw of another Cretaceous Madagascan form, the bothremydid *Kinkonychelys*, is unknown. Its skull had posteriorly widened maxillae, corresponding to posteriorly widened inferior triturating surfaces, which does not correspond to the Upparhatti dentary shape but fits with *Kurmademys.*

The skull of *Sokatra* with rounded rather wide anterior border outline corresponds to a lower jaw with the outline of the reconstructed specimen of Upparhatti (Fig. 7.3E, F, G). The ventral maxilla has a regularly narrow triturating surface that may correspond to that of the Upparhatti dentary, pointing to its possible relationship with this taxon. However, the Upparhatti basisphenoid does not correspond to that of *Sokatra.*

Thus, insofar as they are known, the Upparhatti lower jaw fragments and, by inference, their skull, do not correspond to any known podocnemidid and bothremydid. The jaw fragments do not fit with the double-ridged maxillae of *Sokatra.* They do not belong neither to the Indian *Kurmademys* and *Sankuchemys* and the Madagascan *Kinkonychelys*, nor to the Indian "*Carteremys*" *pisdurensis* and *Carteremys leithii.* However, both *C. leithii* and *Sankuchemys* may have been similar to the Upparhatti form in the presence of a short skull as in many bothremydids, and not an elongated skull as that of "*C.*" *pisdurensis* and of the podocnemidid *Schweboemys* s.l. However, *C. leithii* shell is strongly decorated while no marked decorated shell element is found in Upparhatti except for some elements having basic aquatic ornamentation in places (i.e. scarce light polygones, small dichotomic sulci). Unknown *Sankuchemys* shell cannot be compared. As far as Podocnemididae are concerned, the Upparhatti skull-jaw pattern does not correspond to that of any podocnemidid of Erymnochelyinae from the continental Africa and Africa-Arabian littoral platform, including *Shweboemys* (s.l.) and *Erymnochelys* group, notably of Upper Cretaceous of Ibeceten and Madagascar and living species. Thus, with their short symphysis and wide rounded jaw, the dentaries from Upparhatti indicate the presence of a still unnamed bothremydid in the Upper Cretaceous of India.

Vertebrae: There are 24 more or less complete cervical elements, belonging to at least two forms, one short-necked and one long-necked and three fragmentary thoracic vertebrae (for measurements see Table 7.4).

Table 7.4 Measurements of selected Upparhatti cervical and dorsal vertebrae

Measurements in cm	length	Medial height	% h/l	Fig.	Measurements in cm	length	Medial height	% h/l	Fig.
Cervical vertebrae									
DUGF/T169 long centrum (axis)	1.8	0.5	27%	7.4A	DUGF/T175 (6th?) ventrally partial centrum			ca. 50%	7.4H
DUGF/T168 long median centrum	2.25	0.7	31%	7.4B	DUGF/T165 centrum (2d to 6th)	1.65	0.7	42%	
DUGF/T177 (4–6th) pedicel + linked postzyg	1.5			7.4C,	DUGF/T170 medioposterior partial centrum, similar by its external appearance to DUGF/T168	2.45	1.4	57%	
DUGF/T176 (6–7th) pedicel + separated postzyg	1.4			7.4D	DUGF/T171 medioposterior partial centrum	2			
DUGF/T174 (7th?) centrum	1.7	1.05	61%	7.4E	DUGF/T179 fragment	1.4			
DUGF/T189 (7–8th) partial neural arch	1.7			7.4F	DUGF/T185 fragmentary centrum, short with a protruding crest	1.1	0.9 (and 1.5 at ventral crest)	81%	
DUGF/T178 8th partial centrum	1.65	1.1	66%	7.4G					
Dorsal vertebra									
DUGF/T172 1st centrum	2.3	0.7	30%	7.4I					

Cervical Vertebrae (Fig. 7.4 A–H): DUGF/T169, Fig. 7.4A, long axis centrum; DUGF/T168, Fig. 7.4B, long median centrum; DUGF/T177, Fig. 7.4C, median (4–6th), posterior neural arch pedicel with linked postzygapophyses; DUGF/T176, Fig. 7.4D, posterior (6–7th); DUGF/T174, Fig. 7.4E, posterior centrum (7th?); DUGF/T189, Fig. 7.4F, partial posterior (7–8th), neural arch; DUGF/T178, Fig. 7.4 G, 8th partial; DUGF/T175 Fig. 7.4H, centrum (6th?) ventrally incomplete; DUGF/T165, centrum (2d to 6th); DUGF/T170, medioposterior partial centrum, similar by its external appearance to DUGF/T168; DUGF/T53, fragment of posterior central condyle; DUGF/T171, medioposterior partial centrum; DUGF/T179, fragment; DUGF/T180 to DUGF/T184, 5 fragments of anterior central parts (with a cotyle); DUGF/T185, a fragmentary centrum, short with a protruding crest; DUGF/T186 to DUGF/T188, three fragments of transverse apophyses. Dorsals (thoracic vertebrae) (Fig. 7.4, I–J): First dorsals; DUGF/T172, Fig. 7.4I, 1st dorsal centrum; DUGF/T104, eroded, and DUGF/T120, posterior thoracic vertebrae, narrow and low for their length like DUGF/T172. DUGF/T120, Fig. 7.4J, is possibly a most posterior vertebra, directly covered by the medial junction of two pleurals at the posterior extremity of the carapace.

With one exception, the Upparhatti cervical centra are procoelous. The exception is an axis centrum (DUGF/T169) which is biconvex (Fig. 7.4A). The abundance of procoelous cervical vertebrae, presence of a biconvex atlas, and the lack of any opisthocoelous vertebrae is indicative of a neck with a biconvex axis and procoelous third to 8th vertebrae. This conforms to the Pelomedusoides neck formula (Williams 1950; Broin and De la Fuente 1993). Other characters confirm the pleurodiran attribution of these centra, including the presence of transverse apophyses forming lateral wings, a single elongated process (i.e. postzygapophysis pedicel) for the postzygapophysis, which is elevated at the posterior extremity of the neural arch, and occasionally (in one form) postzygapophyses linked in a semi-circular collar.

Two morphotypes are present in the Upparhatti cervicals and first dorsals, indicating the presence of at least two taxa. These are designated "(cervical) long 'morphotype a'" and "(cervical) short 'morphotype b'". 'Morphotype a' is represented by cervical centra that are longer in relation to the height (height 30 to 50% of length). In this morphotype, the wings (transverse apophyses) are medially positioned, relatively thin (not deep) on their central insertion with oblique borders, as much inclined anteriorly and posteriorly as seen in dorsal or ventral views (Fig. 7.4A, B, J). The centra in this morphotype include axis DUGF/T169, median DUGF/T168, first dorsals DUGF/T172 and DUGF/T170, DUGF/T53, DUGF/T171 and DUGF/T104. 'Morphotype b' is represented by centra that are higher in relation to their length (h/l ca 60–70%). This morphotype is represented by specimens such as DUGF/T174 and DUGF/T175 (Fig. 7.4E, H). In 'morphotype b', the preserved transverse processes are located on the medio-posterior part of the centrum from two-thirds of the centrum length toward the posterior end.

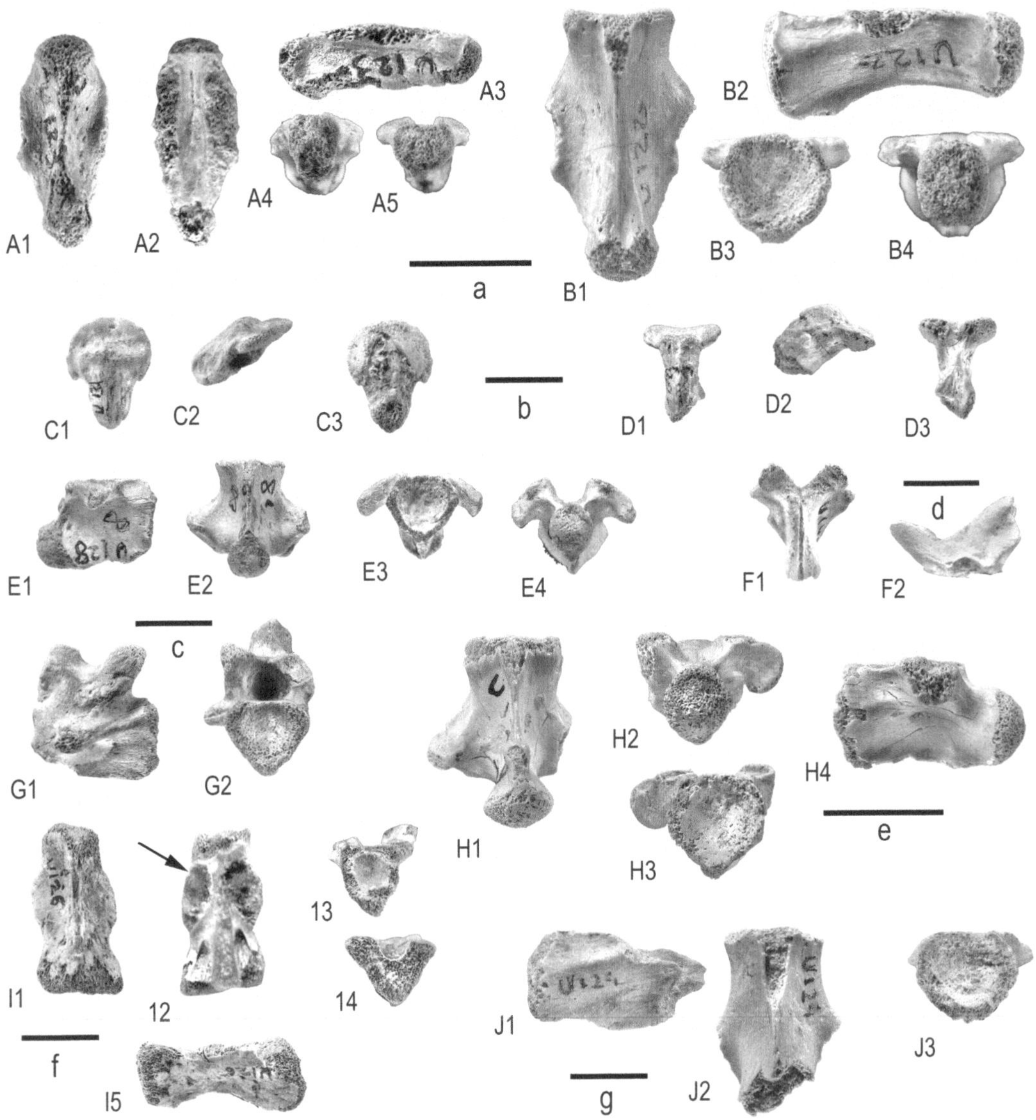

Fig. 7.4 Bothremydidae, Upparhatti, southwestern India, intertrappean beds, Maastrichtian. Fragmentary cervical and dorsal vertebrae, DUGF/T coll. (Work catalog U numbers on the specimens): Bothremydidae gen. et sp. indet., long-necked form a, **A**, **B**, **I** and **J**; cf. Bothremydidae gen. et sp. indet., short-necked form b, **C to H**. **A1–A5**. DUGF/T169, axis centrum, ventral (**A1**), dorsal (**A2**), left lateral (**A3**), anterior (**A4**) and posterior (**A5**) views; **B1-B4**. DUGF/T168, median cervical centrum ($5–6^{th}$), ventral (**B1**), left lateral (**B2**), anterior (**B3**) and posterior (**B4**) views; **C1–C3**. DUGF/T177, median cervical ($4–5^{th}$), pedicel with linked postzygapophyses, dorsal (**C1**), right lateral (**C2**) and ventral (**C3**) views; **D1–D3**. DUGF/T176, posterior cervical ($6–7^{th}$), pedicel with separated postzygapophyses, dorsal (**D1**), right lateral (**D2**) and ventral (**D3**) views; **E1–E4**. DUGF/T174, posterior cervical centrum (7th), right lateral (**E1**), ventral (**E2**), anterior (**E3**) and posterior (**E4**) views; **F1–F2**. DUGF/T189, partial anterior cervical neural arch ($7–8^{th?}$), dorsal (**F1**) and left lateral (**F2**) views; **G1–G2**. DUGF/T178, fragmentary 8th cervical, right lateral (**G1**) and anterior (**G2**) views; **H1–H4**. DUGF/T175, cervical centrum (6th?) ventral (**H1**), posterior (**H2**), anterior (**H3**) and left lateral (**H4**) views; **I1–I5**. DUGF/T172, 1st dorsal centrum, ventral (**I1**), dorsal (**I2**), anterior (**I3**), posterior (**I4**) and left lateral (**I5**) views; **J1–J3**. DUGF/T170, medioposterior partial centrum, left lateral (**J1**), ventral (**J2**) and anterior (**J3**) views; arrow: insertion of the first thoracic rib. Scale bar equals 1 cm, a for **A1–B4**, b for **C1–D3**, c for **E1–E4** and **G1–G2**, d for **F1–F2**, e for **H1–H4**, f for **I1–I5**, and g for **J1–J3**

Also, the protruding wings of these centra are stout, triangular, and their posterior border is nearly transverse while the anterior one is well inclined (Fig. 7.4E, G, H).

The ventral surface of both 'morphotype a' and 'morphotype b' cervicals and the first dorsal centra bear a sharp ventral crest which is partly reduced to completely reduced. This is short and slightly horizontal in the long axis DUGF/T169, short and in the form of a concave bow in the long median cervical DUGF/T168, and the ventral median edge is slightly concave anteriorly in front of the protruding crest in the long first dorsal (incomplete) centrum DUGF/T170. The crest is protruding and medially weakly concave in a short probable 7th cervical DUGF/T174, while in DUGF/T178, a short 8th cervical, it is straight and more protruding. In first long dorsal DUGF/T172, the crest is concave, and it is straight in DUGF/T171. Overall the posterior extremity of the crest is strong in the long median cervical DUGF/T168, protruding and bifurcating in the two small posterior tubercles, visible as separated below the central condyle (Fig. 7.4B1, B4); the tubercles are fused into a single tubercle in the more posterior cervicals DUGF/T175 (Fig. 7.4H1, H2) and DUGF/T178. Thus, below the cervical posterior condyles, the ventral two tubercles may be juxtaposed or well separated. In the long vertebrae from the anterior part of the neck (DUGF/T169, an axis), DUGF/T168 (a mid-cervical), and DUGF/T53 (unidentified position), the separated but juxtaposed tubercles appear as a single very weak process when seen in posterior view (Fig. 7.4A4, B4). In the short cervicals from the posterior part of the neck, such as DUGF/T175, probably a 6th, and DUGF/T164, a 7th, the tubercles are fused in a ventral process, narrower than the condyle, forming a ventral extension of the condyle at the end of the ventral crest. In these centra, the process is visible in posterior view (Fig. 7.4E4).

The anterior cotyles and posterior condyles vary from slightly higher than wide to slightly wider than high. Depending on the morphotype of vertebra and its position on the neck, the cotyles vary from rounded to rather triangular, and the condyles vary from rounded to ovoid (Fig. 7.4). None of the central extremities are clearly widened (transversally stretched in a lenticular shape as in living forms (Fig. 7.6A3, C3) nor clearly narrowed relative to their height.

Prezygapophyses are preserved in two specimens, DUGF/T176 (6th or 7th cervical) and DUGF/T178 (7th?) (Fig. 7.4F, G). These are well separated and curved upwards (Fig. 7.4F1, G1). Both these specimens are examples of short cervicals. A pedicel with postzygapophyses is preserved in two specimens, a medio-posterior vertebra (4th to 6th) DUGF/T174 and the 7th or 8th more posterior vertebra DUGF/T176. From comparison with complete neck in extant taxa, both can be identified as being from cervicals of the short morphology because of their abrupt inclination upward and posteriorly (Fig. 7.4C2, D2) from the anterior base of the neural arch just behind the prezygapophyses. In DUGF/T177, the postzygapophyses facets are linked in a rounded half-circle, the collar, forming a single articulation (Fig. 7.4C1, C3) as in some of the neck vertebrae of South American podocnemidines. The fusion of the facets leaves a rounded very small foramen at the middle part of the collar that is visible dorsally and ventrally (Fig. 7.4C1, C3). While in DUGF/T176 (Fig. 7.4D), the facets are separated, not forming a full or a partial collar and medially open as in most anterior cervicals of South American podocnemidines. Both posterior pedicels DUGF/T176 and DUGF/T177 correspond to anterior partial cervical DUGF/T189.

The first dorsal vertebra is represented by the centrum DUGF/T172 which has an anterior cotyle and a flat posterior surface (Fig. 7.4I). Other fragments of dorsal centra are present (DUGF/T104, DUGF/T120). DUGF/T172 is a centrum relatively long with regard to its width, and in its external appearance it is similar to DUGF/T169, whereas DUGF/T168, also a long centrum, is similar to DUGF/T170.

Comparisons of Vertebrae: The Upparhatti cervical centra vary in length, height, development of the postzygapophyses, and development of the mid-ventral crest and posterior ventral tubercles. In Pleurodira, the height of the vertebral neural arch and the inclination of the pedicel with the postzygapophyses vary according to position in the neck, since the height of the pedicel increases from the axis up to the 8th. The inclination of the pedicel varies also taxonomically according to the relative length and depth of the vertebrae, specific to each taxon, especially when the neural arch and the centrum of the cervical vertebrae are both lowered. However, in extant Pelomedusoides (Pelomedusidae and Podocnemididae), there are several morphotypes of centra with several distinct characters (figures for living taxa in Kasper 1903; Williams 1950; Lapparent Broin and De la Fuente 1993) (MNHN ZA AC coll.) and the axis is biconvex and the cervicals behind it are procoelous. Considering the height and length of primitive cervical vertebrae, particularly during the Triassic (Gaffney 1990; Sterli et al. 2007; pers. obs. in SMNS and UNLaR), the derived states in Pelomedusoides are inferred, the stratigraphically oldest known taxa of this group not having the most primitive states as demonstrated below by the study of their remains.

In Podocnemididae, a gradation in neck length is present that is related to the posterior height of the skull. Primitively (*Proganochelys* [see Gaffney 1990] and *Paleochersis* [see Sterli et al. 2007], the neck vertebrae are short and high (centrum and neural arch), more than in extant forms. But in extant taxa, the cervical height varies with the skull and shell height apparently in relation to feeding habits. The podocnemidine *Peltocephalus* has a higher skull and shorter cervicals (Fig. 7.5A, B) compared to the *Podocnemis* group of the new world. In old world Erymnochelyinae, *Erymnochelys madagascariensis* and known close relatives in its group, the neck is shorter and relatively high (Fig. 7.6A, B), compared to the Eocene *Neochelys*, an erymnochelyine taxon with a less domed skull and shell (Pérez-García and Lapparent de Broin 2015: Fig. 12). In other podocnemidids as in pelomedusids (Fig. 7.6C–E), the cervical vertebrae are also lowered, and they show a reduction of the primitive ventral crest. The primitively well protruding and ventrally straight crest reduces, the ventral median border becomes curved from axis to 6th cervical at least, the crest of the cervicals 7–8 or 8 only remaining protruding.

Cervical - Central Articulations (Joints): Two morphotypes of joints are present in Pelomedusoides: joint morphotype (A), a saddle-shaped one, is derived for podocnemidines (Fig. 7.5A, B), and joint morphotype (B),

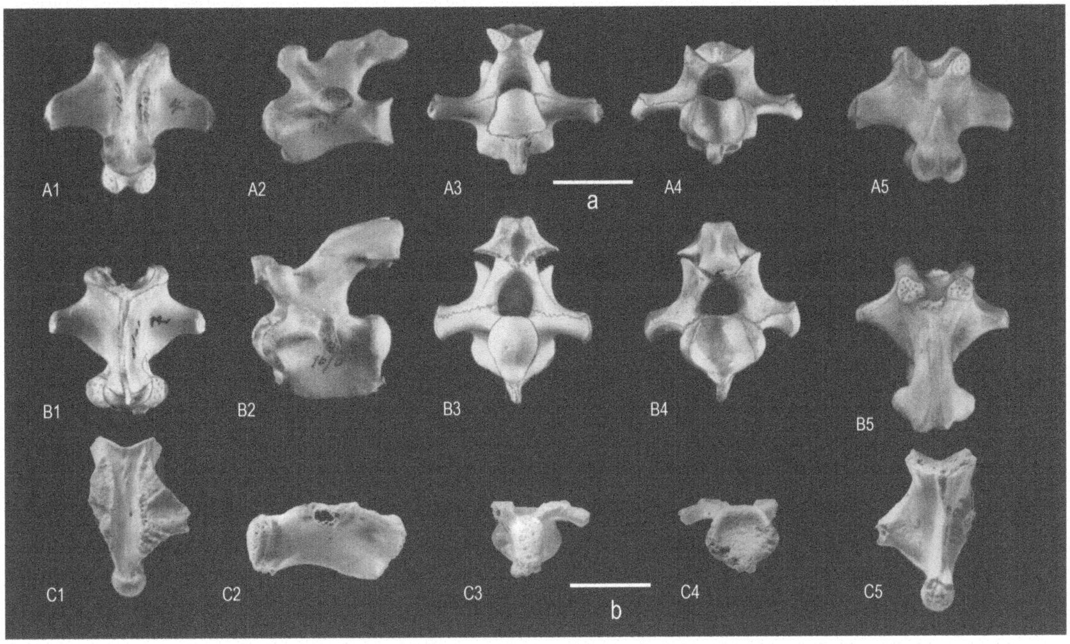

Fig. 7.5 *Peltocephalus dumerilianus*, Guyane, extant, MNHN (ZA AC) Lescure coll., **A1–A5**. 4th cervical vertebra; **B1–B5**. 7th cervical; ventral (**A1, B1**), left lateral (**A2, B2**), posterior (**A3, B3**), anterior (**A4, B4**) and dorsal (**A5, B5**) views. Bothremydidae indet., late Early Miocene (Burdigalian), As Sarrar, Saudi Arabia, loc. 9, MNHNF deposit, no 1081, **C1–C5**. cervical vertebra, 3th to 6th centrum, dorsal (**C1**), right lateral (**C2**), posterior (**C3**), anterior (**C4**) and ventral (**C5**) views. Scale bar equals 1 cm, a for **A1–B5** and b for **C1–C5**

without saddle condyles and with posterior tubercles below the condyle, is basic for Pelomedusoides, (Fig. 7.6A, D for living specimens).

Saddle-shaped joints (morphotype A) are present only in South American Podocnemidinae, both extant and fossil forms closely related with *Podocnemis* and in *Peltocephalus*. In living *Podocnemis* species, saddle-shaped condyles are seen from 2nd to 6th–7th centra, the 7th and 8th condyles being convex and ovoid, respectively and in a bow shape (convex ventrally) (Williams 1950: fig. 11). The centrum is moderately elongated. This joint shape is associated with the reduction of the two posterior tubercles below the condyle up to and including the 8th, so that tubercles are no longer visible in posterior view. The saddle-shaped condyle corresponds to a cotyle in the shape of short, transversally stretched heart. This morphotype of vertebrae is also known in *Podocnemis* relatives *Cambaremys langertoni* França and Langer 2005 (Fig. 2) from the Maastrichtian of Brazil and *Lapparentemys vilavilensis* (Broin 1971) (Broin 1991; Gaffney et al. 1998; Gaffney et al. 2006) from the Paleocene of Bolivia. The entire vertebra, including the neural arch, is rather lowered but not completely flattened, as in extant Pelomedusidae. The ventral crest is medioposteriorly reduced and is concave up to and including the 6th, but straight at the 7th and straight and protruding at the 8th. In Podocnemidinae, cervicals of several taxa bear a postzygapophysis collar composed of more or less fused facets of articulation, from the 2nd cervical, where it is generally incomplete, or the 3rd to the 6th or 7th, depending on each species. When there is incomplete fusion in the anterior neck vertebrae, the two kidney-shaped postzygapophyses are disposed in a half circle outline. The corresponding skull is moderately low, elongated (roundly trapezoid, somewhat pear-shaped), and the posterior emargination is moderate and secondarily reduced. *Bauruemys elegans*, from the Maastrichtian of Brazil, belongs to these skull and cervical morphotypes but differs in that the vertebrae are relatively slightly shorter, the ventral crest is slightly less reduced, there is no known full collar fusion of postzygapophyses and the saddle shape is only weakly developed on the mid-posterior condyles. Rather, corresponding condyles are very slightly more domed medially. Posteriorly, condyles are depressed, as in the saddled vertebra 7 of *Podocnemis expansa*, the cotyles are similarly shaped to above mentioned podocnemidine saddled forms, and the cotyles and condyles have the height and width of the saddle-shaped vertebrae of extant *Podocnemis*. Overall, the condition of "saddle shape joint" is acquired in *B. elegans* by the reduction of the two ventral tubercles below the posterior condyle. The tubercles are weakly drawn in ventral view and hidden in posterior view in *Bauruemys* as in *Podocnemis*. Thus, the cervical vertebrae of this taxon are considered as being of the first "saddle-shaped" morphotype state, acquired just before the full derived state of *Podocnemis* and other cited podocnemidine fossils.

The extant *Peltocephalus* also has saddle-shaped joints and reduced tubercles below the condyles of centra 2–6. But it has no postzygapophysis collar and fusion (Fig. 7.5). The vertebrae are short (for their height), the joints are a little higher than in *Podocnemis*. The anterior median crest is less reduced than in *Podocnemis*, concave in lateral view up to the 6th, barely concave at the 6th, ventrally straight and well protruding in the 7th (Fig. 7.5B) and 8th vertebrae. The contribution of the fused tubercles to the centrum remains only posteriorly visible in the 5th to 7th cervicals as a slight ventral condyle bulge, increasing from the 5th to the 7th making the posterior condyle saddle-shaped above the bulge in the 5th and 6th. In 7th, the condyle is not a saddle but an ovoid-convex integrating the bulge (Fig. 7.5B2, B3), and the tubercles have fully disappeared in the 8th, the condyle being stretched in a convex bow. The higher *Peltocephalus* cervical pattern accompanies a much secondarily covered skull with a protruding snout, wide between eyes but rather pointed, corresponding to a strong lower jaw hook, and roof regularly elevated and domed from the premaxillae to the back of the skull, with respect to all the other cited Podocnemidina, and the carapace is domed (Figs. in Broin 1977; Gaffney 1979, 1990; Gaffney et al. 2011).

Joint morphotype (B) of Pelomedusoides is characterized by the primitive absence of saddle joints and no apparent reduction of the posterior tubercles, as visible in posterior view. It is the basic morphotype present in all the Pelomedusoides with the exception of Podocnemidinae. Elongation of the vertebrae is present or absent according to the skull and shell height as in podocnemidines. Joint morphotype B is present with difference in the elongation process in Pelomedusidae (Fig. 7.6C–E) and, notably, in podocnemidid Erymnochelyinae: *Erymnochelys* (Fig. 7.6A–B) and the old world forms of the Erymnochelyinae known by cervicals. Compared to podocnemidines of the *Podocnemis* group, but as in *Peltocephalus*, *E. madagascariensis* and relatives, the neck vertebral centra are relatively shorter and less lowered, and overall, they have a higher neural arch. In *E. madagascariensis*, below the centrum, there is a very protruding anterior crest, with a ventral inclined border (decreasing in height from the anterior condyle up to the 2/3 length, below the neural arch), from the 2nd up to and including the 6th cervical (Fig. 7.6A2). For their part, the 7th and 8th vertebrae have a ventral crest that is completely protruding and straight from condyle to cotyle, as in one Upparhatti specimen DUGF/T178 (Fig. 7.4G) and as in *Peltocephalus*. In *Erymnochelys*, the tubercles are separated in ventral view in cervicals 2 and 3, then juxtaposed in vertebrae 4 to 6 (Fig.7.6A3), separated at the 7th (Fig. 7.6B3) and fused in the 8th. The differences of *Peltocephalus* (with saddle cervicals without tubercles), in spite of apparently shared height with *Erymnochelys*, explicitly demonstrate the homoplastic acquisitions accompanying rather similarly higher skulls (in Gaffney 1979: Figs. 129 and 131; in Lapparent de Broin and Werner 1998: Figs. 5 and 6) and more domed shells than in other podocnemidids. The skulls are not identically shaped but both have a marked hook, for a similar feeding pattern (similar capture pattern but not similar nutrients - living *Erymnochelys* eating plants, *Peltocephalus* being principally carnivorous), to which are added many other differences in the shell (in particular in scute sulci positions in the anterior plastral lobe) and axial skeleton. In association with the higher vertebrae and shell, the higher *Erymnochelys* skull roof has secondarily little emargination, as in *Peltocephalus*. This feature is also present in *Dacquemys* (Erymnochelyinae) from Fayum, (Williams 1954b; Gaffney et al. 2011). But it has a less high skull, although domed and still more covered, accompanying another feeding pattern as indicated by the strongly crested crushing maxillae (needing powerful means for the lower jaw movement). The muscles insertions are differently distributed and protected in a more covered skull, as known in continental erymnochelyine forms and in some littoral *Shweboemys* subgroup forms. But this does not seem to be an essential condition in other littoral podocnemidoid forms with also powerful maxillae, but with skulls more posteriorly emarginated, as shown in some other members of the *Shweboemys* subgroup and in Bothremydini. The vertebrae of the continental *Dacquemys* and littoral *Shweboemys* s.l. are not preserved (Gaffney et al. 2006; Pérez-García et al. 2017b.) In *E. madagascariensis* and its closer relatives, *Erymnochelys* sp. from the late Miocene of Democratic Republic of Congo (Zaire) (Hirayama 1992) and *Turkanemys* (Wood 2003), the condyles of cervical vertebrae are low, much widened and widely stretched transversally and horizontally, up to the 6th included (Fig. 7.6A3). In the 7th, the condyle is rounded, not widened (Fig. 7.6B3), and the 8th is even more rounded and heart-shaped, because both integrate the posteroventral tubercles (while it is narrowly stretched and arched in a bow in *Peltocephalus* as in *Podocnemis*). No such vertebrae with lowered and widened condyle or with integrated tubercles are present in Upparhatti.

Fused Postzygapophyses: Beside Podocnemidinae, the fusion of the postzygapophyses facets to form a rounded structure named collar, as in Upparhatti DUGF/T178, exists in many pleurodires, including members of the Chelidae (Broin and De la Fuente 1993; De la Fuente et al. 2001; De la Fuente 2003) and various Pelomedusoides (in particular, see figures of extant taxa in Kasper 1903; Williams 1950), but there are no linked postzygapophyses facets in extant Pelomedusidae. Moderately derived in height and length, extant pelomedusid cervicals with relatively low centra and a moderately elongate, arched pedicel rather represent a relatively “moderately-elongate” morphotype of vertebrae. For example, Pelomedusidae *Pelusios* and *Pelomedusa* have moderately lowered neural arch as well as zygapophyses and centrum (Fig. 7.6C) (Broin and De la Fuente 1993, atlas and axis in Kasper 1903): centrum is lowered from the second vertebra, neural arch is lowered from the third; as in

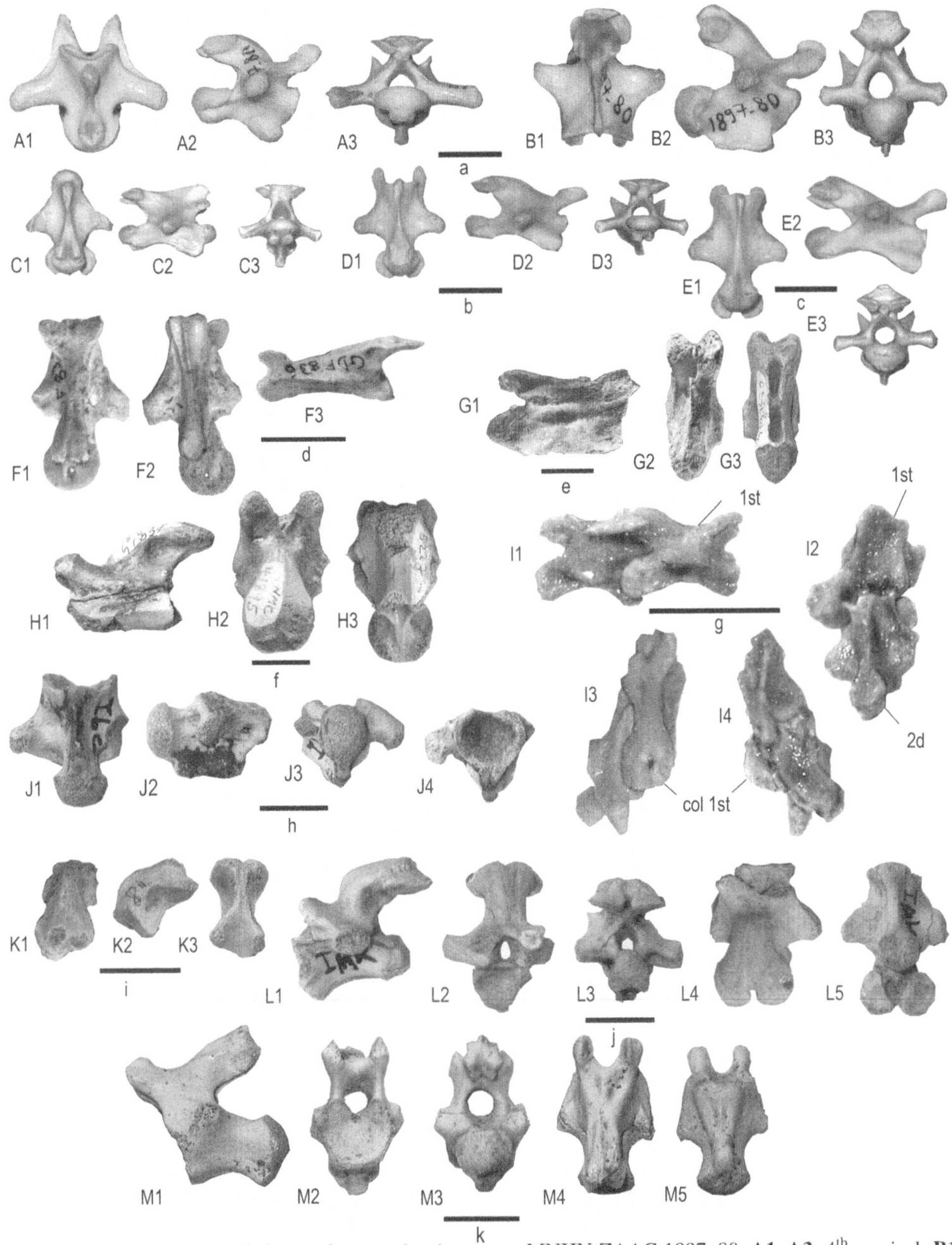

Fig. 7.6 Comparison of vertebrae. *Erymnochelys madagascariensis*, extant, MNHN ZAAC 1897–80, **A1–A3**. 4th cervical; **B1–B3**. 7th cervical; *Pelusios gabonensis*, extant, MNHN ZA AC 1934–373, **C1–C3**. 2nd cervical (axis); **D1–D3**. 4th cervical; **E1–E3**. 7th cervical; ventral (**A1, B1, C1, D1, E1**), right lateral (**A2, B2, C2, D2, E2**) and posterior (**A3, B3, C3, D3, E3**) views. Araripemyidae, *Taquetochelys decorata*, Gadoufaoua (Niger), Late Aptian, MNHNF GDF 836, 3rd to 8th cervical, **F1–F3**. dorsal (**F1**), ventral (**F2**) and left lateral (**F3**) views. Pelomedusoides indet., Hamada of Guir, Kem Kem (Morocco), Albian-Cenomanian boundary: form 1, **G1–G3**. NCM 41976, caudal (?), right lateral (**G1**), dorsal (**G2**) and ventral (**G3**) views; form 2, Pelomedusoides indet., **H1–H3**. NCM 41895, 5–6th cervical, left lateral (**H1**), dorsal (**H2**) and ventral (**H3**) views. *Teneremys lapparenti*, Gadoufaoua (Niger), Late Aptian, MNHNF GDF 918, **I1–I4**. two cervicals between 2nd and 8th, right lateral (**I1**), ventral (**I2**), dorsal (**I3**) and latero-ventral (**I4**) views. Aff. *Erymnochelys* sp., continental Senonian, Ibeceten, Niger, MNHNF: **J1–J4**. IBC 1865, centrum 8th cervical, ventral (**J1**), right lateral (**J2**), posterior (**J3**) and anterior (**J4**) views; **K1–K3**. IBC 1866, 8th cervical, neural postzygapophyses pedicel, dorsal (**K1**), left lateral (**K2**) and ventral (**K3**) views. ?Aff. *Erymnochelys* sp., In Akhamil (Algeria), Late Albian, MNHNF HGN 180, **L1–L5**. 5th cervical, left lateral (**L1**), anterior (**L2**), posterior (**L3**), dorsal (**L4**) and ventral (**L5**) views. Bothremydidae indet g. sp., Sidi Daoui, Oulad Abdoun Basin, (Morocco), Ypresian phosphates, OCP DEK/GE 460, **M1–M5**. possible 6th cervical, left lateral (**M1**), anterior (**M2**), posterior (**M3**), dorsal (**M4**) and ventral (**M5**) views. Abbreviations: 1st, most anterior of the two vertebrae; 2nd, most posterior of the two vertebrae; col 1st; postzygapophyse collar of the most anterior vertebrae. Scale bar equals 1 cm, a for **A1–B3**, b for **C1–D3**, c for **E1–E3**, d for **F1–F3**, e for **G1–G3**, f for **H1–H3**, g for **I1–I4**, h for **J1–J4**, i for **K1–K3**, j for **L1–L5**, and k for **M1–M5**

Podocnemis from the third, the ventral median edge is concave in lateral view with the anterior crest progressively reducing in height from the anterior cotyle up to the mid-length of the centrum from the axis to the 8th included, but without central protruding crest in the 7th and 8th vertebrae, as seen also in araripemydids. However, the centra are longer and less deep in these (Fig. 7.6F3). Wide transversally stretched (not bowed) condyles (and corresponding cotyles), similar to those of *Erymnochelys* are present in *Pelusios* (a little less deep in *Pelomedusa*) and the, separated and protruding posterior tubercles are present up to the 6th included, but fused to the condyle in the 7th and 8th (Fig. 7.6C–E). The postzygapophyses pedicel is elevated and slender with juxtaposed but not fused facets in a collar as in DUGF/T176.

A postzygapophyseal podocnemidine collar is present in several cervicals with fully fused facets, other vertebrae in the neck having facets more or less approximated according to their position in the neck. Various podocnemidines, including *Brasilemys* (Broin 1991; Lapparent de Broin 2000b; França and Langer 2005; Lapparent de Broin et al. 2007) present a collar of fused articular facets as in *Podocnemis* group. In this group, the collar has some partly fused postzygapophyses facets besides fused facets depending on the position of the vertebrae in the neck. Such a collar of fused postzygapophyses exists at least in the following taxa, beside the Podocnemidinae and the Upparhatti specimen DUGF/T177:

(1) In fossil Chelidae such as *Prochelidella portezuelae* De la Fuente, 2003, fossil and extant long-necked species of the *Hydromedusa* group (including *Yaminuechelys*), *Chelodina* group (3rd to 8th) (Broin and De la Fuente 1993; Lapparent de Broin 2000b; De la Fuente et al. 2001), and in such taxa with much lowered vertebrae (centrum and neural arch).

(2) In several Cretaceous Pelomedusoides including (possibly) the basal taxon *Euraxemys* (Gaffney et al. 2006), and (definitively) the long-necked Araripemydidae and *Teneremys lapparenti* Broin, 1980 (Lapparent de Broin et al. 2007), a basic "pelomedusoid" from the late Aptian of Gadoufaoua (Niger).

All the post-atlas cervicals of *Araripemys* from the Aptian-Albian boundary of Araripe, Brasil (Gaffney and Meylan 1991; Meylan and Gaffney 1991; Meylan 1996) have fully fused postzygapophyses. Two isolated araripemydid cervicals with fused postzygapophyses are preserved in the material of *Taquetochelys decorata* Broin, 1980 (MNHNF GDF 836, GDF 837) (Fig. 7.6F) from the late Aptian of Gadoufaoua (Niger). They were first attributed to *Araripemys* (see Lapparent de Broin et al. 2007) and are similar to those of *Laganemys* Sereno and Elshafie, 2013 of the same locality. Actually, *Laganemys* is a junior synonym of *Taquetochelys*, having identical shell elements (personal observation, Pérez-García 2019). The two vertebrae of *Teneremys* are preserved in a rather complete (skull, two vertebrae, shell) specimen MNHNF GDF 918 (Fig. 7.5I). They have a complete collar of fused articular facets. In *Teneremys*, the cervical vertebrae (neural arch as well as zygapophyses and centrum) are more moderately lowered than in Araripemyidae with centra that are also moderately elongated relative to their height and a concave ventral median face in lateral view. In *Teneremys*, in conformity with the moderate height of the cervicals, the shell is less flattened than that of Araripemyidae, and similar to that of *Pelomedusa* from early Miocene to extant (Lapparent de Broin 2008a) except for having some primitive features such as either continuous or posteriorly interrupted neural series, first neural either quadratic or hexagonal (short sided posteriorly), and shorter anterior and posterior shell borders. Thus, this moderately elongate morphotype of vertebrae accompanies a low skull with a strong posterior emargination (extant pelomedusids) or a very low skull with a very strong posterior emargination as in *Teneremys* and araripemydids (Broin 1980; Gaffney and Meylan 1991; Lapparent de Broin et al. 2007; Meylan 1996; Sereno and Elshafie 2013). In all these taxa, the extremity of the pedicel with the collar does not differ much: the whole pedicel with the collar is thinner and flatter in araripemydids, but in *Teneremys* it is shorter and moderately elevated although more elongated and less elevated than in DUGF/T176. In the latter, the pedicel is more elevated than in the preceding ones, and it is as elevated as but a little less compact than in Podocnemidines.

From these comparisons, it appears that in all the morphotypes (short or long-necked), from front to back the relative dimensions and outlines of the cotyles and condyles and the proportions of the ventral crest and its relations with the posterior tubercles vary. It seems that in all the derived Podocnemidoidea, the centrum is ventrally concave anterior to the 7th or 8th vertebra, the last vertebra has a narrower condyle, and a protruding ventral crest is present in the 7th and 8th or only in the 8th. The Podocnemidinae of South America are associated with the presence of saddle joints and reduction of the two posterior ventral tubercles, except in the 7th of *Peltocephalus* where the fused tubercles are posteriorly visible prolonging the crest, but where they are fully reduced in the 8th. No African, European and Indian (Upparhatti) vertebra corresponds to Podocnemidinae. By contrast in Erymnochelyinae and known necks of Bothremydidae, the tubercles are present posteriorly. However, in contrast to Upparhatti vertebrae, *Erymnochelys* and closest relatives have cotyles and condyles that are clearly low, much wider than high than in *Neochelys*, Ibeceten and Upparhatti specimens. Similarly, *Pelusios* contrasts by the tubercles that are more anteriorly placed below the centrum and those of the 8th fused with the condyle or reduced variably depending on the species, and by the condyles that are low and wide and without mid-ventral crest, even on the 8th (Fig. 7.6).

Relationships of Upparhatti Taxa with Erymnochelyines having Short Vertebrae: The fused postzygapophyses collar is not present in any known erymnochelyine. However, this group is particularly important to consider with respect to India, because of presence of Erymnochelyine in the continental Senonian of Ibeceten (Niger) (Broin et al. 1974) and early Maastrichtian of Madagascar (Gaffney and Forster 2003), besides the presence of various bothremydids in the Early Albian-Turonian, Campanian and Maastrichtian of Madagascar (Lapparent de Broin and Werner 1998; Gaffney and Krause 2011) particularly, *Kinkonychelys* (Kurmademydinae) from the Maastrichtian northwestern Mahajanga basin (Gaffney et al. 2009).

At Ibeceten (Fig. 7.6J, K), eight fragmentary vertebrae are known: five cervical centra of intermediate position, a pedicel of a 5–6th cervical, a centrum and a pedicel of the 8th cervical, and one 1st dorsal centrum (MNHNF IBC coll.). All these Senonian Ibeceten fragments are short and conform to those of short-necked Upparhatti and *Erymnochelys* specimens: in the attachment point of the transverse apophyses; the variation in the mid-ventral crest, which ranges from concave to protruding depending on the progressive position of the vertebra in the neck from the 2nd to the 8th; and by the presence of a short and abruptly arched postzygapophyseal pedicel in the fragmentary 8th vertebra (Fig. 7.6K). The centrum of the 8th Ibeceten vertebra (Fig. 7.6J) is similar to that of DUGF/T178, also a 8th (Fig. 7.4G), and DUGF/T174, a 7th (Fig. 7.4E). Both are a little less long than DUGF/T175 (Fig. 7.4H), but the latter is preserved with a slightly eroded ventral crest that appears less protruding than in the two other vertebrae. Overall it is different by its relatively a little longer centrum (height about 50% of the length instead of about 60% in DUGF/T174 and DUGF/T178), but closer to the other shorter morphotype (height about 30% of length). However, all the vertebrae of Ibeceten, as the basisphenoid and all the dentaries, are found together with epi-ento-hyoplastra elements bearing the characteristic scute pattern of the *Erymnochelys* group (with linked gulars posterior to the reduced intergular: see Pérez-García et al. 2017b) (Fig. 7.11A1,A2), a pattern that is not found in the Indian specimens. A moderately short cervical vertebral morphotype is present in the European Eocene erymnochelyine *Neochelys*. *Neochelys* does not correspond to the *Erymnochelys* group by the gular / intergular pattern, but is erymnochelyine by the anterior position of the humeropectoral sulcus and the wide inner carotid canal. The only species of this genus for which cervical vertebrae are described is *N. laurenti* from the early-middleYpresian Saint-Papoul locality (France) (MNHNF and Dominique Teodori coll.). The observed vertebrae of this taxon are represented by a mid-cervical (4th–6th?) and a 7th cervical vertebra (MNHNF SPP31, Pérez-García and Lapparent de Broin 2015: fig. 12). They generally conform to those of *Erymnochelys*, Upparhatti and Ibeceten in particular in the position and shape of the transverse apophyses at mid-length of the vertebra, with oblique anterior border and transversal posterior borders. As the Senonian Ibeceten and Upparhatti specimens, *Neochelys* centra differ from *Erymnochelys* and are closer relatives by having rounded and horizontally stretched posterior condyles and a narrow, approximately quadrangular posterior 7th condyle with a few fused protruding tubercles. The protruding tubercles at the posterior end of the crest are separated on the preserved mid-cervical and fused on the 7th cervical of short cervicals of *Neochelys* as in one Ibeceten specimen (Fig. 7.6J3) and in *Erymnochelys* specimens (see above). The isolated vertebrae of the Upparhatti fossil taxa share most characters with *Erymnochelys* and closest relatives but they are less derived, because these latter alone present the transversally stretched and low condyles. Anyway, the complete vertebral neck of Ibeceten form, *Neochelys* and Upparhatti form is not complete enough to really consider the peculiarities of each taxon in their entirety, although all belong to a short morphotype that is apparently not much derived. Taxonomic relationship of short Upparhatti vertebrae to *Erymnochelys* relatives is ruled out but not to the erymnochelyines although no one erymnochelyine is known with fused postzygapophyses.

Relationships of Upparhatti Vertebrae with Bothremydidae: Little is described and figured about axial skeleton of bothremydid. Fragments of cervicals are preserved in the late Albian *Cearachelys* from Araripe (Brazil). They are neither described nor figured but indicated as having no fused postzygapophyses without referring to their position on the neck (Gaffney et al. 2006: p. 618). Several cervicals (not described and not figured) and one caudal (described but not figured) are preserved in the Paleocene *Taphrosphys sulcatus* (Leidy 1856) (see Gaffney 1975; Zangerl 1947). The relative length of their centra is not given. Fused postzygapophyses are said to be present in this taxon (Gaffney et al. 2006: p. 618) but as they are not figured, the morphology of this species is not comparable with that of other taxa including the Upparhatti specimens.

The only view of a bothremydid neural arch with postzygapohyses fused into a collar is given for "*Bothremys-or-Chedighaii*" in Gaffney et al. (2007). No information on locality and age for this taxon is given but mention of *Bothremys* in the Upper Cretaceous of North America was made in Zangerl (1948), Gaffney and Zangerl (1968) and Gaffney et al. (2006)). This is the first specimen of bothremydid with postzygapophyses which can be compared with the Upparhatti specimen DUGF/T177. One 8th vertebra of *Acleistochelys* Gaffney et al., 2007 from the Paleocene of Mali is represented in the same figures (Gaffney et al. 2007).

It presents a strong, straight and protruding ventral crest as the other examined podocnemidoid taxa, but the postzygapophyseal area is not preserved. Both these bothremydids represent a short cervical morphotype and have the wings (transverse apophyses) positioned well in the middle part of the vertebra and oblique anterior and posterior borders as in Upparhatti long centra (Fig. 7.4A1, B1, J2, I1). Short Upparhatti centra (Fig. 7.4E2, H1) have a less inclined posterior border as in Kem Kem undetermined form 2 vertebra (Fig. 7.6H), and more similar to the transverse or nearly transverse posterior border in other living and fossil forms (Fig. 7.6A1–E1, F1–F2, J1, M4–M5).

In the "*Bothremys*-or-*Chedighaii*" vertebra the postzygapophyses are fused and the wings occupy all the space between the cotyle and the collar, whereas in *Acleistochelys*, the shorter wings occupy only a median part of the centrum between the two extremities. In these features, none of these taxa correspond to Upparhatti specimens DUGF/T175 and DUGF/T176, even if there is a pedicel with a collar in DUGF/T177 as in the "*Bothremys*-or-*Chedighaii*" vertebra. However, the latter is important because it is the first illustrated proof of the presence of a postzygapophyseal collar in Bothremydidae. Figure 7.6M represents a vertebra of an indeterminate bothremydid from the Phosphates of Morocco, from the Ypresian of Sidi Daoui, Oulad Abdoun phosphate basin (OCP DEK/GE 460). From this site, two bothremydid species were described (without associated cervicals), *Ummulisani rutgersensis* Gaffney et al., 2006 and *Bothremys kellyi* Gaffney et al., 2006. However, the presence of other unnamed species cannot be ruled in view of the unstudied indeterminate shell specimens in the OCP collections at Khouribga (Morocco). The vertebra of the latter is stout, and the centrum is short for its height as in the "*Bothremys*-or-*Chedighaii*" specimen and the figured Malian vertebra of *Acleistochelys* (Gaffney et al. 2007: figs. 8D and 9D). The OCP vertebra has a quadrangular cotyle, slightly wider than high, and it is considered as a possible 6th because of the inclination of the neural arch pedicel, the concavity of the ventral median crest in lateral view (Fig. 7.6M1), and the presence of juxtaposed tubercles protruding below the rounded condyle (Fig. 7.6M3). The posterior extremity of the neural arch has a small longitudinal crest above and between the separated postzygapophyses on the top of the pedicel (Fig. 7.6M1, M3). It is a short vertebral morphotype as the two forms described by Gaffney et al. (2007), the above described erymnochelyines and Upparhatti specimens. But it is distinct by the strong forward inclination of the neural arch, thus positioning the prezygapophyses anteriorly in relation to the cotyle front and the transverse posterior border of the wing. It differs from all the cervical elements of Upparhatti. This short and stout vertebra from Moroccan phosphates might belong to one of the two described taxa from the Ypresian level. It may belong to the crushing feeder *B. kelly* Gaffney et al., 2006, better than to *Ummulisani rutgersensis* Gaffney et al., 2006 which was placed in the *Taphrosphys* group by Gaffney et al. (2006). The latter group represents a snapping-cutting feeder by its narrower maxillae without maxillary crests, and it has a long anterior shell border. Therefore, it is inferred to have long and low cervical centra and hence does not represent short vertebral morphotype. Although there are taxa with short cervicals such as OCP DEK/GE 460 in Moroccan phosphates, there is no Upparhatti specimen corresponding to this stout and short Eocene OCP "*Bothremys*-or-*Chedighaii*" and nor to the Paleocene *Acleistochelys* vertebrae. Some remains of cervical vertebrae of the bothremydid *Polysternon atlanticum* Lapparent de Broin and Murelaga, 1996 are known from the late Campanian of Laño (Spain). They consist of fragments of centrum. A centrum possibly from a median position has a cotyle that is slightly higher than wide and slightly triangular. Its condyle is clearly higher than wide and ovoid with the tubercles united below forming a stout extremity of the mid-ventral crest. Both features are similar to those in the posterior vertebra of *Acleistochelys*. The Laño *Polysternon* centra are of a moderately long vertebral morphotype, which corresponds to the moderately long anterior border of the carapace in *Polysternon*. Although clearly anteriorly elongated in relation to primitive Pelomedusoides, the *Polysternon* shell (UM II, coll. Guy Costa) is quadrangular and not as much medially protruding as in the more rounded/ovoid shells of the presumed long-necked taxa, such as those of *Taphrosphys* group (Taphrosphyini part).

Pleurodira with a Possible Elongated Neck from India: "*Carteremys*" *pisdurensis* had a long skull for its width (Jain 1977) that is compatible with long neck vertebrae and its carapace was presented as ovoid (Jain 1986), although surely not as ovoid as in *Taphrosphys* (from Jain 1986: pl. 1). Thus, it might have had also long cervicals. *Carteremys leithii* represents a form with an elongated and anteromedially rounded protruded shell as in taxa with a substantially elongated neck. In extant Pleurodira (with fossil representatives), taxa with a more substantially elongated anterior shell (more or less "quadrangular" or clearly ovoid shell) are represented by chelids of the "*Acanthochelys-Phrynops* s.l." group, *Chelus*, *Hydromedusa* group and *Chelodina* s.l. group, and Pelomedusidae. All of them have moderately to much elongated neck. By its carapace outline, *Taphrosphys* s.l. (Taphroshyini of Gaffney et al. [2006]) ought to be part of the elongated forms. According to Broin, (1988), the African tribe Nigeremydini (Gaffney et al. 2006), new rank (subtribe Nigeremydina Gaffney et al., 2006), has to be separated from the more widely distributed tribe Taphrosphyini (Gaffney et al. 2006), new rank (subtribe Taphrosphyina Gaffney et al., 2006) (Africa, Middle-East, Europe, Americas). The original separation of the *Nigeremys* and

Taphrosphys groups of Broin (1988) is restored here. Nigeremydini, a new rank, is a regional group restricted to the "Trans-Saharan seaway" of Late Cretaceous-Paleocene times. Nigeremydini include the Maastrichtian *Nigeremys* (Bergounioux and Crouzel 1968) from Niger, *Arenila* Lapparent de Broin and Werner, 1998 from Egypt, *Sokotochelys* (Halstead 1979; Walker 1979) from Nigeria (figured but poorly diagnosed and considered as lost), and the Paleocene *Azabbaremys* Gaffney et al., 2001b and *Acleistochelys* from Mali. All are from the littoral coasts of this inner trans-Saharan African sea that connected the Atlantic Guinea Gulf and the North African Tethys margin during the Late Cretaceous and Paleocene time intervals. The diagnosis of the *Nigeremys* group was given in Broin (1988). Its diagnosis is made more precise with the integration of taxa discovered after 1988. The diagnosis includes: the combination of a triangular, longer and stronger anterior skull, a long snout anterior to the choanae with important maxillae in length and robustness, and posteriorly a shorter basicranium with large oval bothremydid fossae. There is some variability according to the depth of the fossae and robustness of the maxillae, and possible individual similarity of one feature with other bothremydid taxa. This combination of features is missing in Taphrosphyini. These characters are not taken into consideration in Gaffney et al. (2006), Romano et al. (2014) and all other papers following Gaffney et al.'s (2006) analysis. They indicate that Nigeremydini was derived independently from the Taphrosphyini s.s. and seem to be closer to more primitive forms such as 'middle' Late Cretaceous taxa of Kem Kem beds (Morocco) (see *Galianemys* in Lapparent de Broin and Werner 1998: fig. 4a). If the Upparhatti basisphenoid and dentaries do not correspond to the skulls of these two tribes, some cervicals might correspond to one of them. Nigeremydini are considered to correspond to a short neck morphotype, as confirmed by *Acleistochelys* vertebra, in contrast to Taphrosphyini and *Carteremys* (by estimation). They correspond to a not ovoid and shorter anterior shell border and the shell is not strongly decorated in contrast to Taphrosphyini shells. This is attested by cf. *Arenila* and all identified and unidentified shell remains which accompany the Nigeremydini in Upper Cretaceous (Maastrichtian) beds of Egypt, Niger, Nigeria and Mali (including the localities Tagnout Chaggeret and Ibeceten, MNHNF coll.) and Paleocene beds of Mali (MNHNF coll. and possibly NHM) (Halstead 1979; Lapparent de Broin 2000a; Lapparent de Broin and Werner 1998; Walker 1979).

As seen above, a long-necked morphotype would appear in some of these taxa (closely related or not) with shells that are anteromedially clearly more elongate and ovoid, such as *Taphrosphys* s.l. (with the most ovoid anterior shell), by estimation in *Elochelys* from the Campanian of France (MHNM) and the Indian *Carteremys* (see the ovoid shells compared by Williams 1953, but without known vertebrae).

Bothremydidae with Elongated Cervicals and Postzygapophyses Collar: Accoding to Gaffney et al. (2006), the presence of collar in *Taphrosphys sulcatus* cannot be confirmed. However, a 7th cervical *Taphrosphys* vertebra is known for *T. congolensis* from the Paleocene of Angola (MRAC coll.) (Wood 1975). The specimen is transversely flattened post mortem so that its height is amplified and its posterior condyle and anterior cotyle were narrowed in relation to their height. However, the height and length proportions of the centrum (height about 25–30% of the length) anteriorly and medially indicate that it belongs to a long-necked morphotype, in agreement with the much medially elongate ovoid anterior shell. The postzygapophyses are well separated and the facets are laterally exposed, oppositely facing, instead of posteriorly facing as in the pedicel of DUGF/T176 (Fig. 7.6D, 6th or 7th). However, as far as the presence of fused postzygapophyses in *T. congolensis* is concerned, it appears possible that the facets of this vertebra are identical to those of the 7th in *Podocnemis* (vertebra posterior to those with fused postzygapophyses). The ventral crest, although partly eroded in *T. congolensis*, is more prominent only posteriorly and the posterior tubercles are fused, elongating the condyle, and in continuity with the mid-ventral crest. If the collar is not observed, the vertebra appears to fit with a long neck form, in agreement with the ovoid much protruding shell of all known species of *Taphrosphys* s.l. group. The centrum seems to have been relatively a little higher than in DUGF/T168 (attributed to the long 'morphotype a'). Other bothremydids with elongated vertebrae exist but no vertebrae with fused postzygapophyses had been described.

From the late early Miocene of As Sarrar (Saudi Arabia), an indeterminate bothremydid centrum, from a median position in the neck was recovered (Fig. 7.5C). This vertebra is considered as bothremydid because it was found with carapace fragments characteristic of bothremydids in having a decoration of dichotomic sulci that is well marked as in *Elochelys perfecta* for example, in proportions and contours of the elements, and because the centrum fits with that of *Taphrosphys congolensis*. It possibly belongs to a contemporaneous and geographically closer bothremydid skull from Ghaba, Oman (Roger et al. 1994: pl. 1, fig. 1). It is a much smaller form than the "*Shweboemys*" (s.l.) elements that are also present in the Saudi Arabian locality with the vertebra. This Ghaba long centrum (Fig. 7.5C) is similar to DUGF/T168 (Fig. 7.4B) of the Upparhatti 'morphotype a' in being slender, having a concave ventral crest which is prominent anteriorly above the juxtaposed tubercles, and the transverse apophyses with both inclined sides positioned in the middle part of the centrum anteroposteriorly (also as in

DUGF/T 169, Fig. 7.4A). This Saudi Arabian centrum appears as relatively a little deeper than DUGF/T 168. However, its condyle is clearly narrow as in DUGF/T169 (Fig. 7.4A1, B1 compared to Fig. 7.5C). On the other hand, among the isolated Arabian elements there is no indication of a very elongated shell: the costal 1 and anterior peripherals are moderately elongated, less than in *Taphrosphys* and *Carteremys*, but the peripheral 1- nuchal are missing, which are the best length markers.

To conclude, in having elongated centrum of 'morphotype a', with protruding posterior tubercles in the median and posterior vertebrae, Upparhatti specimens represent one or two long-necked Bothremydidae species with DUGF/T169 and DUGF/T172 representing one and DUGF/T168 and DUGF/T170 the more robust second one. By comparison with defined taxa, these long vertebrae could be from taxa with an anteriorly elongated carapace (ovoid or long ovo-quadrangular) like that of members of the *Taphrosphys* group, of *Carteremys leithi* and possibly of "*Carteremys*" *pisdurensis*. The Upparhatti peripheral 1 DUGF/T57 (Fig. 7.7A) may belong to such an elongated carapace. However, this is not decorated and there are no shell fragments with the ornamentation typical of *Taphrosphys* and *Carteremys* group, either from western or central India (Table 7.1). This peripheral 1 (DUGF/T57) cannot be compared with the undecorated "*Carteremys*" *pisdurensis* because its anteromedial carapace border is not well preserved and the peripheral 1- nuchal part is only known by a strange reconstruction (Jain 1986: fig. 2). However, knowing that the elongate skull of this taxon is probably associated with at least moderately elongate centra, a moderately elongate anterior carapace border, not obligatorily anteromedially ovoid, is likely (Jain 1986: pl. 1 [2]) even if the reconstruction is not reliable. For this taxon, it was thought (Jain 1986) that it was a member of the *Shweboemys* subgroup. But the presence of this subgroup under the name "*Shweboemys*" *pisdurensis* is definitely not acceptable and, as all other defined species from India, it is now considered as belonging to Bothremydidae (see bothremydid shell characters below). We note that "*Shweboemys*" subgroup differentiated from erymnochelyines on the borders of northern Africa-Arabian littoral platform during the middle–late Eocene, possibly as early as Lutetian and surely by Priabonian (Lapparent de Broin et al. 2018, Zouhri et al. 2017). This littoral group preserves carapaces with a particularly short nuchal border and the length of the vertebrae is unknown. The presence of this Eocene-Oligocene subgroup in the Upper Cretaceous of India would be problematic. Anyway, the referral of "*C.*" *pisdurensis* to Bothremydidae is here postulated, and some relationship of long centra of Upparhatti Bothremydidae indet. with "*Carteremys*" *pisdurensis* neck cannot be disregarded.

In the long-necked form (DUGF/T168), the wings with both oblique borders are positioned in the mid part of the vertebra as in the short "*Bothremys*-or-*Chedighaii*" and the Malian bothremydid, but in the presence of a longer and lower centrum it differs from these two forms. By contrast, the Upparhatti shorter cervicals with poszygapophyse collar, could correspond also to a bothremydid and eventually to one of the other Indian taxa. We do not know which cervicals might belong to *Kurmademys* as it had a relatively moderately elongated shell but its cervicals are unknown. If confirmed for *Taphrosphys*, it will prove the presence of widely spread homoplastic character of fused postzygapophyses in long as well as in short-necked bothremydids.

As far as short-necked form is concerned, the Upparhatti specimens of 'morphotype b' are attributed to Bothremydidae indet. In Upparhatti, there is at least a short form b with fused postzygapophyses (Fig. 7.4C) as in "*Bothremys* or *Chedighaii*", but there is no proof of fused postzygapophyses in the long 'morphotype a'. In the absence of associated shell and vertebrae, and the poor sample of known bothremydid vertebrae in the world, the two morphotypes of cervical vertebrae in the Upparhatti assemblage described here cannot be referred to any of the known taxa. The short vertebrae from Upparhatti have some similarities with the primitive erymnochelyines including the Ibeceten specimens. However, the similarities with Bothremydidae and with the Erymnochelyinae of Ibeceten (already partly differentiated into *Erymnochelys* and the line of its relatives) seem to be only in shared primitive features, inherited from common ancestors that lived prior to the fragmentation of the Gondwana. The morphology of the Upparhatti short vertebrae with a collar of fused postzygapophyses unknown in Erymnochelyinae but known in short-necked Bothremydidae, put the balance of evidence in favor of Bothremydidae. This is supported by the absence of any derived podocnemidine or erymnochelyine feature in all the remaining Upparhatti vertebral material and considering the fact that the dentaries and basisphenoid specimens are attributable to relatively primitive Bothremydidae. It is also indicated by the following postcranial material. Keeping the plesiomorphic features aside, it appears that Bothremydidae alone is present in known material of India. Derived fused postzygapophyses are shared by chelids and continental araripemyids and basic pelomedusoids, placed outside of Podocnemidoidea within Pelomedusoides. These groups are known as part of the northern Gondwana as early as in the Aptian of Niger (Gadoufaoua) and Albian of Brasil (Araripe), much prior to the deposition of Kem Kem beds of Morocco and the Late Cretaceous (Cenomanian-Turonian) transgression. Fused postzygapophyses represent an early derived but homoplastic pleurodiran feature that was present in Pelomedusoides before the separation of Madagascar and

India from each other and from Africa. It is noted that podocnemidines have fused postzygapophyses and saddles joints, and relatively long centra in *Podocnemis* group following the separation of Africa and America. The presence of subfamily Podocnemidinae is not detected by any other skeletal part, neither in the Upparhatti fauna nor in podocnemidoids of any another part of the old world outside South-America, i.e. Africa, Europe, Middle East and Indian subcontinent. Contrary to Gaffney et al. (2011) and Cadena et al. (2010, 2012), we do not accept living *Peltocephalus* as being sister taxon of *Erymnochelys madagascariensis* or the South American late Paleocene *Carbonemys* and Mio-Pliocene *Caninemys* as belonging to Erymnochelyinae. If Erymnochelyinae are not present in South America, we may not find Podocnemidinae in India either. In Upparhatti material, no saddle joints on short or long centra are found to attest the presence of a podocnemidine vertebra. No erymnochelyine vertebra is detected either.

As seen above, some features of the Upparhatti short 'morphotype b' are present in Cretaceous erymnochelyines, i.e the forms which are already characterized as belonging to *Erymnochelys* and relatives (such as the Senonian Ibeceten form). They had not yet become derived by the formation of stretched and low condyles (that characterize the *Erymnochelys* group at least as early as the Miocene, beside the intergular pattern). Other not yet found erymnochelyines had to exist leading to Tertiary European and African taxa (such as *Neochelys*, stereogenyines and other African erymnochelyines) which do not have the *Erymnochelys* plastral gular/intergular scute pattern (Pérez-García et al. 2017b: Fig. 1). In addition to Bothremydidae, certain possible African Pelomedusoides were present as early as Albian-Cenomanian boundary and up to Maastrichtian, in Kem Kem and Sahara areas including Niger and Sudan (Gmira 1995; Lapparent de Broin 2000a) but they are not well identified. As a matter of fact, no erymnochelyine element is preserved in the Upparhatti site nor in any other Indian locality to attest the presence of a possible erymnochelyine in the sub-continent, contrary to the condition in Madagascar. The absence of fused postzygapophyses in the primitive Ibeceten erymnochelyines, contrary to their presence in the short form of Upparhatti vertebrae, does not favor any relationship of the latter with Erymnochelyinae.

To conclude on vertebrae, the taxonomic status of figured and sufficiently preserved cervical elements is proposed as: (1) Bothremydidae indet. 'morphotype a': longer cervical centrum in relation to the height (height on length ca. 27–31%), axis DUGF/T169, median 5th–6th DUGF/T168, and first dorsal DUGF/T170 (incomplete centrum) and DUGF/T172; (2) Bothremydidae indet. 'morphotype b': shorter cervical centrum length in relation to the height (height on length: 42–50% and 61–66%), such as cervicals DUGF/T129 and DUGF/T175 and fragments DUGF/T174, 176, 177, 178, and 189, among which there is one vertebral pedicel with fused postzygapopyses (DUGF/T177, Fig. 7.4C).

Shell elements (Figs. 7.7, 7.8, 7.9, 7.10 and 7.11, Table 7.5): Shell elements are numerous isolated plates and probably come from several individuals although there are not enough plates of each kind to reconstruct a single individual. A few elements are thin (Figs. 7.7A, C, 7.8I) but most of the elements are robust. Thus, based on the thickness of the shell plates, it appears that at least two morphotypes of shells are present. The estimated length of the shells ranged from 15 cm to about 50 cm in length. Other features (buttress scar) indicate the presence a third species.

Carapace

Peripherals (Figs. 7.7, 7.8): Thirty-eight peripherals are preserved documenting most of the peripheral series.

Peripheral 1 DUGF/T57 (Fig. 7.7A) is an example of a thin peripheral plate. It is 0.5 cm thick at the costo-marginal sulcus and thin marginally in longitudinal section. It has a smooth and flat surface that lacks ornamentation and is bright and brown-colored. The sinuous furrows of the scutes are well pronounced. The plate is narrow, longer than wide, measuring 4.2 cm in length medially and 3 cm in width anteriorly. Thus, the anterior shell border is well elongated. Marginal scutes 1 and 2 are both long dorsally, but marginal 2 is slightly longer than marginal 1, this being equal to about half the lateral border of the nuchal; the vertebral 1 is anteriorly nearly as wide as the pleural 1 on the plate, and posteriorly narrower. The 1st marginal is moderately wide, covering a third of the anterior border of peripheral plate 1 in width. Vertebral scute 1 was rather narrow. Its lateral border, on the first peripheral bone, was close to but not crossing, the medial suture with the nuchal. The measurements of peripheral 1 correspond in size to a specimen of *Erymnochelys madagascariensis* that has a carapace length of 35.5 cm, with: anterior border width of 3 cm; medial border length of 4.1 cm; marginal 1 dorsal overlap is 2.2 cm; maximal depth of nuchal border just anterior to the marginal vertebral 1 sulcus is 0.6 cm; thinnest border (posteriorly) is 0.4 cm. The ventral border overlap of the marginals 1 and 2 is very short and shortens from 0.7 cm to 0.4 cm at the nuchal corner.

The presence of long marginal 1 scute differs from members of the *Erymnochelys* group which have shorter first marginals, except in the derived Pliocene *Kenyemys* Wood, 1983 from Africa. But the latter is exceptional in lacking vertebral scute 1 dorsal overlap on the peripheral 1. This marginal also differs from the European *Neochelys* (Broin 1977, Pérez-García and Lapparent de Broin 2013, 2015) and South American *Podocnemis* group (Carvalho et al. 2002: fig. 1[8]; Broin 1991: pl. 1 [1], pl. 2 [2, 8]). Long marginals 1 on peripherals 1 are freuent in Bothremydidae such as *Carteremys leithii*. They are not visible in pictures (photos) of

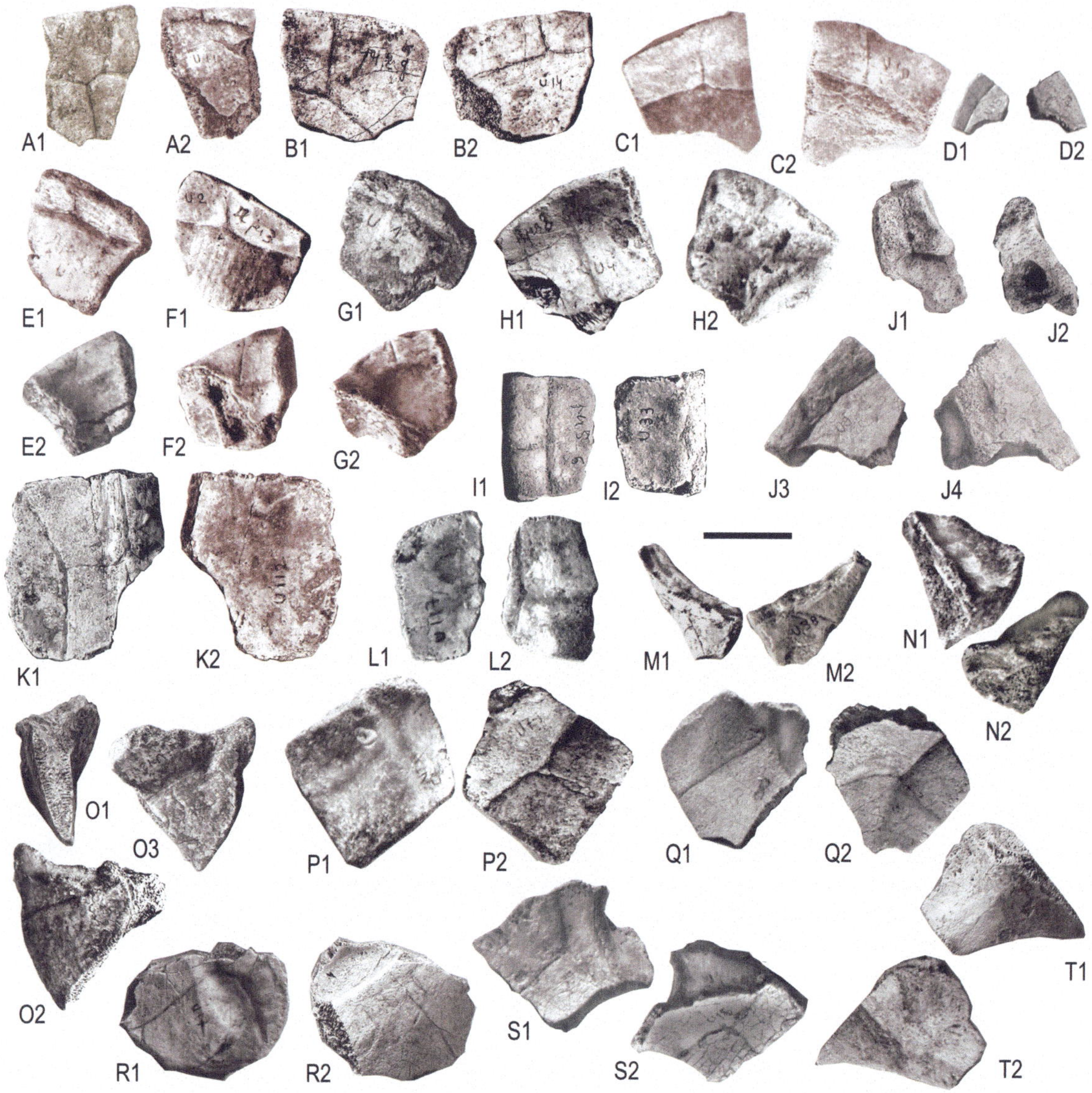

Fig. 7.7 Bothremydidae indet. g. sp., Upper Cretaceous (Maastrichtian) intertrappean beds, Upparhatti, southwestern India, DUGF/T coll. (Work catalog U numbers on the specimens): anterior to posterior peripherals, **A1–A2**. DUGF/T57, left peripheral 1, dorsal (**A1**) and ventral (**A2**) views; **B1–B2**. DUGF/T60, right peripheral 1, dorsal (**B1**) and ventral (**B2**) views; **C1–C2**. DUGF/T56, left peripheral 2, ventral (**C1**) and dorsal (**C2**) views; **D1–D2**. DUGF/T55, left peripheral 4, dorsal (**D1**) and ventral (**D2**) views; **E1–E2**. DUGF/T49, **F1–F2**. DUGF/T48, **G1–G2**. DUGF/T47, right peripherals 3, dorsal (**E1, F1, G1**) and ventral (**E2, F2, G2**) views; **H1–H2**. DUGF/T50, peripheral, extremity of bridge, dorsal (**H1**) and ventral (**H2**) views; **I1–I2**. DUGF/T79, bridge right peripherals 5 or 6, dorsal (**I1**) and ventral (**I2**) views; **J1–J4**. DUGF/T61, left peripheral 4, medial (**J1**), posterior (**J2**), dorsal (**J3**) and ventral (**J4**) views; **K1–K2**. DUGF/T158, **L1–L2**. DUGF/T163, bridge right peripherals 5 or 6, dorsal (**K1, L2**) and ventral (**K2, L1**) views; **M1–M2**. DUGF/T84, right peripheral 7, dorsal (**M1**) and ventral (**M2**) views; **N1–N2**. DUGF/T59, right peripheral 7, postero-dorsal (**N1**) and ventral (**N2**) views; **O1–O3**. DUGF/T58, right peripheral 7, posterior (**O1**), dorsal (**O2**) and ventral (**O3**) views; **P1–P2**. DUGF/T125, **Q1–Q2**. Q, DUGF/T51, right peripherals 8, ventral (**P1, Q1**) and dorsal (**P2, Q2**) views; **R1–R2**. DUGF/T63, left peripheral 8, dorsal (**R1**) and ventral (**R2**) views; **S1–S2**. DUGF/T62, right peripheral 8, ventral (**S1**) and medio-dorsal (**S2**) views; **T1–T2**. DUGF/T54, left peripheral 8, dorsal (**T1**) and ventral (**T2**) views. Scale bar equals 2 cm

Kurmademys but are only shown in by a hypothetical reconstruction. In "*C.*" *pisdurensis* the peripheral 1 is given as wide, but it is only known by a dubious reconstruction. The presence of a long and narrow marginal 1 scute overlap on the peripheral 1, which is little less long than the marginal 2 overlap, in DUGF/T57 agrees with various bothremydids. Thus, for example, the specimen from the Cenomanian of Nazaré (Portugal) attributed to *Algorachelus*, the oldest European bothremydid, and figured in Pérez-García et al. (2017a), has a narrow peripheral 1 with a long marginal overlap, a little less long than the marginal 2 overlap and on a relatively narrow width as in DUGF/T 57. But this specimen presents a very

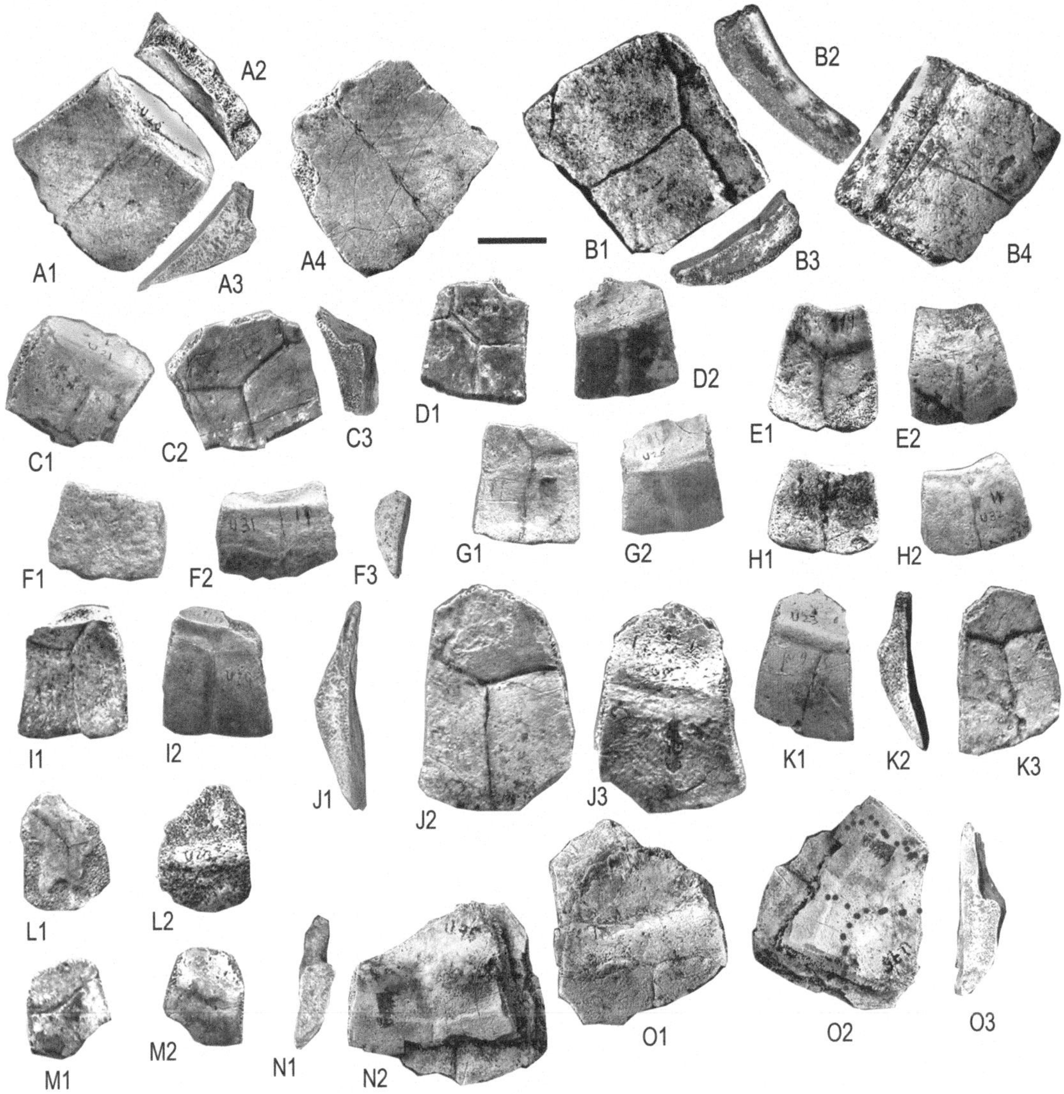

Fig. 7.8 Bothremydidae indet. g. sp., Upper Cretaceous (Maastrichtian) intertrappean beds, Upparhatti, southwestern India, DUGF/T coll. (Work catalog U numbers on the specimens): Posteriormost peripherals, **A1–A4**. DUGF/T66, right peripheral 9 or 10?, ventral (**A1**), anterior (**A2**), medial (A3) and dorsal (A4) views; **B1–B4**. DUGF/T65, left peripheral 9, dorsal (**B1**), medial (**B2**), anterior (**B3**) and ventral (**B4**) views; **C1–C3**. DUGF/T67, fragmentary right peripheral 9, ventral (**C1**), dorsal (**C2**) and posterior (**C3**) views; **D1–D2**. DUGF/T68, fragmentary right peripheral 9, dorsal (**D1**) and ventral (**D2**) views; **E1–E2**. DUGF/T70, right posterior peripheral, dorsal (**E1**) and ventral (**E2**) views; **F1–F3**. DUGF/T77, fragmentary right posterior peripheral, dorsal (**F1**), ventral (**F2**) and anterior (**F3**) views; **G1–G2**. DUGF/T72, left posterior peripheral, dorsal (**G1**) and ventral (**G2**) views; **H1–H2**. DUGF/T78, fragmentary right posterior peripheral, dorsal (**H1**) and ventral (**H2**) views; **I1–I2**. DUGF/T70, right posterior peripheral, dorsal (**I1**) and ventral (**I2**) views; **J1–J3**. DUGF/T64, left peripheral 11, anterior (**J1**), dorsal (**J2**) and ventral (**J3**) views; **K1–K3**. DUGF/T69, right peripheral 11, ventral (**K1**), posterior (**K2**) and dorsal (**K3**) views; **L1–L2**. DUGF/T71, posterior peripheral, dorsal (**L1**) and ventral (**L2**) views; **M1–M2**. DUGF/T74, fragmentary right posterior peripheral, dorsal (**M1**) and ventral (**M2**) views; **N1–N2**. **O1–O3**. DUGF/T121, (catalogue number of one of these specimens is missing) associated left peripheral 11 and pygal: **N1–N2**. fragmentary pygal, right border (**N1**) and ventral (**N2**) view, **O1–O3**. left peripheral 11, ventral (**O1**), dorsal (**O2**) view (partly hidden by pygal, sulci represented by dotted lines) and posterior (**O3**) border. Scale bar equals 2 cm

wide vertebral 1 and a light but well visible decoration of marked polygons on the carapace and on the plastron, not present in Upparhatti plates. The overlap length proportion and the decoration are not appreciable in the *Algorachelus* figured specimens of the type locality in (Pérez-García 2016, 2017; Pérez-García et al. 2017a), but the peripheral 1 is wider with an anteriorly wider vertebral 1 overlap.

Many other bothremydid carapaces have a narrow peripheral 1 with a long marginal overlap: both Afro-American forms (Gaffney et al. 2006, but most are figured by diagrams

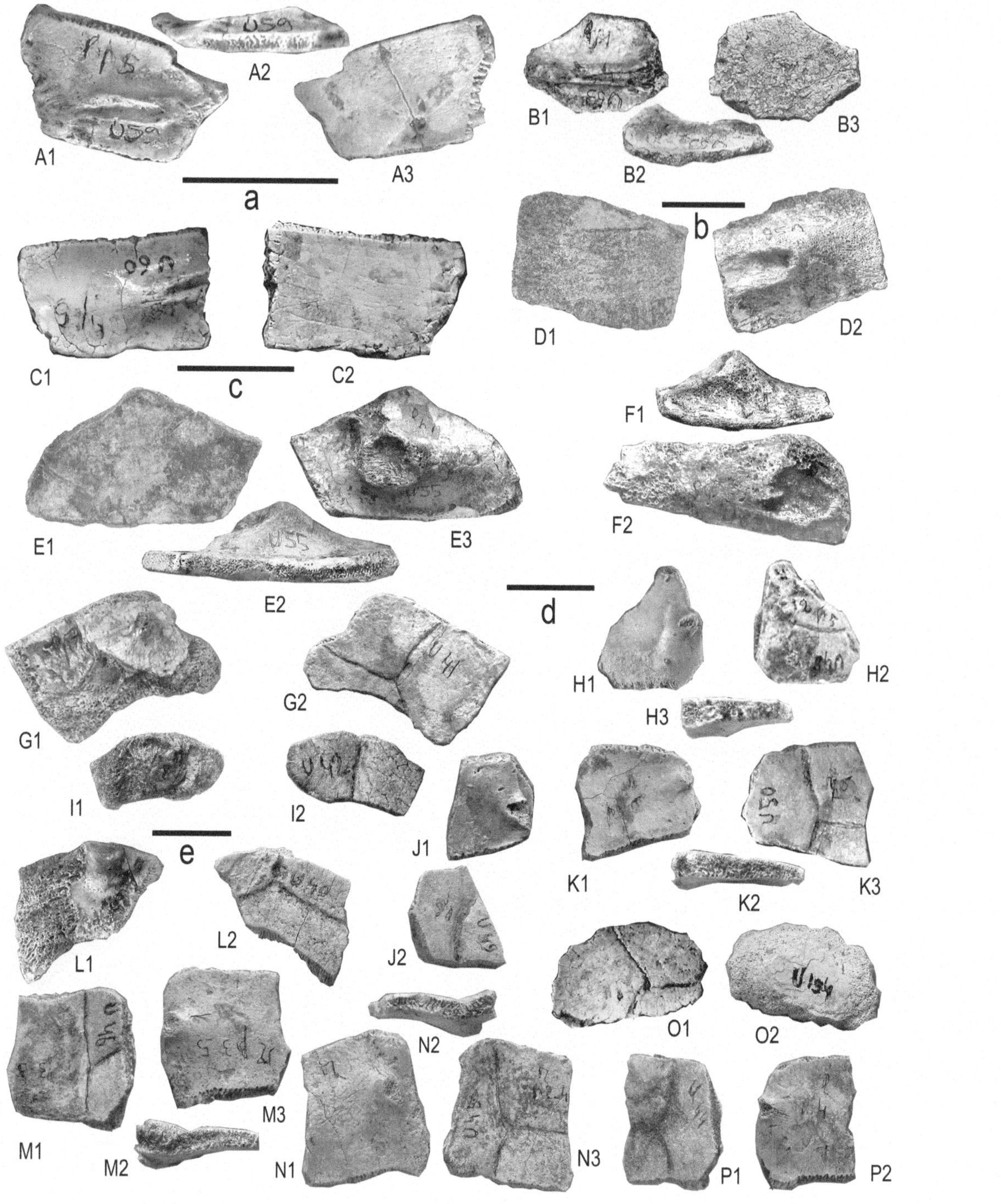

Fig. 7.9 Bothrmydidae, Upper Cretaceous (Maastrichtian) intertrappean beds, Upparhatti, southwestern India, DUGF/T coll. (Work catalog U numbers on the specimens): Fragmentary costals. Bothremydidae indet. g. sp.: **A1–A3**. DUGF/T105, **B1–B3**. DUGF/T99, right costals 5, lateral part, ventral (**A1, B1**), posterior (**A2, B2**) and dorsal (**A3, B3**) views; **C1–C2**. DUGF/T106, right costal 5, lateral part, ventral (**C1**) and dorsal (**C2**) views. Cf. Bothremydidae indet g. sp.: **D1–D2**. DUGF/T102, left costal 5, lateral part, dorsal (**D1**) and ventral (**D2**) views. **E1–E3**. DUGF/T101, right costal 1, dorsal (**E1**), posterior (**E2**) and ventral (**E3**) views; **F1–F2**. DUGF/T129, right costal 1, lateral part, posterior (**F1**) and ventral (**F2**) views; **G1–G2**. DUGF/T87, right costal 7, ventral (**G1**) and dorsal (**G2**) views; **H1–H3**. DUGF/T94, right costal 5, medial part, ventral (**H1**), dorsal (**H2**) and posterior (**H3**) views; **I1–I2**. DUGF/T88, right costal 7, ventral (**I1**) and dorsal (**I2**) views; **J1–J2**. DUGF/T95, right costal 6, medial part, ventral (**J1**) and dorsal (**J2**) views; **K1–K3**. DUGF/T96, right costal 6, medial part, ventral (**K1**), posterior (**K2**) and dorsal (**K3**) views; **L1–L2**. DUGF/T86, right costal 8, ventral (**L1**) and dorsal (**L2**) views; **M1–M3**. DUGF/T92, left costal 3, medial part, dorsal (**M1**), posterior (**M2**), and ventral (**M3**) views; **N1–N3**. DUGF/T9, right costal 2 or 4, medial part, ventral (**N1**), posterior (**N2**), and dorsal (**N3**) views; **O1–O2**. DUGF/T240, right costal 5, medial part, dorsal (**O1**) and ventral (**O2**) views; **P1–P2**. DUGF/T90, left costal 4, dorsal (**P1**) and ventral (**P2**) views. Scale bar equals 2cm, a for **A1–A3**, b for **B1–B3** and **D1–D2**, c for **C1–C2**, d for **E1–F2**, **H1–H3**, **J1–K3**, **M1–P2**, and e for **G1–G2**, **I1–I2**, **L1–L2**

without photographs) and European forms, the Late Cretaceous forms such as *Elochelys perfecta* and *Polysternon*, and the Oligocene "*Neochelys*" *montolivensis* (Roman 1897) (unnamed genus of Bothremydidae with a shell partly similar to that of *Polysternon*) (Nopcsa 1931b; Broin 1977, 1988; Gaffney et al. 2006; Lapparent de Broin et al. 2018). The presence of the marginals 1 and 2 of approximately similar long overlap with the pleural 1 and vertebral 1 scutes approximately equally overlapping the narrow peripheral 1 plate is not a common feature. The peripheral 1 (DUGF/T57) is distinct by its smaller width for its considerable length (as in *Elochelys* and extant pelomedusids) but with a less inclined anterior border. It produces a long anterior carapace border but with less ovoid shell form, elongate transversally all along the anterior border width, as in *Pelusios gabonensis* (but not as in other *Pelusios* species such as *P. adansonii* and *Pelomedusa*) (Broin 1983; Lapparent de Broin 2008a), *Carteremys leithii* and *Taphrosphys*. Among Indian taxa, DUGF/T57 is too narrow for *Carteremys leithii* and also lacks the latter's marked decoration. However, with an elongate anterior shell, it was probably a form with a long neck, corresponding to the long centra of Upparhatti.

Peripheral 2 is represented by DUGF/T 56, a left peripheral (Fig. 7.7C). This element, complete except for a broken lateroposterior corner, is thin, 0.45 cm medially and 0.54 cm thick at the lateral extremity of the margino-costal sulcus and thin towards its marginal edge. The thinness of this element corresponds to the thin peripheral 1 (DUGF/T57) described above, but it is from a larger individual. It is also brown-coloured and smooth and has the same morphology of slightly sinuous scute sulci as in DUGF/T57. Dorsally the marginal 2 overlap is long, corresponding to the long marginal 2 border on peripheral 1. The extent of the overlap of peripheral 2 by marginal 3 is reduced. Ventrally, overlap of peripheral 2 by marginals 2 and 3 is elongated relative to the overlap of the marginals on the peripheral 1. DUGF/T56 contours conform to the following 3rd peripherals of the more robust taxon but the plate does not belong to it because of the longer marginals, as in the peripheral 1.

DUGF/T 60 (Fig. 7.7B) is a partial right peripheral 1 of a slightly larger and relatively wider form than DUGF/T56 and 57. It is complete except for the posterior part of the contact with the costal 1. It has a flat, eroded surface and pale beige-rosy color as are most of the Upparhatti specimens. It is 0.8 cm thick at the medial margino-costal sulcus, 4.9 cm wide at its anterior edge and has a length of about 4.4 cm. The plate is relatively thicker marginally than the previous plates, but its lateral edge is not rounded in cross section. As in the DUGF/T 56 and 57, the marginal scutes are dorsally long and ventrally short. However, the posterior border of marginal 1 slopes toward the nuchal, indicating that the length of marginal 1 is less than that of marginal 2. This is different from DUGF/T57 but similar to the condition in many other Pelomedusoides such as *Elochelys* and *Polysternon*. It is also different from *Erymnochelys* which has a much longer first marginal.

Specimens DUGF/T47-49 (Fig. 7.7E–G) are 3rd peripherals of the robust morphology. Measurements of DUGF/T48 are: length on width, 3.4 cm × 3.9 cm; maximal depth, 1 cm; and length of marginals, 1.2–1.3 cm. These peripherals have much shorter marginals. The portion of these peripherals that is covered by the marginal scutes forms a shorter border that is turned upward, and more or less thickened in longitudinal section. Ventrally, the axillary buttress insertion is present near the posterior edge of the peripheral. DUGF/T48 has well marked rounded ridges corresponding to a vertebral 1 scute radiation. The insertion of the axillary buttress near the posterior border of the peripheral differs from the condition in the podocnemidid *Erymnochelys madagascariensis*, *Turkanemys* Wood, 2003 and *Lapparentemys vilavilensis* (in Broin 1991) in which the insertion of the buttress begins more anteriorly near the boundary between the 2nd and 3rd peripherals. In *Podocnemis lewyana*, *P. sextuberculata* and *Neochelys*, the buttress begins at the mid posterior part of the 3rd peripheral, then expands anteriorly. In *Cambaremys* França and Langer 2005 and *Peltocephalus*, the boundary is just a little posterior to the anterior end of the 3rd peripheral. In bothremydids, where known, the insertion of the axillary buttress and beginning of the bridge varies from the middle to the posterior part of the 3rd peripheral (Lapparent de Broin and Werner 1998: Figs. 8, 9, 10 and 11) in front of the mid lateral border of the costal 1. It is posterior on the 3rd in the *Taphrosphys* group and in *Carteremys leithii*, as in DUGF/T47, DUGF/T48, and DUGF/T49. In the referred specimen of *Carteremys* proposed as a neotype (GSI coll. N° 20337: Williams 1953: pl. 3), the axillary buttress was also posteriorly inserted near the posterior border of the peripheral 3, slightly more posteriorly than in *Polysternon provinciale* and *P. atlanticum*, *Elochelys perfecta* and the undescribed Miocene bothremydid from As Sarrar, Saudi Arabia (coll. loc 6, No. 806, MNHNF provisional deposit). In contrast, the buttress inserts on the 2nd, and nearly the 1st- and 2^{d} boundary, in cf. *Arenila*, a member of the Nigeremydini (*Nigeremys* group in Lapparent de Broin and Werner 1998: Fig. 13). However, the short marginal scute border in relation to the overlap of the plates DUGF/T47, DUGF/T48, and DUGF/T49 by the costal 1, and the plate robustness, are not of shells with an elongate anterior part such as *Elochelys* and *Carteremys leithii*, beside the absence of strong decoration, as seen above. They might belong to a short morphotype as in *Kurmademys*.

The thickness and color of the 4th to 8th bridge peripherals vary depending on the part of the plates. Generally, the plates are brown to pale color with more or less dark parts

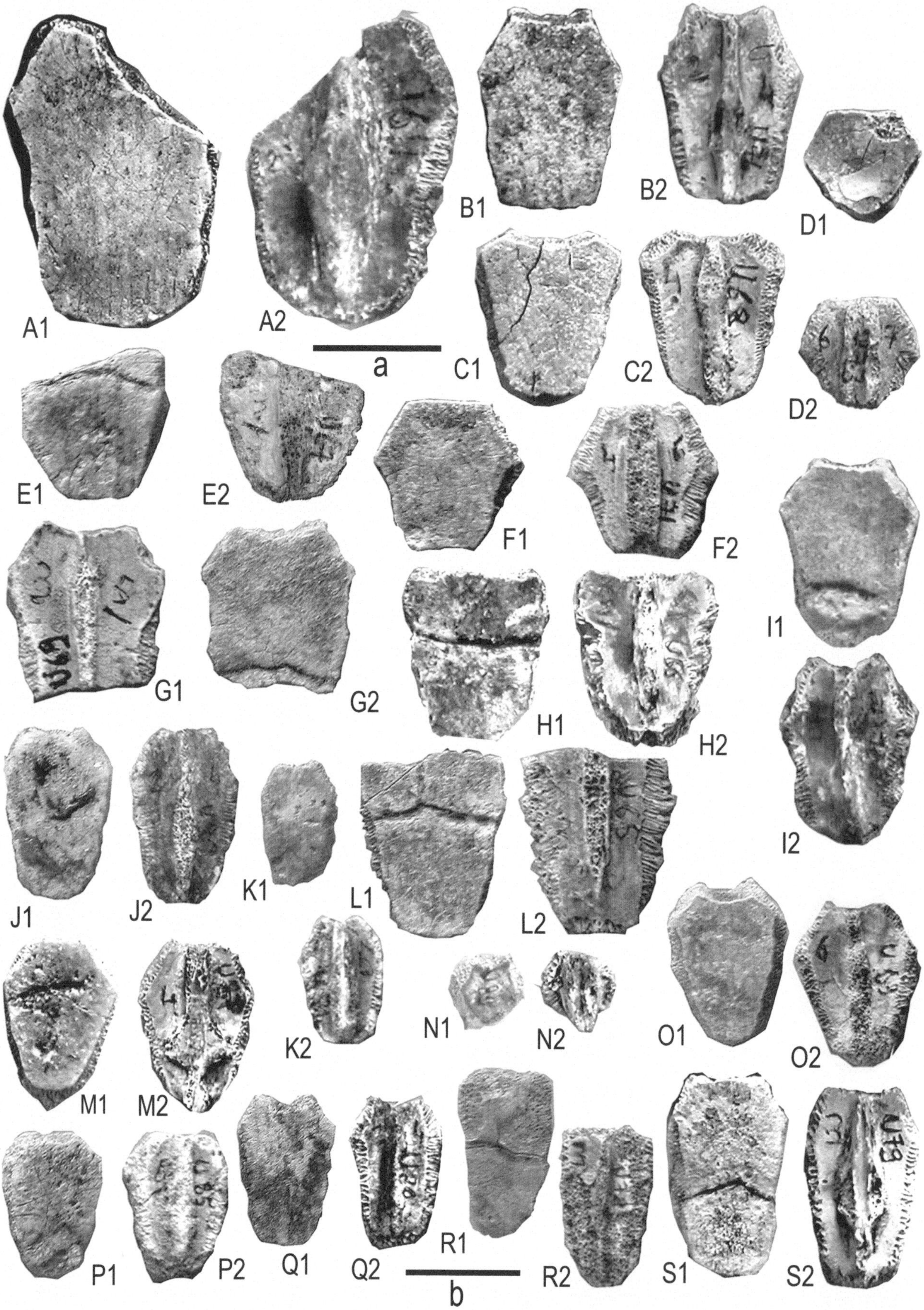
A1
A2
a
B1
B2
C1
C2
D1
D2
E1
E2
F1
F2
G1
G2
H1
H2
I1
I2
J1
J2
K1
K2
L1
L2
M1
M2
N1
N2
O1
O2
P1
P2
Q1
Q2
R1
R2
S1
S2
b

◀ **Fig. 7.10** Bothremydidae indet g. sp., Upper Cretaceous (Maastrichtian) intertrappean beds, Upparhatti, southwestern India, DUGF/T coll. (Work catalog U numbers on the specimens): Neurals. **A1–A2**. DUGF/T110, neural 2, dorsal (**A1**) and ventral (**A2**) views; **B1–B2**. DUGF/T122, neural 2, dorsal (**B1**) and ventral (**B2**) views; **C1–C2**. DUGF/T114, neural 2 or 4, dorsal (**C1**) and ventral (**C2**) views; **D1–D2**. DUGF/T119, last neural (6–7?), dorsal (**D1**) and ventral (**D2**) views; **E1–E2**. DUGF/T113, neural 3, dorsal (**E1**) and ventral (**E2**) views; **F1–F2**. DUGF/T117, neural 4, dorsal (**F1**) and ventral (**F2**) views; **G1–G2**. DUGF/T115, neural 3 (or 5?), dorsal (**G1**) and ventral (**G2**) views; **H1–H2**. DUGF/T112, neural 3 or 5, dorsal (**H1**) and ventral (**H2**) views; **I1–I2**. DUGF/T116, neural 3 or 5, dorsal (**I1**) and ventral (**I2**) views; **J1–J2**. DUGF/T128, neural 2 or 4, dorsal (**J1**) and ventral (**J2**) views; **K1–K2**. DUGF/T130, neural 3 or 5, dorsal (**K1**) and ventral (**K2**) views; **L1–L2**. DUGF/T111, neural 3 or 5, dorsal (**L1**) and ventral (**L2**) views; **M1–M2**. DUGF/T133, neural 2 or 4, dorsal (**M1**) and ventral (**M2**) views; **N1–N2**. DUGF/T135, last neural (6–7?), dorsal (**N1**) and ventral (**N2**) views; **O1–O2**. DUGF/T109, neural 2 or 4, dorsal (**O1**) and ventral (**O2**) views; **P1–P2**. DUGF/T131, neural 3 or 5, dorsal (**P1**) and ventral (**P2**) views; **Q1–Q2**. DUGF/T126, neural 2 or 4, dorsal (**Q1**) and ventral (**Q2**) views; **R1–R2**. DUGF/T124, neural 3 or 5, dorsal (**R1**) and ventral (**R2**) views; **S1–S2**. DUGF/T125, neural 3 or 5, dorsal (**S1**) and ventral (**S2**) views. Scale bars equals 1 cm, a for **A1–E2**, **H1–I2**, **M1–N2**, **S1–S2** and b for **F1–G2**, **J1–L2**, **O1–R2**

and have a smooth surface more or less eroded. Plates DUGF/T51 and DUGF/T63 have the basic fine, distinct decoration of micropores and dichotomic sulci united in small flat polygons, and dorsally have some radiating ridges below the posterior marginal (9th) and the costal 3 (DUGF/T63, Fig. 7.7R).

Peripheral 4 is represented by DUGF/T61 (Fig. 7.7D) and DUGF/T55 (Fig. 7.7J). DUGF/T61 is from a small (juvenile) individual. These peripherals are open for the bridge, showing both the axillary buttress insertion scar (I1) and the bridge internal cavity (I2). These peripherals belong to the more robust taxon as shown by their short rounded marginal border.

Peripherals 5 and 6 are represented by 5 elements. These are formed by two parts, one dorsal part that contacts the costals and a ventral part that contacts the plastron at the bridge. Among these peripherals DUGF/T79 (Fig. 7.7I) and DUGF/T158 (Fig. 7.7K) have a rounded border below the two successive marginals seen in the "strong" peripherals 3 and 4 and the fragmentary plate DUGF/T163 (Fig. 7.7L). The upper part, which is in contact with the costals, has a reduced costal overlap compared to previous plates.

Peripheral 7, which is from the posterior extremity of the bridge, is represented by three partial plates, DUGF/T84 (Fig. 7.7M), DUGF/T59 (Fig. 7.7N) and DUGF/T58 (Fig. 7.7O). The dorsal and ventral parts of the plate become closed, squeezing the inner bridge cavity just in front of its posterior extremity and the inguinal buttress insertion on the edge of the plate. The overlap of the marginal on these peripherals is longer than in peripheral 4. Plates DUGF/T58 and DUGF/T59, which are darker, have their surface incurved and with a notch at the rear border of the sulcus between the two overlapping marginals. This is not present in DUGF/T84, which is pale-rosy, more like the "strong" peripherals 3 and 4 (possibly it belongs to the same individual).

Peripheral 8 is represented by eight specimens. These can be subdivided into three groups. In group 1, which includes right (DUGF/T125) (Fig. 7.7P), left (DUGF/T54) (Fig. 7.7 T) and right (DUGF/T52) peripherals, the bridge extremity and the suture surface for the inguinal buttress insertion in the anteromedial corner of the plate is very short. The ventral face shows a weakly inclined surface without a prominent transverse ridge just medial to the ventral boundary of the marginals and the marginal overlap is long. In group 2, which includes left (DUGF/T63), a 8th (Fig. 7.7R) and right (DUGF/T 62) (Fig. 7.7S), the bridge extremity on the anteromedial corner of the plate is larger, the suture for insertion of the inguinal buttress is longer, and the ventral medial limit of the skin, medial to the edge of the marginals, is marked by a prominent rounded ridge. The space between the ridge and the marginal sulcus is long and the marginal overlap is short. In DUGF/T63 (Fig. 7.7R), there are scute striations becoming plate ridges which are particularly elevated, wide and rounded; the external border is shortened, notched by an injury which has been repaired. Group 3 peripherals include right DUGF/T51 (Fig. 7.7Q). In this peripheral, the extremity of the bridge is as deep as in group 2 peripherals, but the transverse ridge is less marked, and its marginal sulcus is close to the ridge, indicating that the marginal overlap is long, as it is in group 1 peripherals.

In the posterior peripherals and pygal (Figs. 7.7 and 7.8), the thickness varies according to the part of the plates and to the plate size. The color varies from brown to pale color with some more or less dark parts. The surface is smooth and more or less eroded. As in peripheral 8, several morphotypes seem to be present based on the development of three characters. These are:

First, the presence or absence of the prominent ventral transversal rounding that is medial to the visceral boundary of the marginal with the skin, and posterior to a depression in the medial part of the plate. This can be either just a ridge (inclined surface) or an angled (about 90°) narrow lip. The marked rounding, when present, may vary depending on position on the shell border. It may exist in Upparhatti forms without a marked lip but just a ridge, like in extant podocnemidids from South America. Specimens DUGF/T68, 70, 77 and 78 on the one hand and large DUGF/T65 and 121, on the other hand, have a prominent lip posterior to the depression. DUGF/T64, 67 and 69 on the one hand, and

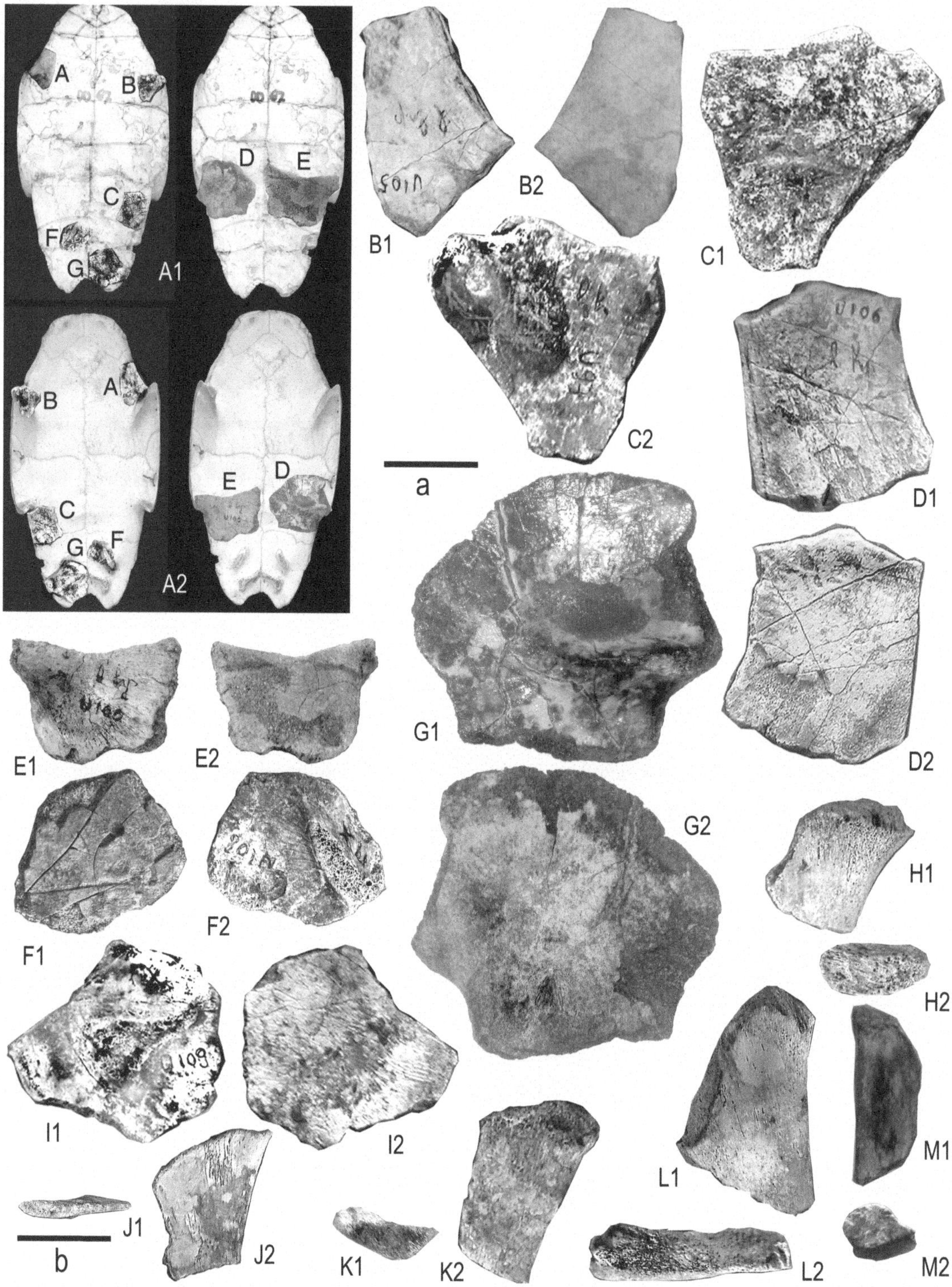

Fig. 7.11 Bothremydidae indet g. sp., Upper Cretaceous (Maastrichtian) intertrappean beds, Upparhatti, southwestern India, DUGF/T coll. (Work catalog U numbers on the specimens): Plastral elements. **A1–A2**. Plastral elements (labelled A to G on ventral and dorsal faces) on the plastron of *Erymnochelys madagascariensis*, MNHN (ZA AC) 1892–494, ventral (**A1**) and dorsal (**A2**) views (not to scale). **B1–B2**. DUGF/T151, lateral part of right hyoplastron, dorsal (**B1**) and ventral (**B2**) views; **C1–C2**. DUGF/T143, posteromedial part of left hyoplastron, dorsal (**C1**) and ventral (**C2**) views; **D1–D2**. DUGF/T152, posterior part of left hypoplastron, dorsal (**D1**) and ventral (**D2**) views; **E1–E2**. DUGF/T146, posterior part of left hypoplastron, dorsal (**E1**) and ventral (**E2**) views; **F1–F2**. DUGF/T154, fragment of right xiphiplastron with pubic suture, ventral (**F1**) and dorsal (**F2**) views; **G1–G2**. DUGF/T237, posterior part of right hypoplastron, dorsal (**G1**) and ventral (**G2**) views; **H1–H2**. DUGF/T138, right posterior hypoplastral buttress, anterior (**H1**) and dorsal (**H2**) views; **I1–I2**. DUGF/T155, fragment of left xiphiplastron with ischiatic suture, dorsal (**I1**) and ventral (**I2**) views; **J1–J2**. DUGF/T139, right posterior hypoplastral buttress, dorsal (**J1**) and anterior (**J2**) views; **K1–K2**. DUGF/T137, right anterior hyoplastral buttress, dorsal (**K1**) and anterior (**K2**) views; **L1–L2**. DUGF/T136, left posterior hypoplastral buttress, posterior (**L1**) and laterodorsal views (**L2**); **M1–M2**. DUGF/T236, right anterior hyoplastral buttress, anterior (**M1**) and dorsal (**M2**) views; Scale bars equals 2 cm for **B1** to **M2**

Table 7.5 Comparison of bothremydid shell elements of *Taphrosphys* spp., *Elochelys perfecta*, *Carteremys* group, *Kurmademys* and "*C.*" *pisdurensis* with two Upparhatti gen. et sp. indet., A, dorsal shell, B, plastron. Abbreviations: ant, anterior, anteriorly. AP, antero-posteriorly; ento, entoplastron; epi, epiplastron; epi hyo sut, epi-hyoplastral suture; HP, humeropectoral sulcus; hyo, hyoplastron; hypo, hypoplastron; med, medially; meso, mesoplastron; mod, moderately; per, peripheral; post, posterior, posteriorly; quadrag, quadrangular; reconst., reconstructed; sut, suture.

Dorsal shell	*T. sulcatus*	*T. congolensis*	*T. ambiguus*	Upparhatti	*Elochelys perfecta*	*Carteremys leithii*	*Carteremys leithii GSI 20237*	*C.* aff. *leithii* Kutch	*Kurmademys*	"*Carteremys*" *pisdurensis*
Dorsal shell-plastral length	ca. 65 cm shell (Gaffney, 1975) – plastron 37 cm (Wood, 1975)	51 cm plastron gives ca. 53 cm for shell (Wood, 1975)	32 cm plastron holotype gives ca. 34 cm shell	form a, shell ca. 32 cm from per Fig. 7.6 A	(2 spec)	t ype: 18 cm (sketch)	20–32 cm	Kutch, 1/2 plastron 16cm for shell ca. 34 cm	shell ca. 27 *cm*	shell 22.5 to 25 cm 25
				form b, shell ca. 20 cm from per Fig. 7.6B	type : 22 cm					
Decoration of polygones	Yes	Yes	Yes	No	Yes	Yes	Yes	Yes	No	No indication
	Much protruding	Much protruding	Much protruding	No	But not protuberant	Protuberant" intricate network of grooves"	protuberant	Much protuberant		
Anterior shell border	Ovoid very narrow and well protruding medially	Ovoid very narrow and well protruding medially	?	Form a somehow ovoid protruding. Form b not protuding	Ovoid, mod narrowed and medially protruding	Barely rounded (sketch, reconst)	Rounded, narrowed and slightly protruding, medially	[Rounded and elongated from the ant lobe]	Quadr-angulo-round	not protruding mod elongated (reconst.)
shell	Widened mid part + posteriorly	Widened mid part + post.	?	?	Widened medially (at the mid line only).	?	Widened mid part + posteriorly	?	Wide, rather quadrang. Not widened at mid part and post	Wide, rather quadrang. Not widened at mid part and post
Nuchal	[Very long, narrow reconst.]	Very long, narrow, V notch	?	Form a long and narrow anteriorly as in *E. perfecta*, from per. 1. Form b, from per 1, not long anteriorly	Long. Ant. mod wide	?	Long mod wide ant	[long : from the ant lobe]	[Moderate long, wide anteriorly]	moderate long and ant not narrow

(continued)

Table 7.5 (continued)

Dorsal shell	*T. sulcatus*	*T. congolensis*	*T. ambiguus*	Upparhatti	*Elochelys perfecta*	*Carteremys leithii*	*Carteremys leithii GSI 20237*	*C.* aff. *leithii* Kutch	*Kurmademys*	*"Carteremys" pisdurensis*
Costal 1	Very long, very narrow ant –medially	Very long, very narrow ant - medially	?	[form a: very long, form b: mod long]	Long mod. narrowed ant – medially	Long	Long well narrowed ant – medially	[long from the ant lobe]	Long mod. narrowed ant - medially	Long, very narrow ant. – medially
Neural 1	Long narrow	Long mod. wide	?	[a: long, narrow; b: mod long and wide	Long wide	?	[Long, narrow]	Long narrow	Mod. long and wide	Long narrow
Plastron	*T. sulcatus*	*T. congolensis*	*T. ambiguus*	*Upparhatti*	*Elochelys perfecta*	*Carteremys leithii lost type and GSI 20337*		*C.* cf. *leithii* Kutch	*Kurmademys*	*"Carteremys" pisdurensis*
Axillary buttress	Short AP med round. Oblique to post border costal 1	Short AP med round. Oblique to post border costal 1	?	Short AP. med round	?	?	At post part costal 1 + per 3 (from dislocation). Per 4 ending at post part costal 1	?	[Very short. Transversal below costal 1] up to at 2/3 border costal 1 and per	[At peripheral 4]
	At per 4 . Per 4 ending at post costal 1	at mid-post per 3. Per 4 ending at post costal 1		?	At mid per 3. Per 4 ending at post costal 1					
Anterior lobe	Trapezoid, wide at base	Trapezoid, wide at base	Trapezoid, wide at base	?	Trapezoid slightly rounded, long and wide	Rounded, well long (not wide)	[Rounded long, from the dorsal shell]	Rounded well long not wide. slight intergular protrusion	?	Rounded slightly trapezoid, well long not wide
Intergular width/ gulars. Length/HP	Wider. Up to HP	Wider. Up to HP	Wider. Up to HP	?	Wider or as wide. Up to HP	Very wide. Up to HP	?	Wider. Up to HP	?	Few wider. Up to HP
Gulars	On epi up to ento suture	On epi up to ento suture	Long up onto ento	?	On epi not up to ento	Narrow Up to ento by a point	?	On epi up to ento	?	Up to ento and onto ento
HP sulcus	Post to epi hyo sut.	Post to epi hyo sut.	Post: to epi hyo sut.	?	Ant to epi-hyo sut.	Post to sut epi-hyo? Oblique	?	Post to epi hyo sut.	? : At least crosses epi-hyo sut. laterally	Oblique. Med at epi-hyo sut. level - Laterally post/ epi-hyo sut>.
Pectoro abdominal sulcus	Close to hyo-hypo sut. Cross meso	Anterior to meso: contact	Close to hyo-hypo sut. Cross meso		Anterior to meso : contact or not	?	?	Anterior to meso [contact]	Well anterior to meso	Anterior to mesoplastra [not contact]
Inguinal buttress	AP short. Wide?	AP mid-long, mid-wide	?	AP (1) short, wide (2) long, narrow (3) mid-long, wide	?	?	?	?	AP mid-long, wide	?

DUGF/T71, 72, 73, 74, 75, 76 and 80 on the other hand, have a flatter lip or barely a ridge with just a slight depression. Some like DUGF/T66 (a 10th) are intermediate with a prominent rounding on one side only, the ridge decreasing towards the other side. In DUGF/T66, there are striations, well-marked both dorsally and ventrally (Fig. 7.8A1, A4) as in peripherals 3 (DUGF/T49) and 8 (DUGF/T63). – Second, the ventral marginal overlap is either long, close to the ridge or lip boundary, or short, with marginal sulcus well separated from the lip or ridge. – Third, the plate may be narrow and long, and rectangular or quadrangular in shape. This is of diagnostic value subject to caution because the posterior plates may be more or less narrow, depending on their position on the posterior border, but in a limited frame of variability. However, some taxa have a shorter posterior border and more quadrangular plates. DUGF/T68 and DUGF/T70 (Fig. 7.8D, I), probably a 9th left and a 11th peripherals, are rather long plates for their width with a long marginal overlap, and the median sulcus of the marginals is close to the ridge or lip. DUGF/T69, a small right 11th peripheral (Fig. 7.8K), and DUGF/T64, a larger left 11th peripheral (Fig. 7.8J), are long and narrow with a transversely inclined ventral ridge not marked by a lip. Some wider more quadrate plates are moderately lipped such as DUGF/T66 (a 10th, Fig. 7.8A) with a shallow lip posteriorly and a thicker lip anteriorly. DUGF/T67, an incomplete right 9th of a quadrate form (Fig. 7.8C) is not as narrow as DUGF/T64 and DUGF/T69; it appears to bear in the anteromedial corner (Fig. 7.8C2, on the right) the sulcus separating the pleural scutes 3 and 4 as in the smaller and narrower 9th DUGF/T68 (Fig. 7.8D, on the left). But the ridge is laterally moderately marked in DUGF/T67, while forming a strong narrow lip in the narrower DUGF/T68, thus both plates possibly belong to two distinct taxa, which agrees with the less quadrangular shape and less depth of DUGF/T68.

Specimen DUGF/T121 is a pygal preserved attached onto the left peripheral 11 (Fig. 7.8N, O). DUGF/T 121 is a rather short plate, quadrangular, with a shorter ventral marginal overlap than in the peripheral 11 (by reduction of the overlap posteriorly) and a longer space between the marginal sulcus and the pronounced lip (by enlargement of the lip). Specimen DUGF/T65 (Fig. 7.8B), a left 9th peripheral, is of the same quadrangular shape and robust type as DUGF/T66, and also slightly ridged dorsally (Fig. 7.8B1) and polygonal ventrally (Fig. 7.8B4). As with small DUGF/T68, it bears the sulcus between pleurals 3 and 4, but DUGF/T68 is a narrow bright plate with a marginal border touching the lip. DUGF/T65 is a wide quadrangular plate with lip and sulcus separated from one another. Specimens DUGF/T65 and DUGF/T 121 (Fig. 7.7B, N–O), which are large quadrangular posterior peripherals with a much more marked angle forming a prominent transverse lip posterior to the depression and with a longer space between the marginal sulcus and the ridge, and a short ventral marginal overlap might belong to the same taxon as the peripherals 8 (DUGF/T62 and 63) (Fig. 7.7R–S) and DUGF/T66 (Fig. 7.8A1), with variability of the lip according to the position, closer to the border posteriorly than anteriorly. These plates do not seem to belong to the same taxa as the long and ventrally flattened 11th plates (DUGF/T64, and DUGF/T69).

DUGF/T68 and DUGF/T70 are also different as specimens DUGF/T77 (Fig. 7.8F), DUGF/T78 (Fig. 7.8H), DUGF/T71 (Fig. 7.8L), DUGF/T72 (Fig. 7.8G), DUGF/T73 (Fig. 7.8E), and DUGF/T74 (Fig. 7.8M); these are small, and more or less complete plates, that are rather short relative to their width. In these peripherals, the ventral lip is either prominent (DUGF/T77 and DUGF/T78) or weak (DUGF/T71-73). Peripheral DUGF/T74 (Fig. 7.8M) has a marginal overlap length and lip size decreasing from anterior to posterior, showing as in other plates the variability of the overlap according to the position but also according to the taxon. All these peripherals show without doubt the presence of several taxa, at least two: surely one narrower and more elongate anteriorly with longer marginal. The second one or ones with shorter marginals (one large and one small): surely one more "quadrangular", wider and dumpy, and possibly another one, slender and also less anteriorly elongate.

Measurements: DUGF/T64: full medial length: 6.2 cm; DUGF/T121, pygal: 4.5 cm; width on depth at lip: DUGF/T64: 3.5 × 1.1 cm; DUGF/T 121: 4.3 × 1.3 cm; posterior width: DUGF/T 64: 4.5 cm, DUGF/T121: 4.9 cm; marginal overlap length, ventral: DUGF/T 64: 3.05 cm; DUGF/T 121: 1.3 cm. Distance: anterior part of lip to posterior border: DUGF/T64: 3.3 cm (inclined ridge), DUGF/T121: 2.95 cm (angled lip).

Costals (Fig. 7.9): Twenty-seven costal elements are preserved, although some are partial and so much eroded that it is not possible to identify their position.

The first costal is best represented by DUGF/T101 (Fig. 7.9E). This element bears the insertion scar of the axillary buttress suture, which is rather wide lateromedially and comes from an anterolateral position, probably from the 3rd peripheral, to end on the 2nd rib where it is bulged with a rounded pit, for the medioventral end of the axillary buttress. This configuration is known in the bothremydid from the early Miocene of Saudi Arabia (loc 6, 803). On the dorsal surface, there is no sign of a vertebral sulcus medially, probably because the surface is too much eroded and because the plate is broken. The second first costal, specimen DUGF/T129 (Fig. 7.9F) is a very short part of the plate and it also bears the suture of the axillary buttress anterolaterally directed and ending in a similar rounded pit as in many podocnemidoids (plastral buttresses are notably illustrated in Lapparent de Broin and Werner [1998]; Lapparent de Broin and Murelaga [1999] [including *Polysternon*];

Carvalho et al. [2002]; França and Langer [2005]; Gaffney et al. [2006], and Cadena [2011]). In addition, twelve isolated axillary buttresses are present. The dorsal part of some axillary buttresses is rounded such as in DUGF/T229 (Fig. 7.11M) which corresponds to the round, spotted scar seen on the costals DUGF/T101 and 129 (Fig. 7.9E and F). Another example such as DUGF/T137 (Fig. 7.11K2), is more ovoid-triangular in shape with a pinched lateral part as in *Polysternon atlanticum*. This buttress does not correspond to any of the few preserved Upparhatti first costals. However, such a scar shape is rather similar, although narrower, to the scar in *Erymnochelys* and other podocnemidids and it is similar to the scar in some other bothremydids. According to Gaffney et al. (2006), the buttress is a straight, inclined and thin scar in *Taphrosphys, Bothremys* or *Chedighaii* and *Kurmademys*. However, it may be a little curved; thus, in *Taphrosphys* spp., it is very close to the junction angle of the peripherals 3–4 with the costal 2. This buttress prolongs the thoracic rib 2 in a long bow, which is close to the common suture of costals 1 and 2 (Lapparent de Broin and Werner 1998).

Vertebral with Pleural Contact Morphotypes: Costals 2 to 8 are mostly represented by fragments. Based on the surfaces of the medial portion of these costals, three vertebro-pleural contact morphotypes can be recognized.

(1) Morphotype 1 is represented by specimens DUGF/T94, a right 3rd or a 5th; DUGF/T96, a right 6th; DUGF/T240, a left 5th (Fig. 7.9H, K and O, respectively). They have a flat surface on each part of the vertebrals sulci, narrow but well-marked, and the neural contact is thick dorsoventrally (see Fig. 7.9H3. and K2). Measurements of right 6th costal DUGF/T96: medial maximal length 2 cm, anterior vertebral medial border 1 cm wide, medial vertebral width at posterior border, 1.3 cm, thickness at neural contact, 0.6 cm, posterior thickness at sulcus vertebrocostal contact, 0.5 cm.

(2) Morphotype 2 is represented by specimens DUGF/T95, right 6th; DUGF/T92, right 3d or 5th; DUGF/T91, right 2nd or 4th; DUGF/T90, left 4th (Fig. 7.9J, M, N and P, respectively). They have a surface angle at the vertebral-pleural sulci and a strong medial thickening at the neural contact that is longitudinally parallel to the neural-costal border. The dorsal vertebral scute surface is therefore elevated relative to the pleural scute portion (see Fig. 7.9M2 and N2). In this morphotype, the vertebrals are narrow, the vertebral-pleural sulcus is close to the medial border of the plate forming narrower vertebral scutes than the first morphotype, and the thickness of the elevated medial part above the rib extremity is distinctive.

3) Morphotype 3 is represented by the plate DUGF/T88, a right 7th costal which bears the iliac scar (Fig. 7.9J). It has two slight bulgings on each side of the vertebral 4-pleural 4 sulcus, the vertebral 4 is not narrowed and there is no particular medial thickening. Costals 7 meet at the midline.

Measurements: 2^{d}–4th costal DUGF/T93: medial maximal length 3.15 cm, anterior and posterior vertebral medial border 0.4 cm and 0.9 cm wide, medial width at sulci between vertebral and costals crossing: 7.5 cm, thickness at neural contact: 0.8 cm, posterior thickness at sulcus vertebrocostal contact, 0.4 cm; 3rd or 5th costal DUGF/T92 (Fig. 7.9N): medial maximal length 3.15 cm, medial border at sulci between vertebrals and costal crossing: 0.9 cm wide, thickness at neural contact: 0.7 cm, posterior thickness at sulcus vertebrocostal contact, 0.5 cm; right 6th costal DUGF/T95 (Fig. 7.9M): medial maximal length 2.2 cm, medial vertebral width at posterior border: 0.7 cm, thickness at neural contact: 0.6 cm, posterior thickness at sulcus between vertebral and costals crossing, 0.3 cm; the anterior vertebral border is incomplete and the costals 3 and 4 had to meet anteriorly in the missing part.

Inguinal Buttresses Morphotypes (Figs. 7.9, 7.11): Costal 5 is represented only by the lateral ends of the bone, which preserve the ventral sutural cavity for the inguinal buttress. Two inguinal buttresses morphotypes are preserved.

Morphotype 1 is represented by DUGF/T10, a left 5th costal (Fig. 7.9D). It has a rounded anteroposteriorly long groove (half the anteroposterior dimension of the plate), but lateromedially short on the elevated bulge extending upon the included rib in the dermal bone. This basic morphology is known in podocnemidoids, podocnemidids as well as bothremydids, with a variable width (medio-laterally) and length (anteroposteriorly) such as in podocnemidines (extants, *Cambaremys* França and Langer 2005).

Morphotype 2 is represented by specimens DUGF/T105, 99 and 106, all right 5th costals (Fig. 7.9A, B and C, respectively). The inguinal buttress has an anteroposteriorly very short scar, i.e. a thin groove with parallel borders extending on a low bulge (as seen in Fig. 7.9A, B). This costal type is known in podocnemidids such as *Podocnemis* in Carvalho et al. (2002) where the thin groove is very wide (from medial to lateral) on the plate and much inclined. It is also known in bothremydids according to illustrations in Gaffney et al. (2006): *Kurmademys*, *Taphrosphys sulcatus*, and "*Bothremys*-or-*Chedighaii*", and four bothremydid specimens from As Sarrar (Saudi Arabia, DAMR, specimens loc 9, Nos. 1061, 1064, 1065, 1108, MNHNF provisional deposits), which are all from different individuals. In *T. congolensis*, the scar is wide mediolaterally but longer anteroposteriorly. In *Carteremys leithii* figured type, the scar is unknown, but in the neotype, it was wide medio-laterally as shown by the elevated crushed plates as seen in dorsal view at this place. The scar is unknown in "*Carteremys*" *pisdurensis*.

There are some isolated inguinal buttresses. Some are robust, such as DUGF/T138 (Fig. 7.11) and have a large (anteromedially elongate) dorsal border with a short rounded mediolateral extremity which may correspond to costals 5 such as DUGF/T102 (Fig. 7.9D). This condition is similar to that in some bothremydids, *Erymnochelys* and other podocnemidoids. Other very thin (anteroposteriorly) inguinal buttresses such as DUGF/T140 and 139 (Fig. 7.11J) correspond to the costals 5 (DUGF/T105, 99 and 106) with thin and wide scars (Fig. 7.9A, B and C).

Costal 7 is represented by DUGF/T87and 88 (Fig. 7.9G, I) which are right costals. In DUGF/T87, the important anterior part of the iliac scar is visible in ventral view. This nearly reaches the anterior border of the costal. The proximal articular surface shows that the costals meet behind sixth neural. Dorsally the junction of vertebrals 4 and 5 is located on the 7th costals. In DUGF/T88, the iliac suture scar extends up to the anterior border of the plate. Dorsally, the vertebral 4—pleural 4 sulcus crosses the plate vertically and the plate proportions indicate another species than DUGF/T87. The junction of the vertebrals 4 and 5 is located posterior to this plate, on the 8th costal. Flat slight polygons cover the plate, which are well visible below the pleural scute. The vertebro-pleural contact morphotypes are different (see above).

Costal 8 is represented by DUGF/T86 (Fig. 7.9L) a right costal. Ventrally the iliac scar crosses the plate. The anterior end of this scar would have been on costal 7 and the posterior end on the pygal as in primitive taxa. As with DUGF/T88, DUGF/T86 is from an individual in which costal 8 is overlain by vertebrals 4 and 5, and pleural 4. However, the iliac scar and the dorsal surface of DUGF/T86 differ from those of DUGF/T 88. In DUGF/T88, the surface is depressed along the area of the vertebro-pleural sulcus, while in DUGF/T86 it is flat along the sulci. In this feature, DUGF/T86 is like DUGF/T87 which is flat at the place where the sulci of vertebrals 4 and 5 and pleural 4 meet.

Although the pygal is not preserved here, as shown by costal 8, the iliac scar overlaps the pygal in this Upparhatti taxon. This is a primitive feature present in the Jurassic *Platychelys* and early Cretaceous *Notoemys* (Cadena Rueda and Gaffney 2005), in most primitive chelids (Lapparent de Broin 2000a; De la Fuente et al. 2001; Lapparent de Broin and De la Fuente 2001), in some indeterminate Pelomedusoides forms (Oliveira and Kellner 2007), some pelomedusids and some bothremydids. However, this character is neither present in *Kurmademys* nor in some Cretaceous or extant podocnemidids, especially in *Erymnochelys* (Gaffney et al. 2006; Lapparent de Broin and Murelaga 1999), but it is present in some other Cretaceous podocnemidids such as in specimens from Peiropolis (Bauru group, Brazil) (Zoological Museum of the USP coll.).

Neurals (Fig. 7.10): Twenty-four neurals are preserved. The first neural is not represented. 23 of these neurals are hexagonal. Of these, nine are neurals 2 or 4, 12 are neurals 3 or 5, and two are neural 6. One neural is pentagonal. This is identified as the 6th or 7th. A great deal of variability is present and two morphological types can be recognized in the neurals 2 to 5, some being wide relative to their length and others narrow, giving the appearance of long neurals.

The neurals lack any polygonal ornamentation but, among the wide neurals, some are longitudinally ridged (DUGF/T110, 122 and 113, Fig. 7.10A1, B1, E1). One of the narrow neurals is also striated (DUGF/T126, Fig. 7.10 Q1). Most of the neurals are thick, but a few of both the relatively wide (DUGF/T111, 115, 117, 118 and 133) and narrow neurals (DUGF/T124 and 126) are thin. It is impossible to determine which of the thicker neurals correspond to costals (Fig. 7.9). The larger neural (DUGF/T110) (Fig. 7.10A), which is a thick-ridged plate, corresponds to a 45–50 cm long carapace. The widest part of the neural is relatively narrow compared to the width of the anterior and posterior ends of the neurals 3 to 5 of "*Carteremys*" *pisdurensis*, and those of *Kurmademys* and *Carteremys*.

Measurements and ratios of width and length of the neurals show great heterogeneity. The narrower neurals, such as DUGF/T128, 130, 124 and 125 (Fig. 7.10J, K and Q to S), which are nearly as narrow anteriorly, medially and posteriorly in relation to their length, are outside the range of variability of "*C.*" *pisdurensis*, *Kurmademys* and *Erymnochelys*. Some of the wide neurals such as DUGF/T110, 122, 114, 111 and 115 (Fig. 7.10A, B, E, F and G) are included in the variability of *Erymnochelys*, but DUGF/T114 and 111 are also in the variability of "*C.*" *pisdurensis* (2nd, 5th). But among the neurals that are narrow for their length, none is as narrow posteriorly as in the equivalent ones in *Kurmademys and* "*C.*" *pisdurensis* (both not especially decorated, except for the possible basic ornamentation) and in *Carteremys* (strongly decorated).

Plastron (Fig. 7.11): The plastral elements described by Kelkar and Gupte (1943) have not been recovered and so few additional elements are preserved in the present collection that a reconstruction of the plastron is not possible. To orient the position of the available pieces, they have been superimposed on a plastron of an extant *Erymnochelys madagascariensis* (Fig. 7.11A). However, it is likely that they are not all conspecific and they are artificially resized, so they are not at the same scale. Moreover, the specimens were not found together.

The posterolateral hyoplastral fragment, DUGF/T151, and the posterior hypoplastral fragment, DUGF/T152 (Fig. 7.11B and D) are both thin and pale coloured. They share the presence of a very narrow dorsal (visceral) border of the pectoral and abdominal scutes, with these being in a depressed area. This is also the case in another more fragmentary hyoplastron (DUGF/T84), which complements the specimen DUGF/T59 and shares the very narrow exposure of the scutes on the visceral surface of the bone, although the exposure becomes much thicker at the process. The anterior lobe of the hyoplastral fragment DUGF/T151 is probably elongated and the dorsal border of the scute is narrow as in podocnemidids such as *Podocnemis lewyana*, *Erymnochelys madagascariensis* (Fig. 7.11A1–A2), and aff. *Erymnochelys* sp. from the Miocene of Namibia (Lapparent de Broin 2003, 2008a, b) and some bothremydids. The posterior hypoplastron DUGF/T152, which is complete up to the xiphiplastral suture but incomplete medially, was probably long relative to its width. It had a straight lateral border that converged slightly posteriorly toward the sagittal line.

Another pale coloured, much smaller hypoplastral fragment, DUGF/T146 (Fig. 7.11E), seems to be nearly complete posteriorly and medially, in the inguinal notch, although it lacks its posteromedial corner and anterolateral part. It is wide relative to its length. It differs from specimen DUGF/T151 and DUGF/T152 described above in having a wider dorsal border of the scute that is situated in a slightly depressed area. The lateral outline of this hyoplastral fragment is slightly rounded, rather than being straight.

Another group of fragments shown in Fig. 7.11, DUGF/T143 (hyoplastral), DUGF/T145 (hypoplastral), DUGF/T155 and 154 (xiphiplastral) (Fig. 7.11C, G, I and F) differ from DUGF/T146 in being of larger size and robustness. This is also the case in a fragment of hypoplastron (DUGF/T145) that is not figured. These hyo- and hypo-plastral fragments also have a narrow scute border as in DUGF/T146, 151 and 152 and *Erymnochelys*. However, the border is relatively thicker and lacks a depressed area. The inguinal notch forms an acute angle, which indicates that the borders of the posterior lobe converged posteromedially as for the posterior lobe of DUGF/T151.

In the preserved xiphiplastron DUGF/T154 (Fig. 7.11F), the pubic scar is ovoid slightly anteriorly pointed and a little elongated. In DUGF/T155 (Fig. 7.11I), the ischiatic scar is triangular, large laterally, elongated and narrow medially, towards the longitudinal median suture. The xiphiplastral point is missing in all the Upparhatti specimens, but it is clear that the anal notch was wide for its length (Fig. 7.11I) and the border of its preserved part was straight with a narrow anal scute dorsal overlap as in DUGF/T152.

The pubic scar is not as elongated as it is in *Taphrosphys*, which has the longest pubic scar known, and is not as wide as in many podocnemidoids. In particular, it is longer relative to its width than in *Erymnochelys* (Fig. 7.11A1–A2) and aff. *Erymnochelys* sp. from the Early Miocene of Namibia (Lapparent de Broin 2003) where it is oval, but it is equal to bothremydids such as *Polysternon* in length and width. The ischiatic scar is not roundly spotted and not close to the anal notch as it is in *Taphrosphys*. It is larger laterally than in *Erymnochelys* (Fig. 7.11A1–A2) and aff. *Erymnochelys* sp. from Namibia. Among bothremydids, the ischiatic scar is similar to that of *Polysternon*. Its anterior part (Fig. 7.11I) is anterior to the anal notch and its posterior lateral end is not prolonged in the xiphiplastral point, as in extant and most fossil podocnemidids (Fig. 7.11A1–A2) and many bothremydids such as *Polysternon*, *Rosasia* and *Kurmademys*. This is a derived condition. The condition is not known in "*C.*" *pisdurensis*. The elongated posterolateral part of the ischiatic scar extending to the xiphiplastral point is present in some Bothremydidae (*Taphrosphys* by the rounded spot, and *Bothremys* or *Chedighaii*).

Robustness and decoration: The Upparhatti plates vary in robustness and ornamentation although without strong taphrosphyine sculpture, or without the marked and raised polygons and dichotomic sulci present everywhere in *Carteremys leithii*. The thin posterior peripheral (DUGF/T70), which is probably a 11th peripheral (Fig. 7.8I), and a stronger, probable 9th peripheral (DUGF/T66), (Fig. 7.8A), have fine parallel sulci becoming dichotomic at the rear, isolating some fine elongated polygons that are not raised. In DUGF/T66, oblique, well visible and fine ridges are also present, as in some extant and fossil podocnemidids (not known up to now for Bothremydidae in the literature). Some of the other plates are covered by small and weak polygons, like the basic condition in extant podocnemidids and various fossil podocnemidoids, but not as well marked as in the strongly ornamented *Carteremys leithii* or *Taphrosphys* among bothremydids. Also, the shape of these pleurodiran polygons differs from larger and wider than long polygons of marine chelonioid forms. Most Upparhatti plates have a blunt relatively eroded surface. Micropores for the microvascularisation of the dermis are visible at places or everywhere on the shell elements.

Some of the robust pieces, particularly neurals and peripherals, have striations formed of fine stripes or ridges radiating from the growth centers of the scutes as in podocnemidids including, for example, *Erymnochelys madagascariensis* and aff. *Erymnochelys* from Ibeceten. This is seen in peripherals, such as DUGF/T48, a 2nd peripheral (Fig. 7.7F), two specimens that are possibly 9th peripherals, DUGF/T66 (Fig. 7.8A) and DUGF/T68 (Fig. 7.8D), one 8th peripheral, DUGF/T 63 (Fig. 7.7R), and neurals such as DUGF/T110, DUGF/T122, and DUGF/T113 (Fig. 7.10A, B and E). The ridges become larger and rounded at their extremities on the border of the scute, below the marginal and peripheral. In aff. *Erymnochelys* from Ibeceten and in *E. madagascariensis*, the ridges also become larger and rounded at their extremity. This is particularly so for the lateral ends of the ridges of the costal scutes at the point where they join the marginals on the peripherals. These striations and ridges have not yet been mentioned and figured in members of the Bothremydidae and they are not described for *Kurmademys* and "*Carteremys*" *pisdurensis*, and also not visible on their pictures. It appears to be a feature of continental freshwater taxa and is known for a long time in Gondwanan Podocnemidoidea, being also present in early-late Cretaceous South American podocnemidids. As such, the basic ornamentation of Upparhatti elements does not allow their placement in a family. However, it eliminates the possibility that these elements are from a member of the littoral *Taphrosphys* group, *Carteremys*, or any other littoral taxon with a strongly decorated shell.

Remarks on compared Indian shell: Upparhatti Shell elements have been particularly compared with the Indian taxa known by shell, i.e. *Carteremys leithii* (Carter 1852) (strongly ornamented by marked polygons and ridges (see in Williams 1953, left costal 1 for example), and *Kurmademys* and "*C.*" *pisdurensis*. In both these taxa the shells are not strongly decorated, as Upparhatti elements.

"*Carteremys*" *pisdurensis* is not shown as ornamented and by this character alone it is not a bothremydid of the strongly decorated groups including *Taphrosphys* and *Carteremys leithii* (known by carapace and plastron in Carter [1852] and by shell without figured plastron in Williams [1953]). In this character, it is similar to Upparhatti shells. The shell of "*Carteremys*" *pisdurensis* was figured by Jain (1986) as ovoid with a problematic reconstruction of nuchal part (for an anteromedial border not well preserved but moderately long). Pictures indicate that it belongs to a new bothremydid genus in the following features: a wide carapace medioanteriorly longer than the anterior lobe, with convex lateral borders (Jain 1986: pl. 1, Fig. 2); a plastron with rather wide lobes, the anteromedially lateral borders of which are not much posteriorly convergent; an anterior round-trapezoid lobe important in the anterior opening (lobe rounded in one specimen and more transverse anteriorly in another specimen); a posterior lobe with rounded anteromedial lateral borders, filling most of the posterior shell opening, but short straighter borders of anal notch; pectoro-abdominal sulcus well separated from the hyo-hypoplastral suture (probably as in *C. leithii*), coming up to the tip of the mesoplastra (anterior to the tip in *Kurmademys*); and neurals moderately long for their width (3rd and 4th a little longer), neurals 1 and 2 not narrowed at their anterior and posterior extremities as are the neurals 3 to 5 at their contacts but not as much as in *Kurmademys* and more in *C. leithii* neotype. Particularly, scute pattern of the anterior lobe shows a long intergular that is wide all along and widely joins the humeropectoral sulcus in a V onto the posterior part of the entoplastron, with the intergular that is wider than the gulars, and those which reach and barely overlap the entoplastron; lateral to the intergular and from their junction, the humeropectoral sulcus is inclined backward and remains posterior to the epihyoplastral suture. This pattern of "*C.*" *pisdurensis* is close to that of *Taphrosphys* (though it differs by the absence of decoration), but the humeropectoral sulcus is completely posterior to the epihyoplastral suture in the latter. The pattern figured for "*C.*" *pisdurensis* seems somehow similar to that of the decorated *C. leithii* (as figured by Carter (1852), by sulci and without the sutures), but in the latter, the wider intergular is figured joining the humeropectoral sulcus only by a point and the gulars are smaller.

The anterior lobe of *Kurmademys* is mostly unknown, and only known is the lateral part of the humeropectoral sulcus, also primitively posterior to the epihyoplastral suture as in "*C.*" *pisdurensis* and as in most of bothremydids, except the European *Polysternon* and *Elochelys*.

Discussion on Upparhatti Shell Elements: In Upparhatti, three taxa are probably represented due to the three buttresses morphotypes and when we consider the proximal end of the costals, the size and proportions of the neurals and peripherals and the ornamentation. Each form was not represented by each category of pieces. Among the peripherals, besides the form with quadrangular peripherals, it is possible there are two narrower (more rectangular) morphotypes for one larger species and one smaller. The two narrower morphotypes have flatter and thinner plates. In the larger one, the peripheral 1 has long marginals 1 and 2, which equally overlap the plate in length. The peripheral 1 of the other narrower morphotype is not preserved. In both narrower forms, the posterior peripherals have a less protruding ventral lip at the medial boundary of the marginal scutes which are longer and closer to the skin boundary ridge, and the 8th peripheral has a shorter area for contact with the bridge. The morphotype of the stronger form has robust and more quadrangular anterior peripherals with a rounded border and with a short marginal overlap (unknown in *Kurmademys*), whereas the posterior peripherals are wide (as in *Kurmademys*). The insertion of the buttress on the 8th peripheral is long, and there are striations and ridges externally on the preserved peripherals and neurals. It is possible that the robust plastral elements with a thicker rounded border (Fig. 7.11C, G F and I) likely correspond to the taxon with the more robust peripherals.

The relative proportions of the marginal scutes overlap, the costals overlap by the buttresses and the shape of the scar buttresses do not differ in the two podocnemidid sexes and thus indicate that several taxa groups are present and there are no marked differences due to sexual dimorphism.

The costals 1 and 5 with a stronger rounded inguinal buttress insertion (more common in Podocnemidoids) may be of the robust morphotype. However, it is uncertain which morphotype the thin and small costal 5 represents. The thin (i.e. relatively short anteroposteriorly) axillary and inguinal buttress scars are both present in some members of *Bothremys* group and in *Kurmademys* (as figured in Gaffney et al. 2006). But thin inguinal buttress scars associated with rounded axillary buttress scars are also present in a South American pre-*Podocnemis* species (Carvalho et al. 2002: Figs. 1 and 2). On the other hand, both rounded axillary and inguinal buttresses, more or less long below the plate, are widely distributed in the Bothremydidae (Gaffney et al. 2006; Lapparent de Broin and Werner 1998: Figs. 7, 8, 10, 11 and pl. 5) as well as in members of the Podocnemididae, such as in *Erymnochelys* and in *Lapparentemys vilavilensis* (Broin 1991). The buttresses morphology is thus considered to be of specific value but in the context of possible homoplasies. However, its relevance to the phylogeny is not yet known. It is not possible to associate each buttress with each plastral element. The axillary buttress with rounded and short extremity of DUGF/T236 is similar to that of podocnemidids (including *Podocnemis* and *Erymnochelys*) as well as many bothremydids such as *Polysternon*, where it corresponds to a rounded axillary buttress scar below the costal 1 as in DUGF/T101 and DUGF/T129. Contrastingly, the thin (anteroposteriorly) mediolaterally wide inguinal buttresses of DUGF/T105 and DUGF/T99 are similar to those of *Kurmademys* (see Gaffney et al. 2006: fig. 257), *Bothremys* or *Chedighaii* (Schmidt, 1940) (in Gaffney et al. 2006, Fig. 264) and *Taphrosphys*; the morphotype represented by DUGF/T105 and DUGF/T99 differs from the thin but mediolaterally narrow morphotype (DUGF/T106).

In Upparhatti, there are three morphotypes of costals 5, two thin (one wide and one narrow) and one rounded, representing three species. In the presence of few costals 1 with axillary buttresses (two fragments with only one morphotype of rounded scar) but three costal morphotypes, the suitable association with other parts of carapace is not clear. In any case, one Upparhatti taxon shares with *Kurmademys* the morphotype of thin inguinal buttresses.

The robust medially angulated upraised costals and narrow vertebrals may be of a robust peripheral morphotype. As far as Upparhatti taxa are considered, they do not belong to *Kurmademys*, *Carteremys* or "*C.*" *pisdurensis*. Also, they do not appear to be the same as any other named form from elsewhere, so they may constitute new taxa. It is uncertain whether the medially flat and relatively thin costals with wider vertebrals are of the morphotype with thin peripherals or represent a third taxon. The two types of costals 7 cannot be associated with any another shell element.

The neurals also indicate the presence of more than one taxon. Probably the wider neurals with radiated striations like in podocnemidids are of the robust morphotype of peripherals with an ornamentation of ridges. The narrow neurals might represent the taxon with narrower peripherals and exclude the wide *Kurmademys* shell (with neurals narrower on their anterior and posterior borders in relation to their width at the posterior angle of the short side). At least no Upparhatti neural belongs to *Kurmademys* for this reason. No one is of the decorated *Carteremys leithii* group or of a member of the *Taphrosphys* group, even though the shape of some narrow and rather long neurals is similar to that of some neurals of *C. leithii*, because they differ in their ornamentation.

On the whole there are three undecorated taxa in the Upparhatti assemblage. In one of the taxa, the anterior border of the shell is not protruding or notched and is moderately elongated as in *Kurmademys*. This state of anterior border is different from that of *Taphrosphys* group and *Carteremys*, which have a medially elongated border as in another Upparhatti species, but differ from the latter in their decoration. The feature of the anterior extremity of the bridge and insertion of the axillary buttress posteriorly on the 3rd peripheral is seen in various Bothremydidae including *Kurmademys*. Finally, the ubiquitous *Taphrosphys* group (Taphrosphyini s.s.) and the Trans-Saharan *Nigeremys* group (Nigeremydini, with a very medial parasagittal axillary process issued from the second peripheral) are surely not represented. The character of the posterior extremity of the bridge and inguinal buttress at the 8th is like in podocnemidids and various bothremydids, and possibly *Kurmademys*, but the presence of a rounded axillary buttresses scar distinguishes it from *Kurmademys*. When known, pleurals 3 and 4 meet on the 9th peripheral as in *Kurmademys* and various other bothremydids and some podocnemidid species, but not on the 8th as in *Erymnochelys*. Vertebrals 4 and 5 meet on the costal 8 as in nearly all the podocnemidoids including *Kurmademys*, or on the costal 7 as in American bothremydini "*Bothremys* or *Chedighaii barberi*" (in Gaffney et al. 2006).

Perhaps except DUGF/T146, plastral elements have narrow scute dorsal borders, which is a primitive pleurodiran feature known in some podocnemidids and in pelomedusids. In bothremydids, the feature is known in *Taphrosphys* (s.l.) *ambiguus* from Mont-Aimé (France), a locality definitely dated as Maastrichtian (Broin 1977; Montenat and Merle 2018), in which the overlap is narrow. The anterior lobe of "*C.*" *pisdurensis* has narrow scute overlap as in DUGF/T151 and 155 but the overlap in posterior lobe is unknown; its hypoplastral border is slightly rounded as in the hypoplastron DUGF/T146. But the latter has a wide scute overlap that is not clear in "*C.*" *pisdurensis*. In *Polysternon provinciale* and *P. atlanticum*, the dorsal overlapping border is widened all along with the widening of the lobes and the lateral borders are rounded as in DUGF/T146. DUGF/T146 is, therefore, closer to *Polysternon* on these two points than to other above compared bothremydids. These two features are not described in other bothremydids but probably the taxa with narrow plastra and a straight border outline also had a narrow dorsal scute overlap as in the other Upparhatti specimens and *Taphrosphys* s.l. *ambiguus*. *Taphrosphys* and *Carteremys* are considered absent in Upparhatti due to the absence of marked polygonal decoration. The possibility of the presence of new taxa with one rather elongated form as in *Carteremys* but not similarly decorated is indicated. Another stronger form, wider as in *Kurmademys* or "*C.*" *pisdurensis*, is present. And possibly the smaller elements of narrower peripherals could belong to a third form.

Girdles and limbs (Fig. 7.12): The specimens of the limbs and girdles are mostly incomplete. Several sizes are represented, from small (from individuals with 20–25 cm long carapaces) to large (from individuals with 40–45 cm long carapaces) and more than one morphotype appear for each element. However, it is not possible to associate the different limb and girdle morphotypes with the carapace and cranial elements or to unite the elements in individual taxa. In interpreting the morphology of the limb and girdle elements, comparisons have been made with the appendicular skeleton of some extant taxa since few podocnemidoid and almost no bothremydid appendicular skeletons are figured.

Anterior girdle: The scapula is represented by twelve fragments of varying size, all preserving the base of the scapular and acromial processes and the glenoid facet (Fig. 7.12M, N and O). From comparison with an extant *Erymnochelys,* the smallest specimens (DUGF/T190 and

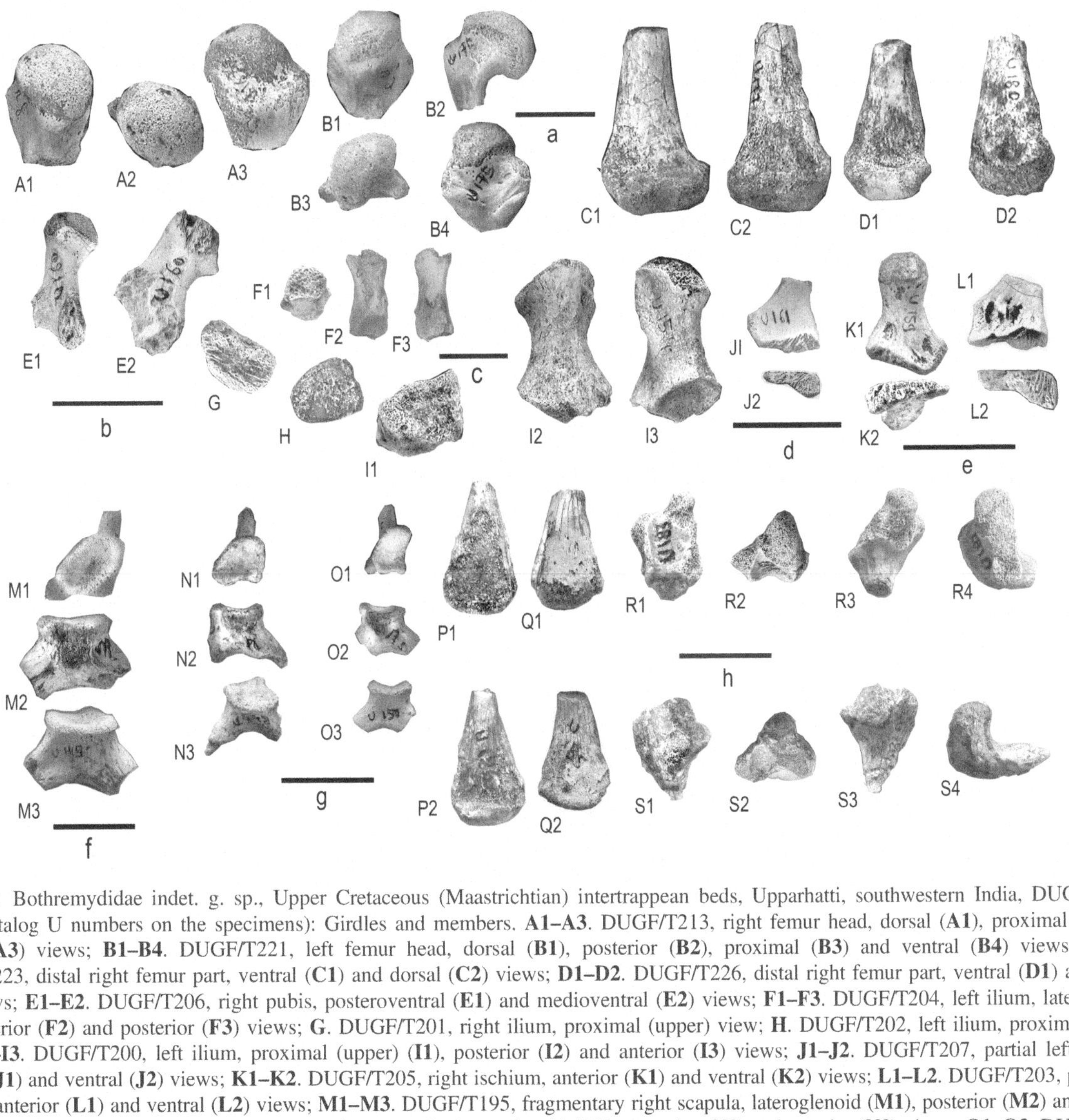

Fig. 7.12 Bothremydidae indet. g. sp., Upper Cretaceous (Maastrichtian) intertrappean beds, Upparhatti, southwestern India, DUGF/T coll. (Work catalog U numbers on the specimens): Girdles and members. **A1–A3**. DUGF/T213, right femur head, dorsal (**A1**), proximal (**A2**) and ventral (**A3**) views; **B1–B4**. DUGF/T221, left femur head, dorsal (**B1**), posterior (**B2**), proximal (**B3**) and ventral (**B4**) views; **C1–C2**. DUGF/T223, distal right femur part, ventral (**C1**) and dorsal (**C2**) views; **D1–D2**. DUGF/T226, distal right femur part, ventral (**D1**) and dorsal (**D2**) views; **E1–E2**. DUGF/T206, right pubis, posteroventral (**E1**) and medioventral (**E2**) views; **F1–F3**. DUGF/T204, left ilium, lateroglenoid (**F1**), anterior (**F2**) and posterior (**F3**) views; **G**. DUGF/T201, right ilium, proximal (upper) view; **H**. DUGF/T202, left ilium, proximal (upper) view; **I1–I3**. DUGF/T200, left ilium, proximal (upper) (**I1**), posterior (**I2**) and anterior (**I3**) views; **J1–J2**. DUGF/T207, partial left ischium, anterior (**J1**) and ventral (**J2**) views; **K1–K2**. DUGF/T205, right ischium, anterior (**K1**) and ventral (**K2**) views; **L1–L2**. DUGF/T203, partial left ischium, anterior (**L1**) and ventral (**L2**) views; **M1–M3**. DUGF/T195, fragmentary right scapula, lateroglenoid (**M1**), posterior (**M2**) and anterior (**M3**) views; **N1–N3**. DUGF/T194, fragmentary right scapula, lateroglenoid (**N1**), posterior (**N2**) and anterior (**N3**) views; **O1–O3**. DUGF/T196, fragmentary right scapula, lateroglenoid (**O1**), posterior (**O2**) and anterior (**O3**) views; **P1–P2**. DUGF/T232, distal right humerus part, dorsal (**P1**) and ventral (**P2**) views; **Q1–Q2**. DUGF/T231, distal left humerus part, dorsal (**Q1**) and ventral (**Q2**) views; **R1–R4**. DUGF/T229, partial left humerus head part, ventral (**R1**), proximal (**R2**), dorsal (**R3**) and medial (**R4**) views; **S1–S4**. DUGF/T230, partial left humerus head part, ventral (**S1**), proximal (**S2**), dorsal (**S3**) and medial (**S4**) views. Scale bars equals 2 cm, a for **A1–D2**, b for **E1–E2**, c for **F1–I3**, d for **J1–J2**, e for **K1–L2**, f for **M1–M3**, g for **N1–O3**, h for **P1–S4**

196), which are 0.8 and 1 cm high at the junction of the scapular and acromial processes, are from an individual with a carapace of about 20–25 cm long. The largest (DUGF/T 251), which is 1.8 cm high at the junction of the scapular and acromial processes, is from an individual with a carapace of about 43–45 cm long. The angle between the dorsal scapula and its ventral acromion is about 95° for DUGF/T194 (Fig. 7.12N3) and about 115° for DUGF/T195 and DUGF/T196 (Fig. 7.12M3 and O3), which are greater than in podocnemidids such as *E. madagascariensis* and *Podocnemis lewyana* and *Cambaremys* (80°) and *Peltocephalus* (90°), and greater than in the bothremydid scapula from Kisalpuri (to be described separately), which is also about 80°. The angle of DUGF/T194 (95°) is greater than that of "*C.*" *pisdurensis* in which it is about 90–92°. The angle of the two others is greater too. Two morphotypes of glenoid facet are also present in both small and large categories. In one, it is shorter and sub-triangular (Fig. 7.12M1 and N1) as in *E. madagascariensis*; and in the other, it is narrower and bean-shaped (Fig. 7.12O1) as in *Peltocephalus*, *P. lewyana* and *P. vogli* (carapaces 15–25 cm long). Here also three species are represented.

Humerus: Seven partial humeri are preserved, two eroded proximal parts (DUGF/T229 and DUGF/T230) (Fig. 7.12R, S) and four distal parts including DUGF/T232 and DUGF/T231 (Fig. 7.12P, Q). All are of approximately the same size and, if complete, would have been about 5 cm long. As with the smaller of the preserved scapulae, they correspond to animals of moderate size, and probably would be from individuals with a 20–25 cm long carapace. They are worn but would have had a rather rounded head (observable at its base) like in *Podocnemis*. One specimen, DUGF/T232 (Fig. 7.12P) consists of the distal part of a humerus that has a wide distal extremity and a particularly narrow shaft anteriorly. This corresponds to a slender proximal part of the second specimen, DUGF/T230 (Fig. 7.12S). The other specimens that preserved the distal portions of the humerus, such as DUGF/T231, have wider shafts relative to their distal extremity. These correspond to the wider proximal part of specimen DUGF/T229 (Fig. 7.12Q, R). The feature of a relatively wider distal humerus is also seen in *Taphrosphys sulcatus* (in Gaffney, 1975), *P. vogli* and *Peltocephalus*. Thus, it is likely that the differences between the two morphotypes are of specific significance, as is the case in terrestrial testudinids (Lapparent de Broin and Antunes 2000). Here, only two morphotypes are represented.

Pelvis: It is represented by 11 fragments of the ilium, two of pubis and five of ischia. All are more or less eroded. DUGF/T200 is the largest and best preserved ilium (Fig. 7.12I). From comparison with *E. madagascariensis*, this would have been from an individual of about 35.5 cm long. The bone is massive, even more so than that of *E. madagascariensis* and much more than that of *Peltocephalus*. The dorsal end of the ilium forms the surface of suture with costals 7 and 8. It is massive and roughly trapezoido-rounded in DUGF/T200 and DUGF/T202 (Fig. 7.12I and H). A smaller fragment, DUGF/T204 is more quadrangular (Fig. 7.12F) and DUGF/T247 is ovo-rectangular (Fig. 7.12G). The acetabular surface of DUGF/T200 is rounded and mediolaterally elongate. DUGF/T204, DUGF/T156, DUGF/T199 (in which the top is missing) and DUGF/T198 have the same acetabular morphology. Specimen DUGF/T197 and DUGF/T247 differ in having a rectangular acetabular surface that is much shorter mediolaterally. Thus, there are three possible ilial morphotypes differing by the shape of the dorsal ends but only two by the shape of the acetabular process (longer vs shorter acetabular surface). The rectangular dorsal surface is closer to that of extant Pelomedusidae while the more trapezoido-rounded end is similar to that of the erymnochelyine *Erymnochelys* and the bothremydid *Polysternon atlanticum* Lapparent de Broin and Murelaga, 1999 (pl. 4, fig. 2). The dorsal iliac suture does not correspond exactly to the scar on the costals because of the presence of cartilage between these surfaces. For example, in *Erymnochelys*, the scar on the costal is rather triangular and more notched than the top of the ilium. In cf. *Arenila krebsi*, the costal scar is rounded when the iliac suture is triangular (Lapparent de Broin and Werner 1998). Thus, we cannot confidentially deduce the shape of sutural surface on the ilium from the scar on the costals. However, the triangular costal iliac surface in *Taphrosphys*, *Kurmademys*, "*Bothremys* or *Chedighaii*" and other bothremydids as illustrated in Gaffney et al. (2006) differs from the Upparhatti trapezoido-rounded morphotype. The pubis is represented by two fragments, DUGF/T247, which is preserved in articulation with the ilium and DUGF/T206, which is nearly complete (Fig. 7.12E). Both are from smaller individuals. The pubis is relatively slender relative to its height compared to the pubis of *Erymnochelys*, with an equidimensional triangular acetabular facet. DUGF/T206 has an articular facet a little eroded but relatively short, rounded at mid length and anteriorly pointed. It might correspond to the scar on the xiphiplastral fragment DUGF/T154 (Fig. 7.11F). This is shorter and less rounded at its extremity than in the extant forms. The pubic scar of *Kurmademys* is small and oval but its figure is stylized (Gaffney et al. 200: Fig. 257). DUGF/T 206 is also not similar to the round shape of the small scar, in "*C.*" *pisdurensis* also stylized by Jain (1986).

Five ischium fragments are present, the most complete of which are DUGF/T207, 111 and 205 (Fig. 7.12J, L and K). These are of relatively small size, corresponding to the size of the pubis and the smaller ilia and probably belong to the same taxon. The preserved acetabular facet of DUGF/T205 is semicircular. The ventral suture for the xiphiplastron is a mediolaterally elongate triangular surface, its anterior border transversely oriented, and the lateral part of the triangle is not posteriorly elongated. The posterior process of the ischium does not extend to the xiphiplastral point, which agrees with the ischiatic scar on the preserved xiphiplastron (DUGF/T155) (Fig. 7.11I).

The morphology of both ischiatic-pubic scars and xiphiplastral sutures are diagnostic and not intraspecifically variable in podocnemidoids, although homoplastic. Here, the pubic and ischiatic sutures of the separated bones correspond to the pubic and ischiatic scars seen (respectively) in the two xiphiplastral fragments (Fig. 7.11F and I), although of larger individuals. They differ from *Taphrosphys* s.l., which has a pubic suture that is narrow and elongated and an ischiatic suture that is in the form of a rounded spot. They also differ from "*Bothremys* or *Chedighaii barberi*", which have a long and narrow pubic scar (anteriorly pointed) and an ischiatic scar in the shape of a right triangle with an incurved hypotenuse extending posterolaterally to the anal notch in the xiphiplastral point (Gaffney et al. 2006). But, as in Upparhatti specimens, the ischiatic scar is not arched and is anterior to the anal notch in "*Bothremys*" *barberi* from New Jersey (in Gaffney and Zangerl 1968: fig. 8). By the elongation of the ischiatic lateral part for prolongation in the xiphiplastral point, and its arched shape, "*Bothremys* or *Chedighaii barberi*" also differs from *Erymnochelys, Shweboemys* subgroup as represented by *Cordichelys antiqua* and *Shweboemys* s.l. sp. from As Sarrar, Saudi Arabia and from *Podocnemis*. This ischiatic prolongation in the xiphiplastral point of "*Bothremys* or *Chedighaii barberi*" is a primitive feature and occurs in primitive chelids (Lapparent de Broin and De la Fuente 2001) and in the dortokid *Ronella* Lapparent de Broin, in Gheerbrant et al. 1999 (see Lapparent de Broin et al. 2004). It is rarely present in some advanced South American podocnemidines because of a particularly large anal notch, as well as in one species of the erymnochelyine *Eocenochelus* Pérez-García et al., 2017b (possibly in some males). In the Saudi Arabian bothremydid, the pubic (oval) and ischiatic scars (not reaching the anal notch transverse line) are small, which recalls those scars identified for bothremydids such as *Kurmademys* and "*C.*" *pisdurensis* in India and those of *Polysternon atlanticum*. In *P. provinciale*, the scars are larger, the pubic scar is moderately oval (not anteriorly narrow as that of DUGF/T154) and the ischiatic scar is large as that of DUGF/T155, but just reaching the transverse line of the anal notch instead of being slightly anterior to this line.

The morphology of the acetabular facet of Upparhatti specimens is similar to that of extant Podocnemididae but differs in the triangular acetabular facet. The morphology of the shaft of the ilium also shows specific differences such as the more robust shafts in small and large ilia (DUGF/T108), pubis (DUGF/T206) and ischium (DUGF/T205), in comparison with *Erymnochelys* and more so in comparison with *Peltocephalus*.

From these comparisons, different podocnemidoid genera may or may not have an identical pubic or ischiatic scar, and *Taphrosphys* only has an autapomorphic ischiatic rounded scar. However, the Upparhatti preserved specimens are not of *Taphrosphys* or *Bothremys* spp., and they are also different from *Kurmademys* and "*C.*" *pisdurensis*.

Femur: Eighteen femora are preserved (Fig. 7.12A to D. Nine of these have the head and trochanters, two are preserved with the base of trochanters and seven preserved only the distal portion of the element. As for humerus, there are two morphotypes. One, for example DUGF/T223 (Fig. 7.12C) is distally relatively wider than the other exemplified by DUGF/T226 (Fig. 7.12D). Differences are also present in the shape of the head. In one, it is wider and rounded as in DUGF/T213 (Fig. 7.12A) and in the other, it is narrower and more elongated, as exemplified by DUGF/T221 (Fig. 7.12B). We hypothesize that the distally wider morphotype of femur, like that of DUGF/T223, is from the same taxon as the distally wider humerus (DUGF/T 232), while the narrower femur DUGF/T226 is from the same taxon as the narrower humerus (DUGF/T229). No particular feature indicates a bothremydid or a podocnemidid affinity.

Discussion

The turtle fauna of Upparhatti is represented by a mixed assemblage consisting of at least three taxa of Podocnemidoidea. Many characters present in the elements from Upparhatti are podocnemidoid, known both in podocnemidids and bothremydids. Many features of the carapace are consistent with the presence of a podocnemidid but the presence of a member of this group cannot be positively demonstrated as the characters are general podocnemidoid plesiomorphies. However, some elements have more typical features, found only in bothremydids, such as the basisphenoid, the lower jaws, some cervical vertebrae, the axillary buttresses insertion posteriorly on the 3rd peripheral and the presence of thin inguinal buttresses. Most plates are comparable with known bothremydids. Thus, the Upparhatti shell material can be included in the Bothremydidae.

The presence of two taxa in the Upparhatti assemblage is demonstrated clearly by variation in the morphology of the cervical vertebrae. One of these, designated "form a" has long cervical centra, and the other, designated "form b," has short cervical centra. As the cervical vertebrae are unknown in the described Indian taxa, it is not possible to compare the Upparhatti "form a" and "form b" with already described taxa. However, "form a" can be attributed to Bothremydidae and might also be an indication to the presence of an elongated form morphologically similar to *Carteremys* or another member of the *Taphrosphys* group, since the members of this group are considered as all having long cervical vertebrae. Even if the presence of a shell elongated anteriorly and with narrower posterior peripherals is proved, as in *Carteremys* or *Taphrosphys*, none of the Upparhatti elements have their marked shell ornamentation. So, it is unlikely that any of these taxa is present in the Upparhatti material. Although the long-necked "form a" is taxonomically distinct, it cannot be positively identified. Another possibility is that there are two species within this long-necked vertebrae form.

"Form b" has similarities with the *Erymnochelys* group as well as with bothremydids and these similarities seems to represent shared primitive podocnemidoid features. It is also provisionally identified as Bothremydidae indet.

Shell, girdle and limb elements occur in two to three distinct morphotypes, agreeing with the vertebral evidence for the presence of two taxa at least in the assemblage. However, none of these elements can be included in established taxa and the possibility that some of the postcranial elements might be those of *Sankuchemys* (the shell of which is unknown) for example, because of its presence with *Carteremys* in Mumbai (Bombay) or a kurmademydine cannot be excluded. The only skull elements in the Upparhatti assemblage are one basisphenoid, and three pieces of a single type of dentary. Features of these elements show that they cannot be of a bothremydid or a podocnemidid known outside India. Dentaries are interpreted as being from a bothremydid, probably similar to the Madagascan *Sokatra* in having a short skull, and a rounded outline, but they differ by a narrow triturating surface. In the latter feature they differ also from *Sankuchemys*, *Carteremys* and *Kurmademys* and Madagascan *Kinkonychelys*. However, it is unknown which postcranial elements are associated with the Upparhatti basisphenoid and jaw elements. As the postcranials of the two Madagascan genera and *Sankuchemys* are unknown, no comparison is possible with them. Despite the problems in identifying the Upparhatti material because of the poor understanding of the postcranial skeleton of many bothremydids, the basiphenoid and probably also the dentaries show that at least one of the Upparhatti bothremydids is definitely distinct from the geographically and stratigraphically closest genera *Sokatra*, *Sankuchemys*, *Kinkonychelys* and *Kurmademys*. Shells and limb girdles show the possibility of three taxa, and approximations, notably, with *Kurmademys* and with "*Carteremys*" *pisdurensis*. However, none of the shell elements can be identified as those of *Kurmademys* or "*Carteremys*" *pisdurensis*, although the possibility of being close to them exists for some of them, provided that they had the same cervical morphotype. Thus, no Upparhatti postcranial can be attributed to any of the established genera of the Indian tribe of Kurmademydini, although knowledge of their postcranial and even the cranial bones remains unknown.

Conclusion

Taxonomic Implications

It is evident from the above discussion that Bothremydidae are a major (or single) component of Indian Late Maastrichtian pleurodire assemblages. There are four defined Indian taxa at specific level but nine to ten are present. At least eight Indian species of Bothremydidae and probably nine are present in the Late Maastrichtian and one in the Early Albian to Middle Turonian strata. Bothremydids are the only group of turtles known to be present in the intertrappean beds (Table 7.1). Three species are included in the *Carteremys* group: *Carteremys leithii* from Worli Hills, Bombay, cf. *Carteremys* aff. *leithii* from Lakshmipur in western India, and cf. *Carteremys* sp. from Kisalpuri in central India (to be described). In addition, *Sankuchemys sethnai* (proposed as Kurmademydini by Gaffney et al. 2009) is present in the Amboli Quarry in Bombay (Mumbai). At least two taxa, and probably three, are present in Upparhatti in southwestern India, by a long-necked form and one or two short-necked forms. Although the taxa from Upparhatti cannot be diagnosed at generic and species level, they are taxonomically distinct from the named taxa. *"Carteremys" pisdurensis* was described from the Upper Cretaceous (Late Maastrichtian) infratrappean beds. In southern India, the Kurmademydine bothremydid *Kurmademys kallamedensis* comes from the Upper Cretaceous Kallamedu Formation, which is also Maastrichtian in age although its age relative to the intertrappean and the infratrappean beds is uncertain. The presence of the Bothremydidae in India earlier in the Late Cretaceous is shown by the Early Albian – Middle Turonian bothremydid from Kunnam. It is too poorly preserved for comparison with the Maastrichtian taxa, but it indicates the ancient presence of the family in India at a time India was not separated from Madagascar and the faunas were less differentiated. Thus, a minimum of nine to ten bothremydid taxa are present in the Cretaceous of India. Although a Paleocene record of pelomedusoids from India has not been recorded so far, it is assumed that the Bothremydidae continue into the Paleocene of the Indian subcontinent as is evident from the presence of "*Podocnemis*" *indica* Lydekker, 1887 in the Paleocene of Salt Range in Pakistan (see Appendix).

"Carteremys" pisdurensis is here identified as a Maastrichtian representative of Bothremydidae. As in nearly all members of the Bothremydidae, a part of its skeleton is

unknown. Despite the poor illustration of the skull, its palatal morphology shows that it is not a podocnemidid "*Shweboemys*" s.l. (*Shweboemys* subgroup) member (contrary to Jain, 1986). As its shell conforms to Bothremydidae, we revert it back to *"Carteremys" pisdurensis* as in Jain (1977), not because it belongs to this genus, but at least it is also a bothremydid, while *Shweboemys* is a podocnemidid. The genus has to be erected keeping in view the fossil material at hand. It is present only in the infratrappean beds (Lameta Formation) of central India, and it is the only identifiable taxon from these beds. Thus, based on these turtle remains, it is concluded that the faunas of the infratrappean and intertrappean beds generically differ, but the whole sample indicates a rich familial diversity and particularly in Upparhatti, with several genera, which is not rare in the basins with Late Cretaceous-Paleogene fossil turtles, such as those of USA (Alabama, New Jersey), Moroccan phosphates and Egypt (Ammonite Hills).

Paleobiogeographical Implications

Since many of the turtles from India are represented primarily by postcranial elements which are not adequately studied in other parts of the world, it is difficult to establish taxonomic relationships on the basis of such elements. However, in the present form, the Indian specimens do not exhibit any close relationship with other geographic regions as they appear to be endemic at the generic level when precisely identified. Even in our present state of understanding they do not exhibit any characteristic feature of the taxa known from other parts of the world, only sharing plesiomorphic or homoplastic characters. Ecological arguments work equally well against or in favour of any biogeographic pattern since many Bothremydidae and some podocnemidids Erymnochelyines are littoral, and thus, separation of the continents does not present a complete barrier to faunal interchange. However, such a dispersal occurs by sea ways along the coasts or between short sea arms if there are no contrary streams, because these turtles still have a freshwater limb pattern and lack typical marine paddles that allow crossing of large and deep sea arms. Some bothremydid forms, like most of podocnemidids, are only freshwater forms that need a terrestrial land connection. But some taxa (*Erymnochelys* relatives) are found either in littoral or in freshwater sites and some others can be found together all along the river (such as *Rosasia* of the Beira littoral area in Viso and Taveiro site) and in its brackish to salted estuary (in Aveiro site) so that they can be considered as euryhaline forms. The Bothremydidae appeared in a continental northern Gondwanan Africa-South America landmass that was still linked or close to the Indo-Madagascar landmass, at the very beginning of the Cretaceous, when Pelomedusoides diversified. During the Early Cretaceous, outside the frontal mid-part of Africa where they abound, southeast to Niger and the closest to Madagascar Pelomedusoides are known only in Nyasaland (Malawi) by *Platycheloides nyasae* Haughton (1928). The latter is a primitive form in relation to bothremydid diversification. But Bothremydidae are present in the Cenomanian of Africa as well as Madagascar (Lapparent de Broin and Werner 1998: tab. 4). Further, at that time bothremydids were already present in India in the Early Albian – Middle Turonian Karai Formation of Kunnam (Ayyasami and Das 1990). The bothremydid group spread both via continental routes and by littoral coastal routes. As both Africa and Madagascar are well separated from India in the Late Cretaceous, bothremydid forms may have moved between these areas either before (continental forms) or from the Cenomanian-Turonian transgression onwards (littoral forms). All the decorated material of the Table 7.1 might share a common African ancestor; but the strongly decorated littoral *Taphrosphys* s. l. (present in Africa and not in India) is not similar to the Indian littoral decorated *Carteremys* in the jaw pattern and ovoid carapace shape, although the plastron is close by the gular / intergular pattern. However, the jaw pattern of *Carteremys* is also present in the unornamented freshwater "*C.*" *pisdurensis* which is not a member of the *Carteremys* group. As far as freshwater forms are concerned, no primitive Pelomedusoides, such as *Platycheloides* and the forms from the early-'middle' Cretaceous Sahelian area including Saharan Continental Intercalaire localities (Nigerian, Sudanese, Kem Kem, in Akhamil and other Aptian-Albian to Early Cenomanian localities) (Broin et al. 1974; Gmira, 1995; Lapparent de Broin 2000a; Zouhri 2017) are present in India and we do not know to which taxon Kunnam bothremydid species belongs. The presence of a podocnemidid Erymnochelyinae in the Maastrichtian of Madagascar (Gaffney et al. 2003) as in the Senonian of Ibeceten in Africa and the absence of any recognized podocnemidid in India (and particularly erymnochelyines) do not favour any privileged relation between India and Madagascar during the Maastrichtian and rather indicates an older time of connection for turtle dispersal between these landmasses. As indicated by their environmental preferences, the turtles indicated the presence of continental forms in India before its separation from Africa and Madagascar.

Faunal connections between Madagascar and India have been hypothesised on the basis of the relationship between the Mastrichtian *Kinkonychelys* from Madagascar and *Kurmademys* of the Kallamedu Formation, India. Further, Maastrichtian *Sankuchemys* of the intertrappean beds of Bombay was also considered as closely related to *Kinkonychelys* and *Kurmademys*, being positioned in the same Kurmademydini tribe (Gaffney et al. 2009). However, the skull morphology of *Kinkonychelys* and *Kurmademys* is relatively primitive in nature and is not similar to that of the poorly preserved skull of *Sankuchemys*. The skull of *Sankuchemys* is much crushed and show few diagnostic characters, but in any case, it is different by its crested and not posteriorly widened maxillae from those of *Kinkonychelys* and *Kurmademys*. *Sankuchemys* does not show the deep oval fossa pterygoidea and the posteriorly widened but uncrested triturating maxillary surfaces present in the skull of *Kinkonychelys* and *Kurmademys*. These two last features could indicate a close relationship between *Kinkonychelys* and *Kurmademys*. But these characters are already present for example in some African 'middle' Cretaceous forms from Kem Kem beds (late Albian-early Cenomanian) and persist as well in some Senonian and Maastrichtian forms of Europe such as *Polysternon* and related forms (Gaffney et al. 2006). Both *Kinkonychelys* and *Kurmademys* are known from deposits dated as Maastrichtian, by which time India and Madagascar were well separated from each other and the Indian subcontinent was at the peak of its isolation. The similarity between the two taxa based on earlier acquired features cannot be used as an evidence for a particular recent relationship. It appears they do not share any autapomorphic characters. They only belong to the same family, sharing characters appearing as possibly acquired before the Maastrichtian and known in the not yet broken Gondwanan taxa. All the more since bothremydids were present in India in the Early Albian – Middle Turonian Karai Formation of Kunnam, the Maastrichtian Indian forms could have evolved from this lineage.

At least, as far as taxonomic relationships of turtles within the framework of Indian endemism are concerned, there is a possibility of similarity for a part of terrestrial Upparhatti specimens with Pisdura and Kallamedu material on one hand, and between the eventually littoral Western and Central Indian material on the other hand.

Based on phylogenetic studies of terrestrial vertebrates, it has been suggested that many Late Cretaceous taxa of South America, India and Madagascar may represent relictual lineages that evolved vicariantly from their ancestral stocks somewhere in Gondwana during the Early and "middle" Cretaceous time (Ali and Krause 2011; Turner and Sertich 2010; Zaher et al. 2011). This hypothesis seems to hold good for the present turtle knowledge as far as Maastrichtian relationships between Africa, Madagascar and India are concerned, as it agrees with the Cretaceous Gondwanan crocodile dispersal (Martin and Lapparent de Broin 2016).

However, recent vertebrate fossil evidences from the Late Cretaceous of India and Madagascar have clearly pointed to a close biogeographic connection between these two land masses in the Maastrichtian. For example, the close phylogenetic relationship between *Indophis fanambinana* Pritchard et al., 2014 from the Late Cretaceous of Madagascar and *I. sahnii* Rage and Prasad, 1992 from India provides evidence for a land connection between India and Madagascar in the Late Cretaceous. *Madtsoia madagascarensis* from Madagascar and *M. pisdurensis* of India also points to the existence of a similar connection between India and Madagascar that favoured faunal interchanges between the two landmasses. The presence of same squamate genera in the vertebrate assemblages of Madagascar and India may be as well attributed to more recent segregation of these landmasses or to a more ancient (pre-Maastrichtian) faunal differentiation dating back to the time of connection with Africa, as indicated by the possible presence of *Madtsoia* aff. *madagascariensis* in Africa as early as in the Senonian of Ibeceten (Niger) (Broin et al. 1974; Rage 1981). The first of these views seems to be supported by recent phylogenetic analysis of abelisaurid dinosaur *Arcovenator* from the Upper Cretaceous (Late Campanian) of southern France (Aix-en-Provence), which showed that *Arcovenator* is a sister taxon to *Majungasaurus* of Madagascar and *Rajasaurus, Rahiolisaurus* and *Indosaurus* of India within the subfamily Majungasaurinae (Tortosa et al. 2014). It was envisaged that these abelisaurid genera could disperse from India/Madagascar into Europe via Africa across the narrow Mediterranean Tethys in the Late Cretaceous. The adapisoriculid mammals with their presence in the Late Cretaceous of India, and Paleocene of Africa (Morocco) and Europe also point to the existence of terrestrial connection between India, Madagascar, Africa and Europe close to the K/Pg boundary (Prasad et al. 2010; Smith et al. 2010; Goswami et al. 2011).

These hypotheses do not fit with the turtle dispersal and distribution between Africa, Europe, India and Madagascar during the Maastrichtian, even in the current imperfect state of knowledge. However, further resolution of the global biogeography of podocnemidoid turtles in general, and the relationships of Indian and African turtles in particular, will require a better understanding of the postcranial skeleton in fossil taxa. Hyperfamily Pelomedusoides phylogenies have principally been established based on skull anatomy. Some recent works, either for Bothremydidae or for Podonemididae, introduced postcranial characters, but missing some important complexes of characters, such as the saddle-shaped central joints which has to be linked with the reduction of the posterior central ventral tubercles, the intergular /gular / humeropectoral sulcus position pattern of *Erymnochelys* group, or the "secondary palate" morphology (presented above, and see Zourhi 2017) (Broin 1988; Broin in Antunes and Broin 1988; Lapparent de Broin and Werner 1998; Lapparent de Broin 2000b; França and Langer 2006; Gaffney et al. 2006, 2009; Cadena et al. 2010; Gaffney et al. 2011; Cadena et al. 2012; Romano et al. 2014; Menegazzo et al. 2015). As the postcranial characters are generally not or less commonly included in data matrices of the cladistic analyses, the comparisons between taxa are not complete. In the present paper, it is shown that how many more characters are present in the postcranial skeleton that might be utilized in taxonomic studies. However, further comparisons are necessary to establish the value of these characters. For this, comparative studies of the postcranial characters that are observable in extant species might help in understanding the variability within a genus and a family and taking into account the locomotion mode and feeding pattern. With this, the relationships between North-African, Madagascar and Indian forms might then be better understood. However, even with the poor material of Upparhatti, besides the lack of attestation of any podocnemidid in India, the relative isolation of continental India seems evident at least for the freshwater forms. Also evident is the absence of littoral *Bothremys* group spp. and *Nigeremys* group spp. The former is known widely from other parts of the world (Africa-Palestine, Europe, North America) but, apparently, it did not reach the Indian subcontinent. The latter, confined to the borders of the Trans-Saharan seaway and African northeastern platform, was also far from Madagascar and India. The possibility of close relationships between the Indian *Carteremys* and the ubiquitous *Taphrosphys* s.l. is not verified either, although they share some shell characters (shell form partially, plastral scute features, ornamentation) but with contradictory skull characters (maxillae shape). Some plastral scute features are shared by the Indian "*C.*" *pisdurensis* and *Taphrosphys* s.l., but they have different skull characters and ornamentation. If a connection between western India, Madagascar and Tethys margins was possible for turtles during the Late Cretaceous, one can ask why genus *Taphrosphys* s.l. is not present in India, on account of its wide dispersal potential along the coasts and its widespread presence in North and South Americas (Gaffney et al. 2006), Europe up to Paris basin (Broin 1977; Lapparent de Broin 2001) and Africa up to Syria (Lapparent de Broin 2000a; Bardet et al. 2000) during the Maastrichtian and Paleogene. If a connection was present between India and Madagascar in the Maastrichtian, one or several bothremydid Indian genera might have been present in Madagascar, and the Madagascan genera *Kinkonychelys* and *Sokatra* might have been present in India. Overall it is important to consider Nigerian or Madagascan *Erymnochelys* relatives, a group known by littoral members, which are not found in India. Perhaps the fossil turtle record is still too poor in India.

Acknowledgements We thank for their hospitality, access to their collections and logistic support: Sunil Bajpai (Indian Institute of Technology, Roorkee, India); Mbarek Amaghzaz and Baâdi Bouya at OCP, Khouribga; the Collection Manager of Paleobiology at CMN, Ottawa; Dale A. Russell, MEAS of NCSU, Charlotte USA; Sandra Chapman, NHM, London; Guy Costa, Montpellier; Dominique Teodori, Toulouse; Dominique Baudet, Paleontology and Wim Van Neer, Zoology, Tervuren MRAC; Ronald Boettchern and Rupert Wild, SMNS, Stuttgart; Robert Jullien at MHNM; Monique Vianey-Liaud and Suzanne Jiquel; at UM II. At MNHN, Paris: ZA AC, Salvador Bailon, Anatomie Comparée, Laure Pierre-Huyet and Roger Bour, Zoologie des Reptiles et Amphibiens; F, Nour Eddine Jalil, Paléontologie des vertébrés). At Museo-Universidad de La Rioja, UNLaR, Argentina, the Direction of collections. At MZUSP, São Paulo, Paulo Emílio Vanzolini. We thank the reviewers Ren Hirayama, Waseda University (Tokyo), Adán Pérez-García, Facultad de Ciencias, UNED (Madrid) and Gustavo Ribeiro de Oliveira, Museu Nacional / UFRJ (Rio de Janeiro) for their useful comments on the manuscript. GVRP acknowledges the financial support from the Science and Engineering Research Board, New Delhi through J. C. National Fellowship for field work in southwestern and central India, and the MNHN "Professeur Invité" grant under which this collaborative work was undertaken.

Appendix: Other Horizons of the Indian Subcontinent (Outside India) from Which a Pleurodiran was Documented

"Podocnemis" indica Lydekker, 1887: cf. Bothremydidae genus indet., Paleocene beds of Nila, Salt Range, Pakistan. In his review, Wood (1970) considered it as a probable "pelomedusid" (i.e. Pelomedusoides) turtle but not related to *Podocnemis* (s.l.), whereas Broin (1987) hypothesized this taxon as a probable member of the podocnemidid *Shweboemys* group, to which *Carteremys* Williams, 1953, with both *Carteremys leithii* and *Carteremys pisdurensis* Jain, 1977 (then considered as *Shweboemys pisdurensis* [Jain, 1977] in Jain [1986]) was previously referred. Here, these two species are defined as Bothremydidae. *Carteremys leithii* is a form with an anteriorly rounded and elongated shell (and consequently probably an elongated neck) having strong decoration. "*Carteremys*" *pisdurensis* is also a bothremydid of an undetermined genus. The anterior shell border is missing in Jain (1986), the carapace having or not having a rounded protrusion but being rather quadrangular and not decorated. For now, its bothremydid group is not defined. *"Podocnemis" indica* is a bothremydid by the rounded anterior shell with protruding border, with a long nuchal and a long medial part of the pleurals 1, unknown in podocnemidids and present in bothremydids such as *Carteremys leithii*. This is in contrast with the quadrangular carapace of *Kurmademys* and most of the specimens of Upparhatti. It possibly presents at least an undifferentiated basic ornamentation as in *Kurmademys* and Upparhatti specimens. But no ornamented decoration is visible on the figure (a distinction from *Carteremys*), nor any mention of decoration is made in the text by Lydekker (1887). It has a tectiform shape and moderately wide vertebrals, that are slightly widened at mid length. Neurals are moderately longer than wide and slightly narrowed at their junction (as frequently in Bothremydidae) as in neurals from Upparhatti, but in contrast to *Carteremys leithii*, "*Carteremys*" *pisdurensis* and *Kurmademys*.

Pleurodira were not found in the early-middle Eocene Kuldana Formation of Pakistan where only Laurasiatic cryptodiran taxa are represented (Broin, 1987).

An Indian fossil species has been recently recognized as being a Pelomedusoides podocnemidid by Ferreira et al. (2018): *Piramys auffenbergi* Prasad, 1974, from the late Miocene of Piram Island, Gujarat. It represents a Stereogenyina Gaffney et al., 2011 partim, subgroup *Shweboemys* nov. in Zouhri et al. (2017) (and see Pérez-García et al. 2017b).

References

Ali, J. R., & Krause, D. W. (2011). Late Cretaceous bioconnections between Indo-Madagascar and Antarctica: Refutation of the Gunnerus Ridge causeway hypothesis. *Journal of Biogeography, 38*, 1855–1872.

Anantharaman, S., Wilson, G. P., Das Sarma, D. C., & Clemens, W. A. (2006). A possible Late Cretaceous "haramiyidan" from India. *Journal of Vertebrate Paleontology, 26*, 488–490.

Andrews, C. W. (1903). On some pleurodiran chelonians from the Eocene of the Fayum, Egypt. *Annals and Magazine of Natural History, S., 7*(11), 115–122.

Andrews, C. W. (1906). *A descriptive Catalog of the Tertiary Vertebrata of the Fayum, Egypt* (pp. I–XXXVII, l–324). London: Trustees of the British Museum (Natural History).

Antunes, M. T., & de Broin, F. (1988). Le Crétacé terminal de Beira Litoral, Portugal: Remarques stratigraphiques et écologiques; étude complémentaire de *Rosasia soutoi* (Chelonii, Bothremydidae). *Ciências da Terra (Universidade Nova de Lisboa), 9*, 153–200.

Arratia, G., Lopez-Arbarello, A., Prasad, G. V. R., Parmar, V., & Kriwet, J. (2004). Late Cretaceous-Paleocene percomorphs (Teleostei) from India—Early radiation of Perciformes. In G. Arratia, M. V. H. Wilson, & R. Cloutier (Eds.), *Recent advances in the origin and early radiation of vertebrates* (pp. 635–663). München: Verlag Dr. Friedrich Pfeil.

Ayyasami, K., & Das, I. (1990). Unusual preservation of a Cretaceous turtle fossil. *Journal of the Geological Society of India, 36*, 519–522.

Bajpai, S., Sahni, A., Jolly, A., & Srinivasan, S. (1990). Kachchh intertrappean biotas: Affinities and correlation. In A. Sahni (Ed.), *Cretaceous event stratigraphy and the correlation of Indian nonmarine strata* (pp. 101–105). Chandigarh: Contributions from the seminar cum workshop International Geological Correlation Programme, 216 and 245.

Bardet, N., Cappetta, H., Pereda-Suberbiola, X., Mouty, M., Al Maleh, A. K., Ahmad, A. M., et al. (2000). The marine vertebrate faunas from the Late Cretaceous of Syria. *Geological Magazine, 137*(3), 269–290.

Bardet, N., Jalil, N. E., Lapparent de Broin, F., Germain, D., Lambert, O., & Amaghzaz, M. (2013). A giant chelonioid turtle from the Late Cretaceous of Morocco with a suction feeding apparatus unique among tetrapods. *PLoS One, 8*(7), e63586. https://doi.org/10.1371/journal.pone.0063586. Supporting Information_S1, Video_S1.

Bardet, N. E., Gheerbrant, E., Noubhani, A., Cappetta, H., Jouve, S., Bourdon, E. et al. (2017). Les Vertébrés des Phosphates crétacés-paléogènes (70.6 – 46.6 Ma) du Maroc. In S. Zouhri (Ed.), La Paléontologie des Vertébrés du Maroc. *Mémoires de la Société géologique de France* (Vol. 180, pp. 355–454).

Baur, G. (1891). Note on some little known American fossil tortoises. *Proceedings of the Academia of Natural Sciences of Philadelphia*, 411–430.

Bergounioux, F. M., & Crouzel, F. (1968). Deux tortues fossiles d'Afrique. *Bulletin de la Société d'Histoire naturelle de Toulouse, 104*(2), 1–2.

Blanford, H. F. (1862). On the Cretaceous and other rocks of the South Arcot and Trichinopoly district, Madras. *Memoirs of the Geological Survey of India, 4*, 1–217.

Boyer, D. M., Prasad, G. V. R., Goswami, A., Krause, D. W., Godinot, M., & Verma, O. (2010). New postcrania of *Deccanolestes* from the Late Cretaceous of India and their bearing on the evolutionary and biogeographic history of euarchontan mammals. *Naturwissenschaften, 97*(4), 365–377.

Bragina, L. G., & Bragin, N. Y. (2013). New data on the Albian—Cenomanian radiolarians from the Karai Formation (South India). *Stratigraphy and Geological Correlation, 21*(5), 515–530.

Broin, F. de (1971). Une espèce nouvelle de Tortue pleurodire (? *Roxochelys vilavilensis* n. sp.) dans le Crétacé supérieur de Bolivie. *Bulletin de la Société géologique de France, S. 7, 13*(3–4), 445–452.

Broin, F. de (1977). Contribution à l'étude des Chéloniens. Chéloniens continentaux du Crétacé et du Tertiaire de France. *Mémoires du Muséum national d'Histoire naturelle, Paris, C, 38*, I-IX, 1–366.

Broin, F. de (1980). Les Tortues de Gadoufaoua (Aptien du Niger): aperçu sur la paléogéographie des Pelomedusidae (Pleurodira). *Mémoires de la Société géologique de France, N.S., 139*, 39–46.

Broin, F. de (1983). Chéloniens, *6*, 211–233. In L. Blanc-Vernet, F. de Broin, E. Buffetaut, H. Camps-Fabrer, P. Carbonel, et al. (Eds.), *Sahara ou Sahel? Quaternaire récent du Bassin de Taoudenni (Mali)* (p. 470). Marseille: N. Petit Maître and J. Riser.

Broin, F. de (1987). Lower vertebrates from the Early-Middle Eocene Kuldana Formation of Kohat (Pakistan). Chelonia. *Contributions from the Museum of Paleontology, University of Michigan, 27(7)*, 169–185.

Broin, F. de (1988). Les Tortues et le Gondwana. Examen des rapports entre le fractionnement du Gondwana et la dispersion géographique des Tortues pleurodires à partir du Crétacé. *Studia Geologica Salmanticensia, Studia Paleocheloniologica, 2*(5), 103–142.

Broin, F. de (1991). Fossil turtles from Bolivia. *Revista Técnica de Yacimientos Petrolíferos Fiscales Bolivianos, 12*(3–4), 509–527.

Broin, F. de, Buffetaut, E., Koeniguer, J.-C., Rage, J.-C., Russell, D., Taquet, P. et al. (1974). La faune de vertébrés continentaux du gisement d'In Beceten (Sénonien du Niger). *Comptes rendus de l'Académie des Sciences, Paris, 279*, 469–472.

Broin, F. de, & De la Fuente, M.S. (1993). Les tortues fossiles d'Argentine. *Annales de Paléontologie, 79*(3), 169–231.

Cadena, E. A. (2011). Potential earliest record of podocnemidoid turtles from the Early Cretaceous (Valanginian) of Colombia. *Journal of Paleontology, 85*(5), 877–881.

Cadena, E. A., Bloch, J. I., & Jaramillo, C. A. (2010). New podocnemidid turtle (Testudines: Pleurodira) from the Middle-Upper Paleocene of South America. *Journal of Vertebrate Paleontology, 30*(2), 367–382.

Cadena, E. A., & Gaffney, E. S. (2005). *Notoemys zapatocaensis*, a New Side-Necked Turtle (Pleurodira: Platychelyidae) from the Early Cretaceous of Colombia. *American Novitates, 3470*, 19 pp.

Cadena, E. A., Ksepka, D. T., Jaramillo, C. A., & Bloch, J. I. (2012). New pelomedusoid turtles from the late Paleocene Cerrejón Formation of Colombia and their implications for phylogeny and body size evolution. *Journal of Systematic Palaeontology, 10*, 313–331.

Carrington da Costa, J. (1940). Um novo Quelónio fossil. *Comunicaçãos dos Serviços Geológicos de Portugal, 21*, 107–125.

Carter, H. J. (1852). Geology of the island of Bombay. *Journal of the Bombay Branch of the Royal Asiatic Society, 21,* 161–215.

Carvalho, P., Bocquentin, J., & de Lapparent de Broin, F. (2002). Une nouvelle espèce de *Podocnemis (*Pleurodira, Podocnemididae) provenant du Néogène de la formation Solimões, Acre, Brésil. *Geobios, 35*(6), 677–686.

Chiplonkar, G. W. (1940). A new species of fossil frog from the intertrappean beds of Worli Hill, Bombay. *Journal of the Bombay Natural History Society, 40*(4), 799–804.

Cope, E. D. (1864). On the limits and relations of the raniformes. *Proceedings of the Academia of Natural Sciences of Philadelphia, 16,* 181–183.

Cope, E. D. (1868). On the origin of genera. *Proceedings of the Academia of Natural Sciences of Philadelphia, 20,* 242–300.

Courtillot, V., Besse, J., Vandamme, D., Montigny, R., Jaeger, J. J., & Cappetta, H. (1986). Deccan flood basalts at the Cretaceous/Tertiary boundary? *Earth and Planetary Science Letters, 80,* 361–374.

Courtillot, V., Féraud, G., Maluski, H., Vandamme, D., Moreau, M. G., & Besse, J. (1988). Deccan flood basalts and the Cretaceous/Tertiary boundary. *Nature, 333,* 843–846.

Cripps, J. A., Widdowson, M., Spicer, R. A., & Jolley, D. W. (2005). Coastal ecosystem responses to late stage Deccan Trap volcanism: The post K-T boundary (Danian) palynofacies of Mumbai (Bombay), west India. *Palaeogeography, Palaeoclimatology, Palaeoecology, 216,* 303–332.

Das Sarma, D. C., Anantharaman, S., Vijayasarathi, G., Nath, T. T., & Rao, C. V. (1995). Paleontological studies for the search of micromammals in the infra- and inter-trappean horizons of Andhra Pradesh. *Records of the Geological Survey of India, 128,* 223.

De la Fuente, M. (2003). Two new pleurodiran turtles from the Portezuelo Formation (Upper Cretaceous) of Northern Patagonia, Argentina. *Journal of Paleontology, 77*(3), 559–575.

De la Fuente, M., Lapparent de Broin, F. de, & Manera de Bianco, T. (2001). The oldest and first nearly complete skeleton of a chelid, of the *Hydromedusa* sub-group (Chelidae, Pleurodira), from the Upper Cretaceous of Patagonia. *Bulletin de la Société géologique de France, Paris, 172*(2), 105–112.

Duncan, R. A., & Pyle, G. (1988). Rapid eruption of Deccan flood basalt at the Cretaceous/Tertiary boundary. *Nature, 331,* 841–843.

Ferreira, G. S., Bandyopadhyay, S., & Joyce, W. G. (2018). A taxonomic reassessment of *Piramys auffenbergi*, a neglected turtle from the late Miocene of Piram Island, Gujarat, India. *PeerJ, 6,* e5938. https://doi.org/10.7717/peerj.5938.

Foote, B. (1876). Geological features of southern Maharashtra and adjacent districts. *Memoirs of the Geological Survey of India, 12*(1), 1–268.

França M. A. G. de, & Langer M. C. (2005). A new freshwater turtle (Reptilia, Pleurodira, Podocnemidae) from the Upper Cretaceous (Maastrichtian) of Minas Gerais, Brazil. *Geodiversitas, 27*(3), 391–411.

França, M. A. G. de, & Langer, M. C. (2006). Phylogenetic relationships of the Bauru group turtles (Late Cretaceous of South-Central Brazil). *Revista Brasileira de Paleontologia, 9*(3), 365–373.

Gaffney, E. S. (1975). A revision of the side-necked turtle *Taphrosphys sulcatus* (Leidy) from the Cretaceous of New Jersey. *American Museum Novitates, 2571,* 1–24.

Gaffney, E. S. (1979). Comparative cranial morphology of recent and fossil turtles. *Bulletin of the American Museum of Natural History, 164*(2), 65–376.

Gaffney, E. S. (1990). The comparative osteology of the Triassic turtle *Proganochelys. Bulletin of the American Museum of Natural History, 194,* 1–263.

Gaffney, E. S., Campbell, K. E., & Wood, R. C. (1998). Pelomedusoid side-necked turtles from Late Miocene sediments in South-Western Amazonia. *American Museum Novitates, 3245,* 1–12.

Gaffney, E. S., Chatterjee, S., & Rudra, D. K. (2001a). *Kurmademys*, a new side-necked turtle (Pelomedusoides: Bothremydidae) from the Late Cretaceous of India. *American Museum Novitates, 3321,* 1–16.

Gaffney, E. S., & Forster, C. A. (2003). Side-necked turtle lower jaws (Podocnemididae, Bothremydidae) from the Late Cretaceous Maevarano Formation of Madagascar. *American Museum Novitates, 3397,* 1–13.

Gaffney, E. S., & Krause, D. W. (2011). *Sokatra*, a new side-necked turtle (Late Cretaceous, Madagascar) and the diversification of the main groups of pelomedusoides. *American Museum Novitates, 3728,* 1–28.

Gaffney, E. S., Krause, D. W., & Zalmout, I. S. (2009). *Kinkonychelys*, a new side-necked turtle (Pelomedusoides: Bothremydidae) from the Late Cretaceous of Madagascar. *American Museum Novitates, 3662,* 1–25.

Gaffney, E. S., & Meylan, P. A. (1991). Primitive pelomedusid turtle. In J. G. Maisey (Ed.), *Santana fossils, an illustrated atlas* (pp. 335–339). Neptune City, N.J.: TFH Publ. Inc.

Gaffney, E. S., Meylan, P. A., Wood, R. G., Simons, E., & de Almeida Campos, D. (2011). Evolution of the side-necked turtles: The family Podocnemididae. *Bulletin of the American Museum of Natural History, 350,* 1–237.

Gaffney, E. S., Moody, R. T. J., & Walker, C. A. (2001b). *Azabbaremys*, a new side-necked turtle (Pelomedusoides: Bothremydidae) from the Paleocene of Mali. *American Museum Novitates, 3320,* 1–16.

Gaffney, E. S., Roberts, E., Sissoko, F., Bouaré, M., Tapanila, L., & O'Leary, M. (2007). *Acleistochelys*, a new side-necked turtle (Pelomedusoides: Bothremydidae) from the Paleocene of Mali. *American Museum Novitates, 3549,* 1–24.

Gaffney, E. S., Sahni, A., Schleich, H., Singh, S. D., & Srivastava, R. (2003). *Sankuchemys*, a new side-necked turtle (Pelomedusoides: Bothremydidae) from the Late Cretaceous of India. *American Museum Novitates, 3405,* 1–10.

Gaffney, E. S., Tong, H., & Meylan, P. A. (2006). Evolution of the side-necked turtles: The families Bothremydidae, Euraxemydidae, and Araripemydidae. *Bulletin of the American Museum of Natural History, 300,* 1–700.

Gaffney, E. S., & Wood, R. C. (2002). *Bairdemys*, a new side-necked turtle (Pelomedusoides: Podocnemididae) from the Miocene of the Caribbean. *American Museum Novitates, 3359,* 1–28.

Gaffney, E. S., & Zangerl, R. (1968). A revision of the Chelonian Genus *Bothremys* (Pleurodira: Pelomedusidae). *Fieldiana Geology, 16*(7), 193–239.

Gautam, J. P., Pandey, B., Pathak, D. B., & Jaitly, A. K. (2015). Recognition of the Early Albian *Douvilleiceras mammillatum* zone in the Cauvery Basin, SE India. *Earth Science India, 8*(IV), 100–111.

Gayet, M., Rage, J.-C., & Rana, R. S. (1984). Nouvelles ichthyofaune et herpétofaune de Gitti Khadan, le plus ancien gisement du Deccan (Crétacé/Paléocène) à microvertébrés. Implications paléobiogéographiques. *Mémoires de la Société géologique de France, N. S., Paris, 147,* 55–65.

Ghevariya, Z. G. (1988). Intertrappean dinosaurian fossils from Anjar area, Kachchh district, Gujarat. *Current Science, 57*(5), 248–251.

Gmira, S. (1995). Etude des Chéloniens fossiles du Maroc. Anatomie. Systématique. Phylogénie. *Cahiers de Paléontologie,* 1–141.

Godinot, M., & Prasad, G. V. R. (1994). Discovery of Cretaceous aboreal eutherians. *Naturwissenschaften, 81,* 79–81.

Goswami, A., Prasad, G. V. R., Benson, R. B., Verma, O., & Flynn, J. J. (2012). New vertebrates from the Late Cretaceous Kallamedu Formation, Cauvery Basin, South India, including a troodontid dinosaur, a gondwanatherian mammal, and a *Simosuchus*-like notosuchian crocodyliform. *Journal Vertebrate Paleontology, 32A,* 102.

Goswami, A., Prasad, G. V. R., Upchurch, P., Boyer, D. M., Seiffert, E. R., Verma, O., et al. (2011). A radiation of arboreal basal eutherian mammals beginning in the Late Cretaceous of India. *Proceedings of the National Academy of Sciences of the United States of America, 108,* 16333–16338.

Goswami, A., Prasad, G. V. R., Verma, O., Flynn, J. J., & Benson, R. B. J. (2013). A troodontid dinosaur from the latest Cretaceous of India. *Nature Communications, 4,* 1703.

Grandidier, A. (1867). Liste des Reptiles nouveaux découverts, en 1866, sur la côte Sud-Ouest de Madagascar. *Revue et magasin de zoologie pure et appliquée, 2*(19), 232–234.

Gray, J. E. (1870). *Supplement to the catalogue of shield reptiles in the collection of the British Museum. Part I. Testudinata* (120 pp.). Trustees of the British Museum, London.

Gray, J. E. (1871). Notice of a fossil Hydraspidae (*Testudo leithii* Carter) from Bombay. *Annals and Magazine of Natural History, London, 4*(8), 339–340.

Haas, G. (1978a). A Cretaceous pleurodire turtle from the surroundings of Jerusalem. *Israel Journal of Zoology, 27,* 20–33.

Haas, G. (1978b). A new turtle of the genus *Podocnemis* from the lower Cenomanian of 'Ein Yabrud. *Israel Journal of Zoology, 27,* 169–175.

Halliday, T. J. D., Cuff, A. R., Prasad, G. V. R., Thanglemmoi, M. S., & Goswami, A. (2016). New record of *Egertonia* (Elopiformes, Phyllodontidae) from the Late Cretaceous of South India. *Papers in Palaeontology, 2*(2), 287–294.

Halstead, B. (1979). A taxonomic note on new fossil turtles. *The Nigerian Field. The International Field Studies Journal of West Africa, Monograph, 1.* (Suppl. to. *44*), 48–49.

Haughton, S. H. (1928). On some reptilian remains from the dinosaur beds of Nyasaland. *Transactions of the Royal Society of South Africa, 16,* 67–75.

Hay, O. P. (1908). The fossil turtles of North America. *Publications of the Carnegie Institution, 75,* 568 pp.

Hirayama, R. (1992). Fossil turtles from the Neogene strata in the Sinda Basin, Eastern Zaire. *African Study Monographs, Suppl., 17,* 49–65.

Hora, S. L. (1938). On some fossil fish scales from the intertrappean beds at Deothan and Kheri, Central Provinces. *Records of the Geological Survey of India, 73,* 267–294.

Jain, S. L. (1977). A new fossil pelomedusid turtle from the Upper Cretaceous Pisdura sediments, Central India. *Journal of the Palaeontological Society of India, (for 1975), 20,* 360–365.

Jain, S. L. (1986). New pelomedusid turtle (Pleurodira: Chelonia) remains from Lameta Formation (Maastrichtian) at Dongargaon, Central India, and a review of pelomedusids from India. *Journal of the Palaeontological Society of India, 31,* 63–75.

Jain, S. L., & Sahni, A. (1983). Some Upper Cretaceous vertebrates from central India and their paleogeographical implications. In H. K. Maheshwari (Ed.), *Cretaceous of India. Indian Association of Palynostratigraphers Symposium* (pp. 66–83). Lucknow.

Kasper, A. (1903). Ueber den Atlas und Epistropheus bei den Pleurodiren Schildkröten. *Arbeiten aus dem Zoologischen Instituten der Universität Wien und der Zoologischen Station in Triest, 14,* 36–49.

Kelkar, K. V., & Gupte, R. B. (1943). Intertrappean beds at Upparhatti. *Journal of the Bombay University, 11*(5), 121–125.

Keller, G., Adatte, T., Bajpai, S., Mohabey, D. M., Widdowson, M., Khosla, et al. (2009). K-T transition in Deccan Traps of Central India marks major marine seaway across India. *Earth and Planetary Science Letters, 282,* 10–23.

Khajuria, C. K., Prasad, G. V. R., & Manhas, B. K. (1994). Paleontological constraints on the age of Deccan Traps, peninsular India. *Newsletter in Stratigraphy, 31,* 21–32.

Khosla, A. (2001). Diagenetic alterations of Late Cretaceous dinosaur eggshell fragments of India. *Gaia, 16,* 45–49.

Khosla, A., Prasad, G. V. R., Verma, O., Jain, A. K., & Sahni, A. (2004). Discovery of a micromammal-yielding Deccan intertrappean site near Kisalpuri, Dindori district, Madhya Pradesh. *Current Science, 87*(3), 380–383.

Khosla, A., & Sahni, A. (1995). Parataxonomic classification of Late Cretaceous dinosaur eggshells from India. *Journal of the Palaeontological Society of India, 40,* 87–102.

Khosla, A., & Sahni, A. (2000). Late Cretaceous (Maastrichtian) ostracods from the Lameta Formation, Jabalpur Cantonment area, Madhya Pradesh, India. *Journal of the Palaeontological Society of India, 45,* 57–78.

Khosla, A., & Sahni, A. (2003). Biodiversity during the Deccan volcanic eruptive episode. *Journal of Asian Earth Sciences, 21,* 895–908.

Khosla, A., Sertich, J. J. W., Prasad, G. V. R., & Verma, O. (2009). Dyrosaurid remains from the intertrappean beds of India and the Late Cretaceous distribution of the Dyrosauridae. *Journal of Vertebrate Paleontology, 29*(4), 1321–1326.

Knight, K. B., Renne, P. R., Halkett, A., & White, N. (2003). 40Ar/39Ar dating of the Rajahmundry Traps, eastern India and their relationship to the Deccan Traps. *Earth and Planetary Science Letters, 208,* 85–99.

Krause, D. W., Prasad, G. V. R., Koenigswald, W. V., Sahni, A., & Grine, F. E. (1997). Cosmopolitanism among Gondwanan Late Cretaceous mammals. *Nature, 390,* 504–507.

Lapparent de Broin, F. de (2000a). African chelonians from the Jurassic to the Present. A preliminary catalog of the African fossil chelonians. *Palaeontologia Africana, 36,* 43–82.

Lapparent de Broin, F. de (2000b). The oldest pre-podocnemidid turtle (Chelonii, Pleurodira), from the early Cretaceous, Ceará state, Brasil, and its environment. *Treballs del Museu de geologia de Barcelona, 9,* 43–95.

Lapparent de Broin, F. de (2001). The European turtle fauna from the Triassic to the Present. *Dumerilia, 4*(3), 155–216.

Lapparent de Broin, F. de (2003). The Miocene chelonians from the southern Namibia. In B. Senut and M. Pickford (Coord.), Faunas from the southern Namibia. *Memoirs of the Geological Survey of Namibia, 19,* 67–102.

Lapparent de Broin, F. de (2004). *Neochelys* sp. (Chelonii, Erymnochelyinae), from Silveirinha, early Eocene, Portugal. *Ciencias da Terra (UNL) (2003), 15,* 117–132.

Lapparent de Broin, F. de (2008a). Miocene Chelonians from Western Namibia. *Memoirs of the Geological Survey of Namibia, 20,* 107–145.

Lapparent de Broin, F. de (2008b). Contributor. In Didier Merle (coord.), *Stratotype Lutétien, Muséum national d'Histoire naturelle, Paris* (288 pp.), Biotope, Mèze; BRGM, Orléans.

Lapparent de Broin, F. de, & Antunes, M. T. (2000). Pleistocene chelonians from Gruta da Figueira Brava (Arrábida, Portugal). Memorias da *Academia* das *Ciencias* de *Lisboa, Classe de Ciências, 38,* 101–112.

Lapparent de Broin, F. de, Bardet, N., Amaghzaz, M., & Meslouh, S. (2014). A strange new chelonioid turtle from the latest Cretaceous phosphates of Morocco. *Comptes Rendus Palevol, 13,* 87–95. https://doi.org/10.1016/j.crpv.2013.07.008 + Electronic Supp.

Lapparent de Broin, F. de, & De la Fuente, M. S. (2001). Oldest World Chelidae (Chelonii, Pleurodira), from the Cretaceous of Patagonia,

Argentina. *Comptes Rendus de l'Académie des Sciences Paris, II, 333*, 463–470.

Lapparent de Broin, F. de, De la Fuente, M. S., & Fernandez, M. (2007). *Notoemys* (Chelonii, Pleurodira), Late Jurassic of Argentina: new examination of the anatomical structure and comparisons. *Revue de Paléobiologie, 26*(2), 99–136.

Lapparent de Broin, F. de, & Murelaga, X. (1996). Une nouvelle faune de Chéloniens dans le Crétacé supérieur européen. *Comptes rendus de l'Académie des Sciences, Paris, IIa, 323*(8), 729–735.

Lapparent de Broin, F. de, & Murelaga, X. (1999). Turtles from the Upper Cretaceous of Laño (Iberian Peninsula). *Estudios del Museo de Cioencias Baturales d'Alava, 14*(1), 135–211.

Lapparent de Broin, F. de, Murelaga Bereikua, X., & Codrea, V. (2004). Presence of Dortokidae (Chelonii, Pleurodira) in the earliest Tertiary of the Jibou Formation, Romania: Paleobiogeographical implications. *Acta Palaeontologica Romaniae, 4*, 203–215.

Lapparent de Broin, F. de, Murelaga, X., Pérez-García, A., Farrés, F., & Altimiras, J. (2018). A new cheloniid turtle from the upper Eocene Ebro basin of the Osona county (Catalonia, Spain) in its Eocene faunistic context. *Fossil Record, 21*, 1–48.

Lapparent de Broin, F. de, Prasad, Guntupalli V. R., Bajpai, S., & Verma, O. (2009). *Chelonian remains from the Upper Cretaceous Deccan Intertrappean Beds of peninsular India: review, new material and comparisons.* In: Gaffney Turtle Symposium (Drumheller, Alberta, 17–18 October 2009), Royal Tyrrell Museum. Abstracts, 95–97.

Lapparent de Broin, F. de, & Werner, C. (1998). New late Cretaceous turtles from the Western Desert, Egypt. *Annales de Palaéontologie, 84*(2), 131–214.

Leidy, J. (1856). Notices of remains of extinct turtles of New Jersey, collected by Prof. Cook, of the State Geological Survey, under the direction of Dr. W. Kitchell. *Proceedings of the Academy of Natural Sciences of Philadelphia, 8*, 303–304.

Leidy, J. (1865). Memoir on the extinct reptiles of the Cretaceous formations of the United States. *Smithsonian Contributions to Knowledge, 14*(6), 1–135.

Linnaeus, C. (1758). *Systema Naturae*, 10th edn (Vol. 1). Stockholm: Laurentius Salvius.

Lourembam, R. S., Prasad, G. V. R., & Grover, P. (2017). Ichthyofauna (Chondrichthyes, Osteichthyes) from the Upper Cretaceous intertrappean beds of Piplanarayanwar, Chhindwara District, Madhya Pradesh, India. *Island Arc, 26*(1), https://doi.org/10.1111/iar.12180.

Loyal, R. S., Mohabey, D. M., Khosla, A., & Sahni, A. (1998). Status and paleobiology of the Late Cretaceous Indian theropods with description of a new theropod eggshell oogenus and oospecies, *Ellipsoolithus khedaensis*, from the Lameta Formation, district Kheda, Gujarat, western India. *Gaia, 15*, 379–387.

Lydekker, R. A. (1887). Eocene chelonians from the Salt-Range. *Palaeontologia Indica, 10*(4), 59–65.

Lydekker, R. (1890). Note on certain vertebrate remains from the Nagpur district. *Records of the Geological Survey of India, 23*, 20–24.

Martin, J. E., & Lapparent de Broin, F. de (2016). A miniature notosuchian with multicuspid teeth from the Cretaceous of Morocco. *Journal of Vertebrate Paleontology, 36*(6), 17 pp. https://doi.org/10.1080/02724634.2016.1211534.

Matheron, P. (1869). Notice sur les reptiles fossiles des dépots fluvio-lacustres crétacés du bassin à lignite de Fuveau. *Mémoires de l'Académie Impériale des Sciences, Belles-Lettres et Arts de Marseille* (pp. 1–39).

Matley, C. A. (1929). The Cretaceous dinosaurs of the Trichinopoly district and the rocks associated with them. *Records of the Geological Survey of India, 61*, 337–349.

Menegazzo, M. C., Bertini, R. J., & Manzini, F. F. (2015). A new turtle from the Upper Cretaceous Bauru Group of Brazil, updated phylogeny and implications for age of the Santo Anastacio Formation. *Journal of South American Earth Sciences, 58*, 18–32.

Meylan, P. A. (1996). Skeletal morphology and relationships of the early Cretaceous side-necked turtle, *Araripemys barretoi* (Testudines: Pelomedusoides: Araripemydidae), from the Santana Formation of Brazil. *Journal of Vertebrate Paleontology, 16*(1), 20–33.

Meylan, P. A. & Gaffney, E. S. (1991). *Araripemys* Price, 1973. In J. G. Maisey (Ed.), *Santana fossils, an illustrated atlas* (pp. 326–334). Neptune, N.J.: T.F.H. Pub.

Młynarski, M. (1976). Testudines, In O. Kuhn (Ed.), *Handbuch der Paläoherpetologie, 7*(7), 1–130. Stuttgart. New York: G. Fischer Verlag.

Mohabey, D. M. (1996). *Pycnodus lametae* (Pycnodontidae), a holostean fish from freshwater Upper Cretaceous Lameta Formation of Maharashtra. *Journal of the Geological Society of India, 47*, 593–598.

Mohabey, D. M., Head, J. J., & Wilson, J. A. (2011). A new species of the snake *Madtsoia* from the Upper Cretaceous of India and its paleobiogeographic implications. *Journal of Vertebrate Paleontology, 31*, 588–595.

Mohabey, D. M., & Udhoji, S. G. (1996). Fauna and flora from Late Cretaceous (Maestrichtian) non-marine Lameta sediments associated with Deccan volcanic episode, Maharashtra: Its relevance to the K-T boundary problem, paleoenvironment, and paleogeography. *Gondwana Geological Magazine, National Symposium Deccan Flood Basalts, Sp., 2*, 349–364.

Montenat, C., & Merle, D. (coord.) (2018). *Stratotype Danien. Muséum national d'Histoire naturelle, Paris.* Patrimoine géologique (Vol. 9, 480 pp).

Muzzy, D. (1856). The geological features of Madurai, Trichinopoly, Tanjore and Poothacotta. *Madras Journal of Literature and Science, New Series, 1*, 90–102.

Nagendra, R., Sathiyamoorthy, P., Pattanayak, S., Reddy, A. N., & Jaiprakash, B. C. (2013). Stratigraphy and paleobathymetric interpretation of the Cretaceous Karai Shale Formation of Uttatur Group, Tamil Nadu, India. *Stratigraphy and Geological Correlation, 21*(7), 675–688.

Noble, G. K. (1930). The fossil frogs of the inter-trappean beds of Bombay, India. *American Museum Novitates, 401*, 1–13.

Nopcsa, F. (1931a). Note préliminaire sur quelques tortues du Danien du Midi de la France. *Annales du Musée d'Histoire Naturelle de Marseille, 22, 6*, 4 pp.

Nopcsa, F. (1931b). Sur des nouveaux restes de Tortues du Danien du Midi de la France. *Bulletin de la Société géologique de France, S.5, 3–4*, 223–235.

Oliveira, G. R., & Kellner, A. W. A. (2007). A new side-necked turtle (Pleurodira, Pelomedusoides) from the Santana Formation (Early Cretaceous), Araripe Basin, Northeastern Brazil. *Zootaxa, 1425*, 53–61.

Parham, J. F., & Pyenson, N. D. (2010). New sea turtle from the Miocene of Peru and the iterative evolution of feeding ecomorphologies since the Cretaceous. *Journal of Paleontology, 84*(2), 231–247.

Pérez-García, A. (2016). A new turtle taxon (Podocnemidoidea, Bothremydidae) reveals the oldest known dispersal event of the crown Pleurodira from Gondwana to Laurasia. *Journal of Systematic Paleontology,* 15(9), 1–23. http://dx.doi.org/10.1080/14772019.2016.1228549

Pérez-García, A. (2017). A new turtle taxon (Podocnemidoidea, Bothremydidae) reveals the oldest known dispersal event of the crown Pleurodira from Gondwana to Laurasia. *Journal of Systematic Palaeontology, 15*(9), 709–731.

Pérez-García, A. (2018a). New information on the Cenomanian bothremydid turtle *Algorachelus* based on new, well-preserved material from Spain. *Fossil Record, 21*, 119–135.

Pérez-García, A. (2018b). New insights on the only bothremydid turtle (Pleurodira) identified in the British record: *Palemys bowerbankii* new combination. *Palaeontologia Electronica,* 21.2.28A 1–12. https://doi.org/10.26879/849.

Pérez-García, A. (2019). Identification of the Lower Cretaceous pleurodiran turtle *Taquetochelys decorata* as the only African araripemydid species. *Comptes Rendus Palevol, 18,* 24–32. https://doi.org/10.1016/j.crpv.2018.04.004.

Pérez-García, A., Antunes, M. T., Barroso-Barcenilla, F., Callapez, P. M., Segura, M., Soares, A. F., et al. (2017a). A bothremydid from the middle Cenomanian of Portugal identified as one of the oldest pleurodiran turtles in Laurasia. *Cretaceous Research, 78,* 61–70.

Pérez-García, A., & de Lapparent de Broin, F. (2013). A new species of *Neochelys* (Chelonii, Podocnemididae) from the Ypresian (early Eocene) of the south of France. *Comptes Rendus Palevol, 12,* 269–277. https://doi.org/10.1016/j.crpv.2013.05.011.

Pérez-García, A., & de Lapparent de Broin, F. (2015). New insights into the anatomy and systematics of "*Papoulemys laurenti*", a representative of *Neochelys* (Chelonii, Podocnemididae) from the early Eocene of the south of France. *Paläontologische Zeitschrift, 89*(4), 901–923.

Pérez-García, A., Lapparent de Broin, F. de, & Murelaga, X. (2017b). The *Erymnochelys* group of turtles (Pleurodira, Podocnemididae) in the Eocene of Europe: New taxa and paleobiogeographical implications. *Palaeontologia Electronica, 20*(1) *14A,* 1–28.

Prasad, G. V. R., & Cappetta, H. (1993). Late Cretaceous selachians from India and the age of Deccan Traps. *Palaeontology, 36,* 231–248.

Prasad, G. V. R., & Godinot, M. (1994). Eutherian tarsal bones from the Late Cretaceous of India. *Journal of Paleontology, 68,* 892–902.

Prasad, G. V. R., Jaeger, J.-J., Sahni, A., Gheerbrant, E., & Khajuria, C. K. (1994). Eutherian mammals from the Upper Cretaceous (Maastrichtian) inter-trappean beds of Naskal, Andhra Pradesh, India. *Journal of Vertebrate Paleontology, 14,* 260–277.

Prasad, G. V. R., & Khajuria, C. K. (1990). A record of microvertebrate fauna from the intertrappean beds of Naskal, Andhra Pradesh. *Journal of the Palaeontological Society of India, 35,* 151–161.

Prasad, G. V. R., & Lapparent de Broin, F. de (2002). Cretaceous crocodile remains from Naskal (India): Comparisons and biogeographic affinities. *Annales de Paléonologie, 88*(1), 19–71.

Prasad, G. V. R., & Rage, J.-C. (1991). A discoglossid frog in the latest Cretaceous Maastrichtian) of India. Further evidence for a terrestrial route between India and Laurasia in the latest Cretaceous. *Comptes rendus de l'Académie des Sciences, Paris, 313,* 273–278.

Prasad, G. V. R., & Rage, J.-C. (1995). Amphibians and squamates from the Maastrichtian of Naskal, India. *Cretaceous Research, 16,* 95–107.

Prasad, G. V. R., & Rage, J.-C. (2004). Fossil frogs (Amphibia: Anura) from the Upper Cretaceous intertrappean beds of Naskal, Andhra Pradesh. *Revue de Paléobiologie, 23*(1), 99–116.

Prasad, G. V. R., & Sahni, A. (1987). Coastal-plain microvertebrate assemblage from the terminal Cretaceous of Asifabad, Andhra Pradesh. *Journal of the Palaeontological Society of India, 32,* 5–19.

Prasad, G. V. R., & Sahni, A. (1988). First Cretaceous mammal from India. *Nature, 332,* 638–640.

Prasad, G. V. R., & Sahni, A. (2014) Vertebrate fauna from the Deccan volcanic province: Response to volcanic activity. In G. Keller & A. C. Kerr (Eds.), *Volcanism, impacts, and mass extinctions: Causes and effects, Geological Society of America, Special Paper* 505, 1–18.

Prasad, G. V. R., & Singh, V. (1991). Microvertebrates from the inter-trappean beds of Rangareddi district, Andhra Pradesh and their biostratigraphic significance. *Bulletin of the Indian Geological Association, 24*(1), 1–20.

Prasad, G. V. R., Verma, O., Flynn, J. J., & Goswami, A. (2013). A new Late Cretaceous vertebrate fauna from the Cauvery basin, South India: Implications for Gondwanan paleobiogeography. *Journal of Vertebrate Paleontology, 33,* 1260–1268.

Prasad, G. V. R., Verma, O., Gheerbrant, E., Goswami, A., Khosla, A., Parmar, V., et al. (2010). First mammal evidence from the Late Cretaceous of India for biotic dispersal between India and Africa at the K/T transition. *Comptes Rendus Palevol, 9,* 63–71.

Prasad, G. V. R., Verma, O., Sahni, A., Krause, D. W., Khosla, A., & Parmar, V. (2007a). A new Late Cretaceous Gondwanatherian mammal from Central India. *Proceedings of the Indian National Science Academy, 73*(1), 17–24.

Prasad, G. V. R., Verma, O., Sahni, A., Parmar, V., & Khosla, A. (2007b). A Cretaceous hoofed mammal from India. *Science, 318,* 937.

Prasad, K. N. (1974). The vertebrate fauna from Piram Island, Gujarat, India. *Memoires of the Geological Survey of India, New Series, 41,* 1–23.

Prasahd, B. (1918). On a new fossil unionid from the intertrappean beds of peninsular India. *Records of the Geological Survey of India, 51* (4), 368–374.

Pritchard, A. C., McCartney, J., Krause, D. W., & Kley, N. J. (2014). New snakes from the Upper Cretaceous (Maastrichtian) Maevarano Formation, Mahajanga Basin, Madagascar. *Journal of Vertebrate Paleontology, 34*(5), 1080–1093.

Rabi, M., & Botfalvai, G. (2006). A new bothremydid (Chelonia: Pleurodira) fossil assemblage from the Late Cretaceous (Santonian) of Hungary—Additional studies in the historical paleobiogeography of Late Cretaceous bothremydid. 4th Annual Meeting of the European Association of Vertebrate Paleontologists. *Hankeriana, 5,* 61–65.

Rage, J.-C. (1981). Les Continents Péri-atlantiques au Crétacé Supérieur: Migrations des Faunes Continentales et Problèmes Paléogéographiques. *Cretaceous Research, 2,* 65–84.

Rage, J.-C., & Prasad, G. V. R. (1992). New snakes from the Late Cretaceous (Maastrichtian) of Naskal, India. *Neues Jahrbuch für Geologie und Paläontologie, Abhandlungen, 187,* 83–97.

Rage, J.-C., Prasad, G. V. R., & Bajpai, S. (2004). Additional snakes from uppermost Cretaceous (Maastrichtian) of India. *Cretaceous Research, 25,* 425–434.

Rage, J.-C., Prasad, G. V. R., Verma, O., Khosla, A., & Parmar, V. (2020). Anuran lissamphibian and squamate reptiles from the Upper Cretaceous (Maastrichtian) Deccan intertrappean sites in Central India, with a review of a lissamphibian and squamate diversity in the northward drifting Indian plate. In G. V. R. Prasad & R. Patnaik (Eds.), *Biological consequences of plate tectonics: New perspectives on post-Gondwana break-up* (pp. 99-121). Cham: Switzerland.

Rai, J., Ramkumar, M., & Sugantha, T. (2013). Calcareous nannofossils from the Ottakovil Formation, Cauvery Basin, South India: Implications on age and Late Cretaceousenvironmental conditions. In M. Ramkumar (Ed.), *On a sustainable future of the earth's natural resources* (pp. 109–122). Berlin: Springer.

Raju, D. S. N., Ravichandran, C. N., Dave, A., Jaiprakash, B. C., & Singh, J. (1991). K/T boundary events in the Cauvery and Krishna-Godavari Basins and the age of Deccan volcanics. *Geoscience Journal, 12,* 177–190.

Rana, R. S. (1989). Alligatorine teeth from the Deccan intertrappean beds near Rangapur, Andhra Pradesh, India: Further evidence of Laurasiatic elements. *Current Science, 59,* 49–51.

Rana, R. S. (2005). Lizard fauna from intertrappean (Late Cretaceous-Early Paleocene) beds of peninsular India. In D. M. Mohabey (Ed.), Indian Non-marine Late cretaceous: Advances and challenges. *Gondwana Geological Magazine, Sp. Vol. 8*, 123–132.

Rana, R. S., & Sati, K. K. (2000). Late Cretaceous-Paleocene crocodilians from the Deccan Trap associated sedimentary sequences of peninsular India. *Journal of the Palaeontological Society of India, 45*, 123–150.

Rana, R. S., & Wilson, G. P. (2003). New Late Cretaceous mammals from the intertrappean beds of Rangapur, India and paleobiogeographic framework. *Acta Palaeontologica Polonica, 48*, 331–348.

Roger, J., Pickford M., Thomas, H., Lapparent de Broin, F. de, Tassy P., Van Neer, W., et al. (1994). Découverte de vertébrés fossiles dans le Miocène de la région du Huqf au Sultanat d'Oman. *Annales de Paléontologie 80*(4), 253–273.

Roman, F. (1897). Recherches stratigraphiques et paléontologiques dans le Bas-Languedoc. *Annales de l'Université de Lyon*, 1–366.

Romano, P. S. R., Gallo, V., Ramos, R. R. C., & Antonioli, L. (2014). *Atolchelys lepida*, a new side-necked turtle from the Early Cretaceous of Brazil and the age of crown Pleurodira. *Biological Letters, 10*, 20140290.

Sahni, A., & Khosla, A. (1994). A Maastrichtian ostracode assemblage (Lameta Formation) from Jabalpur cantonment, Madhya Pradesh, India. *Current Science, 67*, 456–459.

Sahni, A., Kumar, K., Hartenberger, J. L., Jaeger, J.-J., Rage, J.-Cl., Sudre, J., et al. (1982). Microvertébrés nouveaux des Trapps du Deccan (Inde): Mise en évidence d'une voie de communication terrestre probable entre la Laurasie et l'Inde à la limite Crétacé-Tertiaire. *Bulletin de la Société Géologique de France, 24*, 1093–1099.

Sahni, A., Rana, R. S., & Prasad, G. V. R. (1984). SEM study of thin eggshell fragments from the intertrappeans (Cretaceous-Tertiary transition) of Nagpur and Asifabad, peninsular India. *Journal of the Palaeontological Society of India, 29*, 26–33.

Sahni, A., Tandon, S. K., Jolly, A., Bajpai, S., Sood, A., & Srinivasan, S. (1994). Upper Cretaceous dinosaur eggs and nesting sites from the Deccan volcano-sedimentary province of peninsular India. In K. Carpenter, K. F. Hirsch, & J. R. Horner (Eds.), *Dinosaur eggs and babies* (pp. 204–226). Cambridge: Cambridge University Press.

Sahni, A., & Tripathi, A. (1990). Age implications of the Jabalpur Lameta Formation and intertrappean biotas. In A. Sahni (Ed.), *Cretaceous event stratigraphy and correlation of the Indian non-marine strata*. Contributions Seminar cum Workshop (International Geological Correlation Programme 216 & 245), (pp. 35–37). Chandigarh.

Samant, B., & Mohabey, D. M. (2009). Palynoflora from Deccan volcano-sedimentary sequence (Cretaceous-Paleogene transition) of central India: Implications for spatio-temporal correlation. *Journal of Biosciences, 34*, 811–823.

Sánchez-Villagra, M. R., & Winkler, J. D. (2006). Cranial variation in *Bairdemys* turtles (Podocnemididae: Miocene of the Caribbean region) and description of new material from Urumaco, Venezuela. *Journal of Systematic Palaeontology, 4*, 241–253.

Schmidt, K. P. (1940). A new turtle of the genus *Podocnemis* from the Cretaceous of Arkansas. *Field Museum of Natural History-Geology, 8*(1), 1–12.

Sereno, P. C., & El Shafie, J. (2013). *A new long-necked turtle, Laganemys tenerensis (Pleurodira: Araripemydidae), from the Elrhaz Formation (Aptian–Albian) of Niger, Morphology and evolution of turtles* (pp. 215–250). Netherlands: Springer.

Singh, R. S., & Kar, R. K. (2002). Paleocene palynofossils from the Lalitpur intertrappean beds, Uttar Pradesh, India. *Journal of the Geological Society of India, 60*(2), 213–216.

Singh, S. D., & Sahni, A. (1996). Bombay intertrappeans: New data on age and faunal affinities. In J. Pandey, R. J. Azmi, A. Bhandari & A. Dave (Eds.), *Contributions to XV Indian Colloquium Micropalaeontology and Stratigraphy* (pp. 465–469). Dehradun.

Singh, S. D., Sahni, A., Gaffney, E. S., & Schleich, H. H. (1998). Reptilia from the intertrappean beds of Bombay (India). *Veroffentlichungen aus dem Fuhlrott-Museum, 4*, 307–320.

Smith, T., Sigé, B., & De Bast, E. (2010). Euarchontan affinity for the Paleocene Afro-European adapisoriculid mammals and their origin in the late Cretaceous Deccan Traps of India. *Naturwissenschaften, 9*, 7417–7422.

Špinar, Z. V., & Hodrová, M. (1985). New knowledge of the genus *Indobatrachus* (Anura) from the Lower Eocene of India. *Amphibia-Reptilia, 6*, 363–376.

Sterli, J., De la Fuente, M. S., & Rougier, G. W. (2007). Anatomy and relationships of *Paleochersis talampayensis*, a Late Triassic turtle from Argentina. *Palaeontographica, A, 281*, 1–61.

Suárez, J. M. (1969). Um Quelônio da Formaçao-Bauru. Anais do XXIII Congresso Brasileiro de Geologia: 167–176, 7 fig. *Boletim* do Departamento de *Geografia, Faculdade* de *Filosofia, Ciências* e *Letras*, de *Presidente Prudente, 2*, 35–54.

Sukheswala, R. N. (1947). A fossil tortoise (*Testudo leithii*) from the intertrappean of the Worli Hill, Bombay. *Proceedings of the Indian Science Congress, 33rd Session*, Abstract (Vol. 3, p. 97).

Sundaram, R., Henderson, R. A., Ayyasami, K., & Stilwell, J. D. (2001). A lithostratigraphic revision and paleoenvironmental assessment of the Cretaceous System exposed in the onshore Cauvery Basin, southern India. *Cretaceous Research, 22*, 743–762.

Swinton, W. E. (1939). A new fresh-water tortoise from Burma. *Records of the Geological Survey of India, 74*, 548–551.

Thomas, H., Sen, S., Khan, M., Battail, B., & Ligabue, G. C. (1982). The Lower Miocene Fauna of Al-Sarrar (Eastern Province, Saudi Arabia). Chelonia: 117–118. *Atlal, Journal of Saudiarabian Archaeology, 5*(3), 109–136.

Tong, H., & Buffetaut, E. (1996). A new genus and species of pleurodiran turtle from the Cretaceous of southern Morocco. *Neues Jahrbuch für Geologie und Paläontologie Abhandlungen, 199*(1), 133–150.

Tong, H., Gaffney, E. S., & Buffetaut, E. (1998). *Foxemys*, a new side-necked turtle (Bothremydidae: Pelomedusoides) from the Late Cretaceous of France. *American Museum Novitates, 3251*, 1–19.

Tortosa, T., Buffetaut, E., Vialle, N., Dutour, Y., Turini, E., & Cheylan, G. (2014). A new abelisaurid dinosaur from the Late Cretaceous of southern France: Paleobiogeographical implications. *Annales de Paléontologie, 100*, 63–86.

Turner, A. H., & Sertich, J. J. W. (2010). Phylogenetic history of *Simosuchus clarki* (Crocodyliformes: Notosuchia) from the Late Cretaceous of Madagascar. In D. W. Krause & N. J. Kley (Eds.), *Simosuchus clarki (Crocodyliformes: Notosuchia) from the Late Cretaceous of Madagascar* (Vol. 30, pp. 177–236). Society of Vertebrate Paleontology, Supplementary Issue 1: Memoir 10.

Verma, K. K. (1965). Note on a new species of fossil frog from the intertrappean beds of Malabar Hill, Bombay. *Current Science, 34*, 182–183.

Verma, O., Khosla, A., Kaur, J., & Prashanth, M. (2016). Myliobatid and pycnodont fish from the Late Cretaceous of Central India and their paleobiogeographic implications. *Historical Biology, 29*(2), 253–265.

Verma, O., Prasad, G. V. R., Goswami, A., & Parmar, V. (2012). *Ptychodus decurrens* Agassiz (Elasmobranchii: Ptychodontidae) from the Upper Cretaceous of India. *Cretaceous Research, 33*, 183–188.

Vianey-Liaud, M., Jain, S. L., & Sahni, A. (1987). Dinosaur egg shells (Saurischia) from the Late Cretaceous intertrappean and Lameta

Formations (Deccan, India). *Journal of Vertebrate Paleontology, 7,* 408–424.

Walker, C. (1979). New Turtles from the Cretaceous of Sokoto. *The Nigerian Field, the International Field Studies, Journal of West Africa, 1*(Suppl. to 44), 42–48.

Weems, R. E., & Knight, J. L. (2013). A new species of *Bairdemys* (Pelomedusoides: Podocnemididae) from the Oligocene (early Chattian) Chandler Bridge Formation of South Carolina, USA, and its paleobiogeographic implications for the genus. In D. B. Brinkman et al. (Eds.), *Morphology and evolution of turtles* (Vol. 18, pp. 289–303), Vertebrate Palaeobiology and Palaeoanthropology, Dordrecht: Springer.

West, W. D. (1950). General report of the Geological Survey of India for the year 1948. *Records of the Geological Survey of India, 82,* 1–253.

Whatley, R., & Bajpai, S. (2000). A new fauna of Late Cretaceous non-marine Ostracoda from the Deccan intertrappean beds of Lakshmipur, Kachchh (Kutch District), Gujarat, Western India. *Revista Española de Micropaleontologia, 32*(3), 385–409.

Widdowson, M., Pringle, M. S., & Fernandez, O. A. (2000). A post K-T boundary (Early Paleocene) age for feeder dykes, Goa, India. *Journal of Petrology, 41,* 1177–1194.

Williams, E. E. (1950). Variation and selection in the cervical central articulations of living turtles. *Bulletin of the American Museum of Natural History, 94*(9), 505–562.

Williams, E. E. (1953). Fossils and the distribution of chelyid turtles. 1. "*Hydraspis" leithii* (Carter) in the Eocene of India is a pelomedusid. *Breviora, 1*(13), 1–8.

Williams, E. E. (1954a). A key and description of the living species of the genus *Podocnemis* (Sensu Boulenger), Testudines, Pelomedusidae. *Bulletin of the Museum of Comparative Zoology of Harvard, 11*(8), 277–295.

Williams, E. E. (1954b). New or redescribed pelomedusid skulls from the Tertiary of Africa and Asia (Testudines, Pelomedusidae). 1. *Dacquemys paleomorpha*, new genus, new species from the Lower Oligocene of the Fayum, Egypt. *Breviora, 35,* 1–8.

Wilson, J. A., Sereno, P. C., Srivastava, S., Bhatt, D. K., Khosla, A., & Sahni, A. (2003). A new abelisaurid (Dinosauria, Theropoda) from the Lameta Formation (Cretaceous, Maastrichtian) of India. *Contributions from the Museum of Paleontology, University of Michigan, 31*(1), 1–42.

Wilson, G. P., Das Sarma, D. C., & Anantharaman, S. (2007). Late Cretaceous sudamericid Gondwanatherians from India with paleobiogeographic considerations of Gondwanan mammals. *Journal of Vertebrate Paleontology, 27*(2), 521–531.

Wilson, J. A., Mohabey, D. M., Peters, S. E., & Head, J. J. (2010). Predation upon Hatchling Dinosaurs by a new snake from the Late Cretaceous of India. *PLoS Biology, 8*(3), 1–10.

Wood, R. C. (1970). A review of the fossil Pelomedusidae (Testudines, Pleurodira) of Asia. *Breviora, 357,* 1–24.

Wood, R. C. (1975). Redescription of "*Bantuchelys*" *congolensis*, a fossil pelomedusid turtle from the Paleocene of Africa. *Revue de Zoologie africaine, 89*(1), 127–144.

Wood, R. C. (1983) *Kenyemys williamsi*, a fossil pelomedusid turtle from the Pliocene of Kenya. In A. G. J. Rhodin & K. Miyata (Eds.), *Essays in honor of Ernest E. Williams, Advances in herpetology and evolutionary biology* (pp. 74–85). Cambridge, Massachussetts: Museum of Comparative Zoology.

Wood, R. C. (2003). Fossil turtles from Lothagam. In M. G. Leakey & J. M. Harris (Eds.), *Lothagam: The dawn of humanity in eastern Africa* (pp. 115–136). New York: Columbia University Press.

Woodward, A. S. (1908). On some fish-remains from the Lameta Beds at Dongargaon, Central Provinces. *Palaeontologia Indica, N.S, 13,* 1–6.

Yadagiri, P. & Ayyasami, K. (1989). A carnosaurian dinosaur from the Kallamedu Formation (Maastrichtian horizon), Tamilnadu. In M. V. A. Sastry, V. V. Sastry, C. G. K. Ramanujam, H. M. Kapoor, B. R. Jagannatha Rao, P. P. Satsangi, et al. (Eds.), *Symposium on Three Decades of Development in Paleontology and stratigraphy in India. Volume 1. Precambrian to Mesozoic. Geological Society of India Special Publication, 11(1),* 523–528.

Zaher, H., Pol, D., Carvalho, A. B., Nascimento, P. M., Riccomini, C., Larson, P., et al. (2011). A complete skull of an Early Cretaceous sauropod and the evolution of advanced titanosaurs. *PLoS ONE, 6* (2), e16663. https://doi.org/10.1371/journal.pone.0016663.

Zangerl, R. (1947). Redescription of *Taphrosphys olssoni*, a fossil turtle from Peru. *Fieldiana, Geology, 10,* 29–40.

Zangerl, R. (1948). The Vertebrate Fauna of the Selma Formation of Alabama. Part II. The Pleurodiran Turtles. *Fieldiana Geology, Memoirs. 3*(2), 17–56.

Zouhri, S. (2017). La Paléontologie des Vertébrés du Maroc. *Mémoires de la Société Géologique de France, 180,* 1–614.

Zouhri, S., Khalloufi, B., Bourdon, E., de Lapparent de Broin, F., Rage, J.-C., M'Haïdrat, L., et al. (2017). Marine vertebrate fauna from the Late Eocene Samlat Formation of Ad-Dakhla, southwestern Morocco. *Geological Magazine, 155*(7), 1–25, 1596–1620.

Chapter 8
The Age of Dinosaurs in the Land of Gonds

Sankar Chatterjee

Abstract The fossil record of dinosaurs from India provides a highly significant contribution to understanding the origin and evolution of dinosaurs and their paleobiogeographic significance. As India rifted from Gondwana and drifted northwards during the age of dinosaurs, the mobile episode in Indian geology provides a unique opportunity to study the diversity of dinosaurs in time and space. The dinosaurs from the Gondwana and post-Gondwana sediments of India have been collected and studied since their discovery in the 1920s, but the full range of their significance and evolutionary history remained fragmentary. After the independence of India, a renaissance arose in the study of dinosaurs at the Indian Statistical Institute (ISI) under the leadership of Pamela Robinson, as more and more dinosaur skeletons were discovered from different localities. This exploration by ISI paleontologists represented a pivotal moment in the history of vertebrate paleontology in India and became a starting point for a remarkable increase in our knowledge of Triassic, Jurassic, and Cretaceous dinosaur faunas. It inspired a new generation of students working under Ashok Sahni's direction at Panjab University to engage in the Cretaceous research. This paper offers an updated and comprehensive review of the anatomy, systematics, and evolution of Indian dinosaurs within historical, paleobiogeographic, and paleoecologic contexts. The occurrence of Indian dinosaurs is currently restricted to central and southern India, and the record extends across all three Mesozoic periods. It is generally regarded that dinosaurs originated in the Late Triassic Period in Argentina, about 230 million years ago. However, *Alwalkeria*, a theropod discovered in the Lower Maleri Formation of India, was contemporaneous with the oldest Argentinean dinosaurs. Similarly, *Barapasaurus* from the Early Jurassic Kota Formation is considered as one of the oldest, gigantic sauropod dinosaurs with a quadrupedal pose. The Late Triassic and Early Jurassic dinosaurs of India are diverse and document their early radiation. With the breakup of Gondwana, India began to disintegrate and drifted northwards, carrying its dinosaur fauna like a passenger ship, until it collided with the Oman-Kohistan-Ladakh Arc in the Late Cretaceous, forming a biotic corridor to Africa and Europe. The Late Cretaceous dinosaurs from the Lameta Formation, consisting of several species of titanosaurs and abelisaurs, provide intimate documentation of the last 'geologic minutes' before their extinction. Along with dinosaur bones, the largest titanosaurid hatchery is known from the Lameta Formation, extending for more than 1,000 km. Most egg clutches contain about 10 to 12 spherical eggs ranging in diameter from 15 to 20 cm. Surprisingly, these eggs were empty, showing no signs of embryos, perhaps indicating hatching failure during some environmental crisis. At the Cretaceous-Paleogene boundary, India was ground zero for two catastrophic events—the Shiva impact and Deccan volcanism—both linked to dinosaur extinction. The combination of twin asteroid impacts (Chicxulub and Shiva), with prolonged Deccan volcanism created an unprecedented and ultimately catastrophic environmental crisis across the globe, triggering the end-Cretaceous mass extinction.

Keywords Mesozoic • Sauropods • Abelisaurs • Cretaceous-Paleogene boundary • Shiva crater • India

Introduction

The dinosaurs were among the most successful animals ever to live on Earth, representing the glory of evolution and the mystery of extinction. For 166 million years, these reptiles ruled the Earth, flourishing in every continent from

S. Chatterjee (✉)
Department of Geosciences, Museum of Texas Tech University, 3301 4th Street, Lubbock, Texas 79409, USA
e-mail: sankar.chatterjee@ttu.edu

G. V. R. Prasad and R. Patnaik (eds.), *Biological Consequences of Plate Tectonics: New Perspectives on Post-Gondwana Break-up—A Tribute to Ashok Sahni*,
Vertebrate Paleobiology and Paleoanthropology Series, https://doi.org/10.1007/978-3-030-49753-8_8

Antarctica to well beyond the Arctic Circle. To the astonishment and delight of most Indians, dinosaur bones and their nesting sites have been found in different regions of central and southern India. In recent times, India is gaining prominence as one of the important centers for dinosaur radiation. Although numerically and taxonomically less impressive than those of North America, China, Argentina, or Mongolia, the fossil record of India is highly important for understanding the origin and early evolution of certain clades of dinosaurs in Gondwana. There is a fascinating correlation between evolution of Indian dinosaurs and the continental drifting of the Indian plate from Antarctica to Asia over 160 million years. In the Early Cretaceous, as Gondwana began to break up, the Indian subcontinent began its northward journey across the equator carrying its precious dinosaur fauna and its nesting sites like a passenger ship during its collision course to Asia (Chatterjee et al. 2017). In this paper, I describe our current understanding on the origin, evolution, distribution, and paleobiogeography of Indian dinosaurs in three successive periods of the Mesozoic—Triassic, Jurassic, and Cretaceous—and their extinction event at the Cretaceous-Paleogene boundary. Along with the new dinosaur finds and their evolutionary significance, I here provide a historical narrative of the development of vertebrate paleontology in independent India.

Dinosaurs in Gondwana and Post-Gondwana Formations

India was the type area of the Gondwana formations, represented by the basal sequence of Late Paleozoic tillites, followed by fluvial sequences of shale, sandstone, and coal beds containing distinctive seed fern *Glossopteris* (Medlicott 1872). These unusual rocks were named 'Gondwana' meaning 'land of the Gonds,' because this area in central India is the home of the current Gond tribes, who had established a kingdom in the 14th century. The stratigraphic correlation of the Indian Gondwana formations with similar Gondwana rocks across Antarctica, Australia, South Africa, and South America gave birth to the concept of the supercontinent Gondwana and the continental drift theory (Wegener 1915). This ancient supercontinent Gondwana incorporated present-day South America, Africa, Arabia, Madagascar, India, Australia, and Antarctica. It was fully assembled by Neoproterozoic time, some 600 million years ago, and the first stage of its breakup began in the Early Jurassic Period, about 180 million years ago.

The Gondwana basins of peninsular India are traditionally considered extensional rift-basins in older Precambrian basement, which were reactivated in the Late Paleozoic through long-time subsidence, creating accommodation space for infilling Gondwana sediments. The basin-fill succession of Gondwana sediments, about 5-km-thick sequence, is characterized by a complex mosaic of fluvial-lacustrine deposits that range in age from Early Permian through Late Jurassic. The overlying Early Cretaceous Rajmahal/Jabalpur formations, and the Late Cretaceous Lameta/Kallamedu formations, and Deccan Trap are not included within the Gondwana Supergroup since by that time India had separated from Gondwana and was embarked on its northward journey (Chatterjee et al. 2017). The Gondwana basins occur along the extant river valleys and include four depocenters: (1) the Damodar, (2) the Son-Mahanadi, (3) the Satpura, and (4) the Pranhita-Godavari, of which the Pranhita-Godavari Valley strata contain a rich assemblage of early dinosaur fossils.

The fossil record of dinosaurs is biased toward lowland sedimentary environments especially the flood plain deposits. Unlike other parts of the world, the dinosaur faunas from Indian Gondwana and post-Gondwana sediments are limited in diversity through poor sampling and exploration, but are important from an evolutionary perspective. Indian dinosaur fossils hold the key to their early origin in the Late Triassic time, their acquisition of gigantic size in the Jurassic, their nesting behavior, dietary preference, in the Late Cretaceous during India's northward journey and finally their ultimate demise at the end of the Cretaceous.

There are eight successive horizons in the Gondwana and post-Gondwana sequences from which dinosaur faunas are currently known: (1) Upper Triassic (Late Carnian-Early Norian) Lower Maleri Formation); (2) Upper Triassic (Late Norian) Upper Maleri Formation; (3) Upper Triassic (Rhaetian) Lower Dharmaram Formation; (4) Lower Jurassic (Hettangian) Upper Dharmaram Formation; (5) Lower Jurassic (Sinemurian-Toarcian) Kota Formation; (6) Upper Jurassic (Tithonian) Bagra Formation; (7) Upper Cretaceous Nimar Formation (Cenomian-Turonian); and (8) Upper Cretaceous (Maastrichtian) Lameta and Kallamedu formations. Among these, the last two formations represent post-Gondwana sediments deposited following the break-up of Gondwana during the Cretaceous. These dinosaur-bearing horizons of India are shown in Table 8.1. It appears from the Table that India has a rapidly growing dinosaur record from Carnian to Maastrichtian with some major gaps in the Late Jurassic-Early Cretaceous interval (Chatterjee et al., 2017). Using an allegory of the four parts of the day, the Indian dinosaur dynasty is divided here into four episodes: dawn, day, evening, and night—chronologically narrating their origin, radiation, diversification, and finally their demise.

Table 8.1 Composite Gondwana and post-Gondwana formations of India (Dinosaur – bearing formations are shown by asterisk)

Age (Ma)		Era	Period	Epoch	Formation	Plate Tectonic and Global Event
66	POST-GONDWANA			Danian	DECCAN TRAP	END-CRETACEOUS EXTINCTION
		MESOZOIC	Cretaceous	Late	Lameta*/Kallamedu* Nimar*/Chikiala/Gangapur	OKL Arc Corridor Separation of Madagascar from India
				Early	RAJMAHAL TRAP	Separation of India from East Gondwana
					Rajmahal/ Jabalpur	
145	GONDWANA SUPERGROUP		Jurassic	Late	Bagra*	Gondwana Breakup
				Middle	Upper Kota*	
				Early	Lower Kota* Upper Dharmaram*	
201			Triassic	Late	Lower Dharmaram*	END-TRIASSIC EXTINCTION
					Maleri*/Tiki	
				Middle	Yerrapalli / Denwa	
				Early	Panchet / Kamthi	
252		PALEOZOIC	Permian	Late	Kundaram	END-PERMIAN EXTINCTION
					Raniganj	
				Middle	Barren Measures	
					Barakar	Gondwana Coal Formation
				Early	Karharbari	Gondwana Deglaciation
					Talchir	Gondwana Glaciation
300		PRECAMBRIAN				
541						Gondwana Supercontinent
3000			Precambrian		Precambrian basement	

Dawn: The Beginnings of Dinosaurs in the Late Triassic

The Triassic Period began soon after the most catastrophic event in geological history, at the end of the Permian, when life on Earth was almost (95% of species) completely wiped out during the end-Permian mass extinction by an environmental holocaust of a magnitude never seen before. In the aftermaths of the end-Permian extinction, some forms of life rebounded as death created opportunity, and survivors occupied vacant niches. Groups of animals that formerly played minor roles now assumed prominence, and new groups appeared.

The Late Triassic Period, ranged from 235 to 200 million years ago, was a pivotal time in vertebrate evolution - several major groups such as lissamphibians, turtles, lizards, crocodiles, dinosaurs, birds, mammals, and numerous extinct groups of archosauromorphs and archosaurs first appeared in the fossil record. When dinosaurs first originated in the terrestrial ecosystems in the Late Triassic, all the major continental plates were joined into a single supercontinent, Pangea, which slowly drifted northward (Wegener 1915) (Fig. 8.1). As Pangea rifted apart during the Late Jurassic and Cretaceous periods, a vast array of new body plans, structures, and levels of complexity evolved in the dinosaurs because of geographic isolation.

When dinosaurs first appeared in the fossil record during the Carnian stage of the Late Triassic as rare members of the archosaur communities, all the continents were still joined together in Pangea (Wegener 1915). One of the remarkable features of the Triassic was the widespread emergence of continents, followed by the extensive spread of nonmarine deposits, composed largely of redbeds. These redbeds were deposited in a complex river-floodplain-lacustrine system in many parts of the world and have yielded a rich record of the Triassic vertebrate fauna. In the Pangean world, dinosaurs and other tetrapods could migrate to any part of the globe over dry land, so continental Triassic vertebrates are quite similar across Pangea. However, there were both climatic

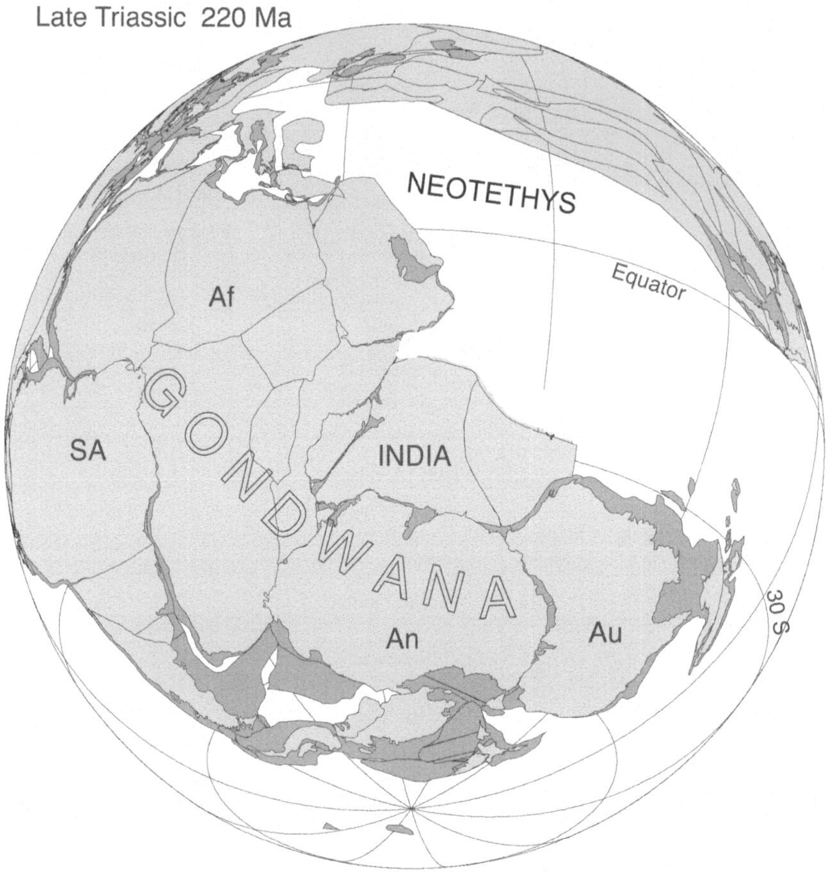

Fig. 8.1 Paleogeographic reconstruction of Gondwana during the Late Triassic (~220 Ma) (modified from Chatterjee and Scotese 1999)

and physical barriers that impede migration of some groups of tetrapods (Fig. 8.2). Climates were warm and uniform, since Pangea was symmetrically disposed above and below the equator, and dominated by monsoonal circulation in the coastal region. The heartlands of the Pangea were hot and dry. There were no ice caps in the polar regions.

Dense forests and swamps covered the land from the equator to high latitudes. The Triassic *Dicroidium* flora includes *Dicroidium*, spore-bearing true ferns, seed-bearing ferns, conifers, horsetails, cycads, and ginkgoes, which would have formed forests and swamps along the water's edge. The Triassic meadows and prairies of lowlands would have been covered in ferns, low cycads, club mosses, and horsetails. The uplands were dominated by coniferous forests. Inside these forests, a bewildering variety of small dinosaurs inhabited the dense understory. Other animals in the scene include several groups of archosaurs including crocodilians, pterosaurs, and birds, as well as frogs, turtles, lizards, and mammals.

In the Triassic Period, archosaurs replaced the therapsids as the dominant component of the terrestrial communities and became highly diversified. Archosaurs are divided into two primary lines, one leading to crocodilians (Crurotarsi), the other line to birds (Ornithodira) (Fig. 8.3A) (Chatterjee 1982a, 2015). The key character for the basal dichotomy of archosaurs seems to be different style of ankle joint. The Crurotarsi, which include phytosaurs, aetosaurs, rauisuchians, ornithischians, and crocodylomorphs, all have a crurotarsal joint. In this group, the astragalus and calcaneum articulate with each other by means of a peg-and-socket joint, allowing rotational movement between them.

The Ornithodira includes pterosaurs, lagosuchids, silesaurids, and dinosaurs (including birds). In this assemblage, the ankle joint is a mesotarsal hinge between the proximal and distal rows of the tarsal bones (Fig. 8.3D, E).

The Late Triassic is often called the 'Dawn of the Age of Dinosaurs,' even though dinosaurs were small and inconspicuous at that time and certainly not yet dominant. For millions of years they were but a small part of the reptilian fauna. These first dinosaurs were no larger than a terrier, with a long neck and a long and stiff tail. They had to compete for survival with larger contemporary reptiles, most notably the crocodile-like phytosaurs and fearsome rauisuchians. By the time early dinosaurs came on the scene, they had already evolved into three major lineages—sauropodomorphs, theropods, and ornithischians—with different dietary behaviors and modes of locomotion.

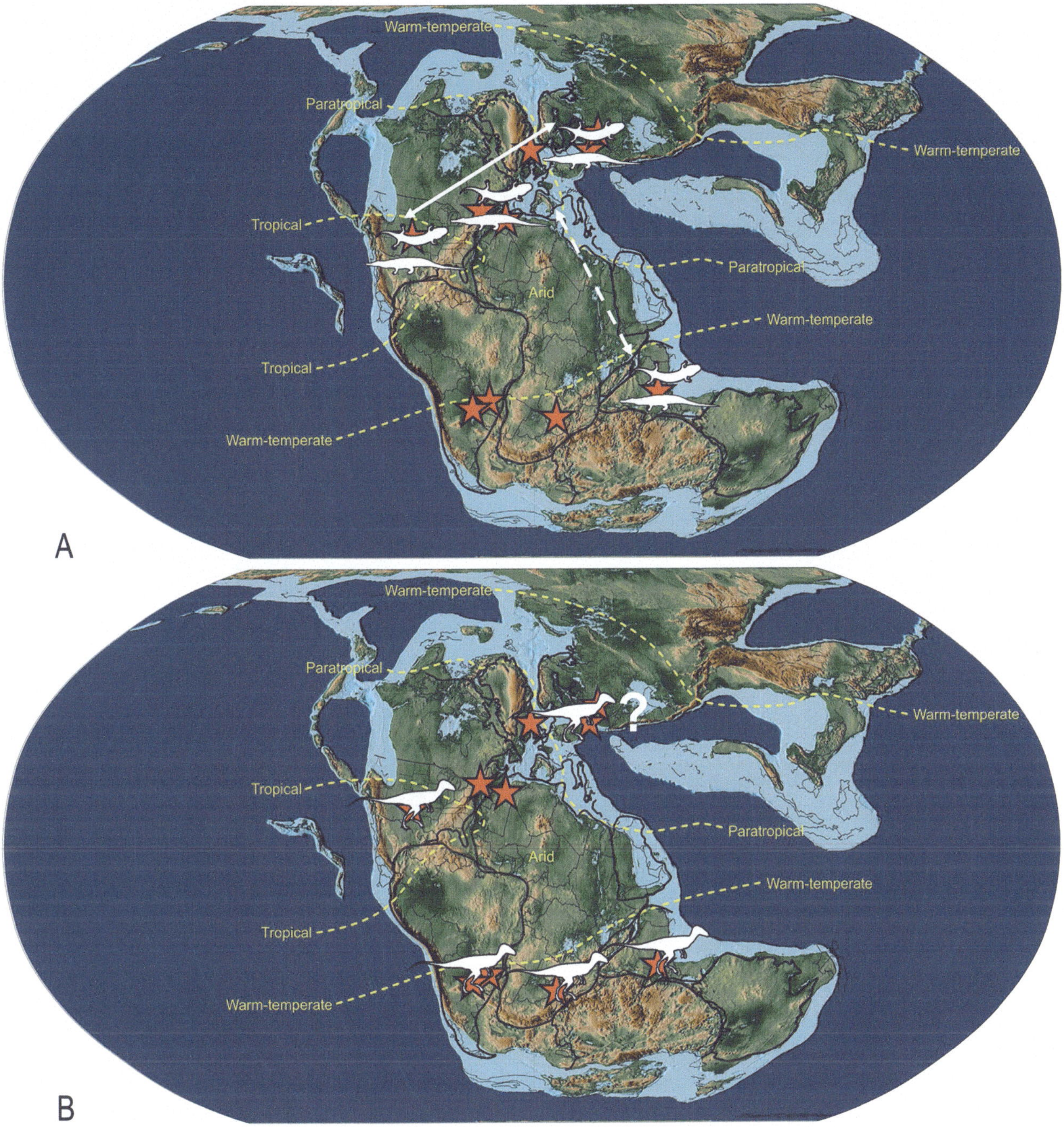

Fig. 8.2 The Late Carnian-Norian paleogeographic map, displaying the radiation of tetrapods; locations of fossil sites of Late Triassic tetrapods are shown by orange stars. **A**, semiaquatic phytosaurs and metoposaurs showing some degree of provincialism; and **B**, early dinosaurs. Among Gondwana continents, only India and Morocco had yielded phytosaur-metoposaur assemblage, which is common in Laurasia. For early radiation of dinosaurs, southern Gondwana such as Argentina and India might be the centers of their origins. Dinosaurs were rare in the early Late Triassic. (courtesy Volkan Sarigul)

The beginning of dinosaur evolution is currently known from the Carnian stage of the Late Triassic (about 230 million years ago) Gondwana beds of Argentina (Ischigualasto Formation), Brazil (Santa Maria Formation), Zimbabwe (Pebbly Arkose Formation) and India (Lower Maleri Formation). By the Carnian stage, dinosaurs had diversified into three major lineages, Ornithischia, Sauropodomorpha, and Theropoda—with different dietary behaviors and modes of locomotion, and by the Norian stage (around 208 million years ago), some dinosaur groups had become species-rich

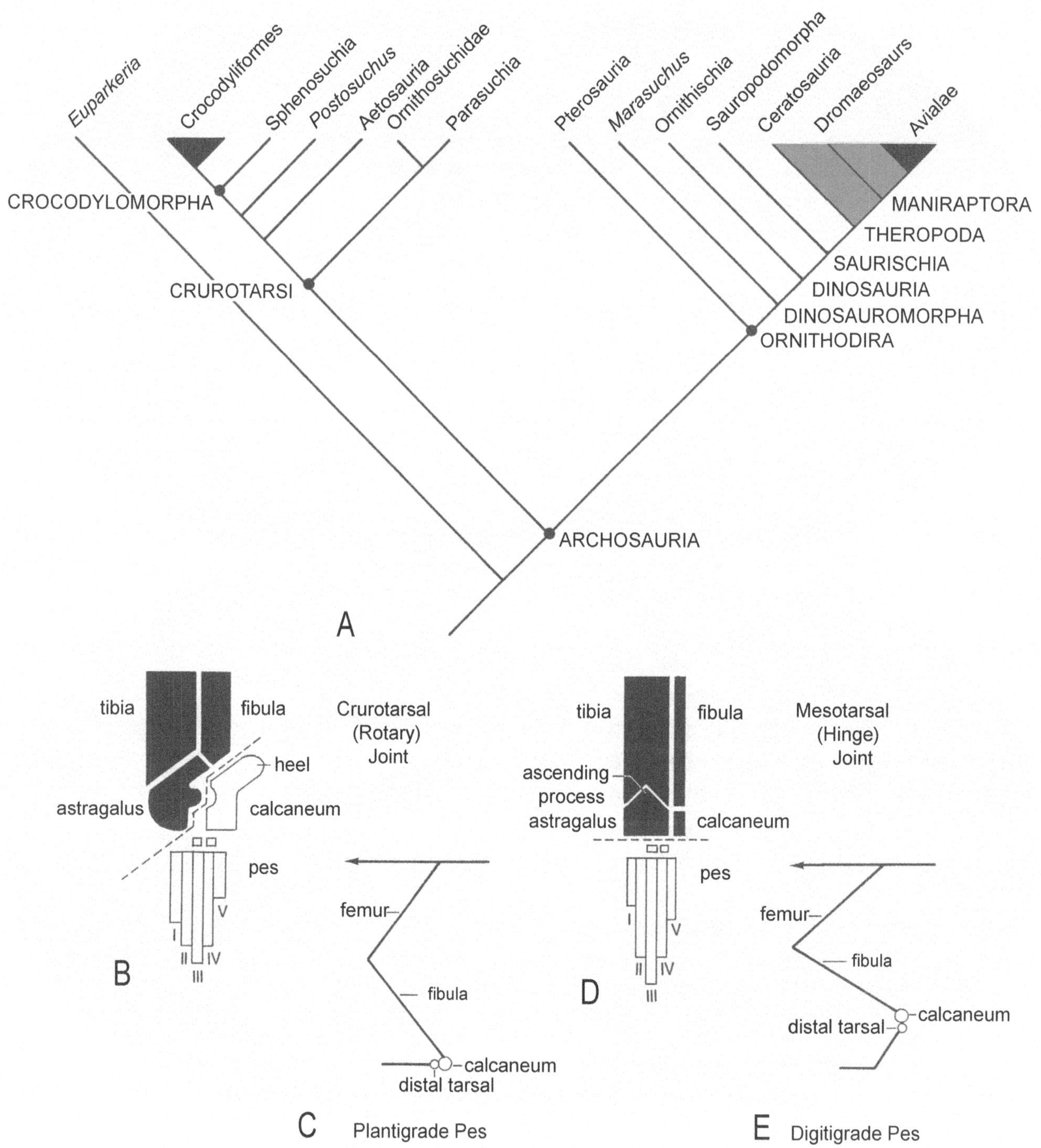

Fig. 8.3 The phylogeny of archosaurs. **A**, cladogram showing the postulated relationships of the basal dichotomy of archosaurs into Crurotarsi (leading to crocodilians) and Ornithodira (leading to birds); **B**, the crurotarsal ankle joint, in which the astragalus forms part of the crus, but the calcaneum, distal tarsals, and combined metatarsals move as a unit on the astragalus and fibula. The peg-and-socket joint between the astragalus and calcaneum allows rotary motion between them; **C**, schematic lateral view of the hindlimb of the Crurotarsi showing the primitive, plantigrade pose; **D**, the characteristic hinge joint of Ornithodira between the proximal and distal rows of the tarsi. The astragalus has a typical ascending process for locking the tibia; the calcaneum is reduced; **E**, schematic lateral view of the hindlimb of the Ornithodira showing improved posture and the digitigrade pes

and numerically abundant. By the end of the Triassic, dinosaurs had also diversified into relatively large, plant-eating forms such as basal sauropodomorphs.

Here I describe the early dinosaurs from India from three successive horizons: a rhynchosaur-dominated Lower Maleri Formation, an archosaur-dominated Upper Maleri, and the

Lower Dharmaram Formation with abundant basal sauropodomorphs but rare theropod members. Similar Late Triassic assemblages occur in the coeval beds of Argentina, Brazil, and Europe, but contrast sharply with North American faunas.

Discovery of Indian Dinosaurs

These are exciting times in vertebrate paleontology, heralding a dinosaur renaissance. So many new discoveries are being made across the globe—from North America, Europe, South America, China, Mongolia, Africa, Australia, and Antarctica; so many new ways of looking at fossils are being devised. The dinosaur fossils of India have much to contribute to the fervor. This is why it is so exciting for me to write this paper. Our story begins with vertebrate collecting in independent India and with the people who did the digging.

In my formative years, I joined the Indian Statistical Institute (ISI) as a Research Scholar in 1964 to work with the British paleontologist Pamela Robinson of University College London, who was a Visiting Professor and scientific advisor at the Geological Studies Unit of ISI. Dr. Robinson was invited through the influence of her mentor and famous geneticist J. B. S. Haldane, who at that time relinquished British citizenship and joined ISI. ISI is a world-renowned academic institution devoted to the research, teaching and application of statistics, mathematics, computer sciences, natural sciences, and social sciences. Dr. Robinson established the Geology unit at ISI and created an integrated research program to foster the study of vertebrate paleontology of the Gondwana formations and disseminate the knowledge (Fig. 8.4A, E). She mentored and influenced other vertebrate paleontologists at the department including Sohan Lal Jain, Tapan Roy Chowdhury, T. S. Kutty, and me. She taught us how to explore new fossils sites by systematic excavation, to record the field data, to map the Gondwana formations, to study the environment of deposition, and to delve into scientific potential, significance, and uniqueness of Indian dinosaurs in paleobiogeography and evolution. Her confidence in our ability ignited our enthusiasm. Sacrificing her own research, she chose to share with us her dream of creating an Indian center for natural history. When I became her Research Associate at the University College, London I got to know Dr. Robinson's vision, dreams, research, scientific rigor, and frustrations more closely. Dr. Robinson began the renewal of vertebrate paleontology in newly independent India. During that time, the Geology Unit of ISI became a vital research center, stimulating an intellectual resurgence for paleontology. Many famous paleontologists from America such as Alfred S. Romer of Harvard University (Fig. 8.4C, E) and Edwin S. Colbert (Fig. 8.4D) from the American Museum of Natural History visited our unit, gave seminars, studied fossil collections, shared their insights, and participated in fossil excavation. It was a golden age of vertebrate paleontology in India. Although most of the ISI paleontologists mentioned above are no longer with us, the tradition is continued today by Saswati Bandopadhopadhya, Dhurjati Sengupta, and subsequently their students, who made, and continue to make important strides in vertebrate paleontology research.

Soon after, two other centers—Geological Survey of India and Panjab University—began the systematic collection of vertebrate fossils from the Gondwana formations, following in the footsteps of ISI. From GSI, P. M. Datta, D. M. Mohabey, Z. G. Ghevariya, P. P. Satsangi, and P. Yadagiri made significant contributions to the discovery of vertebrate fossils. Similarly, my good friend Ashok Sahni would inspire and guide the outstanding young graduate students in Panjab University in vertebrate paleontology (Fig. 8.4B). Some of his students including Guntupalli Prasad, Sunil Bajpai, Ashu Khosla, K. Kumar, R. S. Loyal, and R. S. Rana would continue his legacy in vertebrate paleontology. This paper is a tribute to Ashok Sahni for his great contribution and dedication to Indian vertebrate paleontology.

Lower Maleri Formation (Late Carnian-Early Norian)

The first world of Indian dinosaurs began about 230 million years ago in the Upper Triassic Maleri Formation. Early in their history, dinosaurs were small and inconspicuous in the Late Triassic ecosystems. Even though we know a great deal about early dinosaurs from their fossil record in Argentina, such as *Herrerasaurus* and *Eoraptor*, it is not certain that South America was the exact cradle of their origin because they are also found quite early in the Lower Maleri Formation of India during the same time. Most likely, dinosaurs started in Argentina or India and as they dispersed across Pangea, they were evolving into new species. The Maleri is rich in fossils of many kinds. There are plants, together with an abundance of species of various tetrapods. Most importantly for us, there are dinosaurs. In the Lower Maleri Formation, as in other parts of the world, dinosaurs appear to have been minor players on the ecological stage, but by the Upper Maleri time, they became more prominent and numerous.

The Maleri Formation in the Pranhita Godavari Valley of the Adilabad district, in the northern region of Telengana, was known for tetrapod fossils for almost a century (Table 8.1). Vertebrate fossils were collected in 1866 by Reverend Stephen Hislop of Nagpur, and later on by the

Fig. 8.4 Visiting scientists and resident paleontologists at the Geological Studies Unit of the Indian Statistical Institute in early 1970s. **A**, Pamela Robinson; **B**, Ashok Sahni; **C**, Alfred Sherwood Romer and author; **D**, the author and Ned Colbert; **E**, Geology Museum of the Indian Statistical Institute, Kolkata, during the visit of Alfred Sherwood Romer of Harvard University. At the background, partial skeleton of the Kota sauropod *Barapasaurus* and a mounted skeleton of a Maleri rhynchosaur *Hyperodapedon* (at the center); from left to right: 1, Tapan Roy Chowdhury; 2, Al Romer; 3, the author; 4, Pamela Robinson; 5, Sohan Lal Jain; 6, T. S. Kutty. Other paleontologists in the figure came from the Geological Survey of India and University College, London

members of the Geological Survey of India (GSI), mostly as surface finds. The whole GSI collection of the Maleri fossils was sent to the famous German paleontologist Frederick von Huene of Tübingen University, who had worked with similar fossils from Brazil and East Africa. Huene (1940) described the fragmentary remains of the Maleri vertebrates such as metoposaurs, rhynchosaurs, and phytosaurs in the Memoirs of the Geological Survey of India.

Dr. Pamela Robinson suggested that I begin field work in the Lower Maleri Formation in Adilabad district of Andhra Pradesh, looking there for vertebrate fossils as part of my dissertation project. Huene's memoir was my guide in searching out the old fossil localities. On my first field trip in the Pranhita-Godavari Valley, I was mesmerized by the stark beauty of the landscape of the Maleri Formation. The redbeds there rise – a rainbow spectacle of red, white, yellow, lavender, and blue in the badlands, ablaze with dramatic light and color. Here in these redbeds, in the very first field season, I discovered several associated skeletons of rhynchosaurs, *Hyperodapedon huxleyi*; one such complete skeleton was mounted at the Geology Museum of ISI (Fig. 8.4). During repeated field works in this area, I had the pleasure and thrill of discovering other spectacular skeletons of fossil tetrapods from the Lower Maleri Formation including malerisaurids, phytosaurs, cynodonts, and their relatives, which are currently housed at the Geology Museum of ISI. The most common tetrapods include a metoposaurid tetrapod *Metoposaurus* (Roy Chowdhury, 1965), a protorosaurid *Malerisaurus robinsonae* (named in honor of Dr. Robinson; Chatterjee, 1980), an archosauromorph *Hyperodapedon* (Chatterjee, 1974), a phytosaur *Parasuchus* (Chatterjee, 1978a) (Fig. 8.5), and a cynodont *Exaeretodon statisticae* (named in honor of Indian Statistical Institute; Chatterjee 1982b).

One predatory dinosaur *Alwalkeria maleriensis* is one of the earliest known dinosaurs in the world, lived in the Maleri ecosystem around 230 million years ago (Chatterjee 1987). I named the genus *Alwalkeria* in honor of my former mentor Alick Walker of the University of Newcastle upon Tyne, England, who pioneered and advanced the field of archosaur anatomy. *Alwalkeria* was a small, lightly-built, gracile theropod, like *Eoraptor* of Argentina, and would be about 1.5 m long with an estimated weight of 5 kg. The partial skull of *Alwalkeria* is about 4 cm long, with a subnarial fossa, and the jaws were equipped with sharp, conical heterodont teeth. The hind foot suggests its cursorial adaptation. *Alwalkeria* was relatively small compared to its contemporary archosaurs in the Maleri ecosystems such as phytosaurs and rauisuchians, but fleet-footed. In size and proportion, *Alwalkeria* is comparable with *Eoraptor* of the Ischigualasto Formation of Argentina. The premaxilla is loosely attached to the maxilla but interrupted by a subnarial fossa; and the maxilla shows deep antorbital fossa. Except for some small nipping teeth at the front, all other teeth of *Alwalkeria* were recurved, sharp but unserrated, excellent for capturing small prey and insects. The lower jaw was fused at the front with a symphysis, allowing to move upper and lower teeth against each other in orthal motion. Perhaps *Alwalkeria* fed on carrion or small vertebrates (Fig. 8.6).

In 1970, I wanted to explore the Late Triassic Tiki Formation in the Rewa Basin of the Son-Mahanadi Valley in central India, which is coeval with the Lower Maleri Formation and has yielded typical Maleri fauna such as rhynchosaur-phytosaur-metoposaur assemblage. At that time, the famous vertebrate paleontologist from Harvard University, Professor Alfred Sherwood Romer and his wife Ruth were visiting ISI. Pamela Robinson brought Dr. and Mrs. Romer to my field area, where they stayed for a week to participate in the field work; in the evenings, sitting around the campfire we sampled the local mohua brew, which inspired stories from Romer late into the night. It was a memorable experience to hear Romer narrating his expeditions in South America and Texas and his insights in vertebrate paleontology and evolution. At the time, I was excavating an articulated skeleton of a new rauisichian from the Tiki Formation with Pranab Majumdar. Romer offered valuable suggestion in the field during excavation (Fig. 8.5 B, C). Later I christened the name of this Tiki rauisuchian as *Tikisuchus romeri* in honor of Prof. Romer (Chatterjee and Majumdar 1987). *Tikisuchus* was the top predator in the Tiki and Maleri ecosystems, overshadowing the emerging dinosaurs such as *Alwalkeria*.

Upper Maleri Formation (Late Norian)

My colleague at the Indian Statistical Institute (ISI), T. S. Kutty was a prolific fossil hunter (Fig. 8.4). While I was prospecting the Lower Maleri Formation, Kutty was engaged in mapping and collecting the Upper Maleri and Dharmaram formations in adjacent areas of Adilabad district and there discovered a series of amazing early dinosaur remains. He never got a chance to publish these specimens. Before his retirement, he handed over to me all the dinosaur specimens he had collected for collaborative works. Since spectacular specimens of early dinosaurs are known from the Late Triassic Gondwana beds of Argentina, I invited my Argentinean colleagues Fernando Novas and his student Martin Ezcurra to study and compare Indian specimens with those of Argentina. The detailed study of the Upper Maleri dinosaurs and their phylogenetic relationships filled a major gap in the early history of dinosaur radiation (Novas et al. 2011). The typical Lower Maleri tetrapods such as rhynchosaur *Hyperodapedon*, protorosaur *Malerisaurus*, phytosaur *Parasuchus*, and cynodont *Exaeretodon* apparently

Fig. 8.5 Field photographs in the Late Triassic Maleri and Tiki formations during excavation. **A**, Dhuiya Pradhan, the author, driver, and a local field guide watching two articulated skeletons of the Maleri phytosaur *Parasuchus*. Inside each skeleton of *Parasuchus* in the stomach region, undigested remains of a prey *Malerisaurus* were found; **B**, Al Romer and the author in the Tiki field look at the newly discovered rauisuchian skeleton *Tikisuchus*; **C**, Pamela Robinson, Al Romer, the author, and Ruth Romer discuss on *Tikisuchus*, the largest predator in the Triassic

disappeared in the transition time between the Lower and Upper Maleri formations. The Upper Maleri dinosaur fauna includes three basal sauropodomorph dinosaurs: an unnamed guaibasaurid (ISI R277), a nonplateosaurian *Nambalia roychowdhurii* and a plateosaurian *Jaklapallisaurus asymmetrica*, thus suggesting that India may be an important center

Fig. 8.6 Paleoecology of the Late Triassic tetrapods from the Lower Maleri Formation. At the back, three individuals of *Hyperodapedon* startled by the violent action of gharial-like *Parasuchus*, seizing a *Malerisaurus*, while her mate flees to the right. At the middle, two individuals of theropod dinosaur *Alwalkeria* devouring a baby of *Malerisaurus*. In the water, in the far right, a salamander-like *Metoposaurus* is cautiously watching the predation of *Parasuchus* (after Chatterjee et al. 2017)

for the origin and early radiation of sauropodomorph dinosaurs. The specific name of *N. roychowdhurii* is given in honor of our colleague at ISI, Tapan Roy Chowdhuri (Fig. 8.4). The phylogenetic relationships of the Late Triassic Maleri dinosaurs are shown in a simple cladogram highlighting the Maleri taxa (Fig. 8.7).

Two unnamed basal dinosaurs are fragmentary. ISI R282 is represented by two sacral vertebrae, a proximal caudal, a perforated right ilium, proximal and distal ends of the right pubis, and a probable left pubis. Some elements resemble those of the basal dinosaurs of Argentina such as *Staurikosaurus* and *Herrerasaurus*, other characters are unique to this taxon. ISI R284 is represented by a solitary right ilium, showing some dinosaurian hallmarks such as perforated acetabulum, highly arched dorsal margin of the iliac blade, and the preacetabular process which is strongly compressed transversely. Because of the fragmentary nature, both specimens are difficult to diagnose.

The dinosaurs from the Upper Maleri Formation are dominated by basal sauropodomorphs. The basal sauropodomorphs have been found in all continents including Antarctica and are common in the Norian beds of Argentina (*Coloradosaurus*, *Mussasaurus*, *Riojasaurus*) and Germany (*Sellosaurus*, *Efraasia*, *Plateosaurus*, and *Ruehleia*). ISI R277 is very similar to *Guaibasaurus*, a primitive saurischian dinosaur from the Early Norian Caturrita Formation of Brazil. *Guaibasaurus* is known from a partial skeleton and appears to be more closely related to sauropodomorphs than theropods. It was a bipedal dinosaur around 2 m long, with short, sharply clawed forelimbs but its affinity is controversial. Bonaparte et al. (2007) concluded that a combination of several postcranial features distinguishes *Guaibasaurus* from all other dinosaurs, and it is closely related with *Saturnalia*. The Maleri specimen is represented by several postcranial elements, which compare well with those of *Guaibasaurus* and was allocated in the same family

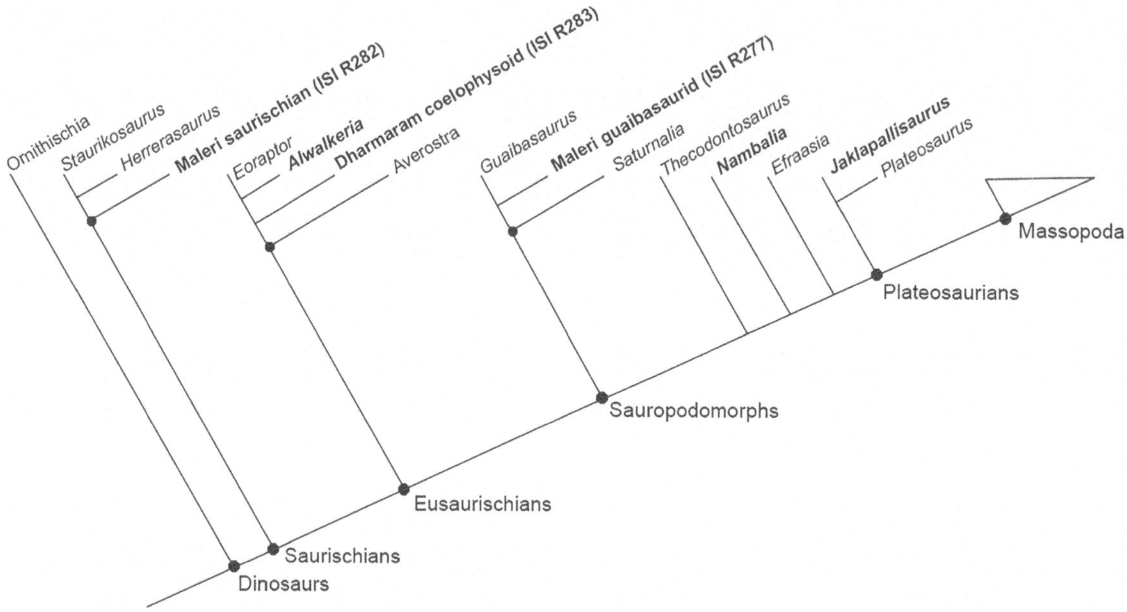

Fig. 8.7 Cladogram showing the origin and early evolution of dinosaurs during the Late Triassic of India. Indian taxa are shown in bold fonts

of Guaibasauridae (Novas e al. 2011). In our phylogenetic analysis, the Maleri guaibasaurid appears to be a stem sauropodomorph and a sister-group of *Nambalia* (Fig. 8.7).

Two distinctive basal sauropodomorphs are represented by *Nambalia* and *Jaklapallisaurus* respectively, which are more derived than *Guaibasaurus*. *Nambalia* is known from partially articulated postcranial material. It is a basal sauropodomorph dinosaur that shows several autapomorphies in femur and astragalus morphology. It is more derived than *Thecodontosaurus* of England but closely related to *Efraasia* of Germany (Fig. 8.7). *Nambalia* was a lightly built, small-sized, herbivorous quadrupedal sauropodomorph, about 3 meters long. *Jaklapallisaurus* is more derived than *Nambalia*, and is a plateosaurian sauropodomorph dinosaur, known from postcranial material and closely resembles *Plateosaurus* from Germany (Novas et al. 2011). *Jaklapallisaurus* was a medium-sized sauropodomorph, lightly built, about 4 meters long. *Nambalia* and *Jaklapallisaurus* were the first large herbivorous dinosaurs in India that probably lived in herds and their size gave them a larger range of plants on which to feed (Fig. 8.8).

Lower Dharmaram Formation (Rhaetian)

The vertebrate fauna of the Lower Dharmaram Formation is represented by pseudosuchians such as aetosaurs and advanced phytosaurs, which died out at the end-Triassic extinction (Kutty and Sengupta 1989). The dinosaurs are represented by fragmentary remains of sauropodomorphs and neotheropods. Some elements of sauropodomorphs (ISI R279, 280, 281) resemble *Jaklapallisaurus* among basal sauropodomorphs. An isolated femur with a strongly inturned head (ISI R283) is reminiscent of neotheropods such as *Coelophysis*. The Dharmaram coelophysoid can be excluded from Averostra because it lacks a proximally well-developed anterior trochanter extending beyond the level of the ventral margin of the femoral head. All neotheropods became extinct at the end of the Triassic Period except for Averostra (Novas et al. 2011).

The Lower Dharmaram Formation yielded the last vestiges of crurotarsan assemblage including phytosaurs (*Nicrosaurus*) and aetosaurs (*Desmatosuchus* and *Paratypothorax*) before their final exit at the end-Triassic extinction. The Triassic-Jurassic boundary lies between the Lower and Upper Dharmaram formations marking the end-Triassic mass extinction event about 201 million years ago.

End-Triassic Mass Extinction Event

At the end-Triassic mass extinction, most of the non-dinosaurs died out, but dinosaurs were unscathed. On land, all archosaurs other than crocodylomorphs and Avemetatarsalia (pterosaur and dinosaurs), some remaining theropods, and many large tetrapods became extinct. This extinction event vacated terrestrial ecological niches, allowing the dinosaurs to assume the dominant role in the Jurassic period. The cause of the end-Triassic extinction is a matter of considerable debate. The extinction event is typically attributed to climatic changes associated with the

Fig. 8.8 Paleoecology of the Late Triassic dinosaur community in the Upper Maleri landscape. In the foreground is a small herd of five predatory dinosaurs similar to *Coelophysis*, three are engaged in dividing a vulnerable prey, while two other wanders in search of further opportunity. In the background left two herbivorous sauropodomorphs *Jaklapallisaurus* emerge from the *Araucarioxylon* forest: one stoops to drink from a pool of water while the other watches with mild interest the conflict in the foreground. On the far right, standing on hind legs, a sauropodomorph *Nambalia* stretches to get at leafy vegetation. Crouched in the center is a small guaibasaurid munching on a fallen conifer tree, inciting the curiosity of the two wandering *Coelophysis* type dinosaurs. On the extreme left two small mammals squat on a river stone

degassing of basaltic flows from the massive volcanism of the Central Atlantic magmatic province (CAMP) during the initial rifting of the Pangea (McHone 2003).

The victims of this end-Triassic mass extinction included therapsids and largely crurotarsan archosaurs such as phytosaurs, aetosaurs, and rauisuchians that moved slowly. It is intriguing why dinosaurs, who kept a low profile during the Late Triassic, survived the end-Triassic extinction and went on to dominate terrestrial fauna during the Jurassic and Cretaceous periods. Certainly, the bone histology of dinosaurs indicates rapid growth, and perhaps high metabolic activity (Padian et al. 2001). During the end-Triassic global warming, food became scarce for all the animals, including the dinosaurs. But early on dinosaurs adapted important anatomical features such as improved posture and gait, and high metabolic activity that may have prepared them during the climatic crisis, giving them greater mobility for long-distance travel in search of food. Due to scarcity of vegetation and decline of prey population, competition for available food resources became severe, and the balance of power shifted from crurotarsan archosaurs to dinosaurs. In this power vacuum, dinosaurs emerged as the winner by default, becoming the great, terrifying giants that ruled the Jurassic and the Cretaceous worlds.

Day: Triumph of the Dinosaurs in the Jurassic

Dinosaurs survived the end-Triassic catastrophe and rebounded with new vigor. Pangea was largely intact during most of the Jurassic Period, allowing dispersal of dinosaurs unhindered across the globe. In the Early Jurassic, dinosaurs came to dominate every niche on land. While sauropodomorphs were still the dominant herbivores, the giant sauropods began to take their place. The descendant of Triassic *Coelophysis*, the Averostra, grew into powerful predators. The Jurassic Period was the time of the titans, particularly during the latter part. Many new dinosaurs emerged in great numbers and varieties, both carnivores and herbivores. Among them were stegosaurs, brachiosaurs, camarasaurs, allosaurs, and many others. Much of the Jurassic was a warm, moist, hothouse world. The warm, wet climate allowed for the proliferation of lush, green gymnosperm vegetation across the world. On land, dinosaurs and flying pterosaurs dominated the ecosystems, and birds made their first appearance. Early mammals were still small but diversified. Towards the end of the Jurassic Period, the supercontinent began to rift. North America rotated toward the northwest, away from Africa, creating the proto-Atlantic Ocean. Because of the breakup of Pangea, the global climate began to change. The interior of the land masses became more humid and seasonal snow and ice frosted the polar regions. In the Southern Hemisphere, India was in the warm temperate region of Gondwana (Chatterjee et al. 2017). Although Pangea was slowly breaking up, many large landmasses remained intact. As a result, dinosaur communities were similar across the globe.

Upper Dharmaram Formation (Early Jurassic, Hettangian)

In the early Jurassic, the basal sauropodomorphs ('prosauropods') were the dominant herbivores. Gondwana was intact during the Early Jurassic Period (Fig. 8.9). The second phase of dinosaurs on the Indian subcontinent, represented by the Jurassic forms, was not all like the first. The dinosaurs were not the same, the ecosystems were not the same. As the Jurassic progressed, the world became wetter, especially at the higher latitudes and the giant sauropods took over. The Upper Dharmaram Formation contained an abundance of sauropodomorph dinosaurs, but none of the Late Triassic crurotarsans such as phytosaurs, aetosaurs, and rauisuchians survived, thus hinting at the victory of dinosaurs by default. The Upper Dharmaram fauna is comparable to that of the Early Jurassic horizons of the Upper Elliot and Clarens Formation of southern Africa, the Kayenta Formation of Arizona, and the Lower Lufeng Formation of China.

The associated fauna of the Upper Dharmaram Formation consisted of a sphenosuchian crocodylomorph, a neotheropod, and two sauropodomorphs (Chatterjee et al. 2017; Kutty et al. 2007). The sphenosuchian crocodylomorphs were completely terrestrial cursors that interacted with the dinosaurs. The Indian sphenosuchian compares well with *Sphenosuchus* of the Elliot Formation of South Africa, *Kayentasuchus* from the Kayenta Formation of Arizona, and *Dibothrosuchus* from the Lower Lufeng Formation of China. Among dinosaurs both theropods and sauropodomorphs are known. Theropod material is fragmentary, but some elements show striking similarities to those of *Dilophosaurus*, a neotheropod from the Kayenta Formation of Arizona. *Dilophosaurus* measured about 7 m long and weighed about 400 kg.

In honor of our ISI colleagues, I and my two coauthors named and described two sauropodomorphs from the Upper Dharmaram Formation: *Pradhania gracilis* and *Lamplughsaura dharmaramensis*, (Kutty et al. 2007). During our field work in the Pranhita-Godavari Valley, gifted fossil collector Dhuiya Pradhan was our outstanding field assistant (Fig. 8.5A). Dhuiya never went to a school and could barely sign his name, yet he could easily identify every bone in the field. I was always impressed with his anatomical knowledge and his native instinct for finding new fossil sites. We named the genus *Pradhania* in honor of Dhuiya Pradhan. *Pradhania* is known from a few skull elements, vertebrae and hand bones. It was a massospondylid sauropodomorph of modest size, about 4 m long and weighing about 1000 kg, closely resembling *Massospondylus* of South Africa, but more derived than *Riojasaurus* of Argentina (Fig. 8.10).

Similarly, we selected the generic name *Lamplughsaura* in honor of late Pamela Lamplugh Robinson of University College London, our mentor at ISI. *Lamplughsaura* is heavier and larger than *Pradhania* and is known from several partial skeletons (Fig. 8.10). It was a large quadrupedal animal growing up to 10 m long, with a small head, long and flexible neck, powerful limbs, and a long tail. Phylogenetic analysis suggests that *Lamplughsaura* is either a basal sauropod or, less likely, a stem sauropodomorph. It bridges a morphological gap between bipedal plateosaurids such as *Pradhania,* and quadrupedal, long-necked sauropods such as *Kotasaurus* and *Barapasaurus*.

In a poetic twist, it was Pamela who first introduced Dhuiya to paleontological research. While excavating a *Lystrosaurus* skeleton from the Early Triassic Panchet Formation, Pamela noticed one of the village boys watching intently from a distance. She asked if he would like to join the dig, to which he readily agreed. That boy grew up to be Dhuiya, one of the finest fossil collectors in India. The naming of these two dinosaurs from the Upper Dharmaram Formation immortalized this chance encounter of a field scientist and a curious boy (Fig. 8.10).

Lower Kota Formation (Sinemurian-Toarcian)

The Kota Formation in the Pranhita-Godavari Valley conformably overlies the fluvial deposits of the Dharmaram Formation and is unconformably overlain by the Early Cretaceous Gangapur Formation. My ISI colleague Dhiraj Rudra had carefully mapped the Lower Kota Formation and interpreted its depositional history. Rudra (1982) divided the Kota Formation into two lithological units, a lower and an

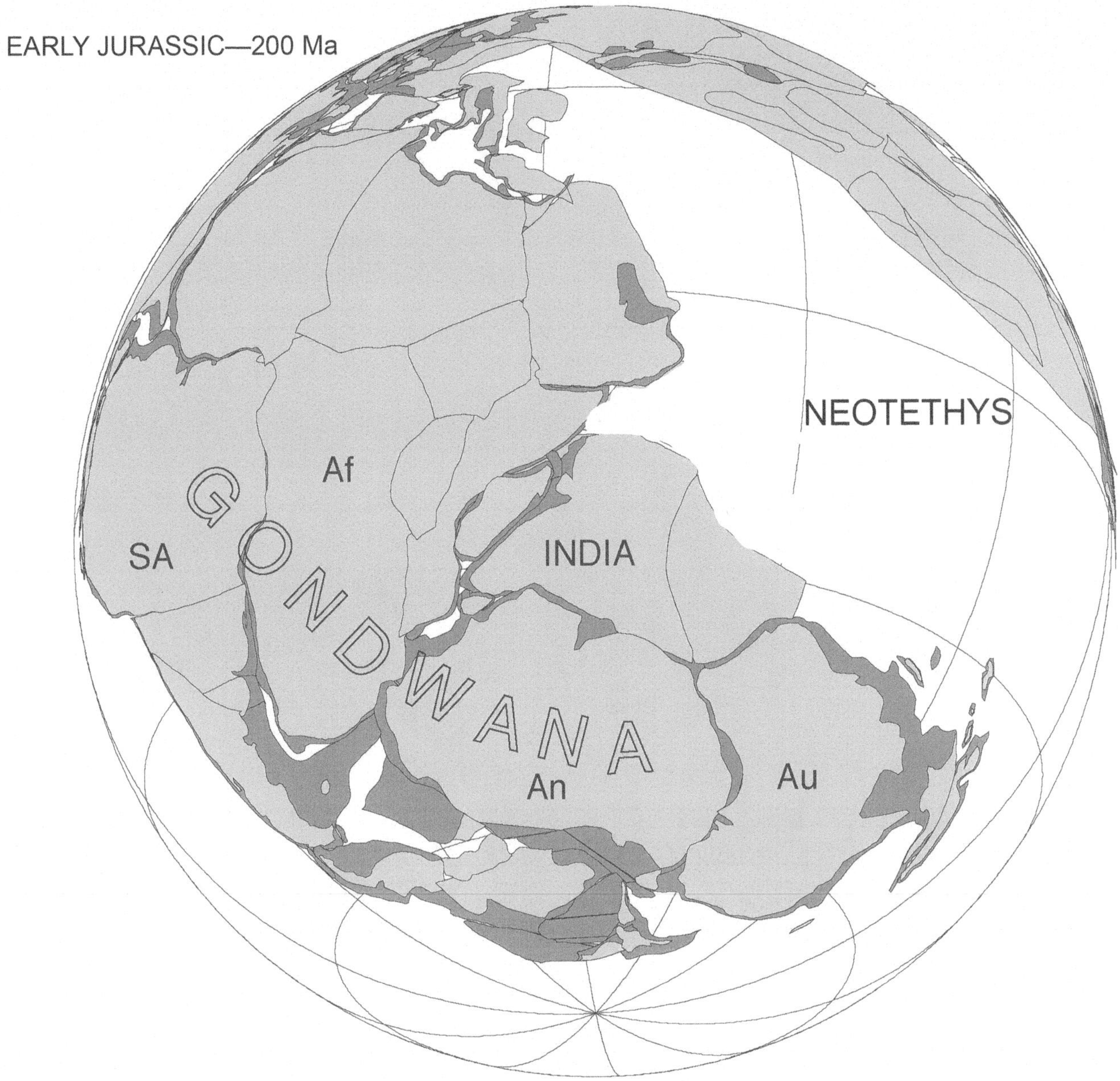

Fig. 8.9 Paleogeographic reconstruction of Gondwana in the Early Jurassic (~200 Ma) (after Chatterjee and Scotese 1999)

upper. The lower unit of Early Jurassic age (Sinemurian through Toarcian), called here the Lower Kota Formation, included 15 to 25-m-thick, hard, compact, poorly sorted, cross-bedded arkosic sandstone, typical of stream channel facies. It hosted a freshwater, lacustrine carbonate deposit interbedded with fluviatile sandstones and mudstones. This sandstone shows a fining upward cycle and grades both laterally and vertically into mudstone of the flood plain deposits that are typical of a meandering river environment. The upper unit, called here the Upper Kota Formation, extends to Middle Jurassic on the basis of palynomorph (Vijaya and Prasad 2001) in age (Aalenian through Callovian), and includes a marl and limestone bed, succeeded by red mudstones of fluvial or flood plain origin. The limestone was deposited in a series of interconnected playa lakes, under a moderately hot climatic regime.

The mudstone of the Lower Kota Formation has yielded a rich assemblage of vertebrates including two early sauropods, three mammals, invertebrates, and charophytes. *Kotasaurus* (Yadagiri 1988, 2001) and *Barapasaurus* (Jain et al. 1975; Bandyopadhyay et al. 2010) represent the early members of two well-known sauropods. The mammalian

Fig. 8.10 Paleoecology of the Early Jurassic sauropodomorph community in the Dharmaram landscape. In the foreground is the solitary plateosaurid *Pradhania* (adult body length ~4 m). In the background is a group of stem sauropod *Lamplughsaura* (adult body length ~10 m) (after Chatterjee et al. 2017)

fauna includes a morganucodont – *Indotherium* Yadagiri, 1984 (=*Indozostrodon* Datta and Das 2001), and a kuehneotherid *Kotatherium* (Datta 1981). These mammalian taxa are small, represented by isolated postcanine teeth.

When I first joined the ISI in 1964, Pamela Robinson invited me to collaborate with other ISI paleontologists Sohan Lal Jain, Tapan Roy Chowdhury, and T. S. Kutty on a recently discovered sauropod dinosaur fauna from the Lower Kota Formation. They discovered a rich layer of monospecific sauropod bone assemblage that had been excavated from an area of 276 m^2 from the Lower Kota Formation; it consisted of approximately 300 postcranial elements and few spoon-shaped teeth of an adult individual near the village of Pochampalli, Sironcha in the Chanda district of Maharashtra. On the basis of 6 left femora, it was estimated that there were at least six individuals. The discovery of the Kota sauropod material ignited the interest in vertebrate paleontology research in India.

The taphonomy of the sauropod bone bed at Sironcha in Maharashtra state is interesting. Apart from one partially associated skeleton, the bones were disarticulated, dispersed, and transported by river current. Mass accumulations of a single species of a sauropod suggest group behavior. The bones of multiple individuals occur in a single bone bed. The absence of delicate elements such as skull material clearly suggests some sort of *post mortem* disturbance during transportation. The limb bones show a preferred NE-SW direction, indicating that the long bones were aligned in response to current flow. However, the state of preservation of the majority of the fossil bones is remarkably good, with very little distortion or abrasion, indicating that the transport of bones was short. The bone bed is reminiscent of a log jam in a river; this is evidenced by the presence of actual logs up to 3 m long along with the bones. The bone assemblage of several sauropod individuals in one place may suggest sudden death due to a catastrophic event such as a flash flood in the flood plain area causing the death of a small herd of sauropods rather than to an attritional event.

The sauropod bones were skillfully prepared by Pranab Majumdar at the Geology unit of ISI for comparative study. Because of the wide publicity of the discovery of the earliest sauropod from India, Edwin H. Colbert from the American Museum of Natural History and his wife Margaret visited our Geology unit in 1964, and went on to Sironcha with the ISI paleontologists to see the Kota sauropod material *in situ*, and to further participate in the digging (Fig. 8.11A, B). Later Colbert wrote a nice account of the Sironcha field trip and the significance of the Kota sauropod in his two memoirs, *A Fossil Hunter's Note* Book (Colbert 1980) and *Digging into the Past* (Colbert 1988).

The first discovery of a nearly complete dinosaur skeleton from the Kota Formation was a major breakthrough in the understanding and appreciating of Mesozoic terrestrial vertebrates from India; it inspired systematic collecting in other unexplored regions. The dinosaur collection from the Lower Kota Formation included six partial skeletons and many disarticulated bones, which were beautifully preserved. I took part in studying and describing the new sauropod material with my senior colleagues, we named the taxon *Barapasaurus tagorei* (Jain et al. 1975). The generic name "barapa" means 'big legged dinosaur' in Bengali (because the femur was about 1.5 m long); the specific name commemorates the centenary birth of India's Nobel-laureate poet Rabindranath Tagore.

Records of Early Jurassic sauropods are quite rare, often known from isolated fragmentary remains. Represented by nearly a complete postcranial material, *Barapasaurus* fills a major gap in our understanding of the origin and early evolution of sauropod dinosaurs. *Barapasaurus* was hailed as one of the oldest sauropod dinosaurs in the world (Jain et al. 1975), and our paper received national and international media coverage. The whole skeleton of *Barapasaurus* was installed in the Geology Museum of ISI by Pranab Majumdar. It was the first mounted dinosaur skeleton in India (Fig. 8.11C).

Sauropod dinosaurs possess a highly specialized set of skeletal adaptations related to their gigantic size, obligate quadrupedalism, graviportal locomotion, and strictly herbivorous diets that accrued over a 140 million-year history. The sauropod body plan is unique among terrestrial tetrapods, with a short and deep trunk combined with a small skull and very long slender neck and tail, and massive columnar limbs. The earliest diverging forms such as *Lamplughsaura*, *Vulcanodon*, *Kotasaurus*, *Shunosaurus*, *Barapasaurus*, and *Jobaria* are successively closer to Neosauropoda, which includes Diplodocoidea and Macronaria (Wilson 2002).

Barapasaurus was the first to have attained gigantic size, with a body length of 14 m, and the first to be fully quadrupedal, with graviportal and columnar limbs (Fig. 8.12). The skeletal elements of *Barapasaurus* show many primitive characters that are absent in later sauropods. Typical sauropod features of *Barapasaurus* are: the quadrupedal posture with columnar limbs, four coalesced sacral vertebrae with a well- developed sacricostal yoke, low deltopectoral crest of the humerus, absence of olecranon process, triradiating proximal end of the ulna with deep radial fossa, ilium with low and subrectangular ischial peduncle, and ischial shaft equal to pubic shaft (Bandyopadhyay et al. 2010). The most diagnostic features of *Barapasaurus* are present in the vertebral column. The posterior dorsal vertebrae are tall and possess slit-like neural arches; the neural spines are not bifurcated. Gigantism and elongation of the neck were major biomechanical problems of sauropods, which was compensated by complex vertebral pneumacity. This problem was overcome in *Barapasaurus*, through pneumatization of the

A B C

Fig. 8.11 Field photographs in the Early Jurassic Kota Formation during visit of Ned Colbert, excavating the *Barapasaurus* skeleton. **A**, Sohan Lal Jain and Ned Colbert; **B**, Tapan Roy Chowdhury and Ned Colbert; **C**. mounted skeleton of *Barapasaurus* at the Geology Museum of Indian Statistical Institute

centra and neural spines of the cervical and dorsal vertebrae, which significantly reduced the body weight (Jain et al. 1975). Dorsal centra are opisthocoelous. In the hindlimb, the femur lacks the lesser trochanter; the fibula has a broad triangular scar for the tibia; and in the ankle, the astragalus has a posterior fossa, divided by a crest. *Barapasaurus,* a descendant from the Late Triassic dinosaurs of modest size, is a prime example of the fact that certain dinosaurs suddenly became giants in the Early Jurassic. *Barapasaurus* is now considered as a basal member of Eusauropoda, and is more derived than *Vulcanodon* of South Africa and *Shunosaurus* of China (Wilson 2002; Chatterjee et al. 2017).

Kotasaurus yamanpalliensis, from the same horizon of the lower Kota Formation, is a basal sauropod, smaller in size (body length ~9 m) and more primitive than *Barapasaurus tagorei*. It is interesting that two sympatric species of early sauropods lived in central India about 180 million years ago (Fig. 8.12). The bone bed of *Kotasaurus* was first

Fig. 8.12 Paleoecology of the Early Jurassic Kota Formation of India. Two basal sauropods, the smaller *Kotasaurus* (front) startled by the arrival of the larger *Barapasaurus* (behind) at the stream. At the lower right foreground, a group of small kuehneotheriid mammal *Kotatherium* look up from their feeding (after Chatterjee et al. 2017)

discovered by my colleague Dhiraj Rudra from the mudstone horizon near the village Yamanpalli, while mapping the Kota Formation. It was however, later excavated and described by P. Yadagiri of Geological Survey of India (Yadagiri 2001). *Kotasaurus* is known essentially from several postcranial skeletons but lacks a skull. It was an obligate quadruped with vertical limbs, and a long neck and tail. A long, low preacetabular process of the ilium, that maintains its dorsoventral width over its entire length, is the diagnostic feature of *Kotasaurus*. The vertebral morphology of *Kotasaurus* is quite different from that of *Barapasaurus*. The vertebrae are massive, without pneumacity. Cervical vertebrae are opisthocoelous, but dorsal vertebrae lack spinal laminae; sacral vertebrae consist of three co-ossified centra; the scapula is tall and has a narrow proximal blade; the humerus is somewhat twisted with a low deltopectoral crest. The femur retains a fourth trochanter. There is a prominent astragalar peg situated anteroventrally.

Upper Kota Formation

The Upper Kota Formation of Middle Jurassic (Vijaya and Prasad 2001) age has yielded a wide variety of small tetrapods including a kayentachelid turtle, two sphenodontids, an iguanian lizard, a pterosaur and seven mammal taxa. A few isolated teeth found in that formation have been attributed to a theropod. A scute and other fragmentary limb bones indicate the presence of a scelidosaurid ornithischian in this formation, but nothing has been diagnosed so far (Chatterjee et al. 2017). Some dental morphotypes of ornithischian and theropod dinosaurs are also known from the upper unit of the Kota Formation (Prasad and Parmar 2020).

Upper Jurassic Bagra Formation

During the Late Jurassic, the vast continent of Pangea started to break up, with narrow Neotethys Ocean dividing it into two landmasses, Laurasia to the north and Gondwana to the south. Gondwana broke apart into two roughly equal parts: West Gondwana comprised South America and Africa, and east Gondwana was composed of Madagascar, India, Australia, and Antarctica (Fig. 8.13). During this period, Earth's climate changed from hot and dry to humid and tropical. On land, dinosaur diversity and scale increased through time, with peaks in the Late Jurassic leading to gigantism. The Late Jurassic formations in many parts of the world have

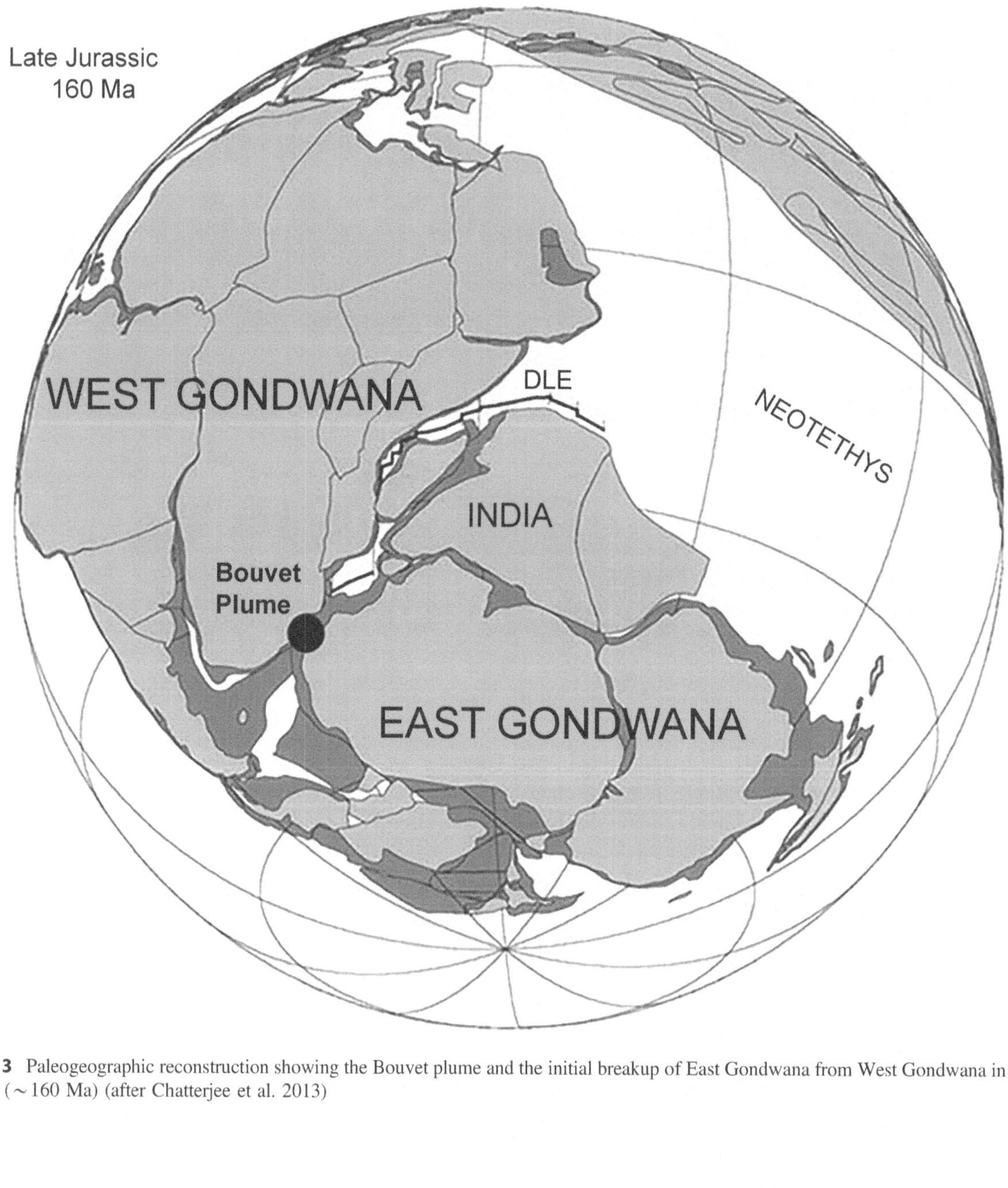

Fig. 8.13 Paleogeographic reconstruction showing the Bouvet plume and the initial breakup of East Gondwana from West Gondwana in the Late Jurassic ($\sim$160 Ma) (after Chatterjee et al. 2013)

produced spectacular dinosaur assemblages including giant sauropods, stegosaurs, allosaurs, and many other forms.

The Late Jurassic sediments remain poorly sampled for dinosaur remains in India. To date, I have found only sauropod remains, probably a titanosaur from the Bagra Formation of Satpura Basin about three decades ago, but the material has yet to be described (Chatterjee and Hotton, 1986).

Origin and Early Evolution of Sauropodomorpha

Sauropodomorpha is a clade of herbivorous dinosaurs that originated during the Late Triassic and were prevalent, both in diversity and giant size, in global terrestrial ecosystems throughout much of the Mesozoic, with at least 175 valid taxa currently known. The earliest members of this group, the basal sauropodomorphs, first appeared in the Ischigualasto Formation of Argentina (Carnian) and achieved a near global distribution by the end of the Triassic. In India, we have several groups of basal sauropodomorphs, as discussed earlier including a Maleri guaibasaurid, a non-plateosaurian *Nambalia*, and the plateosaurian *Jaklapallisaurus* during the Late Triassic. In the Early Jurassic, we encounter a massospondylid sauropodomorph *Pradhania*, and several basal sauropods including *Kotasaurus* and *Barapasaurus* (Fig. 8.14). Thus, the origin and early radiation of sauropodomorphs is nicely documented in the Indian Late-Triassic-Early Jurassic Gondwana sediments. India may be an important center for the origin and early radiation of sauropodomorph dinosaurs.

Evening: Dinosaurs in the Drifting Indian Plate During the Cretaceous

The Cretaceous Period spanned about 80 million years, during which dinosaurs diversified rapidly with the drifting of the continents. During the Early Cretaceous, India began to rift, and drift apart from East Gondwana. The third phase of Indian dinosaurs, in the Late Cretaceous, is quite distinct from the second. It was a different world. During much of the Late Cretaceous, India began to drift northward as an island continent carrying its dinosaur fauna like a passenger ship, until it collided with the Kohistan-Ladakh Arc and incorporated it around 80 million years ago, forming a biotic corridor for dinosaur faunal exchange (Fig. 8.15). The records of dinosaurs from the Post-Gondwana formations of Early Cretaceous sediments of India are essentially blank because of poor sampling rather than for lack of fossil remains. During the Late Cretaceous, dinosaur remains are known from the Nimar Sandstone and the Lameta Formation.

Nimar Sandstone (Cenomanian-Turonian)

In 1996, Ashu Khosla of Panjab University discovered an interesting sauropod bone bed from the Late Cretaceous Nimar Sandstone, in the basal unit of the Bagh Group of Narmada Valley, Central India. Later he and his colleagues described these bones as those of sauropods (Khosla et al. 2003). The material includes limb elements and numerous

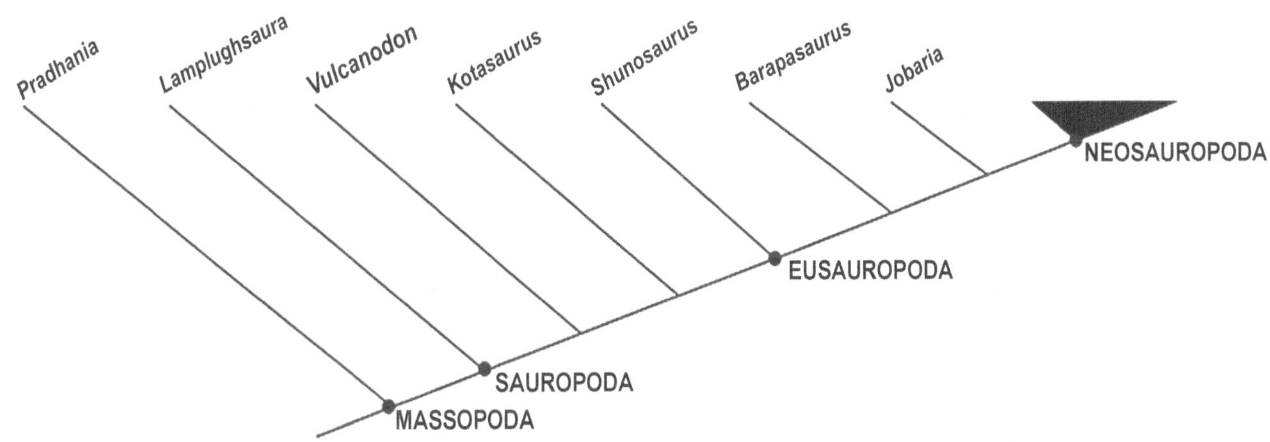

Fig. 8.14 Cladogram showing the origin and early evolution of sauropods during the Early Jurassic of India. Indian taxa are shown by bold fonts (after Chatterjee et al. 2017)

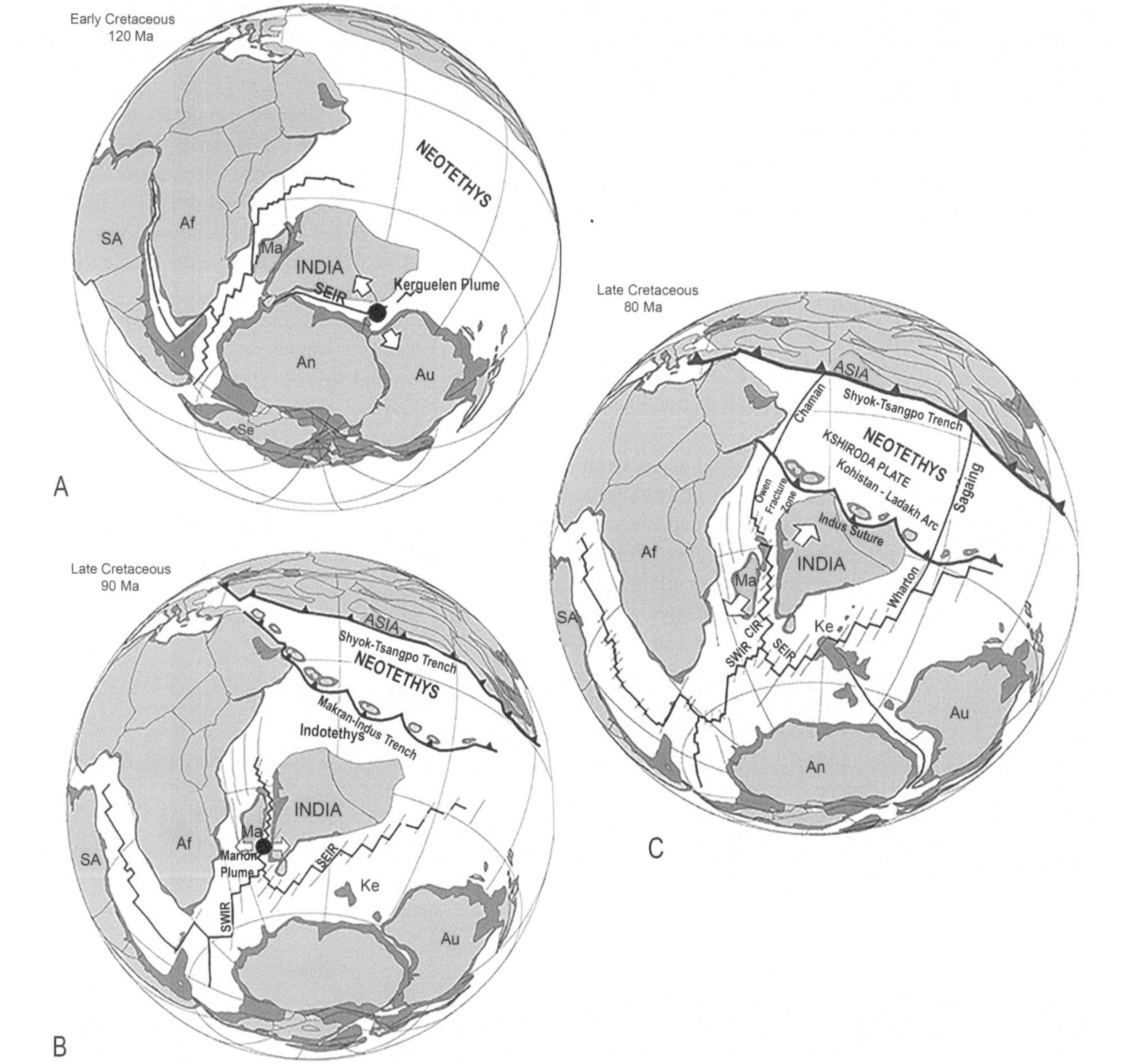

Fig. 8.15 Paleogeographic position of India during the Cretaceous. **A**, the separation of India from Antarctica-Australia in the Early Cretaceous (~ 120 Ma) along the Southeast Indian Ridge; **B**, paleogeographic reconstruction of Gondwana fragments during the Late Cretaceous showing the India-Madagascar rift (~ 90 Ma) and the location of the Marion plume. India became an island continent during a brief period of Late Cretaceous; **C**, Paleogeographic position of India showing collision of the Indian plate with the Oman-Kohistan-Ladakh (OKL) Arc during the Late Cretaceous (~ 80 Ma) time along the Indus Trench. The OKL Arc formed an important corridor for tetrapod migration between India and Africa; other two biotic corridors were: Laxmi Ridge-Seychelles island between India and Madagascar; and emergent Ninetyeast Ridge-Kerguelen Plateau-Antarctica between India and South America (modified from Chatterjee et al. 2013)

fragmentary bones belonging to at least two individuals. Dr. Khosla was kind enough to show me this sauropod material, which looked like a new titanosaur species, perhaps ancestral to the abundant titanosaur remains from the younger Lameta Formation. The Nimar sauropod was a gigantic titanosaur with an estimated length of the femur exceeding >1.5 m. Isolated abelisaurid teeth have also been documented from the green sandstone representing the top most part of the Bagh Group (Prasad et al. 2016).

Lameta Formation (Maastrichtian)

At many places around the fringe of the Deccan volcanic province, fossiliferous sedimentary beds are found associated with the Deccan lava flows. Traditionally, these sedimentary beds are called infratrappean beds (below the basal lava flows, such as the Lameta Formation) or intertrappean beds (interlayered with lava flows), depending on their physical position with respect to the volcanic flows. The Lameta Formation is a thin but extensive horizon that underlies the basal flows of the Deccan volcanic sedimentary sequences. The Lameta sediments were deposited in an arid terrestrial environment, in a meandering channel. Vertebrate fossils are found in over bank deposits and meander cut-off lakes. The vegetation was sparse with shrub cover near the main stream. The Lameta Formation has yielded one of the finest records of dinosaur bones and nesting sites with fossil eggs and eggshells. On the basis of its stratigraphic position, vertebrate assemblages and associated microfossils such as ostracods, pollens, and charophytes, the age of the Lameta Formation is now considered to be Maastrichtian. The type section of the Lameta Formation is the Lameta Ghat on the banks of the Narmada River near Jabalpur, Madhya Pradesh.

What is particularly interesting from historical point of view is that dinosaurs were known from the Lameta Formation more than two decades before Sir Richard Owen coined the term 'Dinosauria' in 1842, the 'fearfully great lizards.' The Lameta Formation near the city of Jabalpur, central India, is important for the discovery of one the earliest, if not the earliest record of dinosaur remains anywhere in the world. In 1917–19, Charles Matley of the Geological Survey of India (GSI) collected a substantial quantity of sauropod dinosaur material from several localities on the western slope of Bara Simla Hill in the Jabalpur Cantonment area. In 1920, Matley collected additional fragmentary sauropod material from localities in Pisdura in Maharasthra. This material formed the basis of a number of taxa described from the Late Cretaceous of India. Much of the collection went to the British Museum, but the sauropod remains were kept at the Indian Museum in Calcutta (hereafter Kolkata), in the repository of the GSI collection. This remarkable discovery of Indian dinosaur material occurred before the first description of the first dinosaur *Megalosaurus* from the Oxford University campus by William Buckland in 1824. In 1922, the legendary and flamboyant paleontologist Barnum Brown of the American Museum of Natural History (Fig. 8.16A), on the steam ship to India met a vivacious young woman, Lilian, who he wooed and wed in a palm-shaded chapel in Kolkata. The newly-wed couple traveled through different parts of India during their honeymoon, collecting fossils on their way. They collected a great variety of Neogene Siwalik mammals at the foothills of the Himalaya. The trip was chronicled by Lilian Brown in her entertaining book, *I Married a Dinosaur* (Brown 1950). Barnum Brown also visited Matley's site at Jabalpur, collected some skull material of theropod dinosaurs, as well as a dermal osteoderm of a titanosaur, and brought back the material to the American Museum of Natural History, but he never reported the discovery. Soon after Brown's excavation, the army officer Captain W. H. Sleeman (celebrated for eradicating India of its centuries-long-terror of *Thugees*) discovered some dinosaur bones in 1828 from the same fossiliferous site of Barnum Brown at Bara Simla Hill, Jabalpur. The first systematic description of Indian dinosaurs was carried out by Lydekker (1877), who described Sleeman's find from Jabalpur, and ereted a new sauropod taxon *Titanosaurus indicus* on the basis of two procoelous distal caudal vertebrae and an imperfectly preserved femur. Later, Frederick von Huene of Tübingen University was invited to collaborate on the Lameta dinosaurs. Huene and Matley (1933) described the Lameta dinosaur material in a monograph. In 1932–33, Matley collected an additional associated partial skeleton of a titanosaur from the Lameta Formation exposed in the adjacent Chota Simla Hill of Jabalpur, which was sent back to the Natural History Museum (then the British Museum of Natural History); the material was recently described by Wilson et al. (2011) as *Jainosaurus* cf. *septentrionalis*.

I became involved with the Lameta dinosaurs in a strange, roundabout way. My discovery of two nearly complete and articulated phytosaur skeletons, along with their stomach contents (Fig. 8.5A), from the Lower Maleri Formation (Chatterjee 1978a) caught the attention of Professor Joseph T. Gregory, the world authority on phytosaurs, and chairman of the Department of Paleontology at the University of California, Berkeley (Fig. 8.16B). Prof. Gregory kindly invited me to his department as a Visiting Professor in early 1976 to teach vertebrate paleontology, and to compare Indian phytosaurs with North American collection at his museum. I eagerly left India for my new assignment, thus beginning my scientific career in the USA. While at Berkeley, a colleague of mine, Robert Long borrowed Barnum Brown's skull material of the Lameta theropods housed at the AMNH, passing it on to me for study during my stay, since the material was originally from India. There was an irony in this situation that would define my future direction of research from Triassic to Cretaceous tetrapods. Studying the Indian dinosaur fossils collected by Barnum Brown during his historical honeymoon trip to Jabalpur, five decades earlier, forced me back home to India many times for extensive research and exploration.

Huene and Matley (1933) identified two large theropods, called carnosaurs at that time, from the Lameta Formation on the basis of skull architecture: *Indosuchus* and *Indosaurus*. I have identified the well-preserved jaw material of the Lameta theropod collected by Barnum Brown, as *Indosuchus* (Chatterjee 1978b). Later, these Lameta theropods were correctly identified as abelisaurs within Neotheropoda, a primitive well-known theropod clade that thrived during the Cretaceous Period in Argentina, Madagascar, and other Gondwana continents (Chatterjee and Rudra 1996). Like most theropods, abelisaurs were predatory bipeds, superficially resembling tyrannosaurs in size and proportion,

Fig. 8.16 Field photographs in the Late Cretaceous Lameta Formation at Bara Simla Hill, Jabalpur. **A**, Barnum Brown of the American Museum of Natural History, who discovered some dinosaur remains from the site in 1921, but later described by Sankar Chatterjee in 1978; **B**, Joseph Gregory, chairman of the Paleontology department at the University of California, Berkeley, who invited Sankar Chatterjee at his department; **C**, Dhiraj Rudra, then head of the Geology unit of Indian Statistical Institute, who accompanied Sankar Chatterjee in the field; **D**, excavation of a titanosaurid *Jainosaurus* at Bara Simla Hill; **E** and **F**, braincase and caudal vertebra of *Jainosaurus*; **G**, Pranab Majumdar and Sankar Chatterjee in front of the camp; **H**, Cretaceous-Paleogene Boundary ejecta layer at Bara Simla Hill, showing evidence of shocked quartz grains in thin section and SEM photographs from this layer (after Chatterjee t al. 2006)

though phylogenetically more primitive. They had stocky hindlimbs and extensive ornamentation in the skull bones.

Another surprise was waiting for me in the USA. While looking at the Barnum Brown collection at AMNH in New York, I discovered an osteoderm of a titanosaur from the Jabalpur site, which was later described by us (D'Emic et al. 2009). My research on the Late Cretaceous Lameta dinosaur material at America would inspire me to visit the Jabalpur site, collect additional dinosaur material from other sites of the Lameta Formation, and study the Deccan volcanism, Shiva crater, and Cretaceous-Paleogene Boundary (K-Pg) mass extinction. If I had never been invited to Berkley I never would have arrived at my current research on Cretaceous dinosaurs, and their extinction.

A Passage to India: Discovery of Cretaceous Dinosaurs

I spent several field seasons in India in collaboration with Dhiraj Rudra, then Head of the Geology unit of the Indian Statistical Institute, to collect the last evolutionary taxa of dinosaurs of India, and study the causes and consequences of their extinction at K-Pg boundary. My wife Sibani accompanied me on several field trips in India. We visited different Lameta sites in Madhya Pradesh and Gujarat and the Kallamedu site in the Cauvery Basin of Tamil Nadu in search of the Maastrichtian dinosaurs. The most productive fossil localities in the Lameta Formation are located near Rahioli village in Balasinor, Jabalpur, and Pisdura regions, traversing a linear E-W stretch of approximately 700 km. During our field work, we discovered remarkable remains of dinosaurs and their eggs in Balasinor and Jabalpur, identified the Shiva crater at the western shelf of India (which is linked to dinosaur extinction), and collected K-Pg boundary samples for meteoritic impacts (Chatterjee and Rudra 1996; Chatterjee et al. 2006; Novas et al. 2010). All these expeditions to India were funded by the National Geographic Society and Smithsonian Institution, and the field logistics were provided by the Indian Statistical Institute. The new dinosaur material from these expeditions is housed at the Geology Museum of Indian Statistical Institute.

The Lameta Dinosaurs

During my field work in the Lameta Formation in 1987–88, my first task was to locate the classical dinosaur fossil site on the western slope of Bara Simla Hill of Jabalpur where Matley, Sleeman, and Brown collected the material. Because of vegetation and monsoon rains, the Bara Simla site remained lost for almost 60 years. It was frequently visited by geology students for field work, their professors proudly reminding them this was where the very first dinosaur was discovered; but nobody knew from exactly what horizon of the Bara Simla Hill the dinosaur fossils had been collected. Dhiraj Rudra (Fig. 8.16C) and Pranab Majumdar (Fig. 8.16G) from ISI, and my wife Sibani assisted me in my search for the original dig. One day, after a futile week- long search, Sibani sat on a rock for our midday lunch break. I suddenly realized that was no rock my wife was sitting on, it was in fact a sauropod femur! The site matched perfectly with Captain Sleeman's 1928 description of the bone bed close to his bungalow, where he found fossil fragments of animals and plants in the sediments below the Deccan Trap.

For many years the dinosaur site was a restricted military area; as a Captain, Sleeman had no problem gaining access, we however had to get formal permission from Colonel A. M. Kapoor of the Gun and Shell Factory to dig. Over the next two weeks, under his military protection, we uncovered the site: first exposing Sibani's lunch-break femur, then some vertebrae, ribs, and a beautiful braincase all belonging to a titanosaur *Jainosaurus* (Chatterjee and Rudra 1996) (Fig. 8.16D, E, F). Our discovery made big headlines in the local newspaper and TV channel, inspiring the local school children and college students to visit the site. To control the crowd and protect the site, Colonel Kapoor posted a soldier as sentry with a rifle. It was quite hilarious to see an armed soldier guarding 65-million-year-old dinosaur bones as a national treasure. On that trip, on the opposite Chota Simla hill, we also discovered a titanosaur nesting site, near the Shiva temple. Here, Matley had collected an associated partial skeleton of a titanosaur *Jainosaurus* during 1932–33, which was sent to the British Museum of Natural History, and described by Wilson et al. (2011).

In the Bara Simla site, above the dinosaur-bearing bed, a 2.7 m- thick sandstone layer has yielded shocked quartz grains in the upper part of a sandstone layer, indicating the close proximity of a crater in India (which was later identified as the Shiva crater) from which thick ejecta could be emplaced ballistically. Shocked quartz grains from the Jabalpur section are coarse, and larger (300–400 μm) than most of the shocked quartz grains reported elsewhere, implying that the proximity of the crater was close by. Such coarse-grained and a thick boundary layer could not be derived as airborne fallout from the Chicxulub impact crater (Chatterjee et al. 2006). In this single site we see both the very last dinosaurs, and the cause of their extinction in the form of Shiva ejecta layer covering the bone bed.

The Lameta Formation has yielded two major groups of Gondwana dinosaurs—titanosaurid sauropods and abelisaurid theropods (Fig. 8.17), as well as a solitary ankylosaur (Huene and Matley1933, Chatterjee and Rudra 1996, Wilson et al. 2005, Novas et al. 2010, Chatterjee et al. 2017). Many of the original dinosaur taxa that were erected by early English paleontologists were based on fragmentary or

Fig. 8.17 Interaction of Late Cretaceous (Maastrichtian) Lameta dinosaurs. In the foreground two individuals of abelisaurid theropod *Rahiolisaurus* are aggressively looking for the opportunity to attack the young or weak individuals among the herd of armored titanosaur *Jainosaurus* at their nesting site. Large adult titanosaurs shielded their young at the center (after Chatterjee and Rudra 1996)

isolated material and their utility for phylogeny and paleobiogeographic analysis has since proven questionable. Recent discoveries of extensive dinosaur material from the Bara Simla quarry of Jabalpur and Rahioli quarry of Gujarat, as well titanosaurid eggs, have provided a more accurate understanding (Chatterjee and Rudra 1996, Novas et al. 2010). This dinosaur material, now housed at the Geology Museum of the Indian Statistical Institute, provides half of the story. The other half derives from the discovery of widespread dinosaur nesting sites from several infratrappean localities in Jabalpur, Kheda, Balasinor, Rahioli, Dohad, and Bagh in the Narmada Valley. These represent one of the most geographically extensive nesting sites known in the world (Prasad and Sahni 2014).

Titanosaurs

Titanosaurs were the most diverse and geographically widespread clade of Cretaceous sauropods in Gondwana, and have been recorded from all of the southern continents including Antarctica. From South America, titanosaurs also migrated to North America during the Late Cretaceous Period through the isthmus of Panama, and are represented by *Alamaosaurus* in the Big Bend National Park of Texas. Titanosaurs were the largest terrestrial animals in Gondwana and the only sauropod that possessed osteoderms. They are regarded as the most advanced of all sauropods, stockier and more robustly built. To date, 40 titanosaurian genera have been discovered all over the world, but the greatest numbers come from Patagonia in Argentina. Fossil evidence provides remarkable anatomical details about osteology, skin impressions, body armor, embryos, eggs, nests, and footprints (Novas 2009).

In India, fragmentary titanosaur remains were discovered from multiple localities in the Lameta Formation (Lydekker 1877, Matley 1923, Huene and Matley 1933). During 1984–1986, my colleague at ISI, Sohan Lal Jain collected a braincase and associated and mostly articulated postcranial material of a titanosaur from the fluvio-lacustrine deposit of the Lameta Formation in the Dongargaon locality of Chandrapur district, Maharasthtra, about 16 km south-east from the well-known Pisdura locality. Bones were concentrated in an excavated site of 184 m^2 area, and the association strongly suggested the presence of a single individual. The braincase from this collection was described by Berman and Jain (1982). Several large sauropod eggs and coprolites have also been known from this locality. Later, Jain and Bandopadhyaya (1997) described the postcranial material as *Titanosaurus colberti*, honoring E. H. Colbert of AMNH, who visited ISI in 1964 and 1977 to see the progress of *Barapasaurus* material and inspire paleontologists. Wilson and Upchurch (2003) replaced the generic name *Titanosaurus* with the new name *Isisaurus*, honoring ISI, because it possessed some unique characters. *Isisaurus colberti* had a short, vertically directed neck and long forelimb, making it considerably different from other sauropods. Cervicals and dorsals are opisthocoelous with well-marked pleurocoels;

caudals are procoelous throughout the series. *Isisaurus* is known from much better remains than most titanosaurs.

With the recent discovery of new cranial and postcranial material, Indian titanosaurs can be grouped into two distinct taxa: *Jainosaurus septentrionalis* (Chatterjee and Rudra 1996, Wilson et al. 2005, 2011) and *Isisaurus colberti* (Jain and Bandyopadhyay 1997). Although *Jainosaurus* and *Isisaurus* coexisted in India, the braincase morphology of these two genera is remarkably different, indicating they were not closely related to each other (Fig. 8.18). Both genera are sympatric and relatively large (~25 m). Our field collection of titanosaur braincase material suggests that *Jainosaurus* remains are dominant in the Bara Simla Hill, but *Isisaurus* material is prevalent in both Dongargaon and Rahioli sites.

There are about 10 species of titanosaurs that possessed osteoderms, which are reported from Brazil, Argentina, Mali, Malawi, Madagascar, India, and Spain. A large ellipsoid osteoderm of a titanosaur from the Bara Simla Hill of Jabalpur, collected by Barnum Brown, is associated with *Jainosaurus* material (D'Emic et al. 2009). *Jainosaurus* appears to be the most basal member of titanosaurs and shows close affinities with other titanosaurs such as *Malawisaurus* of Africa, and *Mendozasaurus* of South America, characterized by the development of scutes. The braincase of *Jainosaurus* resembles that of *Vahiny* from the Late Cretaceous of Madagascar indicating their closely phyletic relationships (Rogers and Wilson 2014). *Jainosaurus* is grouped in the Titanosauridae (Novas 2009).

The osteoderm appears to be absent in associated skeletons of *Isisaurus*, indicating that the taxon was not armored. *Isisaurus*, which appears to be more derived than *Jainosaurus*, has been reported from the contemporary Pab Formation of Pakistan on the basis of a braincase (Wilson et al. 2005). *Isisaurus* shows similarities to *Antarctosaurus* and *Argentinosaurus* of Argentina, *Opisthocoelicoda* of Mongolia, and *Alamosaurus* of North America; these taxa are accommodated in the family Antarctosauridae. In India, the oldest titanosaurian remains may come from the Nimar Sandstone (Cenomian-Turonian) in Narmada Valley (Khosla et al. 2003), or even earlier from the Late Jurassic Bagra Formation (Chatterjee and Hotton 1986). The distribution of titanosaur body fossils is complemented by their abundant footprint record, which hints at a much earlier origin, in the early or Middle Jurassic (Wilson and Upchurch 2005).

The presence of titanosaur osteoderm in India extends the geographic range of armored titanosaurs to India, Africa, Madagascar, South America, and Europe. Titanosaurids were widespread, known from the Late Cretaceous deposits of South America, Australia, North America, Mongolia, and China. Although titanosaurids were dominantly Gondwanan sauropods, they migrated from South America to North America, and to Eurasia via India.

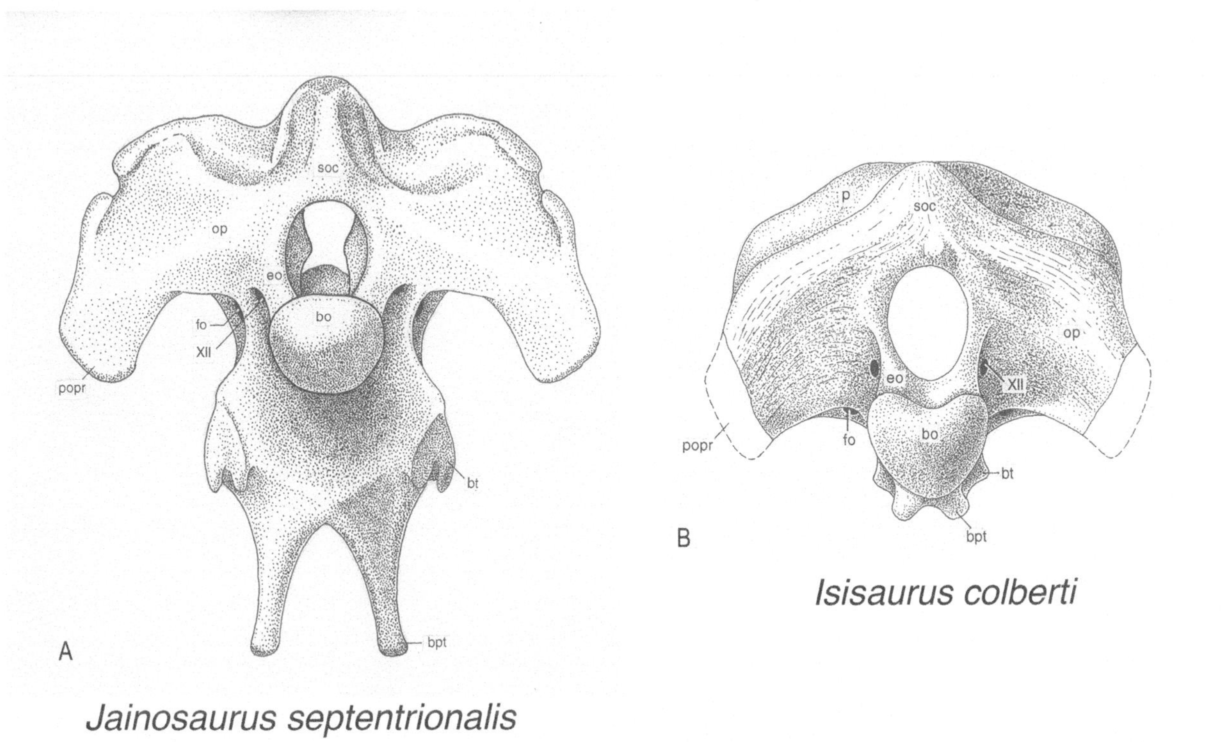

Fig. 8.18 Titanosaurid braincase. **A**, *Jainosaurus* from Jabalpur site; **B**, *Isisaurus* from Rahioli site. The braincase morphology of two taxa is remarkably different, suggesting their different phylogenetic positions in titanosaurid cladogram

Abelisaurs

Like titanosaurs, abelisaurs were distinctive Gondwanan dinosaurs. They were large predatory dinosaurs with a disjunct distribution across the Gondwana landmass during the Cretaceous, and have been recorded from South America, India, Madagascar, and Africa. They are derived members of the ceratosaurian clade, which originated in the Jurassic of Laurasia, but diversified in the Cretaceous of Gondwana. Like other theropods, they were fully bipedal and had forelimbs that were greatly reduced. Diagnostic characters of abelisaurids include anteroposteriorly short and deep premaxialla; dorsoventrally deep snouts at the level of narial openings; frontals dorsoventrally thickened resulting in a dorsal buckling (*Abelisaurus* of Argentina), paired horn-like structures (*Carnotosaurus* of Argentina), or a dome-like prominence (*Majungasaurus* of Madagascar) (Novas 2009). Abelisaurs contain two distinct families: the small and primitive Noasauridae, and the large and more derived Abelisauridae, which were the Gondwanan version of *T. rex*, though highly primitive. Both noasaurids and abelisaurids are known from the Lameta Formation. Novas et al. (2004) reviewed the abelisaurid material from the Lameta Formation originally described by Huene and Matley (1933). They identified *Laevisuchus indicus* as a small noasaurid, related to *Noasaurus* of Argentina and *Masiakasaurus* of Madagascar on the basis of their peculiar cervical vertebrae.

Indosuchus and *Indosaurus* were diagnosed on the basis of isolated cranial elements (Chatterjee 1978b). The skull roof of both genera is smooth without any horn-like protuberance as is the condition of *Abelisaurus* of Argentina. However, the distinctive peg-and-socket articulation between the lacrimal and the frontal bone of *Indosuchus* is very similar to that of *Abelisaurus* (Novas et al. 2004).

In the early 1980s, Suresh Srivastava, a paleontologist from the Geological Survey of India, discovered some disarticulated skeletons of an abelisaurid dinosaur around Temple Hill, near Rahioli village in the Balasinor area of Gujarat. The material was later described by Wilson et al. (2003) as a new taxon, *Rajasaurus narmadensis*. We decided to explore Rahioli site for additional dinosaur material. During our field work in 1995, we discovered a rich graveyard of multispecies, a dinosaur bone bed in the mudstone facies of the Lameta Formation showing disarticulated elements of titanosaurid, abelisaurid, and nodosaurid remains (Fig. 8.19). In this trip, Dhiraj Rudra from ISI and my wife and two sons participated in the excavation. The most remarkable discovery from this quarry represents a nearly complete postcranial skeleton of a large abelisaurid theropod *Rahiolisaurus* (the genus is named after the Rahioli village; Novas et al. 2010) (Fig. 8.19A). In addition, the quarry has yielded bones of a small abelisaur, probably a noasaurid, limbs and girdles of a titanosaur *Isisaurus* associated with a braincase (Fig. 8.19B, C), and several bones of a nodosaurid ankylosaur (Fig. 8.19E, F). Also, we recovered intact titanosaurid eggs in and around this area (Fig. 8.19D). Later, this site was designated as the 'Dinosaur Fossil Park' by the Geological Survey of India and has become a tourist location in Gujarat.

It is likely that *Indosuchus*, *Indosaurus*, and *Rahiolisaurus* may represent the basal members of abelisaurids, similar to *Abelisaurus* of Argentina. From the associated postcranial material, it appears that *Rahiolisaurus* was a gracile and slender-limbed, basal abelisaurid, whereas *Rajasaurus* was a heavy-bodied, stout-limbed form. An adult-sized *Rahiolisaurus* was identified as a large-sized abelisaurid, about 8 m long (Novas et al. 2010). *Rahiolisaurus* had seven co-ossified sacral vertebrae and is more derived than *Majungasaurus* of Madagascar but resembles *Carnotaurus* of Argentina in many postcranial features and shares a close ancestry. The lack of a skull roof precludes conclusions on whether or not *Rahiolisaurus* was a derived member of the horned abelisaurid clade Carnotaurinae.

Rajasaurus, another abelisaur from this site, is known from associated skull and postcranial elements (Wilson et al. 2003). The pelvic and hindlimb elements of *Lametasaurus* were included with *Rajasaurus* because of close morphological similarity. However, the partial skull roof and head crest of *Rajasaurus* is very distinctive and more derived than other Lameta abelisaurs. The skull roof of *Rajasaurus* shows a single frontal horn (Wilson et al. 2003) like that of *Majungasaurus* of Madagascar, whereas *Carnotaurus* of Argentina had developed a pair of bull-like horns. In life, these horns were probably encased in keratin sheath. These horned abelisaurs – *Rajasaurus*, *Majungasaurus* and *Carnotaurus* – are grouped in a highly derived subfamily Carnotaurinae of abelisaurids (Novas 2009). It thus appears that *Rajasaurus* is the most derived member of Indian abelisaurs.

The distribution of closely related abelisaur taxa indicates that dispersal routes between India-Madagascar and India-South America were operative during the Late Cretaceous period via Antarctica-90° East Ridge (Fig. 8.15C). The recent discovery of abelosauroid antecedents from the Early Cretaceous, noasaurids and abelisaurids from the Mid-Cretaceous, and a hornless abelisaurid from early Late Cretaceous of Niger provides evidence for the origin of this group on Africa, thus indicating 'Pan-Gondwana' distribution (Sereno et al. 2004). However, a record of an early abelisaur from Europe indicates more widespread distribution between Laurasia and Gondwana. Tortosa et al. (2014) reported a new abelisaur *Acrovenator* from the late Cretaceous deposits of southern France. Phylogenetically, *Acrovenator* is extremely similar to Indian taxa such as *Rajasaurus*, *Rahiolisaurus*, *Indosaurus* as well as the Madagascan genus *Majungasaurus*, and all grouped in a

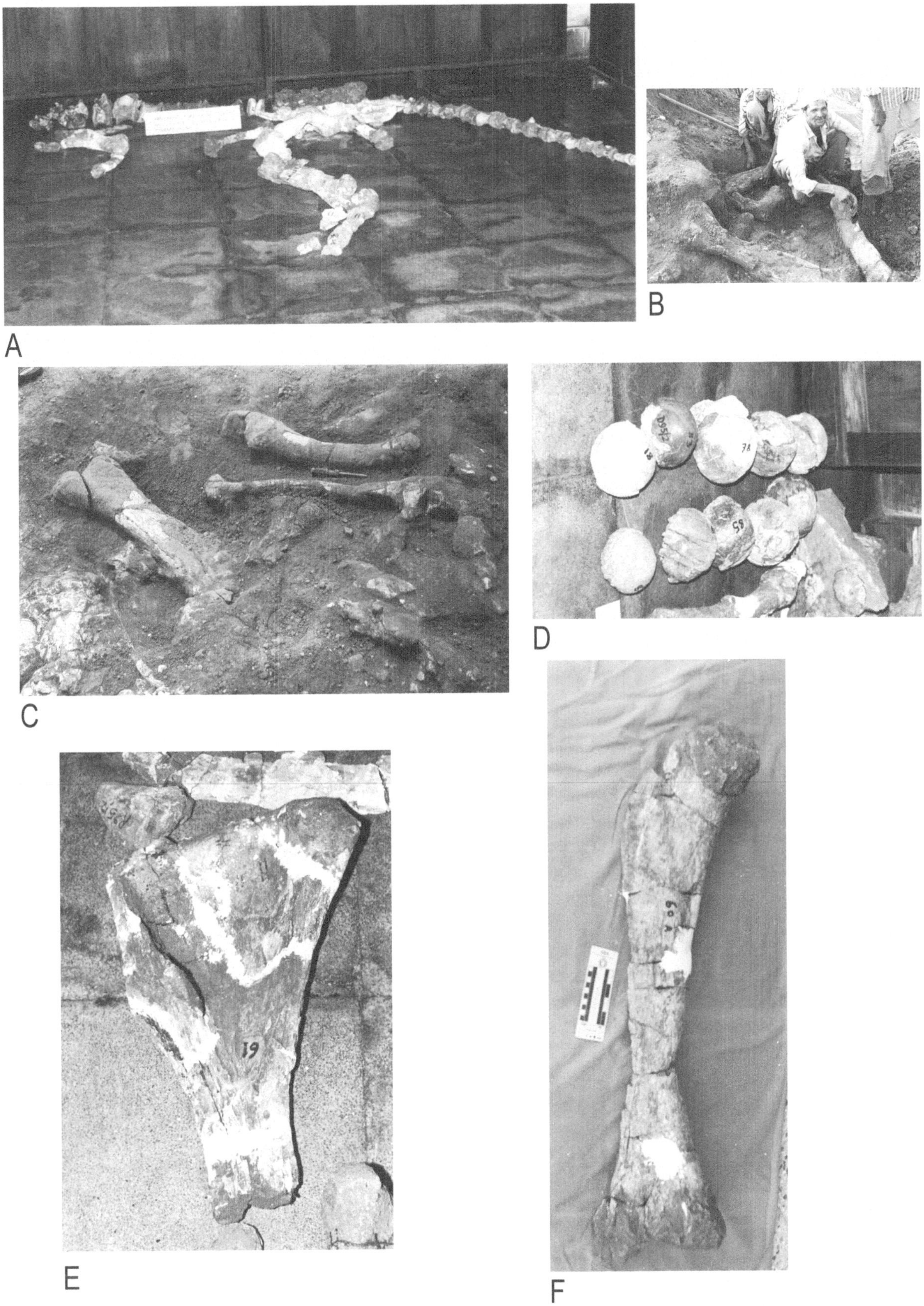

Fig. 8.19 Field photographs of the Late Cretaceous Lameta Formation in Rahioli site. **A**, a nearly complete skeleton of an abelisaurid theropod *Rahiolisaurus*; **B**, excavation of titanosaurid bones; **C**, disarticulated elements of titanosaurid and abelisaurid; **D**, titanosaurid spherical eggs; **E**, proximal end of an ankylosaur humerus; **F**, femur of an ankylosaur

new subfamily Majungsaurinae. Since *Acrovenator* is Late Campanian in age, whereas Indo-Madagascaran abelisaurids are Maastrichtian, it is most likely that post-Cenomanian interchanges occurred among Europe, Africa, India, and Madagascar. Dispersal routes between India and Africa via the OKL Arc, and between India and Madagascar via the Laxmi Ridge-Seychelles remained effective during the Late Cretaceous. To date, no records of abelisaurs have been reported in Antarctica and Australia.

Ankylosaurs

The ankylosaur is a quadrupedal herbivorous group of armored dinosaurs known largely from the Late Cretaceous deposits of all continents with the possible exception of Africa, ranging from the Kimmeridgian to the end of the Maastrichtian. The solitary ornithischian misidentified from the Lameta Formation is *Lametasaurus*, was classified as a stegosaur on the basis of a number of dermal scutes, sacrum containing 5 sacral vertebrae, a pelvis, a tibia, and teeth (Matley 1923). However, *Lametasaurus* is a fossil chimera; most skeletal elements in the original description, especially the hindlimbs and pelvis have been allocated to *Rajasaurus* (Wilson et al. 2003) and are now regarded as a *nomen dubium* (Novas et al. 2010). In 1995, we discovered associated material of an ankylosaur from the Lameta Formation near Rahioli quarry, possibly a nodosaurid, represented by several spines, limbs and girdle bones, indicating its presence in India during the Maastrichtian time (Chatterjee and Rudra 1996). The new material represents isolated vertebrae, scapulocoracoid, humerus, femur, and several pieces of armor such as hollow lateral spikes and solid dorsal scutes typical of ankylosaurs (Fig. 8.19E, F).

Although most ankylosaurs are known from Laurasian landmasses, in recent times, some Gondwana ankylosaurs have been discovered. However, Gondwana dinosaurs were less abundant and less derived than the Laurasian taxa. For example, the recorded ankylosaurs from Gondwana belongs to Nodosauridae; no members of more-derived Ankylosauridae have been discovered in the southern landmasses. For example, *Minmi* from the Early Cretaceous of Australia is a small, basal ankylosaur that might have radiated to other Gondwana landmasses (Molnar 1980). Coria and Saladago (2001) reported the presence of a nodosaurid from the Late Cretaceous Allen Formation of Patagonia. *Antarctopedia* is another basal nodosaurid ankylosaur from the Late Cretaceous of Antarctica (Saladago and Gasparini 2006). The presence of a new nodosaurid from the Lameta Formation indicates that the clade was widespread on the Gondwana landmasses during the Cretaceous.

Kallamedu Formation

During 1998, we explored the Late Cretaceous Kallamedu Formation of Ariyalur in Tamil Nadu in search of the last dinosaurs. We found isolated elements of titanosaurs, theropod teeth, crocodiles, birds, and a beautiful skeleton of a side necked turtle *Kurmademys* (Gaffeny et al. 2001). Here in a stream section, we discovered a 2m-thick, marine oyster-rich, coarse-grained tsunami deposit overlying the continental dinosaur-bearing Kallamedu Formation at the K-Pg boundary. We interpreted the tsunami deposit, linking it to the Shiva impact event (Chatterjee et al. 2006).

Goswami et al. (2013) reported an isolated troodontid tooth serrated with typical coarse-denticles from the Kallamedu Formation of Cauvery Basin, Tamil Nadu. Troodontid is a small-bodied, maniraptoran theropod, predominantly restricted to North America, Europe, and Asia. Discovery of a troodontid from India indicates faunal migration from Laurasia to India via the OKL Arc during the Late Cretaceous time. Abelisaurid teeth and a simosuchid crocodile tooth have also been described from this site indicating faunal exchanges between India and Madagascar during the Late Cretaceous (Prasad et al. 2013).

Dinosaur Eggs and Babies

A fairly extensive record of large, spherical titanosaurid egg clutches, eggshell fragments, and nesting sites have been found in the Lameta Formation can be traced almost continuously along the north of the Narmada Rift from Jabalpur in Madhya Pradesh to Anjar in Gujarat, running over 1,000 km in E-W direction. Both complete and fragmented, these eggs and eggshells are restricted to a thin (3–12 m) lithological unit, a calcareous sandstone of the Lameta Formation, which is interpreted as calcretized paleosol (Sahni and Khosla 1994). These dinosaur nesting sites, comprised of hundreds of nests and thousands of eggs, represent the largest dinosaur hatchery in the world from a single lithological unit (Sahni et al. 1994). Most of these eggs belong to titanosaurids (Fig. 8.19D). These egg-bearing horizons were laid down intermittently across a wide river flood plain during a period when the climate varied between semi-arid and subhumid conditions. The spectacular eruptions of Deccan volcanism began during that time, covering much of central and western India, destroying the habitats of dinosaurs.

Most titanosaurid egg clutches contain about 10 to 12 spherical eggs ranging in diameter from 15 to 20 cm, and were laid down in excavated hollows in linear and circular fashion. The density of clutches suggests that on each occasion, a large number of gregarious females laid eggs at the same time and in the same nesting ground. The sheer size of the adults and proximity of the clutches suggest little or no parental care took place once the eggs were laid – there simply would not have been room for all the parents to tend their young at the same time. The eggs were presumably laid down by two known titanosaurid genera – *Isisaurus* (Pisdura and Balasinor areas) and *Jainosaurus* (Jabalpur area).

The microstructure of the fossil egg shell provides some clue to the producer of the egg. As a rule, large eggs have thick shells and small eggs have thin shells. All these eggs have been grouped under the oogenera *Megaloolithus* Mikhailov 1991 and *Fusioolithus* Fernández and Khosla 2015 and have nine distinct morphostructural units (Khosla and Sahni 1995, Vianey-Liaud et al. 2003). The eggshell is largely made of polycrystalline calcite minerals held in a matrix of collagen fibers. The microscopic structure of the shell in cross section reveals two main layers—an inner mammillary layer and an outer spongy, or palisade layer. The spongy layer is thicker, with calcite crystals arranged in the protein matrix as vertical palisades, separated by minute pore canals. The microscopic structure of the shell in cross section reveals a variation in number of layers among dinosaurs: a single layer is found in sauropods and ornthischians, two distinct layers in theropods, and three layers in birds. Pore canals pass through the shell to permit embryonic gas exchange (Chatterjee 2015). The cross-section of a typical titanosaurid eggshell from the Lameta Formation shows the development of a single layer (basal mammillary layer), separated by pore canals penetrating the thickness of the eggshell, which may vary about 2 to 4 mm in diameter (Sahni et al. 1994). Similar titanosaurid eggs are known from France, Spain, and Patagonia.

Baby dinosaurs have long been a source of fascination to both paleontologists and public. Unlike other dinosaur nesting sites such as Montana, Alberta, Gobi Desert, China, and Argentina, where embryos, hatchlings, and juveniles are common in the nests, no such skeletal remains of baby dinosaurs have yet been recovered from India (except one example) despite such extensive nesting sites. The absence of dinosaur hatchlings and juveniles in India is very unusual and intriguing. Hatchling failure is common in birds, often linked to environmental contamination by pesticides and waste disposal. Perhaps pollution from the Deccan volcanic emission might have contributed largely to the hatchling failure of Indian dinosaurs.

Unlike some dinosaurs, the giant titanosaurids did not care for their eggs. Adults have never been found near nests, so it seems that sauropod hatchlings had to fend themselves from the start, as suggested by their precocial development of the skeleton. And that left the way clear for predators to feast on them as they emerged. There may be one rare example - a partial skeleton of a titanosaurid hatchling, collected from the Dholi-Dungri village, near Balasinor of Gujarat. It is represented by a portion of the left side of the anterior thorax, a partial shoulder girdle and forelimb preserved in anatomical articulation, and the bones are completely ossified (Wilson et al. 2010). An articulated skeleton of a snake *Sanajeh* was found coiled around an egg in a nesting site; the skeleton of a titanosaur hatchling lay nearby (Fig. 8.20). This suggests that *Sanajeh* was a predator caught in the act of devouring a hatchling. Like many early snakes, *Sanajeh* did not have the wide gape seen in boids and pythons and it would have been difficult to crack or swallow titanosaur eggs, which were large, spherical ($\sim$ 14–20 cm diameter), and thick-shelled. Most likely, it was an opportunistic hunter, preying on babies. Titanosaurs in the Lameta ecology such as *Isisaurus* and *Jainosaurus* were large (20–25 m), and were effective deterrents to predators, but their small hatchlings (50 cm), unguarded by parents, were easy prey for stealthy predators like snakes.

Dinosaur Coprolites and Food Preference

The adaptive radiation of dinosaurs resulted in the evolution of a host of species in both the herbivore and carnivore guilds. Identifying the food habits of many dinosaurs are fairly straightforward from their tooth morphology, tooth wear facets, inferred jaw mechanics, as well as general body plan. Sauropods and ornithischians were herbivores, but theropods were carnivores. Coprolite composition can discriminate between herbivores and carnivores. Plant dietary residues are common in coprolites produced by herbivores. On the other hand, inclusions of bone fragments, teeth, fish scales, or mollusk shells provide evidence of carnivory. Additionally, isotopic geochemistry of the preserved stomach contents of some fossils, and coprolites, and the depositional environment in which dinosaurs are found, in association with local plant fossils, can help us to infer their preferred diets.

One of the most novel discoveries was found in the titanosaur diet - phytolith fossils (silicified grasses) were found in titanosaur coprolites from Pisdura, central India. Dinosaur coprolites are rarely preserved in exceptionally favorably conditions because feces are largely composed of soft material. They have been recognized by their familiar fecal shapes, but are highly variable. Matley (1933) discovered a large number of coprolites from the Lameta Formation of Pisdura locality attributed to titanosaur (*Isisaurus*), whose bones were found in the same locality. The large elongated droppings are up to 10 cm wide and 17 cm long, cylindrical, and certainly indicate hefty plant source material. A variety of coprolitic segments have yielded woody

Fig. 8.20 Active nesting sites of the Late Cretaceous Lameta titanosaurs in Rahioli, Gujarat. Life restoration of a large madtsoiid snake *Sanajeh* devouring a newly hatched titanosaur (*Rahiolisaurus*) as the adults are at the distant river. At the background Deccan volcano is erupting (after Chatterjee et al. 2017)

tissue, cuticles, pollen, and water-ingested microorganisms. The plant tissues are mostly of gymnosperm origin. Carbon isotopic ratio ($^{13}C/^{12}C$) and other markers from the coprolites suggest that titanosaurs used C_3 vegetation as its primary diet. On the other hand, gut fermentation may not have been an active mechanism in the digestion process of titanosaurs (Ghosh et al. 2003). Prasad et al. (2005) documented that the grass fossils, in the form of phytoliths, are preserved in the Pisdura coprolites. They concluded that titanosaurs fed indiscriminately on a wide range of C_3 plants including dicotyledons, conifers, palms as well as five taxa of extant grasses (Poaceae).

About one-third of the Earth's vegetative cover comprises savannas, grasslands, and other grass-dominated ecosystems, but the early history of grass is obscure. It is generally believed that dinosaurs did not eat grasses, because the earliest fossil record comes from the Paleocene sediments. The phytoliths from the Pisdura coprolites shed new light on the early origin of grass and suggest that grass covered the Late Cretaceous ecosystem as a green blanket and provided food for herbivorous dinosaurs.

Night: Death of the Dinosaurs

At the Cretaceous-Paleogene (K-Pg) boundary, the Indian plate had moved considerably northward with the subduction of the Kshiroda Plate under Asia, gliding smoothly along two parallel transform faults on its two sides: the Owen-Chaman Fault in the west, and the Wharton Ridge-Sagaing Fault in the east (Fig. 8.21). In this paleobiogeographic setting, India was ground zero for two catastrophes, the Shiva impact and Deccan volcanism that heralded a biotic holocaust across the globe (Chatterjee et al. 2017).

The Cretaceous-Paleogene (K-Pg) boundary mass extinction ($\sim$66 Ma) is one of the most devastating events in the history of life, marking the end of the dinosaur era. The fossil record suggests that about 75% of the plant and animal species inhabiting the Earth perished at the end of the Cretaceous; this occurred over a geologically short period of time. In contrast to the end-Triassic extinction that gave the dinosaurs the advantage to diversify in the Jurassic, the

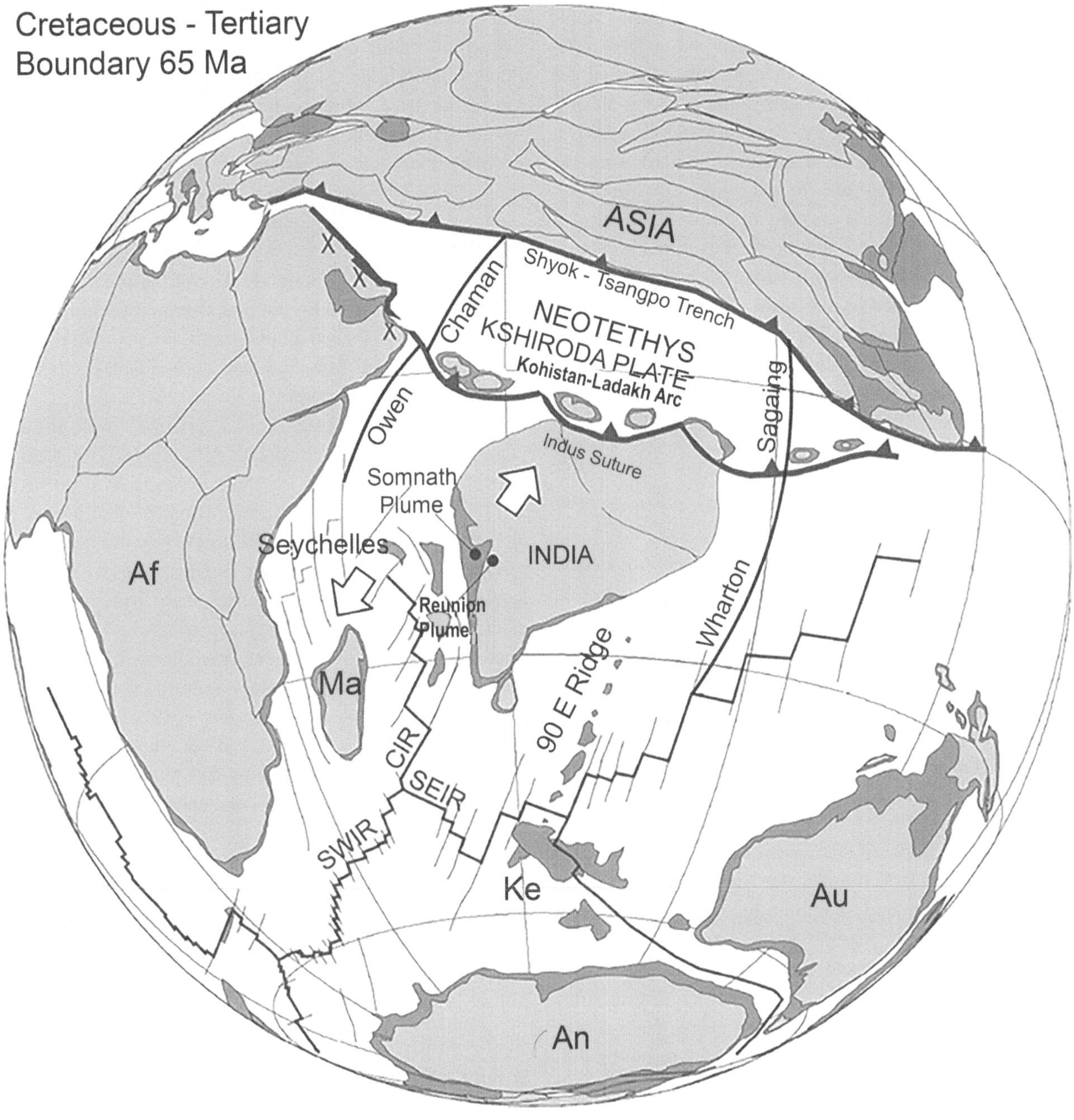

Fig. 8.21 Paleogeographic reconstruction of India in relation to other Gondwana continents during the Cretaceous-Tertiary boundary (~65 Ma). The Oman-Kohistan-Ladakh Arc maintained the biotic link between India and Africa (modified from Chatterjee et al. 2017)

end-Cretaceous extinction finally snuffed out the nonavian dinosaurs, along with the pterosaurs in the sky and the plesiosaurs and mosasaurs in the seas. The extinction also destroyed certain groups of mammals, birds, lizards, insects, and plants on land. In the oceans, ammonites, reef-building rudists, and most species of coccolithosphores and planktonic foraminifera died out. The K-Pg extinction was severe, global, rapid, and selective, affecting biotas of all the major continents and oceans.

The wealth of data accumulated over the past two decades suggests that dinosaurs disappeared suddenly at the K-Pg mass extinction event 66 million years ago (Brusatte et al. 2014). Many hypotheses have been offered over the years to explain dinosaur extinction, but only a few have received serious consideration. The K-Pg extinction was a global event, so we should examine globally effective agents: geographic change, oceanographic change, climatic change, or extraterrestrial events that could cause the environmental crisis. Fossil record of marine plankton and terrestrial plants suggests that, in the last million years or so of the Cretaceous, the warm global climate that prevailed through the period was cooling in fits and spurts: Earth experienced a catastrophic winter at the K-Pg boundary (Scotese 2015). Large amounts of sulfuric acid aerosols, emitted from the Deccan Trap, blocked the sunlight and caused environmental stresses and periodic rapid cooling of Earth. Difficulty in assigning a single cause to end-Cretaceous extinction rests on a series of events that nearly simultaneously occurred: massive volcanism such as the Deccan Traps and the extraterrestrial bolide impacts. The most recent work on the K-Pg extinction has centered on two hypotheses that suggest a violent end to the Cretaceous: twin large asteroid impacts, and a giant volcanic eruption such as Deccan Traps.

Asteroid Impacts

A key breakthrough in the dinosaur extinction debate occurred in 1980, when Alvarez at al. (1980) discovered high concentrations of an iridium anomaly from a 2.5 cm-thick clay layer at the K-Pg boundary marine sections of Gubbio Italy and Stevens Klint in Denmark. As iridium remains abundant in carbonaceous chondrites, the Alvarez team suggested that a giant asteroid, about 10 km in diameter, struck the earth at the time of K-Pg boundary and produced a circummundane dust cloud that blocked the Sun, chilled the planet, and killed the dinosaurs and other organisms. They proposed that the iridium had settled out of a global dust cloud kicked up by the impact of the asteroid. Later, the iridium anomaly has been recognized globally in both continental and marine K-Pg boundary sections. Alvarez group had calculated from the amount of iridium dispersed around the globe that the source crater measures 150 to 200 km in diameter. The impact theory was strengthened by three additional independent forms of evidence in the K-Pg sediments in different parts of the world: (1) shocked quartz, a distinctive signature of an impact event (Bohor et al. 1987); spherules representing droplets of impact melts (Smit 1999); and (3) carbon soot particles resulting from global fires (Wolbach et al. 1988).

Chicxulub Crater, Yucatan Peninsula, Mexico

Locating the K-Pg impact crater would provide the most corroborative evidence of all for the asteroid impact theory, but prior to 1990, no crater of the right age and size has been discovered. Finding a killer 66 million years after the crime was committed is no easy task. Eventually, the K-Pg impact ejecta such as spherules, iridium anomalies, and shocked quartz around the Gulf Coast led to the crime scene in the Chicxulub crater on the northern margin of the Yucatan Peninsula in Mexico. The Chicxulub crater is more than 180 km in diameter, which is buried by 1,100 m of carbonate strata—half on land and half on the sea floor, but the gravity anomaly maps clearly reveal its morphology (Hildebrand et al. 1995). The gravimetric signature of the crater showed a semicircular outline, opening up toward the north-west in a horse-shoe pattern. Recent drill core samples from the crater have confirmed the radiometric date of the impact precisely, coinciding with the K-Pg boundary around 66 million years ago. Moreover, shocked quartz grains, impact melt rocks and breccias recovered from the drill core samples support the impact origin of the crater.

The Chicxulub crater is a pristine, well-preserved peak-ring crater similar to the Schrödinger crater on the Moon, linked to the K-Pg mass extinction event (Kring 2017). The peak ring formed in a matter of minutes. Just after the impact, deep granite bedrock, flowing like a liquid, rebounded into a central tower as tall as 10 km before collapsing into a circular ridge. The sharp iridium peak and ejecta components such as glass spherules and shocked quartz at the thin K-Pg boundary sections in different parts of the globe suggest that the impact-induced catastrophes lasted for extremely short time periods—virtually a geological blink of an eye. The impact caused giant tsunamis in all directions. The emission of dust and particles caused environmental changes close to a nuclear winter. For the past. 30 years, many scientists have claimed the Chicxulub impact to be sole cause for the K-Pg mass extinction, though not without controversy. One of the strongest critics is Gerta Keller, who suggested that Chicxulub impact occurred about 100,000–150,000 years before the K-Pg mass extinction. She suggested that there were multiple impacts at the K-Pg transition (Keller 2014).

Shiva Crater, Mumbai Offshore Basin, India

Half a world away from Chicxulub, in what is now the Mumbai Offshore basin, a giant Shiva impact occurred at the K-Pg boundary, on the western shelf of India-Seychelles continent, which was drifting northward from the breakup of Gondwana (Fig. 8.21). A Ganymed-sized asteroid ($\sim$40 km diameter) striking the Mumbai offshore basin, excavated a large, multiringed complex crater ($\sim$500 km diameter), and extruded fluid ejecta along the downrange of the crater basin (Fig. 8.22A) (Chatterjee and Rudra 1996, Chatterjee et al. 2006, Chatterjee 2015). The Shiva crater (19°45′26″N, 71° 39′46″E) is largely submerged and buried by 2- to 4-km-thick strata of post-impact Tertiary sediments on the western shelf, but the eastern part of the crater rim, the

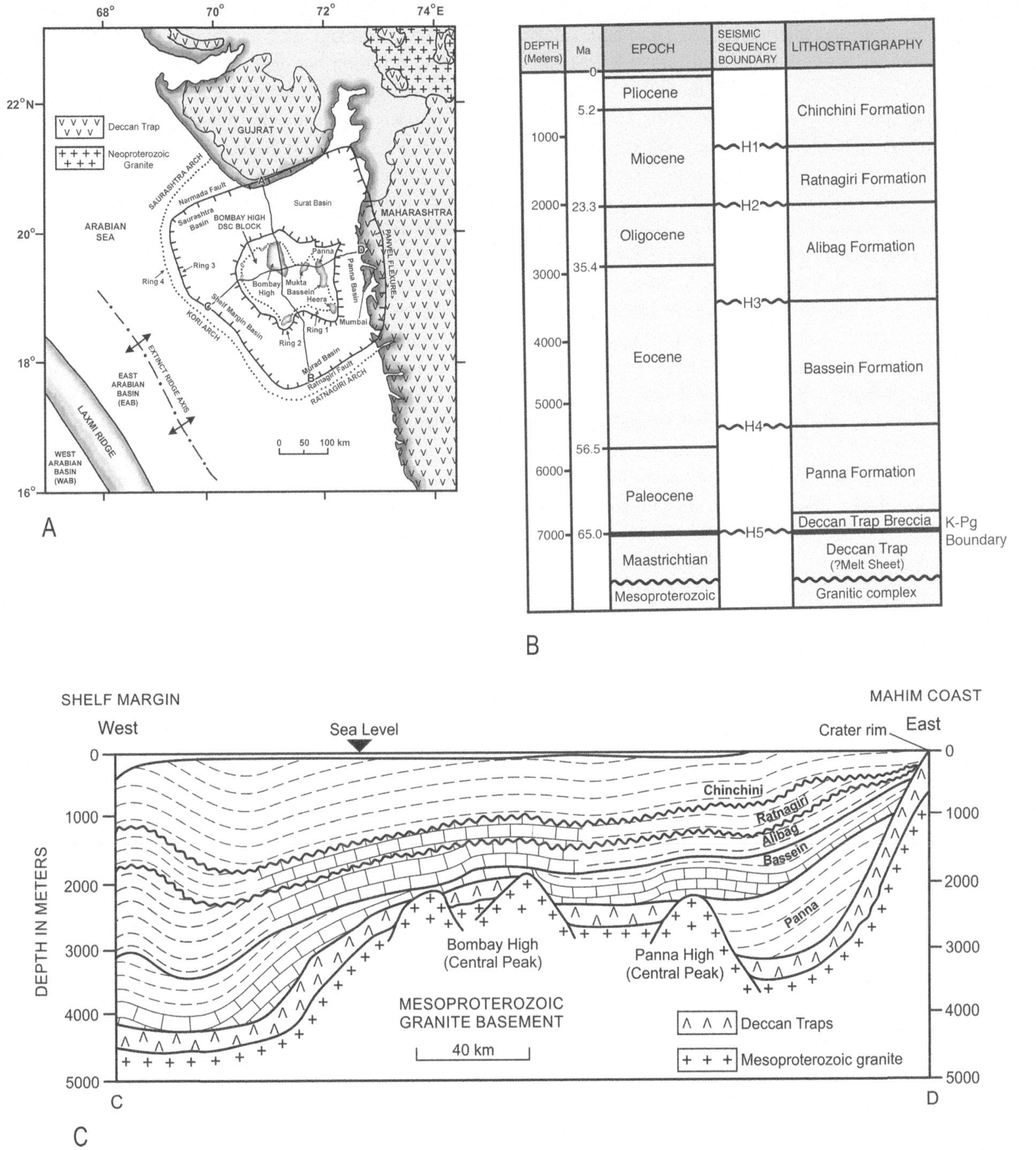

Fig. 8.22 Shiva crater on the western shelf of India. **A**, present day location of Shiva crater at the Mumbai offshore basin. The Shiva crater is a complex peak ring crater, about 500 km across and is buried by 4-km thick, post-impact sediments. The peak ring is about 200 km diameter and consists of several structural highs including Bombay High, Mukta High, Panna-Bassein High, Heera High and several unnamed peaks. The peak ring is the structural trap for oil and gas and is the largest oil field in India. **B**, generalized stratigraphy of the Shiva crater showing the target rocks at the basement, succeeded by the post-impact sediments. **C**, cross-section across the Shiva crater in west-east direction (C–D line in Fig. A), showing the peak rings, which are covered by post-impact sediments (modified from Chatterjee et al. 2006)

Panvel Flexure, comes onshore along the Mumbai coast and remains tectonically and geothermally active, resulting in hundreds of hot springs along this faulted margin (Fig. 8.22B, C). The central region of the crater lies in the Arabian Sea around Bombay High, the largest hydrocarbon field in India, which is located 160 km off the coast of Mumbai. The Oil and Natural Gas Corporation (ONGC) discovered and studied the basin in detail using geophysical and deep drill core data, which are described in several scientific publications and internal reports (Rao and Talukdar 1980, Basu et al.1982, Bhandari and Jain 1984, Mathur and Nair 1993, Zutshi et al. 1993). The ONGC scientists have generously provided some of the drill core samples from the Bombay High and shared many unpublished data with me, which helped us to reconstruct the complex morphology of the subsurface structure of the Shiva crater and confirm its impact origin (Chatterjee et al. 2006).

The Shiva crater has been identified from gravity, seismic and well logs data and buried central peaks (Basu et al. 1982, Bhandari and Jain 1984, Mathur and Nair 1993, Rao and Talukdar 1980, Zutshi et al. 1993, Mehrotra et al. 2010). Gravity data of the Shiva crater show a significant gravity-low-anomaly over the central peaks of the Bombay High region, similar to the pattern of the Chicxulub crater (Hildebrand et al. 1995). Subsurface structure and stratigraphy of the Shiva crater are known primarily from petroleum exploration drill holes, and from seismic stratigraphy by the ONGC. The Shiva structure is a well-preserved, complex peak-ring crater with subsurface mountains, including Bombay High, Mukta High, Panna-Bassein High, Heera High, and several other unnamed peaks, which rise up to 4 km above the crater floor (Fig. 8.22A). The peak-ring consists of highly fractured Mesoproterozoic granite intruding through a veneer of the Deccan Trap, which was rebounded by the bolide impact. The granitic basement rocks of Bombay High field have yielded Rb-Sr isochron age of ~1450 Ma, which is interpreted as the formation and emplacement time of the granite during the Mesoproterozoic (Rathore et al. 2004). However, the radiometric clock of the basement rock was reset to a younger age by the Shiva impact event, but the precise date is not yet available. Most likely, the Proterozoic basement was the target rock for the impact. The peak-ring is about 250 km in diameter and is surrounded by annular trough, and is bordered outside by an elevated and faulted crater rim. The morphology of the Shiva crater closely approximates the Chicxulub Crater of Mexico and bears all the hallmarks of a large complex, peak-ring crater. However, the shape of the Shiva crater is more squarish than circular, like the Barringer crater of Arizona with a tear drop shape in the downrange direction because of the oblique impact. The post-impact sedimentation in the crater basin began with the Early Paleocene Panna Formation and continued through the Pliocene (Fig. 8.22B). These Tertiary strata are nearly horizontal, following the uneven surface of the crater floor. The sediments overlying the peak-ring form an ideal trap for petroleum accumulation and is the most productive oil horizon, where oil and gas is produced primarily from the Miocene reservoirs (Fig. 8.22 C). Exploration for hydrocarbons in the Bombay High field started in the early sixties, and the first oil discovery was made in the Miocene Limestone in 1974. Since then the basin is under active hydrocarbon exploration and exploitation. Ranked 38th worldwide, Shiva has reserves exceeding 4 billion barrels of oil, 24.2 trillion cubic feet of gas, and 0.3 barrels of natural gas liquid (Wandrey 2004). Many impact craters are the most productive hydrocarbon sites. Using K-Ar dating, Rathore et al. (1997) estimated the age of Deccan Trap basement rock around the K-Pg boundary (~66 Ma). The Deccan Trap and the overlying Paleocene Panna Formation together bracket the age of the Shiva crater right at the K-Pg boundary (Fig. 8.22B). Future radiometric dating of the drill core samples will provide additional evidence for the age of the crater.

We can speculate about the size and trajectory of the Shiva bolide. Using a scaling equation (Melosh 1989), the 500-km-diameter Shiva crater was formed by a bolide of 40 km in diameter, about the size of the 'Near Earth Asteroid' (NEA), Ganymed. The shape of the Shiva crater suggests the trajectory of the bolide. Although hypervelocity impacts normally create circular craters, impact from a low angle (~15° from the horizontal) generates an elongated teardrop crater such as the Messier and the Schiller craters on the Moon, Orcus Patera on the Mars, the Rio Cuarto crater in Argentina, and the Shiva crater in India. We speculate the Shiva bolide came from the southwest in a northeasterly direction at a low angle, where the tip of the teardrop indicates the downrange direction. The Shiva bolide flew obliquely from the southwest, over South Africa, and the Arabian Sea, before hitting the western shelf of India at roughly a 15° angle. It created a large oblique crater where it landed and showered India with a particular hard rain of fluid ejecta and dust. As a result, the ejecta was emplaced ballistically along the NE downrange direction. An oblique impact is more devastating than high-angle impact. The "dry target" rock consists of bilayered, rheologically variable Proterozoic granite and Deccan basalt (Chatterjee et al. 2006). The asymmetric distribution of fluid ejecta on the NE side of the Shiva crater indicates the downrange direction (Fig. 8.23A) (Chatterjee et al. 2006). The distribution of ejecta around the crater is more sensitive to the angle of impact and currently serves as the best guide to the obliquity of impacts.

The Shiva impact might have triggered the widespread distribution of the Deccan alkali volcanism that occurred during the late phase of Deccan eruption, because of the synchroneity of two events in space and time. Deccan alkali rocks have been reported from the basement of the Shiva crater (Rathore et al. 1997). Outside the crater, these spectacular volcanic plugs, which are arranged radially in Anjar, Kadi, Jwahar, Phenai Mata, Amba Dongar, Barwaha, Murud and Napsi, are dense with clearly defined zones of gravity-high, which are interpreted as the fluid ejecta of the Shiva impact in the downrange direction (Fig. 8.23A) (Chatterjee and Rudra 1996, Chatterjee et al. 2006). These

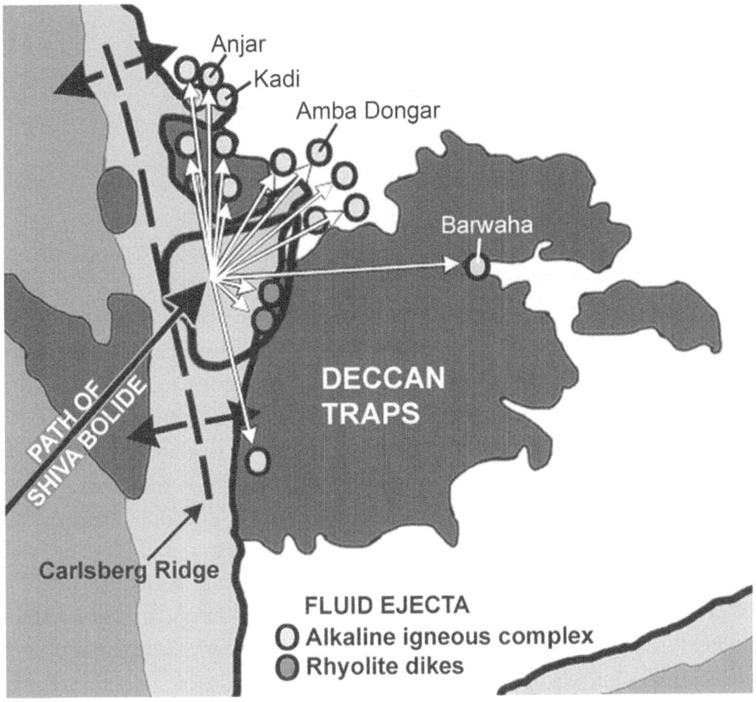

A

B

Fig. 8.23 **A**, radial, asymmetric distribution of fluid ejecta downrange of the Shiva crater; teardrop shape of the crater and asymmetric distribution of melt rocks indicate oblique impact along the NE downrange direction; **B**, Paleogeographic reconstruction showing the initial rift between India and Seychelles in the K-Pg boundary ($\sim$66 Ma). Both the Reunion plume and the Shiva impact have been linked to the separation of Seychelles from India (modified from Chatterjee et al. 2006);

impact melt rocks are rich in iridium, revealing their contamination from the impacting asteroid (Shukla et al. 2001) and showing their consistent ages around 66 Ma (Basu et al. 1993). Most likely, the Deccan alkali rocks represent the melt rocks of bilayered target rocks—Proterozoic granite and Deccan basalt—induced by the Shiva impact (Chatterjee et al. 2006). Drill core samples from the Shiva basin may unravel the origin of the Deccan alkali rocks. Similar lava-like fluid ejecta is common in lunar craters that were emplaced in a downrange to distances as great as a crater radius.

The shock metamorphism of rocks and mineral is an important consequence of hypervelocity impact events. The Proterozoic basement rock of the Shiva crater shows evidence of shock metamorphism, a crucial evidence of the bolide impact event. ONGC drill core samples of the Mesoproterozoic basement rock from the Bombay High show veins of pseudotachylite in the granite, which is usually attributed to bolide impact (Fig. 8.24A, B, C) (Chatterjee et al. 2006). Pseudotachylite is a dark, fine-grained rock that resembles volcanic glass. Both thin sections and SEM images confirm that the composition of the pseudotachylite is pure silica.

The common forms of titanium dioxide (TIO_2) in the Earth's crust are rutile, anatase, and brookite. The first natural occurrence of an unnamed, dense polymorph of TiO_2, attributed to shock metamorphism, was found in the shocked gneisses of the Ries impact crater in Germany (Goresy et al. 2001). Similar shocked rutile has been identified in breccias from the Eocene Chesapeake Bay impact structure in the USA (Jackson et al. 2006). Recently from the ONGC drill core samples of the Proterozoic granite, we have identified in SEM study a shock-induced ultradense polymorph of rutile in the granitic basement rock, which is similar to that of the Ries crater and the Chesapeake Bay impact structure (Fig. 8.24D, E). Experimental data suggests that shocked rutiles are formed at high pressure between 16 and 20 gigapascals, generated by a large body impact. Both pseudotachylite and shocked rutile in the basement rock suggest that the Mesoproterozoic granite was the primary target rock for the Shiva bolide impact.

The Shiva bolide ($\sim$40 km diameter) generated lethal amounts of kinetic energy of 1.45×10^{25} joules. The impact was so powerful that it caused several geodynamic anomalies: it fragmented, sheared, and deformed the lithosphere across the western Indian margin and contributed to major plate reorganization in the Indian Ocean. Geophysical study suggests that the Shiva impact sheared, thinned, and deformed the lithosphere near the Mumbai coast, where the crust-mantle boundary was lifted up more than 50 km, showing unusually high heat flow (Pandey and Agarwal 2001). It initiated rifting between India and Seychelles in the west, creating the Arabian Sea (Chatterjee et al. 2006, Chatterjee et al. 2017).

Killing Mechanism

There is now a general consensus that two asteroids hit the Earth at the end of the Cretaceous, and the twin impacts were a primary cause for the final demise of the dinosaurs, as well as other organisms on land and in the sea (Lerbekmo 2014). The evidence for two large bolide impacts on Earth at the end of the Cretaceous has come from opposite sides of the globe, one on the Yucatan Peninsula of Mexico creating the Chicxulub crater (Hildebrand et al. 1995) and the other on the western shelf of India excavating the Shiva crater (Chatterjee et al. 2006). Another small crater ($\sim$24 km diameter), Boltysh crater at Ukraine, also formed at the K-Pg boundary; the cosmic signatures of three successive impact events at the K-Pg boundary been recorded in Gujarat, India.

The Anjar volcano-sedimentary section in Gujarat consists nine lava flows (F1–F9) four intertrappean beds (Bhandari et al. 1996). The third intertrappean bed, about 6 m thick, contains three closely-spaced iridium and fullerene anomalies at the K-Pg boundary, supporting multiple impact events (Parthasarathy et al. 2002). These three iridium layers are designated as Br-1, Br-2, and Br-2, from top to bottom; they appear to be primary ejecta layers deposited in quick succession from different vaporized meteorite sources from different sites.

We suggested that Br-1 signals the Shiva impact event, Br-2 the Boltysh, and Br-3 may correspond with the Chicxulub impact event. If so, the Anjar section may hold crucial evidence for three distinct episodes of global impact events during the K-Pg transition: Chicxulub in Mexico, Shiva in India, and Boltysh in Ukraine (Chatterjee et al. 2006). Similarly, at the K-Pg section in Oman, there are two distinct iridium anomalies, separated by more than 1 m-thick sediments, which supports the twin impact scenario (Ellwood et al. 2003). The Chicxulub-Shiva one-two impact killer punch, combined with the Deccan volcanism, caused rapid and profound global climate perturbations. Large bolide impacts are disastrous for the biosphere precisely because they exert stupendous energy bursts in extremely short time periods.

The abruptness of the dinosaur extinction suggests a key role for the bolide impacts (Brusatte et al. 2014). The twin impacts would have instantly produced devastating shock waves, a searing global heat pulse, catastrophic environmental effects such as extended darkness, nuclear winter, and acid rain. Impacts would trigger earthquakes of magnitude 13–15 on the Richter scale, generating gigantic mega-tsunami waves that propagated in all directions from

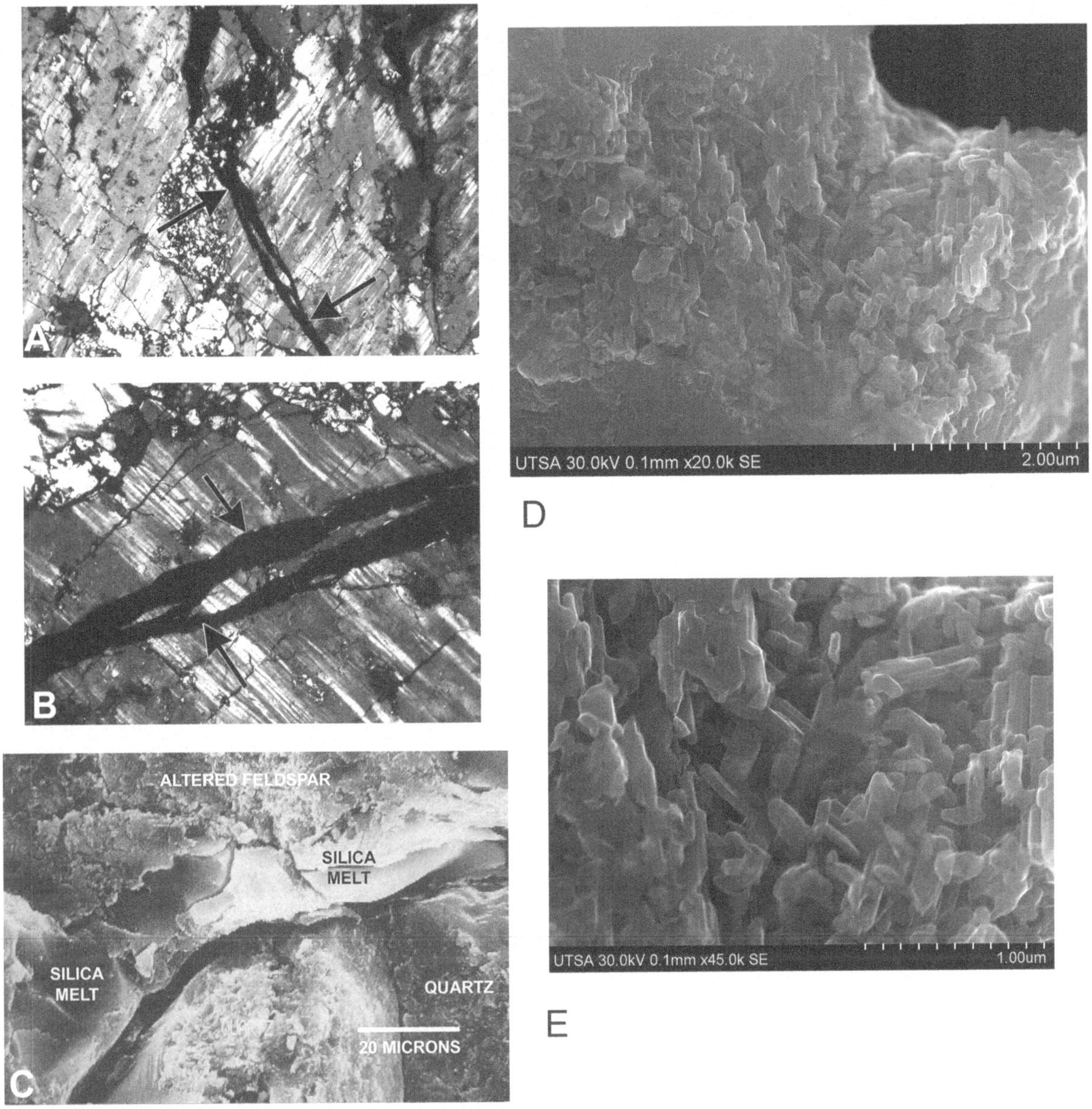

Fig. 8.24 Evidence of shock metamorphism at the basement rock of Mesoproterozoic granite, the target rock of the Shiva impact. A-B, thin section micrographs of basement granite (crossed nicols), showing the vein of pseudotachylite cutting across K-feldspar grain. The granite hosting the pseudotachylite is strongly shock-metamorphosed by the impact; C, SEM photograph of the basement granite showing the highly magnified view of the pseudotachylite vein containing pure silica melt. The impact was so powerful (>100 GPa) that it obliterated the planar deformation features (PDFs) of shocked quartz grain and turned into a melt component; D-E, SEM photographs of the basement granite showing the highly magnified view of shocked rutile (courtesy Necip Guven)

the point of impact and would flood coastal areas. The impacts would have generated more than a 100-million megaton blast (about 10,000 times that of the entire world's nuclear arsenal) and lofted vast amounts of ejecta into the atmosphere that would envelop Earth, block the Sun, and halt photosynthesis (Alvarez 1997). Millions of organisms would die instantly from the direct effect of the impact—shock heating of the atmosphere by the expanding fireball. Most of the world's vegetation would be caught in fireball (Kring and Dudra 2003). The fires consumed Earth's forests, wiped out critical terrestrial habitats and wreaked the base of the food chains. An infrared thermal pulse from a global rain of hot spherules from the K-Pg impacts could be the prime killing agents (Robertson et al. 2004). According to this model, for several hours following the impacts, the entire Earth was scorched with infrared radiation from reentering ejecta that would have killed unsheltered large animals directly and ignited global fires that consumed Earth's forests and their inhabitants. Any animal larger than 25kg was annihilated, very possibly due to insufficiency of oxygen.

Smaller species such as mammals and squamates that lived underground, or freshwater animals such as turtles and crocodiles would have been least vulnerable to heat and fire. Soot and impact ejecta chocked the sky and blocked the Sun, creating a perpetual night that halted photosynthesis and triggered a near-freezing global climate so that plants and phytoplanktons died and food chains collapsed, devastating the biosphere. Since the twin impacts occurred at the coastal region, huge tsunamis produced by the impact destroyed the costal habitat (Chatterjee et al. 2006).

Deccan Volcanism

Although twin impacts played a critical role in the K-Pg mass extinction, global climatic instability, high environmental stress, and acidification of the ocean preceded the K-Pg boundary by ~ 1 million years (Keller 2014). The cause of the precursory climatic perturbation, that pushed some ecosystems to the tipping point, is generally attributed to the early pulses of Deccan volcanism. Deccan volcanism has long been proposed as having a causal relationship in the death of the dinosaurs at the end of the Cretaceous, but the link remains controversial. The Deccan volcanic province extends across approximately one sixth of the subcontinent, encompassing up to 1,000,000 km^2 of eruptive lavas that reach a total thickness of ~ 3000 m near the eruptive center in western India. The main phase of Deccan eruptions initiated $\sim 250{,}000$ years before the Cretaceous-Paleogene boundary (Schoene et al. 2015). Habitat destruction of such an extensive area, by prolonged Deccan volcanism, pushed Maastrichtian dinosaurs to the outer fringe of the volcanic province, causing population declines in many dinosaur species (Fig. 8.25). However, it is clear that the major dinosaur species persisted in India during the second Deccan phase, up until the K-Pg boundary iridium layer in the Anjar section, suggesting that at least in India, Deccan volcanism did not cause any substantial change in dinosaur species richness.

However, Deccan volcanism might have a direct adverse effect on marine organisms. The Deccan Traps have released tens of thousands of petagrams of carbon dioxide and methane, leading to tropical warming and loss of oxygen from the ocean's interior. Temperature-dependent anoxia may explain severity of end-Cretaceous marine mass extinction. Volcanically induced warming is clearly a powerful driver of marine anoxia.

Volcanic emissions led to increasingly toxic environment by releasing mercury into the atmosphere in gaseous form that contaminated water and soil. Mercury emitted into the air eventually settles globally into oceans, rivers and lakes, and is subsequently deposited in sediments. The toxic effects of mercury pollution are deleterious for wildlife and the entire ecosystem. Font et al. (2016) reported a high concentration of mercury (Hg) anomaly (>2 orders of magnitude) at the K-Pg transition in marine sections of France, Italy, Brazil, Argentina, which is linked to the major Deccan eruption episode. There were numerous, pulsed elevations of mercury concentrations in the marine sediments globally during the K-Pg transition. These peaks show that mass extinction coincided with large-scale, episodic volcanism. Pulsing mercury emissions from Deccan volcanism occurring over a million years, might have poisoned the ocean-atmosphere system during the end-Cretaceous crisis. High mercury concentrations polluting an environment reduce the reproductive success and impose hatchling failure of many birds and turtles. In recent times, bird hatchlings are especially at risk from toxic chemicals such as mercury, which accumulates in the eggs in increasing concentration; this may adversely influence embryonic development, or the critical period of hatching, or the first days of the hatched chick's life (Becker et al. 1993). Thousands of dinosaur eggs have been discovered, stretching 1000 km along the Lameta Formation of India, but they seem to be empty; all but one isolated example lack embryos. Did Deccan volcanisms trigger the hatchling failure of dinosaurs in India by mercury-induced poisoning? We don't know but this raises an interesting possibility. Perhaps detection of mercury levels in the titanosaur eggshell may answer this question.

The episodic nature of Deccan volcanism may possibly explain the survival of many freshwater and terrestrial communities in the Deccan volcanic province during the periods of quiescence. Deccan Traps might have contributed to the latest Cretaceous environmental change and biological turnover in the marine realm for a prolonged period, before and after the twin impacts. Enormous amounts of CO_2, SO_2 and other pollutants from the Deccan eruptions possibly caused climate warming and cooling, ocean acidification, and high-stress-environments. The pattern of extinctions seems unequivocal in the rapid demise of planktic foraminifera that is related to the main phase of Deccan eruption. The strong carbonate dissolution effects are observed in foraminifera from the intertrappean and infratrappean sediments of the Krishna-Godavari wells, as well as in the Meghalaya section. This indicates ocean acidification due to volcanic CO_2 and SO_2 resulting in decreased pH of ocean waters (Keller 2014). The Deccan eruption coincided with long-term global warming of 4 °C in the oceans and 8 °C on land, followed by rapid short-term "impact winter" conditions induced by twin impacts (Scotese 2015, Vellekoop et al. 2016).

The environmental crisis induced by the Deccan volcanic emissions would have been intensified a million times by the collision of twin asteroids at the K-Pg boundary. The killing mechanism of marine biota was likely ocean acidification, and mercury poisoning induced by volcanism, resulting in

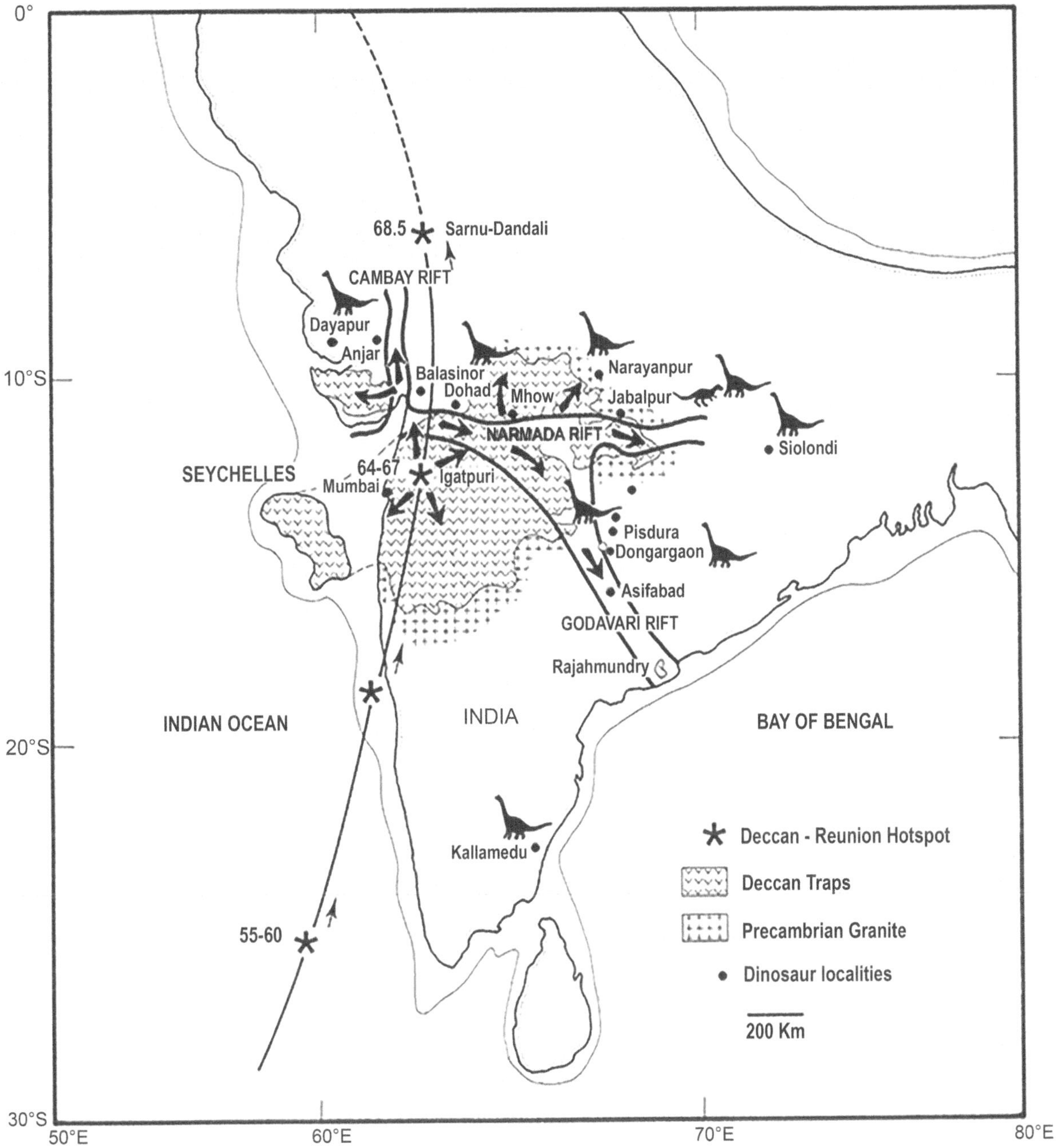

Fig. 8.25 Sketch map showing localities of Maastrichtian dinosaurs around the Deccan volcanic province. The position of India was reconstructed during the Late Cretaceous time (~65 Ma); asterisks indicate the Deccan-Reunion hotspot track; the plume might have started from the Parh Group of volcanics of northeastern Baluchistan, the northernmost manifestation of the Deccan volcanism (shown by dashed line). Dinosaurs were thriving and reproducing along the periphery of the Deccan volcanic province, when the lava was erupting (modified from Chatterjee and Rudra 1996)

the carbonate crisis that led to the collapse of the marine food chain associated with extreme climatic perturbation and anoxic condition. The Deccan volcanic model can better explain the selectivity and the stair-step extinction pattern in marine organisms. Deccan volcanism was an accomplice, but not the main killer, in the massive destruction of life, especially the dinosaurs and other terrestrial organisms. It seems that twin impacts were the proximate cause of the K-Pg extinction, whereas the long-term volcanism contributed heavily to the breakdown of stable ecological communities, and disrupted the biosphere. The K-Pg extinction was a compound crisis, induced by both impact and volcanism. It seems that twin impacts at the K-Pg boundary were the final blow to the already overstressed global ecosystems initiated by the prolonged period of Deccan volcanism (Chatterjee 2015). Two large impacts such as Shiva and Chicxulub in quick succession on the antipodal position, in concert with Deccan eruptions, would have devastating effects globally leading to climatic and environmental catastrophes that wiped out dinosaurs and many organisms at the K-Pg boundary.

Impact-Induced Volcanism

The Deccan volcanism began spewing more lava coincident to the Shiva bolide impact, on the western shelf of India. This was possibly precipitated by the closely proximity of the Shiva impact to the Deccan-Reunion Hotspot, 66 million years ago. A cause-and-effect connection between the bolide impact and Deccan volcanism has been the subject of extensive discussion and speculation. The main phase of Deccan volcanism, which accounts for about 80% of the 3,500-m-thick pile, is reported to have erupted within a short time in C29R, within a 50,000-year period (Schoene et al. 2015). This rapid phase 2 of Deccan eruptions also appeared to coincide with the date for the Shiva impact. At the K-Pg boundary, the normal trickle of Deccan lava eruption became a torrent, as evinced by the thick pile of phase 2 Deccan volcanism. The Shiva impact was so powerful that it not only threw huge clouds of dust into atmosphere, but also rattled the magma chamber of the Earth, causing spectacular Deccan volcanic eruptions. The spatial and temporal coincidence of Deccan volcanism with the Shiva crater led to the suggestion that the main phase of the Deccan Trap might have been triggered by the Shiva impact (Chatterjee and Rudra 1996). Impact could enhance the volcanic activity through decompression melting beneath or near the impact site (Jones et al. 2002). In close geographic proximity of the Shiva crater, the head of the Reunion plume during the K-Pg boundary was positioned near Igatpuri, creating the spectacular Deccan eruption (Fig. 8.23B).

The Deccan volcanic province is one of the largest volcanic eruptions in earth's history, comprising >1.3 million km^3 of erupted lava flows throughout west-central India. Alkali igneous complexes within the Deccan lavas, interpreted as the fluid impact ejecta, straddle the traces of the asteroid impact; this is indicated by a spike in iridium levels. This strengthens the idea that the Shiva impact and the Deccan eruptions happened coincident in geographic proximity and very close together in time (Chatterjee et al. 2006). Perhaps, these two catastrophic events in India were genetically related, i.e., the Shiva impact enhanced the Deccan phase 2 eruption by expansion of the deep-crustal magma chamber, triggering the rapid eruption during the K-Pg boundary (Chatterjee et al. 2017). The spurt in Deccan volcanism at the K-Pg boundary was possibly enhanced by the Shiva impact.

Aftermath of the K-Pg Extinction

Puzzling mass extinctions punctuate the Earth's history. Mass extinction events may be the primary force for major shifts in life's history and important drivers of evolutionary novelty. When a mass extinction strikes, it is not necessarily the most fit fauna that survive, but merely the most fortunate. After each mass extinction event, life rebounds, is revamped, and diversifies. Many dominant groups in the pre-extinction period are eliminated, providing opportunities for surviving lineages to proliferate. Mass extinctions are fundamental game-changers in the history of life, by removing successful incumbents, and by encouraging the meek to proliferate and inherit the Earth.

The K-Pg extinction is a chilling reminder of the fragility of earth's biosphere. Life was devastated but biota that survived the bottleneck effect, inhabited a world in which the slate was wiped clean, and vast and diverse ecological zones were newly available for new-comers for exploration and exploitation. Following the combined cataclysm, evolution in all groups of vertebrates, especially modern birds and placental mammals, was explosive and diverse. The aftermath of twin impacts eliminated dinosaurs from land, and opened up opportunities for placental mammals and primates to evolve and flourish. The K-Pg extinction left an impoverished fauna on the land with the disappearance of nonavian dinosaurs and shallow marine organisms, but a rebound during the early Paleogene once again brought back biodiversity. The demise of the dinosaurs left vast regions virtually devoid of large land animals; these vacant niches were soon occupied with newly evolving and diversifying mammals. It led to the explosive evolution of placentals including primates, our early ancestors.

Mammals evolved into diverse ecophenotypes during the Jurassic and Cretaceous, but compared to extant species were generally small. All mammal groups were severely affected by the end-Cretaceous extinction, particularly metatherians. The survivors proliferated rapidly after the dinosaur extinction, forming diverse mammalian faunas in North America in Early Paleocene. It is tempting to consider that the death of dinosaurs at the end of the Cretaceous lifted a major impediment to the ecological expansion of placental mammals. That view might be true, but the tempo and mode of ecological replacement of dinosaurs by placental mammals was not instantaneous in the Paleocene, but was more gradual. It occurred over several stages, across the 10 million years following the extinction event, eventually culminating in the emergence of the Eocene, when the traumatized ecosystems had finally recovered. The post-K-Pg boundary high stress conditions and delayed recovery can be attributed to last phase 3 of Deccan volcanism during Early Danian. A gradual recovery to larger morphotypes and higher diversity of placentals began only after the Deccan phase 3 ended, and full recovery was restored during the Paleocene-Eocene Thermal Maximum (PETM), when India collided with Asia. At that time, placental mammals underwent an explosive evolution, as recorded in the rich fossils record of the Cambay Shale Formation of Vastan, Gujarat. With the demise of the dinosaurs, mammals evolved a greater variety of forms in the newly vacated niches. We see this in the rapid increase of their body size, and the ecological diversity of placental mammals that are today found in the Cambay Shale Formation. India became the major center for several groups of placental radiation such as bats, whales, artiodactyls, perissodactyls, primates, and lagomorphs, thus strengthening the 'out-of-India' hypothesis (Chatterjee et al. 2017).

Acknowledgements This paper is a story of an itinerant paleontologist who witnessed the birth of vertebrate paleontology in a newly independent India, and actively participated in that development. I would like to take this opportunity to express my sincere thanks and gratitude to the late Pamela Robinson, and the other paleontologists at the Indian Statistical Institute who inspired me to pursue paleontology during my formative years. I thank Dhiraj Rudra for accompanying me on field excursions, and for providing years of much needed logistic support. Necip Guven must be gratefully acknowledged for identifying the shocked rutile from the basement of the Shiva crater. I thank Saswati Bandopadhyay for providing the photograph of *Barapasaurus*. Appreciation goes to Oliver McRae for his accompanying artwork and the careful reading of this manuscript. Many thanks go to Guntupalli Prasad for inviting me to contribute this paper and his careful editing. I thank three anonymous reviewers for providing useful suggestions. This paper is dedicated to my friend Ashok Sahni for his pioneering work in Indian paleontology and stratigraphy. The research was done under the support of Texas Tech University.

References

Alvarez, W. (1997). *T. rex and the Crater of the Doom*. Princeton: Princeton University Press.

Alvarez, L. W., Alvarez, W., Asaro, F., & Michel, H. V. (1980). Extraterrestrial cause for the Cretaceous-Tertiary extinction. *Science, 208*, 1095–1108.

Bandyopadhyay, S., Gillette, D. D., Ray, S., & Sengupta, D. P. (2010). Osteology of *Barapasaurus tagorei* (Dinosauria: Sauropoda) from the early Jurassic of India. *Palaeontology, 53*, 533–569.

Basu, A. R., Renne, P. R., Das Gupta, D. K., Teichmann, F., & Poreda, R. J. (1993). Early and late alkali igneous pulses and a high-^{3}He plume origin for the Deccan flood basalts. *Science, 261*, 902–906.

Basu, D. N., Banerjee, A., & Tamhane, D. M. (1982). Facies distribution and petroleum geology of Bombay Offshore Basin, India. *Journal of Petroleum Geology, 5*, 57–75.

Becker, P. H., Furness, R. W., & Henning, D. (1993). Mercury dynamics in young Common Tern (*Sterna hirundo*) chicks from a polluted environment. *Ecotoxicology, 2*, 33–40.

Berman, D. S., & Jain, S. L. (1982). The braincase of a small sauropod dinosaur (Reptilia: Saurischia) from the Upper Cretaceous Lameta Group, Central India, with review of Lameta Group localities. *Annals of Carnegie Museum, 51*, 405–422.

Bhandari, L. L., & Jain, S. K. (1984). Reservoir geology and its role in the development of L-III reservoir, Bombay high field, India. *Journal of Petroleum Geology, 7*, 27–46.

Bhandari, N., Shukla, P. N., Ghevariya, Z. G., & Sundaram, S. M. (1996). K/T boundary layer in Deccan intertrappeans at Anjar, Kutch. *Geological Society of America Special Paper, 307*, 417–424.

Bohor, B. F., Modreski, P. J., & Foord, E. E. (1987). Shocked quartz in the Cretaceous-Tertiary boundary clay: Evidence for a global distribution. *Science, 224*, 705–799.

Bonaparte, J. F., Brea, G., Schultz, C. L., & Martinelli, A. G. (2007). A new specimen of *Guaibasaurus candelariensis* (basal Saurischia) from the Late Triassic Caturitta Formation of southern Brazil. *Historical Biology, 19*, 73–82.

Brusatte, S. L., Butler, R. J., Barrett, P. M., Carrano, M. T., Evans, D. C., Lloyd, G. T., et al. (2014). The extinction of dinosaurs. *Biological Review, 90*, 628–642.

Chatterjee, S. (1974). A rhynchosaur from the Upper Triassic Maleri Formation of India. *Philosophical Transactions of the Royal Society of London, B276*, 209–261.

Chatterjee, S. (1978a). A primitive parasuchid (phytosaur) reptile from the Upper Triassic Maleri Formation of India. *Palaeontology, 21*, 83–127.

Chatterjee, S. (1978b). *Indosuchus* and *Indosaurus*, Cretaceous carnosaurs from India. *Journal of Paleontology, 52*, 570–580.

Chatterjee, S. (1980). *Malerisaurus*, a new eosuchian reptile from the Upper Triassic Maleri Formation of India. *Philosophical Transactions of the Royal Society of London, B290*, 219–258.

Chatterjee, S. (1982a). Phylogeny and classification of thecodontian reptile. *Nature, 295*, 317–320.

Chatterjee, S. (1982b). A new cynodont reptile from the Triassic of India. *Journal of Paleontology, 56*, 203–214.

Chatterjee, S. (1987). A new theropod dinosaur from India with remarks on the gondwana-laurasia connection in the late triassic. In G. D. McKenzie (Ed.), *Gondwana six: stratigraphy, sedimentology and paleontology. Geophysical Monograph, 41* (pp. 183–189). Washington DC: American Geophysical Union.

Chatterjee, S. (2015). *The rise of birds: 225 million years of evolution*. Baltimore: Johns Hopkins University Press.

Chatterjee, S., Goswami, A., & Scotese, C. R. (2013). The longest voyage: Tectonic, magmatic, and paleoclimate evolution of the Indian plate during its northward flight from Gondwana to Asia. *Gondwana Research, 23*, 238–267.

Chatterjee, S., Guven, N., Yoshinobu, A., & Donofrio, R. (2006). Shiva structure: A possible KT boundary impact crater on the western shelf of India. *Special Publication, Museum of Texas Tech University, 50*, 1–39.

Chatterjee, S., & Hotton, N., III. (1986). The paleoposition of India. *Journal of Southeast Asian Earth Sciences, 1*, 145–189.

Chatterjee, S., & Majumdar, P. K. (1987). *Tikisuchus romeri*, a new rauisuchid reptile from the Late Triassic of India. *Journal of Paleontology, 61*, 787–793.

Chatterjee, S., & Rudra, D. K. (1996). KT events in India: Impact, volcanism, and dinosaur extinction. *Memoirs of the Queensland Museum, 39*, 489–532.

Chatterjee, S., & Scotese, C.R. (1999). The breakup of Gondwana and the evolution and biogeography of the Indian plate. In A. A. Sahni & R. S. Loyal (Eds.), *Gondwana assembly: New issues and perspectives* (pp. 397–425). New Delhi: Indian National Academy of Sciences.

Chatterjee, S., Scotese, C. R., & Bajpai, S. (2017). The restless Indian plate and its epic voyage from Gondwana to Asia: Its tectonic, paleoclimatic, and paleobiogeographic evolution. *Geological Society of America Special Paper, 529*, 1–147.

Colbert, E. H. (1980). *A fossil hunter's notebook*. New York: E. P. Dutton.

Colbert, E. H. (1988). *Digging into the past*. New York: Dembner Books.
Coria, R. A., & Saladgo, L. (2001). South American ankylosaurs. In K. Carpenter & J. I. Kirkland (Eds.), *The armored dinosaurs* (pp. 159–168). Bloomington: Indiana University Press.
Datta, P. M. (1981). The first jurassic mammal from India. *Zoological Journal of the Linnean Society, 73,* 307–312.
Datta, P. M., & Das, D. P. (2001). *Indozostrodon simpsoni*, gen. et sp. nov., an Early Jurassic megazostrodontid mammal from India. *Journal of Vertebrate Paleontology, 21,* 528–534.
D'Emic, M. D., Wilson, J. A., & Chatterjee, S. (2009). The titanosaur (Dinosauria: Sauropoda) osteoderm record: Review and first definitive specimen from India. *Journal of Vertebrate Paleontology, 29,* 165–177.
Ellwood, B. B., MacDonald, W. D., Wheeler, C., & Benoist, S. L. (2003). The K-T boundary in Oman: Identified using magnetic susceptibility field instruments with geochemical confirmation. *Earth and Planetary Science Letters, 206,* 529–540.
Fernández, M. S., & Khosla, A. (2015). Parataxonomic review of the Upper Cretaceous dinosaur eggshells belonging to the oofamily Megaloolithidae from India and Argentina. *Historical Biology, 27* (2), 158-180.
Font, E., Adatte, T., Sial, A. N., de Lacerda, L. D., Keller, G., & Punekar, J. (2016). Mercury anomaly, Deccan volcanism, and the end-Cretaceous mass extinction. *Geology, 44,* 171–174.
Gaffney, E. S., Chatterjee, S., & Rudra, D. K. (2001). *Kurmademys*, a new side-necked turtle (Pelomedusoides: Bothremydidae) from the Late Cretaceous of India. *American Museum Novitates, 3321,* 1–16.
Ghosh, P., Bhattacharya, S. K., Sahni, A., Kar, R. K., Mohabey, D. M., & Ambwani, D. M. (2003). Dinosaur coprolites from the Late Cretaceous (Maastrichtian) Lameta Formation of India: Isotopic and other markers suggesting a C_3 plant diet. *Cretaceous Research, 24,* 743–750.
Goresy, A. E., Chen, M., Dubrovinsky, L., Gillet, P., & Graup, G. (2001). An ultradense polymorph of rutile with seven-coordinated titanium from the Ries crater. *Science, 293,* 1467–1470.
Goswami, A., Prasad, G. V. R., Verma, O., & Benson, R. B. J. (2013). A troodontid dinosaur from the latest Cretaceous of India. *Nature Communications, 4,* 1703. https://doi.org/10.1038/ncomms2716.
Hildebrand, A. R., Pilkington, M., Conors, C. Ortiz-Aleman, & R. E. Chavez. (1995). Size and structure of the Chicxulub crater revealed by horizontal gravity gradients and cenotes. *Nature, 376,* 415–417.
Huene, F. V. (1940). The tetrapod fauna of the Upper Triassic Maleri beds. *Memoirs of the Geological Survey of India, Palaeontologia Indica (NS), 32,* 1–42.
Huene, F. V., & Matley, C. A. (1933). Cretaceous Saurischia and Ornithischia of the central provinces of India. *Memoirs of the Geological Survey of India, Palaeontologia Indica, 21,* 1–74.
Jain, S. L., & Bandopadhyay, S. (1997). New titanosaurid (Dinosauria: Sauropoda) from the Late Cretaceous of Central India. *Journal of Vertebrate Paleontology, 17,* 114–136.
Jain, S. L., Kutty, T. S., Roy Chowdhury, T. K., & Chatterjee, S. (1975). The sauropod dinosaur from the Lower Jurassic Kota Formation of India. *Proceedings of the Royal Society of London, B, 188,* 221–228.
Jackson, J. C., Horton, J. W., Chow, I. M., & Belkin, H. E. (2006). A shock-induced polymorph of anatase and rutile from the chesapeake bay impact structure, Virginia. *U. S. A. American Mineralogist, 91,* 604–608.
Jones, A. P., Price, G. D., Price, N. J., De Carli, P.S., & Clegg, R. A. (2002). Impact induced melting and the development of large igneous provinces. *Earth and Planetary Science Letters, 202,* 551–561.
Keller, G. (2014). Deccan volcanism, the Chicxulub impact, and the end-Cretaceous mass extinction: Coincidence? Cause and effect? *Geological Society of America Special Paper, 505,* 57–89.
Khosla, A., Kapur, V. V., Sereno, P. C., Wilson, J. A., Wilson, G. P., Duthell, D., et al. (2003). First dinosaur remains from the Cenomanian-Turonian Nimar Sandstone (Bagh Beds), district Dhar, Madhya Pradesh, India. *Journal of the Palaeontological Society of India, 48,* 115–127.
Khosla, A., & Sahni, A. (1995). Parataxonomic classification of Late Cretaceous dinosaur eggshells from India. *Journal of the Palaeontological Society of India, 40,* 87–102.
Kring, D.A. (2017). Chicxulub and the exploration of large peak-ring craters through scientific drilling. *GSA Today* 27. https://doi.org/10.1130/GSATG352A.1.
Kring, D. A., & Dudra D. D. (2003). The day the world burned. *Scientific American, 289,* 98–105.
Kutty, T. S., Chatterjee, S., Galton, P. M., & Upchurch, P. (2007). Basal sauropodomorphs (Dinosauria: Saurischia) from the Lower Jurassic of India: Their anatomy and relationships. *Journal of Paleontology, 81,* 1552–1574.
Kutty, S., & Sengupta, D. P. (1989). Late Triassic formations of the Pranhita-Godavari valley and their vertebrate faunal sequence–A reappraisal. *Indian Journal of Earth Sciences, 16,* 189–206.
Lerbekmo, J. F. (2014). The Chicxulub-Shiva extraterrestrial one-two killer punches to Earth 65 million years ago. *Marine and Petroleum Geology, 40,* 203–207.
Lydekker, R. (1877). Notices on new and other Vertebrata from Indian Tertiary and Secondary rocks. *Records of the Geological Survey of India, 10,* 30–43.
Matley, C. A. (1923). Note on an armored dinosaur from the Lameta beds of Jubbulpore. *Records of the Geological Survey of India, 55,* 105–109.
Matley, C. A. (1933). The coprolites of Pisdura, central Provinces. *Records of the Geological Survey of India, 74,* 530–534.
Mathur, R. B., & Nair, K. M. (1993). Exploration of Bombay high offshore basin. In S. K. Biswas (Ed.), *Proceedings of the Second symposium on Petroliferous Basins of India* (pp. 365–396). Dehra Dun, India: Indian Petroleum Publishers.
McHone, J. G. (2003). Volatile emissions from central Atlantic magmatic province basalts: Mass assumptions and environmental consequences. In W. E. Hames, J. G. McHone, P. Renne, & R. Ruppel (Eds.), *The central Atlantic magmatic province: Perspective from the rifted fragments of Pangea* (pp. 241–254). *American Geophysical Union Monograph, 136*.
Medlicott, H. B. (1872). Note on the Lameta or infra-trappean formation of central India. *Records of the Geological Survey of India, 5,* 115–120.
Mehrotra, N. C., Venkatachala, B. S., & Kapoor, P. N. (2010). Palynology in hydrocarbon exploration: High impact palynological studies in western offshore and Krishna-Godavari basins. *Journal of the Geological Society of India, 75,* 364–379.
Melosh, H. J. (1989). *Impact cratering: A geologic process*. New York: Oxford University Press.
Mikhailov, K. E. (1991). Classification of fossil eggshells of amniotic vertebrates. *Acta Palaeontologica Polonica, 36*(2), 193-238.
Molnar, R. E. (1980). An ankylosaur (Ornithischia: Reptilia) from the Lower Cretaceous of southern Queensland. *Memoirs of the Queensland Museum, 20,* 65–75.
Novas, F. E. (2009). *The age of dinosaurs in South America*. Bloomington: Indiana University Press.
Novas, F. E., Agnolin, F. L., & Bandopadhyay, S. (2004). Cretaceous theropods from India: A review of specimens described by Huene and Matley (1933). *Revista del Museo Argentino de Ciencias Naturales "Bernardino Rivadavia", 6,* 67–103.

Novas, F., Chatterjee, S., Rudra, D. K., & Datta, P. M. (2010). *Rahiolisaurus gujaratensis*, n. gen. n. sp., a new abelisaurid theropod from the Late Cretaceous of India. In S. Bandopadhyay (Ed.), *New aspects of Mesozoic biodiversity* (pp. 45–62). Berlin: Springer.

Novas, F. E., Ezcurra, M. D., Chatterjee, S., & Kutty, T. S. (2011). New dinosaur species from the Upper Triassic Upper Maleri and Lower Dharmaram formations of central India. *Earth and Environment Science Transactions of the Royal Society of Edinburgh, 101,* 333–349.

Padian, K., Ricqles, A. J. D., & Horner, J. R. (2001). Dinosaurian growth rates and bird origins. *Nature, 412,* 405–408.

Pandey, O. P., & Agarwal, P. K. (2001). Nature of lithosphere deformation beneath the western continental margin of India. *Journal of the Geological Society of India, 57,* 497–505.

Parthasarathy, G., Bhandari, N., Vairamani, M., & Narasaiah, B. (2002). Natural fullerenes from the Cetaceous-Tertiary boundary at Anjar, Kutch, India. *Geological Society of America Special Paper, 356,* 345–350.

Prasad, G. V. R., & Parmar, V. (2020). First ornithischian and theropod dinosaur teeth from the Middle Jurassic Kota Formation of India: Paleobiogeographic relationaships. In G. V. R. Prasad & R. Patnaik (Eds.), *Biological consequences of plate tectonis: New perspectives on post-Gondwana break-up* (pp. 1–30). Cham: Switzerland.

Prasad, G. V. R., & Sahni, A. (2014). Vertebrate fauna from the Deccan volcanic province: Response to volcanic activity. *Geological Society of America Special Publication, 505,* 193–211.

Prasad, G. V. R., Verma, O., Flynn, J. J., & Goswami, A. (2013). A new Late Cretaceous vertebrate fauna from the Cauvery basin, South India: Implications for Gondwana paleobiogeography. *Journal of Vertebrate Paleontology, 33,* 1260–1268.

Prasad, G. V. R., Verma, V., Grover, P., Priyadashini, R., Sahni, A., & Lourembam, R. S. (2016). Isolated archosaur teeth from the green sandstone capping the Coralline Limestone (Bagh Group) of the Narmada valley: Evidence for the presence of pre-late to late Maastrichtian abelisaurids in India. *Island Arc, 25,* 410–420.

Prasad, V., Stromberg, C. A. E., Alimohammadian, H., & Sahni, A. (2005). Dinosaur coprolites and the early evolution of grasses and grazers. *Science, 310,* 1177–1180.

Rao, B. P., & Talukdar, S. N. (1980). Petroleum geology of Bombay high field, India. *American Association of Petroleum Geologist Memoir, 30,* 487–506.

Rathore, S. S., Vijan, A. R., Singh, M. P., Prabhu, B. N., & Misra, K. N. (1997). K-Ar dating of basement rocks of Bombay offshore basin. *Proceedings of the Second International Petroleum Conference Exhibition, 1,* 583–588.

Rathore, S. S., Vijan, A. R., Singh, M. P., Prabhu, B. N., & Sahu, A. (2004). Isotopic evidence of Middle Proterozoic magmatism from Bombay high field: Implications to crustal evolution of western offshore of India. *Proceedings of the Indian Academy of Science, 113,* 27–36.

Robertson, D. S., McKenna, M., Toon, O. B., Hope, S., & Lillegraven, J. A. (2004). Survival in the first hours of the Cenozoic. *Geological Society of America Bulletin, 116,* 760–763.

Rogers, K. C., & Wilson, J. A. (2014). *Vahiny depereti*, gen. et sp. nov., a new titanosaur (Dinosauria, Sauropoda) from the Upper Cretaceous Maevarano Fomation, Madagascar. *Journal of Vertebrate Paleontology, 34,* 606–617.

Roy Chowdhury, T. K. (1965). A new metoposaurid amphibian from the Upper Triassic Maleri Formation of Central India. *Philosophical Transactions of the Royal Society of London, B250,* 1–52.

Rudra, D. K. (1982). Upper Gondwana stratigraphy and sedimentation in the Pranhita-Godavari valley, India. *Quarterly Journal of the Geological, Mining and Metallurgical Society of India, 54,* 56–79.

Sahni, A., & Khosla, A. (1994). Paleobiological, taphonomical, and paleoenvironmental aspects of Indian Cretaceous sauropod nesting sites. *Gaia, 10,* 215–223.

Sahni, A., Tandon, S. K., Jolly, A., Bajpai, S., Sood, A., & Srinivasan, S. (1994). Upper cretaceous dinosaur eggs and nesting sites from the Deccan volcano-sedimentary province of peninsular India. In K. Carpenter, K. F. Hirsh, & J. R. Horner (Eds.), *Dinosaur eggs and babies* (pp. 204–223). Cambridge: Cambridge University Press.

Saladgo, L., & Gasparini, Z. (2006). Reappraisal of an ankylosaurian dinosaur from the Upper Cretaceous of James Ross Island (Antarctica). *Geodiversitas, 28,* 119–135.

Schoene, B., Samperton, K. M., Eddy, M. P., Keller, G., Adatte, T., Bowring, S. A., et al. (2015). U- Pb geochronology of the Deccan Traps and relation to the end-Cretaceous mass extinction. *Science, 347,* 182–184.

Scotese, C. R. (2015). Some thoughts on global climate change: The transition from Icehouse to Hothouse. Earth History: The Evolution of the Earth System as revealed through Plate Tectonics, Palaeogeography, Paleoclimate, and the Evolution of Life. *PALEOMAP Project, 15e.*

Sereno, P. C., Wilson, J. A., & Conrad, J. L. (2004). New dinosaurs link southern landmasses in the mid-Cretaceous. *Proceedings of the Royal Society of London, B271,* 1325–1330.

Shukla, A. D., Bhandari, N., Kusumgar, S., Shukla, P. N., Ghevariya, Z. G., Gopalan, K., et al. (2001). Geochemistry and magnetostratigraphy of Deccan flows at Anjar, Kutch. *Proceedings of the Indian Academy of Sciences, 110,* 111–132.

Smit, J. (1999). The global stratigraphy of Cretaceous-Tertiary boundary impact ejecta. *Annual Review of Planetary Sciences, 27,* 75–113.

Tortosa, T., Buffetaut, E., Vialle, N., Dutour, Y., Turini, E., & Cheylan, G. (2014). A new abelisaurid from the Late Cretaceous of southern France: Paleobiogeographical implications. *Annales de Paléontologie, 100,* 63–86.

Vellekoop, J., Esmeray-Senlet, S., Miller, K. G., Browning, J. V., Sluji, A., Schootbrugge, B. V., et al. (2016). Evidence for Cretaceous-Paleogene boundary bolide "impact winter" conditions from New Jersey, USA. *Geology, 44,* 619–622.

Vianey-Liaud, M., Khosla, A., & Garcia, G. (2003). Relationships between European and India dinosaur eggs and eggshells of the oofamily Megaloolithidae. *Journal of Vertebrate Paleontology, 23,* 575-585.

Vijaya & Prasad, G. V. R. (2001). Age of the Kota Formation, Pranhita-Godavari valley, India: A palynological approach. *Journal of the Palaeontological Society of India, 46,* 77–93.

Wandrey, C. J. (2004). Bombay geologic province: Eocene to Miocene composite total petroleum system, India. *US Geological Survey Bulletin, 2208-F,* 1–20.

Wegener, A. (1915). *The origin of oceans and continents.* London: Methuen.

Wilson, J. A. (2002). Sauropod dinosaur phylogeny: Critique and phylogenetic analysis. *Zoological Journal of the Linnean Society, 136,* 217–276.

Wilson, J. A., Malkani, M. S., & Gingerich, P. D. (2005). A sauropod braincase from the Pab Formation (Upper Cretaceous, Maastrichtian) of Balochistan, Pakistan. *Gondwana Geological Magazine, 8,* 101–109.

Wilson, J. A., Barrett, P. M., & Carrano, M. T. (2011). An associated partial skeleton of *Jainosaurus* cf. *septentrionalis* (Dinosauria: Sauropoda) from the Late Cretaceous of Chhota Simla, Central India. *Palaeontology, 54,* 981–998.

Wilson, J. A., Mohabey, D. M., Peters, S. E., & Head, J. J. (2010). Predation upon hatchling dinosaurs by a new snake from the Late Cretaceous of India. *PLoS Biology, 8*(3), e1000322. https://doi.org/10.1371/journal.pbio.1000322.

Wilson, J. A., Sereno, P. C., Srivastava, S., Bhatt, D. K., Khosla, A., & Sahni, A. (2003). A new abelisaurid (Dinosauria, Theropoda) from the Lameta Formation (Cretaceous, Maastrichtian) of India. *Contributions from the Museum of Paleontology University of Michigan, 3,* 1–42.

Wilson, J. A., & Upchurch, P. (2003). A revision of *Titanosaurs* Lydekker (Dinosauria – Sauropoda), the first dinosaur genus with a 'Gondwanan' distribution. *Journal of Systematic Palaeontology, 1,* 126–160.

Wolbach, W. S., Gilmore, I., Angers, E., Orth, C. J., & Book, R. R. (1988). Global fire at the Cretaceous-Tertiary boundary. *Nature, 324,* 665–669.

Yadagiri, P. (1984). New symmetrodonts from Kota Formation (Early Jurassic) India. *Journal of the Geological Society of India, 25,* 414–421.

Yadagiri, P. (1988). A new sauropod *Kotasaurus yamanpalliensis* from Lower Jurassic Kota Formation of India. *Records of the Geological Survey of India, 11,* 102–127.

Yadagiri, P. (2001). The osteology of *Kotasaurus yamanpalliensis*, a sauropod dinosaur from the Early Jurassic Kota Formation. *Journal of Vertebrate Paleontology, 21,* 242–252.

Zutshi, P. L., Sood, A., Mohapatra, P., Raman, K. K. V., Dwivedi, A. K., & Srivastava, H. C. (1993). Lithostratigraphy of Indian petroliferous basins. *Documents V, Bombay Offshore Basin*. Dehra Dun, India: Unpublished KDMIPE Report, ONGC.

Chapter 9
Leaving Gondwana: The Changing Position of the Indian Subcontinent in the Global Faunal Network

Thomas J. D. Halliday, Patricia A. Holroyd, Emmanuel Gheerbrant, Guntupalli V. R. Prasad, Agustín Scanferla, Robin M. D. Beck, David W. Krause, and Anjali Goswami

Abstract The paleogeographic history of the Indian subcontinent is unique among Earth's landmasses. From being part of the southern supercontinent Gondwana for most of the Mesozoic, through a period of isolation as a drifting entity in the Late Cretaceous, to colliding with Asia near the Paleocene – Eocene boundary, the Indian subcontinent has been associated with, and dissociated from, a variety of landmasses. This paleogeographic history has been invoked to explain aspects of the subcontinent's modern-day fauna, with a combination of endemic radiations, remnants from Gondwana, and more recent immigrants from Laurasia. Here, network approaches document how vertebrate faunas of the Indian subcontinent, and specifically their relationships to those of other landmasses, changed during the subcontinent's isolation from close faunal relationships with Madagascar and South America in the Late Cretaceous to a more Laurasian fauna most similar to those of Europe by the Eocene.

Keywords Cretaceous • India's Drift • Vertebrate • Faunal Similarity • Laurasia • Gondwana

Introduction

Among Earth's modern landmasses, the Indian subcontinent has arguably had the most complex paleobiogeographic history. Part of the massive southern supercontinent Gondwana for most of the Mesozoic, and particularly closely associated with Madagascar, Africa, Antarctica and Australasia, it sequentially separated from these landmasses and drifted northwards (Norton and Sclater 1979; Powell 1979; Subrahmanya 1998; Ali and Aitchison 2008; Chatterjee et al. 2013, 2017). The precise timing and pattern of separation of the Gondwanan continents is not universally agreed upon, but the Indian subcontinent appears to have entirely separated from Madagascar by 84–92 Ma (Storey et al. 1995, 1997; Torsvik et al. 2000; Yatheesh et al. 2006), and

Electronic supplementary material The online version of this chapter (https://doi.org/10.1007/978-3-030-49753-8_9) contains supplementary material, which is available to authorized users.

T. J. D. Halliday (✉)
School of Geography, Earth, and Environmental Sciences, University of Birmingham, Edgbaston, B15 2TT, UK
e-mail: t.j.d.halliday@bham.ac.uk

T. J. D. Halliday · A. Goswami
Department of Genetics, Evolution, and Environment, UCL, Gower Street, London, WC1E 6BT, UK

P. A. Holroyd
Museum of Paleontology, University of California, Valley Life Sciences Building, 1101, Berkeley, CA 94720, USA

E. Gheerbrant
Centre de Recherche CR2P UMR CNRS 7207 Sorbonne Université, MNHN/CNRS/UPMC, 57 rue Cuvier, CP 38, 75005 Paris, France

G. V. R. Prasad
Department of Geology, Centre for Advanced Studies, University of Delhi, Delhi, 110007, India

A. Scanferla
CONICET-Instituto de Bio y Geociencias, del NOA (IBIGEO), 9 de Julio N° 14 (A4405BBB), Rosario de Lerma, Salta, Argentina

R. M. D. Beck
School of Environment and Life Sciences, University of Salford, The Crescent, Salford, M5 4WT, UK

D. W. Krause
Department of Earth Sciences, Denver Museum of Nature & Science, 2001 Colorado Boulevard, Denver, CO 80205, USA

A. Goswami
Department of Earth Sciences, UCL, Gower Street, London, WC1E 6BT, UK

A. Goswami
Department of Life Sciences, Natural History Museum, Cromwell Road, London, SW9 5DJ, UK

G. V. R. Prasad and R. Patnaik (eds.), *Biological Consequences of Plate Tectonics: New Perspectives on Post-Gondwana Break-up—A Tribute to Ashok Sahni*,
Vertebrate Paleobiology and Paleoanthropology Series, https://doi.org/10.1007/978-3-030-49753-8_9

remained isolated in the Indian Ocean throughout the rest of the Late Cretaceous and all or most of the Paleocene. Near the Paleocene – Eocene boundary, the Indian subcontinent finally collided with mainland Asia (Najman et al. 2010; Gibbons et al. 2015; Hu et al. 2016; Matthews et al. 2016), allowing for the possibility of biotic exchange (Krause and Maas 1990; Briggs 2003; Ali and Aitchison 2008; Bajpai 2009; Klaus et al. 2016). The position of India from the Cretaceous to the Eocene, as well as the location of all sites mentioned in this chapter, are found in Figs. 9.1, 9.2 and 9.3. Despite its increasingly well-documented, though still controversial, geophysical history, there remain certain aspects of the biogeographic history of the Indian subcontinent that render the pattern more complex. Throughout the subcontinent's isolation, oceanic islands and island arcs have been hypothesized to link, directly or indirectly, the Indian subcontinent with Asia, Madagascar, Antarctica, South America, and Africa-Arabia, potentially allowing faunal exchange among these landmasses. The Kerguelen Plateau (Krause et al. 1997; Sampson et al. 1998; Hay et al. 1999), the Chagos-Laccadive Ridge and Seychelles-Mascarene Plateau (Sahni 1984; Ali and Aitchison 2008), the Gunnerus Ridge (Case 2002), and the Kohistan-Ladakh island-arc system (Prasad and Sahni 2009; Chatterjee and Scotese 2010; Chatterjee et al. 2017) have all been invoked as stepping stones for faunal exchange. The subaerial nature and extent of some of these features has been questioned or refuted (Ali and Aitchison 2008; Ali and Krause 2011; Hu et al. 2016; Kapur and Khosla 2016; Smith et al. 2016). Nonetheless, these bio-connection hypotheses provided explanations for the observation that the 'mixed' fauna of the Indian subcontinent, including taxa closely related to those in both Gondwanan and Laurasian landmasses, has persisted since its isolation in the Late Cretaceous and Paleocene (Sahni et al. 1987; Prasad and Sahni 2009; Chatterjee and Scotese 2010; Smith et al. 2016; Chatterjee et al. 2017).

The major Late Cretaceous vertebrate fossil sites of the Indian subcontinent are located in the Deccan Volcanic Province (DVP) of central and western India, the Cauvery Basin of southeastern India, and the Narmada Valley of Central India (see reviews in Prasad 2012; Kapur and Khosla 2016). Within these areas, the major fossiliferous vertebrate-bearing strata include the Lameta Formation and the intertrappean beds of the DVP, the Kallamedu Formation in the Cauvery Basin, and the Bagh beds of the Narmada Valley. All four are dominated by vertebrate microfossils, but fragmentary and fragile larger bones are also represented. Previous work has confirmed that the Lameta and intertrappean beds are extremely similar to one another faunally. The fauna of the Kallamedu Formation, however, more closely resembles that of the Maevarano Formation of Madagascar (Halliday et al. 2017), on the basis of the preservation of taxa such as a *Simosuchus*-like notosuchian crocodyliform (Prasad et al. 2013; Prasad and Goswami 2015), which appears to be absent from the DVP. Other classically 'pan-Gondwanan' taxa such as titanosaurid and abelisaurid dinosaurs, bothremydid turtles, and gondwanatherian mammals are found across the Late Cretaceous of the Indian subcontinent, as well as "madtsoiid"

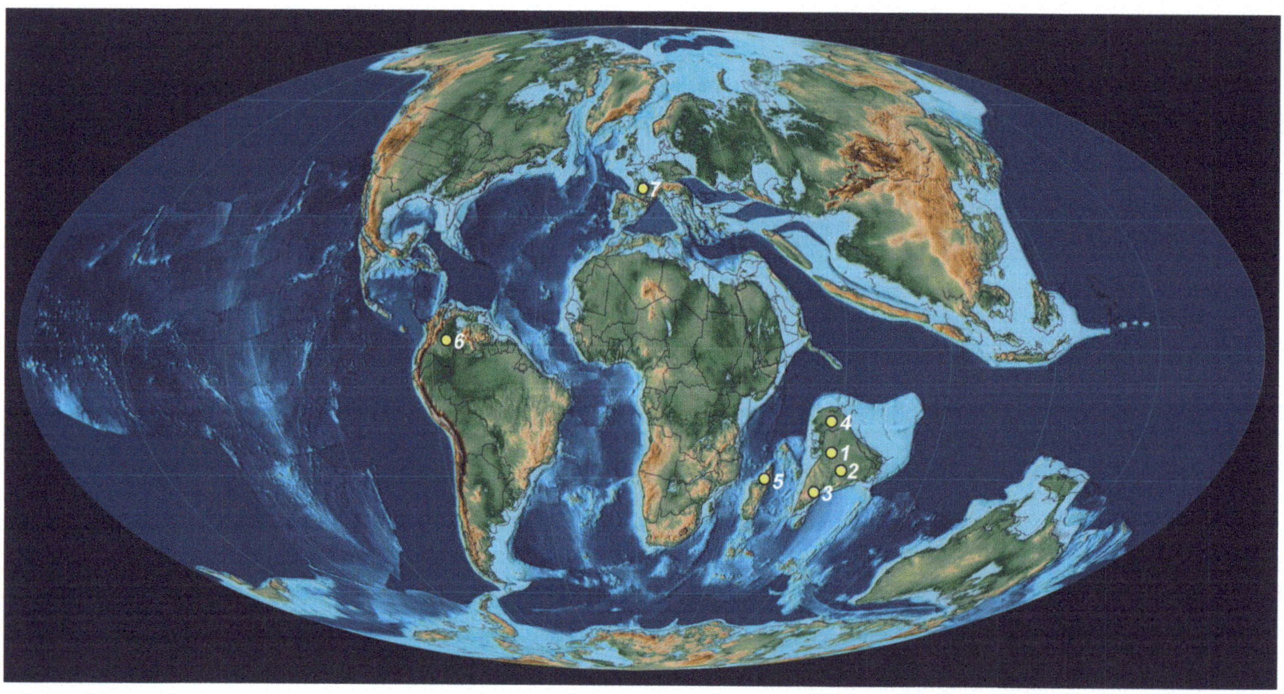

Fig. 9.1 Global paleogeographic map during Late Cretaceous showing the main fossil localities mentioned in the text. 1. Lameta Formation (India), 2. Intertrappean beds (India), 3. Kallamedu Formation (India), 4. Pab Formation (Pakistan), 5. Maevarano Formation (Madagascar), 6. Ortega Formation (Colombia), 7. Rognacian Formation (France)

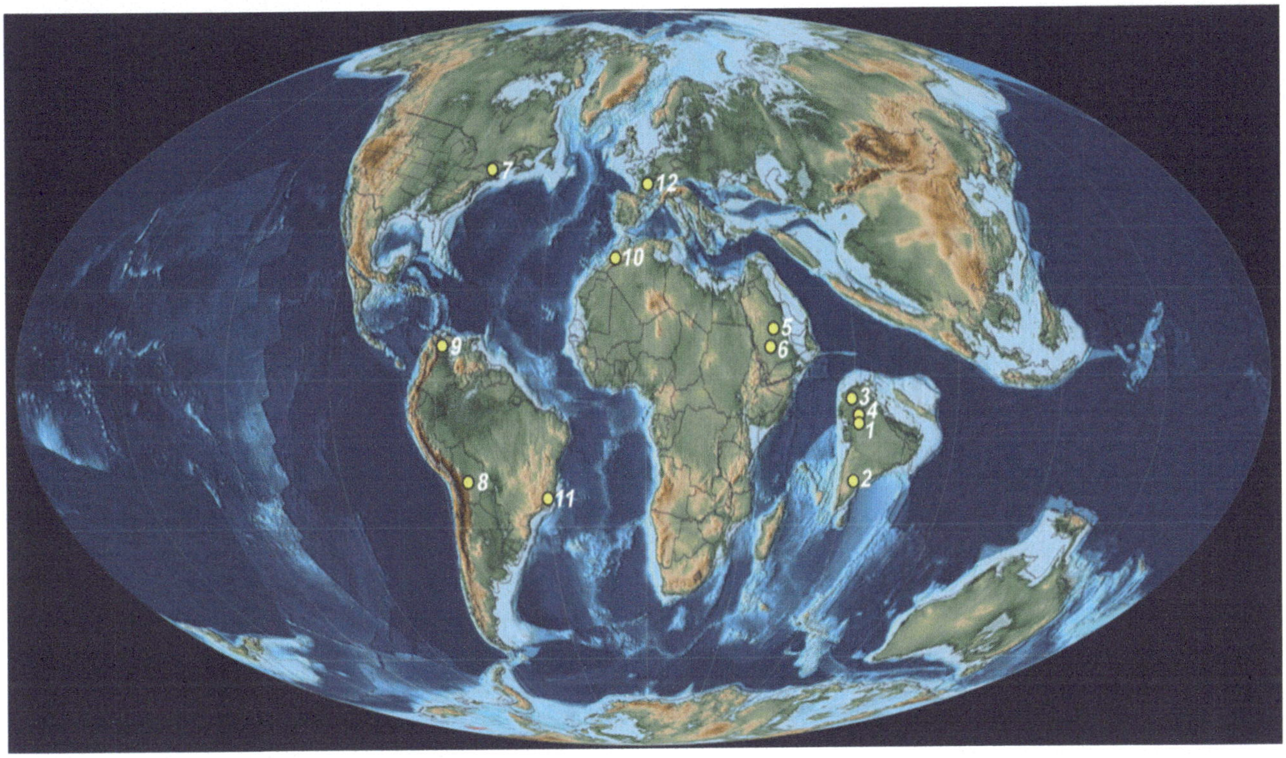

Fig. 9.2 Global paleogeographic map during Paleocene showing the main fossil localities mentioned in the text. 1. Fatehgarh Formation (India), 2. Niniyur Formation (India), 3. Khadro Formation (Pakistan), 4. Akli Formation (India), 5. Aruma Formation (Saudi Arabia), 6. Umm Himar Formation (Saudi Arabia), 7. Hornerstown Formation (United States), 8. Santa Lucía Formation (Bolivia), 9. Cerrejón Formation (Colombia), 10. Jbel Guersif Formation (Morocco), 11. Itaboraí Formation (Brazil), 12. Soissonais Formation (France)

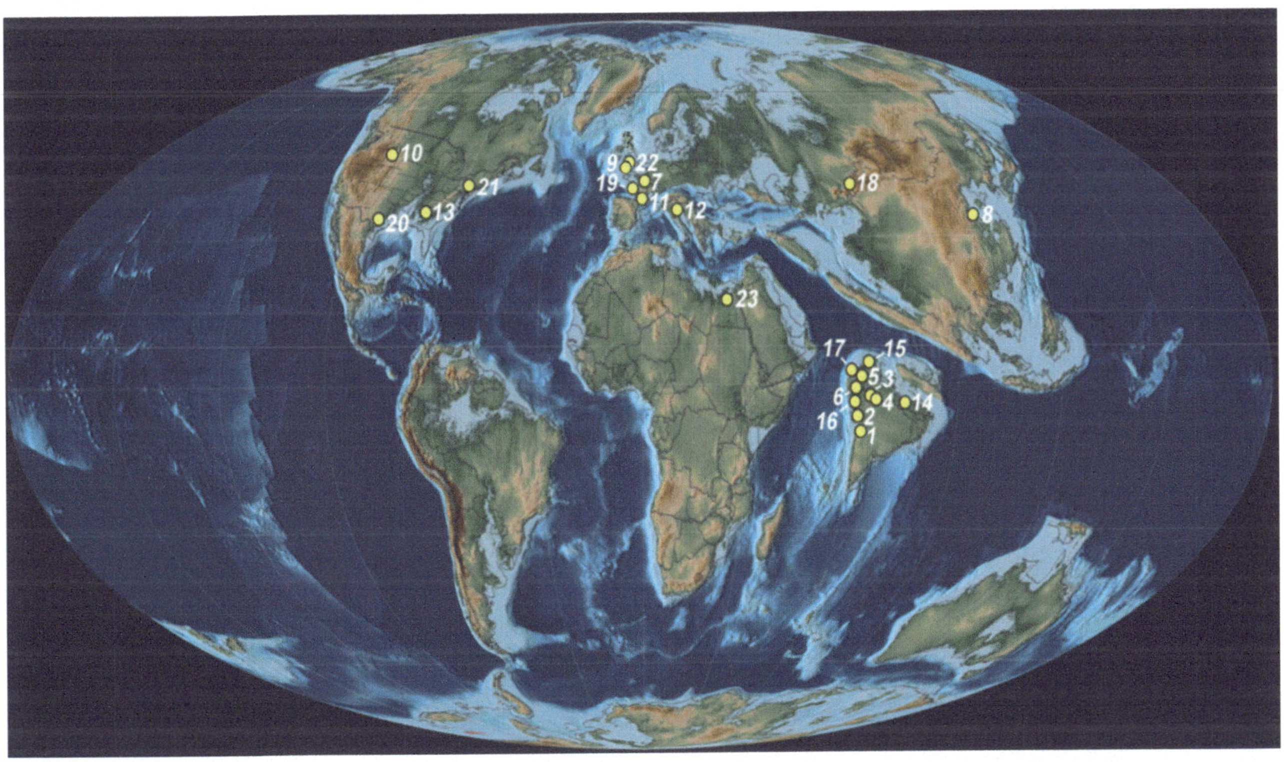

Fig. 9.3 Global paleogeographic map during Eocene showing the main fossil localities mentioned in the text. 1. Cambay Shale (India), 2. Naredi Formation (India), 3. Kapurdi Formation (India), 4. Khuiala Formation (India), 5. Mami Khel clay (Pakistan), 6. Ghazij Formation (Pakistan), 7. Sezanne-Broyes Formation (France), 8. Niushan Formation (China), 9. Suffolk pebble beds, 10. Willwood Formation (United States), 11. Gres d'Assignan (France), 12. Monte Bolca (Italy), 13. Tallahatta Formation (United States), 14. Subathu Formation (India), 15. Kuldana Formation (Pakistan), 16. Harudi Formation (India), 17. Chorgali Formation (Pakistan), 18. Irtysh River (China/Kazakhstan), 19. Grabels (France), 20. Yegua Formation (United States), 21. Shark River Formation (United States), 22. Elmore Formation (UK), 23. Mokattam (Egypt)

snakes (Bonaparte 1999). This last group is known from the Cretaceous and Paleocene of Gondwana, but it is worth mentioning that the identification of some specimens to Madtsoiidae should be treated with caution (Rio and Mannion 2017). Both the DVP and the Kallamedu Formation also contain taxa generally associated with Laurasia, such as a possible troodontid dinosaur (Goswami et al. 2013), phyllodontid fish (Halliday et al. 2016), pelobatid frogs (Sahni et al. 1982), costatan (='discoglossid') frogs (although referral of Indian specimens to these taxa is doubted by Agnolin 2012) and anguimorph lizards (Prasad and Rage 1991, 1995, 2004; Rage et al. 2020) and eutherian mammals (Prasad et al. 1994, 2010; Rana and Wilson 2003; Boyer et al. 2010; Goswami et al. 2011). In addition to the DVP and the Kallamedu Formation, the shallow marine Karai Formation of the Cauvery Basin primarily records marine invertebrates, but shark and ichthyosaur remains are also known (Underwood et al. 2011). Only shark and abelisaurid dinosaur teeth, and fragmentary bones of sauropod dinosaurs have so far been recovered from the Upper Cretaceous Bagh Group of the Narmada Valley (Khosla et al. 2003; Prasad et al. 2016, 2017). Finally, the Pab Formation of Pakistan represents a terrestrial Late Cretaceous ecosystem, also from the Maastrichtian, and is characterized by a dinosaur fauna dominated by titanosaurids, as well as the baurusuchid crocodyliform *Pabwehshi* (Wilson et al. 2001, 2005).

The most productive vertebrate-bearing Paleocene formation of the Indian subcontinent is the nearshore Fatehgarh Formation, which lies in the Barmer Basin of northwestern India (Rana et al. 2006b). The vertebrate fauna recovered from this basin was considered to be of Late Cretaceous age (Mathur et al. 2006), but this estimate has been revised to the early Paleocene on the basis of vertebrate faunal differences with the Deccan intertrappean beds (Rana et al. 2006b). Recently, a yet-to-be-described chondrichthyan fauna has been recovered from the shallow marine Lower Paleocene Niniyur Formation of the Cauvery Basin, indicating close shallow marine connections between India and northwestern Africa and southern Europe (Prasad pers. obs.), but this has not been included in the present study. In addition, the Danian Khadro Formation of Pakistan has yielded probable madtsoiid snake and dyrosaurid crocodylomorph material (Storrs 1986; Rage et al. 2014), while the Upper Paleocene-Eocene Akli Formation of the Barmer Basin preserves a diverse fish fauna (Rana et al. 2005).

In addition, some Deccan intertrappean beds have been proposed to be early Paleocene (Wilson et al. 2016), but these are disputed and are not included in the present analysis. Fossil sites from the early Eocene of the Indian subcontinent are concentrated in shallow marine lignite deposits, and many specimens have been recovered from the associated mines at Vastan and Tadkeshwar (the Cambay Shale Formation) and Panandhro (the Naredi Formation). The Cambay Shale includes several large "ungulate" taxa including endemic forms from the Indian subcontinent, such as cambaytheriids and anthracobunids, and more cosmopolitan families such as palaeoryctids, nyctitheriids, hyaenodontids, notharctids, and leporids, to name but a few (see review of Indian subcontinental early Eocene vertebrates in Chatterjee et al. 2017). The Panandhro lignite mine fauna is dominated by cosmopolitan fish taxa, such as lepisosteids and many chondrichthyans. A taxonomically diverse fish fauna has also been described from the marginal marine to shallow marine deposits associated with lignite seams of the Lower Eocene Akli and Kapurdi formations of the Barmer Basin (Rana et al. 2005, 2006a), and the Khuiala Formation of the Jaisalmer Basin (Kumar et al. 2007) in western India. In the north of India and in Pakistan, the shallow marine Subathu and Kuldana Formations contain some of the earliest cetaceans (Sahni et al. 1981; Kumar and Sahni 1985; Thewissen et al. 2001, 2007), distinguishing them from the sites in western India. Finally, the Mami Khel and Ghazij localities of Pakistan contain several mammalian groups such as dichobunid artiodactyls, chalicotheres, and early primates and carnivorans (e.g., Clyde et al. 2003; Missiaen and Gingerich 2012, 2014).

Middle Eocene sites include the aforementioned Subathu and Kuldana sites, as well as the marine Harudi Formation of Kachchh, which, like the Subathu and Kuldana formations, is dominated by protocetid and remingtonocetid whales, and the Chorgali and Subathu fomations, from which helaletid and raoellid artiodactyls are known (Thewissen et al. 2001, 2020).

Assessing the degree to which transoceanic or island-hopping dispersal during the Indian subcontinent's isolation affected overall composition of its terrestrial and freshwater vertebrate fauna is an important aspect of its paleobiogeographic reconstruction. Several qualitative comparisons have been made between the Indian subcontinent and other landmasses during the Late Cretaceous and later, but few quantitative comparisons exist. Faunal similarity metrics allow quantification of the degree to which sites, formations, or other geographic/geologic divisions share taxa, and have a long history in paleontology from the identification of local ecological differences (e.g., Raup and Crick 1979) to the formation of barriers to migration over geological time (e.g., Cecca 2002). Networks are useful ways of mathematically displaying the presence or absence of relationships among objects. Combining the two to determine the Indian subcontinent's close faunal relations during its long period of isolation will help to quantify and potentially clarify previously described patterns.

When describing a biota as 'cosmopolitan', 'endemic', or using the presence of shared animals or plants to infer general biogeographic relationships, it is important to consider the biota as a whole, and to integrate information from elsewhere. Doing so allows for further testing of important paleobiogeographic scenarios and questions. For instance, in the case under consideration here, does the presence of Laurasian clades in the Late Cretaceous of the Indian subcontinent cause it to be faunally intermediate with respect to the southern and northern continents? Does immigration from Laurasia explain the observed relative lack of similarity between the sites of the DVP on the one hand and both the Cauvery Basin and Madagascar on the other? To answer these and other questions, it is necessary to compare simultaneously the faunal composition of Late Cretaceous and Early Paleogene vertebrate-bearing localities worldwide.

Methods

Faunal Similarity Networks

Networks are an informative way of displaying and analysing the interaction between different pools of data, and have a long and successful history of use in, for example, the social sciences, in which entire journals are dedicated to network models, and epidemiology (e.g., Reis et al. 2007). Networks have been used in evolutionary biology in the context of modularity (Esteve-Altava et al. 2015) and ecology (Blonder et al. 2012), while their mathematical similarity to unrooted phylogenetic trees and cluster hierarchies (both of which are merely special cases of a network graph) has allowed them to be used to address macroevolutionary questions concerning phylogenetic relationships (Bapteste et al. 2013) and paleobiogeography (Kiel 2017; Dunhill et al. 2016). Network methods allow the testing of wider ranges of hypotheses than bifurcating hierarchical clustering methods in paleobiogeography, or bifurcating cladograms in phylogenetic inference. Uncertainty in relationships or similarity can be incorporated into the network, permitting a more nuanced picture of macroevolution to be obtained. In the context of paleobiogeography, use of networks can investigate the degree to which clustering of faunas is based on geography or environment, as has been done by Kiel (2017) on Tethyan brachiopod faunas. More promising, however, is the capability to assess how ecosystems change over time with respect to other sites. Sidor et al. (2013) used an adapted version of the network edge density statistic to compare the homogeneity of terrestrial faunas across the Permian – Triassic boundary. Changes in the values of such network parameters may be able to indicate the growing or waning importance of certain critical faunas, chart the movements of continents, or demonstrate changing global ecosystem dynamics over important geological intervals such as mass extinction events. A modification of this method correcting for phylogeny was introduced by Button et al. (2017) to identify changes in cosmopolitanism in the aftermath of the Permian – Triassic mass extinction. Here, it is worth noting that the networks of Sidor et al. (2013) and Button et al. (2017) are structurally different from those of Kiel (2017). Sidor et al. (2013) represents both taxa and localities as different classes of node, with edges connecting a taxon to a locality if that taxon is represented in that locality. On the other hand, Kiel (2017) treats only localities as nodes, with edges connecting localities with weights proportional to some faunal similarity metric. It is this latter approach that we use here.

Global Patterns

Three network statistics are of potential use when describing the relationships among global faunas. The first is the density statistic, which measures the proportion of all possible edges between nodes that are observed. A network with a density of one is ‘complete’, and has all nodes connected to all other nodes. A network with a density of zero has no edges at all. In the context of a faunal similarity network, a complete network would represent a set of localities, all of which have pairwise similarity scores above the threshold value. Depending on the scale of the analysis, this could be interpreted as evidence that multiple sites are part of the same local ecosystem, or evidence for widespread cosmopolitanism across many taxa. Conversely, a network with no edges would show that all sites are independent from one another, indicating a high degree of endemism, and implying long periods of isolation.

Secondly, node centrality is a family of statistics used to identify highly connected nodes. Betweenness centrality takes the shortest paths between all pairs of nodes, and calculates the number of times those paths include any given node. In the context of faunal similarity indices, a node with a high betweenness centrality would represent a true ‘mixed’ fauna comprising several taxa that are otherwise representative of two or more clusters of localities.

The final network statistic considered here is network modularity. Clusters of nodes that are more linked to one another than to external nodes can be identified using the betweenness statistic as applied to edges rather than nodes. The modularity statistic of a matrix, given the identified clusters, quantifies the degree to which those clusters are truly separate modules.

Here, we employ all three statistics to document the changing position of the Indian subcontinent in a global paleobiogeographic context from the Late Cretaceous through to the end of the Eocene.

Data Collection

Global terrestrial, freshwater, and shallow marine occurrence data for vertebrate families and genera were downloaded from the Paleobiology Database (paleobiodb.org, access date 6th of June, 2017), spanning all time intervals from the Late Cretaceous to the Eocene inclusive. These were then vetted by correcting spellings, merging duplicate names, removing ichnotaxa and ootaxa, ensuring that the familial assignment of each genus was correct, and that formation names were consistent and had complete faunal occurrence lists. The modified Paleobiology Database files are found in Appendix 1. Occurrences were divided into 33 time bins defined by uncertainties in fossil range dates, and were analysed both as separate time bins and as a whole (Table 9.1). Within each analysis, formations with fewer than three families or genera present were excluded to avoid problems with estimating faunal similarity with very small sample sizes. Family- and genus-level data were analyzed separately; for the purposes of this paper, all graphs presented and discussed use the family-level dataset unless otherwise stated. Genus-level data and results can be found in Appendix 1, but because families are more inclusive, discussion of faunal similarity patterns are more informative at that level. Pairwise faunal similarity measures were calculated using the modified Forbes metric (Alroy 2015a), which ranges from zero if no taxa are shared between two sites to one if they are identical, or if one fauna is a subset of the other. The primary advantage of the modified Forbes metric is that it is robust to uneven and biased sampling (Alroy 2015b), a well-known problem in paleontological datasets.

An additional advantage over other methods is that it does not require the data to include numbers of specimens, merely coding a taxon as present or absent in each locality. Thresholds in Forbes metric scores varying from 0.05 to 0.95 were applied to each analysis, in each case treating all values less than the threshold as zero. Undirected weighted graphs were drawn using the igraph (Csardi and Nepusz 2006) package in R (R Development Core Team 2010), using formations as nodes and non-zero Forbes scores as edges. Without thresholding the network at all, the network is invariably extremely dense to the point of being uninformative. Thresholds gradually remove those less connected faunas, leaving only those with some degree of connectivity in the graph. Threshold values in the discussion are typically 0.5 (when the presence of an edge indicates greater similarity than dissimilarity), except when otherwise stated. Nodes were identified that connected to formations on the Indian subcontinent after appropriate thresholds were determined. Clusters were identified using the edge betweenness statistic, and the modularity of the thresholded network calculated based on these clusters. Network edge density was calculated in each time-binned analysis and at each threshold level, to give a measure of 'decay' of density in a given time bin. Betweenness metrics of centrality were

Table 9.1 Time bins into which the data set was divided. Divisions between bins encompass the uncertainties in fossil dating. Fossil occurrences were placed into every bin in which they may have occurred. This approach ensures that faunas with non-overlapping date ranges are not erroneously placed in the same bin for analysis, while allowing flexibility in the sets of faunas compared. Differences in bin duration are not problematic in the context of this analysis as the patterns discussed in this analysis are not sample-size dependent. Bins in bold are those discussed in the manuscript as Late Cretaceous, Early Paleocene, Early Eocene, and Middle Eocene respectively

Bin	Maximum age	Minimum age	Description
1	–	100.5	Earlier than analysis
2	100.5	99.6	Earliest Cenomanian
3	99.6	93.9	Cenomanian
4	93.9	93.5	Earliest Turonian
5	93.5	89.8	Turonian
6	89.8	89.3	Earliest Coniacian
7	89.3	86.3	Coniacian
8	86.3	85.8	Early Santonian
9	85.8	83.6	Late Santonian
10	83.6	83.5	Earliest Campanian
11	83.5	72.1	Campanian
12	72.1	70.6	Early Maastrichtian
13	**70.6**	**66**	**Maastrichtian/Lancian**
14	**66**	**63.3**	**Danian/Puercan**
15	63.3	62.5	Danian/Torrejonian
16	62.5	61.7	Danian/Torrejonian/Peligran
17	61.7	61.6	Danian/Tiffanian
18	61.6	59.2	Selandian/Tiffanian
19	59.2	59	Thanetian/Tiffanian
20	59	58.7	Thanetian/Tiffanian/Itaboraian
21	58.7	57	Thanetian/Tiffanian/ Itaboraian/Gashatan
22	57	56.8	Thanetian/Tiffanian/ Riochican/Gashatan
23	56.8	56	Thanetian/Clarkforkian
24	56	55.8	Ypresian/Clarkforkian
25	**55.8**	**50.3**	**Ypresian/Wasatchian**
26	50.3	48.6	Ypresian/Bridgerian
27	48.6	48	Ypresian/Arshantan/Sharamurunian
28	48	47.8	Ypresian/Mustersan
29	47.8	46.2	Lutetian/Bridgerian
30	**46.2**	**41.3**	**Lutetian/Uintan**
31	41.3	40.4	Bartonian/Uintan
32	40.4	38	Bartonian/Duchesnean
33	38	37.2	Bartonian/Chadronian
34	37.2	33.9	Priabonian/Chadronian

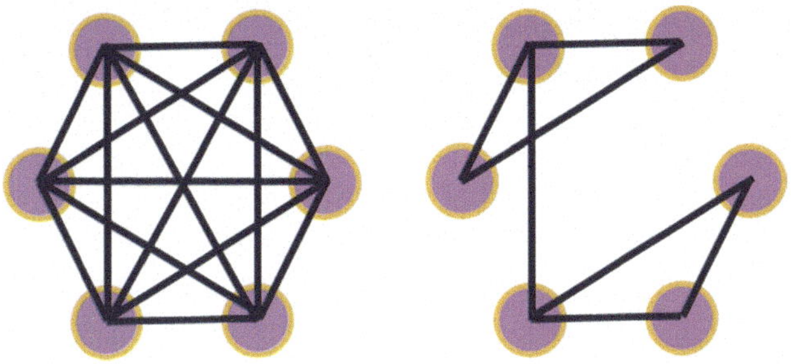

Fig. 9.4 A graph with connections between all nodes (left) is a 'complete' graph, and has a density of one. The density of a graph is calculated as the proportion of the possible connections present. For example, the more structured graph on the right has seven of 15 present, and therefore has a density of 0.47

compared for localities in different continents. The full R code for this analysis, including Alroy's Forbes functions, are found in Appendix 2.

Network density, a measure of the total proportion of the possible edges, is a way of measuring homogeneity among faunas. Dense networks or subnetworks contain a greater proportion of taxa that are found in all or most faunas. For example, a network density of one without thresholding would indicate that all sites shared at least one taxon with all other sites, suggesting a strong degree of connectivity among the ecosystems. Conversely, a network density of zero would indicate that all faunas were exclusively composed of endemic taxa (Fig. 9.4).

Results and Discussion

The Late Cretaceous of the Indian Subcontinent

Four Late Cretaceous (Bin 13, see Table 9.1) formations from subcontinental India are represented by sufficient taxa to be included in the analysis: the Kallamedu Formation of the Cauvery Basin, the Lameta Formation and the intertrappean beds – all from India – and the Pab Formation of western Pakistan. In the Late Maastrichtian, at a threshold of 0.75, the Kallamedu Formation has connections to the Madagascan Maevarano (0.84) and Colombian Ortega (0.76) formations. The Ortega Formation has yielded few fossils, and here is represented by only three families, but the similarity between the Kallamedu and Maevarano formations is based on substantial faunal overlap of a large sample size. The Lameta and intertrappean beds are connected (0.86), but are not similar to any other locality at this threshold, while the Pab Formation is isolated, though most similar to the Kallamedu Formation (0.64). No site on the Indian subcontinent shows high similarity scores with any African site, but at a wider level, Gondwanan sites, including those from Africa and Australia, cluster together, indicating a strong level of cohesion of a "pan-Gondwanan" fauna throughout the Cretaceous.

The Paleocene of the Indian Subcontinent

The early Paleocene (Bin 14, see Table 9.1) vertebrate fossil record of the Indian subcontinent is essentially blank, with the possible exception of a few of the later intertrappean beds, which are thought by some to represent a post-Cretaceous ecosystem (Singh and Kar 2002; Keller et al. 2009; Wilson et al. 2016), as well as the Fatehgarh, Akli, and Khadro formations. The Fatehgarh Formation is entirely represented by fish, and has no connections with any other lower Paleocene formation at a threshold of 0.5. The Akli Formation is most similar to the Aruma Formation of Saudi Arabia (0.54), but otherwise has no similarity to any other late Paleocene fauna at a threshold of 0.5. The Khadro Formation has only yielded a madtsoiid snake and a dyrosaurid crocodylomorph among its vertebrate remains, and as such has too few taxa to be included in this analysis.

The Eocene of the Indian Subcontinent

By the early Eocene (Bin 25, see Table 9.1), the overall signal demonstrating a Gondwanan heritage is gone in all localities on the Indian subcontinent. At a threshold level of 0.6, the Cambay Shale is most similar to the European Ypresian (European Reference Level MP 8–9) localities at Sézanne-Broyes (France) and La Coma (Spain), as well as the Kuldana Formation of Pakistan and the Niushan Formation of China. The Mami Khel Formation has connections with the Suffolk Pebble Beds (UK), the Willwood Formation (USA), and the Ghazij Formation (Pakistan), which is in turn connected to the Ypresian Grès d'Assignan locality in France. The Subathu Formation, which ranges from the late Paleocene to the middle Eocene, is unconnected to any other locality throughout the early Eocene. The fish fauna of the Kapurdi Formation clusters with two other major fish dominated faunas, those of Monte Bolca in Italy, and Tallahatta, USA.

In the middle Eocene (Bin 30, see Table 9.1), the patterns are not qualitatively different from those of the early Eocene. The Subathu, Kuldana, and Chorgali formations of the Indian subcontinent are all similar to one another, with the

Subathu and Kuldana formations displaying some similarity with the Irtysh River Formation of China, and the Kuldana Formation similar to the Grabels Locality (MP 13–14) of France. Separated from these are the Naredi Formation, which displays similarity with the Yegua and Shark River formations of North America, and the Elmore Formation of the UK. The only Gondwanan connection exists between the Harudi Formation of India and the Mokattam Formation of Egypt. This African connection is entirely due to both sites being shallow marine formations dominated by marine mammals – both protocetids and sirenians. Eocene sites of the Indian subcontinent were less similar to one another than they had been in the Cretaceous, and were dominated by Laurasian taxa, with the signal particularly coming from various families of euungulates (the group comprising artiodactyls, perissodactyls, and their stem relatives). The lack of similarity to one another is perhaps driven by a greater environmental heterogeneity. Whereas the sampled Late Cretaceous sites of the Indian subcontinent are mostly fluviolacustrine sites dominated by terrestrial taxa, the early to middle Eocene includes both terrestrial and shallow marine localities.

More broadly, Eocene Gondwanan localities are no longer as closely associated with one another as they were in the Cretaceous, with the exception of a cluster of South American sites that preserve their unique endemic identity. The switch from Cretaceous to Paleogene faunas has coincided with the loss of a unique Gondwanan identity, and greater similarities between north and south.

The Laurasian signal throughout the Eocene of India reflects shared members of families and genera but, at a more qualitative level, comparisons among the families reveal that this Laurasian affinity in terms of euungulates is part of a bigger story concerning globally distributed and closely-related lineages. The one truly endemic family of euungulates in the Eocene of the Indian subcontinent is the cambaytheriids, thought to be most closely related to perissodactyls (Cooper et al. 2014; Rose et al. 2014). This has led some to suggest that perissodactyls originated on the Indian subcontinent, but the plethora of other lineages also considered to be close relatives of perissodactyls, such as the North American phenacodontids (Halliday et al. 2015), the Asian phenacolophids (Mao et al. 2016), and the South American litopterns and notoungulates (Buckley 2015; Welker et al. 2015; Westbury et al. 2017) imply that these various groups may be closely related to one another, and a Paleogene global distribution may be in effect.

Global Faunal Network of the Cretaceous and Paleogene

Globally, and without dividing the data into time bins, faunal clustering at a threshold level of 0.75 initially revealed an unsurprising split between the Cretaceous and Paleogene faunas. Within each era-level cluster, faunas were broadly divided into geographic bins. In the Cretaceous, Gondwanan localities were tightly clustered together, with all sites on the Indian subcontinent – namely the intertrappean beds, Lameta Formation, and the Kallamedu Formation – clustered together to the exclusion of other sites. At a threshold level of 0.75 (Fig. 9.5), the Kallamedu Formation was linked to the Maevarano Formation of Madagascar and the Ortega Formation of Colombia (Ezcurra 2009). It must be noted that the Ortega Formation contains only very fragmentary fossils that cannot be identified below family level (Ezcurra 2009), while only minimally meeting the sample size standard for inclusion in this analysis. In addition, the assignation of the dinosaur remains to the Ortega Formation has been disputed (Jiménez-Velandia 2017). For the purposes of this analysis, as the fossils included here were collected from the same place, the label assigned to that group does not matter, but the small sample size must qualify interpretation of this locality as being closely faunally associated. Comparisons involving smaller numbers of taxa are far more susceptible to change on the discovery of additional material. In addition to the Maevarano and Ortega formations, similarity was found between the Kallamedu Formation and the Rognacian sites of France (Tabuce et al. 2004; Tortosa et al. 2014), on the basis of several otherwise Gondwanan lineages such as lepisosteids, abelisaurids, and bothremydids. Another group contributing to this similarity are titanosaurids, which are most abundant in Gondwana but do have a substantial Laurasian component throughout the Cretaceous. All four sites are Maastrichtian in age. The Cretaceous intertrappean beds and Lameta Formation, and the Paleocene Fatehgarh Formation clustered with one another and with the Cerrejón Formation of Colombia, demonstrating again the close faunal similarities among the Gondwanan continents during the Mesozoic. Close similarities in Late Cretaceous faunas of South America, Madagascar, and India to the exclusion of Africa may represent vicariance, supporting the Africa-first hypothesis (Krause et al. 2006). However, some workers have argued for later dispersals between the Indian subcontinent, Madagascar, Africa and southern Europe close to the Cretaceous – Paleogene boundary, as exemplified by

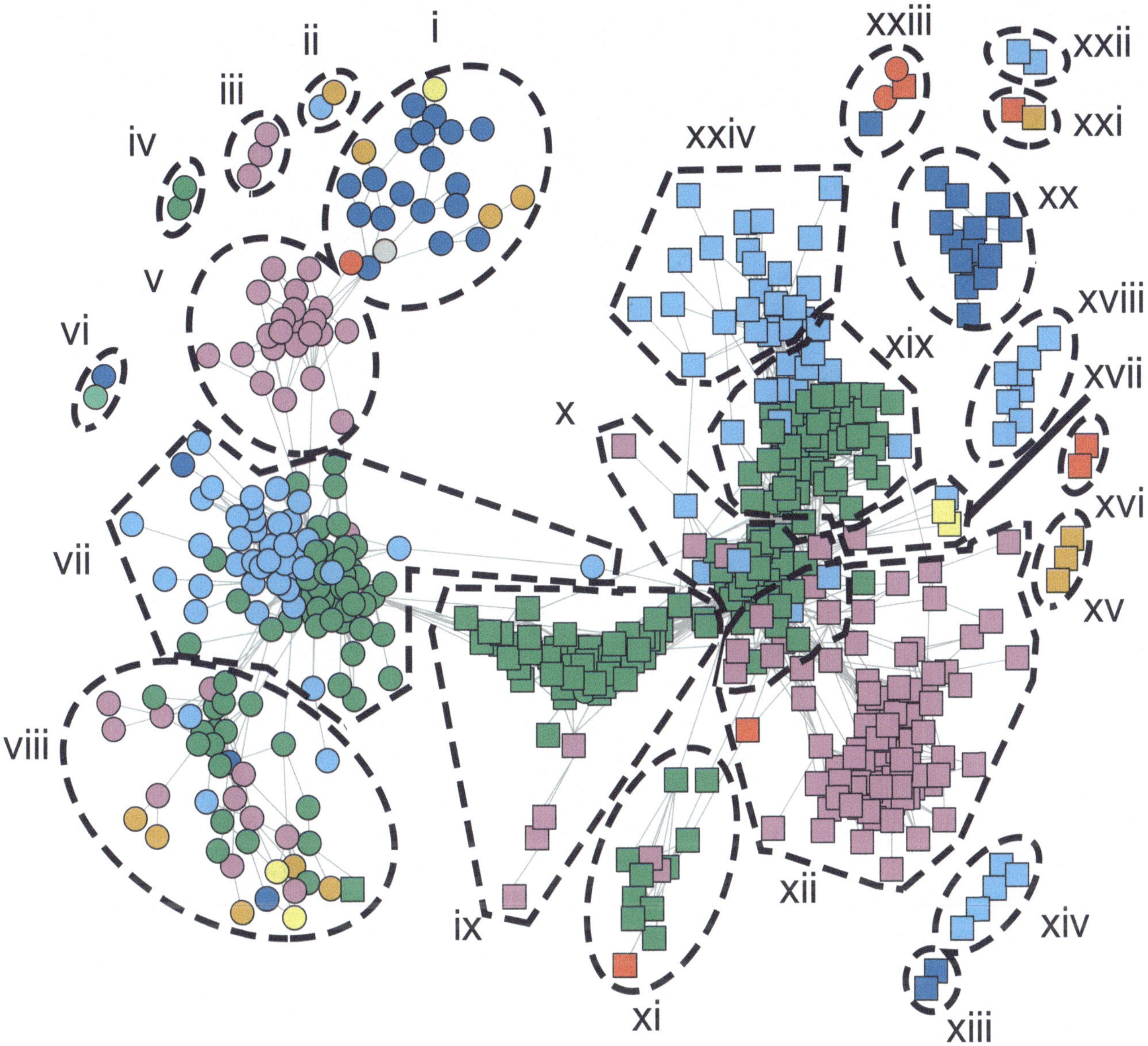

Fig. 9.5 Faunal network of all Cretaceous to Eocene vertebrate sites, thresholded to 0.75 and with all unconnected localities removed. The strongest division is between Cretaceous (circle) and Paleogene (square) sites, followed by geographic divisions based on continents, represented by colors as follows: North America – dark green, Asia – sky blue, South America – navy blue, Europe – pink, the Indian subcontinent – red, Madagascar – grey, Africa – orange, Australia and New Zealand – yellow, Antarctica - pale green. 24 clusters can be identified, as follows: i. Cretaceous Gondwana, ii. the Shendi and Wadi Milk formations, iii. Aquitaine Basin, iv. Turonian-Campanian Utah, v. Cretaceous Europe, vi. Maastrichtian Patagonia and Antarctica, vii. Campanian-Maastrichtian North America and East and Central Asia, viii. Cretaceous global shallow marine sites generally dominated by marine reptiles and sharks, ix. Paleocene North America and Europe, x. Early Eocene North America, Asia, and Europe, xi. Fish-dominated marine faunas of the Eocene, xii. Late Eocene Europe, xiii. Eocene Peru, xiv. Paleocene Mongolia, xv. Paleogene North Africa, xvi. Eocene northern Indian subcontinent, xvii. Eocene Australia, xviii. Paleocene China, xix. Late Eocene North America and Asia, xx. Paleogene Argentina and Brazil, xxi. Marine mammal sites in the Indian subcontinent and Africa, xxii. Wanghudun, xxiii. Cretaceous-Paleogene boundary sites in the Indian subcontinent and South America, xxiv. Eocene Asia. This cluster pattern is very well supported, with a modularity score of 0.76

madtsoiid snakes, majungasaurine abelisaurids, and adapisoriculid mammals (Prasad et al. 2010; Tortosa et al. 2014; Rio and Mannion 2017).

Eocene Gondwanan sites in general, and those from the Indian subcontinent specifically, are more separated, and all are closely connected with Laurasian localities, or entirely separated from the main cluster (Fig. 9.5). Among the former Gondwanan continents in the Cenozoic, it is only South America for which geography is a bigger driver of global faunal affinities than time, with all localities from that continent found as a separate cluster. Whereas clustering in the Cretaceous was driven by geography, with several close connections between temporally divergent localities spanning approximately 30 million years, the Paleogene sites

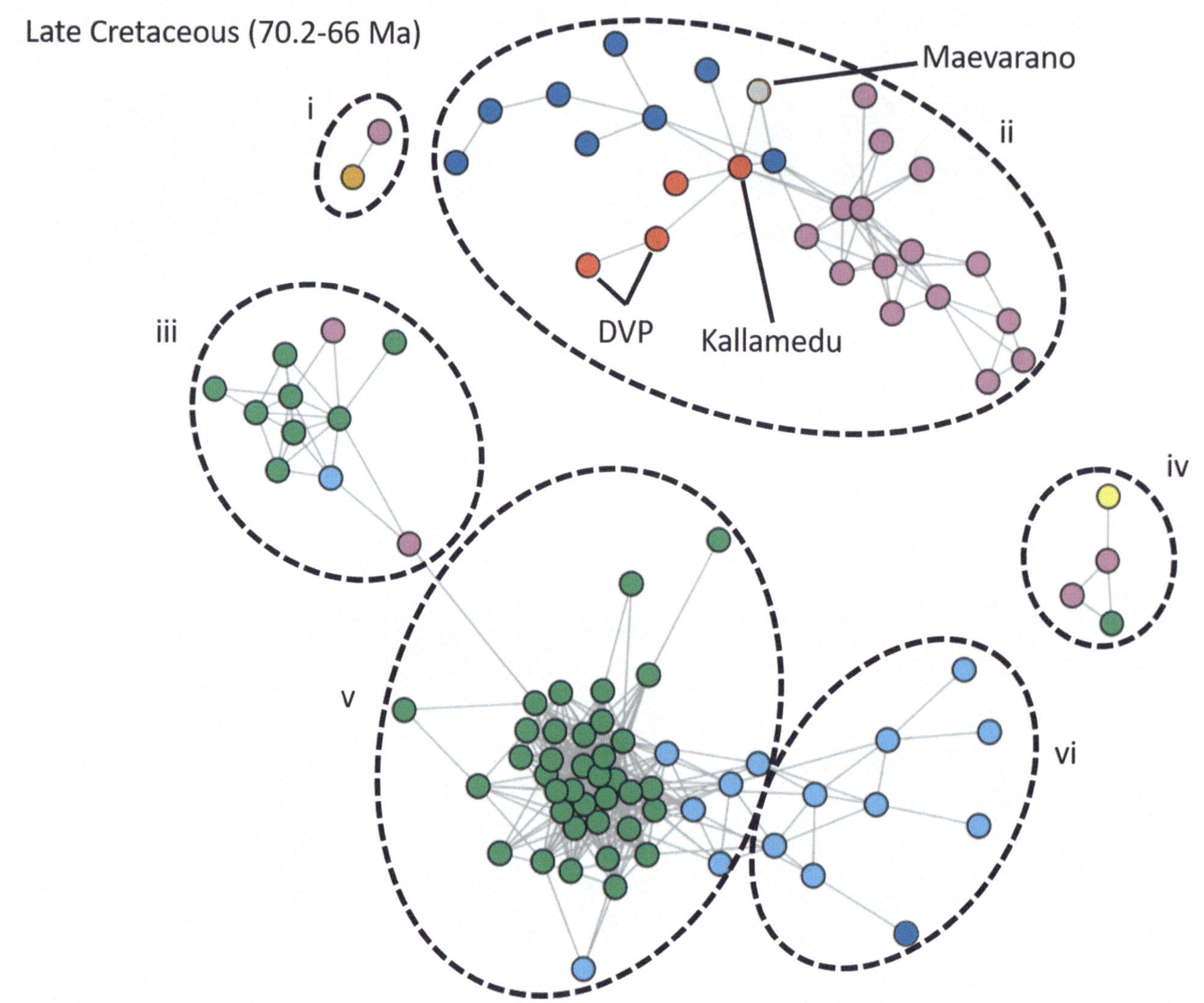

Fig. 9.6 Faunal network of Maastrichtian (Bin 13) vertebrate localities, threshold = 0.75. The network is divided into six clusters by edge betweenness. i. Ahlen and Dukamaje formations of Europe and Africa, ii. Most of Europe and Gondwana, including all localities on the Indian subcontinent, iii. Eastern Laurentia, including Mount Laurel, Marshalltown, Tar Heel formations, concentrating on marine taxa, hence also the Palmyra Phosphates, iv. High latitude localities from Canada, Russia, and New Zealand, v. Terrestrial North America, mostly around the Western Interior Seaway, as well as East Asian localities, vi. Mongolia and China, along with a single South American site – Lago Colhue Huapi. Colors are as in Fig. 9.5

across the same period are divided into at least three major temporal groups – the Paleocene as a whole, the early Eocene, and the middle Eocene. In each temporal group, the abundant North American sites provide the backbone to the network, with connections forward and backward in time, with Asian and European sites connecting to that network in temporally defined clusters. Localities on the Indian subcontinent such as the Naredi Formation and the Cambay Shale are often part of these offshoots, variously close to Laurasian localities or isolated as an unconnected fauna.

When time bins were considered separately, the Maastrichtian global sites showed a very similar pattern to all sites taken at once (Fig. 9.6). The Kallamedu Formation of India was highly connected at a threshold level of 0.75, connecting to the Maevarano Formation of Madagascar, the Pab Formation of Pakistan, and the DVP faunas of India, and both South American and European sites. Europe, for the most part, clusters with Gondwana, recapitulating the Late Cretaceous Eurogondwana concept of Le Loeuff (1991; see also Gheerbrant and Rage 2006; Rabi and Sebők 2015; Rio and Mannion 2017; Ezcurra and Agnolin 2012), while North America and Asia are divided into three major clusters – the southeastern US, a combined East Asia and Western Interior USA cluster, and a China-Mongolia cluster.

In the early Eocene, at a threshold of 0.75, South America and Africa are resolved as fully unconnected clusters, while the Indian subcontinent is part of a major grouping including all North American, European, and approximately half of the

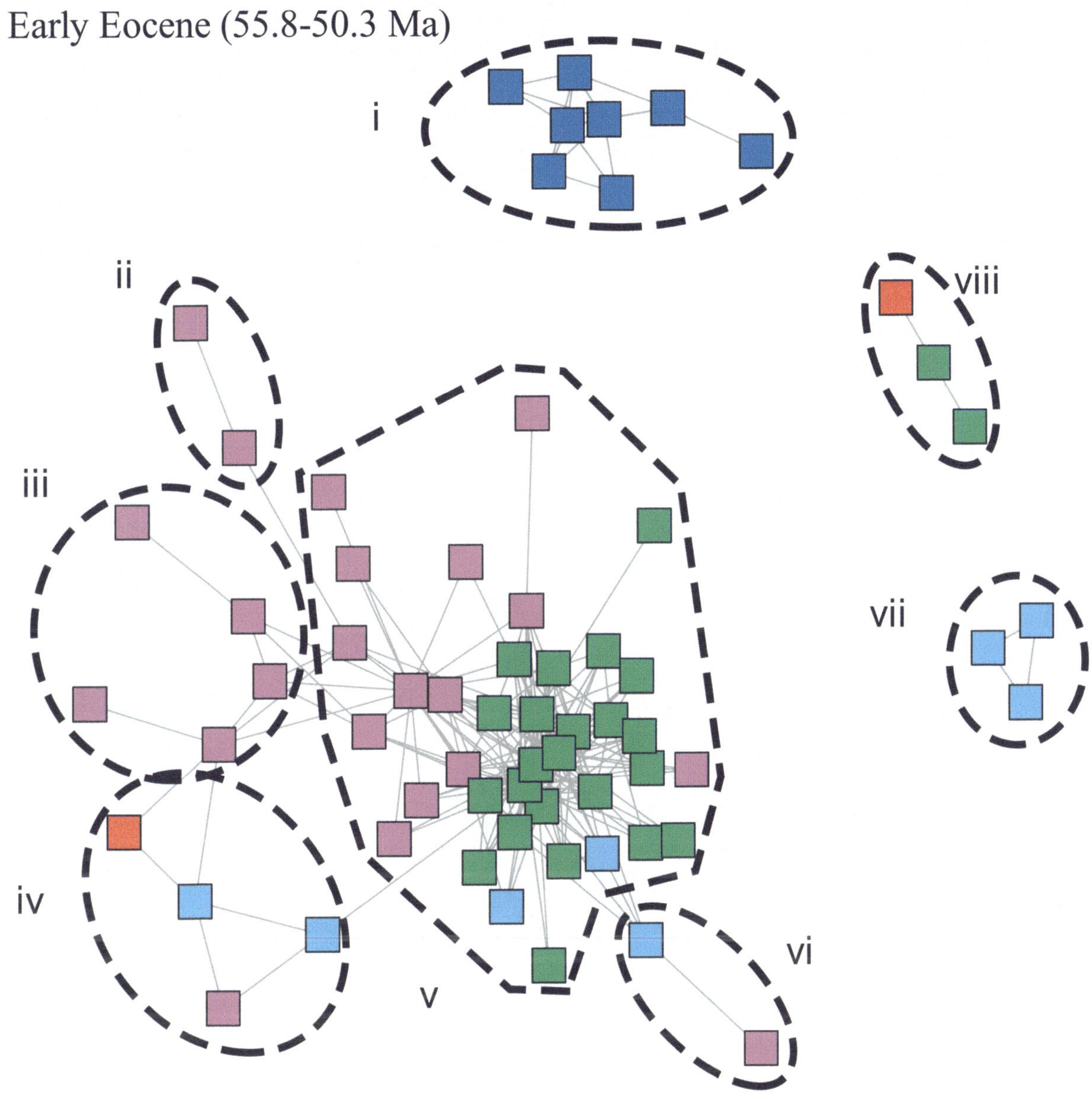

Fig. 9.7 Faunal network of the early Eocene (Bin 25), threshold = 0.75. The network is divided into eight identifiable clusters. i. South America, ii. Grès d'Assignan and La Borie, France, iii. other French Eocene sites, iv. Cambay Shale, connected to Asia and France, v. Large group comprising North America and mostly non-French European sites, vi. Wutu (China) and Clacaires des Montaigues (France), shark-dominated marine sites, vii. East Asia, viii. Marine (Bashi and Nanjemoy [North America], mammal-dominated sites, Kapurdi [India], elasmobranch-dominated marginal marine site). Colors are as in Fig. 9.5

Asian localities (Fig. 9.7), demonstrating strong Laurasian affinities during this time. The closest connections for the Cambay Shale and Subathu formations are through the Niushan Formation of China and the Sézanne-Broyes locality (MP 8–9) in France, which in turn connect more widely to Europe and Asia. This pattern is maintained through to the late Eocene, during which time the faunas of the Indian subcontinent maintain connections with northern rather than southern landmasses (Fig. 9.8).

Density, Centrality, and Modularity

Throughout the time period covered by this analysis, edge density of time-binned global networks at the family level remained high (between 0.15 and 0.6), while edge density at the genus level was low during the Cretaceous and high in the aftermath of the end-Cretaceous mass extinction and for the duration of the Paleocene. Previous observations of the end-Permian mass extinction have suggested a global

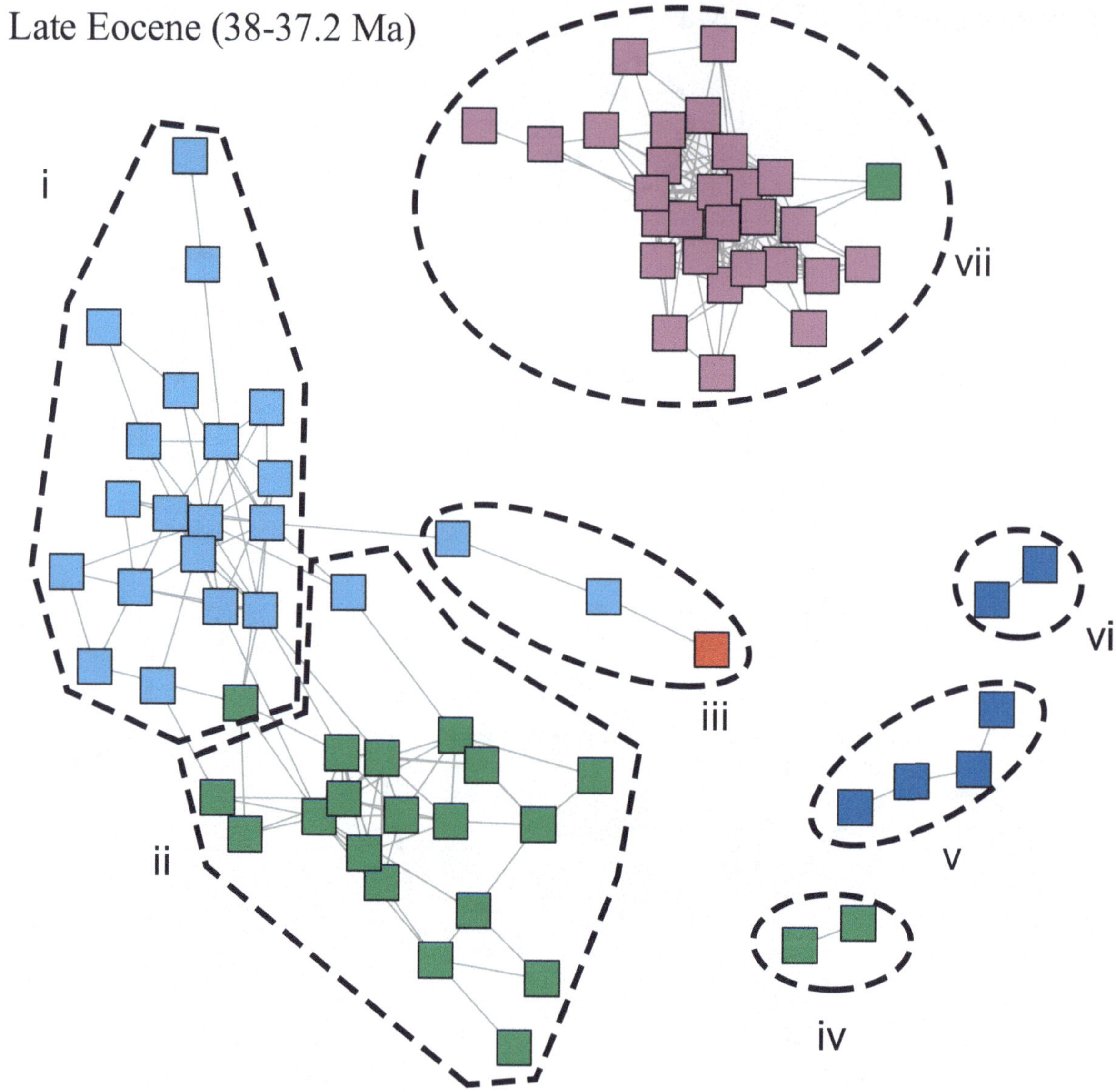

Fig. 9.8 Faunal network of the Late Eocene (Bin 33), threshold = 0.75. The network is divided up into seven identifiable clusters. i. Asia (and the Pruett Formation of the USA), ii. North America (and Changxindian Formation of China), iii. The Kuldana and Iwaki formations and the Shangqinglong fissure fill site, iv. Marine (Jackson and Yazoo formations, USA), v. Argentina, vi. Peru, vii. Europe (and the Wind River Formation of the USA). Colors are as in Fig. 9.5

disaster fauna (Sahney and Benton 2008; Button et al. 2017; although this was not found in the network analysis of Sidor et al. 2013), and as such, we might have expected an increase in network density through the Cretaceous – Paleocene extinction as well. The increase in density observed at the global scale is probably due to a high proportion of Paleocene sites from the Western Interior of North America. With geographically connected and proximate sites, it is expected that some overlap in fauna – and therefore an edge in the network – would be more likely. When a comparison is made at solely the continental level, there is no difference in the edge density between the Late Cretaceous and early Paleocene of North America, suggesting that the apparent global increase is probably a result of spatial bias of fossil localities rather than a result of any identifiable changes in patterns of global faunal heterogeneity. The differences in

Table 9.2 List of faunas of the Indian subcontinent. Betweenness scores are for the Bin indicated at a threshold of 0.5, and are the highest proportion of shortest routes obtained for that formation. Faunas with n/a values are those represented by too few taxa for Forbes faunal similarity calculations, or with no connections to other faunas at a threshold of 0.5, and therefore are not represented in the faunal networks

	Lithologic Unit	Age	Environment	Betweenness (Bin)
Barmer Basin	Kapurdi	Early Eocene	Nearshore to shallow marine	72/3,655 (25)
	Akli	Late Paleocene to Early Eocene	Nearshore terrestrial	0/1,378 (22)
	Fatehgarh	Early Paleocene (Danian)	Nearshore to shallow marine	n/a
Jaisalmer Basin	Khuiala	Early Eocene (Ypresian)	Shallow marine	n/a
Western India	Harudi	Eocene	Lagoonal	0/1,485 (30)
	Naredi	Eocene	Terrestrial to shallow marine	42 (30)
	Intertrappean Beds	Late Cretaceous	Fluvio-lacustrine	17/6,670 (12)
	Lameta	Late Cretaceous	Fluvio-lacustrine	108/6,670 (12)
Cauvery Basin	Kallamedu	Late Maastrichtian	Fluvio-deltaic terrestrial	144/2,301 (13)
Cambay Basin	Cambay Shale	Early Eocene	Lagoonal to shallow marine	28/3,655 (25)
Himalayan Foreland Basin	Subathu	Eocene	Shallow marine to marine	42/4,950 (29)
Kala Chitta Hills	Kuldana	Eocene	Fluvio-deltaic to lagoonal	1,997/5,050 (33)
	Mami Khel	Eocene	Fluvio-deltaic	2/3,403 (26)
	Chorgali	Eocene	Shallow marine	0/1,485 (30)
Lower Indus Basin	Ghazij	Early Eocene	Deltaic to shallow marine	41.5/4,095 (28)
	Khadro	Paleocene	Coastal, shallow marine	n/a
	Pab	Late Cretaceous	Fluvial	25/6,670 (12)

the geographic arrangements of the continents during these two mass extinction events make any direct comparison difficult – reduced migration as a result of oceanic barriers in the Late Cretaceous perhaps making faunal cosmopolitanism more difficult than on Pangaea at the Permian – Triassic boundary.

Betweenness centrality scored at threshold 0.5 revealed that, in the Late Cretaceous, the Kallamedu Formation had among the highest betweenness scores in both Bins 12 and 13, being the sixth most central in the late Maastrichtian as part of 144 of a possible 2,301 shortest paths. The Lameta Formation also had relatively high centrality scores. For the majority of the Eocene, no site on the Indian subcontinent was among the most central localities, but had consistently low, non-zero scores (Table 9.2). The one exception to this pattern is the fauna of the Kuldana Formation for which the error bars extend into Bin 33. Here, the Kuldana Formation acts as a bridge between two large clusters, and thus is part of nearly half of all shortest paths. This high score is not, however, a result of a mixed Gondwanan and Laurasian fauna, but is instead a result of anthracobunid, diacodexid, and omomyid placental mammals. Further, the non-zero scores exhibited by other Cenozoic Indian subcontinent faunas are largely a result of the connection among shallow marine sites rather than a mixed terrestrial fauna. The overall pattern of higher scores in the Late Cretaceous followed by generally lower scores through the Paleogene suggests that the Late Cretaceous vertebrate fauna of the Indian subcontinent can reasonably be said to be a particularly highly 'cosmopolitan' fauna with a mixture of northern and southern clades, linking together sites from Gondwana and Laurasia. The Eocene of the Indian subcontinent, on the other hand, while it contains a number of clades with Gondwanan affinities, is dominated by an influx of Laurasian clades, particularly placental mammals. The Gondwanan component to the Eocene faunas is reduced to being relictual, with little impact on placement in a faunal similarity network.

A Theoretical Paleocene Fauna from the Indian Subcontinent

Although the scarcity of Paleocene rocks from the Indian subcontinent means that its Paleocene fauna may never be well known, the occurrences of taxa in both the Cretaceous and Eocene of the subcontinent is indicative of at least some of the faunal components during this missing interval. Of the 42 families present in the Maastrichtian of the Indian subcontinent, 11 are also found in the Eocene. Some of these are known from those few Paleocene sites in continental India to have yielded vertebrate fossils, such as dyrosaurid crocodylomorphs (Storrs 1986; Prasad and Singh 1991; Rana and Sati 2000; Khosla et al. 2009; Smith et al. 2016) and madtsoiid snakes (Rage et al. 2014). Myliobatid, pycnodont, and eotrigonodontid fish are known from the Cretaceous, Eocene, and also the early Paleocene Fatehgarh Formation (Kumar and Loyal 1987; Rana 1990; Rana et al. 2006b; Smith et al. 2016). In addition to these taxa, there are six families known from Cretaceous and Eocene sites in the Indian subcontinent but absent from the Paleocene, namely ariid (Jain and Sahni 1983; Bajpai and Thewissen 2002) and phyllodontid fish (Rana et al. 2004; Halliday et al. 2016), boid snakes (Jain 1989; Wilson et al. 2010; Smith et al. 2016), crocodylids (Rana and Sati 2000; Bajpai and Thewissen 2002), costatan frogs (Prasad and Rage 1991; Bajpai and Kapur 2008; Folie et al. 2013; Rage et al. 2020), and adapisoriculid mammals (Prasad et al. 2010; Goswami et al. 2011; Kapur et al. 2017). When combined with known vertebrate occurrences from the Paleocene faunas of the Fatehgarh, Akli, and Khadro formations, the inhabitants of a hypothetical Paleocene Indian subcontinent can be placed into the faunal network. This is, of course, extremely speculative: there is the assumption that presence in the Cretaceous and Eocene necessitates presence in the Paleocene, which might not be the case, and the lack of sampling is necessarily restrictive. Nonetheless, theorizing on the composition of otherwise absent faunas allows us to explore possibilities while also producing hypotheses that could be confirmed or rejected by future discoveries.

Here, the 'Theoretical Paleocene Fauna' (TPF) of the Indian subcontinent occupies a place in the network distinct from the majority of North American sites. Largely due to the presence of dyrosaurids, the early Paleocene is strongly similar to the Umm Himar Formation of Saudi Arabia. High similarity scores between TPF and the Umm Himar Formation are partially a result of the low diversity of the latter. Dyrosaurids are also known from the Paleocene phosphate basins of Morocco (Jouve et al. 2005), but other differences between those faunas and TPF result in lower similarity scores. TPF also has some similarity to the fish-dominated marine Hornerstown Formation in New Jersey, USA, as well as the Santa Lucia Formation of Bolivia, which has also yielded both boid and madtsoiid snakes. If included with middle Paleocene localities, TPF shows affinities with the Cerrejón Formation (Colombia) and the Jbel Guersif Formation (Adrar Mgorn localities, Thanetian of Morocco), while late Paleocene sites that cluster with TPF include the Aruma and Itaborai formations, as well as the 'Argiles à Lignites du Soissonais' fauna (MP 7, early Eocene of Rivecourt) in France. As only those taxa present in the Cretaceous and earliest Paleocene are considered for this ersatz Paleocene fauna, it is unsurprising that the majority of high faunal similarities are between localities on the Indian subcontinent and elsewhere in Gondwana. Given, however, that these are taxa that are also present in the Eocene, and which in the Eocene also cluster with Gondwanan sites, this demonstrates the extent to which immigration must be generating the observed similarities between Eocene faunas of the Indian subcontinent and Laurasia.

We found no relationship between familial faunal similarity and distance between latitude-longitudinal sites, contrary to the hypothesis that greater distance would result in lower faunal similarity. The exception to this is during the Paleocene, when the relationship is strong (R^2 of 0.5 or more). This pattern remains true even when looking solely at North America, indicating that oceanic barriers are not the cause, and that there is something fundamentally different about the Paleocene data that results in a different pattern. Although a full exploration of this pattern is beyond the scope of this paper, it may be that Paleocene families are typically younger following radiation after a mass extinction event, and having had less time to disperse, are more likely to have a distance-controlled distribution. Alternatively, the lack of substantial high-latitude dispersal during the Paleocene compared with both the Cretaceous and Eocene (e.g. Beard and Dawson 1999; Martin et al. 2005; Hooker 2015) might have resulted in a reduction in the homogenizing effect of intercontinental migration. Linnaean ranks are arbitrary, and therefore families need not be equivalent among different clades, even where defined. Indeed, this is known to be the case among extant organisms (Avise and Liu 2011). Families are defined solely by morphological differences greater than that of a genus and are a hypothesis of relatedness, but in rapid radiations such as those thought to have occurred in mammals, teleosts, and birds after the end-Cretaceous mass extinction, rates of morphological evolution are expected to have increased substantially (Beck and Lee 2014; Halliday et al. 2019), so ranks in different time periods might not be equivalent. The arbitrariness of Linnaean ranks may also impact the study in differential treatment of taxonomy by different workers. The family concept is little-used among dinosaur workers (Richard Butler pers. comm.), but remains important for distinguishing mammals, where the enigmatic "condylarths" can reliably be divided into more-or-less monophyletic groups (but see De Bast and Smith 2013 for a counterexample) that are defined as families, but whose higher-level relationships are unclear (Halliday et al. 2015). Even with perfect vetting of the data, different practices in the definition of arbitrary taxonomic ranks among paleontologists can impact later interpretation (see Hendricks et al. 2014 for related discussion about non-species ranks in the fossil record). Although ranks are arbitrary, the analyses within each time bin will be largely unaffected. The equivalence of the unit of measurement of 'family' is irrelevant to the calculated similarity, where each example of that unit is defined equivalently throughout the bin. It is in comparing network statistics between time bins that we might have reason for concern, but the pattern exhibited by abelisaurids leads us to have confidence that the Late Cretaceous networks are not substantially different from the Cenozoic networks as a result of substantially different data; the clustering of abelisaurid-bearing faunas of a Madagascar – Indian subcontinent cluster close to Southern Europe mirrors phylogenetic hypotheses about these taxa, resulting in plausible biogeographic scenarios.

Conclusions

As has long been known, the faunal composition of Late Cretaceous and Paleogene sites on the Indian subcontinent is a mixture of taxa with Gondwanan and Laurasian affinities. Faunal network analysis demonstrates that the faunal signal from the Cauvery Basin, the Pab Formation, and the Deccan Volcanic Province links these faunas from the Indian subcontinent most strongly with South American and Madagascan localities, both in pairwise comparisons of faunal similarity and clusters derived from edge betweenness statistics. In these clusters, the Kallamedu Formation falls with the Maevarano Formation of Madagascar, as in previous work, but also with several South American sites in particular – the Ortega Formation, but at lower threshold values, also the Marília and Adamantina Formations, while the sites of the DVP cluster separately. Those few clades present in the Cretaceous of the Indian subcontinent with Laurasian affinities – for instance, alytids, troodontids, phyllodontids, and adapisoriculids – are exceptions to the rule, but are numerous enough to result in the Kallamedu Formation in particular being one of the most faunally connected localities of the Late Cretaceous under the node betweenness statistic.

The Paleocene vertebrate fauna of the Indian subcontinent is poorly known, but those clades that persisted from the Late Cretaceous through the Paleocene and into the Eocene retained their historic ties with Gondwana, even while the signal as a whole shifts in the Eocene to one favouring a Laurasian-dominated fauna, relegating those pan-Gondwanan clades to relictual status within a larger ecosystem. The Laurasian signal is largely driven by shared placental mammalian families that have widespread Laurasian distributions. These similarities reflect significant dispersal events that took place in the early Paleogene among the Indian subcontinent, Asia, Europe, and North America (Krause and Maas 1990; Beard and Dawson 1999; Bossuyt and Milinkovitch 2001; Bowen et al. 2002; Clyde et al. 2003; Smith et al. 2006; Clementz et al. 2011). The subsequent evolution and extinction of clades elsewhere weakened the faunal similarity with other Gondwanan landmasses, and as a result, the Eocene faunas of the Indian subcontinent have lower centrality scores. This metric reflects the lessening degree to which the subcontinent's faunas can be said to be a mixed, cosmopolitan fauna as the Cenozoic progresses, highlighting the importance of intercontinental migration in the development of faunas, and emphasising the biogeographic uniqueness of the diverse fauna on the Indian subcontinent at the end of the Mesozoic.

Author Contributions. TJDH and AG designed the analysis, TJDH wrote the code and conducted the analysis, and all authors screened segments of the data and contributed to the manuscript.

Acknowledgments This work would not have been possible without the dedication of time and resources by several people to the Paleobiology Database, particularly the many volunteers who input the substantial amount of data required for this project, chief among whom are John Alroy, Matthew Carrano, Philip Mannion, Mark Uhen, William Clyde, and Erin Leckey. We thank Richard Butler, David Button, and Philip Mannion for discussion and three anonymous reviewers for their comments. TJDH was funded by Leverhulme Early Career Fellowship ECF-2017-577. AG and TJDH were supported by Leverhulme Trust Grant RPG-2014/364. GVRP acknowledges grant received under J.C. Bose National Fellowship of the Science and Engineering Research Board, New Delhi. DWK acknowledges support from National Science Foundation grant EAR-1664432.

Appendix 1: Data and Bins

This appendix is freely available on Springer Link at https://doi.org/10.1007/978-3-030-49753-8.

Appendix 2: Faunal Network Function

```
#FUNCTION TO PERFORM THRESHOLDED FAUNAL NETWORK ANALYSES IN R

#PARAMETERS REQUIRED - MATRIX, NAME

#DEPENDENT ON IGRAPH PACKAGE



##### THIS IS JOHN ALROY'S FUNCTION FOR CALCULATING FORBES SIMILARITY INDICES ######


forbes<-function(x,y,corrected) {

	if (missing(corrected))  {

		corrected <- T

	}

	if (is.numeric(x) && is.numeric(y) && min(x) == 0 && min(y) == 0 && length(x) == length(y))
	{

		a <- length(which((x * y) > 0))

		b <- length(which(x > 0)) - a

		c <- length(which(y > 0)) - a

	} else	{

		a <- length(na.omit(match(x,y)))

		b <- length(x) - a

		c <- length(y) - a

	}

	n <- a + b + c

	if (corrected == T)	{

		return(a * (n + sqrt(n))/((a + b) * (a + c) + a * sqrt(n) + (b * c)/2))

	} else	{

		return(a * n/((a + b) * (a + c)))

	}

}
```

```
forbesMatrix<-function(x,corrected)	{
	x[is.na(x)] <- 0
	m = matrix(nrow=ncol(x),ncol=ncol(x))
	for (i in 1:ncol(x))	{
		for (j in 1:ncol(x))	{
			m[i,j] = forbes(x[,i],x[,j],corrected)
		}
	}
	return(m)
}

################################################
###########    END ALROY CODE    ###########
################################################


faunal_network<-function(mat, n) {

#READ IN OCCURRENCE MATRIX

m<-read.table(mat, header=T, row.names=1)

#REMOVE COLUMNS (IE SITES) WITH FEWER THAN THREE TAXA

# COMMENT OUT IF YOUR MATRIX IS IN FORMAT COLUMNS=LOCALITIES, ROWS=TAXA
m<-t(m)

m<-m[,colSums(m>0)>=3]
```

```
#CONVERT OCCURRENCE MATRIX INTO ADJACENCY MATRIX

fm<-forbesMatrix(m)
row.names(fm)<-colnames(m)

colnames(fm)<-colnames(m)

#ESTABLISH THRESHOLD VALUES

t<-c(0.05,0.1,0.15,0.2,0.25,0.3,0.35,0.4,0.45,0.5,0.55,0.6,0.65,0.7,0.75,0.8,0.85,0.9,0.95)

#FOR EACH THRESHOLD VALUE

for(i in 1:length(t)) {

#DUMMY MATRIX

fm1<-fm

#THOSE LOWER THAN THE THRESHOLD TO ZERO

fm1[fm1<t[i]]<-0

#CALCULATE GRAPH FROM ADJACENCY MATRIX

fmg<-graph_from_adjacency_matrix(fm1, mode="undirected", weight=T, diag=F)

#WRITE GRAPH TO FILE

write_graph(fmg, file=paste(n, i,".ncol"),format="ncol")

}

}
```

References

Agnolin, F. (2012). A new Calyptocephalellidae (Anura, Neobatrachia) from the Upper Cretaceous of Patagonia, Argentina, with comments on its systematic position. *Studia Geologica Salmanticensia, 48*, 129–178.

Ali, J. R., & Aitchison, J. C. (2008). Gondwana to Asia: Plate tectonics, paleogeography and the biological connectivity of the Indian sub-continent from the Middle Jurassic through latest Eocene (166–35 Ma). *Earth-Science Reviews, 88*(3–4), 145–166. https://doi.org/10.1016/j.earscirev.2008.01.007.

Ali, J. R., & Krause, D. W. (2011). Late Cretaceous bioconnections between Indo-Madagascar and Antarctica: refutation of the Gunnerus Ridge causeway hypothesis. *Journal of Biogeography, 38*(10), 1855–1872. https://doi.org/10.1111/j.1365-2699.2011.02546.x.

Alroy, J. (2015a). A new twist on a very old binary similarity coefficient. *Ecology, 96*(2), 575–586. https://doi.org/10.1890/14-0471.1.

Alroy, J. (2015b). A simple way to improve multivariate analyses of paleoecological data sets. *Paleobiology, 41*(3), 377–386. https://doi.org/10.1017/pab.2014.21.

Avise, J. C., & Liu, J. X. (2011). On the temporal inconsistencies of Linnean taxonomic ranks. *Biological Journal of the Linnean Society, 102*(4), 707–714. https://doi.org/10.1111/j.1095-8312.2011.01624.x.

Bajpai, S. (2009). Biotic perspective of the Deccan volcanism and India-Asia collision: Recent advances. *Indian Academy of Sciences Platinum Jubilee Special: Current Trends in Science*, 505–516.

Bajpai, S., & Kapur, V. V. (2008). Earliest Cenozoic frogs from the Indian subcontinent: Implications for out-of-India hypothesis. *Journal of the Palaeontological Society of India, 53*(1), 65–71.

Bajpai, S., & Thewissen, J. G. M. (2002). Vertebrate fauna from Panandhro Lignite Mine (Lower Eocene), District Kachchh, western India. *Current Science, 82*, 507–509.

Bapteste, E., van Ierse, L., Janke, A., Kelchner, S., Kelk, S., McInerney, J. O., et al. (2013). Networks: Expanding evolutionary thinking. *Trends in Genetics, 29*(8), 439–441. https://doi.org/10.1016/j.tig.2013.05.007.

Beard, K. C., & Dawson, M. R. (1999). Intercontinental dispersal of Holarctic land mammals near the Paleocene/Eocene boundary: Paleogeographic, paleoclimatic and biostratigraphic implications. *Bulletin de la Société Géologique de France, 170*(5), 697–706.

Beck, R. M. D., & Lee, M. S. Y. (2014). Ancient dates or accelerated rates? Morphological clocks and the antiquity of placental mammals. *Proceedings Biological Sciences/The Royal Society, 281* (1793). https://doi.org/10.1098/rspb.2014.1278.

Blonder, B., Wey, T. W., Dornhaus, A., James, R., & Sih, A. (2012). Temporal dynamics and network analysis. *Methods in Ecology and Evolution, 3*(6), 958–972. https://doi.org/10.1111/j.2041-210X.2012.00236.x.

Bonaparte, J. F. (1999). Tetrapod faunas from South America and India: A paleobiogeographic interpretation. *Proceedings of the Indian National Science Academy, 65*(3), 427–437.

Bossuyt, F., & Milinkovitch, M. C. (2001). Amphibians as indicators of early Tertiary "out-of-India" dispersal of vertebrates. *Science, 292* (5514), 93–95. https://doi.org/10.1126/science.1058875.

Bowen, G. J., Clyde, W. C., Koch, P. L., Ting, S. Y., Alroy, J., Tsubamoto, T., et al. (2002). Mammalian dispersal at the Paleocene/Eocene boundary. *Science, 295*(5562), 2062–2065. https://doi.org/10.1126/science.1068700.

Boyer, D. M., Prasad, G. V. R., Krause, D. W., Godinot, M., Goswami, A., Verma, O., et al. (2010). New postcrania of Deccanolestes from the Late Cretaceous of India and their bearing on the evolutionary and biogeographic history of euarchontan mammals. *Naturwissenschaften, 97*(4), 365–377. https://doi.org/10.1007/s00114-010-0648-0.

Briggs, J. C. (2003). The biogeographic and tectonic history of India. *Journal of Biogeography, 30*(3), 381–388.

Buckley, M. (2015). Ancient collagen reveals evolutionary history of the endemic South American 'ungulates'. *Proceedings of the Royal Society B-Biological Sciences, 282*(1806). https://doi.org/10.1098/rspb.2014.2671.

Button, D. J., Lloyd, G. T., Ezcurra, M. D., & Butler, R. J. (2017). Mass extinctions drove increased global faunal cosmopolitanism on the supercontinent Pangaea. *Nature Communications, 8*. https://doi.org/10.1038/s41467-017-00827-7.

Case, J. A. (2002). A new biogeographic model for dispersal of Late Cretaceous vertebrates into Madagascar and India. *Journal of Vertebrate Paleontology, 22*(3, Supplement), 42A.

Cecca, F. (2002). *Paleobiogeography of marine fossil invertebrates: Concepts and methods*. Boca Raton, Florida: CRC Press.

Chatterjee, S., Goswami, A., & Scotese, C. R. (2013). The longest voyage: Tectonic, magmatic, and paleoclimatic evolution of the Indian plate during its northward flight from Gondwana to Asia. *Gondwana Research, 23*(1), 238–267. https://doi.org/10.1016/j.gr.2012.07.001.

Chatterjee, S., & Scotese, C. (2010). The wandering Indian plate and its changing biogeography during the Late Cretaceous-Early Tertiary Period. In S. Bandyopadhyay (Ed.), *New aspects of mesozoic biodiversity* (pp. 105–126).

Chatterjee, S., Scotese, C. R., & Bajpai, S. (2017). The restless Indian plate and its epic voyage from Gondwana to Asia: Its tectonic, paleoclimate, and paleobiogeographic evolution. *The Geological Society of America Special Paper, 529*, 1–147.

Clementz, M., Bajpai, S., Ravikant, V., Thewissen, J. G. M., Saravanan, N., Singh, I. B., et al. (2011). Early Eocene warming events and the timing of terrestrial faunal exchange between India and Asia. *Geology, 39*(1), 15–18. https://doi.org/10.1130/g31585.1.

Clyde, W. C., Khan, I. H., & Gingerich, P. D. (2003). Stratigraphic response and mammalian dispersal during initial India-Asia collision: Evidence from the Ghazij Formation, Balochistan, Pakistan. *Geology, 31*(12), 1097–1100. https://doi.org/10.1130/g19956.1.

Cooper, L. N., Seiffert, E. R., Clementz, M., Madars, S. I., Bajpai, S., Hussain, S. T., et al. (2014). Anthracobunids from the Middle Eocene of India and Pakistan are stem perissodactyls. *PLoS ONE, 9* (10). https://doi.org/10.1371/journal.pone.0109232.

Csardi, G., & Nepusz, T. (2006). The igraph software package for complex network research. *InterJournal, Complex Systems*, 169.

De Bast, E., & Smith, T. (2013). Reassessment of the small arctocyonid *Prolatidens waudruae* from the Early Paleocene of Belgium, and its phylogenetic relationships with ungulate-like mammals. *Journal of Vertebrate Paleontology, 33*(4), 964–976. https://doi.org/10.1080/02724634.2013.747531.

Dunhill, A. M., Bestwick, J., Narey, H., & Sciberras, J. (2016). Dinosaur biogeographical structure and Mesozoic continental fragmentation: A network-based approach. *Journal of Biogeography, 43*(9), 1691–1704. https://doi.org/10.1111/jbi.12766.

Esteve-Altava, B., Diogo, R., Smith, C., Diego, J. C. B., & Rasskin-Gutman, D. (2015). Anatomical networks reveal the musculoskeletal modularity of the human head. *Scientific Reports, 5*. https://doi.org/10.1038/srep08298.

Ezcurra, M. D. (2009). Theropod remains from the uppermost Cretaceous of Colombia and their implications for the paleozoogeography of western Gondwana. *Cretaceous Research, 30*(5), 1339–1344. https://doi.org/10.1016/j.cretres.2009.08.004.

Ezcurra, M. D., & Agnolin, F. L. (2012). A new global paleobiogeographical model for the Late Mesozoic and Early Tertiary. *Systematic Biology, 61*(4), 553–566. https://doi.org/10.1093/sysbio/syr115.

Folie, A., Rana, R. S., Rose, K. D., Sahni, A., Kumar, K., Singh, L., et al. (2013). Early Eocene frogs from Vastan Lignite Mine, Gujarat, India. *Acta Palaeontologica Polonica, 58*(3), 511–524.

Gheerbrant, E., & Rage, J.-C. (2006). Paleobiogeography of Africa: How distinct from Gondwana and Laurasia? *Palaeogeography Palaeoclimatology Palaeoecology, 241*(2), 224–246. https://doi.org/10.1016/j.paleo.2006.03.016.

Gibbons, A. D., Zahirovic, S., Muller, R. D., Whittaker, J. M., & Yatheesh, V. (2015). A tectonic model reconciling evidence for the collisions between India, Eurasia and intra-oceanic arcs of the central-eastern Tethys. *Gondwana Research, 28*(2), 451–492. https://doi.org/10.1016/j.gr.2015.01.001.

Goswami, A., Prasad, G. V. R., Upchurch, P., Boyer, D. M., Seiffert, E. R., Verma, O., et al. (2011). A radiation of arboreal basal eutherian mammals beginning in the Late Cretaceous of India. *Proceedings of the National Academy of Sciences of the United States of America, 108*(39), 16333–16338. https://doi.org/10.1073/pnas.1108723108.

Goswami, A., Prasad, G. V. R., Verma, O., Flynn, J. J., & Benson, R. B. J. (2013). A troodontid dinosaur from the latest Cretaceous of India. *Nature Communications, 4*. https://doi.org/10.1038/ncomms2716.

Halliday, T. J. D., Cuff, A. R., Prasad, G. V. R., Thanglemmoi, M. S., & Goswami, A. (2016). New record of *Egertonia* (Elopiformes, Phyllodontidae) from the Late Cretaceous of South India. *Papers in Palaeontology, 2,* 287–294. https://doi.org/10.1002/spp2.1040.

Halliday, T. J. D., dos Reis, M., Ferguson-Gow, H., Tamuri, A. U., Yang, Z., & Goswami, A. (2019). Rapid morphological evolution in eutherian mammals post-dates the origin of the crown group. *Proceedings of the Royal Society B, 286*, 20182418.

Halliday, T. J. D., Prasad, G. V. R., & Goswami, A. (2017). Faunal similarity in Madagascan and South Indian Late Cretaceous vertebrate faunas. *Palaeogeography Palaeoclimatology Palaeoecology, 468,* 70–75. https://doi.org/10.1016/j.paleo.2016.11.046.

Halliday, T. J. D., Upchurch, P., & Goswami, A. (2015). Resolving the relationships of Paleocene placental mammals. *Biological Reviews*. https://doi.org/10.1111/brv.12242.

Hay, W. W., DeConto, R. M., Wold, C. N., Wilson, K. M., Voigt, S., Schulz, M., et al. (1999). Alternative global Cretaceous paleogeography. *Geological Society of America Special Paper, 332,* 1–48.

Hendricks, J. R., Saupe, E. E., Myers, C. E., Hermsen, E. J., & Allmon, W. D. (2014). The generification of the fossil record. *Paleobiology, 40*(4), 511–528. https://doi.org/10.1666/13076.

Hooker, J. J. (2015). A two-phase Mammalian Dispersal Event across the Paleocene-Eocene transition. *Newsletters on Stratigraphy, 48* (2), 201–220. https://doi.org/10.1127/nos/2015/0060.

Hu, X. M., Garzanti, E., Wang, J. G., Huang, W. T., An, W., & Webb, A. (2016). The timing of India-Asia collision onset – Facts, theories, controversies. *Earth-Science Reviews, 160,* 264–299. https://doi.org/10.1016/j.earscirev.2016.07.014.

Jain, S. L. (1989). Recent dinosaur discoveries in India, including eggshells, nests and coprolites. In D. Gillette & M. Lockley (Eds.), *Dinosaur tracks and traces* (pp. 99–108). New York: Cambridge University Press.

Jain, S. L., & Sahni, A. (1983). Some Upper Cretaceous vertebrates from Central India and their paleogeographical implications. In H. K. Maheswari (Ed.), *Cretaceous of India: Paleoecology, paleogeography, and time boundaries* (pp. 66–83). Lucknow: Indian Association of Palynologists.

Jiménez-Velandia, H. F. (2017). *Revisión de la anatomia y relaciones filogen éticas de la vértebra dorsal de saurópodo proveniente del Cretácico Temprano en cercanías de La Paz, Cesar*. Colombia: Universidad EAFIT.

Jouve, S., Iarochene, M., Bouya, B., & Amaghzaz, M. (2005). A new dyrosaurid crocodyliform from the Paleocene of Morocco and a phylogenetic analysis of Dyrosauridae. *Acta Palaeontologica Polonica, 50*(3), 581–594.

Kapur, V. V., Das, D. P., Bajpai, S., & Prasad, G. V. R. (2017). First mammal of Gondwanan lineage in the early Eocene of India. *Comptes Rendus Palevol, 16*(7), 721–737. https://doi.org/10.1016/j.crpv.2017.01.002.

Kapur, V. V., & Khosla, A. (2016). Late Cretaceous terrestrial biota from India with special reference to vertebrates and their implications for biogeographic connections. In A. Khosla & S. G. Lucas (Eds.), *Cretaceous Period: Biotic diversity and biogeography* (pp. 161–171). New Mexico Museum of Natural History and Science Bulletin.

Keller, G., Adatte, T., Bajpai, S., Mohabey, D. M., Widdowson, M., Khosla, A., et al. (2009). K-T transition in Deccan Traps of central India marks major marine seaway across India. *Earth and Planetary Science Letters, 282*(1–4), 10–23. https://doi.org/10.1016/j.epsl.2009.02.016.

Khosla, A., Kapur, V. V., Sereno, P., Wilson, J. A., Wilson, G. P., Dutheil, D., et al. (2003). First dinosaur remains from the Cenomanian-Turonian Nimar Sandstone (Bagh Beds), district Dhar, Madhya Pradesh, India. *Journal of the Palaeontological Society of India, 48,* 115–127.

Khosla, A., Sertich, J. J. W., Prasad, G. V. R., & Verma, O. (2009). Dyrosaurid remains from the intertrappean beds of India and the Late Cretaceous distribution of Dyrosauridae. *Journal of Vertebrate Paleontology, 29*(4), 1321–1326.

Kiel, S. (2017). Using network analysis to trace the evolution of biogeography through geologic time: A case study. *Geology*, 711–714. https://doi.org/10.1130/G38877.1.

Klaus, S., Morley, R. J., Plath, M., Zhang, Y. P., & Li, J. T. (2016). Biotic interchange between the Indian subcontinent and mainland Asia through time. *Nature Communications, 7*. https://doi.org/10.1038/ncomms12132.

Krause, D. W., & Maas, M. C. (1990). The biogeographic origins of the late Paleocene-early Eocene mammalian immigrants to the western interior of North America. In T. M. Bown & K. D. Rose (Eds.), *Dawn of the Age of Mammals in the Northern part of the Rocky Mountain Interior, North America* (pp. 71–105). Geological Society of America.

Krause, D. W., O'Connor, P. M., Rogers, K. C., Sampson, S. D., Buckley, G. A., & Rogers, R. R. (2006). Late Cretaceous terrestrial vertebrates from Madagascar: Implications for Latin American biogeography. *Annals of the Missouri Botanical Garden, 93*(2), 178–208. https://doi.org/10.3417/0026-6493(2006)93%5b178:lctvfm%5d2.0.co;2.

Krause, D. W., Prasad, G. V. R., von Koenigswald, W., Sahni, A., & Grine, F. E. (1997). Cosmopolitanism among Gondwanan Late Cretaceous mammals. *Nature, 390*(6659), 504–507. https://doi.org/10.1038/37343.

Kumar, K., & Loyal, R. S. (1987). Eocene ichthyofauna from the Subathu Formation, northwestern Himalaya, India. *Journal of the Palaeontological Society of India, 32,* 60–84.

Kumar, K., Rana, R. S., & Singh, H. (2007). Fishes of the Khuiala Formation (Early Eocene) of the Jaisalmer basin, western Rajasthan, India. *Current Science, 93*(4), 553–559.

Kumar, K., & Sahni, A. (1985). Eocene mammals from the Upper Subathu Group, Kashmir Himalaya, India. *Journal of Vertebrate Paleontology, 5,* 153–168.

Le Loeuff, J. (1991). The Campano-Maastrichtian vertebrate faunas from southern Europe and their relationships with other faunas in the world; paleobiogeographical implications. *Cretaceous Research, 12,* 93–114.

Mao, F. Y., Wang, Y. Q., Li, Q., & Jin, X. (2016). New records of archaic ungulates from the Lower Eocene of Sanshui Basin,

Guangdong, China. *Historical Biology, 28*(6), 787–802. https://doi.org/10.1080/08912963.2015.1034120.
Martin, J. E., Case, J. A., Jagt, J. W. M., Schulp, A. S., & Mulder, E. W. A. (2005). A new European marsupial indicates a Late Cretaceous high-latitude transatlantic dispersal route. *Journal of Mammal Evolution, 12,* 495–511.
Mathur, S. K., Mathur, S. C., & Loyal, R. S. (2006). First microvertebrate assemblage from the Fatehgarh Formation (Cretaceous), Barmer district, western Rajasthan. *Journal of the Geological Society of India, 67*(6), 759–769.
Matthews, K. J., Muellner, R. D., & Sandwell, D. T. (2016). Oceanic microplate formation records the onset of India-Eurasia collision. *Earth and Planetary Science Letters, 433,* 204–214. https://doi.org/10.1016/j.epsl.2015.10.040.
Missiaen, P., & Gingerich, P. D. (2012). New early Eocene tapiromorph perissodactyls from the Ghazij Formation of Pakistan, with implications for mammalian biochronology in Asia. *Acta Palaeontologica Polonica, 57*(1), 21–34. https://doi.org/10.4202/app.2010.0093.
Missiaen, P., & Gingerich, P. D. (2014). New basal Perissodactyla (Mammalia) from the Lower Eocene Ghazij Formtion of Pakistan. *Contributions from the Museum of Paleontology, University of Michigan, 32,* 139–160.
Najman, Y., Appel, E., Boudagher-Fadel, M., Bown, P., Carter, A., Garzanti, E., et al. (2010). Timing of India-Asia collision: Geological, biostratigraphic, and paleomagnetic constraints. *Journal of Geophysical Research-Solid Earth, 115.* https://doi.org/10.1029/2010jb007673.
Norton, I. O., & Sclater, J. G. (1979). A model for the evolution of the Indian Ocean and the break-up of the Gondwanaland. *Journal of Geophysical Research, 84*(B12), 6803–6830.
Powell, C. M. (1979). A speculative tectonic history of Pakistan and surroundings: Some constraints from the Indian Ocean. In A. Farah & K. A. Dejong (Eds.), *Geodynamics of Pakistan* (pp. 5–24). Geological Survey of Pakistan.
Prasad, G. V. R. (2012). Vertebrate biodiversity of the Deccan volcanic province of India: A review. *Bulletin de la Société Géologique de France, 183*(6), 597–610.
Prasad, G. V. R., & Goswami, A. (2015). Cretaceous vertebrate faunal diversity and evolution on the Indian raft: Fossil evidence from the Cauvery Basin, south India. In Y. Zhang, S. Z. Wu, & G. Sun (Eds.), *12th Symposium on Mesozoic Terrestrial Ecosystems* (pp. 75–77). Shenyang: China.
Prasad, G. V. R., Jaeger, J.-J., Sahni, A., Gheerbrant, E., & Khajuria, C. K. (1994). Eutherian mammals from the Upper Cretaceous (Maastrichtian) intertrappean beds of Naskal, Andhra Pradesh, India. *Journal of Vertebrate Paleontology, 14*(2), 260–277.
Prasad, G. V. R., & Rage, J.-C. (1991). A discoglossid frog in the latest Cretaceous (Maastrichtian) of India – Further evidence for a terrestrial route between India and Laurasia in the latest Cretaceous. *Comptes rendus de l'Académie des Sciences Série II, 313*(2), 273–278.
Prasad, G. V. R., & Rage, J.-C. (1995). Amphibians and squamates from the Maastrichtian of Naskal, India. *Cretaceous Research, 16* (1), 95–107. https://doi.org/10.1006/cres.1995.1006.
Prasad, G. V. R., & Rage, J.-C. (2004). Fossil frogs (Amphibia: Anura) from the Upper Cretaceous intertrappean beds of Naskal, Andhra Pradesh. *Revue de Paléobiologie, 23,* 99–116.
Prasad, G. V. R., & Sahni, A. (2009). Late Cretaceous continental vertebrate fossil record from India: Paleobiogeographical insights. *Bulletin de la Société Géologique de France, 180*(4), 369–381.
Prasad, G. V. R., & Singh, V. (1991). Microvertebrates from the intertrappean beds of Rangareddi District, Andhra Pradesh & their biostratigraphic significance. *Bulletin of the Indian Geologists' Association, 24,* 1–20.
Prasad, G. V. R., Verma, O., Flynn, J. J., & Goswami, A. (2013). A new Late Cretaceous vertebrate fauna from the Cauvery Basin, South India: Implications for Gondwanan paleobiogeography. *Journal of Vertebrate Paleontology, 33*(6), 1260–1268. https://doi.org/10.1080/02724634.2013.777348.
Prasad, G. V. R., Verma, O., Gheerbrant, E., Goswami, A., Khosla, A., Parmar, V., et al. (2010). First mammal evidence from the Late Cretaceous of India for biotic dispersal between India and Africa at the KT transition. *Comptes Rendus Palevol, 9*(1–2), 63–71. https://doi.org/10.1016/j.crpv.2009.12.003.
Prasad, G. V. R., Verma, V., Grover, P., Priyadarshini, R., Sahni, A., & Lourembam, R. S. (2016). Isolated archosaur teeth from the green sandstone capping the Coralline Limestone (Bagh Group) of the Narmada valley: Evidence for the presence of pre-Late to Late Maastrichtian abelisaurids in India. *Island Arc, 25*(6), 410–420. https://doi.org/10.1111/iar.12142.
Prasad, G. V. R., Verma, V., Sahni, A., Lourembam, R. S., & Rajkumari, P. (2017). Elasmobranch fauna from the upper most part of the Cretaceous Bagh Group, Narmada valley, India. *Island Arc, 26*(5). https://doi.org/10.1111/iar.12200.
R Development Core Team. (2010). 'R: A language and environment for statistical computing'. R Foundation for Statistical Computing, Vienna, Austria. Retrieved from http://www.R-project.org.
Rabi, M., & Sebők, N. (2015). A revised Eurogondwana model: Late Cretaceous notosuchian crocodyliforms and other vertebrate taxa suggest the retention of episodic faunal links between Europe and Gondwana during most of the Cretaceous. *Gondwana Research, 28* (3), 1197–1211. https://doi.org/10.1016/j.gr.2014.09.015.
Rage, J.-C., Metais, G., Bartolini, A., Brohi, I. A., Lashari, R. A., Marivaux, L., et al. (2014). First report of the giant snake *Gigantophis* (Madtsoiidae) from the Paleocene of Pakistan: Paleobiogeographic implications. *Geobios, 47*(3), 147–153. https://doi.org/10.1016/j.geobios.2014.03.004.
Rage, J.-C., Prasad, G. V. R., Verma, O., Khosla, A., & Parmar, V. (2020). Anuran lissamphibian and squamate reptiles from the Upper Cretaceous (Maastrichtian) Deccan intertrappean sites in Central India, with a review of a lissamphibian and squamate diversity in the northward drifting Indian plate. In G. V. R. Prasad & R. Patnaik (Eds.), *Biological consequences of plate tectonics: New perspectives on post-Gondwana break-up* (pp. 99–121). Cham: Switzerland.
Rana, R. S. (1990). Paleontology and paleoecology of the intertrappean (Cretaceous-Tertiary transition) beds of the peninsular India. *Journal of the Palaeontological Society of India, 35,* 105–120.
Rana, R. S., Kumar, K., Loyal, R. S., Sahni, A., Rose, K. D., Mussell, J., et al. (2006a). Selachians from the early Eocene Kapurdi Formation (Fuller's Earth), Barmer district, Rajasthan. *Journal of the Geological Society of India, 67*(4), 509–522.
Rana, R. S., Kumar, K., & Singh, H. (2004). Vertebrate fauna from the subsurface Cambay Shale (Lower Eocene), Vastan Lignite Mine, Gujarat, India. *Current Science, 87*(12), 1726–1733.
Rana, R. S., Kumar, K., & Singh, H. (2006b). Paleocene vertebrate fauna from the Fatehgarh Formation of Barmer District, Rajasthan, Western India. In D. K. Sinha (Ed.), *Micropaleontology: Application in stratigraphy and paleoceanography* (pp. 113–130). New Delhi, India: Narosa Publishing House.

Rana, R. S., Kumar, K., Singh, H., & Rose, K. D. (2005). Lower vertebrates from the Late Paleocene-Earliest Eocene Akli Formation, Giral Lignite Mine, Barmer District, western India. *Current Science, 89*(9), 1606–1613.

Rana, R. S., & Sati, K. K. (2000). Late Cretaceous-Paleocene crocodilians from the Deccan Trap associated sedimentary sequences of peninsular India. *Journal of the Palaeontological Society of India, 45,* 123–136.

Rana, R. S., & Wilson, G. P. (2003). New Late Cretaceous mammals from the intertrappean beds of Rangapur, India and paleobiogeographic framework. *Acta Palaeontologica Polonica, 48*(3), 331–348.

Raup, D. M., & Crick, R. E. (1979). Measurement of faunal similarity in paleontology. *Journal of Paleontology, 53*(5), 1213–1227.

Reis, B. Y., Kohane, I. S., & Mandl, K. D. (2007). An epidemiological network model for disease outbreak detection. *Plos Medicine, 4*(6), 1019–1031. https://doi.org/10.1371/journal.pmed.0040210.

Rio, J. P., & Mannion, P. D. (2017). The osteology of the giant snake *Gigantophis garstini* from the Upper Eocene of North Africa and its bearing on the phylogenetic relationships and biogeography of Madtsoiidae. *Journal of Vertebrate Paleontology, 37*(4). https://doi.org/10.1080/02724634.2017.1347179.

Rose, K. D., Holbrook, L. T., Rana, R. S., Kumar, K., Jones, K. E., Ahrens, H. E., et al. (2014). Early Eocene fossils suggest that the mammalian order Perissodactyla originated in India. *Nature Communications, 5.* https://doi.org/10.1038/ncomms6570.

Sahney, S., & Benton, M. J. (2008). Recovery from the most profound mass extinction of all time. *Proceedings of the Royal Society B-Biological Sciences, 275*(1636), 759–765. https://doi.org/10.1098/rspb.2007.1370.

Sahni, A. (1984). Cretaceous-Paleocene terrestrial faunas of India – Lack of endemism during drifting of the Indian Plate. *Science, 226* (4673), 441–443. https://doi.org/10.1126/science.226.4673.441.

Sahni, A., Bhatia, S. B., Hartenberger, J.-L., Jager, J.-J., Kumar, K., Sudre, J., et al. (1981). Vertebrates from the Subathu Formation and comments on the biogeography of Indian subcontinent during the Early Paleogene. *Bulletin de la Société Géologique de France, 23* (6), 689–695.

Sahni, A., Kumar, K., Hartenberger, J.-L., Jaeger, J.-J., Rage, J.-C., Sudre, J., et al. (1982). Microvertébrés nouveaux des Trapps du Deccan (Inde): Mise en evidence d'une voie de communication terrestre probable entre la Laurasie et l'Inde à la limite Crétacé-Tertiaire. *Bulletin de la Société Géologique de France, 24,* 1093–1099.

Sahni, A., Rana, R. S., & Prasad, G. V. R. (1987). New evidence for paleobiogeographic intercontinental Gondwana relationships based on late Cretaceous-earliest Paleocene coastal faunas from peninsular India. In *Gondwana Six: Stratigraphy, sedimentology, and paleontology* (pp. 207–218). American Geophysical Union. https://doi.org/10.1029/gm041p0207.

Sampson, S. D., Witmer, L. M., Forster, C. A., Krause, D. W., O'Connor, P. M., Dodson, P., et al. (1998). Predatory dinosaur remains from Madagascar: Implications for the Cretaceous biogeography of Gondwana. *Science, 280*(5366), 1048–1051. https://doi.org/10.1126/science.280.5366.1048.

Sidor, C. A., Vilhena, D. A., Angielczyk, K. D., Huttenlocker, A. K., Nesbitt, S. J., Peecook, B. R., et al. (2013). Provincialization of terrestrial faunas following the end-Permian mass extinction. *Proceedings of the National Academy of Sciences of the United States of America, 110*(20), 8129–8133. https://doi.org/10.1073/pnas.1302323110.

Singh, R. S., & Kar, R. K. (2002). Paleocene palynofossils from the Lalitpur intertrappean beds, Uttar Pradesh. *Journal of the Geological Society of India, 60,* 213–216.

Smith, T., Kumar, K., Rana, R. S., Folie, A., Sole, F., Noiret, C., et al. (2016). New early Eocene vertebrate assemblage from western India reveals a mixed fauna of European and Gondwana affinities. *Geoscience Frontiers, 7*(6), 969–1001. https://doi.org/10.1016/j.gsf.2016.05.001.

Smith, T., Rose, K. D., & Gingerich, P. D. (2006). Rapid Asia-Europe-North America geographic dispersal of earliest Eocene primate *Teihardina* during the Paleocene-Eocene Thermal Maximum. *Proceedings of the National Academy of Sciences of the United States of America, 103*(30), 11223–11227. https://doi.org/10.1073/pnas.0511296103.

Storey, M., Mahoney, J. J., & Saunders, A. D. (1997). Cretaceous basalts in Madagascar and the transition between plume and continental lithosphere mantle sources. In J. J. Mahoney, & M. F. Coffin (Eds.), *Large igneous provinces: Continental, oceanic, and planetary flood volcanism.* Washington D.C., USA: American Geophysical Union. https://doi.org/10.1029/gm100p0095.

Storey, M., Mahoney, J. J., Saunders, A. D., Duncan, R. A., Kelley, S. P., & Coffin, M. F. (1995). Timing of hotspot-related volcanism and the breakup of Madagascar and India. *Science, 267*(5199), 852–855. https://doi.org/10.1126/science.267.5199.852.

Storrs, G. W. (1986). A dyrosaurid crocolide (Crocodylia, Mesosuchia) from the Paleocene of Pakistan. *Postilla, 197,* 1–16.

Subrahmanya, K. R. (1998). Tectono-magmatic evolution of the west coast of India. *Gondwana Research, 1*(3–4), 319–327. https://doi.org/10.1016/s1342-937x(05)70847-9.

Tabuce, R., Vianey-Liaud, M., & Garcia, G. (2004). A eutherian mammal in the latest Cretaceous of Vitrolles, southern France. *Acta Palaeontologica Polonica, 49*(3), 347–356.

Thewissen, J. G. M., Cooper, L. N., Clementz, M. T., Bajpai, S., & Tiwari, B. N. (2007). Whales originated from aquatic artiodactyls in the Eocene Epoch of India. *Nature, 450*(7173), 1190–1994. https://doi.org/10.1038/nature06343.

Thewissen, J. G. M., Nanda, A. C., & Bajpai, S. (2020). *Indohyus*, endemic radiation of raoellid artiodactyls in the Eocene of India and Pakistan. In G. V. R. Prasad & R. Patnaik (Eds.), *Biological consequences of plate tectonics: New perspectives on post-Gondwana break-up* (pp. 337–346). Cham: Switzerland.

Thewissen, J. G. M., Williams, E. M., & Hussain, S. T. (2001). Eocene mammal faunas from northern Indo-Pakistan. *Journal of Vertebrate Paleontology, 21*(2), 347–366. https://doi.org/10.1671/0272-4634(2001)021%5b0347:emffni%5d2.0.co;2.

Torsvik, T. H., Tucker, R. D., Ashwal, L. D., Carter, L. M., Jamtveit, B., Vidyadharan, K. T., et al. (2000). Late Cretaceous India-Madagascar fit and timing of break-up related magmatism. *Terra Nova, 12*(5), 220–224. https://doi.org/10.1046/j.1365-3121.2000.00300.x.

Tortosa, T., Buffetaut, E., Vialle, N., Dutour, Y., Turini, E., & Cheylan, G. (2014). A new abelisaurid dinosaur from the Late Cretaceous of southern France: Paleobiogeographical implications. *Annales de Paléontologie, 100*(1), 63–86. https://doi.org/10.1016/j.annpal.2013.10.003.

Underwood, C. J., Goswami, A., Prasad, G. V. R., Verma, O., & Flynn, J. J. (2011). Marine vertebrates from the 'middle' Cretaceous (early Cenomanian) of South India. *Journal of Vertebrate Paleontology, 31*(3), 539–552. https://doi.org/10.1080/02724634.2011.574518.

Welker, F., Collins, M. J., Thomas, J. A., Wadsley, M., Brace, S., Cappellini, E., et al. (2015). Ancient proteins resolve the evolutionary history of Darwin's South American ungulates. *Nature, 522,* 81–84. https://doi.org/10.1038/nature14249.

Westbury, M., Baleka, S., Barlow, A., Hartmann, S., Paijmans, J. L. A., Kramarz, A., et al. (2017). A mitogenomic timetree for Darwin's enigmatic South American mammal *Macrauchenia patachonica. Nature Communications, 8,* 15951. https://doi.org/10.1038/ncomms15951.

Wilson, G. P., Widdowson, M., Anantharaman, S., Das Sarma, D. C., Wilson, J. A., & Renne, P. R. (2016). *New mammalian fossils from the intertrappean beds of the southern part of the Deccan Volcanic Province and the Cretaceous-Paleogene transition in India.* Paper presented at the SVP 76th Annual Meeting, Salt Lake City.

Wilson, J. A., Malkani, M. S., & Gingerich, P. D. (2001). New crocodyliform (Reptilia, Mesoeucrocodylia) from the Upper Cretaceous Pab Formation of Vitakri, Balochisan (Pakistan). *Contributions from the Museum of Paleontology, University of Michigan, 30,* 321–336.

Wilson, J. A., Malkani, M. S., & Gingerich, P. D. (2005). A sauropod braincase from the Pab Formation (Upper Cretaceous, Maastrichtian) of Balochistan, Pakistan. *Gondwana Geology Magazine, 8,* 101–109.

Wilson, J. A., Mohabey, D. M., Peters, S. E., & Head, J. J. (2010). Predation upon hatchling dinosaurs by a new snake from the Late Cretaceous of India. *PLoS Biology, 8*(3). https://doi.org/10.1371/journal.pbio.1000322.

Yatheesh, V., Bhattacharya, G. C., & Mahender, K. (2006). The terrace like feature in the mid-continental slope region off Trivandrum and a plausible model for India-Madagascar juxtaposition in immediate pre-drift scenario. *Gondwana Research, 10*(1–2), 179–185. https://doi.org/10.1016/j.gr.2005.11.021.

Chapter 10
Island Africa and Vertebrate Evolution: A Review of Data and Working Hypotheses

Jean-Claude Rage* and Emmanuel Gheerbrant

Abstract From mid-Cretaceous to Early Miocene, Africa was geographically isolated as a gigantic island, the Island Africa. This considerably impacted local evolution of continental vertebrates. The native fauna of the Island Africa is and was depauperate for poorly known reasons. Although an Island, Africa was intermittently connected to Laurasia by discontinuous routes across the Neo-Tethys. Owing to these routes, various allochthonous groups of Eurasian origin entered Africa where they found favorable evolutionary conditions with relaxed competitive pressure; many of these groups evolved successfully, giving rise to new African clades and important endemic radiations. Thus, Island Africa acted as a secondary cradle, boosting the evolution of stem groups of Eurasian origin. Notably, all Cenozoic African mammals likely derived from stem groups of Laurasian origin. A remarkable number of endemic African lineages evolved successfully in extant African faunas, and even colonized several continents. They survived in particular the Miocene event of the Great Old World Interchange. The earlier faunal relations of Africa and Eurasia across the Tethys likely contributed to the success of the endemic African evolution.

Keywords African endemism • Cretaceous • Dispersals • Eurasia • Vertebrata • Mammalia • Paleogene • Secondary cradle

*Deceased after submission on March 30 2018.

E. Gheerbrant (✉)
Centre de Recherches CR2P UMR CNRS 7207 Sorbonne Université, MNHN/CNRS/UPMC, 57 rue Cuvier, CP 38, 75005 Paris, France
e-mail: emmanuel.gheerbrant@mnhn.fr

Introduction

This work aims to review, based on our current knowledge, some broad questions on the vertebrate evolution in the Arabo-African Island, that could be only briefly mentioned or addressed in our paper Gheerbrant and Rage (2006). The geographical history of Africa is marked by two major events: separation of the Arabo-African plate from other continents during the Mesozoic and connection to Eurasia during the Cenozoic. As a result of these two events, the African continental faunas have evolved in four successive major paleobiogeographical contexts since the Jurassic: (1) Pangean Africa; (2) Gondwanan Africa; (3) insular Arabo-Africa; and (4) Africa in the Old World. We focus here on a critical but still poorly known stage of the African history: the evolution of the Arabo-Africa island - hereafter referred to as *Island Africa* - which overlapped parts of the Mesozoic and Cenozoic, during ca. 80 Ma (Fig. 10.1). The faunal evolution of the Island Africa includes three critical periods in the emergence of modern faunas, which are well documented in Laurasia: The Cretaceous-Paleogene (K-Pg), Paleocene-Eocene and Eocene-Oligocene transitions. These global events more or less affected the endemic African faunas, but their biotic local impact remains much more poorly known than in most other continental provinces. The insular context strongly impacted evolution of the African continental biotas. However, while the process of separation from other continents and contact with Laurasia are relatively well known, the evolutionary history of biotas in the Island Africa remains largely unknown. The question of the evolutionary role of the Island Africa was raised in recent phylogenetic analyses and/or paleontological studies (e.g., Gorscak and O'Connor 2016; Jacobs et al. 2016; Eberle et al. 2017; Elewa 2017).

The main barrier encountered in studying the time interval that corresponds to the existence of Island Africa is the rarity of fossils as compared to other continental provinces,

G. V. R. Prasad and R. Patnaik (eds.), *Biological Consequences of Plate Tectonics: New Perspectives on Post-Gondwana Break-up—A Tribute to Ashok Sahni*,
Vertebrate Paleobiology and Paleoanthropology Series, https://doi.org/10.1007/978-3-030-49753-8_10

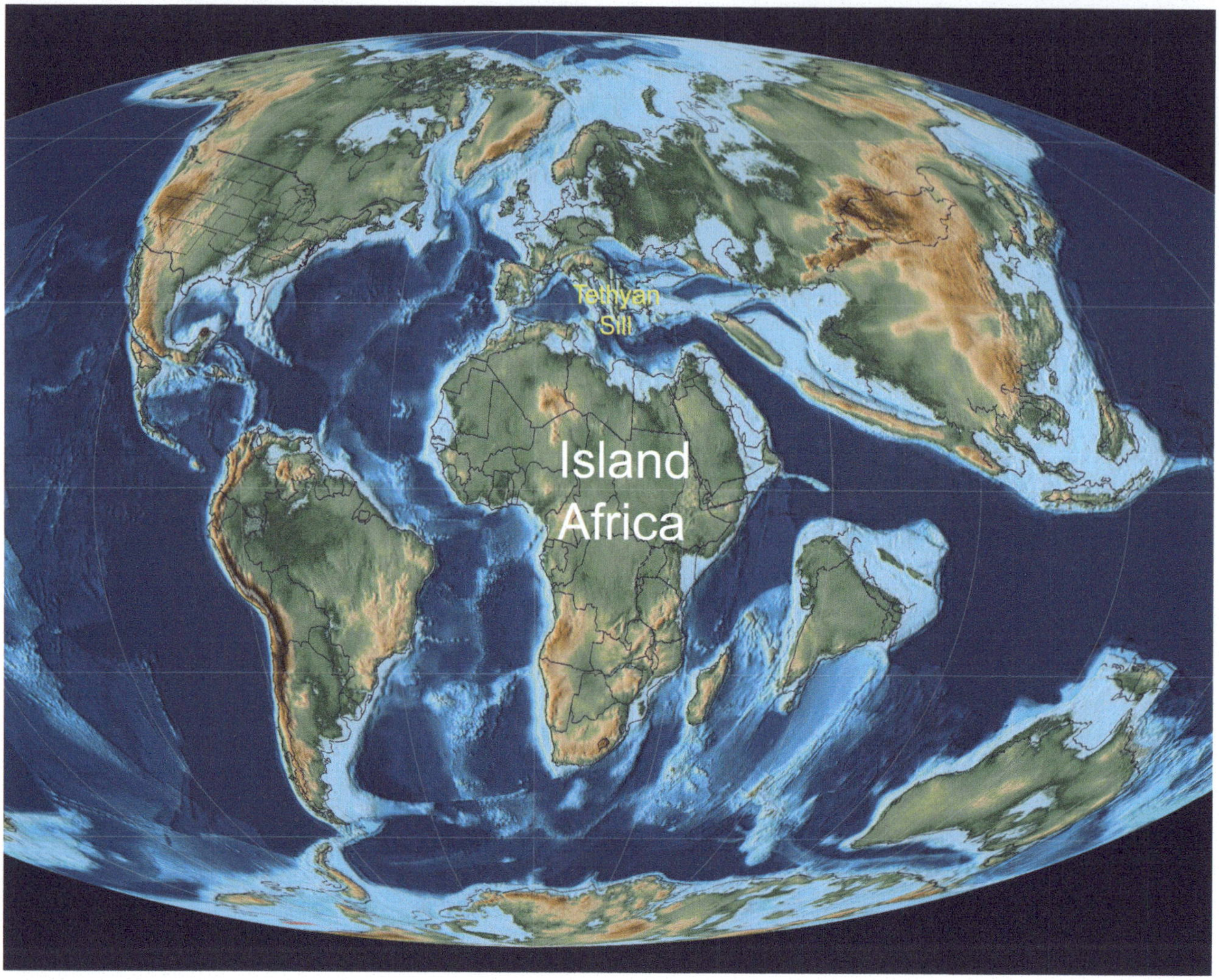

Fig. 10.1 Paleogeographic map for the K-Pg boundary showing the isolated Arabo-African plate forming the *Island Africa*. The Island Africa existed for ca. 90 Ma, from the Aptian/Albian to the Oligocene/Miocene transitions. The Mediterranean Tethyan Sill corresponds to a series of partly emerged continental platforms, such as the Apulian plate, that were located intermediate between Eurasia and Africa across the Tethyan area and that acted as filtering intercontinental dispersal routes for terrestrial vertebrates during episodes of low eustatic levels (e.g., Gheerbrant & Rage 2006). Paleogeographic map from *The PALEOMAP PaleoAtlas* (Scotese 2016)

including Gondwanan ones such as South America. This rarity - sometimes called "African Gap" (O'Connor et al. 2006; Gottfried et al. 2009) - was often regarded as a consequence of the lack of field work in Africa, absences being only non-significant "pseudo-absences" (*sensu* Upchurch et al. 2002). However, field studies - which started early during the first naturalist and mining explorations of the continent - have not been infrequent, far from it, and it now appears that rarity of fossils is implicitly accepted as the natural condition, not an artifact; this is for instance implied by all studies that evidenced relationships between South America and Madagascar-India, and Australia - but not with Africa (see below).

Our objective is not to trace the history of Africa during this period, which is an impossible task for the time being, but to make *a review of the best of our knowledge*, to emphasize the main problems that face us, and to propose hypotheses where possible.

The problems we would like to stress are as follows:

What was the *native* fauna of Africa, that is to say the autochthonous continental fauna that inhabited Africa at the beginning of isolation? What was its evolutionary fate?

How and why was Africa depauperate during isolation? What event(s), if any, could have caused low African paleobiodiversity?

Why were allochthonous forms, i.e., immigrants that entered Island Africa, so successful?

Why did a number of African taxa successfully survived Eurasian immigrants when Africa contacted Eurasia, and even colonized other continents?

We here review and discuss the available data on these questions, following our former discussion in Gheerbrant and Rage (2006).

The Native African Fauna (Fig. 10.2)

The African native fauna is naturally defined as the assemblage of autochthonous taxa present in Africa just after separation from South America. They are vicariant taxa resulting from the splitting of Gondwana (Fig. 10.2). They were present on West Gondwana, i.e., the Africa-South America block that was more precisely named Samafrica by Upchurch (2008), which may be regarded as the starting point in the paleobiogeographical history of the Island Africa. The native taxa of the Island Africa are Samafrican (= West Gondwanan) vicariants, it being understood that the Samafrican stock likely included older, pan-Gondwanan and even Pangean taxa (e.g., some fishes and mammals).

Among the vicariants may be cited various clades that are still living in Africa: freshwater osteichthyans (Lepidosirenidae), amphibians (Pipidae, Atlanticanura sensu Frazão et al. 2015), acrodontan lizards, and pelomedusid chelonians. Several extinct groups found in the Cretaceous of Africa are also evidence of this vicariance event: freshwater fishes (*Mawsonia*; Carvalho and Maysey 2008), some extinct chelonians (Araripemydidae, Euraxemydidae; Gaffney et al. 2006; Sereno and ElShafie 2013), crocodilians (*Araripesuchus*; clades of Pholidosauridae, including the genus *Sarcosuchus*; Turner 2006; Le Loeuff et al. 2010), dinosaurs (Abelisauridae, Carcharodontosauridae, Rebbachisauridae (Sereno et al. 2004; Novas et al. 2005) and mammals (Haramiyida, Eutriconodonta, Dryolestoidea, stem Cladotheria, and probably Gondwanatheria) (see also Gheerbrant and Rage 2006: Table 1). However, it should be noted that the extension of some of these taxa into the Late Cretaceous of Africa remains unknown; this is especially true for the mammals that are poorly documented in the Late Cretaceous of Africa (Table 10.1). Although this list of taxa needs to be confirmed for some of them, and is obviously not exhaustive and may include more, it nevertheless shows that no native African mammal groups survived up to the present and that the above mentioned extinct groups did not survive the Cretaceous in Africa. The only possible exception is raised by the tribosphenidan mammals from the early Cretaceous Anoual-Ksar Metlili locality (Sigogneau-Russell 1991; Kielan-Jaworowska 1992) which could be the stem group of in particular the early Cenozoic African therians. This is the hypothesis of an African Cretaceous origin of the placentals as suggested by Kielan-Jaworowska (1992) and of the afrotherians as suggested by molecular studies (e.g., Dos Reis et al. 2012). However, it should be noted that early tribosphenidans are unknown in other Gondwanan continents, and that the Anoual-Ksar Metlili tribosphenidans are not African natives. Their presence in Africa might indicate a Jurassic/Cretaceous trans-Tethyan dispersal from Eurasia. Mesozoic early tribosphenic mammals are known in South America, Madagascar, India (Parmar et al. 2015) and Australia, but they belong to the convergent Gondwanan clade Australosphenida (Luo et al. 2001). Most remarkably, the Gondwanan clade Australosphenida remains unknown in Africa.

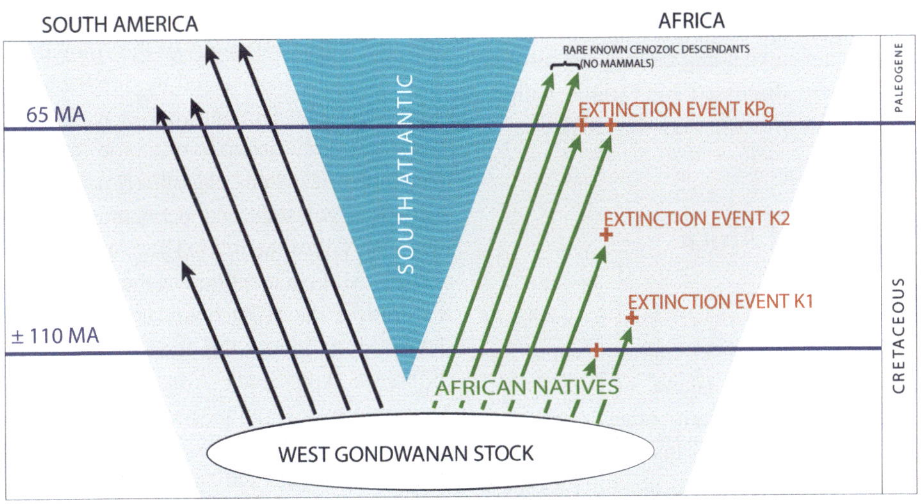

Fig. 10.2 Diagrammatic view of the origin and evolution of the native African fauna. Several vicariant lineages evolved in Africa from the West Gondwanan stock, following the South Atlantic oceanic opening. They constitute the *native African fauna* (autochthonous fauna). The African fossil record documents a poor diversity of the African natives and a depauperate African Island. It raises the hypothesis that the African natives were affected by several significant extinction events during their evolutionary history. One or several extinction events (K1 and K2 in Fig. 10.2) are in particular postulated during the Cretaceous in Africa, in addition to the K-Pg event; a middle Cretaceous extinction event (K1) is especially suggested in relation to the climatic hothouse events reported at this time, when South America and Africa began to drift apart (see text)

Table 10.1 Number of Cretaceous and Paleogene species of vertebrates (all vertebrates, tetrapods, mammals only) known in Island Africa and South America as estimated from the Paleobiology Database (calculation made on the Fossilworks website). Note that, with the exception of mammals in the Early Cretaceous of South America (gap), diversity is lower in Africa than in South America. The tetrapods column shows that, with regard to South America, in Africa the number of species did not markedly increase between the Early and Late Cretaceous (despite probable arrival of immigrants), and that during the Paleogene increase is more marked in Africa (due to immigrants and new AFS; see Fig. 10.5) but the numbers in Africa remain markedly lower. The recorded numbers of species for "vertebrates" (Cretaceous-Paleogene) and "mammals" (Cretaceous) are less significant because of inclusion of fishes and gaps, respectively

	Vertebrates			Tetrapods			Mammals		
	Island Africa	South America	% Island Africa	Island Africa	South America	% Island Africa	Island Africa	South America	% Island Africa
Early Cretaceous	64	289	60.38	55	84	65.47	13	1	(gap in South America)
Late Cretaceous	105	230	45.65	71	207	34.29	0	20	(gap in Africa)
Paleogene	388	581	66.78	323	556	58.09	248	496	50

It is generally acknowledged that the separation between Africa and South America took place by the Aptian-Albian transition (Gheerbrant and Rage 2006), that is about 110 Ma. Recently, Granot and Dyment (2015) suggested a younger age, ca. 100 Ma on geophysical basis. However, the separation of the two continents likely occurred earlier, between 110 and 100 Ma, and possibly close to 103 Ma (Seton et al. 2012; Heine et al. 2013). Anyway, the breakup did not separate abruptly and synchronously everywhere the two areas; for an indeterminate period of time, the separation zone likely acted as a filter route, as suggested by the peculiar environment (presence of amounts of evaporites; Burke and Sengör 1988; Chaboureau et al. 2012). Gorscak and O'Connor (2016), and Gorsack et al. (2017) also showed that the diachronic opening of the South Atlantic from the South to the North (i.e., northward rifting) led to a South–North faunal provinciality of the Cretaceous African faunas. Possible intermittent links between South America and Africa by island chains extending across an already opened South Atlantic were discussed by various authors (see Oliveira et al. 2009; Ezcurra and Agnolín 2012).

The Fauna of the Island Africa

After separation from South America, began a period of African isolation that lasted about 80 million years. The Arabo-African plate was a gigantic island, i.e., the Island Africa, geographically set apart from other continents (Fig. 10.1). Its fossil record confirms strong biotic endemicity, but it also shows that it was not fully isolated biologically.

The Fate of the Native Fauna (Fig. 10.2)

Most of the native taxa did not clearly expand in Africa or even became extinct, leaving a poor fauna with low diversity and low competition pressure in the Island Africa. Among the above-mentioned clades, the Atlanticanura and Acrodonta widely diversified. However, the African component of the Atlanticanura has remained less diversified than representatives in other continents. Acrodontan lizards (today "agamids" plus chamaeleonids) have had a somewhat peculiar history. They were present in Samafrica by the mid-Cretaceous (Apesteguía et al. 2016) but they did not survive at most the Campanian in South America (Simões et al. 2015) whereas they largely diversified in Africa.

Mammals illustrate the unsuccessful evolution of the native fauna. The Mesozoic mammals known in Africa, most from the early Cretaceous of the Anoual-Ksar Metlili locality (Morocco), are represented by rare haramiyidans (incl. family Hahnodontidae: Huttenlocker et al. 2018), a possible gondwanatherian, eutriconodonts, some "symmetrodontans" (stem Trechnotheria), stem cladotheres (non-tribosphenic therians) such as dryolestids and peramurans, and stem tribosphenic therians (Tribosphenida). They are predominated by African natives. All taxa identified as the native African mammals did not survive the K-Pg boundary, even if we do not know if they went extinct at this time or before. This is a noticeable difference with South America where at least two archaic Mesozoic taxa (South American natives) survived the K-Pg boundary into the Paleocene: the peligrotheriid dryolestoids and the sudamericid gondwanatheres. The paleobiogeographical origin of the African Mesozoic native mammals remains poorly understood. Some are possible Gondwanan vicariants (Gondwanatheria, ?Dryolestoidea [see Chornogubsky 2011]). Others may be early Pangean vicariants (Haramiyida, most Eutriconodonta, "symmetrodontans", ?Dryolestoidea, ?Peramura).

The Mesozoic African mammal taxa which might possibly belong to lineages crossing the K-Pg boundary are represented by the tribosphenic therians (tribosphenidans) *Tribotherium* and *Hypomylos* from the Anoual-Ksar Metlili locality (Morocco). They could belong to the stem groups of the Cenozoic eutherian and placentals (Kielan-Jaworowska 1992; Luo et al. 2001; Kielan-Jaworowska et al. 2004), and in particular of the afrotherians known in Africa. However, tribosphenidans are most likely of Laurasian origin (Luo et al. 2001; Kielan-Jaworowska et al. 2004), and are instead not native mammals in Africa. Alternatively, Sigogneau-Russell (1991) and Sigogneau-Russell et al. (2001) postulated the controversial view that the early African tribosphenic mammals such as *Tribotherium* found in Anoual-Ksar Metlili locality may indicate a "Gondwanan" origin of the Tribosphenida (including Theria).

The poor evolutionary success of most of the native African clades, and even the extinctions of several native groups rendered the African native assemblages markedly depauperate in the Late Cretaceous and Paleogene. The causes of such a *depauperate Island Africa* remain poorly understood.

The Island Africa: an Underpopulated Biotic Province?

During isolation, the fauna of Africa was poor and, more specifically, noticeable absences are observed. Several taxa, well represented elsewhere in Gondwanan continents, are lacking in Africa (Gheerbrant and Rage 2006), among them: dipnoans of the *Ptychoceratodus madagascariensis* group and *Atlantoceratodus* (Cione et al. 2007), carnotaurine (*sensu* Sereno et al. 2004) dinosaurs, chelid turtles, non-acrodontan iguanians, boine snakes, and australosphenidan mammals. The absence of the Gondwanan clade australosphenidans is peculiarly remarkable. Aside from animals, the tree *Nothofagus* is a good example of a widely distributed Gondwanan taxon lacking in Africa. Compared to the noticeable cosmopolitanism of faunas of other Gondwanan continents, the Island Africa appears peculiar, i.e., endemic, and depauperate.

An attempt at a quantitative comparison between African faunas and those from other continents provided contrasting results. It faced various problems: taxonomic conceptions different from author to author, presence of several taxa *incertae sedis* and stem groups not included in families, and uncertain stratigraphic positions of localities. However, the estimated numbers of species show that the African fauna was less diverse and less rich than that of South America, which is the last Gondwanan continent to separate from Africa. The relative paucity of African faunas with regard to those of other continents, which is quite well illustrated by the continental distribution and relative number of known Cretaceous and Paleogene vertebrate localities (Fig. 10.3), was already noted in several studies, e.g. Gheerbrant and Rage (2006), Krause et al. (2006), Gorscak et al. (2017). O'Connor et al. (2006) referred to this African pattern as the "African Gap", a notion taken up by Gorscak et al. (2017).

Part of the absences of Gondwanan taxa in Africa might result from selective extinctions (Figs. 10.2, 10.4, 10.5) which impoverished the African fauna more markedly than in the other continents; in addition, paucity was likely increased by the fact that various Gondwanan groups did not enter the Island Africa. A striking illustration of this African pattern is provided by clades that are present in South America, Madagascar and India but that are lacking in intervening Africa (i.e., most of the above cited groups; Krause et al. 1997; Rage 2003; Krause et al. 2006). Similarly, Woodburne and Case (1996), Goin et al. (2016) and Poropat et al. (2016) disclosed faunal affinities between South America and Australia, including Antarctica, but excluding Africa. These paleobiogeographical relationships correspond to the "Africa first" model of Sereno et al. (2004) (see also Krause et al. 2006).

Mammals from the Late Cretaceous offer one example of this African characteristic. Whereas only one undetermined species of mammal (known by one vertebra from the Cenomanian of Libya; Nessov et al. 1998; Rage and Cappetta 2002) is recorded for the whole Late Cretaceous period in Africa, a noticeable diversity of 24 species of Late Cretaceous mammals are found in South America (Rougier et al. 2011), mostly of archaic groups (Gondwanatheria and Dryolestoidea, especially the endemic radiation of Meriodiolestida that includes 16 described species). Even Madagascar has yielded a small diversity of Late Cretaceous endemic mammals in the locality of Berivotra, including one or two species of multituberculates, two gondwanatherians, and one zhelestid eutherian (Averianov et al. 2003; a species initially interpreted as a marsupial by Krause 2001). The Late Cretaceous fauna of India also yielded five eutherians (adapisoriculids, and one "condylarth"), one gondwanatherian, and one haramiyidan (Prasad 2012). As far as large "reptiles" are concerned, the diversity in Africa is not markedly lower than that in South America during the early Late Cretaceous, but it markedly fell during the latest Cretaceous. According to Krause et al. (2006) only two families remained in Africa during the Campanian-Mastrichtian, while 12 or 13 were present in South America during this

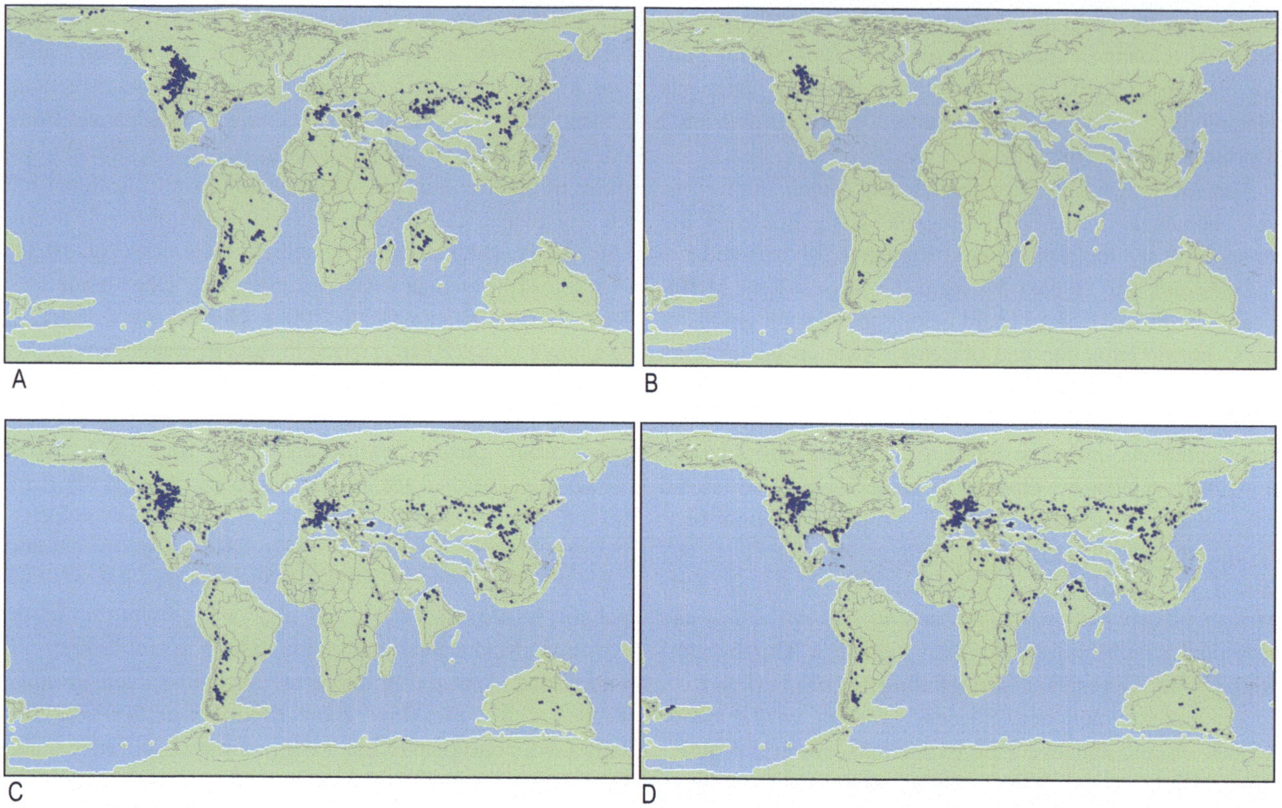

Fig. 10.3 Known Late Cretaceous and Paleogene localities with continental vertebrates. Data from Paleobiology Database and maps reconstructed on the Fossilworks website. A. Late Cretaceous terrestrial vertebrate localities. B. Late Cretaceous mammal localities. C. Paleogene terrestrial vertebrate localities. D. Paleogene mammal localities. The rarity of the Cretaceous and Paleogene vertebrate sites in Africa provide a good picture of the general poor African fossil record. It should be noted that maps A and B do not show (data missing in the Paleobiology Database) the Cenomanian Libyan locality Draa Ubari that yielded the only known late Cretaceous mammal in Africa (Nessov et al. 1998)

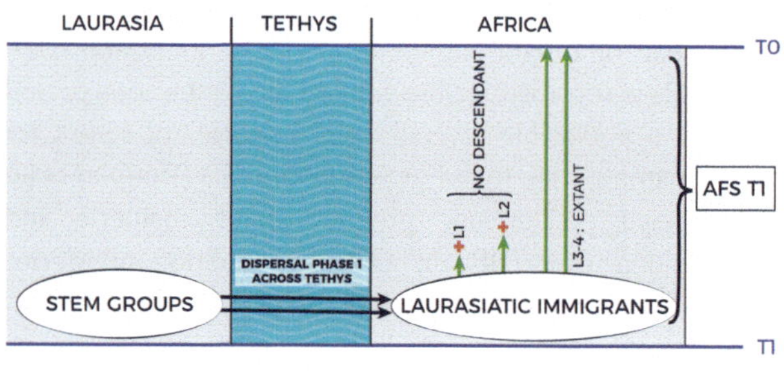

Fig. 10.4 Diagrammatic explanation of an African Faunal Stratum (AFS). At time T1, mammals of Laurasian origin crossed the Tethys and settled in Africa. From the various immigrant stem species several African lineages (L1-4) emerged and evolved. Some of them (L1-2) became extinct, others (L3-4) reach the Present (T0) and are part of extant African faunas. The old Laurasian immigrants and all issued lineages that evolved in Africa made up the AFS (here AFS T1)

time interval (six in Madagascar, nine in India). Small "reptiles" (squamates) cannot be taken into account because of the uncertain age of the rich locality Wadi Abu Hashim (Cenomanian or Maastrichtian). The diversity of amphibians also is low in the Campanian-Maastrichtian of Africa, it includes only one family (Gardner and Rage 2016) whereas at least five families are known from the same interval in South America (one in Madagascar and possibly seven in India; Rage et al. 2020).

However, it is presumed that the comparison of the Late Cretaceous African mammal faunas is biased by strong fossil gaps for this period in the African Island (Table 10.1). The comparison of the Paleogene mammal faunas of the African Island is likely more significant. From the published literature, we estimated that around 400 species of mammals are recorded for the whole Paleogene in Africa (Gheerbrant 2001, and EG's work in progress), which is noticeably low with respect to the Paleogene of North America (e.g., Woodburne 2004) and Europe (e.g., Aguilar and Michaux 1997). Estimation based on the Paleobiology Database (PaleoDB, Fossilworks website) also shows a much lower (half) Paleogene mammal species diversity in Africa than in South America (Table 10.1). In addition, it should be recalled that the number of Paleogene African mammals includes species that originated from Laurasian immigrants (e.g., marsupials, rodents, simiiformes, anthracotheres). Amphibians are also less diverse in Africa than in South America during the Paleogene. Four to six families were

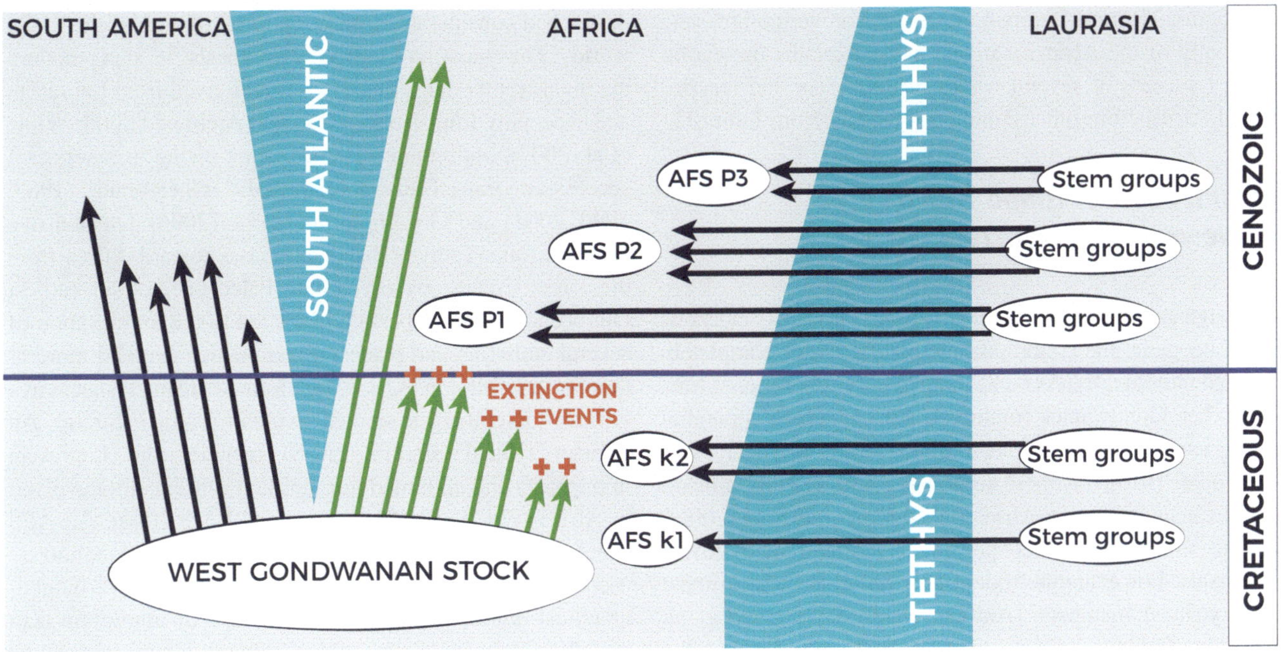

Fig. 10.5 Diagram summarizing the paleobiogeographical processes at the origin of the composition and evolutionary history of the continental fauna in the Island Africa. Several vicariant lineages evolved as the native fauna and were filtered over time by an unknown number extinctions, some being presumed local events (e.g., Late Cretaceous K1 and K2 extinction events in Fig. 10.2) and some being more global (K-Pg event); few of them (none mammal) crossed the K-Pg boundary, which resulted in vacant ecological niches. Aside from the native lineages, a significant number of old Laurasian immigrant lineages, most coming from Eurasia, crossed the Tethys and entered Africa at different times; they took advantage of local conditions and evolved forming successive African Faunal Strata during the Cretaceous (AFS K1-2) and the Paleogene (AFS P1-3). The Paleogene AFS (five to seven identified Paleogene AFS; see text and Gheerbrant and Rage 2006) are much better documented by the African fossil record than the Cretaceous AFS. These AFS evolved more or less successfully. Some include major African radiations that are still present in extant faunas, such as afrotherians, paenungulates (hyraxes and proboscideans), hyaenodonts, catarrhines primates, and hystricognath rodents. Others went extinct either in the Paleogene or in the Neogene (e.g., hyaenodonts). By the end of the Paleogene, the endemic African fauna resulted from the addition of the poor native fauna and taxa issued from the evolution of the different AFS (i.e., the descendants of old Laurasian immigrants). The extant African fauna results from the addition to this endemic fauna of the Eurasian taxa that colonized Africa after the closure of the Tethys at the Oligocene-Miocene transition

reported from South America, whereas three or four are known from Africa. Squamates are the only exception because they appear as diverse in Africa as in South America during the Paleogene; in both continents about ten families were recovered, it being understood that African squamates also likely include Eurasian immigrants.

Looking for the causes of the extinctions of the native African taxa leads to the search of unfavorable environmental conditions, more specifically to unfavorable climatic conditions. Unfortunately, there is no agreement about the climate of that time. Scotese et al. (1999) suggested that a large part of Africa was arid during the opening of South Atlantic (see also Burke and Sengor 1998), but Chaboureau et al. (2012) later reconstructed a wet climate for that time. Holz (2015) reported several episodes of warm greenhouse and "hothouse" at this time. Hothouse episodes were especially observed in the Aptian-Turonian. During the Late Turonian extremely hot temperatures (>40°C) were reported in equatorial Atlantic areas (Bice et al. 2006); the Aptian and Cenomanian are indeed also time of extinction events in continental faunas (Holz 2015). High temperatures also strongly affected plant evolution during mid-Cretaceous (Mays et al. 2017). Therefore, climatic perturbations took place during the Cretaceous, more specifically during the first stages of African isolation. However, the above studies did not identify events that would have affected Africa only and cannot account for the African selective extinctions. Only Russell and Paesler (2003) noted events restricted to Africa. A major perturbation of the environment took place on the continent from the Barremian-Aptian to the Santonian-Campanian and a strong disturbance of the environment in Africa impacted biotas. This event occurred during the separation of Africa from South America and at the times during which the native African fauna should have settled in the continent. However, it is acknowledged that the causes of low faunal density in Africa remain largely unknown.

Anyway, whatever the causes could have been, it clearly appears that the fauna was poor in the Island Africa, while at that time vertebrate faunas markedly developed on other continents. The poor diversity of the African vertebrates, and especially of the African natives, likely explains the evolutionary success of several waves of dispersals that reached Island Africa from other continents, mainly from Laurasia.

Colonization of Island Africa by Newcomers (Figs. 10.4, 10.5)

Although geographically an island during the Late Cretaceous and Paleogene, the Arabo-African plate was not completely isolated biologically. As far as biotic relations between Africa and other Gondwanan continents are concerned, potential filter routes connected Africa to South America, Madagascar and India. However, these routes did not entail significant Gondwanan arrivals in Africa (Gheerbrant and Rage 2006). On the contrary, they likely permitted or helped out of Africa dispersals. For example, rodent (caviomorphs) and primate (plathyrrines) mammals (Antoine et al. 2012; Bond et al. 2015) of African origin reached South America during the Eocene. In addition, based on molecular evidence, Vidal et al. (2008) suggested that amphisbaenian squamates dispersed similarly during the Eocene. Some dispersals, inferred from molecular phylogenies, also occurred between Africa and Madagascar; they involved reptilian taxa during the Late Cretaceous and "reptiles" and mammals (primates, afrotherians) during the Paleogene. Most of them were also likely out-of-Africa dispersals (Gheerbrant and Rage 2006). Dispersals of vertebrates from Africa to India were suggested but are not demonstrated. The dispersal of adapisoriculid mammals from India to Africa (Prasad et al. 2010) appears to be the only argued dispersal from a Gondwanan continent to Africa.

Paradoxically, while relations with other Gondwanan continents were very limited, interchanges with Eurasia were developed. They took place on the Mediterranean Tethyan Sill (Fig. 10.1), a discontinuous terrestrial route on fragments of continental shelves, and were controlled by variations of the sea-level (Gheerbrant and Rage 2006). Excepting the mammals with Laurasian affinities from the earliest Cretaceous of Anoual-Ksar Metlili (Sigogneau-Russell 1991; Sigogneau-Russell et al. 1998; Kielan-Jaworowska et al. 2004), Cretaceous interchanges mainly involved "reptiles" and occurred in both directions, but during the Paleogene the interchanges were dominated by dispersals of mammals toward Africa. Several Paleogene mammals that came from Laurasia evolved as more or less successful endemic lineages in Africa; some of them have become important components of extant faunas, but others went extinct. Conversely, we know only a few examples of Paleogene African mammal taxa that colonized Laurasia such as the hyaenodontids (Solé et al. 2009).

We here focus on mammals because their fossil history is comparatively well-documented and because their history in Africa had considerable impact on the evolution of the living world. The most consensual hypothesis is that modern mammals (eutherians, placentals) are derived from Laurasian and especially Eurasian stem groups (Archibald 2003; Wible et al. 2007), and entered Island Africa owing to a series of successive trans-Tethyan dispersals. Gheerbrant (1987, 1990, 2001) and Gheerbrant and Rage (2006) detected five to seven trans-Tethyan dispersal phases toward Africa from the latest Cretaceous or earliest Paleocene to the earliest Oligocene. These dispersal phases led to the immigration of several eutherian and placental taxa, which evolved more or less successfully in Africa and formed distinct successive African assemblages named African Faunal Strata. An African Faunal Stratum (AFS) includes the Laurasian immigrants and all issued lineages that subsequently evolved in Africa (Fig. 10.4) (Gheerbrant and Rage 2006). An AFS should not be confused with a biogeographical cohort as used by Gheerbrant (2001) and Beard (2016). A biogeographical cohort includes an assemblage of immigrant taxa issued from a single dispersal event and continental source (e.g., Gheerbrant 2001). An example is the biogeographical cohort of immigrant mammals of Asian origin which includes the simiiform primates and the hystricognath and possibly anomaluroid rodents (Marivaux et al. 2017; Coster et al. 2015) that entered Africa by the Middle/Late Eocene (Seiffert 2012; Beard 2016). An AFS can include a biogeographical cohort of immigrants of common geographical origin or different lineages with distinct origins (including those of the same biogeographical cohort).

It should be noted that in addition to the Paleogene AFS, earlier trans-Tethyan dispersals took place during the Mesozoic; the best known example corresponds to Laurasian mammal taxa from the Anoual Ksar Metlili locality such as the gobiconodontids (Sigogneau-Russell 2003) and the tribosphenidans. No other documented Cretaceous Laurasian dispersals in Africa are known by contrast to several reported Cretaceous out of Africa dispersals (e.g., Le Loeuff 1991; Pereda-Suberbiola 2009). As a consequence, the Jurassic and Cretaceous AFS (Fig. 10.5, AFS K1 and K2) remain nearly unknown. For the Late Cretaceous period, at least, this is probably related to the poor Mesozoic fossil record of Africa as mentioned above.

The most successful AFS significantly contributed to the modern fauna of Africa and, for some taxa, to the World fauna. For example, the earliest known placental dispersal that entered Africa (as documented by the Adrar Mgorn and Ouled Abdoun Paleocene faunas) was especially important since it was at the origin of the afrotherians (macroscelideans, tenrecoids, tubulidentates and paenungulates). It corresponds to an Early Paleocene or Late Cretaceous AFS that gave rise to some of the most spectacular endemic African radiations such as that of the hyracoid and proboscidean paenungulates (Fig. 10.6A). This AFS has been therefore prominent in the history of African faunas. In addition to extant taxa, this AFS includes extinct forms such as primitive eutherians (stem placentals: cimolestids, todralestids) (Gheerbrant 1992; Gheerbrant and Rage 2006). Molecular studies suggest a Late Cretaceous age for the afrotherian origin, but the exact age of the earliest African placentals remains a major challenge for current paleontological field research.

Subsequent dispersal phases and related AFS (Fig. 10.5, AFS P1-3) are at the origin of other extant famous endemic African mammal taxa. For instance, they explain the arrival from Asia of the simiiform primates in Africa in the Eocene (e.g., Beard 2004, 2016; Seiffert 2006, 2012), as part of the Lutetian/Bartonian or Bartonian/Priabonian dispersal phase and African Faunal Stratum. The first African simiiforms, from the Late Eocene, are basal to an endemic major African lineage from which evolved the cercopithecoids and hominoids and their stem-groups such as oligopithecids and propliopithecids (Fig. 10.6B) (Seiffert 2012). Earliest African rodents such as the zegdoumyids (Vianey-Liaud et al. 1994; Hartenberger et al. 1997) probably arrived earlier as part of the Thanetian/Ypresian AFS (Gheerbrant and Rage 2006).

A

B

Fig. 10.6 Illustration of some remarkable taxa of the endemic mammal fauna from the Island Africa. A. Reconstruction of the stem proboscidean *Phosphatherium* from the Earliest Eocene (55 Ma) of Morocco. *Phosphatherium* belongs to the oldest confirmed faunal stratum from Africa including placental mammals: the Late Cretaceous/early Paleocene African Faunal Stratum (AFS) which is at the origin of the afrotherians (see text). B. Reconstruction of *Aegyptopithecus*, an early African stem catarrhine simiiform primates known from the Early Oligocene (30 Ma) of the Fayum in Egypt. *Aegyptopithecus* belongs to a later AFS, the Bartonian/Priabonian African Faunal Stratum that leads to colonization of the Island Africa by simiiform primates, following a trans-Tethyan dispersal from Asia (reconstructions by C. Letenneur)

Some of the lineages that evolved from the Paleogene AFS remained strictly endemic (hyracoids, macroscelids, chrysochlorids, anomalurids, catarrhine primates) to the Island Africa before the contact with Eurasia, but others succeeded in dispersing overseas out of the Island Africa (hyaenodontids toward Laurasia, embrithopods in south-eastern Europe and Turkey, stem platyrrhine primates and caviomorph rodents toward South America, tenrecids and lemuriforms toward Madagascar).

Whatever the details of the history of the dispersals, the success of several taxa of Laurasian and mostly Eurasian origin that entered the Island Africa is striking and meaningful. It may be supposed that this flourishing evolutionary history was triggered and subsequently favored by the peculiar environmental conditions, including low diversity of vertebrate assemblages that prevailed in Africa at that time.

Africa as a Secondary Cradle

As stated above, Island Africa was depauperate after separation from South America. The poor diversity of the native fauna likely resulted in low competition pressure, and newcomers might have settled easily and subsequently evolved in favorable African ecosystems with free niches. This corresponds to the standard view about ecological opportunities, according to which an environment in which resources are underexploited, i.e., in vacant niches, favors adaptation of newcomers. Such conditions obviously occur in depauperate areas (Gheerbrant and Rage 2006; Gavrilets and Losos 2009). In addition, Gavrilets and Vose (2005) suggested that larger areas, which was the case of the Island Africa, favor colonization of niches and subsequent diversification of the emerging clade.

Therefore, both theoretical studies and paleontological observations converge to indicate that the Island Africa played the role of a *secondary cradle* for several major clades, including extant ones. Beard (2004) similarly pointed to this key evolutionary role of the insular Africa in his concept of the "African melting pot". Taxa of Laurasian origin entered Africa and found there favorable conditions, with a number of free ecological niches, to evolve locally, giving rise to new adaptive radiations. This process repopulated Africa, adding new taxa to the few native groups.

Aside from mammals, the role of Africa as a *secondary cradle* during isolation appears to have been significant for African testudinid turtles (Hofmeyr et al. 2016), an African unnamed clade of elapoid snakes (Kelly et al. 2009), and perhaps a large unnamed clade of colubrid snakes (McCartney et al. 2014). This list is certainly not exhaustive, but it emphasizes the role of Africa at *high taxonomic levels*.

The End of Isolation of Africa

It is generally acknowledged that massive continental faunal interchanges involving a large diversity of land mammals between Africa and Eurasia began during the Early Miocene, as a consequence of the closure of the Tethys (Cooke 1968; Coryndon and Savage 1973; Bernor et al. 1987; Van der Made 1999; Sen 2013). During the Late Oligocene, taxa that crossed the Tethys seaway were rare and the African mammal fauna remained widely endemic (Kappelman et al. 2003). The few known mammals crossing the Tethys during the Late Oligocene include an elephantoid from the Bugti Hills of Pakistan (Antoine et al. 2003) and a viverrid carnivore from Kenya (Rasmussen and Gutierrez 2009; Leakey et al. 2011). The latest trans-Tethyan interchanges with Africa likely took place in the area of the present contact when the two plates got closer together. This corresponds to the Iranian Route as reported by Gheerbrant and Rage (2006).

The disappearance of the marine barrier (closure of Tethys), replaced by a terrestrial contact between Africa and Eurasia ("*Gomphotherium* landbridge"; Rögl 1999), caused a marked geodispersal (*sensu* Lieberman and Eldredge 1996; Upchurch 2008) in both directions. This event took place during the Burdigalian (Bernor et al. 1987; Van der Made 1999; Mein 1999; Rögl 1999). Van der Made (1999) and Sen (2013) more precisely suggested a date close to 20 Ma for the beginning of these interchanges. This event, which shaped the present mammal fauna of Africa, may be labelled the *Great Old World Interchange* (GOWI; Gheerbrant 2001) by analogy to the younger *Great American Interchange* (Stehli and Webb 1985; Woodburne 2010). It marks the end of isolation and endemicity in Africa, which is the end of the Island Africa.

The subsequent evolution of the endemic African fauna after the *Great Old World Interchange* (GOWI) is characterized by the survival and even success of a number of major African lineages in extant faunas. Among mammals they include the afrotherians, the primates (catarrhines and lemuriforms), and several rodents (hystricognaths and anomaluroids). The extant afrotherians includes six placental orders, among which the endemic African ungulates (hyracoids, proboscideans and sirenians). Some of these endemic lineages widely diversified out of Africa after the GOWI, such as the hyracoids, sirenians, proboscideans, tubulidentates and primates (catarrhines and lemuriforms). This is remarkable with respect to the evolution of other endemic continental faunas which are generally affected by important extinctions. For instance, in South America, which also experimented a long geographical isolation south of the Tethys, none of the ungulate orders (notoungulates, litopterns, astrapotheres, etc.) survived. However, other endemic South American lineages also successfully evolved in extant faunas (marsupials, xenarthrans, platyrrhines primates, caviomorph rodents), and even outside South America such as in North America (opossums) and in Australia (australidelphians, the main extant marsupial radiation). Of note, the platyrrhines primates and caviomorph rodents which represent major South American endemic extant clades are of African origin. Extinctions obviously also strongly affected the endemic African fauna after the closure of Tethys; among mammals this concerns many hyracoids (medium and large sized species), all the embrithopods, several lineages of primates and rodents, hyaenodontids, and ptolemaiids. However, the success of several other African endemic taxa *out of Africa*, especially the primates (catarrhines and lemuriforms), proboscideans, sirenians and hystricognath rodents, is a notable trait of the African evolution. This relative African success might be related in part to the peculiar biogeographical history of Africa. The Island Africa is characterized by repeated trans-Tethyan biotic relations with Laurasia (Fig. 10.5), which resulted in contacts between faunas that sustained an evolutionary competitive and adaptive dynamism among groups then living in Africa. It might account for the survival and success of several old African groups in living faunas.

The trans-Tethyan faunal relations of the Island Africa with Eurasia significantly contributed to the biogeographical pattern of the modern Old World Realm that was definitely established by the closure of Tethys. The modern Old World faunal pattern, which largely results from a mix of African plus Eurasian faunas, has early biogeographical roots dating from the time of the trans-Tethyan dispersals.

Conclusions

From approximately the end of the Early Cretaceous to the Early Miocene, i.e., during about 80 million years, Africa was geographically isolated as a gigantic island. After the beginning of isolation and for poorly known reasons, the native continental fauna of Africa was poor, notably with respect to South America, the last continent linked to Africa; it resulted in the impression of a *depauperate Island Africa*. After the Cretaceous, the three key periods in vertebrate evolution in the Island Africa (Cretaceous-Paleogene (K-Pg), Paleocene-Eocene and Eocene-Oligocene) are either poorly documented (K-Pg event) or not marked by significant faunal turnovers in Africa, as far as we know. For instance, the mammal faunas from the Ouarzazate basin (Morocco) show faunal continuity at the Paleocene-Eocene boundary for several lineages of insectivoran-grade eutherian mammals (Gheerbrant et al. 1998). The Eocene-Oligocene biotic changes in Africa are best documented by the Fayum faunas in Egypt. The Fayum mammal assemblages, and especially the primates, show some changes at the Eocene-Oligocene boundary (e.g., Seiffert 2006, 2007, 2012), but those changes remain moderate with respect to, for instance, the faunal evolution in Europe that is marked at the same time by the "Grande Coupure" faunal revolution. The main changes in the African Paleogene faunas are in fact marked by immigrations of new Laurasian lineages.

Despite geographic isolation, intermittent and discontinuous emerged routes indeed permitted several Laurasian and especially Eurasian continental taxa to enter the Island Africa. Surprisingly, there is very little evidence of immigrations from other Gondwanan continents (Gheerbrant and Rage 2006). In the Island Africa, various Laurasian taxa of mainly Eurasian origin found favorable environmental conditions (low competition pressure, vacant ecological niches) and they settled and evolved successfully locally. Some of them gave rise to remarkable endemic adaptive radiations, which repopulated Africa in addition to the remaining native taxa, and they constitute important clades of the extant African fauna (e.g., afrotherians, catarrhine primates, hystricognath rodents) (Fig. 10.6). Thus, Island Africa acted as a *secondary cradle,* boosting the evolution of stem groups that came from Eurasia. One of the main issues of the peculiar history of Africa is indeed that all Cenozoic African mammals (including living ones) are derived from Laurasian stem groups, that evolved locally in several AFS (Figs. 10.4, 10.5).

The evolutionary history of African faunas is also characterized by the relative success of the endemic taxa, in particular after the *Great Old World Interchange* that followed the establishment of the definite terrestrial contact with Eurasia by the Early Miocene. Not only a significant number of endemic African clades survived up to today, but some of them went out of Africa and spread successfully over other continents. This African evolutionary success was likely sustained by the precocious contacts between faunas of the Island Africa and of Eurasia. These early Eurasian-African trans-Tethyan relations prefigure the paleobiogeographic pattern of the modern Old World fauna. They constitute the peri-Tethyan roots of the modern Old World fauna which was established after the closure of the Tethys.

Acknowledgements Jean-Claude Rage, who initiated this paper, sadly suddenly passed away after submission of our manuscript. The notion of a depauperate native African fauna was an old idea of Jean-Claude that stimulated the writing of this paper. EG dedicates posthumously this work to the honour and memory of our dear colleague and friend, Jean Claude. The authors dedicate this paper to Professor Ashok Sahni whose activity in India has significantly impacted our knowledge of Gondwanan vertebrates. We thank R. Patnaik and G.V.R. Prasad for having invited us to contribute to this volume. We thank the editors and the three reviewers for their very useful comments and corrections which helped to improve our manuscript. Thanks are due also to C. Letenneur (CR2P, MNHN) who made the Figs. 10.2, 10.4, 10.5 and 10.6.

References

Aguilar, J. P., & Michaux, J. (Eds.). (1997). *BiochroM'97. Mémoires et Travaux de l'Ecole Pratique des Hautes Etudes, Institut de Montpellier* 21.

Antoine, P. O., Marivaux, L., Croft, D. A., Billet, G., Ganerød, M., Jaramillo, C., et al. (2012). Middle Eocene rodents from Peruvian Amazonia reveal the pattern and timing of caviomorph origins and biogeography. *Proceedings of the Royal Society B, 279,* 1319–1326.

Antoine, P. O., Welcomme, J. L., Marivaux, L., Baloch, I., Benammi, M., & Tassy, P. (2003). First record of Paleogene Elephantoidea (Mammalia, Proboscidea) from the Bugti Hills of Pakistan. *Journal of Vertebrate Paleontology, 23,* 977–980.

Apesteguía, S., Daza, J. D., Simões, T. R., & Rage, J.-C. (2016). The first iguanian lizard from the Mesozoic of Africa. *Royal Society Open Science, 3,* 160462.

Archibald, J. D. (2003). Timing and biogeography of the eutherian radiation: Fossils and molecules compared. *Molecular Phylogenetics and Evolution, 28,* 350–359.

Averianov, A. O., Archibald, J. D., & Martin, T. (2003). Placental nature of the alleged marsupial from the Cretaceous of Madagascar. *Acta Palaeontologica Polonica, 48,* 149–151.

Beard, K. C. (2004). *The hunt for the dawn monkey: Unearthing the origins of monkeys, apes, and humans.* Berkeley: University of California Press.

Beard, K. C. (2016). Out of Asia: Anthropoid origins and the colonization of Africa. *Annual Review of Anthropology, 45,* 199–213.

Bernor, R. L., Brunet, M., Ginsburg, L., Mein, P., Pickford, M., Rögl, F., et al. (1987). A consideration of some major topics concerning old world Miocene mammalian chronology, migrations and paleogeography. *Geobios, 20,* 431–439.

Bice, K. L., Birgel, D., Meyers, P. A., Dahl, K. A., Hinrichs, K. U., & Norris, R. D. (2006). A multiple proxy and model study of Cretaceous upper ocean temperatures and atmospheric CO2 concentrations. *Paleoceanography, 21,* PA2002.

Bond, M., Tejedor, M. F., Campbell, K. E., Jr., Chornogubsky, L., Novo, N., & Goin, F. (2015). Eocene primates of South America and the African origins of new world monkeys. *Nature, 520,* 538–541.

Burke, K., & Sengör, A. M. C. (1988). Ten metre global sea-level change associated with South Atlantic Aptian salt deposition. *Marine Geology, 83,* 309–312.

Carvalho, M. S. S., & Maysey, J. G. (2008). New occurrence of *Mawsonia* (Sarcopterygii: Actinistia) from the Early Cretaceous of the Sanfranciscana Basin, Minas Gerais, southeastern Brazil. In L.

Cavin, A. Longbottom & M. Richter (Eds.), *Fishes and the break-up of pangaea. Geological society, London, Special publications, 295*, pp. 109–144.

Chaboureau, A. C., Donnadieu, Y., Sepulchre, P., Robin, C., Guillocheau, F., & Rohais, S. (2012). The Aptian evaporites of the South Atlantic: a climatic paradox? *Climate of the Past, 8*, 1047–1058.

Chornogubsky, L. (2011). New remains of the dryolestoid mammal *Leonardus cuspidatus* from the Los Alamitos Formation (Late Cretaceous, Argentina). *Paläontologische Zeitschrift, 85*, 343–350.

Cione, A. L., Gouiric, S., Goin, F., & Poiré, D. (2007). *Atlantoceratodus*, a new genus of lungfish from the Upper Cretaceous of South America and Africa. *Revista del Museo de La Plata, Paleontologia, 10*, 1–12.

Cooke, H. B. S. (1968). The fossil mammal fauna of Africa. *Quarterly Review of Biology, 43*, 234–264.

Coryndon, S. C., & Savage, R. J. G. (1973). The origin and affinities of African mammal faunas. In N. F. Hughes (Ed.), *Organisms and continents through time. Palaeontological Association of London, Special Papers in Palaeontology, 12*, pp. 121–135.

Coster, P. M. C., Beard, K. C., Salem, M. J., Chaimanee, Y., & Jaeger, J.-J. (2015). New fossils from the Paleogene of central Libya illuminate the evolutionary history of endemic African anomaluroid rodents. *Frontiers in Earth Science, 3*(56), 51–55.

Dos Reis, M., Inoue, J., Hasegawa, M., Asher, R. J., Donoghue, P. C., & Yang, Z. (2012). Phylogenomic datasets provide both precision and accuracy in estimating the timescale of placental mammal phylogeny. *Proceedings of the Royal Society B, 279*, 3491.

Eberle, J., Fabrizi, S., Lago, P., & Ahrens, D. (2017). A historical biogeography of megadiverse Sericini—another story "out of Africa"? *Cladistics, 33*, 183–197.

Elewa, A. M. T. (2017). Ostracod provincialism and migration as a response to movements of earth's plates: Cretaceous-Paleogene ostracods of West Africa, North Africa and the Middle East. *Journal of African Earth Sciences, 134*, 92–105.

Ezcurra, M. D., & Agnolín, F. L. (2012). A new global paleobiogeographical model for the late Mesozoic and early Tertiary. *Systematic Biology, 61*, 553e566.

Fossilworks website. Retrieved January 2018, from http://fossilworks.org.

Frazão, A., da Silva, H. R., & de Moraes Russo, C. A. (2015). The Gondwana breakup and the history of the Atlantic and Indian Oceans unveils two new clades for early neobatrachian diversification. *PLoS ONE, 10*, e0143926.

Gaffney, E. S., Tong, H., & Meylan, P. A. (2006). Evolution of the side-necked turtles: The families Bothremydidae, Euraxemydidae, and Araripemydidae. *Bulletin of the American Museum of Natural History, 300*, 1–318.

Gardner, J. D., & Rage, J.-C. (2016). The fossil record of lissamphibians from Africa, Madagascar, and the Arabian plate. In J. D. Gardner & T. Přikryl (Eds.), *Contributions in honour of Zbyněk Roček. Palaeobiodiversity and Palaeoenvironments, 96*, 169–220.

Gavrilets, S., & Losos, J. B. (2009). Adaptive radiation: Contrasting theory with data. *Science, 323*, 732–737.

Gavrilets, S., & Vose, A. (2005). Dynamic patterns of adaptive radiation. *Proceedings of the National Academy of Sciences of the USA, 102*, 18040–18045.

Gheerbrant, E. (1987). Les Vertébrés continentaux de l'Adrar Mgorn (Maroc, Paléocène); une dispersion de mammifères transtéthysienne aux environs de la limite Mesozoïque/Cénozoïque? *Geodinamica Acta, 1*, 233–246.

Gheerbrant, E. (1990). On the early biogeographical history of the African placentals. *Historical Biology, 4*, 107–116.

Gheerbrant, E. (1992). Les mammifères paléocènes du Bassin d'Ouarzazate (Maroc). I. Introduction générale et Palaeoryctidae. *Palaeontographica A, 224*, 67–132.

Gheerbrant, E. (2001). Histoire paléobiogéographique des mammifères paléogènes du domaine arabo-africain (vol. I). Base de données et listes fauniques des mammifères du Paléogène arabo-africain et bibliographie (Vol. II). Rapport d'activités (Vol. III). Mémoire d'Habilitation à Diriger des Recherches, Université Pierre et Marie Curie, Paris.

Gheerbrant, E., & Rage, J.-C. (2006). Paleobiogeography of Africa: How distinct from Gondwana and Laurasia? *Palaeogeography, Palaeoclimatology, Palaeoecology, 241*, 224–246.

Gheerbrant, E., Sudre, J., Sen, S., Abrial, C., Marandat, B., Sigé, B., et al. (1998). Nouvelles données sur les mammifères du Thanétien et de l'Yprésien du bassin d'Ouarzazate (Maroc) et leur contexte stratigraphique. *Palaeovertebrata, 27*, 155–202.

Goin, F. J., Woodburne, M. O., Zimicz, A. N., Martin, G. M., & Chornogubsky, L. (2016). *A brief history of South American metatherians*. Dordrecht: Springer.

Gorscak, E., & O'Connor, P. M. (2016). Time-calibrated models support congruency between Cretaceous continental rifting and titanosaurian evolutionary history. *Biology Letters, 12*, 20151047.

Gorscak, E., O'Connor, P. M., Roberts, E. M., & Stevens, N. J. (2017). The second titanosaurian (Dinosauria: Sauropoda) from the Middle Cretaceous Galula Formation, Southwestern Tanzania, with remarks on African titanosaurian diversity. *Journal of Vertebrate Paleontology, 37*, e1343250.

Gottfried, M. D., Stevens, N. J., Roberts, E. M., O'Connor, P. M., & Chami, R. (2009). A new Cretaceous lungfish from the Rukwa Rift Basin, Tanzania. *African Natural History, 5*, 31–36.

Granot, R., & Dyment, J. (2015). The Cretaceous opening of the South Atlantic Ocean. *Earth and Planetary Science Letters, 414*, 156–163.

Hartenberger, J.-L., Crochet, J. Y., Martinez, C., Feist, M., Godinot, M., Mannai Tayech, B., et al. (1997). Le gisement de mammifères de Chambi (Éocène, Tunisie centrale) dans son contexte géologique. Apport à la connaissance de l'évolution des mammifères en Afrique. *Mémoires et Travaux de l'Ecole Pratique des Hautes Etudes, Institut de Montpellier, 21*, 263–274.

Heine, C., Zoethout, J., & Müller, R. D. (2013). Kinematics of the South Atlantic rift. *Solid Earth, 4*, 215–253. https://doi.org/10.5194/se-4-215-2013.

Hofmeyr, M. D., Vamberger, M., Branch, W., Schleicher, A., & Daniels, S. R. (2016). Tortoise (Reptilia, Testudinidae) radiations in Southern Africa from the Eocene to the present. *Zoologica Scripta, 46*, 389–400.

Holz, M. (2015). Mesozoic paleogeography and paleoclimate–A discussion of the diverse greenhouse and hothouse conditions of an alien world. *Journal of South American Earth Sciences, 61*, 91–107.

Huttenlocker, A. K., Huttenlocker, A. D., Grossnickle, D. M., Kirkland, J. I., Schultz, J. A., & Luo, Z. X. (2018). Late-surviving stem mammal links the lowermost Cretaceous of North America and Gondwana. *Nature, 558*, 108–112.

Jacobs, L. L., Polcyn, M. J., Mateus, O., Schulp, A. S., Gonçalves, A. O., & Morais, M. L. (2016). Post-Gondwana Africa and the vertebrate history of the Angolan Atlantic coast. *Memoirs of Museum Victoria, 74*, 343–362.

Kappelman, J., Rasmussen, D. T., Sanders, W. S., Feseha, M., Bown, T., Copeland, P., et al. (2003). Oligocene mammals from Ethiopia and faunal exchange between Afro-Arabia and Eurasia. *Nature, 426*, 549–552.

Kelly, C. M. R., Barker, N. P., Villet, M. H., & Broadley, D. G. (2009). Phylogeny, biogeography and classification of the snake superfamily Elapoidea: A rapid radiation in the late Eocene. *Cladistics, 25*, 38–63.

Kielan-Jaworowska, Z. (1992). Interrelationships of Mesozoic mammals. *Historical Biology, 6,* 185–201.

Kielan-Jaworowska, Z., Cifelli, R. L., & Luo, Z. X. (2004). *Mammals from the age of dinosaurs*. New York: Columbia University Press.

Krause, D. W. (2001). Fossil molar from a Madagascan marsupial. *Nature, 412,* 497–498.

Krause, D. W., O'Connor, P. M., Curry Rogers, K., Sampson, S. D., Buckley, G. A., & Rogers, R. R. (2006). Late Cretaceous terrestrial vertebrates from Madagascar: Implications for Latin American biogeography. *Annals of the Missouri Botanical Garden, 93,* 178–208.

Krause, D. W., Prasad, G. V. R., Koenigswald, W., Sahni, A., & Grine, F. E. (1997). Cosmopolitanism among Gondwanan Late Cretaceous mammals. *Nature, 390,* 504–507.

Leakey, M., Grossman, A., Gutiérrez, M., & Fleagle, J. G. (2011). Faunal change in the Turkana Basin during the late Oligocene and Miocene. *Evolutionary Anthropology, 20,* 238–253.

Le Loeuff, J. (1991). The Campano-Maastrichtian vertebrate faunas from southern Europe and their relationships with other faunas in the world; paleobiogeographical implications. *Cretaceous Research, 12,* 93–114.

Le Loeuff, J., Métais, E., Dutheil, D. B., Rubino, J. L., Buffetaut, E., Lafont, F., et al. (2010). An Early Cretaceous vertebrate assemblage from the Cabao Formation of NW Libya. *Geological Magazine, 147,* 750–759.

Lieberman, B. S., & Eldredge, N. (1996). Trilobite biogeography in the Middle Devonian: Geological processes and analytical methods. *Paleobiology, 22,* 66–79.

Luo, Z. X., Cifelli, R. L., & Kielan-Jaworowska, Z. (2001). Dual origin of tribosphenic mammals. *Nature, 409,* 53–57.

Marivaux, L., Adnet, S., Benammi, M., Tabuce, R., & Benammi, M. (2017). Anomaluroid rodents from the earliest Oligocene of Dakhla, Morocco, reveal the long-lived and morphologically conservative pattern of the Anomaluridae and Non - Anomaluridae during the Tertiary in Africa. *Journal of Systematic Palaeontology, 15*(7), 539–569. https://doi.org/10.1080/14772019.2016.1206977.

Mays, C., Cantrill, D. J., & Bevitt, J. J. (2017). Polar wildfires and conifer serotiny during the Cretaceous global hothouse. *Geology, 45,* 1119–1122.

McCartney, J. A., Stevens, N. J., & O'Connor, P. M. (2014). The earliest colubroid-dominated snake fauna from Africa: Perspectives from the Late Oligocene Nsungwe Formation of Southwestern Tanzania. *PLoS ONE, 9,* e90415.

Mein, P. (1999). Biochronologie et phases de dispersion chez les vertébrés cénozoïques. *Bulletin de la Société Géologique de France, 170,* 195–204.

Nessov, L. A., Zhegallo, V. I., & Averianov, A. O. (1998). A new locality of Late Cretaceous snakes, mammals and other vertebrates in Africa (western Libya). *Annales de Paléontologie, 84,* 265–274.

Novas, F. E., de Valais, S., Vickers-Rich, P., & Rich, T. (2005). A large Cretaceous theropod from Patagonia, Argentina, and the evolution of carcharodontosaurids. *Naturwissenschaften, 92,* 226–230.

O'Connor, P. M., Gottfried, M. D., Stevens, N. J., Roberts, E. M., Ngasala, S., Kapilima, S., et al. (2006). A new vertebrate fauna from the Cretaceous Red Sandstone Group, Rukwa Rift Basin, southwestern Tanzania. *Journal of African Earth Sciences, 44,* 277–288.

Oliveira, F. B., Molina, E. C., & Marroig, G. (2009). Paleogeography of the South Atlantic: A route for primates and rodents into the new world? In P. A. Garber, A. Estrada, J. C. Bicca-Marques, E. W. Heymann, & K. B. Strier (Eds.), *South American primates: Comparative perspectives in the study of behavior, ecology, and conservation* (pp. 55–68). New York: Springer.

Parmar, V., Prasad, G. V. R., & Kumar, D. (2015). An overview of Jurassic mammalian fauna of India. In Y. Zhang, S. Z. Wu, & G. Sun (Eds.), *12th Symposium on Mesozoic Terrestrial Ecosystems* (pp. 17–20). Shenyang: China.

Pereda-Suberbiola, X. (2009). Biogeographical affinities of Late Cretaceous continental tetrapods of Europe: A review. *Bulletin de la Société géologique de France, 180,* 57–71.

Poropat, S. F., Mannion, P. D., Upchurch, P., Hocknull, S. A., Kear, B. P., Kundrát, M., et al. (2016). New Australian sauropods shed light on Cretaceous dinosaur paleobiogeography. *Scientific Reports, 6,* 34467.

Prasad, G. V. R. (2012). Vertebrate biodiversity of the Deccan volcanic province of India: A review. *Bulletin de la Société géologique de France, 183*(6), 597–610.

Prasad, G. V. R., Verma, O., Gheerbrant, E., Goswami, A., Khosla, A., Parmar, V., et al. (2010). First mammal evidence from the Late Cretaceous of India for biotic dispersal between India and Africa at the KT transition. *Comptes Rendus Palevol, 9,* 63–71.

Rage, J.-C. (2003). Relationships of the Malagasy fauna during the Late Cretaceous: Northern or Southern routes? *Acta Palaeontologica Polonica, 48,* 661–662.

Rage, J.-C., & Cappetta, H. (2002). Vertebrates from the Cenomanian, and the geological age of the Draa Ubari fauna (Libya). *Annales de Paléontologie, 88,* 79–84.

Rage, J.-C., Prasad, G. V. R., Verma, O., Khosla, A., & Parmar, V. (2020). Anuran lissamphibians and squamate reptiles from Upper Cretaceous (Maastrichtian) Deccan intertrappean sites in Central India, with a review of a lissamphibian and squamate diversity in the northward drifting Indian Plate. In G. V. R. Prasad & R. Patnaik (Eds.), *Biological consequences of plate tectonics: New perspectives on* post-Gondwana *break-up* (pp. 99–121). Cham: Springer.

Rasmussen, D. T., & Gutierrez, M. (2009). A mammalian fauna from the late Oligocene of northwestern Kenya. *Palaeontographica A, 288,* 1–52.

Rögl, F. (1999). Mediterranean and Paratethys. Facts and hypotheses of an Oligocene to Miocene paleogeography. *Geologica Carpathica, 50,* 339–349.

Rougier, G., Gaetano, L., Drury, B. R., Colella, R., Gomez, R. O., & Arango, P. N. (2011). A review of the Mesozoic mammalian record of South America. In J. Calvo, J. Porfiri, B. González Riga & D. Dos Santos (Eds.), *Paleontologia y dinosaurios desde America Latina* (pp. 195–214). Mendoza: Ediunc.

Russell, D. A., & Paesler, M. A. (2003). Environments of mid-Cretaceous Saharan dinosaurs. *Cretaceous Research, 24,* 569–588.

Scotese, C. R. (2016). PALEOMAP PaleoAtlas for GPlates and the PaleoData plotter program, PALEOMAP project. http://www.earthbyte.org/paleomap-paleoatlas-for-gplates/.

Scotese, C. R., Boucot, A. J., & McKerrow, W. S. (1999). Gondwanan paleogeography and paleoclimatology. *Journal of African Earth Sciences, 28,* 99–114.

Seiffert, E. R. (2006). Revised age estimates for the later Paleogene mammal faunas of Egypt and Oman. *Proceedings of the National Academy of Sciences, 103,* 5000–5005.

Seiffert, E. R. (2007). Evolution and extinction of Afro-Arabian primates near the Eocene-Oligocene boundary. *Folia Primatologica, 78,* 314–327.

Seiffert, E. R. (2012). Early primate evolution in Afro-Arabia. *Evolutionary Anthropology, 21,* 239–253.

Sen, S. (2013). Dispersal of African mammals in Eurasia during the Cenozoic: Ways and whys. *Geobios, 46,* 159–172.

Sereno, P. C., & ElShafie, S. J. (2013). A new long-necked turtle, *Laganemys tenerensis* (Pleurodira: Araripemydidae), from the Elrhaz Formation (Aptian–Albian) of Niger. In D. Brinkman, P. Holroyd, & J. Gardner (Eds.), *Morphology and evolution of turtles* (pp. 215–250). Dordrecht: Springer.

Sereno, P. C., Wilson, J. A., & Conrad, J. L. (2004). New dinosaurs link southern landmasses in the mid-Cretaceous. *Proceedings of the Royal Society B, 271,* 1325–1330.

Seton, M., Müller, R. D., Zahirovic, S., Gaina, C., Torsvik, T., Shephard, G., et al. (2012). Global continental and ocean basin reconstructions since 200 Ma. *Earth-Science Reviews, 113,* 212–270.

Sigogneau-Russell, D. (1991). Découverte du premier mammifère tribosphénique du Mésozoïque africain. *Comptes rendus de l'Académie des Sciences de Paris II, 313,* 1635–1640.

Sigogneau-Russell, D. (2003). Diversity of triconodont mammals from the Early Cretaceous of North Africa—Affinities of the amphilestids. *Palaeovertebrata, 32,* 27–55.

Sigogneau-Russell, D., Evans, S. E., Levine, J. F., & Russell, D. A. (1998). The Early Cretaceous microvertebrate locality of Anoual, Morocco: A glimpse at the small vertebrate assemblages of Africa, pp. 177–181. In S. G. Lucas, J. I. Kirkland & J. W. Estep (Eds.), *Lower and middle Cretaceous terrestrial ecosystems. New Mexico Museum of Natural History and Science Bulletin, 14,* 177–181.

Sigogneau-Russell, D., Hooker, J. J., & Ensom, P. C. (2001). The oldest tribosphenic mammal from Laurasia (Purbeck Limestone Group, Berriasian, Cretaceous, UK) and its bearing on the 'dual origin' of Tribosphenida. *Comptes rendus de l'Académie des Sciences, 333,* 141–147.

Simões, T. R., Wilner, E., Caldwell, M. W., Weinschütz, L. C., & Kellner, A. W. A. (2015). A stem acrodontan lizard in the Cretaceous of Brazil revises early lizard evolution in Gondwana. *Nature Communications, 6,* 8149.

Solé, F., Gheerbrant, E., Amaghzaz, M., & Bouya, B. (2009). Further evidence of the African antiquity of hyaenodontid ("Creodonta", Mammalia) evolution. *Zoological Journal of the Linnean Society, 156*(4), 827–846.

Stehli, F. G., & Webb, S. D. (Eds.). (1985). *The Great American Biotic Interchange*. New York: Plenum Press.

Turner, A. H. (2006). Osteology and phylogeny of a new species of *Araripesuchus* (Crocodyliformes: Mesoeucrocodylia) from the Late Cretaceous of Madagascar. *Historical Biology, 18,* 255–369.

Upchurch, P. (2008). Gondwanan break-up: Legacies of a lost world? *Trends in Ecology and Evolution, 23,* 229–236.

Upchurch, P., Hunn, C. A., & Norman, D. B. (2002). An analysis of dinosaurian biogeography: Evidence for the existence of vicariance and dispersal patterns caused by geological events. *Proceedings of the Royal Society B, 269,* 613–621.

van der Made, J. (1999). Intercontinental relationship Europe-Africa and the Indian Subcontinent. In G. Rössner & K. Heissig (Eds.), *The Miocene land mammals of Europe* (pp. 457–472). Munich: F. Pfeil.

Vianey-Liaud, M., Jaeger, J.-J., Hartenberger, J.-L., & Mahboubi, M. (1994). Les rongeurs de l'Eocène d'Afrique nord-occidentale (Glib Zegdou (Algérie) et Chambi (Tunisie)) et l'origine des Anomaluridae. *Palaeovertebrata, 23,* 93–118.

Vidal, N., Azvolinsky, A., Cruaud, C., & Hedges, S. B. (2008). Origin of tropical American burrowing reptiles by transatlantic rafting. *Biology Letters, 4,* 115–118.

Wible, J. R., Rougier, G. W., Novacek, M. J., & Asher, R. J. (2007). Cretaceous eutherians and Laurasian origin for placental mammals near the K/T boundary. *Nature, 447,* 1003–1006.

Woodburne, M. O. (Ed.). (2004). *Late Cretaceous and Cenozoic mammals of North America*. New York: Columbia University Press.

Woodburne, M. O. (2010). The great American biotic interchange: Dispersals, tectonics, climate, sea level and holding pens. *Journal of Mammalian Evolution, 17,* 245–264.

Woodburne, M. O., & Case, J. A. (1996). Dispersal, vicariance, and the Late Cretaceous to Early Tertiary land mammal biogeography from South America to Australia. *Journal of Mammalian Evolution, 3,* 121–161.

Chapter 11
Contrasting Biomarker Signatures of Cretaceous and Early Paleogene Coal-Bearing Sediments During the Northward Flight of India

Swagata Paul and Suryendu Dutta

Abstract The northward drift of the Indian plate from Early Cretaceous until it collided with the Eurasian plate in Early Eocene provides a unique opportunity to track its biotic evolution. The distribution of higher plant terpenoids of Early Cretaceous and Eocene coal- and lignite-bearing sequences of Kutch Basin, western India have been investigated to understand the change in floral evolution during the northward journey of the Indian plate. The higher plant terpenoids of studied Early Cretaceous samples are characterized by sesquiterpenoids and high abundance of diterpenoids. Angiosperm-derived triterpenoids are not detected here. Such terpenoid pattern clearly depicts the prevalence of gymnosperm, especially conifer dominated forests during that time in western India. The presence of diterpenoids such as norisopimarane, labdane, *ent*-beyerane, isopimarane, abietane, norabietane and phyllocladane in the aliphatic fraction and 16,17-bisnordehydroabietane, 16,17-bisnorsimonellite, dehydroabietane, 18-norabieta-6,8,11,13-tetraene, 1,2,3,4-tetrahydroretene, simonellite, C_1 simonellite, totarane, sempervirane, retene and 2-methylretene in the aromatic fraction suggest that the Early Cretaceous coal forming plants in western India belonged to Araucariaceae and Podocarpaceae families of conifer. The terpenoids of Eocene lignite-bearing sequences are represented by the significant occurrence of angiosperm-derived triterpenoids and the complete absence of diterpenoids. The angiosperm-derived triterpenoids are amyrin-derivatives and the compounds are des-A-olean-12-ene; A-neo-oleana-3(5),2-diene; olean-2,13(18)-diene; olean-2,18-diene; ursa-2,12-diene; olean-12-ene; olean-18-ene etc. in aliphatic fraction and 24,25–dinoroleana-1,3,5(10),12-tetraene; 24,25,26,27-tetranor-oleana-1,3,5(10),6,8,11,13-heptaene; 24,25,26,27,28-pentanor-ursa-1,3,5(10),6,8,11,13,15,17-nonaene; C-ring cleaved aromatized triterpenoids etc. in aromatic fraction. Variation in biomarker assemblages from Early Cretaceous to Early Paleogene depicts a remarkable shift of floral association during the voyage of Indian continent. The reduction of coniferous vegetation and the proliferation of angiosperms during the Early Paleogene are observed in the present study.

Keywords Early Cretaceous • Eocene • Floral Diversity • Diterpenoids • Conifers • Triterpenoids • Angiosperms

Introduction

After separation from Gondwana, the northward journey of the Indian subcontinent from its original position towards Asia initiated during Early Cretaceous and continued to move as an isolated Gondwanan landmass until it collided with the Eurasian plate in Early Eocene ($\sim$52 Ma) (Scotese 2001, 2014). During its northward journey, Indian plate had covered 9000 km in 160 million years by changing its position from southern Subtropical Arid Climatic Belt to the present day position via Equatorial Belt (Chatterjee et al. 2013). It entered into the humid equatorial region during Paleocene and occupied very close to the present day position during Oligocene ($\sim$30 Ma) (Chatterjee et al. 2013). The aim of the present study is to trace the floral diversity of the Indian continent during its northward journey by analyzing the biomarker assemblages of low-rank coals and associated sediments using gas chromatography-mass spectrometry (GC-MS) technique. Biomarkers or molecular fossils with taxonomic specificity preserve valuable information about the paleochemotaxonomic assignment of their source plants which offers an opportunity to reconstruct the ancient vegetation. Application of chemotaxonomy to the coals and fossil resins has gained considerable attention since the last few decades for elucidating the contemporary floral association (Simoneit et al. 1986; Otto et al. 1997,

S. Paul · S. Dutta (✉)
Department of Earth Sciences, Indian Institute of Technology Bombay, Mumbai, 400076, India
e-mail: s.dutta@iitb.ac.in

G. V. R. Prasad and R. Patnaik (eds.), *Biological Consequences of Plate Tectonics: New Perspectives on Post-Gondwana Break-up—A Tribute to Ashok Sahni*,
Vertebrate Paleobiology and Paleoanthropology Series, https://doi.org/10.1007/978-3-030-49753-8_11

2002; Bechtel et al. 2002, 2005; Stefanova et al. 2005; Dutta et al. 2011b; Shi et al. 2014; Paul et al. 2015; Paul and Dutta 2016 among others). Terpenoids, the widely distributed chemical constituents of plant kingdom, serve as the most important tool in chemotaxonomy. Terpenoids, which are commonly associated with sedimentary rocks and crude oils (geoterpenoids), are the degraded products of their biological precursors of living plant origin (bioterpenoids). The geoterpenoids are widely used for studying paleofloral diversity as they yield valuable evidence about their botanical source by retaining the basic chemical structures of their respective bioterpenoids during diagenesis (Mills et al. 1984; Simoneit 1986; Otto et al. 2002, 2003; Pereira et al. 2009; Dutta et al. 2009, 2011a). Various types of terpenoids viz. monoterpenoids, sesquiterpenoids, diterpenoids and triterpenoids are commonly found in nature and previous studies have suggested that the higher plant triterpenoids are derived from angiosperms, while diterpenoids commonly occur in gymnosperms, especially conifers (Otto and Wilde 2001).

In the present article, we investigate the higher plant terpenoids in Early Cretaceous and Eocene coal- and lignite-bearing sequences from Kutch Basin, western India and relate these terpenoids with particular plant taxa in order to evaluate the contemporary floral association. Here, we also focus on the diagenetic history of preserved terpenoids during the formation of coal from plant biomass.

Material and Methods

Sampling

Early Cretaceous coals and associated shale samples were collected from Guneri mine section (23°47′6″N, 69°50′24″ E), Trambau mine section (23°9′12″N, 69°43′42″E) and Chawad river section (23°27′42″N, 69°13′36″E) of Kutch Basin, western India (Fig. 11.1A). Very thin seams or streaks and pockets of coal were found to be associated with these sections and these unproductive coal seams were quarried for the local consumption. Previous studies suggested that all the three sections belonged to Early Cretaceous Bhuj Formation of Kutch Basin (Prakash 1985; Chakraborty et al. 2011). The lithologs of Trambau and Guneri mines and Chawad river sections are given in Fig. 11.1B–D.

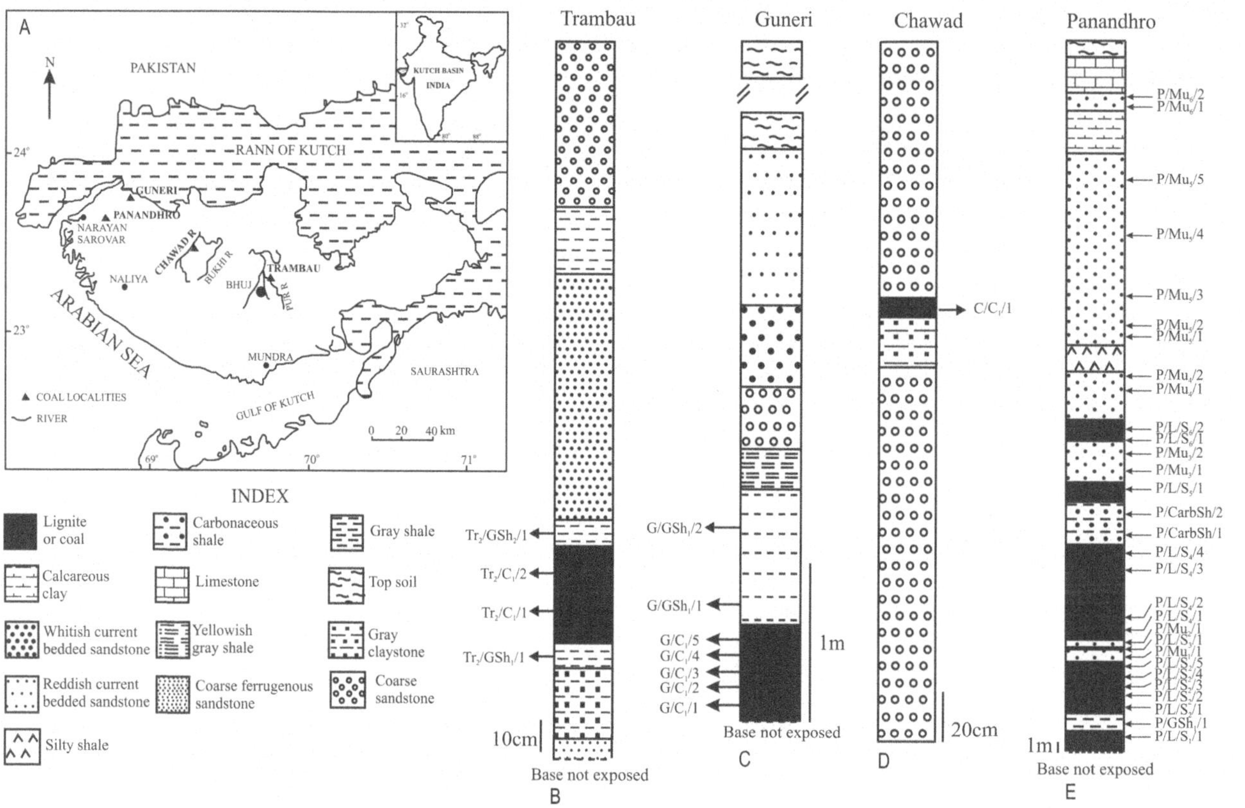

Fig. 11.1 Location map and lithological units of studied Early Cretaceous and Eocene sections from Kutch Basin, western India: (**A**) Geological map of Kutch Basin showing the location of studied sections (Trambau mine, Guneri mine, Chawad river section and Panandhro mine); (**B**) litholog of Trambau coal mine; (**C**) litholog of Guneri coal mine; (**D**) litholog of Chawad river section; (**E**) litholog of Panandhro lignite mine

Eocene samples were obtained from Panandhro lignite mine (23°39′56″N and 68°44′55″E), Kutch Basin, western India (Fig. 11.1A). Based on the microfossil assemblages, the age of the upper part of the mine section was assigned as Middle Eocene (Saraswati et al. 2014). The thickness of the Panandhro mine was about 56 m and 6 lignite seams of 0.5 m to 9 m were exposed in the lower part of the mine. The lignite bands were separated by the intermediate mudrocks, gray and carbonaceous shales. The upper part was represented by the mudstone and limestone horizons. The litho-column of Panandhro mine section is shown in Fig. 11.1E.

Analytical Method

Organic geochemical study has been carried out for all collected Early Cretaceous and Eocene samples. The powdered sediment samples (more than 100 mesh) were sonicated with dichloromethane and methanol (9:1; v:v) solution for bitumen extraction. The extracted product was treated with pentane overnight for asphaltene separation. After asphaltene precipitation, the remaining portion was dried and passed through silica gel (100–200 mesh) for separation of aliphatic and aromatic fractions using organic solvents of different polarity (column chromatography technique). First, *n*-hexane was used to extract aliphatic fraction which was followed by the separation of aromatic portion using a mixture of *n*-hexane and dichloromethane (4:1; v:v). The aliphatic and aromatic fractions were examined by gas chromatography-mass spectrometry (GC-MS) technique. The experimental conditions are described in Paul et al. (2015).

Results

Terpenoid Composition of Early Cretaceous Sediments

Aliphatic Fraction

Terpenoids in aliphatic fraction contain sesquiterpenoids, diterpenoids and bacteria-derived triterpenoids. Partial mass chromatogram at *m/z* 123 of the representative Early Cretaceous sample is characterised by the presence of sesquiterpenoids and diterpenoids (Fig. 11.2). Drimane based sesquiterpenoids (peaks: **2–8**) are detected in the studied coals and associated sediments. Aliphatic

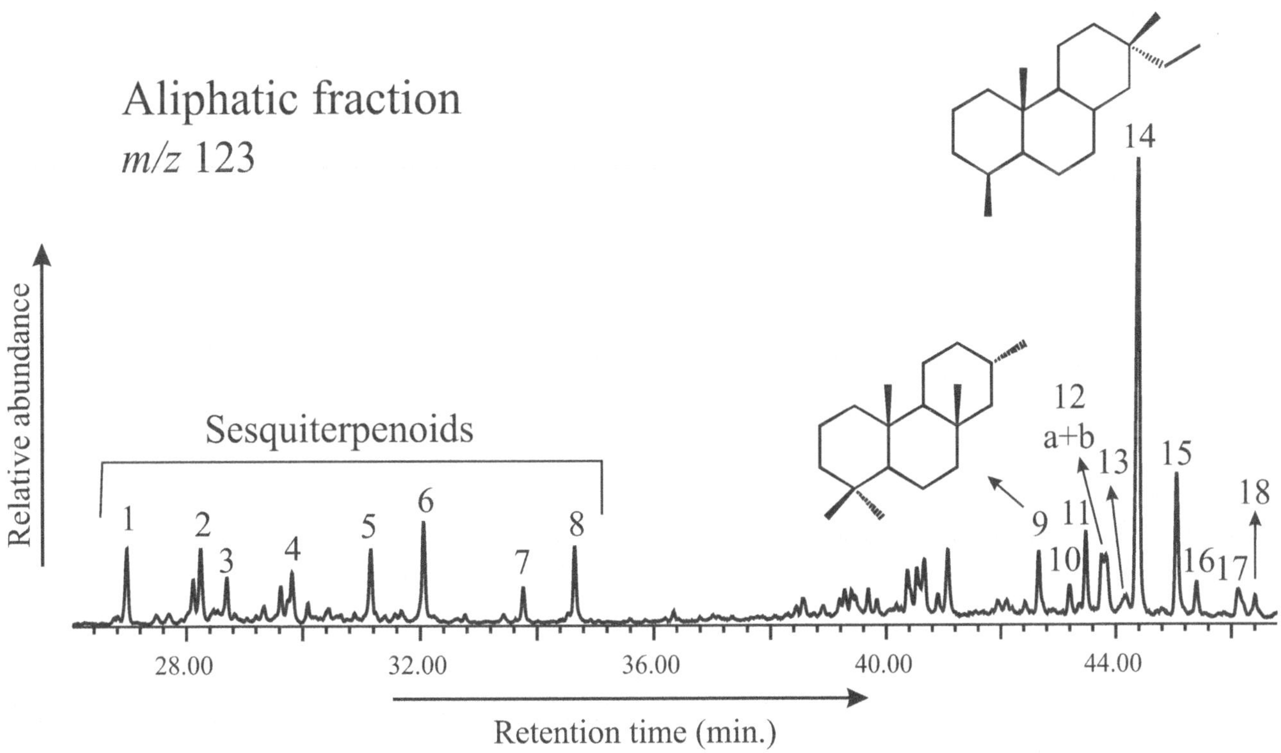

Fig. 11.2 Partial mass chromatogram at *m/z* 123 of aliphatic fraction of the representative Early Cretaceous sample ($Tr_2/C_1/2$) from Kutch Basin, western India is showing the distribution of sesquiterpenoids and diterpenoids. **1**: unknown C_{14} bicyclic sesquiterpenoids; **2**, **3**, **4**: nordrimane isomers; **5**: rearranged drimane; **6**: 8β(H)-drimane; **7**, **8**: homodrimanes; **9**: C_{19} tricyclic terpane; **10**: 8α(H)-labdane; **11**: 4α(H)-19-norisopimarane; **12**: C_{19} tricyclic terpane **(a)** + 8β(H)-labdane **(b)**; **13**: norabietane isomer; **14**: 4β(H)-18-norisopimarane; **15**: *ent*-beyerane; **16**: isopimarane; **17**: abietane; **18**: 16α(H)-phyllocladane

diterpenoids are characterised by 8α(H)- and 8β(H)-labdane (peaks: **10**, **12b**), 4α(H)-19- and 4β(H)-18-norisopimarane (peaks: **11**, **14**), norabietane (peak: **13**), *ent*-beyerane (peak: **15**), isopimarane (peak: **16**), abietane (peak: **17**) and 16α(H)-phyllocladane (peak: **18**). Partial mass chromatogram at *m/z* 191 represents the distribution of bacteria-derived hopanoid class compounds. Higher plant triterpenoids viz. amyrin derivatives are not detected here.

Aromatic Fraction

Diterpenoids and triterpenoids are detected in aromatic fraction of Early Cretaceous samples. The distribution of diterpenoids is given in the Figs. 11.3 and 11.4. The major aromatic diterpenoids are represented by 16,17-bisnordehydroabietane (peak: **19**), 16,17-bisnorsimonellite (peak: **21**), 18-norabieta-6,8,11,13-tetraene (peak: **22**), dehydroabietane (peak: **23**), 1,2,3,4-tetrahydroretene (peak: **24**), simonellite (peak: **26**), diaromatic totarane (peak: **27**), sempervirane (peak: **29**), retene (peak: **30**), C_1 simonellite (peak: **31**) and 2-methylretene (peak: **32**). A-ring opened isohexyl alkylnaphthalenes such as 2-Methyl-1-(4-methyl pentyl)naphthalene (peak: **20**); 6-isopropyl-2-methyl-1-(4-methylpentyl)naphthalene (peak: **25**) and its methylated counterpart 6-isopropyl-2,4-dimethyl-1-(4-methylpentyl) naphthalene (peak: **28**) are also recorded in the analyzed samples. Such compounds are considered to be derived from higher plant diterpenoids in sediments by aromatization, rearrangement and methylation processes (Ellis et al. 1996; Bastow et al. 1999). Aromatized triterpenoids contain cyclized monoaromatic hopanoids or benzohopanes. Two groups of benzohopanes viz. hopanoids cyclized at the C-16 and C-20 positions are found here. The samples lack amyrin-derived aromatized triterpenoids. Benzohopanes are thought to be formed from hopanoid precursor during early diagenesis by the dehydration and cyclization and subsequent degradation of the extended hopane side chain (Ostroukhov et al. 1983; Hussler et al. 1984; Cong et al. 2015).

Terpenoid Composition of Eocene Sediments

Aliphatic Fraction

The major aliphatic triterpenoids including hopanoids and non-hopanoid group of compounds are identified by analyzing the partial mass chromatograms at *m/z* 191 (Fig. 11.5). Diterpenoids are completely absent in both aliphatic and aromatic fractions in the studied Eocene samples. Hopanoid series is characterised by C_{27} to C_{32} hopanes and C_{28} to C_{30} hopenes (Fig. 11.5). Panandhro samples show high abundance of unknown C_{27} to C_{29} terpane. Monounsaturated members of hopane family such as an unidentified C_{30} hopene (peak: **54**); hop-17(21)-ene (peak: **51**); 30-norneohop-13(18)-ene (peak: **47**); 28,30-bisnorhop-13 (18)-ene (peak: **41**) are abundant in the samples. C_{30}-C_{32} Hopanes are represented by αβ-hopane (peak: **53**), ββ-hopane (peak: **58**), αβ-homohopanes (peaks: **56**, **57**), ββ-homohopane (peak: **59**) and ββ-bishomohopane (peak: **60**). Demethylated hopanes (C_{27}-C_{29}) such as ββ-30-norhopane (peak: **55**), βα-30-norhopane (peak: **52**), bisnorhopane (peaks: **42**, **46**), 17α(H)-22,29,30-trisnorhopane (peak: **39**) and 17β(H)-22,29,30-trisnorhopane (peak: **40**) are abundantly present in the samples. Bacteriohopantetrol (C_{35} tetrahydroxibacteriohopane) is considered as precursor of hopanoids. This compound is found in many prokaryotic bacteria in nature (Ourisson et al. 1979; Böcker et al. 2013). Hopane can also be originated from aerobic bacteria, fungi and cryptogams derived 3-desoxyhopanes in minor amount (Bechtel et al. 2001; Böcker et al. 2013). Hopenes are the precursors of hopanes and their presence indicates the immature organic matter (del Rio et al. 1992).

Non-hopanoid triterpenoid group is characterised by the angiosperm-derived higher plant triterpenoids. Aliphatic A-ring degraded triterpenoids and pentacyclic monoene and diene compounds are abundantly detected in the all studied samples and they are majorly related to the oleanane and ursane skeletons (Fig. 11.5). Aliphatic A-ring degraded compounds are identified as mono- and di-unsaturated des-A-olean-12-ene (peak: **33**); des-A-oleana-5(10),13(18)-diene (peak: **34**); des-A-urs-13(18)-ene (peak: **35**) and saturated des-A-lupane (peak: **36**). The non-hopanoid pentacyclic triterpenoids are represented by several diene compounds viz. oleana-2,13(18)-diene (peak: **43**), oleana-2,12-diene (peak: **44**), oleana-2,18-diene (peak: **45**), ursa-2,12-diene (peak: **50**) and monoene compounds such as —olean-12-ene (peak: **48**) and olean-18-ene (peak: **49**). Two peaks of A-ring contracted compounds such as A-neo-oleana-3(5),12-diene (peak: **37**) and A-neo-ursa-3 (5),12-diene (peak: **38**), having prominent fragments at *m/z* 135, 175, 271, 365 are found in all samples in moderate amount (ten Haven et al. 1992a).

Aromatic Fraction

The distribution of aromatic triterpenoids is given in the Fig. 11.6. Monoaromatic triterpenoids are represented by 24,25-dinoroleana(ursa)-1,3,5(10),12-tetraene (peaks: **66**, **67**, **68**, **71b**); dinor-oleana(ursa)-1,3,5(10)-triene (peak: **69**) and unknown A-ring aromatized compound with the base peak and molecular ion at *m/z* 169 and *m/z* 374, respectively (peak: **65**). C-ring cleaved des-A-compounds (peaks: **61**, **62**) having structural similarity with oleanane and lupane skeleton are

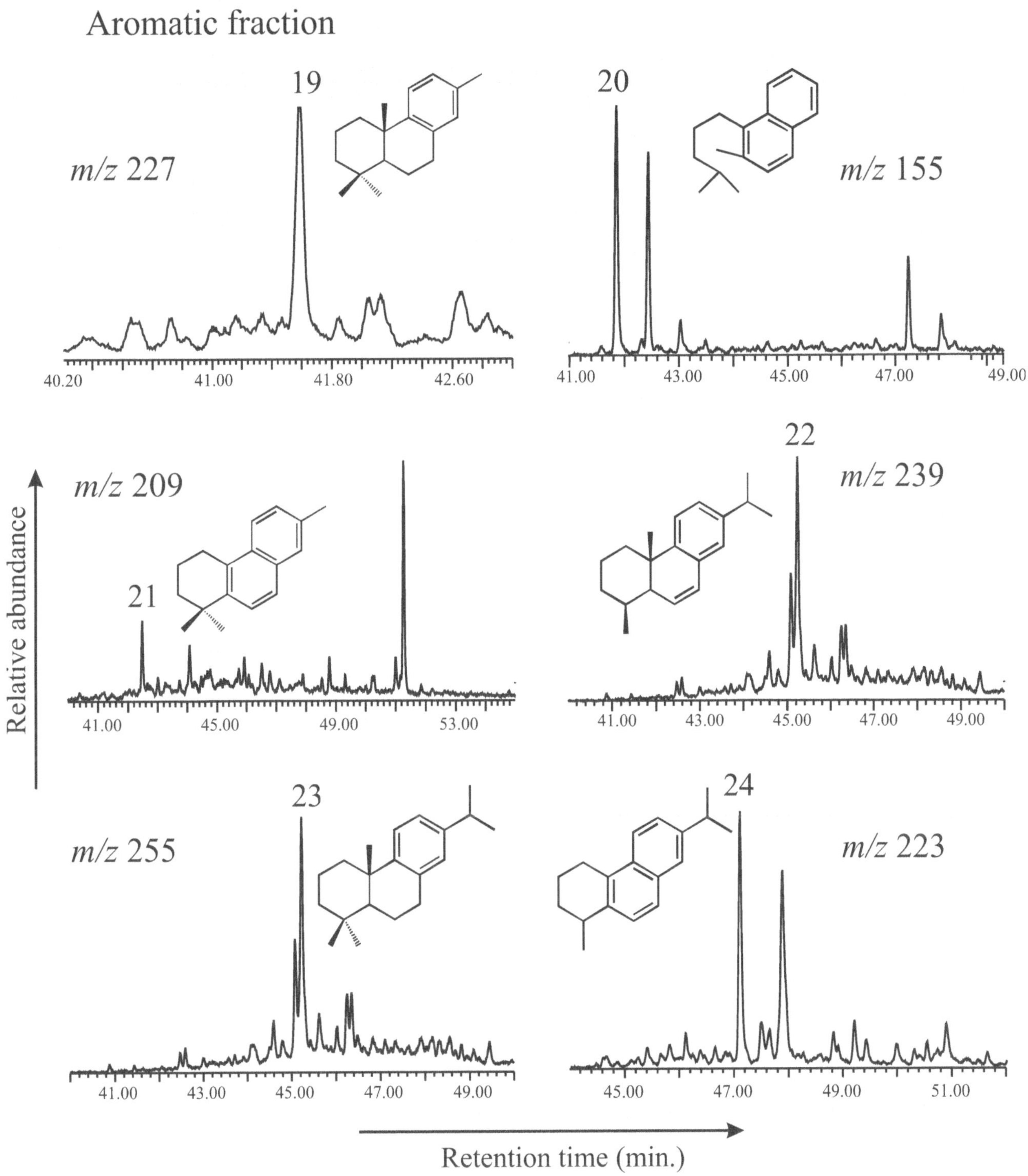

Fig. 11.3 Selected ion chromatograms at *m/z* 227, 155, 209, 239, 255 and 223 of aromatic fraction of the representative Early Cretaceous sample ($G/C_1/5$) from Kutch Basin, western India is showing the distribution of diterpenoids and A-ring opened isohexyl alkylnaphthalenes. **19**: 16,17-bisnordehydroabietane; **20**: 2-Methyl-1-(4-methylpentyl)naphthalene; **21**: 16,17-bisnorsimonellite; **22**: 18-norabieta-6,8,11,13-tetraene; **23**: dehydroabietane; **24**: 1,2,3,4-tetrahydroretene

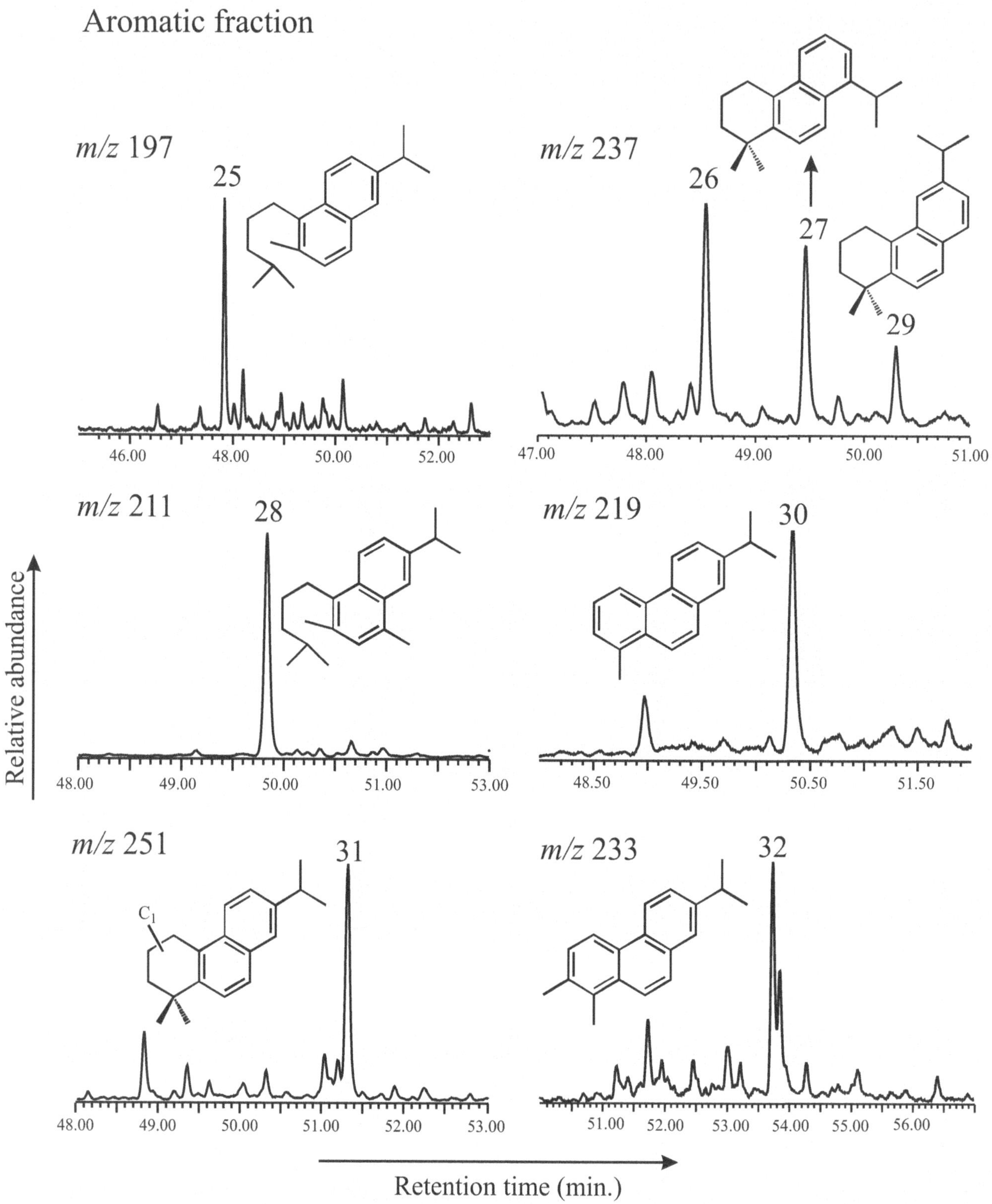

Fig. 11.4 Selected ion chromatograms at *m/z* 197, 237, 211, 219, 251 and 233 of aromatic fraction of the representative Early Cretaceous sample (G/C_1/5) from Kutch Basin, western India is showing the distribution of diterpenoids and A-ring opened isohexyl alkylnaphthalenes. **25**: 6-isopropyl-2-methyl-1-(4-methylpentyl) naphthalene; **26**: simonellite; **27**: diaromatic totarane; **28**: 6–isopropyl-2,4–dimethyl-1-(4-methylpentyl)naphthalene; **29**: sempervirane; **30**: retene; **31**: C_1 simonellite; **32**: 2-methylretene

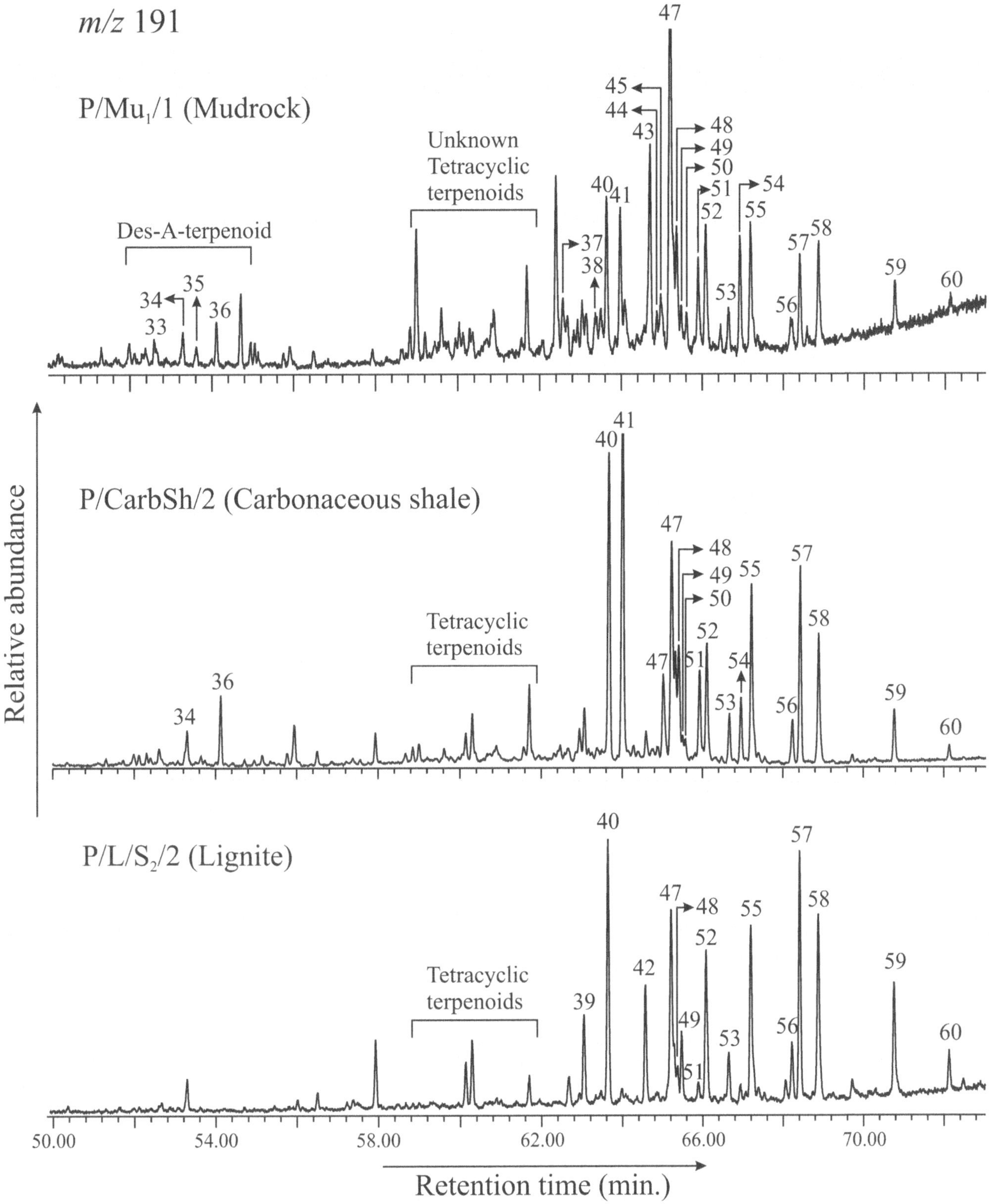

Fig. 11.5 Partial mass chromatograms at *m/z* 191 for the representative mudrock (P/Mu$_1$/1), carbonaceous shale (P/CarbSh/2) and lignite (P/L/S$_2$/2) showing the distribution of triterpenoids from Panandhro lignite mine, Kutch Basin, western India. **33**: des-A-olean-12-ene; **34**: des-A-oleana-5(10),13(18)-diene; **35**: des-A-urs-13(18)-ene; **36**: des-A-lupane; **37**: A-neo-oleana-3(5),12-diene; **38**: A-neo-ursa-3(5),12-diene; **39**: 17α(H)-22,29,30-trisnorhopane; **40**: 17β(H)-22,29,30-trisnorhopane; **41**: 28,30-bisnorhop-13(18)-ene; **42**: bisnorhopane; **43**: oleana-2,13(18)-diene; **44**: oleana-2,12-diene; **45**: oleana-2,18-diene; **46**: Bisnorhopane; **47**: 30-norneohop-13(18)-ene; **48**: olean-12-ene; **49**: olean-18-ene; **50**: ursa-2,12-diene; **51**: hop-17(21)-ene; **52**: βα-30-norhopane; **53**: αβ-hopane; **54**: C_{30} hopene; **55**: ββ-30-norhopane; **56**: αβ-homohopane (22S); **57**: αβ-homohopane (22R); **58**: ββ-hopane; **59**: ββ-homohopane; **60**: ββ-bishomohopane

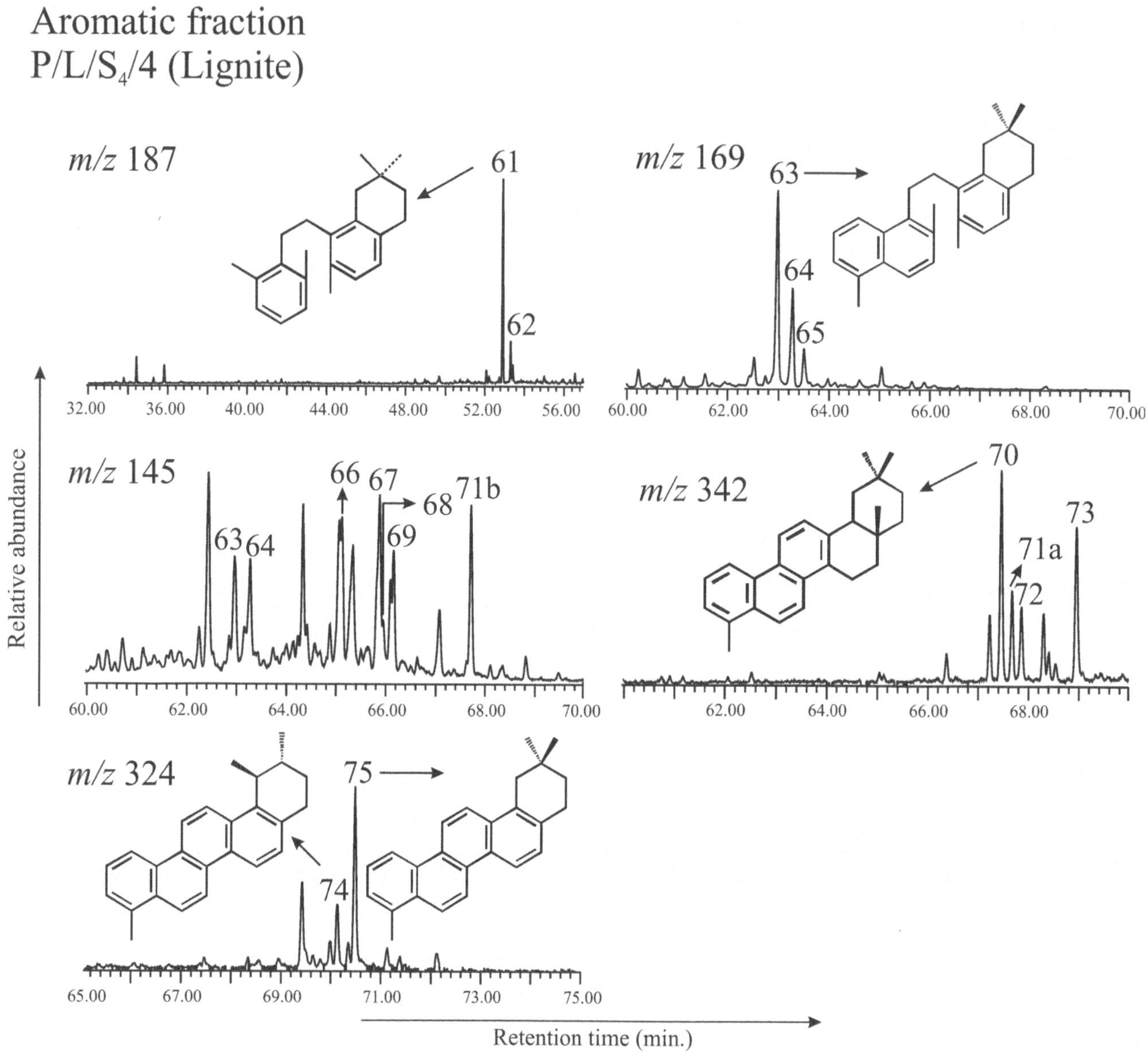

Fig. 11.6 Selected mass chromatogram of *m/z* 187, 145, 169, 342 and 324 for aromatic fraction of representative lignite sample (P/L/S_4/4) from Panandhro lignite mine, Kutch Basin, western India. **61**: diaromatic C-ring cleaved des-A-triterpenoid with oleanane skeleton; **62**: diaromatic C-ring cleaved des-A-triterpenoid with lupane skeleton; **63**: triaromatic C-ring cleaved triterpenoid; **64**: Isomer of **63**; **65**: unknown A-ring aromatized triterpenoid; **66**: 24,25-dinoroleana(ursa)-1,3,5(10),12-tetraene; **67**: 24,25-dinoroleana(ursa)-1,3,5(10),12-tetraene isomer; **68**: 24,25-dinorursa-1,3,5 (10),12-tetraene isomer; **69**: dinor-oleana(ursa)-1,3,5(10)-triene; **70**: 24,25,26,27-tetranor-oleana-1,3,5(10),6,8,11,13-heptaene; **71**: 24,25,26,27-tetranor-ursa-1,3,5(10),6,8,11,13-heptaene **(a)** + 24,25-dinoroleana(ursa)-1,3,5(10),12-tetraene isomer **(b)**; **72**: 24,25,26,27-tetranor-lupa-1,3,5 (10),6,8,11,13-heptaene; **73**: 24,25,26,27-tetranor-ursa-1,3,5(10),6,8,11,13-heptaene isomer; **74**: 24,25,26,27,28-pentanor-ursa-1,3,5 (10),6,8,11,13,15,17-nonaene; **75**: 24,25,26,27,28-pentanor-oleana-1,3,5(10),6,8,11,13,15,17-nonaene

recognized as diaromatic (B- and D-ring aromatized) compounds. Tetracyclic C-ring cleaved compounds with oleanane and ursane structure (peaks: **63, 64**) and pentacyclic 24,25,26,27-tetranor-oleana-1,3,5(10),6,11,13-heptaene (peak: **70**); 24,25,26,27-tetranor-ursa-1,3,5(10),6,8,11,13-heptaene (peaks: **71a, 73**) and 24,25,26,27-tetranor-lupa-1,3,5(10), 6,8,11,13-heptaene (peak: **72**) characterise the triaromatic triterpenoids. The observed pentacyclic triaromatic compounds show their base peak and molecular ion at *m/z* 342 with the prominent fragment at *m/z* 257. 24,25,26,27-Tetranor-lupa-

1,3,5(10),6,8,11,13-heptaene and 24,25,26,27-tetranor-ursa-1,3,5(10),6,8,11,13-heptaene show additional prominent fragments at *m/z* 299 and 327, respectively. 24,25,26,27,28-Pentanor-ursa-1,3,5(10),6,8,11,13,15,17-nonaene (peak: **74**) and 24,25,26,27,28-pentanor-oleana-1,3,5(10),6,8,11,13,15,17-nonaene (peak: **75**) are identified as tetraaromatic triterpenoids.

Discussion

Inferences from Early Cretaceous Terpenoids

The terpenoid compositions of Early Cretaceous coals and associated sediments from Kutch Basin are represented by the significant occurrence of diterpenoids and the absence of higher plant-derived triterpenoids. Such distribution is characteristic for gymnosperm, especially conifer-derived source (Pereira et al. 2009; Widodo et al. 2009). Conifer resins are considered as the source of diterpenoids (Pereira et al. 2009; Widodo et al. 2009). Conifer families viz. Pinaceae, Cupressaceae, Araucariaceae and Podocarpaceae could synthesize resins, among them extant Pinaceae and Araucariaceae produce resins in considerable amount (Langenheim 1995; Dutta et al. 2009, 2011a). In the present study, bicyclic, tricyclic and tetracyclic diterpenoids are recorded. Labdanes (peaks: **10**, **12b**) are found in the aliphatic fractions of the studied samples. Labdane class compounds have limited chemotaxonomic value as they are widely distributed in all modern conifer families (Otto and Wilde 2001; Pereira et al. 2009). The observed tricyclic diterpenoids except diaromatic totarane (peak: **27**) and sempervirane (peak: **29**) belong to regular abietane (peaks: **13**, **16**, **19**, **21**, **23**, **22**, **24**, **26**, **31**, **30**, **32**) and isopimarane (peaks: **11**, **14**, **16**) class of compounds and are the diagenetic derivatives of their biological precursors viz. abietic acid, dehydroabietic acid, dehydroabietol, isopimaric acid and isopimaradiene. The diagenetic pathways of aliphatic and aromatic tricyclic diterpenoids are given in the Fig. 11.7. Such precursors are the common constituents of extant conifer families and therefore, are considered as the non-source specific conifer biomarkers (Pereira et al. 2009). C_{20} Tetracyclic diterpenes are the most likely precursors of tetracyclic diterpanes such as *ent*-beyerane (peak: **15**) and 16α(H)-phyllocladane (peak: **18**) (Noble et al. 1985). The phyllocladene and phyllocladanol are the possible biological source of phyllocladane (Fig. 11.7) while beyer-15-ene (hibaene) would form beyerane by the reduction of their double bond during diagenesis (Noble et al. 1985; Otto et al. 1997). A-ring opened isohexyl alkylnaphthalenes such as 2-methyl-1-(4-methylpentyl)naphthalene (peak: **20**), 6-isopropyl-2-methyl-1-(4-methylpentyl)naphthalene (peak: **25**) and its methylated counterpart 6-isopropyl-2,4-dimethyl-1-(4-methylpentyl)naphthalene (peak: **28**) are detected in the present samples and are considered as the derivatives of diterpenoid precursors (Ellis et al. 1996). Laboratory experiment had shown that the above mentioned isohexyl alkylnaphthalenes were derived from phyllocladene (Ellis et al. 1996) (Fig. 11.7). These tetracyclic diterpenes are abundantly found in the leaf resins of all conifer families (Podocarpaceae, Cupressaceae and Araucariaceae) except Pinaceae (Noble et al. 1985; Otto and Wilde 2001; Pereira et al. 2009). Moreover, modern Pinaceae resins exclusively contain labdane and abietane type diterpenoids and lack phenolic abietanes and tetracyclic diterpanes (Otto and Wilde 2001; Otto and Simoneit 2002; Otto et al. 2002; Hautevelle et al. 2006; Pereira et al. 2009; Dutta et al. 2011a). Labdane derivatives are not detected in the aromatic fractions of the studied samples. Exclusive association of labdane and abietane, however, is not found in the studied samples which suggest the major contribution of Pinaceae into the source vegetation is unlikely. The tricyclic diterpenoids e.g., diaromatic totarane (peak: **27**) and sempervirane (peak: **29**) are possibly the diagenetic products of phenolic abietanes such as totarol and sempervirol, respectively (Otto et al. 1997). These two phenolic abietanes are commonly distributed in modern Cupressaceae and Podocarpaceae families (Otto and Wilde 2001; Cox et al. 2007; Stefanova and Simoneit 2008; Menor-Salván et al. 2010; Charrié-Duhaut et al. 2013). Again, sesquiterpenoids like derivatives of cedrane and cuparene are the proposed biomarkers for Cupressaceae family (Grantham and Douglas 1980; Shi et al. 2014), thus considered as important chemotaxonomic markers. These compounds, however, are not detected in the studied samples, implying that Cupressaceae did not contribute to the biomass. Therefore based on terpenoid assemblages, it can be interpreted that Early Cretaceous coal forming plants in western India belonged to Araucariaceae and/or Podocarpaceae conifer families.

Inferences from Eocene Terpenoids

The terpenoid composition of Eocene lignite-bearing sequence is represented by hopanoids (bacteria-derived triterpenoids) and higher plant triterpenoids. The higher plant triterpenoids are found in both aliphatic and aromatic fractions and are represented by oleanane, ursane and lupane related compounds. The presence of higher plant triterpenoids and the complete absence of diterpenoids are typical for angiosperm dominated sources. The identified triterpenoids are derived from C(3) oxygenated precursor compounds like α- and β-amyrin and lupeol during diagenesis (Trendel et al. 1989; ten Haven et al. 1992a; Jacob et al. 2007). Amyrins are

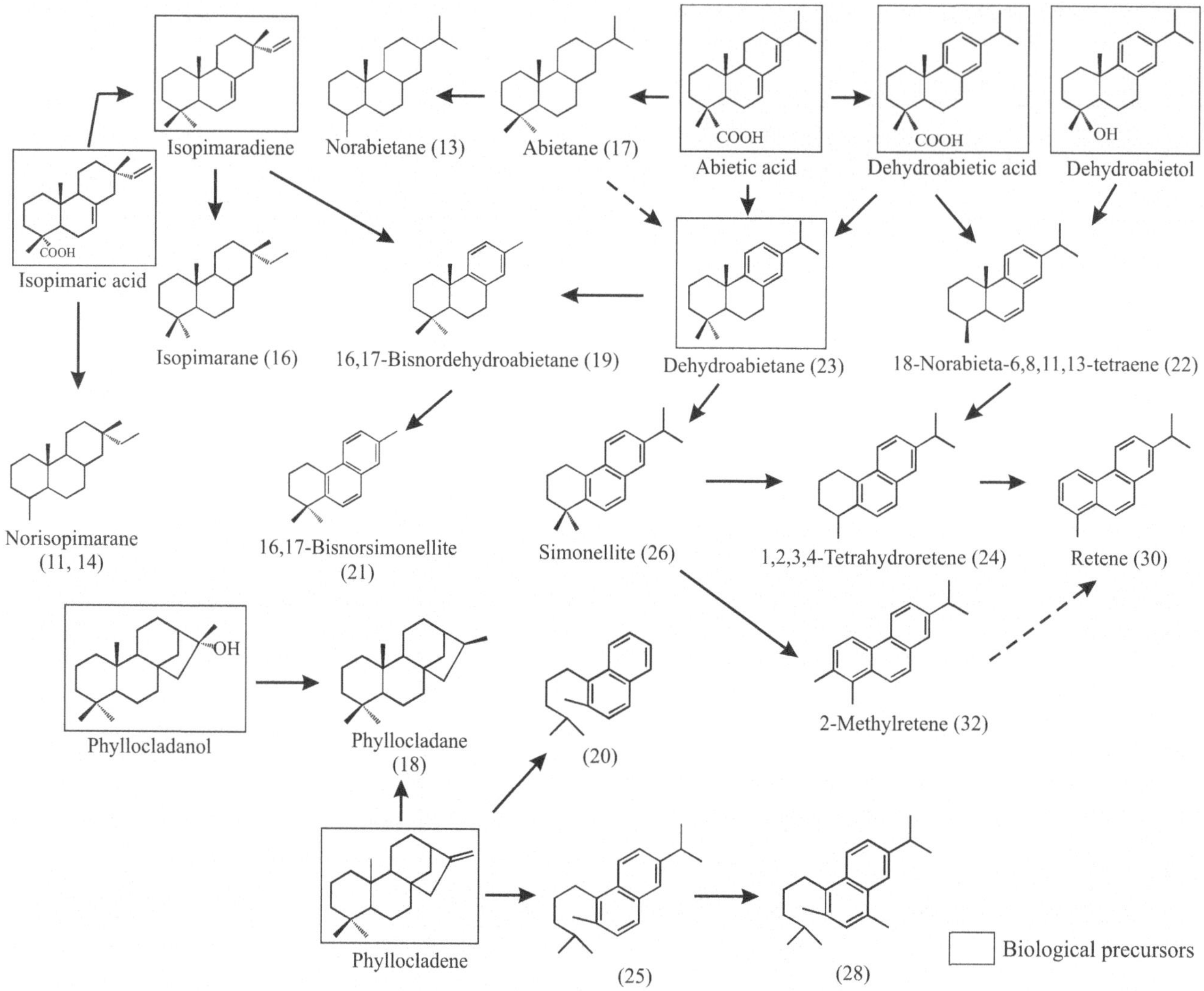

Fig. 11.7 Diagenetic pathways of aliphatic and aromatic diterpenoids in Early Cretaceous samples from Kutch Basin, western India (modified after Ellis et al. 1996; Otto et al. 1997, 2006, 2007; Otto and Simoneit 2001, 2002; Haberer et al. 2006; Hautevelle et al. 2006; Menor-Salván et al. 2010)

the important component of the root, wood, bark and leaf waxes of angiosperms (Bechtel et al. 2001). Loss of this oxygenated function from α- and β-amyrin leads to the formation of triterpenadienes i.e., olean-2,12-diene (peak: **44**) and urs-2,12-diene (peak: **50**), respectively (ten Haven and Rullkötter 1988; Jacob et al. 2007). The monoene compounds (e.g., olean-12-ene; peak: **48**) are originated from its corresponding diene triterpenenoids. Olean-2,13(18)-diene (peak: **43**) is formed due to acid-catalysed double bond rearrangement of oleana-2,12-diene (peak: **44**) (ten Haven et al. 1992a). At the later stage of diagenesis, olean-18-ene (peak: **48**) is formed from olean-12-ene (peak: **49**) by the isomerization process (Rullkötter et al. 1994) and oleanane is formed from olean-18-ene during catagenesis. The absence of oleanane in all studied samples suggests that the samples are within diagenetic stage (Rullkötter et al. 1994). A-ring degraded or des-A-triterpenoids (peaks: **33**, **34**, **35**, **36**) are possibly formed from unsaturated triterpenoids under influence of either microbial degradation or photochemical activity by the formation of A-seco-triterpenoids as intermediate product (Logan and Eglinton 1994; Jacob et al. 2007). The detected aromatized triterpenoids in the present study have previously been reported from very recent to ancient sediments, crude oils and coals (Hazai et al. 1986; Chaffee et al. 1984; Chaffee and Fookes 1988; Wolff et al. 1989; Loureiro and Cardoso 1990; Stout 1992; ten Haven et al. 1992b; Rullkőtter et al. 1994; Yessalina and Suzuki 2005; Jacob et al. 2007). It is widely accepted that these compounds are formed due to the oxidation of 3-oxygenated triterpenoids like α and β-amyrin and lupeol at early part of diagenesis probably mediated by the microbial activities (Wakeham et al. 1980; Wolff et al. 1989; Stout 1992; Rullkőtter et al. 1994). In addition, germanicone (olean-18-en-3-one), a pentacyclic triterpenoid is detected in the polar fraction of the lignites and carbonaceous shales of Panandhro mine, Kutch Basin. The compound is identified by comparing its characteristic fragmentation pattern with published literature (Branco et al. 2004). Earlier, the compound has been found in the stem bark of mangrove plant *Barringtonia racemosa* (Yang et al. 2006). Laboratory experiment suggests that germanicol (olean-18-en-3-ol) on Jones oxidation would produce germanicone (Nair 1991). Germanicol is commonly found in association with the mangrove ecosystem (Koch et al. 2003; Yang et al. 2006; Silva and Madureira 2012). This 3-hydroxy olean-18-ene has been reported in organic matter derived from *Rhizophora mangle* which is a tropical red mangrove tree (Koch et al. 2003). Palynological study suggests the prevalence of *Rhizophoraceae* in adjacent Cambay Basin, western India during the Early Paleogene (Prasad et al. 2009). Therefore, the occurrence of germanicone in the studied samples can be regarded as the contribution from mangrove flora into the organic matter.

Implications for Early Cretaceous and Eocene Vegetation of India

Change in dominant vegetational pattern in Indian continent during its northward journey has been inferred here by the variation in distribution of higher plant terpenoid in Early Cretaceous and Eocene coal and lignite-bearing sequences from Kutch Basin, western India. Paleogeographic model suggests that the journey of the Indian subcontinent initiated during Early Cretaceous. Biomarker signatures reveal that conifers especially Araucariaceae and Podocarpaceae families were the dominant contributors to the Indian vegetation during that time. Previous studies on paleobotanical aspect have also supported the inference (Lakhanpal 1970; Singh 2006; Srivastava 2011). Fossils of these two families have been reported from almost all Late Jurassic-Early Cretaceous localities in India viz. Rajmahal Hills, Bihar; Jabalpur in Madhya Pradesh; Golapilli, Raghavapuram and Vemavaram in Andhra Pradesh; Sriperumbudur, Uttatur and Ramand in Tamil Nadu (Singh 2006). Bose and Banerji (1984) studied the Mesozoic fossil flora from Kutch Basin. The floral assemblage of Bhuj Formation is dominated by pteridospermalean foliage. Other important plant remains are *Sagenopteris*, *Linguifolium*, *Nilssoniopteris*, *Elatocladus–Pagiophyllllm*–and *Araucarites*–like conifer leaves and cones. *Ptilophyllum–Brachyphyllum–and Allocladus* are also common in this formation. During Early Paleogene India entered into the equatorial belt and tropical angiosperms covered the whole continent replacing the Mesozoic conifers (Lakhanpal 1970; Singh 2006). In our present investigation, terpenoid composition of Eocene lignite-bearing sequences from Kutch Basin, western India also indicates the prevalence of dominant angiosperm vegetation during that time. Earlier work on palynological assemblages revealed the occurrence of well-diversified and widely distributed tropical rainforest elements in Early Paleogene lignite-bearing sequences from western India (Prasad et al. 2009; Paul et al. 2015; Sharma and Saraswati 2015). In addition, previous chemotaxonomical studies suggested that Early Paleogene ambers from Kutch and adjacent Cambay Basin were derived from Dipterocarpaceae, a tropical angiosperm family (Dutta et al. 2009; Mallick et al. 2014; Paul et al. 2015) which is the prominent component of the tropical rainforest of Asia in the present day (Ashton 1982, 1988).

Conclusions

A significant difference in higher plant terpenoid assemblages of Early Cretaceous and Eocene coal- and lignite-bearing sequences from Kutch Basin is evident in the present study which suggests remarkable shift of floral association during the voyage of the Indian subcontinent. The terpenoid composition of Early Cretaceous samples is represented by the high abundance of diterpenoids in both aliphatic and aromatic fractions and the absence of higher plant-derived triterpenoids (angiosperm markers). Such terpenoid distribution clearly depicts that the gymnosperm, especially conifer dominated vegetation served as a major source of the organic matter. The detailed diterpenoid distribution suggests that Early Cretaceous conifer forest in western India was dominated by Araucariaceae and Podocarpaceae families. In contrast, the terpenoid composition of Eocene lignite-bearing succession from western India is characterised by the significant occurrence of higher plant triterpenoids in aliphatic and aromatic fractions and complete absence of diterpenoids. Such terpenoid signatures indicate that western India was covered by the angiosperm dominated vegetation during Eocene time under wet and humid climatic conditions. The presence of germanicone suggests that the mangroves contributed significantly into peat-forming organic matter.

Acknowledgements We are grateful to IIT Bombay, India for providing instrument facilities to carry out the research work. We express thanks to Department of Science and Technology, New Delhi (Grant No. DST/SJF/E&ASA-01/2016-17 under Swarnajayanti Fellowships Scheme) for the financial support.

References

Ashton, P. S. (1982). Dipterocarpaceae. In C. G. G. J. van Steenis (Ed.), *Flora Malesiana* (Vol. 9, pp. 237–552)., Series 1: Spermatophyta Hague: Martinus-Nijhoff.

Ashton, P. S. (1988). Dipterocarp biology as a window to the understanding of tropical forest structure. *Annual Review of Ecology and Systematics, 19,* 347–370.

Bastow, T. P., Aarssen, B. G. K. V., Alexander, R., & Kagi, R. I. (1999). Biodegradation of aromatic land-plant biomarkers in some Australian crude oils. *Organic Geochemistry, 30,* 1229–1239.

Bechtel, A., Gruber, W., Sachsenhofer, R. F., Gratzer, R., & Püttmann, W. (2001). Organic geochemical and stable carbon isotopic investigation of coals formed in low-lying and raised mires within the Eastern Alps (Austria). *Organic Geochemistry, 32,* 1289–1310.

Bechtel, A., Sachsenhofer, R. F., Kolcon, I., Gratzer, R., Otto, A., & Püttmann, W. (2002). Organic geochemistry of the Lower Miocene Oberdorf lignite (Styrian Basin, Austria): Its relation to petrography, palynology and the paleoenvironment. *International Journal of Coal Geology, 51,* 31–57.

Bechtel, A., Sachsenhofer, R. F., Zdravkov, A., Kostova, I., & Gratzer, R. (2005). Influence of floral assemblage, facies and diagenesis on petrography and organic geochemistry of the Eocene Bourgas coal and the Miocene Maritza-East lignite (Bulgaria). *Organic Geochemistry, 36,* 1498–1522.

Böcker, J., Littke, R., Hartkopf-Fröder, C., Jasper, K., & Schwarzbauer, J. (2013). Organic geochemistry of Duckmantian (Pennsylvanian) coals from the Ruhr Basin, western Germany. *International Journal of Coal Geology, 107,* 112–126.

Bose, M. N., & Banerji, J. (1984). The fossil flora of Kutch I–Mesozoic-megafossils. *The Palaeobotanist, 33,* 1–189.

Branco, A., Pinto, A. C., & Filho, R. B. (2004). Chemical constituents from *Vellozia graminifolia* (Velloziaceae). *Annals of the Brazilian Academy of Sciences, 76,* 505–518.

Chaffee, A. L., & Fookes, C. J. R. (1988). Polycyclic aromatic hydrocarbons in Australian coals III. Structural elucidation by proton nuclear magnetic resonance spectroscopy. *Organic Geochemistry, 12,* 261–271.

Chaffee, A. L., Strachan, M. G., & Johns, R. B. (1984). Polycyclic aromatic hydrocarbons in Australian coals II: Novel tetracyclic components from Victorian brown coal. *Geochimica et Cosmochimica Acta, 48,* 2037–2043.

Chakraborty, S., Jana, B. N., Bhattacharya, S. K., & Robertson, I. (2011). Carbon isotopic composition of fossil leaves from the Early Cretaceous sediments of western India. *Journal of Earth System Science, 120,* 703–711.

Charrié-Duhaut, A., Porraz, G., Cartwright, C. R., Igreja, M., Connan, J., Poggenpoel, C., et al. (2013). First molecular identification of a hafting adhesive in the Late Howiesons Poort at Diepkloof Rock Shelter (Western Cape, South Africa). *Journal of Archaeological Science, 40,* 3506–3518.

Chatterjee, S., Goswami, A., & Scotese, C. R. (2013). The longest voyage: Tectonic, magmatic, and paleoclimatic evolution of the Indian plate during its northward flight from Gondwana to Asia. *Gondwana Research, 23,* 238–267.

Cong, X.-S., Zong, Z.-M., Li, M., Gao, L., Wei, Z.-H., Li, Y., et al. (2015). Enrichment and identification of cyclized hopanoids from Shengli lignite. *Fuel Processing Technology, 134,* 399–403.

Cox, R. E., Yamamoto, S., Otto, A., & Simoneit, B. R. T. (2007). Oxygenated di- and tricyclic diterpenoids of southern hemisphere conifers. *Biochemical Systematics and Ecology, 35,* 342–362.

del Rio, J. C., Gonzalez-Vila, F. J., & Martin, F. (1992). Variation in the content and distribution of biomarkers in two closely situated peat and lignite deposits. *Organic Geochemistry, 18,* 67–78.

Dutta, S., Mallick, M., Bertram, N., Greenwood, P. F., & Mathews, R. P. (2009). Terpenoid composition and class of Tertiary resins from India. *International Journal of Coal Geology, 80,* 44–50.

Dutta, S., Mallick, M., Kumar, K., Mann, U., & Greenwood, P. F. (2011a). Terpenoid composition and botanical affinity of Cretaceous resins from India and Myanmar. *International Journal of Coal Geology, 85,* 49–55.

Dutta, S., Mathews, R. P., Singh, B. D., Tripathi, S. M., Singh, A., Saraswati, P. K., et al. (2011b). Petrology, palynology and organic geochemistry of Eocene lignite of Matanomadh, Kutch Basin, western India: Implications to depositional environment and hydrocarbon source potential. *International Journal of Coal Geology, 81,* 91–102.

Ellis, L., Singh, R. K., Alexander, R., & Kagi, R. I. (1996). Formation of isohexylalkylaromatic hydrocarbons from aromatization-rearrangement of terpenoids in the sedimentary environment: A new class of biomarker. *Geochimica et Cosmochimica Acta, 60,* 4747–4763.

Grantham, P. J., & Douglas, A. J. (1980). The nature and origin of sesquiterpenoids in some Tertiary fossil resins. *Geochimica et Cosmochimica Acta, 44,* 1801–1810.

Haberer, R. M., Mangelsdorf, K., Wilkes, H., & Horsfield, B. (2006). Occurrence and paleoenvironmental significance of aromatic hydrocarbon biomarkers in Oligocene sediments from the Mallik 5L-38 Gas Hydrate Production Research Well (Canada). *Organic Geochemistry, 37,* 519–538.

Hautevelle, Y., Michels, R., Malartre, F., & Troullier, A. (2006). Vascular plant biomarker as proxies for paleoclimatic changes at the Dogger/Malm transition of the Paris Basin (France). *Organic Geochemistry, 37,* 610–625.

Hazai, I., Alexander, G., Szekely, T., Essinger, B., & Radek, D. (1986). Investigations of hydrocarbons constituents of a young sub-bituminous coal by gas chromatography–mass spectrometry. *Journal of Chromatography, 367,* 117–133.

Hussler, G., Albrecht, P., Ourisson, G., Cesario, M., Guilhem, J., & Pascard, C. (1984). Benzohopanes, a novel family of hexacyclic geomarkers in sediments and petroleums. *Tetrahedron Letters, 25,* 1179–1182.

Jacob, J., Disnar, J.-R., Boussafir, M., Albuquerque, A. L. S., Sifeddine, A., & Turcq, B. (2007). Contrasted distributions of triterpene derivatives in the sediments of Lake Caçó reflect paleoenvironmental changes during the last 20,000 yrs in NE Brazil. *Organic Geochemistry, 38,* 180–197.

Koch, B. P., Rullkötter, J., & Lara, R. J. (2003). Evaluation of triterpenols and sterols as organic matter biomarkers in a mangrove ecosystem in northern Brazil. *Wetlands Ecology and Management, 11,* 257–263.

Lakhanpal, R. N. (1970). Tertiary floras of India and their bearing on the historical geology of the region. *Taxon, 19,* 675–694.

Langenheim, J. H. (1995). Biology of amber-producing trees: Focus on case studies of Hymenaea and Agathis. In K. B. Anderson & J. C. Crelling (Eds.), *Amber, resinite, and fossil resins* (Vol. 617, pp. 1–31). Washington, D. C.: ACS Symposium Series.

Logan, G. A., & Eglinton, G. (1994). Biogeochemistry of the Miocene lacustrine deposit, at Clarkia, northern Idaho, U.S.A. *Organic Geochemistry, 21,* 857–870.

Loureiro, M. R. B., & Cardoso, J. N. (1990). Aromatic hydrocarbons in the Paraiba Valley oil shale. *Organic Geochemistry, 15,* 351–359.

Mallick, M., Dutta, S., & Greenwood, P. F. (2014). Molecular characterization of fossil and extant dammar resin extracts: Insights into diagenetic fate of sesqui- and triterpenoids. *International Journal of Coal Geology, 121,* 129–136.

Menor-Salván, C., Najarro, M., Velasco, F., Rosales, I., Tornos, F., & Simoneit, B. R. T. (2010). Terpenoids in extracts of Lower

Cretaceous ambers from the Basque-Cantabrian Basin (El Soplao, Cantabria, Spain): Paleochemotaxonomic aspects. *Organic Geochemistry, 41,* 1089–1103.

Mills, J. S., White, R., & Gough, L. J. (1984). The chemical composition of Baltic amber. *Chemical Geology, 47,* 15–39.

Nair, S. P. (1991). Chemical examination of some Indian medicinal Plants. Ph. D. Dissertation, University of Kerala.

Noble, R. A., Alexander, R., Kagi, R. I., & Knox, J. (1985). Tetracyclic diterpenoid hydrocarbons in some Australian coals, sediments and crude oils. *Geochimica et Cosmochimica Acta, 49,* 2141–2147.

Ostroukhov, S. B., Aref'yev, O. A., & Petrov, A. A. (1983). Hexacyclic monoaromatic hydrocarbons of petroleum. *Petroleum Chemistry USSR, 23,* 53–60.

Otto, A., Gondokusumo, R., & Simpson, M. J. (2006). Characterisation and quantification of biomarkers from biomass burning at a recent wildfire site in Northern Alberta, Canada. *Applied Geochemistry, 21,* 166–183.

Otto, A., & Simoneit, B. R. T. (2001). Chemosystematics and diagenesis of terpenoids in fossil conifer species and sediment from the Eocene Zeitz Formation, Saxony, Germany. *Geochimica et Cosmochimica Acta, 65,* 3505–3527.

Otto, A., & Simoneit, B. R. T. (2002). Biomarkers of Holocene buried conifer logs from Bella Coola and north Vancouver, British Columbia, Canada. *Organic Geochemistry, 33,* 1241–1251.

Otto, A., Simoneit, B. R. T., & Rember, W. C. (2003). Resin compounds from the seed cones of three fossil conifer species from the Miocene Clarkia flora, Emerald Creek, Idaho, USA, and from related extant species. *Review of Palaeobotany and Palynology, 126,* 225–241.

Otto, A., Simoneit, B. R. T., & Wilde, V. (2007). Terpenoids as chemosystematic markers in selected fossil and extant species of pine (Pinus, Pinaceae). *Botanical Journal of the Linnean Society, 154,* 129–140.

Otto, A., Simoneit, B. R. T., Wilde, V., Kunzmann, L., & Püttmann, W. (2002). Terpenoid composition of three fossil resins from Cretaceous and Tertiary conifers. *Review of Palaeobotany and Palynology, 120,* 203–215.

Otto, A., Walther, H., & Puttmann, W. (1997). Sesqui- and diterpenoid biomarkers preserved in *Taxodium*-rich Oligocene oxbow lake clays, Weisselster basin, Germany. *Organic Geochemistry, 26,* 105–111.

Otto, A., & Wilde, V. (2001). Sesqui-, di- and triterpenoids as chemosystematic markers in extant conifers—A review. *Botanical Review, 67,* 141–238.

Ourisson, G., Albrecht, P., & Rohmer, M. (1979). The hopanoids: Paleo-chemistry and biochemistry of a group of natural products. *Pure Applied Chemistry, 51,* 709–729.

Paul, S., & Dutta, S. (2016). Terpenoid composition of fossil resins from western India: New insight into the occurrence of resin-producing trees in Early Paleogene equatorial rainforest of Asia. *International Journal of Coal Geology, 167,* 65–74.

Paul, S., Sharma, J., Singh, B. D., Saraswati, P. K., & Dutta, S. (2015). Early Eocene equatorial vegetation and depositional environment: Biomarker and palynological evidences from Cambay Basin, western India. *International Journal of Coal Geology, 149,* 77–92.

Pereira, R., Carvalho, I. S., Simoneit, B. R. T., & Azevedo, D. A. (2009). Molecular composition and chemosystematic aspects of Cretaceous amber from the Amazonas, Araripe and Recôncavo basins, Brazil. *Organic Geochemistry, 40,* 863–875.

Prakash, A. (1985). The fossil floras of Kachchh. V. Nature, composition and rank (maturation) of Mesozoic coal. *The Palaeobotanist, 34,* 281–293.

Prasad, V., Farooqui, A., Tripathi, S. K. M., Garg, R., & Thakur, B. (2009). Evidence of Late Paleocene-Early Eocene equatorial rain forest refugia in southern Western Ghats, India. *Journal of Biosciences, 34,* 771–797.

Rullkötter, J., Peakman, T. M., & ten Haven, H. L. (1994). Early diagenesis of terrigenous triterpenoids and its impfications for petroleum geochemistry. *Organic Geochemistry, 21,* 215–233.

Saraswati, P. K., Khalonkar, S., Raju, D. S. N., Dutta, S., & Banerjee, S. (2014). Foraminiferal biostratigraphy of lignite mines of Kutch, India: Age of lignite and fossil vertebrates. *Journal of Palaeogeography, 3,* 90–98.

Scotese, C. R. (2001). *Atlas of earth history*. PALEOMAP Project: University of Texas at Arlington.

Scotese, C. R. (2014). Atlas of Early Cretaceous Paleogeographic Maps, PALEOMAP Atlas for ArcGIS, volume 2, The Cretaceous, Maps 23–31, Mollweide Projection. PALEOMAP Project, Evanston, IL.

Sharma, J., & Saraswati, P. K. (2015). Lignites of Kutch, western India: Dinoflagellate biostratigraphy and paleoclimate. *Revue de Micropaléontologie, 58,* 107–119.

Shi, S., Dutta, S., Paul, S., Wang, B., & Jacques, F. M. B. (2014). Terpenoid composition and botanical origin of Late Cretaceous and Miocene ambers from China. *PLoS One, 9,* 1–10.

Silva, C. A., & Madureira, L. A. S. (2012). Source correlation of biomarkers in a mangrove ecosystem. *Annals of the Brazilian Academy of Sciences, 84,* 589–604.

Simoneit, B. R. T. (1986). Cyclic terpenoids of the geosphere. In R. B. Johns (Ed.), *Biological markers in the sedimentary record* (pp. 43–99). Amsterdam: Elsevier.

Simoneit, B. R. T., Grimalt, J. O., Wang, T. G., Cox, R. E., Hatcher, P. G., & Nissenbaum, A. (1986). Cyclic terpenoids of contemporary resinous plant detritus and of fossil woods, ambers and coals. *Organic Geochemistry, 10,* 877–889.

Singh, V. P. (2006). *Gymnosperms (naked seed plants): Structure and development*. New Delhi: Sarup and Sons.

Srivastava, R. (2011). Indian Upper Cretaceous-Tertiary flora before collision of Indian plate: A reappraisal of central and western Indian flora. *Memoir of Geological Society of India, 77,* 281–292.

Stefanova, M., Markova, K., Marinov, S., & Simoneit, B. R. T. (2005). Molecular indicators for coal-forming vegetation of the Miocene Chukurovo lignite, Bulgaria. *Fuel, 84,* 1830–1838.

Stefanova, M., & Simoneit, B. R. T. (2008). Polar aromatic biomarkers of Miocene-aged Chukurovo resinite and correlation with a progenitor macrofossil. *International Journal of Coal Geology, 75,* 166–174.

Stout, S. (1992). Aliphatic and aromatic triterpenoid hydrocarbons in a Tertiary angiospermous lignite. *Organic Geochemistry, 18,* 51–66.

ten Haven, H. L., Peakman, T. M., & Rullkötter, J. (1992a). Δ2-Triterpenes: Early intermediates in the diagenesis of terrigenous triterpenoids. *Geochimica et Cosmochimica Acta, 56,* 1993–2000.

ten Haven, H. L., Peakman, T. M., & Rullkötter, J. (1992b). Early diagenetic transformation of higher–plant triterpenoids in deep-sea sediments from Baffin Bay. *Geochimica et Cosmochimica Acta, 56,* 2001–2024.

ten Haven, H. L., & Rullkötter, J. (1988). The diagenetic fate of taraxer-14-ene and oleanene isomers. *Geochimica et Cosmochimica Acta, 52,* 2543–2548.

Trendel, J. M., Lohmann, F., Kintzinger, J. P., Albrecht, P., Chiarcni, A., Riche, C., et al. (1989). Identification of des-a-triterpenoid hydrocarbons occurring in surface sediments. *Tetrahedron, 45,* 4457–4470.

Wakeham, S. G., Schaffner, C., & Giger, W. (1980). Polycyclic aromatic hydrocarbons in recent lake sediments–II. Compounds derived from biogenic precursors during early diagenesis. *Geochimica et Cosmochimica Acta, 44,* 415–429.

Widodo, S., Bechtel, A., Anggayana, K., & Püttman, W. (2009). Reconstruction of floral changes during deposition of the Miocene Embalut coal from Kutai Basin, Mahakam Delta, East Kalimantan, Indonesia by use of aromatic hydrocarbon composition and stable carbon isotope ratios of organic matter. *Organic Geochemistry, 40,* 206–218.

Wolff, G. A., Trendel, J. M., & Albrecht, P. (1989). Novel monoaromatic triterpenoid hydrocarbons occurring in sediments. *Tetrahedron, 45,* 6721–6728.

Yang, Y., Deng, Z., Proksch, P., & Lin, W. (2006). Two new 18-en-oleane derivatives from marine mangrove plant, *Barringtonia racemosa. Pharmazie, 61,* 365–366.

Yessalina, S., & Suzuki, N. (2005). Occurrences of sesqui-, di-, and triterpenoid hydrocarbons in the Tertiary oils and sediments from the Pacific side of Japan. *Researches in Organic Geochemistry, 20,* 39–47.

Chapter 12

New Specimens of *Frugivastodon* (Mammalia: Apatotheria) from the Early Eocene of India Confirm Its Apatemyid Status and Elucidate Dispersal of Apatemyidae

Floréal Solé, Eric De Bast, Hélène Legendre, Rajendra S. Rana, Kishor Kumar, Kenneth D. Rose, and Thierry Smith

Abstract We here describe 18 new specimens of the sole apatemyid mammal known outside North America and Europe: *Frugivastodon cristatus* from the early Eocene Cambay Shale Formation of Vastan Lignite Mine, Gujarat, India. This mammal was previously represented by a single isolated lower molar, which hindered the establishment of its relationships among Apatemyidae. The new fossils show that the Indian apatemyid is unique and represents a new morphotype among this family. It is notably characterized by mesiodistally elongated lower molars with a reduced m3, a small hypocone on the upper molars, and a transversely wider M1 than in other apatemyids. The new data supports the inclusion of the enigmatic Uintan *Aethomylos* within Apatemyidae. The Indian *Frugivastodon* and the North American *Aethomylos* might represent a distinct clade of Apatemyidae that originated around the Paleocene-Eocene boundary. A paleobiogeographic analysis suggests that *Frugivastodon* dispersed from Europe into India during the early Ypresian. We also review the dispersal events that characterized the history of Apatemyidae.

F. Solé (✉) · E. De Bast · T. Smith
Directorate Earth and History of Life, Paleobiosphere Evolution, Royal Belgian Institute of Natural Sciences, B-1000 Brussels, Belgium
e-mail: floreal.sole@naturalsciences.be

H. Legendre
Institut Polytechnique LaSalle-Beauvais, Beauvais, France

R. S. Rana
Department of Geology, H.N.B. Garhwal University, Srinagar, 246175, Uttarakhand, India

K. Kumar
Wadia Institute of Himalayan Geology, Dehradun, 248001, Uttarakhand, India

K. D. Rose
Center for Functional Anatomy and Evolution, Johns Hopkins University School of Medicine, Baltimore, MD, USA

Keywords Insectivorous mammals • Eocene • Asia • Europe • North America • Dispersal

Introduction

Apatemyids are small extinct mammals that have been known for approximately 150 years from both Europe (Gervais 1848–1852) and North America (Marsh 1872), but, until recently, almost exclusively from their highly specialized dentition. The dentition of apatemyids is characterized by the presence of an enlarged front tooth (generally considered to be the first incisor) in each quadrant, a reduced number of mostly small teeth between this front tooth and the first molar (the first premolar is lost, the third is greatly reduced or absent, and p2 is enlarged and bizarrely shaped), and rather low-crowned molars with broad, basined talonids (Matthew 1909; West 1973a; Gingerich 1982; Gingerich and Rose 1982).

The first skeletons were described from the lake deposits of Messel, Germany (Lutetian, MP 11 reference-level of the mammalian biochronological scale for the European Paleogene, BiochroM'97), thirty years ago (Koenigswald 1987; Koenigswald and Schierning 1987). Since this initial work, several skulls and complete articulated skeletons have been described from the Paleocene and Eocene of North America and Europe (Koenigswald 1990; Bloch and Boyer 2001; Kalthoff et al. 2004; Koenigswald et al. 2005, 2009; Silcox et al. 2010) A skull of *Sinclairella*, previously described by Jepsen (1934), was unfortunately lost more than 40 years ago (see discussion in Silcox et al. 2010). These mammals display numerous arboreal specializations (e.g., flexible ankle; unusually specialized elongate manual digits; short, deep and laterally compressed terminal phalanges; long bushy tail). More recently, cranial features including the morphology of the turbinate bones, bony labyrinth, and the endocranium of these mammals have been detailed thanks to

G. V. R. Prasad and R. Patnaik (eds.), *Biological Consequences of Plate Tectonics: New Perspectives on Post-Gondwana Break-up—A Tribute to Ashok Sahni*,
Vertebrate Paleobiology and Paleoanthropology Series, https://doi.org/10.1007/978-3-030-49753-8_12

the development of X-ray microtomography (Koenigswald et al. 2009; Armstrong et al. 2011; Silcox et al. 2011).

Apatemyids have been reconstructed as insectivorous mammals with foraging behaviours based on their peculiar features: the enlargement of the incisors and the modification of the second and third digits of the manus into elongate probes (Koenigswald 1987; Koenigswald and Schierning 1987; Bloch and Boyer 2001; Koenigswald et al. 2005). Consequently, paleomammalogists propose, in analogy with the extant lemuroid primate *Daubentonia* and the diprotodont marsupial *Dactylopsila*, that "apatemyids used their enlarged incisors to gouge into bark and rotten wood in search of wood-boring grubs and other insects, which they retrieved with their long, slender fingers" (Rose 2006, p. 104). Silcox et al. (2011) notably hypothesized a relationship between the acquisition of this specialized manual extractive feeding regime and the significant increase of brain size observed when comparing the two presently known endocasts of the Apatemyidae (the late Paleocene-early Eocene *Labidolemur kayi* and the late Eocene *Carcinella sigei*).

The relationships of the Apatemyidae among Mammalia are in dispute, with the most frequently suggested relationships being with primates or 'insectivores' (see notably the summaries of this debate in Jepsen (1934), McKenna (1963a), and Silcox et al. (2010)). The recent phylogenetic analysis performed by Silcox et al. (2010) suggests that apatemyids are members of Euarchontoglires rather than Laurasiatheria, a relationship previously proposed by Stehlin (1916), Heller (1930), and Simpson (1940). Silcox et al. (2010) also concluded that apatemyids do not seem to be a member of any living or fossil order and proposed that these mammals should be placed in their own order, Apatotheria, as originally suggested by Scott and Jepsen (1936).

Some 16 North American species (Gunnell et al. 2008a; Czaplewski and Morgan 2015) and 15 European species (Koenigswald et al. 2009) of apatemyids have been described so far, which are grouped among eleven genera (Fig. 12.1). Apatemyids are known from many fossil localities, but they are almost always represented by fragmentary material and are always rare faunal elements; very few localities contain specimens representing more than a few individuals (Simpson 1954; McKenna 1963a, b; Gingerich and Rose 1982). As a consequence, the morphological variability and the systematics of apatemyids have been subject to numerous debates (see the summaries of McKenna 1963a, and Russell et al. 1979), especially the number of species that can be recognized and included among *Apatemys* (West 1973a; Gingerich and Rose 1982). Moreover, there is little agreement regarding the relationships among apatemyids; the sole phylogenetic analysis that questioned the relationships among Apatemyidae included only the six best represented species (Silcox et al. 2010).

The Apatemyidae have been separated into two distinct subfamilies: Apatemyinae and Unuchiniinae. The latter subfamily includes only the Paleocene genus *Unuchinia*; Van Valen (1967) erected this subfamily because this genus represents a distinctive lineage characterized by the combination of very primitive molars for an apatemyid (molars relatively long anteroposteriorly and with tall trigonid, bladelike paraconid, and paralophid nearly straight in occlusal view) and a relatively advanced p4 (simple, and single-rooted – but *U. dysmathes* has a 2-rooted p4). However, Secord (2008) recently provided arguments that strongly support the inclusion of *Unuchinia* among Apatemyidae.

Apatemyids first appeared in the early Paleocene (Torrejonian; North American Land Mammal Age (=NALMA), To-3) of North America, represented by *Unuchinia* and the somewhat more advanced *Jepsenella* (Simpson 1937, 1940; McKenna 1963a; Szalay 1968; Gunnell 1988; Gunnell et al. 2008a). Apatemyids are recorded on this continent until the latest Oligocene (Arikareean NALMA, Ar-3) (Czaplewski and Morgan 2015). It is likely that the family originated in North America.

Apatemyids were also successful in Europe, where they underwent a moderate radiation (Russell et al. 1979). They appeared in Europe in the earliest Eocene, where they are first recorded in the locality of Dormaal (Belgium; MP7 reference-locality; Teilhard de Chardin 1927; Russell et al. 1979). Hooker (2015) recently proposed that the Dormaal fauna is latest Paleocene in age, which would imply that apatemyids have dispersed during the latest Paleocene in Europe; however, we disagree and here consider this fauna (and all the localities included among biozone PEIB *sensu* Hooker 2015) as earliest Eocene in age. The last occurrence of European apatemyids is from the late Eocene MP17b reference-level (Sigé 1975; Koenigswald et al. 2009; Hooker 2010a), and they are not known in Europe after the 'Grande Coupure.'

The only apatemyid to be discovered in a locality situated outside North America and Europe is *Frugivastodon cristatus* from the early Eocene Cambay Shale Formation in Vastan Mine (Gujarat, India; Ypresian; c. 54.5 Ma; Bajpai et al. 2005: Fig. 1). Unfortunately, this taxon has until now been known from only a single isolated lower molar, and possibly because of the lack of information and uncertainty concerning its systematic position, it is rarely mentioned in the literature.

The Cambay Shale Formation, exposed at the Vastan open cast lignite mine near the Vastan village about 40 km

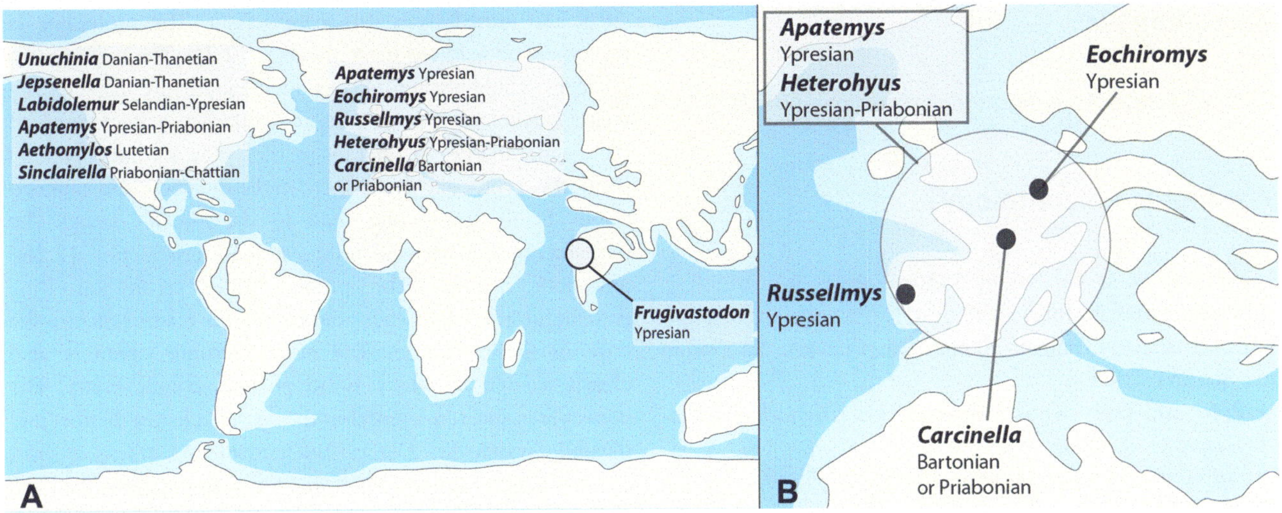

Fig. 12.1 A Geographic and stratigraphic distribution of apatemyids. B focus on European apatemyids. Paleogeographic map adapted from Ron Blakey, Eocene (http://www2.nau.edu/rcb7/050Marect.jpg)

northeast of Surat, Gujarat, western India (Sahni et al. 2006: Fig. 1), is known for its well-preserved diverse fauna of terrestrial mammals as well as other vertebrates (Rana et al. 2004; Bajpai et al. 2005; Rose et al. 2006). The latter include marine and non-marine fishes (Rana et al. 2004; Nolf et al. 2006), the earliest ranid and bombinatorid frogs (Bajpai and Kapur 2008; Folie et al. 2013), agamid lizards (Prasad and Bajpai 2008; Rana et al. 2013), terrestrial and aquatic snakes (Rage et al. 2008), and oldest birds of the Indian subcontinent (Mayr et al. 2007, 2010, 2013). The mammalian fauna of the Cambay Shale Formation is represented by the earliest modern mammals from the Indian subcontinent, including the highest diversity of early bats (Rana et al. 2005; Smith et al. 2007), the earliest lagomorph (Rose et al. 2008), the first Asian ailuravine rodent (Rana et al. 2008), primitive adapoid and omomyid primates (Rose et al. 2009a; Dunn et al. 2016), primitive artiodactyls (Kumar et al. 2010), the first Indian tillodonts (Rose et al. 2009b, 2013), the oldest hyaenodontid in southern Asia (Bajpai et al. 2009; Rana et al. 2015), a primitive tapiroid perissodactyl (Smith et al. 2015), perissodactyl-like cambaytheres (Bajpai et al. 2006; Rose et al. 2014), and the first adapisoriculid from the Eocene of the Indian Subcontinent (Kapur et al. 2017).

We describe in the present paper new fossils from Vastan Lignite Mine that confirm the allocation of *Frugivastodon* to Apatemyidae, hence apatemyids were unambiguously present in southern Asia. Moreover, the new fossils suggest the presence of a new lineage of Apatemyidae in the Eocene, and enable reconstruction of the potential biogeographic origin of this Asian representative.

Material and Methods

The material described here was discovered during several field seasons, between 2004 and 2012, at the Vastan Lignite Mine (Gujarat, India). The specimens were collected by screen-washing using a mesh size of 1 mm. The fossils have been prepared and studied at the RBINS, and are housed at Garhwal University (Srinagar, India) and at Wadia Institute of Himalayan Geology (Dehradun, India).

Abbreviations

H = height; L = length; n = number of specimens; OR = observed range; W = width.

Institutional Abbreviations

AMNH, American Museum of Natural History, New York, USA; CM, Carnegie Museum on Natural History, Pittsburgh, Pennsylvania, USA; GU/RSR/VAS, Garhwal University, Srinagar, India; IITR/SB/VLM, Indian Institute of Technology, Roorkee, India; MNHN.F.MU, Muséum National d'Histoire Naturelle, Mutigny Collection, Paris, France; UCMP, University of California, Museum of Paleontology, Berkeley, California, USA; YPM-PU, Princeton University collection, housed at Yale University Peabody Museum of Natural History, New Haven, Connecticut, USA; WIF/A: Wadia Institute of Himalayan Geology, Dehradun, India.

Measurements

The measurements (in mm) follow Gingerich (1982).

Systematic Paleontology

Apatotheria Scott and Jepsen, 1936.

Apatemyidae Matthew, 1909.

Frugivastodon Bajpai, Kapur, Das, Tiwari, Saravanan and Sharma, 2005.

Type and Only Known Species: *Frugivastodon cristatus* Bajpai et al., 2005.

Frugivastodon cristatus Bajpai et al., 2005.

(Figs. 12.2, 12.3 and 12.4).

Holotype: IITR/SB/VLM 507, left m2.

Referred Specimens: GU/RSR/VAS-533, left M1; GU/RSR/VAS-553, right M1; GU/RSR/VAS-704, right M1; GU/RSR/VAS-1512, left M1; GU/RSR/VAS-1545, right M1; GU/RSR/VAS-1661, left M1; GU/RSR/VAS-1761, right M1; WIF/A 2361, left M1; WIF/A 2362, right M1; GU/RSR/VAS-554, left M2; GU/RSR/VAS-1562, right mandible with m2, m3 and posterior alveoli of m1; GU/RSR/VAS-851, right mandible bearing m2 and alveoli of m1; GU/RSR/VAS-586, right m1; GU/RSR/VAS-1760, right m1; GU/RSR/VAS-587, right m2; GU/RSR/VAS-852, left m2; GU/RSR/VAS-1561, right m2; GU/RSR/VAS-853, right m3.

Type Locality: Vastan Lignite Mine (1–2 m above lower lignite: Lignite Seam 2), about 40 km northeast of Surat, Gujarat, Western India; Cambay Shale Formation, Ypresian, early Eocene.

Emended Diagnosis: Differs from *Apatemys*, *Heterohyus* and *Sinclairella* by having m1 and m2 similar in length but m3 distinctly shorter, and by having higher-crowned molars. Paraconid very reduced on m2 and m3, in contrast to *Unuchinia*, *Labidolemur*, *Apatemys*, *Heterohyus* and *Sinclairella*, in which the paraconid is more developed. Further differs from *Labidolemur*, *Apatemys*, *Heterohyus* and *Sinclairella* in having a smaller hypocone and mesiodistally elongated lower molars. Differs from all other apatemyid genera in having a well-developed postprotocrista and a small metaconule, and transversely wide M1, and from *Aethomylos* in having a small metaconule.

Measurements: Tables 12.1 and 12.2.

Description: Eighteen new specimens (representing 5 loci: m1–m3; M1–M2) are referred to this apatemyid. The specimens, consisting mainly of isolated teeth, represent at least five individuals.

Nine M1s of *Frugivastodon* (Fig. 12.2A–E, H–K) have been discovered in the Vastan Lignite Mine. The paracone is larger and higher than the metacone on the M1, and the two cusps are joined by a thin centrocrista. The protocone is prominent and shifted mesially (Fig. 12.2J). The trigon basin is deep. The paraconule and metaconule are present but weakly developed, corresponding rather to a thickening of the enamel than to distinct cusps (i.e., they are not separated from the protocone by a notch). Both cusps are situated closer to the protocone than to the buccal cusps. The preparaconule crista and premetaconule crista (Fig. 12.2K) are both distinct. The preparaconule crista joins the ectocingulum, while the premetaconule crista reaches the base of the metacone. The postparaconule crista is also clearly visible but there is no postmetaconule crista. The parastyle is short in length but the apex is clearly distinct and mesially projected. The postmetacrista appears as a short crest; it is more transversely oriented than the parastyle. The ectoflexus is variable but shallow. A moderate stylar shelf is present; it is wider labial to the metacone. GU/RSR/VAS-533, GU/RSR/VAS-553, GU/RSR/VAS-1512, and WIF/A 2361 (M1s) display a small but distinct hypocone (Fig. 12.2B, E, I); it is present on a ledge formed by a slight broadening of the postcingulum at the very base of the disto-lingual border of the protocone. The five other M1s have worn (e.g., GU/RSR/VAS-704; Fig. 12.2D) or broken hypocone area. The postcingulum is not continuous with the metacingulum.

Only one M2 can be referred to *Frugivastodon* (Fig. 12.2F, G). The tooth is similar to the M1s in having large paracone and metacone joined by a thin centrocrista, paracone taller than metacone, small paraconule and metaconule, prominent and mesially shifted protocone, deep trigon basin, distinct preparaconule and premetaconule cristae, and a moderate stylar shelf. The M2 differs from the M1s in having a deeper ectoflexus resulting in more labially projecting parastyle and metastyle, and the absence of the postparaconule crista. Unfortunately, the area where the hypocone might be present is damaged; it is thus impossible to determine if a hypocone was present.

All three lower molars appear to be represented in our sample. The m1 and m2 are approximately the same length, whereas m3 is distinguished by being distinctly smaller (Table 12.2). Moreover, the m1s and m2s discovered at Vastan Lignite Mine are conspicuously more transversely compressed and relatively more mesiodistally elongated than the m3s (Fig. 12.3A–C): the mean width/length ratio is 0.53 for the m1 (OR: 0.48–0.58, n = 2) and 0.56 for the m2 (OR: 0.51–0.60, n = 5), while it is 0.69 for the m3 (OR: 0.66–0.72, n = 2). The trigonid is higher than the talonid, as on all the molars. The protoconid is the tallest cusp, the metaconid slightly lower and more distal, and the paraconid low and greatly reduced but nevertheless more distinct on m1 and m2 than on m3. The preprotocristid and paracristid meet at an obtuse angle (nearly a right angle) at the mesiolabial margin of the tooth, where a small, low fourth trigonid cusp

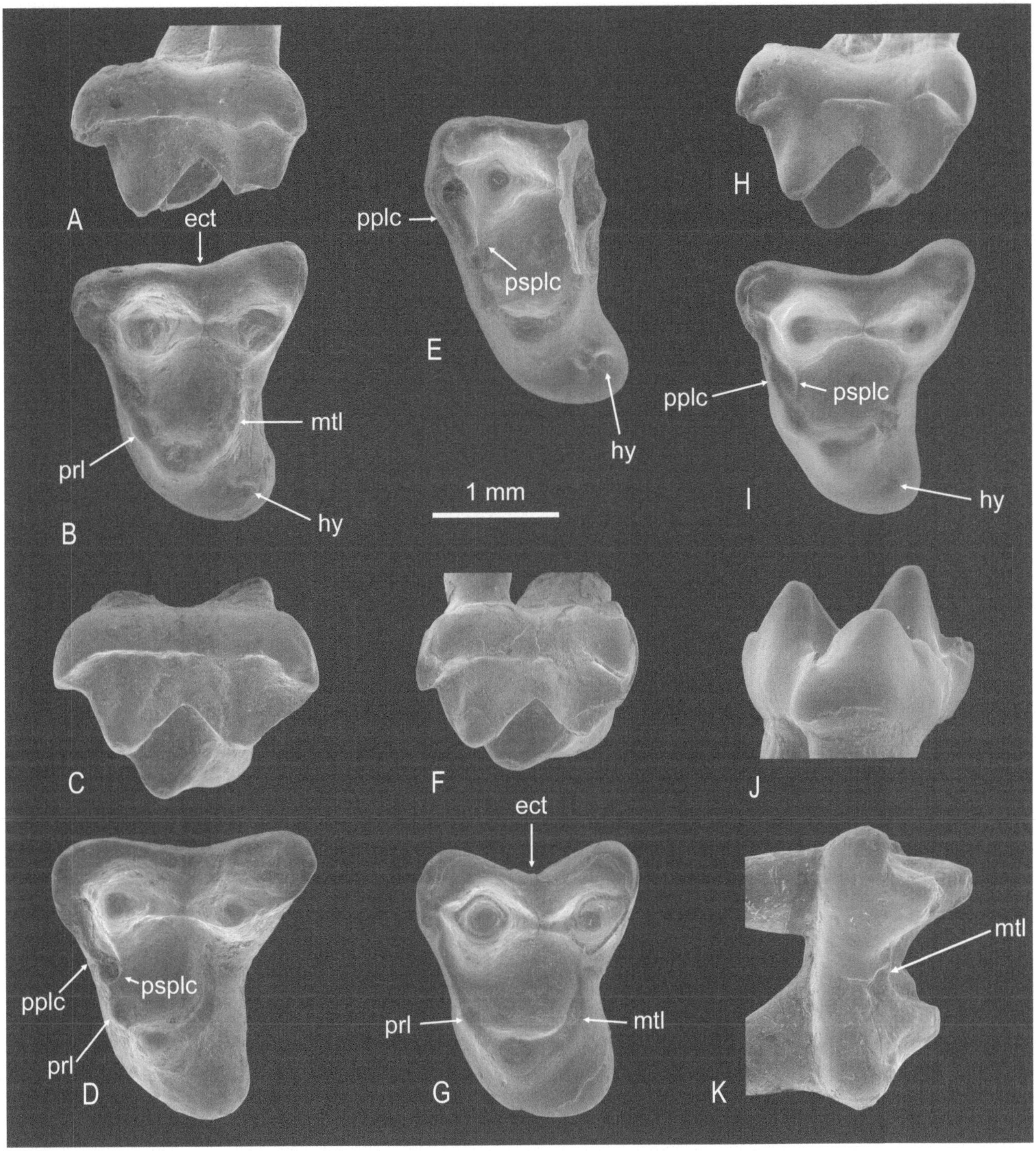

Fig. 12.2 *Frugivastodon cristatus* from Vastan Lignite Mine (early Eocene; India). **A–B**. GU/RSR/VAS-1512, left M1; **A**. labial view; **B**. occlusal view; **C–D**. GU/RSR/VAS-704, right M1 (reversed); **C**. labial view; **D**. occlusal view; **E**. WIF/A 2361, left M1, occlusal view; **F–G**. GU/RSR/VAS-554, left M2; **F**. labial view; **G**. occlusal view; **H–K**. GU/RSR/VAS-553, right M1 (reversed); **H**. labial view; **I**. occlusal view; **J**. lingual view; **K**. distal view; Abbreviations: ect, ectoflexus; hy, hypocone; mtl, metaconule; pplc, preparaconule crista; prl, paraconule; psplc, postparaconule crista

(Fig. 12.3K, L) is present labial to the paraconid and separated from it by a notch. This fourth cusp is diagnostic of apatemyids. Nevertheless, the quadrilateral aspect of the trigonid in occlusal view, typical of other Eocene apatemyids (e.g., *Apatemys*, *Heterohyus*), is less pronounced on the Indian specimens. A premetacristid descends to a notch behind the paraconid and is high enough to lingually enclose the trigonid basin. The talonid basin is enclosed by three

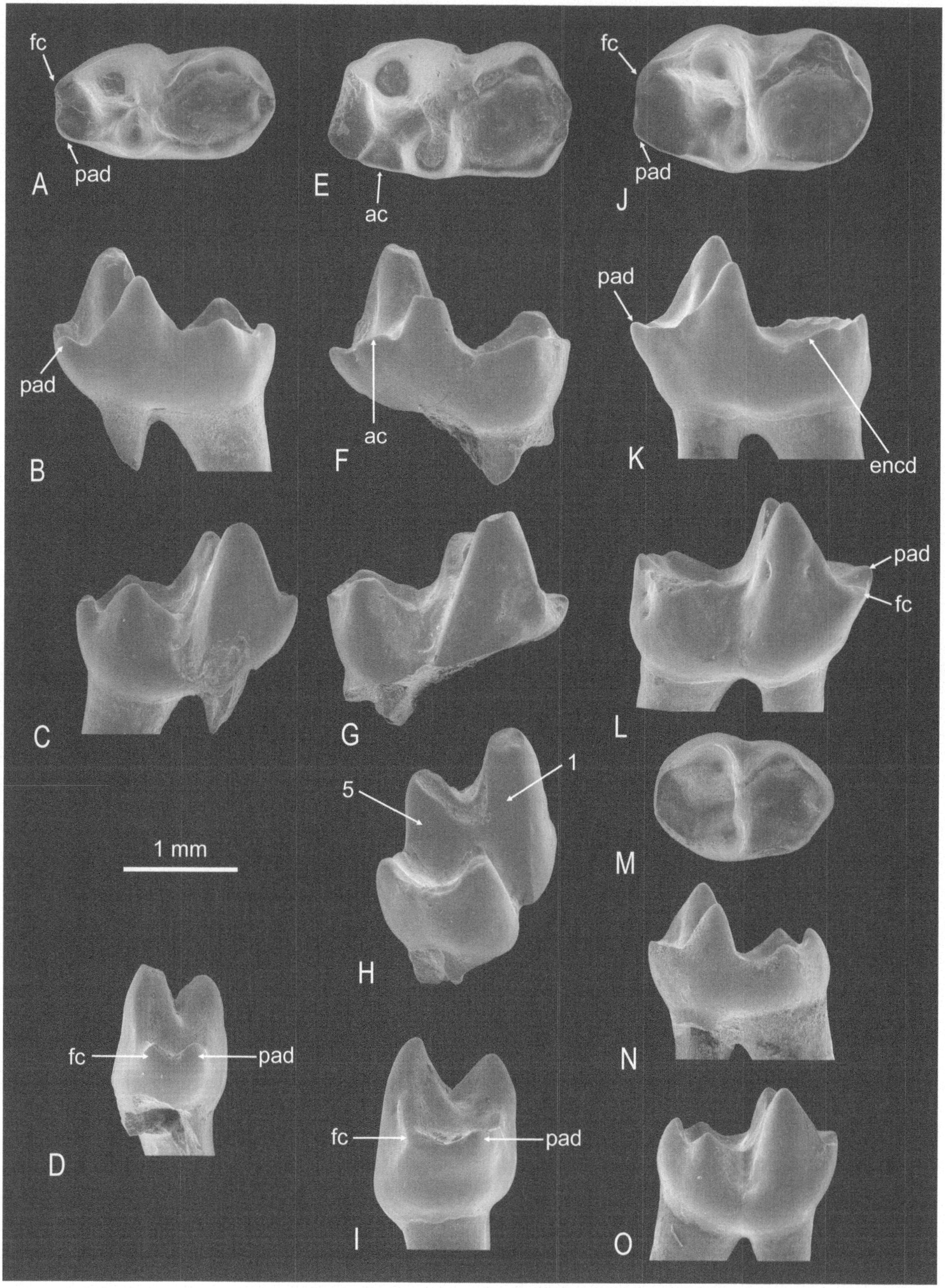

Fig. 12.3 *Frugivastodon cristatus* from Vastan Lignite Mine (early Eocene; India). **A–D**. GU/RSR/VAS-586, right m1; **A**. occlusal view; **B**. lingual view; **C**. labial view; **D**. mesial view; **E–H**. GU/RSR/VAS-1561, right m2; **E**. occlusal view; **F**. lingual view; **G**. labial view; **H**. distal view; **I–L**. GU/RSR/VAS-852, left m2 (reversed); **I**. mesial view; **J**. occlusal view; **K**. lingual view; **L**. labial view; **M–O**. GU/RSR/VAS-853, right m3; **M**. occlusal view; **N**. lingual view; **O**. labial view. Abbreviations: ac, additional cusp?; encd, entocristid; fc, fourth cusp of the trigonid; pad, paraconid. 1 and 5 = indication of the wear facets 1 and 5

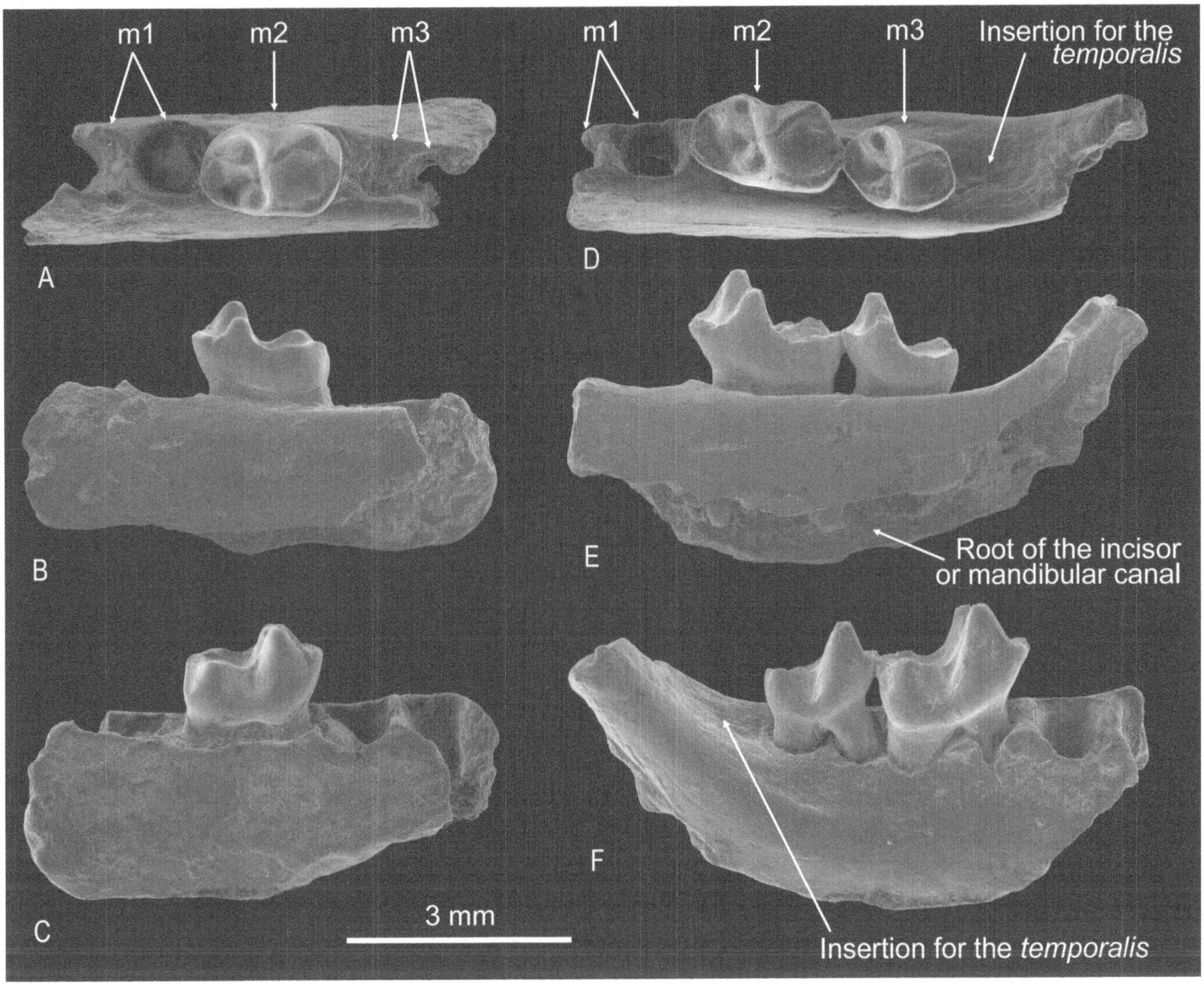

Fig. 12.4 *Frugivastodon cristatus* from Vastan Lignite Mine (early Eocene; India). **A–C**. GU/RSR/VAS-851, right mandible bearing m2 and alveoli of m1; **A**. occlusal view; **B**. lingual view; **C**. labial view; **D–F**. GU/RSR/VAS-1562, right mandible with m2, m3 and posterior alveoli of m1; **D**. occlusal view; **E**. lingual view; **F**. labial view

cusps – hypoconid the tallest, entoconid slightly lower, and a low, posteriorly shifted hypoconulid – and their adjoining crests; however, the distal position of the hypoconid on m1 is only an impression conferred by the narrowness of the talonid. The cusps are situated on the periphery of the crown (i.e., there is no basal inflation). The crests joining these cusps are low, except for the somewhat higher entocristid, which closes the basin lingually. The morphology of m1 is closely similar to that of the m2 except for being narrower and having a more distinct paraconid (Fig. 12.3D, I), a narrow and somewhat extended trigonid and an elongate, basined talonid longer than the trigonid but about the same width, and a deeper notch between the hypoconulid and entoconid. The narrower width of m1 results in lingual and labial cusps that are less separated than on m2 (Fig. 12.3A, J), and consequently a more distinctly notched crest that bridges the paraconid and the fourth cusp and mesially closes the shallow trigonid basin (Fig. 12.3D, I).

Apart from its greater width (and consequent wider separation between lingual and labial cusps; Fig. 12.3I), and the minor differences just noted, m2 is very similar to m1. The trigonid and talonid are approximately equal in length, and the distal margin of the talonid is rounded in occlusal view, somewhat more so than on m1. GU/RSR/VAS-1561 bears a small accessory cuspule (Fig. 12.3E, F) on the lingual part of the trigonid; this cuspule is not present in other specimens.

GU/RSR/VAS-1561 further differs from the other lower molars in having wear facets 1 and 5, on the distal surface of the trigonid (=postvallid), more distinctly offset (Fig. 12.3 H). The cristid obliqua meets and partly ascends the middle

Table 12.1 Measurements (in mm) of all the specimens of *Frugivastodon cristatus* Bajpai et al. 2005 from Vastan Lignite Mine (Ypresian; India)

Specimen	Locus	Length	Width	Height	Ratio L/W	Ratio W/L	Ratio H/L
IITR/SB/VLM 507 (holotype)	m2	1.92	1.03	–	1.86	0.54	–
GU/RSR/VAS-586	m1	1.97	0.95	1.6	2.07	0.48	0.80
GU/RSR/VAS-1760	m1	2.19	1.27	1.6	1.72	0.58	0.73
GU/RSR/VAS-587	m2	2.09	1.11	–	1.88	0.53	–
GU/RSR/VAS-852	m2	2.1	1.18	–	1.78	0.56	–
GU/RSR/VAS-1561	m2	2.16	1.1	1.7	1.96	0.51	0.77
GU/RSR/VAS-851	m2	1.93	1.13	1.4	1.71	0.59	0.74
GU/RSR/VAS-1562	m2	2.09	1.25	1.7	1.67	0.60	0.76
GU/RSR/VAS-1562	m3	1.48	1.07	1.4	1.38	0.72	0.93
GU/RSR/VAS-853	m3	1.56	1.03	1.4	1.51	0.66	0.88
GU/RSR/VAS-533	M1	–	2.51	–	–	–	–
GU/RSR/VAS-1761	M1	–	–	–	–	–	–
WIF/A 2362	M1	–	2.11	–	–	–	–
WIF/A 2361	M1	1.85	2.22	–	0.83	1.20	–
GU/RSR/VAS-1512	M1	1.91	2.07	–	0.92	1.08	–
GU/RSR/VAS-1661	M1	1.94	2.17	–	0.89	1.12	–
GU/RSR/VAS-1545	M1	1.97	2.27	–	0.87	1.15	–
GU/RSR/VAS-704	M1	2	2.24	–	0.89	1.12	–
GU/RSR/VAS-555	M1	1.84	2.21	–	0.83	1.20	–
GU/RSR/VAS-554	M2	1.72	2.08	–	0.83	1.21	–

Table 12.2 Summary measurements (in mm) of *Frugivastodon cristatus* Bajpai et al. 2005 from Vastan Lignite Mine (Ypresian, India)

Locus		N	OR	Mean
M1	L	6	1.84–2	1.92
	W	5	2.07–2.51	2.23
M2	L	1	1.72	–
	W	1	2.08	–
m1	L	2	1.97–2.19	2.08
	W	2	0.95–1.27	1.11
m2	L	3	1.93–2.16	2.07
	W	3	1.10–1.25	1.15
m3	L	2	1.48–1.56	1.52
	W	2	1.03–1.07	1.05

of the postvallid, delimiting facet 1, on the distal surface of the protoconid, from facet 5, on the distal surface of the metaconid, as in some plesiadapiforms (Rose 1975).

The m3 is quite different from m1 and m2 (Figs. 12.3M–O, 12.4D–F). It is short and elliptical in occlusal view and differs from m1 and m2 in having a less distinct paraconid, a shorter paracristid, and a weak or absent fourth cusp. The trigonid is slightly broader than the talonid. The hypoflexid is very shallow. The hypoconulid is relatively more distally located on m3 than on m2, comparable to its position on m1.

The ventral part of the mandible is broken just below the roots of the teeth on the two discovered dentaries (GU/RSR/VAS-1562 and GU/RSR/VAS-851; Fig. 12.4), so its depth cannot be determined. This breakage may result either from the presence of the root of a large incisor – a characteristic of apatemyids – or from the presence of the mandibular canal (Fig. 12.4E). Indeed, as shown by radiographs of a specimen of *Apatemys*, the incisor root may extend back to below m3 in Apatemyidae (Koenigswald et al. 2005). However, the Indian specimens are not adequately preserved to corroborate either hypothesis. The mandible of *Frugivastodon* was probably deep, as usually observed in apatemyids. There is a small fossa distal to the m3 for the insertion of part of the tendon of the temporalis muscle (Fig. 12.4D, F). The mandible is slightly expanded lateral to the m3 in GU/RSR/VAS-1562, but this enlargement is less pronounced in the second mandibular fragment described here (GU/RSR/VAS-851).

Discussion. *Frugivastodon cristatus* was described by Bajpai et al. (2005) based on a single lower molar

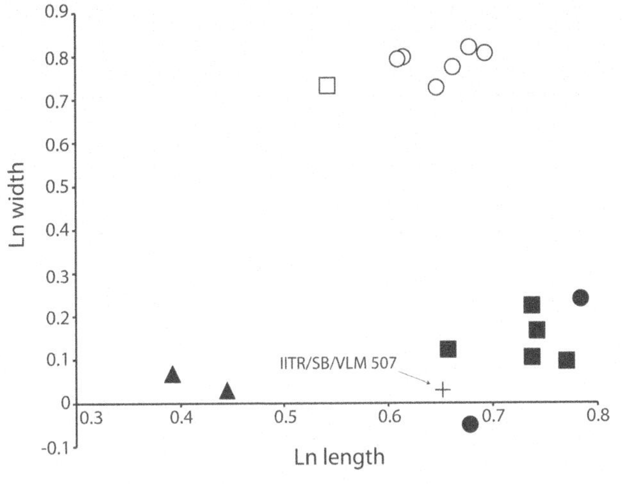

Fig. 12.5 Comparison of the sizes Ln length (mm) × Ln width (mm) of the holotype (IITR/SB/VLM 507) and the m1–m3 and M1–M2 of *Frugivastodon cristatus* described in the present paper. Circles are M1/m1, squares M2/m2, and triangles M3/m3; black symbols correspond to lower teeth, and white figures to upper teeth

(IITR/SB/VLM 507). Our fieldwork in the Cambay Shale Formation of Vastan Lignite Mine (Gujarat, India) has resulted in the discovery of numerous new dental elements for this species, including mandibular fragments and upper molars. The holotype probably corresponds to m2 based on its size (L = 1.92; W = 1.03); but its size is also close to that of one of the two m1s found by our team (Fig. 12.5; Table 12.1). The proportions and the poor development of the paracristid and paraconid, however, support its identification as an m2.

The fossils referred to *Frugivastodon* support the inclusion of this taxon within Apatemyidae based on the following combination of shared features: elongate M1 with the parastyle projecting forward; M2 with prominent styles; weak conules on upper molars; small hypocone present at the distal base of the protocone; lower molar trigonids with a reduced paraconid and a mesiolabial fourth cusp resulting in a quadrate trigonid; absence of labial and lingual cingulids; and rounded talonids on lower molars.

Based on the new fossils we have emended the diagnosis of *Frugivastodon cristatus*, adding further characteristics of the sole apatemyid species found outside Europe and North America. A large part of its molar dentition is now known (Fig. 12.6). Bajpai et al. (2005) considered that *Frugivastodon* differs from the other apatemyids by two major features: the lack of a paraconid and the low talonid cusps on the molars. The molars described and illustrated in the present article show that the paraconid is still present, if small, on the molars; therefore, its absence on the holotype may be explained by (1) the fact that the specimen corresponds to m2, in which the paraconid may be weaker than on m1; and (2) wear or erosion of the specimen. Nor do the new specimens support the presence of talonid cusps lower than in other apatemyids. We believe that the low height of the talonid cusps compared to those of the trigonid on the holotype also results from wear or erosion of the specimen.

One of the most striking features of the new species is the relative size of the lower molars: m1 and m2 are approximately equal in length, but m3 is clearly shorter than m1 and m2 (Table 12.2). This contrasts with the situation observed in many North American and European Eocene apatemyids (*Unuchinia*, *Apatemys*, *Heterohyus*, *Sinclairella*) where the m3 is usually longer than the m1 and m2 (Simpson 1936; West 1973a; Holtzman 1978; Russell et al. 1979) (Fig. 12.7). The three molars are, however, still very close in size in the North American Paleocene apatemyid *Labidolemur major*, the Paleocene-Eocene *Labidolemur kayi*, and the early Eocene *L. serus* and *Apatemys chardini* (McKenna 1963a; West 1972: Table 1; Gingerich and Rose 1982: Table 1; Gingerich 1982: Tables 1–3), and in the European *Apatemys mutiniacus* (Russell et al. 1979: Table 1). This is also the case for the Paleocene *Jepsenella praepropera*, the oldest apatemyid (Szalay 1968: Table 1; West 1973b: Table 1). Paleocene *Unuchinia* differs from other apatemyids in showing increase in molar length from m1 to m3 (Simpson 1936; Holtzman 1978).

Frugivastodon appears primitive relative to other early Eocene apatemyids in the morphology of the teeth. Morphological trends in apatemyids include a tendency for upper molars to expand the hypocone lobe posterolingually and become more squared and relatively narrower transversely; enlargement of the parastyle; reduction of the paraconid on lower molars and enlargement of the fourth trigonid cusp, which tends to extend mesially as far as or even farther than the paraconid (giving the trigonid a peculiar parallelogram aspect in occlusal view); talonid as wide as or wider than the trigonid; and talonid of m3 mesiodistally elongated (McKenna 1963a, b; West 1973a; Russell et al. 1979) (Fig. 12.7). For the majority of these features, *Frugivastodon* is more primitive than other Eocene apatemyids: the M1 is transversely wider and triangular in occlusal view because of the rudimentary hypocone, and it has a small parastyle. On the lower molars, the fourth trigonid cusp is small and not mesially displaced, consequently the trigonid is less oblique and parallelogram-shaped and more triangular in occlusal view than in other early Eocene apatemyids (Fig. 12.7). The talonid and trigonid are of approximately equal width, even on m1; the entocristid is high, closing the talonid basin lingually; the entoconid is almost as high as the hypoconid; and the talonid of m3 is mesiodistally short (Fig. 12.7).

The Ml of apatemyids is typically elongate mesiodistally and narrow transversely compared to M2 and M3 (McKenna 1963a) (Fig. 12.7). M1 of *Frugivastodon*, however, is transversely wide: the width/length (W/L) ratio for M1 is

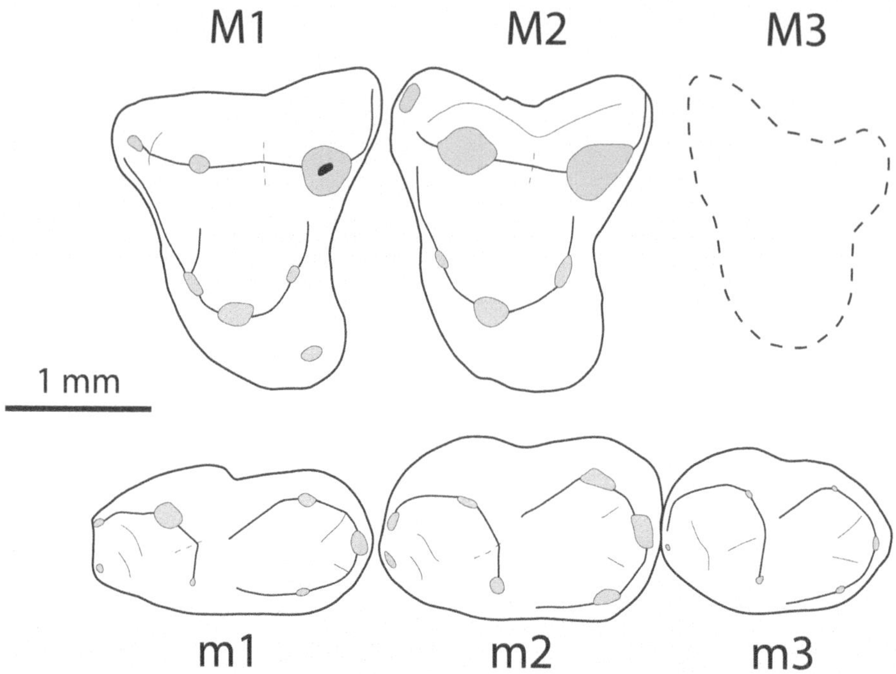

Fig. 12.6 Occlusal sketch of the dentition of *Frugivastodon cristatus*. M1, GU/RSR/VAS-1512; M2, GU/RSR/VAS-554; m1, GU/RSR/VAS-586; m2, GU/RSR/VAS-851; m3, GU/RSR/VAS-853. The outline of M3 is based on a specimen of *Aethomylos* (UCMP 101047) figured by Novacek (1976, Fig. 16D)

Table 12.3 Width/length ratio of the M1 and M2 for several apatemyids. The values presented for *Jepsenella praepropera* are based on Szalay (1968: Table 1), *Labidolemur major* on West (1972), *Labidolemur kayi* on Gingerich (1982: Table 1), *Labidolemur serus* on Gingerich (1982: Table 2), *Apatemys chardini* on Gingerich (1982: Table 3), *Apatemys mutiniacus* on Russell et al. (1979: Table 1), and *Aethomylos simplicidens* on Novacek (1976: Table 6)

Taxa	Width/length ratio	
	M1	M2
Frugivastodon cristatus	1.13 (n = 6)	1.21 (n = 1)
Aethomylos simplicidens	0.98[a] (n = 8)	
Apatemys mutiniacus	0.9 (n = 1)	–
Apatemys chardini	0.89 (n = 1)	1.12 (n = 1)
Labidolemur serus	0.98 (n = 2)	1.25 (n = 2)
Labidolemur kayi	0.95 (n = 1)	1.26 (n = 2)
Labidolemur major	–	1.29 (n = 1)
Jepsenella praepropera	0.96 (n = 1)	1.03 (n = 1)

[a]The deciduous premolars and molars (M1 and M2) were not discriminated by Novacek (1976)

1.15 (OR: 1.08–1.2, n = 6). M2 is also relatively wide (W/L = 1.21, n = 1). The relative width of the M1 of *Frugivastodon* is clearly unusual, while that of the M2 is frequently observed in apatemyids (Table 12.3).

The M1s of *Frugivastodon* are morphologically similar to those provisionally referred by Szalay (1968) to the Paleocene *Jepsenella praepropera* (Fig. 12.7). The similarities listed herein – triangular morphology, small hypocone, presence of both metaconule and paraconule, presence of postparaconule crista – might be regarded as plausible support for Szalay's (1968) allocation of these isolated teeth to *Jepsenella*.

Frugivastodon shares with Eocene apatemyids the reduction of the paraconid on the molars. However, it differs from the Eocene apatemyids *Apatemys* and *Heterohyus* in having higher-crowned molars, as seen in *Labidolemur*, *Jepsenella*, and *Unuchinia* (Table 12.4). The decrease in crown height was considered by McKenna (1963a) as a distinctive trend among apatemyids, and was documented in a limited sample by Gingerich (1982: Table 4). Crown height in *Frugivastodon* is actually greater than in early Eocene *Apatemys* and within the range of the *Labidolemur* species (*L. kayi* ranges from Tiffanian NALMA to early Wasatchian NALMA; *L. serus* is only known in the early

Table 12.4 Relative crown height in mm (protoconid height above base of crown at mesial root/mesiodistal crown length; Gingerich 1982) in Ypresian *Frugivastodon* from Vastan Lignite Mine, Clarkforkian NALMA and Wasatchian *Labidolemur* and *Apatemys* from Clark's Fork and Bighorn basins. The values, except those for *Frugivastodon*, are from Gingerich (1982: Table 4)

Tooth position	Genus and species	N	OR	Mean
m1	*Frugivastodon cristatus*	2	0.73–0.8	0.77
	Apatemys chardini	2	0.63–0.69	0.66
	Labidolemur kayi	6	0.71–0.82	0.76
	Labidolemur serus	2	0.76–0.80	0.78
m2	*Frugivastodon cristatus*	3	0.74–0.77	0.76
	Apatemys chardini	4	0.59–0.67	0.63
	Labidolemur kayi	8	0.74–0.89	0.78
	Labidolemur serus	2	0.77	0.77
m3	*Frugivastodon cristatus*	2	0.88–0.93	0.90
	Apatemys chardini	1	0.59	–
	Labidolemur kayi	5	0.63–0.68	0.66
	Labidolemur serus	1	0.68	–

Wasatchian) (Table 12.4), presumably a plesiomorphic trait. However, because it has been demonstrated that modifications in molar crown height are closely related to diet (e.g., Boyer 2008), one can hypothesize that crown height might be a plastic trait in apatemyids (as generally in mammals) and therefore may not provide strong support for relationships.

In addition, *Frugivastodon cristatus* differs from all other apatemyids in one striking feature: the greater mesiodistal elongation of the lower molars. Because West (1973a) and Russell et al. (1979) hypothesized that there was a progressive lengthening of apatemyid lower molars through the Paleogene, we estimated the length/width (L/W) ratio for various apatemyids from the Paleocene, Eocene, and Oligocene, and for the Indian taxon (Table 12.5). The values calculated for *Frugivastodon* are much higher than those of the other apatemyids in being close to 1.9 for m1 (OR: 1.72–2.07; n = 2) and 1.74 for m2 (OR: 1.67–1.96; n = 5), while the values for these molars are between 1.2 and 1.6 in other apatemyids.

Thus *Frugivastodon* can be confidently assigned to the Apatemyidae, but it differs from contemporary apatemyids in several features that recall the Paleocene species (e.g., relative size of the lower molars, triangular M1 with a small hypocone, M1 transversely wide, high-crowned lower molars) and in displaying elongation of the lower molars and reduction of m3. However, *Unuchinia* differs among Paleocene apatemyids in having m3 longer than m1 and m2. *Frugivastodon* appears clearly less advanced than other Eocene apatemyids in numerous features and seems to belong to a separate lineage of Apatemyidae.

As discussed below (see *Relationships of* Frugivastodon), the closest taxon to *Frugivastodon* could be the enigmatic *Aethomylos simplicidens* from the middle Eocene (Uintan NALMA). *Aethomylos* appears to be confined to the Uintan (Uinta-1) Friars Formation of southern California (Novacek 1976); the record from the Bridgerian Powder Wash locality in Utah (Krishtalka and Stucky 1984) is not certainly referable to this genus. Among the isolated teeth of *Frugivastodon* and *Aethomylos*, are a few narrow molariform teeth (Figs. 12.3A–C; 12.8D–F). Their trigonid is characterized by a low paraconid situated well mesial to the protoconid (though apparently more so in *Aethomylos*, to judge from Novacek's illustration). The talonid is a well-developed, deeply excavated basin, and the hypoconulid is distally located.

Novacek (1976) identified this peculiar tooth (Fig. 12.8 D–F) as either a p4 or dp4 but noted that no apatemyid has a molariform p4. The p4 is simple in known apatemyids, including the earliest taxa such as *Unuchinia*, *Jepsenella*, and *Labidolemur* (Simpson 1929, 1936; West 1972, 1973b), in which it is premolariform and displays only a high rounded trigonid cusp and a single talonid cusp. The morphology of the narrow tooth illustrated by Novacek (1976) (Fig. 12.8D–F), with its apparently extended trigonid, seems to be more consistent with dp4 of primitive placentals generally (e.g., Butler 1952; Zack 2012). The two narrow molariform teeth of *Frugivastodon* (GU/RSR/VAS-586 and GU/RSR/VAS-1760), however, have less extended trigonids and differ from m2 mainly in relative width; cusp positions are otherwise similar. For these reasons we regard them here as m1.

McKenna (1963a) listed the following differences between m1 and m2 of *Labidolemur kayi*: m2 has a relatively much wider trigonid with more separated cusps, a weaker fourth cusp, and a wider talonid. The same distinctions are found when comparing the teeth of *Frugivastodon* here identified as m1s and m2s.

Table 12.5 Length/Width ratio for lower molars of several apatemyids from Europe, North America and India. Values of *H. nanus*, *A. sigogneaui*, *A. mutiniacus*, *A. teilhardi*, *A. uintensis*, *A. "bellus"*, and *Sinclairella Apatemys* and Russell et al. (1979: Table 2). The values for the following taxa have been estimated by ourselves: *Aethomylos simplicidens* from Novacek (1976: Table 5), *Apatemys chardini* from Gingerich (1982: Table 3), *Jepsenella praepropera* (Szalay 1968: Table 1, type AMNH 3592), *Labidolemur kayi* from Gingerich and Rose (1982: Table 1), *Labidolemur serus* (Gingerich 1982: Table 2, holotype UM 69962), *Labidolemur soricoides* from West (1973a: Table 2, m1: holotype AMNH 17400, m2: PU 19992, m3: PU 19938), *Labidolemur major* from West (1972: Table 1), *Unuchinia dysmathes* (from Holtzman 1978: Table 23) and *Frugivastodon cristatus* (present paper); we used the mean values of the length and width when we did not have values for a particular specimen, and we used the maximal width estimated when the values of the trigonid and talonid were both available

	Europe				North America				India			
Eocene		m1	m2	m3		m1	m2	m3		m1	m2	m3
	Heteroyus nanus	1.3	1.2	1.7	*Sinclairella*	1.5	1.4	2	*Frugivastodon cristatus*	1.90 (n = 2)	1.74 (n = 5)	1.45 (n = 2)
	Apatemys sigogneaui	1.3	1.4	1.6	*Apatemys uintensis*	1.4	1.4	1.6				
	Apatemys mutiniacus	1.6	1.4	1.4	*Aethomylos simplicidens*	1.86[a] (n = 2)	1.76 (n = 4)[a]	1.55 (n = 1)				
	Apatemys teilhardi	1.65	1.6	–	*Apatemys "bellus"* (Bridgerian)	1.6	1.4	1.6				
					Apatemys chardini	1.45 (n = 2)	1.46 (mean)	1.55 (n = 1)				
					Labidolemur serus	1.47 (n = 1)	1.41 (n = 1)	–				
Paleocene					*Labidolemur kayi*	1.46 (mean)	1.27 (mean)	1.46 (mean)				
					Labidolemur soricoides	1.55 (n = 1)	1.38 (n = 1)	1.33 (n = 1)				
					Jepsenella praepropera	1.38 (n = 1)	1.27 (n = 1)	1.33 (n = 1)				
					Labidolemur major	1.3 (n = 1)	1.18 (n = 1)	1.25 (n = 1)				
					Unuchinia dysmathes	1.55 (n = 1)	1.5 (n = 1)	1.55 (n = 1)				

[a]Based on the study of *Frugivastodon*, the p4s (or dp4s) of *Aethomylos* (*sensu* Novacek 1976) are here considered as m1s, while the m1s are regarded as m2s

Phylogenetic Analysis

Objectives and Methodology

The systematics and evolution of apatemyids have been controversial, largely because fossils are rare and tend to be very fragmentary; hence authors have focused on the description and validity of species rather than on the phylogenetic relationships among the approximately 30 described species. Silcox et al. (2010) provided the first matrix (240 morphological characters – 68 postcranial, 45 cranial, and 127 dental; 34 taxa) that includes numerous apatemyids. However, this large matrix was built in the attempt to study the broader relationships of these mammals among Placentalia, rather than to study the interrelationships among apatemyids. As a consequence, the matrix only includes six apatemyids: the North American *Jepsenella praepropera*, *Labidolemur kayi*, *Sinclairella dakotensis*, and *Apatemys* sp., and the European *Carcinella sigei*, and *Heterohyus nanus*. The choice of taxa was based on their relative completeness (i.e., presence of skulls and, sometimes, of partial skeletons). This matrix, however, includes the most typical apatemyids: the only Eocene genera that are not represented are the European *Russellmys* and *Eochiromys*, but they are very poorly known and their systematic status is debated (Teilhard de Chardin 1927; Russell et al. 1979; Estravís 1990; Koenigswald et al. 2009). North American Paleocene *Unuchinia* was also excluded from the matrix, but this genus remains poorly known.

In order to test whether *Frugivastodon cristatus* and *Aethomylos simplicidens* are apatemyids, we added these taxa to the matrix of Silcox et al. (2010). Certain similarities shared by *Frugivastodon* and the enigmatic species *Aethomylos simplicidens* from the Uintan NALMA of North America (see *Relationships of* Frugivastodon), suggest that the latter species is a related apatemyid. We performed two analyses: in the first, we added only *F. cristatus* to the matrix, while we added both *F. cristatus* and *A. simplicidens* to the matrix for the second analysis. We performed these two distinct analyses in order to study the impact of *A. simplicidens* on the phylogenetic analysis.

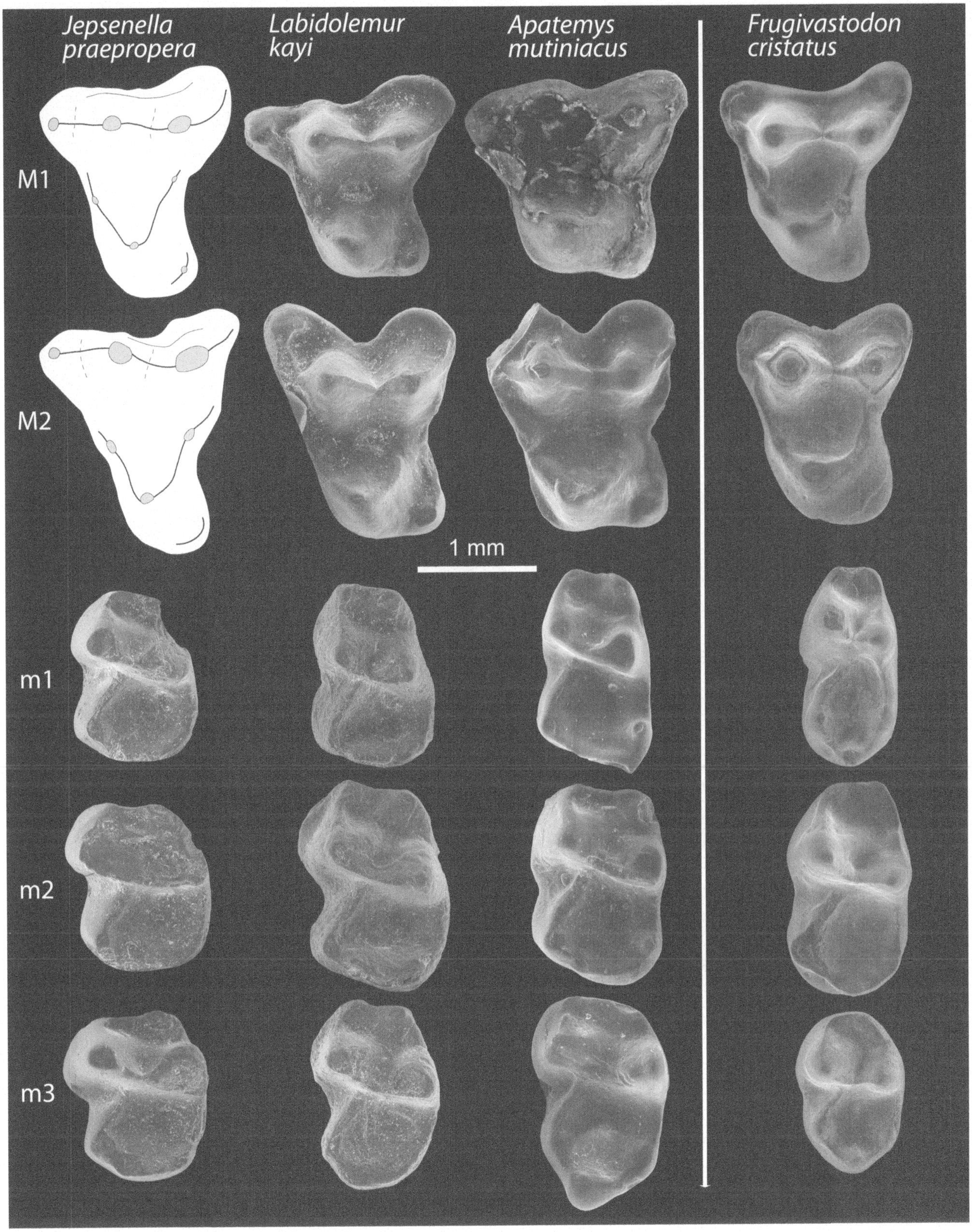

Fig. 12.7 Comparison of the M1–M2 and m1–m3 of several apatemyids from the Paleocene and early Eocene. *Jepsenella praepropera*; M1, AMNH89513 (redrawn from Szalay 1968: Fig. 8A); M2, AMNH89512 (redrawn from Szalay 1968: Fig. 8B); m1–m3, YPM-PU19184 (cast). *Labidolemur kayi*, M1–M2, UM73496 (cast); m1–m3, CM11703 (cast); *Apatemys mutiniacus*; M1, MNHN.F.MU115-L; M2, MNHN.F.MU6680 (cast); m1, MNHN.F.MU6332 (cast); m2, MNHN.F.MU6248 (cast); m3, MNHN.F.MU6024 (cast). *Frugivastodon cristatus*; M1, GU/RSR/VAS-553; M2, GU/RSR/VAS-554; m1, GU/RSR/VAS-586; m2, GU/RSR/VAS-852; m3, GU/RSR/VAS-853. Reversed views: AMNH89512, UM73496, MNHN.F.MU1274, MNHN.F.MU6332, GU/RSR/VAS-553, GU/RSR/VAS-586, GU/RSR/VAS-853

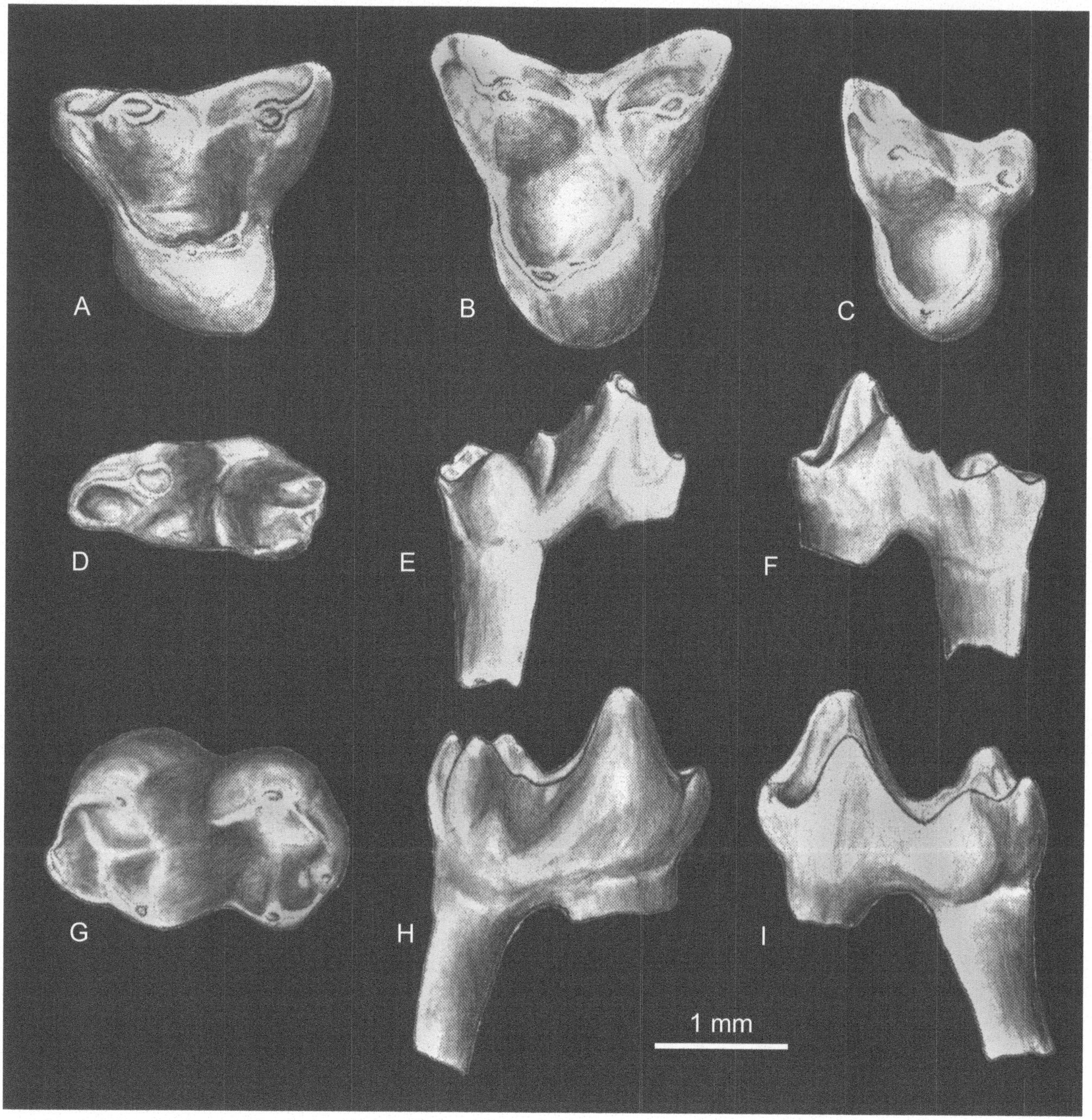

Fig. 12.8 Drawings of several specimens referred to *Aethomylos simplicidens* (early-middle Eocene; North America). **A**. UCMP 99429, right M1?, occlusal view. **B**. UCMP 104178 (holotype), left M2?, occlusal view. **C**. UCMP 101047, left M3, occlusal view. **D–F**. UCMP 101077, left m1; **D**. occlusal view; **E**. labial view; **F**. lingual view. **G–I**. UCMP 101119, right m2; **G**. occlusal view; **H**. labial view; **I**. lingual view. Reversed views: **A**, **D–F** reproduced from Novacek (1976: Figs. 16B–D and 18A–F)

The coding of *Frugivastodon cristatus* is based on the fossils described and illustrated in the present paper. *Aethomylos simplicidens* has been scored only based on the drawings and the description provided by Novacek (1976).

The data matrix was assembled with WinClada 1.00.08 (Nixon 1999, Appendix S1) and the parsimony analysis was performed with TNT 1.5 (Goloboff and Catalano 2016). The search strategy followed that of Spaulding and Flynn (2012). We analysed the data under the 'New Technology search' option, selecting the sectorial search, ratchet and tree-fusing search methods, all with default parameters. Under these settings, replications were run until the minimum length tree was found in 500 separate replicates. The trees generated were then analysed under typical search options (using TBR) to explore fully the tree islands discovered. As in Silcox et al. (2010), all characters were equally weighted and treated as unordered. Results were examined with Winclada 1.00.08 using strict consensus and majority-rule consensus. Bremer support indices were determined using TNT: they have been calculated with TNT for 10 supplementary steps. Bootstrap values were calculated using TNT (10,000 bootstrap replicates).

Results

The first phylogenetic analysis (*Frugivastodon* added to the matrix) yielded three equally parsimonious trees, with a tree length of 1095 steps, CI of 0.29 and RI of 0.52. Such low indices suggest substantial homoplasy. The strict consensus of these three trees is 1101 steps long with a CI = 0.29 and RI = 0.51 (Fig. 12.9). In the strict consensus, two nodes are collapsed. The values of Bremer support index that TNT has found are between one and seven (only one node); three nodes have a Bremer support higher than ten. Seven nodes have a Bootstrap value higher than 50.

The second phylogenetic analysis (*Frugivastodon* and *Aethomylos* added to the matrix) yielded nine equally parsimonious trees, with a tree length of 1099 steps, and CI of 0.29 and RI of 0.52. The strict consensus tree is 1138 steps long with a CI = 0.28 and RI = 0.50 (Fig. 12.10A). In the strict consensus, ten nodes are collapsed. The majority-rule consensus is 1099 steps long (Fig. 12.10B). Its CI and RI equal 0.29 and 0.52, respectively. As in the first analysis, the values of Bremer support index that TNT has found are generally between one and seven, while three nodes have a Bremer support higher than ten (Fig. 12.10A). Seven nodes have a Bootstrap value higher than 50 (Fig. 12.10A).

Discussion

As in Silcox et al. (2010), there are two areas of disagreement among the trees of the first phylogenetic analysis (Fig. 12.9): the relationships of Leptictida and Paleoryctidae to the other major clades and the branching order among the Paleoryctidae (*Pararyctes pattersoni*, *Paleoryctes* sp., *Ottoryctes winkleri*). The relationships among the taxa are the same as those found by Silcox et al. (2010), and they are thus not discussed here. Notably, the addition of *Frugivastodon* to the matrix did not change the relationships among the taxa already analysed.

The node that unites the apatemyids has a Bremer support index of 5 (Fig. 12.9). The Paleocene *Jepsenella praepropera* is the most basal apatemyid; this agrees with the results of Silcox et al. (2010) and previous hypotheses that consider this species as one of the most primitive apatemyids (Simpson 1940; McKenna 1963a; Szalay 1968). *Frugivastodon* is included among Apatemyidae and is the sister taxon to all the other apatemyids in the analysis: *Labidolemur kayi*, *Apatemys* sp., *Carcinella sigei*, *Heterohyus nanus*, and *Sinclairella dakotensis*. This relationship is supported by the following synapomorphies: 159(1) (straight preparacrista on M1); 206(1) (moderate crown height on m1); 214(1) (metaconid positioned well behind the level of the protoconid on m1); 223(1) (mesiolabial projection present on m2); 235(3) (labial cingulid absent on lower molars).

The node that includes *L. kayi*, *Apatemys* sp., *C. sigei*, *H. nanus*, and *S. dakotensis* is only supported by the character 153(1) (M1 more elongate relative to transverse width than M2 or M3). This feature is an unusual character among placental mammals and is therefore phylogenetically highly significant. This group of genera is essentially the subfamily Apatemyinae of McKenna and Bell (1997), excluding *Jepsenella* (Fig. 12.9). The relationships among this clade are similar to those found by Silcox et al. (2010); as a consequence, the addition of *Frugivastodon* to the analysis did not change the relationships among Apatemyinae. *Labidolemur kayi* is the most basal member, while the European *C. sigei* is closely related to the European *H. nanus* and to the North American *S. dakotensis*. Silcox et al. (2010) noted that the relationship between *Carcinella* (from the ?late Eocene of the 'Phosphorites du Quercy') and *Heterohyus+Sinclairella* is poorly supported (two dental characters are related to the P4 in the present analysis). The close relationship between the North American *S. dakotensis* and the European *H. nanus* – a relationship that suggests, as noted by Silcox et al. (2010), that *Sinclairella* may have been a European migrant (see *Geographic origin of* Frugivastodon *and the dispersal of apatemyids*) – is well supported (seven synapomorphies; Bremer support = 2; Bootstrap value = 81) (Fig. 12.9). This relationship is supported by several cranial and postcranial features, as well as numerous dental ones.

The strict consensus of the second phylogenetic analysis, which included *Frugivastodon* and *Aethomylos*, is less resolved than the one produced by the first analysis (Fig. 12.10A). However, the Bremer support and Bootstrap values are very similar to those found in the first analysis; the clades that were well supported in the first analysis are supported in the second.

We found the two areas of disagreement as in the first analysis and Silcox et al. (2010). However, the results of the second analysis display several important differences from that of Silcox et al. (2010). We only list here the differences in the topology with the first analysis and that of Silcox et al. (2010) because the relationships among Euarchontoglires have already been discussed by Silcox et al. (2010). It is worth noting that our phylogenetic analysis does not support Euarchonta (Fig. 12.10), which is a clade that is well supported in molecular analyses (e.g., Springer et al. 2004).

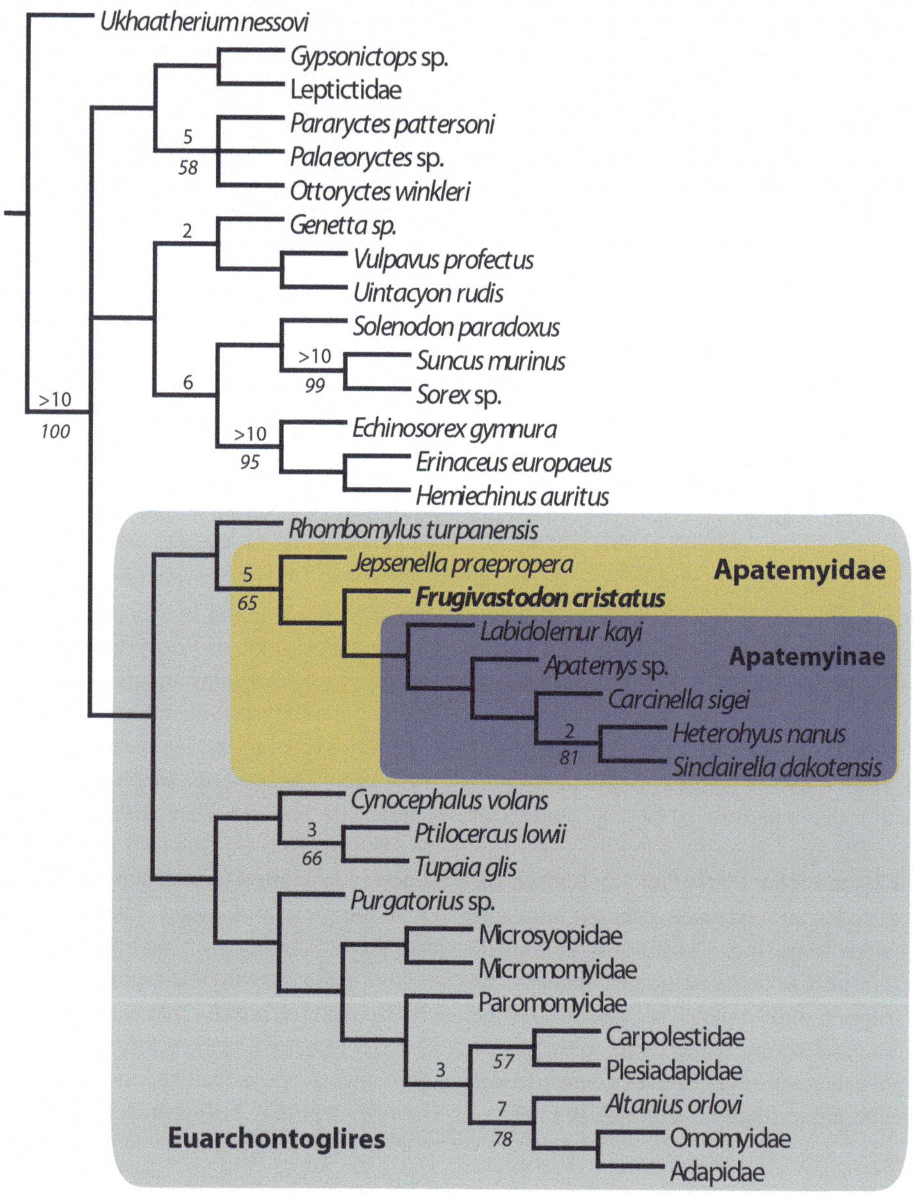

Fig. 12.9 Strict consensus of 3 equally parsimonious trees from the first analysis, adding only *Frugivastodon* to the dataset of Silcox et al. (2010) (L = 1095 steps; CI = 0.29; RI = 0.52): 1101 steps long with a CI = 0.29 and RI = 0.51. Bremer supports (above node) and bootstrap values greater than 50% (below node)

In the strict consensus tree of the second analysis, the majority of the polytomies are localized among the Euarchontoglires (Fig. 12.10A). The primate *Altanius orlovi* is in a polytomy with the Omomyidae and Adapidae (Fig. 12.10A); it was the sister taxon of the two families in the first analysis. However, in the majority-rule consensus of the second analysis, *Altanius orlovi* is more closely related to Omomyidae (Fig. 12.10B).

The relationships of the Paromomyidae in the second analysis are different from those found by the first analysis and by Silcox et al. (2010). In the latter study, the family was the sister taxon of the clade that comprises the Carpolestidae, Plesiadapidae, *Altanius*, Omomyidae, and Adapidae (i.e., Euprimateformes). In our second analysis (majority-rule consensus; Fig. 12.10B), Paromomyidae are closely related to Microsyopidae and Micromomyidae. This relationship, however, seems less probable – based on the current morphological and stratigraphic data – than the relationships found by Silcox et al. (2010) and our first analysis.

In the second analysis there is a lack of resolution concerning the phylogenetic position of the primitive plesiadapiform *Purgatorius* (Fig. 12.10A); its position in our analysis does not support the monophyly of the Primates. *Purgatorius* is indeed more basal among Euarchontoglires

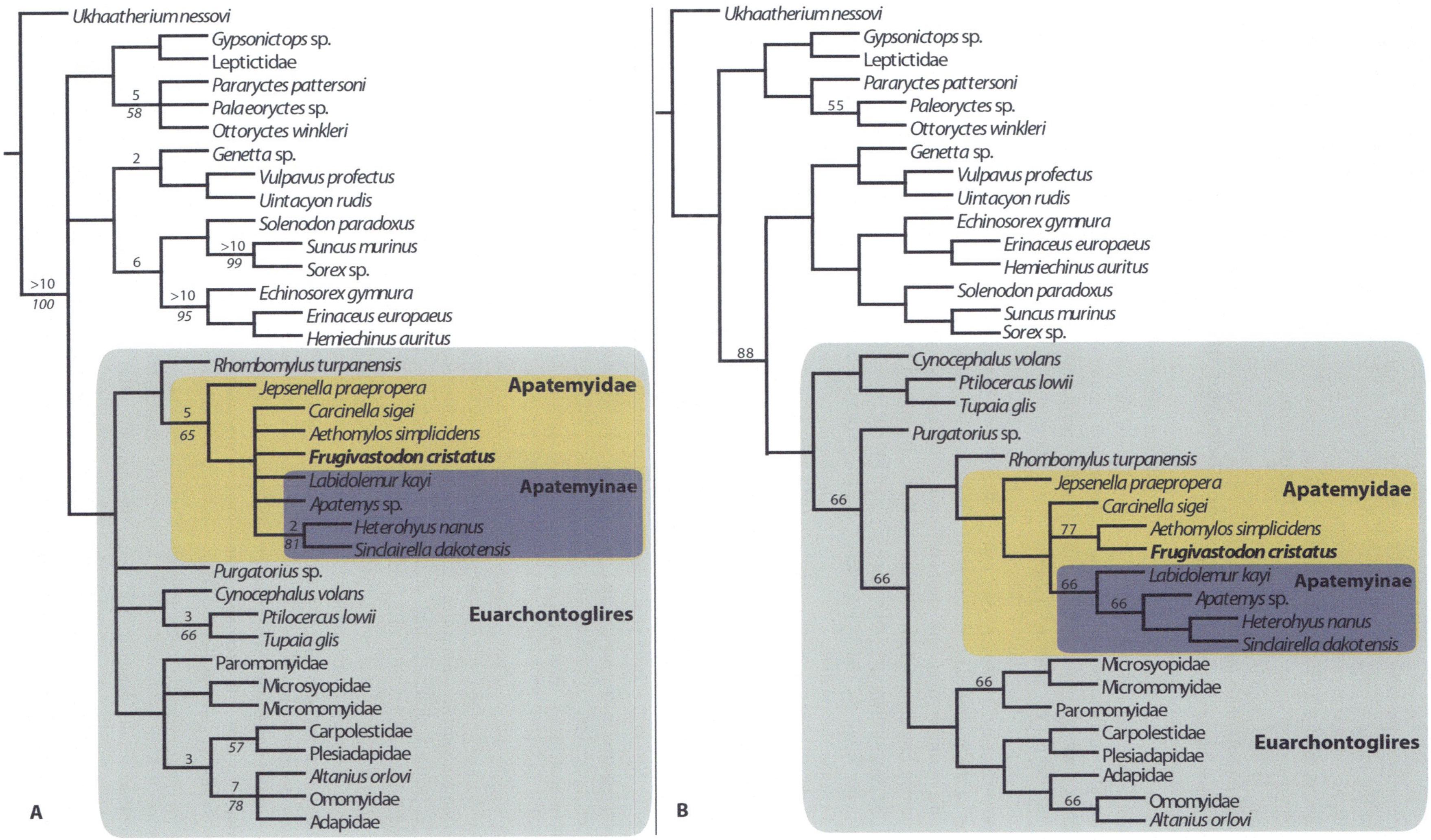

Fig. 12.10 Consensus of 9 equally parsimonious trees from the analysis adding *Frugivastodon* and *Aethomylos* to the dataset of Silcox et al. (2010) (L = 1099 steps; CI = 0.29; RI = 0.52). **A** Strict consensus with Bremer supports (above node) and bootstrap values greater than 50% (below node); 1138 steps long with a CI = 0.28 and RI = 0.50. **B** Majority-rule consensus; 1099 steps long with a CI = 0.29 and RI = 0.52

than in the first analysis and Silcox et al. (2010) (Fig. 12.10B).

Apatemyids and *Rhombomylus turpanensis* are less basal among Euarchontoglires in the majority-rule consensus of the present analysis than in the strict consensus tree of the first analysis and that of Silcox et al. (2010): they are the sister group to Primates (they are even more inclusive than *Purgatorius*) (Fig. 12.10B). In this topology, they are included among Euarchonta. Finally, the Sundatheria (and thus Scandentia) are more basal in the majority-rule consensus than in the first analysis and Silcox et al. (2010).

The relationships listed above, based on our analysis including both *Frugivastodon* and *Aethomylos*, seem less likely than the results of the first analysis and Silcox et al. (2010), based on our current knowledge of the evolution of the Euarchontoglires. Because the present study focused on apatemyids, only further studies will determine if there is merit to any of these rearrangements. However, one can conclude that the inclusion of the fragmentary *Aethomylos* introduces instability into the resulting trees and strongly modifies the relationships among Euarchontoglires. We also performed analyses with the fragmentary possible apatemyids *Eochiromys* and *Russellmys*, but the results were less resolved and the analyses were not informative.

Nevertheless, the second phylogenetic analysis supports allocation of *Aethomylos* to Apatemyidae. The node that unites apatemyids has a Bremer support index of 2 (Fig. 12.10A); it equals five in the first analysis (Fig. 12.9), even though there is little resolution among apatemyid genera. The Paleocene *Jepsenella praepropera* is again the most basal apatemyid (Fig. 12.10). In the strict consensus, only the close relationship between *H. nanus* and *S. dakotensis* is found (Bremer support = 2; Bootstrap value = 81) (Fig. 12.10A). The other apatemyids form a polytomy. In the majority-rule consensus tree (Fig. 12.10B), relationships among some Apatemyidae are resolved. *Apatemys* sp. is closely related to *H. nanus* and *S. dakotensis*. *Labidolemur kayi* is related to *Sinclairella*, *Heterohyus*, and *Apatemys*. This topology is partially similar to that of the first analysis – only *Carcinella* is excluded from this group.

The newly added taxa, *Frugivastodon* and *Aethomylos*, form a monophyletic clade in the majority-rule consensus (Fig. 12.10B). This relationship is supported by the similarities noted above. *Frugivastodon* and *Aethomylos* thus might represent a distinct clade of apatemyids. The position of *Carcinella* is still uncertain. It shares with *Aethomylos* the absence of a hypocone, but some features distinguish *Carcinella* from *Frugivastodon* and *Aethomylos* (see section "Relationships of Frugivastodon").

Therefore, our phylogenetic results indicate that two groups of apatemyids might have evolved independently during the Eocene. The Paleocene age of *Labidolemur kayi* implies that the two groups may have separated during the Paleocene. They both root in Paleocene *Jepsenella*, and as a result, the primitive apatemyid *Jepsenella* should be excluded from the Apatemyinae.

Discussion

Relationships of Frugivastodon

To further assess the affinities of *Frugivastodon*, we compared it to the apatemyids present in Europe and North America at the time it was present in India (c. 54.5 Ma). In North America, two distinct genera are known. *Labidolemur*, first recorded in the Tiffanian, is represented in the early Eocene by *L. kayi* (Cf-1/Wa-6) and *L. serus* (Wa-1/Wa-5). *Apatemys* is represented by *A. pygmaeus* (Wa-0), *A. chardini* (Wa-0/Wa-7), *A. rodens* (Wa-0/Br-3), and *A. bellulus* (Wa-0/Br-3) (Gingerich 1982; Gunnell et al. 2008a; Beard and Dawson 2009).

In Europe, the genus *Apatemys* is known only in the early Eocene, when it was already diverse. Four species are known: *A. teilhardi* at Dormaal (MP7; c. 56 Ma) (Teilhard de Chardin 1927; Russell et al. 1979), *A. sigogneaui* at Pourcy (MP7) and Mutigny and Avenay (MP8+9; Avenay ≈ 51–52 Ma) (Russell et al. 1979), *A. prouti* at Abbey Wood (MP8+9; c. 55.12 Ma) (Hooker 2010b), and *A. mutiniacus* at Mutigny and Avenay (MP8+9) (Russell et al. 1979). A second apatemyid genus, *Heterohyus*, is well diversified in Europe. It is represented in early Eocene localities by indeterminate species: at Silveirinha (MP7), Mutigny, and Avenay (MP8+9), Grauves, and Cuis (MP10) (Russell et al. 1979; Estravís 1992). The genus is far better known in the middle and late Eocene where nine species are presently recorded (Koenigswald et al. 2009: Table 2). In addition to *Apatemys* and *Heterohyus*, there are two poorly known genera in the earliest Eocene of Europe: *Eochiromys landenensis* from Dormaal (two specimens) and *Russellmys denisae* from Silveirinha (one specimen) (Russell et al. 1979).

As discussed in the Systematics section above, the Indian taxon is unusual compared to the typical Eocene apatemyids *Apatemys* and *Heterohyus* due to the relative size and morphology of the molars (M1 transversely elongated and bearing only a small hypocone; high postprotocrista; presence of metaconule; tiny fourth trigonid cusp; paraconid not distally located; trigonid not parallelogram-like; talonid not wider than the trigonid; high entocristid; entoconid only slightly smaller than the hypoconid; reduced m3; talonid mesiodistally short on the m3). As mentioned above, some of the distinctive features of *Frugivastodon* (e.g., morphology of the M1) are more typical of Paleocene apatemyids such as *Labidolemur* and *Jepsenella* than of contemporaneous apatemyids. The Indian taxon notably shares with *Labidolemur* and *Jepsenella* the high crown and the relative size of the molars (primitive features among apatemyids), except the shorter m3. However, the morphology of the Indian fossils clearly distinguishes *Frugivastodon* from *Labidolemur* and *Jepsenella*: the Indian taxon differs from these genera in having more mesiodistally elongated molars, less mesiodistally compressed trigonids, less distally located paraconid, smaller fourth cusp, narrower talonid, higher entocristid, wider M1, and less developed hypocone. *Frugivastodon* is more similar to *Jepsenella* than to *Labidolemur* in having wider, triangular upper molars (a primitive feature). Moreover, *Frugivastodon* and *Jepsenella* possess M1 and M2 that have close L/W ratios – other apatemyids have very different ratios between M1 and M2 (Table 12.3).

Comparison of *Frugivastodon* with the better known apatemyids from the Paleocene and Eocene highlights the numerous unusual features shown by the Indian apatemyid, which suggest that *Frugivastodon* represents a clade separate from that of *Labidolemur*, *Apatemys*, and *Heterohyus*. Among the particular features of *Frugivastodon* is the small hypocone (a primitive feature). This feature is rare among apatemyids, but it characterizes the Paleocene *Jepsenella* (Szalay 1968). It is also found in two European apatemyids known in the Eocene: *Russellmys denisae* from the locality of Silveirinha (Portugal, MP7; Estravís 1990) and *Carcinella sigei* from the 'Phosphorites du Quercy' (France; ?late Eocene; Koenigswald et al. 2009).

Russellmys denisae, from the earliest Eocene of Silveirinha, was described based on a left maxillary fragment bearing the almost intact crown of M2 and bases of M1 and M3. As only one molar is known, the inclusion of this genus among Apatemyidae is problematic (Estravís 1990, 1992). If correctly attributed to this family, it would represent one of the smallest apatemyids recorded in Europe (Estravís 1990; Koenigswald et al. 2009). As in *Frugivastodon*, but unlike other Eocene apatemyids, the hypocone of M2 is either very small or absent; wear makes its presence ambiguous. *Russellmys* also differs from other apatemyids in having a protocone that is less mesially located (resulting in a triangular aspect in occlusal view), a high postprotocrista, and a metaconule. *Frugivastodon* and *Russellmys* share the triangular morphology of the upper molars in occlusal view, the presence of a small metaconule and a distinct postprotocrista (the trigon is distally open in other Eocene apatemyids due to the absence of this structure), and a weak hypocone. The styles are shorter in the Portuguese fossil than in *Frugivastodon* (as well as in all the other apatemyids). Moreover, the area where the hypocone may be situated is somewhat more labially located than in *Frugivastodon* and apatemyids in general – the postcingulid actually extends lingually in apatemyids. As a consequence, the comparison between *Frugivastodon* and *Russellmys* does not evidence similarities between these two taxa, leaving open the question of the systematic status of *Russellmys*.

Estravís (1990) noted similarities between *Russellmys* and *Aethomylos*. Koenigswald et al. (2009, p. 62) followed Estravís (1990) and considered that *Aethomylos* is "almost certainly an apatemyid." *Aethomylos* was initially considered as Insectivora *incertae sedis* because of the peculiar morphology of its upper and lower molars (Novacek 1976; Gunnell et al. 2008b), although Novacek considered the possibility that it could represent an apatemyid. He listed several similarities with apatemyids, including the bowl-shaped structure of the talonid basin, the high entocristid, the morphology of the angular mesiolabial corner of the trigonid (position of the fourth cusp of the trigonid), the low profile of the trigonids, and the more posterior position of the metaconid relative to that of the protoconid (Novacek 1976) (Fig. 12.8). At the same time, however, he noted several important differences: the mesiolabial corner of the trigonid is farther posterior than the paraconid (hence the trigonid lacks the typical parallelogram occlusal outline of Eocene apatemyids), and the upper molars differ in lacking a distinct hypocone (Fig. 12.8). In these differences *Aethomylos* resembles *Frugivastodon*. *Frugivastodon* and *Aethomylos* also share the presence of a distinct postprotocrista that closes the trigon distally, a high entocristid, an entoconid almost as high as the hypoconid, mesiodistally elongated molars (Table 12.5), talonids of similar width to the trigonids (the talonid is wider than the trigonid, at least on the m1, in other Eocene apatemyids), and an m3 shorter than the m1 and m2. Because the teeth of *Aethomylos* are isolated, and upper loci were not identified with confidence, it is premature to reach any conclusions about similarity to *Frugivastodon* in upper molar proportions. The similarities shared by *Frugivastodon* and *Aethomylos* support the conclusions of Estravís (1990) and Koenigswald et al. (2009) that *Aethomylos* is probably an apatemyid. Moreover, the results of the phylogenetic analysis presented herein support a close relationship between *Frugivastodon* and *Aethomylos* despite the great geographic separation between the two taxa (see above; Fig. 12.10B). More complete specimens of both taxa would obviously help to test this hypothesis.

Carcinella was described by Koenigswald et al. (2009) based on a well-preserved skull found in the 'Phosphorites du Quercy'. This small apatemyid is possibly late Eocene in age; its precise locality and age are unknown. It differs from *Heterohyus* (Eocene of Europe) by having a remnant hypocone on M1 and no hypocone on M2 (derived features?). This genus differs from *Frugivastodon* in the loss of the postprotocristid and metaconule (derived features), resulting in a distally open trigon as in most Eocene apatemyids. The protocone area of the M1 and M2 is also reduced compared to *Frugivastodon*. This feature also characterizes the Eocene

Apatemys, *Heterohyus* and *Sinclairella*. The lower dentition of *Carcinella* is unfortunately unknown, which prevents comparison to the lower molars of *Frugivastodon*. However, the large size of the M3 – notably of its protocone – indicates that the m3 was probably mesiodistally elongated, as seen in *Apatemys* and *Heterohyus*, and probably as long as the m1 (a derived feature), rather than reduced as in *Frugivastodon*. Thus, although *Carcinella* seems to be potentially close to *Frugivastodon* based on the absence of a hypocone, the overall morphology of the dentition recalls the apatemyines (*Heterohyus*, *Apatemys*, *Sinclairella*). If *Carcinella* is indeed closely related to *Apatemys* and *Heterohyus*, this would imply that at least two groups of apatemyids independently greatly reduced or lost the hypocone.

Geographic Origin of Frugivastodon *and the Dispersal of Apatemyids*

Apatemyidae first appear in the early Paleocene of North America (Torrejonian, To-3), but are not known in Europe until the earliest Eocene. Consequently, it seems reasonable to postulate that the family originated in North America and dispersed into Europe around the Paleocene-Eocene boundary (e.g., during the PETM, Fig. 12.11). European and North American faunas were very similar (at the genus and family levels) during the earliest Eocene (e.g., Simpson 1947; McKenna 1975; Krause and Maas 1990). Dispersals from North America into Europe have been evidenced for numerous mammalian groups (e.g., Nyctitheriidae, Oxyaenidae, Palaeanodonta, Pantodonta, Rodentia) (Russell 1975; Smith 1996; Lucas 1998; Smith and Smith 2001; Gheerbrant et al. 2005; Solé et al. 2011), whereas several mammals may have followed the opposite direction (e.g., Artiodactyla, Carnivoraformes, Hyaenodonta, Perissodactyla, Primates) (Gingerich 1989; Smith and Smith 2001; Godinot and de Broin 2003; Hooker and Dashzeveg 2003; Smith et al. 2006; Solé et al. 2014). The dispersals that are inferred to have occurred at that time have been recently summarized by Hooker (2015).

The presence of three species referred to *Apatemys* and four or possibly five lineages of apatemyids in the early Eocene of Europe led Russell et al. (1979) to conclude that apatemyids experienced a rapid and important radiation when they arrived in Europe. As a consequence, some paleontologists have also suggested that they were already present in Europe in the Paleocene. Szalay (1968) discussed this possibility based on an upper molar figured by Russell et al. (1966) as *Aboletylestes* sp. that he considered similar to those he referred to *Jepsenella*; however, this isolated tooth is now considered to be a pantolestan, *Nosella europaea* (López-Martínez and Peláez-Campomanes 1999). The absence of apatemyids in the Paleocene fossil record of Europe could result from the scarcity of fossiliferous localities known in Europe during this epoch and the rarity of apatemyids. It has recently been suggested that European mammals from the Paleocene document several episodes of dispersal from North America into Europe (De Bast and Smith 2017).

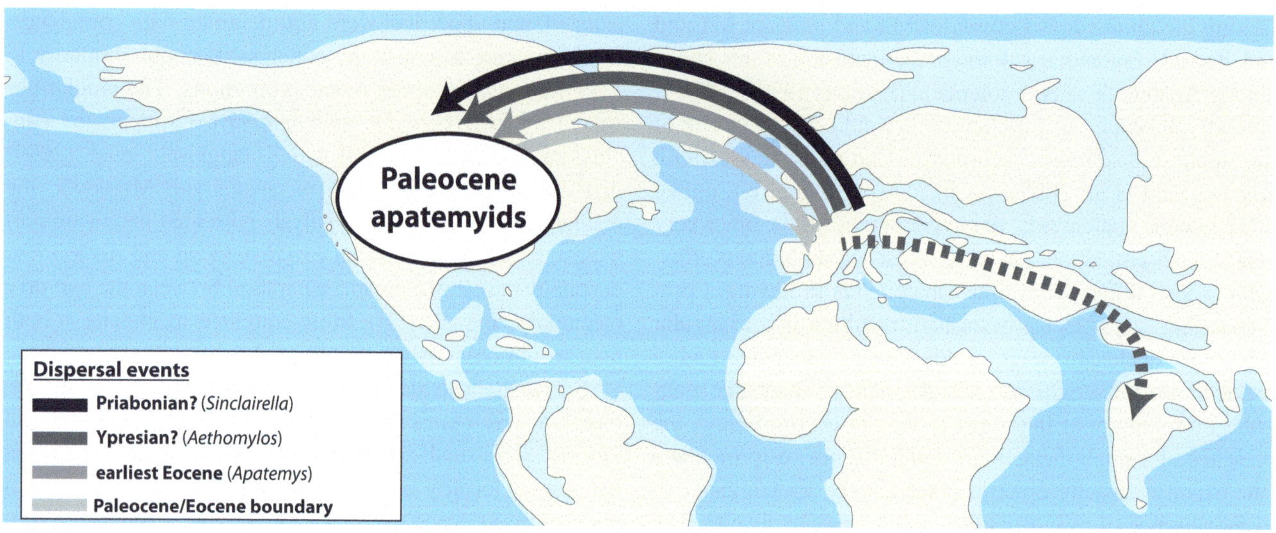

Fig. 12.11 Schematic map showing the hypothetical geographic dispersion of apatemyids. Apatemyidae dispersed between North America and Europe via the Greenland Bridge and Thulean route; they dispersed from Europe to India probably through the Turgai Strait and/or along the Tethyan shore. Oldest apatemyids are recorded in the Paleocene of North America (Torrejonian) (Simpson 1940; McKenna 1963a; Szalay 1968; West 1973b). See text for explanations of the dispersal events. The exact timing of the dispersals of Apatemyidae in India is unknown but probably occurred in the Ypresian. Paleogeographic map adapted from Ron Blakey, Eocene (http://www2.nau.edu/rcb7/050Marect.jpg)

Apatemyids seem to have dispersed between Europe and North America in the earliest Eocene. Gingerich (1982) postulated that *Apatemys chardini* may have dispersed from Europe into North America (Fig. 12.11; earliest Eocene) as part of the Wasatchian faunal cohort that arrived in North America in the earliest Eocene (Gingerich 1989; Woodburne et al. 2009; Hooker 2015). Therefore, European apatemyids may have been closer to the centre of origin of *Apatemys* than North American ones. If *A. chardini* evolved from a European taxon, this would imply rapid radiation of apatemyids in Europe, followed by dispersal back into North America in the earliest Eocene. Unfortunately, the few isolated teeth from the earliest Eocene (Wa-0) of North America (*A. pygmaeus* from the Red Hot local fauna of Mississippi [Beard and Dawson 2009], and *Apatemys* sp. from the southern Bighorn Basin of Wyoming [Strait 2001]) are insufficient to shed light on the geographic origin of *Apatemys*. It may be noted that Sluijs et al. (2014) proposed an age younger than Wa-0 for the Red Hot fauna.

None of the early Eocene apatemyids found in the Belgium, London and Paris basins (=Northwestern European Province) (e.g., Dormaal, Abbey Wood, Mutigny, Avenay) and in the south of France and Iberian Peninsula (=Southwestern European Province) (e.g., Silveirinha, Rians) seem to be related to *Frugivastodon*. Even *Apatemys teilhardi*, the oldest European apatemyid (together with the very fragmentary *Eochiromys*) and the slightly younger *A. mutiniacus*, appear too derived to be closely related to *Frugivastodon* (Fig. 12.7): the difference in width between the (wider) talonid and the trigonid on m1 is greater than in *Frugivastodon*; the position of the entoconid is more distal; the talonid is open lingually; the paraconid is more distally located (giving a pronounced trapezoidal aspect to the trigonid in occlusal view); the M1 is narrower transversely (Table 12.3); and the internal crests, paraconule, metaconule and postparaconule crista are reduced. However, lower molar proportions of *A. teilhardi* are intermediate between those of the late Paleocene-early Eocene apatemyids from North America and *Frugivastodon* (Table 12.5). The oldest European apatemyids might have undergone a mesiodistal elongation of the lower molars similar to *Frugivastodon*.

Because apatemyids were restricted to Europe and North America before the discovery of *Frugivastodon*, the presence of this latter taxon in southern Asia is intriguing. How can we explain the presence of *Frugivastodon* in India? Two hypotheses can be envisaged: (1) apatemyids dispersed from North America into Asia and thus into India via the Bering Land Bridge; (2) they dispersed first from North America into Europe via the Greenland Bridge and Thulean route, and then from Europe to India via land connections across the Turgai Strait and/or along the Tethyan shore – the southern route being more likely. As suggested above, the Indian apatemyid shares mesiodistal elongation of the molars (a possible derived character) with *A. teilhardi* from the earliest Eocene of Europe. Certain other particular features of *Frugivastodon* are shared only with some North American apatemyids from the Paleocene or the Paleocene-Eocene boundary (*Jepsenella*, *Labidolemur*). Since apatemyids are otherwise unknown from Asia, the hypothesis of a direct dispersal from North America into Asia seems less likely. Consequently, present evidence suggests Europe as the potential geographic source of *Frugivastodon*.

The possibility of dispersal from Europe into India is strengthened by the presence of several mammal taxa in the Vastan local fauna (bats, rodents, primates, tillodonts) that strongly suggest faunal connections between Europe and India during the early Eocene (Smith et al. 2007, 2016; Rana et al. 2008; Rose et al. 2009a, b). The direction of this dispersal is, however, still uncertain. By contrast, the Vastan fauna displays only a few similarities to Asian faunas (Rose et al. 2008). The timing of this dispersal is uncertain but it presumably occurred during the early Ypresian or even earlier (Fig. 12.11; Ypresian).

The European endemic genus *Heterohyus*, which is mainly characterized by its incisor morphology (Russell et al. 1979) and the development of the hypocone, is first recorded in Europe at Mutigny (MP8+9), a locality of the Paris Basin that ranges in age between 55.12 Mya (age of Abbey Wood; Hooker 2010b) and 52–51 Mya (age of Avenay; Smith and Smith 2013) – while Vastan age is considered to be around 54.5 Ma (Bajpai et al. 2005; Adatte et al. 2014; Samanta et al. 2016; Smith et al. 2016). As a result, *Heterohyus* and *Frugivastodon* appeared almost at the same time, but in different Laurasian regions. These two endemic lineages can be viewed as representing a phase of local evolution that followed the dispersal that occurred around the Paleocene-Eocene boundary and in the early Ypresian from North America to Laurasia.

As discussed above, certain molar features suggest that *Frugivastodon* could be closely related to North American *Aethomylos*. However, because of the geographic and temporal separation between the two taxa (India *vs* North America, early Eocene *vs* middle Eocene), and the very limited fossil evidence on which their possible relationship is based, the following interpretations are largely speculative. Given the older age of *Frugivastodon*, and some features that might be interpreted as more primitive, it may be postulated that the ancestors of *Aethomylos* could have dispersed from Europe (or from India through Europe) into North America either during the Bridgerian (i.e., during the Ypresian) or around the Brigerian/Uintan boundary (Fig. 12.11; Ypresian?). It is worth noting that the Uintan, like the Bridgerian, experienced only minor immigrations (Woodburne et al. 2009). In the absence of evidence, a dispersal route through Asia and across the Bering land bridge cannot be ruled out.

The beginning of the Bridgerian records a new evolutionary diversification of the mammals, probably related to the Early Eocene Climatic Optimum (EECO) (Woodburne et al. 2009). However, none of the 'new' taxa recorded in the Bridgerian are immigrants according to Woodburne et al. (2009). Once again, because apatemyids, except for *Frugivastodon*, are unknown in Asia, a dispersal from Europe into North America appears to be equally as likely as dispersal directly from Asia to North America across Beringia. Finally, the sudden appearance of *Aethomylos* in the North American fossil record implies that the group that includes *Aethomylos* and *Frugivastodon* may have survived at least until the Uintan.

Another possible dispersal of European apatemyids into North America has been recently discussed by Silcox et al. (2010). As shown by these authors, *Sinclairella dakotensis* and the European *Heterohyus nanus* have a close relationship that suggests *Sinclairella* may have been a European migrant. In this case, the dispersal may have occurred during the Priabonian (*Sinclairella* is first recorded in the Chadronian NALMA, Ch-3) (Fig. 12.11; Priabonian).

The dispersals reconstructed here imply that apatemyids dispersed several times from Europe into North America. Conversely, one can hypothesize that parallel evolution could have occurred in Europe (or India) and North America. As is often the case, more definitive conclusions on the dispersal of apatemyids will require an improved fossil record.

Conclusions

The new fossils discovered at Vastan Lignite Mine confirm the reference of *Frugivastodon* to Apatemyidae. The Apatemyidae were thus present in the Indian subcontinent as well as in Europe and North America. The morphology of the teeth discovered in the early Eocene of India differs from that of most Apatemyidae known from the Eocene of North America and Europe. Similarities between the teeth of *Frugivastodon* and those of *Aethomylos*, from the early and middle Eocene of western North America, strengthen the interpretation that *Aethomylos* is an apatemyid. Further, the resemblances suggest that *Frugivastodon* and *Aethomylos* represent a separate clade of apatemyids, which probably originated during the late Paleocene or, at least, around the Paleocene-Eocene boundary, although where this clade originated is uncertain.

The numerous differences in the dental anatomy of *Frugivastodon* compared to apatemyines (*Labidolemur, Apatemys, Heterohyus, Sinclairella*) indicate that the early Eocene radiation of Apatemyidae was more complex than has been envisioned, but the history and phylogeny of the apatemyids remain poorly understood. Europe appears to have been a crucial area for deciphering the evolution of these insectivorous mammals.

Acknowledgements It is an honor to dedicate this paper to Ashok Sahni, in recognition of his unequalled contributions to the field of vertebrate paleontology in India, and his integral role in establishing our project on the early Eocene vertebrates of Gujarat. We are grateful to G. V. R. Prasad and Rajeev Patniak for inviting us to contribute to this volume honoring Prof. Sahni. We thank other members of our field team who participated in the paleontological expeditions at the Vastan lignite mine, including H. E. Ahrens, R. H. Dunn, A. Folie, F. D. H. Gould, P. Missiaen, H. Singh, L. Singh, and G. McKusick Voegele, and the personnel of the Gujarat Industries Power Company Ltd. for facilitating our excavations. At the RBINS, Julien Cillis assisted during the production of SEM images. We are grateful to Richard Smith for access to his cast collection, and to G. Billet and K. Le Verger (MNHN) for the picture of the specimen MNHN.F.MU115-L. We are also grateful to M. Novacek for allowing us to reproduce his illustrations of *Aethomylos simplicidens*, A. Henrici for information concerning the Carnegie Museum Powder Wash specimens referred to *Aethomylos*, and R. Stucky for information on occurrence of *Aethomylos*. Field work and research was supported by the National Geographic Society (grants 6868-00, 7938-05, 8356-07, 8710-09, 8958-11, and 9240-12 to KDR); the Director, Wadia Institute of Himalayan Geology, Dehradun, India (to KK); and the Federal Science Policy Office of Belgium (BELSPO BR/121/A3/PalEurAfrica to T.S.). In this endeavor, a networking project (BELSPO BL/36/fwi05 to T.S.) funded travels of two Belgian and three Indian researchers to India and Belgium, respectively, to exchange their scientific expertise. Finally, we thank Philip Gingerich, Jerry Hooker, and Mary Silcox for their insightful reviews, which led to significant improvements in the manuscript.

References

Adatte, K. H., Khozyem, H., Spangenberg, J. E., Samant, B., & Keller, G. (2014). Response of terrestrial environment to the Paleocene–Eocene Thermal Maximum (PETM), new insights from India and NE Spain. *Rendiconti Online della Società Geologica Italiana, 31,* 5–6. https://doi.org/10.3301/ROL.2014.17.

Armstrong, S. D., Bloch, J. I., Houde, P., & Silcox, M. T. (2011). Cochlear labyrinth volume in euarchontoglirans: Implications for the evolution of hearing in Primates. *Anatomical Record, 294,* 263–266.

Bajpai, S., & Kapur, V. V. (2008). Earliest Cenozoic frogs from the Indian subcontinent: Implications for out-of-India hypothesis. *Journal of the Palaeontological Society of India, 53,* 65–71.

Bajpai, S., Kapur, V. V., Das, D. P., Tiwari, B. N., Saravanan, N., & Sharma, R. (2005). Early Eocene land mammals from Vastan Lignite Mine, District Surat (Gujarat), western India. *Journal of the Palaeontological Society of India, 50,* 101–113.

Bajpai, S., Kapur, V. V., & Thewissen, J. G. M. (2009). Creodont and condylarth from the Cambay Shale (Early Eocene, ~55–54 Ma), Vastan Lignite Mine, Gujarat, western India. *Journal of the Palaeontological Society of India, 54,* 103–109.

Bajpai, S., Kapur, V. V., Thewissen, J. G. M., Das, D. P., & Tiwari, B. N. (2006). New Early Eocene cambaythere (Perissodactyla, Mammalia) from the Vastan Lignite Mine (Gujarat, India) and an evaluation of cambaythere relationships. *Journal of the Palaeontological Society of India, 51,* 101–110.

Beard, K. C., & Dawson, M. R. (2009). Early Wasatchian mammals of the Red Hot local fauna, uppermost Tuscahoma Formation, Lauderdale County, Mississippi. *Annals of Carnegie Museum, 78,* 193–243. https://doi.org/10.2992/007.078.0301.

Biochrom'97. (1997). Synthèses et corrélations. In J. P. Aguilar, S. Legendre & J. Michaux (Eds.), *Actes du congrés BiochroM'97. Mémoires des travaux EPHE de l'Institut de Montpellier* (Vol. 21, pp. 769–805). Montpellier.

Bloch, J. I., & Boyer, D. M. (2001). Taphonomy of small mammals in freshwater limestones from the Paleocene of the Clarks Fork Basin. *University of Michigan Papers on Paleontology, 33,* 185–198.

Boyer, D. (2008). Relief index of second mandibular molars is a correlate of diet among prosimian primates and other euarchontan mammals. *Journal of Human Evolution, 55,* 1118–1137.

Butler, P. M. (1952). The milk-molars of Perissodactyla, with remarks on molar occlusion. *Proceedings of the Zoological Society of London, 121,* 777–817. https://doi.org/10.1111/j.1096-3642.1952.tb00784.x.

Czaplewski, N. J., & Morgan, G. S. (2015). A late-surviving apatemyid (Mammalia: Apatotheria) from the latest Oligocene of Florida, USA. *PeerJ, 3,* e1509. https://doi.org/10.7717/peerj.1509.

De Bast, E., & Smith, T. (2017). The oldest Cenozoic mammal fauna of Europe: Implication of the Hainin reference fauna for mammalian evolution and dispersals during the Paleocene. *Journal of Systematic Palaeontology, 15,* 741–785. https://doi.org/10.1080/14772019.2016.1237582.

Dunn, R. H., Rose, K. D., Rana, R. S., Kumar, K., Sahni, A., & Smith, T. (2016). New euprimate postcrania from the Early Eocene of Gujarat, India, and the strepsirrhine–haplorhine divergence. *Journal of Human Evolution, 99,* 25–51. https://doi.org/10.1016/j.jhevol.2016.06.006.

Estravís, C. (1990). *Russellmys denisae,* un nouvel Apatémyidé (Mammalia) de Silveirinha, Eocene inférieur du Portugal. *Géobios, 23,* 761–765. https://doi.org/10.1016/s0016-6995(06)80342-6.

Estravís, C. (1992). *Estudos dos mamíferos do Eocenico inferior de Silveirinha (Baixo Mondego).* Ph.D. dissertation, Universidade Nova de Lisboa, Lisbon.

Folie, A., Rana, R. S., Rose, K. D., Sahni, A., Kumar, K., Singh, L., et al. (2013). Early Eocene frogs from Vastan Lignite Mine, Gujarat, India. *Acta Palaeontologica Polonica, 58,* 511–524. https://doi.org/10.4202/app.2011.0063.

Gervais, P. (1848–1852). *Zoologie et paléontologie Françaises.* Paris: Arthus Bertrand.

Gheerbrant, E., Rose, K. D., & Godinot, M. (2005). First palaeanodont [?pholidotan] mammal from the Eocene of Europe. *Acta Palaeontologica Polonica, 50,* 209–218.

Gingerich, P. D. (1982). Studies on Paleocene and early Eocene Apatemyidae (Mammalia, Insectivora). II. *Labidolemur* and *Apatemys* from the early Wasatchian of the Clark's Fork Basin, Wyoming. *Contributions from the Museum of Paleontology, University of Michigan, 26,* 57–69.

Gingerich, P. D. (1989). New earliest Wasatchian mammalian fauna from the Eocene of northwestern Wyoming: Composition and diversity in a rarely sampled high-floodplain assemblage. *University of Michigan Papers on Paleontology, 28,* 1–97.

Gingerich, P. D., & Rose, K. D. (1982). Studies on Paleocene and early Eocene Apatemyidae (Mammalia, Insectivora). *Contributions from the Museum of Paleontology, University of Michigan, 26,* 49–55.

Godinot, M., & Lapparent de Broin, F. de. (2003). Arguments for a mammalian and reptilian dispersal from Asia to Europe during the Paleocene–Eocene boundary interval. *Deinsea, 10,* 255–275.

Goloboff, P. A., & Catalano, S. (2016). TNT version 1.5, including a full implementation of phylogenetic morphometrics. *Cladistics, 32,* 221–238. https://doi.org/10.1111/cla.12160.

Gunnell, G. F. (1988). New species of *Unuchinia* (Mammalia: Insectivora) from the Middle Paleocene of North America. *Journal of Paleontology, 62,* 139–141.

Gunnell, G. F., Bown, T. M., Bloch, J. I., & Boyer, D. M. (2008a). Proteutheria. In C. M. Janis, G. F. Gunnell, & M. D. Uhen (Eds.), *Evolution of Tertiary mammals of North America, Volume 2: Small mammals, xenarthrans, and marine mammals* (pp. 63–81). Cambridge: Cambridge University Press.

Gunnell, G. F., Bown, T. M., Hutchison, J. H., & Bloch, J. I. (2008b). Lipotyphla. In C. M. Janis, G. F. Gunnell & M. D. Uhen (Eds.), *Evolution of Tertiary mammals of North America, Volume 2: Small mammals, xenarthrans, and marine mammals* (pp. 89–125). Cambridge: Cambridge University Press.

Heller, F. (1930). Die Säugetierfauna der mitteleozänen Braunkohle des Geiseltales bei Halle. *Jahrbuch des Halleschen Verbandes, 9,* 13–41.

Holtzman, R. C. (1978). Late Paleocene mammals of the Tongue River Formation, western North Dakota. *North Dakota Geological Survey Report of Investigation, 65,* 1–88.

Hooker, J. J. (2010a). The 'Grande Coupure' in the Hampshire Basin, UK: Taxonomy and stratigraphy of the mammals on either side of this major Paleogene faunal turnover. In J. E. Whitaker & M. B. Hart (Eds.), *Micropaleontology, sedimentary environments and stratigraphy: A tribute to Dennis Curry (1912–2001)* (pp. 147–215). The Micropaleontological Society, Special Publications, The Geological Society, London.

Hooker, J. J. (2010b). The mammal fauna of the early Eocene Blackheath Formation of Abbey Wood, London. *Monograph of the Palaeontographical Society, London, 165,* 1–162. https://doi.org/10.1080/02724634.2013.783489.

Hooker, J. J. (2015). A two-phase mammalian dispersal event across the Paleocene–Eocene transition. *Newsletters on Stratigraphy, 48,* 201–220.

Hooker, J. J., & Dashzeveg, D. (2003). Evidence for direct mammalian faunal interchange between Europe and Asia near the Paleocene–Eocene boundary. *Geological Society of America, Special Paper, 369,* 479–500. https://doi.org/10.1130/0-8137-2369-8.479.

Jepsen, G. L. (1934). A revision of the American Apatemyidae and the description of a new genus, *Sinclairella,* from the White River Oligocene of South Dakota. *Proceedings of the American Philosophical Society, 74,* 287–305.

Kalthoff, D. C., von Koenigswald, W., & Kurz, C. (2004). A new specimen of *Heterohyus nanus* (Apatemyidae, Mammalia) from the Eocene of Messel (Germany) with unusual soft part preservation. *Courier Forschungsinstitut Senckenberg, 252,* 1–12.

Kapur, V. V., Das, D. P., Bajpai, S., & Prasad, G. V. R. (2017). First mammal of Gondwanan lineage in the early Eocene of India. *Comptes Rendus Palevol, 16,* 721–737. https://doi.org/10.1016/j.crpv.2017.01.

von Koenigswald, W. (1987). Apatemyiden-Skelette aus dem Mitteleozän von Messel und ihre paläobiologische Aussage. *Carolinea, 45,* 31–35.

von Koenigswald, W. (1990). Die Paläobiologie der Apatemyiden (Insectivora s.l.) und die Ausdeutung der Skelettfunde von *Heterohyus nanus* aus dem Mitteleozän von Messel bei Darmstadt. *Palaeontographica A, 210,* 41–77.

von Koenigswald, W., Rose, K. D., Grande, L., & Martin, R. D. (2005). First apatemyid skeleton from the lower Eocene Fossil Butte Member, Wyoming (USA), compared to the European apatemyid from Messel, Germany. *Palaeontographica A, 272,* 149–169.

von Koenigswald, W., Ruf, I., & Gingerich, P. D. (2009). Cranial morphology of a new apatemyid, *Carcinella sigei* n. gen. n. sp. (Mammalia, Apatotheria) from the late Eocene of Southern France. *Palaeontographica A, 288,* 53–91.

von Koenigswald, W., & Schierning, H.-P. (1987). The ecological niche of an extinct group of mammals, the early Tertiary apatemyids. *Nature, 326,* 595–596. https://doi.org/10.1038/326595a0.

Krause, D. W., & Maas, M. C. (1990). The biogeographic origin of late Paleocene–early Eocene mammalian immigrants to the western Interior of North America. *Geological Society of America, Special Paper, 243,* 71–105. https://doi.org/10.1130/spe243-p71.

Krishtalka, L., & Stucky, R. K. (1984). Middle Eocene marsupials (Mammalia) from Northeastern Utah and the mammalian fauna from Powder Wash. *Annals of the Carnegie Museum, 53,* 31–45.

Kumar, K., Rose, K. D., Rana, R. S., Singh, L., Smith, T., & Sahni, A. (2010). Early Eocene artiodactyls (Mammalia) from western India. *Journal of Vertebrate Paleontology, 30,* 1245–1274. https://doi.org/10.1080/02724634.2010.483605.

López-Martínez, N., & Peláez-Campomanes, P. (1999). New mammals from south-central Pyrenees (Tremp Formation, Spain) and their bearing on late Paleocene marine-continental correlations. *Bulletin de la Société Géologique de France, 170,* 681–696.

Lucas, S. G. (1998). Fossil mammals and the Paleocene/Eocene series boundary in Europe, North America, and Asia. In M.-P. Aubry, S. G. Lucas & W. A. Berggren (Eds.), *Late Paleocene–early Eocene climatic and biotic events in the marine and terrestrial records* (pp. 451–500). New York: Columbia University Press.

Marsh, O. C. (1872). Preliminary description of new Tertiary mammals, Parts I-IV. *American Journal of Science, 4,* 1–35.

Matthew, W. D. (1909). The Carnivora and Insectivora of the Bridger Basin, Middle Eocene. *Memoirs of the American Museum of Natural History, 9,* 291–567.

Mayr, G., Rana, R. S., Rose, K. D., Sahni, A., Kumar, K., Singh, L., et al. (2010). *Quercypsitta*-like birds from the early Eocene of India (Aves, ?Psittaciformes). *Journal of Vertebrate Paleontology, 30,* 467–478. https://doi.org/10.1080/02724631003617357.

Mayr, G., Rana, R. S., Rose, K. D., Sahni, A., Kumar, K., & Smith, T. (2013). New specimens of the early Eocene bird *Vastanavis* and the interrelationships of stem group Psittaciformes. *Paleontological Journal, 47,* 1308–1314.

Mayr, G., Rana, R. S., Sahni, A., & Smith, T. (2007). Oldest fossil avian remains from the Indian subcontinental plate. *Current Science, 92,* 1266–1269.

McKenna, M. C. (1963a). Primitive Paleocene and Eocene Apatemyidae (Mammalia, Insectivora) and the primate–insectivore boundary. *American Museum Novitates, 2160,* 1–39.

McKenna, M. C. (1963b). Fossil Mammalia from the early Wasatchian four mile Fauna, Eocene of Northwest Colorado. *University of California Publications of Geological Sciences, 37,* 1–130.

McKenna, M. C. (1975). Fossil mammals and early Eocene North Atlantic land continuity. *Annals of the Missouri Botanical Garden, 62,* 335–353. https://doi.org/10.2307/2395200.

McKenna, M. C., & Bell, S. K. (1997). *Classification of mammals above the species level.* New York: Columbia University Press.

Nixon, K. C. (1999). *Winclada (BETA) v. 0.9.9.* Published by the author, Ithaca, New York, USA.

Nolf, D., Rana, R. S., & Singh, H. (2006). Fish otoliths from the Ypresian (early Eocene) of Vastan, Gujarat, India. *Bulletin de l'Institut Royal des Sciences Naturelles de Belgique, 76,* 105–118.

Novacek, M. J. (1976). Insectivora and Proteutheria of the later Eocene (Uintan) of San Diego County, California. *Natural History Museum of Los Angeles County, Contributions in Science, 283,* 1–52.

Prasad, G. V. R., & Bajpai, S. (2008). Agamid lizards from the early Eocene of western India: Oldest Cenozoic lizards from South Asia. *Palaeontologia Electronica, 11*(4a), 1–19.

Rage, J.-C., Folie, A., Rana, R. S., Singh, H., Rose, K. D., & Smith, T. (2008). A diverse snake fauna from the early Eocene of Vastan Lignite Mine, at Gujarat, India. *Acta Palaeontologica Polonica, 53,* 391–403. https://doi.org/10.4202/app.2008.0303.

Rana, R. S., Augé, M., Folie, A., Rose, K. D., Kumar, K., Singh, L., et al. (2013). High diversity of acrodontan lizards in the Early Eocene Vastan Lignite Mine of India. *Geologica Belgica, 16,* 290–301.

Rana, R. S., Kumar, K., Escarguel, G., Sahni, A., Rose, K. D., Smith, T., et al. (2008). An ailuravine rodent from the lower Eocene Cambay Formation at Vastan, western India, and its paleobiogeographic implications. *Acta Palaeontologica Polonica, 53,* 1–14. https://doi.org/10.4202/app.2008.0101.

Rana, R. S., Kumar, K., & Singh, H. (2004). Vertebrate fauna from the subsurface Cambay Shale (Lower Eocene), Vastan Lignite Mine, Gujarat, India. *Current Science, 87,* 1726–1733.

Rana, R. S., Kumar, K., Zack, P. S., Solé, F., Rose, K. D., Missiaen, P., et al. (2015). Craniodental and postcranial morphology of *Indohyaenodon raoi* from the early Eocene of India, and its implications for ecology, phylogeny, and biogeography of hyaenodontid mammals. *Journal of Vertebrate Paleontology, 35,* 1–22. https://doi.org/10.1080/02724634.2015.965308.

Rana, R. S., Singh, H., Sahni, A., Rose, K. D., & Saraswati, P. K. (2005). Early Eocene chiropterans from a new mammalian assemblage (Vastan Lignite Mine, Gujarat, western peninsular margin): Oldest known bats from Asia. *Journal of the Palaeontological Society of India, 50,* 93–100.

Rose, K. D. (1975). The Carpolestidae, early Tertiary primates from North America. *Bulletin of the Museum of Comparative Zoology at Harvard College, 147,* 1–74.

Rose, K. D. (2006). *The beginning of the age of mammals.* John Hopkins: University Press, Baltimore, USA.

Rose, K. D., DeLeon, V. B., Missiaen, P., Rana, R. S., Sahni, A., Singh, L., et al. (2008). Early Eocene lagomorph (Mammalia) from western India and the early diversification of Lagomorpha. *Proceedings of the Royal Society of London, Series B, 275,* 1203–1208. https://doi.org/10.1098/rspb.2007.1661.

Rose, K. D., Holbrook, L. T., Rana, R. S., Kumar, K., Jones, K. E., Ahrens, H. E., et al. (2014). Early Eocene fossils suggest that the mammalian order Perissodactyla originated in India. *Nature Communications, 5*(5570), 1–9. https://doi.org/10.1038/ncomms6570.

Rose, K. D., Kumar, K., Rana, R. S., Sahni, A., & Smith, T. (2013). New hypsodont tillodont (Mammalia, Tillodontia) from the early Eocene of India. *Journal of Paleontology, 87,* 842–853. https://doi.org/10.1666/13-027.

Rose, K. D., Rana, R. S., Sahni, A., Kumar, K., Missiaen, P., Singh, L., et al. (2009a). Early Eocene primates from Gujarat, India. *Journal of Human Evolution, 56,* 366–404. https://doi.org/10.1016/j.jhevol.2009.01.008.

Rose, K. D., Rana, R. S., Sahni, A., Kumar, K., Singh, L., & Smith, T. (2009b). First tillodont from India: Additional evidence for an early Eocene faunal connection between Europe and India? *Acta Palaeontologica Polonica, 54,* 351–355. https://doi.org/10.4202/app.2008.0067.

Rose, K. D., Smith, T., Rana, R. S., Sahni, A., Singh, H., Missiaen, P., et al. (2006). Early Eocene (Ypresian) continental vertebrate assemblage from India, with description of a new anthracobunid (Mammalia, Tethytheria). *Journal of Vertebrate Paleontology, 26,* 219–225. https://doi.org/10.1671/0272-4634.

Russell, D. E. (1975). Paleoecology of the Paleocene–Eocene Transition in Europe. In F. S. Szalay (Ed.), *Approaches to primate paleobiology* (pp. 28–61). *Contributions to Primatology, 5*. Basel: Karger Publishers.

Russell, D. E., Godinot, M., Louis, P., & Savage, D. E. (1979). Apatotheria (Mammalia) de l'Éocène inférieur de France et de Belgique. *Bulletin Museum National d'Histoire Naturelle, Paris, série, 4*(1), 203–242.

Russell, D. E., Louis, P., & Poirier, M. (1966). Gisements nouveaux de la faune Cernaysienne (mammifères paléocènes de France). *Bulletin de la Société Géologique de France, série, 7*(8), 845–856.

Sahni, A., Saraswati, P. K., Rana, R. S., Kumar, K., Singh, H., Alimohammadian, H., et al. (2006). Temporal constraints and depositional paleoenvironments of the Vastan Lignite Sequence, Gujarat: Analogy for the Cambay Shale hydrocarbon source rock. *Indian Journal of Petroleum Geology, 15,* 1–20.

Samanta, A., Bera, M. K., & Sarkar, A. (2016). Climate modulated sequence development in a tropical rift basin during the Late Paleocene to Early Eocene super greenhouse Earth. *Sedimentology, 63,* 917e939. https://doi.org/10.1111/sed.12243.

Scott, W. B., & Jepsen, G. L. (1936). The mammalian fauna of the White River Oligocene. Part I. Insectivora and Carnivora. *Transactions of the American Philosophical Society New Series, 28,* 1–153.

Secord, R. (2008). The Tiffanian land-mammal age (middle and late Paleocene) in the Northern Bighorn Basin, Wyoming. *University of Michigan Papers on Paleontology, 35,* 1–192.

Sigé, B. (1975). Insectivores primitifs de l'Eocène supérieur et de l'Oligocène inférieur d'Europe occidentale; Apatémyidés et Leptictidés. *Colloque International du C.N.R.S., 218,* 653–673.

Silcox, M. T., Bloch, J. I., Boyer, D. M., & Houde, P. (2010). Cranial anatomy of Paleocene and Eocene *Labidolemur kayi* (Mammalia: Apatotheria) and the relationships of the Apatemyidae to other mammals. *Zoological Journal of the Linnean Society, 160,* 773–825. https://doi.org/10.1111/j.1096-3642.2009.00614.x.

Silcox, M. T., Dalmyn, C. K., Hrenchuk, A., Bloch, J. I., Boyer, D. M., & Houde, P. (2011). Endocranial morphology of *Labidolemur kayi* (Apatemyidae, Apatotheria) and its relevance to the study of brain evolution in Euarchontoglires. *Journal of Vertebrate Paleontology, 31,* 1314–1325. https://doi.org/10.1080/039.031.0613.

Simpson, G. G. (1929). A collection of Paleocene mammals from Bear Creek, Montana. *Annals of the Carnegie Museum, 19,* 115–122.

Simpson, G. G. (1936). A new fauna from the Fort Union of Montana. *American Museum of Natural History Novitates, 873,* 1–27.

Simpson, G. G. (1937). *Unuchinia*, new name for *Apator* Simpson, not Semenow. *Journal of Paleontology, 11,* 78.

Simpson, G. G. (1940). Studies on the earliest primates. *Bulletin of the American Museum of Natural History, 77,* 185–212.

Simpson, G. G. (1947). Holarctic mammalian faunas and continental relationships during the Cenozoic. *Bulletin of the Geological Society of America, 58,* 613–688. https://doi.org/10.1130/0016-7606.

Simpson, G. G. (1954). An apatemyid from the early Eocene of New Mexico. *American Museum Novitates, 1654,* 1–4.

Sluijs, A., van Roij, L., Harrington, G. J., Schouten, S., Sessa, J. A., LeVay, L. J., et al. (2014). Warming, euxinia and sea level rise during the Paleocene–Eocene Thermal Maximum on the Gulf Coastal Plain: Implications for ocean oxygenation and nutrient cycling. *Climate of the Past, 10,* 1421–1439.

Smith, T. (1996). *Leptacodon dormaalensis* (Mammalia, Lipotyphla), un nyctithère primitif de la transition Paléocène–Eocène de Belgique. *Belgian Journal of Zoology, 126,* 153–168.

Smith, T., Kumar, K., Rana, R. S., Folie, A., Solé, F., Noiret, C., et al. (2016). New early Eocene vertebrate assemblage from western India reveals a mixed fauna of European and Gondwana affinities. *Geoscience Frontiers, 7,* 969–1001. https://doi.org/10.1016/j.gsf.2016.05.001.

Smith, T., Rana, R. S., Missiaen, P., Rose, K. D., Sahni, A., Singh, H., et al. (2007). High bat (Chiroptera) diversity in the early Eocene of India. *Naturwissenschaften, 94,* 1003–1009. https://doi.org/10.1007/s00114-007-0280-9.

Smith, T., Rose, K. D., & Gingerich, P. D. (2006). Rapid Asia–Europe–North America geographic dispersal of earliest Eocene primate *Teilhardina* during the Paleocene–Eocene Thermal Maximum. *Proceedings of the National Academy of Sciences, 103,* 11223–11227. https://doi.org/10.1073/pnas.0511296103.

Smith, T., & Smith, R. (2001). The creodonts (Mammalia, Ferae) from the Paleocene–Eocene transition in Belgium (Tienen Formation, MP7). *Belgian Journal of Zoology, 131,* 117–136.

Smith, T., & Smith, R. (2013). A land micro-mammal fauna from the early Eocene marine Egem deposits (NP12, Belgium) and the first occurrence of the peradectid marsupial *Armintodelphys* outside North America. *Geologica Belgica, 16,* 302–310.

Smith, T., Solé, F., Missiaen, P., Rana, R. S., Kumar, K., Sahni, A., et al. (2015). First early Eocene tapiroid from India and its implication for the paleobiogeographic origin of perissodactyls. *Palaeovertebrata, 39*(e5), 1–9.

Solé, F., Gheerbrant, E., & Godinot, M. (2011). New data on the Oxyaenidae from the Early Eocene of Europe; biostratigraphic, paleobiogeographic and paleoecologic implications. *Palaeontologia Electronica, 14,* 1–41.

Solé, F., Smith, R., Coillot, T., De Bast, E., & Smith, T. (2014). Dental and tarsal anatomy of '*Miacis*' *latouri* and a phylogenetic analysis of the earliest carnivoraforms (Mammalia, Carnivoramorpha). *Journal of Vertebrate Paleontology, 31,* 1–21. https://doi.org/10.1080/02724634.2013.793195.

Spaulding, M., & Flynn, J. J. (2012). Phylogeny of the Carnivoramorpha: The impact of postcranial characters. *Journal of Systematic Palaeontology, 10,* 653–677. https://doi.org/10.1080/14772019.2011.630681.

Springer, M. S., Stanhope, M. J., Madsen, O., & de Jong, W. W. (2004). Molecules consolidate the placental mammal tree. *Trends in Ecology and Evolution, 19,* 430–437.

Stehlin, H. G. (1916). Die Säugetiere des Schweizerischen Eocaens. *Abhandlungen Schweizerischen Palaontologischen Gesellschaft, 41,* 1299–1552.

Strait, S. G. (2001). New Wa-0 mammalian fauna from Castle Gardens in the Southeastern Bighorn Basin. *University of Michigan Papers on Paleontology, 33,* 127–143.

Szalay, F. S. (1968). Origins of the Apatemyidae (Mammalia, Insectivora). *American Museum Novitates, 2352,* 1–11.

Teilhard de Chardin, P. (1927). Les mammifères de l'Eocène inférieur de la Belgique. *Mémoires du Musée royal d'Histoire naturelle de Belgique, 36,* 1–33.

Van Valen, L. (1967). New Paleocene insectivores and insectivore classification. *Bulletin of the American Museum of Natural History, 135,* 217–284.

West, R. M. (1972). A new late Paleocene apatemyid (Mammalia, Insectivora) from Bison Basin, central Wyoming. *Journal of Paleontology, 46,* 714–718.

West, R. M. (1973a). Review of the North American Eocene and Oligocene Apatemyidae (Mammalia: Insectivora). *Special Publications the Museum Texas Tech University, 3,* 1–42.

West, R. M. (1973b). Antemolar dentition of the Paleocene apatemyid insectivorans *Jepsenella* and *Labidolemur*. *Journal of Mammalogy, 54,* 33–40. https://doi.org/10.2307/1378870.

Woodburne, M. O., Gunnell, G. F., & Stucky, R. K. (2009). Land mammal faunas of North America rise and fall during the Early Eocene Climatic Optimum. *Denver Museum of Nature and Science Annals, 1,* 1–74.

Zack, S. P. (2012). Deciduous dentition of *Didymictis* (Carnivoramorpha: Viverravidae): Implications for the first appearance of "Creodonta". *Journal of Mammalogy, 93,* 808–817. https://doi.org/10.1644/11-mamm-a-245.1.

Chapter 13
Size Variation Amongst the Non-volant Mammals from the Early Eocene Cambay Shale Deposits of Western India: Paleobiogeographic implications

Vivesh V. Kapur

Abstract Within the Indian subcontinent, the oldest Cenozoic record of non-volant mammals comprises various groups including perissodactyls, artiodactyls, primates, lagomorphs, tillodonts, proteutherians, rodents and didelphimorphs. These mammals have been reported from the subsurface horizons of the Cambay Shale which are dated at ~54.5 Ma (Early Eocene). A statistical analysis (at the ordinal level) of size variation(s) in the lower dentitions of these taxa from the Cambay Shale allows their size categorization into three broad groups: small, medium and large. Based on the data on (a) phylogenetic relationships of large perissodactyls (e.g., cambaytheres), (b) an Early Eocene age of the fauna, (c) the timing of collision between the Indian plate and Eurasia, and (d) considering that it was unlikely for the large-sized component (perissodactyls) to disperse across oceanic barriers, the potential presence of a short-lived corridor for dispersal between Indian subcontinent and Eurasia prior to ~54.5 Ma cannot be ruled out. Further, the overall data hint at the presence of cambaythere-like mammals during the late Paleocene of the Indian and/or Eurasian continent(s), emphasizing the need for rigorous reconnoitering for Paleocene mammals (presently unknown) within the Indian subcontinent.

Keywords Vastan lignite mine • Early Eocene • India-Eurasia collision • Mammals • Dispersals • Oceanic barriers

V. V. Kapur (✉)
Birbal Sahni Institute of Palaeosciences, 53, University Road, Lucknow, Uttar Pradesh 226007, India
e-mail: viveshkapur@gmail.com; viveshvir_kapur@bsip.res.in

Introduction

Many previous studies on extant mammals have shown strong correlation between tooth size (particularly for the first lower molar) and body mass (Kay 1975; Legendre and Roth 1988; Bown et al. 1994; Millien and Bovy 2010 and references therein). Previous studies have also linked body mass of organisms (including mammals) to latitude (popularly known as "Bergmann's Rule"), dispersal rates and dispersal mechanisms (Hillebrand 2004 and references therein). Dispersal is a basic strategy for animals' survival, and may occur when the preferred habitat of a population or assemblage becomes uncongenial (e.g., by decline in resource availability) possibly due to climatic or other factors (Clobert et al. 2012; Stenseth and Lidicker 2012; De Queiroz 2014). Faunal dispersal entails three broad mechanisms: sweepstakes (jump dispersal or chance dispersal), diffusion (filter-bridge or island hopping) and corridor migration (Simpson 1940; Pielou 1979; Huggett 1998). Amongst these, jump dispersal is the most difficult whereas dispersal via a terrestrial corridor is considered the least resistant mode. Extant animals (including mammals) show variations in their abilities to disperse over vast oceanic barriers due to physiology, morphology and also size (~body mass) (Teitelbaum et al. 2015; Beard 2016 and references therein). For obvious reasons, volant forms (e.g., chiropterans) find it easier to cross over geographic barriers as opposed to non-volant large mammals. Thus, non-volant animals are given precedence for inferring biogeographic scenarios.

The underlying premise concerning dispersal of non-volant mammals across oceanic barriers (as also considered in the present study) is that mammals of large body size (in general) could not cross ocean barriers because they could not raft or swim, unlike small- to medium-sized mammals. On the contrary, large-sized modern mammals (e.g., elephants, hippopotamuses and deer) have been noted

G. V. R. Prasad and R. Patnaik (eds.), *Biological Consequences of Plate Tectonics: New Perspectives on Post-Gondwana Break-up—A Tribute to Ashok Sahni*,
Vertebrate Paleobiology and Paleoanthropology Series, https://doi.org/10.1007/978-3-030-49753-8_13

as agile swimmers (Johnson 1980; Mazza et al. 2013 and references therein). However, the observed swimming capability of the large mammals is limited. For example: modern elephants (*Elephas maximus*), having a body mass range of 2000–6000 kg (Sukumar 2006), are not known to swim across oceanic barriers greater than 50 km (Johnson 1980), while red deer (*Cervus elaphus*), having a body mass range of 120–240 kg (source: https://en.wikipedia.org/wiki/Red_deer), barely manage to swim distances greater than 3 km (Crouchley et al. 2011; Mazza et al. 2013 and references therein). Further, the scientific community holds varied opinions on the swimming ability of hippopotamuses. Studies on extant hippopotamuses (body mass ~3800 kg; Garland 1983) suggest incapability of these animals for swimming owing to a) a natural tendency to avoid deep waters, b) a daily requirement of large quantities of drinking water for survival, and c) sensitivity to prolonged exposure to sunlight (Eltringham 1999; Fisher et al. 2007; Coughlin and Fish 2009; Mazza et al. 2013). In contrast, hippopotamuses floating/swimming in sea waters have been reported (see Fraedrich 1968 in Van der Geer et al. 2010). Additionally, there is no dearth of literature on fossil dwarf elephants (body mass of ~200 kg; Davies and Lister 2001), and pygmy hippopotamuses (body mass of ~280 kg; Taylor et al. 2013) from oceanic and ocean-like islands worldwide (Herridge 2010 and references therein). The occurrence of dwarf elephants and pygmy hippopotamuses on islands has generally been explained by postulating land-bridge connections between islands and mainland, or as a consequence of chance (sweepstakes mode) dispersal from mainland (Van der Geer et al. 2010 and references therein). However, the latter dispersal mechanism for large mammals has been recently doubted by Mazza et al. (2013). Moreover, considering that small mammals have limited salt-water tolerances (see Van Duzer 2004; Masters et al. 2007), a poor dispersal capability has been argued for small mammals [e.g., *Lepus californicus* (commonly known as the "jack rabbit") weighing ~3.6 Kg] via rafting or sweepstakes (Prescott 1959; Mazza et al. 2013).

While reconciling the biogeographic principles discussed by Matthew (1915, 1939) and Simpson (1940), McKenna (1973) in his article titled "*Sweepstakes, Filters, Corridors, Noah's Arks, and Beached Viking Funeral Ships in Paleogeography*" viewed the Indian subcontinent as a Noah's Ark that permitted one-way dispersal of biota into Eurasia, after the Indian subcontinent collided with Eurasia during the early Paleogene. Paleontological data have limitations owing to the incomplete nature of the fossil record and lack of consensus within the scientific community on the degree of similarity/dissimilarity (in the context of phylogenetic relationships) within various groups of taxa at the familial, generic or species level. Nevertheless, many recent studies have emphasized the importance of size and evolutionary relationships within mammalian groups for testing paleobiogeographic and paleogeographic hypotheses (Perfit and Williams 1989; Ali and Krause 2011; Solé and Smith 2013; Kapur and Khosla 2016, 2019; Klaus et al. 2016) following the pioneering work of George Gaylord Simpson (Simpson 1940), who discussed the migration (via sweepstakes route) of animals between Africa and Madagascar (refer Fig. 6 in Simpson 1940). In the Indian subcontinent's context, Prasad and Sahni (1999) also emphasized the size limitations on the biota involved in biotic exchange, to explain the presence of Laurasian elements within the Indian subcontinent during the Late Cretaceous (Maastrichtian).

Numerous groups of non-volant mammals have been reported from the subsurface Cambay Shale deposits in the western part of India, including perissodactyls, artiodactyls, primates, proteutherians, lagomorphs, tillodonts, rodents and didelphimorphs (Bajpai et al. 2005a, b, c, 2006, 2007a, b, 2008, 2009; Rose et al. 2007, 2009a, b, 2013; Kapur and Bajpai 2015; Smith et al. 2015, 2016; Kapur et al. 2017a, b) (also refer to Table 13.1 and Fig. 13.1, present chapter). This fauna is largely endemic at the familial and/or generic levels but also shows Laurasian affinities. Moreover, a very recent study has shown the presence of a Gondwanan mammal (*Bharatlestes kalami*) in the Early Eocene of India (Kapur et al. 2017a, b). Mammals from the Cambay Shale are the oldest Cenozoic record from the Indian subcontinent and, with a few exceptions e.g., lagomorphs, are known mainly from dental elements. Thus, the data allows for the statistical analysis of size variation(s) in the mammalian assemblage, using the size variation(s) in the dentition (including that for the first lower molar). In the present article, the size variation(s) (at ordinal level) within the oldest Cenozoic record of non-volant mammalian assemblage from the Cambay Shale have been studied, and their age and known evolutionary relationship(s) have been considered to shed some light on the degree of connectivity between the Indian subcontinent and Eurasia. This study further considers the limitations on size for an animal and the possible mechanisms for dispersal during biotic exchange in the context of available data on the timings and dynamics of collision between the Indian plate and Eurasia.

Methodology

Initially, comprehensive data on the non-volant mammals reported from the Cambay Shale were compiled from available literature. For size variations (at ordinal level), only data for the lower dentition (p4-m3) (considering that most of the mammalian finds from Cambay Shale are based on these teeth, [Tables 13.1 and 13.2]) were incorporated and inspected visually by plotting bivariate (length × width)

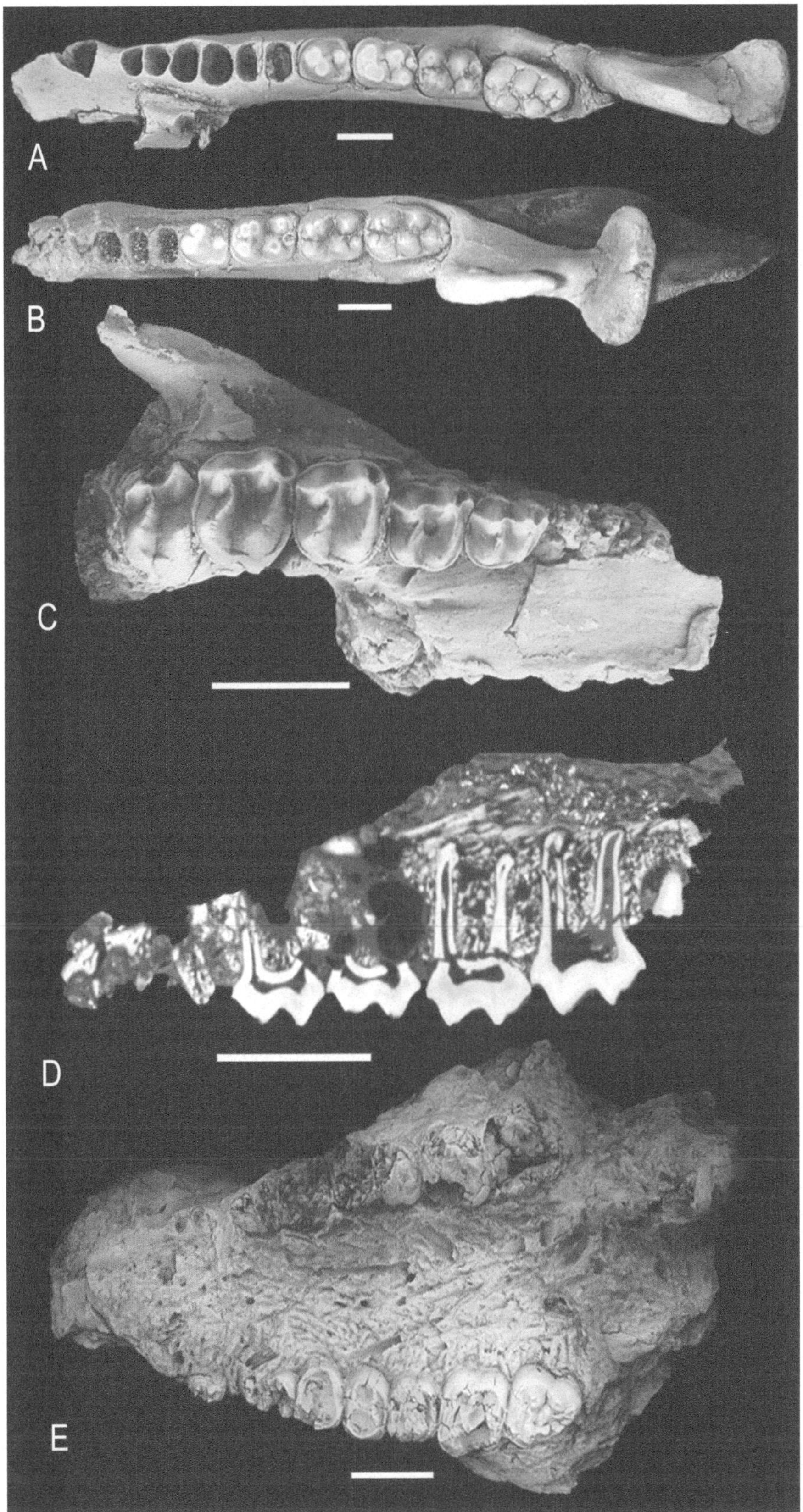

Fig. 13.1 **A.** *Cambaytherium thewissi* (VLM-505), holotype right dentary with *in situ* p4-m3, occlusal view; **B.** *C. thewissi* (VLM-505), holotype left dentary with *in situ* p4-m3, occlusal view; **C.** *Cambaylophus vastanensis* (VLM-760), holotype right maxillary fragment with dp3, dp4, m1, m2 and erupting m3, occulusal view; **D.** Micro-CT image of *Cambaylophus vastanensis* (VLM-760), section view; **E.** *Kalitherium marinus* (VLM-931), holotype rostrum with damaged crowns of right P1 and P2, worn crowns of right P3-M2, and well-preserved crown of right M3, damaged bases of left C-M3, ventral view. Scale bar equals 1 cm for **A–D**; 3 cm for **E**. **A–B** reproduced from Bajpai et al. (2005a), **C** reproduced from Kapur and Bajpai (2015), and **E** reproduced from Bajpai et al. (2006a). All figures reproduced with permission

Table 13.1 Dental material of the non-volant mammals reported from the Early Eocene Cambay Shale, Gujarat, India

Order	Family	Taxon	Dental material and remarks
Perissodactyla (Owen 1848)			
	Cambaytheriidae (Bajpai et al. 2005a)		
		Cambaytherium thewissi (=*Indobune vastanensis*) (Bajpai et al. 2005a)	**Holotype**: VLM 505 (left dentary with p4–m3, right dentary with p4–m3); **Referred material**: VLM 548 (left m1 or m2); VLM 551 (right m1 or m2); VLM 726 (right dentary with p3, m1–m3); VLM 764 (right upper molar); VLM 765 (left upper molar); VLM 767 (right juvenile dentary, with m1 and unerupted p2); VLM 830 (right dentary with poorly preserved alveoli for p1-m3 and fragmentary crown for m2 or m2); VLM 894 (icisor, PI?, right P2 or P3, right P4, and upper molar fragment); VAS 5 (left maxilla with dP3-4, P4 (from crypt), M1); VAS 18 (right M1); VAS 12 (right maxilla with heavily worn M2); VAS 1 (partial left M3); tentatively referred: VAS 10 (upper C); VAS 2 (left p1); VAS 3 (left p2); VAS 4 (talonid of right m1); TAD 9002 (mandibular symphysis with alveoli for left and right incisors, canines, p1, and right p2); TAD 9202 (left maxilla with M1-M3); WIF/A 1200 (mandible with left p3-4, m2, and symphysis); WIF/A 4217 (right maxilla with P4-M2); WIF/A 4219 (right M3); WIF/A 4220 (right M3); WIF/A 4221 (right M3); WIF/A 4222 (right m3); WIF/A 4223 (left m3); WIF/A 4224 (right m2); WIF/A 4232 (right p2?); WIF/A 4233 (left m1); WIF/A 4234 (left dp3). **Remarks**: Data compiled from Bajpai et al. (2005a, 2006a); Rose et al. (2006, 2014); Smith et al. (2016)
		Cambaytherium bidens (Bajpai et al. 2005a)	**Holotype**: VLM 503 (right dentary with c – m3, left dentary with p1–m3); **Referred material**: VLM 502 (left dentary with p1–p3, m1, m3, right dentary with p3–m3, right premaxilla with C1–P1, left C, right P2, left P4, mandibular symphysis, maxilla fragment, nasals, atlas); VLM 520 (m1–m2); VLM 521 (left mandibular fragment with m3 and mx); VLM 532 (left c1, m1–m3, mandibular fragments); VLM 535 (right proximal astragalus) VLM 541 (left C1); VLM 542 (left c1); VLM 543 (p1); VLM 544 (unworn left p4); VLM 545 (unworn left p4); VLM 546 (left mx); VLM 547 (unworn left mx); VLM 558 (unworn left p2–p3 and three unicuspid teeth); VLM 786 (right dentary with p2-m3); VLM 791 (right dentary with m1-m2); VLM 792 (left dentary with p2-m3); VLM 795 (left dentary with m1-m3); VLM 811 (left dentary with p3-m3); VLM 957 (left dentary with p4-m3 and alveoli for i1-p2). **Remarks**: Data compiled from Bajpai et al. (2005a, 2006a)
		Cambaytherium minor (Bajpai et al. 2005a)	**Holotype**: VLM 539 (right mx with anterolabial part missing); **Referred material**: VLM 540 (left mx with anterior trigonid missing); VLM 761 (right dentary with m1-m3); VLM 830 (right dentary with fragmentary m1); VLM 841 (isolated left m1); VLM 895 (fragmentary right m2). **Remarks**: Data compiled from Bajpai et al. (2005a, 2006a)
		Cambaytherium sp. A (Bajpai et al. 2005a)	**Holotype**: VLM 549 (presumed right P2)
		Cambaytherium gracilis (Smith et al. 2015)	**Holotype**: TAD 9001 (mandible with left p3-m3, alveoli for left p1–2 and right p1, and symphysis with alveoli for incisors and canines). WIF/A 4210 (left dentary with p2–m3); WIF/A 4211 (right dentary with p4–m1, m3); WIF/A 4212 (right dentary with m3); WIF/A 4213 (right maxilla with P1–P3); WIF/A 4214 (left maxilla with P4, M2–M3); WIF/A 4215 (left P2); WIF/A 4235 (right p4); WIF/A 4236 (left p4); WIF/A 4237 (right m2); WIF/A 4238 (right dP4); WIF/A 4239 (right P4); WIF/A 4240 (left P4); WIF/A 4241 (left M2); WIF/A 4242 (left M3); WIF/A 4243 (right M3); TAD 9555 (right dentary with m2-m3)
		Kalitherium marinus (Bajpai et al. 2006a)	**Holotype**: VLM 931 (snout without premaxilla, but with orbits and right postorbital process, root of right canine, damaged crowns of right P1 and P2, worn crowns of right P3-M2, and well-preserved crown of right M3, damaged bases of left C-M3)
	Tapiroidea indet (Gill 1872)		

(continued)

Table 13.1 (continued)

Order	Family	Taxon	Dental material and remarks
		Cambaylophus vastanensis (Kapur and Bajpai 2015)	**Holotype**: VLM 760 (right maxillary fragment with alveoli for dP2 and crowns for well preserved dP3–dP4 and M1–M2 and an erupting M3; portion of anterior palate and part of right zygomatic arch)
	Helaletidae? (Osborn 1892)		
		Vastanolophus holbrooki (Smith et al. 2016)	**Holotype**: VAS 323 (right m1); **Referred material**: VAS 1681 (right fragmentary p4)
Artiodactyla (Owen 1848)			
	Dichobunidae (Turner 1849)		
		Gujaratia indica (=*Diacodexis indicus*) (Bajpai et al. 2005a)	**Holotype**: VLM 511 (left dentary with p3–m3). **Referred material**: GU 26 (left p3); GU 27 (left m3); GU 28 (left m2); GU 29 (left m1); GU 31 (left m1); GU 196 (left M1); GU 242 (right DP4); GU 269 (left M1); GU 329 (left p2); GU 437 (left M3); GU 513 (right dp4); GU 647 (left M2); GU 648 (right m1); GU 650 (right P3); GU 668 (left M2); GU 1500 (right DP4); GU 1501 (left M2–3); GU 1620 (left dentary with p4–m2); GU 1621 (right dentary with m2–m3 and alveoli for i3–m1); GU1622 (right dentary with m1–m3 and alveoli for p4). **Remarks**: Data compiled from Bajpai et al. (2005a); Kumar et al. (2010)
		Gujaratia parvus (=*Diacodexis parvus*) (Kumar et al. 2010)	**Holotype**: GU 1568 (right M1); **Referred material**: GU 207 (left m1)
		Diacodexidae indet. (Kumar et al. 2010)	**Referred material**: GU 423 (left dentary fragment with p3 and part of alveolus for p2); GU 1552 (left M2 or M3)
		Artiodcatyla indet. (Kumar et al. 2010)	**Referred material**: GU 30 (left M3); GU 446 (right p4); GU 651 (right p2)
Condylartha (Van Valen 1978)			
	?Arctocyonidae (Giebel 1855)		
		?Arctocyonidae indet. (Bajpai et al. 2009)	**Referred material**: VLM 960 (unworn left upper M1/ or M2/)
Creodonta (Cope 1875)			
	Hyaenodontidae (Leidy 1869)		
		Indohyaenodon raoi (Bajpai et al. 2009)	**Holotype**: VLM 899 (left dentary with alveolus for P/1 and crowns for P/2-M/3, M/1 damaged); **Referred material**: VLM 742 (right dentary with crowns for P/2-P/4, damaged area of M/1 and crowns for M/2-M/3); VLM 782 (right dentary with alveolus for M/1and crowns for P/2-P/4 and M/2-M/3); VLM 926 (left dentary with alveoli for C/1-P/3 and crowns for P/4-M/2, P/4 damaged); GU 321 (isolated left m1); GU 652 (right dentary with p4, worn m2, and alveoli for p2, p3, and canine); GU 767 (right dentary with p4–m3 and alveoli for i2, i3, c, and p1–3); GU 1630 (fragmentary right dentary with p3 and alveoli for c, p2, p4, and m1); GU 1631 (isolated right m1 with lingual half of its talonid broken); GU 1680 (palatal part of a skull with right maxilla having complete P1–M3 and alveolus for canine, and complete left P1); GU 1721 (damaged isolated M1 or M2); TAD 9000 (left dentary with m1); WIF/A 4209 (right dentary with m2). **Remarks**: Data compiled from Bajpai et al. (2009); Rana et al. (2005); Smith et al. (2016)
Tillodontia (Marsh 1875)			
	Esthonychidae (Cope 1883)		
		cf. *Esthonyx* sp. (Rose et al. 2009b)	**Referred material:** VAS 1651 (upper mesial incisor); VAS 1587 (left M3)
		Anthraconyx hypsomylus (Rose et al. 2013)	**Holotype**: VAS 762 (right dentary with dp3-4, m1. m2 in cript)
		Indoesthonyx suratensis (Smith et al. 2016)	**Holotype**: WIF/A 2333 (right dentary with alveoli for i2–m2 and *in situ* m3; **Referred material**: WIF/A 4225 (right m1)
		cf. *Indoesthonyx suratensis* (Smith et al. 2016)	**Referred material**: WIF/A 4226 (left M2); WIF/A 4227 (right M1); WIF/A 4228 (right m2); WIF/A 4229 (left I2); WIF/A 4230 (left i2)

(continued)

Table 13.1 (continued)

Order	Family	Taxon	Dental material and remarks
Primates (Linnaeus 1758)			
	Notharctidae (Trouessart 1879)		
		Marcgodinotius indicus (Bajpai et al. 2005c)	**Holotype**: VLM 800 (right dentary with m2 crown and alveoli for m3, m1-c); **Referred material**: VLM 817 (right dentary with crowns for m2-3 and alveoli for m1-p1 and partial alveolus for canine); VLM 642 (M2 or M3); VAS 7 (left dentary with M1); GU 40 (right P4); GU 41 (left dP4); GU 44 (left M1); GU 45 (left M2); GU 46 (left M3); GU 49 (right M1); GU 50 (right dP4); GU 51 (left M3); GU 52 (right M3); GU 54 (left M1); GU 227 (right dentary with C, P3-4, M2-3, and roots of P1-2); GU 552 (right P2); GU 600 (right M1); GU 611 (right M1); GU 643-GU 646 (left M2s); GU 703 (right P3); GU 727 (right dentary with P4 and M2); GU 743 (right dentary with M2); GU 744 (right dentary with M2 and alveoli of all other teeth); GU 1534 (left M3); GU 1536 (right P4); GU 1538 (left P3); GU 1544 (left P3); GU 1554 (right P4); GU 1575 (left M2); GU 1591 (left M3); GU 1602 (right M2); WIF/A 2334 (right dentary with m2-3, alveoli for i1-m1); WIF/A 4231 (left dentary with p3-m2); **Tentatively referred material**: GU 21 (partial left P4); GU 42 (left dP4); GU 43 (rightM1); GU 47 (left M3); GU 48 (left M3); GU 243 (left dP4); GU 538 (left M3); GU 705 (right dP4); GU 706 (right M3). **Remarks**: Initially, the genus *Marcgodinotius* was described as an adapiform and not assigned to a family (Bajpai et al. 2005c). However, detailed study carried out by Rose et al. (2009) on complimentray dental material led the authors to assign *Marcgodinotius* to the primituve European cercamoniine family Notharctidae. Data compiled from Bajpai et al. (2005c); Rose et al. (2009), Smith et al. (2015)
		Asiadapis cambayensis (=*Suratius robustus*) (Rose et al. 2007)	**Holotype**: VAS 6 (right dentary with p3-m2 and alveoli for c, p2, m3); **Referred material**: VLM 928 (left dentary with antermost alveolus for c1, alveolus for p2, alveoli for m1, m3 and crowns for p3, p4 and m2); VAS 32 (right m3); GU 33 (right dp4); GU 34 (right P4); GU 35 (left p3); GU 36 (right m2); GU 37 (right m2); GU 38 (right p4); GU 330 (right P3); GU 598 (left m3); GU 609 (right M2); GU 610 (right M3); GU 642 (right m2), GU 656 (left maxilla with M1, M2 and M3); GU 700 (right M3); GU 701 (left M2); GU 702 (right M1); GU 745 (left dentary with p3-m2); GU 1505 (right dentary fragment with m3); GU 1537 (right m1); GU 1550 (left M1); GU 1627 (left p4); GU 1649 (right m3). Data compiled from Bajpai et al. (2007b); Rose et al. (2007, 2009a)
		cf. *Asiadapis cambayensis* (Rose et al. 2007, 2009a in Smith et al. 2016)	**Referred material**: TAD 9004 (edentulous right dentary with alveoli for c-m3)
		cf. *Marcgodinotius* (Rose et al. 2009a)	**Referred material**: GU 1619 (left m3)
	Omomyidae (Trouessart 1879)		
		Vastanomys gracilis (Bajpai et al. 2005c)	**Holotype**: VLM 771, a left dentary with crown for m2 and alveoli for c-m1 and m3; **Referred material**: VLM 617 (isolated right upper M1 or M2). **Remarks**: When initially reported, the genus *Vastomys* (Bajpai et al. 2005c) was not assigned to any family. However, omomyid status of *Vastanomys* was recognized in a detailed cladistics analysis (Bajpai et al. 2008). The Omomyid status of *Vastanomys* has been equivocally considered, although, Rose et al. (2009a), doubt the basal position of *Vastanomys* within Omomyidea
		Vastanomys major (Rose et al. 2009a)	**Holotype**: VAS 1541 (right m2)
	Eosiimidae (Beard et al. 1994)		
		Anthrasimias gujaratensis (Bajpai et al. 2008)	**Holotype**: VLM 1137 (left M1); **Referred material**: VLM 1100 (left M2); VLM 1017 (right M3); VLM 1201 (dP4)
	Primates indet. (Linnaeus 1758)		

(continued)

Table 13.1 (continued)

Order	Family	Taxon	Dental material and remarks
		Primates indet. 1 (Bajpai et al. 2005c)	**Referred material**: VLM 710 (isolated right M3)
	Primates indet. (Linnaeus 1758)		
		Primates indet. 2 (Bajpai et al. 2005c)	**Referred material**: VLM 659 (isolated left?deciduous P4)
Didelphimorphia (Gill 1872)			
	Didelphidae (Gray 1821)		
		Indodelphis luoi (Bajpai et al. 2005b)	**Holotype**: VLM 595 (isolated left mx)
Rodentia (Bowdich 1821)			
	Rodentia indet (Bowdich 1821)		
		Anthramys vastani (=*Meldimys musak*) (Bajpai et al. 2007a; also see Rana et al. 2008)	**Holotype**: VLM 1104 (isolated left P4); VAS 609 (isolated right M2); **Paratypes**: VAS 608 (right deciduous P4); VAS 606 (left P4); VAS 449 (right M1); VAS 128 (left M1); VAS 607 (left M1); VAS 450 (right M2); VAS 454 (right M2); VAS 451 (right M3); VAS 452 (right M3); VAS 453 (right M3); VAS 455 (left M3); VAS 455 (left M3); VAS 456 (left M3); VAS 604 (right m1); VAS 605 (left m1); VAS 603 (left m2)
	Chappatimyidae (Hussain et al. 1978)		
		Chappatimyidae gen. et sp. indet. (Bajpai et al. 2007a)	**Referred material**: VLM 581 (left m1 or m2)
Proteutheria (Romer 1966)			
	Paleoryctidae (Winge 1917)		
		Anthraryctes vastanensis (Bajpai et al. 2005a)	**Holotype**: VLM 506 (Isolated M3)
	Cimolestidae (Marsh 1889)		
		Suratilestes gingerichi (Bajpai et al. 2005a)	**Holotype**: VLM 509 (left dentary with alveoli for p2-3, crowns for p4, m1, m2 trigonid, and m3); **Referred material**: VLM 573 (isolated p3)
Apatotheria (Scott and Jepsen 1936)			
	Apatemyidae (Matthew 1909)		
		Frugivastanodon cristatus (Bajpai et al. 2005a)	**Holotype**: VLM 507 (isolated left lower molar)
Insectivora (Cuvier 1817)			
	Vastanidae (Bajpai et al. 2005a)		
		Vastania sahnia (Bajpai et al. 2005a)	**Holotype**: VLM 510 (right dentary with one alveolus for p2, and crowns for p3-m3)
Eutheria			
	Adapisoriculidae (Van Valen 1967)		
		Bharatlestes kalami (=*Indolestes kalamensis*) (Kapur et al. 2017a, b)	**Holotype**: VLM 804 (right dentary with alveoli for p2, p3 and m1; crowns for p4, m2 and m3)
Pantodonta (Cope 1873)			
	cf. Coryphodontidae (Marsh 1876)		
		cf. Coryphodontidae indet (Smith et al. 2016)	**Referred material**: TAD 9204 (right P1); TAD 9501 (fragment of a possible right incisor)

Table 13.2 Dental measurements (length and width, in mm) of known non-volant mammals from the Cambay Shale, Gujarat, western India

Author(s)/ Reference (s)	Taxon	Specimen(s)	Remarks	p4		m1		m2		m3	
				Length (mm)	Width (mm)	Length (mm)	Width (mm)	Length (mm)	Width (mm)	Length (mm)	Width (mm)
Perissodactyla											
Bajpai et al. (2005a)	*Cambaytherium thewissi*	IITR/SB/VLM 505	Holotype right dentary	9.40	7.40	11.30	8.00	11.90	8.60	14.20	9.00
Bajpai et al. (2005a)	*Cambaytherium thewissi*	IITR/SB/VLM 505	Holotype left dentary	9.40	7.40	11.00	7.90	11.40	8.60	13.50	–
Bajpai et al. (2005a)	*Cambaytherium thewissi*	IITR/SB/VLM 548	mX	–	–	–	–	12.20	7.70	–	–
Bajpai et al. (2005a)	*Cambaytherium thewissi*	IITR/SB/VLM 551	mX	–	–	–	–	12.40	7.80	–	–
Rose et al. (2014)	*Cambaytherium thewissi*	Numerous specimens	Refer Supplementary Table 1 of Rose et al. (2014); mean values taken for p4 (9 specimens), m1 (9 specimens), m2 (17 specimens), m3 (13 specimens).	10.98	8.36	11.81	8.23	13.32	9.36	16.69	9.55
Bajpai et al. (2005a)	*Cambaytherium minor*	IITR/SB/VLM 539	Holotype right mX	–	–	–	–	10.50	5.30	–	–
Bajpai et al. (2005a)	*Cambaytherium bidens*	IITR/SB/VLM 503	Holotype right dentary	10.90	9.20	10.80	9.10	13.20	9.70	17.20	10.70
Bajpai et al. (2005a)	*Cambaytherium bidens*	IITR/SB/VLM 503	Holotype left dentary	11.00	8.90	13.00	8.70	13.80	10.00	17.50	10.50
Bajpai et al. (2005a)	*Cambaytherium bidens*	IITR/SB/VLM 502	Holotype right dentary	11.00	8.20	12.00	8.30	13.80	9.60	16.00	9.40
Bajpai et al. (2005a)	*Cambaytherium bidens*	IITR/SB/VLM 502	Holotype left dentary	–	–	–	–	–	–	16.40	9.40
Bajpai et al. (2005a)	*Cambaytherium bidens*	IITR/SB/VLM 520	Left dentary with m1 and m2	–	–	11.40	8.00	13.20	7.90	–	–
Bajpai et al. (2005a)	*Cambaytherium bidens*	IITR/SB/VLM 521	m3	–	–	–	–	–	–	15.80	9.60
Bajpai et al. (2005a)	*Cambaytherium bidens*	IITR/SB/VLM 532	dentary with c1, m2 & m3	–	–	–	–	13.20	10.00	16.50	10.00

(continued)

Table 13.2 (continued)

Author(s)/Reference(s)	Taxon	Specimen(s)	Remarks	p4		m1		m2		m3	
				Length (mm)	Width (mm)	Length (mm)	Width (mm)	Length (mm)	Width (mm)	Length (mm)	Width (mm)
Bajpai et al. (2005a)	*Cambaytherium bidens*	IITR/SB/VLM 544	isolated p4	10.90	7.90	–	–	–	–	–	–
Bajpai et al. (2005a)	*Cambaytherium bidens*	IITR/SB/VLM 545	isolated p4	10.60	7.50	–	–	–	–	–	–
Artiodactyla											
Bajpai et al. (2005a)	*Gujaratia indica* (=*Diacodexis indicus*)	IITR/SB/VLM 511	Left dentary with p1, p2, p3, p4–m3	5.20	3.20	5.00	3.60	5.40	4.10	5.80	4.10
Kumar et al. (2010)	*Gujaratia indica* (=*Diacodexis indicus*)	GU/RSR/VAS 27	isolated m3	–	–	–	–	–	–	6.50	4.10
Kumar et al. (2010)	*Gujaratia indica* (=*Diacodexis indicus*)	GU/RSR/VAS 28	isolated m2	–	–	–	–	5.10	4.10	–	–
Kumar et al. (2010)	*Gujaratia indica* (=*Diacodexis indicus*)	GU/RSR/VAS 29	isolated m1	–	–	4.40	2.80	–	–	–	–
Kumar et al. (2010)	*Gujaratia indica* (=*Diacodexis indicus*)	GU/RSR/VAS 31	isolated m1	–	–	4.40	2.70	–	–	–	–
Kumar et al. (2010)	*Gujaratia indica* (=*Diacodexis indicus*)	GU/RSR/VAS 648	isolated m1	–	–	4.60	3.00	–	–	–	–
Kumar et al. (2010)	*Gujaratia indica* (=*Diacodexis indicus*)	GU/RSR/VAS 1620	p4, m1 & m2	4.70	2.90	4.30	3.10	4.60	3.60	–	–
Kumar et al. (2010)	*Gujaratia indica* (=*Diacodexis indicus*)	GU/RSR/VAS 1621	c, p1, p2, p3, m2-m3; p4-m1 (alveoli only)	–	–	–	–	5.00	4.00	5.50	3.50
Kumar et al. (2010)	*Gujaratia indica* (=*Diacodexis indicus*)	GU/RSR/VAS 1622	p4 (alveoli); m1-m3	–	–	4.30	3.10	4.30	3.60	5.50	3.60
Kumar et al. (2010)	*Gujaratia indica* (=*Diacodexis indicus*)	GU/RSR/VAS 207	isolated m1	–	–	3.40	2.40	–	–	–	–
Kumar et al. (2010)	Artiodactyla indet.	GU/RSR/VAS 446	isolated p4	5.00	2.70	–	–	–	–	–	–

(continued)

Table 13.2 (continued)

Author(s)/ Reference (s)	Taxon	Specimen(s)	Remarks	p4		m1		m2		m3	
				Length (mm)	Width (mm)	Length (mm)	Width (mm)	Length (mm)	Width (mm)	Length (mm)	Width (mm)
Creodonta											
Bajpai et al. (2009)	*Indohyaenodon raoi*	IITR/SB/VLM 899	Holotype, left dentary with alveolus for p1; crowns for p2-m3	5.80	2.40	4.60	2.60	5.80	3.60	6.90	4.30
Bajpai et al. (2009)	*Indohyaenodon raoi*	IITR/SB/VLM 926	left dentary with alveoli for c1 to p3; crowns for p4 to m2; p4 damaged	4.90	2.10	4.60	2.60	6.20	3.70	–	–
Bajpai et al. (2009)	*Indohyaenodon raoi*	IITR/SB/VLM 742	right dentary with crowns for p2 to p4; crowns for m2 & m3; m1 damaged	5.20	2.40	–	–	6.00	2.80	–	–
Bajpai et al. (2009)	*Indohyaenodon raoi*	IITR/SB/VLM 782	right dentary with alveolus for m1; crowns for p2 to p4; crowns for m2 & m3	–	–	–	–	5.90	3.70	7.40	4.30
Rana et al. (2015)	*Indohyaenodon raoi*	GU/RSR/VAS 321	isolated m1	–	–	5.50	3.20	–	–	–	–
Rana et al. (2015)	*Indohyaenodon raoi*	GU/RSR/VAS 652	dentary with c1, p2, p3, p4 and m2	6.10	3.00	–	–	6.70	3.50	–	–
Rana et al. (2015)	*Indohyaenodon raoi*	GU/RSR/VAS 767	dentary with c1,p1, p2, p3, p4, m1, m2 and m3	5.40	2.60	5.40	2.70	6.00	3.30	6.30	4.20
Rana et al. (2015)	*Indohyaenodon raoi*	GU/RSR/VAS 1630	dentary with p2, p3, p4, m1	–	–	4.80	2.30	–	–	–	–
Rana et al. (2015)	*Indohyaenodon raoi*	GU/RSR/VAS 1631	isolated m1	–	–	5.40	3.00	–	–	–	–
Tillodontia											
Rose et al. (2013)	*Anthraconyx hypsomylus*	GU/RSR/VAS 762	Holotype, right dentary with dp3, dp4 and m1	–	–	6.20	4.90	–	–	–	–
Primates											
Bajpai et al. (2005c)	*Marcgodinotius indicus*	IITR/SB/VLM 800	Holotype right dentary with alveoli for c1, p1, p2, p3, p4, m1 and m3; crown for m2	–	–	–	–	2.23	1.73	–	–
Bajpai et al. (2005c)	*Marcgodinotius indicus*	IITR/SB/VLM 817	right dentary with partial alveolus for c1; alveoli for p1, p2, p3, p4; crowns for m2 & m3	–	–	–	–	–	–	2.22	1.50
Bajpai et al. (2005c)	*Marcgodinotius indicus*	IITR/SB/VLM 642	isolated right upper molar (M2 or M3); lost during photography	–	–	–	–	–	–	–	–
Rose et al. (2009a)	*Marcgodinotius indicus*	GU/RSR/VAS 7	isolated lower m1	–	–	2.40	1.65	–	–	–	–
Rose et al. (2009a)	*Marcgodinotius indicus*	GU/RSR/VAS 40	isolated lower p4	–	–	2.25	1.30	–	–	–	–

(continued)

Table 13.2 (continued)

Author(s)/ Reference (s)	Taxon	Specimen(s)	Remarks	p4		m1		m2		m3	
				Length (mm)	Width (mm)	Length (mm)	Width (mm)	Length (mm)	Width (mm)	Length (mm)	Width (mm)
Rose et al. (2009a)	*Marcgodinotius indicus*	GU/RSR/VAS 44	isolated lower m1	–	–	2.20	1.40	–	–	–	–
Rose et al. (2009a)	*Marcgodinotius indicus*	GU/RSR/VAS 45	isolated lower molar m2	–	–	–	–	2.35	1.75	–	–
Rose et al. (2009a)	*Marcgodinotius indicus*	GU/RSR/VAS 46	isolated lower molar m3	–	–	–	–	–	–	2.30	1.50
Rose et al. (2009a)	*Marcgodinotius indicus*	GU/RSR/VAS 49	isolated lower molar m1	–	–	2.40	1.55	–	–	–	–
Rose et al. (2009a)	*Marcgodinotius indicus*	GU/RSR/VAS 51	isolated lower molar m3	–	–	–	–	–	–	2.25	1.50
Rose et al. (2009a)	*Marcgodinotius indicus*	GU/RSR/VAS 52	isolated lower molar m3	–	–	–	–	–	–	2.40	1.40
Rose et al. (2009a)	*Marcgodinotius indicus*	GU/RSR/VAS 54	isolated lower molar m1	–	–	2.30	1.55	–	–	–	–
Rose et al. (2009a)	*Marcgodinotius indicus*	GU/RSR/VAS 227	dentary with p3, p4, m2 & m3	2.30	1.40	–	–	2.30	1.75	2.40	1.60
Rose et al. (2009a)	*Marcgodinotius indicus*	GU/RSR/VAS 600	isolated lower molar m1	–	–	2.30	1.55	–	–	–	–
Rose et al. (2009a)	*Marcgodinotius indicus*	GU/RSR/VAS 611	ioslated lower molar m1	–	–	2.20	1.40	–	–	–	–
Rose et al. (2009a)	*Marcgodinotius indicus*	GU/RSR/VAS 643	isolated lower molar m2	–	–	–	–	2.15	1.55	–	–
Rose et al. (2009a)	*Marcgodinotius indicus*	GU/RSR/VAS 644	isolated lower molar m2	–	–	–	–	2.20	1.50	–	–
Rose et al. (2009a)	*Marcgodinotius indicus*	GU/RSR/VAS 645	isolated lower molar m2	–	–	–	–	2.40	1.70	–	–
Rose et al. (2009a)	*Marcgodinotius indicus*	GU/RSR/VAS 646	isolated lower molar m2	–	–	–	–	2.15	1.55	–	–
Rose et al. (2009a)	*Marcgodinotius indicus*	GU/RSR/VAS 727	dentary with lower premolar p4 & molar m2	2.00	1.35	–	–	2.20	1.70	–	–
Rose et al. (2009a)	*Marcgodinotius indicus*	GU/RSR/VAS 743	dentary with molar m2	–	–	–	–	2.20	1.60	–	–
Rose et al. (2009a)	*Marcgodinotius indicus*	GU/RSR/VAS 744	dentary with molar m2	–	–	–	–	2.30	1.60	–	–
Rose et al. (2009a)	*Marcgodinotius indicus*	GU/RSR/VAS 1534	isolated lower molar m3	–	–	–	–	–	–	2.30	1.40
Rose et al. (2009a)	*Marcgodinotius indicus*	GU/RSR/VAS 1536	isolated lower premolar p4	2.30	1.25	–	-	–	–	–	–
Rose et al. (2009a)	*Marcgodinotius indicus*	GU/RSR/VAS 1554	isolated lower premolar p4	2.20	1.30	–	–	–	–	–	–

(continued)

Table 13.2 (continued)

Author(s)/ Reference (s)	Taxon	Specimen(s)	Remarks	p4		m1		m2		m3	
				Length (mm)	Width (mm)	Length (mm)	Width (mm)	Length (mm)	Width (mm)	Length (mm)	Width (mm)
Rose et al. (2009a)	*Marcgodinotius indicus*	GU/RSR/VAS 1575	isolated lower molar m2	–	–	–	–	2.30	1.60	–	–
Rose et al. (2009a)	*Marcgodinotius indicus*	GU/RSR/VAS 1591	isolated lower m3	–	–	–	–	–	–	2.35	1.35
Rose et al. (2009a)	*Marcgodinotius indicus*	GU/RSR/VAS 1602	isolated lower molar m2	–	–	–	–	2.30	1.60	–	–
Rose et al. (2009a)	cf. *Marcgodinotius*	GU/RSR/VAS 1619	isolated lower molar m3	–	–	–	–	–	–	1.75	1.05
Rose et al. (2009a)	*Vastanomys gracilus*	IITR/SB/VLM 771	Holotype left dentary with alveoli for c1, p1, p2, p3, p4, m1 and m3; crown for m2	–	–	–	–	2.05	1.81	–	–
Rose et al. (2009a)	*Vastanomys major*	GU/RSR/VAS 1541	Holotype isolated lower molar m2	–	–	–	–	2.80	2.45	–	–
Bajpai et al. (2007b)	*Asiadapis cambayensis* (=*Suratius robustus*)	IITR/SB/VLM 928	Holotype; Left dentary with alveoli for c1, p2, m1, m3; crowns for p3, p4 and m2	2.54	1.92	–	–	2.70	2.40	–	–
Rose et al. (2007)	*Asiadapis cambayensis* (=*Suratius robustus*)	GU/RSR/VAS 6	Holotype right dentary with alveoli for c, p2 and m3; crowns for p3, p4, m1, m2	2.60	1.70	2.85	2.00	2.90	2.15	–	–
Rose et al. (2009a)	*Asiadapis cambayensis* (=*Suratius robustus*)	GU/RSR/VAS 32	isolated lower molar m3	–	–	–	–	–	–	2.80	1.75
Rose et al. (2009a)	*Asiadapis cambayensis* (=*Suratius robustus*)	GU/RSR/VAS 36	isolated lower molar m2	–	–	–	–	2.65	2.10	–	–
Rose et al. (2009a)	*Asiadapis cambayensis* (=*Suratius robustus*)	GU/RSR/VAS 37	isolated lower molar m2	–	–	–	–	2.85	2.35	–	-
Rose et al. (2009a)	*Asiadapis cambayensis* (=*Suratius robustus*)	GU/RSR/VAS 38	isolated lower premolar p4	2.90	1.70	–	–	–	–	–	–
Rose et al. (2009a)	*Asiadapis cambayensis* (=*Suratius robustus*)	GU/RSR/VAS 598	isolated lower molar m3	–	–	–	–	–	–	3.40	2.05
			isolated lower molar m2	–	–	–	–	2.95	2.30	–	–

(continued)

Table 13.2 (continued)

Author(s)/ Reference (s)	Taxon	Specimen(s)	Remarks	p4		m1		m2		m3	
				Length (mm)	Width (mm)	Length (mm)	Width (mm)	Length (mm)	Width (mm)	Length (mm)	Width (mm)
Rose et al. (2009a)	*Asiadapis cambayensis* (=*Suratius robustus*)	GU/RSR/VAS 642									
Rose et al. (2009a)	*Asiadapis cambayensis* (=*Suratius robustus*)	GU/RSR/VAS 745	lower dentary with p3, p4, m1, m2	3.00	1.80	3.10	2.20	3.10	2.40	–	–
Rose et al. (2009a)	*Asiadapis cambayensis* (=*Suratius robustus*)	GU/RSR/VAS 1505	isolated lower molar m3	–	–	–	–	–	–	3.25	1.90
Rose et al. (2009a)	*Asiadapis cambayensis* (=*Suratius robustus*)	GU/RSR/VAS 1537	isolated lower molar m1	–	–	3.00	2.00	–	–	–	–
Rose et al. (2009a)	*Asiadapis cambayensis* (=*Suratius robustus*)	GU/RSR/VAS 1627	isolated lower premolar p4	2.70	1.65	–	–	–	–	–	–
Rose et al. (2009a)	*Asiadapis cambayensis* (=*Suratius robustus*)	GU/RSR/VAS 1649	isolated lower m3	–	–	–	–	–	–	3.30	2.05
Proteutheria											
Bajpai et al. (2005a)	*Suratilestes gingerichi*	IITR/SB/VLM 509	Holotype; left dentary with alveoli for p2-3, crowns for p4, m1, m2 trigonid, and m3	1.16	0.60	1.33	0.90	1.26	0.96	1.33	0.93
Bajpai et al. (2005a)	*Suratilestes gingerichi*	IITR/SB/VLM 573	isolated lower p3 (not included in sheet)	–	–	–	–	–	–	–	–
Bajpai et al. (2005a)	*Frugivastanodon cristatus*	IITR/SB/VLM 507	Holotype; isolated left lower molar (m1 or m2); considered as m1 in this sheet	–	–	3.50	2.10	-	–	–	–
Bajpai et al. (2005a)	*Vastania sahnia*	IITR/SB/VLM 510	Holotype; right dentary with one alveolus for p2, and crowns for p3-m3	1.00	0.61	1.41	0.86	1.43	0.88	1.33	0.77
Rodentia											
Rana et al. 2008 (also see Bajpai	*Anthramys vastani*	GU/RSR/VAS 605	isolated lower molar m1	–	–	2.20	1.80	–	–	–	–

(continued)

Table 13.2 (continued)

Author(s)/ Reference (s)	Taxon	Specimen(s)	Remarks	p4		m1		m2		m3	
				Length (mm)	Width (mm)	Length (mm)	Width (mm)	Length (mm)	Width (mm)	Length (mm)	Width (mm)
et al. 2007a)	*(=Meldimys musak)*										
Rana et al. 2008 (also see Bajpai et al. 2007a)	*Anthramys vastani (=Meldimys musak)*	GU/RSR/VAS 604	isolated lower molar m2	–	–	–	–	2.30	1.95	–	–
Rana et al. 2008 (also see Bajpai et al. 2007a)	*Anthramys vastani (=Meldimys musak)*	GU/RSR/VAS 603	isolated lower molar m3	–	–	–	–	–	–	2.35	2.00
Didelphimorphia											
Bajpai et al. (2005b)	*Indodelphis luoi*	IITR/SB/VLM 595	Holotype; Isolated left lower molar (considered here as an m1)	–	–	1.59	0.71	–	–	–	–

graphs. Further, for statistical analysis, box plots were drawn and one-way ANOVA tests were conducted on the data on dental area (length × width) for individual mammalian teeth (p4-m3) provided in the published literature. As many previous studies have shown strong correlation between first lower molar (m1) size and body size of mammals (Kay 1975; Legendre and Roth 1988; Bown et al. 1994; Millien and Bovy 2010 and references therein), emphasis was placed on the statistical analysis carried out on the first lower molars. Further, the available data on age, evolutionary relationship (s) of Cambay Shale non-volant mammals (particularly perissodactyls) and the timings and dynamics of the India-Eurasia collision were considered for drawing inferences.

Timing of India-Eurasia Collision

The northward drift of the Indian subcontinent, after its separation from Madagascar around 90 Ma (Bardintzeff et al. 2010 and references therein) culminated with the subcontinent colliding with Eurasia. However, the timing of initiation and final collision is intensely debated, with estimates ranging from ~70 Ma to ~34 Ma (Patriat and Achache 1984; Garzanti and Van Haver 1988; Willems et al. 1996; Yin and Harrison 2000; Besse and Coutillot 2002; Guillot et al. 2003; Ali and Aitchison 2004; Ding et al. 2005; Yin 2006; Aitchison et al. 2007; Cai et al. 2011; DeCelles et al. 2014; Hu et al. 2015, 2016). The commonly accepted ~55–50 Ma age for the initial collision event has been constrained mainly on the basis of changes in the rate of northern drift of the Indian subcontinent (Molnar and Tapponnier 1975; Powell et al. 1988; Dewey et al. 1989; Klootwijk et al. 1992; Acton 1999; Besse and Courtillot 2002), pre-collision size of the Indian plate (Treloar and Coward 1991; Guillot et al. 2003) and geological evidence from the suture zone between the northern margin of the Indian plate and the southern margin of the Asian plate (Garzanti and Van Haver 1988; Pognante and Spencer 1991; Tonarini et al. 1993; Beck et al. 1995; Searle et al. 1997; Guillot et al. 1997; De Sigoyer et al. 2000; Najman et al. 2010; Henderson et al. 2010; Zhang et al. 2012; Zhu et al. 2015). However, a thorough review of previous studies and availability of new datasets from geological and paleomagnetic studies, consideration of pre-collision size of the Indian subcontinent, popularly referred to as "Greater India", its motion history, the reviewed position of stable Eurasia and the southern margin of Asia, structure of the Neo-Tethys Ocean and the inferred age of the youngest marine sedimentation in the Tethyan region, led Aitchison et al. (2007) to propose a much younger age (~34 Ma) for the initial continent-continent collision. The variable age constraints for the Indian subcontinent and Eurasia collision event are tied to the use of a variety of approaches (seismology, structural geology, geomagnetism, geochemistry and others), leading to various connotations of the term "collision onset" used in the extensive literature available on the subject (see Zhu et al. 2005 and Hu et al. 2016). In addition, the inferred west to east propagation of the collision, also reflects diachroneity (Rowley 1996; also see Ding et al. 2017). In the context of the present article, "collision onset" is the time at which the biological entities (e.g., terrestrial mammals) start showing taxonomic similarity across the two continents in the fossil record. I believe that the independent nature of the biotic evidence discussed in the present article does not contravene the definition of the "collision onset" in a tectonic framework (*sensu* Hu et al. 2015).

A Brief Historical Background of Biogeographically Important Vertebrate Fossil Record (Including Mammals) from the Cretaceous-Paleogene (K-Pg) Interval of India

In the context of the present article, it is worth providing readers with a brief historical background of the significant (Late Cretaceous) mammalian record from the Indian subcontinent and the state of affairs that led to the recovery of Late Cretaceous and Early Eocene mammals from the subcontinent. Large dinosaurian bones (e.g., *Titanosaurus*) have been known from the Late Cretaceous (Maastrichtian) of the Indian subcontinent since the 19th century (Sleeman 1844; Lydekker 1877). Employment of surface prospecting technique(s) augmented the fossil data on dinosaurian remains in the form of cranial and postcranial elements, eggs and coprolites (Chatterjee 1978; Carrano et al. 2010 and references therein). The collection of large-sized remains by surface prospecting had a bias as it excluded the small-sized faunal remains. Techniques for prospecting for microvertebrates had been successfully employed in North America by the middle part of the 20th century (Hibbard 1949); however, it was in the late 1960's that Professor Ashok Sahni introduced to the Indian scientific community the bulk screen-washing maceration techniques now employed for the recovery of microvertebrates from Late Cretaceous sedimentary sequences of peninsular India (Sahni 1968). In the years that followed, unswerving efforts to apply these techniques to the Late Cretaceous (Maastrichtian) Deccan associated sedimentary deposits by the Indian paleontological community (led by Ashok Sahni) resulted in the recovery of numerous microvertebrate fossils belonging to fishes, frogs, lizards, snakes, turtles, dinosaurian egg shells, dinosaurian dental elements and mammalian dentitions (see Kapur et al. 2019; Kapur and Khosla 2019 and references therein). Prasad and Sahni (1988) reported the first mammal from the Late Cretaceous (Maastrichtian) intertrappean deposits of Naskal, southern India. Since then, the mammalian data from the Maastrichtian of India have also grown both numerically and taxonomically, represented by at least three groups: adapisoriculids, gondwanatherians and possible haramiyids (Prasad et al. 2007, 2010 and references therein). Overall, the vertebrate fauna from the Maastrichtian of India is an admixture of Laurasian and Gondwanan components. With respect to the Laurasian component, Prasad and Sahni (1999) postulated biotic exchange via an island arc system (Kohistan-Dras-Ladakh island arc system) north of the Indian subcontinent and emphasized the size constraints of the biota that was involved in biotic exchange. A detailed review of the Late Cretaceous (Maastrichtian) fauna from the Indian subcontinent by Kapur and Khosla (2016) reiterated the viewpoint (based on geophysical evidence) expressed by Aitchison et al.'s (2007) that no clear land connections existed between India and its surrounding landmasses during this time interval. It is noteworthy that the inferences on the degree of connectivity surrounding between India and its landmasses, particularly based on biotic evidence, consider vertebrate-yielding intertrappean deposits to be Maastrichtian in age (as also widely reported in the available literature). However, there are a few intertrappean localities (e.g. Papro, Lalitpur district, Uttar Pradesh State, north India) that have been assigned a Paleocene age based on palynological data (Singh and Kar 2002); these are, unfortunately, devoid of mammals. In view of the recent data on the timing and episodic nature of the Deccan volcanism (Keller et al. 2012 and references therein), a detailed age evaluation (based on vertebrate remains) of the numerous intertrappean sites located on the fringes of the Deccan Volcanic Province has been lately provided by Kapur and Khosla (2019). Regrettably, no definite Paleocene mammals have yet been recorded from the subcontinent.

About ten or slightly more years ago, the Paleogene mammalian record was confined to the Eocene Harudi Formation of Kutch and the coeval Subathu Formation in northwestern Himalaya (Bajpai 2009 and references therein). However, detailed works on primitive cetacean (whale) fossils from the Indian subcontinent (Bajpai and Gingerich 1998; also see Bajpai and Thewissen 2014), set the stage for unraveling the bigger role the Indian subcontinent may have played in terms of the origin and early evolution of these mammals that once thrived on land but presently occupy exclusively marine realms. Detailed paleontological studies on fossil sirenians (sea cows) from the Paleogene succession of Kutch region, Gujarat, western India have also generated a lot of interest regarding the origin and early diversification of these large plant-eating mammals (Bajpai et al. 2006b; Thewissen and Bajpai 2009).

About 20 years ago, the timing was perfect for reconnoitering fossil vertebrates in the Early Eocene lignite-associated subsurface shale deposits, as India had upped its lignite exploration in the western states of Gujarat and Rajasthan, thereby exposing the subsurface Eocene horizons (largely lignite-associated shales). Initial efforts yielded vertebrates (including mammalian remains) from the Panandhro Lignite Mine, Kutch District, Gujarat (Bajpai and Thewissen 2002). Lower vertebrate fossils (mainly fish remains) were also reported from the lignite mines of Rajasthan (Kumar et al. 2005; Rana et al. 2005). In 2001, borehole samples of the Cambay Shale Formation, Surat District, Gujarat yielded numerous fish ear bones (otoliths) (Samant and Bajpai 2001). Subsequently (from 2002 to 2006), as part of a Government of India funded project, the author (VVK) pursuing a doctorate degree at Indian Institute of Technology Roorkee (IITR, India) reconnoitered for fossil vertebrates in the subsurface lignite-associated sedimentary deposits of Cambay Basin in Gujarat. These efforts resulted in the discovery of the world's oldest otoliths belonging to gobiid fish from Cambay Shale subsurface deposits in the open-pit Vastan Lignite Mine, Surat District, Gujarat (Bajpai and Kapur 2004). In the very next year, the first Early Eocene land mammals were reported from the Cambay Shale and included artiodactyls, perissodactyls, primates and proteutherians (Bajpai et al. 2005a). This report not only remedied the paucity of the mammalian fossil record from the Early Eocene of India, but also helped to a certain extent in developing some understanding of the origin and early evolutionary history of modern mammalian orders (particularly perissodactyls) in the years that followed.

Age of the Cambay Shale Mammals

The age of the Cambay Shale mammals has been deciphered mainly on the basis of detailed studies carried out in the open pit "Vastan Lignite Mine–North Pit" (21° 25′ 47″ N and 73° 07′ 30″ E), situated about 65 km northeast of the city of Surat and close (about 3 km) to the small village "Nani Naroli" in Surat District, Gujarat (see Kapur et al. 2017a: Fig. 3). Initially, the age of the Cambay Shale mammals was considered to be late Early Eocene (middle Ypresian), which was based on the larger benthic foraminifer *Nummulites burdigalensis* recovered about 15–20 m above the mammal-yielding level (Sahni et al. 2006). A more reliable basal Eocene age for the mammalian horizon was provided by the record of an age-diagnostic dinoflagellate assemblage (Garg et al. 2008). Subsequently, a pronounced, negative carbon isotopic shift at a stratigraphic level over 20 m above the mammal horizon, correlating to the Second Eocene Thermal Maximum (ETM2), was earmarked. The combined data on $\delta^{13}C$, $^{87}Sr/^{86}Sr$ and dinoflagellates (Clementz et al. 2011) provided an age older than the ETM2 datum (~53.7 Ma) and slightly younger than the Paleocene-Eocene Thermal Maximum (PETM, ~55.8 Ma) for the mammal-yielding level. As such, the Cambay Shale mammalian record provides the oldest known Cenozoic record from India as well as from South Asia. Within the Indian subcontinent, the Eocene lignite-associated sediments of the Ghazij Formation (Pakistan) are also known for mammals that have been reported from multiple horizons (Thewissen et al. 2001 and references therein). However, the paleontological and stratigraphic (bio-, chemo- and magneto-) data suggest that the Ghazij mammals are younger than the Cambay Shale mammals (Gingerich et al. 1997, 2001; Clyde et al. 2003; Missiaen and Gingerich 2012; Kapur and Bajpai 2015).

Size Variation Amongst the Non-volant Mammalian Groups Known from the Cambay Shale

As numerous dental remains (including lower p4-m3) are known for a variety of mammalian groups from the Cambay Shale, we can examine and categorize the size variation(s) amongst these groups. In the context of the present article and for reasons cited earlier, data on volant forms (e.g., chiropterans) known from the Cambay Shale (Bajpai et al. 2005a; Smith et al. 2007) have been excluded from the present study. Bivariate plots (length × width) for p4, m1, m2 and m3 for individual groups of non-volant Cambay Shale mammals were constructed using the data provided in Table 13.2 and are shown in Fig. 13.2A–D. Visual interpretation within these plots suggests the presence of three broad categories of non-volant mammals within the Cambay Shale.

Further, using the data provided on tooth area (length × width) for p4-m3 in Table 13.2, the Minimum, First Quartile, Second Quartile (Median), Third Quartile and Maximum values were also calculated for the p4-m3 for perissodactyls, artiodactyls, creodonts, primates and proteutherians known from the Cambay Shale (refer Table 13.3). These data have been graphically represented as box plots (Fig. 13.3A–D).

Box plots also allow size categorization for Cambay Shale mammals into three broad groups: small, medium and large. The median ranges considered for this grouping in the present article are as follows: small mammals (p4: 0.61–5.40 mm^2; m1: 1.20–7.35 mm^2; m2: 1.21–7.44 mm^2; m3: 1.02–6.97 mm^2), medium mammals (p4: 10.29–18.30 mm^2; m1: 8.16–18.00 mm^2; m2: 15.48–23.45 mm^2; m3: 19.25–31.82 mm^2) and large mammals (p4: 69.56–100.28 mm^2; m1: 86.90–113.10 mm^2; m2: 55.65–138.00 mm^2; m3: 127.80–184.04 mm^2) (also refer to Table 13.3 and Fig. 13.3).

For further inferences on size variation(s), one-way ANOVA test was conducted that allowed comparisons of the means (tooth area: length × width) between individual mammalian groups (at ordinal level) to determine whether any of these means are statistically significantly different from each of the mammalian groups. The results for the one-way ANOVA test are summarized in Table 13.4 while the dataset is provided in Supplementary Data S1.

The results of the one-way ANOVA test confirm the visual interpretations above and demonstrate that the tooth areas for the Cambay Shale perissodactyls are statistically significantly different ($p < 0.05$) from those of artiodactyls, creodonts, tillodonts, primates, proteutherians, rodents and didelphimorphs. Emphasizing the results for the first lower molar (m1), the tooth areas for artiodactyls, creodonts and tillodonts are statistically similar ($p > 0.05$). However, additional tooth specimens are warranted for tillodonts as they are here represented by a single specimen (no. GU/RSR/VAS762). The tooth areas for both artiodactyls and creodonts are statistically significantly different ($p < 0.05$) compared to the tooth areas of primates, proteutherians, rodents and didelphimorphs. Numerically, the dataset on p4-m3 tooth areas for proteutherians, rodents and dildelphimorphs is small (owing to the dearth of specimens), but it is statistically similar ($p > 0.05$) to that of primates (represented by a numerically rich dataset). Thus, considering that the tooth size [particularly first lower molar (m1)] is strongly correlated with body size in mammals, the one-way ANOVA test confirms the three broad size categorizations (mentioned earlier) for the non-volant mammalian assemblage from the Cambay Shale. The small-sized mammals are represented by primates, proteutherians, rodents, and didelphimorphs, the medium-sized mammals consist of artiodactyls, creodonts, and tillodonts, while the large-sized component is represented only by perissodactyls.

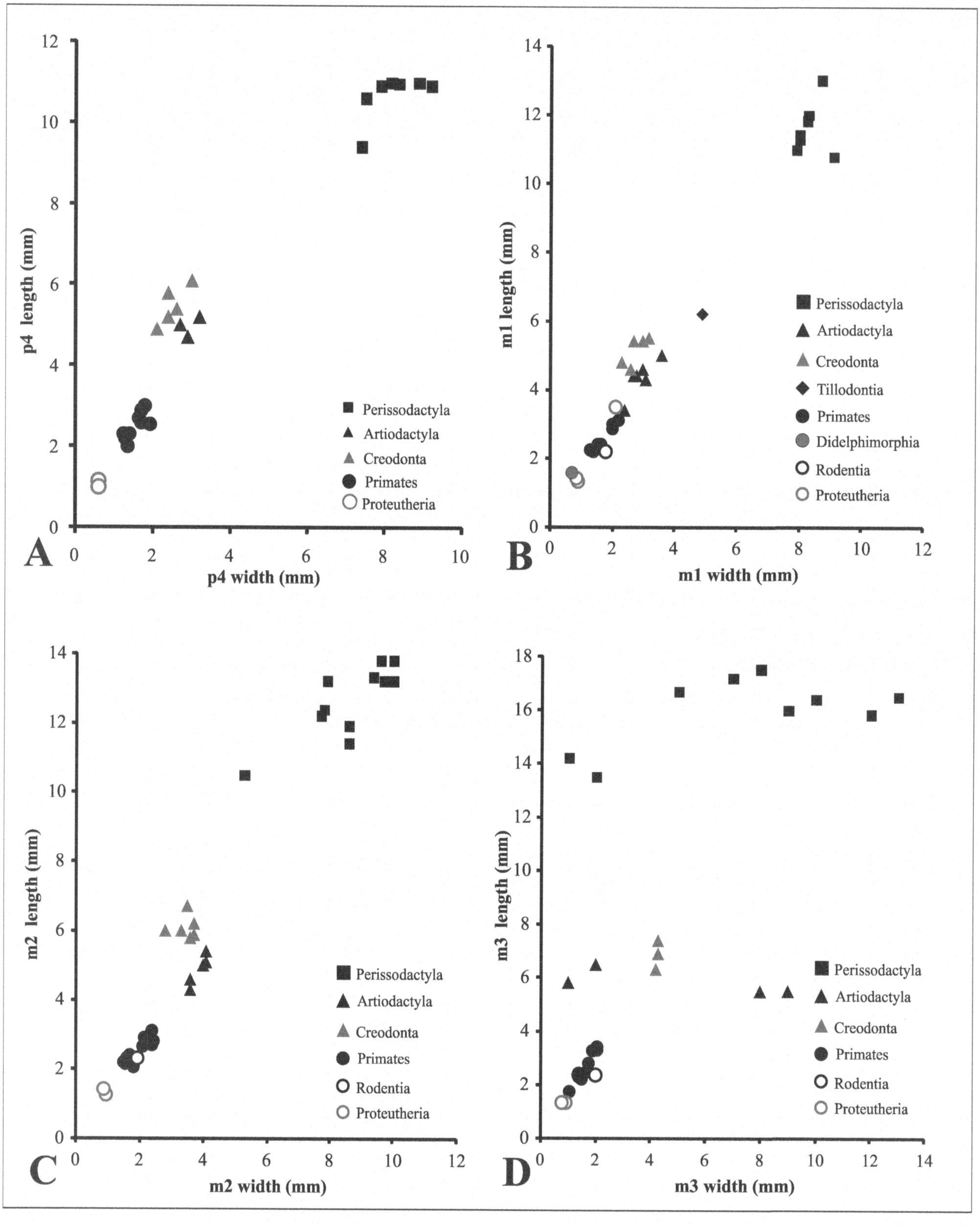

Fig. 13.2 Bivariate plots (width × length) for the dental elements (p4-m3) of the non-volant mammalian assemblage known from the Cambay Shale, Gujarat, western India. (**A**) p4, (**B**) m1, (**C**) m2 and (**D**) m3. Compiled from dataset provided in Table 13.2

Table 13.3 Minimum, First Quartile, Medium (Second Quartile), Third Quartile and Maximum values (length × width, in mm sq.) calculated for the p4, m1, m2 and m3 from the dataset on the Cambay Shale mammals (Gujarat, western India) provided in Table 13.2

p4	Perissodactyla	Artiodactyla	Creodonta	Primates	Proteutheria
Minimum	69.56	13.50	10.29	2.70	0.61
1st Quartile (Q1)	77.02	13.57	12.48	2.88	0.63
Median (Q2)	88.16	13.63	13.92	4.42	0.65
3rd Quartile (Q3)	93.32	15.14	14.04	4.88	0.67
Maximum	100.28	16.64	18.30	5.40	0.70
m1					
Minimum	86.90	8.16	11.04	2.93	1.20
1st Quartile (Q1)	90.80	12.10	11.96	3.20	1.20
Median (Q2)	97.20	13.33	13.27	3.64	1.21
3rd Quartile (Q3)	98.94	13.57	15.80	5.27	4.28
Maximum	113.10	18.00	17.60	6.82	7.35
m2					
Minimum	55.65	15.48	16.80	3.30	1.21
1st Quartile (Q1)	97.38	16.56	20.07	3.68	1.22
Median (Q2)	104.28	20.00	21.36	3.94	1.23
3rd Quartile (Q3)	130.02	20.91	22.66	6.30	1.25
Maximum	138.00	22.14	23.45	7.44	1.26
m3					
Minimum	127.80	19.25	26.46	1.84	1.02
1st Quartile (Q1)	151.36	19.66	28.07	3.30	1.08
Median (Q2)	156.77	21.79	29.67	3.41	1.13
3rd Quartile (Q3)	169.69	24.50	30.75	5.22	1.18
Maximum	184.04	26.65	31.82	6.97	1.24

Evolutionary Relationship(s) of the Large-Sized Component (Perissodactyls) from the Cambay Shale

Interestingly, the fossil data from the northern continents suggests sudden appearance of modern orders of mammals namely: artiodactyls, perissodactyls, and primates (APP taxa), close to the Paleocene-Eocene (P-E) transition (Gingerich 2006). Krause and Maas (1990) proposed the "Out-of-India" hypothesis and viewed the Indian subcontinent as a Noah's Ark where APP taxa originated and subsequently ferried into Eurasia after the subcontinent collided with Eurasia at/close to the P-E interval. Intriguingly, the evolutionary relationship(s) of Cambay Shale perissodactyls with known perissodactyls from other regions (particularly from Eurasia) becomes significant if we consider that these large-sized animals were unable to cross oceanic barriers surrounding the Indian subcontinent. The scientific community has been debating the basal evolutionary relationship(s) within perissodactyls, with many previous studies linking perissodactyl origins to an archaic group of Paleocene-Eocene ungulate mammals, the phenacodontid 'condylarths', based on the oldest and most primitive genus of the group, *Tetraclaenodon*, from the Paleocene of North America (Radinsky 1966; Van Valen 1978; Thewissen and Domning 1992). However, the monophyly of 'Phenacodontidae' has also been debated in the literature (Penkrot et al. 2008; Kondrashov and Lucas 2012). Nevertheless, a well-known bunodont ungulate, the phenacolophid genus *Radinskya* from the late Paleocene of China (McKenna et al. 1989), has also been considered significant to ascertain the origin or early evolution of perissodactyls (Rose 1996; Beard 1998; Hooker 2005; Holbrook 2014; also see Mao et al. 2016 and references therein). It should be noted here that Springer et al. (2011) emphasized the molecular data and provided the evidence for an Asian origin of perissodactyls. Nevertheless, these studies brought about a shift in the viewpoint of the geographic center of origin of perissodactyls from North America towards the Asian continent. Regarding the earlier proposal for an Indian origin of perissodactyls, Krause and Maas's (1990) "Out of India" hypothesis had support only in the form of endemic bunodont ungulates known from the Eocene of the subcontinent that included quettacyonine arctocyonids (Gingerich et al. 1997) and anthracobunid tethytheres (Dehm and Oettingen-Spielberg 1958; West 1980, 1983; Wells and Gingerich 1983; Kumar 1991). However, the reports of numerous fossil remains of the endemic perissodactyl family Cambaytheriidae from the Early Eocene of India (Bajpai et al. 2005a, 2006a; Bajpai 2009) allowed comparisons with the already known bunodont ungulates

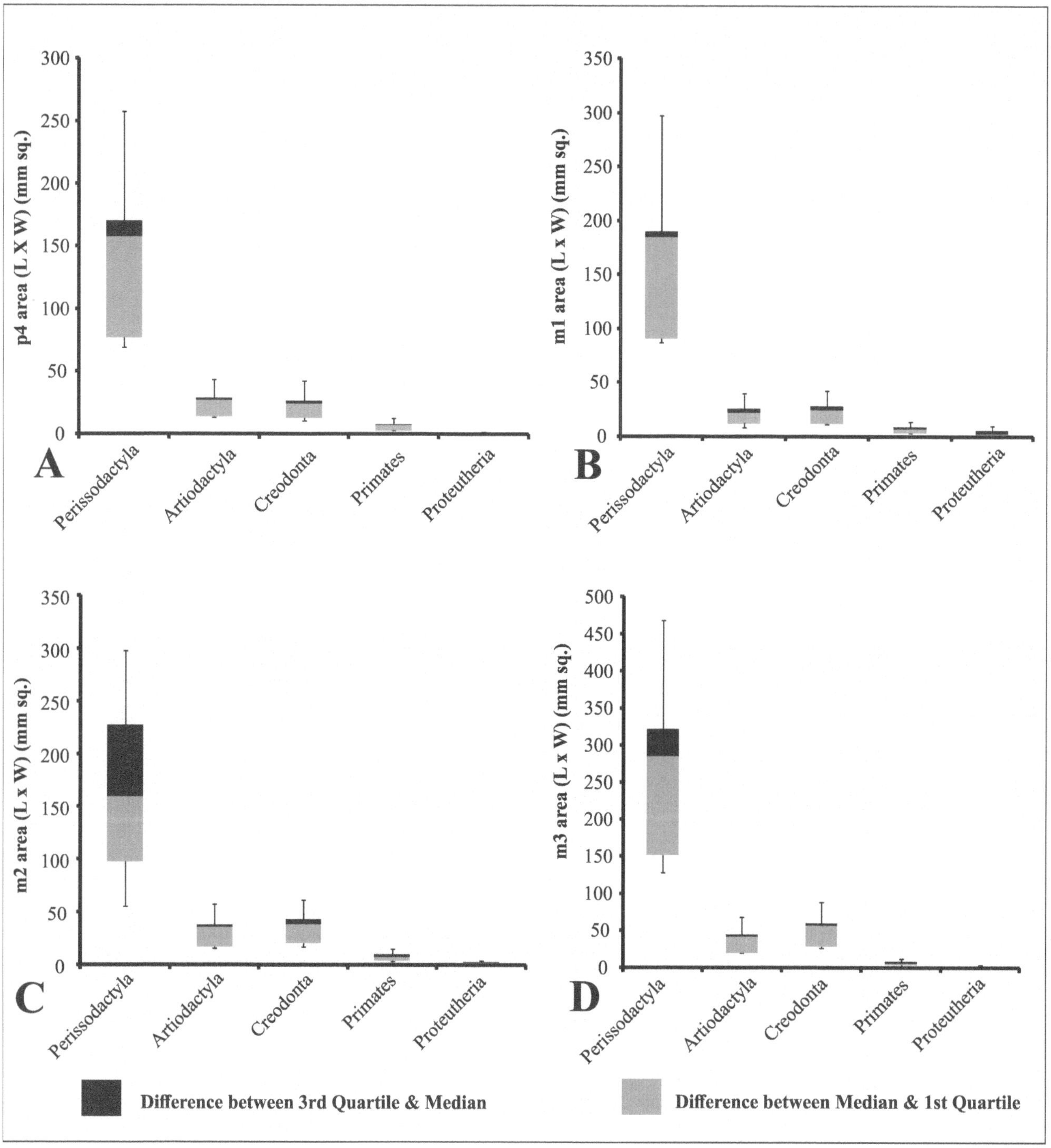

Fig. 13.3 Box plots as graphical representation of the data from Table 13.3, showing size variations among non-volant mammals from the Cambay Shale, Gujarat, western India. **A** for p4; **B** for m1; **C** for m2; and **D** for m3. *Note* Lack of box-plots for the p4, m1, m2 and m3 for tillodonts, rodents and didelphimorphs due to absence or dearth of data for these animals from Cambay Shale

from the Eocene of the region, and further, supported Krause and Maas's (1990) viewpoint on the origin of perissodactyls within the Indian subcontinent. The endemic perissodactyl family Cambaytheriidae from the Cambay Shale is presently represented by two genera, *Cambaytherium* and *Kalitherium* (Bajpai et al. 2005a, 2006a). Recently, Rose et al. (2014) included the long-known genus *Nakusia* from the younger Ghazij Formation of Pakistan (Ginsburg et al. 1999) within Cambaytheriidae. It is noteworthy that the "archaic perissodactyl" status of cambaytheres was initially questioned on the basis of a study that linked cambaytheres to anthracobunids and quettacyonid condylarths (Rose et al. 2006). The initial

Table 13.4 Summary of one-way ANOVA test conducted for area (length × width, in mm sq.) of premolar (p4) and molars (m1, m2, and m3) for known non-volant mammalian orders: Perissodactyla, Artiodactyla, Creodonta, Tillodonta, Primates, Proteutheria, Rodentia, and Didelphimorphia from the Cambay Shale, Gujarat, western India. $<10^{-3}$ and $<10^{-6}$ represents statistically different tooth areas ($p < 0.05$) while $>10^{-3}$ represents statistically similar tooth areas ($p > 0.05$)

Premolar (p4)	Perissodactyla (n = 8)	Artiodactyla (n = 3)	Creodonta (n = 5)	Tillodontia (n = 0)	Primates (n = 9)	Proteutheria (n = 2)	Rodentia (n = 0)	Didelphimorphia (n = 0)
Perissodactyla	–	$<10^{-6}$	$<10^{-6}$	–	$<10^{-6}$	$<10^{-3}$	–	–
Artiodactyla	$<10^{-6}$	–	**$>10^{-3}$**	–	$<10^{-6}$	$<10^{-3}$	–	–
Creodonta	$<10^{-6}$	**$>10^{-3}$**	–	–	$<10^{-6}$	$<10^{-3}$	–	–
Tillodontia	–	–	–	–	–	–	–	–
Primates	$<10^{-6}$	$<10^{-6}$	$<10^{-6}$	–	–	$<10^{-3}$	–	–
Didelphimorphia	–	–	–	–	–	–	–	–
Rodentia	–	–	–	–	–	–	–	–
Protutheria	$<10^{-3}$	$<10^{-3}$	$<10^{-3}$	–	$<10^{-3}$	–	–	–
Molar (m1)	Perissodactyla (n = 7)	Artiodactyla (n = 7)	Creodonta (n = 6)	Tillodontia (n = 1)	Primates (n = 10)	Proteutheria (n = 3)	Rodentia (n = 1)	Didelphimorphia (n = 1)
Perissodactyla	–	$<10^{-6}$	$<10^{-6}$	$<10^{-3}$	$<10^{-6}$	$<10^{-6}$	$<10^{-3}$	$<10^{-3}$
Artiodactyla	$<10^{-6}$	–	**$>10^{-3}$**	$<10^{-3}$	$<10^{-6}$	$<10^{-3}$	$<10^{-3}$	$<10^{-3}$
Creodonta	$<10^{-6}$	**$>10^{-3}$**	–	$<10^{-3}$	$<10^{-6}$	$<10^{-3}$	$<10^{-3}$	$<10^{-3}$
Tillodontia	$<10^{-3}$	$<10^{-3}$	$<10^{-3}$	–	$<10^{-6}$	$<10^{-3}$	–	–
Primates	$<10^{-6}$	$<10^{-6}$	$<10^{-6}$	$<10^{-6}$	–	**$>10^{-3}$**	**$>10^{-3}$**	**$>10^{-3}$**
Didelphimorphia	$<10^{-3}$	$<10^{-3}$	$<10^{-3}$	–	**$>10^{-3}$**	**$>10^{-3}$**	–	–
Rodentia	$<10^{-3}$	$<10^{-3}$	$<10^{-3}$	–	**$>10^{-3}$**	**$>10^{-3}$**	–	–
Protutheria	$<10^{-6}$	$<10^{-3}$	$<10^{-3}$	$<10^{-3}$	**$>10^{-3}$**	–	**$>10^{-3}$**	**$>10^{-3}$**
Molar (m2)	Perissodactyla (n = 11)	Artiodactyla (n = 5)	Creodonta (n = 6)	Tillodontia (n = 0)	Primates (n = 20)	Proteutheria (n = 2)	Rodentia (n = 1)	Didelphimorphia (n = 0)
Perissodactyla	–	$<10^{-6}$	$<10^{-6}$	–	$<10^{-6}$	$<10^{-3}$	$<10^{-3}$	–
Artiodactyla	$<10^{-6}$	–	**$>10^{-3}$**	–	$<10^{-6}$	$<10^{-3}$	$<10^{-3}$	–
Creodonta	$<10^{-6}$	**$>10^{-3}$**	–	–	$<10^{-6}$	$<10^{-3}$	$<10^{-3}$	–
Tillodontia	–	–	–	–	–	–	–	–
Primates	$<10^{-6}$	$<10^{-6}$	$<10^{-6}$	–	–	$<10^{-3}$	**$>10^{-3}$**	–
Didelphimorphia	–	–	–	–	–	–	–	–
Rodentia	$<10^{-3}$	$<10^{-3}$	$<10^{-3}$	–	**$>10^{-3}$**	$<10^{-3}$	–	–
Protutheria	$<10^{-3}$	$<10^{-3}$	$<10^{-3}$	–	$<10^{-3}$	–	$<10^{-3}$	–
Molar (m3)	Perissodactyla (n = 8)	Artiodactyla (n = 4)	Creodonta (n = 3)	Tillodontia (n = 0)	Primates (n = 12)	Proteutheria (n = 2)	Rodentia (n = 1)	Didelphimorphia (n = 0)
Perissodactyla	–	$<10^{-6}$	$<10^{-6}$	–	$<10^{-6}$	$<10^{-6}$	$<10^{-3}$	–
Artiodactyla	$<10^{-6}$	–	$<10^{-3}$	–	$<10^{-6}$	$<10^{-3}$	$<10^{-3}$	–
Creodonta	$<10^{-6}$	$<10^{-3}$	–	–	$<10^{-6}$	$<10^{-3}$	$<10^{-3}$	–
Tillodontia	–	–	–	–	–	–	–	–
Primates	$<10^{-6}$	$<10^{-6}$	$<10^{-6}$	–	–	$<10^{-3}$	**$>10^{-3}$**	–
Didelphimorphia	–	–	–	–	–	–	–	–
Rodentia	$<10^{-3}$	$<10^{-3}$	$<10^{-3}$	–	**$>10^{-3}$**	$<10^{-3}$	–	–
Protutheria	$<10^{-6}$	$<10^{-3}$	$<10^{-3}$	–	$<10^{-3}$	–	$<10^{-3}$	–

recognition of the perissodactyl status of cambaytheres (Bajpai et al. 2005a, 2006a; Bajpai 2009) was acknowledged when independent detailed phylogenetic analysis suggested a stem group position for cambaytheres within perissodactyls (Cooper et al. 2014; Rose et al. 2014). It is worth mentioning here that cambaytheres along with the late Paleocene Asian *Radinskya* and *Hallensia* were consistently placed close to the anthracobunids as stem perissodactyls in the phylogenetic analysis by Cooper et al. (2014). However, in a separate independent analysis by Rose et al. (2014) cambaytheres were always found to be positioned at the base of the perissodactyl stem with *Radinskya* sometimes occurring in a sister relationship with the perissodactyl-anthracobunid clade in the phylogenetic analysis. These studies further led to the advancement of opinions on (a) the close relationships of anthracobunids to perissodactyls rather than to proboscideans or sirenians and (b) the origin of perissodactyls within the Indian subcontinent.

The tapiromorphs *Cambaylophus vastanensis* and *Vastanolophus holbrooki* represent a second group of perissodactyls from the Cambay Shale (Kapur and Bajpai 2015; Smith et al. 2015). These finds are the oldest records of tapiromorphs from South Asia, if we consider the earliest Eocene (~55.5 Ma) *Orientolophus* from the upper part of the Lingcha Formation, Henyang Province, China as the oldest record (Ting et al. 2011; Kapur and Bajpai 2015 and references therein). *Cambaylophus* is known from a maxillary fragment with *in situ* deciduous premolars and permanent molars, while *Vastanolophus* is known from a fragmentary isolated premolar and an isolated lower molar. As a result, a direct comparison between these two taxa is lacking at present. Initial phylogenetic analysis suggested a close relationship of *Cambaylophus* with the Asian *Orientolophus* (Kapur and Bajpai 2015) and has subsequently been supported in a recent study on perissodactyls by Bai et al. (2018). However, considering the opinions on the origin of perissodactyls within the Indian subcontinent (Cooper et al. 2014; Rose et al. 2014, also see Bajpai 2009; Krause and Maas 1990), the interrelationships between Early Eocene Indo-Pakistan and Asian tapiromorph finds still need to be evaluated rigorously. Tapiromorphs are considered as an important component in the Eocene of Asia and have allowed the division of the Early Eocene Bumbanian Asian Land Mammal Age (ALMA) into three interval zones namely: *Orientolophus*, *Homogalax*, and *Heptodon* interval zones (Ting 1998). Missiaen and Gingerich (2012) had earlier proposed the presence of two separate tapiromorph lineages within the Indian subcontinent during the Eocene and attempted a biochronological correlation between the Indo–Pakistan and Asian Eocene land mammals. These authors correlated the Upper Ghazij Formation with the middle to late Bumbanian ALMA. In addition, the *Cambaylophus* find assisted in suggesting that the Cambay Shale mammals represent the early Bumbanian ALMA and are clearly older than mammals (including tapiromorphs) known from the Upper Ghazij Formation, Pakistan (Kapur and Bajpai 2015 and references therein). The age assignment for the helaletid *Vastanolophus*, together with its primitive dental morphology as compared to Early Eocene (~53 Ma) helaletids from America, allowed Smith et al. (2015) to propose an Asian origin for this group. Very recently, Bai et al. (2018) have suggested that the four groups (i.e., equids, ceratomorphs, ancyclopods, and brontotheres) representing perissodactyls most likely diverged close to 56 Ma and quite possibly within the Paleocene epoch.

Considering the phylogenetic relationships and evidence on the origin of perissodactyls within the Indian subcontinent, it is intriguing that Cambay perissodactyls are still younger (as discussed earlier) as compared to the record of this group from Eurasia. This certainly raises the possibility of the presence of older (late Paleocene) cambaythere-like mammals within India and/or Eurasia. If this is correct, then considering that it was unlikely for the large-sized perissodactyls to disperse across large oceanic barriers (as discussed earlier), the possibility of a short-lived corridor for dispersal between Indian subcontinent and Eurasia before ~54.5 Ma cannot be ruled out.

Discussion

It is evident from this study of size variation in the lower dentitions of known non-volant Cambay Shale mammals that they can be broadly categorized into three groups: small, medium, and large. Considering that the size variability within the non-volant Cambay Shale mammalian assemblage has implications for the dispersal mechanism used by individual groups of animals for biotic exchanges between India and Eurasia, it is quite likely that the small- and medium-sized mammals employed either or both active (chance dispersal and island hopping mechanisms) and passive modes for dispersal. Further, if we take into account the Early Eocene (~54.5 Ma) age for the large-sized component (perissodactyls), and their close evolutionary relationships with fauna from Eurasia, and considering that it was unlikely for the large-sized component (perissodactyls) to disperse across large oceanic barriers, the question arises: did a short-lived corridor for dispersal between Indian subcontinent and Eurasia existed before ~54.5 Ma?

Interestingly, the literature is rife with examples pointing to the significance of the late Paleocene time interval in the context of the Indian subcontinent-Eurasia collision. Most geophysical models suggest that the northward moving Indian plate suddenly slowed down from ~21.1 cm/year to ~9.5 cm/year during the late Paleocene (~57 Ma) (Acton 1999; Klootwijk et al. 1992). Ali and Aitchison (2008), citing Acton (1999), identified a "57 Ma event" to mark the collision between the Indian subcontinent and the Neotethyan intra-oceanic arc system. Further, these authors proposed initial terrestrial migrations occurring between India and Asia close to ~57 Ma, as a result of a glancing connection between the northeastern portion of the Indian subcontinent and Sumatra. Hu et al. (2015), while providing a robust chronology based on combined data from detrital zircons and biostratigraphy (using radiolarians and nannofossils) from the suture zone (Sangdanlin area, Central Himalaya), suggested Asian-derived detritus was deposited onto the India Plate within this region as a consequence of "collision onset" between the two landmasses during the late

Paleocene. Very recently, Klaus et al. (2016), while providing a phylogeographical meta-analysis taking into account the maximum dispersal events, have shown that a long-term terrestrial dispersal corridor between the Indian subcontinent and Asia came into existence only close to ~44 Ma and favoured trans-oceanic dispersal for biotic components prior to ~44 Ma (Fig. 13.4). However, Klaus et al. (2016) marked 'estimated change points' within the ~60–55 Ma interval, are based on faunal data including crustaceans, amphibians, and reptiles, but not mammals. Interestingly, the scientific literature is also flooded with reports on the presence of foraminiferal limestones (carbonate platforms) along the northern margin of the Indian subcontinent that were deposited during the late Paleocene under shallow marine conditions (Afzal et al. 2005; Tewari et al. 2010 and references therein) and also the presence of an island-arc system north of the Indian subcontinent (Clift et al. 2002 and references therein). Further, Haq et al. (1987) recognized major sea-level drops during the late Paleocene (Fig. 13.4) that would certainly have resulted in intermittent exposure of the shallow carbonate platforms north of the Indian subcontinent. Intriguingly, previous studies have documented abrupt changes from marine to continental settings during the late Paleocene, as evident from the presence of red beds, volcanic arenites and coal-bearing sandstones in the northern part of the Indian subcontinent (Searle et al. 1997; Garg and Ateequzzaman 2000). It is quite likely that the intermittent exposure of carbonate platforms together with the volcanic island arcs served as temporary corridors for dispersal across the Tethys Sea close to the Paleocene-Eocene interval (i.e., ~55–56 Ma) or within Late Paleocene (i.e., prior to 56 Ma) (Fig. 13.4).

Regarding the origin and subsequent dispersal of perissodactyls, a Central American origin with subsequent dispersal to Eurasia and other continents was initially envisaged by Radinsky (1966) on the assumption that perissodactyls may have evolved from the phenacodontids possibly thriving during the late Paleocene in Central America (also see Gingerich 1976; Van Valen 1978) (Fig. 13.5). Later, Gingerich (1986, 1989) proposed an African origin for perissodactyls with subsequent dispersal to Asia and Europe and finally into North America. Gingerich's proposal was based on a) an equatorial position of the African landmass, b) sudden appearance of hyaenodontid creodonts, artiodactyls, primates and perissodactyls in the early Paleogene fossil record of North America, and c) absence of precursor fauna (for the above groups including perissodactyls) in the late Paleocene of North America (Fig. 13.5). Beard (1998) envisaged an "East of Eden" model i.e., an Asian origin of perissodactyls with subsequent trans-Beringian dispersal towards North America considering the paucity of evidence for precursor fauna in North American fossil record (Fig. 13.5). Recently, Cooper et al. (2014) and Rose et al. (2014) based on phylogenetic analysis contemplated perissodactyl origins within the Indian subcontinent supporting Krause and Maas's proposal for an "Out-of-India" dispersal (discussed earlier) (Fig. 13.5). These independent phylogenetic analyses suggest a close relationship between the late Paleocene Asian taxon *Radinskya* and stem perissodactyls (including cambaytheres) (as discussed earlier). It should be noted here that initially *Radinskya* was considered as a phenacolophid (McKenna et al. 1989); however, this taxon was later considered as the most primitive member of Altungulata (McKenna and Bell 1997; Gheerbrant et al. 2005) with a close relationship to perissodactyls (Holbrook 2014 and references therein). In addition, Holbrook (2014) considered the exclusion of *Radinskya* from the family Phenocolophidae. This interpretation has recently received support from Mao et al. (2016). Nonetheless, as pointed out earlier *Radinskya* is an important component from Asia in the context of biogeogeographic dispersals out-of and/or into- the continent. Moreover, given the presence of older perissodactyls (compared to the ones from Cambay Shale) within Eurasia, inter-relationships between Indian and Eurasian perissodactyls, and the evidence for the origin of perissodactyls within the Indian subcontinent, the presence of cambaythere-like mammals within the late Paleocene of the Indian subcontinent and/or Eurasia can also not be ruled out at this stage. This certainly warrants rigorous reconnoitering of the Indian subcontinent for Paleocene mammals, which are presently unknown in the fossil record.

Caveats

It should be considered that the data presented here are based only on the lower dental elements (p4, m1, m2 and m3). Also, the known non-volant mammalian assemblage from the Cambay Shale includes groups for which there is a dearth of lower dental elements (e.g., tillodonts, proteutherians, rodents, and didelphimorphs) or for which the lower dentition is not yet recorded (e.g., lagomorphs are known from post-cranial elements while a condylarth is known only by an isolated upper molar). Moreover, many mammalian finds from the Cambay Shale are housed in various laboratories and have not yet been studied. There is also a need to build a consensus among paleontologists on the taxonomic status and evolutionary relationships of a few mammalian finds from the Cambay Shale. Nevertheless, the data presented here are significant, and future paleontological endeavors will augment the dataset numerically and very

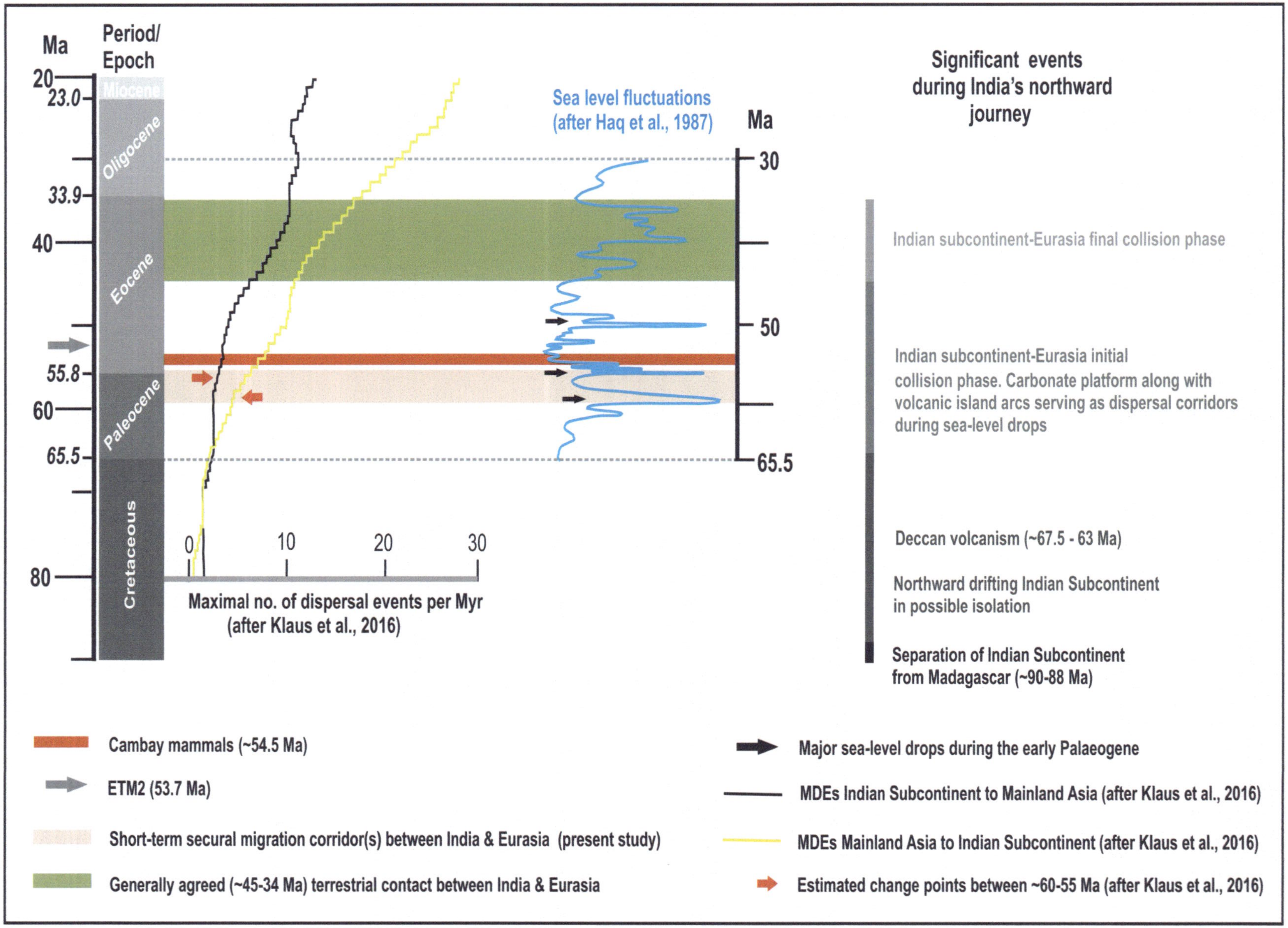

Fig. 13.4 Important events close to the Indian subcontinent-Eurasia collision phase. *Note* ETM2 (Second Eocene Thermal Maximum); MDE (Maximum Dispersal Events). The Cretaceous-Paleogene (K-Pg) boundary has been marked following Gradstein et al. (2012); however, subsequent works (e.g., Renne et al. 2013) place the K-Pg boundary at ~66 Ma

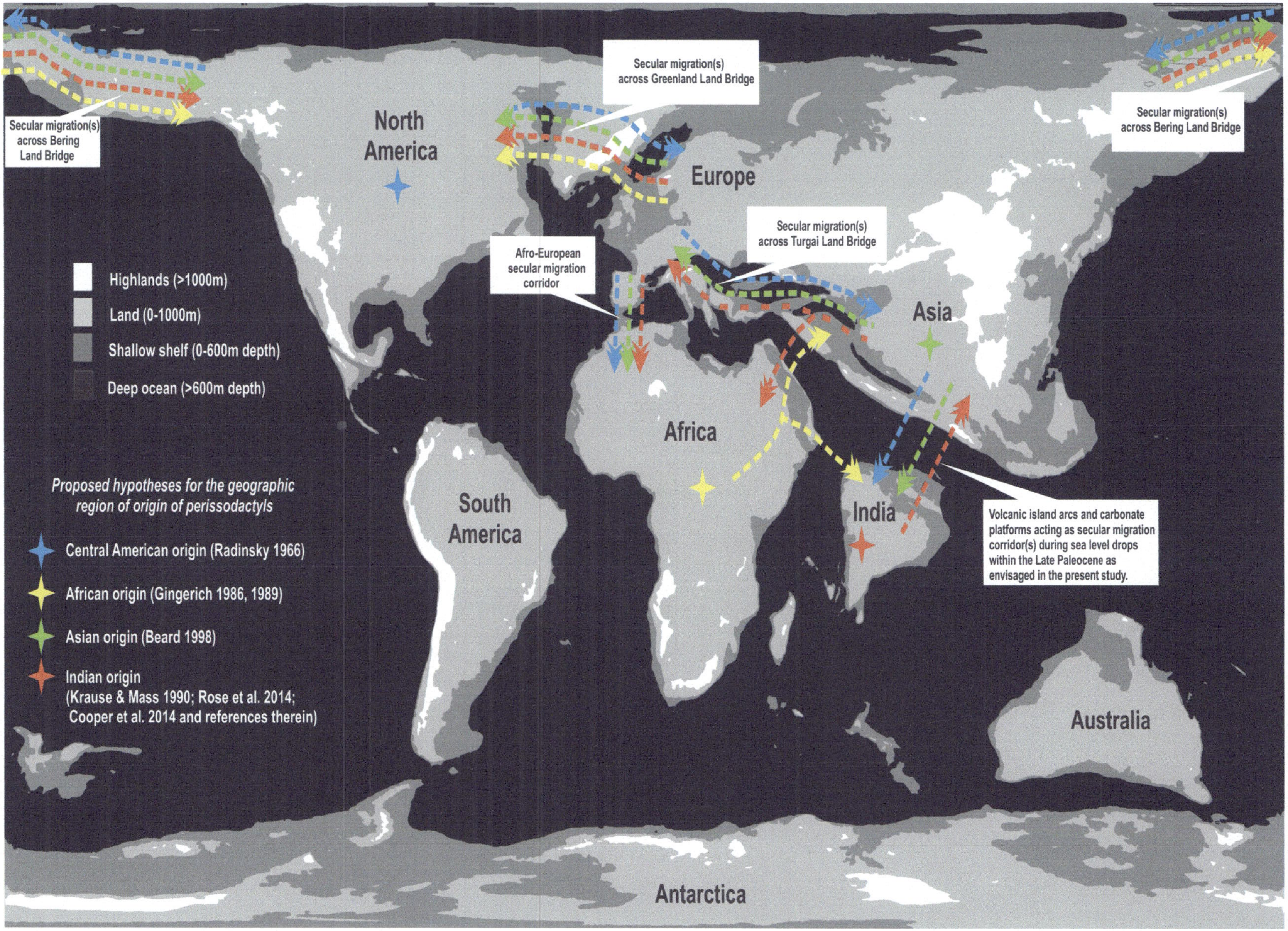

Fig. 13.5 Early Paleogene paleogeographic setup of the globe depicting the variously proposed hypotheses for the geographic region of origin for perissodactyls. *Note* Star marks the region of origin while arrow marks the direction of migration. Map modified after Scotese (2016), with permission

likely taxonomically, consequently affecting the resolution of the size groupings. In a similar scenario, the timings and the nature of the Indian subcontinent-Eurasia collision has also been continuously debated in the literature for the last five decades or more. Future efforts by the geoscience community incorporating independent biotic evidence, collaborative and interdisciplinary research should help us to better understand the dynamics of biotic exchanges across the Indian subcontinent and Eurasia during the early Paleogene.

Acknowledgements At the onset, I would like to thank Prof. G.V.R. Prasad and Prof. Rajeev Patnaik for the invitation to contribute to this volume. I would also like to acknowledge the Director (Birbal Sahni Institute of Palaeosciences, India) for constant encouragement and providing permission to contribute this research article. The author acknowledges funding support from BSIP in the form of in-house projects 2.17 (2017–2019) and 3.9 (2019–2021). Dr. Sebastian Klaus (Goethe Universität, Germany), Prof. M. P. Singh (University of Lucknow, India), Prof. Christopher R. Scotese (Director, PEOMAP Project, Illinois, USA) and Dr. (Mrs) Rajni Tewari (BSIP, Lucknow, India) kindly provided permission(s) to reproduce, reuse and redraw figure(s) for this article. I would also like to thank Prof. Daryl Paul Domning (Howard University, Washington DC, USA) for valuable discussions on the manuscript and also for providing grammatical enhancements. Prof. I.B. Singh (Lucknow, India) and Dr. Rahul Garg (Lucknow, India) are also acknowledged for insightful discussions. I also thank staff members of the Gujarat Industries Power Company Limited (GIPCL) and Gujarat Mineral Development Corporation (GMDC) for providing necessary permission(s) and extending logistic support during numerous field excursions and help in many other ways. I gratefully acknowledge Dr. David Krause (Denver Museum of Nature and Science, Colorado, USA) and two anonymous reviewers for a critical review of the manuscript and for their constructive and insightful comments which helped me in improving the quality of the manuscript.

Appendix

Data on the area (length × width, in mm sq.) of premolar (p4) and molars (m1, m2, and m3) measurements for known non-volant mammalian orders: Perissodactyla, Artiodactyla, Creodonta, Tillodonta, Primates, Proteutheria, Rodentia, and Didelphimorphia from the Cambay Shale, Gujarat, western India that has been used in the present study for conducting the ANOVA test

Data p4							
Perissodactyla	Artiodactyla	Creodonta	Primates	Proteutheria	Tillodontia	Didelphimorphia	Rodentia
69.56	13.50	10.29	2.70	0.61	NIL	NIL	NIL
69.56	13.63	12.48	2.86	0.70	–	–	–
79.50	16.64	13.92	2.88	–	–	–	–
86.11	–	14.04	3.22	–	–	–	–
90.20	–	18.30	4.42	–	–	–	–
91.79	–	–	4.46	–	–	–	–
97.90	–	–	4.88	–	–	–	–
100.28	–	–	4.93	–	–	–	–
–	–	–	5.40	–	–	–	–
Data m1							
Perissodactyla	Artiodactyla	Creodonta	Tillodontia	Primates	Proteutheria	Rodentia	Didelphimorphia
90.40	18.00	11.96	30.38	3.96	1.20	3.96	1.13
86.90	12.32	11.96	–	2.93	7.35	–	–
97.20	11.88	17.60	–	3.08	1.21	–	–
98.28	13.80	14.58	–	3.72	–	–	–
113.10	13.33	11.04	–	3.57	–	–	–
99.60	13.33	16.20	–	3.57	–	–	–
91.20	8.16	–	–	3.08	–	–	–
–	–	–	–	5.70	–	–	–
–	–	–	–	6.82	-	–	–
–	–	-	–	6.00	–	–	–
Data m2							
Perissodactyla	Artiodactyla	Creodonta	Primates	Rodentia	Proteutheria	Tillodontia	Didelphimorphia
55.65	15.48	16.80	3.30	4.49	1.21	NIL	NIL
93.94	16.56	19.80	3.33	–	1.26	–	–
96.72	20.00	20.88	3.33	–	–	–	–
98.04	20.91	21.83	3.52	–	–	–	–
102.34	22.14	22.94	3.68	–	–	–	–

(continued)

Data m2							
Perissodactyla	Artiodactyla	Creodonta	Primates	Rodentia	Proteutheria	Tillodontia	Didelphimorphia
104.28	–	23.45	3.68	–	–	–	–
124.68	–	–	3.68	–	–	–	–
128.04	–	–	3.71	–	–	–	–
132.00	–	–	3.74	–	–	–	–
132.48	–	–	3.86	–	–	–	–
138.00	–	–	4.03	–	–	–	–
–	–	–	4.08	–	–	–	–
–	–	–	4.11	–	–	–	–
–	–	–	5.57	–	–	–	–
–	–	–	6.24	–	–	–	–
–	–	–	6.48	–	–	–	–
–	–	–	6.70	–	–	–	–
–	–	–	6.79	–	–	–	–
–	–	–	6.86	–	–	–	–
Data m3							
Perissodactyla	Artiodactyla	Creodonta	Primates	Rodentia	Proteutheria	Tillodontia	Didelphimorphia
127.80	19.25	26.46	1.84	4.70	1.02	NIL	NIL
150.40	19.80	29.67	3.17	–	1.24	–	–
151.68	23.78	31.82	3.22	–	–	–	–
154.16	26.65	–	3.33	–	–	–	–
159.39	–	–	3.36	–	–	–	–
165.00	–	–	3.38	–	–	–	–
183.75	–	–	3.45	–	–	–	–
184.04	–	–	3.84	–	–	–	–
–	–	–	4.90	–	–	–	–
–	–	–	6.18	–	–	–	–
–	–	–	6.77	–	–	–	–
–	–	–	6.97	–	–	–	–

References

Acton, G. D. (1999). Apparent polar wander of India since the Cretaceous with implications for regional tectonics and true polar wander. The Indian subcontinent and Gondwana: A paleomagnetic and rock magnetic perspective. *Memoir Geological Society of India, 44*, 129–175.

Afzal, J., Khan, R. F., Khan, S. N., Alam, S., & Jalal, M. (2005). Foraminiferal biostratigraphy and paleoenvironments of the paleocene lockhart limestone from Kotal Pass, Kohat, Northern Pakistan. *Pakistan Journal of Hydrocarbon Research, 15*, 9–23.

Aitchison, J. C., Ali, J. R., & Davis, A. M. (2007). When and where did India and Asia collide? *Journal of Geophysical Research*. https://doi.org/10.1029/2006JB004706.

Ali, J. R., & Aitchison, J. C. (2004). Problem of positioning Paleogene Eurasia: A review, efforts to resolve the issue, implications for the India–Asia collision. *Monograph-American Geophysical Union, 149*, 23–35.

Ali, J. R., & Aitchison, J. C. (2008). Gondwana to Asia: Plate tectonics, paleogeography and the biological connectivity of the Indian sub-continent from the Middle Jurassic through latest Eocene (166–35 Ma). *Earth Science Reviews, 88*(3–4), 145–166.

Ali, J. R., & Krause, D. W. (2011). Late Cretaceous bioconnections between Indo-Madagascar and Antarctica: Refutation of the Gunner vs. Ridge causeway hypothesis. *Journal of Biogeography, 38*, 1855–1872.

Bai, B., Wang, Y.-Q., & Meng, J. (2018). The divergence and dispersal of early perissodactyls as evidenced by Early Eocene equids from Asia. *Communications Biology*. https://doi.org/10.1038/s42003-018-0116-5.

Bajpai, S. (2009). Biotic perspective of the Deccan volcanism and India–Asia collision: Recent advances. *Current trends in Science, Platinum Jubilee Special Publication, Indian Academy of Sciences*, 505–516.

Bajpai, S., Das, D. P., Kapur, V. V., Tiwari, B. N., & Srivastava, S. S. (2007a). Early Eocene rodents (Mammalia) from Vastan Lignite Mine, Gujarat, India. *Gondwana Geological Magazine, 22*(2), 91–95.

Bajpai, S., & Gingerich, P. D. (1998). A new Eocene Archaeocete (Mammalia, Cetacea) from India and the time of the origin of whales. *Proceedings of the National Academy of Sciences, USA, 95*, 15464–15468.

Bajpai, S., & Kapur, V. V. (2004). Oldest known gobiids from Vastan Lignite Mine (Early Eocene), district Surat, Gujarat. *Current Science, 87*(4), 433–435.

Bajpai, S., Kapur, V. V., Das, D. P., & Tiwari, B. N. (2007b). New early Eocene primate (Mammalia) from Vastan Lignite Mine, district Surat (Gujarat), western India. *Journal of the Palaeontological Society of India, 52*(2), 231–234.

Bajpai, S., Kapur, V. V., Das, D. P., Tiwari, B. N., Saravanan, N., & Sharma, R. (2005a). Early Eocene land mammals from Vastan Lignite Mine, district Surat (Gujarat), western India. *Journal of the Palaeontological Society of India, 50*(1), 101–113.

Bajpai, S., Kapur, V. V., & Thewissen, J. G. M. (2009). Creodont and Condylarth from the Cambay Shale (Early Eocene, ~55–54 Ma), Vastan Lignite Mine, Gujarat, Western India. *Journal of the Palaeontological Society of India, 54*(1), 103–109.

Bajpai, S., Kapur, V. V., Thewissen, J. G. M., Das, D. P., & Tiwari, B. N. (2006a). New early Eocene cambaythere (Perissodactyla, Mammalia) from the Vastan Lignite Mine (Gujarat, India) and an evaluation of cambaythere relationships. *Journal of the Palaeontological Society of India, 51*(1), 101–110.

Bajpai, S., Kapur, V. V., Thewissen, J. G. M., Tiwari, B. N., & Das, D. P. (2005b). First fossil marsupials from India: Early Eocene *Indodelphis* n.gen. and *Jaegeria* n.gen from Vastan lignite mine, district Surat, Gujarat. *Journal of the Palaeontological Society of India, 50*(1), 147–151.

Bajpai, S., Kapur, V. V., Thewissen, J. G. M., Tiwari, B. N., Das, D. P., Sharma, R., et al. (2005c). Early Eocene primates from Vastan lignite mine, Gujarat, western India. *Journal of the Palaeontological Society of India, 50*(2), 43–54.

Bajpai, S., Kay, R. F., Williams, B. A., Das, D. P., Kapur, V. V., & Tiwari, B. N. (2008). The oldest Asian record of Anthropoidea. *Proceedings of the National Academy of Sciences of the USA, 105,* 11093–11098.

Bajpai, S., & Thewissen, J. G. M. (2002). Vertebrate fauna from Panandhro lignite field (Lower Eocene), district Kachchh, western India. *Current Science, 82,* 507–509.

Bajpai, S., & Thewissen, J. G. M. (2014). Protocetid cetaceans (Mammalia) from the Eocene of India. *Palaeontologia Electronica, 17*(3), 34A, 19.

Bajpai, S., Thewissen, J. G. M., Kapur, V. V., Tiwari, B. N., & Sahni, A. (2006b). Eocene and Oligocene sirenians (Mammalia) from Kachchh, India. *Journal of Vertebrate Paleontology, 26,* 400–410.

Bardintzeff, J. M., Liégeois, J. P., Bonin, B., Bellon, H., & Rasamimanana, G. (2010). Madagascar volcanic provinces linked to the Gondwana break-up: Geochemical and isotopic evidences for contrasting mantle sources. *Gondwana Research, 18,* 295–314.

Beard, K. C. (1998). East of Eden: Asia as an important center of taxonomic origination in mammalian evolution. *Bulletin of the Carnegie Museum of Natural History, 34,* 5–39.

Beard, K. C. (2016). Out of Asia: Anthropoid origins and the colonization of Africa. *Annual Review of Anthropology, 45,* 199–213.

Beck, R. A., Burbank, D. W., Sercombe, W. J., Riley, G. W., Barndt, J. K., Berry, J. R., et al. (1995). Stratigraphic evidence for an early collision between Northwest India and Asia. *Nature, 373*(6509), 55–58.

Besse, J., & Courtillot, V. (2002). Apparent and true polar wander and the geometry of the geomagnetic field over the last 200 Myr. *Journal of Geophysical Research, 107*(B11), 2300. https://doi.org/10.1029/2000JB000050.

Bown, T. M., Holroyd, P. A., & Rose, K. D. (1994). Mammal extinctions, body size, and paleotemperature. *Proceedings of the National Academy of Sciences, USA, 91,* 10403–10406.

Cai, F. L., Ding, L., & Yue, Y. H. (2011). Provenance analysis of upper Cretaceous strata in the Tethys Himalaya, southern Tibet: Implications for timing of India-Asia collision. *Earth and Planetary Science Letters, 305*(1–2), 195–206.

Carrano, M. T., Wilson, J. A., & Barrett, P. M. (2010). The history of dinosaur collecting in Central India. In R. T. J. Moody, E. Buffetaut, D. Naish, & D. M. Martill (Eds.), *Dinosaurs and other extinct saurians: A historical perspective* (pp. 1828–1947). London: Geological Society Special Publications.

Chatterjee, S. (1978). *Indosuchus* and *Indosaurus*, Cretaceous carnosaurs from India. *Journal of Paleontology, 52,* 570–580.

Clementz, M., Bajpai, S., Ravikant, V., Thewissen, J. G. M., Saravanan, N., Singh, I. B., et al. (2011). Early Eocene warming events and the timing of terrestrial faunal exchange between India and Asia. *Geology, 39,* 15–18.

Clift, P. D., Hannigan, R., Blusztajn, J., & Draut, A. E. (2002). Geochemical evolution of the Dras-Kohistan Arc during collision with Eurasia; evidence from the Ladakh Himalaya, India. *Island Arc, 11,* 255–273.

Clobert, J., Baguette, M., Benton, T. G., & Bullock, J. M. (2012). *Dispersal ecology and evolution*. United Kingdom: Oxford University Press.

Clyde, W. C., Khan, I. H., & Gingerich, P. D. (2003). Stratigraphic response and mammalian dispersal during initial India Asia collision: Evidence from the Ghazij Formation, Balochistan, Pakistan. *Geology, 31,* 1097–1100.

Cooper, L. N., Seiffert, E. R., Clementz, M., Madar, S. I., Bajpai, S., Hussain, S. T., et al. (2014). Anthracobunids from the Middle Eocene of India and Pakistan are stem perissodactyls. *PLoS ONE, 9* (10), e109232. https://doi.org/10.1371/journal.pone.0109232.

Coughlin, B. L., & Fish, F. E. (2009). *Hippopotamus* underwater locomotion: Reduced gravity movements for a massive mammal. *Journal of Mammalogy, 90,* 675–679.

Crouchley, D., Nugent, G., & Edge, K.-A. (2011). Removal of red deer (*Cervus elaphas*) from Anchor and Secretary Islands, Fiordland, New Zealand. In C. R. Veitch, M. N. Clout, & D. R. Towns (Eds.), *Island invasives: Eradication and management* (pp. 422–425). Gland, Switzerland: IUCN.

DeCelles, P., Kapp, P., Gehrels, G., & Ding, L. (2014). Paleocene – Eocene foreland basin evolution in the Himalaya of southern Tibet and Nepal: Implications for the age of initial India-Asia collision. *Tectonics, 33*(5), 824–849.

Dehm, R., & Oettingen-Spielberg, T Zu. (1958). Palaeontologische und geologische Untersuchungen im Teriar von Pakistan, 2. Die Mitteleocaecen Saeugetiere von Ganda Kas bei Basal in Nordwest-Pakistan. *Bayerische Akademie der wissenschaften, Mathematisch-Physikalische Klasse, 91,* 1–54.

Davies, P., & Lister, A. M. (2001). *Palaeoloxodon cypriotes*, the dwarf elephant of Cyprus: Size and scaling comparisons with *P. falconeri* (Sicily-Malta) and mainland *P. antiques*. The World of Elephants - International congress, Rome.

De Queiroz, A. (2014). *The monkey's voyage: How improbable journeys shaped the history of life*. New York: Basic Books.

De Sigoyer, J., Chavagnac, V., Blichert-Toft, J., Villa, I. M., Luais, B., Guillot, S., et al. (2000). Dating the Indian continental subduction and collisional thickening in the northwest Himalaya: Multichronology of the Tso Morari eclogites. *Geology, 28*(6), 487–490.

Dewey, J. F., Cande, S., & Pitman, W. C., III. (1989). Tectonic evolution of the India/Eurasia collision zone. *Eclogae Geologicae Helvetiae, 82*(3), 717–734.

Ding, L., Kapp, P., & Wan, X. Q. (2005). Paleocene – Eocene record of ophiolite obduction and initial India-Asia collision, south central Tibet. *Tectonics, 24*(3). https://doi.org/10.1029/2004tc001729.

Ding, L., Maksatbek, S., Cai, F. L., Wang, H. Q., Song, P. P., Ji, W. Q., et al. (2017). Processes of initial collision and suturing between India and Asia. *Science China Earth Sciences, 60,* 635–651.

Eltringham, S. K. (1999). *The Hippos*. London: Poyser.

Fisher, R. E., Scott, K. M., & Naples, V. L. (2007). Forelimb myology of the pygmy *Hippopotamus* (*Choeropsis liberiensis*). *Anatomical Record, 290,* 673–693.

Garg, R., & Ateequzzaman, K. (2000). Dinoflagellate cysts from the Lakadong Sandstone, Cherrapunji area: Biostratigraphical and palaeoenvironmental significance and relevance to sea level changes in the Upper Paleocene of the Khasi Hills, South Shillong Plateau, India. *The Palaeoboanist, 49,* 461–484.

Garg, R., Ateequzzaman, K., Prasad, V., Tripathi, S. K. M., Singh, I. B., Jauhri, A. K., et al. (2008). Age-diagnostic dinoflagellate cysts from the lignite-bearing sediments of the Vastan lignite mine, Surat

District, Gujarat, Western India. *Journal of the Palaeontological Society of India, 53,* 99–105.

Garland, T. J. (1983). The relation between maximal running speed and body mass in terrestrial mammals. *Journal of Zoology, London, 199,* 157–170.

Garzanti, E., & Van Haver, T. (1988). The Indus clastics: Forearc basin sedimentation in the Ladakh Himalaya (India). *Sedimentary Geology, 59,* 237–249.

Gheerbrant, E., Domning, D. P., & Tassy, P. (2005). Paenungulata (Sirenia, Proboscidea, Hyracoidea, and relatives). In K. D. Rose & D. J. Archibald (Eds.), *The rise of placental mammals: Origins and relationships of the major extant clades* (pp. 84–105). Baltimore (MD): Johns Hopkins University Press.

Gingerich, P. D. (1976). Cranial anatomy and early evolution of early Tertiary Plesiadapidae (Mammalia, Primates). *University of Michigan Papers Paleontology, 15,* 1–140.

Gingerich, P. D. (1986). Early Eocene *Cantius torresi*-oldest primate of modern aspect from North America. *Nature, 319,* 319–321.

Gingerich, P. D. (1989). New earliest Wasatchian mammalian fauna from the Eocene of northwestern Wyoming: Composition and diversity in a rarely sampled high-floodplain assemblage. *University of Michigan Papers on Paleontology, 28,* 1–97.

Gingerich, P. D. (2006). Environment and evolution through the Paleocene-Eocene thermal maximum. *Trends in Ecology & Evolution, 21,* 246–253.

Gingerich, P. D., Abbas, S. G., & Arie, M. (1997). Early Eocene *Quettacyon parachai* (Condylarthra) from the Ghazij Formation of Baluchistan (Pakistan): Oldest Cenozoic land-mammal from South Asia. *Journal of Vertebrate Paleontology, 17,* 629–637.

Gingerich, P. D., Arif, M., Khan, I. H., Ul-Haq, M., Bloch, J. I., Clyde, W. C., et al. (2001). Gandhera Quarry, a unique mammalian faunal assemblage from the Early Eocene of Baluchistan (Pakistan). In G. F. Gunnell (Ed.), *Eocene biodiversity: Unusual occurrences and rarely sampled habitats* (pp. 251–262). New York: Kluwer Academic/Plenum Publishers.

Ginsburg, L., Durrani, K. H., Kassi, A. M., & Welcomme, J. L. (1999). Discovery of a new Anthracobunidae (Tethytheria, Mammalia) from the lower Eocene lignite of the Kach-Harnai area in Baluchistan (Pakistan). *Comptes Rendus Académie des Sciences Paris, Sciences Terre Planètes, 328,* 209–213.

Gradstein, F. M., Ogg, J. G., Schmitz, M. D., & Ogg, G. M. (2012). *The geological time scale 2012*. Amsterdam: Elsevier.

Guillot, S., de Sigoyer, J., Lardeaux, J. M., & Mascle, G. (1997). Eclogitic metasediments from the Tso Morari area (Ladakh, Himalaya): Evidence for continental subduction during India–Asia convergence. *Contributions to Mineralogy and Petrology, 128* (2-3), 197–212.

Guillot, S., Garzanti, E., Baratoux, D., Marquer, D., Maheo, G., & de Sigoyer, J. (2003). Reconstructing the total shortening history of the NW Himalaya. *Geochemistry Geophysics Geosystems, 4,* 1064.

Haq, B. U., Hardenbol, J., & Vail, P. R. (1987). *Mesozoic and Cenozoic chronostratigraphy and cycles of sea-level change*. Sea-level changes: An integrated Approach, SEPM Special Publication no. 42.

Henderson, A. L., Najman, Y., Parrish, R., BouDagher Fadel, M., Barford, D., Garzanti, E., et al. (2010). Geology of the Cenozoic Indus Basin sedimentary rocks: Paleoenvironmental interpretation of sedimentation from the western Himalaya during the early phases of India-Eurasia collision. *Tectonics, 29*(6). https://doi.org/10.1029/2009tc002651.

Herridge, V. L. (2010). *Dwarf elephants on Mediterranean islands: A natural experiment in parallel evolution*. Ph.D. Dissertation, University College London.

Hibbard, C. W. (1949). Techniques in collecting microvertebrate fossils. *Contributions from the Museum of Paleontology, University of Michigan, 8*(2), 7–19.

Hillebrand, H. (2004). On the generality of the latitudinal diversity gradient. *The American Naturalist, 163*(2), 192–211.

Holbrook, L. T. (2014). On the skull of *Radinskya* (Mammalia) and its phylogenetic position. *Journal of Vertebrate Paleontology, 34*(5), 1203–1215.

Hooker, J. J. (2005). Perissodactyla. In K. D. Rose & J. D. Archibald (Eds.), *The rise of placental mammals* (pp. 199–214). Baltimore and London: The Johns Hopkins University Press.

Huggett, R. J. (1998). *Fundamentals of biogeography*. London: Routledge.

Hu, X., Garzanti, E., Moore, T., & Raffi, I. (2015). Direct stratigraphic dating of India-Asia collision onset at the Selandian (middle Paleocene, 59 ± 1 Ma). *Geology, 43*(10), 859–862.

Hu, X., Garzanti, E., Wang, J., Huang, W., An, W., & Webb, A. (2016). The timing of India-Asia collision onset: Facts, theories, controversies. *Earth Science Reviews, 160,* 264–299.

Johnson, D. L. (1980). Problems in the land vertebrate zoogeography of certain islands and the swimming powers of elephants. *Journal of Biogeography, 7*(4), 383–398.

Kapur, V. V., & Bajpai, S. (2015). Oldest South Asian tapiromorph (Perissodactyla, Mammalia) from the Cambay Shale Formation, western India, with comments on its phylogenetic position and biogeographic implications. *The Palaeobotanist, 64,* 95–103.

Kapur, V. V., Das, D. P., Bajpai, S., & Prasad, G. V. R. (2017a). First mammal of Gondwanan lineage in the Early Eocene of India. *Comptes Rendus Palevol, 16,* 721–737.

Kapur, V. V., Das, D. P., Bajpai, S., & Prasad, G. V. R. (2017b). Corrigendum to "First mammal of Gondwanan lineage in the Early Eocene of India". *Comptes Rendus Palevol, 16,* 820.

Kapur, V. V., & Khosla, A. (2016). Late Cretaceous Terrestrial Biota from India with special reference to vertebrates and their implications for biogeographic connections. In A. Khosla & S. G. Lucas (Eds.), *Cretaceous period: Biotic diversity and biogeography* (pp. 161–172). New Mexico Museum of Natural History and Science Bulletin.

Kapur, V. V., & Khosla, A. (2019). Faunal elements from the Deccan volcano-sedimentary sequences of India: A reappraisal of biostratigraphic, palaeoecologic, and palaeobiogeographic aspects. *Geological Journal, 54,* 2797–2828. https://doi.org/10.1002/gj.3379.

Kapur, V. V., Khosla, A., Tiwari, N. (2019). Palaeoenvironmental and palaeobiogeographical implications of the microfossil assemblage from the Late Cretaceous intertrappean beds of Manawar area, District Dhar, Madhya Pradesh, Central India. *Historical Biology, 31* (9), 1145–1160. https://doi.org/10.1080/08912963.2018.1425408.

Kay, R. F. (1975). The functional adaptations of Primate molar teeth. *American Journal of Physical Anthropology, 43,* 195–216.

Keller, G., Adatte, T., Bhowmick, P. K., Upadhyay, H., Dave, A., Reddy, A. N., et al. (2012). Nature and timing of extinctions in Cretaceous-Tertiary planktic foraminifera preserved in Deccan intertrappean sediments of the Krishna-Godavari Basin, India. *Earth and Planetary Science Letters, 341,* 211–221.

Klaus, S., Morley, R. J., Plath, M., Zhang, Ya-Ping, & Li, Jia-Tang. (2016). Biotic interchange between the Indian subcontinent and mainland Asia through time. *Nature Communications, 7,* 12132.

Klootwijk, C. T., Gee, J. S., Peirce, J. W., Smith, G. M., & McFadden, P. L. (1992). An early India-Eurasia contact: Paleomagnetic constraints from Ninety-East Ridge, ODP Leg 121. *Geology, 20,* 395–398.

Kondrashov, P. E., & Lucas, S. G. (2012). Nearly complete skeleton of *Tetraclaenodon* (Mammalia, Phenacodontidae) from the Early Paleocene of New Mexico: Morpho-functional analysis. *Journal of Paleontology, 86*(1), 25–43.

Kumar, K. (1991). *Anthracobune aijiensis* nov. sp. (Mammalia: Proboscidea) from the Subathu Formation, Eocene from NW Himalaya, India. *Geobios, 24,* 221–239.

Krause, D. W., & Maas, M. C. (1990). The biogeographic origins of Late Paleocene – Early Eocene mammalian immigrants to the western interior of North America. *Geological Society of America Special Paper, 243,* 71–105.

Kumar, K., Rana, R. S., & Paliwal, B. S. (2005). Osteoglossid and Lepisosteid fish remains from the Paleocene Palana Formation, Rajasthan, India. *Palaeontology, 48*(6), 1187–1209.

Kumar, K., Rose, K. D., Rana, R. S., Singh, L., Smith, T., & Sahni, A. (2010). Early Eocene artiodactyls (Mammalia) from western India. *Journal of Vertebrate Paleontology, 30*(4), 1245–1274.

Legendre, S., & Roth, C. (1988). Correlation of carnassial tooth size and body weight in Recent carnivores (Mammalia). *Historical Biology, 1,* 85–98.

Lydekker, R. (1877). Notices of new and other Vertebrata from Indian Tertiary and Secondary rocks. *Records of the Geological Survey of India, 10,* 30–43.

Mao, F.-Y., Wang, Y.-Q., Li, Q., & Jin, X. (2016). New records of archaic ungulates from the Lower Eocene of Sanshui Basin, Guangdong, China. *Historical Biology, 28*(6), 787–802.

Masters, J. C., Lovegrove, B. G., & de Wit, M. J. (2007). Eyes wide shut: Can hypometabolism really explain the primate colonization of Madagascar? *Journal of Biogeography, 34,* 31–37.

Matthew, W. D. (1915). Climate and evolution. *Annals of the New York Academy of Sciences, 24,* 171–318.

Matthew, W. D. (1939). *Climate and evolution* (2nd ed.). New York: New York Academy of Sciences.

Mazza, P. A. P., Lovari, S., Masini, F., Masseti, M., & Rustioni, M. (2013). A multidisciplinary approach to the analysis of multifactorial land mammal colonization of islands. *Bioscience, 63*(12), 939–951.

McKenna, M. C. (1973). Sweepstakes, filters, corridors, Noah's arks, and beached viking funeral ships in paleogeography. In D. H. Tarling & S. K. Runcorn (Eds.), *Implications of continental drift to the earth sciences* (Vol. 1, pp. 295–308). New York: Academic Press.

McKenna, M. C., & Bell, S. K. (1997). *Classification of mammals above the species level.* New York: Columbia University Press.

McKenna, M. C., Chow, M., Ting, S., & Luo, Z. (1989). *Radinskya yupingae*, a perissodactyl-like mammal from the late Paleocene of southern China. In D. R. Prothero & R. M. Schoch (Eds.), *The evolution of perissodactyls* (pp. 24–36). New York: Oxford University Press.

Millien, V., & Bovy, H. (2010). When teeth and bones disagree: Body mass estimation of a giant extinct rodent. *Journal of Mammalogy, 91*(1), 11–18.

Missiaen, P., & Gingerich, P. D. (2012). New Early Eocene tapiromorph perissodactyls from the Ghazij Formation of Pakistan, with implications for mammalian biochronology in Asia. *Acta Palaeontologica Polonica, 57,* 21–34.

Molnar, P., & Tapponnier, P. (1975). Cenozoic tectonics of Asia: Effects of a continental collision. *Science, 189,* 419–426.

Najman, Y., Appel, E., Boudagher-Fadel, M., Bown, P., Carter, A., Garzanti, E., et al. (2010). Timing of India-Asia collision: Geological, biostratigraphic, and paleomagnetic constraints. *Journal of Geopysical Research Solid Earth, 115*(B12), B12416. https://doi.org/10.1029/2010JB007673.

Patriat, P., & Achache, J. (1984). India – Eurasia collision chronology has implications for crustal shortening and driving mechanism of plates. *Nature, 311,* 615–621.

Penkrot, T. A., Zack, S. P., Rose, K. D., & Bloch, J. I. (2008). Postcranial morphology of *Apheliscus* and *Haplomylus* (Condylarthra, Apheliscidae): Evidence for a Paleocene Holarctic origin of Macroscelidea. In E. J. Sargis, & M. Dagosto (Eds.), *Mammalian evolutionary morphology: A tribute to Frederick S. Szalay* (pp. 73–106). Dordrecht: Springer Science.

Perfit, M. R., & Williams, E. E. (1989). Geological constraints and biological retrodictions in the evolution of the Caribbean Sea and its islands. In C. A. Woods (Ed.), *Biogeography of the West Indies: Past, present, and future* (pp. 47–102). Gainesville, FL: Sandhill Crane.

Pielou, E. C. (1979). *Biogeography.* New York: Wiley.

Pognante, U., & Spencer, D. A. (1991). First report of eclogites from the Himalayan belt, Kaghan valley (northern Pakistan). *European Journal of Mineralogy, 3,* 613–618.

Powell, C. M., Roots, S. R., & Veevers, J. J. (1988). Pre-breakup continental extension in eastern Gondwanaland and the early opening of the eastern Indian Ocean. *Tectonophysics, 155,* 261–283.

Prasad, G. V. R., & Sahni, A. (1988). First Cretaceous mammal from India. *Nature, 332*(6164), 638–640.

Prasad, G. V. R., & Sahni, A. (1999). Were there size constraints on biotic exchanges during the northward drift of the Indian plate? *Proceedings of the Indian National Science Academy, 65A*(3), 377–396.

Prasad, G. V. R., Verma, O., Gheerbrant, E., Goswami, A., Khosla, A., Parmar, V., et al. (2010). First mammal evidence from the Late Cretaceous of India for biotic dispersal between India and Africa at the KT transition. *Comptes Rendus Palevol, 9,* 63–71.

Prasad, G. V. R., Verma, O., Sahni, A., Parmar, V., & Khosla, A. (2007). A Cretaceous hoofed mammal from India. *Science, 318,* 937.

Prescott, J. H. (1959). Rafting of jack rabbit on kelp. *Journal of Mammalogy, 40,* 443–444.

Radinsky, L. B. (1966). The adaptive radiation of the phenacodontid condylarths and the origin of the Perissodactyla. *Evolution, 20,* 408–417.

Rana, R. S., Kumar, K., Singh, H., & Rose, K. D. (2005). Lower vertebrates from the Late Paleocene-Earliest Eocene Akli Formation, Giral Lignite Mine, Barmer District, western India. *Current Science, 89,* 1606–1613.

Renne, P. R., Deino, A. L., Hilgen, F. J., Kuiper, K. F., Mark, D. F., Mitchell, W. S., III, et al. (2013). Time scales of critical events around the Cretaceous – Paleogene boundary. *Science, 339,* 684–687.

Rose, K. D. (1996). Skeleton of Early Eocene *Homogalax* and the origin of Perissodactyla. *Palaeovertebrata, 25,* 243–260.

Rose, K. D., Holbrook, L. T., Rana, R. S., Kumar, K., Jones, K. E., Ahrens, H. E., et al. (2014). Early Eocene fossils suggest that the mammalian order Perissodactyla originated in India. *Nature Communications, 5*(5570), 1–9.

Rose, K. D., Kumar, K., Rana, R. S., Sahni, A., & Smith, T. (2013). New hypsodont tillodont (Mammalia, Tillodontia) from the Early Eocene of India. *Journal of Paleontology, 87*(5), 842–853.

Rose, K. D., Rana, R. S., Sahni, A., & Smith, T. (2007). A new adapoid primate from the Early Eocene of India. *Contributions from the Museum of Paleontology, University of Michigan, 31*(14), 379–385.

Rose, K. D., Rana, R. S., Sahni, A., Kumar, K., Missiaen, P., Singh, L., et al. (2009a). Early Eocene primates from Gujarat, India. *Journal of Human Evolution, 56,* 366–404.

Rose, K. D., Rana, R. S., Sahni, A., Kumar, K., Singh, L., & Smith, T. (2009b). First tillodont from India: Additional evidence for an Early Eocene faunal connection between Europe and India? *Acta Palaeontologica Polonica, 54,* 351–355.

Rose, K. D., Smith, T., Rana, R. S., Sahni, A., Singh, H., Missiaen, P., et al. (2006). Early Eocene (Ypresian) continental vertebrate assemblage from India, with description of a new anthracobunid (Mammalia, Tethytheria). *Journal of Vertebrate Paleontology, 26,* 219–225.

Rowley, D. B. (1996). Age of initiation of collision between India and Asia; a review of stratigraphic data. *Earth and Planetary Science Letters, 145,* 1–13.

Sahni, A. (1968). Techniques in prospecting for terrestrial microvertebrates. *Journal of the Palaeontological Society of India, 13*(5), 38–43.

Sahni, A., Saraswati, P. K., Rana, R. S., Kumar, K., Singh, H., Alimohammadian, H., et al. (2006). Temporal constraints and depositional paleoenvironments of the Vastan Lignite Sequence, Gujarat: Analogy for the Cambay Shale hydrocarbon source rock. *Indian Journal of Petroleum Geology, 15,* 1–20.

Samant, B., & Bajpai, S. (2001). Fish otoliths from the subsurface Cambay shale (Lower Eocene), Surat lignite field, India. *Current Science, 81*(7), 758–759.

Scotese, C. R. (2016). Continental flooding and orography. https://www.youtube.com/watch?v=y-Qh1Zp9WoM&t=31s.

Searle, M., Corfield, R. I., Stephenson, B., & McCarron, J. (1997). Structure of the North Indian continental margin in the Ladakh-Zanskar Himalayas: Implications for the timing of obduction of the Spontang ophiolite, India-Asia collision and deformation events in the Himalaya. *Geological Magazine, 134*(3), 297–316.

Simpson, G. G. (1940). Mammals and land bridges. *Journal of the Washington Academy of Science, 30,* 137–163.

Singh, R. S., & Kar, R. K. (2002). Paleocene palynofossils from the Lalitpur Intertrappean beds, Uttar Pradesh, India. *Journal of the Geological Society of India, 60,* 213–216.

Sleeman, W. H. (1844). *Rambles and recollections of an Indian official.* London: I. J. Hatchard and Sons.

Smith, T., Kumar, K., Rana, R. S., Folie, A., Solé, F., Noiret, C., et al. (2016). New Early Eocene vertebrate assemblage from western India reveals a mixed fauna of European and Gondwanan affinities. *Geoscience Frontiers.* https://doi.org/10.1016/j.gsf.2016.05.001.

Smith, T., Rana, R. S., Missiaen, P., Rose, K. D., Sahni, A., Singh, H., et al. (2007). Highest diversity of earliest bats in the Early Eocene of India. *Naturwissenschaften, 94,* 1003–1009.

Smith, T., Solé, F., Missiaen, P., Rana, R. S., Kumar, K., Sahni, A., et al. (2015). First Early Eocene tapiroid from India and its implication for the paleobiogeographic origin of perissodactyls. *Palaeovertebrata, 39*(2), e5.

Solé, F., & Smith, T. (2013). Dispersals of placental carnivorous mammals (Carnivoramorpha, Oxyaenodonta & Hyaenodontida) near the Paleocene – Eocene boundary: A climatic and almost worldwide story. *Geologica Belgica, 16*(4), 254–261.

Springer, M. S., Meredith, R. W., Janecka, J. E., & Murphy, W. J. (2011). The historical biogeography of Mammalia. *Philosophical Translations of the Royal Society B, 366,* 2478–2502.

Stenseth, N. C., & Lidicker, W. Z. (2012). *Animal dispersal.* London: Chapman & Hall.

Sukumar, R. (2006). A brief review of the status, distribution and biology of wild Asian elephants. *International Zoo Yearbook, 40,* 1–8.

Taylor, L. A., Rudd, J., Hummel, J., Clauss, M., & Schwitzer, C. (2013). Weight loss in pygmy hippos (*Choeropsis liberiensis*). In B. Steck (Ed.), *International studbook for the year 2012: Pygmy Hippopotamus* (pp. 20–25). Basel: Zoo Basel.

Teitelbaum, C. S., Fagan, W. F., Fleming, C. H., Dressler, G., Calabrese, J. M., Leimgruber, P., et al. (2015). How far to go? Determination of migration distance in land mammals. *Ecology Letters, 18,* 545–552.

Tewari, V. C., Kumar, K., Lokho, K., & Siva Siddaiah, N. (2010). Lakadong limestone: Paleocene-Eocene boundary carbonate sedimentation in Meghalaya, northeastern India. *Current Science, 98*(1), 88–95.

Thewissen, J. G. M., & Bajpai, S. (2009). A new Miocene sirenian from Kutch, India. *Acta Palaeontologica Polonica, 54*(1), 7–13.

Thewissen, J. G. M., & Domning, D. P. (1992). The role of the phenacodontids in the origin of the modern orders of ungulate mammals. *Journal of Vertebrate Paleontology, 12,* 494–504.

Thewissen, J. G. M., Williams, E. M., & Hussain, S. T. (2001). Eocene mammal faunas from northern Indo-Pakistan. *Journal of Vertebrate Paleontology, 21,* 347–366.

Ting, S. (1998). Paleocene and early Eocene land mammal ages of Asia. *Bulletin of the Carnegie Museum of Natural History, 34,* 124–147.

Ting, S., Tong, Y., Clyde, W. C., Koch, P. L., Meng, J., Wang, Y., et al. (2011). Asian early Paleogene chronology and mammalian faunal turnover events. *Vertebrata PalAsiatica, 49,* 1–28.

Tonarini, S., Villa, I. M., Oberli, F., Meier, M., Spencer, D. A., Pognante, U., et al. (1993). Eocene age of eclogite metamorphism in Pakistan Himalaya: Implications for India-Eurasia collision. *Terra Nova, 5,* 13–20.

Treloar, P. J., & Coward, M. P. (1991). Indian plate motion and shape: Constraints on the geometry of the Himalayan orogen. *Tectonophysics, 191,* 189–198.

Van der Geer, A., Lyras, G., de Vos, J., & Dermitzakis, M. (2010). *Evolution of Island mammals.* West Sussex: Wiley.

Van Duzer, C. (2004). *Floating Islands: A global bibliography.* California: Cantor Press.

Van Valen, L. (1978). The beginning of the age of mammals. *Evolutionary Theory, 4,* 103–121.

Wells, N. A., & Gingerich, P. D. (1983). Review of Eocene Anthracobunidae (Mammalia:Proboscide) with a new genus and species, *Jozaria palustris*, from the Kuldana Formation of Kohat (Pakistan). *Contributions from the Museum of Paleontology, University of Michigan, 26,* 117–139.

West, R. M. (1980). Middle Eocene large mammal assemblage with Tethyan affinities, Ganda Kas region, Pakistan. *Journal of Paleontology, 54,* 508–533.

West, R. M. (1983). South Asian Middle Eocene moeritheres (Mammalia: Tethytheria). *Annals of the Carnegie Museum of Natural History, 52,* 359–373.

Willems, H., Zhou, Z., Zhang, B., & Grafe, K. U. (1996). Stratigraphy of the upper Cretaceous and lower Tertiary Strata in the Tethyan Himalayas of Tibet (Tingri area, China). *Geologische Rundschau, 85*(4), 723–754.

Yin, A. (2006). Cenozoic tectonic evolution of the Himalayan orogen as constrained by along-strike variation of structural geometry, exhumation history, and foreland sedimentation. *Earth Science Reviews, 76*(1–2), 1–131.

Yin, A., & Harrison, M. T. (2000). Geologic evolution of the Himalayan-Tibetan orogen. *Annual Review of Earth and Planetary Sciences, 28,* 211–280.

Zhang, Q. H., Willems, H., Ding, L., Gräfe, K.-U., & Appel, E. (2012). Initial India-Asia continental collision and foreland basin evolution in the Tethyan Himalaya of Tibet: Evidence from stratigraphy and paleontology. *Journal of Geology, 120*(2), 175–189.

Zhu, B., Kidd, W. S. F., Rowley, D. B., Currie, B. S., & Shafique, N. (2005). Age of initiation of the India-Asia collision in the east-central Himalaya. *The Journal of Geology, 113*(3), 265–285.

Zhu, D.-C., Wang, Q., Zhao, Z.-D., Chung, S.-L., Cawood, P. A., Niu, Y., et al. (2015). Magmetic record of India-Asia collision. *Scientific Reports, 5,* 14289. https://doi.org/10.1038/srep14289(2015).

Chapter 14
Indohyus, Endemic Radiation of Raoellid Artiodactyls in the Eocene of India and Pakistan

J. G. M. Thewissen, A. C. Nanda, and Sunil Bajpai

Abstract The South Asian artiodactyl genus *Indohyus* contains two species: *I. indirae* and *I. major*. *Indohyus* is the senior synonym of several other genera: *Raoella*, *Kunmunella*, *Metkatius* and *Haqueina* (in part). The dental morphology of a large collection of *I. indirae* from the locality Sindkhatudi in Jammu & Kashmir is described here. While morphological variation in teeth is significant, there are no discrete features that could be used to recognize more than one species.

Keywords Dental Morphology • Evolution • Mammalia • Paleogene • Tertiary • Vertebrate Paleontology

Introduction

In the Eocene, several families of small artiodactyls occurred on the Indian Subcontinent, including the families Diacodexeidae (Thewissen et al. 1983; Kumar 1998; Bajpai et al. 2005; Kumar et al. 2010), Helohyidae (West 1980), and Raoellidae. While diacodexids and helohyids are widely distributed across the world (Russell and Zhai 1987; Theodor et al. 2007), raoellids are abundant only in middle Eocene of Pakistan and India. Orliac and Ducrocq (2011) reviewed the published records of raoellids from outside of India and Pakistan, and found it to be unlikely that any of these are raoellids, except for a single molar from Shanghuang in China.

Raoellid specimens were first described by Pilgrim (1940) before the partition of India and Pakistan, and he named *Khirtharia dayi*. Dehm and Oettingen Spielberg (1958) published additional specimens from the Kala Chitta Hills of northern Pakistan. *Khirtharia* is a form with low cusps on its bunodont molars. In this, it is similar to *Bunodentus*, a raoellid described by Ranga Rao (1972), and the two genera were synonymized by Sahni et al. (1981).

A separate radiation of raoellids is lophodont. Ranga Rao (1971) described the genus *Indohyus* as a ?choeropotamid from the Kalakot region of Kashmir. Several other genera of more lophodont raoellids were subsequently also described from other localities, including some in Pakistan.

Collecting efforts in the Kalakot region yielded many Eocene mammalian fossils (Ranga Rao 1971, 1972; Sahni and Khare 1972, 1973; Kumar and Sahni 1985). One of its localities, now called Sindkhatudi, was worked extensively by Ranga Rao (1971, 1972). The vast majority of fossils, more than 90%, pertains to *Indohyus*, but there are some fossils of *Bunodentus*, primates, small tapiroids, rhinocerotoids, and anthracobunid perissodactyls. Recent work on this fauna has mostly involved *Indohyus* (Thewissen et al. 2007; Cooper et al. 2011), a genus of particular importance because it is thought to be the sister group to cetaceans (Thewissen et al. 2007; Geisler and Theodor 2009; Uhen 2010; Gatesy et al. 2013).

Materials and Methods

Specimens described here were collected by Mr. Anne Ranga Rao at a locality he discovered in 1970, and which is now referred to as Sindkhatudi (Kumar 2000). Fossils here are found in a mud-supported bone bed, the matrix of which is purple and grey mudstone, with some silty parts, partly

J. G. M. Thewissen (✉)
Department of Anatomy and Neurobiology, Northeast Ohio Medical University, Rootstown, Ohio, USA
e-mail: thewisse@neomed.edu

A. C. Nanda
Wadia Institute of Himalayan Geology, Dehradun, 248001, India

S. Bajpai
Department of Earth Sciences, Indian Institute of Technology Roorkee, Roorkee, 247664, India

G. V. R. Prasad and R. Patnaik (eds.), *Biological Consequences of Plate Tectonics: New Perspectives on Post-Gondwana Break-up—A Tribute to Ashok Sahni*,
Vertebrate Paleobiology and Paleoanthropology Series, https://doi.org/10.1007/978-3-030-49753-8_14

indurated by calcium carbonate. These rocks contain hundreds of isolated bones, mostly pertaining to *Indohyus*. Ranga Rao excavated the entire locality and moved fossiliferous blocks of sediment to his residence on 180/3 Rajpur Road in Dehradun, where they still reside. Fossils were extracted from a number of these blocks until 1999, when Ranga Rao died at the age of 69. After the death of his wife, Dr. Friedlinde Obergfell, in 2007, specimens from the collection were curated and fossils extracted from existing blocks in the lab of Thewissen. The former residence of Ranga Rao and Dr. Obergfell is now the repository for the Ranga Rao collection and most specimens are available for study by appointment with A. C. Nanda.

Acronyms

GSP-UM, Geological Survey of Pakistan, University of Michigan, Ann Arbor, U.S.A.; H-GSP, Howard University, Geological Survey of Pakistan, Washington, D.C., U.S.A.; LUVP, Lucknow University, Vertebrate Palaeontology, Lucknow, India; ONG, Oil and Natural Gas Commission, Dehra Dun, India; RR, Ranga Rao collection, Dehra Dun, India; VPL,Vertebrate Palaeontology Laboratory, Panjab University, Chandigarh, India.

Systematic Paleontology

Class Mammalia
Order Artiodactyla
Family Raoellidae Sahni et al., 1981
Genus *Indohyus* Ranga Rao, 1971
Indohyus indirae Ranga Rao, 1971
(Figs. 14.1–14.2)

Indohyus indirae Ranga Rao, 1971: 126, pl. I, II, III, IV, V.1, and VI.1; Thewissen et al., 1987: 263, Fig. 6A–E.

Indohyus (?) kalakotensis Ranga Rao, 1971: 131, pl. V.2 and VI.2.

Raoella dograi Sahni and Khare, 1972 (for 1971): 47, Fig. 3, pl. 1.3–4.

Kunmunella rajauriensis Sahni and Khare, 1972 (for 1971): 50, Fig. 4, pl. 1.5.

Haqueina haquei Sahni and Khare, 1973 (for 1972): 45, Fig. 13–14, pl. 5.5–8, 6.1–2.

Mesonychidae indet. West, 1980: 512, Fig. 1.

Metkatius kashmiriensis Kumar and Sahni, 1985: 157, Fig. 3F, 4A–C.

Kunmunella kalakotensis Kumar and Sahni, 1985: 160, Fig. 4I–M, 5A–C.

Kunmunella transversa Kumar and Sahni, 1985: 162, Fig. 5D–F; Thewissen et al., 2001a: 358, Fig. 5E–G.

Indohyus Thewissen et al., 2007, Fig. 1a–h, k; Cooper et al., 2011, Fig. 1–6, 7.1–7.13.

Holotype: ONG/K/1, left mandible with p3-m3 and alveolus of p2.

Paratypes: ONG/K/2 (p1-3); ONG/K/3 (c1-p4); ONG/K/4 (m3 with additional alveoli); ONG/K/5 (p3, m2-3); ONG/K/6 (mandibular fragment).

Referred Specimens: From the Kalakot area in Jammu & Kashmir (J&K): ONG/K/7 (M1-3, holotype of *I.(?) kalakotensis*); LUVP 15003 (P4-M3, holotype of *Raoella dograi*); LUVP 15004 (M3, holotype of *Kunmunella rajauriensis*), LUVP 15014, 15015; VPL/K 562 (skeletal fragments, poor teeth, holotype of *Metkatius kashmiriensis*); VPL/K/ 502 (m3); VPL/K/ 524 (M2); VPL/K/ 525 (palate); VPL/K/ 527 (P4-M3); VPL/K/ 531 (M2-3); VPL/K/532 (palate); VPL/K/ 572 (m3); VPL/K/526 (P4-M3, holotype of *Kunmunella transversa*); RR 4 (endentulous mandible); RR 17 (mx); RR 18 (D4); RR 80 (mandible with alveoli for i-p1, and crowns p2-3); RR 93 (Mx); RR 95 (mx); RR 100 (p2-m3); RR 135 (m1-3); RR 147 (p2-m3); RR 150 (m1-2); RR 151 (m3); RR 152 (m1-3); RR 153 (P2, p4-m3); RR 159 (edentulous mandible); RR 160 (m3); RR 163 (edentulous symphysis); RR 201 (mx); RR 202 (mx); RR 207 (skull with left P2-M3, right P4-M3); RR 208 (skull with left P1-4, M2-3, right P1-M3); RR 212 (m3); RR 252 (m3); RR 253 (m1-3, unerupted p2-4); RR 255 (p2-m3); RR 254 (mandible with alveoli for c-p4, crowns for m1-3); RR 262 (juv. skull with D3-4 and M1-2); RR 289 (D4); RR 300 (edentulous mandible); RR 310 (m1-3); RR 314 (m2-3); RR 315 (m1-3); RR 320 (p3-m2); RR 321 (p2-m3); RR 323 (mandible with alveoli for i1-p1, and crowns for p2-m3); RR 324 (mandible with alveoli for i1-p1, and crowns for p2-m1); RR 324 (p2-m1); RR 325 (p4-m3); RR 326 (m2-3); RR 327 (p2-m2); RR 328 (p4-m3); RR 329 (p3-m3); RR 330 (upper jaw with I3-P2); RR 331 (premaxilla with I3-P2); RR 365 (m1-3); RR 366 (m3); RR 367 (m1-3); RR 368 (p2-m3); RR 369 (p3-m3); RR 370 (p3-m1); RR 400 (m2-3); RR 401 (m1-3); RR 407 (D4); RR 408 (D4); RR 409 (d4); RR 412 (p4-m1, M3); RR 414 (D4); RR 415 (d3); RR 416 (mandible with i1, c, d3-4, m1); RR 419 (P3-M1); RR 420 (P3-M1); RR 422 (p4-m3); RR 528 (juvenile rostrum with erupted D2, D4-M2, and unerupted I2-C1, P3-P4, M3); RR 542 (c, p2-m3); RR 587 (D3-4); RR 588 (D3-4, M1-2); RR 602 (premaxilla and maxilla with alveoli for incisors and canines, and poorly preserved crowns for P2-M1); RR 608 (m3); RR 609 (m2); RR 616 (P4-M1); RR 617 (m2-3); RR 621 (P4-M3); RR 622 (m3); RR 627 (D3-4, M1-2); RR 630 (P3-M2); RR 633 (m3); RR 634 (m2-3); RR 636 (px, m3); RR 637 (m3); RR 638 (m1-3); RR 642 (p4-m2); RR 647 (m1-3); RR 648 (m1-2); RR 650 (D4); RR 712 (m1-2); RR 717 (p4-m3); RR 718 (m2-3); RR 735 (d4); RR 736 (D3-4); RR 743 (mx); RR 748 (mx);

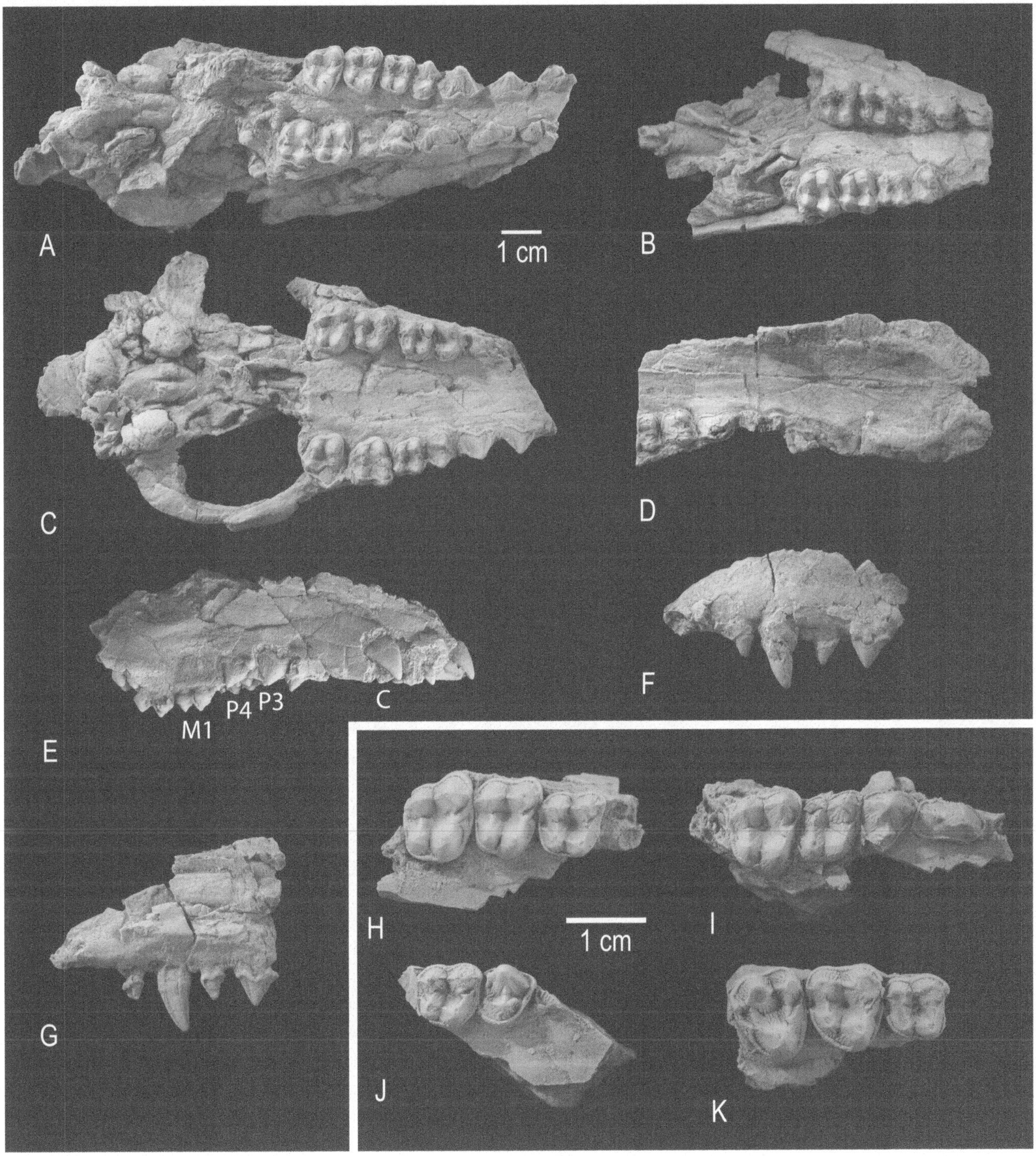

Fig. 14.1 *Indohyus indirae*. **A.** Skull in ventral view (RR 208); **B.** palate in ventral view (RR 209); **C.** skull in ventral view (RR 207); **D.** rostrum in ventral view (RR 602); **E.** rostrum in lateral view (RR 528); **F–G.** premaxilla and maxilla in lateral (RR 331) and medial view (RR 330); **H.** right M1-3 (RR 402); **I.** right P3-M2 (RR 630); **J.** maxilla in occlusal view (RR 616); **K.** maxilla in occlusal view (RR 402). Photography by David Waugh

From the Ganda Kas Area in Punjab (including H-GSP locality 62): H-GSP 1024 (p4).

From Gali Jhagir in Punjab: H-GSP 97184 (M2); H-GSP 97187 (p2-m3).

From Chorlakki in NWFP: GSP-UM 102 (m2), 696 (m2 trigonid).

Type Locality and Horizon: From the basal Murree (Kalakot Zone) near Sind (33°14′30″:74°22′25″) two and half miles west-north-west of Kalakot, a hundred feet above the Kalakot-Rajauri road. "All specimens are collected from the middle of a 20 feet thick mauve colored claystone immediately overlying the Subathu Formation" as quoted

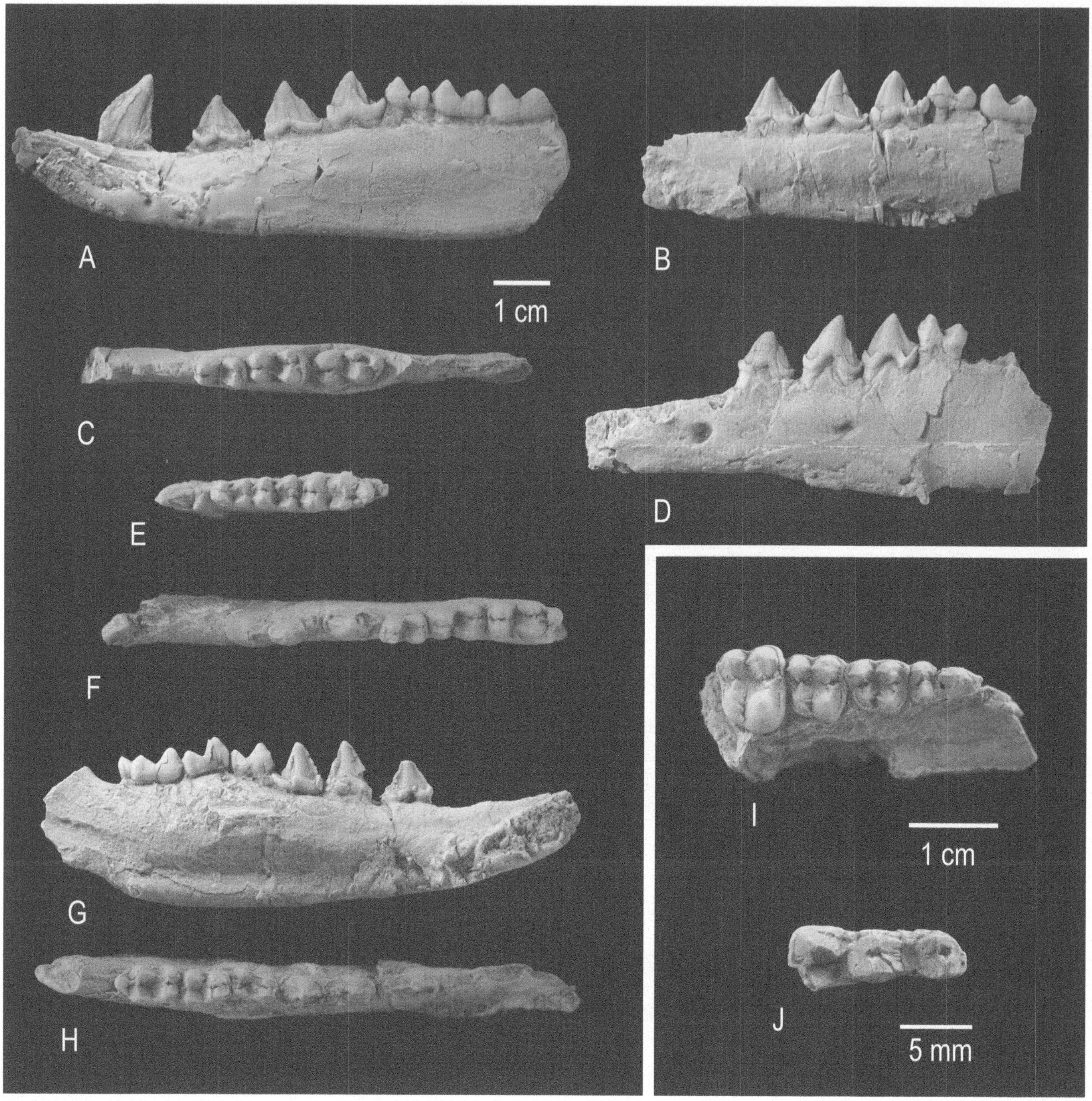

Fig. 14.2 *Indohyus indirae*. **A.** Mandible in lingual view, p1 not preserved, but its alveolus is (RR 542); **B.** mandible in lingual view (RR 327); **C.** mandible in occlusal view (RR 367); **D.** mandible in labial view (RR 324); **E.** mandible in occlusal view (RR 325); **F.** mandible in occlusal view (RR 369); **G–H.** mandible in lingual and occlusal view (RR 323); **I.** maxilla with D3-4, M1-2 (RR 588), **J.** d4 (RR 409). Photography by David Waugh

from Ranga Rao (1972). This locality is now usually referred to as Sindkhatudi and was described in some detail by Nanda and Kumar (1999), and is generally held to be included in the Subathu Formation (Kumar and Sahni 1985). The locality is on the right bank of the Jigni Tawi River, near km stone 98, 100 m upstream from a bridge near Sindkhatudi village, 9 km from Kalakot on Kalakot-Rajauri road, Rajauri District, Jammu and Kashmir, India (Stop 16 of Traverse Route 1 of Nanda and Kumar 1999).

Distribution: Jammu & Kashmir (India), Punjab (Pakistan), and North-West Frontier Province (Pakistan). The majority of specimens of *Indohyus* have been found in the Kalakot region of J&K. Specimens are rare in the collections from the Ganda Kas area of Punjab (West 1980), the Gali Jhagir area of Punjab (Thewissen et al. 2001a) and at Chorlakki in North-West Frontier Province (NWFP, Thewissen et al. 1987).

Age: Lutetian, middle Eocene (Kumar 2000).

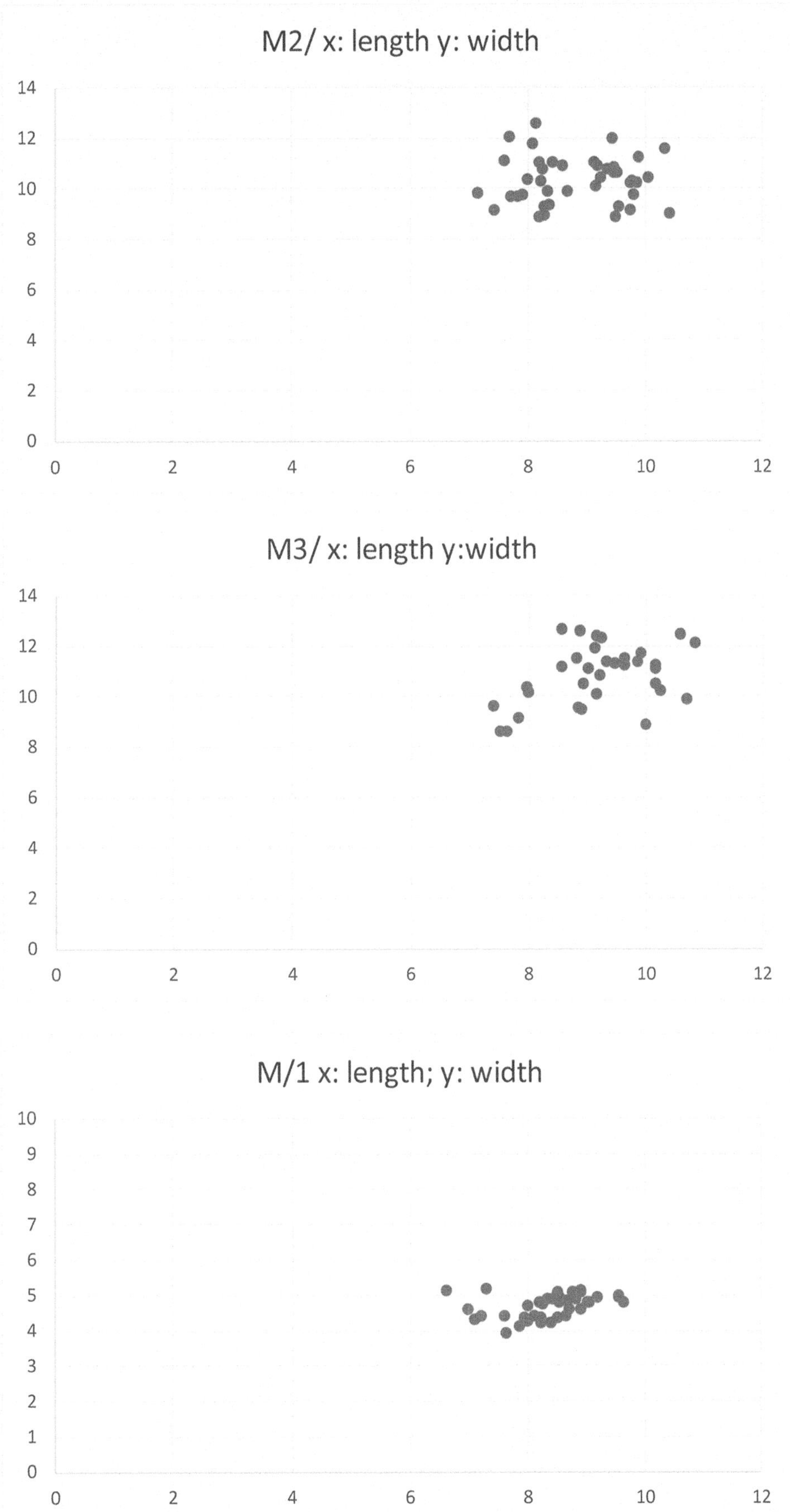

Fig. 14.3 Scatter plots of length and width of M2, M3, and M1 for *Indohyus indirae* from Sindkhatudi. Axes in mm scale

Table 14.1 Means for cheek teeth of *Indohyus* from Sindkhatudi (in mm)

Indohyus		P2		P3		P4		M1		M2		M3	
		L	W	L	W	L	W	L	W	L	W	L	W
Upper	Number	4	4	6	6	9	9	18	17	25	25	18	19
	Mean	9.97	4.12	9.61	4.72	7.32	7.88	8.49	8.74	8.82	10.1	9.19	10.19
Lower	Number	14	14	20	20	27	27	41	43	40	44	41	45
	Mean	9.61	3.84	9.87	3.9	9.43	4.45	8.32	4.63	9.18	5.62	12.51	5.95

Description: The dental formula of *Indohyus* is 3.1.4.3/3.1.4.3. Means for sizes of teeth are presented in Table 14.1. Alveoli for the upper incisors are preserved in RR 330, 528, and 602 and these are lined up in a parasagittal plane, not a coronal plane (Fig. 14.1D). The alveoli are widely spaced, and the alveolus for I2 is the largest of the three. The crown for an unerupted I2 is preserved in RR 528, this tooth has a high, pointed crown and is somewhat flat linguo-labially, but the cingulum is not preserved (Fig. 14.1E). I3 has a low crown, less than half as high as I2, with crests descending from its pointed tip (Fig. 14.1F–G). A strong cingulum surrounds the tooth. The upper canine is slightly curved and flat linguo-labially, with sharp crests descending rostral and caudal sided, and lacking a cingulum (RR 330 and 331). The canine is much larger than the incisors and first premolar. It has a curved crown with no cingulum. There is no evidence of sexual dimorphism in the canine, or elsewhere.

The P1 and P2 are similar in shape, but p2 is larger and more elongate rostrocaudally (Fig. 14.1A, E). These premolars are triangular in labial view, with two roots, that are closely approximated in P1. The main cusp, the paracone, is located on the rostral part of the tooth, and strong crests descend from it rostrally and caudally. The caudal of these crests in some specimens bears a weak cusp behind the paracone. Weak styles maybe present where these crests touch the cingulum. The enamel is crenulated, and a strong cingulum surrounds the tooth. The cingulum of P1 is expanded caudo-lingually.

The P3 is triangular in labial view, and has a more symmetrical crown than P2, with two roots (Fig. 14.1A, B, I, and J). There is one major cusp, and crests descend from it rostrally and caudally. A small cusp occurs on the rostral crest near the paracone in some specimens. The caudal crest is crenulated, and crenulations sometimes line up to appear like a second, parallel crest (RR 630). The caudal cingulum is occasionally extended lingually (RR 419). The cingulum is strong, crenulated, and is expanded caudo-lingually.

The P4 bears a large paracone and also a large protocone, and has three roots. In labial view, this tooth is dominated by the triangular paracone, which is similar in height to P3 and much higher than the paracone of M1. Crests descend the paracone rostrally and caudally, and the caudal crest bears a consistent crenulation that could be interpreted as the metacone. The paracone is placed somewhat posterior to the middle of the crown, hence the crest on its rostral surface is less steep than the crest on its caudal surface. The cingulum is strong and crenulated and surrounds the entire tooth. The cingulum bends occlusally where the preparacrista reaches it, but there is no parastyle. The protocone is located on the protocone lobe, and this cusp is immediately lingual to the paracone. The protocone of P4 is similar in height to the protocone of M1. The labial side of the protocone of P4 is connected to the base of the paracone by a weak crest. This crest arches rostrally in some specimens (RR 207), but not in others (RR 630), whereas in still others there are several crests, to the paracone, to the rostral cingulum, and to the caudal cingulum (RR 616). The enamel is crenulated, and some of these crenulations are so strong that they could be interpreted as variable cusps.

The D3 has three large cusps (RR 588, Fig. 14.2I). The rostral most cusp, the paracone, is the largest, and is located on a narrow part of the tooth. The paracone is triangular in labial view, and narrow linguo-labially. The caudal part of the tooth has two cusps, both most circular in circumference than the paracone. The two caudal cusps are located directly linguolabial to each other, and connected by a weak crest. The labial of these, the metacone, has a weak labial cingulum. The D4 is similar in shape to the upper molars (Fig. 14.2I), although its length is proportionally greater than its width.

The upper molars of *Indohyus* are quadritubercular with a crest connecting protocone to paracone and also connecting hypocone to metacone (Fig. 14.1A–C, H–K). While these two crests are the strongest on the tooth, they vary among specimens, making some teeth more lophodont in nature (RR 618) and others more bunodont (RR 625). In general, smaller individuals tend to have stronger crests and larger individuals tend to have more bunodont cusps. Weak crests connect paracone to metacone. The protocone is always larger than the posterolingual cusp, here called the hypocone, even though it may instead be homologous to a metaconule (Theodor et al. 2007). Occasionally, in *Indohyus*, the hypocone is missing on M3 (RR 402). A small paraconule is located on the paracrista, and a weak preparacrista descends to the rostral base of the paracone from the paraconule. A weak preparaconule crista and postmetaconule crista may be present, ending, respectively in a small parastyle and metastyle. The cingulum is usually complete, but not strong. In some specimens the cingulum is interrupted, for instance, lingual to the protocone (RR 618).

No lower incisors are preserved in their alveoli, but RR 323 and 324 preserve their alveoli. The alveolus for i2 is slightly larger than those for the other two incisors. All three alveoli are closely spaced, suggesting that their crowns touched. Several small isolated unicuspid teeth have been recovered at Sindkhatudi (RR 53, 110, 729) and these may or may not represent *Indohyus* incisors. However, no small spatulate teeth have been found here, so, since the taxon is very common at this locality, it is unlikely that the lower incisors were spatulate. A single specimen preserves the lower canine in situ (RR 542, Fig. 14.2A). This tooth is single-rooted but remarkably long rostro-caudally, resembling the triangular outline of a lower premolar in labial view, although with a somewhat higher tip. The cingulum is weak in some areas and absent in others, whereas lower p2-4 have stronger cingula.

The first lower premolar is single rooted and separated by small diastemata from the canine and p2. No p1 is preserved in its alveolus. p2, p3, and p4 are similar in size and shape (Fig. 14.2A, B, D, F–H). They are two-rooted teeth, triangular in labial view, with a strong lingual cingulum, and weak to absent labial cingulum. The tip of the tooth, the protoconid, is placed near the center of the tooth, and crown height is greater than tooth length. The postprotocristid descends to the cingulum caudally, and a crenulation is located on this crest occasionally (RR 324). Occasionally, a second crest parallels the postprotocristid, without reaching the top of the protoconid. The caudal cingulum is sometimes broad, especially in p4, but never bears a cusp. In most specimens, the crown of p3 is higher than that of p2 and p4, although the difference is minor. The labial side of these premolars is convex rostro-caudally. The lingual side, in most specimens, shows two depressions, one just caudal to the preprotocrista, and one just rostral to the postprotocrista. This causes the rostro-caudal profile of this tooth to be concave-convex-concave.

Several specimens of d4 are preserved (Fig. 14.2J). This tooth is elongate, and bears three rows of two cusps from rostral to caudal. The rostral pair of cusps is similar in height, and these cusps are connected by two weak crests. The labial of these is the paraconid. Each of these cusps is connected to the cusp caudal to them by a crest, and these crests surround a distinct basin, larger in extent than the talonid basin. The lingual cusp of the middle pair is the largest cusp of all: the metaconid. Labial to it, is a smaller cusp, the protoconid. Caudal to the middle pair of cusps is the talonid basin, which is flanked by two crests, the labial of which is stronger, the cristid obliqua, which connects the center of the caudal face of the trigonid to the hypoconid, the larger of the two caudal cusps. It is connected to the entoconid by a crest on which a small hypoconulid is located.

The three lower molars are similar in shape (Fig. 14.2A–H), except that m3 has a hypoconulid lobe with a large cusp, unlike m1-2. The trigonid bears two cusps in close approximation and the metaconid is taller than the protoconid. Overall, the molars are lower than the premolars. The paracristid sweeps low lingually from the protoconid to an area rostral to the metaconid, ascending the metaconid slightly. The metacristid is absent or weak, connecting the caudal side of protoconid and metaconid by a low curving crest. Steep tooth wear on the rostral side of protoconid and metaconid can result in the formation of a straight crest that connects the proto- and metacristid on their cranial side, giving the impression of a protolophid. The talonid basin is broad. In most specimens, it is flanked by a high and straight hypocristid, resembling a hypolophid, but in a smaller number of specimens the hypocristid is interrupted where it is crossed by a valley that connects the talonid basin to the caudal side of the tooth (RR 367, 368). This difference is not consistently associated with other features in which the lower molars vary. The cristid obliqua is straight, reaching the trigonid just posterior to the protoconid, and a weaker crests extends rostrally from the entoconid. The hypoconid and entoconid are similar in height, but the hypoconid has a broader base. A small hypoconulid may be present on the caudal surface of the hypocristid on M1-2, but it does not reach the apex of that crest. The hypoconulid of m3 is always present, but varies in size. A crest connects the lingual side of the m3 hypoconulid to the caudal side of the entoconid, and occasionally a low crest extends straight rostrally from the hypoconulid. Weak cingula may be present.

The postcranial osteology of *Indohyus* was described by Cooper et al. (2011). The cranial anatomy of *Indohyus* has not been described in detail although figures of skulls have been published (Thewissen et al. 2007, 2009) and characters scored are widely used in cladistics data matrices (Thewissen et al. 2007; Geisler and Theodor 2009; Uhen 2010; Gatesy et al. 2013).

Discussion: The synonymy of the many taxa here included in *Indohyus indirae* is widely accepted (Kumar and Sahni 1985; Russell and Zhai 1987; Thewissen et al. 1987; Theodor et al. 2007). Thewissen et al. (2001a) referred two specimens from Gali Jhagir in Punjab to *Kunmunella transversa*, citing the difference in proportions of the M2 and M3 as the most significant difference between *Kunmunella* and *Indohyus*, with the latter genus having narrower upper molars. Plotting length versus width for these two teeth (Fig. 14.3) does document significant variation in this regard. The M2 plot appears to show two clusters of teeth with similar width, but distinguished by length smaller and greater than 8. 7 mm. This distinction is less clear for the M3 plot, and jaws with teeth that are in the shorter cluster for M2, are not consistently in the shorter cluster of the M3 plot. Indeed, in other important metric features, such as m1 dimensions, all specimens form a tight cluster (Fig. 14.3). At present, it appears best to attribute these specimens to the same species: *Indohyus indirae*.

This does not negate the fact that dental variation in *Indohyus* is significant, and a great size range is present, probably more significant than the normal variation in most species. Indeed, the taphonomy of the Sindkhatudi locality may be relevant here. This locality is a bone bed with disarticulated bones only, suggesting that disarticulation occurred before burial. As such, this locality may represent a time-averaged assemblage. If *Indohyus* was a quickly evolving lineage, it is possible that this locality combined specimens that lived hundreds of thousands of years apart, enough time to allow for significant anagenesis. Indeed, remarkable variation also occurs in non-metric part of *Indohyus*' teeth. The size of the hypocone is more variable among the specimens in our sample than the size of other cusps. In one extreme examples, RR 402 (Fig. 14.1K), the hypocone in M1-2 is unremarkable in size, but the cusp is absent altogether in M3.

Indohyus major Thewissen et al., 1987

Holotype: GSP-UM 1435, trigonid of a lower molar.

Type Locality and Horizon: Middle Eocene of Chorlakki, NWFP, Pakistan (Thewissen et al. 1987)

Distribution and Age: Middle Eocene of type locality.

Discussion: The type specimen of *Indohyus major* is approximately twice as large, in linear dimensions, as other *Indohyus*. No additional specimens have ever been recovered after its initial description. A single fragmentary m3 from China matches the type specimen in size (Orliac and Ducrocq 2011), who also reassigned this species to the genus ?*Khirtharia*. Not having studied the specimen, we are unable to fully assess its shape, but believe that it may represent a more basal, diacodexid or dichobunid artiodactyl.

Discussion

The Eocene of all northern hemisphere continents abounds with small, bunodont artiodactyls. The most plesiomorphic of these are the diacodexeids (Thewissen et al. 1983; Thewissen et al. 1987; Bajpai et al. 2005; Kumar et al. 2010). Diacodexeidae and Raoellidae are the common artiodactyls of early-middle Eocene of the Indian Subcontinent. The molar morphology of these two families allows easy for distinction: Diacodexeidae lack a hypocone, whereas Raoellidae have that cusp. *Khirtharia* (which includes *Bunodentus*) has low, blunt molar cusps and small premolars, whereas *Indohyus* has high, sharp cusps and large, trenchant premolars. With such a significant difference in molar morphology, it is debatable whether these should be included in the same family, but this assessment will have to await a more comprehensive study of *Khirtharia*.

Diacodexeids are known from all northern continents, and have been used to assess arguments related to intercontinental migrations and the timing of the collision between Indian and Asian continents, although the great morphological differences between diacodexeids from the Indian, European, and North American continents may suggest that these are not very closely related. Indeed, the genus *Diacodexis* has been considered to not occur in Indo-Pakistan, and the Indian subcontinent specimens have been referred to *Gujaratia*. Similar forms classified as dichobunids, have been referred from the part of Asia that is not the Indian Subcontinent (Tsubamoto et al. 2013, Métais et al. 2007).

Raoellids cannot be used for paleogeographic studies as they only occur on the Indian (and maybe the Asian) subcontinent. Most studies find that cetaceans are the sister group to raoellids (Thewissen et al. 2007; Geisler and Theodor 2009; Uhen 2010; Gatesy et al. 2013), and cetaceans are generally thought to have originated on the Indian continent. The sister group of this combined clade could be some of the diacodexeids (Spaulding et al. 2009) or possibly Hippopotamidae (Gatesy et al. 2013). If the record of raoellids from China (Orliac and Ducrocq 2011) is incorrect, then raoellids were not involved in the intercontinental migrations of the middle Eocene, and instead were endemic to the Indian subcontinent. If anything, it might suggest that the Indian subcontinent went through a period of relative isolation.

Indohyus is of great importance in paleontology, as most authors have found it to be the sistergroup to cetaceans. In its dentition too, there are some synapomorphies with basal cetaceans, in particular pakicetids. These include the elongation of the rostrum with upper incisors not touching and implanted rostro-caudally (Fig. 14.1D). Lower incisors of diacodexeids are spatulate and their crowns touch (Thewissen et al. 1983). This is unlikely to be the case in *Indohyus*, where no positively identifiable lower incisors are known, but where, among the many isolated teeth, there are none with spatulate crowns. It is thus likely that lower incisors were pointed, similar to pakicetids (Thewissen and Hussain 1998). *Indohyus* also has large, triangular (in labial view) upper and lower premolars. Indeed, the size is striking and led West (1980) to mistake a premolar of *Indohyus* for a mesonychid tooth.

In molar morphology, *Indohyus* does not compare well to pakicetids, which have large paracones, protocones, protoconids, and hypoconids, and very reduced to absent metacones, hypocones, metaconids, entoconids, conules, cristae, and basins. Such reduction of crown morphology extends into the premolars: whereas most bunodont Eocene artiodactyls have a protocone on P4, Eocene cetaceans do not. On the other hand, *Indohyus* has quadritubercular, bunodont teeth with well-developed basins.

Taken together, the differences in dental morphology point to differences in diet or food processing function between early and middle Eocene artiodactyls, *Indohyus*, and middle Eocene cetaceans. Isotopic and gross morphological evidence indicate that these cetaceans were freshwater meat eaters, while *Indohyus* ate land plants, although it lived in and near water (Thewissen et al. 2001b, 2007, 2011). Evidence from tooth wear (Thewissen et al. 2011) indicates that *Indohyus* used its molars to process food differently from generalized artiodactyls, and, in some regards, more similar to Eocene cetaceans. It is likely then that the diet of *Indohyus* included plant food of particular and unusual texture or consistency that required food processing methods uncommon in other early artiodactyls. At this point, there is no clarity on what these may be.

References

Bajpai, S., Kapur, V. V., Das, D. P., Tiwari, B. N., Saravanan, N., & Sharma, R. (2005). Early Eocene land mammals from Vastan lignite mine, district Surat (Gujarat), Western India. *Journal of the Palaeontological Society of India, 50,* 101–113.

Cooper, L. N., Thewissen, J. G. M., Bajpai, S., & Tiwari, B. N. (2011). Postcranial morphology and locomotion of the Eocene raoellid *Indohyus* (Artiodactyla, Mammalia). *Historical Biology*. https://doi.org/10.1080/08912963.2011.624184.

Dehm, R., & Spielberg, T. O. (1958). Paleontologische und geologische Untersuchungen Im Tertiaer von Pakistan. 2. Die mitteleocaenenSauegetiere von Ganda Kasbei Basal in Nord-West Pakistan. *AbhandlungenBayerischenAkademie der Wissenschaftenmathematisch -naturwissenschaftlicheKlasse. neueFolge, 91,* 1–53.

Gatesy, J., Geisler, J. H., Chang, J., Buell, C., Berta, A., Meredith, R. W., et al. (2013). A phylogenetic blueprint for a modern whale. *Molecular Phylogenetics and Evolution, 66,* 479–506.

Geisler, J. H., & Theodor, J. M. (2009). *Hippopotamus* and whale phylogeny. *Nature, 458,* E1–4.

Kumar, K. (1998). Tertiary vertebrates from the pre-Siwalik Himalayan foreland basin, Northwest India. Himalayan Foreland Basin, Workshop, University of Jammu, March 16–19, 1998, Abstracts 33–36.

Kumar, K. (2000). Correlation of continental Eocene vertebrate localities of the Indian Subcontinent. *Himalayan Geology, 21,* 63–85.

Kumar, K., Rose, K. D., Rana, R. S., Singh, L., Smith, T., & Sahni, A. (2010). Early Eocene artiodactyls (Mammalia) from western India. *Journal of Vertebrate Paleontology, 30,* 1245–1274.

Kumar, K., & Sahni, A. (1985). Eocene mammals form the Upper Subathu Group, Kashmir Himalaya, India. *Journal of Vertebrate Paleontology, 5,* 153–168.

Métais, G., Naing Soe, A., Marivaux, L., & Beard, K. C. (2007). Artiodactyls from the Pondaung Formation (Myanmar): New data and reevaluation of the South Asian faunal Province during the middle Eocene. *Naturwissenschaften, 94,* 759–768.

Nanda, A. C., & Kumar, K. (1999). Excursion guide on the Himalayan foreland basin (Jammu-Kalakot-Udhampur sector). *Wadia Institute of Himalayan Geology, Special Publication, 2,* 85 pp.

Orliac, M. J., & Ducrocq, S. (2011). Eocene raoellids (Mammalia, Cetartiodactyla) outside the Indian subcontinent: Paleogeographical implications. *Geological Magazine*. https://doi.org/10.1017/S0016756811000586.

Pilgrim, G. E. (1940). Middle Eocene mammals from northwest India. *Proceedings of the Zoological Society London, 110,* 124–152.

Ranga Rao, A. (1971). New mammals from Murree (Kalakot Zone) of the Himalayan foot hills near Kalakot, Jammu & Kashmir State, India. *Journal of the Geological Society of India, 12,* 125–134.

Ranga Rao, A. (1972). New mammalian genera and species from the Kalakot zone of Himalayan foot hills near Kalakot, Jammu & Kashmir State, India. *Special Paper, Directorate of Geology, Oil & Natural Gas Commission, 1,* 1–22.

Russell, D. E., & Zhai, R. J. (1987). The Paleogene of Asia: Mammals and stratigraphy. *Memoires de la Musém National d'Histoire. Naturelle, Paris, Science de la Terre, 52,* 1–488.

Sahni, A., Bhatia, S. B., Hartenberger, J.-L., Jaeger, J.-J., Kumar, K., Sudre, J., et al. (1981). Vertebrates from the Subathu Formation and comments on the biogeography of Indian subcontinent during the early Paleogene. *Bulletin de la Societé Geologique de France, 23,* 689–695.

Sahni, A., & Khare, S. K. (1972) (for 1971). Three new Eocene mammals from Rajauri District, Jammu and Kashmir. *Journal of the Palaeontological Society of India, 16,* 41–53.

Sahni, A., & Khare, S. K. (1973) (for 1972). Additional Eocene mammals from the Subathu Formation of Jammu and Kashmir. *Journal of the Palaeontological Society of India, 17,* 31–49.

Spaulding, M., O'Leary, M. A., & Gatesy, J. (2009). Relationships of Cetacea (Artiodactyla) among mammals: Increased taxon sampling alters interpretations of key fossils and character evolution. *PLoS One, 4*(9). https://doi.org/10.1371/journal.pone.0007062.

Theodor, J., Erfurt, J., & Métais, G. (2007). The earliest artiodactyls, Diacodexeidae, Dichobunidae, Homacodontidae, Leptochoeridae, and Raoelllidae. In D. R. Prothero & S. E. Foss (Eds.), *The evolution of artiodactyls* (pp. 32–58). Baltimore: Johns Hopkins Univ.

Thewissen, J. G. M., Cooper, L. N., Clementz, M. T., Bajpai, S., & Tiwari, B. N. (2007). Whales originated from aquatic artiodactyls in the Eocene epoch of India. *Nature, 450,* 1190–1195.

Thewissen, J. G. M., Cooper, L. N., George, J. C., & Bajpai, S. (2009). From land to water: The origin of whales, dolphins, and porpoises. *Evolution: Education and Outreach, 2,* 272–288.

Thewissen, J. G. M., Clementz, M. T., Sensor, J. D., & Bajpai, S. (2011). Evolution of dental wear and diet during the origin of whales. *Paleobiology, 37,* 655–669.

Thewissen, J. G. M., Gingerich, P. D., & Russell, R. E. (1987). Artiodactyla and Perissodactyla (Mammalia) form the early-middle Eocene Kuldana Formation of Kohat (Pakistan). *Contributions of the Museum of Paleontology, University of Michigan, 27,* 247–274.

Thewissen, J. G. M., & Hussain, S. T. (1998). Systematic review of the Pakicetidae, Early and middle Eocene Cetacea (Mammalia) from Pakistan and India. *Bulletin of the Carnegie Museum, 34,* 220–238.

Thewissen, J. G. M., Russell, D. E., Gingerich, P. D., & Hussain, S. T. (1983). A new artiodactyl (Mammalia) from the Eocene of North-West Pakistan. Dentition and classification. *Proceedings of the Koninklijke Nederlandse Akademie van Wetenschappen, Series B, 86,* 153–180.

Thewissen, J. G. M., Williams, E. M., & Hussain, S. T. (2001a). Eocene mammal faunas from northern Indo-Pakistan. *Journal of Vertebrate Paleontology, 21,* 347–366.

Thewissen, J. G. M., Williams, E. M., Roe, L. J., & Hussain, S. T. (2001b). Skeletons of terrestrial cetaceans and the relationship of whales to artiodactyls. *Nature, 413,* 277–281.

Tsubamoto, T., Egi, N., Takai, M., Htike, T.-, Thein, Z.-M.-M. (2013). A new genus and species of bunodont artiodactyl from the Eocene Pondaung Formation, Myanmar. *Paleontological Research, 17,* 297–311.

Uhen, M. D. (2010). The origin(s) of whales. *Annual Review of Earth and Planetary Sciences, 38,* 189–219.

West, R. M. (1980). A middle Eocene large mammal assemblage with Tethyan affinities, Ganda Kas region, Pakistan. *Journal of Paleontology, 54,* 508–533.

Chapter 15
The Murine Cradle

Lawrence J. Flynn, Yuri Kimura, and Louis L. Jacobs

Abstract The unique geological, geophysical, and tectonic setting of the Indian Subcontinent created a theater of evolution that provided a relatively insulated refuge for origination and early diversification of many Mesozoic and early Cenozoic age vertebrate groups. This phenomenon continued into the Miocene Epoch in association with developing physical barriers: mountains, major river systems, and oceanic borders. The terrestrial barriers were semi-permeable filters that allowed differential passage of select organisms, and endemic evolution of certain clades. Muroid rodents of modern grade are a group that evolved by the earliest Miocene, in part endemically after they appeared on the subcontinent. One endemic muroid lineage led to the Murinae, extant mice and rats, and the semi-isolated physical setting of India and Pakistan promoted local evolution of murines during the middle Miocene, prior to their dispersion throughout Eurasia and beyond around ten million years ago. Here we note dental morphological features that appeared early in the evolution of Murinae to document the divergence of two major clades leading to extant murines. Their earliest presumed members are morphologically similar in older samples before clear taxonomic separation of molar phenotypes.

Keywords Murinae • Miocene • Indian Subcontinent • Siwalik • Taxonomy • Biogeography

L. J. Flynn (✉)
Department of Human Evolutionary Biology, Harvard University, Cambridge, MA 02138, USA
e-mail: ljflynn@fas.harvard.edu

Y. Kimura
Department of Geology and Paleontology, National Museum of Nature and Science, 4-1-1 Amakubo, Tsukuba, Ibaraki 305-0005, Japan

L. L. Jacobs
Roy M. Huffington Department of Earth Sciences, Southern Methodist University, Dallas, TX 75275, USA

Introduction

The special geological, geophysical, and tectonic setting of the Indian Subcontinent limited immigration and dispersal but promoted evolution of endemic vertebrate groups during its Mesozoic passage northward, prior to docking with the southern margin of Asia (Jacobs et al. 2011; Chatterjee et al. 2017). Subsequent uplift along the northern margin of the once island continent, as its leading edge subducted under what is now Tibet, contributed to the high terrain of the Tibetan Plateau, and importantly the spectacular mountain ranges bounding it. The Himalaya-Hindu Kush and other ranges hem in the subcontinent along its northern limits. Mighty river systems crossing Indo-Pakistan, with the tropical forests of Myanmar and Thailand, further isolate the subcontinent on the east. To the west, mountainous and arid terrain and its extensive coastline to the south complete the physical barriers around the Indo-Pakistan peninsula. The northern highlands deliver clastic wedges of terrestrial deposits to the south via the major river systems, yielding near-continuous sedimentation and consecutive fossiliferous horizons that span the Neogene of the Cenozoic Era. The most continuous series of superposed fossil horizons, the Miocene Siwalik deposits of the Potwar Plateau, Pakistan (coupled with Zinda Pir and Manchar sequences to the southwest), is complemented in India by Lower and Middle Siwalik outcroppings near Jammu, Nurpur and Haritalyangar, and by the vast Upper Siwaliks near Chandigarh (Flynn et al. 2013; Patnaik 2013; Nanda 2015). The dense biotic record of the Indian Subcontinent spans much of the Neogene; contrasts in its paleofauna with that of other biogeographic regions are due, in part, to isolation.

Among the groups of small mammals that show endemism to the Indian Subcontinent, true mice provide a striking example of local evolution prior to a phase of dispersion of nearly global scale. Mice, formally Subfamily Murinae, are derived from early Miocene muroid rodents,

G. V. R. Prasad and R. Patnaik (eds.), *Biological Consequences of Plate Tectonics: New Perspectives on Post-Gondwana Break-up—A Tribute to Ashok Sahni*,
Vertebrate Paleobiology and Paleoanthropology Series, https://doi.org/10.1007/978-3-030-49753-8_15

some of which have a fossil record in the early Miocene of the Indian Subcontinent (de Bruijn et al. 1981; Lindsay 2017). Murinae is a species-rich group of nearly 600 species in 130 genera (Steppan 2017), the core of Family Muridae. There are four other crown group subfamilies (see, e.g., Steppan and Schenk 2017), mainly African in distribution except for Gerbillinae, which are distributed throughout Eurasia as well as Africa. The gerbil fossil record dates from the early Miocene and is represented in the Siwaliks. As sister taxon to gerbils, Murinae originated in early Miocene time (Aghová et al. 2018) and were represented by successive stem taxa *Potwarmus* and *Antemus*. Murines began to diversify by the end of the middle Miocene, and spread throughout the Old World by 10 Ma (Sen 2003 notes *Progonomys* somewhat less than 10 Ma in Turkey; van Dam et al. 2001 date *Progonomys* in Spain at 9.8 Ma [current time scale]: López-Antoñanzas et al. 2019 at near 11 Ma in Lebanon; see also Kimura et al. 2017).

Early field work in collaboration with the Geological Survey of Pakistan (GSP) concentrated on the Potwar Plateau, where Jacobs (1977) recovered primitive fossil murine rodents in the middle Miocene type area of the Chinji Formation, at the time a surprisingly old find. Murines were then considered to have evolved rapidly from an early late Miocene murid stock in the Mediterranean region represented by *Progonomys* Schaub, 1938, so discovery of middle Miocene mice in Pakistan was a surprise. Jacobs (1978) outlined murid evolution in the middle and late Miocene of South Asia. In the following we develop the thesis that much (not all) of early mouse evolution is sampled by and recorded in the cradle of the Indian Subcontinent, until Murinae burst onto the global stage before 10 Ma.

Previous Work

When *Antemus chinjiensis* Jacobs, 1977 was named, very few rodent fossils from the Siwaliks had been described. *Antemus* opened the door to a new world for micromammals – a semi-isolated Miocene world – a theater of evolution for unique lineages in the island-like setting of the Indian Subcontinent.

Antemus was a murine by definition of its crown pattern, but clearly a primitive murine. It was more primitive than the universally-accepted stem murine *Progonomys* Schaub, 1938 in that its synapomorphic lingual cusps were weakly connected or unconnected to the adjacent anterocone and protocone. *Antemus* turned out to be quite common in localities dating to 13.8–12.7 Ma. Lindsay (1988) extended work on the Potwar Plateau to older horizons (early middle Miocene) and described yet-more-primitive muroids. *Potwarmus primitivus* (Wessels et al. 1982) was smaller than *Antemus*, quite common at localities dating to ∼14 Ma, and distinguished by lack of connections of lingual cusps on upper molars and its simple, cricetid-like m1. The middle Miocene Chinji Formation preserved a stem murine lineage represented by the *Potwarmus-Antemus* succession, preceding more derived *Progonomys*-like mice at the top of the formation.

Late middle and early late Miocene murines were derived with respect to *Antemus*. Jacobs and Downs (1994) and Jacobs and Flynn (2005) noted material of about 12.4 Ma as derived with respect to *Antemus* in more stable, but low connections to lingual cusps in upper molars. Murine samples from single localities appeared to be homogeneous and represent single species. Younger murid samples dating to 11.2 Ma, however, presented considerable morphological variation that Kimura et al. (2016) thought might represent more than one genus.

Early exploratory work on localities dating to about 9.2 Ma had clearly revealed two dominant mice in distinct genera: *Progonomys debruijni* Jacobs, 1978 and *Karnimata darwini* Jacobs, 1978. More recent work by Kimura (Kimura et al. 2015, 2016, 2017) defined and specified the ages of older forerunners: *Progonomys hussaini* Cheema et al., 2000 and *Karnimata fejfari* Kimura et al., 2017. These extended the history of the genera to older localities dating to 10.5 Ma. Can *Progonomys* and *Karnimata* be distinguished in older localities of about 11.2 Ma as Kimura et al. (2016) hypothesized? Kimura et al. (2017) considered assemblages dating to 11.2 Ma to include "*Progonomys* cf. *P. hussaini*" and "?*Karnimata* sp." What is the pattern of emergence and stabilization of phenotypes in this case of small mammal evolution? We suggest that one approach to this problem lies in analysis of large samples of early murids now available from single sites.

Herein we examine murine assemblages that predate 10.5 Ma and find that the *Progonomys* and *Karnimata* lineages appear to diverge deeper in time from a single population with inconsistently separable molar morphology, showing great variation. Our interpretation leads us to propose early (>11 Ma) occurence of *Progonomys* and *Karnimata* M1 phenotypes within murine assemblages that show variation but inconsistent species distinction of many specimens. The emergence of some *Progonomys* and *Karnimata* dental features prior to the definitive divergence of those lineages implies ancestral murines (∼11.2 Ma) in which genomic discriminants of distinct morphology (particularly M1 shape) had yet to fully diverge.

If *Progonomys* and *Karnimata* molar morphology can be distinguished in assemblages as old as 11.2 Ma, can recognition of those lineages be pushed back further? Kimura et al. (2016, 2017) postulated 'no', and thought that the 11.2 Ma murines were preceded by a variable murine designated "cf. *Progonomys* spp." We revisit this below.

Materials and Methods

Our systematic work on early murines is based on isolated molars from fossiliferous sediment retrieved by wet screening. We focus on six small mammal localities at the top of the Chinji Formation and low in the Nagri Formation of the Potwar Siwaliks. These occur in a narrow time span, 11.6–11.2 Ma. Localities of the Potwar Plateau are constrained stratigraphically by superposition relative to each other, with ages generally resolved on the scale of 100,000 years for the interval of approximately 18–6 Ma (Barry et al. 2013). Stratigraphic sections are matched to the geomagnetic polarity time scale by magnetostratigraphy as discussed by Barry et al. (2002, 2013), and in the present case correlated to magnetic chrons preceding Chron C5n.2n. Ages for individual localities are interpolated between chron reversal dates (Hilgen et al. 2012) by stratigraphic thickness. Fossils studied here come from sites Y504 (11.64 Ma), Y83 (11.6 Ma), Y76 and Y809 (11.4 Ma), all high in the Chinji Formation, and from low in the Nagri Formation, Y791 (11.23 Ma) and Y797 (11.18 Ma). Age estimates reflect close spacing in time; they are latest middle Miocene to early late Miocene, using the top of chron C5r.2n (11.59 Ma) to signal the beginning of the late Miocene.

Field teams processed bulk samples of fossiliferous sediment from each locality by washing in water in tandem screen-bottom boxes. Dried residue was rewashed after soaking in kerosene to reduce the volume of sediment. Lighter sand and minerals were separated by flotation with a heavy liquid (sodium polytungstate), and the concentrate was sorted by microscope. Specimens are catalogued in the data base originally labeled Yale–Geological Survey of Pakistan (YGSP), the Potwar Plateau localities being designated with a "Y" prefix, and are currently housed in the Human Evolutionary Biology department at Harvard University.

The murine fossils are isolated cheek teeth, representing three upper molar (M1–M3) and three lower molar (m1–m3) loci. We used morphological details of the crowns and root structure to recognize species. Upper and lower case, respectively, are used for teeth from upper and lower jaws. Descriptions utilize standard Cope-Osborn cusp terminology (Fig. 15.1). The figure is modified from Kimura et al. (2013), and shows also the widely used t1–t9 terminology. Molar length and width were measured using a Keyence VHX-1000 digital microscope, with major axis (length) parallel to the tooth row, and width perpendicular to that axis; maximum dimensions were recorded in mm. The Leo 1450 scanning electron microscope was used to view and image isolated murine molars at a magnification of 90× and working distance of 14 mm. All images were made in the variable pressure mode, using pre-purified nitrogen at a pressure of 15–25 Pa, and a four-quadrant backscatter electron detector. The electron beam was set at 20 kV and a probe current of 193 pA.

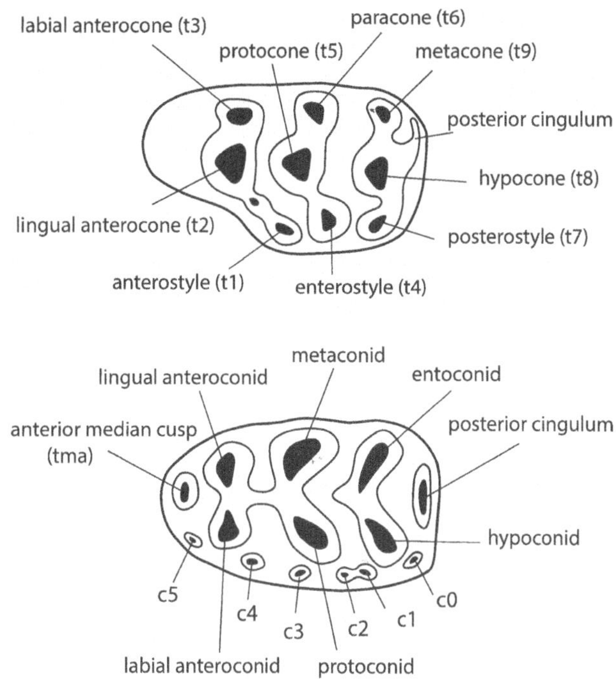

Fig. 15.1 Molar terminology used in this study, after Kimura et al. (2017). Left M1 figured above, and left m1 below

Systematics

Family Muridae Illiger, 1811

Subfamily Murinae Illiger, 1811

Progonomys morganae Kimura et al., 2017

Referred Material: left M1 YGSP 34963; left M2 YGSP 33649 and 33672; right m2 YGSP 33558; left and right m1 YGSP 54059 and 54049, right m1 33670 (Fig. 15.2A–D).

Localities and Ages: 34963 from Y791, 33649, 33670 and 33672 from Y504, 54049 and 54059 from Y809; 33558 from Y76; 11.6–11.2 Ma (Fig. 15.2, Table 15.1), lower part of the Nagri Formation and upper level of the Chinji Formation, Potwar Plateau, Pakistan.

Description: Previously reported from the Nagri Formation, additional specimens are recorded from older levels of the Nagri Formation and upper levels of the Chinji Formation. The M1 referred to this species is larger than the holotype YGSP 33180 (see Kimura et al. 2017), with length to width ratio of about 1.75. The M1 lacks a deep anterolingual flexus

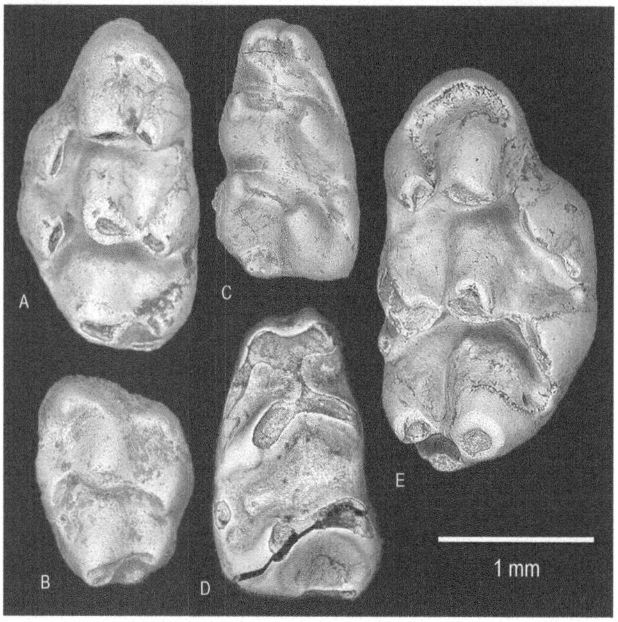

Fig. 15.2 *P. morganae* (**A–D**), **A.** M1 YGSP 34963 from Y791; **B.** M2 33672 from Y504; **C.** m1 33670 from Y504; **D.** m1 54059 from Y809. Unnamed Genus C, **E.** M1 YGSP 36822 fromY797

(inflection of the medial wall of the tooth). Its lingual anterocone is central and the small labial anterocone is appressed to it. There is a weak precingulum. The anterostyle is posterior and laterally pinched, with its major axis nearly antero-posterior. The enterostyle is isolated posterolingual to the protocone. The paracone is only slightly posterior to the protocone.

The protocone and hypocone are the major cusps of M2. The paracone is somewhat posterior to the protocone and larger than the small metacone. The anterostyle is large and isolated; the labial anterocone is small and low. The enterostyle is large, nearly isolated from the protocone in one of two specimens. In one of two, the paracone and metacone are closely positioned with a low marginal ridge between them. A low, thin posterior cingulum is preserved in one specimen.

The 6-cusped lower first molar is narrow, with the lingual cusps slightly anterior to the labal cusps. The double anteroconids are twinned and positioned close to each other, and there is no anterocentral cusp (tma). A low, median connection (anterior mure) from the anteroconids runs posteriorly to the protoconid-metaconid junction. The

Table 15.1 Comparative molar dimensions (mm) for *Progonomys morganae*, M1 of "indeterminate *Progonomys/Karnimata*" grade (labeled according to dominant morphotype), and pre-*Progonomys* molars

	Locality	YGSP number	Taxon name	Length	Width	Height[a]
M1	791	34963	*Progonomys morganae*	1.77	1.01	0.72
m1	809	54059	*Progonomys morganae*	1.45	0.88	–
m1	809	54049	*Progonomys morganae*	1.35	0.85	–
m1	504	33670	*Progonomys morganae*	1.48	0.92	–
m2	76	33558	*Progonomys morganae*	1.05	0.91	–
M2	504	33649	*Progonomys morganae*	1.13	0.95	–
M2	504	33672	*Progonomys morganae*	1.15	1.05	–
M1	791	34965	*Progonomys* cf. *P. hussaini*	1.71	1.06	0.62
M1	791	34976	*Progonomys* cf. *P. hussaini*	1.94	1.14	0.62
M1	797	33869	*Progonomys* cf. *P. hussaini*	1.8	1.12	0.74
M1	797	36820	*Progonomys* cf. *P. hussaini*	1.77	1.05	–
M1	797	36821	*Progonomys* cf. *P. hussaini*	1.86	1.1	0.74
M1	797	33878	*Progonomys* cf. *P. hussaini*	1.81	1.07	0.67
M1	797	36819	*Progonomys* cf. *P. hussaini*	1.72	1.05	0.62
M1	797	36826	*Progonomys* cf. *P. hussaini*	1.8	1.1	0.69
M1	797	36823	*Progonomys* cf. *P. hussaini*	1.68	1.09	0.72
M1	791	34973	*Karnimata* cf. *K. fejfari*	1.9	1.19	0.76
M1	791	34975	*Karnimata* cf. *K. fejfari*	1.92	1.23	0.66
M1	797	36824	*Karnimata* cf. *K. fejfari*	2.06	1.32	0.53
M1	797	33864	*Karnimata* cf. *K. fejfari*	1.9	1.21	0.71
M1	797	33876	*Karnimata* cf. *K. fejfari*	1.93	1.21	0.65
M1	797	33852	*Karnimata* cf. *K. fejfari*	1.93	1.22	0.73
M1	797	33866	*Karnimata* cf. *K. fejfari*	1.95	1.22	0.69
M1	797	36822	New Genus C, unnamed	2.23	1.25	0.78
M1	76	33567	"Pre-*Progonomys*" cf. *K. fejfari*	1.87	1.12	0.58
M1	76	33588	"Pre-*Progonomys*" cf. *P. hussaini*	1.93	1.13	0.73
M1	76	33587	"Pre-*Progonomys*" cf. *P. hussaini*	1.68	1.01	0.49

(continued)

Table 15.1 (continued)

	Locality	YGSP number	Taxon name	Length	Width	Height[a]
M1	76	33571	"Pre-*Progonomys*"	1.71	1.08	0.56
M1	76	33572	"Pre-*Progonomys*"	1.79	1.16	0.67
M1	76	33580	"Pre-*Progonomys*"	1.83	1.14	0.66
M1	76	33584	"Pre-*Progonomys*"	1.75	1.13	0.56
M1	76	33586	"Pre-*Progonomys*"	1.69	1.13	0.59
M1	76	33591	"Pre-*Progonomys*"	1.75	1.11	0.62
M2	76	33755	"Pre-*Progonomys*"	1.08	0.92	–
M2	504	33664	"Pre-*Progonomys*"	1.21	1.17	–
M3	76	33788	"Pre-*Progonomys*"	0.64	0.68	–
M3	76	33786	"Pre-*Progonomys*"	0.69	0.77	–
m1	809	54087	"Pre-*Progonomys*"	1.83	1.13	–
m1	809	54051	"Pre-*Progonomys*"	1.73	1.14	–
m1	504	33647	"Pre-*Progonomys*"	1.86	1.04	–
m1	76	33777	"Pre-*Progonomys*"	1.65	1.05	–
m2	504	33654	"Pre-*Progonomys*"	1.3	1.11	–
m2	76	33561	"Pre-*Progonomys*"	1.27	1.1	–
m3	809	54043	"Pre-*Progonomys*"	0.92	0.83	–
m3	809	54044	"Pre-*Progonomys*"	1.06	1.01	–
m3	809	54047	"Pre-*Progonomys*"	1.06	0.98	–

[a]For M1, cervix to fresh or slightly worn tip of hypocone

hypoconid-entoconid pair of cusps is slightly larger than the protoconid-metaconid, and a low cingulum extends posterior to the entoconid. The labial cingulum is weakly developed, with a small cuspule (c1) anterior to the hypoconid. The single m2 has low major cusps and a weak labial cingulum without cuspules, only a minor thickening of enamel near the hypoconid.

Comment: *Progonomys morganae* Kimura et al., 2017 was based on a dozen early late Miocene teeth from the Nagri Formation of the Potwar Plateau. Specimens are small in size, but similar to other *Progonomys* in their elongate first molars, although the anterolingual margin of the M1 is unusual in lack of a strong inflection (the anterior lobe is not strongly asymmetrical). In the original description, one specimen was recognized from Y797, low in the Nagri Formation. Here we note more material from Y797 and 791, and describe specimens from upper levels of the Chinji Formation, localities Y504, Y809, and Y76, extending the time range back to 11.6 Ma.

Kimura et al. (2017) considered *P. morganae* to be a close relative (congener) of *Progonomys hussaini*, the latter being a basal member of the lineage leading to extant Murini. Either species of *Progonomys* would indicate differentiation of the Murini and its sister, the Arvicanthini. The early record of *P. morganae* would push back this split – or call into question its genus level identification.

Indeterminate Grade: *Progonomys* Schaub, 1938 – *Karnimata* Jacobs, 1978

(cf. *Progonomys hussaini* Cheema et al., 2000 – cf. *Karnimata fejfari* Kimura et al., 2017)

Referred Material: Samples from these localities of ~ 11.2 Ma comprise about 146 isolated teeth of this grade but 12 teeth represent *P. morganae* or unnamed species.

Localities and Ages: Y791 and Y797, both low in the Nagri Formation, Potwar Plateau, Pakistan, 11.23 and 11.18 Ma, respectively, early late Miocene.

Description: The molars display the primitive murine plan of extra lingual cusps forming chevrons in upper molars, cusps in opposition on lower molars, reduction of longitudinal connections, and lower first molars with a double anteroconid. The anteroconids, twinned with the protoconid-metaconid, comprise four cusps that dominate the mesial end of m1, but the four are not united in the X-pattern characteristic of later murines. Molars are low-crowned with distinct cusps; crests connecting them are variable in height. Relative positions of cusps vary across the samples. Root form is primitive, two on lower molars, three on upper molars.

The greatest observed dental variation occurs at the M1 position (Figs. 15.3 and 15.4). Upper first molars vary slightly in size (Table 15.1), isolation of the enterostyle, and

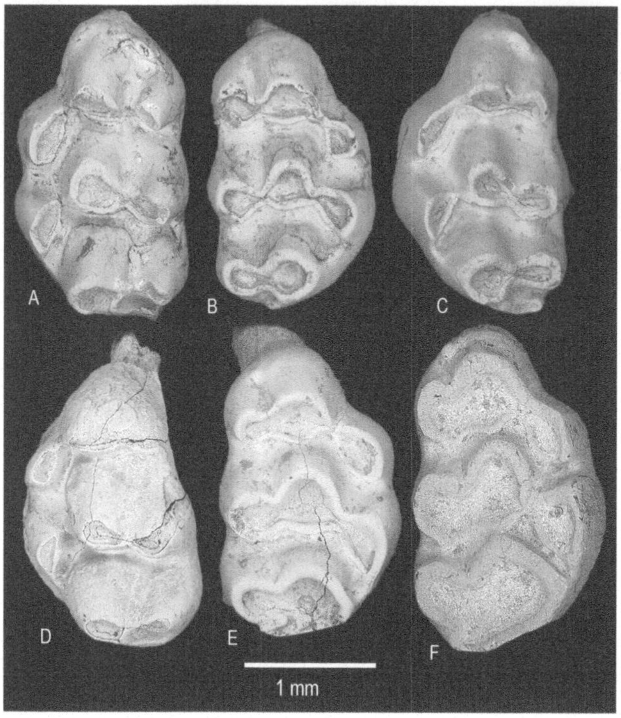

Fig. 15.3 M1 variation: Indeterminate *Progonomys/Karnimata* grade showing *Karnimata* cf. *K. fejfari* morphology. **A.** YGSP 34973 from Y 791; **B.** YGSP 33567 from Y76; **C.** YGSP34975 from Y791; **D.** YGSP 33864 from Y797; **E.** YGSP 33876 from Y797; **F.** YGSP 36824 from Y797

importantly in relative position and shape of the anterostyle. The latter feature, particularly, distinguishes the late Miocene Potwar Plateau lineages of *Progonomys* and *Karnimata*. Whereas *Karnimata* has a rounded anterostyle of anterior location (and therefore close in position to the lingual anterocone), *Progonomys* shows a more posterior, laterally compressed cusp with longer crest joining the lingual anterocone.

Of 26 well-preserved M1 from these two sites, most display *Karnimata* or *Progonomys* morphology. Several, two in particular, have bulbous cusps weakly joined into chevrons. The distinction between the *Karnimata* and *Progonomys* forms is subtle, with intermediate morphologies in evidence. The *Karnimata*-like specimens (Fig. 15.3A–F) tend to be larger and more bunodont (Table 15.1). The labial anterocone is moderately distinct and paired with the lingual anterocone. In most teeth, a precingulum shelf is weakly developed. The anterostyle tends to be anterior in position, close to and connected by a short crest to the lingual anterocone. The enterostyle is weakly connected to the protocone and generally well separated from the hypocone.

The *Progonomys*-like M1 are smaller in average size, with cusps more strongly connected except YGSP 36820, which has an isolated anterostyle (Fig. 15.4A). The labial anterocone is distinctly smaller than the lingual anterocone and positioned close to it (Fig. 15.4F, G). A weak precingulum occurs in a minority of specimens (Fig. 15.4D, H). The anterostyle is posterior to the anterocones, usually laterally compressed, and joined to the lingual anterocone by a narrow, elongated crest. The enterostyle is joined weakly to the protocone and generally shows a low ridge running toward the hypocone.

Lower first molars show similar variation (Fig. 15.5E–H). They are generally of modern grade, meaning opposed, twinned hypoconid-entoconid, protoconid-metaconid, and paired anteroconids. Lingual cusps are slightly anterior to their pair. The double anteroconids are usually well developed and divergent, weakly joined posteriorly to the protoconid-metaconid by a single ridge of variable position. In one m1, the lingual anteroconid is poorly developed. Most specimens show no anterocentral cusp (tma), but have a well developed posterior cingulum. The labial cingulum is moderately developed, with usually a single cuspule located anterior to the hypoconid. Lower first molars, unlike other teeth, show moderate bimodality in size, which might be used at this tooth locus to recognize species (the small specimen, Fig. 15.5H, may represent *Progonomys*).

Differences among second and third molars are so slight that identifications based on size alone would be unreliable. In shape, M2 narrows strongly posteriorly, the metacone closely positioned to the hypocone. A short, thin posterior cingulum is present. The enterostyle is weakly connected, or isolated from, the protocone, and shows a low marginal crest running toward the hypocone. The paracone is large and somewhat posterior to the protocone. The anterostyle is large and the labal anterocone is much smaller but shows wear (Fig. 15.5A–C). Lower m2 is bilophodont and narrows posteriorly. It has a prominent anterolateral crescent (=labial anteroconid). The posterior cingulum is strong. The labial cingulum is well developed and bears one or two cuspules.

Comment: Both *Karnimata* and *Progonomys* M1 phenotypes occur in the 11.2 Ma samples from Y791 and Y797, but apparently as end members of broadly distributed morphological states. Other molar loci show no consistent morphological variation that can be interpreted to associate those teeth with the M1 variants. Although many but not all M1 can be separated based on morphology, the collection as a whole does not display correlated variation, and is not obviously bimodal in size or morphology.

The bulk of the mouse collection from low in the Nagri Formation, about 11.2 Ma, presents similar molars with subtle morphological differences that obscure confident recognition to the species level. Previously (Jacobs and Flynn 2005) we considered the sites Y791 and 797 to include early species of both *Karnimata* and *Progonomys*. This may, indeed, be the case as certain teeth, particularly end members on the spectrum of M1 morphology, appear to represent one or the other genus. Yet, identifying all teeth in

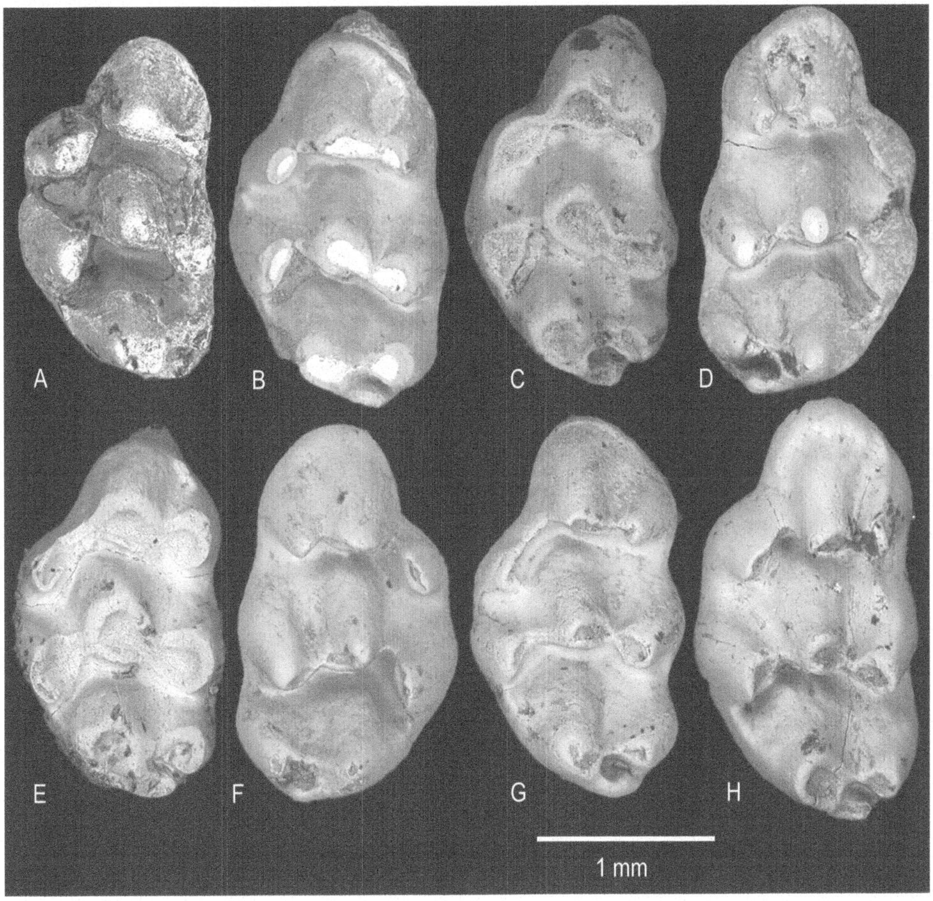

Fig. 15.4 M1 variation: "Indeterminate *Progonomys/Karnimata*" grade showing *Progonomys* cf. *P. hussaini* morphology. **A.** YGSP 36820 fromY797; **B.** YGSP 36826 from Y797; **C.** YGSP 36819 from Y797; **D.** YGSP 36821 from Y797; **E.** YGSP 34965 from Y791; **F.** YGSP 33869 from Y797; **G.** YGSP 33878 from Y797; **H.** YGSP 33588 from Y76

the collection would be challenging because some similar M1 would be classified with question, and molars from other dental positions present very few differences to guide identification. The morphological and size ranges in all features overlap.

Further, these assemblages differ from both *Karnimata fejfari* and *Progonomys hussaini*. Lingual cusps of M1 and M2 are not firmly united in chevrons, anteroconids of m1 generally have a weak posterior connection, the *Karnimata* phenotype typically is not as large as *K. fejfari*, and the anterostyle of *Progonomys* M1 is pinched laterally less frequently than in *P. hussaini*. We consider that these samples do not include *Karnimata fejfari* and *Progonomys hussaini*, but likely record an early phase in the separation of those species. Summarizing, some of the specimens can be recognized to represent the *Karnimata* and *Progonomys* lineages as two species of similar morphology. Alternatively, a single murine is sampled, one that embraces broad morphological variation (see also Kimura et al. 2016). We use as a grade rubric "Indeterminate *Progonomys/Karnimata*" for these samples of ~11.2 Ma.

Unnamed "Genus C", new species at Y797

Material and Age: One M1, YGSP 36822 from Y797, ~11.2 Ma, low Nagri Formation (Fig. 15.2E).

Description: The single specimen attributed to this taxon is distinctive in large size and morphology (Table 15.1); it is relatively high crowned. The M1 is strongly asymmetrical in outline, the double anterocone being extended anteriorly and preceded by a strong anterior cingulum. The anterostyle is compressed diagonally and located well posterior to the anterocones, but connected to them by a thin crest. The enterostyle is also diagonal, well posterior to the protocone, but connected by thin crests to both the protocone and the hypocone. The metacone is opposite and joined to the larger hypocone. The posterior cingulum is prominent. YGSP 36822 displays accessory cuspules between anterostyle and enterostyle, and between paracone and metacone.

Comment: This distinctive molar represents an unnamed murine genus other than *Karnimata* or *Progonomys*. It is the largest (L, W = 2.23, 1.25) murine specimen in the collection,

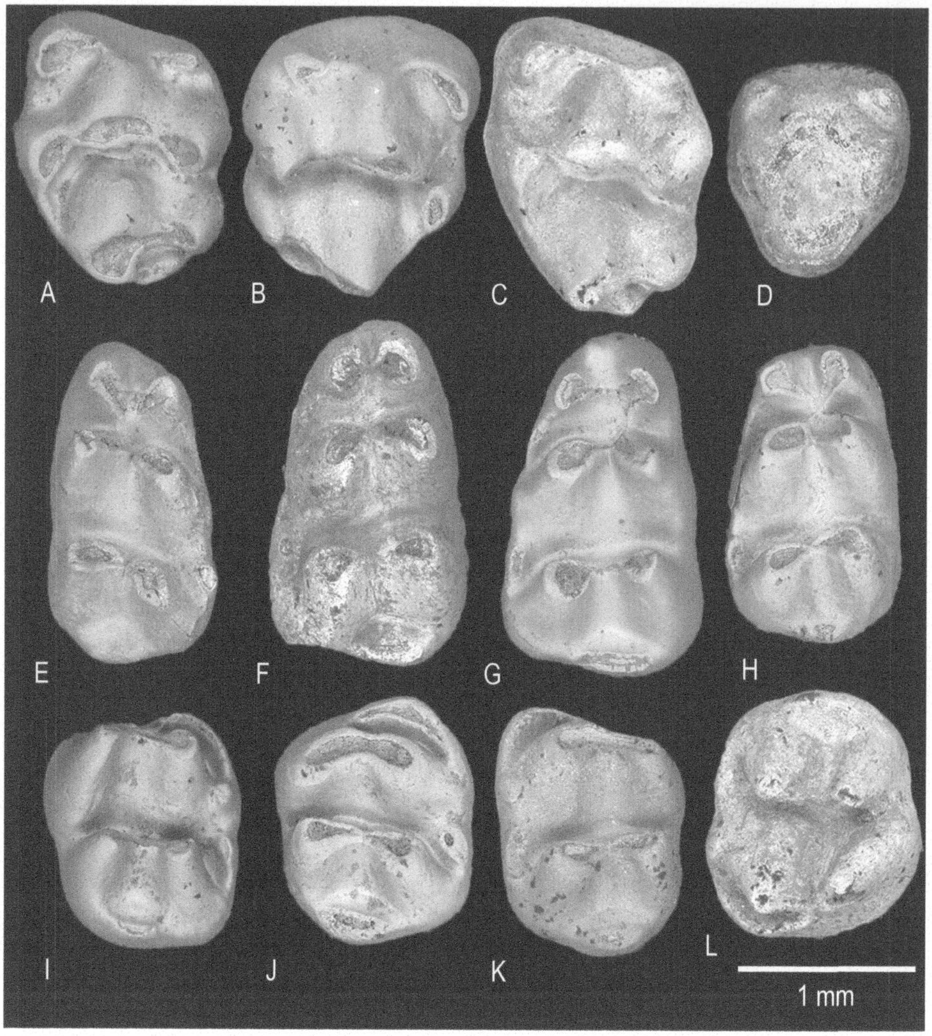

Fig. 15.5 "Indeterminate *Progonomys/Karnimata*" grade molars (*Karnimata* cf. *K. fejfari* or *Progonomys* cf. *P. hussaini*). **A.** M2 YGSP 33845; **B.** M2 36846; **C.** M2 36875; **D.** M3 36885; **E.** m1 36860; **F.** m1 36859; **G.** m1 34971; **H.** m1 34960; **I.** m2 33847; **J.** m2 36871; **K.** m2 36836; **L.** m2 36885. All from locality Y797, except **G** and **H** from Y791; anterior upward, except **L** anterior downward

larger than any cf. *Karnimata* sp. Its length/width ratio is near 1.8. The anterior cingulum and accessory cuspules are unique. The importance of this tooth is that it records murine diversity at 11.2 Ma that was not apparent previously.

Unnamed "Genus D", new species from Y809

Distinctive Features: Large murine with relatively broad molars, M1 exceeding 2.0 mm in length, mean m1 length = 1.8 mm. Anterior chevron on M1 is blunt (short), not attenuated anteriorly, with the anterostyle anterior in position relative to the anterocones.

Distinctive Specimen: YGSP 54088 (Fig. 15.6A), slightly worn right M1 broken posteriorly through the metacone and hypocone, posterior cingulum missing, from Y809, upper Chinji Formation of the Potwar Plateau, early late Miocene, 11.4 Ma.

Additional Material and Measurements: Nineteen additional specimens (Table 15.2) from the upper Chinji and the Nagri formations, sites Y504, Y83, Y809, Y76, Y791, Y797, Y259, Y450, 11.6–10.2 Ma (Y259 and Y450 are in the upper part of the Nagri Formation).

Description: M1 is represented by three specimens, each broken posteriorly. This molar (and M2) is broad, and although broken, its length-width ratio as preserved is less than 1.5. Cusps are strongly inclined and chevrons are widely separated. The anterolingual margin of M1 is slightly inflected so that the molar is not as asymmetrical as in other murines from the same localities. Correspondingly, the anterostyle is shifted only somewhat to the posterior. M1 is particularly broadened at the second chevron, where the paracone and enterostyle bulge widely. The enterostyle is connected firmly to the protocone. The precingulum, weak

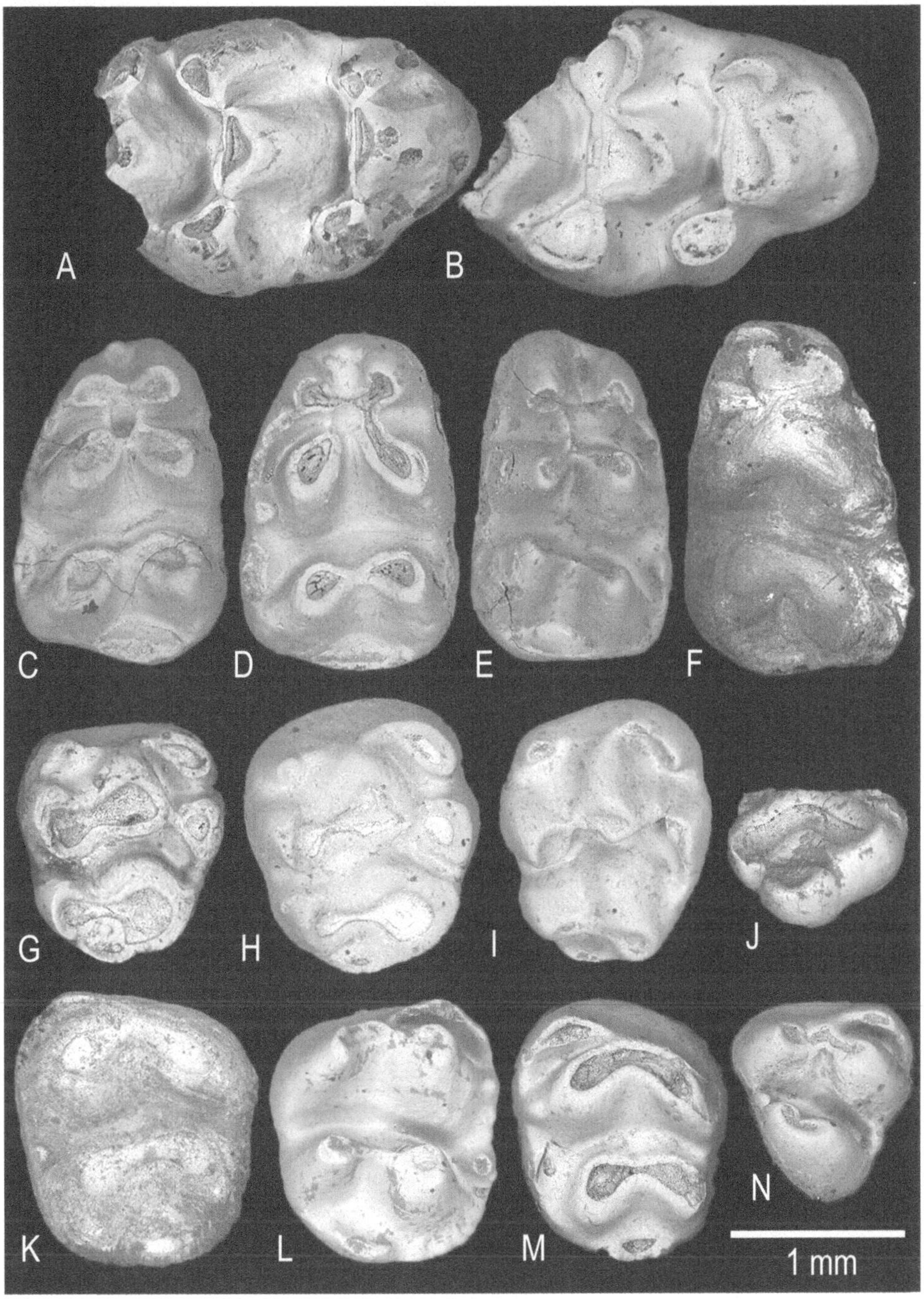

Fig. 15.6 Genus D, new species from Y809. **A.** M1 YGSP 54088 from Y809; **B.** M1 54093 from Y450; **C.** m1 33799 from Y76, **D.** m1 34989 from Y791; **E.** m1 33669 from Y504; **F.** m1 36866 from Y797; **G.** M2 53816 from Y83; **H.** M2 36840 from Y797; **I.** M2 33662 from Y504; **J.** M3 33790 from Y76; **K.** m2 36868 from Y797; **L.** m2 36662 from Y504; **M.** m2 34991 from Y791; **N.** m3 54048 from Y809

in YGSP 54088, is stronger in the larger YGSP 54093 (Fig. 15.6B) which, from site Y450, is more than one million years younger.

M2 is represented by several specimens that recall M1 in their large size, especially width (Fig. 15.6H). M2 has a prominent anterostyle and small labial anterocone. The enterostyle is firmly united with the protocone. The posterior chevron of M2, although one pair of cusps, is broad relative to other murines of the same age. M3 (Fig. 15.6J), assigned to this species on the basis of large size, has both the anterostyle and small labial anterocone.

Lower first molars are large and usually oval in outline, with convex sides. The anteroconids splay widely, with room between them anteriorly, 1 of 4 specimens showing a

Table 15.2 Molar lengths and widths (mm) for robust murine, Genus D, new species at Y809

Locus	Locality	YGSP number	Length	Width
M1	450	54093	2.3+	1.63
M1	259	34003	Broken	~ 1.4
M2	259	34191	1.48	1.45
M2	797	34839	1.51	1.37
m2	797	34991	1.33	1.18
m1	797	36866	1.84	1.14
m1	791	34989	1.77	1.16
M2	797	36832	1.45	1.27
M2	797	36840	1.44	1.28
m2	797	36868	1.47	1.37
m2	76	33544	1.33	1.12
m2	76	33545	1.42	1.29
m3	76	33546	1.02	0.98
M3	76	33790	0.9	0.97
m1	76	33799	1.78	1.21
M1	809	54088	2+	1.36
m3	809	54084	1.02	0.96
M2	83	53816	1.3	1.15
M2	504	33662	1.46	1.21
m1	504	33669	1.79	1.15

small median accessory cusp (tma, Fig. 15.6C). The connection posterior to the anteroconids is variable, but usually stronger between the lingual anteroconid and metaconid. The labial cingulum is moderately developed, strong anterior to the protoconid and always with a c1 cuspule anterior to the hypoconid. The lower second molar is wider than m1 on average. It has a small to crescentic labial anteroconid. The labial cingulum is well developed, and bares a strong accessory cuspule anterior to the hypoconid (Fig. 15.6L). Of the two large m3, one has a crescentic labial anteroconid and a small accessory cuspule between the protoconid and hypoconid (Fig. 15.6N); in the other m3, the labial anteroconid is a low crest, and there is no accessory cuspule.

Comment: This murine indicates further earliest late Miocene diversity in the Potwar murine assemblages. All three mice, Unnamed Genus D, *Progonomys morganae* and Unnamed Genus C, are uncommon, minorities in the early murine assemblages with respect to *Progonomys-Karnimata*. Genus D represents a murine of body size larger than the common murines, with more robust and broad molars, but smaller than Genus C. M1 of Genus D is *Karnimata*-like in anterior position of the anterostyle, and m1 has widely spaced anteroconids and a developed labial cingulum. Specimen 54093 from locality Y450 is an outlier in both younger age (10.2 Ma) and larger size.

Indeterminate Grade: More plesiomorphic than *Progonomys*.

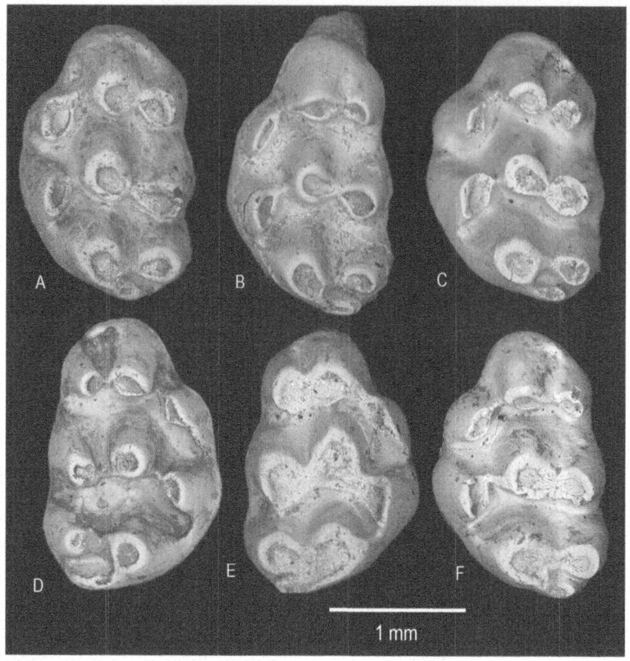

Fig. 15.7 M1 variation in "Pre-*Progonomys*" grade murines: variation of locality Y76. **A.**YGSP 33571, **B.** 33591, **C.** 33572, **D.** 33580, **E.** 33584, **F.** 33586

Referred Material: Of 222 isolated teeth from these localities dated to 11.6–11.4 Ma, by far the majority (204) represents a primitive murine that is less derived than *Progonomys* or *Karnimata*. All molar positions are represented in the material (Table 15.1, Figs. 15.7 and 15.8).

Localities and ages: Y76 and Y809 (11.4 Ma) and Y83 (11.6 Ma) and Y 504 (11.64 Ma), all high in the Chinji Formation, near in age to the middle-late Miocene boundary, Potwar Plateau, Pakistan.

Description: Although "murine" in form, the structural grade is more plesiomorphic than that of *Progonomys* or *Karnimata*. Cusps are bunodont, not strongly inclined, and the upper molars have weakly connected lingual cusps. M1 is oval in outline, broad at the second chevron, and blunt anteriorly with unextended lingual anterocone and a minor anterolingual indentation. Specimens are smaller than later *Karnimata*, overlapping in size with *Progonomys* cf. *P. hussaini*. The length-width ratio of M1 is low, about 1.6 (see Table 15.1) and the m1 has a comparable ratio (but see slender YGSP 33647, Fig. 15.8C). Cusps of the double anterocone, the protocone-paracone pair, and the hypocone-metacone pair are positioned closely, as if the lingual cusps were casually attached to the M1 crown. The anterostyle of M1 is weakly connected to the lingual anterocone (Fig. 15.7A) and pinched on some specimens. The enterostyle is weakly joined to the protocone (or unconnected until advanced wear). The enterostyle is isolated from the hypocone. The posterior cingulum is short and blunt.

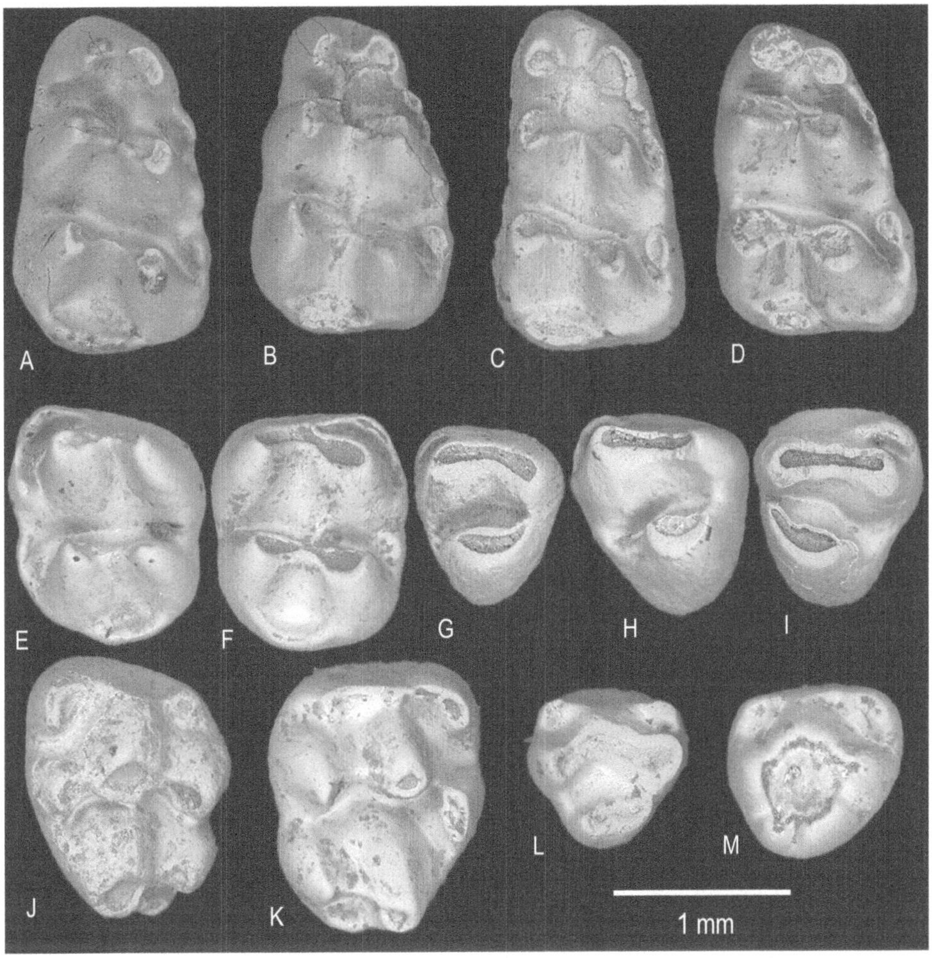

Fig. 15.8 Molars of "Pre-*Progonomys*" grade. (**A–D**, m1): **A.** YGSP 54087 from Y809, **B.** 54051 from Y809, **C.** 33647 from, Y504, **D.** 33777 from Y76; (**E–F**, m2): **E.** 33654 from Y504, **F.** 33561 from Y76; (**G–I**, m3): **G.** 54043 from Y 809, **H.** 54044 from Y809, **I.** 54047 from Y809; (**J–K**, M2): **J.** 33664 from Y504; **K.** 33755 from Y76; (**L–M**, M3): **L.** 33788 from Y76, **M.** 33786 from Y76

M2 narrows posteriorly (Fig. 15.8J, K). Its anterostyle is modest, with a weak crest running toward the anterior wall of the protocone. The labial anterocone is small. The enterostyle is weakly joined to the protocone, isolated from the hypocone. M3 (Fig. 15.8L, M) has one chevron and a single posterior cusp. Both the anterostyle and labial anterocone are small.

Some variation in m1 is noted above. Generally oval in outline, one specimen (Fig. 15.8C) is relatively longer, with straighter sides. Lingual cusps of m1 are somewhat anterior to labial cusps. The twin anteroconids are generally well separated; no tma is evident. The anteroconids have a variable connection to the following metaconid-protoconid pair of cusps, two of four (Fig. 15.8A, D) showing almost no connection and the other two with a thin ridge running toward the labal side of the metaconid. The well-developed labial cingulum bears multiple cuspules. The posterior cingulum is an isolated oval.

Four cusps dominate m2 in a bilophodont pattern. The tooth is slightly narrower posteriorly than anteriorly. The posterior cingulum is a low shelf; the labial cingulum is variable in strength and presence of cuspules. The labial anteroconid is a thin crest. The m3 is also bilophodont, but cusps in the first loph are indistinct, and the second loph is one paisley-shaped cusp. There is variation in size and strength of the crescentic labial anteroconid.

Comment: The great majority of murine specimens from upper levels of the Chinji Formation considered here appear to represent a single species displaying considerable variation. Over 200 specimens include molars from all dental positions. Eighteen other teeth from these localities are assigned above as *Progonomys morganae* or Genus D. A few specimens (like Fig. 15.7A) present some (but not all) characters reminiscent of *Progonomys* or *Karnimata*, but they are clearly less derived overall than any other late Miocene genus. Cusps are arranged in chevrons, but are weakly connected, molars have low length-width ratios, and the cusps are blunt and not strongly inclined. Previously we included these samples too simplistically as *Progonomys hussaini* (e.g., Jacobs and Flynn 2005), or as a variation considered a bulbous "new genus" in Kimura et al. (2016, 2017). This merely obscured the reality that samples from Chinji Formation localities Y504, 83, 76, 809 represent a grade of murine evolution more primitive than *Progonomys* or *Karnimata* or other early genera, such as *Parapodemus*. This stem murine at 11.6–11.4 Ma represents a grade of development preceding the early Indo-Pakistan radiation of modern murines (Kimura et al. 2017) that was well established by 10.5 Ma, one million years later. Informally we use the term "*Pre-Progonomys* spp." for specimens of this murine grade.

Great Morphological Variation of Early Murines

In the early middle Miocene of the Indian Subcontinent the characteristic Siwalik microfaunal composition was established, including dominance of murine rodents. Stem Murinae *Potwarmus* and *Antemus* (16–13 Ma) were abundant in small mammal assemblages of the lower part of the Chinji Formation. Subsequent faunas were dominated by murines as well, those between 11 and 8 Ma comprising two common lineages, the *Progonomys* and the *Karnimata* groups (Jacobs 1978; Kimura et al. 2017). Successive members of these genera document evolutionary trends (Kimura et al. 2013, 2015) endemic to the Indian Subcontinent. At least two other murines co-occurred during any time interval with the common species, but in low abundance.

Between 12 and 11 Ma the pattern was similar: a dominant murine accompanied by at least two uncommon species. Sites low in the Nagri Formation, 11.2 Ma, record dominant "Indeterminate *Progonomys/Karnimata*" grade (although some M1 variants can be considered to evidence the roots of *Progonomys* cf. *P. hussaini* or of *Karnimata* cf. *K. fejfari*), and the uncommon *P. morganae*, Genus C, and Genus D. Older sites from upper levels of the Chinji Formation, 11.6–11.4 Ma preserve an abundant "Pre-*Progonomys*" murine stock (instead of *Progonomys* or *Karnimata*) along side uncommon mouse species.

The large murine samples of 11.2 Ma in the Potwar Siwaliks are taken to represent either distinct species, some individuals of which are so similar in morphology that they cannot be distinguished with confidence, or a murine stock containing individuals displaying variously *Progonomys*-like or *Karnimata*-like morphology, from which those generic lineages evolved. The Y791-797 assemblages ($\sim$11.2 Ma) may record the initial differentiation of *Karnimata* and *Progonomys* soon after their split and before strong morphological differences were established (see Kimura et al. [2016] for quantitative analysis) or some time before their split when morphological variants characterized the parent of the daughter species. This implies a process of speciation measured on the time scale of hundreds of thousands of years.

The density through time of the murine fossil record is remarkable and affords a mode of research on terrestrial vertebrates that is rarely achieved. The well dated samples distinguished on the scale of 10^5 years or better can be used to date cladogenic events, which bear on the pattern of speciation in murines. Coupled with the level of endemism in faunas of the Potwar Plateau that is a result of the physical barriers at the margins of the subcontinent, the evolutionary events seen in Siwalik murine samples represent *in situ* microevolution of lineages leading to extant species. Kimura et al. (2015, 2016, 2017) place Potwar fossil rodents into lineages leading to extant clusters of species, generally crown tribes.

Early Murine Evolution in the Semi-isolated Cradle of the Indian Subcontinent

Our study (Kimura et al. 2017) specifically discussed the early paleontological records of three tribes, and specified a calibration point for Apodemini, including the oldest stratigraphic occurrence of *Parapodemus* at 10.5 Ma. That fossil record can be related to a timetree generated by calculated rates of molecular evolution for an array of murids (Fig. 15.9). Lecompte et al. (2008) built a comprehensive phylogeny based on mitochondrial and nuclear genes. Tribal relationships of their clades are stable and duplicated in other analyses such as Steppan and Schenk (2017), but revised by Aghová et al. (2018). The Lecompte et al. (2008) timetree of successive splitting events, calibrated at the base of the Murinae by the fossil age of *Antemus*, agrees with the Potwar succession of fossil lineages (Fig. 15.9). The early records of Murini, Arvicanthini, and Apodemini at 10.1–10.5 Ma fall very near the node dates calculated by Lecompte et al. (2008) and Aghová et al. (2018). Note that localities Y791 and Y797 at 11.2 Ma already present the

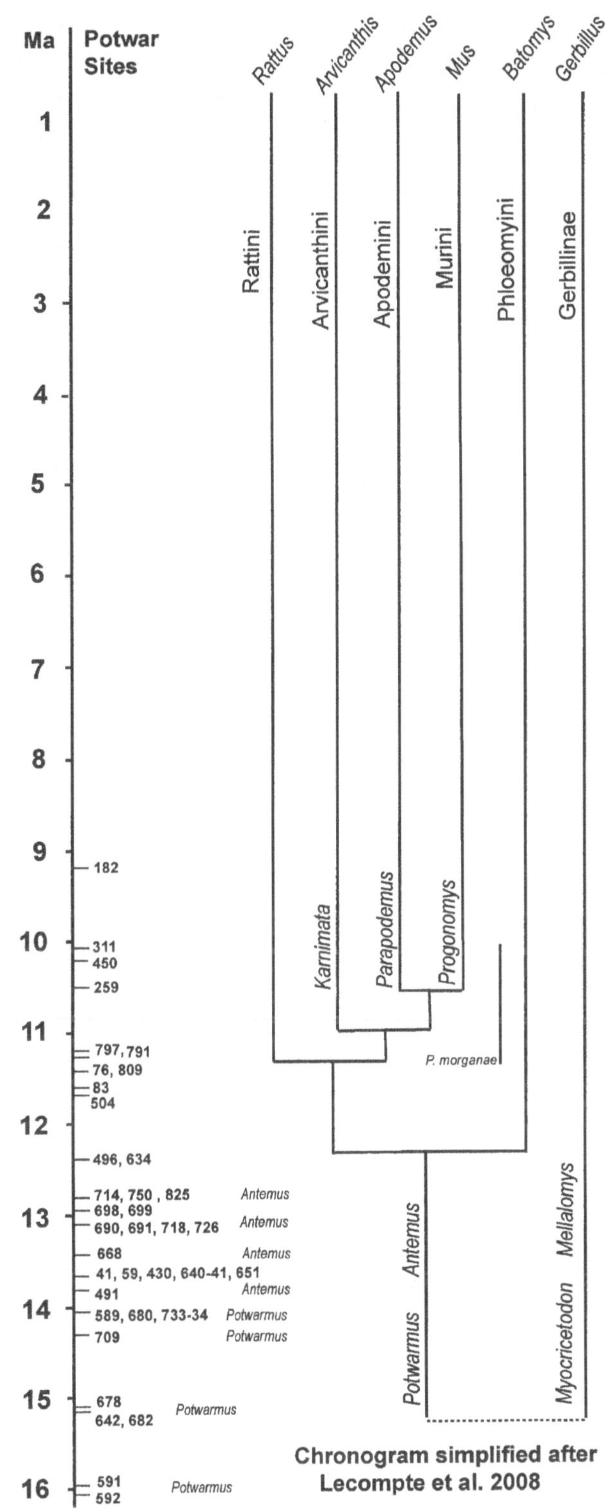

◄ **Fig. 15.9** Dendrogram of relationships of selected murine tribes (Subfamily Murinae) and Subfamily Gerbillinae based on molecular analyses (Lecompte et al. 2008; Steppan and Schenk 2017; Aghová et al. 2018). Here a simplified branching pattern of Lecompte et al. (2008) is drawn to maintain relationships proportional to relative molecular divergence. The dendrogram is calibrated in time, given an approximate date for emergence of crown Murinae based on the appearance by 14 Ma of Siwalik *Antemus*. The timetree places the differentiation of *Karnimata* and *Progonomys-Parapodemus* by 11 Ma. Fossils from our sequence of localities (adjacent to the left of the time line) confirm this and suggest that the differentiation of these groups started somewhat before the time predicted. Younger than 10 Ma (e.g., Y182 at 9.4 Ma) Siwalik localities yield derived species of *Progonomys, Karnimata, Parapodemus* and other murines. Potwar sites dating to 16–13 Ma contain stem gerbils *Myocricetodon* and *Mellalomys*, and the successive stem murine genera *Potwarmus* and *Antemus*. Fossil levels that yield the stem murines *Potwarmus* or *Antemus* are specified in small type on the left. *Antemus* persists in Siwalik sites to about 13 Ma. The younger localities Y496 and Y634 yield murines that are derived with respect to *Antemus* but primitive with respect to *Progonomys*. "Pre-*Progonomys*" morphology is observed at still younger sites dating to 11.6–11.4 Ma. The range of *Progonomys morganae*, an early record for the genus, is indicated separately

roots of Arvicanthini and Murini+Apodemini in the form of *Karnimata* cf. *K. fejfari* and *Progonomys* cf. *P. hussaini*. Also (Fig. 15.9), the Murini and Apodemini are more closely related than either is to Arvicanthini; as such, an early *Progonomys*-like mouse (>11 Ma) would represent a stem species for those combined tribes. Some dental elements older still at Y76, 11.4 Ma, arguably show that phenotypes later characteristic of distinct *Karnimata* and *Progonomys* lineages were already then present in stem Murinae. The fossil record suggests that the divergence of these groups is older than the Lecompte et al. (2008) estimate, in agreement with Aghová et al. (2018).

Assuming that both proportionality of the murine tree based on molecular data and its calibration, based largely on the fossil appearance of *Potwarmus* and *Antemus*, are approximately correct, the combined dataset suggests that the *Mus-Rattus* divergence would have occurred by 12 Ma (see Fabre et al. 2013; Steppan and Schenk 2017). This estimate is in general agreement with the estimates of Jacobs and Downs (1994) and Jacobs and Flynn (2005) – although the time scales used by those authors have been supplanted by that of Hilgen et al. (2012) – but now our interpretation is based on much richer and more resolved fossil data. The Lecompte et al. (2008) timetree also projects divergence of basal crown Murinae (represented by *Batomys*) at 12–13 Ma. We interpret this estimate to suggest dispersion of stem murines by this time to Southeast Asia, where Phloemyini occur. Finally, murines and gerbils are considered to have diverged by 16 million years ago. Paleontological evidence argues that primitive gerbils, represented by *Myocricetodon*, date to deep in the early Miocene, well before 16 Ma (de Bruijn et al. 1981; Wessels 1996; Theocaropoulos 2000; Lindsay and Flynn 2015).

The Lecompte et al. (2008) murine dendrogram uses the appearance of *Antemus* at 14 Ma as a calibration point. The effect of an older calibration point, e.g., the appearance of the older stem genus *Potwarmus* (18–16 Ma) would be to extend all nodes proportionally in time. For example, the *Mus-Rattus* split may project to ~13 Ma.

Interpretation

Potwar fossil murine assemblages of around 9.2 Ma (e.g., locality Y182, Fig. 15.9) were dominated by two common mice, *Progonomys debruijni* and *Karnimata darwini* (Jacobs 1978). These were distinct morphologically, the former being smaller, and were accompanied by the less common species *Parapodemus hariensis* (also see Patnaik 2020) and a large unnamed species of *Karnimata*. Older assemblages dated to 10.1–10.5 Ma had different species recorded in abundance: *Progonomys hussaini* and *Karnimata fejfari*. These were less distinctive morphologically and closer in size than their younger congeners; while still abundant, they were recovered with uncommon *Parapodemus badgleyae* and *Progonomys morganae*.

Yet older assemblages at 11.2 Ma also show this pattern: *Progonomys* and *Karnimata* comprising a rich collection of fossils accompanied by *P. morganae* and large forms (Genus C, Genus D). The *Progonomys* and *Karnimata* morphs resemble their successors, but are still less distinct than are the species of 10.5–10.1 Ma. They are close in size to each other, and many individuals lack the derived features of either, so that many teeth cannot be identified with certainty. This array of murine fossils conceivably represents a mix of the two lineages – or one interbreeding population with variation that later became characteristic of the two successor lineages. Samples of this age capture murines at a time in their history either when they are diverging or just after their split and before definitive morphologies became characteristic of the distinct genera.

The Potwar murine record presents a rare case in paleontology that offers a distant view of a speciation event – the dynamics of a lineage split. We perceive a long process of phenotypic divergence among mice, one that conceivably involved sympatric speciation and hybridization admixture (see Arnold et al. 2015). Possibly the ultimate differentiation of these murine tribes was tied to partitioning of resources by different stem lineages, which Kimura et al. (2013) argue was reflected in tooth morphology.

The argument presented here begins with a premise: the isolated nature of the Indian Subcontinent that promoted endemic evolution of various terrestrial vertebrates in the Mesozoic and early Cenozoic is reflected and continues in Miocene age Siwalik mammals. Docking of the Indian Plate with Asia built a subcontinent isolated by high mountain ranges to the north, large rivers and subtropical habitat on the east, and rough desertic terrain on the west. Fossil sites of the Potwar Plateau repeatedly sample evolving muroid lineages characteristic of the Siwalik biogeographic subregion. The unique setting plus a wealth of well-dated fossil assemblages capture a significant speciation event – the splitting of extant tribes Murini and Arvicanthini.

We argue that this event was completed before the 10.5 Ma estimate that was previously apparent (Kimura et al. 2016, 2017), and was probably under way before 11.2 Ma, when early representatives of the genera *Progonomys* and *Karnimata* (our indeterminate grade *Progonomys-Karnimata*) are seen. Some yet older specimens (a minority at 11.4 Ma) show *Progonomys* or *Karnimata*-like morphology, which we suggest reflects variation in a common ancestor – early morphological evidence of genomic divergence that later became characteristic of successor genera (see Kimura et al. 2016). It is, of course, possible that the 11.4 Ma assemblages include cryptic (hard to recognize) species of *Progonomys* and *Karnimata*.

The rich Potwar micromammal record sheds light on some questions and raises more. It illustrates the antiquity of tribe level rodent origins, but forces review of recognition of those events as reflected in dental morphology. It cannot yet diagnose stem taxa, such as the last common ancestor of *Progonomys* and *Parapodemus* (Fig. 15.9). It does show occurrence of daughter phenotypes in assemblages preceding distinct samples, and reframes systematics of basal lineages like *Progonomys morganae*.

We see the 11.6–11.2 Ma Potwar micromammals as capturing, fortuitously, the split between early Murini (+Apodemini) and Arvicanthini. These samples catch the process of splitting speciation at the time that these lineages diverged in the cradle of the Indian Subcontinent. They suggest that the process of genomic divergence and subsequent population splitting in mammals is a process measured on the scale of hundreds of thousands of years – a million in some cases.

Acknowledgments We are pleased to contribute to this volume in honor of Professor Ashok Sahni. Professor Sahni has encouraged diverse lines of research throughout his career, all the while promoting international collaborations. We much appreciate his example in fostering scientific advancement. Some of the ideas developed here took root when our friend I. U. Cheema discovered the hypodigm of *Progonomys hussaini*. We acknowledge the help of all field collaborators of the Geological Survey of Pakistan and the Pakistan Museum of Natural History, and from various institutions in the United States, especially our small mammal colleague, Will Downs, whose efforts made this study possible. Roy Beavers of Southern Methodist

University skillfully produced the SEM images for this paper. Evaluations by David Pilbeam and two reviewers stimulated improvement of the study, and we thank the editors of this volume for their help and encouragement. Financial support was provided by JSPS KAKENHI Grant Number JP15H06884 (Grant-in-Aid for Young Scientists Start-up) and JSPS KAKENHI Grant Number 18K13650. NSF grant EAR 0958178 to D. Pilbeam, C. Badgley and T. Cerling supported a portion of the research.

References

Aghová, T., Kimura, Y., Bryja, J., Dobigny, G., Granjon, L. & Kergoat, G. J. (2018). Fossils know it best: Using a new set of fossil calibrations to improve the temporal phylogenetic framework of murid rodents (Rodentia: Muridae). *Molecular Phylogenetics and Evolution, 128*, 98–111.

Arnold, M. L., Brothers, A. N., Hamlin, J. A. P., Taylor, S. J., & Martin, N. H. (2015). Divergence-with gene-flow—What humans and other mammals got up to. In N. Gauntier (Ed.), *Reticulate evolution* (pp. 255–295). Interdisciplinary Evolution Research 3. Cham: Springer.

Barry, J. C., Behrensmeyer, A. K., Badgley, C. E., Flynn, L. J., Peltonen, H., Cheema, I. U., et al. (2013). The Neogene Siwaliks of the Potwar Plateau, Pakistan. In X. Wang, L. J. Flynn, & M. Fortelius (Eds.), *Fossil mammals of Asia: Neogene biostratigraphy and chronology* (pp. 373–398). New York: Columbia University Press.

Barry, J. C., Morgan, M. E., Flynn, L. J., Pilbeam, D., Behrensmeyer, A. K., Raza, S. M., et al. (2002). Faunal and environmental change in the Late Miocene Siwaliks of northern Pakistan. *Paleobiology, 28*, 1–71.

Chatterjee, S., Scotese, C. R., & Bajpai, S. (2017). The restless Indian Plate and its epic voyage from Gondwana to Asia: Its tectonic, paleoclimatic, and paleobiogeographic evolution. *Geological Society of America Special Paper, 529*, 1–147.

Cheema, I. U., Raza, S. M., Flynn, L. J., Rajpar, A. R., & Tomida, Y. (2000). Miocene small mammals from Jalalpur, Pakistan, and their biochronologic implications. *Bulletin National Science Museum, Tokyo, C26*, 57–77.

de Bruijn, H., Hussain, S. Taseer, & Leinders, J. J. M. (1981). Fossil rodents from the Murree Formation near Banda Daud Shah, Kohat, Pakistan. *Proceedings of the Koninklijke Nederlandse Akademie van Wetenschappen, B84*, 71–99.

Fabre, P.-H., Pagès, M., Musser, G. G., Fitriana, Y. S., Fjeldså, J., Jennings, A., et al. (2013). A new genus of rodent from Wallacea (Rodentia: Muridae: Murinae: Rattini), and its implication for biogeography and Indo-Pacific Rattini systematics. *Zoological Journal of the Linnaean Society, 169*, 408–447.

Flynn, L. J., Lindsay, E. H., Pilbeam, D., Raza, S. M., Morgan, M. E., Barry, J. C., et al. (2013). The Siwaliks and Neogene evolutionary biology in South Asia. In X. Wang, L. J. Flynn, & M. Fortelius (Eds.), *Fossil mammals of Asia: Neogene biostratigraphy and chronology* (pp. 353–372). New York: Columbia University Press.

Hilgen, F. J., Lourens, L. J., & van Dam, J. A. (2012). The Neogene period. In F. M. Gradstein, J. G. Ogg, M. S. Schmitz, & G. M. Ogg (Eds.), *The geologic time scale 2012* (Vol. 2, pp. 923–978). Amsterdam: Elsevier.

Jacobs, L. L. (1977). A new genus of murid rodent from the Miocene of Pakistan and comments on the origin of the Muridae. *PaleoBios, 25*, 1–11.

Jacobs, L. L. (1978). Fossil Rodents (Rhizomyidae and Muridae) from Neogene Siwalik deposits, Pakistan. *Museum of Northern Arizona Press, Bulletin, 52*, 1–103.

Jacobs, L. L., & Downs, W. R. (1994). The evolution of murine rodents in Asia. In Y. Tomida, C.-K. Li & T. Setoguhi (Eds.), *Rodent and lagomorph families of Asian origins and diversification* (pp. 149–156). Tokyo: National Science Museum Monograph 8.

Jacobs, L. L., & Flynn, L. J. (2005). Of mice ... again: The Siwalik rodent record, murine distribution, and molecular clocks. In D. E. Lieberman, R. J. Smith, & J. Kelley (Eds.), *Interpreting the past: Essays on human, primate, and mammal evolution* (pp. 63–80). Boston: Brill Academic Publishers Inc.

Jacobs, L. L., Strganac, C., & Scotese, C. (2011). Plate motions, Gondwana dinosaurs, Noah's arks, beached Viking funeral ships, ghost ships, and landspans. *Anais da Academia Brasileira de Ciencias, 83*(1), 3–22.

Kimura, Y., Flynn, L. J., & Jacobs, L. L. (2016). A paleontological case study for species delimitation in diverging fossil lineages. *Historical Biology, 28*(1–2), 189–198.

Kimura, Y., Flynn, L. J., & Jacobs, L. L. (2017). Early Late Miocene murine rodents from the upper part of the Nagri Formation, Siwalik Group, Pakistan, with a new fossil calibration point for the Tribe Apodemurini (*Apodemus/Tokudaia*). *Fossil Imprint, 73*(1–2), 197–212.

Kimura, Y., Hawkins, M. T. R., McDonough, M. M., Jacobs, L. L., & Flynn, L. J. (2015). Corrected placement of *Mus-Rattus* fossil calibration forces precision in the molecular tree of rodents. *Scientific Reports, 5*, 14444, 9 p. https://doi.org/10.1038/srep14444.

Kimura, Y., Jacobs, L. L., & Flynn, L. J. (2013). Lineage-specific responses of tooth shape in murine rodents (Murinae, Rodentia) to Late Miocene dietary change in the Siwaliks of Pakistan. *PLoS ONE, 8*(10), e76070. https://doi.org/10.1371/journal.pone.0076070.

Lecompte, E., Aplin, K., Denys, C., Catzeflis, F., Chades, M., & Chevret, P. (2008). Phylogeny and biogeography of African Murinae based on mitochondorial and nuclear gene sequences, with a new tribal classification of the subfamily. *BMC Evolutionary Biology, 8, 199*. https://doi.org/10.1186/1471-2148-8-199. Retrieved from http://www.biomedcentral.com/1471-2148/8/199.

Lindsay, E. H. (1988). Cricetid rodents from Siwalik deposits near Chinji Village Part I: Megacricetodontinae, Myocricetodontinae, and Dendromurinae. *Palaeovertebrata, 18*, 95–154.

Lindsay, E. H. (2017). *Democricetodon fejfari* sp. nov. and replacement of Cricetidae by Muridae in Siwalik deposits of Pakistan. *Fossil Imprint 73(3–4)*, 445–453.

Lindsay, E. H., & Flynn, L. J. (2015). Late Oligocene and Early Miocene Muroidea of the Zinda Pir Dome. *Historical Biology, 28*, 215–236.

López-Antoñanzas, R., Renaud, S., Pelaez-Campomanes, P., Azar, D., Kachacha, G., & Knoll, F. (2019). First Levantine fossil murines shed new light on the earliest intercontinental dispersal of mice. *Scientific Reports, 9*, 11874. https://doi.org/10.1038/s41598-019-47894-y.

Nanda, A. C. (2015). *Siwalik mammalian faunas of the Himalayan Foothills*. Dehradun: Wadia Institute of Himalayan Geology.

Patnaik, R. (2013). Indian Neogene Siwalik mammalian biostratigraphy: An overview. In X. Wang, L. J. Flynn, & M. Fortelius (Eds.), *Fossil mammals of Asia: Neogene biostratigraphy and chronology* (pp. 423–443). New York: Columbia University Press.

Patnaik, R. (2020). New data on the Siwalik murines, rhizomyines and ctenodactylines (Rodentia) from the Indian subcontinent. In G. V. R. Prasad & R. Patnaik (Eds.), *Biological consequences of plate tectonics: New perspectives on post-Gondwana break-up* (pp. 363–391). Cham: Springer.

Schaub, S. (1938). Tertiäre und Quartäre Murinae. *Abhandlungen des Schweizerischen Paläontologische Gesellschaft, 61*, 1–38.

Sen, S. (2003). Muridae and Gerbillidae (Rodentia). In M. Fortelius, J. Kappelman, S. Sen, & R. L. Bernor (Eds.), *Geology and*

Paleontology of the Miocene Sinap Formation, Turkey (pp. 125–140). New York: Columbia University Press.
Steppan, S. J. (2017). Tree of Life Web Project. Retrieved from http://tolweb.org/tree?group=Murinae&contgroup=Muroidea.
Steppan, S. J., & Schenk, J. J. (2017). Muroid rodent phylogenetics: 900-species tree reveals increasing diversification rates. *PLoS ONE, 12*(8), e0183070.
Theocaropoulos, C. D. (2000). Late Oligocene – Middle Miocene *Democricetodon*, *Spanocricetodon*, and *Karydomys* n. gen. from the eastern Mediterranean area. *Gaia, 8,* 1–103.
van Dam, J. A., Alcalá, L., Alonso Zara, A., Calvo, J. A., Garcés, M., & Krijgsman, W. (2001). The upper Miocene mammal record from the Teruel-Alfambra región (Spain). The MN system and continental stage/age concepts discussed. *Journal of Vertebrate Paleontology, 21,* 367–385.
Wessels, W. (1996). Myocricetodontinae from the Miocene of Pakistan. *Proceedings Koninklijke Nederlandse Akademie van Wetenschappen, 99,* 253–312.
Wessels, W., de Bruijn, H., de Hussain, S. T., & Leinders, J. J. M. (1982). Fossil rodents from the Chinji Formation, Banda Daud Shah, Kohat, Pakistan. *Proceedings Koninklijke Nederlandse Akademie van Wetenschappen, B85,* 337–364.

Chapter 16
New Data on the Siwalik Murines, Rhizomyines and Ctenodactylines (Rodentia) from the Indian Subcontinent

Rajeev Patnaik

Abstract In the light of new discoveries, previously known Siwalik murid, rhizomyid and ctenodactylid rodents are revised. Murines that are described and discussed here include *Progonomys debruijni, Karnimata darwini* and *Parapodemus hariensis* from the Late Miocene of Haritalyangar and *Mus* cf. *M. pahari* and *Hadromys* sp. from the Pliocene and Early Pleistocene localities near Chandigarh, respectively. Additional spalacids cf. *Kanisamys* sp., *Miorhizomys harii* and a ctenodactylid *Sayimys sivalensis* have also been described from the Miocene ape locality of Haritalyangar, Himachal Pradesh. A PAUP based phylogenetic analysis of cranial characters of *Progonomys debruijni* and various species belonging to the genus *Mus* reveal that the subgenus *Mus* diverged first, followed by *Pyromys*, whereas, *Coelomys* and *Nannomys* were last to separate. A phylogenetic analysis of the spalacids in the matrix places cf. *Kanisamys* sp. as a basal clade comprising the Siwalik *Protachyoryctes tatroti, Eicoryctes kaulialensis* and the African tachyoryctines. The murids (*Progonomys, Karnimata, Parapodemus, Mus* and *Hadromys*), rhizomyines and ctenodactylids provide an excellent example of range expansion-contraction and long distance dispersals in the Miocene and Plio-Pleistocene.

R. Patnaik (✉)
Centre for Advanced Study in Geology, Panjab University, Chandigarh, 160014, India
e-mail: rajeevpatnaik@gmail.com

Keywords Siwalik • Murinae • Rodentia • Phylogeny • Taxonomy • Biogeography

Introduction

Professor Ashok Sahni was among the earliest investigators of fossil rodents from the Indian sub-Himalayas (Sahni and Khare 1973; Sahni and Srivastava 1976; Sahni et al. 1980; Nanda and Sahni 1998) and is still actively involved in recent rodent finds from lignite mines of Vastan, Gujarat, India (Sahni et al. 2006; Rana et al. 2008). His bench mark papers on enamel microstructure of rodents (Sahni 1980, 1985) such as murids (rats and mice), spalacids (the bamboo rats) and ctenodactylids (gundis) are widely cited. As a tribute to his excellent contribution and keeping the legacy alive, the present paper describes some new murid, spalacid and ctenodactylid rodents from the famous Miocene ape site of Haritalyangar and from Pliocene deposits exposed near Chandigarh, India. Some murine rodents described earlier by Vasishat (1985) are revised. The paper also discusses phylogeny and paleobiogeography of these rodents in the light of emerging hypotheses.

Previous Work

Muridae

The exact place of the origin of the family Muridae, whether in the Afrotropical or in the Indomalayan region is not resolved yet (Aghová et al. 2018). However, there is clear indication that the most diverse mammalian subfamily today, the Murinae with 584 species and 129 genera (Musser and Carleton 2005) had its origin in the Indomalayan region (Aghová et al. 2018; Kimura et al. 2015; Patnaik 2014). The

G. V. R. Prasad and R. Patnaik (eds.), *Biological Consequences of Plate Tectonics: New Perspectives on Post-Gondwana Break-up—A Tribute to Ashok Sahni*,
Vertebrate Paleobiology and Paleoanthropology Series, https://doi.org/10.1007/978-3-030-49753-8_16

deep rooted fossil record of murines in Indomalaya supports this view (Kimura et al. 2015; Patnaik 2014; Chaimanee 1998). In the Indian subcontinent the Neogene and Quaternary Siwalik deposits house a continuous record of murine dental remains, the oldest being *Antemus* that appears around 13.8 Ma in the Chinji Formation (Jacobs 1978; Jacobs and Flynn 2005), followed by the appearance of cf. *Progonomys*, at ~12.2 Ma (Jacobs and Flynn 2005). *Karnimata* lineage started to separate from the *Progonomys* one by 11.2 Ma (Kimura et al. 2016). At ~10.5 Ma *Progonomys hussaini, P. morganae, Karnimata fejfari* and *Parapodemus badgleyae* appear (Kimura et al. 2017). By ~9.2 Ma *Progonomys debruijni, Karnimata darwini, Parapodemus hariensis* and a large *Karnimata* emerge (Kimura et al. 2017). From the famous Miocene ape locality, Haritalyangar, India, Vasishat (1985) reported a partial cranium, which he assigned to a new species *Progonomys choprai.* Besides describing another new species, *Parapodemus hariensis*, he erected a new genus and a species *Siwalkomus nagrii* based on complete upper and lower dentitions. *Parapodemus* has also been reported from Middle Siwalik sediments exposed near Mohand, India (Tiwari and Bhandari 2014) and Khaur, Pakistan (Jacobs 1978). By ~8.1 Ma *Parapelomys* evolves on one hand from the *Karnimata* lineage and *Mus* emerges from the *Progonomys* lineage on the other (Jacobs and Flynn 2005). Recently this divergence has been revised (Kimura et al. 2017). The Indian fossils of *Mus* are known to occur in ~3.5–3 Ma deposits (Patnaik et al. 2017). Also between ~3.5 and ~2.5 Ma modern murine taxa such as *Golunda, Cremnomys* and *Millardia* appear (Patnaik 1997; Patnaik et al. 2017; Gupta and Prasad 2001). At ~3 Ma *Bandicota* makes its first appearance (Patnaik 2001; Patnaik et al. 2017), which is followed by *Nesokia* appearing sometime around 2 Ma (Raghavan 1990). *Hadromys* occurs around 3 Ma in Pakistan (Cheema et al. 1997; Musser 1987). *Dilatomys* and *Hadromys* are also known from the Indian Siwaliks from ~3.5 Ma and ~2 Ma deposits, respectively (Patnaik 2001, 1997).

Spalacidae

Today, the rhizomyines are confined to south Asia and east Africa (Gogolevskaya et al. 2010). They have a fully subterranean way of life. The Asian bamboo rats include three species of *Rhizomys* and one species of *Cannomys*. The African mole rats *Tachyoryctes*, include two extant species. Both the Asian *Rhizomys* (*Brachyrhizomys*) and African *Tachyoryctes* have a long fossil record since the Late Miocene. However, *Cannomys* fossils are yet to be discovered. Since the Late Oligocene, over 30 fossil species belonging to 11 genera (*Prokanisamys, Kanisamys, Rhizomyides, Protachyoryctes, Tachyoryctes, Eicooryctes, Miorhizomys, Anepsirhizomys, Rhizomys* (*Brachyrhizomys*), *Pronakalimys* and *Nakalimys* have been recorded from China, Thailand, India, Pakistan, Afghanistan and Africa (Black 1972; Flynn 1982, 1984, 1986; Vasishat 1985; López-Antoñanzas et al. 2013, 2015a).

Ctenodactylidae

Today, gundis are confined to the rocky terrain of North and East Africa, but their fossil record is widespread in Arabia, Anatolia, South and East Asia since the Oligocene. *Prosayimys flynni* comes from the Early Miocene (c. 20 Ma) of the Chitarwata Formation (Zinda Pir Dome, Pakistan) (Baskin 1996). *Sayimys obliquidens* (Bohlin 1946) is probably of late Early Miocene age reported from Taben-buluk (Gansu, China) (Wang et al. 2003). Early Middle Miocene *Sayimys baskini* occurs in the Kamlial Formation of Pakistan (López-Antoñanzas and Sen 2003). *Sayimys intermedius* comes from the Middle Miocene Hofuf Formation at Al Jadidah, Saudi Arabia (Sen and Thomas 1979), from Tayma, north-western Saudi Arabia (López-Antoñanzas and Sen 2004) and from the Lower Manchar Formation of Pakistan (De Bruijn et al. 1989). *Sayimys* cf. *S. intermedius* is known from the Vihowa and Kamlial formations, Early–Middle Miocene of Pakistan Baskin (1996) and from Paşalar, Turkey (Flynn and Jacobs 1990; Peláez-Campomanes and Daams 2002). López-Antoñanzas et al. (2004) have proposed the inclusion of *S. perplexus*, *S. chinjiensis* and *S. badauni* in the wide ranging species, *S. sivalensis. Sayimys sivalensis* comes from Lower Manchar (De Bruijn et al. 1989), Kamlial (Baskin 1996), and Chinji (Hinton 1933; Hussain et al. 1977; Munthe 1980; Dehm et al. 1982; Wessels et al. 1982; Baskin 1996) formations of Pakistan and in India from Chinji (Vasishat 1985), Nagri (Wood 1937; Prasad 1970; Vasishat 1978, 1985) and Tatrot equivalent formations (*S. badauni* of Vasishat 1985).

Material and Methods

Murid, rhizomyid and ctenodactylid rodents collected from well dated (9.2–9.1 Ma) Late Miocene Siwalik deposits exposed around Haritalyangar (see Pillans et al. 2005) are described herein. Late Pliocene and Early Pleistocene Siwaliks of Chandigarh and Kala Amb areas have also yielded few new rodents after screen washing of some sediments. Dilute glacial acetic acid was used to dissolve matrix followed by thorough rinsing and screening under tap water. The residue was sorted under Leica S8APO microscope and photographed. Measurements have been taken under stereozoom microscope using a micrometer. Few images are taken using JEOL 6490 SEM housed at the Department of Geology, Panjab University.

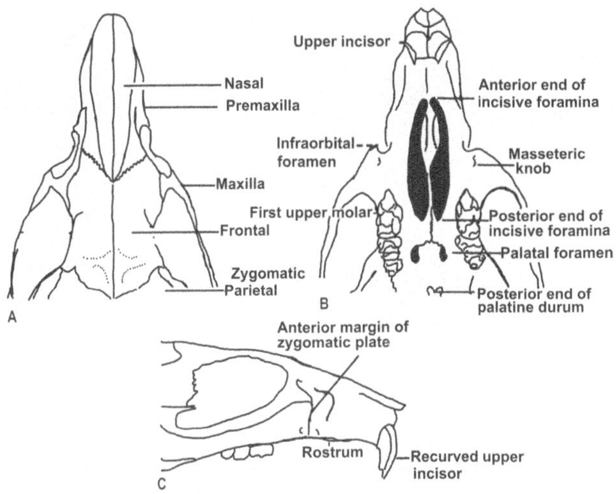

Fig. 16.1 Line drawings showing dorsal (**A**), ventral (**B**) and lateral (**C**) views of murine anterior skull portion (modified from Gerasimov et al. 1990)

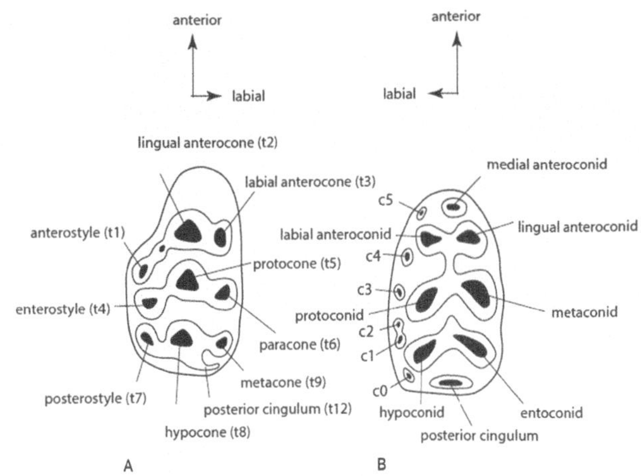

Fig. 16.2 Line drawings of idealized murine left first molars, M1 left (**A**) and m1 right (**B**). Standard Cope-Osborn terminology is used for the cusps, with equivalent t1 to t12 terminology and specialized cingulum cusp names for m1 (modified from Kimura et al. 2017)

The murine rodents described by Vasishat (1985) from Haritalyangar are also revisited in the light of recent work by Kimura et al. (2016, 2017). Haritalyangar sites were tied to paleomagnetically dated section of Pillans et al. (2005) and the age of Late Pliocene sites have recently been revised based on biochronology (Patnaik et al. 2017). The data on rodent fossil occurrences, come from the literature (Badgley et al. 2008; Barry et al. 2002; Barry et al. 2013; Flynn et al. 2013; Flynn et al. 2014; Flynn and Wessels 2013; López-Antoñanzas et al. 2004; 2013; 2015a, b; 2016; Patnaik 2013, 2016). Institutional abbreviations: VPL (Vertebrate Paleontology Laboratory), PUA (Panjab University Anthropology). PAUP (version 4.0b10, Swofford 2002), Phylogenetic Analyses Using Parsimony. Murine cranial and dental terminologies (Figs. 16.1 and 16.2) are from Gerasimov et al. (1990) and Kimura et al. (2017), respectively.

Systematics

Family Muridae Illiger, 1811

Subfamily Murinae Illiger, 1811

Genus *Progonomys* Schaub, 1938

Progonomys choprai Vasishat, 1985 = *Progonomys debruijni* Jacobs, 1978, in part

Holotype: PUA 71-48

Type Locality: Haritalyangar.

Referred Material: Partial skull with premaxilla and a part of the maxilla preserved.

Description: The partial cranium shows slight post-mortem distortion. It is moderately built, has a broad and short incisive foramen (Fig. 16.3A). The posterior end of the incisive foramina reaches the level of the anterior root of the first upper molar. In the lateral view the partial cranium makes a gentle concave arc. The zygomatic plate is squarish in shape. The anterior border of the zygomatic plate (Fig. 16.3D) is straight, rises vertically, then is swept back in an arc at the moderately deep zygomatic notch. The plate is set out laterally and a large masseteric knob is present (Fig. 16.3D). Incisors are re-curved (opisthodont) and not notched. The rostrum is short and deep.

The M1 (Fig. 16.4) is longer than wide. The cusps are strongly inclined (Fig. 16.4B and C). The anterostyle (t1) is displaced well posterior to the lingual anterocone (t2). The labial anterocone (t3) is distinctly smaller than the lingual anterocone (t2). The anterior lobe of the molar is moderate in length and is quite asymmetrical, with an inflection at the anterostyle (t1). The cusp t1 is laterally compressed. The precingulum is weakly developed and bears no cuspule. The enterostyle (t4) is posterior in position and is strongly connected to the protocone (t5). The paracone (t6) is little posterior to the protocone (t5). The metacone (t9) is as large as the cusp t6 and it is strongly inclined posteriorly (Fig. 16.4). The posterior cingulum is distinct and is attached to the hypocone (t8).

Comparisons: Murine skulls in the fossil record are extremely rare. The only other mouse skull from Siwaliks, that of *Mus linnaeusi* from ~2 Ma Pinjor Formation exposed near Chandigarh (Patnaik et al. 1996; Patnaik 1997) is known.

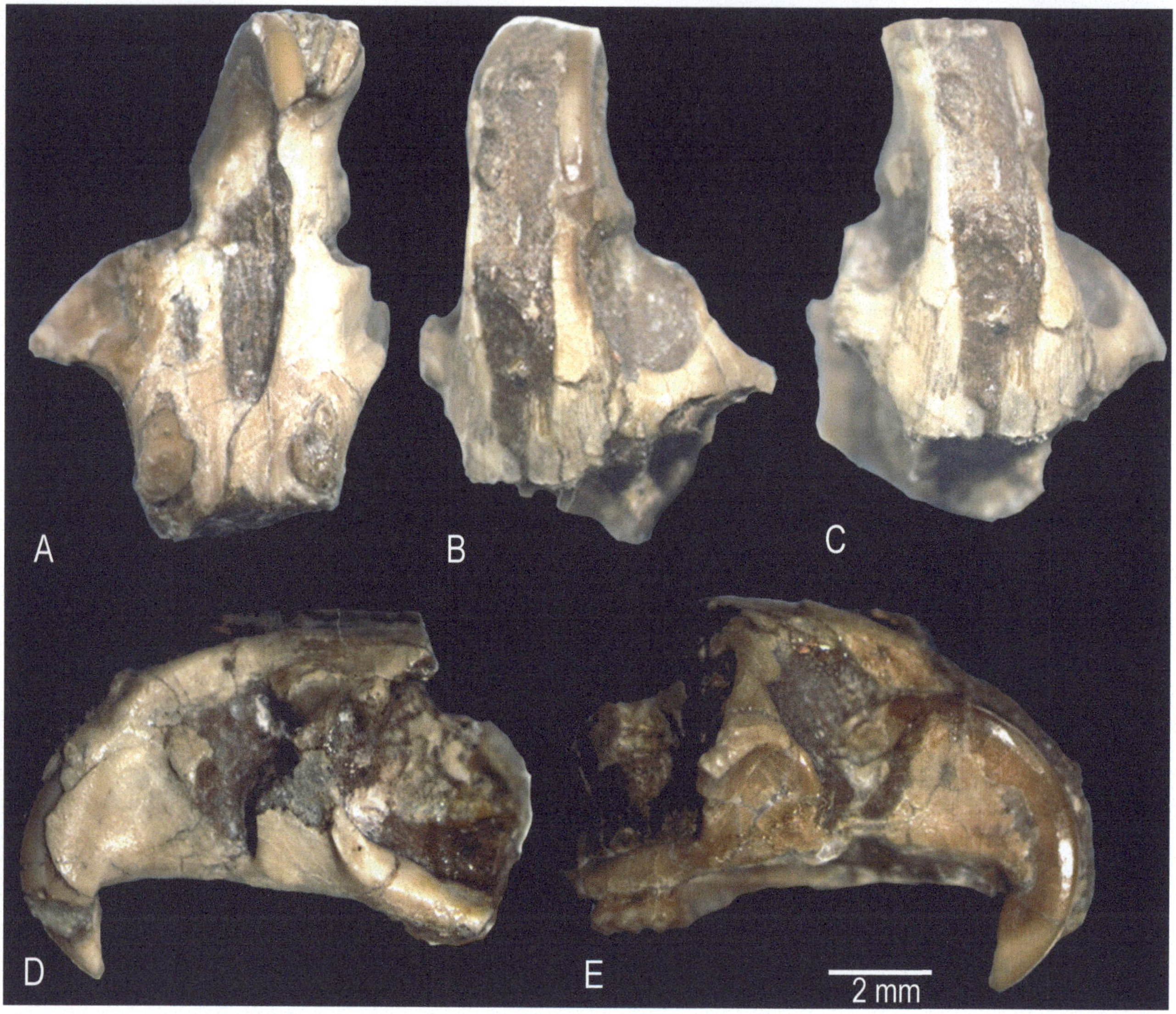

Fig. 16.3 *Progonomys debruijni* partial cranium Nagri Formation, Haritalyangar (PUA 71-48, *Progonomys choprai* Vasishat, 1985). **A**. ventral view; **B**. dorsal view tilted little left laterally; **C**. dorsal view tilted little right laterally; **D**. left lateral view and **E**. right lateral view

The partial skull of *Progonomys* is similar to the front part of *Mus linnaeusi* cranium in having moderate size incisive foramina terminating at the level of the anterior root of the first upper molar, a short and deep rostrum, a moderately deep zygomatic notch, anterior margin of the zygomatic plate straight, and presence of a masseteric knob. The latter species differ from the former in having orthodont upper incisors with apical notch, an accessory cuspule on the anterior margin of the first upper molar, t1 strongly displaced posteriorly (Fig. 16.4A).

Progonomys is considered to be the ancestral stock that gave rise to several taxa belonging to Malacomyni, Apodemurini, Murini, and Praomyini tribes (Kimura et al. 2015; Aghová et al. 2018). The present partial skull provides an opportunity to identify some of the ancestral/generalised characters of murines. Presence of recurved or opisthodont incisors can be treated as an ancestral character (also see Musser 1981). Orthodont and proodont incisors were perhaps later (derived) developments. Another primitive feature could be the absence of subapical notch on the upper incisors and so could be the presence of moderate sized incisive foramina that stops just at the anterior root of the first upper molar. A moderate zygomatic notch, a straight anterior margin of zygomatic plate, a short rostrum and an absence of supraorbital ridges may also be considered as more generalised characters. On these lines *Malacomys longipes* shares some primitive characters with *Progonomys debruijni* such as the presence of opisthodont upper incisors lacking a subapical notch. Other characters on *Malacomys longipes* such as presence of very short incisive foramina, deep zygomatic notch and supraorbital ridges are considered as derived (see Chevret et al. 2003). The apodemurini

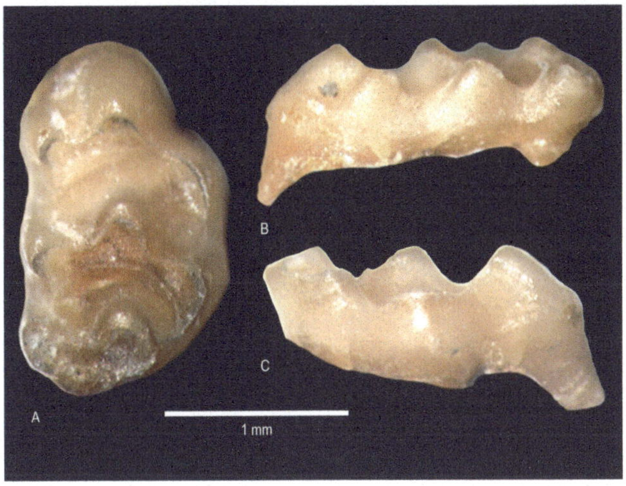

Fig. 16.4 *Progonomys debruijni* right M1 enlarged (PUA 71-48, *Progonomys choprai* Vasishat, 1985). **A**. occlusal view; **B**. lingual view; **C**. labial view

Apodemus flavicollis and the praomyini *Praomys jacksoni* also share some of the primitive skull characters with *Progonomys debruijni* such as the presence of opisthodont upper incisors, short rostrum, moderate zygomatic notch and straight anterior margin of zygomatic plate. However, these species are little derived in having incisive foramina penetrating the anterior region of the first upper molars.

Discussion: The genus *Mus* comprises four subgenera *Mus, Coelomys*, *Nannomys* and *Pyromys*, characterized by morphological and genetic traits (Marshall 1977, 1988; Musser and Newcomb 1983). The subgenus *Mus* contains nine species mainly from Eurasia, *Coelomys* (shrew mice) has five Asiatic species, *Pyromys* (spiny mice) has five Asiatic species and *Nannomys* (pygmy mice) has 19 African species. The systematics of mice (genus *Mus*) has been studied using morphology, biochemistry and molecular biology (Misonne 1969; Marshall 1977, 1981; Bonhomme et al. 1982, 1984; She et al. 1990; Catzeflis and Denys 1992; Lundrigan and Tucker 1994; Sourrouille et al. 1995; Chevret et al. 2003).

Progonomys debruijni partial cranium is quite alike to the skull of *Mus linnaeusi* which belongs to the subgenus *Mus* (Patnaik 1997). Affinity of *Mus linnaeusi* to the subgenus *Mus* is confirmed when compared to the various morphological features of skulls of the modern species belonging to the subgenus. For example skulls of *M. cervicolor, M. cooki, M. caroli, M. musculus*, *M. spicilegus* and *M. spretus* all belonging to the subgenus *Mus* have moderate sized incisive foramina; a wide inter-orbital width less than 4 mm, opisthodont or orthodont upper incisors (Patnaik 1997).

The PUA 71-48 M1 is very close in size, shape and cusp pattern to those of *Progonomys debruijni* described from Pakistani Siwalik deposits of similar age (Jacobs 1978; Kimura et al. 2017). This molar is smaller than those of the older species *Progonomys hussaini* (~10.5–10.1 Ma) of Siwalik reported by Cheema et al. (2000) (see Kimura et al. 2017), with cusps being more inclined. This molar is similar in size to those of a new species recently described as *Progonomys morganae* Kimura et al. (2017) from ~10.5 to 10.1 Ma Siwaliks of Pakistan, but the latter species however is slender and has cusps less inclined.

Genus *Karnimata* Jacobs, 1978

Karnimata darwini Jacobs, 1978

Type Locality: Y 182A of Dhok Pathan Formation dated to 9.2 Ma, Potwar, Pakistan.

Referred Material: 1 maxilla with all molars and 1 maxilla with M2 and M3 preserved (PUA 18-75), earlier referred to as *Siwalkomus nagrii* by Vasishat (1985) from Nagri sediments, Haritalyangar; New Material, 1 maxilla with M2 and M3 preserved (VPL/RP-HM1) and two mandibles (VPL/RP-HM2 and VPL/RP-HM3) with all molars preserved from HD1 site dated to 8.85 Ma (Pillans et al. 2005).

Description: M1 (Fig. 16.5A) is longer than wide (L/W ratio = 1.56) (Table 16.1). The molar is partially damaged and is glued, thus appears a little longer than it actually is.

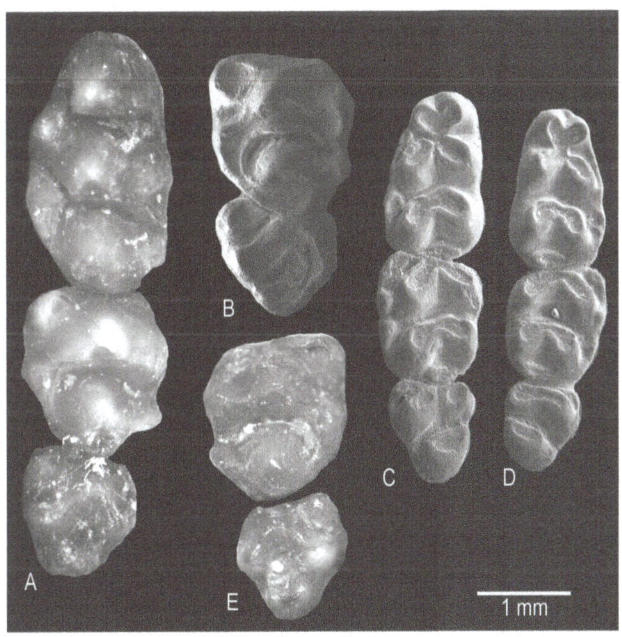

Fig. 16.5 *Karnimata darwini*. **A**. occlusal view of left M1, M2, and M3 (PUA 18-75, *Siwalkomus nagrii*, Vasishat 1985); **B**. Occlusal view of right M2 and M3 (VPL/RP-HM1); **C**. Occlusal view of left M2 and M3 (PUA 18-75, *Siwalkomus nagrii*, Vasishat 1985); **D**. Occlusal view of left m1, m2, and m3 (VPL/RP-HM2); **E**. Occlusal view of left m1, m2, and m3 (VPL/RP-HM3)

Table 16.1 Dimensions of molar specimens studied here

Specimen no.	Taxa	Element	Length in mm	Width in mm
PUA 71-48	*Progonomys debruijni*	M1	1.84	1.15
PUA 18-75	*Karnimata darwini*	M1	2.72	1.59
PUA 18-75	*Karnimata darwini*	M2	1.68	1.59
PUA 18-75	*Karnimata darwini*	M3	1.31	1.22
VPL/RP-HM1	*Karnimata darwini*	M2	1.59	1.54
VPL/RP-HM1	*Karnimata darwini*	M3	1.13	1.31
PUA 18-75	*Karnimata darwini*	M2	1.77	1.59
PUA 18-75	*Karnimata darwini*	M3	1.31	1.31
VPL/RP-HM2	*Karnimata darwini*	m1	1.77	1.09
VPL/RP-HM2	*Karnimata darwini*	m2	1.36	1.18
VPL/RP-HM2	*Karnimata darwini*	m3	0.90	1.00
VPL/RP-HM3	*Karnimata darwini*	m1	1.72	1.09
VPL/RP-HM3	*Karnimata darwini*	m2	1.27	1.09
VPL/RP-HM3	*Karnimata darwini*	m3	0.81	0.81
PUA 71-22	*Parapodemus hariensis*	m1	2.26	1.50
PUA 71-22	*Parapodemus hariensis*	m2	1.53	1.42
PUA 71-22	*Parapodemus hariensis*	m3	1.15	1.30
PUA 18-75	*Parapodemus hariensis*	m1	2.23	1.42
PUA 18-75	*Parapodemus hariensis*	m2	1.53	1.42
PUA 18-75	*Parapodemus hariensis*	m3	1.15	1.15
VPL/RP-KM1	*Mus* cf. *M. pahari*	M1	2.16	1.26
VPL/RP-KM2	*Mus* cf. *M. pahari*	m2	1.2	1.00
VPL/RP-KM3	*Mus* cf. *M. pahari*	m3	0.92	0.96
VPL/RP-NM4	*Hadromys* sp.	M1	2.58	1.8
VPL/RP-HR1	cf. *Kanisamys* sp.	m1	2.70	2.10
VPL/RP-HR1	cf. *Kanisamys* sp.	m2	2.60	2.40
VPL/RP-HR1	cf. *Kanisamys* sp.	m3	3.12	2.30
VPL/RP-HR2	cf. *Kanisamys* sp.	m1	2.70	2.10
VPL/RP-HR2	cf. *Kanisamys* sp.	m2	–	2.5
VPL/RP-BR3	*Miorhizomys harii*	M2	4.15	4.15
VPL/RP-BR2	*Miorhizomys harii*	M2	4.15	4.15
VPL/RP-BR1	*Miorhizomys harii*	m1	5.12	3.33
VPL/RP-BR1	*Miorhizomys harii*	m2	4.16	4.41
VPL/RP-BR1	*Miorhizomys harii*	m3	5.38	4.30
VPL/RP-HCT-3	*Sayimys sivalensis*	m1	2.50	2.10
VPL/RP-HCT-3	*Sayimys sivalensis*	m2	2.72	2.9
VPL/RP-HCT-3	*Sayimys sivalensis*	m3	2.72	3.00

Molar cusp inclination is less than that of *Progonomys*. M1 has eight major cusps, three in each of the first two chevrons. In the first chevron, the anterostyle (t1), is round and vertical in orientation, displaced a little posterior to a much larger lingual anterocone (t2). The former is weakly linked to the latter. The labial anterocone (t3) is a small cusp lying at the level of t2. In the second chevron, enterostyle (t4) is placed posteriorly to a large protocone (t5) and smaller paracone (t6). The enterostyle has a faint posterior spur joining the hypocone (t8) low on its base. The paracone (t6) is isolated from the metacone (t9) and there is no hint of stephanodonty. A short posterior cingulum is well developed.

M2 is similar to M1 in being robust with round cusps, that are hardly inclined. The major chevron comprising t4–t5–t6 dominates the tooth, but the t1 is a prominent cusp located anterolingually. Also, a small t3 is always present. The t4 has a posterior contact with the base of the t8. The t9 is small and separated from the t6. No posterior cingulum is visible. M3 has a triangular occlusal outline with a prominent t1, faint trace of t3, t4–t5–t6 join to form an arc. Two of the three specimens show separate presence of t8 and t9, but in one of these, the cusps are fused.

Lower molars are robust having protoconid-metaconid and hypoconid-entoconid pairs of cusps. The twin cusps of anteroconid are smaller than posterior cusps (protoconid and metaconid) and there is a narrow shelf anterior to them, which could be the trace of a medial anteroconid. Posterior cingulum is oval in shape. The labial cingulum is well developed, and bears accessory cusp c1, c3. Posterior lower molars are similar to m1 in robustness. Their opposite cusp pairs are broad. The outline of m2 is more like a trapezoid with its labial side longer than the lingual side (Fig. 16.5D). The hypoconid-entoconid cusp pair is nearly as broad as the protoconid-metaconid pair. The posterior cingulum is distinct. The labial cingulum is well-developed, bears c1, and c3. The m3 is triangular in shape with a wide anterior portion matching the width of the adjacent m2. It is a small tooth with single, teardrop shaped posterior cusp situated somewhat lingual to the tooth midline. There is a small labial anteroconid. The labial cingulum is absent. There is no posterior cingulum.

Discussion: *Progonomys* has a more posterior anterostyle which tends to be laterally compressed, unlike the anterior, conical anterostyle of *Karnimata* (Kimura et al. 2017). The molars of *Karnimata darwini* are somewhat larger than those of *K. fejfari* (Kimura et al. 2017) reported from an older horizon (10.1 Ma). *Karnimata* represents the earliest clear indication of a major split in the murid evolution, lying at the base of the *Mus-Arvicanthis* split (Kimura et al. 2015). Early *Karnimata* as old as ~11 Ma is not easily distinguishable from early *Progonomys*. Closer to 10 Ma, *Karnimata fejfari* encompasses a recognizable, early step in arvicanthine evolution. *Karnimata* survived until the end of the Miocene in the Indian subcontinent, and until the early Pliocene in Afghanistan (Brandy 1979). *Karnimata* has been thought to occur in the late Miocene of Shanxi and Inner Mongolia, China, but Qiu and Li (2016) distinguish this material as a new genus *Karnimatoides*. Mein et al. (1993) pointed out that the holotype of the Siwalik *Karnimata darwini* might represent the European species named as *Progonomys woelferi* Bachmayer and Wilson (1979). Storch and Ni (2002) however rejected this point of view due to the anterior position of the anterostyle is unlike *Progonomys,* but typical of *Karnimata*. Recently, Kimura et al. (2013a, 2015) emphasized that *K. darwini* is encased in a lineage separate from *Progonomys*. Further, they opined that *Karnimata* is the most recent common ancestor of the tribes Arvicanthini-Otomyini-Millardini.

Genus: *Parapodemus* Schaub, 1938

Parapodemus hariensis Vasishat (1985)

Holotype: PUA 70-13.

Type Locality: Hari Scarp, Nagri Formation, Siwalik Group.

Referred Material: One mandible assigned to *Siwalkomus nagrii* (PUA 18-75) by Vasishat (1985).

Description: The teeth are robust. On m1, the anteroconids and a small medial anteroconid form a chevron that is as wide as the second chevron (protoconid and metaconid) on the m1. The first two chevrons are isolated and they are not connected by a mure. The third chevron has hypoconid and entoconid strongly connected at the middle. The m2 is rectangular having a small labial anteroconid. The lingual cusps are gently inclined labially. The labial cusps are inclined antero-lingually. Labial cingulum is well developed with a large circular c1. Posterior cingulum is oval in outline. The m3 is subtriangular, lacks a labial anteroconid and a labial cingulum. The fused hypoconid-entoconid forms a wide chevron extending from labial to lingual side.

Discussion: Jacobs (1978) described *Parapodemus* from Y 182A of Dhok Pathan Formation dated to 9.2 Ma along with *Karnimata darwini* and *Progonomys debruijni*. However, he did not assign it to any species. Vasishat (1985) found a similar large murid from Haritalyangar and named it *Parapodemus*

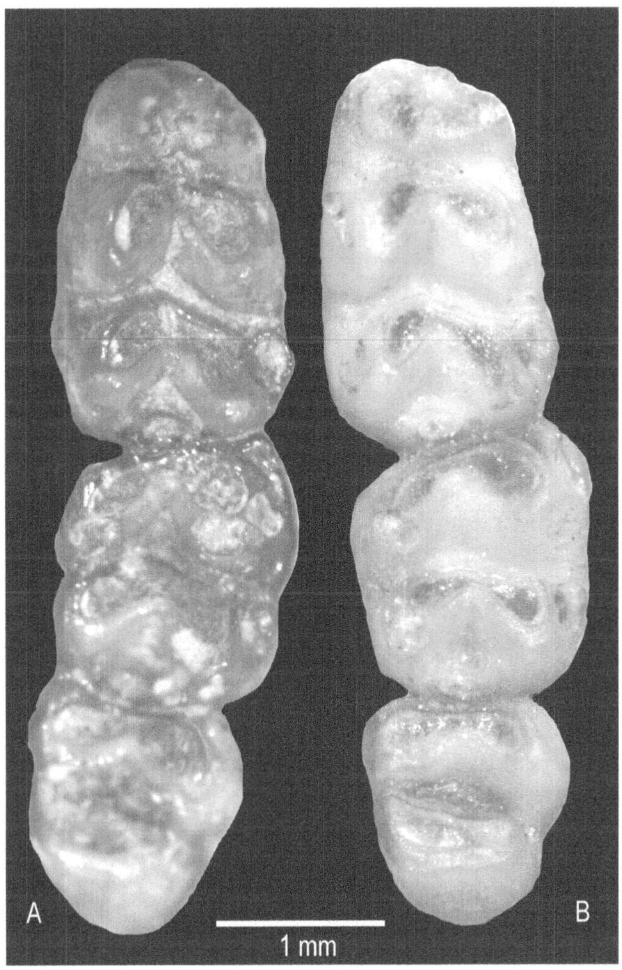

Fig. 16.6 *Parapodemus hariensis*. **A**. occlusal view of right m1, m2, and m3 (PUA 71-22, Vasishat 1985); **B**. occlusal view of right m1, m2, and m3 (PUA 18-75, *Siwalkomus nagrii*, Vasishat 1985)

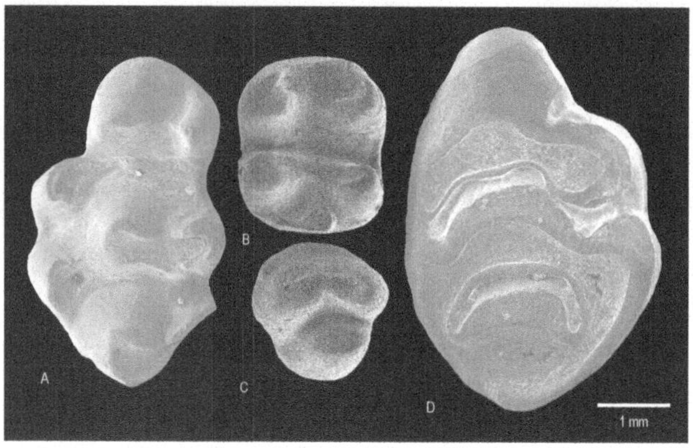

Fig. 16.7 *Mus* cf. *M. pahari.* **A**. occlusal view of left M1 (VPL/RP-KM1); **B**. occlusal view of right m2 (VPL/RP-KM2); **C**. occlusal view of left m3 (VPL/RP-KM3); *Hadromys* sp., **D**. occlusal view of right M1(VPL/RP-NM4)

hariensis (Fig. 16.6A). *Parapodemus hariensis* lower molars can easily be distinguished from those of contemporary *Karnimata darwini* and *Progonomys debruijni* being much larger in size with a large medial anteroconid. Recently, Kimura et al. (2017) recovered upper molars of a more primitive *Parapodemus* from the older Nagri Formation in the Potwar region and have named these as *Parapodemus badgleyae* sp.nov.

Genus *Mus* Schaub, 1938

Mus pahari Thomas, 1916

Mus (*Coelomys*) cf. *M. pahari*

Holotype: Left M1 (VPL/RP-KM1)

Type Locality: Kanthro (Himachal Pradesh) ~3 Ma.

Referred Material: Right m2 (VPL/RP-KM2) and left m3 (VPL/RP-KM3).

Description: The left M1 (Fig. 16.7) is longer than wide. The cusps are weakly inclined posteriorly (Fig. 16.7A). The central cusps are wider. The anterostyle (t1) is displaced well posterior, separated by a wide valley and makes a right angle to the lingual anterocone (t2). It is almost vertical. The labial anterocone (t3) is distinctly smaller than the lingual anterocone (t2) and lies at the same level. The anterior lobe of the molar is short. The enterostyle (t4) is posterior in position and is strongly connected to the protocone (t5). The posterostyle (t6) is little posterior to the protocone. There appears a faint crest between t4 and the hypocone (t8). The metacone (t9) is broken in this specimen.

The right m2 is rectangular in outline and is elongated. Cusps are less inclined anteriorly. A small labial anteroconid is present. The labial cusps are placed little distal to the lingual ones. The posterior cingulum is small, low, triangular in outline and lies close to the hypoconid (Fig. 16.7B).

The left m3 is triangular in outline and has a faint labial anteroconid. The cusps are nearly vertical (Fig. 16.7C).

Comparisons: The M1 (L = 2.16 mm and W = 1.26 mm) (Table 16.1) closely recalls those of *Mus* (*Coelomys*) *pahari* (Lenght range: 1.92–2.22; Width range: 1.15–1.27 mm) reported from the Middle Pleistocene upper layer of Snake Cave locality, Thailand (Chaimanee 1998: plate 11, Figs. 13–16). In fact, the two taxa have a vertical t1 making a deep valley and a right angle with the t2, a short anterior lobe of the t2 and a crest between the cusps t4 and t8. The cusps on the M1 of the extant *Mus pahari* are, however, little posteriorly inclined (including the t1). The crest between t4 and t8 is also weaker compared to that in the fossils from the Snake Cave locality, Thailand (Chaimanee 1998).

Discussion: Presently *Mus pahari* is distributed in the northeast India (Assam and Sikkim), South China, Vietnam, Laos and Thailand. The fossil record is poorly known. Beside the findings from the Middle Pleistocene of Thailand, *Mus pahari* is only known by a couple of Pleistocene mandibles (~100,000 yrs) from China (Zheng 1993). The present specimens come from a site Kanthro, in Himachal Pradesh that has recently been biostratigraphically dated to ~3 Ma (Patnaik et al. 2017). This makes *Mus* (*Coelomys*) cf. *M. pahari* the oldest known species belonging to the subgenus *Coelomys*.

Genus *Hadromys* Thomas, 1911

Hadromys humei Thomas, 1886

Hadromys sp.

Type Locality: Nadah (Haryana), 2 Ma.

Referred Material: M1 (VPL/RP-NM4).

Description: The right M1 is oval shaped (Fig. 16.7D). The hypsodont tooth is heavily worn showing strong connection among the cusps of various chevrons. The central cusps are wider. The labial cusps are worn. The anterostyle (t1) is displaced little posterior to the lingual anterocone (t2). It is almost vertical. The labial anterocone (t3) is smaller than lingual anterocone (t2) and lies a little posterior to it. The anterior lobe of the molar is short. The enterostyle (t4) is posterior in position and is strongly connected to the protocone (t5). The posterostyle (t6) is little posterior to the protocone. The t4 is connected to the hypocone (t8). The metacone (t9) is tiny and merges with t8. The tooth has five roots.

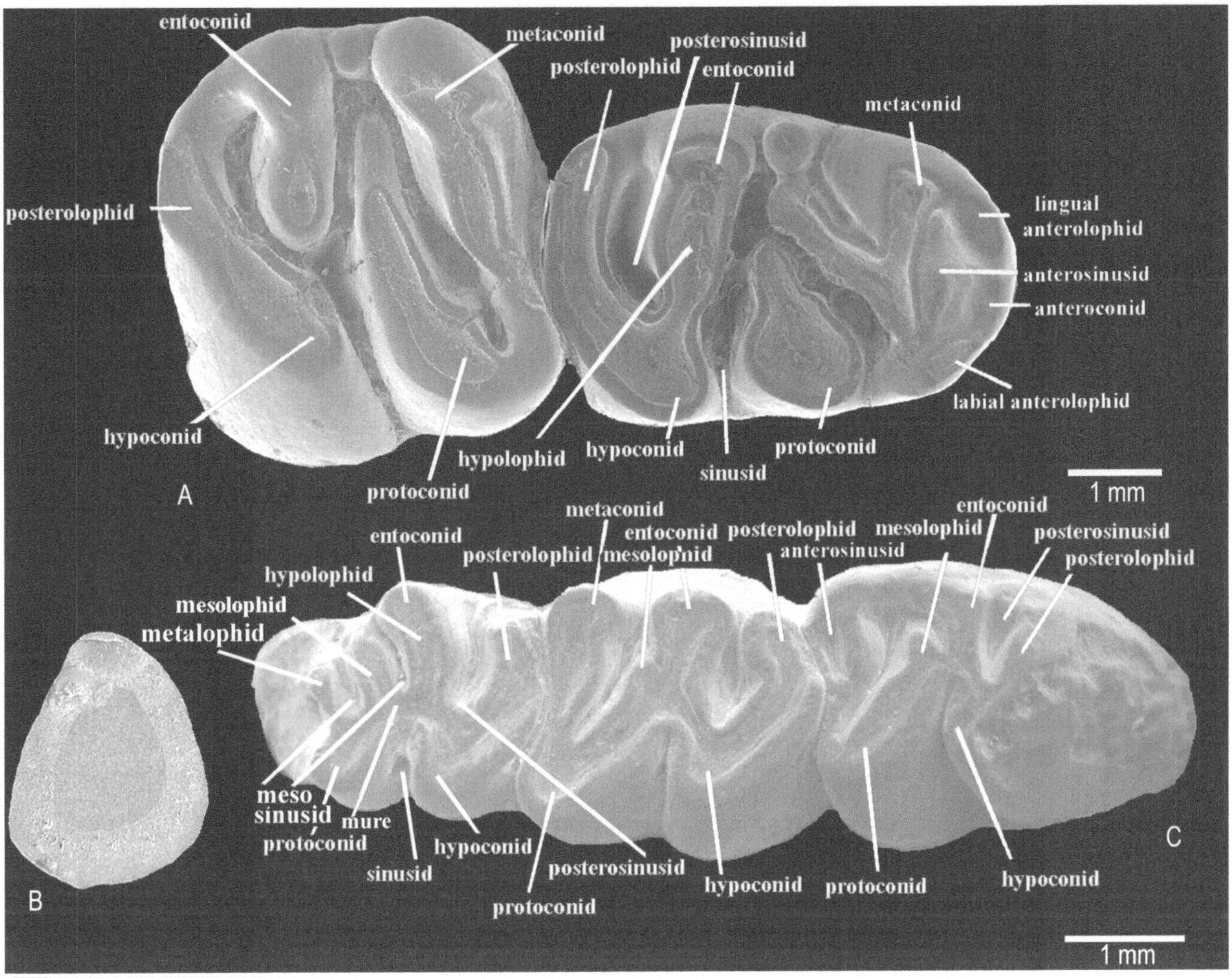

Fig. 16.8 *Miorhizomys harii*. **A**. occlusal view of right m1 and m2 (VPL/RP-BR1); cf. *Kanisamys* sp. **B**. Transverse section of lower incisor (VPL/RP-HR1); **C**. occlusal view of m1, m2, and m3 (VPL/RP-HR1). Dental terminology of Mein and Freudenthal (1971), with some adjustments (López-Antoñanzas et al 2013; Flynn 1982)

Comparisons: The present M1 with length = 2.58 and width = 1.8 mm (Table 16.1) is slightly smaller than those of the extant *Hadromys humei* (Length range 2.46–3.02 width 1.92–2.18 mm, see Chaimanee 1998, Table 3.4). However, its length falls well within the range of fossil *Hadromys humei* from Thailand. Like in the extant taxa the present M1 shows t1 placed slightly distal to t2. The t3 is also little distal to t2 forming an arcuate first chevron. Similarly, t4 and t6 are little distal to t5. M1 of *Hadromys loujacobsi* (Musser 1987) from the Pleistocene Siwalik deposits of Pakistan differs from the present specimen in being slightly larger, more hypsodont, having cusps of first and second chevron placed in an almost straight line, a rather larger and columnar t9 and four roots.

Family: Spalacidae Gray, 1821

Subfamily: Rhizomyinae, Winge, 1887

Tribe: Tachyoryctini Miller and Gidley, 1918

cf. *Kanisamys* sp. indet.

Type Locality: Hari Devi Temple Dhok Pathan Formation, Siwalik Group, age 9.2 Ma.

Referred Material: VPL/RP-HR1, left lower dentition with m1 incomplete (Figs. 16.8B, C and 16.9A–C). VPL/RP-HR2 (Fig. 16.9D, E, F), right mandibular fragment with incomplete m1 and m2.

Diagnosis: The specimens can be assigned to the tribe Tachyoryctini based on a dentary that is slender and shallow, moderately hypsodont teeth (Fig. 16.9), a long, extended masseteric crest inflated below m2, a teardrop shaped cross-section of lower incisors with a longitudinal ridge that runs near the midline of the labial surface.

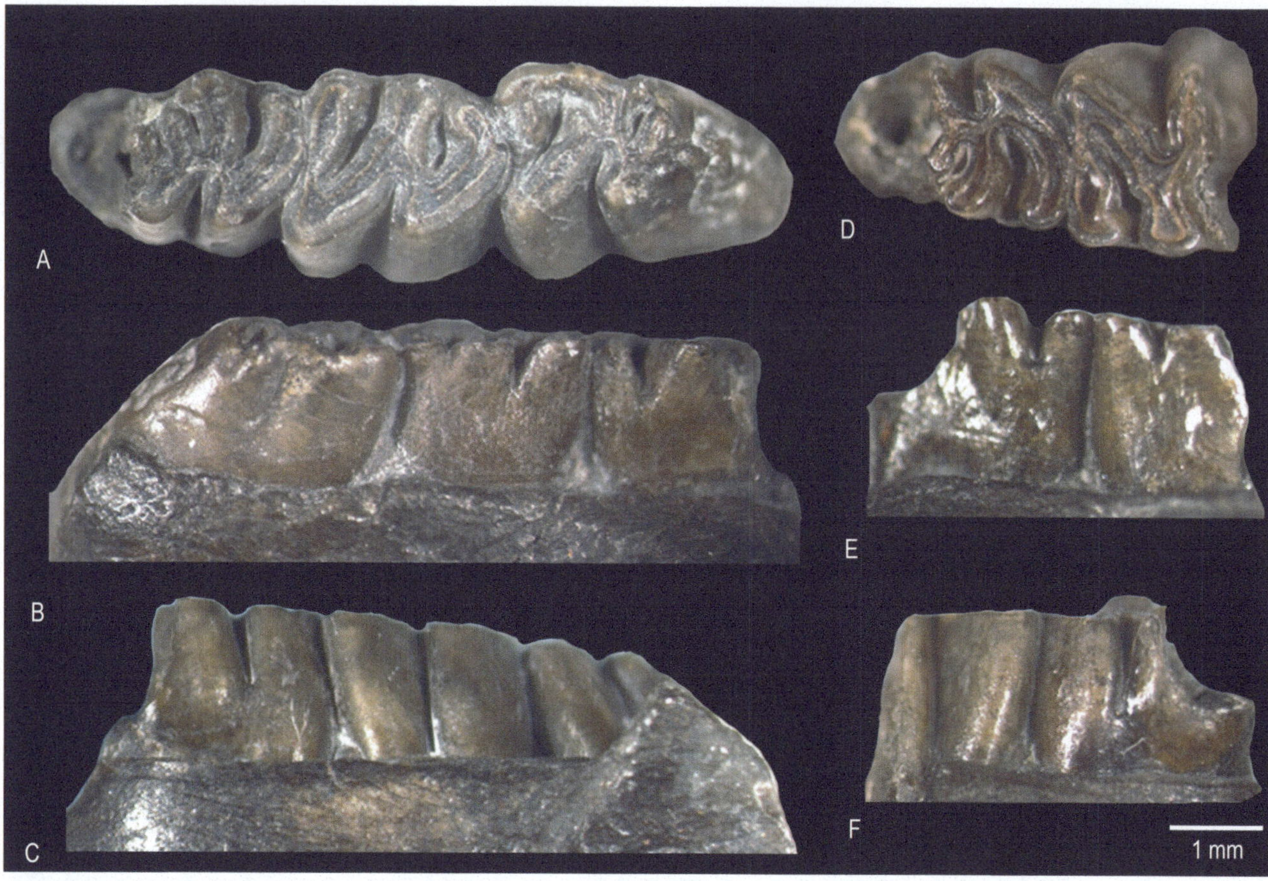

Fig. 16.9 Cf. *Kanisamys* sp. **A-C**. Occlusal (**A**), lingual (**B**), and labial (**C**) views of m1, m2, and m3 (VPL/RP-HR1). **D-F**. Occlusal (**D**), lingual (**E**), and labial (**F**) views of m1 and m2 (VPL/RP-HR2)

Description: The occlusal outline of the m1 tooth is oval, apparently narrower anteriorly. The anterolophid is not preserved on the two specimens nor is the metaconid, however the metalophid is partly preserved in one of the two specimens (Fig. 16.8C). Entoconid-hypolophid makes a long loph and so does the posterolophid. The mesolophid is short, distinct and is separated from the hypolophid. The m1 clearly shows a posterior metalophid (metaconid broken)-protoconid connection. The mesosinusid and the posterosinusid are fairly deep and open. The labial sinusid is relatively shallower. A mure is distinct. The posterolophid is a long continuation of the bent hypoconid.

The occlusal outline of the m2 tooth is rectangular (L = 2.6 and W = 2.4; RI = 0.43) with the posterior part slightly narrower than the anterior one. The main lophids are the anterolophid–metalophid, hypolophid and posterolophid. The protoconid is v-shaped and its posterior arm is broad and oblique. The hypolophid is strong and directed lingually from the short mure. The mesolophid is very reduced and is appressed to the hypolophid. The labial sinusid is fairly deep and directed posterolingually.

The m3 is oval in occlusal outline with the posterior side rounded and slightly narrower than the anterior one. They have the same three main lophids. Anterolophid and metalophid are differentiated and enclose a shallow anterior enamel lake, anterosinusid. The labial sinusid is posterolingually directed and moderately deep. The protoconid is v-shaped with a short and transverse anterior arm and an oblique and broad posterior one, from which the short mesolophid expands. Posterosinusid is triangular in shape and separates hypolophid from posterolophid.

Comparisons: *Kanisamys indicus* found in older (17–11 Ma) sediments is smaller, lower crowned and less lophodont than the present specimens. The same is true when the present specimen is compared with *K. potwarensis*. Moreover, the dentition of *Kanisamys indicus* is characterized by having labial anterolophid, protosinusid, long and well separated mesolophid on m2–m3 and double longitudinal ridges on incisors, further differentiating it from the present taxa. The lower molars of the coeval *Kanisamys sivalensis* and *Kanisamys nagrii* are closer to those of the present taxa as far as

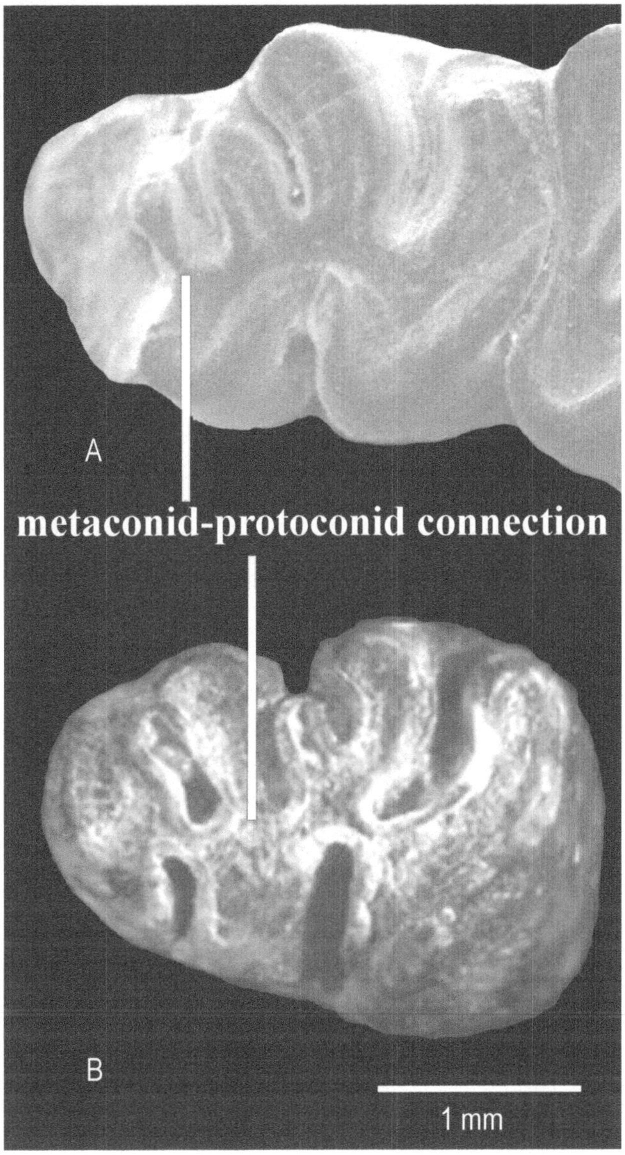

Fig. 16.10 Cf. *Kanisamys* sp. **A**. occlusal view of left m1 (VPL/RP-HR1); *Protachyoryctes makooka*, **B**. occlusal view of left m1 (ASK-VP-3/287, After López-Antoñanzas and Wesselman [2013])

overall morphology and size are concerned. However, the mesolophid in the present m2 is much reduced and the masseteric crest inflated below m2 when comparing with *K. sivalensis* and *Kanisamys nagrii*. *K. sivalensis* m1 shows an anterior metaconid-protoconid connection unlike in the present m1, which appears to show a posterior metalophid-protoconid connection, metaconid being missing. This connection in the present specimen is similar to that seen in the African fossil tachyoryctines such as *Tachyoryctes makooka* (Fig. 16.10). The anterior mesosinusid of the present m1 clearly separates the metalophid and mesolophid as is the case in *Tachyoryctes makooka* (Wesselman et al. 2009; López-Antoñanzas and Wesselman 2013). However, our taxon has the molars clearly smaller than the African taxon. The m3 is strongly appressed with m2 and its anterior border is a little concave. This inflection also offsets the protoconid-metaconid connection on this tooth a little posteriorly. The larger tachyoryctine from the Siwaliks *Protachyoryctes tatroti* Hinton, 1933 comes from the Latest Miocene (6.9–6.5 Ma) Dhok Pathan sediments instead of Tatrot Formation (Flynn 1981, 1982), as the name suggest. Compared to the present m2, that of *Protachyoryctes tatroti* is larger (Table 16.2) and its m1 also shows an anterior connection between the protoconid and metaconid, like that in the m1 of *Kanisamys sivalensis*. Another tachyoryctine from the Late Miocene is *Eicooryctes kaulialensis* (Flynn 1982), which has larger and more hypsodont teeth when compared with our specimens. The m1 of *Eicooryctes kaulialensis* is similar to the present m1 in having a posterior protoconid and metaconid connection, very reduced mesolophid on m2 and m3. The present specimens show close affinity to the genus *Kanisamys*as they share several derived characters with *Kanisamys sivalensis* and *K. nagrii*. However, there are some major differences that do not allow the present specimens to be assigned to either of these contemporary species. For the time being they are considered cf. *Kanisamys* sp. untill more specimens are recovered.

Tribe Rhizomyini Winge, 1887

Miorhizomys Flynn, 2009

Miorhizomys harii (Prasad, 1970)

Type Locality: Baron (31°31′78.0″N:76°40′.66.0″E) Dhok Pathan Formation, Latest Miocene.

Referred Material: VPL/RP-BR1, right lower dentition with incisor. VPL/RP-BR2, right M2 and VPL/RP-BR3, left M2 (Fig. 16.11).

Description: The dentary is shallow and teeth are moderately hypsodont. The teeth are at a very early wear. M2 is square in occlusal outline (L = 4.15; W = 4.15 mm). The protocone is v shaped and it is attached to the anteroloph that runs anterolabially, and joins posterolabially the protoloph, which in turn leads to the paracone on the labial side. The lingual re-entrant goes all the way to the labial side and there is no mure. This could be due to a very early stage of wear (for wear stages in rhizomyines see López-Antoñanzas 2012a, b and López-Antoñanzas 2013a, b, c). The hypocone is concave labio-posteriorly, attached to mesoloph that forms its anterior arm and touches the posteroloph, posteriorly.

Table 16.2 Dental measurements of the present rhizomyids and those described by Flynn (1981)

Specimen No.	Taxa	Element	Length in mm	Width in mm	Reference
VPL/RP-HR1	cf. *Kanisamys* sp.	m1	2.70	2.10	This work
VPL/RP-HR1	cf. *Kanisamys* sp.	m2	2.60	2.40	This work
VPL/RP-HR1	cf. *Kanisamys* sp.	m3	3.12	2.30	This work
VPL/RP-HR2	cf. *Kanisamys* sp.	m1	2.70	2.10	This work
VPL/RP-HR2	cf. *Kanisamys* sp.	m2	–	2.5	This work
VPL/RP-BR3	*Miorhizomys harii*	M2	4.15	4.15	This work
VPL/RP-BR2	*Miorhizomys harii*	M2	4.15	4.15	This work
VPL/RP-BR1	*Miorhizomys harii*	m1	5.12	3.33	This work
VPL/RP-BR1	*Miorhizomys harii*	m2	4.16	4.41	This work
VPL/RP-BR1	*Miorhizomys harii*	m3	5.38	4.30	This work
YGSP 8096	*Kanisamys indicus*	m2	2.38	2.25	Flynn, 1981
YGSP8097	*Kanisamys indicus*	m2	2.31	2.25	Flynn, 1981
YPM 13810	*Kanisamys indicus*	M1	2.38	1.88	Flynn, 1981
YGSP 8247	*Kanisamys indicus*	M1	2.38	1.81	Flynn, 1981
YGSP 8377	*Kanisamys indicus*	M1	2.31	1.75	Flynn, 1981
YGSP 8388	*Kanisamys indicus*	M1	2.40	1.75	Flynn, 1981
YPM 13810	*Kanisamys indicus*	M2	2.25	2.13	Flynn, 1981
GSI D271	*Kanisamys indicus*	M2	2.31	2.06	Flynn, 1981
YGSP 8372	*Kanisamys indicus*	M2	2.44	2.13	Flynn, 1981
YGSP 8105	*Kanisamys indicus*	M2	2.63	2.19	Flynn, 1981
YPM 13810	*Kanisamys indicus*	M3	2.44	1.94	Flynn, 1981
GSI D271	*Kanisamys indicus*	M3	2.44	1.88	Flynn, 1981
YGSP 8109	*Kanisamys indicus*	M3	2.56	2.00	Flynn, 1981
YGSP 8111	*Kanisamys indicus*	M3	2.31	2.00	Flynn, 1981
YGSP 8112	*Kanisamys indicus*	M3	2.44	1.94	Flynn, 1981
YGSP 8220	*Kanisamys indicus*	M3	2.56	2.06	Flynn, 1981
YGSP 8248	*Kanisamys indicus*	M3	2.38	1.94	Flynn, 1981
YGSP 8117	*Kanisamys potwarensis*	M2	2.56	2.56	Flynn, 1981
YGSP 8374	*Kanisamys potwarensis*	M2	2.44	2.69	Flynn, 1981
YGSP 8088	*Kanisamys potwarensis*	M2	2.50	2.81	Flynn, 1981
YGSP 8104	*Kanisamys potwarensis*	m2	2.50	2.50	Flynn, 1981
YGSP 8106	*Kanisamys potwarensis*	m2	2.56	2.44	Flynn, 1981
YGSP 8118	*Kanisamys potwarensis*	m2	2.56	2.50	Flynn, 1981
YGSP 8113	*Kanisamys potwarensis*	m3	2.94	2.25	Flynn, 1981
YGSP 8114	*Kanisamys potwarensis*	m3	2.88	2.25	Flynn, 1981
YGSP 8119	*Kanisamys potwarensis*	m3	2.88	2.38	Flynn, 1981
YGSP 8120	*Kanisamys potwarensis*	m3	3.06	2.44	Flynn, 1981
YGSP 8379	*Kanisamys potwarensis*	m3	2.94	2.31	Flynn, 1981
YGSP 8370	*Kanisamys nagrii*	M1	3.2	2.2	Flynn, 1981
AMNH 30000	*Kanisamys nagrii*	M1	2.7	2.0	Flynn, 1981
AMNH 30000	*Kanisamys nagrii*	M2	2.4	1.9	Flynn, 1981
YGSP 8309	*Kanisamys nagrii*	M2	2.5	2.0+	Flynn, 1981
YGSP 8124	*Kanisamys nagrii*	M2	2.3	2.3	Flynn, 1981
AMNH 39323	*Kanisamys nagrii*	m1	2.8	2.1	Flynn, 1981
GSI 18086	*Kanisamys nagrii*	m2	2.4	2.5	Flynn, 1981
AMNH 30000	*Kanisamys nagrii*	m2	2.5	2.6	Flynn, 1981
AMNH 39323	*Kanisamys nagrii*	m3	2.5	2.3	Flynn, 1981
GSI 18086	*Kanisamys nagrii*	m3	2.7	2.3	Flynn, 1981
Y182	*Kanisamys sivalensis*	M1	2.8	2.3	Flynn, 1981
Y182	*Kanisamys sivalensis*	M2	2.5	2.4	Flynn, 1981
Y182	*Kanisamys sivalensis*	M3	2.4	2.4	Flynn, 1981
Y182	*Kanisamys sivalensis*	m1	2.8	2.1	Flynn, 1981
Y182	*Kanisamys sivalensis*	m2	2.6	2.6	Flynn, 1981

(continued)

Table 16.2 (continued)

Specimen No.	Taxa	Element	Length in mm	Width in mm	Reference
Y182	*Kanisamys sivalensis*	m3	3.0	2.5	Flynn, 1981
GSI D272	*Kanisamys sivalensis*	m1	3.56	3.06	Flynn, 1981
GSI D272	*Kanisamys sivalensis*	m2	3.56	3.75	Flynn, 1981
GSI D272	*Kanisamys sivalensis*	m3	4.00	3.06	Flynn, 1981
YGSP 8360	*Protachyoryctes tatroti*	m1	–	3.00	Flynn, 1981
YGSP 8360	*Protachyoryctes tatroti*	m2	3.63	3.94	Flynn, 1981
YGSP 8360	*Protachyoryctes tatroti*	m3	4.28	3.38	Flynn, 1981
YGSP 15336 (right)	*Protachyoryctes tatroti*	m1	3.88	3.19	Flynn, 1981
YGSP 15336 (right)	*Protachyoryctes tatroti*	m2	3.69	3.63	Flynn, 1981
YGSP 15336 (right)	*Protachyoryctes tatroti*	m3	4.06	3.38	Flynn, 1981
YGSP 15336 (left)	*Protachyoryctes tatroti*	m1	3.88	3.13	Flynn, 1981
YGSP 15336 (left)	*Protachyoryctes tatroti*	m2	3.50	3.44	Flynn, 1981
YGSP 15336 (left)	*Protachyoryctes tatroti*	m3	3.75	3.19	Flynn, 1981
YGSP 15320 (left)	*Eicooryctes kauliaensis*	M1	2.75	2.75	Flynn, 1981
YGSP 15320 (right)	*Eicooryctes kauliaensis*	M1	2.75	2.5+	Flynn, 1981
YGSP 15320 (left)	*Eicooryctes kauliaensis*	M2	3.00	2.75	Flynn, 1981
YGSP 15320 (right)	*Eicooryctes kauliaensis*	M2	3.13	2.7+	Flynn, 1981
YGSP 15308	*Eicooryctes kauliaensis*	M2	3.00	2.88	Flynn, 1981
YGSP 15307	*Eicooryctes kauliaensis*	m1	3.13	2.56	Flynn, 1981
YGSP 15309	*Eicooryctes kauliaensis*	m1	3.0+	2.4+	Flynn, 1981
YGSP 15307	*Eicooryctes kauliaensis*	m2	3.13	3.19	Flynn, 1981
YGSP 15309	*Eicooryctes kauliaensis*	m2	2.94	3.0+	Flynn, 1981
YGSP 15335	*Eicooryctes kauliaensis*	m2	2.81	3.13	Flynn, 1981
YGSP 15335	*Eicooryctes kauliaensis*	m3	3.00	2.69	Flynn, 1981
GSI D97	*Rhizomyides sivalensis*	m2	5.46	5.06	Flynn, 1981
GSI D97	*Rhizomyides sivalensis*	m3	6.13	4.50	Flynn, 1981
GSI D276	*Rhizomyides sivalensis*	m2	5.25	4.31	Flynn, 1981
GSI D276	*Rhizomyides sivalensis*	m3	5.44	4.31	Flynn, 1981
YGSP 15319	*Rhizomyides sivalensis*	m2	4.00	4.94	Flynn, 1981
YGSP 15319	*Rhizomyides sivalensis*	m3	5.38	4.63	Flynn, 1981
DP 394	*Rhizomyides sivalensis*	m2	–	–	Flynn, 1981
DP 394	*Rhizomyides sivalensis*	m3	5.00	4.00	Flynn, 1981
YGSP 8121	*Miorhizomys punjabiensis*	M1	2.94	2.41	Flynn, 1981
YGSP 8123	*Miorhizomys punjabiensis*	M2	2.72	2.75	Flynn, 1981
YGSP 8216	*Miorhizomys punjabiensis*	M2	2.63	2.22	Flynn, 1981
YGSP 8217	*Miorhizomys punjabiensis*	M2	2.78	2.75	Flynn, 1981
YGSP 8125	*Miorhizomys punjabiensis*	m2	2.78	2.75	Flynn, 1981
YGSP 8218	*Miorhizomys punjabiensis*	m2	2.63	2.75	Flynn, 1981
AMNH 19762	*Miorhizomys punjabiensis*	m2	2.75	2.75	Flynn, 1981
GSI D287	*Miorhizomys punjabiensis*	m2	2.63	2.63	Flynn, 1981
AMNH 19762	*Miorhizomys punjabiensis*	m3	3.06	2.56	Flynn, 1981
GSI D287	*Miorhizomys punjabiensis*	m3	3.13	2.44	Flynn, 1981
YGSP 8362	*Miorhizomys nagrii*	M1	3.49	3.49	Flynn, 1981
YGSP 8362	*Miorhizomys nagrii*	M2	3.62	3.65	Flynn, 1981
YGSP 8362	*Miorhizomys nagrii*	M3	2.63	3.14	Flynn, 1981
YGSP 8362	*Miorhizomys nagrii*	m1	4.19	3.12	Flynn, 1981
YGSP 8362	*Miorhizomys nagrii*	m2	3.74	4.00	Flynn, 1981
YGSP 8362	*Miorhizomys nagrii*	m3	3.95	3.48	Flynn, 1981
YGSP 15560	*Miorhizomys blacki*	m2	3.81	4.25	Flynn, 1981
YGSP 15560	*Miorhizomys blacki*	m3	4.13	3.19	Flynn, 1981
YGSP 5494	*Miorhizomys micrus*	m1	–	2.31	Flynn, 1981
YGSP 5494	*Miorhizomys micrus*	m2	3.38	3.19	Flynn, 1981
YGSP 5494	*Miorhizomys micrus*	m3	3.50	2.88	Flynn, 1981

(continued)

Table 16.2 (continued)

Specimen No.	Taxa	Element	Length in mm	Width in mm	Reference
YGSP 8366	*Miorhizomys pilgrimi*	m2	5.25	4.63	Flynn, 1981
YGSP 8366	*Miorhizomys pilgrimi*	m3	5.38	4.56	Flynn, 1981
YGSP 182	*Miorhizomys pilgrimi*	m2	5.25	4.38	Flynn, 1981
YGSP 182	*Miorhizomys pilgrimi*	m3	5.82	4.19	Flynn, 1981
GSI D278	*Miorhizomys pilgrimi*	m2	5.06	5.00	Flynn, 1981
GSI D278	*Miorhizomys pilgrimi*	m3	6.31	4.81	Flynn, 1981
GSI HT6	*Miorhizomys pilgrimi*	m2	5.38	4.75	Flynn, 1981
GSI HT6	*Miorhizomys pilgrimi*	m3	6.00	4.89	Flynn, 1981
YGSP 4810 (right)	*Miorhizomys tetracharax*	M1	3.56	3.94	Flynn, 1981
YGSP 4810 (left)	*Miorhizomys tetracharax*	M1	3.81	4.44	Flynn, 1981
YGSP 4810	*Miorhizomys tetracharax*	M2	4.44	4.63	Flynn, 1981
YGSP 4810 (right)	*Miorhizomys tetracharax*	M3	3.94	3.88	Flynn, 1981
YGSP 4810 (left)	*Miorhizomys tetracharax*	M3	3.69	4.25	Flynn, 1981
YGSP 8170	*Miorhizomys tetracharax*	M3	3.63	5.00	Flynn, 1981
YGSP 39325	*Miorhizomys tetracharax*	m1	4.69	3.31	Flynn, 1981
YGSP 39325	*Miorhizomys tetracharax*	m2	4.63	4.25	Flynn, 1981
YGSP 4810	*Miorhizomys tetracharax*	m2	4.44	4.63	Flynn, 1981
YGSP 9762	*Miorhizomys tetracharax*	m2	3.81	4.0+	Flynn, 1981
YGSP 8169	*Miorhizomys tetracharax*	m2	4.00	4.13	Flynn, 1981
YGSP 39325	*Miorhizomys tetracharax*	m3	5.44	4.13	Flynn, 1981
YGSP 4810	*Miorhizomys tetracharax*	m3	5.19	4.25	Flynn, 1981
YGSP 9762	*Miorhizomys tetracharax*	m3	4.81	3.7+	Flynn, 1981
YGSP 4053	*Miorhizomys choristos*	m1	3.5+	3.1	Flynn, 1981
YGSP 4053	*Miorhizomys choristos*	m2	4.44	4.38	Flynn, 1981
YGSP 4053	*Miorhizomys choristos*	m3	5.22	4.25	Flynn, 1981
DP 678	*Anepsirhizomys opdykei*	m1	6.2	4.7	Flynn, 1981
DP 678	*Anepsirhizomys opdykei*	m2	6.7	5.3	Flynn, 1981

Fig. 16.11 *Miorhizomys harii*. **A**. occlusal view of left M2 (VPL/RP-BR2); **B**. occlusal view of right M2 (VPL/RP-BR3); **C, D, E**, occlusal, labial and lingual views of right m1, m2, and m3 (VPL/RP-BR1)

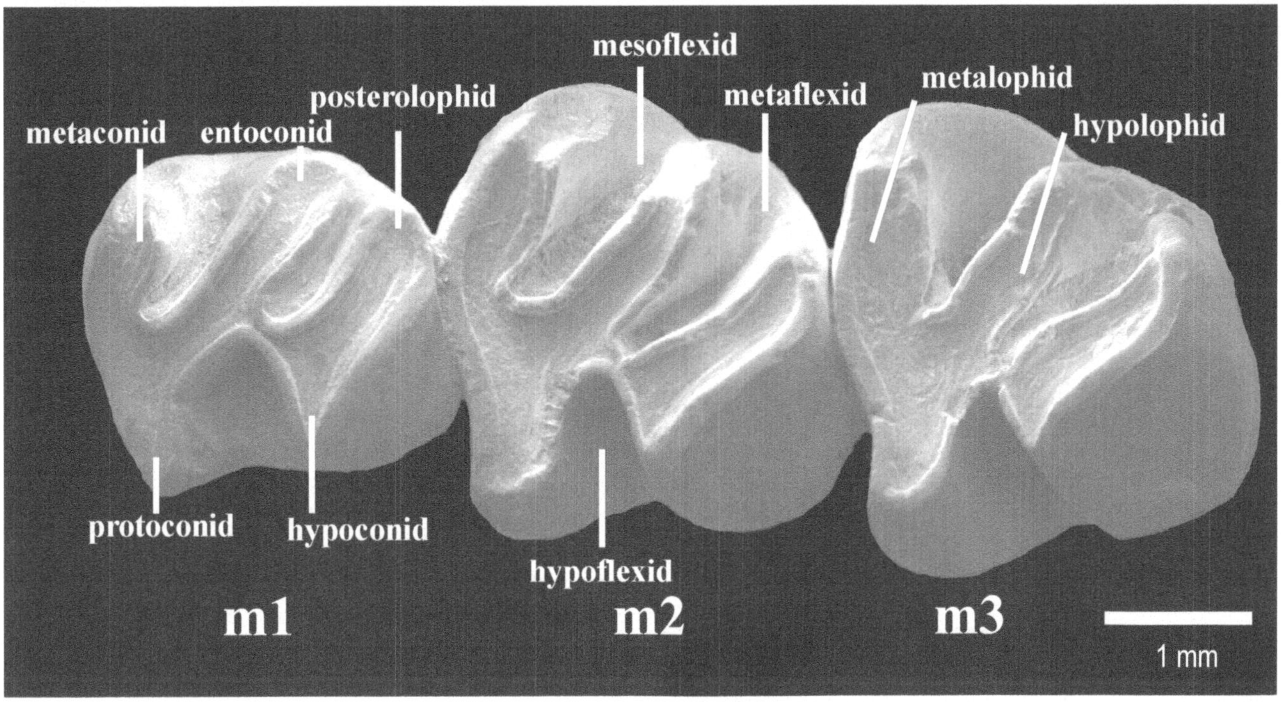

Fig. 16.12 *Sayimys sivalensis*. Occlusal view of right m1, m2, and m3 (VPL/RP-HCT-3). Dental terminology of Baskin (1996) with some adjustments (López-Antoñanzas and Knoll 2011)

The lower incisor is chisel-like, broad and flat anteriorly. It is devoid of any longitudinal ridges.

The occlusal outline of the m1 is oval, narrower anteriorly. The anterolophid is distinct. The anteroconid is however not discernible. Metalophid is straight and terminates lingually at metaconid. Protononid is large, attached to the mesolophid that stops before reaching the lingual border due to early wear (see López-Antoñanzas 2012a, b and López-Antoñanzas 2013a, b, c). The labial re-entrant runs all the way to the lingual side. A trace of mure is visible deep into the re-entrant. The hypoconid is v shaped, its anterior arm comprises hypolophid-entoconid and the posterior arm makes a long anteriorly concave posterolophid. The posterior metaconid-protoconid connection is absent. The posterosinusid is deep and open.

The occlusal outline of the m2 tooth is square (L = 4.16 and W = 4.41). There are four main lophids (anterolophid + metalophid, mesolophid, hypolophid and posterolophid). The protoconid is semicircular and its anterior arm (anterolophid + metalophid) is broad and oblique, oriented posterolabially. Its posterior arm forms the mesolophid which stops before reaching the lingual side. The hypoconid is v-shaped and is not attached to the hypolophid antero-lingually. It is strongly attached to posterolophid. Due to early wear the mure is not visible.

The m3 is longer than wide and is little larger than the m2 (L = 5.38; W = 4.30). It is not fully erupted and is unworn. However, the three main lophids, the anterolophid, mesolophid and hypolophid + posterolophid are distinct. The labial anterolophid is absent, but lingual anterolophid is present. No mure is visible. The mesolophid is long and separated from hypolophid. The posterosinusid is absent. Entoconid is fused with the hypolophid + posterolophid. Protosinusid is absent and the anterosinusid is present.

Comparisons: The teeth are large and in shape and size resemble those of the early tachyoryctine *Rhizomyides* and the late rhizomyine *Miorhizomys*. However, the present taxa differs from *Rhizomyides* in having a shorter m2, a masseteric crest uninflated, shallower dentary, and lower incisor without any longitudinal ridges. It is similar to *Miorhizomys* in having a lower incisor that is flat, without longitudinal ridges, a constricted mure on m1m2, in lacking an extended and inflated masseteric crest and a posterior metaconod-protoconid connection on m1. Among the species of *Miorhizomys* the present specimens are close to those of the Late Miocene *Miorhizomys pilgrimi, Miorhizomys tetracharax* and *Miorhizomys choristos. Anepsirhizomys pinjoricus* comes very close to the present taxa in having several characters in common and differing in being larger and more hypsodont. The present specimens match *M. harri* in size and morphology reported by Prasad (1970).

Family: Ctenodactylidae Zittel, 1893

Genus: Sayimys Wood, 1937

Type species: *Sayimys perplexus* Wood, 1937.

Sayimys sivalensis (Hinton, 1933)

Sayimys sivalensis
Synonyms: *Sayimys perplexus* (PUA 70-22, 70-11) and *S. badauni* (PUA 74-70, 74-73, 74-74).

Type Locality: HD2. 9.1 Ma.

Referred Material: VPL/RP-HCT-3 right lower dentition (Figs. 16.12 and 16.13).

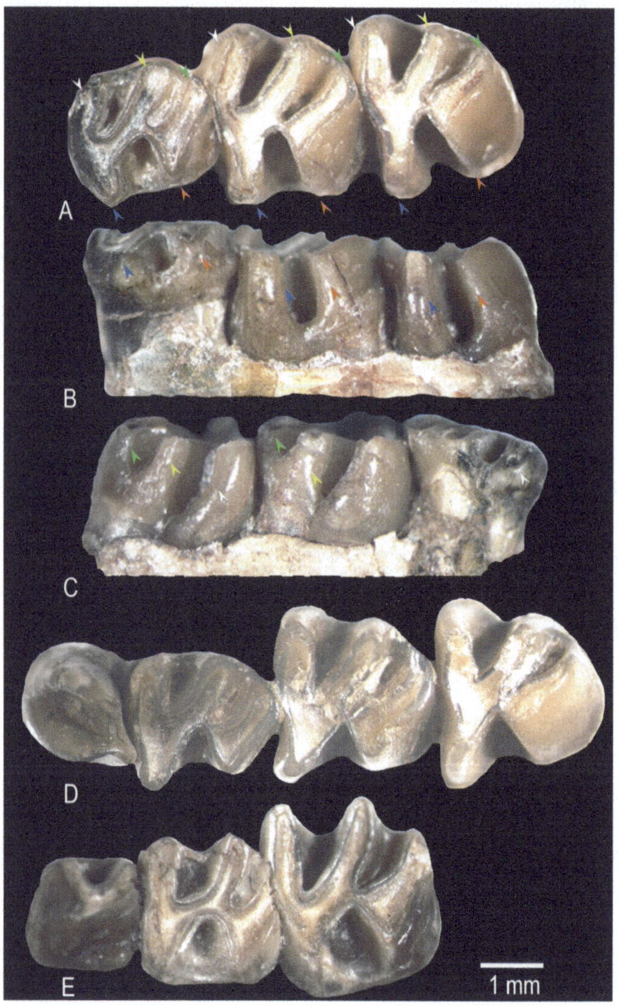

Fig. 16.13 *Sayimys sivalensis*. **A-C**. Occlusal (**A**), lingual (**B**), and labial (**C**) views of right m1, m2, and m3 (VPL/RP-HCT-3). *Sayimys perplexus* **D**. occlusal view of p4, m1, m2 and m3 (PUA-70-22) and **E** *Sayimys badauni* left p4, m1, and m2 (PUA-74-70)

Description: The present mandible is broken ventrally and the incisor is not preserved. The teeth are moderately hypsodont. The m1 is rectangular in occlusal outline. Its anterior margin is convex. The protoconid is v shaped and its anterior arm comprises the metalophid terminating at metaconid. It is much larger than the hypoconid and extends more labially. Till midway the metalophid runs antero-lingually to the midline and then it turns postero-lingually. Its posterior arm or hypolophid runs obliquely postero-lingually terminating at entoconid. The hypolophid is longer than the metalophid and posterolophid. The hypoconid is also v shaped and has a posterior arm posterolophid that runs obliquely almost parallel to the hypolophid. The hypoflexid is an inverted 'u' in shape, very deep and extends up to the middle of the tooth. On the lingual side, the mesoflexid is deeper and longer compared to the metaflexid (Fig. 16.13A, C). A posterior shelf or cingulum is present on the posterolabial margin (Fig. 16.13B, C).

The m2 is triangular in occlusal outline and its anterior margin is 's' shaped. Protoconid is much larger than the hypoconid and extends more labially. As in m1 the metalophid bends midway posteriorly. Again, mesoflexid is deeper than metaflexid but both are of almost equal length. Unlike in m1 the metaflexid is wider, becomes narrower towards the middle of the tooth and thus hypolophid and posterolophid are not parallel to each other. Also, the posterolophid is more oblique and bent more posterolingually compared to the hypolophid. Hypoflexid is very deep and it appears an inverted 'v' in shape (Fig. 16.13A, B). The posterior cingulum is distinct on the posterolabial margin (Fig. 16.13B, C).

The m3 is very similar to m2, but its straight metalophid runs across m3 rather than bending posteriorly. The hypolohid is oblique and so is the posterolophid. The mesoflexid is slightly shorter than the metaflexid. A faint trace of posterior cingulum is present.

Comparison: Compared to the long ranging (Middle Miocene-Late Miocene) and well sampled Siwalik form *S. sivalensis* the present m1m2 are a little larger in size, their metalophids are bent posteriorly. However, the protoconid and metaconid height (1.8 mm) falls very much within the domain of the latter (see Fig. 5B, López-Antoñanzas and Sen 2003). *S. perplexus* illustrated and described (PUA 70-22; m2 L = 2.62; W = 2.77 mm) from 600 m east of Haritalyangar by Vasishat (1985) are very similar to those of *S. sivalensis* in having their anterior margins straight with metalophids running across the tooth (re-illustrated, Fig. 16.13D). Vasishat (1985) described a new species *S. badauni* from a younger locality (Badaun), which he claimed to be of Pliocene age. *S. badauni*'s highly worn lower molars (Vasishat 1985: PUA 74-70, pl.24, Figs. a–c) although a little larger in size (m2 L = 2.90; W = 3.00 mm) appear very similar to those of *S. sivalensis* in having their anterior margins straight with metalophids running across the tooth (re-illustrated, Fig. 16.13E). When plotted on a graph (López-Antoñanzas et al. 2016: Fig. 3) showing length and width of m2 of various *Sayimys* species both *S. perplexus* and *S. badauni* of Vasishat (1985) fell close to the largest *S. sivalensis* m2 of López-Antoñanzas et al. (2016) and *S. sivalensis* m2 described here.

Discussion: The type material of *S. perplexus* Wood, 1937 comes from Nagri Formation exposed at Haritalyangar, the origin of the present specimen. Based on a much larger sample, Munthe (1980) considered *S. perplexus* as a junior synonym of *S. sivalensis* (Hinton 1933), with which López-Antoñanzas and Sen (2003), López-Antoñanzas and Knoll (2011) agree. Munthe (1980) observed a wide variation in the molar morphology of *S. sivalensis* population that included the differential pattern demonstrated by *S. sivalensis* from older levels and younger *S. perplexus*. However, Black (1972), De Bruijn et al. (1989) and Baskin (1996) considered, *S. perplexus* to be valid. Black (1972) argued that *S. perplexus* is larger, metaconid-entoconid are separated by a wider gap, and the hypolophid is more oblique compared to that in *S. sivalensis*. Referring to Munthe (1980), López-Antoñanzas and Sen (2003) found these characters to subsume within the variation of *S. sivalensis* and are not diagnostic enough for erecting a separate species. On similar lines *S. badauni* was also included within *S. sivalensis* (López-Antoñanzas et al. 2004; López-Antoñanzas and Knoll 2011). Black (1972) also mentioned in his diagnosis of *S. perplexus* that the presence of an "anterior loph on m1 and m2 bent at midpoint rather than running straight across tooth" setting this species apart from *S. sivalensis*. Munthe (1980) clearly mentions on page 22 that "since the lingual part of the metalophid slopes first anteriorly and then posteriorly as wear proceeds through the crown, this metaconid-protoconid relationship depends in part on stage of wear". Interestingly, this variation in metalophid position can be clearly seen on the present specimen itself (Fig. 16.13A and C, white arrows). Not only this, the hypolophid also slopes, first posteriorly and then anteriorly (Fig. 16.13A and C, yellow arrows), whereas the posterolophid gently slopes antero-lingually (Fig. 16.13A and C, green arrows). Labially the protoconid very gently slopes posteriorly (Fig. 16.13A and B, blue arrows), whereas the hypoconid slopes posteriorly and at the very end slopes anteriorly (Fig. 16.13A and B, red arrows). The m3 is less worn compared to m2, which in turn is less worn than the m1. The metalophid on m3 is straight, whereas it becomes more oblique as we move from m2 to m1. On similar lines one would expect the variability in the position of hypolophid, posterolophid on the lingual side and protoconid and hypoconid on the labial side depending upon the stage of wear. Therefore, it is quite convincing to allocate the specimens described as *S. perplexus* (Wood 1937; Prasad 1970; Vasishat 1985) and *S. badauni* (Vasishat 1985) to *S. sivalensis*. The age of the rocks exposed around Badaun, is controversial. Prasad (1971) is of the view that Tatrot-equivalent beds are exposed there, whereas Johnson and Vondra (1972) think that there is a break after Dhok Pathan in this region which in turn is overlain by Boulder Conglomerates. Until better age control is available for this site, *S. badauni* (*S. sivalensis*) should be assigned to latest Miocene, as is *Miorhizomys harii* from Baron (a site close to Badaun).

Phylogenetic Analyses

Combining generalised and derived characters identified by Musser (1981), Catzeflis and Denys (1992), Chevret et al. (2003) and those observed herein on the partial cranium of *Progonomys debruijni* a set of character states are listed below for some of the species belonging to various subgenera of *Mus*.

Characters

1. Upper incisors: opisthodont (0); orthodont (1); proodont (2).
2. Upper incisors with subapical notch: absent (0); present (1).
3. Incisive foramina: moderate in size, stops right at the anterior root of the first upper molar (0); penetrates between the anterior root of M1 or between root 2 of M1 (1); short, stops much before it reaches the anterior part of the first upper molar (2).
4. Zygomatic notch: moderate (0); deep (1); absent or barely evident (2).
5. Anterior margin of Zygomatic plate: straight (0); curved (1).
6. Supraorbital ridges: absent (0); present over posterior region of orbit, but temporal ridges absent (1); distinct, extending from frontals to parietals (2).
7. Rostrum: short and deep (0); long and narrow (1).
8. Anterior lobe of M1: normal (0); elongated with an accessory cusp (1); very short (2).
9. Position of t1: displaced well posterior to t2 (0); displaced slightly posterior to t2 (1).
10. Inclination of t1: inclined (0); vertical (1).
11. Posterior cingulum on M1: present (0); absent (1).
12. M3: moderate in size (0); reduced but larger than t8 and t9 of M2 (1); very reduced, smaller than t8 and t9 of M2 (2).
13. Suture between maxillary and palatine: level with anterior of M2 or at the junction between M1 and M2 (0);

Table 16.3 Matrix of character codings used in the analysis of relationships among *Progonomys debruijni* and various species of *Mus*. *Progonomys debruijni* has been used here as the out-group taxon

S. no	Characters taxa	1	2	3	4	5	6	7	8	9	10	11	12	13
1	*Progonomys debruijni*	0	0	0	0	0	0	0	0	0	0	0	?	0
2	*Mus linnaeusi*	1	1	0	0	0	0	0	0	0	0	1	1	0
3	*Mus (Nannomys) setulosus*	1	1	0	0	1	1	0	1	0	0	1	2	1
4	*Mus (Nannomys) minutoides*	0	1	0	1	1	1	0	1	0	0	1	2	1
5	*Mus (Nannomys) mathayi*	1	1	0	1	1	1	0	1	0	0	1	2	1
6	*Mus (Pyromys) saxicola*	1	0	1	1	1	2	0	2	1	0	1	0	0
7	*Mus (Pyromys) shortridgei*	1	1	0	1	1	2	0	1	1	0	1	0	0
8	*Mus (Coelomys) pahari*	1	0	2	1	0	1	1	0	1	1	1	2	1
9	*Mus (Coelomys) mayori*	0	0	2	1	1	2	1	0	1	1	1	2	2
10	*Mus (Coelomys) crociduroides*	0	0	2	2	1	0	1	0	1	1	1	2	2
11	*Mus musculus*	1	1	0	1	0	1	0	0	0	0	1	0	0
12	*Mus cooki*	1	1	0	1	0	1	0	0	0	0	1	1	0
13	*Mus caroli*	1	1	0	1	0	1	0	0	0	0	1	1	0
14	*Mus cervicolor*	0	1	1	1	0	1	0	0	0	0	1	1	0
15	*Mus spretus*	1	1	0	1	0	1	0	1	0	0	1	1	0
16	*Mus spicilegus*	¿	1	0	1	0	1	0	1	0	0	1	0	0

level with posterior of M1 (1); level with mid-region of M1 (2).

Based on the distribution of the above character states among various species of *Mus* (Table 16.3), with *Progonomys debruijni* being the out group, a cladistic analysis using PAUP was run to asses how various subgenera of *Mus* are related to *Progonomys debruijni*.

Bootstrap method with heuristic search:
Number of bootstrap replicates = 1000
Starting seed = 1926734987
Optimality criterion = parsimony
Character-status summary:
Of 13 total characters:
All characters are of type 'ord' (Wagner)
All characters have equal weight
1 character is parsimony-uninformative
Number of parsimony-informative characters = 12
Starting tree(s) obtained via stepwise addition
Addition sequence: simple (reference taxon = Prog)
Number of trees held at each step during stepwise addition = 1
Branch-swapping algorithm: tree-bisection-reconnection (TBR)
Steepest descent option not in effect
Initial 'MaxTrees' setting = 100
Branches collapsed (creating polytomies) if maximum branch length is zero
'MulTrees' option in effect
Topological constraints not enforced
Trees are unrooted
1000 bootstrap replicates completed
Note: Effectiveness of search may have been diminished due to tree-buffer overflow.
Time used = 34.63 sec
Heuristic search settings:
Optimality criterion = parsimony
Character-status summary:
Of 13 total characters:
All characters are of type 'ord' (Wagner)
All characters have equal weight
1 character is parsimony-uninformative
Number of parsimony-informative characters = 12
Starting tree(s) obtained via stepwise addition
Addition sequence: random
Number of replicates = 10
Starting seed = 101993044
Number of trees held at each step during stepwise addition = 1
Branch-swapping algorithm: tree-bisection-reconnection (TBR)
Steepest descent option not in effect
Initial 'MaxTrees' setting = 100
Branches collapsed (creating polytomies) if maximum branch length is zero
'MulTrees' option in effect
Topological constraints not enforced
Trees are unrooted
'MaxTrees' limit (100) hit while swapping on tree #8 (score = 35)
Search terminated prematurely (no room to store new trees)
100 trees retained
Time used = 1.60 sec
100 trees converted from unrooted to rooted.
Character-state optimization: Accelerated transformation (ACCTRAN)
AncStates = "standard"
Tree number 1 (rooted using default outgroup)
Tree length = 35
Consistency index (CI) = 0.5429
Homoplasy index (HI) = 0.4571
CI excluding uninformative characters = 0.5294
HI excluding uninformative characters = 0.4706
Retention index (RI) = 0.7193
Rescaled consistency index (RC) = 0.3905

Both 50% majority rule consensus and Bootstrap 50% majority rule consensus trees support the monophyly of the

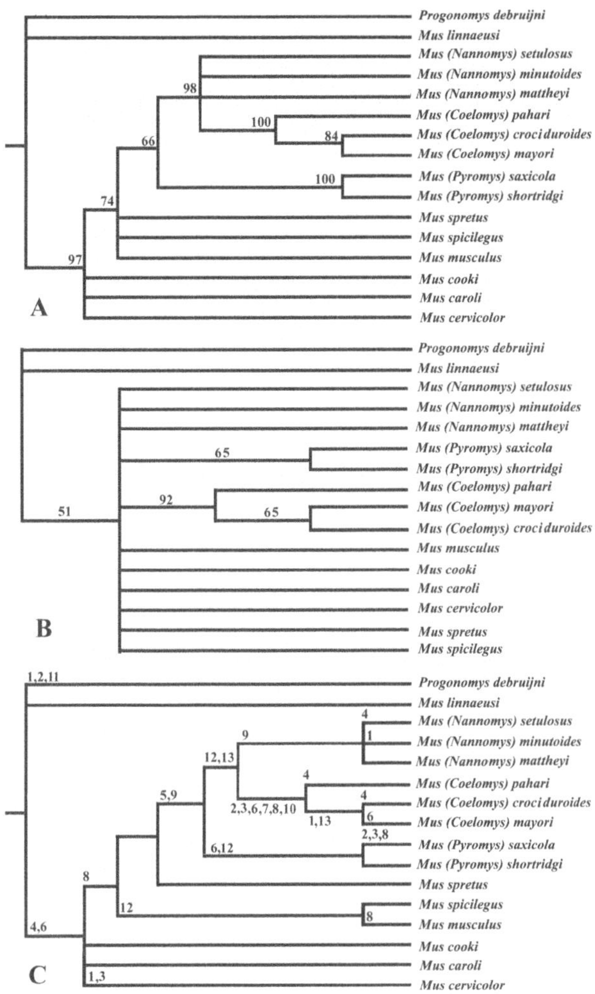

Fig. 16.14 A, 50% majority rule consensus of 96 trees with frequency of occurrences shown above the branch.; B, Bootstrap 50% majority rule consensus tree, with bootstrap support values shown above the branches; C, Heuristic search consensus tree, with branches supported by various characters. Total number of rearrangements tried = 14641. Score of best tree(s) found = 35. Number of trees retained = 100. Tree length = 35. Consistency index (CI) = 0.5429. Homoplasy index (HI) = 0.4571. Retention index (RI) = 0.7193. Rescaled consistency index (RC) = 0.3905

genus *Mus* and its separation into four subgenera; the Eurasian *Mus* (*M. cervicolor, M. cooki, M. caroli, M. musculus, M. spicilegus* and *M. spretus*), African *Nannomys* (*M. setulosus*, *M. mattheyi* and *M. minutoides*) and the Asian *Coelomys* (*M. pahari*, *M. mayori* and *M. crociduroides*) and *Pyromys* (*M. saxicola* and *M. shortridgi*) (Fig. 16.14A and B). The Eurasian subgenus *Mus* appears to be deeply rooted and it seems to be the first to diverge from a *Progonomys*-like ancestor, followed by the split of *Pyromys* from a clade comprising *Nannomys* and *Coelomys*. The fossil species *Mus linnaeusi* was first to diverge from the *Progonomys* lineage. This makes sense as the oldest *Mus* known is ~8 Ma (Kimura et al. 2015). *Mus auctor* appears later at ~6.5 Ma (see Kimura et al. 2015) but has a rather generalised molar morphology (Patnaik 2014) similar to that seen in the subgenus *Mus* and potentially could be an ancestor to *Mus linnaeusi*. *Mus linnaeusi* has been placed in the subgenus *Mus* owing to its short rostrum, moderate sized incisive foramina that penetrates at the anterior root of M1 and opisthodont upper incisors (Patnaik 1997). Next to diverge are the Asiatic taxa *M. cervicolor*, *M. caroli* and *M. cooki* followed by the Eurasiatic *M. spicilegus, M. spretus* and *M. musculus*. The next clade comprises the Asian *Pyromys*, the African *Nannomys* (*M. setulosus*, *M. mattheyi* and *M. minutoides*) and *Coelomys* (*M. pahari*, *M. mayori* and *M. Crociduroides*) supported by a frequency of 98%. The clade *Mus* (*Pyromys*) *saxicola* and *Mus* (*Pyromys*) *shortridgi* occur together with a frequency of 100%. There is a 100% support for a monophyletic group of the subgenus *Coelomys* (*M. pahari*, *M. mayori* and *M. Crociduroides*). The bootstrap consensus tree performed with a heuristic search (see Fig. 16.15C) also shows strong support for clades of *Pyromys* and *Coelomys*.

The cladogram obtained from the heuristic search supports the overall monophyly of the genus *Mus*, consistent

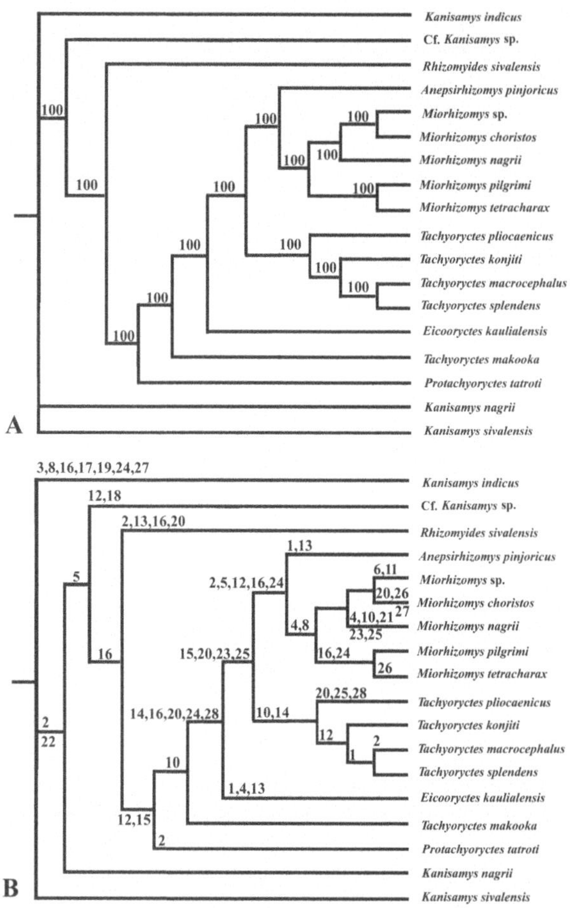

Fig. 16.15 **A**, 50% Majority-rule consensus of 2 trees with frequency of occurrences shown above the branch; **B**, Heuristic search consensus tree, with branches supported by various characters. Total number of rearrangements tried = 55808. Score of best tree(s) found = 68. Number of trees retained = 2. Tree length = 68. Consistency index (CI) = 0.5735. Homoplasy index (HI) = 0.4754. Retention index (RI) = 0.7100. Rescaled consistency index (RC) = 0.4072

with studies based on allozymes (Bonhomme et al. 1984), DNA–DNA hybridization (She et al. 1990; Chevret et al. 2003) and DNA sequencing (Lundrigan et al. 2002; Chevret et al. 2003). The tree topology, however, contrasts with that proposed by Chevret et al. (2003) on morphological and molecular grounds. Chevret et al. (2003: Figs. 1–3 and 5; 2005: Fig. 3) obtained trees where *Coelomys* or *Nannomys* are commonly first to split and the subgenus *Mus* usually the last. Veyrunes et al. (2006) reconstructed an ancestral karyotype for the genus *Mus* with *Coelomys* to be the earliest branch of the *Mus* radiation, followed by the divergence of *Nannomys* from *Pyromys* and *Mus*. The present topology is similar to that of a recent DNA based phylogenetic study on murids, where, Aghová et al. (2018) found the subgenus *Mus* to split first, while *Nannomys* is the last offshoot. Their cladogram differs from the present one having *Coelomys* to split next followed by *Pyromys*.

Spalacids

Recently López-Antoñanzas et al. (2013) carried out a comprehensive cladistic analysis of the spalacids. The characters they identified on mandibles, lower molars and incisors have been used here to place the present taxa in a phylogenetic tree.

Characters

1. Hypsodonty: Crown height on at least one side of the tooth greater than length or width of the tooth. This character has been scored as (0) absent; (1) moderate; (2) high (Table 16.4).
2. Size: Length of the m2: (0) 1.5–2.5 mm; (1) 2.5–3.5 mm; (2) 3.5–4.5 mm; (3) 4.5–5.5 mm; (4) >5.5 mm.
3. Lophodonty: (0) weak; (1) moderate; (2) high.
4. Masseteric crest: (0) with long anterior extension; (1) with short anterior extension; (2) without anterior extension.
5. Lower masseteric crest: (0) inflated under m2; (1) not inflated.
6. Dentary depth: (0) shallow; (1) moderate; (2) deep.
7. Infraorbital foramen: (0) with ventral slit; (1) with abbreviated ventral slit; (2) without ventral slit.
8. Longitudinal ornamentation on the lower incisor: (0) double ridge; (1) single ridge; (2) absent.
9. Anteroconid on the m1: (0) discernible); (1) absent (fused with the anterolophid).
10. Mure on the m1: (0) present; (1) constricted; (2) absent.
11. Labial anterolophid on the m1: (0) present; (1) absent.
12. Mesolophid on the m1: (0) long and complete and separated from the hypolophid; (1) short and well separated from the hypolophid; (2) short and migrated towards the hypolophid; (3) absent; (4) a long continuation of the protoconid.
13. Posterior protoconid–metaconid connection on the m1: (0) present; (1) absent.
14. Anterosinusid on the m1: (0) present; (1) absent.
15. Mure on the m2: (0) present; (1) constricted; (2) absent.
16. Mesolophid on the m2: (0) long and complete and separated from the hypolophid; (1) short and well separated from the hypolophid; (2) short and migrated towards the hypolophid; (3) absent; (4) a long continuation of the protoconid.
17. Labial anterolophid on the m2: (0) present; (1) absent.
18. Hypolophid on the m2: (0) not isolated; (1) isolated.
19. Protosinusid on the m2: (0) present; (1) absent.
20. Anterosinusid on the m2: (0) present; (1) absent.

Table 16.4 Matrix of character codings used in the analysis of relationships among Siwalik and African spalacids. *Kanisamys indicus* has been used here as the out-group taxa

Characters taxa	1	2	3	4	5	6	7	8	9	10	11	12	13	14	15	16	17	18	19	20	21	22	23	24	25	26	27	28
Kanisamys indicus	1	0	1	0	1	0	?	0	1	0	0.1	0	0.1	0	0	0	0	0	0	0	1	0	0	0	0	0	0	0
Kanisamys sivalensis	1	1	2	0	1	0	1	1	1	0	0.1	0.2	1	0	0	2	1	0	1	0	1	1	0	2	0	0	1	0.1
Kanisamys nagrii	1	0.1	2	0	1	0	1	1	1	0	0.1	0.2	1	0	0	2	1	0	1	0	1	0.1	0	2	0	0	1	0.1
cf. *Kanisamys* sp.	1	1	2	0	0	0	?	1	?	0	?	1	?	?	0	2	1	0	1	0	1	1	0	2	0	0	1	0
Rhizomyides sivalensis	1	3	2	0	0	1	?	1	1	0	1	0	0	0	0	0	1	0	1	1	1	1	0	2	0	0	1	0
Protachyoryctes tatroti	1	2	2	0	0	1	?	1	1	0	1	2	1	0	1	2	1	0	1	0	1	1	0	2	0	0	1	0
Tachyoryctes makooka	1	1.2	2	?	?	?	1	?	1	1	1	2	1	0	1	2	1	0	1	0	1	1	0	2	0	0	1	0
Eicooryctes kaulialensis	2	1	2	1	0	1	?	1	1	1	1	2	0	1	1	2.3	1	0	1	1	1	1	0	3	0	0	1	1
Tachyoryctes pliocaenicus	1	0.1	2	0	0	1	1	1	1	2	1	2	1	0	2	2.3	1	0	1	0	1	1	1	3	1	0	1	0
Tachyoryctes konjiti	1	1	2	?	?	?	?	1	1	2	1	3	1	0	2	3	1	0	1	1	1	1	1	3	2	0	1	1
Tachyoryctes macrocephalus	2	2	2	0	0	1	1	1	1	2	1	3	1	0	2	3	1	0	1	1	1	1	1	3	2	0	1	1
Tachyoryctes splendens	2	0.1	2	0	0	1	1	1	1	2	1	3	1	0	2	3	1	0	1	1	1	1	1	3	2	0	1	1
Miorhizomys harii	1	2	2	2	1	0	?	2	1	1	0	0	1	0	1	0	1	0	1	1	1	1	1	0	2	0	1	0
Miorhizomys pilgrimi	1	3	2	2	1	1	1	2	?	?	?	?	?	?	2	4	1	0	1	1	1	1	1	4	1.2	0	1	1
Miorhizomys nagrii	1	1.2	2	1	1	1	1	2	1	0	1	0.1	1	1	0	0	1	0	1	1	0	1	0	0.2	1	0	1	1
Miorhizomys choristos	2	2.3	2	2	1	1	1	2	1	1	1	0	1	0	1	0	1	0	1	0	1	1	1	4	2	1	0	0
Miorhizomys tetracharax	2	2.3	2	2	1	1	1	2	1	1	1	0.1	1	1	2	4	1	0	1	1	1	1	1	4	2	1	1	1
Anepsirhizomys pinjoricus	2	3	2	0	?	1	?	1	1	?	1	0	0	1	2	0	1	0	1	1	1	1	1	0	2	0	1	?

21. m3: (0) reduced (shorter than the m2); (1) enlarged (equal or larger than the m2).
22. Labial anterolophid on the m3: (0) present; (1) absent.
23. Mure on the m3: (0) present; (1) absent.
24. Mesolophid on the m3: (0) long and complete, separated from the hypolophid; (1) short, well separated from the hypolophid; (2) short, migrated towards the hypolophid; (3) absent; (4) a long continuation of the protoconid.
25. Posterosinusid on the m3: (0) large; (1) small; (2) absent (posterolophid fused with hypolophid).
26. Entoconid on the m3: (0) not isolated; (1) isolated.
27. Protosinusid on the m3: (0) present; (1) absent.
28. Anterosinusid on the m3: (0) present; (1) absent.

Kanisamys indicus was chosen as the outgroup because it is the oldest *Kanisamys* species recovered and is basal to all the later tachyoryctines (both Asian and African) and rhizomyines. Due to the lack of *a priori* information, all characters were unordered and equally weighted, a procedure followed by López-Antoñanzas et al. (2013).

Processing of file "C:\Users\Rajeev\Desktop\SringerChapter \Rhizo.nex"
begins...
Data matrix has 18 taxa, 28 characters
Valid character-state symbols: 01234
Missing data identified by '?'
Processing of file "C:\Users\Rajeev\Desktop\SringerChapter \Rhizo.nex"
completed.
Bootstrap method with heuristic search:
Number of bootstrap replicates = 100
Starting seed = 263066449
Optimality criterion = parsimony
Character-status summary:
Of 28 total characters:
All characters are of type 'unord'
All characters have equal weight
2 characters are constant
7 variable characters are parsimony-uninformative
Number of parsimony-informative characters = 19
Multistate taxa interpreted as uncertainty
Starting tree(s) obtained via stepwise addition
Addition sequence: simple (reference taxon = KanI)
Number of trees held at each step during stepwise addition = 1
Branch-swapping algorithm: tree-bisection-reconnection (TBR)
Steepest descent option not in effect
Initial 'MaxTrees' setting = 100
Branches collapsed (creating polytomies) if maximum branch length is zero
'MulTrees' option in effect
Topological constraints not enforced
Trees are unrooted
100 bootstrap replicates completed
Note: Effectiveness of search may have been diminished due to tree-buffer
overflow.
Time used = 12.56 sec Heuristic search settings:
Optimality criterion = parsimony
Character-status summary:
Of 28 total characters:
All characters are of type 'unord'
All characters have equal weight
2 characters are constant
7 variable characters are parsimony-uninformative
Number of parsimony-informative characters = 19
Multistate taxa interpreted as uncertainty
Starting tree(s) obtained via stepwise addition
Addition sequence: simple (reference taxon = KanI)
Number of trees held at each step during stepwise addition = 1
Branch-swapping algorithm: tree-bisection-reconnection (TBR)
Steepest descent option not in effect
Initial 'MaxTrees' setting = 100
Branches collapsed (creating polytomies) if maximum branch length is zero
'MulTrees' option in effect
Topological constraints not enforced
Trees are unrooted
Heuristic search completed
Total number of rearrangements tried = 55808
Score of best tree(s) found = 68
Number of trees retained = 2
Time used = 0.05 sec 2 trees converted from unrooted to rooted.
Tree description:
Optimality criterion = parsimony
Character-status summary:
Of 28 total characters:
All characters are of type 'unord'
All characters have equal weight
2 characters are constant
7 variable characters are parsimony-uninformative
Number of parsimony-informative characters = 19
Multistate taxa interpreted as uncertainty
Character-state optimization: Accelerated transformation (ACCTRAN)
AncStates = "standard"
Tree number 1 (rooted using default outgroup)
Tree length = 68
Consistency index (CI) = 0.5735
Homoplasy index (HI) = 0.4265
CI excluding uninformative characters = 0.5246
HI excluding uninformative characters = 0.4754
Retention index (RI) = 0.7100
Rescaled consistency index (RC) = 0.4072

The 50% majority rule consensus as well as heuristic search trees (Fig. 16.15A and B) support cf. *Kanisamys* sp. as derived from the *Kanisamys* lineage and basal to later tachyoryctines and rhizomyines. The topology of the cladogram indicates that the large taxa of *Rhizomyides*, separated early, followed by a split of *Protachyoryctes tatroti*, then, *Tachyoryctes* (*Protachyoryctes*) *makooka,* which in turn is basal to *Eicooryctes kaulialensis*. Next to diverge are the African *Tachyoryctes pliocaenicus, Tachyoryctes konjiti, Tachyoryctes macrocephalus* and *Tachyoryctes splendens,* forming a monophyletic clade supported by 100% frequency. The younger genus *Anepsirhizomys* groups with a clade of *Miorhizomys*, also supported by 100% frequency, suggesting that this genus is a rhizomyin.

The present *Miorhizomys harii* is nested among the various contemporary *Miorhizomys* species, a monophyletic clade supported by 100% frequency.

As per López-Antoñanzas et al. (2013) the genus *Kanisamys* is not monophyletic and the present topology agrees. Flynn (1982, 1984, 1986) suggested that *Kanisamys nagrii* and *Kanisamys sivalensis* are more derived than *K. indicus* and that *K. indicus* would be a direct ancestor of *Kanisamys sivalensis* (Black 1972; Flynn 1982), which is also substantiated by the present results. An unidirectional dispersal event from southern Asia to Africa at the origin of *Tachyoryctes* (*Protachyoryctes*) *makooka* inferred between 6 and 8 Ma (López-Antoñanzas et al. 2013), followed by a second that gave rise to *Tachyoryctes pliocaenicus, Tachyoryctes konjiti, Tachyoryctes macrocephalus* and *Tachyoryctes splendens* is well supported here.

Rhizomyides sivalensis separates from *Miorhizomy* early in the geological history as observed by López-Antoñanzas et al. (2013). Results obtained by López-Antoñanzas et al. (2013, 2015a) suggest that the genus *Miorhizomys* does not appear to be monophyletic. In the present cladogram, however, it appears monophyletic. Flynn et al. (1990) were of the opinion that *Anepsirhizomys* can be considered as a rhizomyine as well as a tachyoryctine. According to the results of this analysis, *Anepsirhizomys pinjoricus* shares a clade with the rhizomyine *Miorhizomys*. Uncertainities reflect the low number of characters and taxa used in the present analysis.

Paleobiogeography

The tectonic connection between Africa and Eurasia in the Early Miocene (23.0–16.0 Ma), led to formation of a landbridge popularly known as the '*Gomphotherium landbridge*' (Rögl 1999) (Fig. 16.16). Several mammals such as proboscideans, hyracoids, apes and aardvarks took advantage to disperse from Africa to Eurasia and rhinos, ruminants, carnivores, lagomorphs and several rodents from Eurasia to Africa across this bridge, time and again (Sen 2013; Patnaik 2016; Gilbert et al. 2020). Early in the Miocene (20–18 Ma) the early murids (ex. *Potwarmus*, see López-Antoñanzas 2009) colonized both Eurasia and North Africa (Aghová et al. 2018). The reopening of the Mediterranean-Indo-Pacific seaway broke this connection leading to a separate development of Afrotropical groups (Deomyinae, Gerbillinae and Lophiomyinae) and Indomalayan (Murinae) rodent clades (Aghová et al. 2018). Early murines were definitely forest dwellers and browsers (Flynn 2003; Kimura et al. 2013b; Patnaik 2015). Later forms however, became more adapted to open habitat and relied on grass, and insects (Kimura et al. 2013b; Patnaik 2015). At the beginning of the Serravallian *ca.* 13.8 Ma ('Parathethys Salinity Crisis'; Rögl 1999) the connection was again established coinciding with a global cooling (Miller et al. 1991), leading to vegetation shift and seasonal aridification. This allowed repeated dispersals of murine rodents that preferred open habitat from Asia to both Africa and Eurasia, during the Late Miocene. Murine fossil records provide clear evidence for connections between the Indomalaya, the Palearctic, and the Afrotropics. Among them, the dispersal of *Progonomys* is well known. Specimens assigned to near *Progonomys* appear around 11.2 Ma in the Siwaliks (Kimura et al. 2107; Flynn et al. 2020). *Progonomys* has been recovered from many Indomalayan Middle Miocene localities (Jacobs and Flynn 2005). Sometime between 11 and 10 Ma *Progonomys* disperses westward to the Mediterranean area appearing in Egypt (Heissig 1982), Algeria (Wessels 2009), France (Mein et al. 1993) and Spain (Weerd 1976). This is followed by a northeastward dispersal into China (Qiu et al. 2004; Flynn and Wessels 2013). Most likely *Progonomys hussaini* from the subcontinent gave rise to the European *P. cathalai* Schaub (1938) and then to the Chinese *Progonomys sinensis* Qiu et al. (2004) (Kimura et al. 2017). Early *Progonomys* carbon isotope values indicate a C3 diet (Kimura et al. 2013b) most likely a seasonal forest. Dispersal would have been promoted if seasonal forests across South Asia, Europe, China and North Africa was were widespread between 11 and 10 Ma. Recently, Kimura et al. (2017) described a new species *Parapodemus badgleyae* from an older (10.5 Ma) level relative to European *Parapodemus* yielding sites and suggested that *Parapodemus* also dispersed westward from South Asia, giving rise to the European forms, as did *Progonomys* (Kimura et al. 2017). Later, lineages introduced *Progonomys, Karnimata* and *Parapelomys* to the Late Miocene of East Africa (Geraads 2001; Winkler 2002) from Pakistan (Jacobs and Flynn 2005). The Messinian Salinity Crisis *ca.* 6 Ma during the Late Miocene (Hsü 1978) probably led to another faunal interchange of murid taxa between Africa, Asia and Western Palearctic (Benammi et al. 1996; Sabatier 1982; Sen 1977, 1983; Winkler 2002). Splitting from the Asian subgenera of *Mus*, the African subgenus *Nannomys* (genus *Mus*), perhaps colonized the Afrotropics *ca.* 4.5 Ma (Tugen Hills, Kenya) (Bryja et al. 2014). It has been found that beside Late Miocene *Mus* sp. and *Mus auctor,* dental remains have also been found from the Pliocene and Pleistocene of the Indian subcontinent (see Patnaik 2014 for a review). However, their assignment to various subgenera is difficult based solely on the dental features. Therefore assignment of fossil *Mus* from East Africa to the subgenus *Nannomys* solely based on dental remains is difficult (also see views of Catzeflis and Denys 1992). In southeast Asia, Chaimanee (1998) discovered Plio-Pleistocene taxa that she assigned to the extant subgenera *Mus, Coelomys* and *Pyromys*. Chaimanee (1998)

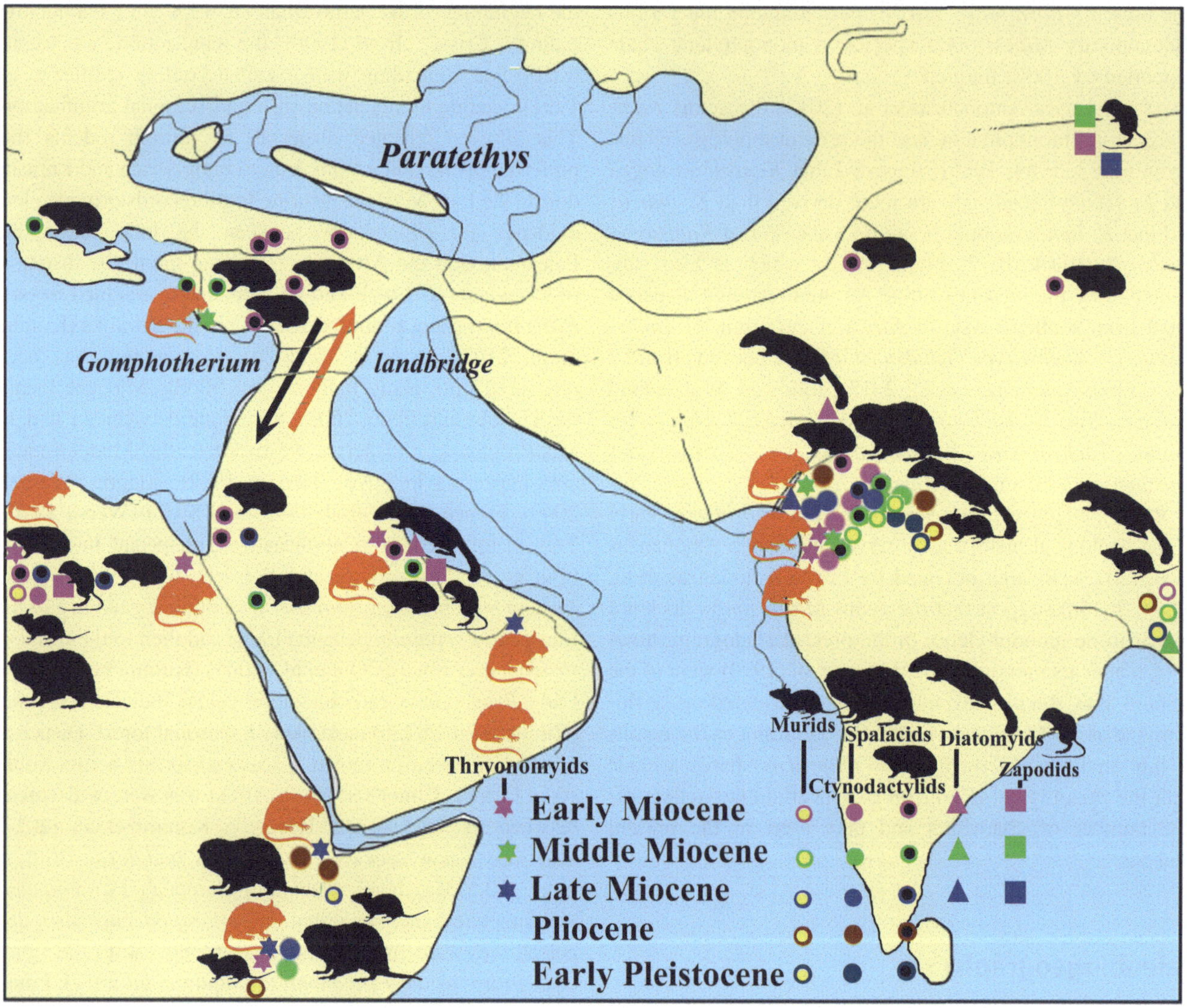

Fig. 16.16 The "Gomphotherium landbridge" formed during the early Miocene (map based on Rögl 1999), Early, Middle, Late Miocene, Pliocene and early Pleistocene sites of murids, spalacids, ctenodactylids and diatomyids and direction of faunal exchanges between Africa and Eurasia (data source: López-Antoñanzas 2009, 2011; López-Antoñanzas and Sen 2004, 2006, 2010; López-Antoñanzas and Knoll 2011; López-Antoñanzas et al. 2004, 2013, 2015a, b, 2016; Flynn and Wessels 2013)

distinguished *Mus* (*Pyromys*) *shortridgei* occurring at several sites in Thailand ranging from Late Pliocene to recent. The oldest record of *Mus* (*Coelomys*) *pahari* thus far is Middle Pleistocene. The members of the subgenus *Mus*, *Mus* (*Mus*) *caroli* and *cooki* date back to the Late Pliocene and *Mus* (*Mus*) *cervicolor* has been reported from the Middle Pleistocene (Chaimanee 1998). If we follow the present topology and a split between *Nannomys* and *Coelomys* that is supposed to have occurred around 6 Ma, the separation of the European members of the subgenus *Mus* and Asian members of the subgenus *Pyromys* would have to be pushed further back in time. A Late Pliocene (~ 3 Ma) occurrence of the present *Mus* cf. *Mus* (*Coelomys*) *pahari* supports an early date of split between *Nannomys* and *Coelomys*. The house mouse *Mus musculus* that originated in the Indian subcontinent or west Asia, perhaps offers one of the most fascinating example of long distance dispersal (Suzuki et al. 2013). *Hadromys* had a wider distribution earlier, Potwar in the west during the Pliocene and Pleistocene to Thailand in the South East during the Pleistocene. It is now confined to a small area in Manipur and provides an example of range contraction, rather than long distance dispersal.

Like the murids, rhizomyines and ctenodactylids illustrate long distance dispersal of rodents consistent with geological and ecological evidence indicating the establishment of an Early Miocene corridor between Afro-Arabia and Eurasia. The presence of *Prokanisamys* sp. from Jebel Zelten (Libya) (Wessels et al. 2003, 2008) provides evidence of an Early Miocene migration or range expansion of the members of this family. The Asian bamboo rats are subtropical forest dwellers and they require moist conditions to thrive (López-Antoñanzas et al. 2015a). Again supporting the presence of moist, subtropical forest from South Asia to North Africa in the Early Miocene. The second dispersal may have occurred around Middle Miocene co-inciding with Middle Miocene Climatic Optimum involving *Pronakalimys* and the third around the Late Miocene involving *Nakalimys* (López-Antoñanzas et al. 2013: Fig. 9). This is also suggestive of existence of a continuous forest between East Africa to Asia. The fourth and fifth dispersals of *Protachyoryctes makooka* (López-Antoñanzas re-allocated *Tachyoryctes makooka* to *Protachyoryctes makooka*) and the African Tachyoryctini might have taken place during the latest Miocene, co-inciding with the spread of grasslands and arid conditions. The African Tachyoryctines with hypsodont teeth are adapted to open habitat. Even if the ctenodactylines are restricted to Africa today, much of their early evolution took place in Asia. Living African gundis are adapted to open conditions but their Miocene ancestors were most likely forest dwellers (Flynn 2003). During the Miocene, ctenodactylines underwent a wide geographical expansion, reaching Sardinia at the beginning of the Early Miocene and the north of Africa possibly in the Middle Miocene, through the Arabian Peninsula (López-Antoñanzas and Knoll 2011) (Fig. 16.16). *S. sivalensis* (*S. perplexus* and *S. badauni* included in *S. sivalensis*) were the last ctenodactylines to survive in Asia. Diatomyids represented by a single species *Laonastes aenigmamus* today, occurs in evergreen forest of Vietman, occupying small area (150 km^2) and prefers crevasses in limestones (Nguyen et al. 2014). Their fossils have been found in south and east Asia and Arabia. They also took advantage of the corridor to reach Saudi Arabia in the Early Miocene (López-Antoñanzas 2011) (Fig. 16.16) as did the Jumping mice Zapodidae, dispersing to Arabia and North Africa (López-Antoñanzas and Sen 2006) (Fig. 16.16). The cane rats (Thryonomyids) that are confined to Africa today, took the opposite route to reach South Asia.

Conclusions

New murine records, *Mus* cf. *M. pahari* from the Pliocene, additional murine *Hadromys* sp. from the Pleistocene and *Karnimata darwini* from Late Miocene are described. Murines described earlier by Vasishat (1985) are reviwed, *Progonomys choprai* is allocated to *Progonomys debruijni*; *Siwalkomus nagrii* to *Karnimata darwini* and the presence of *Parapodemus hariensis* is confirmed. Additional spalacids cf. *Kanisamys* sp., *Miorhizomys harii* and a ctenodactylid *Sayimys sivalensis* have also been added for the Late Miocene. *Sayimys perplexus* and *Sayimys badauni* (Vasishat 1985) have been allocated to *Sayimys sivalensis* as suggested earlier (López-Antoñanzas and Knoll 2011). The age of the Badaun site is not resolved yet and pending further work it should be treated as belonging to the latest Miocene. A PAUP based phylogenetic analysis of cranial characters of *Progonomys debruijni* and species belonging to the genus *Mus* suggest that the subgenus *Mus* separated first, followed by the split of *Pyromys*, and finally that of *Coelomys* and *Nannomys*. Phylogenetic analysis of spalacids reveal that cf. *Kanisamys* sp. was basal to the Siwalik *Protachyoryctes tatroti*, *Eicoryctes kaulialensis* and the African tachyoryctines. The Early Miocene collision of Afro-Arabian and Eurasian Plates and the presence of subtropical moist forest most likely facilitated long distance dispersals or range expansion of early murids, rhizomyines and ctenodactylids. Middle Miocene seasonal forests, followed by Late Miocene drier conditions and spread of grasslands most likely helped the later forms that were adapted to more open conditions to disperse in both the directions.

Acknowledgements The author is grateful to Dr. Kewal Krishan and Dr. Gayatri Padmanadhan of Anthropology, Panjab University to allow access to some of the rodents collected by Prof. R. N. Vasishat. The author thanks Drs. L. J. Flynn and R. López-Antoñanzas for their help with preliminary identification of the rhizomyids and ctenodactylid. The author extends his sincere thanks to the three reviwers for their critical comments and constructive suggestions. The author acknowledges the financial support from MoES (MoES/P.O. (Geoscience)/46/2015) and SERB (HRR/2018/000063), New Delhi.

References

Aghová, T., Kimura, Y., Bryja, J.,Gauthier, D., Granjon, L., & Kergoat, G. J. (2018). Fossils know it best: Using a new set of fossil calibrations to improve the temporal phylogenetic framework of murid rodents (Rodentia: Muridae). *Molecular Phylogenetics and Evolution, 128,* 98–111.

Bachmayer, F., & Wilson, R. W. (1979). A third contribution to the fossil small mammal Fauna of Kohfidisch (Burgenland), Austria. *Annalen des Naturhistorischen Museums in Wien, 83,* 351–386.

Badgley, C., Barry, J. C., Morgan, M. E., Nelson, S. V., Behrensmeyer, A. K., Cerling, T. E., et al. (2008). Ecological changes in Miocene mammalian record show impact of prolonged climatic forcing. *Proceedings of the National Academy of Sciences, 105,* 12145–12149.

Barry, J. C., Behrensmeyer, A. K., Badgley, C., Flynn, L. J., Peltonen, H., Cheema, et al. (2013). The Neogene Siwaliks of the Potwar Plateau, Pakistan. In X. Wang, L. J. Flynn, & M. Fortelius (Eds.), *Fossil mammals of Asia: Neogene biostratigraphy and chronology* (pp. 373–399). New York, Columbia University Press.

Barry, J. C., Morgan, M. E., Flynn, L. J., Pilbeam, D., Behrensmeyer, A. K., Raza, et al. (2002). Faunal and environmental change in the late Miocene Siwaliks of northern Pakistan. *Paleobiology Memoirs, 3,* 1–71.

Baskin, J. A. (1996). Systematic revision of Ctenodactylidae (Mammalia, Rodentia) from the Miocene of Pakistan. *Palaeovertebrata, 25,* 1–49.

Benammi, M., Calvo, M., Prévot, M., & Jaeger, J.-J. (1996). Magnetostratigraphy and paleontology of Ait Kandoula Basin (High Atlas, Morocco) and the African-European late Miocene terrestrial fauna exchanges. *Earth and Planetary Science Letters, 145,* 15–29.

Black, C. C. (1972). Review of fossil rodents from the Neogene Siwalik Beds of India and Pakistan. *Palaeontology, 15,* 238–266.

Bohlin, B. (1946). The fossil mammals from the tertiary deposits of Tabenbuluk, Western Kansu, Part II: Simplicidentata, Carnivora, Artiodactyla, Perissodactyla, and Primates. *Palaeontologica Sinica, C 8 b,* 1–259.

Bonhomme, F., Catalan, J., Britton-Davidian, J., Chapman, V. M., Moriwaki, K., Nevo, E., et al. (1984). Biochemical diversity and evolution in the genus *Mus*. *Biochemical Genetics, 22,* 275–303.

Bonhomme, F., Catalan, J., Gautun, J. C., Petter, F., & Thaler, L. (1982). Caractérisation biochimique de souris africaines référables au sous-genre *Nannomys* Peters, 1876. *Mammalia, 46,* 110–113.

Brandy, L. D. (1979): Rongeurs nouveaux du Néogène d'Afghanistan. *Comptes Rendus de l'Académie des Sciences, Paris, Sér. D, 289,* 81–83.

Bryja, J., Mikula, O. E., Umbera, R., Meheretu, Y., Aghová, T., Lavrenchenko, L. A., et al. (2014). Pan-African phylogeny of *Mus* (subgenus *Nannomys*) reveals one of the most successful mammal radiations in Africa. *BMC Evolutionary Biology, 14,* 256.

Catzeflis, F. M., & Denys, C. (1992). The African *Nannomys* (Muridae), an early offshoot from the *Mus* lineage – Evidence from scnDNA hybridization and compared morphology. *Israel Journal of Zoology, 38,* 219–231.

Chaimanee, Y. (1998). Plio-Pleistocene rodents of Thailand. *Thai Studies in Biodiversity, 3,* 1–303.

Cheema, I. U., Raza, S. M., & Flynn, L. J. (1997). Note on Pliocene Small Mammals from the Mirpur district, Azad Kashmir, Pakistan. *Geobios, 30*(1), 115–119.

Cheema, I. U., Raza, S. M., Flynn L. J., Rajpar, A. R., & Tomida Y. (2000). Miocene small mammals from Jalalpur, Pakistan, and their biochronologic implications. *Bulletin of the National Museum, Tokyo, Japan, 26*(1–2), 57–77.

Chevret, P., Jenkins, P., & Catzeflis, F. (2003). Evolutionary systematics of the Indian mouse *Mus famulus* Bonhote, 1898: molecular (DNA/DNA hybridization and 12S rRNA sequences) and morphological evidences. *Zoological Journal of the Linnean Society, 137,* 385–401.

Chevret, P., Veyrunes, F., & Britton-Davidian, J. (2005). Molecular phylogeny of the genus *Mus* (Rodentia: Murinae) based on mitochondrial and nuclear data. *Biological Journal of the Linnean Society, 84,* 417–427.

De Bruijn, H., Boon, E., & Hussain, T. (1989). Evolutionary trends in *Sayimys* (Ctenodactylidae, Rodentia) from the Lower Manchar Formation (Sind, Pakistan). *Koninklijke Nederlandse Akademie van Wetenschappen Proceedings B, 92,* 191–214.

Dehm, R., Jacobs, L., Wessels, W., de Bruijn, H., & Hussain, S. T. (1982). Fossil Rodents from the type area of the Chinji Formation, Siwalik Group, Pakistan. *Koninklijke Nederlandse Akademie van Wetenschappen Proceedings B, 85,* 259–263.

Flynn, L. J. (1981). Biostratigraphy and systematics of Siwalik Rhizomyidae (Rodentia). Ph.D. Dissertation, The University of Arizona.

Flynn, L. J. (1982). Systematic revision of Siwalik Rhizomyidae (Rodentia). *Geobios, 15,* 327–389.

Flynn, L. J. (1984). Mosaic evolution in a family of fossorial rodents. In E. Buffetaut, J. M. Mazin, & E. Salmon (Eds.), *Actes du Symposium paléontologique Georges Cuvier* (pp. 185–195). Ville de Montbéliard: Montbéliard.

Flynn, L. J. (1986). Species longevity, stasis, and stair steps in rhizomyid rodents. *University of Wyoming Contribution to Geology, Special Paper, 3,* 273–285.

Flynn, L. J. (2003). Small mammal indicators of forest paleoenvironment in the Siwalik deposits of the Potwar Plateau, Pakistan. In W. F. Reumer & W. Wessels (Eds.), *Distribution and migration of Tertiary mammals in Eurasia* (pp. 183–196).

Flynn, L. J. (2009). The antiquity of *Rhizomys* and independent acquisition of fossorial traits in subterranean muroids. *Bulletin of the American Museum Natural History, 331,* 128–156.

Flynn, L. J., & Jacobs, L. L. (1990). Preliminary analysis of Miocene small mammals from Paşalar, Turkey. *Journal of Human Evolution, 19,* 423–436.

Flynn, L. J., Kimura, Y., & Jacobs, L. J. (2020). The murine cradle. In G. V. R. Prasad & R. Patnaik (Eds.), *Biological consequences of plate tectonics: New perspectives on post-Gondwana break-up* (pp. 347–362). Cham: Springer.

Flynn, L. J., Lindsay, E. H., Pilbeam, D., Raza, S. M., Morgan, M. E., Barry, J. C., et al. (2013). The Siwaliks and Neogene evolutionary biology in South Asia. In X. Wang, L. J. Flynn & M. Fortelius (Eds.), *Fossil mammals of Asia: Neogene biostratigraphy and chronology* (pp. 353–372). New York: Columbia University Press.

Flynn, L. J., Morgan, M. E., Pilbeam, D., & Barry, J. (2014). "Endemism" relative to space, time, and taxonomic level. *Annales Zoologici Fennici, 51,* 245–258.

Flynn, L. J., Sahni, A., Jaeger, J.-J., Singh, B., & Bhatia, S. B. (1990). Additional fossil rodents from the Siwalik Beds of India. *Proc.Kon. Ned. Akad. Wetensch. 93,* 7–20.

Flynn, L. J., & Wessels, W. (2013). Paleobiogeography and South Asian small mammals: Neogene latitudinal faunal variation. In X. Wang, L. J. Flynn & M. Fortelius (Eds.), *Fossil mammals of Asia: Neogene biostratigraphy and chronology* (pp. 445–460). New York: Columbia University Press.

Geraads, D. (2001). Rongeurs du Miocene superieur de Chorora (Ethiopie): Dendromuridae, Muridae et conclusions. *Palaeovertebrata, 30,* 89–109.

Gerasimov, S. N. H., Mihailova, V., Auffray, J. C., & Bonhomme, F. (1990). Morphometric step-wise discriminant analysis of the five genetically determined European taxa of the genus *Mus*. *Biological Journal of the Linnean Society, 41*(1–3), 47–64.

Gilbert, C. C., Pugh, K. D., & Fleagle, J. G. (2020). Dispersal of Miocene hominoids (and pliopithecoids) from Africa to Eurasia in light of changing tectonics and climate. In G. V. R. Prasad & R. Patnaik (Eds.), *Biological consequences of plate tectonics: New perspectives on post Gondwana break-up* (pp. 393–412). Cham: Springer.

Gogolevskaya, I. K., Veniaminova, N. A., & Kramerov, D. A. (2010). Nucleotide sequences of B1 SINE and 4.5SI RNA support a close relationship of zokors to blind mole rats (Spalacinae) and bamboo rats (Rhizomyinae). *Gene, 460,* 30–38.

Gupta, S. S., & Prasad, G. V. R. (2001). Micromammals from the Upper Siwalik subgroup of the Jammu region, Jammu and Kashmir State, India: Some constraints on age. *Neues Jahrbuch für Geologie und Paläontologie Abhandlungen, 220*(2), 153–157.

Heissig, K. (1982). Kleinsäuger aus einer obermiozänen (Vallesium) Karstfüllung Ägyptens. *Mitteilungen der Bayerischen Staatssammlung für Paläontolgie und Hisoriche Geologie, 22,* 97–101.

Hinton, M. A. C. (1933). Diagnoses of new genera and species of rodents from Indian Tertiary deposits. *Annals and Magazine of Natural History Series, 10*(12), 620–622.

Hsü, K. J. (1978). Stratigraphy of the lacustrine sedimentation in the Black Sea. *Initial Reports of the DSDP, 42B*, 509–524.

Hussain, S. T., Munthe, J., Shah, S. M. I., West, R. M., & Lukacs, J. R. (1977). The Daud Khel Local Fauna: A preliminary report on a Neogene vertebrate assemblage from the trans-Indus Siwaliks, Pakistan. *Milwaukee Public Museum, Contributions in Biology and Geology, 16*, 1–16.

Jacobs, L. L. (1978). Fossil rodents (Rhizomyidae and Muridae) from Neogene Siwalk deposits, Pakistan. *Museum of North Arizona Bulletin, 52*, 1–103.

Jacobs, L. L., & Flynn, J. J. (2005). Of mice ... again: The Siwalik rodent record, murine distribution and molecular clocks. In D. E. Leiberman, R. J. Smith, & J. Kelley (Eds.), *Interpreting the past: Essays on human, primate, and mammal evolution in honor of David Pilbeam* (pp. 63–80). Boston: Brill Academic Publishers Inc.

Johnson, G. D., & Vondra, C. F. (1972). Siwalik sediments in a portion of the Punjab Reentrant: The sequence at Haritalyangar, District Bilaspur. H.P. *Himalayan Geology, 2*, 118–143.

Kimura, Y., Flynn, L. J., & Jacobs, L. L. (2016). A paleontological case study for species delimitation in diverging fossil lineages. *Historical Biology, 28*(1–2), 189–198.

Kimura, Y., Flynn, L. J., & Jacobs, L. L. (2017). Early Late Miocene Murine Rodents from the upper part of the Nagri Formation, Siwalik group, Pakistan, with a new fossil calibration point for the Tribe Apodemurini (*Apodemus/Tokudaia*). *Fossil Imprint, 73*(1–2), 197–212.

Kimura, Y., Hawkins, M. T. R., McDonough, M. M., Jacobs, L. L., & Flynn, L. J. (2015). Corrected placement of *Mus-Rattus* fossil calibration forces precision in the molecular tree of rodents. *Scientific Reports, 5*, 14444 (9 pp.) https://doi.org/10.1038/srep14444.

Kimura, Y., Jacobs, L. L., Cerling, T. E., Uno, K. T., Ferguson, K. M., Flynn, L. J., et al. (2013b). Fossil mice and rats show isotopic evidence of niche partitioning and change in dental ecomorphology related to dietary shift in Late Miocene of Pakistan. *PLoS ONE, 8* (8), e69308 (10 pp.). https://doi.org/10.1371/journal.pone.0069308.

Kimura, Y., Jacobs, L. L., & Flynn, L. J. (2013a). Lineage-specific responses of tooth shape in murine rodents (Murinae, Rodentia) to Late Miocene dietary change in the Siwaliks of Pakistan. *PLoS ONE, 8*(10), e76070 (12 pp.). https://doi.org/10.1371/journal.pone.0076070.

López-Antoñanzas, R. (2009). First *Potwarmus* from the Miocene of Saudi Arabia and the early phylogeny of murines (Rodentia: Muroidea). *Zoological Journal of Linnean Society, 156*, 664–679.

López-Antoñanzas, R. (2011). First diatomyid out of Asia. *Naturwissenschaften, 98*, 117–123.

López-Antoñanzas, R. (2012a). Dental morphology of the Lesser Bamboo Rat, *Cannomys badius* (Rodentia, Spalacidae). *Zookeys, 228*, 69–75.

López-Antoñanzas, R. (2012b). Dental morphology and wear pattern in *Tachyoryctes* (Spalacidae, Rodentia). *Mammalia, 76*, 309–314.

López-Antoñanzas, R. (2013a). Dental wear pattern in *Rhizomys* (Rodentia, Spalacidae). *Anatomical Record, 296*, 272.

López-Antoñanzas, R. (2013b). Dental morphology and wear pattern of bamboo rats, genus *Rhizomys* Gray, 1831(Rodentia: Spalacidae). *Zoologischer Anzeiger, 253*, 85–92.

López-Antoñanzas, R. (2013c). The tetralophodont pattern of the upper molars in juvenile *Cannomys badius* (Rodentia, Spalacidae). *Mammalia, 77*, 457–460.

López-Antoñanzas, R., Flynn, L. J., & Knoll, F. (2013). A comprehensive phylogeny of extinct and extant Rhizomyinae (Rodentia): evidence for multiple intercontinental dispersals. *Cladistics, 29*, 247–273.

López-Antoñanzas, R., Gutkin, V., Rabinovich, R., Calvo, R., & Grossman, A. (2016). A Transitional Gundi (Rodentia: Ctenodactylidae) from the Miocene of Israel. *PLoS ONE, 11*(4), e0151804. https://doi.org/10.1371/journal.pone.0151804.

López-Antoñanzas, R., & Knoll, F. (2011). A comprehensive phylogeny of the gundis (Ctenodactylinae, Ctenodactylidae). *Journal of Systematic Palaeontology, 9*, 379–398.

Lopez-Antonanzas, R., Knoll, F., Maksoud, S., & Azar, D. (2015b). First Miocene rodent from Lebanon provides the 'missing link' between Asian and African gundis (Rodentia: Ctenodactylidae). *Scientific Reports, 5*, 12871. https://doi.org/10.1038/srep12871.

Lopez-Antonanzas, R., Knoll, F., Wan, S., & Flynn, L. J. (2015a). Causal evidence between monsoon and evolution of rhizomyine rodents. *Scientific Reports, 5*, 9008. https://doi.org/10.1038/srep09008l.

López-Antoñanzas, R., & Sen, S. (2003). Systematic revision of Mio-Pliocene Ctenodactylidae (Mammalia, Rodentia) from the Indian subcontinent. *Eclogae Geologicae Helvetiae, 96*, 521–529.

López-Antoñanzas, R., & Sen, S. (2004). Ctenodactylids from the Lower and Middle Miocene of Saudi Arabia. *Palaeontology, 47*, 1477–1494.

López-Antoñanzas, R., & Sen, S. (2006). New jumping mouse (Rodentia, Zapodidae) from the Lower Miocene of Saudi Arabia and a contribution to the knowledge of zapodid phylogeny. *Journal of Vertebrate Paleontology, 26*, 170–181.

López Antoñanzas, R., Sen, S., & Koufos, G. D. (2005). Presence of a ctenodactylid (Rodentia) in the Keramaria Formation (Middle Miocene) of Chios Island (Greece). *Geobios, 38*, 113–126.

López-Antoñanzas, R., Sen, S., & Saraç, G. (2004). A new large ctenodactylid species from the Lower Miocene of Turkey. *Journal of Vertebrate Paleontology, 24*, 676–688.

López-Antoñanzas, R., & Wesselman, H. B. (2013). *Tachyoryctes makooka* (Tachyoryctini, Spalacidae, Rodentia) and its bearing on the phylogeny of the Tachyoryctini. *Palaeontology, 56*(1), 157–166.

Lundrigan, B. L., Jansa, S. A., & Tucker, P. K. (2002). Phylogenetic relationships in the genus *Mus*, based on paternally, maternally, and biparentally inherited characters. *Systematic Biology, 51*, 410–431.

Lundrigan, B. L., & Tucker, P. K. (1994). Tracing paternal ancestry in mice, using the Y-linked, sex-determining locus Sry. *Molecular Biology and Evolution, 11*, 483–492.

Marshall, J. T. (1977). A synopsis of Asian species of Mus (Rodentia, Muridae). *Bulletin of the American Museum Natural History, 158*, 173–220.

Marshall, J. T. (1981). Taxonomy. In H. L. Foster, J. D. Small & J. G. Fox (Eds.), *The mouse in biomedical research* (pp. 17–26). New York: Academic Press.

Marshall, J. T. (1988). Family Muridae: Rats and mice. In B. Legakul & J. A. McNeely (Eds.), *Mammals of Thailand* (pp. 397–487). Bangkok: Association for the Conservation of Wildlife.

Mein, P., & Freudenthal, M. (1971). Les Cricetidae (Mammalia, Rodentia) du Néogène moyen du Vieux-Collonges. *Partie 1, Le genre Cricetodon Lartet, 1851. Scripta Geologica, 5*, 1–51.

Mein, P., Martín-Suárez, E., & Agustí, J. (1993). *Progonomys* Schaub, 1938 and *Huerzelerimys* gen. nov. (Rodentia); their evolution in Western Europe. *Scripta Geologica, 103*, 41–64.

Miller, G. S., & Gidley, J. W. (1918). Synopsis of the supergeneric groups of rodents. *Journal of Washington Academy Science, 8*, 431–448.

Miller, K. G., Wright, J. D., & Fairbanks, R. G. (1991). Unlocking the ice house: Oligocene–Miocene oxygen isotopes, eustasy, and margin erosion. *Journal of Geophysical Research, 96*, 6829–6848. https://doi.org/10.1029/90jb02015.

Misonne, X. (1969). African and Indo-Australian Muridae: evolutionary trends. *Musee Royal de l'Afrique Centrale, Tervuren, Belgique Annals, Serie 8, Sciences Zoologiques, 172*, 1–177.

Munthe, J. (1980). Rodents of the Miocene Daud Khel Local Fauna, Mianwali District, Pakistan. Part 1. Sciuridae, Gliridae, Ctenodactylidae, and Rhyzomyidae. *Milwaukee Public Museum, Contributions in Biology and Geology, 34*, 1–36.

Musser, G. G. (1981). Results of the Archbold expeditions. No. 105. Notes on systematics of Indo-Malayan murid rodents, and descriptions of new genera and species from Ceylon, Sulawesi, and the Philippines. *Bulletin of the American of Museum Natural History, 168*(3), 225–334.

Musser, G. G. (1987). The occurrence of *Hadromys* (Rodentia: Muridae) in early Pleistocene Siwalik strata in northern Pakistan and its bearing on biogeographic affinities between Indian and northeastern African murine faunas. *American Museum Novitates, 2883*, 1–36.

Musser, G. G., & Carleton M. D. (2005). Superfamily Muroidea. In D. E. Wilson & D. A. M. Reeder (Eds.), *Mammal species of the world: A taxonomic and geographic reference* (pp. 894–1531). Baltimore: The Johns Hopkins University Press.

Musser, G. G., & Newcomb, C. (1983). Malaysian murids and the giant rat of Sumatra. *Bulletin of the American Museum Natural History, 174*(4), 327–598.

Nanda, A. C., & Sahni, A. (1998). Ctenodactyloid rodent assemblage from Kargil Formation, Ladakh Molasse Group: Age and paleobiogeographic implications for the Indian subcontinent in the Oligo-Miocene. *Geobios, 31*, 533–544.

Nguyen, D. X., Nguyen, N. X., Nguyen, D. D., Dinh, T. H., Le, D. T., & Dinh, D. H. (2014). Distribution and habitat of the Laotian rock rat *Laonastes aenigmamus* Jenkins, Kilpatrick, Robinson & Timmins, 2005 (Rodentia: Diatomyidae) in Vietnam. *Biodiversity Data Journal, 2*, e4188.

Patnaik, R. (1997). New murids and gerbillids (Rodentia, Mammalia) from Pliocene Siwalik sediments of India. *Palaeovertebrata, 26*(1–4), 129–165.

Patnaik, R. (2001). Late Pliocene micromammals from Tatrot Formation (Upper Siwaliks) exposed near village Saketi, Himachal Pradesh, India. *Palaeontographica, 261*, 55–81.

Patnaik, R. (2013). Indian Neogene Siwalik mammalian biostratigraphy: An overview. In X. Wang, L. J. Flynn & M. Fortelius (Eds.), *Fossil mammals of Asia: Neogene biostratigraphy and chronology* (pp. 423–443). New York: Columbia University Press.

Patnaik, R. (2014). Phylogeny of Siwalik murine rodents: Implications for *Mus-Rattus* divergence time. *Journal of the Palaeontological Society of India, 59*, 15–28.

Patnaik, R. (2015). Diet and habitat changes among Siwalik herbivorous mammals in response to Neogene and Quaternary climate changes: An appraisal. *Quaternary International, 371*, 232–243.

Patnaik, R. (2016). Neogene–Quaternary mammalian paleobiogeography of the Indian subcontinent: An appraisal. *Comptes Rendus Palevol, 919*, 1–14.

Patnaik, R., Auffray, J.-C., Jeager, J.-J., & Sahni, A. (1996). House mouse ancestor from late Pliocene Siwalik sediments of India. *Comptes Rendus de l'academie des Sciences, Sciences de la Vie, 319*, 431–434.

Patnaik, R., Kotla, S., Singh, P., Singla, A., & Kaur, J. (2017). A new murid rodent assemblage from Upper Siwaliks, Himachal Pradesh, India: Biostratigraphic and phylogenetic implications. *Journal of Asian Earth Sciences*. http://dx.doi.org/10.1016/j.jseaes.2017.05.030.

Peláez-Campomanes, P., & Daams, R. (2002). Middle Miocene rodents from Paşalar, Anatolia, Turkey. *Acta Palaeontologica Polonica, 47*, 125–132.

Pillans, B., Williams, M., Cameron, D., Patnaik, R., Hogarth, J., Sahni, A., Sharma, J. C., Williams, F., & Bernor, R. (2005). Revised correlation of the Haritalyangar magnetostratigraphy, Indian Siwaliks: Implications for the age of the Miocene hominids *Indopithecus* and *Sivapithecus*, with a note on a new hominid tooth. *Journal of Human Evolution, 48*, 507–515.

Prasad, K. N. (1970). The vertebrate fauna from the Siwalik beds of Haritalyangar, Himachal Pradesh, India. *Palaeontologia Indica, 39*, 1–56.

Prasad, K. N. (1971). A note on the geology of Bilaspur, Haritalyangar area, Himachal Pradesh. *Records of the Geological Survey of India, 92*, 72–81.

Qiu, Z.-D., & Li, Q. (2016). Neogene rodents from central Nei Mongol, China. *Palaeontologia Sinica, 198*, (C) 30, i–xi, 1–684.

Qiu, Z.-D., Zheng, S.-H., & Zhang, Z.-Q. (2004). Murids from the Late Miocene Bahe Formation, Lantian, Shanxi. *Vertebrata PalAsiatica, 42*, 67–76.

Raghavan, P. (1990). New records of microfossil assemblages from the basal Pinjor Formation at Panchkula, Haryana (India). *Bulletin Indian Geologists' Association, 23*(1), 29–37.

Rana, R. S., Kumar, K., Escarguel, G., Sahni, A., Rose, K. D., Smith, T., et al. (2008). An ailuravine rodent from the Lower Eocene Cambay Formation at Vastan, western India, and its paleobiogeographic implications. *Acta Palaeontologica Polonica, 53*(1), 1–14.

Rögl, F. (1999). Mediterranean and Paratethys. Facts and hypotheses of an Oligocene to Miocene paleogeography (short overview). *Geologica Carpathica, 50*, 339–349.

Sabatier, M. (1982). Les rongeurs du site Pliocène à Hominidés de Hadar (Ethiopie). *Palaeovertebrata, 12*, 1–56.

Sahni, A. (1980). SEM studies of Eocene and Siwalik rodent enamels. *Geoscience Journal, 1*, 21–30.

Sahni, A. (1985). Enamel structure of early mammals and its role in evaluating relationship among rodents. In W. P. Luckett & J.-L. Hartenberger (Eds.), *Evolutionary relationships among rodents, a multidisciplinary analysis* (pp. 133–150). New York: Plenum Press.

Sahni, A., Bhatia, S. B., Hartenberger, J.-L., Jaeger, J.-J., Kumar, K., Sudre, J., et al. (1980). Vertebrates from the Subathu Formation and comments on the biogeography of Indian subcontinent during the early Paleogene. *Bulletin Geological Society France, 23*, 689–695.

Sahni, A., & Khare, S. K. (1973). Additional Eocene mammals from the Subathu Formation of Jammu and Kashmir. *Journal of the Palaeontological Society of India, 17*, 31–49.

Sahni, A., Saraswati, P. K., Rana, R. S., Kumar, K., Singh, H., Alimohammadian, H., et al. (2006). Temporal constraints and depositional paleoenvironments of the Vastan lignite sequence, Gujarat: Analogy for the Cambay Shale hydrocarbon source rock. *Indian Journal of Petroleum Geology, 15*, 1–20.

Sahni, A., & Srivastava, V. C. (1976). Eocene rodents and associated reptiles from the Subathu Formation of northwestern India. *Journal of Paleontology, 50*, 922–928.

Schaub, S. (1938). Tertiäre und Quartäre Murinae. *Abhandlungen des Schweizerischen Paläontologische Gesellschaft, 61*, 1–38.

Sen, S. (1977). La faune de rongeurs Pliocenes de Calta (Ankara, Turquie). *Bulletin du Museum national d'Histoire naturelle, 465*, 1–171.

Sen, S. (1983). Rongeurs et lagomorphes du gisement pliocène de Pul-e Charkhi, bassin de Kabul, Afghanistan. *Bulletin du Museum National d'Histoire naturelle, 5*, 33–74.

Sen, S. (2013). Dispersal of African mammals in Eurasia during the Cenozoic: Ways and whys. *Geobios, 46*, 159–172.

Sen, S., & Thomas, H. (1979). Découverte de rongeurs dans le Miocène moyen de la Formation Hofuf (Province du Hasa, Arabie Saoudite). *Compte rendu sommaire des séances de la Société Géologique de France, 1*, 34–37.

She, J.-X., Bonhomme, F., Boursot, P., Thaler, L., & Catzeflis, F. (1990). Molecular phylogenies in the genus *Mus:* comparative analysis of electrophoretic, scnDNA hybridization, and mtDNA RFLP data. *Biological Journal of the Linnean Society, 41*, 83–103.

Sourrouille, P., Hänni, C., Ruedi, M., & Catzeflis, F. M. (1995). Molecular systematics of *Mus crociduroides* an endemic mouse of Sumatra (Muridae, Rodentia). *Mammalia, 59,* 91–102.

Storch, G., & Ni, X.-J. (2002). New late Miocene murids from China (Mammalia, Rodentia). *Géobios, 35,* 515–521.

Suzuki, H., Nunome, M., Kinoshita, G., Aplin, K. P., Vogel, P., Kryukov, A. P., et al. (2013). Evolutionary and dispersal history of Eurasian house mice *Mus musculus* clarified by more extensive geographic sampling of mitochondrial DNA. *Heredity, 111,* 375–390.

Swofford, D. L. (2002). *PAUP: Phylogenetic analysis using parsimony (and other methods), Version 4.0b10.* Sunderland: Sinauer Associates.

Tiwari, B. N., & Bhandari, A. (2014). Exotic murid rodent *Parapodemus* sp. from Middle Siwalik of Mohand, district Saharanpur, as harbinger of micropaleontological studies in the region. *Himalayan Geology, 35*(1), 70–74.

Vasishat, R. N. (1985). *Antecedents of Early Man in Northwestern India: Palaeontological and Palaeoecological Evidences.* New Delhi: Inter-India Publications.

Vasishat, R. N. (1978). First record of maxillary dentition of *S. perplexus* (Ctenodactylidae) from the Indian Siwaliks. *Current Science, 47,* 859–860.

Veyrunes, F, Dobigny, G, Fengtang, Y, O'Brien, P. C., Catalan, J., et al. (2006). Phylogenomics of the genus *Mus* (Rodentia; Muridae): extensive genome repatterning is not restricted to the house mouse. *Proceedings of the Royal Society Biological Sciences, 273,* 2925–2934.

Wang, B. Y., Qiu, Z. X., Wang, X. M., Xie, G., Xie, J. Y., Downs, W., et al. (2003). Cenozoic stratigraphy in Danghe area, Gansu Province, and uplift of Tibetan Plateau. *Vertebrata PalAsiatica, 41,* 66–75.

Weerd, A. D. (1976). *Rodent faunas of the Mio-Pliocene continental sediments of the Teruel-Alfambra region, Spain.* Ph.D. Dissertation, Utrecht University.

Wesselman, H. B., Black, M. T., & Asnake, M. (2009). Small mammals (pp. 105–133). In Y. Haile-Selassie & G. Woldegabriel (Eds.), *Ardipithecus kadabba: Late Miocene evidence from the Middle Awash, Ethiopia (The Middle Awash Series)* (p. 664). Berkeley: University of California Press.

Wessels, W. (2009). Miocene rodent evolution and migration. Muroidea from Pakistan, Turkey and Northern Africa. Geologica Ultraiectina 307. Ph.D. Dissertation, Departement Aardwetenschappen.

Wessels, W., de Bruijn, H., Hussain, S. T., & Leinders, J. (1982). Fossil rodents from the Chinji Formation, Banda Daud Shah, Kohat, Pakistan. *Proceedings of the Koninklijke Nederlandse Akademie, 85,* 337–364.

Wessels, W., Fejfar, O., Peláez-Campomanes, P., Meulen, A. van der, & de Bruijn, H. (2003). Miocene small mammals from Jebel Zelten, Libya. In N. López-Martínez, P. Peláez-Campomanes, & M. Hernández Fernández (Eds.), *Surrounding fossil mammals: Dating, evolution and paleoenvironment* (pp. 699–715). Coloquios de Paleontología, Volumen Extraordinario 1, en homenaje al Dr. Remmert Daams.

Wessels, W., Fejfar, O., Peláez-Campomanes, P., Van Der Meulen, A., de Bruijn, H., & El-Arnauti, A. (2008). The age of the small mammal faunas from Jabal Zaltan, Libya. *Garyounis Scientific Bulletin, Special Issue, 5,* 129–138.

Winge, H. (1887). *Jordfundne og nulevende gnavere (Rodentia) fra Lagoa Santa, Minas Geraes, Brasilien.* Kjøbenhavn: F. Dreyer.

Winkler, A. J. (2002). Neogene paleobiogeography and East African paleoenvironments: Contributions from the Tugen Hills rodents and lagomorphs. *Journal of Human Evolution, 42,* 237–256.

Wood, A. E. (1937). Fossil rodents from the Siwalik beds of India. *American Journal of Science, 34,* 64–76.

Zittel, K. A. (1893). Handbuch der Paleontologie. I. Abt. Paleozologie. IV Band. Vertebrata (Mammalia). Munchen.

Zheng, S. (1993). *Quaternary rodents of Sichuan-Guizhou area, China.* Beijing: Science Press.

Chapter 17
Dispersal of Miocene Hominoids (and Pliopithecoids) from Africa to Eurasia in Light of Changing Tectonics and Climate

Christopher C. Gilbert, Kelsey D. Pugh, and John G. Fleagle

Abstract This chapter examines hypothesized dispersal events between Africa and Eurasia involving non-cercopithecoid catarrhines, particularly hominoid apes, and reviews the tectonic and climatic events that may have had a role in shaping them. All available evidence points to hominoid origins in Africa by the latest Oligocene, and the earliest evidence for apes outside of Africa in Eurasia occurs at ~17–16 Ma following a tectonic event at ~19 Ma that resulted in a landbridge between these continents through the Arabian Peninsula. Following their initial dispersal into Eurasia, the estimated number of subsequent dispersals between Africa, Asia, and Europe is dependent on the hypothesized phylogeny of these fossil apes. Here, we examined several recent phylogenetic hypotheses that suggest anywhere between one and four hominoid dispersal events between Africa and Eurasia, and a minimum of zero to two ape dispersals between Europe and Asia. The arrival of pongines, and possibly other apes, in Asia most likely occurred during or right after the Middle Miocene Climatic Optimum (~17–15 Ma), and the extinction of many Asian taxa was probably driven in part by Himalayan tectonic uplift in the Late Miocene (~9–8 Ma), which changed climatic patterns and resulted in the loss of preferred hominoid rainforest habitat. Similarly, climate change in Europe resulting in the loss of preferred habitat almost certainly played a role in European ape extinction.

Keywords Hominoid • Asia • Europe • Biogeography • Pongine • Fossil

Introduction

Fossil apes represent some of the best known faunal elements from the Eurasian Miocene, capturing paleontologists' imaginations since their discovery and recognition in the 1800s (e.g., Falconer and Cautley 1837; Lartet 1856; Gervais 1872; Lydekker 1879). While apes persist only in tropical African and Southeast Asian environments today, the fossil record indicates they were once dispersed across Eurasia, and that all Eurasian apes are, broadly speaking, immigrants from Africa. The earliest fossil evidence of apes comes from the late Oligocene, in ~25.2 Ma deposits in the Rukwa Rift, Tanzania (Stevens et al. 2013) as well as ~24–27 Ma deposits in northern Kenya (Leakey et al. 1995; Rasmussen and Guitierrez 2009). These fragmentary fossils are followed shortly by a radiation of widely-recognized stem hominoids in the early Miocene of East Africa (e.g., Leakey and Leakey 1986a, b; Ward et al. 1991; Begun et al. 1997; Gebo et al. 1997; Senut et al. 2000; Stevens et al. 2013; McNulty et al. 2015; Russo 2016; Nengo et al. 2017; see also Rossie and MacLatchy 2006; Fleagle 2013; Begun 2015). Fossil apes first appear in Eurasia in the latest early Miocene to middle Miocene, ~17–13.5 Ma in Germany, Slovakia, and Turkey, and they are dentally derived compared to the early Miocene

C. C. Gilbert (✉)
Department of Anthropology, Hunter College of the City University of New York, 695 Park Avenue, New York, NY 10065, USA
e-mail: cgilbert@hunter.cuny.edu

C. C. Gilbert
Graduate Center of the City University of New York, 365 Fifth Avenue, New York, NY 10016, USA

C. C. Gilbert · K. D. Pugh
New York Consortium in Evolutionary Primatology, New York, NY, USA

C. C. Gilbert
Division of Paleontology, American Museum of Natural History, Central Park West at 79th Street, New York, NY 10024, USA

K. D. Pugh
Division of Anthropology, American Museum of Natural History, Central Park West at 79th Street, New York, NY 10024, USA

J. G. Fleagle
Department of Anatomical Sciences, Stony Brook University, Stony Brook, NY 11794, USA

G. V. R. Prasad and R. Patnaik (eds.), *Biological Consequences of Plate Tectonics: New Perspectives on Post-Gondwana Break-up—A Tribute to Ashok Sahni*, Vertebrate Paleobiology and Paleoanthropology Series, https://doi.org/10.1007/978-3-030-49753-8_17

East African radiation, instead displaying numerous similarities with contemporaneous African apes such as thick enamel and reduced cingula (Alpagut et al. 1990; Martin and Andrews 1993; Begun 2000; Heizmann and Begun 2001; Andrews and Kelley 2007; Kelley et al. 2008; Begun et al. 2012; Begun 2015). Thus, the morphology and first appearance of Eurasian apes clearly point towards their African origins (e.g., Andrews and Tobien 1977; Harrison 2005; Alba 2012; Begun et al. 2012; Fleagle 2013; Begun 2015).

Once there, apes became widely dispersed throughout Eurasia during the rest of the Miocene epoch, from Spain in the west to China in the east (Begun 2002; Kelley 2002; Alba 2012), until many hominoids became extinct in the late Miocene. Tectonics and climate have clearly played a large role in primate evolution, greatly affecting their biogeographic distributions both past and present (Fleagle and Gilbert 2006). Apes are no exception, and this chapter broadly examines the role that these factors played in facilitating and shaping the dispersal of apes (and the contemporaneous stem catarrhine group known as pliopithecoids) across Eurasia during the early-middle part of the Miocene epoch and their eventual extinction in many localities during the late Miocene. Specifically, we ask the following questions regarding the dispersal of Miocene apes from Africa to Eurasia: (1) When did the first Eurasian apes appear and what tectonic and climatic events influenced their initial dispersal out of Africa?; (2) Was there a single ape dispersal event out of Africa or were there numerous dispersals into Eurasia?; (3) Are pongines (i.e., orangutans and their fossil relatives) the result of a dispersal event from Europe or Africa, and what tectonic and/or climatic events influenced their appearance in Asia?; and (4) What tectonic and climatic events are responsible for the extinction of many Eurasian apes, resulting in their limited distribution in Southeast Asia today?

Hominoid Systematics

Before delving into Miocene hominoid biogeography, a brief discussion of hominoid and, more broadly, catarrhine systematics is warranted (see Table 17.1 and Fig. 17.1).

In this paper, we consider the Proconsulidae, Afropithecidae and Nyanzapithecidae to be early (stem) hominoids (Fig. 17.1), composed of some combination of the following taxa: *Proconsul*, *Ekembo*, *Kalepithecus*, *Afropithecus*, *Morotopithecus*, *Heliopithecus*, *Otavipithecus*, *Nyanzapithecus*, *Rangwapithecus*, *Mabokopithecus*, *Rukwapithecus*, *Turkanapithecus*, and possibly *Kamoyapithecus*, *Equatorius*, *Nacholapithecus*, *Samburupithecus*, and *Oreopithecus* (see also Harrison 2005, 2010; Alba 2012; Begun et al. 2012; Begun 2015; Fleagle 2013; Nengo et al. 2017 for systematic debates and discussions). In general, these taxa display primitive dental features such as broad upper premolars/molars and extensive molar cingula, combined with postcranial features associated with a more pronograde or more primitive anti-pronograde form of locomotion not seen in extant apes (with notable exceptions including *Oreopithecus* and *Morotopithecus*, which display features associated with orthogrady) (Leakey et al. 1988; Rose 1983, 1993; Gebo et al. 1997; Walker 1997; Ward 1998; Ishida et al. 2004; Nakatsukasa and Kunimatsu 2009; Alba 2012; Fleagle 2013; Begun 2015). In addition, compared to most stem catarrhines, some or all of these taxa appear to share a few derived features linking them with crown apes, where preserved: tubular ectotympanics (also seen in *Saadanius* and all crown catarrhines), relatively large canines, relatively broad anterior palates, premaxillae-maxillae sutures that do not extend significantly above the nasal aperture, nasal bones that do not extend greatly above the frontal-maxillary suture, and relatively large body size (e.g., Rae 1997, 1999, 2004; Zalmout et al. 2010). The phylogenetic affinities of East African fossils known collectively as dendropithecoids (or dendropithecids, i.e., *Dendropithecus*, *Micropithecus*, and *Simiolus*; *Limnopithecus* is also traditionally included in this group, but more recently classified as *incertae sedis*; see Harrison 2002, 2010; Fleagle 2013), often identified in the early literature as "small-bodied apes" (e.g., Le Gros Clark and Leakey 1951; Andrews 1978; Fleagle and Simons 1978; Leakey and Leakey 1986b), are considered here to be uncertain relative to crown catarrhines (cercopithecoids and definitive hominoids), and it is entirely possible that they are a paraphyletic assemblage of stem catarrhines and/or stem hominoids (see also Andrews 1978; Harrison 1988; Rae 1999, 2004; Rossie and MacLatchy 2006; Nengo et al. 2017; Rossie and Hill 2018). More data, especially from cranial regions such as the basicranium (e.g., ectotympanic), are necessary to definitively assess the position of dendropithecoids relative to the crown catarrhine node.

Among more derived hominoids, the position of *Kenyapithecus* and *Griphopithecus* (i.e., Kenyapithecidae; possibly also including *Equatorius* and *Nacholapithecus*) relative to crown apes is here considered to be equivocal (Fig. 17.1), but all indications are that they are most likely close to the split between hylobatids and hominids (*Kenyapithecus*, in particular; e.g., see Pickford 1985, 1986; Andrews and Martin 1987; Harrison 1992, 2002, 2005, 2010; Andrews 1996; Ward et al. 1999; Ward and Duren 2002; Kelley et al. 2002, 2008; Alba 2012; Begun 2013, 2015; Pugh 2020). Dryopithecinae is a European hominid clade composed of genera such as *Dryopithecus*, *Pierolapithecus*, *Hispanopithecus*, *Anoiapithecus*, and *Rudapithecus*; the Anatolian/Greek genera *Ouranopithecus* and *Graecopithecus* may or may not be a part of this clade as well (Alba 2012; Begun 2015; Pugh 2020). In general, dryopithecines

Table 17.1 Taxa discussed in the text by age range and geographic distribution

Taxon	Classification	Geographic region	FAD (Ma)	LAD (Ma)	References
Rukwapithecus	Hominoidea (Nyanzapithecidae)	Africa	~25.2		Stevens et al. (2013)
Kamoyapithecus	Hominoidea (*incertae sedis*)	Africa	~24–27		Leakey et al. (1995), Rasmussen and Guitierrez (2009)
Proconsul	Hominoidea (Proconsulidae)	Africa	21	19	Bishop et al. (1969), Harrison and Andrews (2009)
Ekembo	Hominoidea (Proconsulidae)	Africa	~20	~17	Peppe et al. (2009), McCollum et al. (2013)
Lomorupithecus	Pliopithecoidea	Africa	~20–19		Rossie and MacLatchy (2006)
Platodontopithecus	Pliopithecoidea	Asia	~18–17		Harrison and Gu (1999)
Dionysopithecus	Pliopithecoidea	Asia	~18–17		Harrison and Gu (1999)
Afropithecus	Hominoidea (Afropithecidae)	Africa	~17.5–16.8		Leakey and Leakey (1986a), McDougall and Brown (2009)
cf. Griphopithecus	Hominoidea (Kenyapithecidae)	Europe	~17–16		Heizmann and Begun (2001), Böhme et al. (2011)
Kamlial/ Manchar teeth	*incertae sedis*	Asia	~17–16		Barry et al. (1986), Bernor et al. (1988)
Griphopithecus	Hominoidea (Kenyapithecidae)	Europe, Western Asia	~16 or after 14 (See text for details)		Heizmann and Begun (2001), Begun et al. (2003)
Kenyapithecus	Hominoidea (Kenyapithecidae)	Africa, Western Asia	~16 or after 14 [Paşalar]; 13.5–14 [Fort Ternan]		Bishop et al. (1969), Heizmann and Begun (2001), Begun et al. (2003), Pickford et al. (2006), Casanovas-Vilar et al. (2011)
Equatorius	Hominoidea (*incertae sedis*)	Africa	15.5	~14	Ward et al. (1999)
Nacholapithecus	Hominoidea (*incertae sedis*)	Africa	15		Nakatsukasa and Kunimatsu (2009)
Kapi	Hominoidea (Hylobatidae)	Asia	13.8–12.5		Gilbert et al. (2017, 2020)
Khoratpithecus	Hominoidea (Ponginae)	Asia	~12.4	~9–7	Chaimanee et al. (2004), Coster et al. (2010), Jaeger et al. (2011)
Sivapithecus	Hominoidea (Ponginae)	Asia	~12.7	8.5 (8.1?)	Barry et al. (2013), Pillans et al. (2005), Patnaik and Cameron (2007), Patnaik (2008)
Lufengpithecus	Hominoidea (*incertae sedis*)	Asia	~10	~6	Harrison et al. (2002), Qi et al. (2006), Kelley and Gao (2012)
Ankarapithecus	Hominoidea (Ponginae)	Western Asia	~9.8		Alpagut et al. (1996)
Ouranopithecus	Hominoidea	Europe, Western Asia	~9.6	~8.11–7.64	Koufos et al. (2016), Güleç et al. (2007), Begun (2015), Kaya et al. (2016)
Krishnapithecus	Pliopithecoidea	Asia	~9		Sankhyan et al. (2017)
Yuanmoupithecus	Hominoidea (Hylobatidae)	Asia	~9–7		Harrison (2016)
Indopithecus	Hominoidea (Ponginae)	Asia	~8.85–8.6		Pillans et al. (2005)
Oreopithecus	Hominoidea (*incertae sedis*)	Europe	8.1	6.7	De Miguel and Rook (2018)
?Dryopithecus wuduensis	*incertae sedis*	Asia	~8–6		Xue and Delson (1989)
Laccopithecus	Pliopithecoidea	Asia	~7–6		Harrison (2005)
Gigantopithecus	Hominoidea (Ponginae)	Asia	2	0.32–0.4	Zhang et al. (2014)
Pongo	Hominoidea (Ponginae)	Asia	Early Pleistocene	Present	Harrison et al. (2014)
Bunopithecus	Hominoidea (Hylobatidae)	Asia	Early-Middle Pleistocene		Matthew and Granger (1923), Ortiz et al. (2015)
Junzi	Hominoidea (Hylobatidae)	Asia	Holocene		Turvey et al. (2018)

Notes FAD = First Appearance Datum, LAD = Last Appearance Datum. Numbers in parentheses represent estimates from less secure specimens. When only one date or set of dates are given, the taxon is only known from a single dated locality

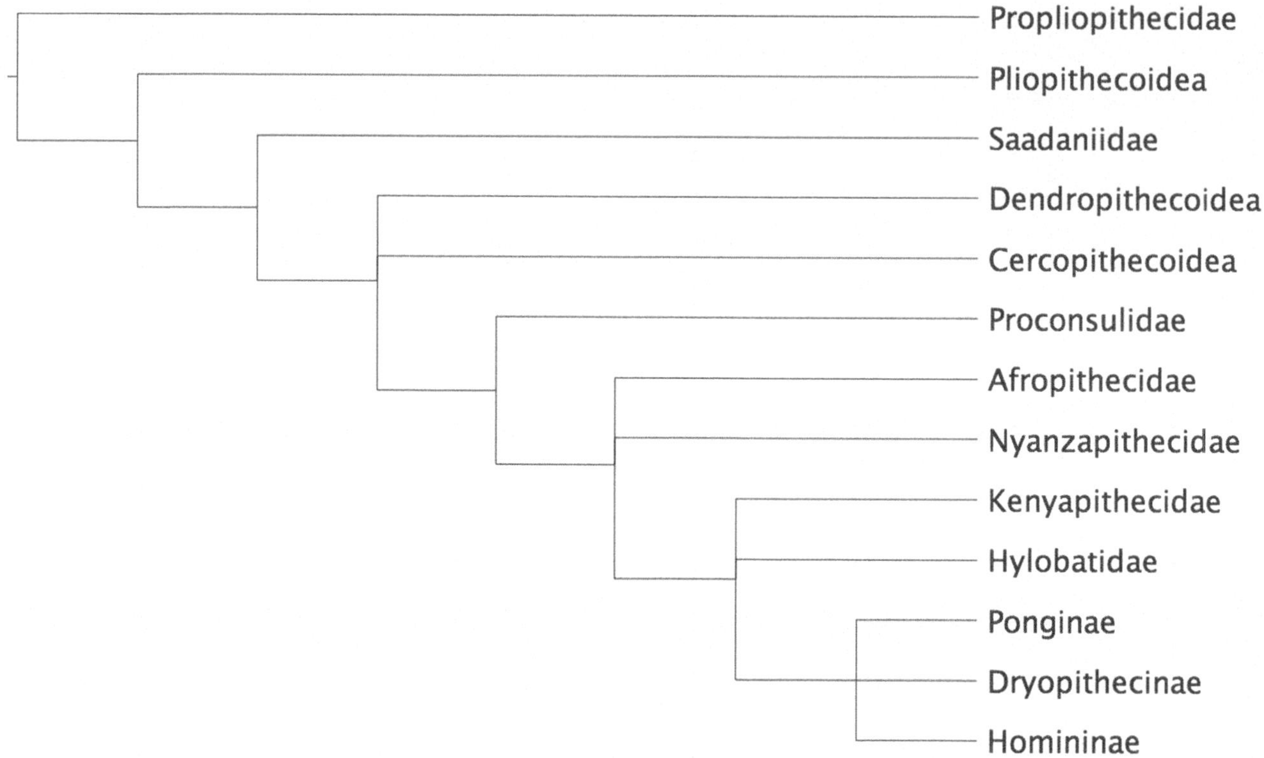

Fig. 17.1 Hypothesized phylogenetic tree of broad-level catarrhine taxa discussed in this paper. For further details on taxonomy used in this paper, see text and Table 17.1. Note that if *Oreopithecus* is included with the nyanzapithecids, then Oreopithecidae is the correct family name for the group

show more derived, extant hominid-like cranial features, molars and premolars with reduced cingula, and many possess postcranial adaptations related to suspensory or anti-pronograde locomotor behavior (reviewed in Alba 2012; Fleagle 2013; Begun 2015).

Hylobatidae is composed of the living gibbons (*Hylobates*, *Hoolock*, *Nomascus*) and siamang (*Symphalangus*), along with the late Miocene *Yuanmoupithecus*, Pleistocene *Bunopithecus*, and Holocene *Junzi* (Table 17.1). There are few clear craniodental synapomorphies uniting the hylobatids to the exclusion of all other catarrhine groups, particularly in the facial region, which may be secondarily derived compared to hominids and most stem hominoids (e.g., Rae 1999, 2004; Nengo et al. 2017). Even so, some derived craniodental features linking all living hylobatids have been suggested, including low sexual dimorphism in canine (and body) size, high crowned canines in both males and females with strong lateral splay, fused premolar roots (upper and lower), relatively simple molars with reduced cingula and bunodont, peripherally placed cusps resulting in large occlusal basins (seen in all crown hominoids and some stem catarrhines to varying degrees), M3s equal to or smaller in size compared to the M1s, relatively shallow mandibular corpora, a mandibular symphysis lacking a simian shelf, and an ectotympanic tube that curves anteriorly (Harrison 2016; Nengo et al. 2017). Extant hylobatids are most distinctive in their postcranial skeleton, which is highly derived for their unique locomotor repertoire including high incidences of ricochetal brachiation (e.g., see Schultz 1933a, b; Andrews and Groves 1975; Larson 1998; Harrison 2016). As discussed below, there are very few possible fossil hylobatids, known only from partial jaws and isolated teeth (Matthew and Granger 1923; Ortiz et al. 2015; Harrison 2016; Gilbert et al. 2020).

The Ponginae include the living orangutans (*Pongo*) and their fossil relatives, which are here considered to include the genera *Sivapithecus*, *Ankarapithecus*, *Khoratpithecus*, *Gigantopithecus*, and *Indopithecus* (see also Begun and Güleç 1998; Kelley 2002; Chaimanee et al. 2004; Jaeger et al. 2011; Begun 2015; Pugh 2020). These taxa are united by derived craniofacial (facial dishing, narrow interorbital distance, supraorbital costae, lack of the anterior digastric groove, etc.) and/or dental features (thick enamel, molar enamel "wrinkling", etc.) resembling extant *Pongo* (e.g., see Andrews and Tekkaya 1980; Andrews and Cronin 1982; Ward and Pilbeam 1983; Ward and Kimbel 1983; Brown and Ward 1988; Olejniczak et al. 2008; Begun 2015). *Lufengpithecus*, often considered to be a pongine (e.g., Kelley 2002; Alba 2012; Begun 2015), is here considered to be *incertae sedis* and may be polyphyletic (see Kelley and Gao 2012; Pugh 2020).

Homininae includes *Gorilla, Pan, Homo,* and their fossil relatives (e.g., *Sahelanthropus, Ardipithecus, Orrorin, Australopithecus, Paranthropus*, etc.). It is possible that genera such as *Ouranopithecus*, *Nakalipithecus*, and *Chororapithecus* are more closely related to Homininae than to other apes (e.g., see de Bonis et al. 1990; Dean and Delson 1992; Kunimatsu et al. 2007; Suwa et al. 2007; Begun et al. 2012; Pugh 2020), but these taxa are probably best left as *incertae sedis* for the time being, especially *Chororapithecus*, which is represented by only a handful of teeth. The hominines generally display crania with pronounced supraorbital tori and sulci, broad interorbital regions, squared orbits, klinorhynchy, and an overlapping and stepped nasal fossa (e.g., see Ward and Kimbel 1983; Shea 1985; Begun et al. 2012; Begun 2015).

The First Eurasian Apes: Early Catarrhine Dispersals Out of Africa in the Miocene

As discussed above, the earliest accepted hominoids are found in the Late Oligocene through early Miocene of East Africa. In addition to hominoids, most evidence indicates that other major advanced catarrhine clades, including pliopithecoids (or pliopithecids) and cercopithecoids (Old World monkeys), also originated in Africa (e.g., Benefit 1993, 1999; Benefit and McCrossin 1997; Harrison and Gu 1999; Benefit and McCrossin 2002; Harrison 2005; Rossie and MacLatchy 2006; Miller et al. 2009; Fleagle 2013). Early stem catarrhines are found in the Late Eocene of North Africa (e.g., Fleagle and Simons 1982; Simons 1987, 1989; Simons and Rasmussen 1996; Simons et al. 1999, 2007; Seiffert et al. 2000, 2010; Seiffert and Simons 2001; Begun 2002; Harrison 2005; Seiffert 2012; Fleagle 2013), and are best known from the classic Late Eocene-Early Oligocene quarries of the Fayum, Egypt (e.g., Seiffert et al. 2010; Fleagle 2013; see Fig. 17.1). Molecular data suggest an even earlier origin for the Catarrhini, perhaps as early as ~51–38 Ma (Chatterjee et al. 2009; Perelman et al. 2011; Pozzi et al. 2014), and crown catarrhines, Cercopithecoidea and Hominoidea, are estimated to have diverged between ~38 and 20 Ma (Steiper and Young 2006; Perelman et al. 2011; Pozzi et al. 2014). Evidence of such an early crown catarrhine divergence in the fossil record is currently quite limited, however, aside from several ~27–24 Ma discoveries in the East African Rift Valley. *Rukwapithecus* and *Nsungwepithecus* (~25 Ma) are the oldest representatives of the hominoid and cercopithecoid lineages respectively, and although very fragmentary, ~27–24 Ma *Kamoyapithecus* shares a few dental features with early Miocene hominoids (particularly *Afropithecus*) as well (see Table 17.1; Madden 1980; Leakey et al. 1995; Rasmussen and Guitierrez 2009; Stevens et al. 2013).

In addition to a scanty record of crown catarrhine groups, advanced stem catarrhine groups (e.g., saadaniids, pliopithecoids, and possibly dendropithecoids) are also poorly known for much of the middle and late Oligocene (Harrison 2005). Pliopithecoids are unknown from Africa with the possible exception of *Lomorupithecus* from the ~20–19 Ma deposits of Napak, Uganda (Rossie and MacLatchy 2006). However, pliopithecoids almost certainly arose in Africa based on the distribution of other stem catarrhine groups. As with all stem taxa, *Lomorupithecus* does not possess the full complement of traits that define later pliopithecoids, so there is some disagreement about its inclusion in the superfamily. For example, Harrison (2010, 2013) prefers to include it in Dendropithecoidea with other small-bodied catarrhines of the early Miocene of East Africa. Furthermore, Harrison and Gu (1999) have suggested that stem pliopithecoids and small-bodied proconsuloids (Harrison and Gu's broad definition of proconsuloids includes almost all non-cercopithecoid catarrhines from Africa at this time) would have been quite similar morphologically, making it very difficult to differentiate stem taxa representing these two clades.

Thus, the available fossil evidence strongly suggests that early catarrhine evolution took place almost entirely (if not exclusively) in Africa. During the time of catarrhine origins and early evolution, Africa remained an island continent, separated from Eurasia by the Tethys Seaway (Seiffert 2012; Begun et al. 2012). Members of these early catarrhine lineages evolved in Africa *in situ* until the early Miocene when major tectonic events occurred (e.g., Steininger et al. 1985; Rögl 1998; van der Made 1999; Harzhauser et al. 2007; Nargolwalla 2009; Begun et al. 2012). During the early Miocene, the Afro-Arabian plate made contact with Eurasia after nearly 100 million years of separation, and fluctuating sea levels resulted in intermittent landbridges that permitted faunal interchange between Africa and Eurasia (Rögl 1998; Koufos et al. 2003; Harzhauser et al. 2007). The first long-lasting connection between Afro-Arabia and Eurasia occurred around 19 Ma, closing off the Tethys Seaway and sparking a large faunal exchange between the continents shortly thereafter (Rögl 1998, 1999; Harzhauser et al. 2007). This semi-permanent connection is often referred to as the "*Gomphotherium* Landbridge" after the proboscidean group that was among the first to disperse from Africa into Eurasia. However, there is fossil evidence from Indo-Pakistan suggesting that occasional connections between Afro-Arabia and Asia may predate the establishment of this landbridge by a few million years (~23–20.5 Ma; Barry et al. 1985; Thomas 1985; Bernor et al. 1987; Welcomme et al. 2001;

Koufos et al. 2003; Harzhauser et al. 2007; Flynn et al. 2013; Antoine et al. 2013; Patnaik 2016; for possible Oligocene connection, see Antoine et al. 2003). In addition to proboscideans (both gomphotheres and deinotheres), numerous other mammalian groups dispersed from Africa into Eurasia, including creodonts (*Hyainailouros*), hyracoids (cf. *Geniohyus*), anthracotheres (*Brachyodus*), possibly some rodents (Pedetidae, Thryonomyidae) and tragulids (*Dorcatherium*), and, of interest here, catarrhine primates (van der Made 1999; Flynn and Wessels 2013; Patnaik 2016). Following an initial wave of African immigrants likely including proboscideans, hyracoids, creodonts, anthracotheres, and possibly *Dorcatherium* and thryonomyid rodents, catarrhines and pedetid rodents appear in Eurasia shortly thereafter (van der Made 1999; Flynn and Wessels 2013; Patnaik 2016, 2020).

The oldest catarrhine fossils found outside of Afro-Arabia are *Platodontopithecus jianghuaiensis* and *Dionysopithecus shuangouensis*, from ~18 to 17 Ma deposits at Sihong, Jiangsu Province, China (Li 1978; Gu and Lin 1983; Harrison and Gu 1999; Harrison 2005). On the basis of a number of features shared among many stem catarrhines, these taxa have been attributed to several different groups in the past, including dendropithecoid “apes” (e.g., Fleagle 1988; Bernor et al. 1988). However, they are now generally agreed to belong to the broader Eurasian pliopithecoid radiation on the basis of a number of shared derived features, such as mesiodistal waisting of the incisor crowns, the lack of a strongly developed lingual pillar on the I^1, a relatively short and steep honing flange on P_3, the mesial position of the buccal molar cusps relative to the lingual cusps, resulting in oblique crests between them, and the presence of a pliopithecine triangle on the lower molars, particularly m2–m3 (Harrison and Gu 1999; Begun 2002).

After their first appearance in China, pliopithecoids become more common and widely distributed in Europe shortly thereafter, by 17–16 Ma during the European mammal zone MN 5 (Hürzeler 1954; Zapfe 1958, 1960; Ginsburg 1975; Ginsburg and Mein 1980; Harrison and Gu 1999; Begun 2002). Harrison and Gu (1999) further hypothesize a subsequent branching event in Europe resulting in the crouzeliine subfamily, followed by a dispersal back eastward to China between ~15 and 8 Ma involving *Laccopithecus* and/or its ancestor(s). The ~9 Ma pliopithecoid from Haritalyangar in the Indian Siwaliks, *Krishnapithecus*, may be an immigrant from the European pliopithecoid stock, but its affinities to earlier Asian dionysopithecines are unclear as well (Sankhyan et al. 2017). In any case, there are potentially multiple pliopithecoid dispersal events between Europe and Asia between ~15 and 8 Ma, although analyses on South Asian Miocene fauna suggest that periods of migration before ~13 Ma and after ~11.6 Ma seem most likely (Badgley et al. 2016).

The earliest evidence of hominoids in Eurasia occurs in ~17–16 Ma deposits at the site of Engelsweis in Germany (Heizmann and Begun 2001; Böhme et al. 2011; Casanovas-Vilar et al. 2011), coinciding with the possible arrival of pliopithecoids from Asia. A worn partial M^3 (ENG. 4/1), found in 1973 and described in detail in 2001, was considered too fragmentary to confidently assign to an existing taxon or to base the erection of a new taxon (Heizmann and Begun 2001). Until additional specimens are recovered, it has been conservatively assigned to cf. *Griphopithecus* to reflect similarities in enamel thickness with *Griphopithecus* specimens from Turkey and Slovakia. This specimen predates other fossil representatives of *Griphopithecus*, and other evidence of hominoids in Europe, by 1–3 Myr (discussion of ages of next oldest sites below), and as a result, the biogeographic hypotheses that stem from it remain controversial and unlikely to be resolved without additional fossils of similar antiquity. However, taken at face value, this information suggests that hominoids dispersed out of Africa slightly later than the pliopithecoids and, in contrast to the apparent Asian dispersal route taken by the earliest pliopithecoids, the earliest hominoids possibly took a more westerly route into Europe first, or in combination with a simultaneous dispersal eastward into Asia.

The Western Asian/Anatolian sites of Paşalar and Çandır are the next most ancient hominoid bearing localities in Eurasia, bearing fossils of two possible stem hominid taxa, *Griphopithecus alpani* and *Kenyapithecus kizili*. Unlike Engelsweis, Paşalar is relatively productive and the fossils discovered there are taxonomically informative (Andrews and Tobien 1977; Alpagut et al. 1990; Kelley et al. 2008). A single fossil ape specimen, the type specimen of *Gr. alpani*, has also been recovered from the similarly aged site of Çandır (Tekkaya 1974). Similarities in craniodental morphology have been noted between the Turkish species and *Kenyapithecus* and *Equatorius* from East Africa, indicating continuity between these regions (Andrews et al. 1996; Begun 2000; Andrews and Kelley 2007; Kelley et al. 2008). The ages of these sites have been difficult to ascertain, and as a result are somewhat controversial; it remains unclear if they are younger or older than the transgression (i.e., expansion) of the Tethys Sea during the Langhian (16.5–14.9 Ma) that once again prevented dispersal between Afro-Arabia and Eurasia (Bernor and Tobien 1990; Andrews et al. 1996; Rögl 1998; Begun et al. 2003; 2012; Andrews and Kelley 2007; Casanovas-Vilar et al. 2011). Begun and colleagues (2003, 2012) argue that these sites are roughly contemporaneous with, but slightly younger than, Engelsweis (before the Langhian transgression), while others have suggested that they date to after the transgression event at ~14 Ma (Bernor and Tobien 1990; Casanovas-Vilar et al. 2011). Whether hominoids entered Eurasia before or after this transgression is essential to understanding the initial dispersal and evolution of the broader Eurasian ape radiation. By ~12.5–11 Ma, hominoids are spread throughout both Europe and Asia (Begun 2002; Kelley 2002; Alba 2012), and whether the dispersals into these continents are virtually simultaneous or successive following the initial European dispersal represented by Engelsweis remains unclear. In any case, a dispersal eastward into Asia must have occurred by >16–13 Ma, and this event may have been contemporaneous with dispersals of crouzeliine pliopithecoids and hylobatids as well (see also biogeographic analyses and Figs. 17.2, 17.3 and 17.4).

Within hominoids, hylobatid origins are still frustratingly unclear. The fossil record is virtually devoid of any hylobatid fossils until *Yuanmoupithecus xiaoyuan* at ~9–7 Ma in southern China, a taxon that is only known from a handful of teeth displaying numerous gibbon-like features such as fused premolar roots, low-crowned molars with rounded corners and peripherally placed bunodont cusps, lower molars with a shallow and wide talonid basin, relatively small lingual cusps linked by a crest, and marked reduction of the buccal cingulum (Pan 2006; Harrison et al. 2008; Harrison 2016). Because hylobatids are estimated to have diverged from hominids by ~20 Ma (e.g., Perelman et al. 2011) and hominids are known to have dispersed into Asia by at least ~13 Ma (as documented by the earliest *Sivapithecus* specimens in the Lower Siwaliks), it is possible if not probable that hylobatids were present in Asia before *Yuanmoupithecus* at ~9–7 Ma, but given their scarcity in the fossil record, they are rare and/or poorly preserved. As recently reviewed by Harrison (2016), numerous specimens have been suggested as possible hylobatid ancestors over the years, but many of these fossils have since been demonstrated to be pliopithecoids (e.g., *Laccopithecus*, *Krishnapithecus*, and *Dionysopithecus*). The recently described ~11.6 Ma *Pliobates* was argued to be a stem hominoid close to hylobatid origins based on derived aspects of its postcranium, particularly the absence of an entepicondylar foramen and lack of a capitular tail on the distal humerus, along with some subtle ulnar-carpal morphologies as well (e.g., a relatively small triquetrum, distal ulnar and capitate articulations, see Alba et al. 2015a). However, some extant platyrrhines (including *Pliobates*-sized atelids) have also lost the entepicondylar foramen (e.g., Gebo 1993; Rose 1997), and atelids also share other postcranial convergences with hominoids and *Pliobates* (e.g., radial head shape, mediolaterally narrow capitate head, see Larson 1998), making the significance of these features in *Pliobates* ambiguous. Instead, *Pliobates* shares numerous primitive features with unambiguous stem catarrhines (particularly pliopithecoids) and dendropithecids: broad premolars and molars, fairly extensive molar cingulae, relatively short snouts, projecting infraorbital rims, a posteriorly divergent palate, and a relatively long ulnar styloid process. Most importantly, *Pliobates* shares a very notable derived feature only found in pliopithecoids among all known catarrhine taxa: a partially ossified tubular ectotympanic. All crown catarrhines possess a fully ossified ectotympanic, and this morphology evolved by at least the middle-late Oligocene as evidenced by *Saadanius* (Zalmout et al. 2010). Moreover, *Pliobates* also lacks a very significant postcranial synapomorphy seen in the distal humerus of nearly all commonly accepted fossil and extant hominoid taxa, namely the development of a spool-shaped trochlea associated with both medial and lateral trochlear keels (e.g., Napier and Davis 1959; Rose 1983, 1988, 1997; Larson 1998; Benefit and McCrossin 2015). Thus, given the available evidence, in our view *Pliobates* is most likely a derived, late-occurring pliopithecoid or other stem catarrhine taxon that has converged on some interesting but minor details of extant hominoid elbows and wrists (e.g., see also Benefit and McCrossin 2015; Nengo et al. 2017).

A few other intriguing specimens from ~17 to 16 Ma Kamlial Formation deposits (Lower Siwaliks) on the Potwar Plateau and the similar-aged Manchar Formation of southern Pakistan have been mentioned as possible stem hylobatids and/or also linked to the early Miocene small-bodied catarrhine group from Africa often referred to as dendropithecoids (e.g., see Barry et al. 1986; Bernor et al. 1988; Harrison and Gu 1999; Fleagle 2013; Harrison 2016). These teeth do not display any distinctive pliopithecoid features, although they do not display any hylobatid synapomorphies either. Most recently a 2D morphometric analysis of molar crown shape demonstrated that these specimens share the greatest affinities with stem catarrhine taxa, although which early catarrhine group still remains unclear (Gilbert et al. 2020). More interesting, however, is a recently discovered hylobatid-sized m3 from Chinji-aged (~13.8-12.5 Ma) deposits near Ramnagar, northern India (*Kapi ramnagarensis*; see Gilbert et al. 2017; Gilbert et al. 2020). A 2D morphometric analysis of m3 crown shape demonstrates that this specimen plots exclusvely in crown hominoid morphospace, closest to extant hylobatids (Gilbert et al. 2020). Additional fossils are necessary to more confidently confirm or refute the hylobatid affinities of the Ramnagar specimen, but it is currently the most probable candidate for a stem hylobatid in the fossil record before ~9 Ma.

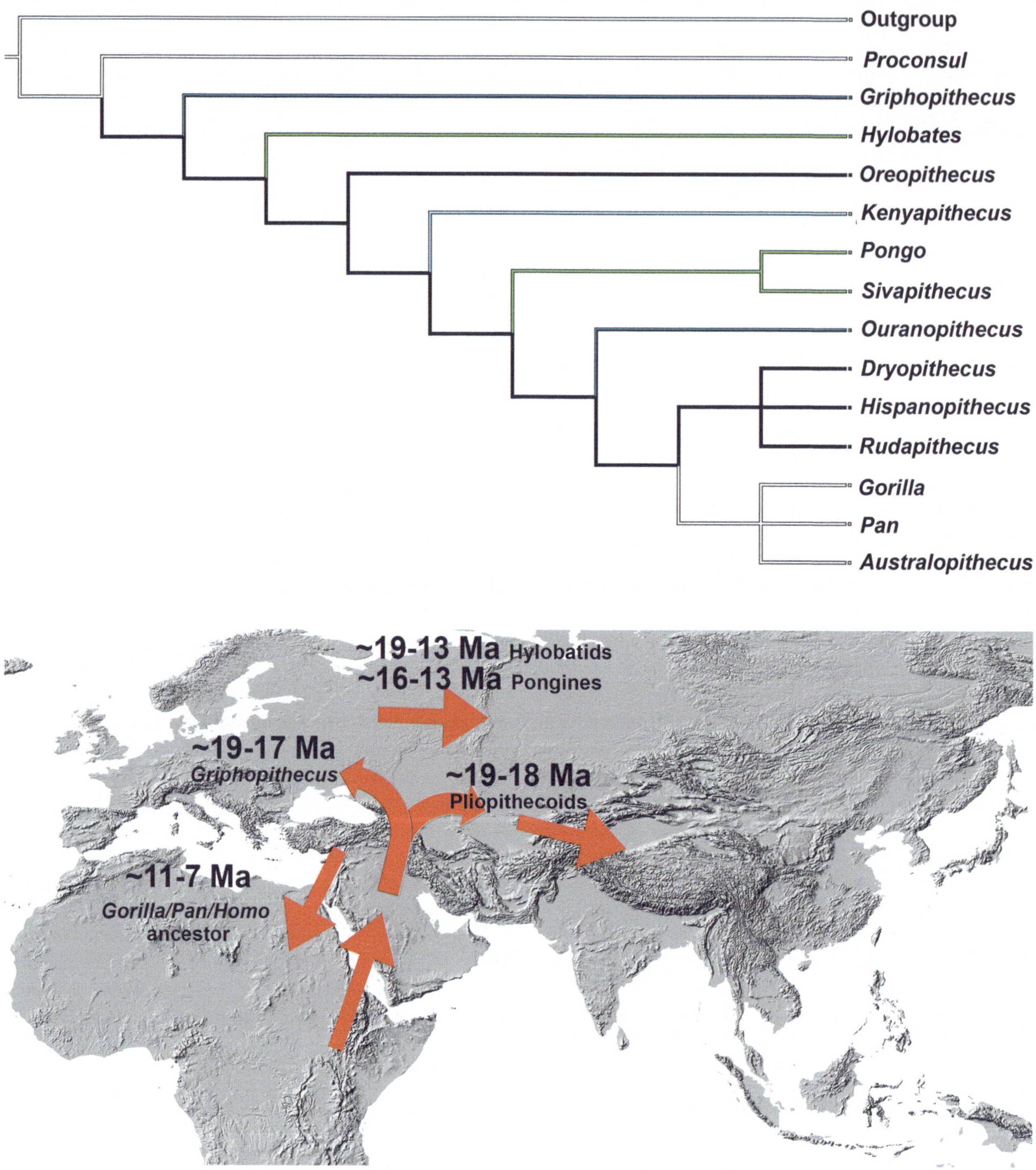

Fig. 17.2 Top panel: Hypothesized phylogenetic tree from Begun et al. (2012) with biogeographic character states mapped on. Color changes along branches indicate a change in character state and an implied biogeographic dispersal event. White = Africa, Blue = Western Asia/Anatolia, Green = Asia, Black = Europe. Bottom panel: Illustration of the main dispersal events implied by the Begun et al. (2012) cladogram. The earliest estimated dispersal dates are bracketed by the ~19 Ma landbridge and the FAD for each group. See text for further details and see Begun et al. (2012) for the taxonomy used in this cladogram

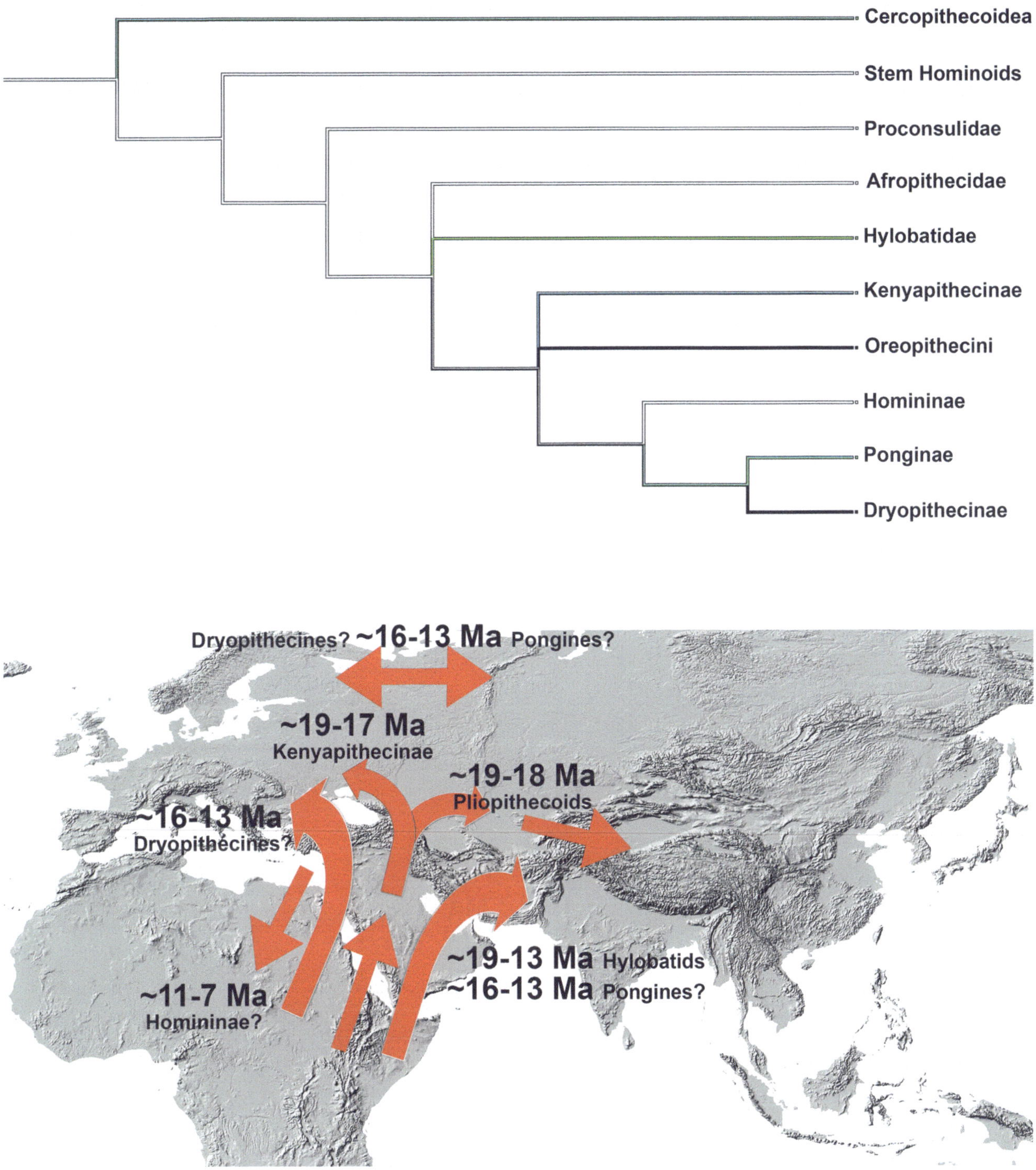

Fig. 17.3 Top panel: Hypothesized phylogenetic tree from Alba (2012) with biogeographic character states mapped on. Color changes along branches indicate a change in character state and an implied biogeographic dispersal event. White = Africa, Blue = Western Asia/Anatolia, Green = Asia, Black = Europe. Bottom panel: Illustration of the main dispersal events implied by the Alba (2012) cladogram. The earliest estimated dispersal dates are bracketed by the ~ 19 Ma landbridge and the FAD for each group. See text for further details and see Alba (2012) for the taxonomy used in this cladogram

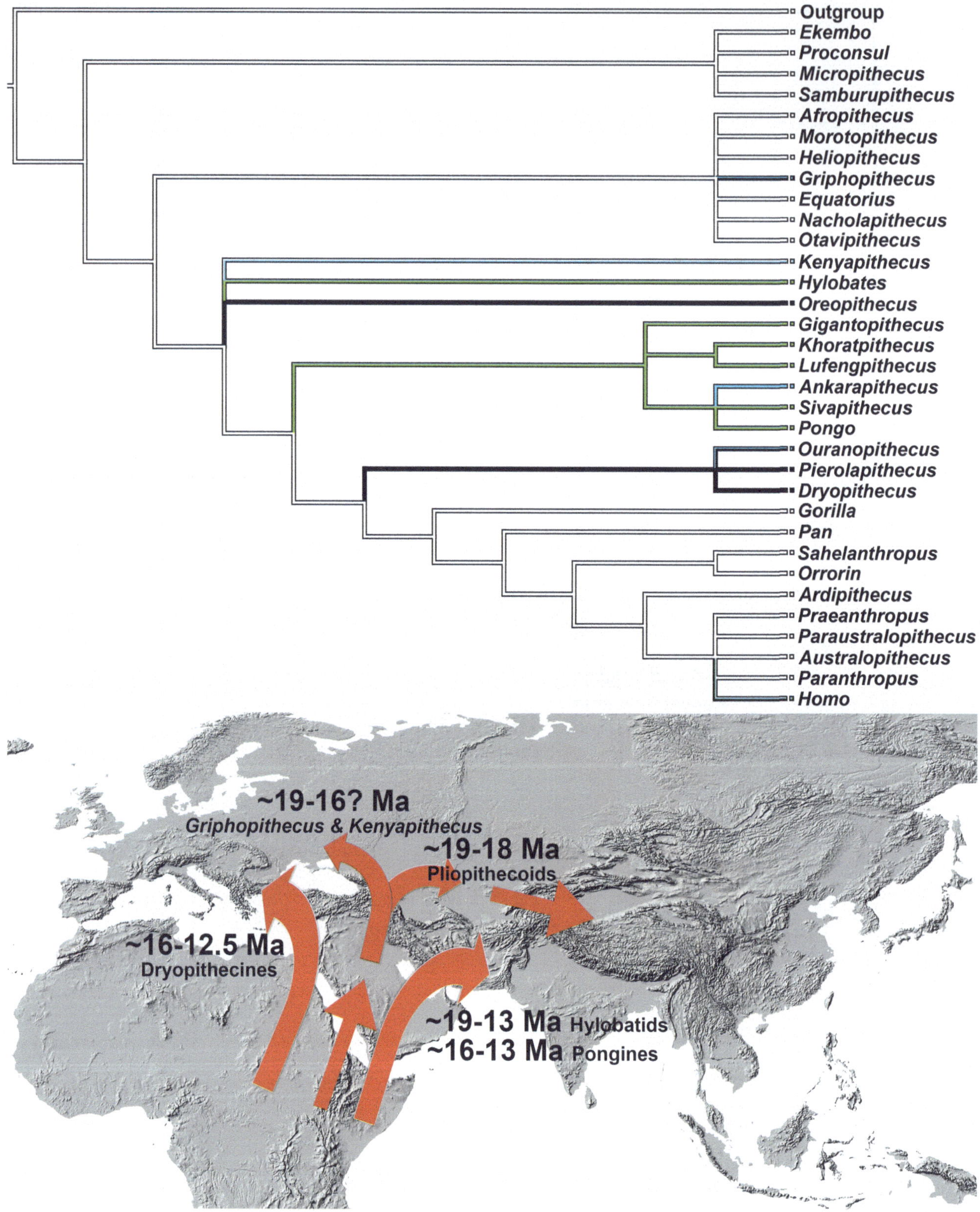

Fig. 17.4 Top panel: Hypothesized phylogenetic tree from Begun (2015) with biogeographic character states mapped on. Color changes along branches indicate a change in character state and an implied biogeographic dispersal event. White = Africa, Blue = Western Asia/Anatolia, Green = Asia, Black = Europe. Bottom panel: Illustration of the main dispersal events implied by the Begun (2015) cladogram. The earliest estimated dispersal dates are bracketed by the ~19 Ma landbridge and the FAD for each group. See text for further details and see Begun (2015) for the taxonomy used in this cladogram

How Many Transcontinental Ape Dispersal Events Occurred and Between Which Continents? Biogeographic Links to Phylogeny, Tectonics, and Climate

Further insight into the timing and geographic dispersal of Eurasian hominoid taxa depends largely on the inferred phylogenetic relationships of the taxa involved. For example, using parsimony to analyze dispersal patterns (the "progression rule", see Hennig 1966; Nelson and Platnick 1981; Strait and Wood 1999), we took three recent phylogenetic hypotheses derived from review papers on ape evolution by Begun et al. (2012), Begun (2015), and Alba (2012), and treated biogeography as an unordered character using four states for each operational taxonomic unit: 0 = Africa, 1 = Western Asia/Anatolia, 2 = Asia, 3 = Europe (see Table 17.1 for geographic regions and Appendix for character state designations by taxon). This biogeographic character was then mapped on each tree using the "Trace Character History" function in Mesquite v. 3.5 (Maddison and Maddison 2018). Wherever a character state change is noted on a tree, a hypothetical dispersal event between biogeographic areas was inferred (see Figs. 17.2, 17.3 and 17.4).

Before discussing the results of such a thought experiment, it is important to recognize the limitations of the fossil record and concomitant limitations on attempts to derive more fine-grained biogeographical details from the preserved evidence. As Maxwell et al. (2018) demonstrate, many patterns in primate and human evolution can be driven by geological sampling and chance as opposed to actual events, so we regard the following analyses as preliminary; obviously, any new fossil discoveries in unexpected areas or time periods could greatly change interpretations of hominoid biogeography. Nonetheless, based on recent phylogenetic hypotheses, some patterns emerge.

First, the relationships suggested by Begun et al. (2012) hypothesize two taxonomic dispersals from Europe into Asia, namely the hylobatids and pongines (see Fig. 17.2). Potentially, both groups could have dispersed during the same time period, but the ancestors of both hylobatids and pongines are most parsimoniously hypothesized to have migrated from Europe rather than originating in Africa or *in situ* from a common ancestor in Asia. Notably, such a biogeographic scenario is at odds with molecular divergence dates that suggest a divergence between hylobatids and great apes ~20 Ma (e.g., Perelman et al. 2011; Pozzi et al. 2014), before any apes are found in Eurasia.

In contrast to the Begun et al. (2012) hypothesis, the relationships favored by Alba (2012) suggest that the hylobatid lineage dispersed to Asia from Africa while the pongine lineage migrated to Asia independently from either Africa or Europe (Fig. 17.3). While both groups could have again dispersed into Asia during the same time period, the Alba (2012) phylogenetic arrangement suggests that they potentially did so from different geographic regions (Fig. 17.3). The Alba (2012) tree also suggests that Dryopithecines could have dispersed into Europe from either Africa or Western Asia/Asia, and that Homininae could have evolved in Africa or dispersed to Africa from Europe (Fig. 17.3).

Finally, the tree from Begun (2015; see also Begun et al. 2003) suggests that both hylobatids and pongines dispersed independently from Africa (Fig. 17.4). Again, the dispersal of these two lineages could have taken place at roughly the same time and as part of a larger faunal migration event (see above), but the overall indication is that the ancestors of both hylobatids and pongines are immigrants and that neither group evolved from earlier Eurasian groups *in situ*. Despite the implication of multiple dispersal events from Africa to Asia, however, it is worth pointing out that there is currently no evidence of "ancestral" hylobatids or pongines in the African fossil record.

In addition to the scenarios laid out above, analyses of two juvenile cranial specimens found in China and assigned to *Lufengpithecus* have been used to suggest a more complicated Asian dispersal pattern for hominids during the Miocene (Harrison et al. 2002; Kelley 2002; Kelley and Gao 2012). *Lufengpithecus* is typically assigned to Ponginae on the basis of geography, a few adult cranial features (supraorbital costae, anteriorly oriented zygomatics, pear-shaped nasal fossa, etc.) and especially postcanine dental similarities to *Pongo* (Schwartz 1990; Begun 2015), however the described juvenile crania were found to lack synapomorphies of pongines, and, rather, their morphology is reported to be consistent with primitive or stem hominids unaffiliated with any extant clade (Kelley and Gao 2012; Ji et al. 2013). Plesiomorphic hominid morphology in China would necessitate at least two independent dispersals of hominids into Asia (Kelley and Gao 2012). A partial mandible assigned to *Dryopithecus* (*?D. wuduensis*; Xue and Delson 1989; see also Harrison 2005) may provide additional evidence for a dispersal of non-pongine hominids into Asia. Assessing the affinities of specimens currently assigned to *Lufengpithecus* more fully is necessary to evaluate this hypothesis; these juvenile crania were recently included in a species level phylogenetic analysis, and results suggested that *Lufengpithecus* is a polyphyletic genus (Pugh 2020). However, adult cranial specimens are necessary to confirm this result.

So what conditions during what time periods would have made such dispersals possible? As outlined above, on the basis of the existing fossil record, the timing for the pongine dispersal to Asia can be estimated to between ~16 and 13 Ma, and for hylobatids anytime between ~19 and 13 Ma is possible. Because other African and Eurasian faunal immigrants appear in the South Asian Siwaliks between ~21 and 18 Ma (e.g., tragulids, bovids, muroid rodents, etc.; see Barry et al. 1985, 1995; Friedman et al. 1992; Downing et al. 1993), it is tempting to speculate that primates were most likely a part of these faunal migrations as well, perhaps including all three major groups discussed in this paper: pliopithecoids, hylobatids, and hominids (pongines, more specifically). While this is certainly one possibility, the current fossil record only supports dionysopithecine pliopithecoids as being possible members of this initial wave (or successive waves) into Asia from Africa (see above). Because both hylobatids and pongines show up later in the Asian record, it seems more likely that they are a part of a different migration event or events from Europe or Africa into Asia, perhaps one(s) that is/are linked to the Miocene Climatic Optimum between ~17 and 15 Ma (Böhme 2003) and the spread of subtropical woodland or forested environments (e.g., Andrews and Kelley 2007).

Another possibility, more in line with the current pongine and hylobatid fossil record, is that the hominoid dispersal event into Asia took place after the Miocene Climatic Optimum, during a period of time when there was faunal diversification in the Siwaliks perhaps associated with global cooling and periods of climatic instability (Barry et al. 1995). This instability, in fact, may have promoted the migration of apes into new geographic regions. In any case, the earliest definitive apes currently known in the Lower Siwaliks are ~13 Ma and found in areas with subtropical to tropical closed or semi-closed woodlands and forested environments, among others (e.g., Flynn et al. 2016, 2020; Singh et al. 2018). It is therefore likely that, some time before 13 Ma, a forested corridor allowed hominoid dispersal from Europe or Africa via Western Asia (see also Badgley et al. 2016).

Late Miocene Eurasian Extinction and its Relationship to Tectonics and Climate

The continued presence of forested mosaics throughout the middle Miocene and into the late Miocene allowed hominoids to survive in South, East and Southeast Asia for millions of years. In South Asia, the climate notably changed following a major tectonic uplift phase associated with the Himalayas and Tibetan Plateau ~9–8 Ma (e.g., Zhisheng et al. 2001), resulting in a significant drop in rainfall, a transition to a more savannah-like environment, and the extinction and turnover of many mammalian species between ~8.5 and 6.0 Ma, including hominoid primates (Pillans et al. 2005; Badgley et al. 2008; Patnaik 2008, 2016). This same ~9–8 Ma uplifting event also intensified the East Asian monsoon pattern (Zhisheng et al. 2001), although hominoid primates such as *Lufengpithecus* were apparently able to persist in subtropical forested or wooded environments there until the terminal Miocene (Harrison et al. 2002; Chang et al. 2015). Likewise, the pongine *Khoratpithecus* persisted in tropical forest environments in Southeast Asia until the late Miocene (Chaimanee et al. 2003, 2006; Sepulchre 2003). *Gigantopithecus*, fossil *Pongo*, and hylobatids survived at higher latitudes until even more recently in East Asian Pleistocene forest environments, but as these forests and other resources began to disappear due to continued climate change and/or possible hominin activity in the Middle-Late Pleistocene through the Holocene, many of these hominoids disappeared as well (Harrison et al. 2014; Takai et al. 2014; Zhang et al. 2015; Ortiz et al. 2015; Bocherans et al. 2017; Turvey et al. 2018).

Similar to the situation in Asia, apes radiated in Europe until the climate changed and environments started to dry out, transitioning from subtropical to deciduous forests and woodlands during the Mid-Vallesian Crisis ~9.6 Ma (Agusti et al. 2003; Begun et al. 2012). This climatic event is hypothesized to be the result of significant changes in deep ocean circulation patterns (Agusti et al. 2003). Also similar to the situation in Asia, however, a few hominoids persisted beyond this climate change event. Several ape taxa (e.g., *Ouranopithecus*, *Oreopithecus*) with masticatory and/or postcranial specializations persisted in southern Europe and Western Asia until about ~6.7 Ma (Spassov et al. 2012; Suwa et al. 2016; Böhme et al. 2017; De Miguel and Rook 2018) before they appear to go extinct (e.g., Ungar and Kay 1995; Harrison and Rook 1997; Güleç et al. 2007). In the case of *Oreopithecus*, isolation in an island environment probably also helped to extend its temporal range beyond that seen for other European hominoids (Rook et al. 1996; Harrison and Rook 1997; Harrison et al. 2002).

In addition to climate change, the late Miocene also bore witness to the arrival of a new group of catarrhine primates in Eurasia, the cercopithecoid monkeys. The first monkeys, represented by the colobine *Mesopithecus*, are found at sites securely dated to ~8.5–5.3 Ma in the Mediterranean region as well as Western Asia, possibly Pakistan, and China (Barry 1987; Delson 1994; Ji et al. 2013; Koufos 2013; Alba et al. 2015b). The first cercopithecines are found in Europe ~6.0–5.3 Ma and extend all the way to China by the early Pliocene (Delson 1980, 1996; Delson et al. 2000; Köhler et al. 2000; Alba et al. 2014). In combination with the climate changes outlined above, possible competition from these new catarrhine immigrants with specialized teeth (e.g., bilophodont molars), multichambered stomachs (colobines), faster generation times, and, in many cases, more generalized and flexible locomotor adaptations, may have played a role in Eurasian hominoid extinction as well (Delson 1975; Andrews 1981; Lovejoy 1981; Jablonski et al. 2000).

Summary and Conclusions

As this brief review indicates, the dispersals of apes into Eurasia during the Miocene, and their eventual extinction in many regions, are ultimately linked to major tectonic events combined with regional and global climatic conditions. At the beginning of this chapter, we asked four broad questions: (1) When did the first Eurasian apes appear and what tectonic and climatic events aided their initial dispersal out of Africa?; (2) Was there a single ape dispersal event out of Africa or were there numerous dispersals into Eurasia?; (3) Are pongines (i.e., orangutans and their fossil relatives) the result of a dispersal event from Europe or Africa, and what tectonic and/or climatic events influenced their appearance in Asia?; and (4) What tectonic and climate events are responsible for the extinction of many Eurasian apes, resulting in their limited distribution in Southeast Asia today? In response, current evidence suggests:

(1) The early Miocene is associated with a tectonic event that had broad impact on mammalian faunal exchange between Eurasia and Africa, namely a land-bridge connection between these continents through the Arabian Peninsula ~22–19 Ma. Pliopithecoids were likely included in the waves of mammalian dispersals associated with this tectonic event, and hominoids likely soon after, by at least ~17–16 Ma in Europe and Anatolia as evidenced by fossils allocated to *Griphopithecus* and *Kenyapithecus*. Several million years after *Griphopithecus* and *Kenyapithecus* are found in Anatolia, we begin to find evidence of apes of a more modern aspect in both Europe and Asia about the same time in the late middle Miocene (~13–12 Ma). In Europe, there is a large radiation of apes throughout this period.

(2) Biogeographic hypotheses of Miocene ape dispersal into Eurasia are heavily dependent on underlying phylogenetic inferences. Thus, uncertainty about the patterns of hominoid dispersals through Eurasia mirrors uncertainty about hominoid evolutionary relationships. Recent hypotheses of fossil hominoid phylogeny suggest anywhere between one and four dispersals from Africa into Eurasia. Within Eurasia, anywhere between zero and two dispersal events involving hylobatids and pongines are generally hypothesized between Europe and Asia, again depending on the phylogenetic scenario preferred.

(3) In Asia, pongines are found beginning in the middle Miocene (e.g., *Sivapithecus*, *Khoratpithecus*) with some fossil taxa persisting until the Pleistocene (e.g., *Gigantopithecus*). Similar to the situation with hominoids more broadly, the origin and dispersal of the orangutan (pongine) clade in Asia from either Africa or Europe depends on the inferred phylogenetic relationships. However, it seems highly likely that the pongine clade is an immigrant group to Asia during the middle Miocene and that the group did not evolve in Asia *in situ* from earlier Asian taxa. The dispersal of pongines to Asia may have been facilitated by the Miocene Climatic Optimum, as the first appearance of pongines is consistent with a dispersal during or immediately following this climatic event.

(4) During the late Miocene, tectonic events in Asia leading to a period of Himalayan and Tibetan uplift ~9–8 Ma resulted in climatic changes that led to the extinction of hominoids in many parts of the continent. A similar change in climate associated with changing oceanic currents occurred in the late Miocene of Europe, also leading to the extinction of many apes there. Thus, it seems clear that the late Miocene extinction of hominoids across large portions of Eurasia is tightly linked to the loss of tropical and subtropical forests and woodland environments. The later Pleistocene-Holocene extinctions of many hylobatids, fossil *Pongo*, and *Gigantopithecus* in the apparent forest refugia of eastern and southeastern Asia may be the result of both natural and anthropogenic pressures on preferred hominoid habitat and resources. Where tropical forests persist, apes survive in the late Miocene through to the Pleistocene and, of course, to present day.

Acknowledgments We thank Rajeev Patnaik for inviting us to contribute to this volume and Prof. Sahni for his many contributions to geology and paleontology. Three anonymous reviewers provided helpful comments that greatly improved an earlier version of this manuscript. Luci Betti-Nash provided the maps for Figs. 17.2–17.4.

Appendix

Character state matrices used to infer biogeographic dispersals in Figs. 17.2, 17.3 and 17.4.

Begun et al. (2012) **matrix**	
Taxon	*Character state(s)*
Outgroup	0
Proconsul	0
Griphopithecus	1, 3
Hylobates	2
Oreopithecus	3
Kenyapithecus	0, 1
Pongo	2
Sivapithecus	2
Ouranopithecus	1, 3
Dryopithecus	3
Hispanopithecus	3
Rudapithecus	3
Gorilla	0
Pan	0
Australopithecus	0
Alba (2012) **matrix**	
Taxon	*Character state(s)*
Cercopithecoidea	0, 1, 2, 3
Stem Hominoidea	0
Proconsulidae	0
Afropithecidae	0
Hylobatidae	2
Kenyapithecidae	0, 1, 3
Oreopithecini	3
Homininae	0
Ponginae	1, 3
Begun (2015) **matrix**	
Taxon	*Character state(s)*
Outgroup	0
Ekembo	0
Proconsul	0
Micropithecus	0
Samburupithecus	0
Afropithecus	0
Morotopithecus	0
Heliopithecus	0
Griphopithecus	1, 3
Equatorius	0
Nacholapithecus	0
Otaviptihecus	0
Kenyapithecus	0, 1
Hylobates	2
Oreopithecus	3
Gigantopithecus	2
Khoratpithecus	2
Lufengpithecus	2
Ankarapithecus	1
Sivapithecus	2
Pongo	2
Ouranopithecus	1, 3
Pierolapithecus	3
Dryopithecus	3
Gorilla	0
Pan	0
Sahelanthropus	0
Orrorin	0
Ardipithecus	0
Praeanthropus	0
Paraustralopihtecus	0
Australopithecus	0
Paranthropus	0
Homo	0, 1, 2, 3

Notes 0 = Africa, 1 = W. Asia, 2 = Asia, 3 = Europe

References

Agusti, J., Sanz de Siria, A., & Garces, M. (2003). Explaining the end of the hominoid experiment in Europe. *Journal of Human Evolution, 45,* 145–153.

Alba, D. M. (2012). Fossil apes from the Vallès-Penedès basin. *Evolutionary Anthropology, 21,* 254–269.

Alba, D. M., Delson, E., Carnevale, G., Colombero, S., Delfino, M., Giuntelli, P., et al. (2014). First joint record of *Mesopithecus* and cf. *Macaca* in the Miocene of Europe. *Journal of Human Evolution, 67,* 1–18.

Alba, D. M., Almecija, S., DeMiguel, D., Fortuny, J., de los Rios, M.P., Pina, M., et al. (2015a). Miocene small-bodied ape from Eurasia sheds light on hominoid evolution. *Science, 350.* https://doi.org/10.1126/science.aab2625.

Alba, D. M., Montoya, P., Pina, M., Rook, L., Abella, J., Morales, J., et al. (2015b). First record of *Mesopithecus* (Cercopithecidae, Colobinae) from the Miocene of the Iberian Peninsula. *Journal of Human Evolution, 88,* 1–14.

Alpagut, B., Andrews, P., & Martin, L. (1990). New hominoid specimens from the Middle Miocene site at Pasalar, Turkey. *Journal of Human Evolution, 19,* 397–422.

Alpagut, B., Andrews, P., Fortelius, M., Kappelman, J., Temizsoy, I., Celebi, H., et al. (1996). A new specimen of *Ankarapithecus meteai* from the Sinap Formation of central Anatolia. *Nature, 382,* 349–351.

Andrews, P. (1978). A revision of the Miocene Hominoidea of East Africa. *Bulletin of the British Museum (Natural History), London, 30,* 85–224.

Andrews, P. (1981). Species diversity and diet in monkeys and apes during the Miocene. In C. B. Stringer (Ed.), *Aspects of human evolution* (pp. 25–61). London: Taylor and Francis.

Andrews, P. (1996). Paleoecology and hominoid paleoenvironments. *Biological Reviews, 71,* 257–300.

Andrews, P., & Cronin, J. E. (1982). The relationships of *Sivapithecus* and *Ramapithecus* and the evolution of the orang-utan. *Nature, 297,* 541–546.

Andrews, P., & Groves, C. (1975). Gibbons and brachiation. *Gibbon Siamang, 4,* 167–218.

Andrews, P., Harrison, T., Delson, E., Bernor, R. L., & Martin, L. (1996). Distribution and biochronology of European and Southwest Asian Miocene catarrhines. In R. L. Bernor, V. Fahlbusch & H.-W. Mittman (Eds.), *The evolution of Western Eurasian Neogene mammal faunas* (pp. 168–207). New York: Columbia University Press.

Andrews, P., & Kelley, J. (2007). Middle Miocene dispersals of apes. *Folia Primatologia, 78,* 328–343.

Andrews, P., & Martin, L. (1987). Cladistic relationships of extant and fossil hominoids. *Journal of Human Evolution, 16,* 101–118.

Andrews, P., & Tekkaya, I. (1980). A revision of the Turkish Miocene hominoid *Sivapithecus meteai*. *Palaeontology, 23,* 85–95.

Andrews, P., & Tobien, H. (1977). New Miocene locality in Turkey with evidence on the origin of *Ramapithecus* and *Sivapithecus*. *Nature, 268,* 699–701.

Antoine, P.-O., Metais, G., Orliac, M. J., Crochet, J.-Y., Flynn, L. J., Marivaux, L., et al. (2013). Mammalian Neogene biostratigraphy of the Sulaiman province, Pakistan. In X. Wang, L. J. Flynn & M. Fortelius (Eds.), *Fossil mammals of Asia: Neogene biostratigraphy and chronology* (pp. 400–422). New York: University Press.

Antoine, P.-O., Welcomme, J.-L., Marivaux, L., Baloch, I., Benammi, M., & Tassy, P. (2003). First record of Paleogene Elephantoidea (Mammalia, Proboscidea) from the Bugti Hills of Pakistan. *Journal of Vertebrate Paleontology, 23,* 977–980.

Badgley, C., Barry, J. C., Morgan, M. E., Nelson, S. V., Behrensmeyer, A. K., Cerling, T. E., et al. (2008). Ecological changes in Miocene mammalian record show impact of prolonged climatic forcing. *Proceedings of the National Academy of Sciences of the USA, 105,* 12145–12149.

Badgley, C., Domingo, M. S., Barry, J. C., Morgan, M. E., Flynn, L. J., & Pilbeam, D. (2016). Continental gateways and the dynamics of mammalian faunas. *Comptes Rendus Palevol, 15,* 763–779.

Barry, J. C. (1987). The history and chronology of Siwalik cercopithecids. *Human Evolution, 2,* 47–58.

Barry, J. C., Behrensmeyer, A. K., Badgley, C. E., Flynn, L. J., Peltonen, H., Cheema, I. U., et al. (2013). The Neogene Siwaliks of the Potwar plateau, Pakistan. In X. Wang, L. J. Flynn & M. Fortelius (Eds.), *Fossil mammals of Asia: Neogene biostratigraphy and chronology* (pp. 373–399). New York: Columbia University Press.

Barry, J. C., Jacobs, L. L., & Kelley, J. (1986). An early middle Mocene catarrhine from Pakistan with comments on the dispersal of catarrhines into Eurasia. *Journal of Human Evolution, 15,* 501–508.

Barry, J. C., Johnson, N. M., Raza, S. M., & Jacobs, L. L. (1985). Neogene mammalian faunal change in southern Asia: Correlations with climatic, tectonic, and eustatic events. *Geology, 13,* 637–640.

Barry, J. C., Morgan, M. E., Flynn, L. J., Pilbeam, D., Jacobs, L. L., Lindsay, E. H., et al. (1995). Patterns of faunal turnover and diversity in Neogene Siwaliks of Northern Pakistan. *Palaeogeography, Palaeoclimatology, Palaeoecology, 115,* 209–226.

Begun, D. R. (2000). Technical comments: Middle Miocene hominoid origins. *Science, 287,* 2375a.

Begun, D. R. (2002). The pliopithecoidea. In W. C. Hartwig (Ed.), *The primate fossil record* (pp. 221–240). Cambridge: Cambridge University Press.

Begun, D. R. (2015). Fossil record of Miocene hominoids. In W. Henke & I. Tattersall (Eds.), *Handbook of paleoanthropology* (pp. 1261–1332). Berlin: Springer.

Begun, D. R., & Güleç, E. (1998). Restoration of the type and palate of *Ankarapithecus meteai*: Taxonomic and phylogenetic implications. *American Journal of Physical Anthropology, 105,* 279–314.

Begun, D. R., Güleç, E., & Geraads, D. (2003). Dispersal patterns of Eurasian hominoids: Implications from Turkey. *Deinsea, 10,* 23–40.

Begun, D. R., Nargolwalla, M. C., & Kordos, L. (2012). European Miocene hominids and the origin of the African ape and human clade. *Evolutionary Anthropology, 21,* 10–23.

Begun, D. R., Ward, C. V., & Rose, M. D. (1997). Events in hominoid evolution. In D. R. Begun, C. V. Ward & M. D. Rose (Eds.), *Function, phylogeny and fossils: Miocene hominoid origins and adaptations* (pp. 389–415). New York: Plenum Press.

Benefit, B. R. (1993). The permanent dentition and phylogenetic position of *Victoriapithecus* from Maboko Island, Kenya. *Journal of Human Evolution, 25,* 83–172.

Benefit, B. R. (1999). *Victoriapithecus*: The key to Old World monkey and catarrhine origins. *Evolutionary Anthropology, 8,* 155–174.

Benefit, B. R., & McCrossin, M. L. (1997). Earliest known Old World monkey skull. *Nature, 388,* 368–371.

Benefit, B. R., & McCrossin, M. L. (2002). The Victoriapithecidae, Cercopithecoidea. In W. C. Hartwig (Ed.), *The primate fossil record* (pp. 241–253). Cambridge: Cambridge University Press.

Benefit, B. R., & McCrossin, M. L. (2015). A window into ape evolution. *Science, 350,* 515–516.

Bernor, R., Brunet, M., Ginsberg, L., Mein, P., Pickford, M., Rögl, F., et al. (1987). A consideration of some major topics concerning Old World Miocene Mammalian chronology, migrations and paleogeography. *Geobios, 20,* 431–439.

Bernor, R., Flynn, L. J., Harrison, T., Hussain, S. T., & Kelley, J. (1988). *Dionysopithecus* from southern Pakistan and the biochronology and biogeography of early Eurasian catarrhines. *Journal of Human Evolution, 17,* 339–358.

Bernor, R. L., & Tobien, H. (1990). The mammalian geochronology and biogeography of Paşalar (Middle Miocene, Turkey). *Journal of Human Evolution, 19,* 551–568.

Bishop, W. W., Miller, J. A., & Fitch, F. J. (1969). New potassium-argon age determinations relevant to the Miocene fossil mammal sequence in East Africa. *American Journal of Science, 267,* 669–699.

Bocherans, H., Schrenk, F., Chaimanee, Y., Kullmer, O., Morike, D., Pushkina, D., et al. (2017). Flexibility of diet and habitat in Pleistocene of South Asian mammals: Implications for the fate of the giant fossil ape *Gigantopithecus*. *Quaternary International, 434A,* 148–155.

Böhme, M. (2003). The Miocene climatic optimum: Evidence from ectothermic vertebrates of Central Europe. *Palaeogeography, Palaeoclimatology, Palaeoecology, 195,* 389–401.

Böhme, M., Aziz, H. A., Prieto, J., Bachtadse, V., & Schweigert, G. (2011). Bio-magnetostratigraphy and environment of the oldest Eurasian hominoid from the Early Miocene of Engelweis (Germany). *Journal of Human Evolution, 61,* 332–339.

Böhme, M., Spassov, N., Ebner, M., Geraads, D., Hristova, L., Kirscher, U., et al. (2017). Messinian age and savannah environment of the possible hominin *Graecopithecus* from Europe. *PLoS ONE, 12,* e0177347. https://doi.org/10.1371/journal.pone.0177347.

de Bonis, L., Bouvrain, G., Geraads, D., & Koufos, G. (1990). New hominid skull material from the late Miocene of Macedonia in Northern Greece. *Nature, 345,* 712.

Brown, B., & Ward, S. (1988). Basicranial and facial topography in *Pongo* and *Sivapithecus*. In J. H. Schwartz (Ed.), *Ourang-utan biology* (pp. 247–260). New York: Oxford University Press.

Casanovas-Vilar, I., Alba, D. M., Garcés, M., Robles, J. M., & Moyà-Solà, S. (2011). Updated chronology for the Miocene hominoid radiation in Western Eurasia. *Proceedings of the National Academy of Sciences of the USA, 108,* 5554–5559.

Chaimanee, Y., Jolly, D., Benammi, M., Tafforeau, P., Duzer, D., Moussa, I., et al. (2003). A Middle Miocene hominoid from Thailand and orangutan origins. *Nature, 422,* 61.

Chaimanee, Y., Suteethorn, V., Jintasakul, P., Vidthayanon, C., Marandat, B., & Jaeger, J.-J. (2004). A new orang-utan relative from the Late Miocene of Thailand. *Nature, 427,* 439–441.

Chaimanee, Y., Yamee, C., Tian, P., Khaowiset, K., Marandat, B., Tafforeau, P., et al. (2006). *Khoratpithecus piriyai*, a late Miocene hominoid of Thailand. *American Journal of Physical Anthropology, 131,* 311–323.

Chang, L., Guo, Z., Deng, C., Wu, H., Ji, X., Zhao, Y., et al. (2015). Pollen evidence of the paleoenvironments of *Lufengpithecus lufengensis* in the Zhaotong Basin, southeastern margin of the Tibetan Plateau. *Palaeogeography, Palaeoclimatology, Palaeoecology, 435,* 95–104.

Chatterjee, H. J., Ho, S. Y. W., Barnes, I., & Groves, C. (2009). Estimating the phylogeny and divergence times of primates using a supermatrix approach. *BMC Evolutionary Biology, 9,* 259.

Coster, P., Benammi, M., Chaimanee, Y., Yamee, C., Chavasseau, O., Emonet, E.-G., et al. (2010). A complete magnetic-polarity stratigraphy of the Miocene continental deposits of Mae Moh Basin, northern Thailand, and a reassessment of the age of hominoid-bearing localities in northern Thailand. *Geological Society of America Bulletin, 122,* 1180–1191.

Dean, D., & Delson, E. (1992). Second gorilla or third chimp? *Nature, 359,* 676.

Delson, E. (1975). Evolutionary history of the Cercopithecidae. In F. Szalay (Ed.), *Approaches to Primate Paleobiology, Contributions to Primatology*, vol. 5 (pp. 167–217). Basel: Karger.

Delson, E. (1980). Fossil macaques, phyletic relationships and a scenario of deployment. In D. G. Lindburg (Ed.), *The macaques: Studies in ecology, behavior and evolution* (pp. 10–30). New York: Van Nostrand.

Delson, E. (1994). Evolutionary history of the colobine monkeys in paleoenvironmental perspective. In A. G. Davies & J. H. Oates (Eds.), *Colobine monkeys: Their ecology, behaviour, and evolution* (pp. 11–43). Cambridge: Cambridge University Press.

Delson, E. (1996). The oldest monkeys in Asia. In O. Takenaka (Ed.), *Abstracts, international symposium: Evolution of Asian Primates* (p. 40). Inuyama: Primate Research Institute.

Delson, E., Terranova, C. J., Jungers, W. L., Sargis, E. J., Jablonski, N. G., & Dechow, P. C. (2000). Body mass in Cercopithecidae (Primates, Mammalia): Estimation and scaling in extinct and extant taxa. *Anthropological Papers of the American Museum of Natural History, 8,* 1–159.

De Miguel, D., & Rook, L. (2018). Understanding climate's influence on the extinction of *Oreopithecus* (late Miocene, Tusco-Sardinian paleobioprovince, Italy). *Journal of Human Evolution, 116,* 14–26.

Downing, K. F., Lindsay, E. H., Downs, W. R., & Speyer, S. E. (1993). Lithostratigraphy and vertebrate biostratigraphy of the early Miocene Himalayan foreland, Zinda Pir Dome, Pakistan. *Sedimentary Geology, 87,* 25–37.

Falconer, H., & Cautley, P. T. (1837). On additional fossil species of the order Quadrumana from the Sewalik Hills. *Journal of the Asiatic Society of Bengal, 6,* 354–360.

Fleagle, J. G. (1988). *Primate adaptation and evolution* (1st ed.). New York: Academic Press.

Fleagle, J. G. (2013). *Primate adaptation and evolution* (3rd ed.). Amsterdam: Academic Press.

Fleagle, J. G., & Gilbert, C. C. (2006). Biogeography and the primate fossil record: The role of tectonics, climate, and chance. In S. Lehman & J. G. Fleagle (Eds.), *Primate biogeography* (pp. 375–418). New York: Springer.

Fleagle, J. G., & Simons, E. L. (1978). *Micropithecus clarki*, a small Ape from the Miocene of Uganda. *American Journal of Physical Anthropology, 49,* 427–440.

Fleagle, J. G., & Simons, E. L. (1982). The humerus of *Aegyptopithecus zeuxis*: A primitive anthropoid. *American Journal of Physical Anthropology, 59,* 175–193.

Flynn, L. J., Lindsay, E. H., Pilbeam, D., Raza, S. M., Morgan, M. E., Barry, J. C., et al. (2013). The Siwaliks and Neogene evolutionary biology in South Asia. In X. Wang, L. J. Flynn, & M. Fortelius (Eds.), *Fossil mammals of Asia: Neogene biostratigraphy and chronology* (pp. 353–372). New York: Columbia University Press.

Flynn, L. J., & Wessels, W. (2013). Paleobiogeography and South Asian small mammals: Neogene latitudinal faunal variation. In X. Wang, L. J. Flynn, & M. Fortelius (Eds.), *Fossil mammals of Asia: Neogene biostratigraphy and chronology* (pp. 445–460). New York: Columbia University Press.

Flynn, L. J., Pilbeam, D., Barry, J. C., Morgan, M. E., & Raza, S. M. (2016). Siwalik synopsis: A long stratigraphic sequence for the later Cenozoic of South Asia. *Comptes Rendus Palevol, 15,* 877–887.

Flynn, L. J., Kimura, Y., & Jacobs, L. J. (2020). The Murine Cradle. In G. V. R. Prasad & R. Patnaik (Eds.), *Biological consequences of plate tectonics: New perspectives on post-Gondwana break-up* (pp. 347–362). Cham: Springer.

Friedman, R., Gee, J., Tauxe, L., Downing, K., & Lindsay, E. (1992). The magnetostratigraphy of the Chitarwata and lower Vihowa formations of the Dera Ghazi Khan area, Pakistan. *Sedimentary Geology, 8,* 253–268.

Gebo, D. L. (1993). Postcranial anatomy and locomotor adaptation in early African anthropoids. In D. L. Gebo (Ed.), *Postcranial adaptation in nonhuman primates* (pp. 220–234). DeKalb: Northern Illinois University Press.

Gebo, D. L., MacLatchy, L., Kityo, R., Deino, A., Kingston, J., & Pilbeam, D. (1997). A hominoid genus from the early Miocene of Uganda. *Science, 276,* 401–404.

Gervais, P. (1872). Sur un singe fossile, d'espece non encore decrite, qui a ete decouvert au Monte Bamboli. *Comptes rendus de l'Académie des Sciences, 74,* 1217–1223.

Gilbert, C. C., Pugh, K. D., Patel, B. A., Campisano, C. J., Singh, N. P., Fleagle, J. G., et al. (2017). New small bodied hominoid from Lower Siwalik deposits surrounding Ramnagar (Jammu and Kashmir), India (abstract). *Journal of Vertebrate Paleontology, Suppl. 37,* 118.

Gilbert, C. C., Ortiz, A., Patel, B. A., Singh, N. P., Campisano, C. J., Fleagle, J. G., Pugh, K. D., and Patnaik, R. (2020). New Middle Miocene ape (Primates: Hylobatidae) from Ramnagar, India fills major gaps in the hominoid fossil record. *Proceedings of the Royal Society B.*

Ginsburg, L. (1975). Les Pliopitheques des faluns helvetiens de la Touraine et de l'Anjou. *Colloq. Int. CNRS Paris, 218,* 877–886.

Ginsburg, L., & Mein, P. (1980). *Crouzelia rhodanica*, nouvelle espèce de Primate catarhinien, et essai sur la position systématique des Pliopithecidae.

Gu, Y., & Lin, Y. (1983). First discovery of *Dryopithecus* in East China. *Acta Anthropologica Sinica, 2,* 305–314.

Güleç, E. S., Sevim, A., Pehlevan, C., & Kaya, F. (2007). A new great ape from the late Miocene of Turkey. *Anthropological Science, 115,* 153–158.

Harrison, T. (1988). A taxonomic revision of the small catarrhine primates from the early Miocene of East Africa. *Folia Primatologia, 50,* 59–108.

Harrison, T. (1992). A reassessment of the taxonomic and phylogenetic affinities of the fossil catarrhines from Fort Ternan, Kenya. *Primates, 33,* 501–522.

Harrison, T. (2002). Late Oligocene to middle Miocene catarrhines from Afro-Arabia. In W. C. Hartwig (Ed.), *The primate fossil record* (pp. 311–338). Cambridge: Cambridge University Press.

Harrison, T. (2005). The zoogeographic and phylogenetic relationships of early catarrhine primates in Asia. *Anthropological Science, 113*, 43–51.

Harrison, T. (2010). Dendropithecoidea, Proconsuloidea, and Hominoidea (Catarrhini, Primates). In L. Werdelin & W. J. Sanders (Eds.), *Cenozoic mammals of Africa* (pp. 429–469). Berkeley: University of California Press.

Harrison, T. (2013). Catarrhine origins. In D. R. Begun (Ed.), *A companion to paleoanthropology* (pp. 376–396). West Sussex: Wiley-Blackwell.

Harrison, T. (2016). The fossil record and evolutionary history of hylobatids. In U. H. Reichard, H. Hirohisa & C. Barelli (Eds.), *Evolution of gibbons and siamang: Phylogeny, morphology and cognition* (pp. 91–110). New York: Springer.

Harrison, T., & Andrews, P. (2009). The anatomy and systematic position of the early Miocene proconsulid from Meswa Bridge, Kenya. *Journal of Human Evolution, 56*, 479–496.

Harrison, T., & Gu, Y. (1999). Taxonomy and phylogenetic relationships of early Miocene catarrhines from Sihong, China. *Journal of Human Evolution, 37*, 225–277.

Harrison, T., Ji, X., & Zheng, L. (2008). Renewed investigations at the Late Miocene hominoid locality of Leilao, Yunnan, China. *American Journal of Physical Anthropology, 135*(S46), 113.

Harrison, T., Jin, C., Zhang, Y., Wang, Y., & Zhu, M. (2014). Fossil *Pongo* from the Early Pleistocene *Gigantopithecus* fauna of Chongzuo, Guangxi, southern China. *Quaternary International, 354*, 59–67.

Harrison, T., & Rook, L. (1997). Enigmatic anthropoid or misunderstood ape? The phylogenetic status of *Oreopithecus bambolii* reconsidered. In D. R. Begun, C. V. Ward & M. D. Rose (Eds.), *Function, phylogeny and fossils: Miocene hominoid evolution and adaptations* (pp. 327–362). New York: Plenum Press.

Harrison, T., Xueping, J., & Su, D. (2002). On the systematic status of the late Neogene hominoids from Yunnan Province, China. *Journal of Human Evolution, 43*, 207–227.

Harzhauser, M., Kroh, A., Mandic, O., Piller, W. E., Göhlich, U., Reuter, M., et al. (2007). Biogeographic responses to geodynamics: A key study all around the Oligo-Miocene Tethyan seaway. *Zoologischer Anzeiger, 246*, 241–256.

Heizmann, E., & Begun, D. R. (2001). The oldest European hominoid. *Journal of Human Evolution, 41*, 465–481.

Hennig, W. (1966). *Phylogenetic Systematics. D. Dwight Davis and Rainer Zangerl, tr.* Urbana: University of Illinois Press.

Hürzeler, J. (1954). Contribution à l'odontologie et à la phylogenèse du genre *Pliopithecus* Gervais. *Annales de Paléontologie, 40*, 5–63.

Ishida, H., Kunimatsu, Y., Takano, T., Nakano, Y., & Nakatsukasa, M. (2004). *Nacholapithecus* skeleton from the Middle Miocene of Kenya. *Journal of Human Evolution, 46*, 69–103.

Jablonski, N. G., Whitfort, M. J., Roberts-Smith, N., & Qinqi, X. (2000). The influence of life history and diet on the distribution of catarrhine primates during the Pleistocene in eastern Asia. *Journal of Human Evolution, 39*, 131–157.

Jaeger, J.-J., Soe, A. N., Chavasseau, O., Coster, P., Emonet, E.-G., Guy, F., et al. (2011). First hominoid from the Late Miocene of the Irrawaddy Formation (Myanmar). *PLoS ONE, 6*, e17065. https://doi.org/10.1371/journal.pone.0017065.

Ji, X., Jablonski, N. G., Su, D. F., Deng, C., Flynn, L. J., You, Y., et al. (2013). Juvenile hominoid cranium from the terminal Miocene of Yunnan, China. *Chinese Science Bulletin, 58*, 3771–3779.

Kaya, F., Kaymakçi, N., Bibi, F., Eronen, J. T., Pehlevan, C., Erkman, A. C., et al. (2016). Magnetostratigraphy and paleoecology of the hominid-bearing locality Corakyerler, Tuglu Formation (Cankiri Basin, Central Anatolia). *Journal Vertebrate Paleontology, 36*. https://doi.org/10.1080/02724634.2015.1071710.

Kelley, J. (2002). The hominoid radiation in Asia. In W. C. Hartwig (Ed.), *The primate fossil record* (pp. 369–384). Cambridge: Cambridge University Press.

Kelley, J., Ward, S., Brown, B., Hill, A., & Duren, D. L. (2002). Dental remains of *Equatorius africanus* from Kipsaramon, Tugen Hills, Baringo district, Kenya. *Journal of Human Evolution, 42*, 39–62.

Kelley, J., Andrews, P., & Alpagut, B. (2008). A new hominoid species from the middle Miocene site of Pasalar, Turkey. *Journal of Human Evolution, 54*, 455–479.

Kelley, J., & Gao, F. (2012). Juvenile hominoid cranium from the late Miocene of southern China and hominoid diversity in Asia. *Proceedings of the National Academy of Sciences of the USA, 109*, 6882–6885.

Köhler, M., Moyà-Solà, S., & Alba, D. M. (2000). *Macaca* (Primates, Cercopithecidae) from the late Miocene of Spain. *Journal of Human Evolution, 38*, 447–452.

Koufos, G. D. (2013). Neogene mammal biostratigraphy and chronology of Greece. In X. Wang, M. Fortelius, & L. J. Flynn (Eds.), *Neogene terrestrial mammalian biostratigraphy and chronology in Asia* (pp. 595–621). New York: Columbia University Press.

Koufos, G. D., de Bonis, L., & Kugiumtzis, D. (2016). New material of the hominoid *Ouranopithecus macedoniensis* from the Late Miocene of the Axios Valley (Macedonia, Greece) with some remarks on its sexual dimorphism. *Folia Primatologia, 87*, 94–122.

Koufos, G. D., Zouros, N., & Mourouzidou, O. (2003). *Prodeinotherium bavaricum* (Proboscidea, Mammalia) from Lesvos island, Greece; the appearance of deinotheres in the Eastern Mediterranean. *Geobios, 36*, 305–315.

Kunimatsu, Y., Nakatsukasa, M., Sawada, Y., Sakai, T., Hyodo, M., Hyodo, H., et al. (2007). A new Late Miocene great ape from Kenya and its implications for the origins of African great apes and humans. *Proceedings of the National Academy of Sciences of the USA, 104*, 19220–19225.

Larson, S. G. (1998). Parallel evolution in the hominoid trunk and forelimb. *Evolutionary Anthropology, 6*, 87–99.

Lartet, E. (1856). Note sur un grand singe fossile qui se rattache au groupe des singes superieurs. *Comptes rendus de l'Académie des Sciences, Paris, 43*, 219–223.

Leakey, M. G., Ungar, P. S., & Walker, A. (1995). A new genus of large primate from the Late Oligocene of Lothidok, Turkana district, Kenya. *Journal of Human Evolution, 28*, 519–531.

Leakey, R. E., & Leakey, M. G. (1986a). A new Miocene hominoid from Kenya. *Nature, 324*, 143–146.

Leakey, R. E., & Leakey, M. G. (1986b). A second new Miocene hominoid from Kenya. *Nature, 324*, 146–148.

Leakey, R. E., Leakey, M. G., & Walker, A. (1988). Morphology of *Turkanapithecus kalakolensis* from Kenya. *American Journal of Physical Anthropology, 76*, 277–288.

Le Gros Clark, W. E., & Leakey, L. S. B. (1951). The Miocene Hominoidea of East Africa. Fossil mammals of Africa. *British Museum (Natural History), London, 1*, 1–117.

Li, C. (1978). A Miocene gibbon-like primate from Shihhung, Kiangsu Province. *Vertebrata PalAsiatica, 16*, 187–192.

Lovejoy, C. O. (1981). The origin of man. *Science, 211*, 341–350.

Lydekker, R. (1879). Further notices of Siwalik Mammalia. *Records of the Geological Survey of India, 12*, 33–52.

Maddison, W. S., & Maddison, D. R. (2018). Mequite: A modular system for evolutionary analysis. Version 3.5. http://mesquiteproject.org.

Martin, L. B., & Andrews, P. (1993). Species recognition in Middle Miocene hominoids. In W. H. Kimbel & L. B. Martin (Eds.), *Species, species concepts, and primate evolution* (pp. 393–427). New York: Plenum Press.

Matthew, W. D., & Granger, W. (1923). New fossil mammals from the Pliocene of Szechuan, China. *Bulletin of the American Museum of Natural History, 48,* 568–598.

Maxwell, S. J., Hopley, P. J., Upchurch, P., & Soligo, C. (2018). Sporadic sampling, not climatic forcing, drives observed early hominin diversity. *Proceedings of the National Academy of Sciences of the USA, 115,* 4891–4896.

McDougall, I., & Brown, F. H. (2009). Timing of volcanism and evolution of the northern Kenya Rift. *Geological Magazine, 146,* 34–47.

Madden, C. T. (1980). New *Proconsul* (*Xenopithecus*) from the Miocene of Kenya. *Primates, 21,* 241–252.

McCollum, M. S., Peppe, D. J., McNulty, K. P., Dunsworth, H. M., Harcourt-Smith, W. E. H., & Andrews, A. L. (2013). Magnetostratigraphy of the early Miocene Hiwegi Formation (Rusinga Island, Lake Victoria, Kenya). *Geological Society of America Abstracts with Programs, 44,* 241.

McNulty, K. P., Begun, D. R., Kelley, J., Manthi, F. K., & Mbua, E. N. (2015). A systematic revision of *Proconsul* with the description of a new genus of early Miocene hominoid. *Journal of Human Evolution, 84,* 42–61.

Miller, E. R., Benefit, B. R., McCrossin, M. L., Plavcan, J. M., Leakey, M. G., & El-Barkooky, A. N. (2009). Systematics of early and middle Miocene Old World monkeys. *Journal of Human Evolution, 57,* 195–211.

Nakatsukasa, M., & Kunimatsu, Y. (2009). *Nacholapithecus* and its importance for understanding hominoid evolution. *Evolutionary Anthropology, 18,* 103–119.

Napier, J. R., & Davis, P. R. (1959). The forelimb skeleton and associated remains of *Proconsul africanus*. *Fossil Mammals of Africa, 16,* 1–69.

Nargolwalla, M. C. (2009). Eurasian middle and late Miocene hominoid paleobiogeography and the geographic origins of the Homininae. Ph.D. Dissertation, University of Toronto.

Nelson, G., & Platnick, N. (1981). *Systematics and biogeography: Cladistics and vicariance*. New York: Columbia University Press.

Nengo, I., Tafforeau, P., Gilbert, C. C., Fleagle, J. G., Miller, E., Feibel, C., et al. (2017). New infant cranium from the African Miocene sheds light on ape evolution. *Nature, 548,* 169–174.

Olejniczak, A. J., Smith, T. M., Wang, W., Potts, R., Ciochon, R., Kullmer, O., et al. (2008). Molar enamel thickness and dentine horn height in *Gigantopithecus blacki*. *American Journal of Physical Anthropology, 135,* 85–91.

Ortiz, A., Pilbrow, V., Villamil, C. I., Korsgaard, J. G., Bailey, S. E., & Harrison, T. (2015). The taxonomic and phylogenetic affinities of *Bunopithecus sericus*, a fossil hylobatid from the Pleistocene of China. *PLoS ONE, 10,* e0131206. https://doi.org/10.1371/journal.pone.0131206.

Pan, Y. (2006). Primates Linnaeus, 1758. In G. Qi & W. Dong (Eds.), *Lufengpithecus hudienensis site* (pp. 131–148). Beijing: Science Press.

Patnaik, R. (2008). Revisiting Haritalyangar, the Late Miocene Ape locality of India. In J. G. Fleagle & C. C. Gilbert (Eds.), *Elwyn Simons: A search for origins* (pp. 197–210). New York: Springer.

Patnaik, R. (2016). Neogene-Quaternary mammalian paleobiogeography of the Indian Subcontinent: An appraisal. *Comptes Rendus Palevol, 15,* 889–902.

Patnaik, R. (2020). New data on the Siwalik murines, rhizomyines and ctenodactylines (Rodentia) from the Indian subcontinent. In G. V. R. Prasad & R. Patnaik (Eds.), *Biological consequences of plate tectonics: New perspectives on post-Gondwana break-up* (pp. 363–391). Cham: Springer.

Patnaik, R., & Cameron, D. (2007). On new dates and ecology of recent hominoid finds from Haritalyangar area, Himachal Pradesh. In A. R. Sankhyan & V. R. Rao (Eds.), *Human origins, genome and people of India: Genomic, paleontological, and archaeological perspectives* (pp. 112–119). New Delhi: Allied Publishers.

Peppe, D. J., McNulty, K. P., Cote, S. M., Harcourt-Smith, W. E., Dunsworth, H. M., & Van Couvering, J. A. (2009). Stratigraphic interpretation of the Kulu Formation (Early Miocene, Rusinga Island, Kenya) and its implications for primate evolution. *Journal of Human Evolution, 56,* 447–461.

Perelman, P., Johnson, W. E., Roos, C., Seuanez, H. N., Horvath, J. E., Moreira, M. A. M., et al. (2011). A molecular phylogeny of living primates. *PLoS Genetics, 7,* e1001342. https://doi.org/10.1371/journal.pgen.1001342.

Pickford, M. (1985). A new look at *Kenyapithecus* based on recent discoveries in Western Kenya. *Journal of Human Evolution, 14,* 113–143.

Pickford, M. (1986). Hominoids from the Miocene of East Africa and the phyletic position of *Kenyapithecus*. *Zeitschrift für Morphologie und Anthropologie, 76,* 117–130.

Pickford, M., Sawada, Y., Tayama, R., Matsuda, Y. K., Itaya, T., Hyodo, H., et al. (2006). Refinement of the age of the Middle Miocene Fort Ternan Beds, Western Kenya, and its implications for Old World biochronology. *Comptes Rendus Geoscience, 338*(8), 545–555.

Pillans, B., Williams, M., Cameron, D., Patnaik, R., Hogarth, J., Sahni, A., et al. (2005). Revised correlation of the Haritalyangar magnetostratigraphy, Indian Siwaliks: Implications for the age of the Miocene hominids *Indopithecus* and *Sivapithecus*, with a note on a new hominid tooth. *Journal of Human Evolution, 48,* 507–515.

Pozzi, L., Hodgson, J. A., Burrell, A. S., Sterner, K. N., Raaum, R. L., & Disotell, T. R. (2014). Primate phylogenetic relationships and divergence dates inferred from complete mitochondrial genomes. *Molecular Phylogenetics Evolution, 75,* 165–183.

Pugh, K. D. (2020). The phylogenetic relationships of Middle-Late Miocene apes: Implications for early human evolution. Ph.D. Dissertation, The Graduate Center of the City University of New York.

Qi, G., Dong, W., Zheng, L., Zhao, L., Gao, F., Yue, L., et al. (2006). Taxonomy, age and environment status of the Yuanmou hominoids. *Chinese Science Bulletin, 51,* 704–712.

Rae, T. C. (1997). The early evolution of the hominoid face. In D. R. Begun, C. V. Ward & M. D. Rose (Eds.), *Function, phylogeny, and fossils: Miocene hominoid evolution and adaptations* (pp. 59–77). New York: Plenum Press.

Rae, T. C. (1999). Mosaic evolution in the origin of the Hominoidea. *Folia Primatologia, 70,* 125–135.

Rae, T. C. (2004). Miocene hominoid craniofacial morphology and the emergence of great apes. *Annals of Anatomy, 186,* 417–421.

Rasmussen, D. T., & Guitierrez, M. (2009). A mammalian fauna from the Late Oligocene of Northwestern Kenya. *Palaeontographica Abt. A, 288,* 1–52.

Rögl, F. (1998). Paleogeographic considerations for Mediterranean and Paratethys seaways (Oligocene to Miocene). *Annalen des Naturhistorischen Museums in Wien, 99,* 279–310.

Rögl, F. (1999). Mediterranean and Paratethys. Facts and hypotheses of an Oligocene to Miocene paleogeography (short overview). *Geologica Carpathica, 50,* 339–349.

Rook, L., Harrison, T., & Engesser, B. (1996). The taxonomic status and biochronological implications of new finds of *Oreopithecus* from Baccinello (Tuscany, Italy). *Journal of Human Evolution, 30,* 3–27.

Rose, M. D. (1983). Miocene hominoid postcranial morphology: Monkey-like, ape-like, neither, or both? In R. L. Ciochon & R. S. Corruccini (Eds.), *New interpretations of ape and human ancestry* (pp. 405–417). New York: Plenum Press.

Rose, M. D. (1988). Another look at the anthropoid elbow. *Journal of Human Evolution, 171,* 193–224.

Rose, M. D. (1993). Locomotor anatomy of Miocene hominoids. In D. L. Gebo (Ed.), *Postcranial adaptation in nonhuman primates* (pp. 252–272). DeKalb: Northern Illinois University Press.

Rose, M. D. (1997). Functional and phylogenetic features of the forelimb in Miocene hominoids. In D. R. Begun, C. V. Ward & M. D. Rose (Eds.), *Function, phylogeny, and fossils: Miocene hominoid evolution and adaptations* (pp. 79–100). New York: Plenum Press.

Rossie, J. B., & MacLatchy, L. (2006). A new pliopithecid genus from the early Miocene of Uganda. *Journal of Human Evolution, 50,* 568–586.

Rossie, J. B., & Hill, A. (2018). A new species of *Simiolus* from the middle Miocene of the Tugen Hills, Kenya. *Journal of Human Evolution, 125,* 50–58.

Russo, G. A. (2016). Comparative sacral morphology and the reconstructed tail lengths of five extinct primates: *Proconsul heseloni*, *Epipliopithecus vindobonensis*, *Archaeolmur adwardsi*, *Megaladapis grandidieri*, and *Paleopropithecus kelyus*. *Journal of Human Evolution, 90,* 135–162.

Sankhyan, A. R., Kelley, J., & Harrison, T. (2017). A highly derived pliopithecid from the Late Miocene of Haritalyangar, India. *Journal of Human Evolution, 105,* 1–12.

Shea, B. T. (1985). On aspects of skull form in African apes and orangutans, with implications for hominoid evolution. *American Journal of Physical Anthropology, 68,* 329–342.

Shultz, A. (1933a). Observations on the growth, classification and evolutionary specialization of gibbons and siamangs. *Human Biology, 5,* 212–255.

Shultz, A. (1933b). Observations on the growth, classification and evolutionary specialization of gibbons and siamang (concluded). *Human Biology, 5,* 385–428.

Schwartz, J. H. (1990). *Lufengpithecus* and its potential relationship to an orang-utan clade. *Journal of Human Evolution, 19,* 591–605.

Seiffert, E. R. (2012). Early primate evolution in Afro-Arabia. *Evolutionary Anthropology, 21,* 239–253.

Seiffert, E. R., & Simons, E. L. (2001). Astragalar morphology of late Eocene anthropoids from the Fayum Depression (Egypt) and the origin of catarrhine primates. *Journal of Human Evolution, 41,* 577–606.

Seiffert, E. R., Simons, E. L., & Fleagle, J. G. (2000). Anthropoid humeri from the late Eocene of Egypt. *Proceedings of the National Academy of Sciences of the USA, 97,* 10062–10067.

Seiffert, E. R., Simons, E. L., Fleagle, J. G., & Godinot, M. (2010). Paleogene anthropoids. In L. Werdelin & W. J. Sanders (Eds.), *Cenozoic mammals of Africa* (pp. 369–391). Berkeley: University of California Press.

Senut, B., Pickford, M., Gommery, D., & Kunimatsu, Y. (2000). A new genus of Early Miocene hominoid from East Africa: *Ugandapithecus major* (Le Gros Clark & Leakey, 1950). *Comptes rendus de l'Académie des Sciences, Paris, II A, 3,* 227–233.

Sepulchre, P. (2003). *Paleoenvironments d'Asie du Sud-Est: apports de l'analyse palynologique de deux sites Miocene de Thailande.* Montpellier: Diplome d'Etudes Approfondies, Universite Montpellier II.

Simons, E. L. (1987). New faces of *Aegyptopithecus* from the Oligocene of Egypt. *Journal of Human Evolution, 16,* 273–289.

Simons, E. L. (1989). Description of two genera and species of late Eocene Anthropoidea from Egypt. *Proceedings of the National Academy of Sciences of the USA, 86,* 9956–9960.

Simons, E. L., & Rasmussen, D. T. (1996). Skull of *Catopithecus browni*, an early Tertiary catarrhine. *American Journal of Physical Anthropology, 100,* 261–292.

Simons, E. L., Plavcan, J. M., & Fleagle, J. G. (1999). Canine sexual dimorphism in Egyptian Eocene anthropoid primates: *Catopithecus* and *Proteopithecus*. *Proceedings of the National Academy of Sciences of the USA, 96,* 2559–2562.

Simons, E. L., Seiffert, E. R., Ryan, T. M., & Attia, Y. (2007). A remarkable female cranium of the Oligocene anthropoid *Aegyptopithecus zeuxis* (Catarrhini, Propliopithecidae). *Proceedings of the National Academy of Sciences of the USA, 104,* 8731–8736.

Singh, N. P., Gilbert, C. C., Patel, B. A., & Patnaik, R. (2018). The taphonomy and paleoecology of the middle Miocene hominoid locality of Ramnagar (Jammu and Kashmir, India). *Journal of Asian Earth Sciences.* https://doi.org/10.1016/j.jseaes.2018.02.020.

Spassov, N., Geraads, D., Hristova, L., Markov, G. N., Merceron, G., Tzankov, T., et al. (2012). A hominid tooth from Bulgaria: The last pre-human hominid of continental Europe. *Journal of Human Evolution, 62,* 138–145.

Steininger, F. F., Rabeder G., & Rögl F. (1985). Land mammal distribution in the Mediterranean Neogene: A consequence of geokinematic and climatic events. In *Geological evolution of the Mediterranean Basin* (pp. 559–571). New York: Springer.

Steiper, M. E., & Young, N. M. (2006). Primate molecular divergence dates. *Molecular Phylogenetics and Evolution, 41,* 384–394.

Stevens, N. J., Seiffert, E. R., O'Connor, P. M., Roberts, E. M., Schmitz, M. D., Krause, C., et al. (2013). Paleontological evidence for an Oligocene divergence between Old World monkeys and apes. *Nature, 497,* 611–614.

Strait, D. S., Wood, B. A. 1999. Early hominid biogeography. Proceedings of the National Academy of Sciences, USA 96, 9196–9200.

Suwa, G., Kono, R. T., Katoh, S., Asfaw, B., & Beyene, Y. (2007). A new species of great ape from the late Miocene Epoch in Ethiopia. *Nature, 448,* 921.

Suwa, G., Kunimatsu, Y., Ataabadi, M. M., Orak, Z., Sasaki, T., & Fortelius, M. (2016). The first hominoid from the Maragheh Formation, Iran. *Palaeobiodiversity and Palaeoenvironments, 96,* 373–381.

Takai, M., Zhang, Y., Kono, R. T., & Jin, C. (2014). Changes in the composition of the Pleistocene primate fauna in southern China. *Quaternary International, 354,* 75–85.

Tekkaya, I. (1974). A new species of Tortonian anthropoids (Primates, Mammalia) from Anatolia. *Bulletin of the Mineral Research and Exploration, Turkey, 83,* 1–11.

Thomas, H. (1985). The Early and Middle Miocene land connection of the Afro-Arabian plateau and Asia: A major event for hominoid dispersal? In E. Delson (Ed.), *Ancestors: The hard evidence* (pp. 42–50). New York: Allan R. Liss.

Turvey, S. T., Bruun, K., Ortiz, A., Hansford, J., Hu, S., Ding, Y., et al. (2018). New genus of extinct Holocene gibbon associated with humans in Imperial China. *Science, 360,* 1346–1349.

Ungar, P. S., & Kay, R. F. (1995). The dietary adaptations of European Miocene catarrhines. *Proceedings of the National Academy of Sciences of the USA, 92,* 5479–5481.

van der Made, J. (1999). Intercontinental relationship Europe-Africa and the Indian subcontinent. In G. E. Rossner & K. Heissig (Eds.), *The Miocene land mammals of Europe* (pp. 457–472). Munchen: Friedrich Pfeil.

Walker, A. (1997). *Proconsul*: Function and phylogeny. In D. R. Begun, C. V. Ward, & M. D. Rose (Eds.), *Function, phylogeny, and fossils: Miocene hominoid evolution and adaptations* (pp. 209–224). New York: Plenum Press.

Ward, C. V. (1998). *Afropithecus*, *Proconsul*, and the primitive hominoid skeleton. In E. Strasser, J. G. Fleagle, A. Rosenberger, & H. McHenry (Eds.), *Primate locomotion: Recent advances* (pp. 337–352). New York: Plenum Press.

Ward, C. V., Walker, A. C., & Teaford, M. F. (1991). *Proconsul* did not have a tail. *Journal of Human Evolution, 21,* 215–220.

Ward, S. C., & Duren, D. L. (2002). Middle and late Miocene African hominoids. In W. C. Hartwig (Ed.), *The primate fossil record* (pp. 385–397). Cambidge: Cambridge University Press.

Ward, S. C., & Kimbel, W. H. (1983). Subnasal alveolar morphology and the systemic position of *Sivapithecus*. *American Journal of Physical Anthropology, 61,* 157–171.

Ward, S. C., & Pilbeam, D. R. (1983). Maxillofacial morphology of Miocene hominoids from Africa and Indo-Pakistan. In R. L. Corruccini & R. S. Ciochon (Eds.), *New interpretations of ape and human ancestry* (pp. 211–238). New York: Plenum Press.

Ward, S., Brown, B., Hill, A., Kelley, J., & Downs, W. (1999). *Equatorius*: A new hominoid genus from the middle Miocene of Kenya. *Science, 285,* 1382–1386.

Welcomme, J.-L., Benammi, M., Crochet, J.-Y., Marivaux, L., Metais, G., Antoine, P.-O., et al. (2001). Himalayan forelands: Paleontological evidence for Oligocene detrital deposits in the Bugti Hills (Balochistan, Pakistan). *Geological Magazine, 138,* 397–405.

Xue, X.-X., & Delson, E. (1989). A new species of *Dryopithecus* from Gansu, China. *Kexue tongbao* (English edition), *34,* 223–229.

Zalmout, I. S., Sanders, W. J., MacLatchy, L. M., Gunnell, G. F., Al-Mufarreh, Y. A., Ali, M. A., et al. (2010). New Oligocene primate from Saudi Arabia and the divergence of apes and Old World monkeys. *Nature, 466,* 360–364.

Zapfe, H. (1958). The skeleton of *Pliopithecus* (*Epipliopithecus*) *vindobonensis* Zapfe & Hürzeler. *American Journal of Physical Anthropology, 16,* 441–455.

Zapfe, H. (1960). Die primatenfunde aus der miozänen Spaltenfüllung von Neudorf an der March (Dévinská Nová Ves), Tsechoslovakei. Mit anhang: Der Primatenfund aus dem Miozän von Klein Hadersdorf in Niederosterreich. *Abbandlungen der Schweizerische Paläontologischen Gesellschaft, 78,* 1–293.

Zhang, Y., Jin, C., Cai, Y., Kono, R., Wang, W., Wang, Y., et al. (2014). New 400–320 ka *Gigantopithecus blacki* remains from hejiang cave, chongzuo City, guangxi, South China. *Quaternary International, 354,* 35–45.

Zhang, Y., Kono, R. T., Wang, W., Harrison, T., Takai, M., Ciochon, R. L., et al. (2015). Evolutionary trend in dental size *in Gigantopithecus blacki* revisited. *Journal of Human Evolution, 83,* 91–100.

Zhisheng, A., Kutzbach, J. E., Prell, W. L., & Porter, S. C. (2001). Evolution of Asian monsoons and phased uplift of the Himalaya-Tibetan plateau since Late Miocene times. *Nature, 411,* 62.

Index

Note: Page numbers followed by *f* and *t* indicates figures and tables, respectively

G. V. R. Prasad and R. Patnaik (eds.), *Biological Consequences of Plate Tectonics: New Perspectives on Post-Gondwana Break-up—A Tribute to Ashok Sahni*, Vertebrate Paleobiology and Paleoanthropology Series, https://doi.org/10.1007/978-3-030-49753-8

D

E

L

M

N

O

The manufacturer's authorised representative in the EU is Springer Nature Customer Service Centre GmbH, Europaplatz 3, 69115 Heidelberg, Germany. If you have any concerns regarding our products, please contact ProductSafety@springernature.com

Printed and bound by CPI Group (UK) Ltd, Croydon, CR0 4YY

26/09/2026

02222876-0001